CW00722077

Handbook of Fruits and Fruit Processing

Handbook of Fruits and Fruit Processing

Editor
Y. H. Hui

Associate Editors
József Barta, M. Pilar Cano, Todd W. Gusek,
Jiwan S. Sidhu, and Nirmal Sinha

Blackwell Publishing Professional
2121 State Avenue, Ames, Iowa 50014, USA

Orders: 1-800-862-6657
Office: 1-515-292-0140
Fax: 1-515-292-3348
Web site: www.blackwellprofessional.com

Blackwell Publishing Ltd
9600 Garsington Road, Oxford OX4 2DQ, UK
Tel.: +44 (0)1865 776868

Blackwell Publishing Asia
550 Swanston Street, Carlton, Victoria 3053, Australia
Tel.: +61 (0)3 8359 1011

First edition, 2006

Library of Congress Cataloging-in-Publication Data

Handbook of fruits and fruit processing / editor, Y.H. Hui; associate editors, J´ozsef Barta . . . [et al.].— 1st ed.
p. cm.
Includes index.
ISBN-13: 978-0-8138-1981-5 (alk. paper)
ISBN-10: 0-8138-1981-4 (alk. paper)
1. Food industry and trade. 2. Fruit—Processing. I. Hui, Y. H. (Yiu H.) II. Barta, J´ozsef.

TP370.H264 2006
664 .8—dc22

2005013055

The last digit is the print number: 9 8 7 6 5 4 3 2

Contents

Part III Commodity Processing

Contributors

Rafael Alique (Chapter 20)
Instituto del Frío (CSIC)
C/José Antonio Novais n° 10
28040 Madrid, Spain
Phone: +34915492300

Jesús Alonso (Chapter 20)
Instituto del Frío (CSIC)
C/José Antonio Novais n° 10
28040 Madrid, Spain
Phone: +34915492300
E-mail: jalonso@if.csic.es

Husam Al-Omariah (Chapter 15)
Biotechnology Department
Kuwait Institute for Scientific Research
P.O. Box 24885, 13109-Safat, Kuwait

Sameer Al-Zenki (Chapter 15)
Biotechnology Department
Kuwait Institute for Scientific Research
P.O. Box 24885, 13109-Safat, Kuwait
Phone: (965)-483-6100
Fax: (965)-483-4670
E-mail: szenki@kisr.edu.kw

Begoña De Ancos (Chapters 2, 4)
Department of Plant Foods Science
and Technology, Instituto del Frío
Consejo Superior de Investigaciones
Científicas (CSIC) Ciudad Universitaria
E-28040 Madrid, Spain
E-mail: ancos@if.csic.es

Csaba Balla (Chapter 7)
Corvinus University of Budapest, Faculty of
Food Science, Department of Refrigeration
and Livestock Products Technology
Hungary 1118, Budapest, Ménesi út 45
Phone: 36-1-482-6064
Fax: 36-1-482-6321
E-mail: csaba.balla@uni-corvinus.hu

József Barta, Ph.D. (Chapters 5, 14)
Head of the Department
Corvinus University of Budapest
Faculty of Food Science
Department of Food Preservation
Budapest, Ménesi út 45
Hungary 1118
Phone: 36-1-482-6212
Fax: 36-1-482-6327
E-mail: jozsef.barta@uni-corvinus.hu

A.S. Bawa (Chapters 9, 11)
Fruits and Vegetables Technology
Defence Food Research Laboratory
Siddarthanagar, Mysore-570 011, India
Phone: 0821-247-3783
Fax: 0821-247-3468
E-mail: dfoodlab@sancharnet.in

N. R. Bhat (Chapter 24)
Arid Land Agriculture Department
Kuwait Institute for Scientific Research
P.O. Box 24885, 13109-Safat, Kuwait
E-mail: nbhat@safat.kisr.edu.kw

M. Pilar Cano, Ph.D. (Chapters 2, 4)
Director
Instituto del Frío-CSIC
C/Jose Antonio Novais, 10
Ciudad Universitaria
28040-Madrid, Spain
Phone: 34-91-5492300
Fax: 34-91-5493627
E-mail: pcano@if.csic.es

Zsuzsanna Cserhalmi (Chapter 6)
Central Food Research Institute
Hungary 1022 Budapest, Hermann O. u. 15
Phone: 36-1-214-1248
Fax: 36-1-355-8928
E-mail: zs.cserhalmi@cfri.hu

Sonia De Pascual-Teresa (Chapters 2, 4)
Department of Plant Foods Science
and Technology, Instituto del Frío
Consejo Superior de Investigaciones
Científicas (CSIC) Ciudad Universitaria
E-28040 Madrid, Spain
E-mail: soniapt@if.csic.es

B. B. Desai (Chapter 24)
Arid Land Agriculture Department
Kuwait Institute for Scientific Research
P.O. Box 24885, 13109-Safat, Kuwait

József Farkas (Chapter 7)
Corvinus University of Budapest
Faculty of Food Science, Department
of Refrigeration and Livestock Products
Technology and Central Food Research Institute
Hungary 1118, Budapest, Ménesi út 45
and, 1022, Budapest, Hermann O. u. 15
Phone: 36-1-482-6303
Fax: 36-1-482-6321
E-mail: j.farkas@cfri.hu

Raquel de Pinho Ferreira Guiné (Chapter 28)
Associate Professor
Department of Food Engineering
ESAV, Polytechnic Institute of Viseu
Campus Politécnico, Repeses
3504-510 Viseu, Portugal
E-mail: raquelguine@esav.ipv.pt

Beatriz Gandul-Rojas (Chapter 26)
Group of Chemistry and Biochemistry
of Pigments. Food Biotechnology Department
Instituto de la Grasa (CSIC).
Av. Padre García Tejero 4, 41012
Sevilla, Spain

Kristen K. Girard (Chapter 21)
Principal Scientist
Ocean Spray Cranberries, Inc.
Ingredients
1 Ocean Spray Dr.
Middleboro MA 02349, USA
E-mail: kgirard@oceanspray.com

Rajinder P. Gupta (Chapter 1)
Department of Microbiology,
College of Basic Sciences and Humanities
Punjab Agricultural University
Ludhiana-141004, India
rpguptag@rediffmail.com

Todd W. Gusek, Ph.D.
Principal Scientist, Central Research
Cargill, Inc.
PO Box 5699
Minneapolis, MN 55440, USA
Phone: (952)742-6523
Fax: (952)742-4925
E-mail: todd_gusek@cargill.com

M. Hamdi (Chapter 34)
Director, Department of Biochemical and Chemical
Engineering Microbial and Food Processes
Higher School of Food Industries
National Institute of Applied Sciences
and Technology. BP: 676. 1080 Tunisia
Phone: 216-98-326675
Fax: 216-71-704-329
E-mail: moktar.hamdi@insat.rnu.tn

Emoke Horváth-Kerkai (Chapter 12)
Corvinus University of Budapest, Faculty
of Food Science, Department of
Food Preservation Hungary 1118
Budapest, Ménesi út 45.
Phone: 36-1-482-6035
Fax: 36-1-482-6327
E-mail: emoke.kerkai@uni-corvinus.hu

Y. H. Hui, Ph.D.
Senior Scientist
Science Technology System
P.O. Box 1374
West Sacramento, CA 95691, USA
Phone: 916-372-2655
Fax: 916-372-2690
E-mail: yhhui@aol.com

Manuel Jarén-Galán (Chapter 30)
Group of Chemistry and Biochemistry
of Pigments. Food Biotechnology Department
Instituto de la Grasa (CSIC)
Av. Padre García Tejero 4, 41012
Sevilla, Spain

Anu Kalia (Chapter 1)
Department of Microbiology,
College of Basic Sciences and Humanities
Punjab Agricultural University
Ludhiana-141004, India
kaliaanu@rediffmail.com

Imre Körmendy (Chapter 3)
Corvinus University of Budapest,
Faculty of Food Science, Department
of Food Preservation Hungary 1118
Budapest, Ménesi út 45
Phone: 36-1-482-6212
Fax: 36-1-482-6327
E-mail: imre.kormendy@uni-corvinus.hu

Olga Martín-Belloso (Chapters 8, 23)
Department of Food Technology, University
of Lleida Av. Alcalde Rovira Roure, 191. 25198
Lleida, Spain
Phone: +34-973-702-593
Fax: +34-973-702-596
E-mail: omartin@tecal.udl.es

Angel Robert Marsellés-Fontanet (Chapter 23)
Department of Food Technology, University
of Lleida Av. Alcalde Rovira Roure, 191. 25198
Lleida, Spain
Phone: +34 973 702 593
Fax: +34 973 702 596
E-mail: rmarselles@tecal.udl.es

M. Isabel Mínguez-Mosquera (Chapters 26, 30)
Group of Chemistry and Biochemistry
of Pigments. Food Biotechnology Department
Instituto de la Grasa (CSIC)
Av. Padre García Tejero 4, 41012
Sevilla, Spain
Phone: +34954691054
Fax: +34954691262
E-mail: minguez@cica.es.

Kuldip Singh Minhas (Chapter 19)
Professor
Food Science and Technology
Punjab Agricultural University
Ludhiana, Punjab, India
Phone: 0161-2401960 Extn. 305

Judit Monspart-Sényi (Chapter 10)
Corvinus University of Budapest, Faculty
of Food Science, Department of Food Preservation
Hungary 1118, Budapest, Ménesi út 45
Phone: 36-1-482-6037
Fax: 36-1-482-6327
E-mail: judit.senyi@uni-corvinus.hu

Lillian G. Occeña-Po (Chapter 33)
Department of Food Science and Human Nutrition
Michigan State University
East Lansing, MI 48824, USA
Phone: 517-432-7022
Fax: 517-353-8963
E-mail: occena@msu.edu

Gemma Oms-Oliu (Chapter 8)
Department of Food Technology, University of
Lleida Av. Alcalde Rovira Roure, 191. 25198
Lleida, Spain
Phone: +34-973-702-593
Fax: +34-973-702-596
E-mail: goms@tecal.udl.es

Györgyi Pátkai (Chapter 13)
Corvinus University of Budapest, Faculty
of Food Science, Department of Food Preservation
Hungary 1118, Budapest, Ménesi út 45
Phone: 36-1-482-6212
Fax: 36-1-482-6327
E-mail: gyorgyi.patkai@uni-corvinus.hu

Antonio Pérez-Gálvez (Chapter 30)
Group of Chemistry and Biochemistry
of Pigments, Food Biotechnology Department
Instituto de la Grasa (CSIC).
Av. Padre García Tejero 4, 41012,
Sevilla, Spain

Isak S. Pretorius (Chapter 25)
The Australian Wine Research Institute
PO Box 197, Glen Osmond
Adelaide, SA 5064
Australia
Phone: +61-8-83036835
Fax: +61-8-83036601
E-mail: Sakkie.Pretorius@awri.com.au

P.S. Raju (Chapter 9)
Fruits and Vegetables Technology
Defence Food Research Laboratory
Siddarthanagar, Mysore-570 011, India
Phone: 0821-247-3783
Fax: 0821-247-3468
E-mail: dfoodlab@sancharnet.in

María Jesús Rodrigo (Chapter 18)
Instituto de Agroquímica y Tecnología
de Alimentos (CSIC). Apartado Postal 73
46100 Burjasot, Valencia, Spain

Concepción Sánchez-Moreno (Chapters 2, 4)
Department of Plant Foods Science and
Technology, Instituto del Frío, Consejo Superior
de Investigaciones Científicas (CSIC)
Ciudad Universitaria, E-28040 Madrid, Spain
E-mail: csanchezm@if.csic.es

Kulwant S. Sandhu (Chapter 19)
Sr. Veg. Technologist (KSS)
Department of Food Science and Technology
Punjab Agricultural University
Ludhiana - 141 004, Punjab, India
Phone: 0161-2405257, 2401960 extn. 8478
(KSS)
E-mail: ptc@satyam.net.in

Jiwan S. Sidhu, Ph.D. (Chapters 22, 32)
Professor, Department of Family Science
College for Women, Kuwait University
P.O. Box 5969, Safat-13060, Kuwait
Phone: (965)-254-0100 extn. 3307
Fax: (965)-251-3929
E-mails: jsidhu@cfw.kuniv.edu;
jiwansidhu2001@yahoo.com

Muhammad Siddiq (Chapters 17, 27, 29)
Food Processing Specialist
Department of Food Science & Human Nutrition
Michigan State University
East Lansing, MI 48824, USA
Phone: 517-355-8474
Fax: 517-353-8963
E-mail: siddiq@msu.edu

Nirmal Sinha, Ph.D. (Chapters 16, 21, 31)
VP, Research and Development
Graceland Fruit, Inc.
1123 Main Street
Frankfort, MI 49635, USA
Phone: 231-352-7181
Fax: 231-352-4711
E-mail: nsinha@gracelandfruit.com

Robert Soliva-Fortuny (Chapter 8)
Department of Food Technology, University
of Lleida Av. Alcalde Rovira Roure, 191. 25198
Lleida, Spain
Phone: +34-973-702-593
Fax: +34-973-702-596
E-mail: rsoliva@tecal.udl.es

Mónika Stéger-Máté (Chapter 35)
Corvinus University of Budapest, Faculty
of Food Science, Department of Food Preservation
Hungary 1118, Budapest, Ménesi út 45
Phone: 36-1-482-6034
Fax: 36-1-482-6327
E-mail: monika.stegernemate@uni-corvinus.hu

M. K. Suleiman (Chapter 24)
Arid Land Agriculture Department
Kuwait Institute for Scientific Research
P.O. Box 24885, 13109-Safat, Kuwait

H.S. Vibhakara (Chapter 11)
Fruits and Vegetables Technology
Defence Food Research Laboratory
Siddarthanagar, Mysore-570 011, India
Phone: 0821-247-3949
Fax: 0821-247-3468

Lorenzo Zacarías (Chapter 18)
Instituto de Agroquímica y Tecnología
de Alimentos (CSIC). Apartado Postal 73
46100 Burjasot, Valencia, Spain
Phone: 34 963900022
Fax: 34 963636301
E-mail: lzacarias@iata.csic.es or
cielor@iata.csic.es

Preface

In the past 30 years, several professional reference books on fruits and fruit processing have been published. The senior editor of this volume was part of an editorial team that published a two-volume text on the subject in the mid-nineties.

It may not be appropriate for us to state the advantages of our book over the others available in the market, especially in contents; however, each professional reference treatise has its strengths. The decision is left to the readers to determine which title best suits their requirement.

This book presents the processing of fruits from four perspectives: scientific basis; manufacturing and engineering principles; production techniques; and processing of individual fruits.

Part I presents up-to-date information on the fundamental aspects and processing technology for fruits and fruit products, covering:

- Microbiology
- Nutrition
- Heat treatment
- Freezing
- Drying
- New technology: pulsed electric fields
- Minimal processing
- Fresh-cut fruits
- Additives
- Waste management

Part II covers the manufacturing aspects of processed fruit products:

- Jams and jellies
- Fruit beverages
- Fruit as an ingredient
- A fruit processing plant
- Sanitation and safety in a fruit processing plant

Part III is from the commodity processing perspective, covering important groups of fruits such as:

- Apples
- Apricots
- Citrus fruits and juices
- Sweet cherries
- Cranberries, blueberries, currants, and gooseberries
- Date fruits
- Grapes and raisins, including juices and wine
- Olives
- Peaches and nectarines
- Pears
- Plums and Prunes
- Red pepper fruits
- Strawberries and raspberries
- Tropical fruits (guava, lychee, papaya, banana, mango, and passion fruit)

Although many topical subjects are included in our text, we do not claim that the coverage is comprehensive. This work is the result of the combined efforts of nearly fifty professionals from industry, government, and academia. They represent eight countries with diverse expertise and backgrounds in the discipline of fruit science and technology. An international editorial team of six members from four countries led these experts. Each contributor or editor was responsible for researching and reviewing subjects of immense depth, breadth, and complexity. Care and attention were paramount to ensure technical accuracy for each topic. In sum, this volume is unique in many respects. It is our sincere hope and belief that it will serve as an essential reference on fruits and fruit processing for professionals in government, industry, and academia.

We wish to thank all the contributors for sharing their expertise throughout our journey. We also thank the reviewers for giving their valuable comments on improving the contents of each chapter. All these professionals are the ones who made this book possible. We trust that you will benefit from the fruits of their labor.

We know firsthand the challenges in developing a book of this scope. What follows are the difficulties in producing the book. We thank the editorial and production teams at Blackwell Publishing and TechBooks, Inc. for their time, effort, advice, and expertise. You are the best judges of the quality of this work.

Y. H. Hui
J. Barta
M. P. Cano
T. W. Gusek
J. S. Sidhu
N. Sinha

Part I
Processing Technology

1
Fruit Microbiology

Anu Kalia and Rajinder P. Gupta

INTRODUCTION

Microbiology is the science that deals with the study of microscopic critters inhabiting planet earth, and of living organisms residing in earth. "Microbe" as a general term features amalgam of a variety of diverse microorganisms with the range spanning from electronmicroscopical cyrstallizable viruses, nucleoid–bearing unicellular prokaryotic bacteria to eukaryotic multicellular fungi and protists. The omnipresent feature of microbes advocates their unquestionable presence on external surface of plant and plant products, particularly skin of fruits and vegetables.

Fruits and vegetables are vital to our health and well being, as they are furnished with essential vitamins, minerals, fiber, and other health-promoting phytochemicals. The present health-conscious generation prefers a diet exhibiting low calories and low fat/sodium contents. A great importance of intake of fruits everyday has been found to half the risk of developing cancer and may also reduce the risk of heart disease, diabetes, stroke, obesity, birth defects, cataract, osteoporosis, and many more to count. Over the past 20 years, the consumption of fresh fruits and vegetables in industrialized countries has increased. However, this has also hiked the chances of outbreaks of food poisoning and food infections related to consumption of fresh fruits and uncooked vegetable salads. Many workers have described the changes that may contribute to the increase in diseases associated with the consumption of raw fruits and vegetables in industrialized countries and foods in general (Hedberg et al., 1994; Altekruse and Swerdlow, 1996; Altekruse et al., 1997; Potter et al., 1997; Bean et al., 1997). A healthy fruit surface harbors diverse range of microbes, which may be the normal microflora, or the microbes inoculated during the processing of fresh produce (Hanklin and Lacy, 1992; Nguyen and Carlin, 1994). However, the microflora could be plant pathogens, opportunistic pathogens, or non-plant pathogenic species. According to Center for Disease Control and Prevention (CDCP), among the number of documented outbreaks of human infections associated with consumption of raw fruits, vegetables, and unpasteurized fruit juices, more than 50% of outbreaks occur with unidentified etiological agents. These new outbreaks of fresh-produce-related food poisoning include major outbreaks by tiny culprits as *Escherichia coli* 0157:H7, *Salmonella*, *Shigella*, *Cyclospora*, Hepatitis A virus, Norwalk disease virus, on a variety of fruits as cantaloupes, apples, raspberries, and other fruits. Erickson and Kornacki (2003) have even advocated the appearance of *Bacillus anthracis* as a potential food contaminant. Factors include globalization of the food supply, inadvertent introduction of pathogens into new geographical areas (Frost et al., 1995; Kapperud et al., 1995), the development of new virulence factors by microorganisms, decreases in immunity among certain segments of the population, and changes in eating habits.

NORMAL MICROFLORA OF FRESH FRUITS

Fresh fruits have an external toughness, may be water proof, wax-coated protective covering, or skin that functions as a barrier for entry of most plant pathogenic microbes. The skin, however, harbors a variety of microbes and so the normal microflora of fruits is varied and includes both bacteria and fungi (Hanklin and Lacy, 1992). These microbes get associated with fruits, since a variety of sources such as the blowing air, composted soil, insects as *Drosophila melanogaster* or the fruit fly inoculate the skin/outer surface with a variety of Gram-negative bacteria (predominantly *Pseudomonas*, *Erwinia*, *Lactobacillus*). Likewise, hand-picking the fresh produce inoculates the fruit surfaces with *Staphylococcus*. Contact with soil, especially partially processed compost or manure, adds diverse human pathogenic microbes generally of the fecal-oral type including the *Enterobacter*, *Shigella*, *Salmonella*, *E. coli* 0157:H7, *Bacillus cereus*, as well as certain viruses such as Hepatitis A Virus, Rotavirus, and Norwalk disease viruses that are transmitted by consumption of raw fruits. Normal fungal microflora of fruits includes molds such as *Rhizopus*, *Aspergillus*, *Penicillum*, *Eurotium*, *Wallemia*, while the yeasts such as *Saccharomyces*, *Zygosaccharomyces*, *Hanseniaspora*, *Candida*, *Debaryomyces*, and *Pichia* sp. *are* predominant. These microbes are restrained to remain outside on fruit surfaces as long as the skins are healthy and intact. Any cuts or bruises that appear during the postharvest processing operations allow their entry to the less protected internal soft tissue.

NORMAL MICROFLORA OF PROCESSED FRUIT PRODUCTS

Postharvest processing methods include diverse range of physical and chemical treatments to enhance the shelf life of fresh produce. The minimally processed fresh-cut fruits remain in a raw fresh state without freezing or thermal processing, or addition of preservatives or food additives, and may be eaten raw or conveniently cooked and consumed. These minimally processed fruits are washed, diced, peeled, trimmed, and packed, which lead to the removal of fruit's natural cuticle, letting easy access by outer true or opportunistic normal microflora to the internal disrupted tissues abrassed during processing. Gorny and Kader (1996) observed that pear slices cut with a freshly sharpened knife retained visual quality longer than the fruits cut with a dull hand-slicer.

Rinsing of fresh produce with contaminated water or reusing processed water adds *E. coli* 0157:H7, *Enterobacter*, *Shigella*, *Salmonella* sp., *Vibrio chloreae*, *Cryptosporidium parvum*, *Giardia lamblia*, *Cyclospora caytanensis*, *Toxiplasma gondii*, and other causative agents of foodborne illnesses in humans, thus increasing the microbial load of the fresh produce that undergo further processing including addition of undesirable pathogens from the crop.

Fruits processed as fruit concentrates, jellies, jams, preserves, and syrups have reduced water activity (a_w) achieved by sufficient sugar addition and heating at 60–82°C, that kills most of xerotolerant fungi as well as restrains the growth of bacteria. Thus, the normal microflora of such diligently processed fruit products may include highly osmophilic yeasts and certain endospore-forming *Clostridium*, *Bacillus* sp. that withstand canning procedures. Similar flora may appear for processed and pasteurized fruit juices and nectars that loose most vegetative bacteria, yeasts, and molds while retaining heat-resistant ascospores or sclerotia producing *Paecilomyces* sp., *Aspergillus* sp., and *Penicillum* sp. (Splittstoesser, 1991). Recently, Walls and Chuyate (2000) reported the occurrence of *Alicyclobacillus acidoterrestris*, an endospore-forming bacteria in pasteurized orange and apple juices.

FACTORS AFFECTING MICROBIAL GROWTH

Fruits are composed of polysaccharides, sugars, organic acids, vitamins, minerals which function as eminent food reservoirs or substrates dictating the kind of microorganisms that will flourish and perpetuate in the presence of specific microflora and specific environmental prevailing conditions. Hence, one can predict the development of microflora on the basis of substrate characteristics. Fresh fruits exhibit the presence of mixed populations, and growth rate of each microbial type depends upon an array of factors that govern/influence the appearance of dominating population, which include the following.

INTRINSIC FACTORS

These imply the parameters that are an inherent part of the plant tissues (Mossel and Ingram, 1955) and thus are characteristics of the growth substrates that include the following.

Hydrogen Ion Concentration (pH)

Microbial cells lack the ability to adjust their internal pH, hence are affected by change in pH, so could grow best at pH values around neutral. Bacteria exhibit a narrow pH range with pathogenic bacteria being the most fastidious; however, yeasts and molds are more acid-tolerant than bacteria. Fruits possess more acidic pH (<4.4) favoring growth of yeasts and molds. Microbes, in general, experience increased lag and generation times at either extremes of the optimum pH range, which is usually quite narrow. The small fluctuations in pH have elaborate impact on microbial growth rates, and the pH changes become more profound if the substrate has low buffering capabilities leading to rapid changes in response to metabolites produced by microorganisms during fermentation (Table 1.1).

Adverse pH affects the functioning of respiring microbial enzymes and the transport of nutrients into the cell. The intracellular pH of microbial cytoplasm remains reasonably constant due to relative impermeability of cell membrane to hydrogen (H^+) and hydroxyl (OH^-) ions as key cellular compounds such as ATP and DNA require neutrality (Brown, 1964). The pH changes also affect the morphology of some microbes as *Penicillum chrysogenum* that show decreased length of hyphae at pH above 6.0. Corlett and Brown (1980) observed varying ability of organic acids as microbial growth inhibitors in relation to pH changes.

Table 1.1. Approximate pH Values of Some Fresh Fruits

Fruits	pH Values	Fruits	pH Values
Apples	2.9–3.3	Limes	1.8–2.0
Bananas	4.5–4.7	Melons	6.3–6.7
Grapefruit	3.4–4.5	Figs	4.6
Watermelons	5.2–5.6	Plums	2.8–4.6
Oranges	3.6–4.3		

Source: Adapted from Jay (1992).

Water Activity (Moisture Requirement)

Water is a universal constituent required by all the living cells, and microbes are no exceptions but the exact amount of water required for growth of microorganisms varies. Hence, several preservation methods involve drying or desiccation of the produce (Worbo and Padilla-Zakour, 1999). The water requirement of microbes is defined as water activity (a_w) or ratio of water vapor pressure of food substrate to that of vapor pressure of pure water at same temperature

$$a_w = \frac{p}{p_o},$$

where p is the vapor pressure of the solution and p_o is the vapor pressure of the solvent.

Christian (1963) related water activity to relative humidity as (Table 1.2)

$$\text{RH} = 100 \times a_w.$$

Thus, the relative humidity of a food corresponding to a lower a_w tends to dry the surface and vice versa. In general, most fresh produce has a_w value above 0.99 which is sufficient for the growth of both bacteria and molds; however, bacteria, particularly Gram-negative, are more stringent regarding a_w changes, while molds could grow at a_w as low as 0.80. The lowest range of permeable a_w values for halophilic bacteria, xerophilic fungi, and osmophilic yeasts is 0.75–0.61. Morris (1962) elaborated the interaction of a_w values with temperature and nutrition and observed that at optimum temperature, range of a_w values remain wide, while lowering/narrowing a_w values reduces growth and multiplication of microbes, and nutritive properties of substrate increase the range of a_w over which microorganisms can survive (Fig. 1.1).

Hence, each microbe has its own characteristic a_w range and optimum for growth and multiplication which are affected by temperature, pH, oxygen availability, and nutritive properties of substrate as well as the presence of organic acids or other secondary metabolites performing inhibitory action, thus narrowing the a_w range that culminates in decreased yield of cells and increased lag phase for growth,

Table 1.2. Lower Limit a_w Values of Certain Microorganisms

Bacteria	Minimum a_w Values	Fungi	Minimum a_w Values
Pseudomonas	0.97	*Mucor*	0.62 (0.94)
E. coli	0.96	*Rhizopus*	0.62
Staphylococcus aureus	0.86	*Botyritis*	0.62
Bacillus subtilis	0.95	*Aspergillus*	0.85
Clostridium botulinum	0.93	*Penicillum*	0.95
Enterobacter aerogenes	0.94		

Source: Adapted from Jay (1992).

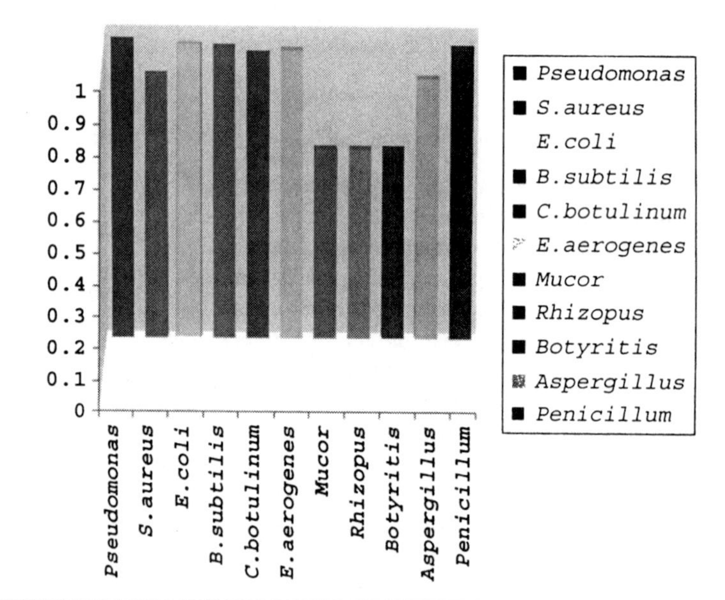

Figure 1.1. Graphical representation of a_W values of various microbes.

and results in decreased growth rate and size of final population (Wodzinsky and Frazier, 1961). Lowering of water activity builds up stress and exerts adverse influence on all vital metabolic activities that require aqueous environment. Charlang and Horowitz (1974) observed the appearance of non-lethal alterations in cell membrane permeability of *Neurospora crassa* cells resulting in loss of various essential molecules, as the dynamic cell membrane should remain in fluid state.

The exception to normal a_w requirements are basically the halophilic bacteria that grow under low a_w values by accumulating potassium ions in the cell (Csonka, 1989), while osmophilic yeasts concentrate polyols as osmoregulators and enzyme protectors (Sperber, 1983). Brown (1976) reported proline accumulation in response to low a_w in halotolerant *Staphylococcus aureus* strains. Xerotolerant fungi accumulate polyhydric alcohols (Troller, 1986). Microbes thus attempt to compensate for increased stress by accumulating compatible solutes.

Redox Potential/Redox Poising Capacity

The type of microbial growth depends upon oxidation and reduction power of the substrate. The oxidation–reduction potential of a substrate may be defined as the ease with which the substrate loses or gains electrons and, in turn, gets oxidized or reduced, respectively. Aerobic microbes require oxidized (positive Eh values) substrates for growth and it is reverse for the anaerobes (Walden and Hentges, 1975). The fruits contain sugars and ascorbic acid for maintaining the reduced conditions, though plant foods tend to have positive values (300–400 mV). Hence, aerobic bacteria and molds most commonly spoil fruits and fruit products. The O/R potential of food can be determined by

- Characteristic pH of food
- Poising capacity
- Oxygen tension of the atmosphere
- Atmospheric access of food

Poising capacity could be defined as the extent to which a food resists externally effected changes in pH that depend on the concentration of oxidizing or reducing compounds in the substrate. The capacity alters the ability of the living tissues to metabolize oxygen at specifically low Eh values that exist in the vacuum-packed foods. Aerobic microbes include bacilli, micrococci, pseudomonas, and actinobacters, and require positive Eh values, while anaerobes such as clostridia and bacteriodes require negative Eh values. However, most yeast and molds are aerobic and few tend to be facultative anaerobes. In the presence of limited amounts of oxygen, aerobic or facultative microbes may produce incompletely oxidized organic acids. Processing procedures such as heating or pasteurization, particularly of fruit juices, make microbes devoid of reducing substances, but favorable for the growth of yeasts.

Available Nutrients

Fruits as substrate act as a reservoir of sugars (source of energy), water, minerals, vitamins, and other growth-promoting factors, while the protein content or nitrogen source appears to be little less in fruits. Carbohydrates include sugars or other carbon sources that act as sources of energy because breakage of bonds or oxidation of these compounds helps in the formation of energy currency of cell or ATP.

Microorganisms have varied nutrient requirements, which are influenced by other conditions such as temperature, pH and Eh values. The microbes become more demanding at decreased temperatures, while under optimum temperature conditions, nutrients control the microbial growth only when present in limiting quantities. Thus, microorganisms that grow on a product become the best-suited by exploiting the product, as pectinolytic bacteria such as *Erwinia cartovora*, *Pseudomonas* sp., or pectinolytic molds grow best on fruits and vegetables.

Nitrogen requirement is usually fulfilled by proteolysis of protein present in substrate and the use of amino acids, nucleotides, certain polysaccharides, and fats under usual microbe-specific conditions. The accessory food substances or vitamins are to be furnished by substrate since microorganisms are unable to synthesize essential vitamins. In general, Gram-positive bacteria are least synthetic and require supply of certain vitamins before growth, while Gram-negative bacteria and molds are relatively independent and could synthesize most of the vitamins. Thus, these two groups of microbes grow profusely on foods relatively low in B-complex vitamins such as fruits under the influence of usual low pH and positive Eh values.

Each microbe has a definite range of food requirements, with some species having wide range and ability to grow on a variety of substrates, while others having narrow range and fastidious requirement allowing growth on limited substrates.

Antimicrobial Factors

Certain naturally occurring substances in substrate (food) work against the microbes, thus maintaining stability of food; however, these are directed toward a specific group of microorganism and have weak activity. Song et al. (1996) reported that the presence of aroma precursor Hexal readily gets converted to aroma volatiles *in vivo* by fresh-cut apple slices. Hexal acts as antibrowning agent as well as inhibits growth of molds, yeasts, mesophilic and psychrotropic bacteria (Lanciotti et al., 1999). Hexanal and (E)-Hexenal in modified atmosphere packaging (MAP) of sliced apples reduce spoilage microbe populations (Corbo *et al.*, 2000).

Spices contain essential oils such as eugenol (clove), allicin (garlic), cinnamic aldehyde and eugenol (cinnamon), allyl isothiocynate (mustard), eugenol and thymol (sage), thymol and isothymol (oregano) that have antimicrobial activity (Shelef, 1983). Buta and Molin (1998) observed reduction in mold growth on fresh-cut peppers by exogenous application of methyl jasmonate.

The antimicrobial compounds may originally be present in food, added purposely or developed by associated microbial growth, or by processing methods. Certain antifungal compounds applied to fruits include benomyl, biphenyl, and other phenylic compounds that exist in small quantities as by-product of phenol synthesis pathways. Beuchat (1976) observed that essential oils of oregano, thyme, and sassafras have bacteriocidal activity, at 100 ppm, to *V. parahaemolyticus* in broth, while cinnamon and clove oils at 200–300 ppm inhibit growth and aflatoxin production by *Aspergillus parasiticus* (Bullerman et al., 1977). The hydroxy-cinnamic acid derivatives as *p*-coumaric, ferulic, caffeic, and chlorogenic acids and benzoic acid in cranberries have antibacterial and antifungal activities and are present in most plant products including fruits.

Extrinsic Factor

Extrinsic factors include parameters imposed from the external environment encountered during storage that affect food, and the microbes that tend to develop on it. These factors include the following.

Temperature of Storage

Microbes grow over a wide range of temperature, and change in temperature at both extremes lengthens the generation time and lag periods. The range is quite wide from −34°C to 90°C, and according to range microbes could be grouped as follows.

Psychrotrophs. These microorganisms grow well at 7°C or below 7°C with the optima ranging from 20°C to 30°C. For example, *Lactobacillus*, *Micrococcus*, *Pseudomonas*, *Enterococcus*, *Psychrobacter*, *Rhodotorula*, *Candida* and *Saccharomyces* (yeasts), *Mucor*, *Penicillum*, *Rhizopus* (molds) and *Clostridium botulinum*, *Listeria monocytogenes*, *Yersinia enterocolitica*, *Bacillus cereus* (pathogenic psychrotrophs). The group of microbes that grow from −10°C to 20°C with the optima of 10–20°C are included as *Psychrophiles* and include certain overlapping genera mentioned above.

Mesophiles. These include microbes growing best between 20°C and 45°C with optimum range of 30–40°C. For example, *Enterococcus faecalis*, *Streptococcus*, *Staphylococcus*, and *Leuconostoc*.

Thermophiles. Microbes that grow well above 45°C with the optima ranging between 55°C and 65°C and with maximum of above 60–85°C are known as thermotolerant thermophiles. For example, *Thermus* sp. (extreme thermophile), *Bacillus sternothermophilus*, *Bacillus coagulans*, *Clostridium thermosaccharolyticum* are endospore-forming thermotolerants and grow between 40°C and 60°C and create major problems in the canning industry.

Thermotrophs. This group includes microbes similar to mesophiles but grows at slightly higher temperature optima and includes pathogenic bacteria in foods. For example, *Salmonella*, *Shigella*, enterovirulent *E. coli*, *Campylobacter*, toxigenic *Bacillus cereus*, *Staphylococcus aureus*, and *Clostridium perfringens*. There exists a relation of temperature to growth rate of microorganisms between minimum and maximum temperature range by (Ratowsky et al., 1982)

$$\sqrt{r} = b(T - T_0),$$

where r is the growth rate, b is the slope of regression line, and T_0 is the conceptual temperature of no metabolic significance.

Relative Humidity of Environment

Success of a storage temperature depends on the relative humidity of the environment surrounding the food. Thus, relative humidity affects a_w within a processed food and microbial growth at surfaces. A low a_w food kept at high R.H. value tends to pick up moisture until the establishment of equilibrium, and foods with high a_w lose moisture in a low-humidity environment. Fruits and vegetables undergo a variety of surface growth by yeasts and molds as well as bacteria, and thus are liable to spoilage during storage at low R.H. conditions. However, this practice may cause certain undesirable attributes such as firmness and texture loss of the climacteric (perishable) fruits calling for the need of altered gas compositions to retard surface spoilage without lowering R.H. values.

Modified Atmosphere Storage

Altering the gaseous composition of the environment that retards the surface spoilage without reducing humidity includes the general practice of increasing CO_2 (to 10%) and is referred as "controlled or modified atmosphere" (MA). MA retards senescence, lowers respiration rates, and slows the rate of tissue softening or texture loss (Rattanapanone and Watada, 2000; Wright and Kader 1997a; Qi et al., 1999). MA storage has been employed for fruits (apples and pears) with CO_2 applied mechanically or as dry ice, and this retards fungal rotting of fruits probably by acting as competitive inhibitor of ethylene action (Gil et al., 1998; Wright and Kader 1997b).

The inhibitory effect increases with decrease in temperature due to increase in solubility of CO_2 at lower temperatures (Bett et al., 2001). Elevated CO_2 levels are generally more microbiostatic than microbiocidal probably due to the phenomena of catabolite repression. However, an alternative to CO_2 application includes the use of ozone gas at a few ppm concentration that acts as ethylene antagonist as well as a strong oxidizer that retards microbial growth. Sarig et al. (1996) and Palou et al. (2002) reported control of postharvest decay of table grapes caused by *Rhizopus stolonifera*. A similar report on effect of ozone and storage temperature on postharvest diseases of carrots was observed by Liew and Prange (1994). In general, gaseous ozone introduction to postharvest storage facilities or refrigerated shipping and temporary storage containers is reported to be optimal at cooler temperatures and high relative humidity (85–95%) (Graham, 1997). The most reproducible benefits of such storage are substantial reduction of spore production on the surface of infected produce and the exclusion of secondary spread from infected to adjacent produce (Kim et al., 1999; Khadre and Yousef, 2001).

Ozone treatment has been reported to induce production of natural plant defense response compounds involved in postharvest decay resistance. Ozone destruction of ethylene in air filtration systems has been linked to extended storage life of diverse ethylene-sensitive commodities.

Implicit Factors

Implicit factors include the parameters depending on developing microflora. The microorganisms while growing in food may produce one or more inhibitory substances such as acids, alcohols, peroxides, and antibiotics that check the growth of other microorganisms.

General Interference

This phenomena works when competition occurs between one population of microbes and another regarding the supply of the same nutrients. Normal microflora of fresh produce helps prevent the colonization of pathogens and succeeds in overcoming the contaminant number by overgrowth and efficient utilization of available resources.

Production of Inhibitory Substances

Some microbes can produce inhibitory substances and appear as better competitors for nutrient supply. The inhibitory substances may include "bacteriocins," the commonest being "nisin" produced by certain strains of *Lactobacillus lactis*, which is heat stable, attached by digestive enzymes, labile and non-toxic for human consumption, and is quite targeted toward inhibition of a narrow spectrum of microbes. Other bacteriocins produced by lactic acid bacteria include lactococcins, lacticins, lactacins, diplococcin, sakacins, acidophilocins, pediocins, and leuconosins. As an inhibitor of spore-forming *Clostridium* spp., which cause cheese blowing due to undesirable gas production, nisin was the first bacteriocin produced by lactic acid bacteria to be isolated and approved for use in cheese spreads. Although mostly active against Gram-positive bacteria, bacteriocins can be microbiocidal under certain conditions, even toward Gram-negative bacteria and yeasts, provided that their cell walls have been sensitized to their action. The antimicrobial action of nisin and of similar bacteriocins is believed to involve cell membrane depolarization leading to leakage of cellular components and to loss of electrical potential across the membrane. *Propioniobacterium* produces propionic acid that has inhibitory effect on other bacteria. Certain microorganisms may produce wide spectrum antimicrobial substances or secondary metabolites capable of killing or inhibiting wide range of microbes called "antibiotics." However, growth of one kind of microbe could lead to lowering of pH of substrate, making the environment unsuitable for other microbes to grow, while organic acid production or hydrogen peroxide formation could also interfere with the growth of background microbial population (Jay, 1992).

Biofilm Formation

Most of the Gram-negative bacteria exhibit quorum sensing or the cell-to-cell communication phenomena that leads to the formation of a multicellular structure in the life of a unicellular prokaryote that provides protection to bacterial species from the deleterious environment by precipitation. Adoption of biofilm formation involves release of autoinducers, particularly called the *N*-acyl homoserine lactones that either activate or repress the target genes involved in biofilm formation (Surette et al., 1999). Quorum sensing has a profound role in food safety in association with behavior of bacteria in food matrix and regulates prime events such as spore germination, biofilm formation on surfaces (Frank, 2000b), and virulence factor production. Cells in biofilm are more resistant to heat, chemicals, and sanitizers due to diffusional barrier created by biomatrix as well as very slow growth rates of cells in biofilms (Costerton, 1995). Morris et al. (1997) have reported certain methods for observing microbial biofilms directly on leaf surfaces and also to recover the constituent microbes for isolation of cultivable microorganisms. Thus, biofilm formation has been emerging as a challenge for the decontamination techniques routinely used in the food and beverage industries, and requires the advent of new revolutionary methods for decontamination or the modification of the older techniques in vision of the current scenario (Frank, 2000a).

FACTORS AFFECTING MICROBIAL QUALITY AND FRUIT SPOILAGE

From quality standpoint, the fresh fruits and the processed fruit products should possess certain characteristics such as fresh-like appearance, taste, aroma, and flavor that should be preserved during storage. Thus, if the primary quality attributes of produce remain unoffended, the shelf-life characteristics lengthen. As discussed before, fruits possess normal microflora as well as the microflora that is added during the handling and postharvest processing of fruits, though harsh treatments during processing can kill or inhibit certain or most of the microflora while letting specific types to become predominant and prevail in the finished product. A variety of factors that affect the microbial quality of fruits include the following.

PREHARVEST FACTORS

These factors basically involve production practices that have tremendous explicit effect on the microbial quality of fruits. Management practices can affect product quality since stressed produce or mechanical injuries permit microbial contamination. Mold growth and decay on winter squash caused by *Rhizoctoina* result from fruits lying on the ground. Food safety begins in field as a number of foodborne disease outbreaks have potential sources in field that contaminate the fresh produce such as the use of partially treated manure, irrigation with livestock-used farm pond water, or storage near roosting birds (Trevor, 1997). Wallace et al. (1997) reported the presence of verocytotoxin producing *E. coli* O157:H7 from wild birds.

POSTHARVEST HANDLING AND PROCESSING

Improper or harsh handling of produce causes skin breaks, bruises, or lesions leading to increased chances of microbial damage. Handlers picking fresh produce with skin lesions could potentially transfer the causative agent to other fruits. The postharvest rots are most prevalent in fruits, particularly the damaged or bruised ones (Sanderson and Spotts, 1995; Bachmann and Earles, 2000). The processing methods involve the use of temperature, moisture content, and ethylene control, thus include the extrinsic parameters discussed earlier.

FRUIT SPOILAGE

The fruit spoilage is manifested as any kind of physical change in color or flavor/aroma of the product that is deteriorated by microflora that affects the cellulose or pectin content of cell walls which, in turn, is the fundamental material to maintain the structural integrity of any horticultural product. Fresh fruits possess more effective defense tactics including the thicker epidermal tissue and relatively higher concentration of antimicrobial organic acids. The higher water activity, higher sugar content, and more acidic pH (<4.4) of fresh fruits favor the growth of xerotolerant fungi or osmophilic yeasts. Lamikarna et al. (2000) have reported bacterial spoilage in neutral pH fruits.

Normal microflora of fruits is diverse and includes bacteria such as *Pseudomonas*, *Erwinia*, *Enterobacter*, and *Lactobacillus* sp. (Pao and Petracek, 1997), and a variety of yeasts and molds. These microbes remain adhered to outer skin of fruits and come from several sources such as air, soil, compost, and insect infestation. Brackett (1987) reported inoculation of *Rhizopus* sp. spores by egg laying in ruptured epidermal fissures of fruits by *Drosophila melanogaster* or the common fruit fly. The microbial load of the fresh produce could be reduced by rinsing with water (Splittstoesser, 1987). However, the source and quality of water dictate the potential for human pathogen contamination upon contact with the harvested produce.

Lund and Snowdon (2000) reported certain common molds to be involved in fruit spoilage such as *Penicillum* sp., *Aspergillus* sp., *Eurotium* sp., *Alternaria* sp., *Cladosporium* sp., and *Botrytis cinerea* of fresh and dried fruits (Fig. 1.2), while certain molds producing heat-resistant ascospores or sclerotia such as *Paecilomyces fulvus*, *P. niveus*, *Aspergillus fischeri*, *Penicillum vermiculatum*, *and P. dangeardii* were observed to cause spoilage of thermally processed fruits or the fruit products exhibiting characteristic production of off-flavors, visible mold growth, starch and pectin solubilization, and fruit texture breakdown (Beuchat and Pitt, 1992; Splittstoesser, 1991).

Figure 1.2. Degradation of fruit texture due to growth of cellulase/pectinase-producing bacteria followed by fungal growth.

Fruit safety risks could be increased by certain spoilage types that create microenvironments suitable for the growth of human pathogens as the primary spoilage by one group of phytopathogens produces substances required for nurturing growth and development of human pathogens. Wade and Beuchat (2003) have well documented the crucial role of proteolytic fungi and the associated implications on the changes in pH of the pericarp of the decayed and damaged raw fruits in survival and growth of various foodborne pathogens. *Botrytis* or *Rhizopus* spoilage of fruits could help create environment for the proliferation of *Salmonella enterica serovar typhimurium* (Wells and Butterfield, 1997), while Dingman (2000) observed the growth of *E. coli* 0157:H7 in bruised apple tissues. Similar reports of Riordan et al. (2000) and Conway et al. (2000) depicted the impact of prior mold contamination of wounded apples by *Penicillum expansum* and *Glomerella cingulata* on survival of *E. coli* 0157:H7 and *Listeria monocytogenes.*

Technically, the fresh produce deteriorating microflora is diverse and remains on surface skin of fruits, and the basis of invasion process could be of two types.

True Pathogens

These microbes possess ability to actively infect plant tissues as they produce one or several kinds of cellulytic or pectinolytic and other degradative enzymes to overcome tough and impervious outer covering of fruits which acts as the first and the foremost effective external protective system, thus causing active invasion and active spoilage in fruits. The degradative enzyme brigade includes the following.

Pectinases

These enzymes depolymerize the pectin, which is a polymer of α-1, 4-linked D-galactopyranosyluronic acid units interspersed with 1, 2-linked rhamnopyranose units. On the basis of site and type of reaction on the pectin polymer, pectinases are of three main types, i.e., pectin methyl esterases produced by *Botrytis cinerea, Monilinia fructicola, Penicillum citrinum, and Erwinia cartovora* (Cheeson, 1980), polygalacturonase, and pectin lyase.

Cellulases

Several types of cellulase enzymes attack the native cellulose and cleave the cross-linkage between β-D-glucose into shorter chains. Cellulases contribute toward tissue softening and maceration as well as yield glucose, making it available to opportunistic microflora.

Proteases

These enzymes degrade the protein content of fresh produce giving simpler units of polypeptides, i.e., amino acids. The action of proteases is limiting in fruit spoilage as fruits are not rich in proteins.

Phosphatidases

These enzymes cleave the phosphorylated compounds present in cell cytoplasm and the energy released is utilized to cope with the increased respiration rates.

Dehydrogenases

These enzymes dehydrogenate the compounds, thus increasing the amount of reduced products that may lead to increased fermentation reaction under microaerobic/anaerobic conditions.

Opportunistic Pathogens

These microorganisms lack the degradative enzyme brigade and thus gain access only when the normal plant product defense system weakens, which is the situation of mechanical injury or cuticular damage caused by the insect infestation or by natural openings present on the surface of the fresh produce. Thus,

Figure 1.3. Growth of *Aspergillus* on surface of apple fruits visible due to formation of spores.

an opportunistic pathogen slips in through the damage caused by biotic and abiotic stresses on the produce and generally involves movement via natural gateways as the lenticels, stomata, hydathodes, or the other pores/lesions caused by insect infestation or invasion by true pathogens. Damage of the product during harvesting or by postharvest processing techniques and equipments enables opportunistic microflora to invade the internal unarmed tissue and causes spoilage (Fig. 1.3).

Hence, spoilage connotes any physical change in color, taste, flavor, texture, or aroma caused by microbial growth in fruit/fruit product, thereby resulting in product that becomes unacceptable for human consumption (Fig. 1.4).

MODES OF FRUIT SPOILAGE

Fruit spoilage occurs as a result of relatively strong interdependent abiotic and biotic stresses posed particularly during the postharvest handling of produce (Fig. 1.5). Harvested fruits continue to respire by utilizing the stored available sugars and adjunct organic acids culminating to significant increase in stress-related/stress-induced carbon dioxide and ethylene production that leads to rapid senescence (Brecht, 1995). Moreover, postharvest processing that involves washing, rinsing, peeling, and other treatments result in major protective epidermal tissue damage and disruption which in turn leads to unsheathing of the vacuole-sequestered enzymes and related substrates and their amalgamation with the cytoplasmic contents. Cutting/dicing increases the a_w and surface area as well as stress-induced ethylene production which accelerates the water loss, while the sugar availability promptly invites enhanced microbial invasion and rapid growth (Wiley, 1994; Watada and Qi, 1999). The physiological state of fruit also determines the pattern of spoilage to be followed as with increase in age/maturity, the normal defense tactics of the plant produce deteriorates. Harvested produce loses water by transpiration, thus gets dehydrated, followed by climacteric ripening, enzymatic discoloration of cut surfaces to senescence, thus increasing possibilities of damage by microflora (Fig. 1.6). Harsh handling and ill-maintained equipment during processing lead to increased damage or

Figure 1.4. Fungal hyphae and spores of *Aspergillus niger* on guava fruits.

Figure 1.5. White hyphal mass of *Aspergillus fumigatus* on surface of orange fruit.

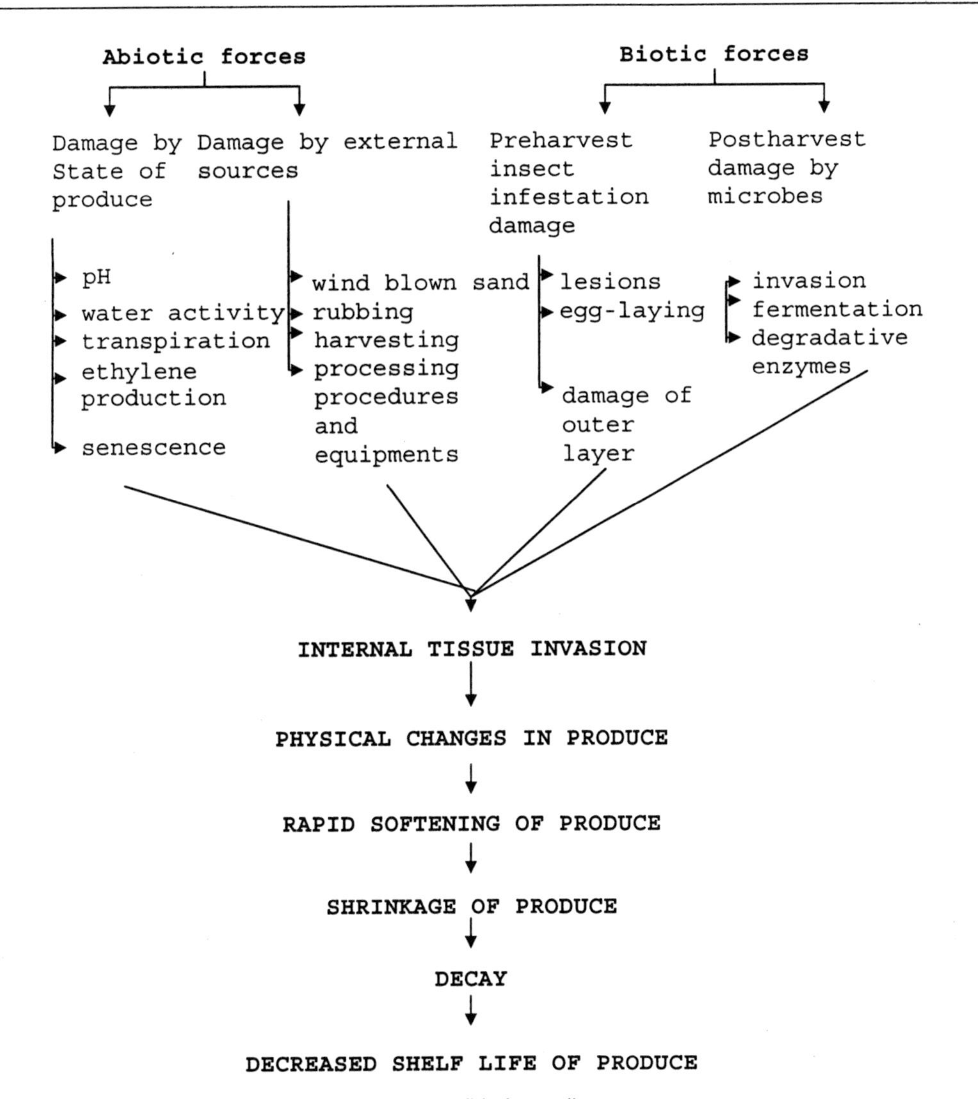

Figure 1.6. Modes of fruit spoilage and factors responsible for spoilage.

removal of the outer cuticle leading to tissue disruption that provokes stress-induced increased respiration and microbial decay (Gorny and Kader, 1996). Spanier et al. (1998) reported the development of off-flavors in fresh-cut pineapple that appeared undamaged physically, in lower portion of container kept at 4°C for 7–10 days. Walls and Chuyate (2000) reported survival of acid- and heat-tolerant *Alicyclobacillus acidoterrestris* that produces 2-methoxy phenol or guaiacol imparting phenolic off-flavor in pasteurized orange and apple juices. Jay (1992) reported osmophilic yeasts to be associated primarily with the spoilage of cut fruits due to their ability to grow faster than the molds and this usually includes the genera such as *Cryptococcus*, *Rhodotorula*, and *Saccharomyces* sp. in fresh fruits, and *Zygosaccharomyces rouxii*, *Hanseniaspora*, *Candida*, *Debaryomyces*, and *Pichia* sp. in dried fruits.

Thus, senescence and spoilage depend on product type, abiotic factors, and microbes involved in deterioration process, and it is convenient to describe spoilage on the basis of visible symptoms. Thus, a customary approach is to name the spoilage type by symptomatological appearance such as soft rot or black rot. However, this definitely results in discrepancy in ascertaining the causal pathogen of spoilage and this ambiguity could be overruled by classifying on the basis of causal microbe such as *Rhizopus* rot, *Cladosporium* rot, etc.

METHODS TO EVALUATE MICROBIAL QUALITY

Food quality and safety are ensured by analysis of food for the presence of microbes, and such microbial analyses are routinely performed as quarantine/regulatory procedures. The methods employed for adjudging the quality of food include an array of microbiological to biochemical assays to ascertain the acceptability or unacceptability of a food product for human consumption or a processing/handling practice that needs to be followed. Thus, enumerating the microbial load of the produce could help in determining the quality as well as the related safety aspects of product and effectiveness of the processing technique employed to kill spoilage microbes.

Microbiological methods for pathogen identification primarily involve conventional cultural techniques of growing microbes on culture media and observing the ability to form viable countable colonies showing characteristic growth on such media as well as the direct microscopic methods for various groups of microbes.

Hence, microbiological criteria are specifically employed to assess:

- Safety of food
- Shelf life of perishable products
- Suitability of food or ingredient for specific purpose
- Adherence to general manufacturing practices

The routine culturing techniques require longer time to obtain results. To overcome this hurdle, nowadays, use of indicator organisms that provide rapid, simple, and reliable information without the requirement of isolation and identification of specific pathogens is performed. However, such tests could be used as the presumptive ones with the confirmation provided by a battery of biochemical tests, and may include specialized serological typing also (Swaminathan and Feng, 1994). The microbiological techniques could be summarized as follows.

Conventional Techniques

Direct Microscopic Count

This method involves the microscopic examination for evaluating the viable or non-viable microbes present in a given sample. This method ushers little value for the determination of microbiological status of a food sample as usually total cell counts exceed 10^5 cfu per g or ml of the sample. New variations of microscopes render researchers the capability to predict the presence of pathogens on the surfaces of fruits clinging or attached to internal surfaces. Confocal scanning laser microscopy has been reported to show the presence of *E. coli* 0157:H7 on surfaces and internal structures of apple (Burnett et al., 2000).

Drawbacks: This technique suffers from a major drawback of not providing the types of bacteria present in the sample as well as it does not differentiate between the normal microflora and the pathogen-causing spoilage.

Aerobic Plate Counts (APC) or Total Plate Counts (TPC)

It is the most practical approach to determine the presence of cultivatable microbes in a sampled food product having ability to spoil food. This technique, thus, reveals the total number of microbes in a food product under a particular set of incubation temperature, time, or culture media and can be used to preferentially screen out a specific group of microbes, thereby, helping in determining the utility of food or food ingredient added for specific purpose. However, the APC of the refrigerated fruits/fruit products indicate utensil or equipment conditions prevailing during storage and distribution of the product.

Drawbacks: Though APC bacterial count is the most practical and easy technique, it suffers from certain inherent drawbacks as listed below:

- It provides the viable cell count that does not reflect the quality of raw material used for processing.
- It is unable to record the extent of quality loss at low count levels.
- It provides negligible information regarding organoleptic quality that is degraded at low counts.
- It requires scrupulous researcher to interpret APC results.

Certain variations to APC method are now available to classify according to the types of microbes as molds, yeasts, or thermophilic spore counts. These counts are basically used for microbiological analysis of the canned fruits/fruit products.

1. *Howard Mold Count*. This technique involves the enumeration of molds in products such as the canned fruits and provides the inclusion of the moldy material.
2. *Yeasts and Mold Counts*. The high sugar products such as fruit drinks or fruit beverages are prone to contamination and overgrowth by yeasts and molds more than the bacterial counterparts and thus enumeration of these microbes gives the presumptive glimpse of the microbiological status of the product. A similar kind of count involves the heat-resistant mold count providing the presence of molds such as *Aspergillus fischeri* and *Byssochlamys fulva* in heat-processed fruit products such as the fruit concentrates.
3. *Thermophilic Spore Count*. The technique again advocates the presence of spore-forming bacteria as the major contaminants of canned fruits, fruit beverages, and fruit juices that are being thermally processed by pasteurization and thus specifically enriches the spore-forming genera.

New Methods for Rapid Analysis

The physical characteristics of food result in non-uniform distribution of microbes and thus such a non-uniform homogenate results in inconsistent presence of specific pathogen providing non-reproducible results following the analysis of the same sample. Thus, the drawbacks of the conventional microbiological analysis criteria are:

- Requirement of the selective or enrichment media for isolation of foodborne pathogen suffers from involvement of several days to provide results.
- Normal microflora interferes with the isolation and identification protocols of low infectious dose and low number pathogens that may be sub-lethally injured during the accomplishment of a variety of processing procedures employed. These microorganisms that exist in state of shock after vigorous heat/chemical/radiation treatments need specific enriched culture media to overcome the shock (Jiang and Doyle, 2003). Thus, unless the injured cells could resuscitate, they could be easily outgrown by other bacteria in the sample. Zhao and Doyle (2001) have reported the use of a universal pre-enrichment broth for growth of heat-injured pathogens in food.

Hence, these rapid methods shorten the assay time by a simple modification of conventional methods or may also involve an array of molecular assay formats and diverse technologies that are quite specific and more sensitive (Mermelstein et al., 2002). Some of the assays involved in the rapid enumeration of pathogens in food samples are as follows.

Modification of Conventional Techniques

- *Miniaturized Biochemical Assays*: The use of certain biochemical test kits for identification of pure cultures of bacterial isolates delivers results in less than 1 day with high accuracy of 90–99% comparable to conventional techniques making the procedure simpler, cost- and performance-effective (Hartman et al., 1992).
- *Modified Process/Specialized Media*: Use of petrifilms (Curiale et al., 1991) and hydrophobic grid membrane filters eliminates the need for media preparation, thus economizes storage and incubation space as well as simplifies disposal after analysis while the use of chromogenic (ONPG/X-gal) or fluorogenic (MUG/GUD) substances provides quick measure of specific enzyme activities to quickly ascertain the presence of a specific microbe, and the bioluminescence assays provide quick assessment of direct live cell counts with sensitivity to provide results with low counts within few minutes.

DNA-Based Assays

Use of DNA probes technically fishes out the target gene sequence specific to a particular pathogenic microbe in the concoction of sample DNA obtained from the food sample with unique sensitivity and reproducibility, and has been developed for detection of most of the foodborne pathogens (Guo et al., 2000; Feng et al., 1996; Lampel et al., 1992; Saiki et al., 1988; Schaad et al., 1995). However, if the target DNA contains several targets, then PCR assays can be used in a multiplex format that ensures the elimination of culturing steps prior to producing the results (Chen and Griffith, 2000; Hill, 1996; Jones and Bej, 1994). PCR protocols can detect very small number/few cells of particular pathogens and have been successfully developed for various fastidious/uncultivatable pathogens (Guo et al., 2000, 2002). DNA fingerprinting methods are the most recent ones for the detection of pathogens in fresh produce and a semi-automated fluorescent AFLP technique for genomic typing of *E. coli* 0157:H7 has been developed (Zhao et al., 2000). Another report of occurrence of *Acidovorax avenae* subsp. *citrulli* in watermelon seeds has been provided by Walcott and Gitaitis (2000).

Antibody-Based Assays

These include the classical agglutination assays as well as the immunodiffusion techniques that are rather simple, quick, and useful methods for confirmation of microbial isolates from food sample but possess low sensitivity. Hence, the new immunological protocols hail the use of ELISA (basic sandwich ELISA method) scoring high sensitivity (Candish, 1991) and immunoprecipitation techniques that provide the results within few minutes as these are automated requiring less manual expertise.

Other Techniques

These rather unconventional methods involve the use of immunomagnetic separations, chromatographic detection of certain organic acids produced by the pathogen during growth and recent techniques as the flow cytometery for deciphering the survival and growth of human fecal-oral pathogens in raw produce. Orr et al. (2000) have detected the presence of *Alicyclobacillus acidoterrestris* in apple juice by sensory and chromatographic analysis of compounds produced by bacteria. The magnetic separation technique is now being employed in both clinical and food microbiology (Olsvik et al., 1994; Safarik and Safarikova, 1999; Bennett et al., 1996). Jung et al. (2003) have used immunomagnetic separation technique in conjunction with flow cytometery to detect the presence of *Listeria monocytogenes* in food.

MAINTAINING MICROBIAL QUALITY OF FRUITS

The microbial quality of fruits or fruit products needs to be maintained at various levels of processing and packaging. Production practices have a tremendous effect on the quality of fruits at harvest, on postharvest quality, and on shelf life. Cultivar or fruit variety, abiotic or environmental factors such as soil type, temperature, frost, and rainy weather at harvest may adversely affect the storage life and quality of produce. Fresh produce that has been stressed by too much or too little water, high rates of nitrogen application, or mechanical injury (scrapes, bruises, abrasions) is particularly susceptible to postharvest diseases. Mold decay on winter squash, caused by the fungus *Rhizoctonia*, results from the fruits lying on the ground, which can be alleviated by using mulch. Evidently, food safety also begins in the field, and should be of special concern, since a number of outbreaks of foodborne illnesses have been traced to the contamination of produce in the field. Management practices such as unscrupulous picking and harsh handling of the fresh produce markedly affect the quality of fruits (Beaulieu et al., 1999). Crops destined for storage should be as free as possible from skin breaks, bruises, spots, rots, decay, and other deterioration. Bruises and other mechanical damage not only affect appearance, but also provide entrance to the decay organisms as well. Postharvest rots are more prevalent in fruits that are bruised or otherwise damaged. Moreover, mechanical injury also increases moisture loss that may hike up to 400% in a single badly bruised apple.

POSTHARVEST AND STORAGE CONSIDERATIONS

The fresh produce once harvested has to be stored for shipment and this is the critical period that exhibits most of the loss regarding microbial decay and spoilage of produce. The extrinsic factors governing microbial growth play an important role during this critical period and involve temperature and water activity.

Temperature

Temperature is the single most important factor in maintaining fruit quality after harvest. Refrigerated storage retards the following elements of deterioration in perishable crops:

- Aging due to ripening, softening, and textural/color changes
- Undesirable metabolic changes and respiratory heat production
- Moisture loss/wilting
- Spoilage due to invasion by bacteria/fungi/yeasts

Refrigeration controls the respiration rate of crop, which is evil enough as this generates heat due to oxidation of sugars, fats, and proteins in the cells resulting in loss of these stored food reserves leading to decreased food value, loss of flavor, loss of saleable weight, and more rapid deterioration. Recent work of Sharma et al. (2001) has provided the insight about the fate of *Salmonellae* in calcium-supplemented orange juice at refrigeration temperature. Since the respiration rate of fruits strongly determines their transit and postharvest life, a constant cold temperature maintained over a span of storage period decreases the deterioration; however, the produce has to be precooled to relieve the field heat (heat held from sun and ambient temperature) by an array of methods such as room cooling, forced air cooling, vacuum cooling, hydrocooling, and top or ice cooling.

However, during refrigeration certain fruits having higher water content get injured over a time period (chilling injury) but store best at 45–55°F. The effect of chilling injury may be cumulative in some crops with the appearance of chilling symptoms becoming evident as pitting or other skin blemishes, internal discoloration, or failure to ripen. Fruits such as muskmelons, peppers, winter squash, tomatoes, and watermelons are moderately sensitive to chilling injury, but if tomatoes, squash, and peppers are over-chilled, then they may particularly become more susceptible to decay by fungal genera such as by *Alternaria*.

Regulation of Water Activity

Transpiration rates are determined by the moisture content of the air, which is usually expressed as relative humidity. Water loss at low R.H. values can severely degrade quality since sugar-rich perishable fruits such as grapes may shatter loose from clusters due to drying out of their stems and this would decrease the aesthetic value of the product as well as saleable weight loss culminating in reduced profits. Thus, the relative humidity of the storage unit directly influences water loss in fresh produce. Most fruit and vegetable crops retain better quality at higher relative humidity (80–95%) maintaining saleable weight, appearance, nutritional quality and flavor, and reduction in wilting, softening, and juiciness but it encourages disease growth. This situation could be overruled by storage at cool temperatures but stringent sanitary preventative protocols have to be enforced. Unfortunately, refrigeration inevitably extracts moisture from fruit surfaces, thus necessitating the use of proper packaging.

Control of Respiration and Ethylene Production

Ethylene, a natural phytohormone, produced by some fruits upon ripening promotes additional ripening of produce exposed to it (Gorny et al., 1999). Damaged or diseased apples produce high levels of ethylene and stimulate the other apples to ripen too quickly, making them more susceptible to diseases. Ethylene "producers" such as apple, apricot, avocado, ripening banana, cantaloupe, honeydew melon, ripe kiwifruit, nectarine, papaya, passionfruit, peach, pear, persimmon, plantain, plum, prune, quince, and tomato show decreased quality, and reduced shelf life with appearance of specific symptoms of injury (Gorny et al., 2000, 2002). Respiration-induced ethylene production causes:

- Softening and development of off-flavor in watermelons
- Increased ripening and softening of mature green tomatoes
- Shattering of raspberries and blackberries.

Packaging

This process is crucial in preventing contamination by microbes as it avoids inward movement of light and air, thus keeping produce dry/moist and this prevents any changes in the textural integrity of produce along with convenient division of the produce in suitable portions needed for transportation, handling, and sale.

Vacuum Packaging. Elimination of air from a gas-impermeable bag in which food product has been placed and sealed reduces the pressure inside the bag, thus creating vacuum. While continuous respiration of the microbes present in/on the food product leads to exhaustion of available oxygen with a respective increase in carbon dioxide level that troubles the execution of biochemical processes and related microbial enzymes, the cells fail to survive the hiked gaseous changes.

Hyperbaric Packaging. High pressure processing (HPP) or high hydrostatic pressure (HHP) or ultra high pressure (UHP) processing subjects liquid and solid foods, with or without packaging, to pressures between 100 and 800 MPa at higher temperatures that relatively increases microbial inactivation. Water activity and pH are among the critical process factors in the inactivation of microbes by HPP. Temperatures ranging from 194°F to 230°F (90–110°C) in conjunction with pressures of 500–700 MPa have been used to inactivate spore-forming bacteria such as *Clostridium botulinum* (Patterson et al., 1995).

Storage of fruit product under low pressure and temperature conditions at high relative humidity reduces the oxygen availability. Thus, during the storage and transportation of various commodities, their compatibility regarding temperature, relative humidity, atmosphere (oxygen and carbon dioxide), protection from odors, and protection from ethylene requirements must be considered.

Edible Film Packaging. This is rather a new packaging advancement regarding fresh or minimally processed fruits as these edible coatings and films extend the shelf life by creating a modified atmosphere and preventing water loss (Ahvenainen, 1996; Baldwin et al., 1995a, b; Nisperos and Baldwin, 1996). Cereal biopolymers such as proteins and polysaccharides are attractive raw materials for use as materials in packaging applications as these are inexpensive, easily processable, thermoplastically originating from renewable resources, edible, and biodegradable, and possess good mechanical properties, thus functioning as excellent gas and grease barriers (Stading et al., 2001; Baldwin et al., 1996; Arvanitoyannis and Blanshard, 1994). Ghaouth et al. (1991) reported effects of chitosan coatings on storability and quality of fresh strawberries.

Sanitizing Agents

Halogenated Sanitizers

1. *Chlorine.* Chlorine has been used to treat drinking water, wastewater, as well as to sanitize food processing equipments and surfaces in processing environments (Botzenhart et al., 1993). Sodium hypochlorite solution Clorox^{TM} or dry, powdered calcium hypochlorite at 50–200 ppm concentration and an exposure time of 1–2 min can be used in hydrocooling or wash water as a disinfectant as it forms hypochlorous acid which is the active species required to perform the microbicidal action (Hendrix, 1991). Norwood and Gilmour (2000) have reported the growth and resistance of *Listeria monocytogenes* to sodium hypochlorite in steady-state multispecies biofilm. However, this antimicrobial action is reduced by a variety of abiotic factors such as temperature, light, and presence of soil and organic debris (Combrink and Grobbelaar, 1984; Folsom and Frank, 2000). A careful inspection and monitoring of wash water should be performed periodically with a monitoring kit.
2. *Bromine.* Bromine alone is not as effective as chlorine but shows an additive or synergistic increase in antimicrobial action upon use with other halogenated compounds, particularly chlorine.
3. *Iodine.* Aqueous iodine solutions and iodophors could be used to sanitize the processing equipments and surfaces and possess greater antimicrobial action range affecting yeasts and molds, reducing vegetative bacterial cells at very low concentrations and lower exposure times (Odlaug, 1981). Moreover, readily water-soluble iodophors have little corrosive action and are not skin irritants.

Ozonation. Ozone is a powerful disinfectant and has long been used to sanitize drinking water, swimming pools, and industrial wastewater. The dump tanks used for fruit precooling could be sanitized by using ozone treatment, as it is an efficient natural species to destroy foodborne pathogens as well as spoilage-causing microbes (Kim et al., 1999), but for certain fresh products as blackberries, ozonation treatment may lead to development of or increase in amount of anthocyanin pigment content (Barth et al., 1995). Kim et al. (2000) have reported the impact of use of electrolyzed oxidizing and chemically modified water on various types of foodborne pathogens.

Hydrogen Peroxide. Hydrogen peroxide could also be used as a disinfectant. Concentrations of 0.5% or less are effective for inhibiting development of postharvest decay caused by a number of fungi. Hydrogen peroxide has a low toxicity rating and is generally recognized as having little potential for environmental damage (Sapers and Simmons, 1998). The use of lactic acid dippings along with the treatment by hydrogen peroxide could lead to inactivation of *E. coli* O157:H7, *S. enteridtis*, and *Listeria monocytogenes* on apples, oranges, and tomatoes (Venkitanarayanan et al., 2002).

Use of certain antibacterial solutions could also help in decreasing the bacterial load. McWatters et al. (2002) reported the consumer acceptance of raw apples treated with antibacterial solution used routinely in household.

Irradiation. Non-ionizing ultraviolet radiations could be used for surface sterilization of food-handling utensils, as these rays do not penetrate foods (Worobo, 2000), while ionizing gamma radiations (Chervin and Boisseau, 1994) that penetrate well, oxidize sensitive cellular constituents (radappertization), and thus require moistening of produce to produce peroxides. Gamma irradiation has been used for the decontamination of a range of products such as fresh produce including fruits and vegetables as well as certain other spoilage labile fresh products such as seafoods and meat (Gunes et al., 2000). A recent report has provided the information regarding marketing of irradiated strawberries for consumption in the United States (Marcotte, 1992).

FRUIT SAFETY

Fruit safety is related to an amalgam of unprecedented agronomical procedures that while accomplishment culminate toward elimination of various human pathogenic species present on the fruits (Meng and Doyle, 2002). Several incidences of transmission of infection by consumption of raw fruits and vegetables have been documented such as *Salmonella typhi* infection by consuming a variety of fresh products (Sanchez et al., 2002; Pixley, 1913), *Salmonella* and *E. coli* in fruit juices as well as certain parasitic helminths primarily *Fasciola hepatica, Fasciolopsis buski* have been observed to encyst on plants and cause human illnesses. Recently, viruses following the fecal-oral route as Hepatitis A virus and Norwalk disease virus have been observed to be associated with consumption of raw fruits such as raspberries, strawberries, and melons.

ASSOCIATED PATHOGENS AND SOURCES OF CONTAMINATION

A healthy fruit surface may get contaminated during the long route of processing and storage dramatically including diverse external sources such as environmental factors, water used, processing equipments, or procedures performed. Bacteria such as *Clostridium botulinum, Bacillus cereus,* and *Listeria monocytogenes* are normal inhabitants of soil, whereas *Salmonella, Shigella, E. coli,* and *Campylobacter* are resident microflora of the rumen of ruminant animals and stomachs of human beings that could potentially contaminate raw fruits and vegetables through contact with feces, sewage, untreated irrigation water, or surface water, while viruses of the fecal-oral route and parasites in form of cysts of liver flukes, tapeworms, and *Giaradia lamblia* contaminate produce by contact with sewage, feces, and irrigation water (Mead et al., 1999; King et al., 2000; Buck et al., 2003). Food pathogens such as *Clostridium, Yersinia,* and *Listeria* can potentially develop on minimally processed fruits and vegetables under refrigerated or high-moisture conditions (Doyle, 2000a, b, c; Meng and Doyle, 2002). Beuchat (2002) has reviewed several ecological factors that influence the growth and survival of human pathogens in raw fruits and vegetables (Table 1.3).

PREHARVEST SOURCES OF CONTAMINATION

Environmental contamination: Human pathogens may enter produce through various pathways or natural structures such as stem, stem scars, or calyx of certain produce (Zhuang et al., 1995), or through damaged surface parts such as wounds, cuts, splits, and punctures caused during maturation by insect infestation (Michailides and Spotts, 1990; Beuchat, 1996; Olsen, 1998; Janisiewicz et al., 1999; Iwasa et al., 1999; Shere et al., 1998; Wallace et al., 1997), damage caused by sand storms and hail/frost (Hill and Faville, 1951; Hill and Wenzel, 1963), damage occurred during the harvesting of fruits (Carballo et al., 1994; Sugar and Spotts, 1993; Wells and Butterfield, 1997), and damage occurred during processing procedures or equipments utilized.

CONTAMINATION DURING POSTHARVEST PROCESSING

Waterborne Contamination

Processing procedures such as rinsing and washing with contaminated water may contribute toward the microbial contamination of fruits and vegetables (Petracek et al., 1998; Buchanan et al ., 1999). Thus, water if not potable could act as source of an array of human pathogenic microbes such as *E. coli* 0157:H7, *Salmonella* sp., *Vibrio chloerae, Shigella* sp., *Cryptosporidium parvum, Giardia lamblia, Cyclospora cayetanensis,* Norwalk disease virus, and Hepatitis A virus.

Cross Contamination

Cross contamination of products can occur from processing equipment and the environment. Eisenberg and Cichowicz (1977) noted that tomato and pineapple products can become contaminated with the mold *Geotrichum candidum,* while the same organism was observed to contaminate orange and grapefruit juices, apples, and ciders (Senkel et al., 1999) indicating a kind of cross contamination during processing operations. Any pathogen internalized in the fruit must survive there to cause illness afterwards but the very survival depends on the physical and chemical attributes of fruit, postharvest processes, and consumer use (Burnett and Beuchat, 2000; 2001). *Salmonella*

Table 1.3. Types of Fungal Spoilage of Fruits

Product	Type of Spoilage	Mold Involved
Citrus fruits		
Oranges	Blue rot	*Penicillum italicum*
Tomatoes, citrus fruits	Sour rot	*Geotrichum candidum*
Citrus fruits	Green mold rot	*Penicillum digitatum*
Citrus fruits	*Alternaria* rot	*Alternaria* sp.
Citrus fruits	Stem end rot	*Phomopsis citri, Diplodia natalensis, Alternaria citri*
Peaches/apricots		
Peaches	Brown rot	*Monilinia fructicola*
Peaches	Pink rot	*Trichothecium* sp.
Peaches, apricots	Black mold rot	*Aspergillus niger*
Peaches, cherries	*Cladosporium rot*	*Cladosporium herbarum*
Apples/pears		
Apples	Soft rot	*Penicillum expansum*
Apples, pears	Lenticel rot	*Cryptosporidium malicorticus, Phylctanea vagabuna*
Apples, pears	Black spot/scab	*Venturia inaequalis*
Apples, pears	Brown rot	*Monilinia fructigena*
Pears	Erwinia rot	*Erwinia cartovora*
Bananas		
Bananas	Bitter rot (Anthracnose)	*Colletotrichum musae*
Bananas	Finger rot	*Pestalozzia, Fusarium, Gleosporium* sp.
Bananas	Crown rot	*Ceratocystis paradoxa, Fusarium roseum; Colletotrichum musae, Verticillium theobromae*
Other fruits		
Pineapples	Pineapple black rot	*Ceratocystis paradoxa*
Water melons	Anthracnose	*C. lagenarium*
Strawberries	Grey mold rot	*Botrytis cinerea*
Grapes	Grey mold rot	*Botrytis cinerea*

Source: Adapted from Jay (1992).

can grow rapidly on cut surfaces of cantaloupe, watermelon, and honeydew melon held at room temperature (Golden et al., 1993), while *E. coli* 0157:H7 can grow in ground apples stored at various temperatures (Fisher and Golden, 1998) and in apple juice at 4°C (Miller and Kaspar, 1994; Fratamico et al., 1997; Splittstoesser et al., 1996), in orange juice at 4°C (Fratamico et al., 1997), on surface of citrus fruits (Pao and Brown, 1998). *Aerobacter, Xanthomonas*, and *Achromobacter* can grow inside the citrus fruits (Hill and Faville, 1951), while *Leuconostoc* and *Lactobacillus* in orange juice and *Listeria* in orange juice at 4°C (Parish and Higgins, 1989). Sometimes these human pathogens could be traced to contamination. These findings indicate the accomplishment of favorable environment for the survival and growth of human pathogenic microbes in/on the fresh produce, thus alerting the empowerment of strict safety and hygiene practices during storage and processing of produce as well as the regulation of stringent quarantine measures for rapid detection and identification of these microbes in processed products.

SAFETY AND SANITATION

Sanitation is of great concern as it protects produce against postharvest diseases as well as protects consumers from foodborne illnesses caused by an array of human pathogens residing in the intestines of ruminants and humans that can get transmitted via the fecal-oral route such as *E. coli* 0157:H7, *Salmonella, Cryptosporidium, Hepatitis A virus*, and *Cyclospora* by contamination of fruits and vegetables. Disinfection of produce by chlorination (Zhao et al., 2001), use of hydrogen peroxide, ozonation, use of quaternary ammonium salts in wash water can help to prevent both postharvest and foodborne diseases. Effectiveness of disinfectant depends on the nature of the cells as well as the characteristics of fruit tissues and juices. Han et al. (2002) reported the inactivation of *E. coli* 0157:H7 on green peppers by ozone gas treatment. Earlier a similar report on the effect of ozone and storage temperature on postharvest diseases of carrots was provided by Liew and Prange (1994). Castro et al. (1993) have reported the use of rather unusual technique of pulsed electric fields for inactivation of microbes in foods.

HEALTH IMPLICATIONS

Human pathogens such as enteric bacteria and viruses cause illnesses exhibiting initial symptoms such as diarrhea, nausea, vomiting, altered peristaltic movement of the intestine, fever that may debilitate patient's health and could aggravate toward certain advanced complications or group of ailments/syndrome sometimes resulting in death of vulnerable age/immunocompromised patients.

E. coli: *E. coli* O157:H7 causes abdominal cramps and watery diarrhea/bloody diarrhea (hemorrhagic colitis) along with fever and vomiting and the incidence recovery within 10 days. However, infection of *E. coli* 0157:H7 in young children and elderly patients results in life-threatening complications as hemolytic uremic syndrome (HUS), which is characterized by acute renal failure, hemolytic anemia, and thrombocytopenia.

Salmonella enteritidis/S. typhimurium: The symptoms share the similarity to *E. coli* infection along with abdominal pain and cholera-like disease and subsides within 2–4 days or may result in prolonged enteritis with passage of mucus and pus in feces and typhoidal speticaemic fever.

Shigella sp.: This bacterium causes shigellosis/bacillary dysentery upon ingestion at very less infective dose by forming shiga toxin and produces inflammation of intestine, capillary thrombosis leading to transverse ulceration or bacteremia manifested as bloody, mucoid scanty feces with tenesmus, fever, and vomiting. HUS may also appear as a rare complication in certain cases.

FUTURE PERSPECTIVES

The future era in fruit microbiology connotes the advent of fully automatized packaging, detection, and status analyzers that would ensure stringent regulations to minimize losses by spoilage and transmission of foodborne human pathogens. The development of new techniques of film coatings for fresh produce involving the use of yeasts and lysozyme combinations to fight against rot-causing microbes keeps the fruits fresh for longer periods or the advent of transgenic fruits which act as vehicles for various diseases such as cancer, Hepatitis, etc. have revolutionized the very idea of consuming fruits. Consumption of cranberry juice was observed to prevent recurrent urinary-tract infections in women (Stapleton, 1999; Henig and Leahy 2000; Howell, 2002). Bacteriocins have been long noticed as potential inhibitors or cidal agents against sensitive microorganisms under certain conditions, but these in foods may cause moderate antimicrobial activity followed by microbial growth, which may indicate development of resistance, application of inadequate quantities of bacteriocin, or its inability to find all cell microenvironments to inactivate the target microorganism. The potential for commercial use of bacteriocins may be enhanced when they are used in multihurdle preservation systems. The use of robots for faster inspection and screening of produce during processing enlarges the horizons for instant food inspection, while with the use of DNA-based techniques, the whole scenario of conventional isolation and cumbersome identification protocols has speeded up to rapid enumeration of very low infective dose pathogens, with even the detection of presence of several uncultivatable microbes. The future techniques presently available as research trials would not only detect the microbes but also eradicate them or the toxic chemicals produced by using tiny molecule/protein-coated computer chips. Thus, future scenario holds the possibilities of better product shelf life and little risk regarding the consumption of fresh fruits or their processed products.

REFERENCES

Ahvenainen, R. 1996. New approaches in improving the shelf life of minimally processed fruits and vegetables. Trends in Food Science and Technology 7:179–186.

Altekruse, S.F., Cohen, M.L. and Swerdlow, D.L. 1997. Emerging foodborne diseases. Emerging Infectious Diseases 3:285–293.

Altekruse, S.F. and Swerdlow, D.L. 1996. The changing epidemiology of foodborne diseases. American Journal of Medical Science 311:23–29.

Arvanitoyannis, M. and Blanshard, J.M. 1994. Study of diffusion and permeation of gases in undrawn and unaxially drawn films made from potato and rice

starch conditioned at different relative humidities. Carbohydrate Polymers 24:1–15.
Bachmann, J. and Earles, R. 2000. Postharvest Handling of Fruit and Vegetables by ATTRA Ozark Mountains at the University of Arkansas in Fayetteville at P.O. Box 3657, Fayetteville, AR 72702 NCAT Agriculture Specialists (August 2000) cited from http://attra.ncat.org/attra-pub/PDF/postharvest.pdf
Baldwin, E.A., Nisperos-Carriedo, M.O. and Baker, R.A. 1995a. Edible coatings for lightly processed fruits and vegetables. Horticulture Science 30:35–38.
Baldwin, E.A., Nisperos-Carriedo, M.O. and Baker, R.A. 1995b. Use of edible coatings to preserve quality of lightly (and slightly) processed products. CRC Critical Reviews in Food Science and Nutrition 35:509–524.
Baldwin, E.A., Nisperos-Carriedo, M.O., Chen, X. and Hagenmaier, R.D. 1996. Improving the storage life of cut apple and potato with edible coating. Postharvest Biology and Technology 9:151–163.
Barth, M.M., Zhou, C., Mercier, J. and Payne, F.A. 1995. Ozone storage effects on anthocyanin content and fungal growth in blackberries. Journal of Food Science 60:1286–1288.
Bean, N.H., Goulding, J.S., Daniels, M.T. and Angulo, F.J. 1997. Surveillance for foodborne disease outbreaks: United States 1988–1992. Journal of Food Protection 60:1265–1286.
Beaulieu, J.C., Bett, K.L., Champagne, E.T., Ingram, D.A. and Miller, J.A. 1999. Flavor, sensory and postharvest evaluations of commercial- versus tree-ripe fresh-cut 'Bounty' peaches. Horticulture Science 34:504.
Bennett, A.R., MacPhee, S. and Bett, R.P. 1996. The isolation and detection of *Escherichia coli* O157 by use of immunomagnetic separation and immunoassay procedures. Letters in Applied Microbiology 22:237–243.
Bett, K.L., Ingram, D.A., Grimm, C.C., Lloyd, S.W., Spanier, A.M., Miller, J.M., Gross, K.C., Baldwin, E.A. and Vinyard, B.T. 2001. Flavor of fresh-cut 'Gala' apples in modified atmosphere packaging as affected by storage time. Journal of Food Quality 24:141–156.
Beuchat, L.R. 1976. Sensitivity of *Vibrio parahaemolyticus* to spices and organic acids. Journal of Food Science 41:899–902.
Beuchat, L.R. 1996. Pathogenic microorganisms associated with fresh produce. Journal of Food Protection 59:204–216.
Beuchat, L.R. 2002. Ecological factors influencing survival and growth of human pathogens on raw fruits and vegetables. Microbes Infection 4:413–423.
Beuchat, L.R. and Pitt, J.I. 1992. Detection and enumeration of heat-resistant molds. In: Compendium of Methods for the Microbiological Examination of Foods, Vanderzant C, Splittoesser DF (eds.), American Public Health Assoc., Washington, DC, pp. 251–263.
Botzenhart, K., Tarcson, G.M. and Ostruschka, M. 1993. Inactivation of bacteria and coliphages by ozone and chlorine dioxide in a continuous flow reactor. Water Science Technology 27:363–370.
Brackett, R.E. 1987. Microbiological consequences of minimally processed fruits and vegetables. Journal of Food Quality 10:195–206.
Brecht, J.K. 1995. Physiology of lightly processed fruits and vegetables. HortScience 30:18–22.
Brown, A.D. 1964. Aspects of bacterial response to the ionic environment. Bacteriological Reviews 28:296–329.
Brown, A.D. 1976. Microbial water stress. Bacteriological Reviews 40:803–846.
Buchanan, R.L., Edelson, S.G., Miller, R.L. and Sapers, G.M. 1999. Contamination of intact apples after immersion in an aqueous environment containing *Escherichia coli* O157:H7. Journal of Food Protection 62(5):444–450.
Buck, J.W., Walcott, R.R. and Beuchat, L.R. 2003. Recent trends in microbiological safety of fruits and vegetables. Online Plant Health Progress doi:10.1094/PHP-2003-0121-01-RV.
Bullerman, L.B., Lieu, F.Y. and Seier, S.A. 1977. Inhibition of growth and aflatoxin production by cinnamon and clove oils, cinnamic aldehyde and eugenol. Journal of Food Science 42:1107–1109.
Burnett, S.L. and Beuchat, L.R. 2000. Human pathogens associated with raw produce and unpasteurized juices, and difficulties in decontamination. Journal of Industrial Microbiology and Biotechnology 25:281–287.
Burnett, S.L. and Beuchat, L.R. 2001. Human pathogens associated with raw produce and unpasteurized juices, and difficulties in contamination. Journal of Industrial Microbiology and Biotechnology 27:104–110.
Burnett, S.L., Chen, J. and Beuchat, L.R. 2000. Attachment of *Escherichia coli* O157:H7 to the surfaces and internal structures of apples as detected by confocal scanning laser microscopy. Applied and Environment Microbiology 66:4679–4687.
Buta, J.G. and Molin, H.E. 1998. Methyl jasmonate extends shelf life and reduces microbial contamination of fresh cut celery and peppers. Journal of Agriculture Food Chemistry 46:1253–1256.
Candish, A.A.G. 1991. Immunological methods in food microbiology. Food Microbiology 8:1–14.

Carballo, S.J., Blankenship, S.M., Sanders, D.C., Ritchie, D.F. and Boyette, M.D. 1994. Comparison of packing systems for injury and bacterial soft rot on bell pepper fruit. Horticulture Technology 4(3):269–272.

Castro, A.J., Barbosa-Canovas, G.V. and Swanson, B.G. 1993. Microbial inactivation of foods by pulsed electric fields. Journal of Food Processing and Preservation 17:47–73.

Charlang, G. and Horowitz, N.H. 1974. Membrane permeability and the loss of germination factor from *Neurospora crassa* at low water activities. Journal Bacteriology 117:261–264.

Cheeson, A. 1980. Maceration in relation to the post-harvest handling and processing of plant material. Journal of Applied Bacteriology 48:1–45.

Chen, J. and Griffiths M.W. 2001. Detection of Salmonella and simultaneous detection of Salmonella and Shiga-like toxin-producing *Escherichia coli* using the magnetic capture hybridization polymerase chain reaction. Letters in Applied Microbiology 32(1): 7–11.

Chervin, C. and Boisseau, P. 1994. Quality maintenance of "ready-to-eat" shredded carrots by gamma irradiation. Journal of Food Science 59:359–361.

Christian, J.H.B. 1963. Water activity and growth of microorganisms. In: Recent Advances in Food Science, Leitch JM, Rhodes DN (eds.), Vol. 3, Butterworths, London, pp:248–255.

Combrink, J.C. and Grobbelaar, C.J. 1984. Influence of temperature and chlorine treatments on post-harvest decay of apples caused by *Mucor piriformis*. South Africa Department of Agriculture and Water Supply, Technical Communication No. 192:19–20.

Conway, W.S., Leverentz, B., Saftner, R.A., Janisiewicz, W.J., Sams, C.E. and Leblanc, E. 2000. Survival and growth of *Listeria monocytogenes* on fresh-cut apple slices and its interaction with *Glomerella cingulata* and *Penicillium expansum.* Plant Diseases 84:177–181.

Corbo, M.R., Lanciotti, R., Gardini, R., Sinigaglia, R. and Guerzoni, M.E. 2000. Effects of hexanal, trans-2-hexenal, and storage temperature on shelf-life of fresh sliced apples. Journal of Agriculture Food Chemistry 48:2401–2408.

Corlett, D.A. Jr. and Brown, M.H. 1980. pH and acidity. In: Microbial Ecology of Foods, Vol.1 ICMSF, Academic Press, New York, pp:92–111.

Costerton, J.W. 1995. Overview of microbial biofilms. Journal of Industrial Microbiology 15:137–140.

Csonka, L.N. 1989. Physiological and genetic responses of bacteria to osmotic stress. Microbiological Reviews 51:121–147.

Curiale, M.S., Sons, T., Mclver, D., McAllister, J.S., Halsey, B., Rhodes, D. and Fox, T.L. 1991. Dry rehydrated film for enumeration of total coliforms and *Escherichia coli* in foods: Collaborative study. Journal Association of Analytical Chemistry 74:635–648.

Dingman, D.W. 2000. Growth of *Escherichia coli* O157:H7 in bruised apple (*Malus domestica*) tissue as influenced by cultivar, date of harvest and source. Applied and Environment Microbiology 66:1077–1083.

Doyle, M. 2000a. Food safety issues arising at food production in a global market. Journal Agribusiness 18:129–133.

Doyle, M.P. 2000b. Reducing foodborne disease. Food Technology 54(11):130.

Doyle, M.P. 2000c. Reducing foodborne disease: What are the priorities? Nutrition 16:647–649.

Eisenberg, W.V. and Cichowicz, S.M. 1977. Machinery mold-indicator organism in food. Food Technology 31:52–56.

Erickson, M.C. and Kornacki, J.L. 2003. *Bacillus anthracis*: Current knowledge in relation to contamination of food. Journal of Food Protection 66:691–699.

Feng, P., Lampel, K.A. and Hill, W.E. 1996. Developments in food technology: Applications and economic and regulatory considerations. In: Nucleic Acid Analysis: Principles and Bioapplications, Dangler CA, Osburn B (eds.), Wiley and Sons, New York, NY, pp.203–229.

Fisher, T.L. and Golden, D.A. 1998. Fate of *Escherichia coli* O157:H7 in ground apples used in cider production. Journal of Food Protection 61(10):1372–1374.

Folsom, J.P. and Frank, J.F. 2000. Heat inactivation of *Escherichia coli* O157:H7 in apple juice exposed to chlorine. Journal of Food Protection 63:1021–1025.

Frank, J.F. 2000a. Control of biofilms in the food and beverage industry. In: Industrial Biofouling, Walker J, Surman S, Hass JH (eds.), John Wiley and Sons, London, pp.205–224.

Frank, J.F. 2000b. Microbial attachment to food and food contact surfaces. Advances in Food Nutrition Research 43:319–370.

Fratamico, P.M., Deng, M.Y., Strobaugh, T.P. and Palumbo, S.A. 1997. Construction and characterization of *Escherichia coli* O157:H7 strains expressing firefly luciferase and green fluorescent protein and their use in survival studies. Journal of Food Protection 60(10):1167–1173.

Frost, J.A., McEvoy, B., Bentley, C.A., Andersson, Y. and Rowe, A. 1995. An outbreak of *Shigella sonnei* infection associated with consumption of iceberg lettuce. Emerging Infectious Diseases 1:26–29.

Ghaouth, A.E., Arul, J., Pannampalam, R. and Boulet, M. 1991. Chitosan coating effect on the storability and quality of fresh strawberries. Journal of Food Science 56:1618–1620.

Gil, M.I., Gorny, J.R. and Kader, K.A. 1998. Responses of 'Fuji' apple slices to ascorbic acid treatments and low oxygen atmospheres. Horticulture Science 33:305–309.

Golden, D.A., Rhodehamel, E.J. and Kautter, D.A. 1993. Growth of *Salmonella* spp. in cantaloupe, watermelon, and honeydew melons. Journal of Food Protection 56(3):194–196.

Gorny, J.R., Cifuentes, R.A., Hess-Pierce, B. and Kader, A.A. 2000. Quality changes in fresh-cut pear slices as affected by cultivar, ripeness stage, fruit size, and storage regime. Journal of Food Science 65:541–544.

Gorny, J.R., Hess-Pierce, B., Cifuentes, R.A. and Kader, A.A. 2002. Quality changes in fresh-cut pear slices as affected by controlled atmospheres and chemical preservatives. Postharvest Biology and Technology 24(3):271–278.

Gorny, J.R., Hess-Pierce, B. and Kader, A.A. 1999. Quality changes in fresh-cut peach and nectarine slices as affected by cultivar, storage atmosphere and chemical treatments. Journal of Food Science 64:429–432.

Gorny, J.R. and Kader, A.A. 1996. Fresh-cut fruit products. Fresh-cut products: Maintaining quality and safety. UCD Postharvest Horticulture Series 10(14):1–14.

Graham, D.M. 1997. Use of ozone for food-processing. Food Technology 51:72–75.

Gunes, G., Watkins, C.B. and Hotchkiss, J.H. 2000. Effects of irradiation on respiration and ethylene production of apple slices. Journal Science Food Agriculture 80:1169–1175.

Guo, X., Chen, J., Beuchat, L.R. and Brackett, R.E. 2000. PCR detection of *Salmonella enterica* serotype Montevideo in and on tomatoes using primers derived from *hil*A. Applied and Environment Microbiology 66:5248–5252.

Guo, X., Van Iersel, M.W., Chen, J., Brackett, R.E. and Beuchat, L.R. 2002. Evidence of association of salmonellae with tomato plants grown hydroponically in inoculated nutrient solution. Applied and Environment Microbiology 68:3639–3643.

Han, Y., Floros, J., Linton, R., Nielsen, S. and Nelson, P. 2002. Response surface modeling for the inactivation of *Escherichia coli* O157:H7 on green peppers (*Capsicum annum*) by ozone gas treatment. Journal of Food Microbiology and Safety 67:1188–1199.

Hanklin, L. and Lacy, G.H. 1992. Pectinolytic microorganisms. In: Compendium of Methods for the Microbiological Examination of Foods, Vanderzant C, Splittstoesser DF (eds.), American Public Health Assoc., Washington DC, pp.176–183.

Hartman, P.A., Swaminathan, B., Curiale, M.S., Firstenberg-Eden, R., Sharpe, A.N., Cox, N.A., Fung, Y.C. and Goldschmidt, M.C. 1992. Rapid methods and automation. In: Compedium of Methods for the Microbiological Examination of Foods, Vanderzant C, Splittstoesser DF (eds.), American Public Health Association, Washington, DC, pp.665–746.

Hedberg, C.W., MacDonald, K.L. and Osterholm, M.T. 1994. Changing epidemiology of food-borne disease: A Minnesota perspective. Clinical and Infectious Diseases 18:671–682.

Hendrix, F.F. Jr. 1991. Removal of sooty blotch and flyspeck from apple fruit with a chlorine dip. Plant Diseases 75(7):742–743.

Henig, Y.S. and Leahy, M.M. 2000. Cranberry juice and urinary-tract health: Science supports folklore. Nutrition 16:684–687.

Hill, E.C. and Faville, L.W. 1951. Studies on the artificial infection of oranges with acid tolerant bacteria. Proceedings of Florida State Horticulture Society 64:174–177.

Hill, E.C. and Wenzel, F.W. 1963. Microbial populations of frozen oranges. Proceedings of Florida State Horticulture Society 76:276–281.

Hill, W.E. 1996. The polymerase chain reaction: Application for the detection of food-borne pathogens. Critical Reviews Food Science and Nutrition 36:123–173.

Howell, A. 2002. Tablets may stop UTIs: Do cranberry pills work as well as juice in preventing urinary tract infections? Prevention May:70.

Iwasa, M., Sou-Ichi, M., Asakura, H., Hideaski, K. and Morimoto, Y. 1999. Detection of *Escherichia coli* O157:H7 from *Musca domestica* (Diptera: Muscidae) at a cattle farm in Japan. Journal of Medical Entomology 36(1):108–112.

Janisiewicz, W.J., Conway, W.S., Brown, M.W., Sapers, G.M., Fratamico, P. and Buchanan, R.L. 1999. Fate of Escherichia coli O157:H7 on fresh-cut apple tissue and its potential for transmission by fruit flies. Applied Environment Microbiology 65(1):1–5.

Jay, J.M. 1992. Spoilage of fruits and vegetables. In: Modern Food Microbiology (4th ed.), Chapman and Hall, New York, pp.187–198.

Jiang, X.P. and Doyle, M.P. 2002. Optimizing enrichment culture conditions for detecting *Helicobacter pylori* in foods. Journal of Food Protection 65:1949–1954.

Jones, D.D. and Bej, A.K. 1994. Detection of food borne microbial pathogens using PCR methods. In: PCR Technology: Current Innovations, Griffin HG, Griffin AM (eds.), CRC Press, Boca Raton, FL, pp.341–365.

Jung, Y.S., Frank, J.F. and Brackett, R.E. 2003. Evaluation of antibodies for immunomagnetic separation combined with flow cytometry detection of *Listeria monocytogenes*. Journal of Food Protection 66:1283–1287.

Kapperud, G., Rorvik, L.M., Hasseltvedt, V., Hoiby, E.A., Iverson, B.G., Staveland, K., Johnson, G., Leitao, J., Herikstad, H., Andersson, Y., Langeland, Y., Gondrosen, B. and Lassen, J. 1995. Outbreak of *Shigella sonnei* infection traced to imported iceberg lettuce. Journal of Clinical Microbiology 33(3):609–614.

Khadre, M. and Yousef, A. 2001. Sporicidal action of ozone and hydrogen peroxide: A comparative study. International Journal of Food Microbiology 71:131–138.

Kim, C., Hung, Y.C. and Brackett, R.E. 2000. Efficacy of electrolyzed oxidizing (EO) and chemically modified water on different types of foodborne pathogens. International Journal of Food Microbiology 61:199–207.

Kim, J., Yousef, A. and Dave, S. 1999. Application of ozone for enhancing the microbiological safety and quality of foods: A Review. Journal Food Protection 62:1071–1087.

King, J.C., Black, R.E., Doyle, M.P., Fritsche, K.L., Hallbrook, B.H., Levander, O.A., Meydani, S.N., Walker, W.A. and Woteki, C.E. 2000. Foodborne illnesses and nutritional status: A statement from an American Society for Nutritional Sciences Working Group. Journal of Nutrition 130:2613–2617.

Lamikarna, O., Chen, J.C., Banks, D. and Hunter, P.A. 2000. Biochemical and microbial changes during storage of minimally processed cantaloupe. Journal of Agriculture Food Chemistry 48:5955–5961.

Lampel, K.A., Feng, P. and Hill, W.E. 1992. Gene probes used in food microbiology. In: Molecular Approaches to Improving Food Safety, Bhatnagar D, Cleveland TE (eds.), Van Nostrand Reinhold, New York, pp.151–188.

Lanciotti, R., Corbo, M.R., Gardini, F., Sinigerglia, M. and Guerzoni, M.E. 1999. Effect of hexanal on the shelf life of fresh apple slices. Journal of Agriculture Food Chemistry 47:4769–4776.

Liew, C.L. and Prange, R.K. 1994. Effect of ozone and storage temperature on postharvest diseases and physiology of carrots (*Daucus carota L.*). Journal of American Society of Horticultural Sciences 119:563–567.

Lund, B.M. and Snowdon, A.L. 2000. Fresh and processed fruits. In: The Microbiological Safety and Quality of Food, Lund BM, Baird-Parker TC, Gould GW (eds.), Vol. I, Aspen Publication, Gaithersburg, MD, pp.738–758.

Marcotte, M. 1992. Irradiated strawberries enter the U.S. market. Food Technology 46(5):80–86.

McWatters, K.H., Doyle, M.P., Walker, S.L., Rimal, A.P. and Venkitanarayanan, K. 2002. Consumer acceptance of raw apples treated with an antibacterial solution designed for home use. Journal of Food Protection 65:106–110.

Mead, P.S., Slutsker, L., Dietz, V., McGaig, L.F., Bresee, J.S., Shapiro, C., Griffin, P.M. and Tauxe, R.V. 1999. Food-related illness and death in the United States. Emerging Infectious Diseases 5:607–625.

Meng, J.H. and Doyle, M.P. 2002. Introduction: Microbiological food safety. Microbes Infection 4:395–397.

Mermelstein, N.H., Fennema, O.R., Batt, C.A., Goff, H.D., Griffiths, M.W., Hoover, D.G., Hsieh, F.H., Juneja, V.K., Kroger, M., Lund, D.B., Miller, D.D., Min, D.B., Murphy, P.A., Palumbo, S.A., Rao, M.A., Ryser, E.T., Schneeman, B.O., Singh, H., Stone, H., Whiting, R., Wu, J.S.B., Yousef, A.E., BeMiller, J.N., Dennis, C., Doyle, M.P., Escher, F.E., Klaenhammer, T., Knorr, D., Kokini, J.L., Iwaoka, W., Chism, G.W., Dong, F.M., Hartel, R., Reitmeier, C., Schmidt, S.J. and Wrolstad, R.E. 2002. Food research trends-2003 and beyond. Food Technology 56(12):30.

Michailides, T.J. and Spotts, R.A. 1990. Transmission of *Mucor piriformis* to fruit of *Prunus persica* by *Carpophilus* spp. and *Drosophila melanogaster*. Plant Diseases 74:287–291.

Miller, L.G. and Kaspar, C.W. 1994. *Escherichia coli* O157:H7 acid tolerance and survival in apple cider. Journal of Food Protection 57(6):460–464.

Morris, C.E., Monier, J.E. and Jacques, M.A. 1997. Methods for observing microbial biofilms directly on leaf surfaces and recovering them for isolation of culturable microorganisms. Applied Environment Microbiology 63:1570–1576.

Morris, E.O. 1962. Effect of environment on microorganisms. In: Recent Advances in Food Science, Hawthorn J, Leitch JM (eds.), Vol. 1, Butterworths, London, pp:24–36.

Mossel, D.A.A. and Ingram, M. 1955. The physiology of the microbial spoilage of foods. Journal of Applied Bacteriology 18:232–268.

Nguyen-the, C. and Carlin, F. 1994. The microbiology of minimally processed fresh fruits and vegetables. Critical Reviews Food Science and Nutrition 34:371–401.

Nisperos, M.O. and Baldwin, E.A. 1996. Edible coatings for whole and minimally processed fruits and vegetables. Food Australia 48(1):27–31.

Norwood, D.E. and Gilmour, A. 2000. The growth and resistance to sodium hypochlorite of *Listeria monocytogenes* in a steady-state multispecies biofilm. Journal of Applied Microbiology 88:512–520.

Odlaug, T.E. 1981. Antimicrobial activity of halogens. Journal of Food Protection 44:608–613.

Olsen, A.R. 1998. Regulatory action criteria for filth and other extraneous materials. III. Review of flies and foodborne enteric disease. Regulatory Toxicology and Pharmacology 28:199–211.

Olsvik, O., Popovic, T., Skjerve, E., Cudjoe, K.S., Hornes, E., Ugelstad, J. and Uhlen, M. 1994. Magnetic separation techniques in diagnostic microbiology. Clinical Microbiological Reviews 7:43–54.

Orr, R.V., Shewfelt, R.L., Huang, C.J., Tefera, S. and Beuchat, L.R. 2000. Detection of guaiacol produced by *Alicyclobacillus acidoterrestris* in apple juice by sensory and chromatographic analyses, and comparison with spore and vegetative cell populations. Journal of Food Protection 63:1517–1522.

Palou, L., Crisosto, C., Smilanick, J., Adaskaveg, J. and Zoffoli, J. 2002. Effects of continuous ozone exposure on decay development and physiological responses of peaches and table grapes in cold storage. Postharvest Biology and Technology 24:39–48.

Pao, S. and Brown, G.E. 1998. Reduction of microorganisms on citrus fruit surfaces during packinghouse procedures. Journal of Food Protection 61(7):903–906.

Pao, S. and Petracek, P.D. 1997. Shelf life extension of peeled oranges by citric acid treatment. Food Microbiology 14:485–491.

Parish, M.E. and Higgins, D.P. 1989. Survival of *Listeria monocytogenes* in low pH model broth systems. Journal of Food Protection 52(3):144–147.

Patterson, M.F., Quinn, M., Simpson, R. and Gilmour, A. 1995. Sensitivity of vegetative pathogens to high hydrostatic pressure treatment in phosphate buffered saline and foods. Journal of Food Protection 58(5):524–529.

Petracek, P.D., Kelsey, D.F. and Davis, C. 1998. Response of citrus fruit to high-pressure washing. Journal of American Society of Horticulture Science 123(4):661–667.

Pixley, C. 1913. Typhoid fever from uncooked vegetables. New York Medical Journal 98:328.

Potter, M.E., Motarjemi, Y. and Käferstein, F.K. 1997. Emerging foodborne diseases. World Health, January-February pp.16–17.

Qi, L., Wu, T. and Watada, A.E. 1999. Quality changes of fresh-cut honeydew melons during controlled-atmosphere storage. Journal of Food Quality 22:513–521.

Ratowsky, D.A., Olley, J., McMeekin, T.A. and Ball, A. 1982. Relationship between temperature and growth rate of bacterial cultures. Journal of Applied Bacteriology 44:97–106.

Rattanapanone, N. and Watada, A.E. 2000. Respiration rate and respiratory quotient of fresh-cut mango (*Magnifera indica* L.) in low oxygen atmosphere. Proceedings of 6th International Symposium Mango 509:471–478.

Riordan, D.C.R., Sapers, G.M. and Annous, B.A. 2000. The survival of *Escherichia coli* O157:H7 in the presence of *Penicillium expansum* and *Glomerella cingulata* in wounds on apple surfaces. Journal of Food Protection 63:1637–1642.

Safarik, I. and Safarikova, M. 1999. Use of magnetic techniques for the isolation of cells. Journal of Chromatography B 722:33–53.

Saiki, R.K., Gelfand, D.H., Stoffel, S., Scharf, S.J., Higuchi, R., Horn, G.T., Mullis, K.B. and Erlich, H.A. 1988. Primer-directed enzymatic amplification of DNA with a thermostable DNA polymerase. Science 239:487–491.

Sanchez, S., Hofacre, C.L., Lee, M.D., Maurer, J.J. and Doyle, M.P. 2002. Animal sources of salmonellosis in humans. Journal of American Veterinary Medical Association 221:492–497.

Sanderson, P.G. and Spotts, R.A. 1995. Postharvest decay of winter pear and apple fruit caused by species of *Penicillium.* Phytopathology 85(1):103–110.

Sapers, G.M. and Simmons, G.F. 1998. Hydrogen peroxide disinfection of minimally processed fruits and vegetables. Food Technology 52:48–52.

Sarig, P., Zahavi, T., Zutkhi, Y., Yannai, S., Lisker, N. and Ben-Arie, R. 1996. Ozone for control of post-harvest decay of table grapes caused by *Rhizopus stolonifer.* Physiological and Molecular Plant Pathology 48:403–415.

Schaad, N.W., Cheong, S.S., Tamaki, S., Hatziloukas, E. and Panopuolos, N.J. 1995. A combined biological and enzymatic amplification (BIO-PCR) technique to detect *Pseudomonas syringae* pv. *phaseolicola* in bean seed extracts. Phytopathology 85:243–248.

Senkel, I.A., Henderson, R.A., Jolbitado, B. and Meng, J. 1999. Use of hazard analysis critical control point and alternative treatments in the production of apple cider. Journal of Food Protection 62(7):778–785.

Sharma, M., Beuchat, L.R., Doyle, M.P. and Chen, J. 2001. Fate of Salmonellae in calcium-supplemented orange juice at refrigeration temperature. Journal of Food Protectection 64:2053–2057.

Shelef, L.A. 1983. Antimicrobial effects of spices. Journal of Food Safety 6:29–44.

Shere, J.A., Bartlett, K.J. and Kaspar, C.W. 1998. Longitudinal study of *Escherichia coli* O157:H7 dissemination on four dairy farms in Wisconsin. Applied Environment Microbiology 64(4):1390–1399.

Song, J., Leepipattanawit, R., Deng, W. and Beaudry, R.M. 1996. Hexanal vapor is a natural, metabolizable fungicide: Inhibition of fungal activity and enhancement of aroma biosynthesis in apple slices. Journal of American Society of Horticulture Science 121:937–942.

Spanier, A.M., Flores, M., James, J., Lloyd, S. and Miller, J.A. 1998. Fresh cut pineapple (*Ananas sp.)* flavour: Effect of storage. In: Development of Food Science. Food Flavours: Formation, Analysis and Packaging Influences, Contis ET, Ho CT, Mussinan CJ, Parliament TH, Shahidi F, Spanier AM (eds.), Elsevier Science Amsterdam, pp.331–343.

Sperber, W.H. 1983. Influence of water activity on food borne bacteria—a review. Journal of Food Protection 46:142–150.

Splittstoesser, D.F. 1987. Fruits and vegetable products. In: Fruit and Beverage Mycology, Beuchat LR (ed.), 2nd edition, Van Nostrand Reinhold, NY, pp.101–128.

Splittstoesser, D.F. 1991. Fungi of importance in processed fruits, In: Handbook of Applied Mycology, Arora DK, Mukerji KG, Marth EH (eds.), Marcel Dekker, Inc., New York, pp.201–219.

Splittstoesser, D.F., McLellan, M.R. and Churey, J.J. 1996. Heat resistance of *Escherichia coli* O157:H7 in apple juice. Journal of Food Protection 59(3):226–229.

Stading, M., Rindlav, W.A. and Gatenholm, P. 2001. Humidity-induced structural transitions in amylose and amylopectin films. Carbohydrate Polymers 45:209–217.

Stapleton, A. 1999. Prevention of recurrent urinary-tract infections in women. Lancet 353:7–8.

Sugar, D. and Spotts, R.A. 1993. The importance of wounds in infection of pear fruit by *Phialophora malorum* and the role of hydrostatic pressure in spore penetration of wounds. Phytopathology 83:1083–1086.

Surette, M.G., Miller, M.B. and Bassler, B.L. 1999. Quorum sensing in *E. coli*, *S. typhimurium* and *Vibrio harveyi*: A new family of genes responsible for autoinducer production. Proceedings of National Academy Sciences USA 96:1639–1644.

Swaminathan, B. and Feng, P. 1994. Rapid detection of food-borne pathogenic bacteria. Annual Review of Microbiology 48:401–426.

Trevor, S. 1997. Microbial food safety: An emerging challenge for small-scale growers. Small Farm News June–July:7–10.

Troller, J.A. 1986. Water relations of food borne bacterial pathogens—an updated review. Journal of Food Protection 49:656–670.

Venkitanarayanan, K.S., Lin, C.M., Bailey, H. and Doyle, M.P. 2002. Inactivation of *Escherichia coli* O157:H7, *Salmonella* enteritidis, and *Listeria monocytogenes* on apples, oranges, and tomatoes by lactic acid with hydrogen peroxide. Journal of Food Protection 65:100–105.

Wade, W.I.N. and Beuchat, L.R. 2003. Proteolytic fungi isolated from decayed and damaged raw tomatoes and implications associated with changes in pericarp pH favorable for survival and growth of foodborne pathogens. Journal of Food Protection 66:911–917.

Walcott, R.R. and Gitaitis, R.D. 2000. Detection of *Acidovorax avenae* subsp. *citrulli* in watermelon seed using immunomagnetic separation and the polymerase chain reaction. Plant Diseases 84:470–474.

Walden, W.C. and Hentges, D.J. 1975. Differential effects of oxygen and oxidation-reduction potential on multiplication of three species of anaerobic intestinal bacteria. Applied Microbiology 30:781–785.

Wallace, J.S., Cheasty, T. and Jones, K. 1997. Isolation of vero cytotoxin-producing *Escherichia coli* O157 from wild birds. Journal of Applied Microbiology 82:399–404.

Walls, I. and Chuyate, R. 2000. Spoilage of fruit juices by *Alicyclobacillus acidoterrestris*. Food Australia 52:286–288.

Watada, A.E. and Qi, L. 1999. Quality of fresh-cut produce. Postharvest Biology and Technology 15:201–205.

Wells, J.M. and Butterfield, J.E. 1997. *Salmonella* contamination associated with bacterial soft rot of fresh fruits and vegetables in the marketplace. Plant Diseases 81:867–872.

Wiley, R.C. 1994. Minimally processed refrigerated fruits and vegetables. Chapman and Hall, London, UK.

Wodzinsky, R.J. and Frazier, W.C. 1961. Moisturer requirements of bacteria II influence of temperature, pH and maleate concentrations on requirements of

Aerobacter aerogenes. Journal of Bacteriology 81:353–358.

Worbo, R. and Padilla-Zakour, O. 1999. Food safety and you. Venture Winter 99 1(4):1–4.

Worobo, R.W. 2000. Efficacy of the CiderSure 3500 Ultraviolet Light Unit in Apple Cider. Cornell University, Department of Food Science and Technology, Ithaca, NY, p.1–6.

Wright, K.P. and Kader, A.E. 1997a. Effect of slicing and controlled atmosphere storage on the ascorbate content and quality of strawberries and persimmons. Postharvest Biology and Technology 10:39–48.

Wright, K.P. and Kader, A.E. 1997b. Effect of controlled atmosphere storage on the quality and carotenoid content of sliced persimmons and peaches. Postharvest Biology and Technology 10:89–97.

Zhao, S.H., Mitchell, S.E., Meng, J.H., Kresovich, S., Doyle, M.P., Dean, R.E., Casa, A.M. and Weller, J.W. 2000. Genomic typing of *Escherichia coli* O157:H7 by semi-automated fluorescent AFLP analysis. Microbes Infection 2:107–113.

Zhao, T. and Doyle, M.P. 2001. Evaluation of universal pre-enrichment broth for growth of heat-injured pathogens. Journal of Food Protection 64:1751–1755.

Zhao, T., Doyle, M.P., Zhao, P., Blake, P. and Wu, F.M. 2001. Chlorine inactivation of *Escherichia coli* O157:H7 in water. Journal of Food Protection 64:1607–1609.

Zhuang, R.Y., Beuchat, L.R. and Angulo, F.J. 1995. Fate of *Salmonella montevideo* on and in raw tomatoes as affected by temperature and treatment with chlorine. Applied Environment Microbiology 61(6):2127–2131.

2
Nutritional Values of Fruits

Concepción Sánchez-Moreno, Sonia De Pascual-Teresa, Begoña De Ancos, and M. Pilar Cano

INTRODUCTION

Nutrient is defined as a substance obtained from food and used in the body to promote growth, maintenance, and repair of body tissues, or simply as a substance that provides nourishment.

Broadly speaking, nutrients are classified into two groups, namely macronutrients (also called energy-producing nutrients or energy-yielding nutrients) and micronutrients (which are characterized by their essentiality for human health and the low quantities in which they need to be ingested). Energy-producing nutrients include carbohydrates, fats, and proteins. Micronutrients often refer to vitamins and minerals.

Phytochemicals, also called bioactive compounds, are substances present in foods in low levels that may have a role in health maintenance in humans.

Fruits have proved to be essential for a balanced diet. This is believed to be mainly due to their content of vitamins, fibers, and phytochemicals, the latter being responsible in part for the antioxidant properties of fruits and foods of fruit origin.

Manufacturing processes are changing the nutritional properties of some foods. For instance, partial hydrogenation of vegetable oil results in the formation of *trans*-fatty acids, and heat treatment of protein solutions in an alkali environment results in the formation of lysinoalanine. Both of these have been shown to have detrimental health effects. On the other hand, some nutrients and bioactive compounds that are naturally present in fruits may undergo transformations during food processing that neither decrease their nutritional value nor bioactive value but may increase it by favoring their absorption and metabolism in the human body.

In general, vitamins, minerals, water, and fibers are considered to be the main nutrients contributed by fruits to a balanced diet, and thus special attention should be addressed to this group of nutrients (Villarino-Rodríguez et al., 2003).

In this chapter, we will present what we consider to be the contribution of fruits to human nutrition in order to understand how the different processing methods used in the food industry may modify their contents, structure, and biological activity in humans.

MACRONUTRIENTS

WATER

Water plays two fundamental roles as a nutrient: (1) protective and regulatory, by being a substrate of biological reactions or acting as the matrix or vehicle in which those reactions take place, and (2) an essential role as the temperature and pH regulator in the human body. Water also has a plastic function through the maintenance of the cell and tissue integrity. Around two-thirds of the human body is composed of water, and in general, the higher the metabolic activity of a given tissue, the higher its percentage of water.

Most of the body water is found within three body compartments: (1) intracellular fluid, which contains approximately 70% water, (2) extracellular fluid, which is the interstitial fluid, and (3) blood plasma. These two compartments contain ~27% water. The body controls the amount of water in each compartment by controlling the ion concentrations in those compartments. Therefore, gains or losses of electrolytes are usually followed by shifts of fluid to restore osmotic equilibrium.

Although a low intake of water has been associated with some chronic diseases, this evidence is insufficient to establish water intake recommendations. Instead, an adequate intake of water has been set by the Food and Nutrition Board of the Institute of Medicine in the United States, to prevent deleterious effects of dehydration. This adequate intake of total water is 3.7 l for men and 2.7 l for women. Fluids should represent 81% of the total intake, and water contained in foods represent the other 19% (IM, 2004).

The body has three sources of water: (1) ingested water and beverages, including fruit juices, (2) the water content of solid foods, and (3) metabolic water. Fruits have a high percentage of water that ranges from 70% to 95% of the eatable part of the fruit (see Table 2.1). For this reason, they are, together with vegetables, a very good source of water in the diet within the solid foods. The content of water in a fruit may be greatly affected by the processing technology, and in fact, some technologies used to increase the shelf life of fruits do so through the reduction of their water content. It is important to bear in mind that the water content of a fruit also changes during maturation, therefore the optimum degree of maturation of a fruit for a given processing technology may be different than for another processing technology. This will also affect the water content in the final product.

Carbohydrates

Energy is required for all body processes, growth, and physical activity. Carbohydrates are the main source of energy in the human diet. The energy produced from carbohydrate metabolism may be used directly to cover the immediate energy needs or be transformed into an energy deposit in the body in the form of fat. Carbohydrates also have a regulatory function, for instance, by selecting the microflora present in the intestines. Fructose has been known to increase plasma urate levels due to rapid fructokinase-mediated metabolism to fructose 1-phosphate. This increase in plasma urate levels seems to cause an

Table 2.1. Fruit Composition (Grams per 100 g of Edible Portion)

Fruit	Water	Carbohydrates	Protein	Fat	Fiber
Apple	86	12.0	0.3	T_r	2.0
Apricot	88	9.5	0.8	T_r	2.1
Avocado	79	5.9	1.5	12	1.8
Banana	75	20.0	1.2	0.3	3.4
Cherry	80	17.0	1.3	0.3	1.2
Grape	82	16.1	0.6	T_r	0.9
Guava	82	15.7	1.1	0.4	5.3
Kiwi fruit	84	9.1	1.0	0.4	2.1
Mango	84	15.0	0.6	0.2	1.0
Melon	92	6.0	0.1	T_r	1.0
Orange	87	10.6	1.0	T_r	1.8
Papaya	89	9.8	0.6	0.1	1.8
Peach	89	9.0	0.6	T_r	1.4
Pear	86	11.5	0.3	T_r	2.1
Pineapple	84	12.0	1.2	T_r	1.2
Plum	84	9.6	0.8	T_r	2.2
Raspberry	86	11.9	1.2	0.6	6.5
Strawberry	91	5.1	0.7	0.3	2.2
Watermelon	93	8.0	1.0	T_r	0.6

Source: Moreiras et al. (2001).

increase in plasma antioxidant capacity in humans (Lotito and Frei, 2004).

In general, the carbohydrates are classified into three groups: monosaccharides, oligosaccharides, and polysaccharides. Monosaccharides include pentoses (arabinose, xylose, and ribose) and hexoses (glucose, fructose, and galactose). Oligosaccharides are sucrose, maltose, lactose, raffinose, and stachyose. Polysaccharides include starch (composed of amylose and amylopectine, both polymers of glucose), glycogen, and other polysaccharides, which form part of fiber which we will review in the following section.

The recommended dietary allowance (RDA) for carbohydrates is 130 g/day, except in the cases of pregnancy (when it is 175 g/day) and lactation (210 g/day). With respect to the total energy consumed per day, carbohydrates should represent 45–65% (IM, 2002).

After water, carbohydrates are the main component of fruits and vegetables and represent more than 90% of their dry matter. The main monosaccharides are glucose and fructose. Their concentration may change depending on the degree of maturation of the fruit. The relative abundance of glucose and fructose also changes from one fruit to another (Table 2.2). For instance, in peaches, plums, and apricots, there is more glucose than fructose and the opposite occurs in the case of apples or pears. Other monosaccharides, such as galactose, arabinose, and xylose, are present in minimal amounts in some fruits, especially orange, lemon, or grapefruit. Fruits such as plums, pears, and cherries also contain the sugar alcohol sorbitol, which acts as a laxative because of osmotic transfer of water into the bowel.

Sucrose is the most abundant oligosaccharide in fruits; however, there are others such as maltose, melibiose, raffinose, or stachyose that have been described in grapes, and 1-kestose in bananas. Other oligosaccharides are rare in fruits. Starch is present in very low amounts in fruits, since its concentration decreases during maturation. The only exception is banana that may have concentrations of starch higher than 3% (Belitz and Grosch, 1997).

During food processing, carbohydrates are mainly involved in two kinds of reactions: on heating they darken in color or caramelize, and some of them combine with proteins to give dark colors known as the browning reaction.

Fiber

Fiber is often referred to as unavailable carbohydrate. This definition has been a controversy for years. Fiber is a generic term that includes those plant constituents that are resistant to digestion by secretions of the human gastrointestinal tract. Therefore, dietary fiber does not have a defined composition, but varies with the type of foodstuff. Perhaps we can say that fiber may not be a carbohydrate and it may be available.

Fiber has mainly a regulatory function in the human body. The role of fiber in human health has been the subject of many studies in the last 30 years. In most of these studies, the results have suggested important roles of fiber in maintaining human health.

Table 2.2. Sugar Contents of Fruits (Grams per 100 g of Edible Portion)

Fruit	Fructose	Glucose	Sucrose	Maltose	Total Sugar
Apple	5.6	1.8	2.6	–	10.0
Apricot	0.4	1.9	4.4	–	6.7
Avocado	0.1	0.1	–	–	0.2
Banana	2.9	2.4	5.9	–	11.3
Cherry	6.1	5.5	–	–	11.6
Grapefruit	1.6	1.5	2.3	0.1	5.7
Grape	6.7	6.0	0.0	0.0	12.9
Mango	3.8	0.6	8.2	–	12.7
Orange	2.0	1.8	4.4	–	8.3
Peach	4.0	4.5	0.2	–	8.7
Pear	5.3	4.2	1.2	–	10.7
Plum	3.2	5.1	0.1	0.1	8.6
Strawberry	2.3	2.6	1.3	–	6.2
Watermelon	2.7	0.6	2.8	–	6.2

Source: Belitz and Grosch (1997) and Li et al. (2002).

The role of fiber in human health is mainly protective against disease, e.g., diseases of the gastrointestinal tract, circulation related diseases and metabolic diseases (Saura-Calixto, 1987).

The major components of dietary fiber are the polysaccharides celluloses, hemicelluloses, pectins, gums, and mucilages. Lignin is the other component that is included in most definitions of fiber but it is not a carbohydrate.

Fiber may be classified as water soluble and insoluble. Gums, mucilages, some hemicelluloses, and pectins are part of the soluble fiber. Celluloses, hemicelluloses, and lignins are insoluble fibers. Fruits are good sources of both classes of fibers, especially soluble fiber. Fiber, together with vitamins, is the main nutritional reason for using fruits for a balanced diet.

There are several fiber-associated substances that are found in fruit fiber, which may have some nutritional interest. Among them are phytates, saponins, tannins, lectins, and enzyme inhibitors. Saponins, which are mainly present in some tropical fruits, may enhance the binding of bile acids to fiber and reduce cholesterol absorption. Tannins are polyphenolic compounds widely distributed in fruits, which can bind proteins and metals and reduce their absorption. Lectins, which are present in bananas and some berries, are glycoproteins that can bind specific sugars and affect the absorption of other nutrients.

The RDA for fiber is 25–30 g/day, depending on age and sex, except in the case of children from 1 to 3 years, in which case it is 19 g/day.

Dietary fiber is present in fruits in amounts that may be as high as 7% of the eatable part of the fruit (see Table 2.1). Within fiber, the most common components in fruits are celluloses, hemicelluloses, and pectins. Pectins are important in the technological process, since they may be deeply modified and this modification not only has an influence on the nutritional value of the final food, but also has an impact on the texture and palatability of the product.

Fats

Fat has three important roles as a nutrient: it is a highly concentrated source of energy, it serves as a carrier for fat-soluble vitamins and there are some fatty acids that are essential nutrients that can only be ingested with fat. Fat also serves as a carrier for some of the bioactive compounds present in fruits such as phytoestrogens and carotenoids that are lypophylic.

Fatty acids are also needed to form cell structures and to act as precursors of prostaglandins. Fatty acids are part of triglycerides, which are the principle form in which fat occurs. Fatty acids may occur naturally with various chain lengths and different numbers of double bonds. They may be saturated (butyric, caproic, caprylic, capric, lauric, palmitic, stearic, and myristic acids), monounsaturated (oleic and palmitoleic acid), and polyunsaturated (linoleic, linolenic, and arachidonic acids) also known as PUFAs. Linoleic and linolenic acids cannot be synthesized in the body and are known as essential fatty acids. They are needed to build and repair cell structures, such as the cell wall and, notably, tissues in the central nervous system, and to form the raw material for prostaglandin production. Inflammatory and other chronic diseases are noted for exhibiting a deficiency of polyunsaturated fatty acids in the bloodstream. Fatty acids that contain double carbon bonds can exist in either of two geometrically isomeric forms: *cis* and *trans*. *Trans*-fatty acids are produced in the hydrogenation process in the food industry and may play a role in atherosclerotic vascular disease (Sardesai, 1998).

In general, fat should represent between 20% and 35% of the total energy consumed per day in order to reduce risk of chronic disease while providing intakes of essential nutrients. This fat should include 10–14 g/day of linoleic acid and 1.2–1.6 g/day of linolenic acid.

Fat content in fruits is in general very low (see Table 2.1). However, in cherimoya (1%) and avocado (12–16%), the lipid levels are higher. In avocado, the most abundant fatty acids are palmitic, palmitoleic, stearic, oleic, linoleic, and linolenic acids, but the amounts may change a lot with the variety, maturity, processing, and storage conditions (Ansorena-Artieda, 2000).

Proteins

The importance of protein in the diet is primarily to act as a source of amino acids, some of which are essential because the human body cannot synthesize them. From the 20 amino acids that are part of the structure of proteins, almost half of them are considered to be essential, including isoleucine, leucine, lysine, methionine, phenylalanine, threonine, tryptophan, and valine. The RDA for proteins is 34–56 g/day, depending on age and sex, and in the case of pregnancy and lactation, it is 71 g/day. With respect to the total energy consumed per day, carbohydrates (proteins) should represent 10–35%.

Proteins are essential structural components of all cells and are needed by the human body to build and repair tissues, for the synthesis of enzymes, hormones, and others. They are also involved in the immune system, coagulation, etc. Therefore, proteins play both regulatory and plastic roles in the human body.

Proteins are made up of a long chain of amino acids, sometimes modified by the addition of heme, sugars, or phosphates. Proteins have primary, secondary, tertiary, and quaternary structures, all of which may be essential for the protein to be active. The primary structure of a protein is its amino acid sequence and the disulphide bridges, i.e., all covalent connections in a protein. The secondary structure is the way a small part, spatially near in the linear sequence of a protein, folds up into α-helix or β-pleated sheets. The tertiary structure is the way the secondary structures fold onto themselves to form a protein or a subunit of a more complex protein. The quaternary structure is the arrangement of polypeptide subunits within complex proteins made up of two or more subunits, sometimes associated with non-proteic groups. Food processing may affect these four structures in many ways, thus modifying the activity of the protein and also its nutritional value. Amino acids and proteins containing lysine or arginine as their terminal amino acids are also involved in the Maillard reactions that have a nutritional and sensory impact on processed foods.

Nitrogenated compounds are present in fruits in low percentages (0.1–1.5%). From a quantitative point of view, fruits are not a good source of proteins, however, in general berries are a better source than the rest of the fruits. Cherimoya and avocado also present higher levels of proteins than other fruits (Torija-Isasa and Cámara-Hurtado, 1999).

There are some free amino acids that may be characteristic of a certain fruit. This is the case of proline which is characteristic of oranges but cannot be found in strawberries or bananas.

MICRONUTRIENTS

VITAMINS

Thirteen vitamins have been discovered to date, and each has a specific function. Vitamins must be supplied in adequate amounts via the diet in order to meet requirements. Scientists are interested in determining the optimal levels of intake for these micronutrients in order to achieve maximum health benefit and the best physical and mental performance.

Vitamin C

Antioxidants have important roles in cell function and have been implicated in processes that have their origins in oxidative stress, including vascular processes, inflammatory damage, and cancer. L-Ascorbic acid (L-AA, vitamin C, ascorbate) is the most effective and least toxic antioxidant. Vitamin C may also contribute to the maintenance of a healthy vasculature and to a reduction in atherogenesis through the regulation of collagen synthesis, prostacyclin production, and nitric oxide (Davey et al., 2000; Sánchez-Moreno et al., 2003a, b). The second US National Health and Nutrition Examination Survey reported that a low intake of vitamin C is associated with blood concentrations of vitamin C = 0.3 mg/dl, whereas blood concentrations in well-nourished persons fluctuate between 0.8 and 1.3 mg/dl. An increase in intake of vitamin C is associated with health status (Simon et al., 2001).

Vitamin C is an essential nutrient for humans; unlike most mammals, we cannot synthesize vitamin C, and therefore must acquire it from the diet. For adults, dietary needs are met by a minimum intake of 60 mg/day. However, the preventative functions of vitamin C in aging related diseases provide compelling arguments for an increase in dietary intakes and RDAs. Men and women who consumed four daily vegetable and fruit servings had mean vitamin C intakes of 75 and 77 mg, respectively. Men and women who consumed five daily vegetable and fruit servings averaged 87 and 90 mg vitamin C, respectively (Taylor et al., 2000).

The primary contributors to daily vitamin intake are fruit juices (21% of total), whereas all fruits together contributed nearly 45% of total vitamin C intake. Relatively high amounts of vitamin C are found in strawberries and citrus fruits, although the availability of vitamin C within these food sources will be influenced by numerous factors. Virtually all of the vitamin C in Western diets is derived from fruits and vegetables. In general, fruits tend to be the best food sources of the vitamin. Especially rich sources of vitamin C are blackcurrant (200 mg/100 g), strawberry (60 mg/100 g), and the citrus fruits (30–50 mg/100 g). Not all fruits contain such levels, and apples, pears, and plums represent only a very modest source of vitamin C (3–5 mg/100 g). However, much fruit is

Table 2.3. Vitamin Content of Fruits (Value per 100 g of Edible Portion)

Fruit	Vitamin C (mg)	Vitamin E (mg) (α-tocopherol)	Vitamin A (μg RAE)	Thiamin (mg)	Riboflavin (mg)	Niacin (mg)	Pyridoxine (mg)	Folate (μg)
Apple	4.6	0.18	3	0.017	0.026	0.091	0.041	3
Apricot	10.0	0.89	96	0.030	0.040	0.600	0.054	9
Avocado	10.0	2.07	7	0.067	0.130	1.738	0.257	58
Banana	8.7	0.10	3	0.031	0.073	0.665	0.367	20
Cherry	7.0	0.07	3	0.027	0.033	0.154	0.049	4
Grape	10.8	0.19	3	0.069	0.070	0.188	0.086	2
Guava	183.5	0.73	31	0.050	0.050	1.200	0.143	14
Kiwi fruit	75.0	–	9	0.020	0.050	0.500	–	–
Orange	53.2	0.18	11	0.087	0.040	0.282	0.060	30
Papaya	61.8	0.73	55	0.027	0.032	0.338	0.019	38
Passion fruit	30.0	0.02	64	0.000	0.130	1.500	0.100	14
Peach	6.6	0.73	16	0.024	0.031	0.806	0.025	4
Pear	4.2	0.12	1	0.012	0.025	0.157	0.028	7
Pineapple	36.2	0.02	3	0.079	0.031	0.489	0.110	15
Plum	9.5	0.26	17	0.028	0.026	0.417	0.029	5
Raspberry	26.2	0.87	2	0.032	0.038	0.598	0.055	21
Strawberry	58.8	0.29	1	0.024	0.022	0.386	0.047	24

Source: USDA (2004).
Note: RAE—retinol activity equivalents.

eaten raw and the low pH of fruits stabilizes the vitamin during storage (Davey et al., 2000).

A summary of the average vitamin C content of certain fruits (mg per 100 g of edible portion) is given in Table 2.3.

Vitamin E

Vitamin E is the generic term for a family of related compounds known as tocopherols and tocotrienols. Naturally occurring structures include four tocopherols (α-, β-, γ-, and δ-) and four tocotrienols (α-, β-, γ-, and δ-). Of the eight naturally occurring forms of α-tocopherol (*RRR*-, *RSR*-, *RRS*-, *RSS*-, *SRR*-, *SSR*-, *SRS*-, and *SSS*-), only one form, *RRR*-α-tocopherol, is maintained in human plasma and therefore is the active form of vitamin E (Trumbo et al., 2003).

α-Tocopherol is the predominant tocopherol form found naturally in foods, except in vegetable oils and nuts, which may contain high proportions of γ-tocopherol (Bramley et al., 2000).

The vitamin E activity of tocopherols is frequently calculated in international units (IU), with 1 IU defined as the biological activity of 1 mg *all-rac*-α-tocopheryl acetate. Recently, the US National Research Council has suggested that vitamin E activity could be expressed as *RRR*-α-tocopherol equivalents (α-TE). One α-TE is defined as the biological activity of 1 mg *RRR*-α-tocopherol. One IU is equal to 0.67 α-TE (Brigelius-Flohé et al., 2002).

Recent research evidences the role of vitamin E in reducing the risk of developing degenerative disease. This role is suggested on the hypothesis that preventing free radical-mediated tissue damage (e.g., to cellular lipids, proteins, or DNA) may play a key role in delaying the pathogenesis of a variety of degenerative diseases (Bramley et al., 2000; Sánchez-Moreno et al., 2003b).

There is some controversy about the optimum range of vitamin E intake for associated health benefits. Some authors recommend intakes of 130–150 IU/day or about 10 times the US Food and Nutritional Board (15 mg/day) on the basis of the protection in relation to cardiovascular disease. Other authors indicate that the optimal plasma α-tocopherol concentration for protection against cardiovascular disease and cancer is >30 mmol/l at common plasma lipid concentrations. A daily dietary intake of only about 15–30 mg α-tocopherol would be sufficient to maintain this plasma level, an amount that could be obtained from the diet (Bramley et al., 2000).

The richest sources of vitamin E are vegetable oils and the products made from them, followed by bread and bakery products and nuts. Vegetables and fruits contain little amount of vitamin E (Bramley et al., 2000).

Table 2.3 shows the range of concentrations (mg per 100 g of edible portion) of vitamin E (α-tocopherol) from certain fruits.

Vitamin B-1, B-2, B-3, B-6, Folate

Thiamin (vitamin B-1), riboflavin (vitamin B-2), niacin (vitamin B-3), and pyridoxine (vitamin B-6), are used as coenzymes in all parts of the body. They participate in the metabolism of fats, carbohydrates, and proteins. They are important for the structure and function of the nervous system (IM, 1998; ASNS, 2004; Lukaski, 2004).

Thiamin diphosphate is the active form of thiamin. It serves as a cofactor for several enzymes involved in carbohydrate catabolism. The suggested intake for thiamin is 1.15 g/day. Thiamin requirement depends on energy intake, thus the suggested RDA is 0.5 mg/1000 kcal.

Riboflavin is required for oxidative energy production. Because riboflavin is found in a variety of foods, either from animal or vegetable origin, riboflavin deficiency is uncommon in Western countries. Recommendations for riboflavin intake are based on energy intake. It is suggested that an intake of 0.6 mg/1000 kcal will meet the needs of most healthy adults. The current RDA is 1.2 g/day.

Niacin (nicotinic acid and nicotinamide). Nicotinamide is a precursor of nicotinamide adenine (NAD), nucleotide, and nicotinamide adenine dinucleotide phosphate (NADP), in which the nicotinamide moiety acts as electron acceptor or hydrogen donor, respectively, in many biological redox reactions. The RDA is expressed in milligram niacin equivalents (NE) in which 1 mg NE = 1 mg niacin or 60 mg tryptophan. For individuals above 13 years of age, the RDA is 16 mg NE/day for males and 14 mg NE/day for females.

The chemical name of vitamin B-6 is pyridoxine hydrochloride. Other forms of vitamin B-6 include pyridoxal, and pyridoxamine. Vitamin B-6 is one of the most versatile enzyme cofactors. Vitamin B-6 in the form of pyridoxal phosphate acts as a cofactor for transferases, transaminases, and decarboxylases, used in transformations of amino acids. The RDA for vitamin B-6 is 1.6 mg/day.

Folate is an essential vitamin that is also known as folic acid and folacin. The metabolic role of folate is as an acceptor and donor of one-carbon units in a variety of reactions involved in amino acid and nucleotide metabolism. The RDA for folate is 400 μg/day. Excellent food sources of folate from fruits (>55 μg/day) include citrus fruits and juices.

Table 2.3 shows the range of concentrations (amount per 100 g of edible portion) of thiamin, riboflavin, niacin, pyridoxine, and folate from selected fruits.

Minerals

An adequate intake of minerals is essential for a high nutritional quality of the diet, and it also contributes to the prevention of chronic nutrition related diseases. However, even in Western societies, intake of some minerals such as calcium, iron, and zinc is often marginal in particular population groups e.g., small children or female adolescents, while the intake of sodium or magnesium, reach or exceed the recommendations.

Table 2.4 shows the mineral content (amount per 100 g of edible portion) from certain fruits.

Iron

Iron (Fe) is an essential nutrient that carries oxygen and forms part of the oxygen-carrying proteins, hemoglobin in red blood cells and myoglobin in muscle. It is also a necessary component of various enzymes. Body iron is concentrated in the storage forms, ferritin and hemosiderin, in bone marrow, liver, and spleen. Body iron stores can usually be estimated from the amount of ferritin protein in serum. Transferrin protein in the blood transports and delivers iron to cells (Lukaski, 2004).

The body normally regulates iron absorption in order to replace the obligatory iron losses of about 1–1.5 mg/day. The RDAs for iron are 10 mg for men over 10 years and for women over 50 years, and 15 mg for 11- 50-year-old females (ASNS, 2004).

Non-heme iron is the source of iron in the diet from plant foods. The absorption of non-heme iron is strongly influenced by dietary components, which bind iron in the intestinal lumen. Non-heme iron absorption is usually from 1% to 20%. The main inhibitory substances are phytic acid from cereal grains and legumes such as soy, and polyphenol compounds from beverages such as tea and coffee. The main enhancers of iron absorption are ascorbic acid from fruits and vegetables, and the partially digested peptides from muscle tissues (Frossard et al., 2000; Lukaski, 2004).

Table 2.4. Mineral Content of Fruits (Value per 100 g of Edible Portion)

Fruit	Fe (mg)	Ca (mg)	P (mg)	Mg (mg)	K (mg)	Na (mg)	Zn (mg)	Cu (mg)	Se (μg)
Apple	0.12	6	11	5	107	1	0.04	0.027	0.0
Apricot	0.39	13	23	10	259	1	0.20	0.078	0.1
Avocado	0.55	12	52	29	485	7	0.64	0.190	0.4
Banana	0.26	5	22	27	358	1	0.15	0.078	1.0
Cherry	0.36	13	21	11	222	0	0.07	0.060	0.0
Grape	0.36	10	20	7	191	2	0.07	0.127	0.1
Guava	0.31	20	25	10	284	3	0.23	0.103	0.6
Kiwi fruit	0.41	26	40	30	332	5	–	–	–
Orange	0.10	40	14	10	181	0	0.07	0.045	0.5
Papaya	0.10	24	5	10	257	3	0.07	0.016	0.6
Passion fruit	1.60	12	68	29	348	28	0.10	0.086	0.6
Peach	0.25	6	20	9	190	0	0.17	0.068	0.11
Pear	0.17	9	11	7	119	1	0.10	0.082	0.1
Pineapple	0.28	13	8	12	115	1	0.10	0.099	0.1
Plum	0.17	6	16	7	157	0	0.10	0.057	0.0
Raspberry	0.69	25	29	22	151	1	0.42	0.090	0.2
Strawberry	0.42	16	24	13	153	1	0.14	0.048	0.4

Source: USDA (2004).

Calcium

Calcium (Ca) is the most common mineral in the human body. Calcium is a nutrient in the news because adequate intakes are an important determinant of bone health and reduced risk of fracture or osteoporosis (Frossard et al., 2000).

Approximately 99% of total body calcium is in the skeleton and teeth, and 1% is in the blood and soft tissues. Calcium has the following major biological functions: (a) structural as stores in the skeleton, (b) electrophysiological—carries a charge during an action potential across membranes, (c) intracellular regulator, and (d) as a cofactor for extracellular enzymes and regulatory proteins (Frossard et al., 2000; ASNS, 2004).

The dietary recommendations vary with age. An amount of 1300 mg/day for individuals aged 9–18 years, 1000 mg/day for individuals aged 19–50 years, and 1200 mg/day for individuals over the age of 51 years. The recommended upper level of calcium is 2500 mg/day (IM, 1997; ASNS, 2004).

Calcium is present in variable amounts in all the foods and water we consume, although vegetables are one of the main sources. Of course, dairy products are excellent sources of calcium.

Phosphorus

Phosphorus (P) is an essential mineral that is found in all cells within the body. The body of the human adult contains about 400–500 g. The greatest amount of body phosphorus can be found primarily in bone (85%) and muscle (14%). Phosphorus is primarily found as phosphate (PO_4^{2-}). The nucleic acids—deoxyribonucleic acid (DNA) and ribonucleic acid (RNA)—are polymers based on phosphate ester monomers. The high-energy phosphate bond of ATP is the major energy currency of living organisms. Cell membranes are composed largely of phospholipids. The inorganic constituents of bone are primarily a calcium phosphate salt. The metabolism of all major metabolic substrates depends on the functioning of phosphorus as a cofactor in a variety of enzymes and as the principal reservoir for metabolic energy (ASNS, 2004).

The RDAs for phosphorus (mg/day) are based on life stage groups. Among others, for youth 9–18 years, the RDA is 1250 mg, which indicates the higher need for phosphorus during the adolescent growth. Adults 19 years and older have a RDA of 700 mg (IM, 1997; ASNS, 2004).

Magnesium

Magnesium (Mg) is the fourth most abundant cation in the body, with 60% in the bone and 40% distributed equally between muscle and non-muscular soft tissue. Only 1% of magnesium is extracellular. Magnesium has an important role in at least 300 fundamental enzymatic reactions, including the transfer of phosphate groups, the acylation of coenzyme A in the initiation of fatty acid oxidation, and the hydrolysis of phosphate and pyrophosphate. In addition, it has a key role in neurotransmission and immune function. Magnesium acts as a calcium antagonist and interacts with nutrients, such as potassium, vitamin B-6, and boron (Lukaski, 2004; ASNS, 2004).

The RDA, from the US Food and Nutrition Board, vary according to age and sex. The RDAs for magnesium are 320 and 420 mg/day for women and men (adults over 30 years), respectively (IM, 1997; ASNS, 2004).

Potassium

Potassium (K) in the form of K^+ is the most essential cation of the cells. Its high intracellular concentration is regulated by the cell membrane through the sodium–potassium pump. Most of the total body potassium is found in muscle tissue (ASNS, 2004).

The estimated minimum requirement for potassium for adolescents and adults is 2000 mg or 50 mEq/day. The usual dietary intake for adults is about 100 mEq/day. Most foods contain potassium. The best food sources are fruits, vegetables, and juices (IM, 2004; ASNS, 2004).

Sodium

Sodium (Na) is the predominant cation in extracellular fluid and its concentration is under tight homeostatic control. Excess dietary sodium is excreted in the urine. Sodium acts in consort with potassium to maintain proper body water distribution and blood pressure. Sodium is also important in maintaining the proper acid–base balance and in the transmission of nerve impulses (ASNS, 2004).

The RDAs for sodium ranges from 120 mg/day for infants to 500 mg/day for adults and children above 10 years. Recommendations for the maximum amount of sodium that can be incorporated into a healthy diet range from 2400 to 3000 mg/day. The current recommendation for the general healthy population to reduce sodium intake has been a matter of debate in the scientific community (Kumanyika and Cutler, 1997; IM, 2004; ASNS, 2004).

Zinc

Zinc (Zn) acts as a stabilizer of the structures of membranes and cellular components. Its biochemical function is as an essential component of a large number of zinc-dependent enzymes, particularly in the synthesis and degradation of carbohydrates, lipids, proteins, and nucleic acids. Zinc also plays a major role in gene expression (Frossard et al., 2000; Lukaski, 2004).

The RDAs for zinc are 8 and 11 mg/day for women and men, respectively (ASNS, 2004).

Copper

Copper (Cu) is utilized by most cells as a component of enzymes that are involved in energy production (cytochrome oxidase), and in the protection of cells from free radical damage (superoxide dismutase). Copper is also involved with an enzyme that strengthens connective tissue (lysyl oxidase) and in brain neurotransmitters (dopamine hydroxylase) (ASNS, 2004).

The estimated safe and adequate intake for copper is 1.5–3.0 mg/day (ASNS, 2004).

Selenium

Selenium (Se) is an essential trace element that functions as a component of enzymes involved in antioxidant protection and thyroid hormone metabolism (ASNS, 2004).

The RDAs are 70 μg/day for adult males, and 55 μg/day for adult females. Foods of low protein content, including most fruits and vegetables, provide little selenium. Food selenium is absorbed with efficiencies of 60–80% (ASNS, 2004).

BIOACTIVE COMPOUNDS

CAROTENOIDS

Carotenoids are lipid-soluble plant pigments common in photosynthetic plants. The term carotenoid summarizes a class of structurally related pigments, mainly found in plants. At present, more than 600 different carotenoids have been identified, although only about two dozens are regularly consumed by humans. The most prominent member of this group is β-carotene. Most carotenoids are structurally arranged as two substituted or unsubstituted ionone rings separated by four isoprene units containing nine conjugated double bonds, such as α- and β-carotene, lutein, and zeaxanthin, and α- and β-cryptoxanthin (Goodwin and Merce, 1983; Van den Berg et al., 2000). These carotenoids, along with lycopene, an acylic biosynthetic precursor of β-carotene, are most commonly consumed and are most prevalent in human plasma (Castenmiller and West, 1998).

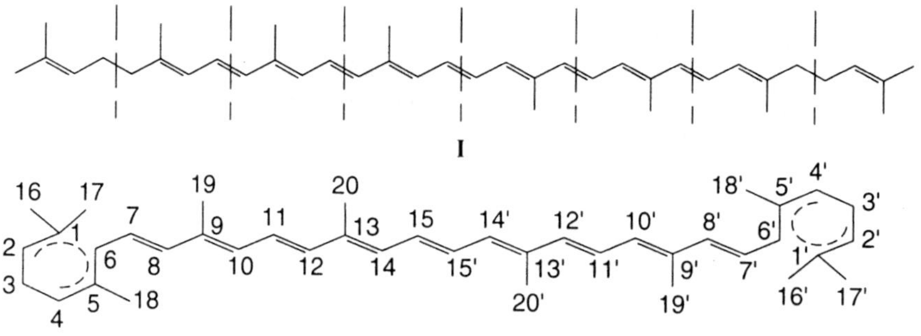

I

II

Figure 2.1. Structure and numbering of the carotenoid carbon skeleton. (*Source*: Shahidi et al., 1998.)

All carotenoids can be derived from an acyclic C40H56 unit by hydrogenation, dehydrogenation, cyclization and/or oxidation reactions (Fig. 2.1). All specific names are based on the stem name carotene, which corresponds to the structure and numbering in Figure 2.1 (Shahidi et al., 1998).

The system of conjugated double bonds influences their physical, biochemical, and chemical properties. Based on their composition, carotenoids are subdivided into two groups. Those contain only carbon and hydrogen atoms, which are collectively assigned as carotenes, e.g., β-carotene, α-carotene, and lycopene. The majority of natural carotenoids contain at least one oxygen function, such as keto, hydroxy, or epoxy groups, and are referred to as xanthophylls or oxocarotenoids. In their natural sources, carotenoids mainly occur in the *all-trans* configuration (Goodwin and Merce, 1983; Van den Berg et al., 2000).

Carotenoid pigments are of physiological interest in human nutrition, since some of them are vitamin A precursors, especially β-carotene. α-Carotene, and α- and β-cryptoxanthin possess provitamin A activity, but to a lesser extent than β-carotene. On the basis of epidemiological studies, diet rich in fruits and vegetables containing carotenoids is suggested to protect against degenerative diseases such as cancer, cardiovascular diseases, and macular degeneration. Recent clinical trials on supplemental β-carotene have reported a lack of protection against degenerative diseases. Much of the evidence has supported the hypothesis that lipid oxidation or oxidative stress is the underlying mechanism in such diseases. To date carotenoids are known to act as antioxidants *in vitro*. In addition to quenching of singlet oxygen, carotenoids may react with radical species either by addition reactions or through electron transfer reactions, which results in the formation of the carotenoid radical cation (Canfield et al., 1992; Sies and Krinsky, 1995; Van den Berg et al., 2000; Sánchez-Moreno et al., 2003c).

Carotenoid intake assessment has been shown to be complicated mainly because of the inconsistencies in food composition tables and databases. Thus, there is a need for more information about individual carotenoids. The estimated dietary intake of carotenoids in Western countries is in the range of 9.5–16.1 mg/day. To ensure the intake of a sufficient quantity of antioxidants, the human diet, which realistically contains 100–500 g/day of fruit and vegetables, should contain a high proportion of carotenoid-rich products. No formal diet recommendation for carotenoids has yet been established, but some experts suggest intake of 5–6 mg/day, which is about twice the average daily U.S. intake. In the case of vitamin A, for adult human males, the RDA is 1000 μg retinyl Eq/day, and for adult females, 800 μg retinyl Eq/day (O'Neill et al., 2001; Trumbo et al., 2003).

Citrus fruits are the major source of β-cryptoxanthin in the Western diet. The major fruit contributors to the carotenoid intake in Western diets are orange (β-cryptoxanthin and zeaxanthin), tangerine (β-cryptoxanthin), peach (β-cryptoxanthin and zeaxanthin), watermelon (lycopene), and banana (α-carotene). Other relatively minor contributors are kiwi fruit, lemon, apple, pear, apricot, cherry, melon, strawberry, and grape (Granado et al., 1996; O'Neill et al., 2001).

FLAVONOIDS

Flavonoids are the most common and widely distributed group of plant phenolics. Over 5000 different flavonoids have been described to date and they are classified into at least 10 chemical groups. Among

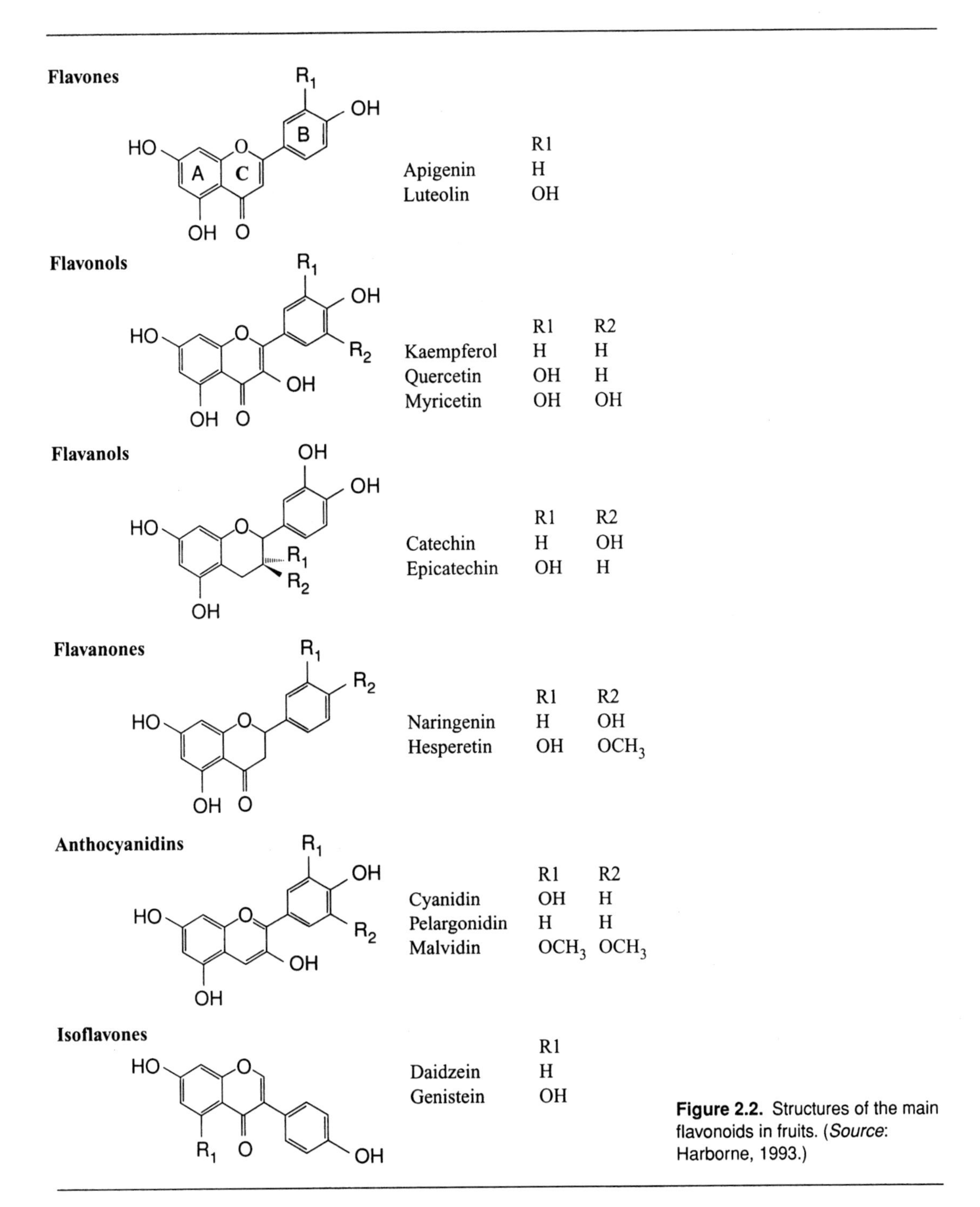

Figure 2.2. Structures of the main flavonoids in fruits. (*Source*: Harborne, 1993.)

them, flavones, flavonols, flavanols, flavanones, anthocyanins, and isoflavones are particularly common in fruits (Fig. 2.2). The most-studied members of these groups are included in Table 2.5, along with some of their fruit sources (Bravo, 1998).

Numerous epidemiological studies support the concept that regular consumption of foods and beverages rich in antioxidant flavonoids is associated with a decreased risk of cardiovascular disease mortality. There is also scientific evidence that flavonoids may

Table 2.5. Classification of Flavonoids and Their Presence in Fruits

Subclasses	Flavonoids	Fruits
Flavones	Apigenin, luteolin	Apples, blueberries, grapefruit, grapes, oranges
Flavonols	Quercetin, kaempferol, myricetin	Apples, berries, plums
Flavanols	Catechin, epicatechin, epigallocatechin gallate	Apples, berries, grapes, plums
Flavanones	Hesperetin, naringenin	Citrus fruits
Anthocyanins	Cyanidin, pelargonidin, malvidin	Berries, grapes
Isoflavones	Genistein, daidzein	Currants, passion fruit

Source: De Pascual-Teresa et al. (2000) and Franke et al. (2004).

protect against some cancers. It has been shown in the past that flavonoid content and structure may change with technological processes increasing or decreasing their contents and biological activity (García-Alonso et al., 2004).

Most of the existing flavonoids in fruits have shown antioxidant activity in *in vitro* studies, and almost all the fruits that have been screened for their antioxidant activity have shown to a lower or higher extent some antioxidant and radical scavenger activity.

Other biological activities of flavonoids seem to be independent of their antioxidant activity. This is the case of the oestrogen-like activity showed by isoflavones. Isoflavones have also shown an effect on total and HDL cholesterol levels in blood.

Anthocyanins have shown to be effective in decreasing capillary permeability and fragility and also have anti-inflammatory and anti-oedema activities.

Flavonols inhibit COX-2 activity and thus may play a role in the prevention of inflammatory diseases and cancer (De Pascual-Teresa et al., 2004).

Factors like modification on the flavonoid structure or substitution by different sugars or acids may deeply affect the biological activity of flavonoids and in this sense different processing of the fruits may also influence their beneficial properties for human health.

PHYTOSTEROLS

Plant-based foods contain a large number of plant sterols, also called phytosterols, as minor lipid components. Plant sterols have been reported to include over 250 different sterols and related compounds. The most common sterols in fruits are β-sitosterol, and its 22-dehydro analogue stigmasterol, campesterol and avenasterol (4-desmethylsterols). Chemical structures of these sterols are similar to cholesterol differing in the side chain (Fig. 2.3). β-Sitosterol and stigmasterol have ethyl groups at C-24, and campesterol has a methyl group at the same position. Plant sterols can exist as free plant sterols, and as bound conjugates: esterified plant sterols (C-16 and C-18

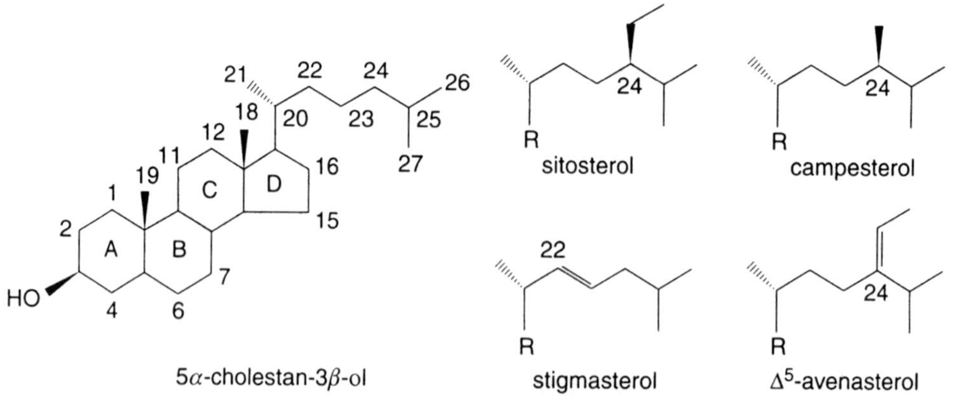

Figure 2.3. Structures of cholesterol (5α-cholestan-3β-ol), sitosterol, campesterol, stigmasterol, and Δ^5-avenasterol. (*Source*: Piironen et al., 2003.)

fatty acid esters, and phenolic esters), plant steryl glycosides (β-D-glucose), and acylated plant steryl glycosides (esterified at the 6-hydroxy group of the sugar moiety). All of these forms are integrated into plant cell membranes (Piironen et al., 2000, 2003).

Plant sterols are not endogenously synthesized in humans, therefore, are derived from the diet entering the body only via intestinal absorption. Since plant sterols competitively inhibit cholesterol intestinal uptake, a major metabolic effect of dietary plant sterols is the inhibition of absorption and subsequent compensatory stimulation of the synthesis of cholesterol. The ultimate effect is the lowering of serum cholesterol owing to the enhanced elimination of cholesterol in stools. Consequently, the higher the dietary intake of plant sterols from the diet, the lower is the cholesterol absorption and the lower is the serum cholesterol level (Ling and Jones, 1995; De Jong et al., 2003; Trautwein et al., 2003).

The usual human diet contains currently around 145–405 mg/day of plant sterols. Dietary intake values depend on type of food intake. Intakes, especially that of β-sitosterol, are increased two- to threefold in vegetarians. For healthy humans, the absorption rate of plant sterols is usually less than 5% of dietary levels. Serum sterol levels of around 350–270 μg/dl in non-vegetarians have been observed (Ling and Jones, 1995; Piironen et al., 2000).

Vegetables and fruits are generally not regarded to be as good a source of sterols as cereals or vegetable oils. The plant sterol content in a food may vary depending on many factors, such as genetic background, growing conditions, tissue maturity, and postharvest changes (Piironen et al., 2000). There are scarce data available on the content of plant sterols in the edible portion of fruits (Wiehrauch and Gardner, 1978; Morton et al., 1995). Recently, the fruits more commonly consumed in Finland have been analyzed. Total sterols ranged from 6 mg/100 g (red currant) to 22 mg/100 g (lingonberry) of fresh weight, in all fruits, except avocado, which contained significantly more sterols, 75 mg/100 g. The content on dry weight basis was above 100 mg/100 g in most products. Peels and seeds were shown to contain more sterols than edible parts (Piironen et al., 2003). In Sweden, the range of plant sterol for 14 fruits is 1.3–44 mg/100 g (fresh weight), only passion fruit contains more than 30 mg/100 g (Normen et al., 1999). Among the fruits found in both reports, orange shows the highest plant sterol content, and banana the lowest. In all the items analyzed, β-sitosterol occurred at the highest concentrations, followed by campesterol or stigmasterol. Detectable amounts of five-saturated plant stanols, sitostanol, and campestanol, were found in specific fruits such as pineapple.

REFERENCES

Ansorena-Artieda D. 2000. Frutas y Frutos Secos. In: Astiasarán I, Martinez A (Eds), Alimentos, Composición y Propiedades. McGraw-Hill International, New York, pp. 191–211.

ASNS (American Society for Nutritional Sciences). 2004. http://www.nutrition.org (accessed 2004).

Belitz HD, Grosch W (Eds). 1997. Química de los alimentos. Acribia S.A., Zaragoza.

Bramley M, Elmadfa I, Kafatos A, Kelly FJ, Manios Y, Roxborough HE, Schuch W, Sheehy PJA, Wagner KH. 2000. Vitamin E. Journal of the Science of Food and Agriculture 80:913–938.

Bravo L. 1998. Polyphenols: chemistry, dietary sources, metabolism, and nutritional significance. Nutrition Reviews 56:317–333.

Brigelius-Flohé R, Kelly FJ, Salonen JT, Neuzil J, Zingg JM, Azzi A. 2002. The European perspective on vitamin E: current knowledge and future research. The American Journal of Clinical Nutrition 76:703–716.

Canfield IM, Forage JW, Valenzuela JG. 1992. Carotenoids as cellular antioxidants. Proceeding of the Society of Experimental Biology and Medicine 200:260–265.

Castenmiller JJM, West CE. 1998. Bioavailability and bioconversion of carotenoids. Annual Review of Nutrition 18:19–38.

Davey MW, Montagu MV, Inze D, Sanmartin M, Kanellis A, Smirnoff N, Benzie IJJ, Strain JJ, Favell D, Fletcher J. 2000. Plant L-ascorbic acid: chemistry, function, metabolism, bioavailability and effects of processing. Journal of the Science of Food and Agriculture 80:825–860.

De Jong A, Plat J, Mensink RP. 2003. Metabolic effects of plant sterols and stanols. The Journal of Nutritional Biochemistry 14:362–369.

De Pascual-Teresa S, Johnston KL, DuPont MS, O'Leary KA, Needs PW, Morgan LM, Clifford MN, Bao YP, Williamson G. 2004. Quercetin metabolites regulate cyclooxygenase-2 transcription in human lymphocytes ex vivo but not in vivo. The Journal of Nutrition 134:552–557.

De Pascual-Teresa S, Santos-Buelga C, Rivas-Gonzalo JC. 2000. Quantitative analysis of flavan-3-ols in Spanish foodstuffs and beverages. Journal of Agricultural and Food Chemistry 48:5331–5337.

Franke AA, Custer LJ, Arakaki C, Murphy SP. 2004. Vitamin C and flavonoid levels of fruits and vegetables consumed in Hawaii. Journal of Food Composition and Analysis 17:1–35.

Frossard E, Bucher M, Machler F, Mozafar A, Hurrell R. 2000. Potential for increasing the content and bioavailability of Fe, Zn and Ca in plants for human nutrition. Journal of the Science of Food and Agriculture 80:861–879.

Garcia-Alonso M, De Pascual-Teresa S, Santos-Buelga C, Rivas-Gonzalo JC. 2004. Evaluation of the antioxidant properties of fruits. Food Chemistry 84:13–18.

Goodwin TW, Merce EI. 1983. Introduction to Plant Biochemistry. Pergamon Press Ltd., London.

Granado F, Olmedilla B, Blanco I, Rojas-Hidalgo E. 1996. Major fruit and vegetables contributors to the main serum carotenoids in Spanish diet. European Journal of Clinical Nutrition 50:246–250.

Harborne JB. 1993. The Flavonoids. Advance in Research Since 1986. Chapman & Hall, London.

IM (Institute of Medicine). 1997. Committee on the Scientific Evaluation of Dietary Reference Intakes. In: Dietary Reference Intakes for Calcium, Phosphorus, Magnesium, Vitamin D, and Fluoride. National Academy Press, Washington, DC.

IM (Institute of Medicine). 1998. Committee on the Scientific Evaluation of Dietary Reference Intakes. In: Dietary Reference Intakes for Thiamin, Riboflavin, Niacin, Vitamin B6, Folate, Vitamin B12, Pantothenic Acid, Biotin, and Choline. National Academy Press, Washington, DC.

IM (Institute of Medicine). 2002. Food and Nutrition Board. In: Dietary Reference Intakes for Energy, Carbohydrate, Fiber, Fat, Fatty Acids, Cholesterol, Protein, and Amino Acids (Macronutrients). National Academy Press, Washington, DC.

IM (Institute of Medicine). 2004. Food and Nutrition Board. In: Dietary Reference Intakes for Water, Potassium, Sodium, Chloride, and Sulfate. National Academy Press, Washington, DC.

Kumanyika SK, Cutler JA. 1997. Dietary sodium reduction. Is there cause for concern? Journal of the American College of Nutrition 16:192–203.

Li BW, Andrews KW, Pehrsson PR. 2002. Individual sugars, soluble, and insoluble dietary fibre contents of 70 high consumption foods. Journal of Food Composition and Analysis 15:715–723.

Ling WH, Jones PJH. 1995. Dietary phytosterols: a review of metabolism, benefits, and side effects. Life Sciences 57:195–206.

Lotito SB, Frei B. 2004. The increase in human plasma antioxidant capacity after apple consumption is due to the metabolic effect of fructose on urate, not apple-derived antioxidant flavonoids. Free Radical Biology and Medicine 37:251–258.

Lukaski HC. 2004. Vitamin and mineral status: effects on physical performance. Nutrition 20:632–644.

Moreiras O, Carbajal A, Cabrera L, Cuadrado C. 2001. Tablas de Composición de los alimentos. Ediciones Pirámide (Grupo Anaya), Madrid.

Morton GM, Lee SM, Buss DH, Lawrence P. 1995. Intakes and major dietary sources of cholesterol and phytosterols in the British diet. Journal of Human Nutrition and Dietetics 8:429–440.

Normen L, Johnsson M, Andersson H, Van Gameren Y, Dutta P. 1999. Plant sterols in vegetables and fruits commonly consumed in Sweden. European Journal of Clinical Nutrition 38:84–89.

O'Neill ME, Carroll Y, Corridan B, Olmedilla B, Granado F, Blanco I, Berg H, Van-den Hininger I, Rousell AM, Chopra M, Southon S, Thurnham DI. 2001. A European carotenoid database to assess carotenoid intakes and its use in a five-country comparative study. British Journal of Nutrition 85:499–507.

Piironen V, Lindsay DG, Miettinen TA, Toivo J, Lampi A-M. 2000. Plant sterols: biosynthesis, biological function and their importance to human nutrition. Journal of the Science of Food and Agriculture 80:939–966.

Piironen V, Toivo J, Puupponen-Pimia R, Lampi A-M. 2003. Plant sterols in vegetables, fruits and berries. Journal of the Science of Food and Agriculture 83:330–337.

Sánchez-Moreno C, Cano MP, De Ancos B, Plaza L, Olmedilla B, Granado F, Martín A. 2003a. High-pressurized orange juice consumption affects plasma vitamin C, antioxidative status and inflammatory markers in healthy humans. The Journal of Nutrition 133:2204–2209.

Sánchez-Moreno C, Cano MP, De Ancos B, Plaza L, Olmedilla B, Granado F, Martín A. 2003b. Effect of orange juice intake on vitamin C concentrations and biomarkers of antioxidant status in humans. The American Journal of Clinical Nutrition 78:454–460.

Sánchez-Moreno C, Plaza L, De Ancos B, Cano MP. 2003c. Quantitative bioactive compounds assessment and their relative contribution to the antioxidant capacity of commercial orange juices. Journal of the Science of Food and Agriculture 83:430–439.

Sardesai VM. 1998. Introduction to Clinical Nutrition. Marcel Dekker Inc., New York.

Saura-Calixto F. 1987. Dietary fibre complex in a sample rich in condensed tannins and uronic acid. Food Chemistry 23:95–106.

Shahidi F, Metusalach, Brown JA. 1998. Carotenoid pigment in seafoods and aquaculture. Critical Reviews in Food Science and Nutrition 38:1–69.

Sies H, Krinsky NI. 1995. Antioxidant vitamins and β-carotene in disease prevention. The American Journal of Clinical Nutrition 62S:1299S–1540S.
Simon JA, Hudes ES, Tice JA. 2001. Relation of serum ascorbic acid to mortality among US adults. Journal of the American College of Nutrition 20:255–263.
Taylor JS, Hamp JS, Johnston CS. 2000. Low intakes of vegetables and fruits, especially citrus fruits, lead to inadequate vitamin C intakes among adults. European Journal of Clinical Nutrition 54:573–578.
Torija-Isasa ME, Cámara-Hurtado MM. 1999. Hortalizas, verduras y frutas. In: Hernández-Rodriguez M, Sastre-Gallego A (Eds), Tratado de Nutrición. Díaz de Santos, Madrid, pp. 413–423.
Trautwein EA, Guus SM, Duchateau JE, Lin Y, Melnikov SM, Molhuizen HOF, Ntanios FY. 2003. Proposed mechanisms of cholesterol-lowering action of plant sterols. European Journal of Lipid Science and Technology 105:171–185.
Trumbo PR, Yates AA, Schlicker-Renfro S, Suitor C. 2003. Dietary reference intakes: revised nutritional equivalents for folate, vitamin E and provitamin A carotenoids. Journal of Food Composition and Analysis 16:379–382.
USDA. 2004. National Nutrient Database for Standard Reference, Release 16-1.
Van den Berg H, Faulks R, Granado F, Hirschberg J, Olmedilla B, Sandmann G, Southon S, Stahl W. 2000. The potential for the improvement of carotenoid levels in foods and the likely systemic effects. Journal of the Science of Food and Agriculture 80:880–912.
Villarino-Rodríguez A, García-Fernández MC, Garcia-Arias MT. 2003. In: García-Arias MT, García-Fernández MC (Eds), Frutas y Hortalizas en Nutrición y Dietética. Universidad de León. Secretariado de publicaciones y medios audiovisuales, León, pp. 353–366.
Wiehrauch JL, Gardner JM. 1978. Sterols content of food of plant origin. Journal of the American Dietetic Association 73:39–47.

3
Fruit Processing: Principles of Heat Treatment

Imre Körmendy

REVIEW OF HEAT TREATMENT PROCESSES AND BASIC IDEAS OF SAFETY

SURVEY OF INDUSTRIAL PROCESSES

The greatest quantity of processed fruit is preserved by heat treatment. A wide range of practical and theoretical knowledge is needed for an industrial process, as Figure 3.1 illustrates. Food processing involves the fields of microbiology, plant biology, thermophysics, food rheology and chemistry, packaging technique, unit operations, reactor techniques, construction and materials science, machinery, and electrophysics.

The most important factors (besides the nature of the raw material and type of product) for constructing a plant are as follows:

- type and size of the container (e.g., from small cans up to large tanks), and
- mode of heat treatment (e.g., batch or continuous pasteurization of closed containers; full aseptic process or some combination; temperature and pressure above or below 100°C and absolute pressure of 100 kPa).

BACKGROUND OF MICROBIAL SAFETY

Early perceptions (Ball and Olson, 1957) initiated the use of first-order (exponential) inactivation kinetics for microbes, including constants as D, D_r, T_r, z and calculation of the heat treatment equivalent (F-value). These early ideas continue to be followed. Meanwhile, concepts have been refined with the availability of computer programs designed to calculate "risk analysis" for food safety in industrial food processing.

The two safety aspects are health protection and control of spoilage. The critical (cold) point or zone is the location in the food, where the maximum concentration of surviving microorganisms is expected. Calculating and measuring the concentration of surviving microorganisms in the cold zone is an important

Receptacle ↔	**Apparatus** ↔	**Transport processes in food materials** ↔	**Quality attribute changes**
Containers Cans Glass jars Pouches Boxes	**Heat treatment of filled and closed containers** Batch process Continuous process	*Heat conduction* *Natural convection* *Forced convection* Fluid mechanics (Newtonian and non-Newtonian)	*Inactivation and multiplication of microbes* *Changes of chem. concentrations and enzyme activities* *Sensory and physical attributes*
Tanks Portable Fixed	**Flow-through type (aseptic and quasi-aseptic processes)**		*Kinetics theory* *Reactor technics*
Load and deformation of receptacles	*Design of heat exchangers*	*Ohmic and dielectric heating*	Extreme values and averages. H. treatment equivalents

Figure 3.1. Co-operation between science and technology for achievements in heat treatment processes.

part of health protection and food safety. Table 3.1 summarizes the cold zones in the food processing industry. Only surviving pathogens are involved in the "cold point" calculations.

Control of spoilage means that the number of nonpathogenic survivors which can multiply in the sterilized food is limited by the process parameters, so that the ratio of spoiled cans (s) is very low. For example, if one in 10,000 cans ($s = 10^{-4}$) is spoiled, and one can initially contains $N_0V = 10^3$ harmful microbes (V is the can volume, N_0 is the initial concentration of microbes), then a pasteurization process is needed which reduces $10^3/10^{-4} = 10^7$ microbes to one survivor. Assuming first-order kinetics [see Eq. 3.5] and decimal reduction time $D = 0.7$ min at 70°C reference temperature, the necessary pasteurization equivalent (see later) is

$$P = (\log 10^7) \times 0.7 = 7 \times 0.7 = 4.9 \text{ min.}$$

The cold zone is (in most cases) near the center of a can, but it can shift toward the surface when filling and closing of hot food, because the surface zone cools down first.

Liquid food leaving a flow-through type sterilizer is a mixture of elements (including microbes) having different residence time periods. No distinction can be made between maximum and average concentrations of surviving pathogens in this case. However, when food pieces are dispersed in a liquid, the maximum survivor concentration will be expected *in the center of the greatest piece with the shortest residence time*.

Most fruit products belong to the groups of medium- and high-acid foods (pH < 4). Typical pathogenic bacteria are the *Salmonella* and *Staphylococcus* species, while lactic acid producing bacteria (*Lactobacillus*, *Leuconostoc*), though inhibiting growth of pathogens, can cause spoilage. Yeasts and

Table 3.1. Location of the Critical Zone or Critical Sample When Inactivating Microbes

Process Characteristics	Location of the Critical Zone or Sample
Food in container:	Central zone in the container
Heat conduction	
Natural convection	
Forced convection	
Natural or forced convection + food pieces	Central zone in the container + the core of food piece
Flow-through type, full aseptic:	Calculation of the average of survivors' concentration at the exit
Liquid food or puree	
Liquid food or puree + food pieces	Calculation of survivors' concentration in the core of the largest piece with shortest residence time
Flow-through type heating + hot filled containers (quasi-aseptic):	Surface zone in the container (after cooling)
Liquid food or puree	
Liquid food or puree + food pieces	Surface zone of liquid food or in the largest piece with shortest residence time at the container wall

molds can also be harmful. The heat-resistant mold *Byssochlamys fulva* cannot be destroyed by temperatures under 100°C (Stumbo, 1973; Ramaswamy and Abbatemarco, 1996).

Contrary to sterilization (pH > 4.5, T > 100°C), no generally accepted reference temperature exists for pasteurization, nor agreement on which organisms are dangerous. Even the criteria for safe shelf-life (time, temperature, etc.) may be uncertain. However, pathogenic species must not survive and $N/N_0 = 10^{-8}$ reduction of either pathogenic or other species causing spoilage would do.

HEAT TREATMENT EQUIPMENT

CLASSIFICATION

The major factors to consider are:

1. Whether the food is pasteurized after filling individual containers or in bulk before filling (full aseptic and quasi-aseptic processes).
2. Type, size, and material of the container or tank.
3. Highest retort temperature under or above 100°C and pressure equal to or above atmospheric pressure.
4. Operational character, batch or continuous operation.
5. Physical background of heating and cooling, considering both the equipment and food material. A great variety of applications can be found. Besides steam, hot-, and cold-water applications, new methods include combustion heating of cans or ohmic heating in aseptic processes. Steam injection and infusion into viscous purees and evaporation cooling have also been adopted (as well as microwave applications).

BATCH-TYPE PASTEURIZERS AND STERILIZERS FOR "FOOD IN CONTAINER" TREATMENT

Open pasteurization tanks, filled with water, are heated by steam injection and cooled by cold water. Racks holding containers are lifted in and out from above by a traveling overhead crane. Heating and cooling in the same tank in one cycle is uneconomical. However, steam and water consumption can be decreased by modifications (e.g., hot water reservoirs).

Horizontal retorts are favored by plants where different products are processed in small or medium volumes. A wide variety of construction is available, usually with the following features (see Fig. 3.2).

Container holding racks are carried into the retort and fixed to a metal frame, which can be rotated at variable speed to increase heat transfer. The retort door with bayonet-lock cannot be opened under inside overpressure. An insulated upper reservoir serves as storage for hot water at the end of the heating period. Automatic control provides for uniform repetition of sterilization cycles. Temperatures and heat treatment equivalent are registered. Heating is provided by steam (injection or heat exchanger), cooling by water.

Vertical retorts had been used up to the second half of the last century.

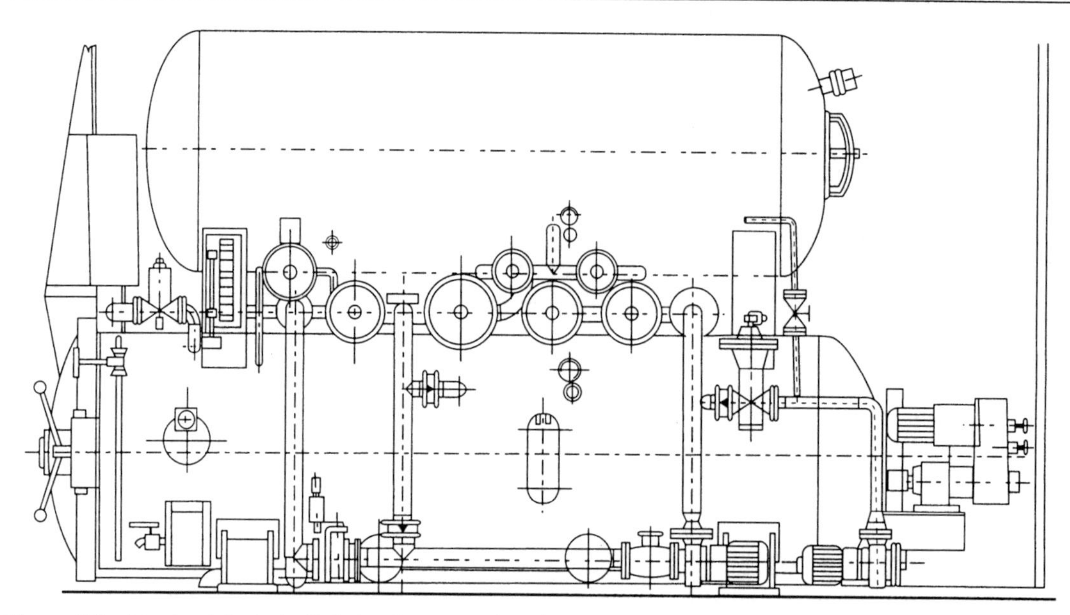

Figure 3.2. A horizontal retort with hot water reservoir and mechanism for the rotation of containers.

Continuous Pasteurizers and Sterilizers for "Food in Container" Treatment

Continuous operation is advantageous in those plants where large volumes are processed for a long period. The specific energy and water consumption of a continuous apparatus would be less than in its batch-type equivalent.

In tunnel pasteurizers horizontal conveyors carry containers through insulated heating and cooling sections. Hot- and cold-water spray, sometimes combined with water baths, would be applied in counter-current flow to container travel. Atmospheric steam is also used (see Fig. 3.3).

In combustion heated pasteurizers, cylindrical cans are rolled above gas-burners along guide-paths (Rao and Anantheswaran, 1988).

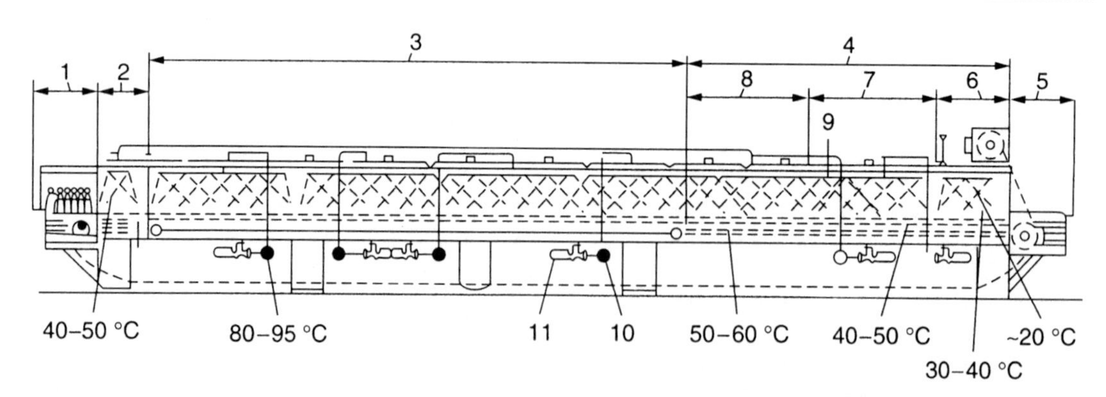

Figure 3.3. Tunnel pasteurizer with hot- and cold-water spray: (1) container feed, (2) section for preheating, (3) maximum temperature zone, (4) zone of counter-current cooling, (5) container discharge, (6) cold-water section, (7) tepid water section, (8) medium hot water cooling section, (9) spray-nozzles, (10) water pump, and (11) filter to the pump (altogether six units).

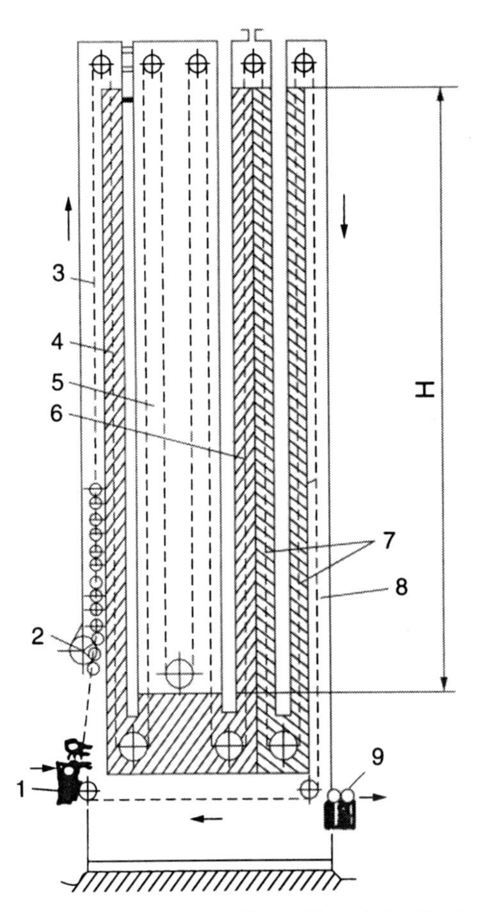

Figure 3.4. Hydrostatic sterilizer: (1) container feed, (2) container holding shell, (3) conveyor, (4) water column for heating, (5) room under steam overpressure, (6) water column for cooling at decreasing pressure, (7) U-shaped cooling bath, and (8) discharge section of the conveyor.

Hydrostatic sterilizers are the best energy and water saving devices with safe operational characteristics. Containers enter and leave the heating chamber through hydrostatic columns (see Fig. 3.4).

The hydrostatic pressure at the bottom level of a column balances the chamber's overpressure:

$$\Delta p = H\rho g. \tag{3.1}$$

For example, if column height ($H = 24$ m) balances a chamber pressure ($\Delta\ p = 230$ kPa), then the saturated steam temperature is $T = 125°\text{C}$. Such sterilizers protrude from the plant building as a (insulated) tower.

It is possible to reduce the height of the sterilizer by applying serially connected lower columns on both the inlet and outlet sides. Such construction needs special control systems on both sides. Hungarian construction with the commercial name "Hunister" works with six 4 m high columns (equivalent to a 24 m high unit) and can be placed into a processing hall of about 8 m inner height (Schmied et al., 1968; Pátkai et al., 1990).

Pasteurizers and sterilizers with a helical path and can-moving reel (commercial name: "Sterilmatic") are popular in the United States. Heating and cooling units are serially connected in the necessary number. Cans rotate or glide along the helical path. Cylindrical units are equipped with feeding and discharge devices, and special rotating valves serve for units under inside overpressure. The horizontal arrangement is favorable. The long production and maintenance praxis of machinery counterbalance the drawbacks of somewhat complicated mechanisms.

Statements for Both Batch-Type and Continuous Apparatus

The output of an apparatus, i.e., the number of containers pasteurized in unit time (Q) can be calculated by the formula:

$$Q = \frac{\Phi W}{V t_m}, \tag{3.2}$$

W is the inside volume filled with containers, heat transfer mediums, and the transport mechanism. The symbol t_m denotes the total treatment time, i.e., cycle period for batch type and total residence time for continuous operation. V is the volume of a single container. Φ is the compactness ratio, i.e., the volume of all containers per inside volume. Greater Φ and short t_m are advantageous and mean better compactness and heat transfer intensity (including the use of elevated temperatures).

The process diagram (see Fig. 3.5) presents ambient and container temperatures and pressures depending on treatment time ($0 \le t \le t_m$) for a sterilizer. Treatment time is the time needed for the progress of the container in a continuous unit. Instrumentation enables quick creation of such diagrams.

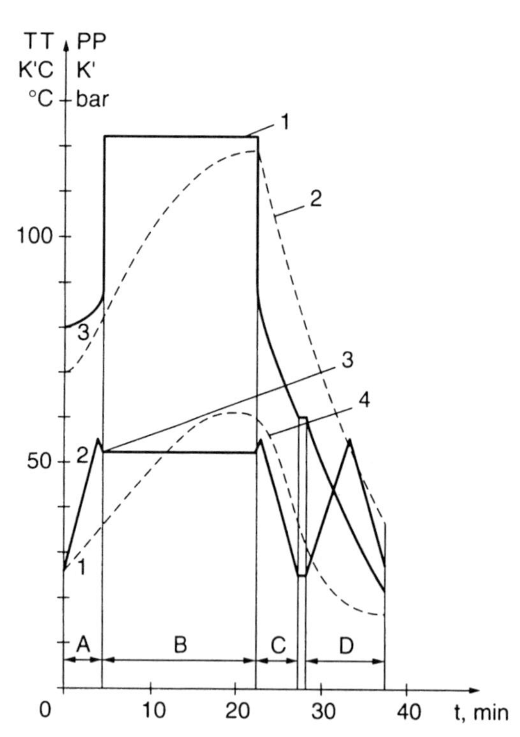

Figure 3.5. Process diagram of a hydrostatic sterilizer: (1) ambient temperature (T_K), (2) central temperature in the food (T_C), (3) ambient pressure (p_K), and (4) pressure in the container (p). (**A–D**) Heating (rising temperature), constant temperature, first cooling section, and second cooling section, respectively.

Table 3.2 presents specific energy and water requirements of a heat treatment apparatus, as these make a considerable contribution to the total consumption of a plant. Reduction can be achieved by heat recuperation and water reuse (filtration, disinfection, etc.).

Flow-Through Type Pasteurizers and Sterilizers

Typical flow-through type equipment consists of a pump which propels liquid food through heating, constant high temperature, and cooling units for (aseptic) filling and sealing (see Fig. 3.6).

Low viscosity liquids are apt to be moved through tubular or plate heat exchangers. Fruit pulps, purees, and other comminuted fruits (containing occasionally dispersed particles) would be processed in units provided with mixers and forwarding devices like scraped surface heat exchangers.

Well-designed equipment consumes energy and water with the same low specific values as hydrostatic sterilizers (see Table 3.2). Pulpy, fibrous juices and concentrates belong to the pseudoplastic or plastic category of non-Newtonian fluids. In addition to flow resistance and heat transfer calculations, the results of (chemical) reactor techniques should be adopted for quality attribute change calculations, including the inactivation of enzymes and microbes. Such concepts as residence time distribution of food elements, macro- and micro-mixing, etc., are involved here. Special problems arise from undesirable deposits and burning on the food side of heat transmission walls.

Table 3.2. Specific Consumption of Steam (400 kPa, Saturated) and Water (About 20°C) of Heat Treatment Processes

Pasteurization or Sterilization Process (Notes)	Equipment (Notes)	Specific Consumption (kg/kg)	
		Steam	Water
“Food in container” treatment (values are related to the mass of food + container)	Tank pasteurizer, vertical retort (without heat recuperation)	0.4–0.55	4–8
	Horizontal retort (with heat recuperation)	0.2–0.36	2–4
	Tunnel pasteurizer	0.15–0.20	1.5–2
	Hydrostatic sterilizer	0.08–0.12	1.2–2
Flow-through type treatment (values are related to the mass of food)	Tubular and plate apparatus (without heat recuperation)	0.12–0.18	–
	Tubular and plate apparatus (with heat recuperation)	0.06–0.12	–

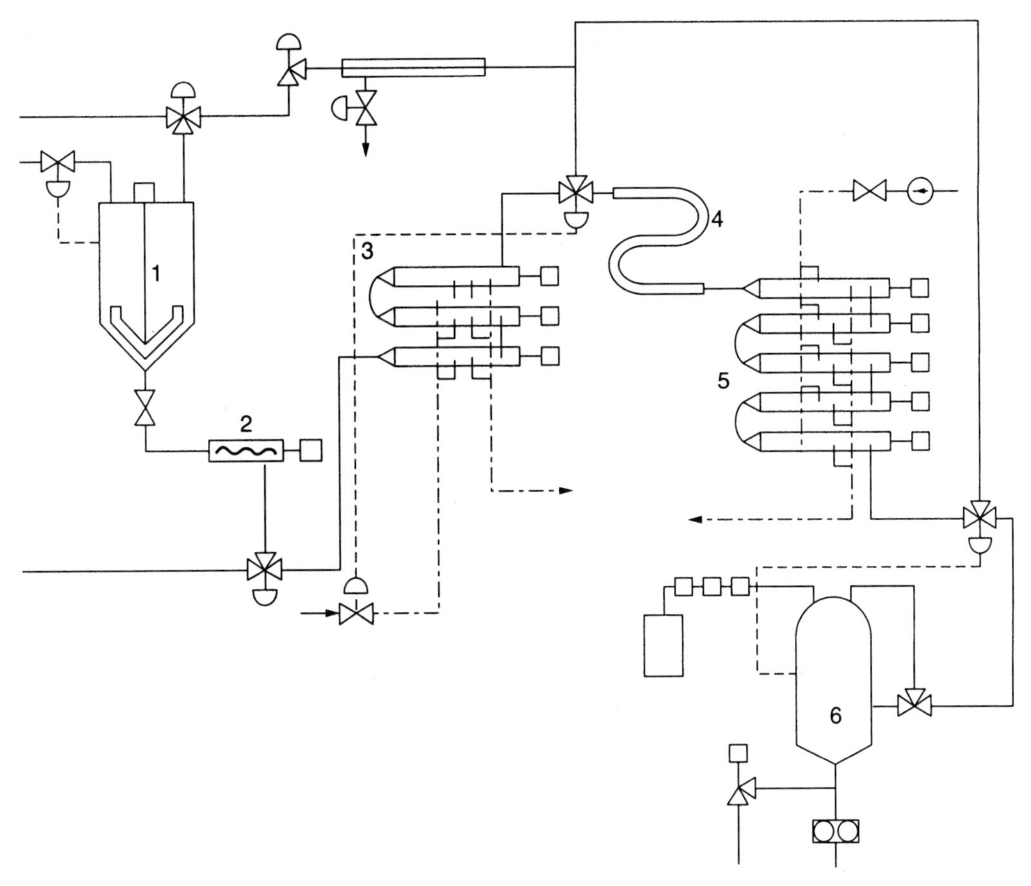

Figure 3.6. Flow-through type pasteurization: (1) feed tank, (2) pump, (3) scarped surface heat exchangers, (4) isolated tube for keeping the food at constant temperature, (5) scarped surface cooling units, and (6) aseptic tank for pasteurized food.

HEAT PROPAGATION UNDER HEAT TREATMENT CONDITIONS

Heat Conduction in Food Holding Containers

Experience has shown that heat propagation in many food materials under the circumstance of pasteurization can be calculated using the principles of conduction. All food products might be treated as conductive, in which no major convective currents develop during heat treatment. Small local movements from density differences or induced vibrations increase the apparent thermal diffusivity. As a consequence, the best way to measure thermophysical constants is by the "in plant" method. Results of "in plant" measurements are often 10–20% higher than respective data from the literature (Pátkai et al., 1990). The calculation of time-dependent temperatures in food is based on the differential equations of unsteady-state heat conduction with initial and boundary conditions. The application of the Duhamel theory is also needed in case of time-dependent variation of the retort temperature (Geankoplis, 1978; Carslaw and Jaeger, 1980).

While analytical solutions are limited to a few simplified tasks, a wide range of industrial problems can be solved using methods of finite differences and finite elements. Figure 3.7 illustrates the elementary annuli (and cylinders) of a cylindrical can. Figure 3.8 illustrates a time-dependent change in ambient temperature and the approximation using step-wise variation. Both figures explain a finite difference method, where differential quotients have been substituted by quotients of suitably small differences (Körmendy, 1987).

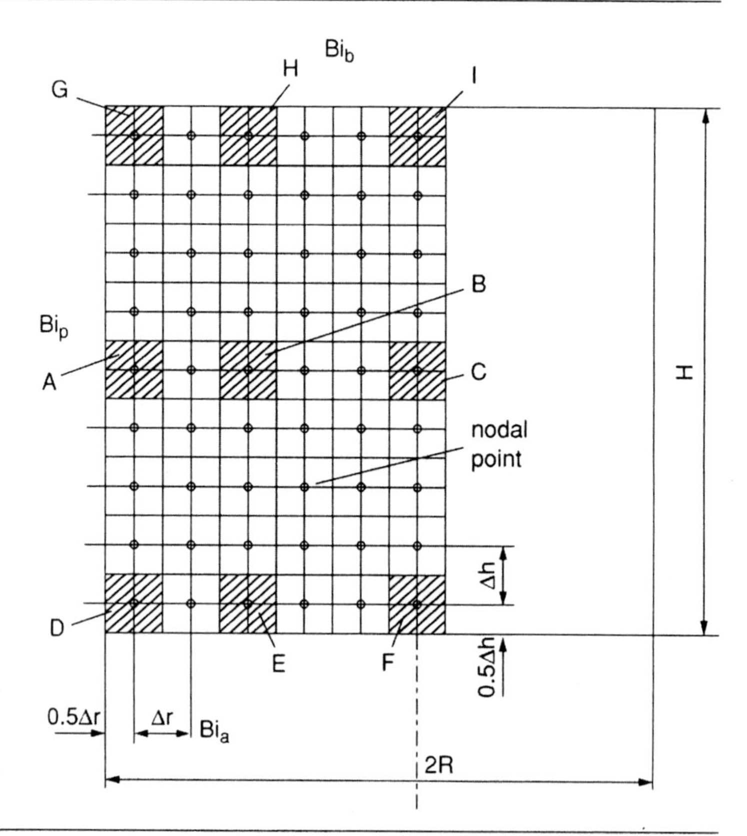

Figure 3.7. Geometry belonging to a finite difference method. (**A–I**) Elementary annuli and cylinders. Bi_a, Bi_b, Bi_p: Biot numbers (bottom, cover, and jacket, respectively).

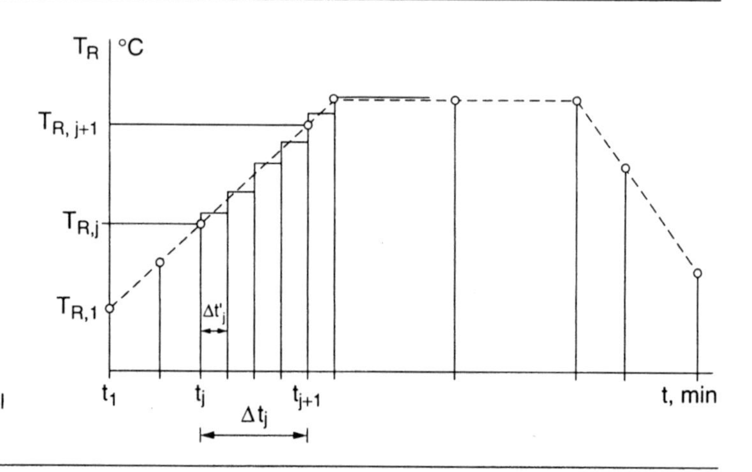

Figure 3.8. Retort temperature (T_R) and its step-wise approximation in a finite difference calculating system. Hollow circles illustrate input data (temperature vs time). Time intervals (Δt_j) are divided into sufficiently small equal time steps ($\Delta t'_j$).

Natural and Forced Convection Heating of Food Holding Containers

Natural convection heat transfer inside containers is based on fluid circulation induced by temperature and density differences. The phenomenon is typical for low viscosity liquids. Temperature differences are the greatest at the container wall, while the central bulk is of near uniform temperature. Practical calculations are based on a heat balance including the mean temperature of food and ambient temperature. The relation $\mathrm{Nu} = C\,(\mathrm{Gr} \times \mathrm{Pr})^m$ between dimensionless terms (groups) can be applied for the "from wall to food" heat transfer. Treatment time is divided into consecutive intervals to enable the use of temperature-dependent physical constants for computer analysis. The result is the time-dependent average temperature in the container (Körmendy, 1987).

The expression of forced convection would be applied to all those achievements, where the effect of mixing is gained by mechanical energy input into a fluid, slurry, or paste. The rate of heat flow may be increased considerably by mixing the food in a container. Actual accomplishment of mixing is done by rotating or tilting the containers. Machines often rotate the containers as a consequence of conveyance. Some hydrostatic sterilizers incorporate periodic tilting. The rotational speed or tilting rhythm depends on the conveyance speed in this equipment.

A number of rotational speeds are available in horizontal retorts. Frictional, gravitational, and inertial forces and their (periodic) variation acting on rotating food elements jointly influence mixing. The volume of the headspace also influences the mixing effect (Körmendy, 1991). An optimum speed exists, because at high rotational speed the mixing effect decreases as the food reaches a new equilibrium state under the overwhelming centrifugal (inertial) force (Eisner, 1970).

Many products contain fruit pieces in a syrup or juice. Heat is transferred from the container wall into the fluid constituent by convection, while the fruit pieces are heated by conduction. Time-dependent temperatures are obtained by using equations for convection and conduction, including initial and boundary conditions (Bimbenet and Duquenoy, 1974).

Heat transfer coefficients have been developed through extensive research. Results are heat transfer coefficients from container wall to liquid food and from liquid food to pieces of food, depending on container geometry, rotational speed, product, and heating and cooling programs. Coefficients take into consideration the relationship between dimensionless terms (Bi, Gr, Nu, Pr, Re, St, We), geometrical proportions, and temperature and viscosity ratios (Rao and Anantheswaran, 1988; Rao et al., 1985; Sablani and Ramaswamy, 1995; Akterian, 1995).

The usual methods based on the determination of the values: f_h, j_h, f_c, j_c (Ramaswamy and Abbatemarco, 1996) are adequate for convective heat transfer calculations (if proper simulation has been used in case of forced convection). For conductive heating, however, it seems advisable to use the previous values for the evaluation of the relevant thermophysical constants and enter the latter values into a computer program that calculates with the help of finite differences.

Heat Transfer in Flow-Through Type Heat Treatment Units

When liquid food (including non-Newtonian slurries and purees) is pumped through channels (tubular, annular, and plate-type heat exchangers), three major flow types would form: sliding (peristaltic), laminar, or turbulent. Heat transfer calculations are available for all three types of publications from the unit operations field (Geankoplis, 1978; Gröber et al., 1963). Relationships between dimensionless terms are applied as previously described.

A number of publications are available on scraped surface heat exchangers, including heat transfer characteristics (Geankoplis, 1978).

LOAD AND DEFORMATION OF CONTAINERS UNDER HEAT TREATMENT CONDITIONS

Containers and Their Characteristic Damages

The main load on containers under heat treatment conditions originate from the difference in ambient and inside pressure. The deformation or damage that occurs from this load depends on the geometry and material of the container, the design of closure, and the headspace volume.

Metallic containers display reversible deformation at small loads and permanent deformations at higher loads (see Fig. 3.9). Excess inside overpressure causes permanent bulging at the end plates of a cylindrical can, while outside overpressure might indent the jacket.

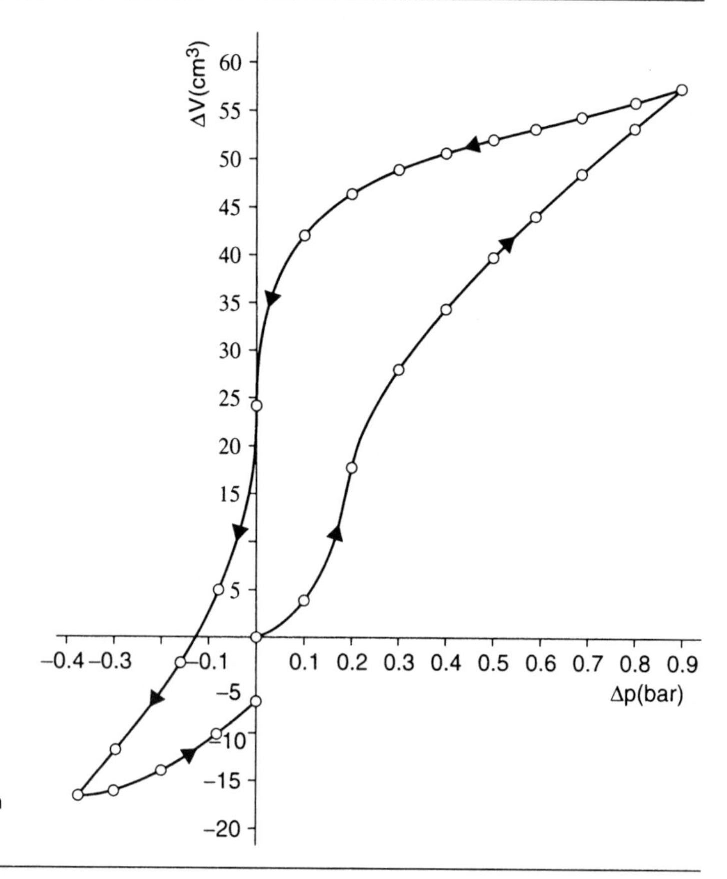

Figure 3.9. Volume change (ΔV) versus pressure difference (Δp) relation of a tinplate can.

Rigid containers like glass jars and bottles undergo very small (and resilient) deformation. Excess inside overpressure can open or cast down the cover. The temporary opening of the cover effects air exhaust or food loss, depending on the position (vertical or horizontal) of the jar.

Plastic bags are susceptible to large deformations, while inside overpressure easily rips the bags. Table 3.3 includes critical loads for some types of containers.

Deformation Versus Load Calculations

Two basic equations are used to calculate the inside pressure of a container, the respective pressure difference, and container volume:

$$p = p_{\mathrm{f}} + \frac{RT_{\mathrm{G}}}{V_{\mathrm{G}}} \sum_{i=1}^{n} \frac{m_{\mathrm{G}i}}{M_i} = p_{\mathrm{f}} + \frac{V_{\mathrm{G0}} T_{\mathrm{G}}}{V_{\mathrm{G}} T_{\mathrm{G0}}} (p_0 - p_{\mathrm{f0}}) \quad (3.3)$$

and

$$V_{\mathrm{G}} = V_{\mathrm{C0}}\,[1 + \alpha_{\mathrm{VC}}\,(T_{\mathrm{C}} - T_{\mathrm{C0}})] + \Delta V - V_{\mathrm{f0}} \times [1 + \alpha_{\mathrm{Vf}}\,(T_{\mathrm{f}} - T_{\mathrm{f0}})]\,. \quad (3.4)$$

Additional relationships and data are needed as time-dependent variations of ambient pressure and temperature, temperatures of the container wall, headspace, and food (average). Initial values (at the moment of closing) include volumes, pressures, and temperatures. According to Equation 3.3, the inside pressure of the container is the sum of the vapor pressure of food and of the partial pressure of gas components. According to Equation 3.4, the actual headspace volume (V_{G}) can be calculated by adding the container volume (increased exclusively by heat expansion) to the pressure difference induced volumetric container deformation (ΔV), and subtracting the volume of food expanded by heat. The measured relationship between ΔV and the pressure difference (Δp, e.g., see Fig. 3.9) is also required for the formula (Körmendy and Ferenczy, 1989; Körmendy et al., 1994, 1995).

Table 3.3. Critical Loads of Tinplate Cans and Glass Jars

	Container				Critical load (kPa)		
Material	V_0	d	j	ep	Δp_1	Δp_2	Notes
Tinplate can	860	99	0.18 0.22 0.24	0.18 0.22 0.24	103 138 159	−52 −60 −67	$\Delta p1$: bulging of end plates
	3200–4760	153	0.28–0.30	0.3–0.32	75–95	−43–−53	Δp_2: indentation at the jacket
	6830	160	0.28	0.3	78	−51	
Glass jar + twist-off lid of tinplate	–	79	–	0.24	107	–	Δp_1: opening of the lid

Note: V_0—volume (cm^3); j—jacket thickness (mm); d—can diameter or lid diameter for jars (mm); ep–end plate thickness or lid plate thickness (mm).

QUALITY ATTRIBUTE CHANGES

ATTRIBUTE KINETICS

Besides the concentration of favorable and unfavorable constituents, enzyme activity, sensory attributes, related physical properties, etc., the concentration of surviving microbes can also be regarded as a quality attribute. Namely, similar descriptive kinetic methods, concepts of extreme values and averages are used for changes in microbial concentration.

The typical method to measure variations of time-dependent attributes in food is by laboratory testing at a number of different constant temperatures, providing for all major conditions of (industrial) heat treatment. The next step is the fitting of expertly chosen relationships to data, and the simultaneous evaluation of kinetic constants.

Heat inactivation of microbes belongs to the population change dynamics field. Since early years (about 1920), first-order equations had been used, meanwhile, a number of other equations were also fitted to experimental results (Casolari, 1988; Körmendy and Körmendy, 1997). The method that shows promise is based on the distribution of lethal time of the individual microbes combined with the two parametric Weibull distribution (Peleg and Penchina, 2000; Körmendy and Mészáros, 1998). The Weibull distribution is useful for fitting to diverse types of time-dependent inactivation courses by selecting appropriate constants.

The generally used first-order (exponential) equations are

$$N = N_0 \times 10^{-t/D}, \quad (3.5)$$

and

$$D = D_r \times 10^{-(T-T_r)/z}. \quad (3.6)$$

The first equation is valid for constant temperature, the second one describes the variation of the decimal reduction time (D) versus temperature (T). D_r, T_r are arbitrary (though expedient) reference values. Time-dependent variation of the logarithm of the concentration of surviving microbes is linear. The value of z is the temperature increment effecting the decimal reduction of D. Naturally, D is not the decimal reduction time in non-exponential relationships.

Reaction theory principles should be used to determine changes in time-dependent chemical concentrations (Levenspiel, 1972; Froment and Bischoff, 1990). However, mostly fitted relationships are used instead of more exact calculations. Kinetic constants are: rate constant (k, or rate constants), energy of activation (E_a), reference temperature (T_r), and reference rate constant (k_r). The rate constant is independent of the initial concentration if the kinetic equation is based on a sound chemical background, otherwise a fitted rate constant might only be valid for a fixed initial concentration (see more details in the publication of Körmendy and Mészáros, 1998).

Empirical (i.e., fitted) equations describe the time-dependent variation of sensory attributes and of related physical properties (color, consistency, etc.). *Concentration based attributes* follow a linear mixing law, i.e., the attribute intensity of a mixture of different volumes and intensities is the weighted mean of component intensities. This evident rule is not valid for *sensory attributes* (see Körmendy, 1994; for food color measurements).

Attribute Intensity Versus Time-dependent Temperature in Food Holding Containers

Temperature always varies during a heat treatment process (see Fig. 3.5). No general procedure exists for calculating time-dependent attribute intensity at variable temperature from constant temperature experiments. Notwithstanding, a few useful methods have been developed since about 1920 and more are expected in the future.

The equivalent sterilization time (F) at constant reference temperature (T_r) induces the same lethal effect as the actual time-dependent temperature variation:

$$F = \int_0^{t_m} 10^{\frac{T(t)-T_r}{z}} \, dt. \qquad (3.7)$$

The previous integral was used later for pasteurization (P), enzyme inactivation (E), chemical and sensory attribute variation (C), replacing the symbol F by P, E, C (cooking value), respectively. Equation 3.7 had been derived originally for first-order (i.e., exponential) destruction. It could be proved later that Equation 3.7 is applicable in all those cases, where attribute intensity at constant temperature depends only on t/D or kt. There are methods for other variable temperature changes (Körmendy and Körmendy, 1997; Peleg and Penchina, 2000). Equations 3.6 and 3.7 undergo modifications, when the z-value is replaced by the energy of activation (Hendrickx et al., 1995). The attribute intensity at the end of a process is easily available by substituting T_r, D_r (or k_r) and the equivalent time (F, P, E, C) into the constant temperature intensity versus time relation [e.g., into Eq. 3.5]. Computerized calculation should provide three intensities in a container: the maximum, the minimum, and the average (see "Background of Microbial Safety" and Table 3.1). The "cold point" of a vertically positioned container is near the bottom, at a distance less than 25% of the container height, in case of natural convection.

Calculation Methods for Flow-Through Type Apparatus

The residence time is uniform for all food elements in case of in-container pasteurization. As a contrast,

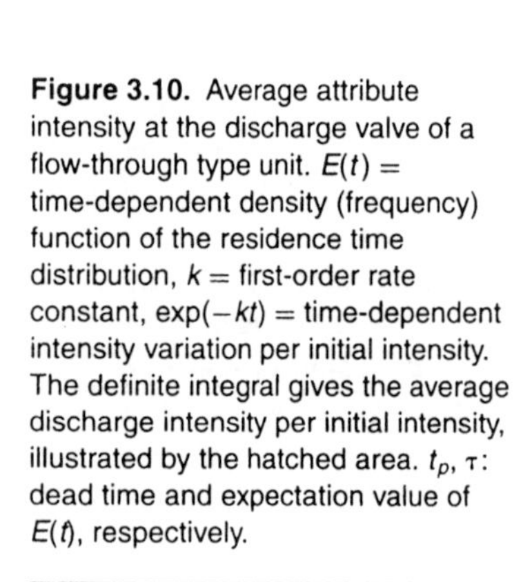

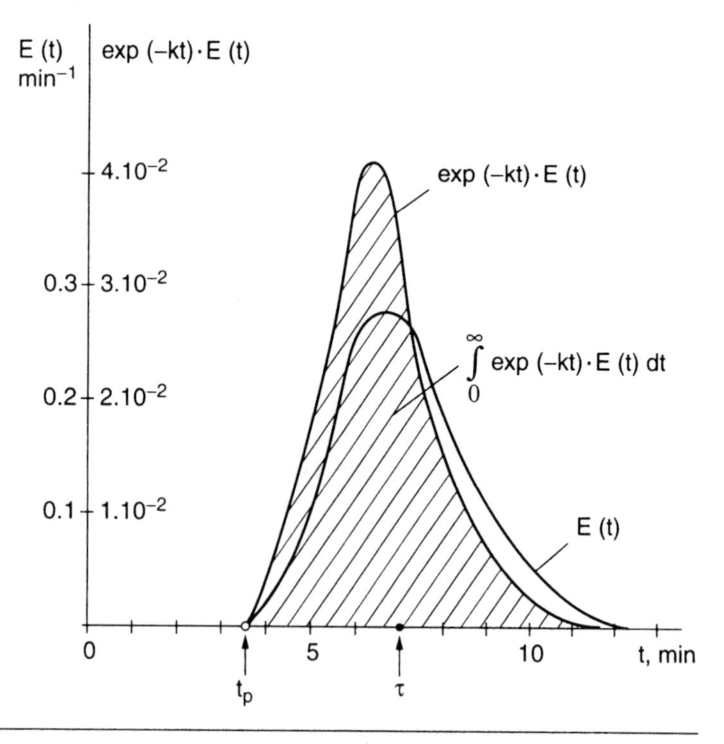

Figure 3.10. Average attribute intensity at the discharge valve of a flow-through type unit. $E(t)$ = time-dependent density (frequency) function of the residence time distribution, k = first-order rate constant, $\exp(-kt)$ = time-dependent intensity variation per initial intensity. The definite integral gives the average discharge intensity per initial intensity, illustrated by the hatched area. t_p, τ: dead time and expectation value of $E(t)$, respectively.

food elements reside for different time intervals in a unit of a flow-through type apparatus, and a distribution function characterizes residence time. A sample at the discharge port of a unit is a mixture of food elements of different residence time intervals. It has been proved for liquid food that the average of the concentration of surviving microbes at the exit can be calculated according to the equation:

$$\overline{N} = N_0 \int_0^{\infty} E(t) \times 10^{-t/D}\,dt, \qquad (3.8)$$

if food temperature is constant (suspended particles are small enough too) and exponential inactivation law exists [see Eq. 3.5]. Equation 3.8 is based on macromixing principles (Levenspiel, 1972) and can be easily converted for other inactivation kinetics. Figure 3.10 demonstrates the essence of calculation useful for a constant temperature unit.

No definite solution exists for a variable temperature unit, presumably the approximation of Bateson (1971) will be useful in the future. As a consequence of his idea, an average temperature ($\overline{T}$) can be assessed for a variable temperature unit and the pertaining value: $\overline{D}$ substituted into Equation 3.8. The heat treatment equivalent for average attribute intensity ($\overline{F}$) is now:

$$\overline{F} = -\log \frac{\overline{N}}{N_0}. \qquad (3.9)$$

The overall equivalent of an apparatus with serially connected units (e.g., heating, constant temperature, and cooling) is the sum of the individual units equivalents (Körmendy, 1994, 1996).

LIST OF SYMBOLS

- Bi — Biot number
- C — dimensionless constant
- C — cooking value (min)
- D — decimal reduction time or time constant (min)
- $\overline{D}$ — average of D (min)
- E — enzyme inactivation value (min)
- E_a — energy of activation (kJ/kmol)
- $E(t)$ — density function of the residence time distribution (earlier: frequency distribution, min^{-1})
- f_c, f_h — cooling and heating rate indexes: time needed for the decimal reduction of the difference between outside and inside temperatures (min)
- F — sterilization or lethality value (equivalent, min)
- $\overline{F}$ — average of F
- g — gravitational constant (m/s^2)
- Gr — Grashof number
- H — water column height (m)
- j_c, j_h — cooling rate and heating rate lag factors
- k — rate constant (for first-order kinetics, min^{-1})
- m — dimensionless exponent
- m_G — mass of a gas component (kg)
- M — molar weight of a gas component (kg/kmol)
- n — number of gas components
- N — concentration of living or surviving microbes (cm^{-3})
- $\overline{N}$ — average of N (cm^{-3})
- Nu — Nusselt number
- p — absolute pressure in a container (kPa)
- P — pasteurization value (equivalent, min)
- Pr — Prandtl number
- Q — number of containers pasteurized in unit time (min^{-1})
- R — universal gas constant [kJ/(kmol K)]
- Re — Reynolds number
- s — spoilage ratio
- St — Stanton number
- t — time (min)
- t_m — total treatment time (min)
- T — temperature (K, °C)
- V — volume in connection with a container (m^3)
- W — inside volume of an apparatus (m^3)
- We — Weber number
- z — temperature increment for the decimal reduction of D (K, °C)
- α_v — volumetric heat expansion coefficient (K^{-1}, $°C^{-1}$)
- Δp — pressure difference, inside minus outside pressure (Pa, kPa)
- ΔV — volume change of a container due to mechanical load (m^3)
- ρ — density of water (kg/m^3)
- Φ — compactness ratio

Indexes: C = container, f = food, G = headspace, i = serial number of gas components, 0 = initial value, r = reference value.

REFERENCES

Akterian, S. G. 1995. Numerical simulation of unsteady heat transfer in canned mushrooms in brine during sterilization. J. Food Eng. 25:45.

Ball, C. O. and Olson, F. C. W. 1957. Sterilization in Food Technology. McGraw-Hill, New York.

Bateson, R. N. 1971. The effect of age distribution on aseptic processing. Chem. Eng. Prog. Syst. Ser. 108 67:44.

Bimbenet, J. J. and Duquenoy, A. 1974. Simulation mathematique de phenomenes interessant les industries alimentaires. I. Transfer de chaleur au cours de la sterilisation. Ind. Aliment. et Agricoles 91:359.

Carslaw, H. S. and Jaeger, J. C. 1980. Conduction of Heat in Solids. Oxford University Press, UK.

Casolari, A. 1988. Microbial death. In: Bazin, M. J. and Prosser, J. I. (Eds), Physiological Models in Microbiology, vol. II. CRC Press, Boca Raton, FL:1–44.

Eisner, M. 1970. Einführung in die Technik und Technologie der Rotations Sterilisation. Günter Hempel, Braunschweig, Germany.

Froment, G. F. and Bischoff, K. B. 1990. Chemical Reactor Analysis and Design. Wiley and Sons, New York.

Geankoplis, C. J. 1978. Transport Processes and Unit Operations. Allyn and Bacon, Boston, MA.

Gröber, H., Erk, S. and Grigull, U. 1963. Die Grundgesetze der Wärmeübertragung. Springer, Göttingen, Germany.

Hendrickx, M., Maesmans, G., De Cordt, S., Noroha, J. and Van Loey, A. 1995. Evaluation of the integrated time–temperature effect in thermal processing of foods. Crit. Rev. Food Sci. Nutr. 35:231.

Körmendy, I. 1987. Outline of a system for the selection of the optimum sterilization process for canned foods. I. Calculation methods. Acta Aliment. 16:3.

Körmendy, I. 1991. Thermal processes, para 5.1.1. to 5.1.4. In: Szenes, E. and Oláh, M. (Eds), Konzervipari Kézikönyv (Handbook of Canning). Integra-Projekt Kft., Budapest, Hungary:163–188 (in Hungarian).

Körmendy, I. 1994. Variation of quality attributes in serially connected independent food reactor units. Part 1. Chem. Eng. Proc. 33:61.

Körmendy, I. 1996. Variation of quality attributes in serially connected independent food reactor units. Part 2. Chem. Eng. Proc. 35:265.

Körmendy, I. and Ferenczy, I. 1989. Results of measurements of deformation versus load relations in tin cans. Acta Aliment. 18:333.

Körmendy, I., Koncz, L. and Sárközi, I. 1994. Deformation versus load relations of tinplate cans. Interpolation between extreme loading cycles. Acta Aliment. 23:267.

Körmendy, I., Koncz, L. and Sárközi, I. 1995. Measured and calculated pressures and pressure differences of tinplate cans under sterilization conditions. Acta Aliment. 24:3.

Körmendy, I. and Körmendy, L. 1997. Considerations for calculating heat inactivation processes when semilogarithmic thermal inactivation models are non linear. J. Food Sci. 34:33.

Körmendy, I. and Mészáros, L. 1998. Modelling of quality attribute variations in food under heat treatment conditions. Important aspects. In: Proceedings of the Third Karlsruhe Nutrition Symposium, Part 2. Karlsruhe, Germany:312.

Levenspiel, O. 1972. Chemical Reaction Engineering. Wiley and Sons, New York.

Pátkai, G., Körmendy, I. and Erdélyi, M. 1990. Outline of a system for the selection of the optimum sterilization process for canned foods. Part II. The determination of heat transfer coefficients and heat conductivities in some industrial equipments for canned products. Acta Aliment. 19:305.

Peleg, M. and Penchina, C.M. 2000. Modelling microbial survival during exposure to a lethal agent with varying intensity. Crit. Rev. Food Sci. Nutr. 40:159.

Ramaswamy, H. S. and Abbatemarco, C. 1996. Thermal processing of fruits. In: Somogyi, L. P., Ramaswamy, H. S. and Hui, Y. H. (Eds), Processing Fruits: Science and Technology, vol. 1. Technomic, Lancaster, CA:25–66.

Rao, M. A. and Anantheswaran, R. C. 1988. Convective heat transfer to fluid foods in cans. Adv. Food Res. 32:39.

Rao, M. A., Cooley, H. J., Anantheswaran, R. C. and Ennis, R. W. 1985. Convective heat transfer to canned liquid foods in a Steritort. J. Food Sci. 50:150.

Sablani, S. S. and Ramaswamy, H. S. 1995. Fluid to particle heat transfer coefficients in cans during end-over-end processing. Lebensm.-Wiss. u.-Technol. 28:56.

Schmied, J., Ott, J. and Körmendy, I. 1968. Hydrostatic sterilizer with automatic temperature control. Hungarian Licence No.: KO-2206, No. of Registration: 158292. Budapest, Hungary.

Stumbo, C. R. 1973. Thermobacteriology in Food Processing. Academic Press, New York.

4
Fruit Freezing Principles

Begóna De Ancos, Concepción Sánchez-Moreno, Sonia De Pascual-Teresa, and M. P. Cano

INTRODUCTION

Freezing is one of the best methods for long-term storage of fruits. Freezing preserves the original color, flavor, and nutritive value of most fruits. Fresh fruits, when harvested, continue to undergo chemical, biochemical, and physical changes, which can cause deterioration reactions such as senescence, enzymatic decay, chemical decay, and microbial growth. The freezing process reduces the rate of these degradation reactions and inhibits the microbiological activity. However, it should be recognized that a number of physical, chemical, and biochemical reactions can still occur and many will be accentuated when recommended conditions of handling, production, and storage are not maintained. Although few microorganisms grow below −10°C, it should be recognized that freezing and frozen storage is not a reliable biocide. The production of safe frozen fruits requires the same maximum attention to good manufacturing practices (GMP) and hazard analysis critical control points (HACCP) principles as those used in fresh products. The quality of the frozen fruits is very dependent on other factors such as the type of fruit, varietal characteristics, stage of maturity, pretreatments, type of pack, and the rate of freezing. The freezing process reduces the fruit temperature to a storage level (−18°C) and maintaining this temperature allows the preservation of the frozen product for 1 year or more. Fruits are frozen in different shapes and styles: whole, halves, slices, cubes, in sugar syrup, with dry sugar, with no sugar added, or as juices, purees, or concentrates, depending on the industrial end-use. The influence of freezing, frozen storage, and thawing on fruit quality has been extensively reviewed (Skrede, 1996; Reid, 1996; Hui et al., 2004). The objective of this chapter is to describe the main principles of manufacturing and processing of frozen fruits (selection of raw material, pretreatments, packaging, freezing process, and frozen storage) and review the current topics on the sensorial and nutritional quality and safety of frozen fruits.

FREEZING PRINCIPLES

The freezing process reduces food temperature until its thermal center (food location with the highest temperature at the end of freezing) reaches −18°C, with the consequent crystallization of water, the main component of plant tissues. Water in fruit and fruit products constitute 85–90% of their total composition. From a physical point of view, vegetable and animal tissues can be considered as a dilute aqueous solution, which is the natural medium where chemical and biochemical cellular reactions take place and microorganisms grow. Crystallization of water during freezing reduces water activity (a_w) in these tissues

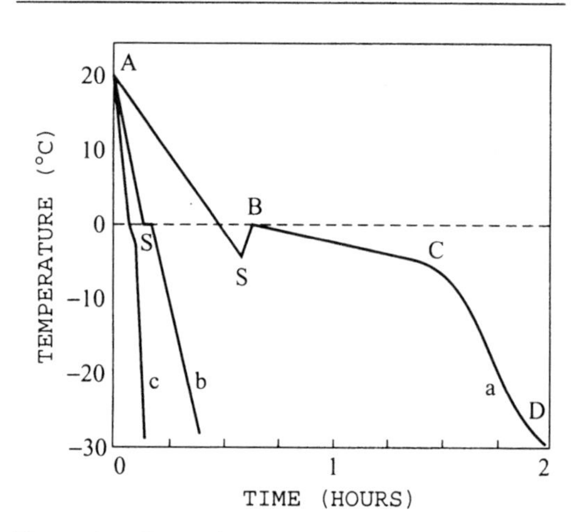

Figure 4.1. Typical freezing curves of foods at different rates: (a) very slow; (b) fast; and (c) very fast (Fennema, 1976).

and consequently produces a decline in chemical and biochemical reactions and microbial growth. Freezing also involves the use of low temperatures and reactions take place at slower rates as temperature is reduced. The study of temperature changes during freezing is basic to an understanding of how products are processed. Figure 4.1 shows typical freezing curves at different freezing rates. When the product is cooling down to 0°C, ice begins to develop (see section A–S, Fig. 4.1). The exact temperature for the formation of first ice crystal depends on the type of product and is a consequence of the constituents concentration independent of water content; for example, fruits with high water content (≈90%) have a freezing point below −2°C or −3°C, while meat with less water content (≈70%) has a freezing point of −1°C; the main difference being the high sugar and organic acid concentration in fruits. Ice formation takes place after the product reaches a temperature below its freezing point (−5°C to −9°C) for only a few seconds. This process is known as supercooling (position S in Fig. 4.1). After that, due to heat release during the first ice formation, the temperature increases until the freezing point is reached (position B in Fig. 4.1). Section B–C in Fig. 4.1 corresponds to the freezing of most of the tissue water at a temperature that is practically constant, with a negative slope from a decline of the freezing point due to solute concentration. The increase of solute concentration as freezing progresses causes the unfrozen portion to undergo marked changes in such physical properties as ionic strength, pH, and viscosity. This increases the risk of enzymatic and chemical reactions, e.g., enzymatic browning or oxidation–reduction, with adverse effects on frozen fruit quality. A short B–C section increases the quality of frozen fruit. This means that a fast rate freezing produces a better quality frozen fruit (see curves b and c of Fig. 4.1). Section C–D corresponds with the cooling of the product until the storage temperature, with an important increase of solute concentration in the unfrozen portion. Below −40°C, new ice formed is undetected. Up to 10% of the water can be unfrozen, mainly joined to protein or polysaccharide macromolecular structures that take part in the physical and biochemical reactions. In frozen foods the relationship between the frozen water and the residual solution is dependent on the temperature and the initial solute concentration. The presence of ice, and an increase in solute concentration, has a significant effect on the reactions and state of the fruit matrix. The concentration of the solute increases as freezing progresses; and thus, solute concentration of the unfrozen matrix can leach out of the cellular structures causing loss of turgor and internal damage. Solute-induced damage can occur whether freezing is fast or slow, and cryoprotectants, such as sugars, are usually added to aqueous solution to reduce the cell damage. (Reid, 1996; Rahman, 1999).

Freezing Rate

Controlling the freezing rate is an important aspect of reducing cell damage, which causes important quality losses in frozen fruits. Three types of cell damage due to freezing have been reviewed:

- solute-induced damage
- osmotic damage
- structural damage.

Although solute-induced damage is present in fast and slow freezing processes, it can be minimized by slow speed. Osmotic and structural damages are dependent on the rate of freezing.

Freezing rate is the speed at which the freezing front goes from the outside to the inside of the product, and depends on the freezing system used (mechanical or cryogenic), the initial temperature of the product, the size and form of the package, and the type of product. The freezing process (as a function of the rate) can be defined as follows (IIR, 1986):

- Slow, 1 cm/h
- Semiquick, 1–5 cm/h
- Quick, 5–10 cm/h
- Very quick, 10 cm/h.

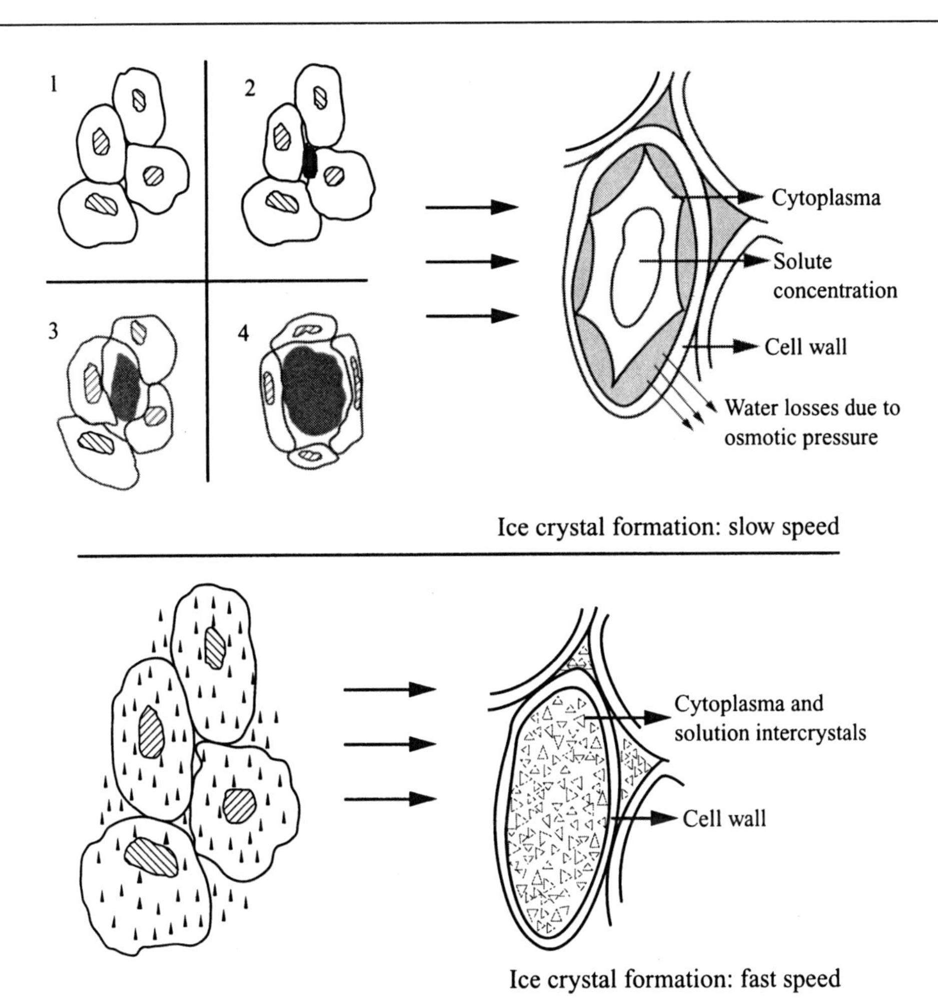

Figure 4.2. Ice crystal formation in plant tissues at slow speed (Fig. 4.2, up) and at fast speed (Fig. 4.2, down).

Generally, quick freezing produces better quality frozen fruits. Rates between 5 and 10 cm/h for "individual quick freezing" is an efficient way to obtain individual frozen fruits with high quality. The rate of freezing is very important in plant tissues because it determines the size, form, and status of the ice crystals, factors that affect cell wall integrity. If the rate of freezing is very slow, large ice crystals are formed slowly in the outer of cells and water from the cells migrate out by osmotic pressure (Fig. 4.2, up). Then, the cellular membranes are damaged during thawing, and the consequence of migration is an important drip loss.

Also, in slow cooling, large sharp ice crystals are formed and may cause damage to delicate organelle and membrane structure of the cell. As a consequence enzymatic systems and their substrates may be released, leading to different effects such as off-flavors and color and textural changes, etc. These effects can be prevented by applying prefreezing treatments like the addition of chemicals or by blanching, a heat treatment that denatures the enzymes. In a rapid rate freezing process, small size and round ice crystals increase at the same time, both inside and outside of the cell, and structural and osmotic damages are minimal (Fig. 4.2, down). Although fast freezing is better than slow freezing in fruit and vegetable products, the importance of freezing speed is sometimes misleading. The initial advantage obtained by fast freezing can be lost during storage due to recrystallization as a consequence of temperature fluctuations. Also, some products, such as whole fruits, will crack if they are exposed to extremely low temperature. This is due to volume expansion, internal stress, and the contraction and expansion phenomenon (Reid, 1996; Rahman, 1999).

FACTORS AFFECTING FROZEN FRUIT QUALITY

The freezing of fruits slows down, but does not stop, the physical, chemical, and biochemical reactions that produce their deterioration. There is a slow progressive change in sensorial and nutritional quality during frozen storage that becomes noticeable after a period of time. Safe, high-quality frozen fruits with maximum nutritional values can be produced if diligent controls are maintained at all times. These include temperature control, extended quality shelf life, microbiological safety, and the retention of nutrients.

Two principles dominate the control of quality and safety in frozen foods: product-process-package factors (PPP) and time-temperature-tolerance factors (TTT). PPP factors need to be considered at an early stage in the production of frozen fruits and they are the bases of commercial success of the product. The PPP factors are as follows:

- *Product*: High-quality frozen food requires high-quality raw materials and ingredients.
- *Process*: The speed and effectiveness of the freezing operations and the use of additional processes (blanching, etc.).
- *Package*: Packaging offering physical and chemical barriers.

TTT factors maintain the quality and safety during storage. TTT concepts refer to the relationship between storage temperature and storage life. For different foods, different mechanisms govern the rate of quality degradation and the most successful way of determining practical storage life is to subject the food to long-term storage at different temperatures. TTT relationships predict the effects of changing or fluctuating temperatures on quality shelf life (IIR, 1986).

Safe, high-quality frozen fruits with maximum nutritional values can be produced if the directions given below are followed:

- selection of suitable product for freezing
- PPP factors
- knowledge of the effect of freezing, frozen storage, and thawing on the fruit tissues that causes physical, chemical, and biochemical changes
- stability of frozen fruits (TTT factors)
- thawing
- microbiological quality and safety of frozen fruits.

SELECTION OF SUITABLE PRODUCT FOR FREEZING

High-quality frozen fruit requires high-quality raw material. Generally, quality cannot be gained from processing, but it certainly can be lost. Fruits are best when frozen fully ripe but still firm and at the peak of quality, with a pleasing color, texture, flavor, and maximum nutritional value. Great differences of frozen fruit quality exist between fruit varieties and cultivars based on chemical, biochemical, and physical characteristics that determine the sensorial and nutritional quality. Differences in cell wall structure, enzyme activity, amounts of pigments, sugars, organic acids, volatile compounds, vitamins C, A, and E, and other components are factors that affect the differences in sensorial and nutritional quality of raw fruits. Freezing potential of fruit varieties or cultivars are evaluated with practical trials after freezing, frozen storage, and thawing of the fruit products. The suitability of varieties or cultivars for freezing can be studied on the basis of physical (texture and color), physical–chemical (pH, acidity, and soluble solids), chemical (volatile, pigments, and polyphenol compounds), nutritional (vitamins and dietary fiber content), and sensorial aspects (firmness, color, and taste). These kinds of studies have been done with different fruits such as kiwi (Cano and Marín, 1992; Cano et al., 1993a), mango (Marín et al., 1992; Cano and Marín, 1995), pineapple (Bartolomé et al., 1996a, b, c), papaya (Cano et al., 1996a; Lobo and Cano, 1998), raspberry (De Ancos et al., 1999, 2000a, b; González et al., 2002), strawberry (Castro et al., 2002), and other fruits. Another criterion for selection of suitable variety or cultivar can be the enzymatic systems activity (polyphenoloxidase, peroxidase, lipoxygenase, etc.), in raw fruit and during freezing and frozen storage. Employing varieties with low enzymatic activities could reduce the development of browning, off-flavors and off-odors, and color and textural changes (Cano et al., 1990b, 1996b, 1998; González et al., 2000).

Harvesting fruits at optimum level for freezing purposes is difficult. The need for efficient production often implies the use of mechanical harvesting at a time when the fruit has reached an acceptable maturity level to avoid mechanical damage. Postharvest techniques allow the storage of unripe climateric fruits at specific atmosphere, temperature, and humidity conditions until they reach proper maturity levels to be frozen (Cano et al., 1990a, b; Marín et al., 1992). Nonclimateric fruits (strawberries, raspberries, etc.) are harvested, preferably when fully ripe but still firm, cooled immediately after picking, and frozen as soon as possible (Gónzalez et al., 2002). However, the quality advantages of immediate freezing could not be detected after a long frozen storage period (6–12 months) (Plocharski, 1989).

PREPARING, PRETREATMENTS, AND PACKAGING

Successful freezing should retain the initial quality present in the raw fruit selected for freeze processing according to freshness, suitability varietal for freezing, and sensorial and nutritional characteristics. Retaining this quality level prior to freezing is a factor of major importance to obtain high-quality frozen fruits.

Preparing

Fruits must be prepared before freezing according to the frozen fruit end-use. Washing, rinsing, sorting, peeling, and cutting the fruits are not specific steps for frozen fruits; these are preparatory operations similar to other types of processing but must be carried out quickly and with great care to avoid damaging the fragile fruit tissue. Peeling, stone removal, and cutting in cubes, slices, or halves are usually mechanical operations. Decreasing the size of the product before freezing results in a faster freezing and consequently a better frozen fruit quality. For economical factors, certain fruits like peaches, apricots, and plums are frozen whole immediately after harvesting and peeling; stone removal and cutting is done after a partial thawing.

Consumption of fruit juices and nectars has increased in the world due to recommendations for better nutrition and healthier diets. Fruits and fruit juices meet these recommendations. Nectars and fruit juices can be manufactured with fresh fruit but with frozen fruit higher yields are obtained.

At present, frozen juices represent an important segment of the international drink industry. Preparing fruit for frozen juice requires different steps: pressing, clarification, heat treatment, and concentration. Also purees and pulps represent an important ingredient for the manufacturing industries for dairy products, cakes, ice-creams, jellies, and jams (Chen, 1993).

Pretreatments

The importance of enzyme content to fruit quality has been extensively reviewed (Philippon and Rouet-Mayer, 1984; Browleader et al., 1999; Robinson and Eskin, 1991; Friedman, 1996). Enzymes, namely polyphenoloxidase (PPO), peroxidase (POD), lipoxygenase (LOX), catalase (CAT), and pectinmethylesterase (PME) are involved in the fast deterioration of fruit during postharvest handling and processing. Enzymes not inactivated before freezing can produce off-flavors, off-odors, color changes, development of brown color, and loss of vitamin C and softness during frozen storage and thawing. Water blanching is the most common method for inactivating vegetable enzymes (Fellows, 2000). It causes denaturation and therefore, inactivation of the enzymes that also causes destruction of thermosensitive nutrients and losses of water-soluble compounds such as sugar, minerals, and water-soluble vitamins. Blanching is rarely used for fruits because they are usually consumed raw and heat treatment causes important textural changes. An alternative to blanching fruit is to use ingredients and chemical compounds that have the same effect as blanching.

Blanching. Heat treatment to inactivate vegetable enzymes can be applied by immersion in hot water, by steam blanching or by microwave blanching. Hot water blanching is usually done between 75°C and 95°C for 1–10 min, depending on the size of the vegetable pieces. Hot water blanching also removes tissue air and reduces the occurrence of undesirable oxidation reactions during freezing and frozen storage. Steam blanching reduces the water-soluble compounds losses and is more energy efficient than water blanching. Of all the enzymes involved in vegetable quality losses during processing, POD and CAT seem to be the more heat stable, and thus could be used as an index of adequate blanching. Generally, a quality blanched vegetable product permits some POD and CAT activity. Complete POD inactivation indicates overblanching. Blanching also helps to destroy microorganisms on the surface of the vegetable. Blanching destroys semipermeability of cell membranes and removes cell turgor. Reduced turgor is perceived as softness and lack of crispness and juiciness. These are some of the most important sensorial characteristics of eating fruit. Although loss of tissue firmness in blanched frozen fruits after thawing indicates that blanching is not a good pretreatment for the majority of the fruits, some results have been interesting (Reid, 1996). Hot water blanching peeled bananas prior to slicing, freezing, and frozen storage produced complete PPO and POD inactivation and a product with acceptable sensorial quality (Cano et al., 1990a). Microwave blanching has not been an effective pretreatment for banana slices (Cano et al., 1990b) but interesting results have been obtained with frozen banana purees (Cano et al., 1997)

Addition of Chemical Compounds. Substitutes for thermal blanching have been tested with different enzymatic inhibitors. They are mainly antibrowning additives such as sulfiting agents (sulfur dioxide or inorganic sulfites salts) and ascorbic acid, which are applied by dipping or soaking the fruit in different solutions before freezing (Skrede, 1996). Enzymatic browning involving the enzyme PPO is the principal cause of fruit quality losses during postharvest and processing. PPO catalyzes the oxidation of mono- and orthodiphenols to quinones, which can cyclize, undergo further oxidation, and polymerize to form brown pigments or react with amino acids and proteins that enhance the brown color produced (Fig. 4.3).

The proposed mechanisms of antibrowning additives that inhibit enzymatic browning are (1) direct inhibition of the enzyme; (2) interaction with intermediates in the browning process to prevent the reaction leading to the formation of brown pigments; or (3) to act as reducing agents promoting the reverse reaction of the quinone back to the original phenols (Fig. 4.3). (Friedman, 1996; Ashie et al., 1996). Other acid treatments such as dipping in citric acid or hydrochloric acid solution (1%) could be a commercial pretreatment for browning control and quality maintenance of frozen litchi fruit (Yueming-Jiang et al., 2004). Although all the fruits contain polyphenolic compounds, some fruits as peaches, apricots, plums, prunes, cherries, bananas, apples, and pears show a greater tendency to develop browning very quickly during processing. Research efforts have been done to develop new natural antibrowning agents in order to replace sulfites, the most powerful and cheapest product until now, but they cause adverse health effects in some asthmatics. In this framework, maillard reaction products have been recognized as a strong apple PPO inhibitor (Billaud et al., 2004). Also, some frozen fruits like apples and cherimoya are pretreated by dipping its slices in sodium chloride solutions (0.1–0.5%) in combination with ascorbic or citric acid, in order to remove intracellular air and reduce oxidative reactions (Reid, 1996; Mastrocola et al., 1998).

Fruit texture is greatly changed by freezing, frozen storage, and thawing. Fruits have thin-walled cells rich in pectin substances, in particular in the middle lamella between cells, and with a large proportion of intracellular water, which can freeze resulting in cell damage. Freezing–thawing also accelerates the release of pectin, producing de-esterification of pectins and softens the fruit tissue. Optimum freezing rate reduces tissue softening and drip loss, and the addition of calcium ions prior to freezing increases the firmness of fruit after thawing. These ions fortify the fruit by changing the pectin structure. Calcium maintains the cell wall structure in fruits by interacting with the pectic acid in the cell walls to form calcium pectate.

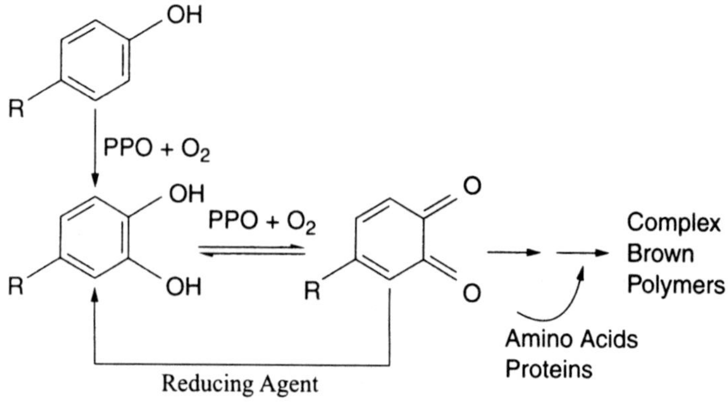

Figure 4.3. Enzyme-catalyzed initiation of browning by PPO showing the point of attack by reducing agents.

Dipping in calcium chloride solution (0.18% Ca) or pectin solution (0.3%) improves the quality of frozen and thawed strawberries (Suutarinen et al., 2000).

Osmotic Dehydration: Addition of Sugars and Syrups. Dipping fruits in dry sugar or syrups is a traditional pretreatment to preserve color, flavor, texture, and vitamin C content and to prevent the browning of freezing–thawing fruits. Sugar or syrups are used as cryoprotectants by taking out the fruit cell water by osmosis and excluding oxygen from the tissues. Partial removal of water before freezing might reduce the freezable water content and decrease ice crystal damage, making the frozen fruit stable. Therefore, minor damage to cellular membranes occurs and oxidative reactions and enzymatic degradation reactions are minimized. The process of dehydration before freezing is known as *dehydrofreezing* (Fito and Chiralt, 1995; Robbers et al., 1997; Bing and Da-Wen, 2002). During osmotic dehydration, the water flows from the fruit to the osmotic solution, while osmotic solute is transferred from the solution into the product, providing an important tool to impregnate the fruit with protective solutes or functional additives. Syrup is considered a better protecting agent than dry sugar. Dry sugars are recommended for fruits, such as sliced peaches, strawberries, figs, grapes, cherries, etc., that produce enough fruit juice to dissolve the sugar. Dipping fruit, whole or cut, in syrup allows a better protection than dry sugar because the sugar solution is introduced inside the fruit. Syrup concentrations between 20% and 65% are generally employed, although 40% syrup is enough for the majority of the fruits. Sucrose is the osmotic agent most suitable for fruits although other substances, including sucrose, glucose, fructose, lactose, L-lysine, glycerol, polyols, maltodextrin, starch syrup, or combinations of these solutes can be used (Bing and Da-Wen, 2002; Zhao and Xie, 2004). Osmotic dehydration is carried out at atmospheric pressure or under vacuum. Among developments in osmotic treatments, vacuum impregnation may be the newest. The exchange of partial freezable water for an external solution is promoted by pressure, producing different structural changes and lower treatment time than osmotic dehydration at atmospheric pressure. Successful applications of dehydrofreezing and vacuum impregnation on fruits have been recently reviewed (Zhao and Xie, 2004). Great color, flavor, and vitamin C retention have been achieved in frozen–thawed strawberries, raspberries, and other types of berries treated with a 20% or 40% syrup concentration before freezing and long-term frozen storage between 6 months and 3 years (Skrede, 1996). The effects of dehydrofreezing process on the quality of kiwi, strawberry, melon, and apples have been reported (Garrote and Bertone, 1989; Tregunno and Goff, 1996; Spiazzi et al., 1998; Talens et al., 2002, 2003). The quality and texture of dehydrofrozen and thawed fruit has been improved by using osmotic solutions in combination with ascorbic acid solution (antibrowning treatment) and/or calcium chloride or pectin solutions (Skrede, 1996; Suutarinen et al., 2000; Talens et al., 2002, 2003; Zhao and Xie, 2004). Another important factor contributing to fruit quality improvement is vacuum impregnation, which is useful in introducing functional ingredients into the fruit tissue structure, conveniently modifying their original composition for development of new frozen products enriched with minerals, vitamins, or other physiologically active nutritional components (Zhao and Xie, 2004).

Packaging

Packaging of frozen fruits plays a key role in protecting the product from air and oxygen that produce oxidative degradation, from contamination by external sources, and from damage during passage from the food producer to the consumer. Package barrier properties protect the frozen fruit from ingress of oxygen, light, and water vapour, each of which can result in deterioration of colors, oxidation of lipids and unsaturated fats, denaturation of proteins, degradation of ascorbic acid, and a general loss of characteristic sensory and nutritional qualities. Similarly, barrier properties protect against the loss of moisture from the frozen food to the external environment to avoid external dehydration or "freezer burn" and weight loss. The primary function of food packaging is to protect the food from external hazards. In addition, packaging materials should have a high heat transfer rate to facilitate rapid freezing. Also, the package material should not affect the food in any way, as indicated by European Directives on food contact materials, including migration limits (EC Directives 1990, 1997) and the Code of Federal Regulations in the United States regarding food contact substances (CFR 2004). A wide range of materials has been used for packaging of frozen fruits, including plastic, metals, and paper/cardboard, or polyethylene bags. Laminates can provide a combination of "ideal" package properties.

Table 4.1. Relative Oxygen and Water Vapour Permeabilities of Some Food Packaging Materials (References Values Measured at 23°C and 85% RH)

Package Material	Relative Permeability	
	Oxygen (ml m^{-2} day^{-1} atm^{-1})	Water Vapour (g m^{-2} day^{-1})
Aluminum	<50 (Very high barrier)	<10 (very high barrier)
Ethylene vinyl acetate (EVOH)	<50 (Very high barrier)	variable
Polyester (PET)	50–200 (High barrier)	10–30 (high barrier)
Polycarbonate (PC)	200–5000 (Low barrier)	100–200 (medium barrier)
Polyethylene (PE)		
High density (HDPE)	200–5000 (Low barrier)	<10 (very high barrier)
Low density (LDPE)	5000–10,000 (Very low barrier)	10–30 (high barrier)
Polypropylene (PP)	200–5000 (Low barrier)	10–30 (high barrier)

Source: Atmosphere Controle 2000 (http://atmosphere-controle.fr/permeability.html).

Board and paper packages are often laminated with synthetic plastics to improve the barrier properties. Table 4.1 shows some comparisons of barrier properties for a range of common package materials. Fruit products can be packaged before freezing (fruits with sugar or syrup, purees and juices concentrated or not) or after freezing (whole or cut fruits). The importance of packaging material to the stability of frozen fruits has been reviewed (Skrede, 1996). In general, quality differences (pigment content, ascorbic acid retention, color, and consistency) between frozen products packaged in different types of packages are mainly detected after a long period of frozen storage (>3 months) and at temperatures over −18°C.

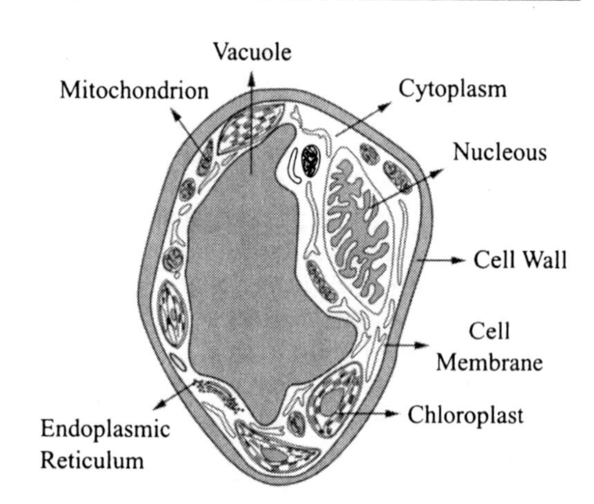

Figure 4.4. Cross-section of a plant cell.

Effect of Freezing, Frozen Storage, and Thawing on Fruit Tissues: Physical, Chemical, and Biochemical Changes

Plant Cell Structure

Understanding the effect of freezing on fruit requires a short review of plant cell structure. A relationship between cell structure properties and freezing cell damage has been extensively reviewed (Reid, 1996; Skrede, 1996). Plant cells are surrounded by a membrane and interspersed with extensive membrane systems that structure the interior of the cell into numerous compartments. The plasmalemma or plasma membrane encloses the plasma of the cell and is the interface between the cell and the extracellular surroundings. Contrary to animal cells, plant cells are almost always surrounded by a cell wall and many of them contain a special group of organelles inside: the plastids (chloroplasts, leucoplasts, amyloplasts, or chromoplasts) (Fig. 4.4).

An important property of the plant cell is its extensive vacuole. It is located in the center of the cell and makes up the largest part of the cells volume and is responsible for the turgor. It helps to maintain the high osmotic pressure of the cell and the content of different compounds in the cell, among which are inorganic ions, organic acids, sugars, amino acids, lipids, oligosaccharides, tannins, anthocyanins, flavonoids, and more. Vacuoles are surrounded by a special type of membrane, the tonoplast. The cell wall of plants consists of several stacked cellulose microfibrils embedded in a polysaccharide matrix able to store water thereby increasing the cell volume (hydration and absorbtion). According to their capacity to bind or store water, the polysaccharides involved in the matrix can be classified as follows: pectin>hemicellulose>cellulose>lignin.

Pectins are mainly polygalacturonic acids with differing degrees of G-galactosyl, L-arabinosyl or L-rhanmosyl residue and are predominant in the middle lamella, the layer between cells. The de-esterification process of pectin is related to the softness of fruit tissues during ripening and processing.

Physical Changes and Quality

Volume Expansion. The first factor that produces mechanical damage to the cell is the volume expansion due to the formation of ice that affects the integrity of cell membrane.

Recrystallization. Ice crystals can change the quality of frozen fruits in different ways. First, the speed of freezing affects frozen–thawing fruit quality. Slow speed freezing produces large and sharp ice crystals that can produce mechanical damage to the fragile plants cell membranes, causing the cell organelles to collapse and lose their contents (sugars, vitamins, pigments, volatile compounds, phenol, enzymes, etc.) and a breakdown of the pectin fraction in the cell wall which affects fruit tissue texture. During frozen storage, retail-display, or the carry-home period, fluctuations in product temperature produces ice recrystallization that affects the number, size, form, and position of the ice crystal formed during freezing. Frequent large fluctuation produces partial fusion of ice and the reforming of large and irregular ice crystals that can damage cellular membranes and produce a freeze-dried product, allowing sublimed or evaporated water to escape.

Sublimation: Freezer Burn. The sublimation of the ice may occur during frozen storage if the packaging product is unsuitable. Moisture loss by evaporation from the surface of the product leads to "freezer burn," which is recognized as a light-colored zone on the surface of the product. Dehydration of the product can be avoided by improving the type of package, increasing humidity, and decreasing the storage temperature.

The recrystallization and freezer burn dehydration increase with temperature fluctuations, but the harmful effect of these two processes on frozen fruit quality can be decreased by lowering the storage temperature below −18°C (IIR, 1996)

Chemical and Biochemical Changes and Quality

The chemical and biochemical reactions related to sensorial and nutritional quality changes of fruits are delayed but not completely stopped at subzero temperature. Quality changes, such as loss of the original fruit color or browning, developing off-odour and off-taste, texture changes, and oxidation of ascorbic acid, are the main changes caused by chemical and biochemical mechanisms that affect fruit quality. Also, pH changes in fruit tissues detected during freezing and frozen storage can be a consequence of these degradation reactions.

Color Changes. Color is the most important quality characteristic of fruits because it is the first attribute perceived by the consumers and is the basis for judging the product acceptability. The most important color changes in fruits are related to chemical, biochemical, and physicochemical mechanisms: (a) breakdown of cellular chloroplasts and chromoplasts, (b) changes in natural pigments (chlorophylls, carotenoids, and anthocyanins), and (c) development of enzymatic browning.

Mechanical damage (ice crystals and volume expansion) caused by the freezing process can disintegrate the fragile membrane of chloroplasts and chromoplasts, releasing chlorophylls and carotenoids, and facilitating their oxidative or enzymatic degradation. Also, volume expansion increases the loss of anthocyanins by lixiviation due to disruption of cell vacuoles.

(i) *Chlorophylls.* Chlorophylls are the green pigment of vegetables and fruits, and their structures are composed of tetrapyrroles with a magnesium ion at their center. Freezing and frozen storage of green vegetables and fruits cause a green color loss due to degradation of chlorophylls (a and b) and transformation in pheophytins, which transfers a brownish color to the plant product (Cano, 1996). One example is kiwi-fruit slices that show a decrease in chlorophyll concentration between 40% and 60%, depending on cultivar, after freezing and frozen storage at −20°C for 300 days (Cano et al., 1993a). Different mechanisms can cause chlorophyll degradation; loss of Mg due to heat and/or acid, which transforms chlorophylls into pheophytins; or loss of the phytol group through the action of the enzyme chlorophyllase (EC 3.1.1.14), which transforms chlorophyll into pheophorbide. Loss of the carbomethoxy group may

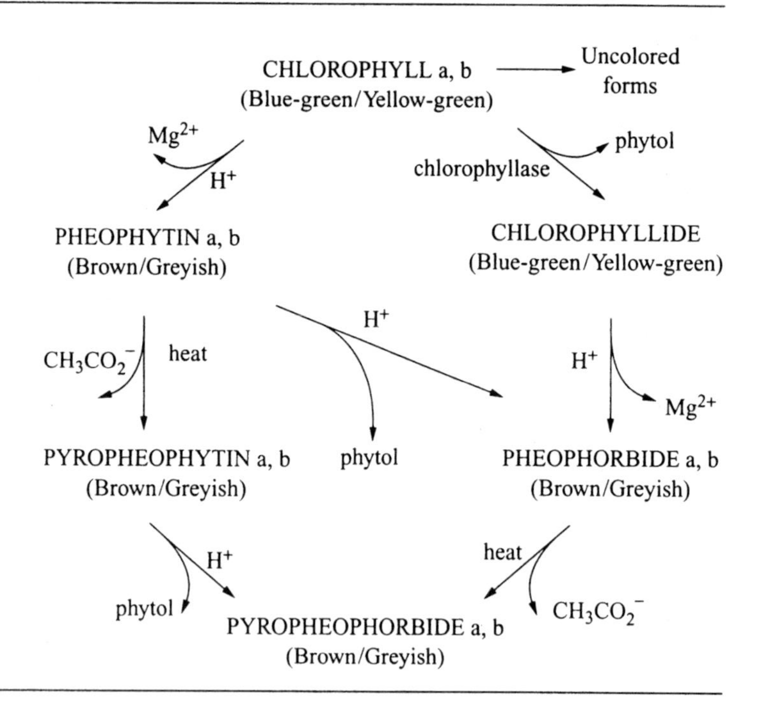

Figure 4.5. Pathways of chlorophyll degradation.

also occur and pyropheophytin and pyropheophorbide can be formed (Fig. 4.5.) (Heaton et al., 1996).

Acids, temperature, light, oxygen, and enzymes easily destroy the chlorophylls. Thus, blanching (temperature/time), storage (temperature/time), and acidity are the important factors to be controlled during processing in order to preserve chlorophylls. Other chlorophyll degradation mechanism can cause degradation by the action of peroxides, formed in the fruit tissue due to the oxidation reaction of polyunsaturated fatty acids catalyzed by the enzyme LOX. An important quality parameter employed to determine the shelf life of frozen green fruits is the formation of pheophytins from chlorophylls. As different types of enzymes can be involved in chlorophyll degradation (LOX, POD, and chlorophyllase), blanching and addition of inorganic salts such as sodium or potassium chloride and sodium or potassium sulphate are efficient treatments to preserve green color (IIR, 1986; Cano and Marín, 1992; Cano et al., 1993a, b).

(ii) *Carotenoids.* Carotenoids are among the most abundant pigment in plant products and are responsible for the yellow, orange, and red color of most of the fruits. All of them are tetraterpenes and contain 40 carbon atoms in eight isoprenes residues. β-carotene and lutein are the carotenoids present in most of the fruits. Important sources of these pigments are as follows (Figure 4.6):

- β-cryptoxanthin: oranges
- lycopene: tomatoes, watermelon, papaya and persimmon
- α-carotene: banana and avocado
- zeaxanthin: orange and peach

Carotenoids are affected by pH, enzymatic activity, light, and oxidation associated with the conjugated double bond system. The chemical changes occurring in carotenoids during processing have been reviewed by several authors (Simpson, 1986; Rodriguez-Amaya, 1997). The main degradation reaction that damages carotenoid compounds is isomerization. Most plants appear to produce mainly *trans* forms of carotenoids but with increased temperature, the presence of light, and catalysts such as acids, isomerization to the *cis* forms increases, and the biological activity is dramatically reduced. However, heat treatments of products rich in carotenoids reduce the degradation of carotenoids because of the inactivation of enzymes LOX and POD. Blanching fruits before freezing could be efficient in the preservation of carotenoids due to enzyme inactivation. Although most carotenoids are heat resistant, some carotenoids, such as epoxycarotenoids, could be

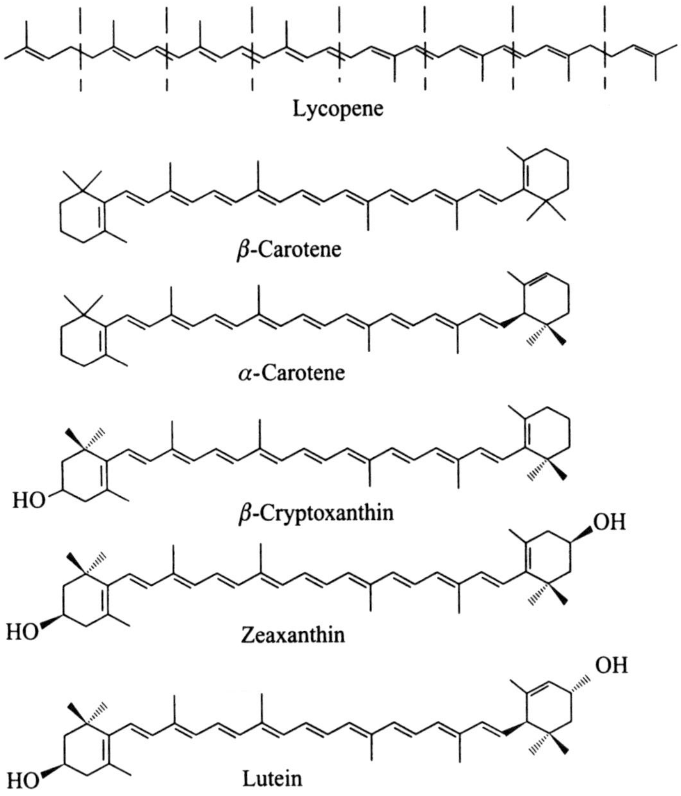

Figure 4.6. Structure of more frequent carotenoids present in fruits.

affected. Carotenoids are fat-soluble pigments and breakdown of chromoplasts, by heat treatment or mechanical damage, improves their extraction with organic solvents and bioavailability but not their loss by lixiviation (Hof et al., 2000). Freezing without protector pretreatment slightly decreases total carotenoid concentration (20%) of some fruits rich in carotenoids, such as mango and papaya. But after 12 months of frozen storage at −18°C, an important decrease of total carotenoid concentration (between 40% and 65%) occurred, although the carotenoid profile was unchanged (Cano and De Ancos, 1994; Cano et al., 1996b). Similar results have been found with frozen tomato cubes. A pronounced stability of total carotenoids, β-carotene, and lycopene was recorded up to the 3rd month of storage. But after 12 months of storage at −20°C, the losses of carotenoids reached 36%, of β-carotene 51%, and of lycopene 48% (Lisiewska and Kmiecik, 2000). Freezing and frozen storage could affect the carotenoid structure and concentration depending on the type of fruit and cultivar (pH, fats, antioxidants, etc.) and the processing conditions (temperature, time, light, oxygen, etc.) (Simpson, 1986; Rodriguez-Amaya, 1997).

(iii) *Anthocyanins.* Anthocyanins are one class of flavonoid compounds, which are widely distributed plant polyphenols, and are responsible for the pink, red, purple, or blue hue of a great number of fruits (grape, plum, strawberry, raspberry, blackberry, cherry, and other types of berries). They are water-soluble flavonoid derivatives, which can be glycosylated and acylated. The effect of freezing, frozen storage, and thawing in different fruits rich in anthocyanins pigments have been reviewed by Skrede (1996). Anthocyanins in cherry fruit underwent pronounced degradation during storage at −23°C (87% after 6 months), but they are relatively stable at −70°C storage (Chaovanalikt and Wrolstad, 2004). But in raspberry fruit, the stability of anthocyanins to freezing and frozen storage depends on the seasonal period of harvest. Spring cultivars were practically unaffected by freezing and frozen storage for 1 year at −20°C, but autumn cultivars showed a decreasing trend in total anthocyanin content (4–17%)(De Ancos et al., 2000b). In general, the freezing process does not affect the level of anthocyanins in raspberry fruit (De Ancos, 2000; Mullen et al., 2002). Authors explain degradation of anthocyanins during frozen storage by different chemical or biochemical mechanisms. Anthocyanins are water-soluble pigments located in the vacuoles of cell and are easily lost by lixiviation when the cell membranes break down. Also oxidation can play an important role in anthocyanin degradation catalyzed by light. PPO and POD enzymatic activities have been related to anthocyanin degradation. Thus, frozen–thawed cherry discoloration disappeared when the fruits were blanched before freezing. The changes in pH during processing can affect anthocyanin stability. Maintenance of red fruit requires an acid medium (pH < 3.5). The flavylium cation structure of anthocyanins transfers a red color to the fruit. But an increase in pH value produces a change from red to blue until the product is colorless, a consequence of transforming flavylium cation into a neutral structure (Fig. 4.7).

The loss of characteristic red color can also be produced by formation of the anthocyanin complex with different products present in the fruit matrix: ascorbic acid, acetaldehyde, proteins, leucoanthocyanins, phenols, quinones, metals (Fe^{3+} and Al^{3+}), hydrogen peroxide, etc. (Escribano-Bailon et al., 1996).

(iv) *Enzymatic Browning.* Browning usually occurs in certain fruits during handling, processing, and storage. Browning in fruit is caused by enzymatic oxidation of phenolic compounds by PPO(EC 1.10. 3.1) (Martínez-Whitaker, 1995). PPO catalyzes either one or two reactions involving molecular oxygen. The first type of reaction is hydroxylation of monophenols, leading to formation of o-hydroxy compounds. The second type of reaction is oxidation of o-hydroxy compounds to quinones that are transformed into polymeric brown pigment (Fig. 4.3). Freezing, frozen storage, and thawing of fruits, like mangoes, peaches, bananas, apples, apricots, etc., quickly develop color changes that result in nonreversible browning or darkening of the tissues. Freezing does not inactivate enzymes; however, some enzyme activity is slowed during frozen storage (Cano et al., 1998). Browning by PPO can be prevented by the addition of sulfites, ascorbic acid, citric acid, cysteine, and others. The addition of antibrowning agents has been discussed in the pretreatments section. Selection of varieties with low PPO activity could help to control browning in

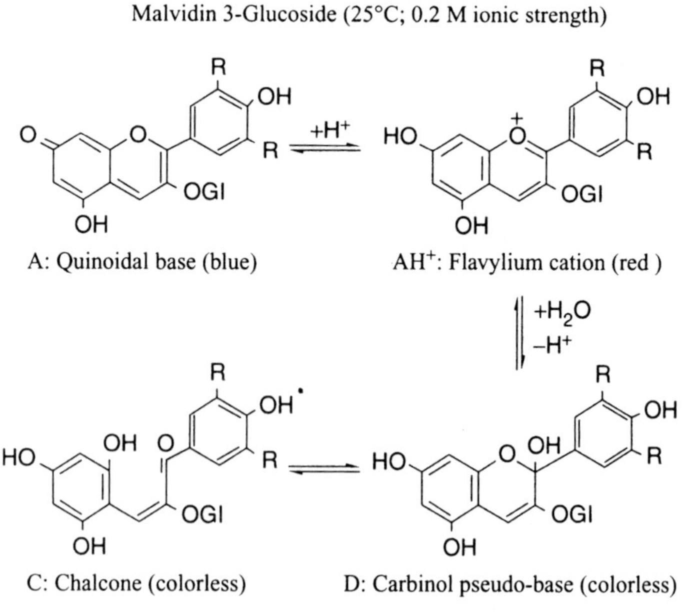

Malvidin 3-Glucoside (25°C; 0.2 M ionic strength)

A: Quinoidal base (blue) ⇌ ($+H^+$) AH+: Flavylium cation (red)

⇅ ($+H_2O$, $-H^+$)

C: Chalcone (colorless) ⇌ D: Carbinol pseudo-base (colorless)

Figure 4.7. Effect of pH on anthocyanins.

frozen–thawed fruits (Cano et al., 1996b; Cano et al., 1998).

Flavor and Aroma Changes. Volatile compounds forming the fruit flavor (alcohols, esters, aldehydes, ketones, acids, furans, terpenes, etc.) are produced through metabolic pathways during harvest, postharvest, and storage and depend on many factors related to species, variety, and type of processing. Although freezing is the best way to preserve fruit aroma (Skrede, 1996), frozen storage and thawing can modify the natural fresh aroma of some fruits such as strawberries (Larsen and Poll, 1995), but other fruits like kiwi (Talens et al., 2003) or raspberry fruits (De Ancos, 2000) do not significantly modify the aroma profile. Freezing, frozen storage, and thawing affect fruits volatile profile in different ways depending on the type of fruit and variety.

Instead of being destroyed during freezing, some enzymes are released. This can cause cell disruption and is one factor in the development of off-flavors and off-odors in plant products during frozen storage. Blanching is the main tool used to inactivate enzymes before freezing, but most fruits suffer important textural changes when blanched. POD enzyme activity has been related to the presence of different volatile compounds such as hexanal, which is produced during lipid oxidation and confers an unpleasant odor to the frozen–thawed product. Cell structure disruption during freezing and frozen storage favors an increase in or preserved important enzymatic POD activity levels in different thawed fruits [mango (Marín et al., 1992) and papaya (Cano et al., 1998)]. It is important to select the suitable fruit varieties for freezing, based on high volatile compounds concentration and low enzymatic activity, to obtain high-quality frozen fruit.

Textural Changes. Texture of frozen fruits is dependent on chemical and biochemical modifications of the cell wall and middle lamella components (pectins, hemicelluloses, and celluloses). Freezing causes severe texture loss due to the cryoconcentration phenomena, which can induce cell wall degradation and a decrease in liquid retention. The size and location of ice crystals cause cell membrane rupture that promotes enzyme and/or chemical activity and contributes to mechanical damage in cell wall material. The influence of freezing rate on tissue integrity, texture, and drip loss has been reviewed by different authors (Skrede, 1996; Cano, 1996; Reid, 1996). Ice recrystallization also leads to greater damage during frozen storage (Reid, 1996). Pectin is an important component of fruit cell wall. In fact, a decrease of the pectin fraction during freezing and frozen storage has been related to a reduction of firmness in different fruits (Lisiewska and Kmiecik, 2000).

Nutritional and Antioxidant Status Changes. Consumption of fruits is related to a good nutritional status and contributes to the prevention of degenerative processes, particularly the lowering of the incidence and mortality rate of cancer and cardiovascular disease (Steinmetz and Potter, 1996; Tibble et al., 1998; Willcox et al., 2003). Nutritional compounds found in fruits are vitamins, sugars, minerals, proteins, and fats. Fruits are the main dietary source of vitamins C, A, and E, which are indispensable for human life. The protective effect of a rich fruit diet has been attributed to certain bioactive compounds with antioxidant and antimutagenic properties. Vitamins A, and C, carotenoids, and phenolics are the main bioactive compounds that contribute to the antioxidant characteristics of fruits (Rice-Evans et al., 1996; Miller and Rice-Evans 1997; Boileau et al., 1999; Gardner et al., 2000). Retention of the nutritional and antioxidant value of fruit is the main goal of all the processing methods, and freezing and frozen storage can be one of the less destructive methods in terms of long-storage periods.

(i) *Vitamin C.* Freezing processes have only a slight effect on the initial vitamin C content of fruit (Cano and Marín, 1992; Marín et al., 1992; De Ancos, 2000). The destruction of vitamin C (ascorbic acid) occurs during freezing and frozen storage, and this parameter has been employed to limit the frozen storage period of frozen fruit. The main cause of loss of vitamin C is the action of the enzyme ascorbate oxidase. If pretreatments or freezing processes do not destroy this enzyme, it is continuously active during the frozen storage. Vitamin C degradation depends on different factors, such as time–temperature conditions, type of fruit, variety, pretreatments, type of package, freezing process, etc. (Skrede, 1996). Thus as the frozen storage temperature decreases, higher vitamin C retention is achieved for different fruits like berries, citrus, tomato, etc. (Skrede, 1996; Lisiewska and Kmiecik, 2000). Also, significantly different vitamin C retention values have been achieved between varieties of fruits such as raspberry (De Ancos, 2000), mango (Marín et al., 1992), and kiwi (Cano and

Marín, 1992), which were frozen and stored under the same conditions. Vitamin C stability in freezing and frozen storage of strawberries seems to be more dependent on storage temperature than on the type of freezing process. Nonstatistical differences were observed between strawberries processed by fast rate freezing (at −20°C) and quick rate freezing (at −50°C to −100°C), but great loss was shown between strawberries stored at −18°C and −24°C (Sahari et al., 2004).

(ii) *Provitamin A and Antioxidant Carotenoids.* Some carotenoids, like β-carotene, α-carotene, and β-cryptoxanthin, are recognized as precursors of vitamin A. These provitamin A carotenoids, in addition to lycopene and lutein, constitute the group of antioxidant carotenoids. The prevailing opinion is that freezing and frozen storage do not prevent degradation of carotenoids. The content of β-carotene, and consequently the provitamin A value, was decreased during frozen storage of mango (Marín et al., 1992), kiwi (Cano and Marín, 1992), papaya (Cano, 1996), and tomato (Lisiewska and Kmiecik, 2000). The losses were mainly due to the activity of enzymes (POD, LOX, and CAT), particularly during frozen storage in an oxygen environment. Lycopene, a characteristic carotenoid in tomato fruit, has been recognized as a powerful antioxidant (Rao and Agarwal, 1999; Lavelli et al., 2000). After 3 months of frozen storage (−20°C and −30°C), great stability of lycopene was recorded. After this period, slow losses occurred, the rate being faster at the higher storage temperature. After 12 months at −20°C and −30°C, the lycopene content was 48% and 26%, respectively, lower than that in the raw material (Lisiewska and Kmiecik, 2000). Other authors have reported an increase in the extraction of lycopene after 1 month of frozen storage, although after 3 and 6 months the loss of lycopene concentration was significantly higher than 40% (Urbanyi and Horti, 1989). Papaya fruit could be an important source of lycopene, but freezing and frozen storage at −20°C during 12 months produced a significant loss of lycopene concentration (34%) in frozen papaya slices (Cano, 1996).

Further discussion on the effect of freezing, frozen storage, and thawing on carotenoid stability are included in the section of color changes.

(iii) *Phenolic Compounds.* The freezing process does not modify either total phenolic content or ellagic acid concentration in raspberry fruit. There is an increasing interest in ellagic acid, a dimeric derivative of gallic acid, due to its anticarcinogenic and antioxidant effects. Although frozen storage produces a slight decrease in ellagic acid content because of PPO enzyme activity, frozen storage is a good methodology to preserve phenolic compounds during long-term periods (De Ancos, 2000).

(iv) *Antioxidant Capacity.* Radical scavenging capacity, a measure of the antioxidant capacity of fruit extracts, was not affected by freezing and long-term frozen storage (De Ancos, 2000).

(v) *Dietary Fiber.* Comparative studies on dietary fiber content between fresh fruit pulp and the corresponding frozen fruit pulp have shown that frozen fruit pulp has lower fiber content than fresh fruit pulp. Freezing and frozen storage induced significant dietary fiber losses ranging from 18% for mango to 50% for other fruits like guava (Salgado et al., 1999).

Stability of Frozen Fruit

Physical, physicochemical, chemical, and biochemical changes that occur in frozen fruit during the storage period lead to a gradual, cumulative and irreversible loss of quality that limits the storage life of frozen fruit. Temperature and length of storage time are the principal factors that limit the frozen storage period of fruit and are known as TTT factors. In general, lower storage temperatures lead to longer storage life. TTT data for each fruit was determined by different quality analysis of samples of the same product, identically processed, and stored at different temperatures in the range of −10°C to −40°C. At certain intervals of frozen storage, sample quality was analyzed. Sensorial analysis, loss of vitamin C, and changes of chlorophylls to pheophytin, or other types of pigment degradation, are the quality analyses used to determine the storage life of the frozen fruit. On the basis of TTT data, different terms have been established to determine the suitable frozen storage life. “High-Quality Life” has been defined as the storage period quality of a frozen product compared to a similar quality of a product just frozen. After this time, frozen fruits are still suitable for consumption, and a second term has been defined as “Practical Storage Life” or the storage time period that provides frozen foods suitable for human consumption. Table 4.2 shows the “Practical Storage Life” for different frozen fruits stored at −12°C, −18°C, and −24°C. Fruit frozen with sugar or syrup added is more sensitive to an increase in frozen storage temperature because they freeze at lower temperature than fruit frozen without sugar. Thus, strawberries without sugar stored at −12°C have longer “Practical Storage Life” (5 months) than fruit with

Table 4.2. "Practical Storage Life" at Different Frozen Storage Temperatures (In Months)

Fruit	−12°C	−18°C	−24°C
Strawberry/raspberry/peach	5	24	>24
(Strawberry/raspberry/peach) + sugar	3	24	>24
Apricot/cherry	4	18	>24
(Apricot/cherry) + sugar	3	18	>24
Fruit juice (concentrated)	–	24	>24

Source: Institute of International Refrigeration (IIR, 1986).

sugar (3 months). These times for suitable storage were obtained on the basis of high-quality raw products, processing in suitable conditions, and without temperature fluctuations during frozen storage. Increasing and fluctuating temperature may occur during transport and retail display. Temperature fluctuations shorten the storage life of frozen foods because of accelerated degradation reactions and increased quality loss (IIR, 1986; Cano, 1996).

Thawing

The quality of the original fruit, preserved by freezing, is retained by quick thawing at low temperature in controlled conditions. During incorrect thawing, chemical and physical damage and microorganism contamination can also occur. Fruit products exhibit large losses of ascorbic acid (up to 40%) and color changes when thawed for an unusually long period, e.g., 24 h at room temperature. Good results in terms of vitamin C and anthocyanins retention (90%) were achieved by thawing small frozen fruits such as bilberry, raspberry, black currant, red currant, and strawberry at room temperature (18–20°C/6–7 h), in a refrigerator (2–4°C/18 h), or in a microwave oven. Color and ascorbic acid retention of fruit was equally affected by thawing temperature and time. Thorough thawing must be determined by taking into account the size of the fruit and/or the type of packaging (Kmiecik et al., 1995).

Microbiological Quality and Safety of Frozen Fruits

Fruit microflora are dominated by spoilage yeast, moulds, and bacteria, but occasionally the presence of pathogenic bacteria, parasites, and viruses capable of causing human infections has also been documented. Fruits can become contaminated with pathogenic microorganisms while growing in fields, orchards, vineyards, or greenhouses, or during harvesting, postharvest handling, processing, distribution, and food preparation (Beuchat, 2002). Freezing halts the activities of spoilage microorganisms in foods but can also preserve some microorganisms for long periods of time. During the freezing process, microbial growth can occur when freezing does not take place rapidly due to increasing temperature or fluctuations during frozen storage, transport or retail display (greater than −18°C), and during slow thawing. Frozen foods have an excellent overall safety record. However, the few outbreaks of food-borne illness associated with frozen foods indicate that some, but not all, human pathogenic microorganisms are killed by freezing processes. Outbreaks associated with frozen foods have been reviewed by Lund (2000). Freezing does not destroy *Clostridium botulinum*, the spoilage organism that causes the greatest problems in plant food processing. However, *C. botulinum* will not grow and produce botulin toxin (a poison) at frozen storage temperature below −18°C or low pH of fruit. Acid media of fruit is a protective factor against microorganism growth. The effect of freezing and frozen storage on the microbiology of some frozen fruit products has been reviewed (Skrede, 1996). Although spoilage microorganisms are not a great problem in frozen fruit and fruit juices, some outbreaks and illnesses associated with frozen food consumption have been due to fruit products. Cases of hepatitis A that were produced by frozen raspberries in the United Kingdom (Reid and Robinson, 1987) and frozen strawberries in the United States (DHHS, 1997) have been referenced. When thawing frozen food, it is important to remember that if the raw product is contaminated and freezing does not totally destroy spoilage and pathogenic microorganisms, as the temperature of food rises, there may be microorganism growths, mainly on the surface of the product. To preserve safety in frozen fruits, recommended temperature requirements exist for each stage of the cold chain. It is recommended that frozen fruits be maintained at −18°C or colder, although exceptions are allowed during brief periods as during transportation or local distribution (−15°C). Retail display cabinets should be at −18°C and never warmer than −12°C (IIR, 1986).

Freezing effects on different types of microorganisms have been recently studied (Archer, 2004). Yeast, moulds, viruses, bacteria, and protozoa are affected in different ways by freezing, frozen storage, and thawing cycles. Although Gram-negative bacteria (*Salmonella* spp. *Escherichia coli*, etc.) are more susceptible to freezing than Gram-positive ones (*Listeria monocytogenes, Staphylococcus aureus*, etc.), the nature of the food can change the survival of some former organisms. Freezing kills microorganisms by physical and chemical mechanisms, and factors related to freezing parameters (ice formation, rate of cooling, temperature/time of storage, etc.), or food matrix composition and nutritional status, or phase of growth determine the survival of the microorganism (Lund, 2000). Several mechanisms have been proposed to explain the damage caused to microorganisms by freezing. Cellular damage caused by internal or external large ice crystals and increase of external or internal solute concentration are some of the mechanisms proposed. Better understanding of the interactions between physical and chemical changes in the microorganism cell and food matrix during freezing, frozen storage, and thawing processes could lead to the designing of safe freezing processes where the microorganisms, if they are present, would not survive. For spoilage and pathogenic microorganisms, the freezing process becomes an important hurdle to overcome (Archer, 2004).

Legislation

Special rules for frozen food safety regulations have not been adopted by either the United States or the "European Communities (EC)" authorities. Frozen food is regulated by the general rules for food processing safety. Codex Alimentarius Commission adopted special rules for frozen foods–Recommended International Code of Practice for Processing and Handling of Quick Frozen Food. The Commission recognized not only temperature as the main consideration to maintain frozen food quality (Codex Alimentarius, 1976) but also other factors. The production of safe frozen food requires maximum attention to GMP and HACCP principles in all the production chain, from raw material (farm) to consumer freezer (table), and to all the steps in between. In the United States, the minimum sanitary and processing requirements for producing safe and wholesome food are an important part of regulatory control over the safety of the nation's food supply and are ruled by GMP (FDA, 2004). HACCP and Application Guidelines were first adopted by The National Advisory Committee on Microbiological Criteria For Foods (NACMCF) for astronauts (1970), seafood (1995), low-acid canned food and juice industry (2002–2004). Other food companies, including frozen foods, already use the HACCP system in their manufacturing processes (NACMCF, 1997). The Codex Alimentarius also recommended a HACCP-based approach to enhance food safety (Codex Alimentarius, 1999).

The central goal of The European Commission on food safety policy is to ensure a high level of protection for human health and consumer interests in relation to food. The Commission's guiding principle, primarily set out in its White Paper on Food Safety, is to apply an integrated approach from farm to table covering all sectors of the food chain, feed production, primary production, food processing, storage, transport, and retail sale. The establishment of the European Food Safety Authority (EFSA) was one of the key measures contained in the Commission's White Paper on Food Safety. EFSA is the keystone of European Communities (EC, 2002) risk assessment regarding food.

FREEZING METHODS

The rate of freezing and the formation of small ice crystals in freezing are critical to reduce tissue damage and drip loss in fruit thawing. Different types of freezing systems are designed for foods. The selection of suitable freezing systems is dependent on the type of product, the quality of frozen product, desire, and economical reasons. Freezing systems are divided according to the material of the heat-transmission medium (Rahman, 1999):

1. Freezing by contact with cooled solid or plate freezing: The product is placed between metal plates and then adjusted by pressure. This method is used for block or regular form products.
2. Freezing by contact with cooled liquid or immersion freezing: The fluids usually used are sodium chloride solutions, glycol and glycerol solutions, and alcohol solutions.
3. Freezing with a cooled gas in cabinet or air-blast freezing: Air-blast freezing allows quick freezing by flowing cold air (−40°C) at relatively high speed between 2.5 and 5 m/s.

4. Cryogenic freezing: Food is frozen by direct contact with liquefied gases, nitrogen and carbon dioxide. Nitrogen boils at −195.8°C and the surrounding food temperature reaches temperatures below −60°C. This is a very fast method of freezing and the rapid formation of ice crystals reduces the damage caused by cell rupture, preserving sensorial and nutritional characteristics. Cryogenic freezing is recommended for cubes, slices, medium or small whole fruits but is not appropriate for whole medium and large fruits such as prunes, peaches, etc., due to the risk of crushing.

FUTURE PERSPECTIVES

Irradiation. Ionizing radiation has been used as a safe and effective method for eliminating bacterial pathogens from different foods and disinfecting fruit, vegetables, and juices. The application of low-dose (<3 kGy) irradiation to a variety of frozen plant foods to eliminate human pathogens has been studied. The amount of ionizing radiation necessary to reduce the bacterial population increases with decreasing temperature. Significant softening was achieved at −20°C, but textural changes were not shown when lower ionization doses were employed at higher temperatures (−5°C) (Sommers et al., 2004).

High Pressure. The quality of frozen/thawed product is closely related to freezing and thawing processes (Cano, 1996). The rate of freezing and the formation of small ice crystals in freezing are critical to minimize tissue damage and drip loss in thawing. Several reports have studied the use of high pressure at subzero temperature (Bing and Da-Wen, 2002; LeBail et al., 2002). The physical state of food can be changed by the external manipulation of pressure and temperature according to the water phase diagram. The main advantage of high-pressure freezing is that when pressure is released, a high supercooling can be obtained, and as a result the ice-nucleation rate is greatly increased and the initial formation of ice is instantaneous and homogeneous throughout the whole volume. The use of high pressure facilitates supercooling, promotes uniform and rapid ice nucleation and growth, and produces small size crystals, resulting in a significant improvement of product quality (LeBail et al., 2002; Bing and Da-Wen, 2002).

From a structural point of view, damage to cells during processing is diminished due to the small size of ice crystals, resulting in a significant improvement of product quality. These advantages have been tested with different fruit tissues. Fruit tissues were frozen under pressure. Peach and mango were also cooled under pressure (200 MPa) to −20°C without ice formation, and then the pressure was released to 0.1 MPa. By scanning electron microscope , it was observed that the cells of fruits frozen under pressure were less damaged compared to those frozen using traditional freezing process, including cryogenic freezing (Otero et al., 2000).

High-Pressure Thawing. Thawing occurs more slowly than freezing. During thawing, chemical and physical damage can occur, as well as microorganism contamination that can reduce the quality of the frozen/thawed product. From a textural point of view, an incorrect thawing can produce an excessive softening of the plant tissue. A quick thawing at low temperature to avoid rising temperature could help in assuring the food quality. High-pressure thawing would be a new application of high-pressure freezing. Recent studies showed that high-pressure thawing can preserve food quality and reduce the necessary thawing time. High-pressure thawing was more effective in texture improvement than was atmospheric pressure thawing (Bing and Da-Wen, 2002).

ACKNOWLEDGMENTS

This work was financed by the Spanish national research projects, AGL-2002-04059-C02-02 and AGL-2003-09138-C04-01, from Ministry of Science and Technology and the Spanish project 07G/0053/2003 from Consejería de Educación, Comunidad Autónoma de Madrid.

REFERENCES

Archer, L.D. 2004. Freezing: an underutilized food safety technology? International Journal of Food Microbiology 90:127–138.

Ashie, I.N.A., Simpson, B.K., Smith, J.P. 1996. Mechanisms for controlling enzymatic reactions in foods. Critical Reviews in Food Science and Nutrition 36(1):1–30.

Bartolomé, A.P., Ruperez, P., Fúster, C. 1996a. Freezing rate and frozen storage effects on color and sensory characteristics of pineapple fruit slices. Journal of Food Science 61:154.

Bartolomé, A.P., Ruperez, P., Fúster, C. 1996b. Changes in soluble sugars of two pineapple fruit cultivars during frozen storage. Food Chemistry 56:163.

Bartolomé, A.P., Ruperez, P., Fúster, C. 1996c. Non-volatile organic acids, pH and tritable acidity changes in pineapple fruit slices during frozen storage. Journal of Science of Food and Agriculture 70:475

Beuchat, R.L. 2002. Ecological factors influencing survival and growth of human pathogens on raw fruits and vegetables. Microbes and Infection 4:413–423.

Billaud, C., Brum-Mérimée, S., Louarme, L. 2004. Effect of glutathione and maillard reaction products prepared from glucose or fructose with gluthathione on polyphenoloxidase from apple-II. Kinetic study and mechanism of inhibition. Food Chemistry 84:223–233.

Bing, L., Da-Wen, S. 2002. Novel methods for rapid freezing and thawing of foods: a review. Journal of Food Engineering 54(3):175–182.

Boileau, T.W.M., Moore, A.C., Erdman, J.W. 1999. "Carotenoids and vitamin A." In Antioxidant Status, Diet, Nutrition and Health, edited by Papas, M. pp. 133–158. New York: CRC Press.

Browleader, M.D., Jackson, P.D., Mobasheri, A.T., Pantelides, S.T., Sumar, S.T., Trevan, M.T., Dey, P.M. 1999. Molecular aspects of cell wall modifications during fruit ripening. Critical Reviews in Food Science and Nutrition 39(2):149–164.

Cano, M.P. 1996. "Vegetables". In Freezing Effects on Food Quality, edited by Lester E. Jeremiah, pp. 247–297. New York: Marcel Dekker Inc.

Cano, M.P., De Ancos, B. 1994. Carotenoids and carotenoid esters composition in mango fruit as influenced by processing method. Journal of Agricultural and Food Chemistry 42:2737–2742.

Cano, M.P., De Ancos, B, Lobo, M.G. 1996a. Effects of freezing and canning of papaya slices on their carotenoid composition. Zeistchirft fuer Lebensmittel-Untersuchung und Forchung 202:270–284.

Cano, M.P., Lobo, M.G., De Ancos, B., Galeazzi, M.A. 1996b. Polyphenol oxidase from Spanish hermaphrodite and female papaya fruits (*Carica papaya* cv Sunrise, Solo group). Journal of Agricultural Food and Chemistry 44:3075–3079.

Cano, M.P., De Ancos, B., Lobo, M.G., Santos, M. 1997. Improvement of frozen banana (*Musa cavendishii*, cv Enana) color by blanching: relationship among browning, phenols and polyphenol oxidase and peroxidase activities. Zeistchirft fuer Lebensmittel Untersuchung und Forschung. 204:60–65.

Cano, M.P., Fúster, C., Marín, M.A. 1993a. Freezing preservation of four Spanish kiwi fruit cultivars (*Actinidia chinensis*, Planch): chemical aspects. Zeitschirift fuer Lebensmittel-Untersuchung und Forschung 195:142–146.

Cano, M.P., Marín, M.A., De Ancos, B. 1993b. Pigment and color stability of frozen kiwi-fruit slices during prolonged storage. Zeistchirft fuer Lebensmittel Untersuchung und Forschung 197:346–352.

Cano, M.P., Lobo, M.G., De Ancos, B. 1998. Peroxidase and polyphenol oxidase in long-term frozen stored papaya slices. Differences among hermaphrodite and female fruits. Journal of Food Science and Agriculture 76:135–141.

Cano, M.P., Marín, M.A. 1992. Pigment composition and color of frozen and canned kiwi fruit slices. Journal of the Agricultural and Food Chemistry 40:2121–2146.

Cano, M.P., Marín, M.A. 1995. Effects of freezing preservation on dietary fibre content of mango (*Mangifera indica* L.) fruit. European Journal of Clinical Nutrition 49(suppl 3):S257-S260.

Cano, M.P., Marín, M.A., Fúster, C. 1990a. Freezing of banana slices. Influence of maturity level and thermal treatment prior to freezing. Journal of Food Science 55(4):1070–1072.

Cano, M.P., Marín, M.A., Fúster, C. 1990b. Effects of some thermal treatments on polyphenoloxidase and peroxidase activities of banana (*Musa cavendishii*, var enana). Journal of Science of Food and Agriculture 51:223–231.

Castro, I., Goncalves, O., Teixera, J.A., Vicente, A.A. 2002. Comparative study of Selva and Camarosa strawberries for the commerical market. Journal of Food Science 67(6):2132–2137.

Chaovanalikt A., Wrolstad, R.E. 2004. Anthocyanins and polyphenolic composition of fresh and processed cherries. Journal of Food Science 69:FCT73-FCT83.

Chen, C.S. 1993. "Physicochemical principles for the concentration and freezing of fruit juices." In Fruit Juice Processing Technology, edited by Steven Nagy, Chin Shu Chen, Philip E. Shaw. Auburndale, FL: Agscience, Inc.

Code Federal Regulation (CFR). 2004. Threshold of Regulation for substance Used in Food contact Articles. Code Federal Regulation, 21 CFR 170.39 (http://gpoaccess.gov/cfr.index.html).

Codex Alimentarius. 1976. Recommended International Code of Practice For The Processing and Handling of Quick Frozen Foods (CAC/RCP 8–1976) (http://www.codexalimentarius.net/search/searchindex.do).

Codex Alimentarius. 1999. Recommended Internatio-nal Code of Practice General Principles of Food Hygiene (CAC/RCP 1–1969, Rev.–1997.

Amd–1999) (http://www.codexalimentarius.net/search/searchindex.do).

De Ancos, B., González, E.M., Cano, M.P. 1999. Differentiation of raspberry varieties according to anthocyanin composition. Zeistchirft fuer Lebensmittel Untersuchung und Forschung 208:33–38.

De Ancos, B., González, E.M., Cano, M.P. 2000a. Ellagic acid, vitamin C and total phenolic contents and radical scavenging capacity affected by freezing and frozen storage in raspberry fruit. Journal of Agricultural Food and Chemistry 48:4565–4570.

De Ancos, B., Ibañez, E., Reglero, G., Cano, M.P. 2000b. Frozen storage effects on anthocyanins and volatile compounds of raspberry fruit. Journal of Agricultural Food and Chemistry 48:873–879.

Department of Health and Human Services (DHHS), Center for Disease Control and Prevention. 1997. Hepatitis A associated with consumption of frozen strawberries. Michigan, March 1997. Morbidity Mortality Weekly Report 46, 288–289.

Escribano-Bailón, T., Dangles, O., Brouillard, R. 1996. Coupling reactions between flavylium ions and catechin. Phytochemistry 41:1583–1592.

European Commission (EC) Directive 90/128/EEC, Plastic Materials and Articles Intended to Come into Contact With Foodstuffs. Official Journal of The European Communities L75, 21.03.1990.

European Commission (EC) Directive 2002/72/EC, Amending for the Directive 82/711/EEC Laying Down the Basic Rules Necessary for Testing Migration of the Constituents of Plastic Materials and Articles Intended to Come With Foodstuffs. Official Journal of The European Communities L220, 15.08.2002.

European Parliament and Council (EC) Regulation 178/2002. 2002. General principles and requirements of food law, establishing the European Food Safety Authority and laying down procedures in matter of food. Official Journal of The European Communities L31, 01.02.2002. (http://europa.eu.int/comm/food/index and (http://www.efsa.eu.int).

Food and Drug Administration (FDA). 2004. Good manufacturing practices (GMPs) for the 21st Century Food are Published in Title 21 of the Code of Federal Regulations, Part 110 (21 CFR 110 Processing). http://www.cfsan.fda.gov/~dms/gmp.toc.html

Fellows, P.J. 2000. Food Processing Technology. Boston: CRC Press.

Fennema, O.R. 1976. The U.S. frozen food industry 1776–1976. Journal of Food Technology 30(6):56–61, 68.

Fito, P., Chiralt, A. 1995. "An update on vacuum osmotic dehydration" In Food Preservation by Moisture Control: Fundamentals and Applications, edited by Barbosa-Cánova, G., Welti-Chanes, J., pp. 351–372. Lancaster, PA: Technomic Publishing Co.

Friedman, M. 1996. Food browning and its prevention: an overview. Journal of Agricultural and Food Chemistry 44:631–653.

Gardner, P.T., White, T.A.C., McPhail, D.B., Duthie, G.G. 2000. The relative contributions of vitamin C, carotenoids and phenolics to the antioxidant potential of fruit juices. Food Chemistry 68:471–474.

Garrote, R.L., Bertone, R.A. 1989. Osmotic concentration at low temperature of frozen strawberry halves. Effect of glycerol, glucose and sucrose solution on exudate loss during thawing. Food Science and Technology 22:264–267.

González, E.M., De Ancos, B., Cano, M.P. 2000. Partial characterization of peroxidase and polyphenoloxidase activities in blackberry fruits. Journal of Agricultural and Food Chemistry 48(11):5459–5464.

González, E.M., De Ancos, B., Cano, M.P. 2002. Preservation of raspberry fruits by freezing: physical, physico-chemical and sensory aspects. European Food Research and Technology 215:497–503.

Heaton, J.W., Lencki, R.W., Marangoni, A.G. 1996. Kinetic model for chlorophyll degradation in green tissue. Journal of Agricultural and Food Chemistry 44:399–402.

Hof, V.H.K.H, Boer, C.J., Tijburg, L.V.M., Lucius, B.R.H.M., Zijp, I., West, C.E., Hautvast, J.G.A.J., Westrate, J.A. 2000. Carotenoid bioavailability in humans from tomatoes processed in different ways determinated from the carotenoid response in the triglyceride-rich lipoprotein fraction of plasma after a single consumption and in plasma after four days of consumption. Journal of Nutrition 130:1189–1196.

Hui Y.H., Cornillon P., Guerrero I., Lim M., Murrel K.D., Nip Wai-Kit. 2004. Hand Book of Frozen Foods. New York: Marcel Dekker Inc.

Institute International of Refrigeration (IIR). 1986. Recommendations for the Processing and Handling of Frozen Foods. 3rd edition. Ed. Institute International of Refrigeration. París.

Kmiecik, W., Jaworska, G., Budnik, A. 1995. Effect of thawing on the quality of small fruit frozen products. Roczniki Panstwowego Zakladu Higieny 46 (2): 135–143.

Larsen, M., Poll, L. 1995. Changes in the composition of aromatic compounds and other quality parameters of strawberries during freezing and thawing. Zeistchirft fuer Lebensmittel Untersuchung und Forschung 201:275–277.

Lavelli, V., Peri, C., Rizzolo, A. 2000. Antioxidant activity of tomato products as studied by model reactions using xanthine oxidase, myeloperosidase, and copper-induced lipid peroxidation. Journal of Agricultural and Food Chemistry 48:1442–1448.

LeBail, A., Chevalier, D., Mussa, D.M., Ghoul, M. 2002. High pressure freezing and thawing of foods: a review. International Journal of Refrigeration 25:504–513.

Lisiewska, Z., Kmiecik, W. 2000. Effect of storage period and temperature on the chemical composition and organolectic quality of frozen tomato cubes. Food Chemistry 70:167–173.

Lobo, G., Cano M. Pilar. 1998. Preservation of hermaphrodite and female papaya fruits (*Carica papaya* L. cv sunrise, solo group) by freezing: physical, physico-chemical and sensorial aspects. Zeistchirft fuer Lebensmittel Untersuchung und-Forschung A 206:343–349.

Lund, B.M. 2000. "Freezing." In The Microbiological Safety and Quality of Food, edited by Lund, B.M., Baird-Parker, T.C., Goudl, G.W., vol I, pp. 122–145. Gaithersburg, MD: Aspen Publishers.

Marín, M.A., Cano, M.P., Fúster, C. 1992. Freezing preservation of four Spanish mango cultivars (*Mangifera indica*, L.): chemical and biochemical aspects. Zeistchirft fuer Lebensmittel Untersuchung und Forschung 194:566–569.

Martiner, M.V., Whitaker, J.R. 1995. The biochemistry and control of enzymatic browning. Trends in Food Science and Technology 6:195–200.

Mastrocola, D., Manzocco, L., Poiana, M. 1998. Prevention of enzymatic browning during freezing, storage and thawing of cherimoya (*Annona Cherimola*, Mill) derivatives. Italian Journal of Food Science 10(3):207–215.

Miller, N.J., Rice-Evans, C.A. 1997. The relative contributions of ascorbic acid and phenolic antioxidants to the total antioxidant activity of orange and apple fruit juices and blackcurrant drink. Food Chemistry 60:331–337.

Mullen, W., Stewart, A.J., Lean, M.E.J., Gardner, P., Duthie, G.G., Grozier, A. 2002. Effect of freezing and storage on the phenolics, ellagitannins, flavonoids, and antioxidant capacity of red raspberry. Journal of Agricultural and Food Chemistry 50(8):5197–5201.

National Advisory Committee on Microbiological Criteria For Foods (NACMCF). 1997. Hazard Analysis and Critical Control Point Principles and Application Guidelines http://vm.cfsan.fda.gov/~comm/nacmcfp.html).

Otero, L., Martino, M., Zaritzky, N., Solas, M., Sanz, P.D. 2000. Preservation of microstructure in peach and mango during high-pressure-shift freezing. Journal of Food Science 65(3):466–470.

Philippon, J., Rouet-Mayer, M.A. 1984. Blanching and quality of frozen vegetables and fruit. Review I. Introduction and enzymatic aspects. International Journal of Refrigeration 7(6):384–388.

Plocharski, W. 1989. Strawberries quality of fruits, their storage life and suitability for processing. Part VI. Quality of fruit frozen immediately after picking or frozen after cold storage under controlled atmosphere conditions. Fruit Science and Rep.(Skierniewice) 16(3):127. The reference journal is:

Rahman M. Sahafiur. 1999. "Food preservation by freezing." In Handbook of Food Preservation, edited by Rahman M. Sahafiur, p. 259. New York: Marcel Dekker Inc.

Rao, A.V., Agarwal, S. 1999. Role of lycopene as antioxidant carotenoid in the prevention of chronic disease: a review. Nutrition Reviews 19:305–323.

Reid, D. 1996. "Fruit freezing." In Processing Fruits: Science and Technology. Vol.1. Biology, Principles and Application, edited by Somogoyi, L.P., Ramaswamy, H.S., Hui, Y.H., p. 169. Lancaster, PA: Technomic Publishing.

Reid, T.M.S., Robinson, H.G. 1987. Frozen raspberries and hepatitits A. Epidemiology and Infection 98:109–112.

Rice-Evans, C., Miller, N.J., Paganga, G. 1996. Structure-antioxidant activity relationships of flavonoids and phenolic acids. Free Radical Biology Medical 20:933–956.

Robbers, M., Sing, R.P., Cunha, L.M. 1997. Osmotic-convective dehydrofreezing process for drying kiwifruit. Journal of Food Science 62(5):1039–1042.

Robinson, D.S., Eskin, N.A.M. 1991. Oxidative Enzymes in Foods. London: Elsevier.

Rodriguez-Amaya, D.B. 1997. Carotenoids and Food Preparation: The Retention of Provitamin A Carotenoids in Prepared, Processed and Stored Foods. Washington, DC: Agency for International Development, OMNI/USAID.

Sahari, M.A., Boostani, M., Hamidi, E.Z. 2004. Effect of low temperature on the ascorbic acid content and quality characteristics of frozen strawberry. Food Chemistry 86:357–363.

Salgado, S.M., Guerra, N.B., Melo-Filho, A.B. 1999. Frozen fruit pulps: effects of the processing on dietary fiber contents. Revista de Nitricao 12(3):303–308.

Simpson, K.L. 1986. "Chemical changes in natural food pigments." In Chemical Changes in Food

During Processing, edited by Finley, J.W., pp. 409–441. Westport, CT: AVI.
Skrede, G. 1996. "Fruits". In Freezing Effects on Food Quality, edited by Lester, E.J., pp. 183–245. New York: Marcel Dekker Inc.
Sommers, C., Fan, X., Niemira, B., Rajkowski, K. 2004. Irradiation of ready-to-eat foods at USDA'S Easter regional research center-2003 update. Radiation Physics and Chemistry 71:509–512.
Spiazzi, E.A., Raggio, Z.I., Bignono, K.A., Mascheroni, R.H. 1998. Experiments on dehydrofreezing of fruits and vegetables mass transfer and quality factors. Advances in the Refrigeration Systems, Food Technologies and Cold Chain, IIF/IIR 6:401–408.
Steinmetz, K.A., Potter, J.D. 1996. Vegetables, fruit and cancer prevention: a review. Journal American Diet Association 53:536–543.
Suutarinen, J., Heiska, K., Moss. P., Autio, K. 2000. The effects of calcium chloride and sucrose pre-freezing treatments on the structure of strawberry tissues. Lensmittel Wissensachaft und Technologie 33:89–102.
Talens, P., Escriche, I., Martínez-Navarret, N., Chiralt, A. 2002. Study of the influence of osmotic dehydration and freezing on the volatile profile of strawberries. Journal of Food Science 67(5):1648–1653.
Talens, P., Escriche, I., Martínez-Navarret, N., Chiralt, A. 2003. Influence of osmotic dehydration and freezing on the volatile profile of kiwi fruit. Food Research International 36:635–642.
Tibble, D.L., Benson, J., Curtin, K., Ma, K.-N., Schaeffer, D., Potter, J.D. 1998. Further evidence of the cardiovascular benefits of diets enriched in carotenoids. American Journal of Clinical Nutrition 68:521–522.
Tregunno, N.B., Goff, H.D. 1996. Osmodehydrofreezing of apples: structural and textural effects. Food Research International. 29:471–479.
Urbanyi, G., Horti, K. 1989. Color and carotenoid content of quick-frozen tomato cubes during frozen storage. Acta Alimentaria 18:247–267.
Willcox, J.K., Catignani, G.L., Lazarus, S. 2003. Tomatoes and caridovascular health. Critical Reviews in Food Science and Technology 43: 1–18.
Yueming-Jiang, Yuebiao-Li, Jianrong-Li. 2004. Browning control, shelf live extension and quality maintenance of frozen litchi fruit by hydrochloric acid. Journal of Food Engineering 63(2):147–151.
Zhao, Y., Xie, J. 2004. Practical applications of vacuum impregnation in fruit and vegetable processing. Food Science and Technology 15:434–451.

5
Fruit Drying Principles

József Barta

FRUIT DRYING

Fruit drying has a long tradition. Inhabitants living close to the Mediterranean Sea and in the Near East traded fruits that had been dried in the open sun. Dried fruit is a delicacy, because of the nutritive value (66–90% carbohydrate) and shelf life. For example, inhabitants of hillside villages isolated from the outside world by snow ate diets consisting primarily of seeds and dried fruits.

Today, the production of dried fruits is widespread. Nearly half of the dried fruits in the international market are raisins, followed by dates, prunes, figs, apricots, peaches, apples, pears, and other fruits. Significant amounts of sour cherries, cherries, pineapples, and bananas are also dried. The selection of fruit for drying depends on local circumstances and customs. For example, in the Middle East, lemons with thin peel are dried whole. The taste and aroma are preserved in the brownish inner fleshy part, which remains soft. In the United States, blackberries, cowberries (ligonberries), and grapes are dried, while in Spain, red grapes are dried.

Apricots, dates, plums, and tropical fruits are dried in the sun in several countries, while apples, pears, prunes, and peaches are dried by artificial means. Dryers with natural air ventilation were used in the 19th century in California for apple drying or to finish products that have previously been dried by the sun. Fruits dried in the sun or in dryers with natural air ventilation are referred to as "evaporated fruit," while fruits dried in dryers with artificial ventilation are described as "dehydrated fruits."

Fruits can be dried whole, in halves, or as slices, or alternatively can be chopped after drying. The residual moisture content varies from small (3–8%) to large (16–18%) amounts, according to the type of fruit. Significant amounts are packaged in small portions (200–1000 g) in manufacturing plants. Often, countries importing dried fruit repackage it to meet the needs of consumers and large kitchens. Fruit mixtures are widely consumed both in the United States and Europe. Fruit is packed as a mixture or each component is packed separately in a transparent, appealing packaging. Well-known components are round slices of apples, apricots and peaches, pear halves, prunes, sour cherries, and dates. Often, walnuts and almonds are also added to the mixture.

Dried fruit is widely used by the confectionery, baking, and sweets industries. Soup manufacturing plants use dried fruits in the various sauces, garnishments, puddings, and ice powders, and food for infants and children. Dried fruits are used in various teas, e.g., rose hips, and by the distilling industry (dried prunes, apricots). Applications include fruit powders processed from juices or pulps that dissolve quickly. The development of the fruit powders was possible through processing, which preserves color and flavor (vacuum drying, lyophilization, and swelling). Artificial drying made it economically possible to use raw materials at competitive prices and of high quality; examples are apples, prunes, and rose hips. Various milling procedures make it possible to dry highly valuable berries with soft flesh (strawberries, raspberries) and mature stone-fruits (apricots, peaches) (Burits and Berki, 1974).

STATE OF WATER IN FRUITS

Fruit drying involves removing water in different forms (both free and bound) and different amounts. The amount and manner of water removal change the structure of fruit depending on the type of bonding, and also determine the character of the reconstituted dried material. Among the various bonding forms of water, the strongest is the chemical, physicochemical bonding, followed by adsorption, osmotic, micro- and macro-capillary, and, finally, rehydration (Imre, 1974). During drying, the weakest bound water is removed first; removing moisture by breaking stronger bonds requires energy. Removal of free water does not change the character of the material in either the dried or rehydrated states. Significantly higher energy and special procedures are required to remove bound water, i.e., to decompose the higher bonding energies (Ginzburg, 1968, 1976).

EQUILIBRIUM STATES

By putting wet material into a closed space, water molecules change to the gaseous state forming a mixture of air and water vapor. At the same time, the molecules of the water vapor adsorb on the surface of the material by moistening it. After a given time, the number of molecules adsorbed on the surface of the material and the number of molecules that change to the gaseous state becomes equal. At this time, there is a state of equilibrium between the gaseous atmosphere in the space and the solid material. The state of the gaseous atmosphere can be characterized by its water activity, which is the ratio of the partial pressure of water vapor to the saturated partial pressure. The equilibrium relative humidity of the material can be determined from the water activity:

$$a_W = \frac{p_1}{p_2}, \quad (5.1)$$

where a_w is the water activity, p_1 is the partial pressure of water in the food, and p_2 is the saturated vapor pressure of water at the same temperature.

The moisture content characterizes the state of the material, i.e., its water content expressed in kg related to 1 kg of dried material. The equilibrium between the atmosphere and the wet material is highly affected by temperature. Knowledge of the correlation between the various factors influencing equilibrium has primary importance for drying technology, since the air moistening state determines the final moisture content being reached at the drying temperature. The relationship among the three features of the state makes it possible to have three types of planar representations.

- The sorption isotherm represents the function of the moisture content of the material with the water activity at constant temperature.
- The sorption isobar is the function of the temperature and the moisture content of the material at the equilibrium relative humidity (ERH).
- The sorption isostheta represents the water activity as a function of the temperature at constant moisture content of the material.

Drying technology uses the sorption isotherms most frequently. Determination of the sorption isotherms is done by actual or theoretical measurements. For representation of isotherms, several empirical correlations were proposed (e.g., Halsey, 1948; Henderson, 1952; Chung and Pfost, 1967). Several authors dealt with correlations of sorption data for fruits. Requirements necessary for a good correlation can be found in numerous reports (Ratti et al., 1989; Crapiste and Rotstein, 1986; Guggenheim, 1966; Iglesias and Chirife, 1976; Pfost et al., 1976; Thompson, 1972). Several methods exist for determination of the sorption isotherms of fruits by measurements. One method consists of the measurement of the relative moisture content and temperature at equilibrium by placing the material with a known moisture content into a closed air space. The measured relative moisture content is equal to the equilibrium relative vapor content. Air moisture values measured at the same temperature but different levels of moisture gives sorption isotherms for a specific fruit. Other methods of measuring sorption isotherms make use of air space with a constant relative vapor content established either by cooling and heating of the saturated air or by salt crystal solutions in a desiccator. The moisture content of the material put into the air space becomes constant, reaching a state of equilibrium. The relative vapor content and the moisture content of the material measured at the

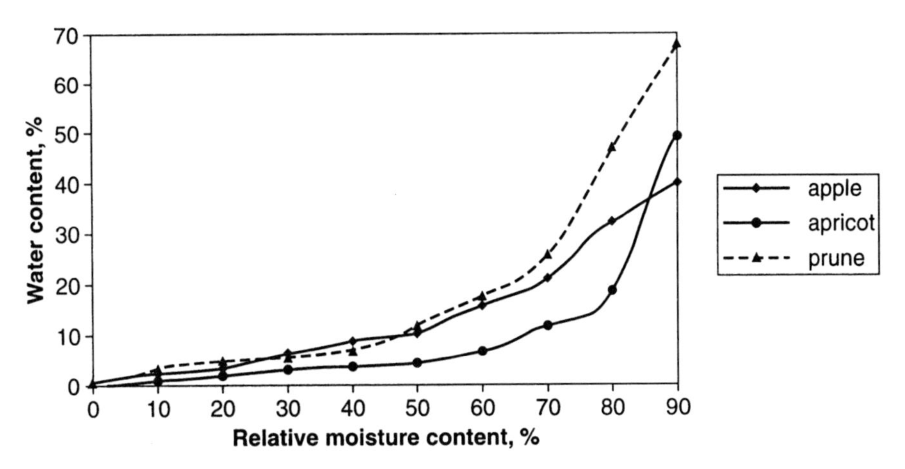

Figure 5.1. Sorption isotherms of some fruits.

temperature of the air space will be one point on the sorption isotherm (Jowitt et al., 1983; Mazza, 1984; Wolf and Jung, 1985). Figure 5.1 shows sorption isotherms of some fruit.

Knowledge of the sorption isotherms of a material is of primary importance from a practical point of view (Wolf et al., 1985). To ensure a product with the required moisture content, sorption isotherms are used to determine the state of the air (temperature, relative vapor content) (Shatadal and Jayas, 1992). The temperature and relative vapor content predict the remoistening and deterioration of a dried product with a given moisture content during storage. Further, it has great importance in selecting the drying procedure and predicting dryability, the binding strength of the moisture, and the shelf life of the fruit (Maroulis et al., 1988). If on the sorption isotherm, low moisture content relates to high a_w value, the material is highly hygroscopic, and drying can be done only with care in a climate-controlled space or in vacuum. Drying procedures with dry air can also be used for fruit having higher moisture content, e.g., fruit at a low a_w value. The design of any process in which the transfer of heat is involved requires knowledge of density as well as thermal properties of fruits being processed. Properties of fruits are discussed by Lewis (1987), Lozano et al. (1979), and Constenla et al. (1989). Empirical equations are proposed for modeling using density and thermal properties of the fruits being processed (Heldman, 1975; Choi and Okos, 1986; Singh and Mannapperuma, 1990; Singh, 1992).

PRINCIPLES OF WATER REMOVAL

Drying a moist material and decreasing the water activity mean evaporation of bound water from inside the solid material into the atmosphere. Breaking water bonds, releasing, and transferring heat connected to phase change require energy. Drying can be done with different types of drying energy: convective (warm air), contact (cooled surface), radiative (infrared rays), and excitation (microwave) energies. With convective drying, the heated air low in moisture content meets the wet material and as a result, the moisture moves onto the surface of the material and then into the drying air. Tasks of the warm air are to transfer heat to the material being dried to establish the drying potential and to transfer moisture into the air. For contact drying, the heat expanded by conduction from the cooled surface of the material evaporates the moisture. With infrared drying, the heat spreads from a radiating body—which can be a spot lamp, a piece of heated metal, or ceramic—directly to the material being dried. This method can be well-applied using vacuum drying for very small or chopped material (Szabó, 1987). For heat exchange by excitation, materials consisting of highly polarized molecules absorb the energy of excitation, resulting in heat necessary for drying the material. Using this method, liquids, pastes, and highly milled materials can be handled quickly and without a deterioration of the product. Vacuum drying can be used for heat-sensitive materials with low moisture content. In a vacuum with no transferring medium, convective heat exchange cannot be applied.

Moisture Transport in Solid Material

The phenomenon of drying is similar regardless of the drying method. This section deals with convective drying, the most widely used method in the fruit processing industry. The wet material (fruit) is placed in an air space with relative moisture content lower than the ERH of the material; moisture is transferred from the solid material (fruit) into the drying medium (air space).

Mass flow of the moisture (q_m in kg/s) is $q_m = \beta_y(Y_s - Y_g)\ A$, where β_y is the material exchange factor at the gaseous side (kg/m^{2}s), Y_s, Y_g are the absolute vapor content of the air at the surface of the material and in the air, respectively, (kg/kg), A is the surface area (m^2).

Simultaneously, the moisture content of the material is decreased. The water moves from the solid (fruit) and changes to vapor either inside or on the surface of the solid material. This vapor moves to the surface and goes into the air. In certain materials, such as gels, moisture transport is caused by diffusion flow of the water in the given material. This diffusion flow is initiated by the moisture difference of the material (Barta et al., 1990). Most foods are capillary-colloidal porous materials in which simultaneous liquid–vapor transport can occur. The character and direction of this transport depend on the texture, shape, and relationship of capillaries and pores. The vapor produced by water evaporation in the capillary-porous structure flows by diffusion to the surface. The so-called Knudsen flow in the micro-capillaries can be several orders of magnitude larger than Poiseuille flow in macro-capillaries. In foods, the conduction form of energy mentioned above occurs together with diffusion and moisture transport that is a function of the type of material and circumstances. For industrial calculations, the various forms of water transport can be handled together by means of an effective apparent diffusion parameter. Using the average apparent diffusion parameter (D_e, m^2/s) the mass flow (q_m, kg/s) of the moisture is in a stationary state:

$$q_m = c_s D_e \frac{dX}{dz} A, \tag{5.2}$$

where A is the surface area perpendicular to the direction of the moisture transport (m^2), c_s is the concentration of the solid material (kg/m^3), z is the length in direction of the moisture transport (m), and X is the moisture content of the material, i.e., the amount of water related to 1 kg dry material.

The relationship above can be derived from Fick's law. In a non-stationary state, the material equation written for water results in a second order, non-linear, parabolic differential equation, which can be given together with the initial and boundary conditions (e.g., material exchange on the surface). The moisture distribution along the length can be determined at an arbitrary drying period (Gion, 1986, 1988; Körmendy, 1985; Mohr, 1984).

Drying Procedure

At steady-state conditions (constant temperature, air flow rate, and air moisture content), the experimental results of drying are plotted by time. Generally, the moisture content (X) related to dry material is shown as a function of time (t). This is presented in Figure 5.2.

This plot shows a typical case where the moisture from the solid material evaporates first from the moisture layer on the surface and decreases

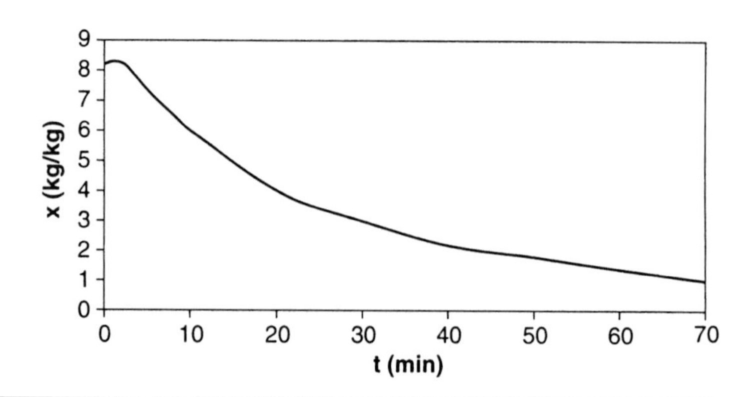

Figure 5.2. Drying curve of a wet material.

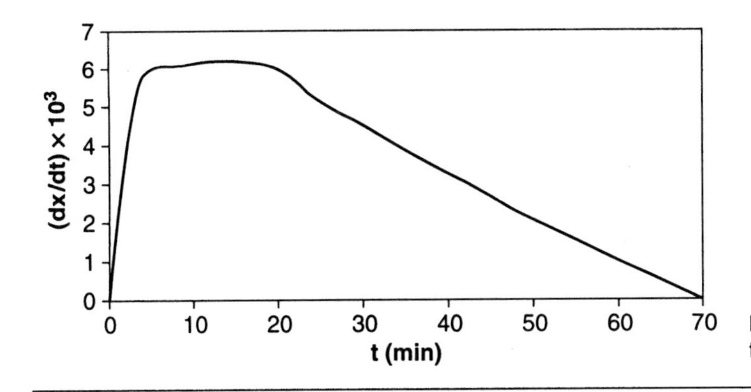

Figure 5.3. Drying rate curve as a function of time.

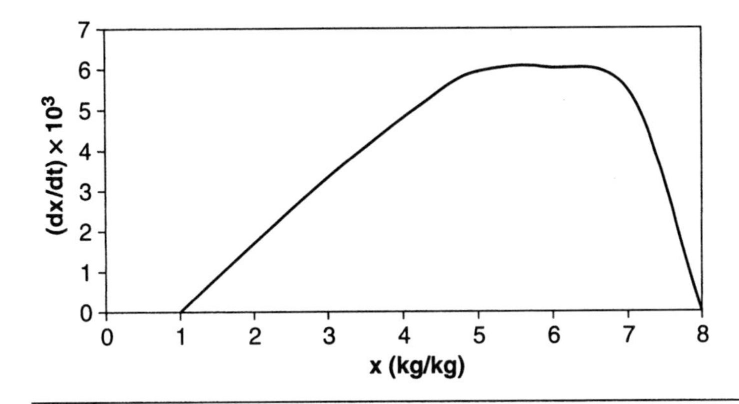

Figure 5.4. Drying rate curve as a function of the moisture content.

continuously until water evaporates from the inside of the solid material. It can be seen in the figure that variations in the drying rate depend on time and moisture content of the fruit product. This change can be seen better if the drying curve is differentiated and a drying flow rate curve is derived. Drying rate can be presented as a function either of the drying period (Fig. 5.3), or the moisture content of the material (Fig. 5.4).

Curves for the drying rate and drying flow rate can be divided into several parts. These parts are the result of the inner mechanism of drying and of changes occurring during drying. In the first step of drying, temperature equalization and moisture transport occur. In the next step, which is the constant rate period, there is a constant moisture flow to the surface, therefore, the surface is always wet. The average moisture measured at drying of the surface is the so-called critical moisture content. The drying rate decreases after reaching the critical moisture content. Drying stops and the drying rate becomes equal to zero when the average moisture content reaches the equilibrium moisture content related to the relative vapor content of the air. Figure 5.5 shows a curve for the average temperature of the material.

At the initial period of drying, the temperature of the material reaches the temperature of a wet thermometer. The temperature does not change in the constant rate period until reaching the critical moisture content. The temperature of the material increases in the falling rate period and becomes equal to the temperature of the drying air when drying stops. Dimensions of the material being dried are of primary importance in drying technology (Figure 5.6).

Linear variation of the size of the material changes the drying period to the second power. Increasing the drying temperature, and therefore the drying rate, the drying period is shortened and the capacity of the equipment is raised. This method is useful only in the constant rate period because the higher temperature of drying air does not result in significant increases in the temperature of the material. The increase in the drying rate is hindered by some stresses

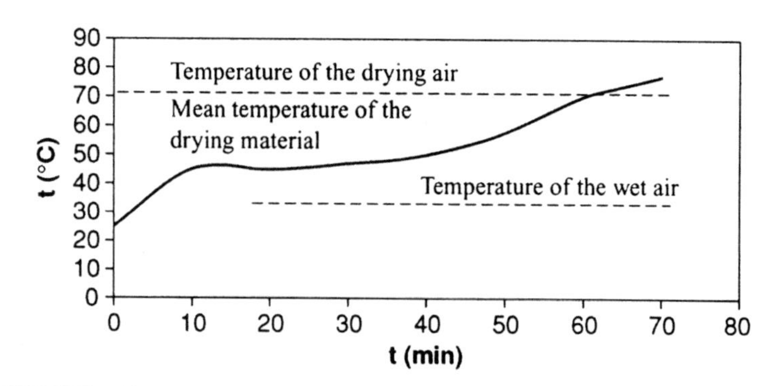

Figure 5.5. Average temperature of drying material as a function of time.

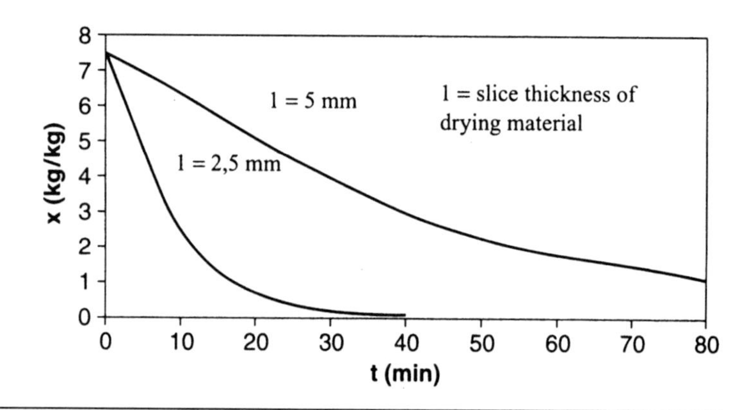

Figure 5.6. Effect of the size of material on the drying process.

in the material, by precipitation of the solute salts on the surface and crust formation. The intensive air ventilation enhances the moisture transport to the surface; however, it can result in crust formation (Barta et al., 1990). Managing the drying process takes into account the following aspects:

- high temperature and intensive air ventilation at the initial period
- mechanical removal of the surface moisture layer
- temperatures ensuring a low drying rate for a long period at the end of drying, "quiet" ventilation.

Effect of the Drying Air Characteristics on the Drying Process

It is necessary to know the factors determining the quality of the finished product, which can help to establish the parameters of the drying procedure. The concept of optimum drying must include the concept of economy, or the optimal application of heat used for drying. The external factors influencing drying are the following: temperature, moisture content, flow rate, direction of the drying air, and drying period. These factors must fit the properties of the material being dried (variety, water content, dimensions) and the methods of preparation (Kilpatrick et al., 1955; Lazar and Farkas, 1971; Lozano et al., 1983; Van Arsdel, 1973; Ratti, 1991). The most important factor is the *temperature of the applied air*. Under the same conditions, the lower the temperature used for drying, the better the quality of the product. Since an increase in temperature increases the rate of some chemical reactions and changes the original composition and properties, it is advisable to limit fruit temperatures to 60°C during the period of falling rate. However, in the constant rate period, where the mass transfer rate is in equilibrium with the heat transfer, 80–85°C can be achieved. The temperature of the drying air is generally 20–25°C higher than that of the material. In the case of drying at excessively high temperatures, the food temperature increases due to the warm air. Then the evaporation of the surface moisture occurs so quickly that there is not enough time to remove moisture from the inside. In this case, undesirable changes occur on the surface of the material. Internal water evaporation becomes more and more difficult except at higher temperatures. Because of the very high temperature, the material becomes too dry and fragile. Thicker plant sections dried in this way develop a hard and strong crust while the inside remains wet, contributing to a product that deteriorates easily. The crust formation also causes shrinkage inappropriate to the amount of water removed, and a resistance to the forces directed inward. The relatively soft inner parts move to the outer crust and the product becomes hollow (Lozano et al., 1983; Barta et al., 1990). The moisture gradient between the inner and outer parts can be described as a counter flowing concentration gradient. The originally uniform distribution of solutes is more diluted inside than on the surface layer, which has less moisture. This may cause an inward diffusion of solutes and undesirable reactions. During drying, the temperature used must match the physical and chemical properties of the material being dried. Generally, the starting temperature is high, gradually decreased, and then held constant for a longer period. The *moisture content of the air* used for drying is also an important factor in the drying process. Vapor absorptivity of the air is a function of the temperature (e.g., it is doubled by increasing the temperature by 15°C). Under the same conditions (temperature, flowing rate, etc.), the more moisture that is absorbed by the drying air, the less the moisture content at the inlet of the dryer. Therefore, drying becomes more economical if the outgoing air removes as much moisture as possible. The drying rate—as was mentioned earlier—can be controlled by the appropriate temperature setting. In order to prevent deterioration, the air moving out of the dryer is mixed with fresh air at a ratio suitable for re-application. Therefore, the same, climate-independent air is used (dryers do not work generally in climatized rooms) with appropriately chosen relative moisture content. In this way, at least partly, the heat of the used air is also utilized. The *flowing rate of the air* used for drying is important both for its appropriate usage and for the quality of the finished product. The higher the air temperature and its decrease, the faster the drying if there is no crust formation. The flow rate of air must be set, so both roles of air can be fulfilled (to evaporate the moisture on the surface and to remove the internal moisture). Both must occur at a given temperature interval. At 1 m/s of air flowing rate, the drying rate is double that of air at rest and at a 2 m/s flow rate, drying is about three times quicker. The flowing rate of the air is hindered by the capacity of the food for fluidization. At high air velocities, e.g., above 10 m/s, the drying air can take the material along. *The flowing direction of the air* can also vary in dryers. If the fresh air entering the drier meets the raw material first, then the airflow and the drying are called direct airflow and direct drying, respectively. This is less economical, but not harmful. If the fresh air comes in contact with the driest material and the used air with the newest material (or the wettest solid), the flowing of air and drying are called counter flowing air and drying, respectively. Direct flow is most commonly applied for drying products that are more sensitive. The finished product is the most sensitive to further drying. When the moisture content of the material being dried is lower, deterioration caused by heat occurs more easily. Drying air is less harmful if it has the highest possible vapor content and the lowest temperature. These properties are features of used air; therefore, it is desirable that the most used air is directed to the finished product. If the flowing direction of the air is perpendicular to the direction of the motion of the material, the flowing and drying are called cross-flowing and drying, respectively. The drying period is determined both by the raw material and the drying air according to the viewpoints discussed above. The upper limit of the drying period is a function of the economical factors (more efficient usage of the dryer) and the risk of deterioration of the material being dried. The drying period for vegetables and fruits cannot be longer than 6–7 h and moreover, with appropriate controls cannot exceed 3 h. Of course, this is highly dependent on the type of dryer. The lower limit of the drying period is influenced by the character, species, and quality of the product. Taking into account the risk of drying too fast, a period shorter than 2 h is not advisable for a standard dryer. Shape, surface area, and thickness have a significant effect on the drying rate. The thinner the material, the larger the surface area, and the faster the movement of moisture from the inside. This happens in such a way that the drying rate curve reaches the breaking point only at the end of drying. Therefore, sensitive materials can be dried at a higher temperature in thin-bedded layers, because the drying period is shorter. This can be done in a thin-layered columnar dryer. The very small sphere-like drops of spray drying are highly effective.

PRINCIPLES OF THE TECHNOLOGICAL CALCULATIONS

The basis of the technological calculations is the law of mass preservation—the sum of masses of materials coming into the dryer is equal to the masses leaving the dryer. For a continuous process, calculations must be done with mass flows of the materials (expressed in kg/s). Drying does not change the mass of the dry material. The incoming and outgoing masses or mass flows are equal. Therefore, the air leaving the dryer takes with it an amount of water equal to the water evaporated from the material being dried. If we study the mass flow and the moisture content of the drying air, we need to add the evaporated water mass flow to that of the air entering the dryer. The drier the air coming into the equipment, the more moisture can be absorbed by it, therefore lower specific air consumption is necessary (the amount of the air needed to evaporate 1 kg water). The mass flow of air is also influenced by the size of the conducting flue and the velocity of the air. As an average, the temperature of the used air decreases by about 20°C during drying and about six times more air is required for the evaporation of water than for its transfer. Drying temperature determines the amount of heat given to the material, therefore the degree of evaporation that affects the structure and quality of the material (e.g., deterioration caused by heat). Theoretically, the heat flow is enough to evaporate the water in food; however, this is not the case in practice. During drying, not only water, but also food and the component parts of the drying equipment must be heated to the drying temperature. Extra heat is necessary to compensate the heat losses, which are always present. Examples are the heat contributed by the equipment together with the warm air and food product. Heat losses can be decreased by insulating the equipment and by using the outgoing air. Drying efficiency is the ratio of the theoretical heat flow to the one used in practice and depends mainly on the type of dryer and quality of the material being dried. However, the temperature, moisture content, and velocity of the drying air also influence the drying period. Calculation of the efficiency is given by the following mathematical formulas. The heat flow needed to warm the material (Φ_m) is

$$\Phi_m = cq_w \Delta T (\text{kJ/s}),$$

where c is the average specific heat of the material (kJ/kg K), q_w is the average mass flow of the material (kg/s), and ΔT is the temperature difference in the material (K).

The heat flow taken for evaporation (Φ_e) is

$$\Phi_e = q_w \Delta H (\text{kJ/s}) \quad \Delta H > r$$
$$\Delta H = r + H'$$

where r is the heat needed to evaporate 1 kg water at the average temperature of the material (kJ/kg), ΔH is the specific enthalpy change (kJ/kg), H' is the specific binding energy of the water (kJ/kg), and q_w is the mass flow of the water being evaporated (kg/s).

The useful heat consumption (Φ_n) is given by

$$\Phi_u = \Phi_m + \Phi_e$$

and the total heat consumption (Φ_t) by

$$\Phi_t = \Phi_u + \Phi_l$$

where Φ_l is the heat loss parameter.

The total efficiency of drying (η) is

$$\eta = \eta_d \eta_{eq}$$

where η_d is the efficiency of drying

$$\eta_d = \frac{\Phi_u}{\Phi_t} = \frac{\Phi_m + \Phi_e}{\Phi_t} \tag{5.3}$$

and

η_{eq} is the efficiency of the drier

$$\eta_{eq} = \frac{\Phi_l}{\Phi_t}. \tag{5.4}$$

DRYERS

The following literary sources can be used for designing a dryer: Lazar and Farkas, 1971; Lozano et al., 1983; Van Arsdel et al., 1973; Ratti, 1991; Suzuki et al., 1976; Crank, 1967; Gion and Barta, 1998; Kudra, 2001; Mujumdar, 2000, 2004; Oetjen 2004. For dielectric drying, the following references are recommended: Schiffmann, 1987; Nelson, 1983; Mohsenin, 1984; Ratti and Mujumdar, 1995; Ginsburg, 1969; Sandu, 1986. To evaluate drying efficiency it is necessary to know the capacity of the dryer and the production ratio of the food being dried (the ratio of the dried finished product to the raw material). The aim of dehydration is the best usage of the properties of the raw material and capacity of the dryer to produce the specified quality dried product. The ideal is when the relationship among changes in heat, moisture content of the air, and volume are matched to changes in water leaving the material. These conditions can be accomplished (including the efficiency) by special equipment suitable for large-scale manufacturing. Assumptions are difficult to make due to variability of the equipment and materials and the milling sizes. Therefore, there is no universal equipment. Each type has an optimum working range determining the type of product that can be processed efficiently. Dehydration can be done under various conditions in different forms; therefore, dryers with various configurations are made (Kudra, 2001). Their classification can be done from several points of view.

According to the working mode of the equipment:

- batch
- continuous.

According to the pressure in the dryers:

- atmospheric dryers
- vacuum dryers (their spread is apparent now).

According to the heat exchange:

- convection heat
- radiation heat
- conduction (contact) heat
- combined
- dielectric.

According to the relationship between the motion of the drying medium and the material:

- direct flow
- counter flow
- cross-flow
- combined flow.

Menon and Mujumdar (1987) gave a detailed scheme for the classification from another point of view. For drying coarse products such as fruits, convection dryers are mainly used. The heat exchange medium is the warm air, and the material being dried is placed evenly on thin-layered perforated sheets, screens, or a grid. Pieces of the material change positioning during the movement of the trays. Due to the low air velocity, the material dries as a stationary layer. Generally, at an air velocity of 1–3 m/s, air does not carry particles. Production of coarse dried products is done in tray, tunnel, or bend dryers and often vacuum dryers are used. Drying for shorter periods and at lower temperature preserves the product. It is a reversible method. Spray drying technology is widely used for drying liquids and food pulp. Drying as a technological process can be controlled at several steps. The controlling possibilities of dryers are summarized in Table 5.1.

The main requirement for rehydrating a dried product is that, upon placing the dried product in water, the amount of water absorbed is the same as the water content in the original, non-dried food. The rehydrated product must also have, as much as possible, the shape, consistency, color, taste, and odor of the original food. Therefore, the dried product must be well dehydrated. A study of rehydration can be done by putting the dried fruit in distilled water for a given period (1–2 h) or by cooking in salty water for 10–20 min. The water absorptivity factor or re-hydration index (W_r) is the ratio of the mass of the dehydrated product (m_r) to the one of the dry product (m_o) (Barta et al., 1990).

$$W_r = \frac{m_r}{m_o} = \frac{U_r + 1}{U_o + 1}, \tag{5.5}$$

where U_r is the water content of the rehydrated product and U_o is the water content of the dried product being studied (both in mass ratio).

In addition to the remoistening index, the dehydrated product is evaluated by sensory and objective methods (its consistency, flavor, smell, and color). Important requirements are the storability, cleanliness—and if required—the size (uniformity).

DRYING TECHNOLOGY OF FRUITS

Drying technology consists of three types of processes:

- Preliminary, preparative procedures
- Drying procedure
- Processing after drying, secondary treatments.

These are closely related. Failure of a process or a type of it cannot be repaired or only in a way which is not economic.

PREPARATIVE PROCEDURES

The aims of the preparative procedures are to ensure the introduction of the product to the dryer free of foreign material and microbial impurities. Production costs of dried product are highly dependent on the preparative procedures, especially in processing of chopped dried products.

Cleaning

Raw materials arriving at a plant are accepted on the basis of a suitable standard by doing quantitative

Table 5.1. Controlling Possibilities of Dryers

Sensory Features	Controlled Features						
	Heat Introduction	Air Supply	Moving Rate of the Conveyor or the Tray	Air Recirculation	Solid Supply Rate	Mass of the Outgoing Air	Product Recirculation
Temperature of the incoming air	*	*	–	*	–	–	–
Temperature of the outgoing air	*	*	–	–	*	*	*
Mass flow rate of the air	*	*	–	*	*	*	–
Temperature of the product	*	*	*	*	*	–	*
Moisture content of the product	*	*	*	*	*	–	*
Heat decay of the product	*	*	*	*	*	–	*
Thickness of the solid medium	–	–	*	–	*	–	–
Holding period	–	–	*	–	*	–	*
Air velocity	–	*	–	*	–	*	–
Evaporation rate	*	*	–	–	*	–	–

Source: Barta et al. (1990).

Note: Mark * present in the meeting points of some columns and rows means that the measurement in the given row can be applied for characterizing of the feature in the column.

and qualitative studies. The incoming fruit must be cleaned to remove impurities since impurities damage the quality.

Washing

Washing can be done before cleaning, after cleaning, and after secondary cleaning. It is important that washing achieves these aims:

- Be effective, removing impurities adhered on the surface.
- Do not cause unnecessary losses in the dry solid and postinfection (frequent change of water is required).
- Do not break and crush the fruit.
- Have economic water usage.

Only water acceptable for drinking is allowed to be used for washing fruits.

Slicing

Slicing affects the shape of the finished product, its uniform residual moisture content, and apparent density. During slicing, the main requirement is the use of a non-destructive, smooth slicing surface and uniform thickness of the slices. The blades of the slicing machine must be protected from alien materials (stone, wood, metal) and set, sharpened, and changed frequently.

DRYING PROCEDURE

The appropriate spreading influences the uniform warming of the fruit, loading the dryer, and air distributions as well as residual moisture content. Thickness of the spreading is influenced by the manner and quality of chopping even for the same type of product.

Excess water must be removed from the fruit by a vibrating sieve before being loaded into the dryer. By removing excess water, the load on the dryer decreases significantly. The product must be dried to the required moisture level by taking into account parameters influencing the drying.

SECONDARY PROCEDURES

The material leaving the dryer is heterogeneous, differing in size, color, cleanliness, and moisture content (e.g., besides the appropriate cube sizes there are smaller pieces, too).

Finishing procedures can be placed into two groups:

- Procedures ensuring the main requirements:
  a. Classification by size
  b. Classification by color, cleanliness, moisture content
  c. Air separating/classifying
- Improving processes which increase the variety and value of the products:
  a. Mashing, cutting, granulation
  b. Fine classification into some given fractions
  c. Powder grinding, sieving
  d. Mixing, producing mixtures

Packing

Dried products cooled to room temperature are packed as required by consumers. Requirements for packing materials are the following: protect the product against moisture, oxygen, light, alien odors, bruising, impurities, and insects. The finer the size of the product and the higher the surface area, the more it is influenced by damaging effects. Packaging is suitable if the material, closure, dimensions, strength, labeling, and aesthetic appearance meet marketing specifications, if it is non-toxic and ensures the safety and quality of the product for a preset time.

Storage

At storage the following must be ensured:

- Safety of the wrapping
- Legibility of the label
- Appropriate storage temperature (conditioned, cooled air space)
- Appropriate turning at the store
- Exclusion of risk for contamination
- Eligibility of the specification of food law
- Appropriate clean environment
- Extermination of rodents and insects.

SPECIFIC REFERENCES

Barta, J., Vukov, K. and Gion, G. 1990. Dehydration by drying. In: Konzervtechnológia, I. (Canning Technology.) (in Hungarian), Eds. Körmendy, I. and Török, Sz., Budapest, Hungary, University of Horticulture and Food Industry, Faculty of Food Science, Press, 482–521.

Burits, O. and Berki, B. 1974. Zöldség- és gyümölcsszárítás. (Drying of Vegetables and Fruits.)

(in Hungarian) Budapest, Hungary, Mezőgazdasági Kiadó, 231–241.

Choi, Y. and Okos, M. R. 1986. Effects of temperature and composition on the thermal properties of foods. In: Food Engineering and Process Applications Vol. 1: Transport Phenomena, Eds. Le Maguer, M. and Jelen, P., England, Elsevier Applied Science Publishers Ltd., 93–101.

Chung, D. S. and Pfost, H. B. 1967. Adsorption and desorption of water vapor by cereal grains and their products. *Trans. A.S.A.E.* 10:549.

Constenla, D. T., Lozano, J. E. and Crapiste, G. H. 1989. Thermophysical properties of clarified apple juice as a function of concentration and temperature. *J. Food Sci.* 54(3):663–668.

Crank, J. 1967. The Mathematics of Diffusion. London, Oxford University Press, 55–123.

Crapiste, G. H. and Rotstein, E. 1986. Sorptional equilibrium at changing moisture contents. In: Drying of Solids, Ed. Mujumdar, A. S., Wiley Eastern Ltd., New Delhi, India, 41–47.

Ginzburg, A. Sz. 1968. Szárítás az élelmiszeriparban. (Drying in Food Industry) Budapest, Hungary, Műszaki Könyvkiadó, 39–47.

Ginzburg, A. Sz. 1969. Application of Infrared Radiation in Food Processing. Chemical and Process Engineering Series, London, Leonard Hill.

Ginzburg, A. Sz. 1976. Élelmiszerek szárításelméletének és technikájának alapjai. (Principles of Drying Theory and Techniques of Foods.) (in Hungarian) Budapest, Hungary, Mezőgazdasági Kiadó, 28–51.

Gion, B. 1986. Simulation of food drying. (in Hungarian) *Élelmezési Ipar.* 40(3):110–125.

Gion, B. 1988. Simulation of drying technological processes on IBM-AT computer for potato and Jerusalem artichoke roots. (in Hungarian) *Élelmezési Ipar.* 42(6):220–224.

Gion, B. and Barta, J. 1998. Processing of dried cubes and flour from Jerusalem Artichoke. *J. Food Phys.* 9:15–22.

Guggenheim, E. A. 1966. Applications of Statistical Mechanics. Oxford, Clarendon Press, 186–206.

Halsey, G. 1948. Physical adsorption on non-uniform surfaces. *J. Chem. Phys.* 16:931.

Heldman, D. R. 1975. Food Process Engineering. Westport, CT, The Avi Publishing Company Inc., 96–103.

Henderson, S. M. 1952. A basic concept of equilibrium moisture. *Agr. Eng.* 33:29.

Iglesias, H. A. and Chirife, J. 1976. Prediction of the effect of temperature on water sorption isotherms of food materials. *J. Food Technol.* 11:109.

Imre, L. 1974. Szárítási kézikönyv. (Handbook of Drying.) (in Hungarian) Budapest, Hungary, Műszaki Könyvkiadó, 39–59.

Jowitt, R., et al. 1983. Physical properties of foods. London, Applied Science Publishers, 50–127.

Kilpatrick, P. W., Lowe, E. and Van Arsdel, W. B. 1955. Tunnel dehydrators for fruit and vegetables. *Adv. Food Res.* 50:385.

Körmendy, I. 1985. Heat-and material transports in canning technological procedures II. *Élelmezési Ipar.* 39(12):463–470.

Lazar, M. E. and Farkas, D. F. 1971. The centrifugal fluidized bed. 2. Drying studies on pieces-form foods. *J. Food Sci.* 36:315.

Lewis, M. J. 1987. Physical Properties of Foods and Food Processing Systems. Chichester, England, Ellis Horwood, 210–295.

Lozano, J. E., Rotstein, E. and Urbicain, M. J. 1983. Shrinkage, porosity and bulk density of foodstuffs at changing moisture contents. *J. Food Sci.* 51:113.

Lozano, J. E., Urbicain, M. J. and Rotstein, E. 1979. Thermal conductivity of apples as a function of moisture content. *J. Food Sci.* 44(1):198–199.

Maroulis, Z. B., Tsami, E., Marinos-Kouris, D. and Saravacos, G. D. 1988. Application of the GAB model to the moisture sorption isotherms for dried fruits. *J. Food Eng.* 7:63–78.

Mazza, G. 1984. Sorption isotherms and drying rates of Jerusalem Artichoke. *J. Food Sci.* 49: 384.

Menon, A. S. and Mujumdar, A. S. 1987. Drying of solids: Principles, classification, and selection of dryers. In: Handbook of Industrial Drying, Ed. Mujumdar, A. S., First edition, New York, NY, Marcel Dekker Inc., 3–46.

Mohr, K. 1984. Oualitätserhalt der Frocknung durch Computersimulation. *Lebensmittelindustrie* 34(4):150–151.

Mohsenin, N. N. 1984. Electromagnetic Radiation Properties of Foods and Agricultural Products. New York, NY, Gordon and Breach Science Publishers, 25–30.

Nelson, S. O. 1983. Dielectric properties of some fresh fruits and vegetables at frequencies of 2.45 to 22 GHz. *Trans. A.S.A.E.* 26:613.

Pfost, H. B., Mauer, S. G., Chung, D. S. and Milliken, G. A. 1976. Summarizing and reporting equilibrium moisture data for grains. A.S.A.E. Paper 76–3520. 1976. A.S.A.E Meeting, St. Joseph, MI.

Ratti, C. 1991. Design of dryers for vegetable and fruit products. Ph.D. Thesis. Universidad Nacional del Sur, (in Spanish) Bahia Blanca, Argentina.

Ratti, C., Crapiste, G. H. and Rotstein, E. 1989. A new water sorption equilibrium expression for solids foods based on thermodynamics considerations. *J. Food Sci.* 54(3):738–742.

Ratti, C. and Mujumdar, A. S. 1995. Infrared drying. In: Handbook of Industrial Drying, Ed. Mujumdar, A. S., Second edition, New York, NY, Marcel Dekker Inc.

Sandu, C. 1986. Infrared radiative drying in food engineering: A process analysis. *Biotechnol. Prog.* 2(3):109–119.

Schiffman, R. F. 1987. Microwave and dielectric drying. In: Handbook of Industrial Drying, Ed. Mujumdar, A. S., First edition, New York NY, Marcel Dekker Inc., 327–356.

Shatadal, P. and Jayas, D. S. 1992. Sorption isotherms of foods, In: Drying of Solids, Ed. Mujumdar, A. S., New York, NY, International Science Publisher., 433–448.

Singh, R. P. 1992. Heating and cooling processes for foods. In: Handbook of Food Engineering, Eds. Heldman, D. R. and Lund, D. B., New York, NY, Marcel Dekker Inc., 247–255.

Singh, R. P. and Mannapperuma, J. D. 1990. Developments in food freezing. In: Biotechnology and Food Process Engineering, Eds. Schwartzberg, H. G. and Rao, M. A., New York, NY, Marcel Dekker Inc., 309–329.

Suzuki, K., Kubota, K., Hasegawa, T. and Hosaka, H. 1976. Shrinkage in dehydration of root vegetables. *J. Food Sci.* 41:1189.

Szabó, Z. 1987. Drying. In: Élelmiszeripari műveletek és gépek. (Procedures and Machines of Food industry.) (in Hungarian), Eds. Szabó, Z., Csury, I. and Hidegkuti, Gy. Budapest, Hungary, Mezőgazdasági Kiadó, 491–556.

Thompson, T. L. 1972. Temporary storage of high-moisture shelled corn using continuous aeration. *Trans. A.S.A.E.* 15:333.

Van Arsdel, W. B., Copley, M. J. and Morgan, A. I. 1973. *Food Dehydration*. Second edition, Westport, CT, Avi Publishing Co., 50–139.

Wolf, W. and Jung, G. 1985. Wasserdampfsorptionsdaten für die Lebensmitteltrocknung. *Zeitschrift für Lebensmitteltechnologie* 36(2):36–38.

Wolf, W., Spiess, W. E. L. and Jung, G. 1985. Sorption Isotherms and Water Activity of Food Materials. New York, NY, Elsevier, 1–5.

GENERAL REFERENCES

Kudra, T. 2001. Advanced Drying Technologies. Marcel Dekker, New York, 265–303.

Mujumdar, A. S. 2000. Drying Technology in Agriculture and Food Sciences. Science Pub. Inc., New Hampshire, USA, 313 pp.

Mujumdar, A. S. 2004 Dehydration of Products of Biological Origin. Science Publishers, Inc., New Hampshire, USA, 541 pp.

Oetjen, G. W. 2004. Freeze-Drying. John Wiley & Sons, Weinheim, Germany, 407 pp.

6
Non-Thermal Pasteurization of Fruit Juice Using High Voltage Pulsed Electric Fields

Zsuzsanna Cserhalmi

INTRODUCTION

The pulsed electric field (PEF) process is a new and innovative non-thermal minimal processing technology that is used as an alternative preservation process for fruit juices. The aim of this technology is to inactivate microorganisms and to decrease the activity of enzymes in order to increase the shelf life of food products without undesirable heat and chemical effects.

The theoretical basis of PEF technology is the use of an external electric field to destabilize cell membranes and form one or more pores in them. PEF technology applies high voltage pulses (generally 20–80 kV/cm) for very short time (μs to ms), producing PEFs between two electrodes. This technique is very similar to electroporation, used in cell biology and genetic manipulation of cells. But, in the case of foods, the applied pulses are shorter and much more intense. The aim of the application of high voltage pulses to foods is not only to disrupt temporarily the cell membranes of microorganisms. However, in this process, the microorganisms are also killed or their numbers are drastically decreased by irreversible disruption of cell membranes.

MECHANISMS OF INACTIVATION BY PEF

The mechanism of inactivation of microorganisms exposed to PEFs has not been fully clarified yet. The most commonly accepted theory is based on the dielectric breakdown of cell membranes resulting in changes in membrane structure and permeability which occur at a critical breakdown voltage (Kinosita and Tsong, 1977a, b; Zimmermann et al., 1974; Coster and Zimmermann, 1975; Castro et al., 1993). It is suggested that an external electric field induces a transmembrane potential over the cell membrane that is larger than the normal potential of the cell. When the overall membrane potential reaches a critical value, rupture takes place. According to Zimmermann (1986), this value is around 0.7–1.1 V for most cell membranes. As shown in Figure 6.1, the cell membrane can be regarded as a capacitor filled with dielectric materials with a low dielectric constant. Accordingly, free charges can be accumulated on both sides of membrane surfaces (Fig. 6.1a). These charges occur by the application of an external electric field and increase the potential difference across the membrane, known as the transmembrane potential. The accumulated charges on two sides of the membrane are opposite and attract each other resulting in membrane compression, which leads to reduction of membrane thickness (Fig. 6.1b). An elastic or viscoelastic restoring force stands opposite to the increased compression. Since electric

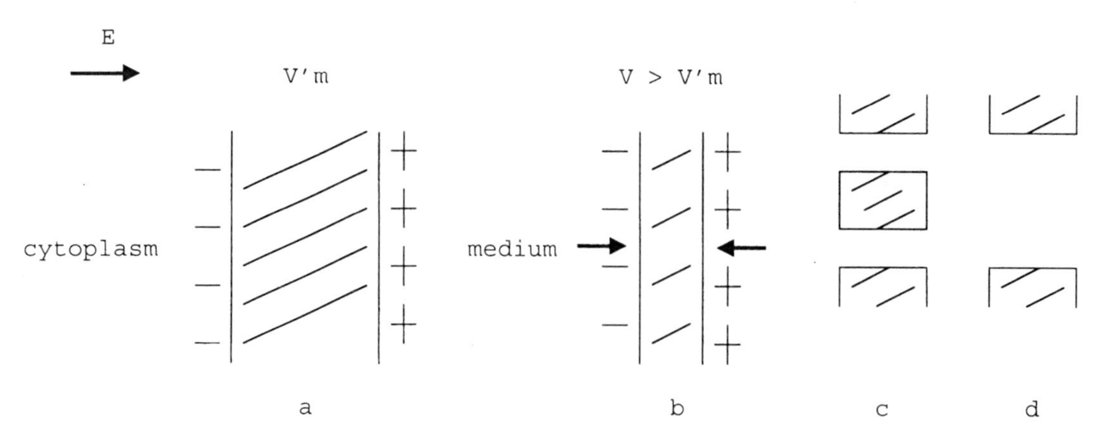

Figure 6.1. Schematic diagram of dielectric breakdown of cell membrane (according to Zimmermann, 1986). (a) cell membrane with potential V'_m, (b) membrane compression, (c) reversible pore formation, and (d) irreversible pore formation.

compression increases more rapidly with decreasing membrane thickness than the elastic restoring compression forces, the cell membrane ruptures (Fig. 6.1c). It occurs if the critical breakdown voltage is reached by a further increase in the external field strength (Zimmermann, 1986).

The change in membrane permeability can be reversible or irreversible, depending on the external electric field strength. When it is equal to or only slightly exceeds the critical value, the increase in membrane permeability is reversible and the change in cell membrane can be recovered within a few seconds after the treatment. In the food industry, when the aim of PEF application is to kill microorganisms in order to extend the shelf life of food products, the magnitude of the electric field strength has to be carefully taken into consideration.

The PEF-induced transmembrane potential (V_m) depends on the electric field strength and on the cell size (Hülsheger et al., 1983):

$$V_m = FrE,$$

where F is the shape factor, r is the cell radius (μm), and E is the electric field strength (V/μm). With the same magnitude of external electric field, the induced transmembrane potential is greater in a larger cell. By this means, the larger cells are more sensitive to PEF than smaller ones. The shape factor is 1.5 for spherical cells. For non-spherical cells, Zimmermann et al. (1974) derived a mathematical equation. The equation is based on the assumption that the cell shape consists of a cylinder with two hemispheres at each end. In this case, the shape factor is

$$F = L(1 - 0.33d),$$

where L is the length of cylinder and d is the diameter.

The shape factor (F) and the critical field strength for four microorganisms are presented in Table 6.1 (Aronsson, 2002).

The electrical breakdown of biological cell membranes can be explained by mechanisms other than dielectric breakdown. These were comprehensively summarized, e.g., by Barbosa-Cánovas et al. (1998a).

Table 6.1. Shape Factor and Critical Electric Field Strength of Microorganisms

	Shape Factor		Critical Electric Field Strength (kV/cm)	
Microorganism	Minimum	Maximum	Minimum	Maximum
Escherichia coli	1.09	1.23	12.2	14.8
Listeria innocua	1.09	1.37	36.6	36.7
Leuconostoc mesenteroides	1.24	1.31	23.0	30.4
Saccharomyces cerevisiae	1.07	1.31	2.2	18.7

According to these authors, the breakdown can occur for the following possible reasons: (1) threshold transmembrane potential and compression of cell membranes, (2) viscoelastic properties of cell membranes, (3) fluid mosaic arrangement of lipids and proteins in cell membranes, (4) structural defects in cell membranes, and (5) colloid osmotic swelling. The basis of these changes was explained by those who studied the electrical breakdown of biological membranes based on model systems such as liposomes, planar bilayers, and phospholipid vesicles.

PEF SYSTEM

A PEF processing system consists of five main components such as a power source, high voltage generator, capacitor bank, switch, and a treatment chamber to which the voltage, current, and temperature controlling, sample handling, and packaging systems can be connected (Fig. 6.2). The power source charges the capacitor bank and a switch is used to discharge the energy from the capacitor bank across the food that is held in a treatment chamber.

A laboratory scale continuous flow PEF system (OSU-4B) designed and constructed at the Ohio State University in the United States is presented in Figure 6.3. This equipment is able to generate both bipolar and unipolar square wave pulses. A syringe pump system consisting of four syringes (syringe A for starter, B for the sample, C for the treated sample, and D for waste) is integrated with the system and run, using a total of 60 ml of sample. The sample is pumped through six PEF treatment chambers connected in series containing stainless steel electrodes with a gap of 0.29 cm (Fig. 6.4). The flow rate of the system is up to 2 ml/s. After treated at each pair of chambers, the sample is cooled via attached coils that are submerged in a refrigerated water bath, set or controlled at the treatment temperature. The pre- and post-treatment temperatures in each pair of chambers are monitored by thermocouples attached to the exit of the chamber pair. The system is able to generate pulse durations up to 10 μs. However, a minimum pulse duration of 2 μs is recommended to obtain a good waveform. The electric field strength (E), one of the most important factors influencing microbial inactivation, is calculated by

$$E = U/d,$$

where U is the peak voltage of the applied voltage (V) and d is the distance between the electrodes (cm). The maximum output voltage and current are 10 kV and 50 A, respectively. The maximum pulse frequency of the system is 5000 Hz, but too high a frequency causes unnecessary heating of the sample. Therefore, its usage must be avoided. The total treatment time (t) can be calculated from

number of pulses received in each chamber (n_p),
number of chambers (n),
pulse duration (τ) [$t = n_p n \tau$].

One of the most important parts of a PEF system is the treatment chamber that consists of two carbon or metal electrodes. Stainless steel is generally applied but other metals may be preferable to reduce electrochemical attack (Barsotti et al., 1999). Parallel plates, parallel wires, concentric cylinders, and a rod-plate are the possible electrode configurations discussed by Hofmann (1989). Parallel plates are the simplest in design and produce the most uniform distribution of electric field (Jeyamkondan et al., 1999). Numerous types of static and continuous flow treatment chambers are known which are named after the designers. Examples include Sale and Hamilton (1967),

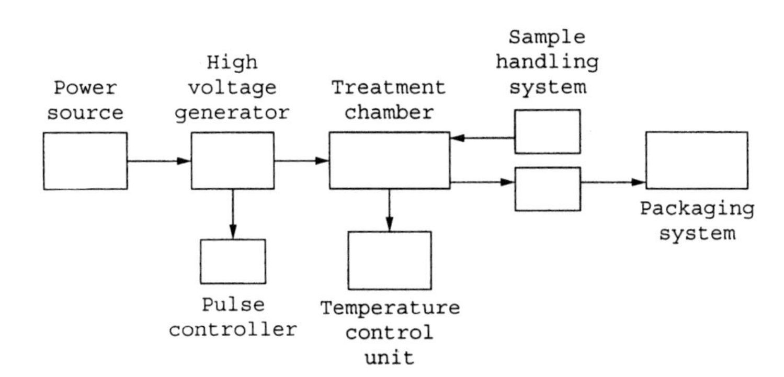

Figure 6.2. PEF system.

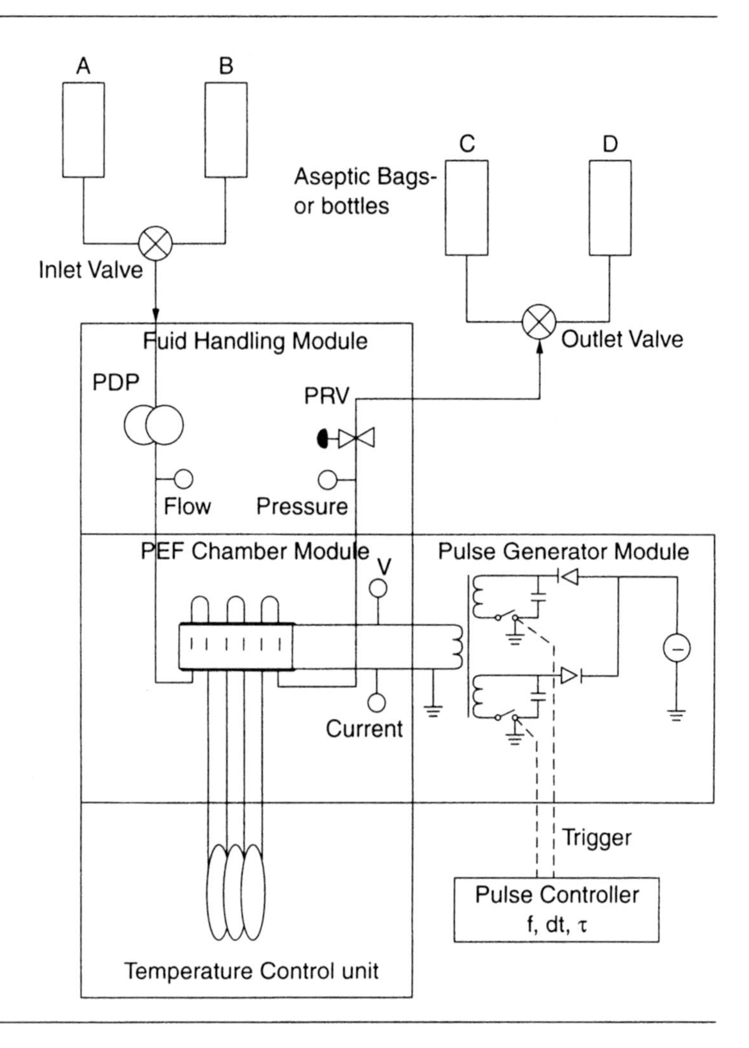

Figure 6.3. OSU-4B PEF system.

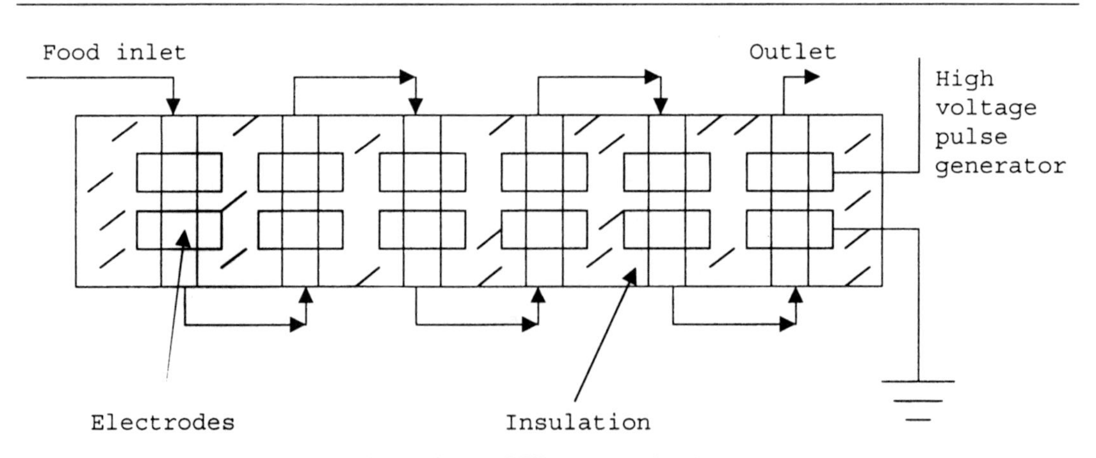

Figure 6.4. Schematic configuration of a continuous PEF treatment chamber.

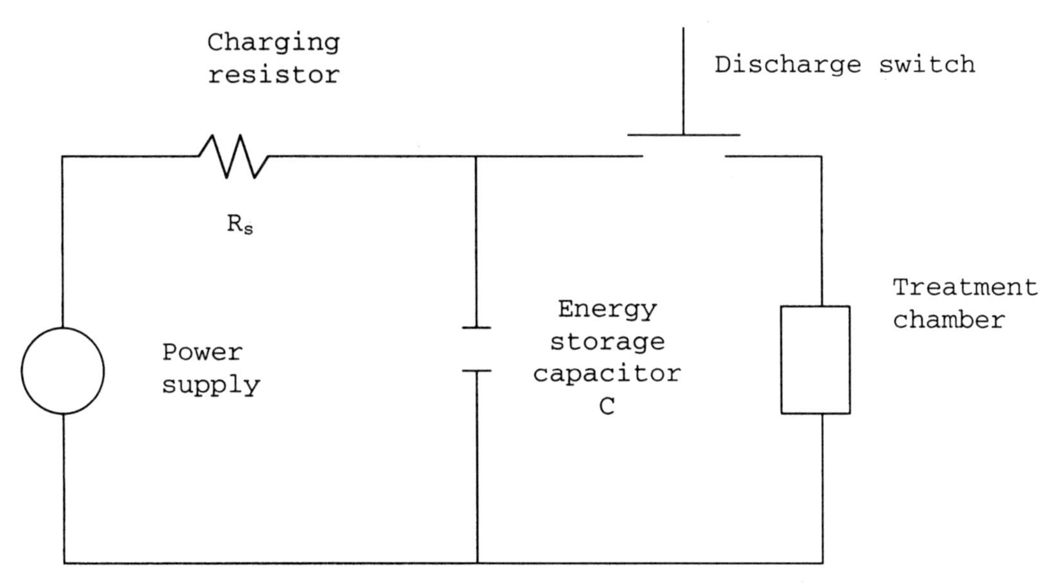

Figure 6.5. Simplified circuit for exponential decay pulse generation.

Dunn and Pearlman (1987), Grahl et al. (1992) static treatment chamber, and Dunn and Pearlman (1987) continuous flow treatment chamber. For a comprehensive summary of the different chambers, see Zhang et al. (1995) and Barbosa-Cánovas et al. (1998b). Also, commercially available PEF systems for food processing are reviewed by Barsotti et al. (1999).

The treatment chambers are designed to avoid dielectric breakdown of foods, which occurs when the applied electric field strength exceeds the dielectric strength of the food, as indicated by a spark.

In the treatment chamber, electric pulses go through the food samples. Depending on the specific electronic system, the pulses may be exponential decay, square wave, oscillatory, or bipolar. The exponential decay and square wave pulses are the most commonly applied waveforms in the PEF technology. Although, exponential decay pulses are easier to obtain, their efficiency has failed to produce the desired effect. In the electrical circuit, a power supply charges a capacitor bank connected in series with a charging resistor (Rs) (Fig. 6.5). When a trigger signal is applied, the charge stored in the capacitor flows through the food in the treatment chamber. Exponential decay pulse rises rapidly to a maximum value and decays slowly to zero (Fig. 6.6). Consequently, the pulses have a long tail with a low electric field, resulting in excess heat generated in the food without bactericidal effect. A typical exponential decay pulse can be characterized by a 1–3 μs pulse width and 10–100 kV/cm peak electric field strength. The square pulse waveform maintains a peak voltage for a longer time than the exponential decay pulses (Fig. 6.7). Consequently, it is more energy efficient and requires less cooling effort (Zhang et al., 1995). The generation of square pulses is complex and usually involves a pulse-forming network with an array of capacitors, inductors, and solid-state switching devices (Fig. 6.8). The high cost involved prohibits practical application of this system.

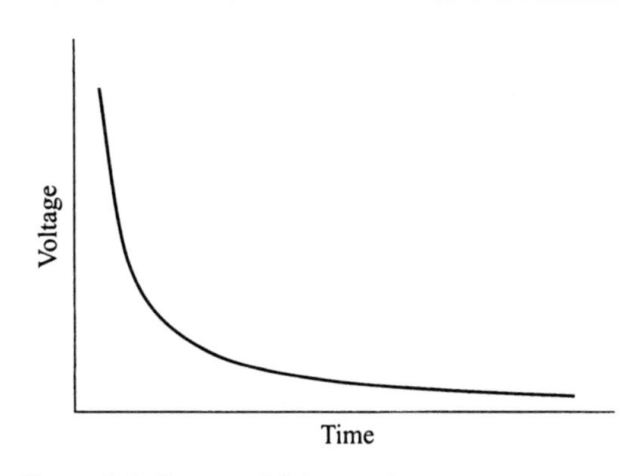

Figure 6.6. Exponential decay pulse.

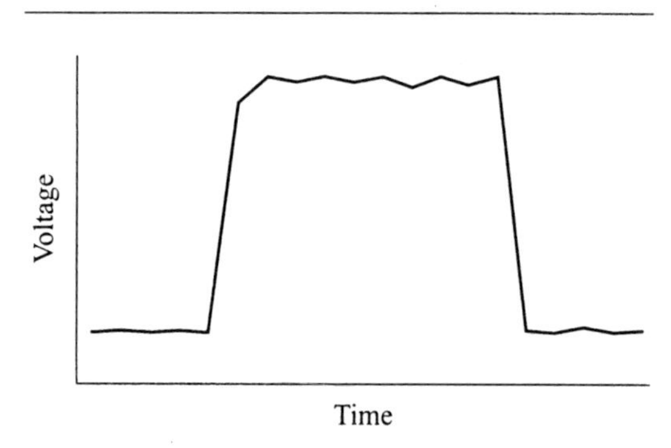

Figure 6.7. Square wave pulse.

During PEF processing for food pasteurization, the following main operating parameters should be considered: voltage (2–50 kV), electrode gap distance (0.2–7 cm), electric field strength (1–100 kV/cm), number of pulses (1–120), pulse width (1 μs–10 ms), pulse frequency (1–1000 Hz), shape of pulses, electrical characteristics, and in the case of a continuous treatment chamber, the flow rate of liquid food. A required residence time (T_r) of sample in a continuous treatment chamber is needed to choose the flow rate correctly (T_r = volume of one chamber/flow rate).

In order to understand the effect of electric fields on foods, it is very important to know the electrical properties of a food item, especially the resistivity. Barsotti et al. (1999) showed that food resistivity, the reciprocal of conductivity, ranges from 0.4 to 100 Ω·m. The electrical resistivity of a food sample has an effect on the whole circuit that is related to the efficiency of energy transfer and temperature. A change in conductivity affects the pulse energy [W(J/ml)] given by the following formula:

$$W = \sigma \tau E^2,$$

where σ is the conductivity (S/cm), τ is the pulse length (μs), and E is the electric field strength (V/μm).

When τ and E are constants, W depends only on σ and, as a result, the application of the same number of pulses generates higher energy input when σ is higher (Wouters et al., 2001a). The other problem for food systems with high conductivity in relation to PEF is the generation of excessive heat. A high conductivity of foods results in low field strengths and excessive production of heat through the generation of current. A low conductivity results in a more effective PEF treatment (Jayaram et al., 1993; Wouters et al., 1999). Lowering the conductivity of food increases the difference between the conductivity of the medium and the microbial cytoplasm and weakens the membrane structure due to an increased flow of ionic substances across the membrane (Jayaram et al., 1993).

EFFECT OF PEF ON FOOD PRESERVATION

EFFECT OF PEF ON MICROORGANISMS

Sale and Hamilton (1967) were the first to report the effect of PEF on microorganisms. *Escherichia coli, Staphylococcus aureus, Micrococcus lysodeikticus, Sarcina lutea, Bacillus subtilis, Bacillus cereus, Bacillus megatherium, Clostidium welchii, Saccharomyces cerevisiae*, and *Candida utilis* were suspended in neutral sodium chloride solution and exposed to a 5–25 kV/cm electric field strength that was applied as a series of direct current pulses from 2 to 20 μs. The authors demonstrated that microbial inactivation was non-thermal and it happened by irreversible loss of membrane function. The sensitivity of microorganisms to this treatment was different

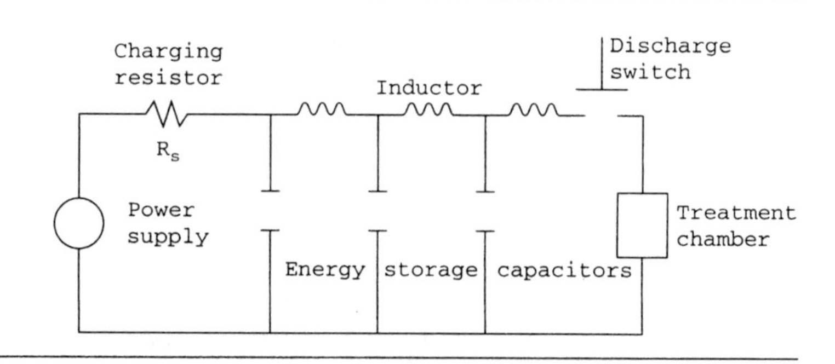

Figure 6.8. Simplified circuit for square wave pulse generation.

depending on the electric field strength and the total treatment time and was also dependent on the concentration and nature of the cells treated. Yeasts were more sensitive to treatment than vegetative bacteria.

Numerous scientific papers and reviews (Knorr et al., 1994; Pothakamury et al., 1996; Barbosa-Cánovas et al., 1996; Barsotti and Cheftel, 1999; Wouters et al., 1999; Raso et al., 2000) deal with the effect of PEF on microorganisms. The inactivation of *Saccharomyces cerevisiae*, *Escherichia coli*, *Listeria monocytogenes*, *Listeria innocua*, *Staphylococcus aureus*, *Salmonella* spp., *Bacillus subtilis*, and *Bacillus cereues* has been extensively studied. Some results are summarized in Table 6.2 for yeast and bacteria suspended in various media or liquid food.

It is generally accepted that yeasts are more sensitive to PEF treatment than vegetative bacteria and, within the bacteria, Gram-negative bacteria are more susceptible to the treatment. It has been proposed that cell morphology affects inactivation of bacteria. With thicker and more rigid cell wall, the microorganisms are more resistant to PEF. Moreover, the membrane potential theory states that induced transmembrane potential is correlated with cell size and shape (Hülsheger et al., 1983).

Considerable work has been reported on the inactivation of *Saccharomyces cerevisiae* in fruit juices, milk, and yoghurt (Matsumoto et al., 1991; Grahl et al., 1992; Zhang et al., 1994a, b; Harrison et al., 1997). Six log cycles reduction in microbial viability could be achieved with 16 pulses for *Saccharomyces cerevisiae* in potato dextrose broth at a treatment electric field of 40 kV/cm (Zhang et al., 1994c). Qin et al. (1995) reported a six decimal inactivation of inoculated *Saccharomyces cerevisiae* in apple juice applied ten times 2.5 μs pulses of 35 kV/cm, or two times 2.5 μs pulses of 50 kV/cm. Aronsson et al. (2001) obtained a 6 log cycles reduction of *Saccharomyces cerevisiae* when it was exposed to 30 kV/cm and 20 pulses with 4 μs pulse duration.

Zhang et al. (1994d) reported a three decimal reduction of *Escherichia coli* treated by 20 pulses at 25 kV/cm. A 4 log cycles reduction of *Escherichia coli* was achieved in model foods such as simulated milk ultrafiltrate, with a peak electric field strength of 16 kV/cm and 60 pulses and a pulse width ranging between 200 and 300 μs (Pothakamury et al., 1995a). The application of 10 pulses at an intensity of 41 kV/cm and at 37°C induced a reduction of 2.3 log cycles for *Escherichia coli* suspended in milk and 3.3 log cycles when suspended in phosphate buffer. After 63 pulses, the reduction reached approximately 4.5 log cycles (Dutreux et al., 2000). Aronsson (2002) reported a 5.4 log cycles reduction of *Escherichia coli* when the latter was exposed to 30 kV/cm and 20 pulses with 4 μs duration.

The Gram-positive bacteria *Listeria monocytogenes* and *Listeria innocua* are less sensitive to PEF treatment than Gram-negative bacteria. Possibly, the more rigid and thicker cell wall of Gram-positive bacteria constitutes a protection to PEF (Aronsson, 2002). Aronsson et al. (2001) studied the effect of PEF treatment on four organisms (*Escherichia coli, Listeria innocua, Leuconostoc mesenteroides, and Saccharomyces cerevisiae*) and found that the most resistant organisms were *Listeria innocua* and *Leuconostoc mesentroides* with only a 3 log cycles reduction when compared with a 6 and 5.4 log cycles reduction in the case of *Saccharomyces cerevisiae* and *Escherichia coli*, respectively. Also, Reina et al. (1998) reported a 3 log cycles reduction of *Listeria monocytogenes* by PEF suspended in milk.

Jeantet et al. (1999) could achieve a 3.5 log cycles reduction of *Salmonella enteritidis* by PEF treatment in diaultrafiltered egg white. The effect of electric field intensity (20–35 kV/cm), pulse frequency (100–900 Hz), pulse number (2–8), temperature (4–30°C), pH (7–9), and inoculum size (10^3–10^7cfu/ml) were also tested. A 4.5, 5 $\geq$, and 6 log cycles reductions were obtained in the case of *Staphylococcus aureus* exposed to 16, 60, and 40 kV/cm, respectively (Pothakamury et al., 1995a; Qin et al., 1998; Zhang et al., 1994c).

Pothakamury et al. (1995b) reported a 4 to 5 log cycles reduction of *Bacillus subtilis* with a peak electric field strength of 16 kV/cm and 50 pulses. The rate of inactivation of cells increased with an increase in electric field intensity. Heinz et al. (1999) found that inactivation of *Bacillus subtilis* was a linear function of field strength. The reduction in microbial count of *Bacillus cereus* cells was hardly more than 1 log cycle using 62 pulses at 20 kV/cm and 50 pulses at 16.7 kV/cm, respectively (Cserhalmi et al., 2002; Pol et al., 2000). The inactivation rate could be increased by the use of bacteriocins or irradiation (Pol et al., 2000; Beczner et al., 2001).

It is generally known that bacterial spores are resistant to different treatments such as an electric field. However, after germination, they become sensitive to PEF treatment (Barbosa-Cánovas et al., 1998b,

Table 6.2. Effect of PEF Processing on Microorganisms

Microorganism	Treatment Conditions	Media	Effects	Reference
Saccharomyces cerevisiae	25–38 kV/cm, 20 pulses, 100 μs pulse duration, $T < 63°C$	Yogurt	2 log reduction	Dunn and Pearlman (1987)
	6.7 kV/cm, 5 pulses, 20 μs pulse duration, $T < 50°C$	Orange juice	5 log reduction	Grahl et al. (1992)
	40 kV/cm, 16 exp. decay pulses, 3 μs pulse width, $T = 15°C$	Potato dextrose agar	6 log reduction	Zhang et al. (1994c)
	36 kV/cm, 10 pulses, 2.5 μs pulse duration, $T < 34°C$	Apple juice	6 log reduction	Qin et al. (1995)
	40 kV/cm, 64 exp. decay pulses, 4 μs pulse width, $T = 15°C$	Apple juice	3.3 log reduction	Harrison et al. (1997)
	25, 30, and 35 kV/cm, 20 and 40 square wave pulses, 2 and 4 μs pulse duration, $T \leq 61°C$, $f = 250$ and 500 Hz	NTM model medium	6.5 log reduction	Aronsson et al. (2001)
	10–28 kV/cm, 25–62 bipolar pulses, 2 μs pulse width, $T = 23–30°C$	Apple juice	3.3–3.9 log reduction	Cserhalmi et al. (2002)
Escherichia coli	18–30 kV/cm, bipolar pulses, 6, 5, 4, and 3 μs pulse duration, 86–172 μs treatment time, $T < 35°C$	Apple juice	5 log reduction	Evrendilek et al. (1999)
	25, 30, and 35 kV/cm, 20 and 40 square wave pulses, 2 and 4 μs pulse duration, $T \leq 61°C$, $f = 250$ and 500 Hz	NTM model medium	5.4 log reduction	Aronsson et al. (2001)
	60, 70, 80 kV/cm, 30 pulses, $T = 20°C$, 30°C, and 32°C	Apple cider	5.35 log reduction	Iu et al. (2001)
	10–30 kV/cm, 48 bipolar pulses, 3 μs pulse width, $T < 35°C$	0.1% NaCl (0.2 S/m) solution	4.5 log reduction	Unal et al. (2001)

Listeria monocytogenes	25 and 35 kV/cm, square wave pulses, 1.5 μs pulse duration, $T = 25°C$, $f = 1.7$ Hz	Milk	1–4 log reduction	Reina et al. (1998)
	10–30 kV/cm, 48 bipolar pulses, 3 μs pulse width, $T < 35°C$	0.1%NaCl (0.2 S/m) solution	1.1 log reduction	Unal et al. (2001)
Listeria innocua	30 kV/cm, 3–3.9 μs pulse width, $T = 49.5°C$	Phosphate buffer (pH 4.0, 0.5 S/m at 40°C	≥6.3 log reduction	Wouters et al. (1999)
	25, 30, and 35 kV/cm, 20 and 40 square wave pulses, 2 and 4 μs pulse duration, $T \leq 61°C$, $f = 250$ and 500 Hz	NTM model medium	7 log reduction	Aronsson et al. (2001)
Staphylococcus aureus	16 kV/cm, 60 exp. decay pulses, 200–300 μs pulse width, $T < T_{lethal}$	Simulated milk ultrafiltrate (SMUF)	4–5 log reduction	Pothakamury et al. (1995a)
Salmonella enteritidis	20–35 kV/cm, 2–8 square wave pulses, $T = 4$–30°C, $f = 100$–900 Hz	Diaultrafiltered egg white	3.5 log reduction	Jeantet et al. (1999)
Salmonella dublin	15–40 kV/cm, 12 μs treatment time, 1 μs pulse duration, $T = 10$–50°C	Skim milk	4 log reduction	Sensoy et al. (1997)
Salmonella typhimurium	10–20 kV/cm, 2000–10,000 pulses, $T < 40°C$	Distilled water, tris-maleate buffer (10 nM, pH 7.4), model beef broth	4 log reduction	Simpson et al. (1999)
Bacillus subtilis	20, 35, and 50 kV/cm, 15 and 30 exp. decay pulses, 2 μs pulse width, $T < 25°C$	0.10–0.15% NaCl solution	≥3.4 log reduction	Marquez et al. (1997)
Bacillus cereus	35 and 50 kV/cm, 30 and 50 exp. decay pulses, 2 μs pulse width, $T < 25°C$	0.10–0.15% NaCl solution	≥5 log reduction	
	10–28 kV/cm, 26–62 bipolar pulses, 2 μs pulse width, $T = 23$–30°C	0.15% NaCl solution	<0.5 log reduction	Cserhalmi et al. (2002)

Barsotti and Cheftel, 1999; Pol et al., 2001a, b). The electric fields do not induce the germination of bacterial spores. They can be reached by other physical or chemical treatments. Hamilton and Sale (1967) and later other authors (Grahl et al., 1992; Cserhalmi et al., 2002) found that the spores of *Bacillus cereus* were very resistant to PEF treatment. Marquez et al. (1997), however, were able to decrease the number of *Bacillus subtilis* and *Bacillus cereus* spores suspended in 0.15% sodium chloride by using PEF. The 30 pulses applied resulted in a 3.4 log cycles reduction of *Bacillus subtilis* spores, while 50 pulses gave a 5 log cycles reduction of *Bacillus cereues* spores. Jin et al. (2001) could also inactivate *Bacillus subtilis* spores by PEF treatment. They inactivated 98% of the spores with a maximum PEF treatment (40 kV/cm for 3500 μs). The scanning electron microscopy (SEM) micrographs showed that after the treatment, the spores shrank and many wrinkles formed on their surfaces. The shrinkage and wrinkles may indicate that the spores were inactivated.

Although these contradictory results can be explained because of different PEF conditions, more research is needed for clarification.

Microbial inactivation by PEF has been influenced by various factors such as process parameters (electric field strength, pulse number, pulse width, pulse shape, processing temperature), microbial characteristics (type, growth stage), and product properties (pH, conductivity). Numerous publications have demonstrated their effects on the rate of microbial inactivation (Vega-Mercado et al., 1997; Pothakamury et al., 1996; Alvarez et al., 2000, 2002; Aronsson and Ronner, 2001; Wouters et al., 2001b; Picart et al., 2002; Cserhalmi et al., 2002; Cserhalmi and Beczner, 2003). In general, increasing the intensity of process parameters increases microbial inactivation. Since Gram-positive and Gram-negative bacteria are more resistant to PEF treatment than yeasts, it has been proposed that cell size and shape may influence the inactivation kinetics. It also has been demonstrated that bacterial cells in the stationary growth phase are more resistant to PEF treatment than they are during the exponential growth phase. Physical and chemical properties of the products in which microorganisms are present significantly influence the rate of microbial inactivation. For example, a treatment medium with lower pH and conductivity is more effective for the inactivation of microorganisms. It should be understood, however, that changing any single parameter can affect microbial inactivation (Wouters et al., 2001b).

Many studies have shown that PEFs cause several morphological changes in microorganisms (Harrison et al., 1997; Wouters and Smelt, 1997; Barbosa-Cánovas et al., 1998b; Shin and Pyun, 1999). Pothakamury et al. (1997) found that after PEF treatment, the cell surface of *Staphylococcus aureus* was rough, the cell wall was broken, and the cytoplasmic contents were leaking out of the cell. Transmission electron microscopy (TEM) pictures of PEF-treated *Escherichia coli* cells have showed ruptured cell walls and leakage of the cytoplasmic contents of cells (Dutreux et al., 2000). Significant differences could be observed between the structure of PEF-treated and PEF-untreated *Saccharomyces cerevisiae* cells studied by SEM or TEM. Surface roughening, budding, and elongation could be observed (Barbosa-Cánovas et al., 1998b; Harrison et al., 1997). Morphological changes were observed in *Listeria innocua* cells after PEF treatment (Calderon-Miranda et al., 1999). By increasing the electric field strength (30–50 kV/cm), the cell wall of *Listeria innocua* lost smoothness and uniformity and at 40 kV/cm or higher electric field strength, extracellular material was observed in the surroundings of cells. Surface roughness of cell walls, clumping of cytoplasm, and cytoplasmic disruption also were observed. TEM results of Rowan et al. (2000) showed ruptured cell walls of *Bacillus cereus* cells with leakage of intracellular contents or cellular debris after 6000 pulses at 30 kV/cm.

Effect of PEF on Enzymes

Compared to the amount of information on the effects of PEF treatment on microorganisms, there is limited information on the inactivation of enzymes and the results are very contradictory. According to the results summarized in Table 6.3, the sensitivity of enzymes to PEF treatment is very different from that of microorganisms. The process parameters, the character of medium, and the structure of the enzyme significantly influence the enzymatic inactivation (Yeom and Zhang, 2001; Van Loey et al., 2002). For example, the activity of alpha-amylase from *Bacillus licheniformis* decreased significantly after 30 exponential decay pulses at 26 kV/cm. After PEF treatment, the remaining activity was only 15% (Ho et al., 1997). However, PEF treatment of alkaline phosphatase resulted in inconsistent results ranging from 5% to 96% inactivation (Ho et al., 1997; Barbosa-Cánovas et al., 1998b; Castro, 1994; Castro et al., 2001). In contradiction, Grahl and Märkl (1996)

Table 6.3. Effects of PEF Processing on Enzymes

Enzyme	Treatment Conditions	Media	Effects	Enzyme Origin	Reference
Amylase	4–26 kV/cm, 30 exp. decay pulses, 2 μs pulse width, $T = 20°C$, $f = 0.5$ Hz	Distilled water	70–85% Inactivation	*Bacillus licheniformis*	Ho et al. (1997)
Alkaline phosphatase	13.2 kV/cm, 70 pulses	Raw milk	96% Inactivation	Milk	Barbosa-Cánovas et al. (1996)
	21.5 kV/cm, 1–1000 exp. decay pulses	Raw milk	No inactivation	Milk	Grahl and Märkl (1996)
	18.8 kV/cm, 400 μs pulse duration, 70 pulses, $T = 22°C$	Raw milk	60% Inactivation	Bovine milk	Barbosa-Cánovas et al. (1998b) and Castro (1994)
		2% Fat and skim milk	60% and 65% Inactivation		
	80 kV/cm, 30 exp. decay pulses, 2 μs pulse duration, $T = 20°C$, $f = 0.5$ Hz	Buffer (pH 9.8), +1 M diethanolamine, +0.5 mM MgCl	5% Inactivation	Bovine intestinal mucosa	Ho et al. (1997)
Glucose oxidase	50 kV/cm, 30 exp. decay pulses, 2 μs pulse width, $T = 20°C$, $f = 0.5$ Hz	50 mM sodium acetate	75% Inactivation	*Aspergillus niger*	
Lipase	21.5 kV/cm, 1–1000 exp. decay pulses	Raw milk	65% Inactivation	Milk	Grahl and Märkl (1996)
	88 kV/cm, 30 exp. decay pulses, 2 μs pulse duration, $T = 20°C$, $f = 0.5$ Hz	Distilled water	85% Inactivation	Wheat germ	Ho et al. (1997)
Lipoxygenase	2.5–20 kV/cm, 1 μs pulse width, 100–400 pulses, $f = 1$ Hz	Green pea juice	No inactivation	Green pea	Van Loey et al. (2002)
	10, 20, and 30 kV/cm, 1–1000 pulses, 5 and 40 μs pulse width, $f = 1$ and 100 Hz	Distilled water	64% Inactivation		

Continued

Table 6.3. *Continued*

Enzyme	Treatment Conditions	Media	Effects	Enzyme Origin	Reference
Papain	20–50 kV/cm, 200–500 square wave pulses, 4 μs pulse width, $T < 35°C$, $f = 1.5$ kHz	Activated papain solution (1 mM EDTA)	5% Inactivation immediately after PEF treatment, 85% inactivation after 24 h storage at 4°C	Papaya	Yeom et al. (1999)
Pectinmethyl esterase	35 kV/cm, 42 pulses, 1.4 μs pulse width, $T_{inlet} = 24°C$, $T_{outlet} = 60°C$, $f = 600$ Hz	100% Orange juice	88% Inactivation	Valencia orange	Yeom et al. (2000)
	10–30 kV/cm, 1–1000 pulses, 5 and 40 μs pulse width, $f = 1$ and 100 Hz	100% Orange juice and distilled water	<10% Inactivation	Orange	Van Loey et al. (2002)
Peroxidase	21.5 kV/cm, 1–1000 exp. decay pulses	Raw milk	25% Inactivation	Milk	Grahl and Märkl (1996)
	73.3 kV/cm, 30 exp. decay pulses, 2 μs pulse width, $T = 20°C$, $f = 0.5$ Hz	100 mM potassium phosphate	30% Inactivation	Soybean	Ho et al. (1997)
Polyphenol oxidase	24 kV/cm, 20–100 exp. decay pulse, $T < 25°C$	Distilled water	97% Inactivation	Apple	Giner et al. (1997)
			70% Inactivation	Peach	
			62% Inactivation	Pear	
	50 kV/cm, 30 exp. decay pulses, 2 μs pulse width, $T = 20°C$, $f = 0.5$ Hz	50 mM potassium phosphate	40% Inactivation	Mushroom	Ho et al. (1997)
	31 kV/cm, 1000 pulses, 1μ-40 μs pulse duration, $f = 1$ Hz	100% Apple juice	No inactivation	Apple	Van Loey et al. (2002)
		Distilled water	10% Inactivation		
Plasmin	15, 30, or 45 kV/cm, 10–50 pulses, 2 μs pulse duration, $T = 15°C$, $f = 0.1$ Hz	Simulated milk ultrafiltrate (SMUF)	90% Inactivation	Bovine milk	Vega-Mercado et al. (1995a, b)
Protease	14–18 kV/cm, 20–98 pulses, 2 μs pulse duration, $T = 50°C$, $f = 2$ Hz	Skim milk	40–80% Inactivation	*Pseudomonas fluorescens M3/6*	

who applied exponential decay pulses at 21.5 kV/cm could not detect any decrease in the activity of alkaline phosphatase in raw milk.

Ho et al. (1997) studied the effect of 30 exponential decay pulses at 50 kV/cm on glucose oxidase from *Aspergillus niger*. PEF treatment induced inactivation was 75%.

Grahl and Märkl (1996) and Ho et al. (1997) observed 65% and 85% loss in lipase activity by applying exponential decay pulses at 21.5 and at 88 kV/cm, respectively.

Yeom et al. (1999) studied the inactivation of papain which was treated by 200–500 square wave pulses of 4 μs at 20–50 kV/cm. The effect of treatment was not perceptible immediately after the treatment. But, after 24 and 48 h of storage at 4°C, significant papain inactivation was observed. After 24 h of storage, the treatment resulted in approximately 85% decrease in papain activity.

Studying the effect of PEF treatment on the quality of Valencia orange juice, Yeom et al. (2000) investigated the effect of treatment on pectinmethylesterase activity. After PEF treatment at 35 kV/cm, the remaining activity of pectinmethylesterase was only 12%. In this study and probably others as well, the process temperature may have played a great part in the inactivation of enzymes since the outlet temperature after the treatment chamber was 60°C. Van Loey et al. (2002) studied the inactivation of pectinmethylesterase dissolved in distilled water and in real food systems. After PEF treatment (10–30 kV/cm, 5 and 40 μs pulse width, 1 and 100 Hz frequency, 1–1000 pulses), pectinmethylesterase activity was less than 10%. Moreover, the PEF treatment of orange juice resulted mostly in an increased enzyme activity. According to the authors, it probably happened because of cell permeabilization and the release of intracellular pectinmethylesterase. Similar to pectinmethylesterase, peroxidase and polyphenoloxidase, dissolved in distilled water, were also resistant to PEF treatment. In contradiction to the results of Van Loey et al. (2002), Giner et al. (1997) obtained 62–97% inactivation of polyphenoloxidase after PEF treatment at 24 kV/cm. A moderate 25–30% reduction in peroxidase activity was obtained by applying PEF treatment at 21.5 kV/cm and at 73.3 kV/cm, respectively (Grahl and Märkl, 1996; Ho et al., 1997).

Vega-Mercado et al. (1995a, b) studied the effect of PEF treatment on plasmin in simulated milk. Applying 50 pulses at 45 kV/cm decreased plasmin activity by 90%. The same authors investigated the inactivation of an extracellular protease from *Pseudomonas fluorescens* by PEF treatment. An 80% reduction in enzyme activity was obtained after PEF treatment with 20 pulses at 18 kV/cm. At lower electric field strength (15, 14 kV/cm), but higher number of pulses (98, 32), 60% and 40% reduction were achieved, respectively.

The above results confirm that the effects of PEF treatment on enzyme activity remain contradictory and more research is needed for clarification.

Effect of PEF on Food Quality

Much of the focus in PEF research is on its effects on microorganisms and enzymes in order to develop a new non-thermal electrical pasteurization process. But this work would be not complete without knowledge of the effects of the PEF process on physical, chemical, and sensory properties of food products. It is generally accepted that the application of high voltage pulses leads to significant inactivation of microorganisms, while causing little changes in food quality as compared to traditional thermal pasteurization. Food products minimally processed by PEF are expected to retain fresh, physical, chemical, and nutritional characteristics, and may possess an extended refrigerated shelf life (Zárate-Rodriguez et al., 2000).

It must be emphasized that foods are typically degassed prior to PEF to avoid the dielectric breakdown of them. The dielectric breakdown is often caused by gas bubbles because the electric field is enhanced within these bubbles. It is therefore recommended to avoid or reduce the presence or formation of such bubbles. The prior deaeration of foods is useful, especially when these products are subjected to a turbulent flow in the continuous treatment chamber (Barsotti et al., 1999). However, such prior deaeration can result in the loss of volatiles and hence quality (Jia et al., 1999).

Ortega-Rivas et al. (1998) investigated the effect of high voltage pulses on soluble solids (Brix°), pH, acidity, and color of apple juice. After applying 2, 4, 8, and 16 pulses at increasing (50, 58, and 66 kV/cm) electric field strength, quality attributes were unchanged except for color. The PEF-treated juices became paler as a function of applied electric field strength. Jin and Zhang (1999), studying the effect of PEF treatment on quality of cranberry juice, found that the overall volatile profile of the sample was not affected by the treatment. No significant difference in color was observed between the untreated and treated samples.

PEF treatment of a protein-fortified fruit beverage by Sharma et al. (1998) resulted in less protein denaturation and less loss of vitamin C in comparison to heat treatment. Also, the PEF-processed beverage maintained its natural orange juice like color better than the heat-treated one which developed a slightly whitish color. Jia et al. (1999) compared the effects of PEF on flavor compounds of orange juice with the effects of heat processing. The average losses of flavor compounds in orange juice processed at 30 kV/cm for 240 or 480 μs were 3% and 9%, respectively, compared to 22% for heat processing. According to the authors, the flavor loss was mainly due to vacuum degassing in the PEF process. The same research group investigated the change in vitamin C and color of fresh orange juice due to PEF treatment (Qiu et al., 1998). The PEF treatment eliminated 99.9% of the microflora, yet caused little loss of vitamin C and color compared with heat pasteurization. Similar results were obtained by Yeom et al. (2000) who also studied the effects of the PEF process on orange juice. At 35 kV/cm, PEF-treated orange juice retained greater amounts of vitamin C and five representative flavor compounds than the heat-pasteurized orange juice during storage at 4°C. The PEF-treated orange juice had a lower browning index, higher whiteness, and higher hue angle values than the heat-pasteurized juice during storage at 4°C. Brix° and pH values were not significantly affected by treatment. Evrendilek et al. (2000), studying the microbial safety and shelf life of apple juice and cider processed by PEF, found that PEF not only extended the shelf life of fresh apple juice and cider, but also maintained a fresh flavor. Their results at 4°C and 22°C storage temperature indicated that PEF processing was significantly effective in retaining vitamin C content in samples. Both PEF and heat processing of apple cider did not affect color stability during storage. Freshly pressed raw apple juice was PEF-treated and investigated by Harrison et al. (2001). They found that the PEF-treated apple juice could be stored at 4°C for 1 month with little or no change in its volatile flavor compounds.

According to Mok and Lee (2000), PEF sterilization of rice wine resulted in superior quality and less change in color and sensory properties than rice wine sterilized by heat. Similarly, the results for PEF treatment of various vegetable and fruit juices result in little or no change in physical and chemical properties of samples (Table 6.4) (Lechner and Cserhalmi, 2004; Cserhalmi et al., 2004a, b).

APPLICATION

One of the most important advantages of PEF treatment is minimal heat production during treatment, yet sufficient microbial inactivation. The shelf life of PEF-treated food products can be significantly extended without undesirable changes in physical, chemical, and sensory properties. Owing to this property, the PEF technology might be a potential alternative preservation process to heat pasteurization. Numerous liquid foods and beverages, including orange, apple, cranberry and peach juices, pea soup, beaten eggs, and skim milk, have been successfully processed with PEF by several research groups (Vega-Mercado et al., 1997; Qin et al., 1995; Barbosa-Cánovas et al., 1998b; Bendicho et al., 2002; Góngora-Nieto et al., 2003). Krupp Maschinentechnik GmbH in Hamburg, Germany, was one of the first that successfully applied PEF to fluid foods such as orange juice and milk (Castro et al., 1993). Pure Pulse Technologies, Inc., that was discontinued more than 2 years ago, has patented the PEF preservation of some fluid foods such as dairy products, fruit juices, and liquid eggs (Dunn and Pearlman, 1987). Washington State University has achieved prominent success in PEF treatment chamber design and construction and in the application of PEF preservation technology to orange juice, apple juice, milk, eggs, and green pea soups (Barbosa-Cánovas et al., 2000). Ohio State University plays an important role in PEF equipment design and in pilot scale application of it for food preservation such as apple juice and cider, orange, and cranberry juices. Berlin University of Technology has applied high electric field pulses, especially for the permeabilization of plant membranes. The aim of their work is to aid and optimize the release of valuable metabolites such as pigments, flavors, bioactive components, and enzymes from plant cells and tissues and to affect mass and heat transfer for subsequent unit operations (dehydration, extraction, freezing) to improve fruit and vegetable juice quality (Angersbach and Knorr, 1998).

FUTURE OF PEF

In spite of a considerable volume of work documenting the effects of PEF treatment on microorganisms, enzymes, and the quality of treated food products, additional systematic research is needed to assure the safety of foods processed by PEF technology. Researchers worldwide have many ongoing projects

Table 6.4. Physical and Chemical Properties of Fruit Juices

Samples	Treatment	$T_{treatment}$(°C)	pH	Brix°	Conductivity (mS)	Viscosity (cP)	NEBI	DE[a]	Organic Acid Content		
									Malic acid	Citric acid	Ascorbic acid
Carrot	Untreated	–	6.32	0.13	1.11	0.94	0.20	2.31	–	–	–
	Treated	30	6.16	0.13	1.13	0.94	0.17		–	–	–
Red beet	Untreated	–	6.08	2.00	1.80	1.02	0.39	0.58	–	–	–
	Treated	33	5.96	2.00	1.73	1.15	0.41		–	–	–
Pumpkin	Untreated	–	6.21	0.23	1.41	0.94	0.21	2.42	–	–	–
	Treated	33	6.19	0.20	1.50	1.11	0.18		–	–	–
Apple	Untreated	–	3.39	11.50	1.32	–	0.33	0.59	3.52	0.31	–
	Treated	34	3.40	11.50	1.35	–	0.36***		3.34	0.30	–
Strawberry	Untreated	–	3.25	4.00	1.84	–	0.50	1.42	219.00	10.19	–
	Treated	38	3.20	4.00	1.91	–	0.51**		211.00	10.01	–
Raspberry	Untreated	–	3.01	3.50	1.83	–	0.69	0.38	0.70	20.89	110.00
	Treated	37	2.98	3.50	1.86	–	0.70		0.69	20.69	97.85
Blackberry	Untreated	–	3.19	6.50	1.53	–	0.71	0.17	6.44	0.54	–
	Treated	35	3.20	6.50	1.57	–	0.73*		6.12	0.50	–
Sour cherry	Untreated	–	3.17	5.36	1.50	–	0.11	0.58	23.08	0.32	–
	Treated	33	3.18	5.83	1.51	–	0.11		22.19	0.29	–
Plum	Untreated	–	3.35	7.47	1.49	–	0.18	0.81	6.72	0.01	–
	Treated	39	3.34	7.42	1.50	–	0.18		6.50	0.01	–
Grapefruit	Untreated	–	3.05	5.10	1.23	1.36	0.11	0.52	0.24	11.91	362.31
	Treated	33	3.09	5.11	1.25	1.48	0.11		0.25	11.91	339.92
Lemon	Untreated	–	2.43	1.16	1.20	0.88	0.09	0.59	2.13	52.38	313.51
	Treated	32	2.42	1.00**	1.23	1.15*	0.10**		2.56	55.62	129.61***
Orange	Untreated	–	3.63	4.53	1.20	1.4	0.11	0.47	1.14	8.21	328.21
	Treated	31	3.61	4.30*	1.28	1.07**	0.11		1.11	8.05	318.41
Tangerine	Untreated	–	3.41	6.50	1.50	1.15	0.11	2.44	0.63	6.91	155.61
	Treated	34	3.43*	6.00	1.50	1.19	0.11		0.69	7.15	157.82

Note: PEF treatment: 28 kV/cm, 50 pulses, 2 μs pulse duration.

* Significant difference p = 0.05%, **p = 0.01%, ***p = 0.001%.

[a] DE (total color difference) = $\sqrt{(\Delta L)^2 + (\Delta a)^2 + (\Delta b)^2}$. DE = 0.0–0.5, not noticeable; DE = 0.5–1.5, slightly noticeable; DE = 1.5–3.0, noticeable; DE = 3.0–6.0, well visible; DE = 6.0–12.0, great.

with a focus on a better understanding of PEF and the answer to industrial and regulatory demands for the commercialization of this method. Major efforts are needed to define:

- inactivation kinetics and processing conditions that allow a more reliable comparison among different systems
- effects of the process on different food components
- flow hydrodynamics
- treatment uniformity

For PEF to become a viable food industry application, both the capital equipment costs and the energy consumption of the process must be reduced. According to Vega-Mercado et al. (1999), the initial investment costs may be between $450,000 and $2 million depending on the scale of the operation.

The short- and long-term research works are comprehensively summarized by Barbosa-Cánovas et al. (2000):

- Confirmation of the mechanisms of microbial and enzyme inactivation.
- Identification of the pathogens of concern most resistant to PEF.
- Identification of the surrogate microorganisms for the pathogens of concern.
- Development of validation methods to ensure microbiological effectiveness.
- Development and evaluation of kinetic models that take into consideration the critical factors influencing inactivation.
- Optimization and control critical process factors.
- Standardization and development of effective methods for monitoring consistent delivery of a specified treatment.
- Treatment chamber design uniformity and processing capacity.
- Identification and application of electrode materials for longer operation time and lower metal migration.
- Process system design, evaluation, and cost reduction.

REFERENCES

Alvarez, I., Raso, J., Palop, A., Sala, F. J. 2000. Influence of different factors on the inactivation of *Salmonella senftenberg* by pulsed electric fields. International Journal of Food Microbiology 55:143–146.

Alvarez, L., Pagan, R., Raso, J., Condon, S. 2002. Environmental factors influencing the inactivation of *Listeria monocytogenes* by pulsed electric fields. Letter on Applied Microbiology 35(6):489–493.

Angersbach, A., Knorr, D. 1998. Impact of high electric field pulses on plant membrane permeabilization. Trends in Food Science and Technology 9:185–191.

Aronsson, K. 2002. Inactivation, cell injury and growth of microorganisms exposed to pulsed electric fields using continuous process. Ph.D. Thesis, Chalmers University of Technology, Department of Food Science, Göteborg, Sweden.

Aronsson, K., Lindgren, M., Johansson, B. R., Rönner, U. 2001. Inactivation of microorganisms using pulsed electric fields: the influence of process on *Escherichia coli*, *Listeria innocua*, *Leuconostoc mesenteroides* and *Saccharomyces ceevisiae*. Innovative Food Science and Emerging Technologies 2:41–54.

Aronsson, K., Rönner, V. 2001. Influence of pH, water activity and temperature on the inactivation of *Escherichia coli* and *Saccharomyces cerevisiae* by pulsed electric fields. Innovative Food Science and Emerging Technologies (2):105–112.

Barbos-Cánovas, G. V., Qin, B. L., Swanson, B. G. 1996. "Biological effects induces by pulsed electric fields of high intensity." In Tecnologias avanzadas en esterilizacion y seguridad de alimentos y otros productos, edited by Rodrigo, M., Martinez, A., Fiszman, S. M., Rodrigo, C., Mateu, A. pp. 151–165. Grafo Impresores, SL, Spain.

Barbosa-Cánovas, G. V., Góngora-Nieto, M. M., Swanson, B. G. 1998a. Nonthermal electric methods in food preservation. Food Science Technology Internal 4(5):363–370.

Barbosa-Cánovas, G. V., Pierson, M. D., Zhang, Q. H., Schaffner, D. W. 2000. Pulsed electric fields. Journal of Food Science; Special Supplement: Kinetics of microbial inactivation for alternative food processing technologies, pp. 65–79.

Barbosa-Cánovas, G. V., Pothakamury, U. R., Palou, E., Swanson, B. G. 1998b. "Biological effects and applications of pulsed electric fields for the preservation of foods." In Nonthermal Preservation of Foods pp. 73–112. Marcel Dekker Inc., New York.

Barsotti, L., Cheftel, J. C. 1999. Food processing by pulsed electric fields: II Biological aspects. Food Reviews International 15(2):181–213.

Barsotti, L., Merle, P., Cheftel, J. C. 1999. Food processing by pulsed electric fields. I. Physical aspects. Food Reviews International 15(2): 163–180.

Beczner, J., Vidács, I., Cserhalmi, Zs. 2001. Combined effect of pulsed electric field, heat treatment and irradiation on some bacteria. 11th World Congress of Food Science and Technology, 22–27 April, Seoul, Korea, Abstracts, p. 179.

Bendicho, S., Barbosa-Cánovas, G. V., Martín, O. 2002. Milk processing by high intensity pulsed electric fields. Trends in Food Science and Technology 13:195–204.

Calderon-Miranda, M. L., Barbosa-Cánovas, G. V., Swanson, B. G. 1999. Inactivation of *Listeria innocua* in skim milk by pulsed electric fields and nisin. International Journal of Food Microbiology 51:19–30.

Castro, A. J. 1994. Pulsed electric field modification of activity and denaturation of alkaline phosphatase. Ph.D. Thesis. Washington State University, Pullman, WA.

Castro, A. J., Barbosa-Cánova, G. V., Swanson, B. G. 1993. Microbial inactivation of foods by pulsed electric fields. Journal of Food Processing and Preservation 17(1):47–73.

Castro, A. J., Swanson, B. G., Barbosa-Cánovas, G. V., Zhang, Q. H. 2001. "Pulsed electric field modification of milk alkaline phosphatase activity." In Pulsed electric fields in food processing. Fundamental aspects and applications, edited by Barbosa-Cánovas, G. V., Zhang, Q. H. pp. 65–82. Technomic Publishing Company Inc., Lancaster.

Coster, H. G. L., Zimmermann, U. 1975. The mechanism of electrical breakdown in the membranes of *Valonia utricularis.* Journal of Membrane Biology 22:73–90.

Cserhalmi, Zs., Beczner, J. 2003. Effect of processing conditions on inactivation of *Saccharomyces cerevisiae* by pulsed electric fields. Workshop on Nonthermal food preservation, 7–11 September, Wageningen, Proceedings, pp. 1–2.

Cserhalmi, Zs., Mészáros, L., Sass-Kiss, Á., Tóth, M. 2004a. Study of fruit juices treated by new preservation techniques. International Congress on Engineering and Food, 7–11 March, Montpellier, Abstracts, p. 83., Proceedings, pp. 180–186.

Cserhalmi, Zs., Sass, Á., Tóth, M., Lechner, N. 2004b. Study of pulsed electric field treated citrus juices. 2nd Central European Congress on Food, 26–28 April, Budapest, Abstracts, p. 251.

Cserhalmi, Zs., Vidács, I., Beczner, J., Czukor, B. 2002. Inactivation of *Saccaharomyces cerevisiae* and *Bacillus cereus* by pulsed electric fields technology. Innovative Food Science and Emerging Technologies 3:41–45.

Dunn, J. E., Pearlman, J. S. 1987. Methods and apparatus for extending the shelf-life of fluid products. US patent 4695472.

Dutreux, N., Notermans, S., Wijtzes, T., Góngora-Nieto, M. M., Barbosa-Cánovas, G. V., Swanson, B. G. 2000. Pulsed electric fields inactivation of attached and free-living *Escherichia coli* and *Listeria innocua* under several conditions. International Journal of Food Microbiology 54:91–98.

Evrendilek, G. A., Zhang, Q. H., Richter, E. R. 1999. Inactivation of *Eschesichia coli* 0157:H7 and *Eshesichia coli* 8739 in apple juice by pulsed electric field. Journal of Food Protection 62:793–796.

Evrendilek, G. A., Jin, Z. T., Ruhlman, K. T., Qin, X., Zhang, Q. H., Richter, E. R. 2000. Microbial safety and shelf-life of apple juice and cider processed by bench and pilot scale PEF systems. Innovative Food Science and Emerging Technologies 1:77–86.

Giner, J., Rauret-Arino, A., Barbosa-Cánovas, G. V., Martin-Belloso, O. 1997. Inactivation of polyphenoloxidase by pulsed electric fields. In Proceedings IFT Annual Meeting, p. 19. Orlando, USA.

Góngora-Nieto, M. M., Pedro, P. D., Swanson, B. G., Barbosa-Cánovas, G. V. 2003. Energy analysis of liquid whole egg pasteurized by pulsed electric fields. Journal of Food Engineering 57(3):209–216.

Grahl, T., Märkl, H. 1996. Killing of microorganisms by pulsed electric fields. Applied Microbiology and Biotechnology 45:148–157.

Grahl, T., Sitzmann, W., Märkl, H. 1992. Killing of microorganisms in fluid media by high-voltage pulses. 10th Dechema Biotechnology Conference Series 5B, pp. 675–678. Verlagsgesellschaft, Hamburg, Germany.

Hamilton, W. A., Sale, A. J. H. 1967. Effects of high electric field on microorganisms II. Mechanism of action of the lethal effect. Biochimica et Biophysica Acta 148:789–800.

Harrison, S. L., Barbosa-Cánovas, G. V., Swanson, B. G. 1997. *Saccharomyces cerevisiae* structural changes induced by pulsed electric field treatment. Lebensmittel-Wissenschaft und Technologie 30:236–240.

Harrison, S. L., Chang, F. J., Boylston, T., Barbosa-Cánovas, G. V., Swanson, B. G. 2001. "Shelf stability sensory analysis and volatile flavour profile of raw apple juice after pulsed electric field, high hydrostatic pressure, or heat exchanger processing." In Pulsed electric fields in food processing. Fundamental aspects and applications, edited by Barbosa-Cánovas, G. V., Zhang, Q. H.

pp. 241–257. Technomic Publishing Company Inc., Lancaster.

Heinz, V., Phillips, S. T., Zenker, M., Knorr, D. 1999. Inactivation of *Bacillus subtilis* by high intensity pulsed electric fields under close to isothermal conditions. Food Biotechnology 13:155–168.

Ho, S. Y., Mittal, G. S., Cross, J. D. 1997. Effects of high electric pulses on the activity of selected enzymes. Journal of Food Engineering 31:69–84.

Hofmann, G. A. 1989. "Cells in electric fields, physical and practical electronic aspects of electro cell fusion and electroporation." In Electroporation and Electrofusion in Cell Biology, edited by Neuman. E., Sowers, A. E, Jordan, C. A. Plenum Press, New York.

Hülsheger, H., Potel, J., Niemann, E. G. 1983. Electric field effects on bacteria and yeast cells. Radiation and Environmental Biophysics 22:149–162.

Iu, J., Mittal, G. S., Griffiths, M. W. 2001. Reduction in levels of *Escherichia coli O157:H7* in apple cider by pulsed electric fields. Journal of Food Protection 64(7):964–969.

Jayaram, S., Castle, G. S. P., Argyrios, M. 1993. The effects of high field DC pulse and liquid medium conductivity on survivability of *Lactobacillus brevis*. Applied Microbiology and Biotechnology 40:117–122.

Jeantet, R., Baron, F., Nau, F., Roignant, M., Brulé, G. 1999. High intensity pulsed electric fields applied to egg white: effect on *Salmonella enteridis* inactivation and protein denaturation. Journal of Food Protection 62:1381–1386.

Jeyamkondan, S., Jayas, D. S., Holley, R. A. 1999. Pulsed electric field processing of foods: a review. Journal of Food Protection 62:1088–1096.

Jia, M., Zhang, Q. H., Min, D. B. 1999. Pulsed electric field processing effects on flavour compounds and microorganisms of orange juice. Food Chemistry 65(4):445–451.

Jin, Z. T., Su, Y., Tuhela, L., Zhang, Q. H., Sastry, S. K., Yousef, A. E. 2001. "Inactivation of *Bacillus subtilis* spores using high voltage pulsed electric fields." In Pulsed electric fields in food processing. Fundamental aspects and applications, edited by Barbosa-Cánovas, G. V., Zhang, Q. H. pp. 167–181. Technomic Publishing Company Inc., Lancaster.

Jin, Z. T., Zhang, Q. H. 1999. Pulsed electric field inactivation of microorganisms and preservation of quality of cranberry juice. Journal of Food Processing and Preservation 23(6):481–497.

Kinosita, K. Jr., Tsong, T. Y. 1977a. Hemolysis of human erythrocytes by a transient electric field. Proceeding of the National Academy of Sciences United States of America 74:1923–1927.

Kinosita, K. Jr., Tsong, T. Y. 1977b. Voltage-induced pore formation and hemolysis of human erythrocytes. Biochimica et Biophysica Acta 471:227–242.

Knorr, D,. Geulen, M., Grahl, T., Sitzmann, W. 1994. Food application of high electric field pulses. Trends in Food Science and Technology 5:71–75.

Lechner, N., Cserhalmi, Zs. 2004. Pulsed electric field (PEF) processing effects on physical and chemical properties of vegetable juices. Fosare seminar series 4 "Novel (mild) preservation technologies in relation to food safety", 22–23 January, Brussels, p. 1.

Marquez, V. O., Mittal, G. S., Griffiths, M. W. 1997. Destruction and inhibition of bacterial spores by high voltage electric field. Journal of Food Science 62:399–401, 409.

Matsumoto, Y., Satake, T., Shioji, N., Sakuma, A. 1991. Inactivation of microorganisms by pulsed high voltage applications. Conference Record of IEEE Industrial Applications Society Annual Meeting, pp. 652–659.

Mok, C., Lee, S. 2000. Sterilization of yakju (rice wine) on a serial multiple electrode pulsed electric field treatment system. Korean Journal of Food Science and Technology 32:356–362.

Ortega-Rivas, E., Zárate-Rodríguez, E., Barbosa-Cánovas, G. V. 1998. Apple juice pasteurization using ultrafiltration and pulsed electric fields. Trans Institution of Chemical Engineers, Part C 76(C4):193–198.

Picart, L., Dumay, E., Cheftel, J. C. 2002. Inactivation of *Listeria innocua* in dairy fluids by pulsed electric fields: influence of electric parameters and food composition. Innovative Food Science and Emerging Technologies 3(4):357–369.

Pol, I. E., Mastwijk, H. C., Bartels, P. V., Smid, E. J. 2000. Pulsed electric field treatment enhances the bactericidal action of nisin against *Bacillus cereus*. Applied and Environmental Microbiology 66:428–430.

Pol, I. E., Masteijk, H. C., Slump, R. A., Popa, M. E., Smid, E. J. 2001a. Influence of food matrix on inactivation of *Bacillus cereus* by combinations of nisin, pulsed electric field treatment, and carvacrol. Journal of Food Protection 64:1018–1021.

Pol, I. E., van Arendonk, W. G. C., Mastwijk, H. C., Krommer, J., Smid, E. J., Moezelaar, R. 2001b. Sensitivities of germination spores and carvacrol-adapted vegetative cells and spores of *Bacillus cereus* to nisin and pulsed-electric-field treatment. Applied and Environmental Microbiology 67:1693–1699.

Pothakamury, U. R., Barbosa-Cánovas, G. V., Swanson, B. G., Spence, K. D. 1997. Ultrastructural changes in *Staphylococcus aureus* treated with pulsed electric fields. Food Science and Technology International 3:113–121.

Pothakamury, U. R., Monsalve-Gonzalez, A., Barbosa-Cánovas, G. V., Swanson, B. G. 1995a. Inactivation of *Escherichia coli* and *Staphylococcus aureus* in model foods by pulsed electric field technology. Food Research International 28:167–171.

Pothakamury, U. R., Monsalve-González, A., Barbosa-Cánovas, G. V., Swanson, B. G. 1995b. High voltage pulsed electric field inactivation of *Bacillus subtilis* and *Lactobacillus delbrueckii.* Spanish Journal of Food Science and Technology 35:101–107.

Pothakamury, U. R., Vega, H., Zhang, Q., Barbosa-Cánovas, G. V., Swanson, B. G. 1996. Effect of growth stage and processing temperature on the inactivation of *Escherichia coli* by pulsed electric fields. Journal of Food Protection 59:1167–1171.

Qin, B. L., Barbosa-Cánovas, G. V., Swanson, B. G., Pedrow, P. D. 1998. Inactivating microorganisms using a pulsed electric field continuous treatment system. IEEE Transactions on Industry Applications 34(1):43–50.

Qin, B. L., Pothakamury, U. R., Vega-Mecado, H., Martion, O., Barbosa-Cánovas, G. V., Swanson, B. G. 1995. Food pasteurization using high intensity pulsed electric fields. Food Technology 12:57–99.

Qiu, X., Sharma, S., Tuhela, L., Jia, M., Zhang, Q. H. 1998. An integrated PEF pilot plant for continuous nonthermal pasteurization of fresh orange juice. Transaction of the American Society of Agricultural Engineers 41(4):1069–1074.

Raso, J., Álvarez, I., Cond6n, S., Sala Trepat, F. J. 2000. Predicting inactivation of *Salmonella senftenberg* by pulsed electric fields. Innovative Food Science and Emerging Technologies 1: 21–29.

Reina, L. D., Jin, Z. T., Zhang, Q. H., Yousef, A. E. 1998. Inactivation of *Listeria monocytogenes* in milk by pulsed electric field. Journal of Food Protection 61(9):1203–1206.

Rowan, N. J., MacGragor, S. J., Anderson, J. G., Fouracre, R. A., Farish, O. 2000. Pulsed electric inactivation of diarrhoeagenic *Bacillus cereus* through irreversible electroporation. Letters in Applied Microbiology 31:110–114.

Sale, A. J. H., Hamilton, W. A. 1967. Effects of high electric fields on microorganisms, I. Killing of bacteria and yeasts. Biochimica et Biophysica Acta 148:781–788.

Sensoy, I., Zhang, Q. H., Sastry, S. K. 1997. Inactivation kinetics of Salmonella Dublin by pulsed electric field. Journal of Food Process Engineering 20:367–381.

Sharma, S. K., Zhang, Q. H., Chism, G. W., Tuhela-Reuning, L., Yousef, A. E. 1998. Developing a protein fortified fruit beverage processed with pulsed electric field treatment. Journal of Food Quality 21:459–473.

Shin, H. H., Pyun, Y. 1999. Cell damage and recovery characteristics of *Lactobacillus plantarum* by high voltage pulsed electric fields treatment. Food Science and Biotechnology 8(4):261–266.

Simpson, R. K., Whittington, R., Earnshaw, R. G., Russel, N. J. 1999. Pulsed high electric field causes 'all or nothing' membrane damage on *Listeria monocytogenes* and *Salmonella typhimurium,* but membrane H^+–ATPase is not a primary target. International Journal of Food Microbiology 48:1–10.

Unal, R., Kim, J. G., Yousef, A. E. 2001. Inactivation of *Escherichia coli O157:H7*, *Listeria monocytogenes*, and *Lactobacillus leichmannii* by combinations of ozone and pulsed electric field. Journal of Food Protection 64(6):777–782.

Van Loey, A., Verachter, B., Hendrickx, M. 2002. Effect of high electric field pulses on enzymes. Trends in Food Science and Technology 12:94–102.

Vega-Mercado, H., Powers, J. R., Barbosa-Cánovas, G. V., Swanson, B. G. 1995a. Plasmin inactivation with pulsed electric fields. Journal of Food Science 60(5):1143.

Vega-Mercado, H., Powers, J. R., Barbosa-Cánovas, G. V., Swanson, B. G., Luedecke, L. 1995b. Inactivation of protease from *Pseudomonas fluorescens* M3/6 using high voltage pulsed electric fields. IFT 1995. Annual Meeting. Book of Abstracts, paper no. 89-3. p. 267.

Vega-Mercado, H., G6ngora-Nieto, M. M., Barbosa-Cánovas, G. V., Swanson, B. G. 1999. "Nonthermal preservation of liquid foods using pulsed electric fields." In Handbook of Food Preservation, edited by Rahman, M. S. Chapter 17. Marcel Dekker, Inc., New York, 478–520.

Vega-Mercado, H., Martín-Belloso, O., Qin, B. L., Chang, F. J., Gángora-Nieto, M. M., Barbosa-Cánovas, G. V., Swanson, B. G. 1997. Non-thermal food preservation: Pulsed electric fields. Trends in Food Science and Technology 8:151–157.

Wouters, P. C., Alvarez, J., Raso, J. 2001a. Critical factors determining inactivation kinetics by pulsed electric field food processing. Trends in Food Science and Technology 12:112–121.

Wouters, P. C., Bos, A. P., Ueckert, J. 2001b. Membrane permeabilisation in relation to inactivation kinetics of *Lactobacillus* species due to pulsed electric fields. Applied and Environmental Microbiology 67:3092–3101.

Wouters, P. C., Dutreux, N., Smelt, J. P. P. M., Lelieveld, H. L. M. 1999. Effects of pulsed electric fields on inactivation kinetics of *Listeria innocua*. Applied and Environmental Microbiology 65:5364–5371.

Wouters, P. C., Smelt, J. P. P. M. 1997. Inactivation of microorganisms with pulsed electric fields: potential for food preservation. Food Biotechnology 11:193–229.

Yeom, H. W., Streaker, C. B., Zhang, Q. H., Min, D. B. 2000. Effects of pulsed electric fields on the quality of orange juice and comparison with heat pasteurization. Journal of Agricultural and Food Chemistry 48:4597–4605.

Yeom, H. W., Zhang, Q. H. 2001. "Enzymatic inactivation by pulsed electric fields: A review." In Pulsed electric fields in food processing, Fundamental aspects and applications, edited by Barbosa-Cánovas, G. V., Zhang, Q. H. pp. 65–82. Technomic Publishing Company Inc. Lancaster.

Yeom, H. W., Zhang, Q. H., Dunne, C. P. 1999. Inactivation of papain by pulsed electric fields in a continuous system. Food Chemistry 67:53–59.

Zárate-Rodriguez, E., Ortega-Rivas, E., Barbosa-Cánovas, G. V. 2000. Quality changes in apple juice as related to nonthermal processing. Journal of Food Quality 23:337–349.

Zhang, Q. H., Barbosa-Cánovas, G. V., Swanson, B. G. 1994b. Engineering aspects of pulsed electric field pasteurization. Journal of Food Engineering 25:268–281.

Zhang, Q. H., Chang, F. J., Barbosa-Cánovas, G. V., Swanson, B. G. 1994c. Inactivation of microorganisms in a semisolid model food using high voltage pulsed electric fields. Lebensmittel-Wissenschaft und Technologie 27:538–543.

Zhang, Q. H., Monsalve-Gonzuález, A., Qin, B., Barbosa-Cánovas, G. V., Swanson, B. G. 1994a. Inactivation of *Saccharomyces cerevisiae* by square wave and exponential-decay pulsed electric field. Journal of Food Protection and Engineering 17:469–478.

Zhang, Q. H, Qin, B. L., Barbosa-Cánovas, G. V., Swanson, B. G. 1994d Inactivation of *Escherichia coli* for food pasteurization by high-intensity short-duration pulsed electric fields. Journal of Food Protection and Preservation 19:103–118.

Zhang, Q. H., Qin, B. L., Barbosa-Cánovas, G. V., Swanson, B. G. 1995. Inactivation of *Escherichia coli* for pasteurization by high-strength pulsed electric fields. Journal of Food Preservation 19:103–118.

Zimmermann, U. 1986. Electrical breakdown, electropermeabilisation and electrofusion. Review of Physiological Biochemistry and Pharmacology 105:176–256.

Zimmermann, U., Pilwat, G., Riemann, F. 1974. Dielectric breakdown on cell membranes. Biophysical Journal 14:881–889.

7

Minimally Processed Fruits and Fruit Products and Their Microbiological Safety

Csaba Balla and József Farkas

MINIMALLY PROCESSED FOODS

TERMINOLOGY AND CLASSIFICATION

Consumer demands for convenient but fresh and healthy foods are driving the food industries to apply new and mild preservation techniques, which satisfy the increasing market demands for fewer preservatives, higher nutritive value, and fresh sensory attributes. Traditional preservation technologies and techniques highly affect the appearance, sensorial characters, and the nutritional value. "Minimal processing techniques have emerged to meet this challenge of replacing traditional methods of preservation whilst retaining nutritional and sensory quality" (Ohlsson and Bengtsson, 2002).

Several terminologies are used for the definition "minimal processing." According to Huis in't Veld (1996), minimal processes are those which "minimally influence the quality characteristics of a food whilst, at the same time, giving the food sufficient shelf life during storage and distribution." Fellows (2000) considers that "minimally process technologies are techniques that preserve foods but also retain to a greater extent their nutritional quality and sensory characteristics by reducing the reliance on heat as the main preservative action."

"Minimally processed" is an equivocal term that is applied to such different types of products as precut, prepackaged fresh produce or fresh meat for short refrigerated storage, and mildly cooked or pasteurized foods (meals or meal components) that can be stored under refrigeration for more than 1 week.

The original purpose of minimal processing was to minimize the heat treatment (thermal processing) used by traditional thermal techniques to reduce the quality loss that has been caused by long- and high-temperature treatments. The target to reduce the quality loss and to extend the shelf life of food including fruit and fruits products forced researchers to develop the nonconventional heat treatment to achieve better balance between preservation and quality and to develop new techniques that extend the shelf life of the products.

However, minimal processing does not inactivate completely all microorganisms present in the raw material. Thus, the microbiological safety during the shelf life of minimally processed foods depends largely on

1. an appropriate refrigerated storage and distribution, which prevents the growth of hazardous microorganisms and
2. restriction of "use-by" time period.

In addition, intrinsic hurdles to microbial growth may contribute to the safety of these products either originating from the raw material or introduced during processing [e.g., low pH, reduced water activity, natural antimicrobials, and modified atmosphere packaging (MAP)].

The aim of this chapter is to give a short review about the minimal processing techniques used for extending the shelf life and quality of fruits and fruit products and the microbiological safety of such products.

There are several systems to classify the methods of preservation used for minimal processing technology. The traditional classification is based on the methods and main effects of the preservation. Ohlsson and Bengtsson (2002) distinguish these different groups:

- thermal methods, with subgroups classified according to the heating system of the food: minimal processing by thermal conduction, convection and radiation, aseptic and semiaseptic processing, sous–vide processing, infrared heating, electric voltage heating, electric resistance/ohmic heating, high-frequency and radio frequency heating, microwave heating, and inductive electrical heating;
- nonthermal methods: irradiation, high-pressure processing, pulsed discharge of high capacitor, pulsed white light, ultraviolet light, laser light, pulsed electric field (PEF), oscillating magnetic field, ultrasound, pulse power system, air ion bombardment;
- MAP;
- active and intelligent packaging; and
- use of natural food preservatives.

The main effects of using traditional thermal method for fruits preservation are as follows: the loss of fresh appearance, destruction of the respiration pathways that serve as the energy for the life of the cell, and the irreversible change of some components, such as protein coagulation, starch shrivelling, texture softening, formation of aroma compounds, loss of vitamin and minerals, or formation of some thermal reaction components. The desirable or nondesirable effects of heat treatments depend on the exposure time and temperature. Minimal processing uses mild heating to avoid nondesirable effects or quality change of the processed food.

The bases of nonthermal methods of minimal processing are different, and only few methods are sufficiently effective in inactivating microorganisms or enzymes that play the main role in the deterioration. Most of them need combination with other preservation methods such as refrigeration and MAP for the prolongation of shelf life and maintenance of quality and safety.

As fruits are living products with active respiration even after harvesting, it is possible to classify the minimal process technologies used for processing of fruits and fruit products:

1. minimal processing technologies with active respiration of fruits and its products and
2. minimal processing technologies for processed fruit products that destroy the respiratory activities of fruits during processing.

The other chapters of this book deal with the basic concept of preservation systems and methods used in preservation or minimal processing. Therefore, in this short chapter, we would like to discuss only the MAP of fresh fruits and fruit products, covering theories, concepts, and practices.

Principles of MAP of Minimally Processed Fruits

Fresh fruits continue to respire after harvesting. This means that the carbohydrate content of the fruits as the substrate of respiration will be involved in the oxidation process. The products will be CO_2, water, and, of course, energy that appears as heat production. The rate of respiration is highly affected by a temperature involving Arrhenius kinetics (Q_{10} values are about 2–3), and as balanced reaction it is affected by the concentration of O_2 and CO_2. The Q_{10} means how many times the respiration rate changes if temperature increases by 10°C. The value of it serves information about the temperature sensitivity of fruit metabolism like respiration. Q_{10} values are of the order of 3 for chemical reactions. The O_2 involved in the oxidation of $NADH^+$ and $FADH^+$ determines the direction of the pathway. The CO_2 level affects the activity of the enzyme decarboxylase responsible for the decarboxylation process of the organic acids in Krebs cycle. Respiration is a balanced reaction. A reduction in oxygen and an increase in carbon dioxide will reduce the rate of respiration, thus extending the shelf life of the product.

To prolong the storage life of fresh fruits and vegetables, controlled atmosphere (CA) storage is frequently used. The basic CA effect on biochemical reactions can also be used to extend the shelf life of processed and ready-to-use fruit products. These products are often peeled and sliced and they are preferred as fruit dishes by consumers. The technique that provides CA condition for this ready-to-use fruit dishes is usually MAP.

Opposite to the nonrespiring foods that are packaged often without any O_2, respiring products such as freshly peeled and sliced fruits require O_2 to produce enough energy to maintain their integrity and quality (aerobic respiration). When these products are packaged (to meet the requirement of safety of ready-to-use food), the key factor in controlling respiration is the gas atmosphere at a certain temperature, especially the concentrations of oxygen and carbon dioxide.

When the fruit tissues respire, they take up oxygen and release carbon dioxide. The increased carbon dioxide and decreased oxygen cause a reduction of the respiration rate of the fruit tissue. This reduces the energy available for chemical changes that occur in fruits and vegetables, resulting in slower rates of ripening and prolonged preripening storability of produce. In the case of ripening fruits, the MAP is used to control the ripening. In the case of plant products that are marketed in an immature form to retard the senescence of cut fruits, MAP may be used to prevent browning of the cut surface when fruits are sliced (Stiles, 1991).

In fruits, the concentrations of both CO_2 and O_2 are important. The reduced O_2 concentration between 20% and 2% level decreases the rate of respiration. The effect is higher at lower oxygen concentration, according to a negative exponential funtion curve. Some oxygen needs to be present to prevent an anaerobic environment, which can be physiologically harmful to tissue metabolism. The critical O_2 concentration that initiates anaerobic respiration depends on the respiratory activity of fruit tissue. Each produce has its own tolerance of oxygen and response to lowering of O_2 level. At the critical O_2 level, the anaerobic respiration starts producing acetaldehyde and alcohol, which poison the tissue, cause physiological disorders or death, and lead to quality loss of perishable fruit. The total absence of O_2 results in off-flavor development and softening. The effect of reduced O_2 level in delaying the ripening of fruits can be reversed by using ethylene, which is the natural hormone of ripening. A reduced O_2 level decreases the biosynthesis of ethylene. According to Burg and Burg (1967), the ethylene receptor contains a metal ion and, when it is in its oxidated state, the binding of ethylene is enhanced. However, the effect of hypoxia on ethylene biosynthesis and action may be indirect. It suppresses the effects of hypoxia on the induction of 1-amino-cyclopropane-1-carboxylic acid (ACC) synthase and/or synthesis of transducer of ethylene action (Wiley, 1994).

The effect of CO_2 on senescence is unclear. It is suggested that CO_2 is a competitive inhibitor of ethylene. Experiments indicate that CO_2 may indeed diminish the action of ethylene, provided the concentration of the latter is less than 1 μl/l (Solomos, 1994). Fidler et al. (1973) observed, in the case of apple, that CO_2 enhances the inhibitory effects of low O_2 on respiration in the concentration range of 1–27% CO_2. It is also supposed that CO_2 is responsible for the ACC oxidase action (Kuai and Dilley, 1992). The tolerance against high CO_2 level is different in different fruits. Strawberries can tolerate high CO_2 level up to 20%, peach up to 10–15%, but McIntosh apples are damaged by about 3% CO_2 level.

The inhibitory effect of CO_2 on growth and metabolism of microorganisms was pointed out by Dixon and Kell (1989). This can be exploited in the preservation of refrigerated food. Carbon dioxide hydrates and dissociates in the water content of food. In most food systems, this involves the following equilibrium (Eq. 7.1) because the pH values are less than eight.

$$CO_2 + H_2O \leftrightarrow H_2CO_3 \leftrightarrow HCO_3^- + H^+. \qquad (7.1)$$

The concentration of CO_2 in solution depends on its partial pressure in the gas phase, temperature, and pH. As pH increases above 8.0, carbonate ions are formed and the equilibrium is shifted further to right. Detailed antimicrobial mechanisms of CO_2 are not fully understood but theorized as follows (Yuan, 2003):

- When CO_2 comes in contact with membrane proteins, it changes the ionic charges of the cell membrane, which can then interrupt the transport of specific ions needed for maintaining homeostasis in the cytoplasm.
- CO_2 permeates the membrane and reacts with water in the cytoplasm (see earlier). The hydrogen ions formed acidify the inside of the cell and the organism requires cellular energy to pump the protons back out. This added energy requirement creates a burden on the cells, thereby inhibiting their growth.
- CO_2 plays a role by either inducing or repressing the synthesis of some cytoplasmic enzymes.

The possibility of the growth of psychrotrophic pathogens such as *Listeria monocytogenes* is substantially less in MAP fruit products than that in higher pH vegetables or in food of animal origin packed

Table 7.1. Permeability Characteristics of Some Plastic Film (Dégen 2002)

		Transmission Rate (g/m^2/day at 38°C and 90% RH)		
Film Type	Density (g/cm^3)	Water Vapor	O_2	CO_2
Low density polyethylene (LDPE)	0.92	15–22	6000–8000	540,000
High density polyethylene (HDPE)	0.96	2	2100	7000
Polypropylene (PP)	0.88–0.90	3–10	1600–5300	4000–9000
Polyvinylchlorid (PVC)	1.25–1.45	40–100	300–8000	1500
Polystyrene (PS)	1.05	100–150	2200–6500	10,000–24,000
Polyvinylidine chloride (PVDC)	1.68	3	16	0

with high partial pressure of carbon dioxide, but it may be higher in prepared salads packed with only low concentration of carbon dioxide (Molin, 2000).

Important parameters influencing the shelf life of freshly prepared produce are the respiration rate, storage temperature, relative humidity (RH), initial microbial load, packaging film and equipment, filling weight in the package, volume and area, light, etc. The respiration rate is affected by product type and size, degree of preparation, product variety and growing conditions, maturity and tissue type, atmospheric composition, and temperature (Kader et al., 1989).

For minimally processed food products, there is a great effect of peeling and slicing on the tissue metabolism. The observed changes include a rise in respiration, DNA and RNA synthesis, including new enzymes, membrane degradation, and the appearance of novel mRNA (Laties, 1978). Investigation has shown that slicing induces a three- to fivefold rise in respiration over that of the parent plant organ. With aging, there is a further two- to threefold increase in respiration (Laties, 1978). Numerous experiments have shown that the nature of both respiratory substrates and pathways changes with aging of slices (Wiley, 1994). The increased CO_2 production of wounded or sliced fruits is further increased by microorganisms living on the surface with good condition for propagation.

Modified Atmosphere Packaging of Fresh Minimally Processed Fruit

The food package must protect and contain the product from the place and time of preparation and manufacture to the point of consumption. The packaging materials have additional task in MAP technology. The factors that affect MAP–induced atmosphere within the package are respiration rate, mass of product in the package, the optimal gas concentration of the fruit in the pack, the free gas volume in the pack, etc. (Geeson et. al., 1985; Hong and Gross, 2001). The other factors are the packaging film factors, such as permeability of film used for packaging and the surface and mass ratio in the package.

The permeability of film for CO_2 and O_2 are mainly determined by the material and the thickness of the film, the area of the surface, and the gas concentration difference between the outside and inside of the package. Since the respiring minimally processed (MPR) products utilize considerable oxygen, suitable plastic film for this type of MPR should have a relatively high oxygen permeability to avoid an oxygen-depleted atmosphere within the package. The gas transmission rates for some polymeric film used for MAP are shown in Table 7.1.

Plastic films used for MAP of fresh fruits need to have relatively high permeability to O_2 and CO_2. Most films have higher permeability to CO_2 than O_2 because of the solubility of CO_2 in polymer. It is suggested that the permeability of film used for MAP of respiring products need to be five times higher for CO_2 than that for O_2 to result in adequate O_2 intakes and CO_2 outputs on the surface. Because an O_2 concentration of less than 10–12% has an effect on respiration of fruits products, it can be proposed that the effective O_2 concentration should be below 5–6% to prolong the shelf life of the product. At 2–3% O_2, there is a high risk of anoxia. The maximum tolerable CO_2 concentrations for many fresh products are in the range of 2–5%.

Active packaging system uses O_2 and CO_2 absorbers in the sealed pack. Some of the absorbers being used are calcium hydroxide, activated charcoal, and sometimes magnesium oxide. Sometimes C_2H_4 absorbers like potassium permanganate are used to prevent the accelerating respiring effect of ethylene.

Minimal Pretreatment Processing

The minimal process operations should be valuable adjuncts to MAP for successful extension of the shelf life of fruit commodities. These operations include washing to remove the nondesirable materials from the surface (soil, insects, pesticides, etc.) and cool the produce, trimming to remove unsound tissue, separating inedible portions from desirable edible segments, cutting edible tissue into suitable shapes and sizes, cooling and temperature conditioning, and applying food additives for pH adjustment, microbial control, oxidative reaction control, and texture modification.

The choice of washing method of fruits is dependent on the purpose of washing and the delicacy of the tissue. The wash water should have a low microbial count and a suitable temperature for effective cleaning. Chlorine is the most widely used among the sanitizing agents in wash water available for fresh produce. However, the most that can be expected at permitted concentrations is a 1- to 2-log population reduction and the reaction of chlorine with organic residues can form potentially mutagenic or carcinogenic reaction products (Sapers, 2003).

Tissue shearing involves trimming, pitting, peeling, and coring. Such shearing actions cause decompartmentalization of cellular components and the bruising of tissue near the shear surfaces. Both decompartmentalization and bruising of tissue lead to oxidative reactions such as the enzymatic browning reaction with the consequence of product darkening and off-flavor development. The presence of cut surfaces with a consequent release of nutrients, the absence of treatments enables to ensure the microbial stability, the active metabolism of fruit tissue, and the confinement of final product enhance the growth extent of the naturally occurring microbial population in minimally processed fruits. The low-acid fruits such as cut melon and tropical fruits can also favor the proliferation of pathogenic species such as *Salmonella* spp. and enteropathogenic *Escherichia coli* up to the infective threshold.

Another problem that necessitates further physiological and biochemical studies is that cut fruit products can lose rapidly their typical flavor and develop loss of freshness even in refrigerated storage, due to changes in some volatile aroma compounds (Lamikanra and Richard, 2002).

Cooling and temperature conditioning are used to decrease the respiratory activities and control the propagation of microorganisms. Cooling methods for removing the thermal energy include forced-air cooling, hydro-air-cooling, hydro-cooling, ice contact cooling, and vacuum cooling, selected according to fruit characteristics.

Tropical and subtropical fruits and some temperate fruits are susceptible to a low-temperature tissue disorder called chilling injury. Generally, the lowest injury-safe temperatures for fruits and vegetables are in the range of 5–13°C.

Spoilage control is one of the most important factors of MAP fruits and fruit products. For fruits with pH values below 4.5 yeast, lactic acid bacteria, and fungi are the major contributors to spoilage. Chlorine in wash water is effective at a level of 10–200 ppm to inactivate molds, yeasts, and bacteria. For effective bacterial control, the chlorinated water must have a pH of 6 or lower. Sometimes hot water is used before peeling to reduce the microbial load on the surface. The temperature used is between 45 and 62°C, but some fruits such as berries are sensitive to hot water.

Food additives diffused or infused into fruit tissues are used in reducing the pH, modifying the textural attributes, inhibiting microbial growth and preventing discoloration. Some fruits may have pH values above pH 4.6 and under MAP conditions, spoilage and possibly human pathogen growth may occur. Generally, as the pH of a cut produce decreases, there is less growth of spoilage and human pathogens. Lemon juice, citric acid, or ascorbic acid can be added to cut fruits for pH adjustment prior to MA packaging. Calcium in plant tissues is involved in the delaying of senescence, reducing respiration, decreasing ethylene production, increasing tissue firmness, and preventing enzymatic browning. The increase in tissue firmness with an elevation of tissue calcium is caused by the interaction of the calcium ions with pectin polysaccharides in both the middle lamellae and parenchyma cell walls. Control of browning is important to the feasibility of fruit processing. Until recently, the most commonly used agents to prevent enzymatic and nonenzymatic browning were sulfites which were multifunctional; besides inhibition of browning they could control the growth of microorganisms, acted as antioxidants and carried out other technical functions. However, they were corrosive, destructive to some nutrients, and could produce softening and off-flavors. The use of sulfiting agents to prevent the browning of fresh-cut fruits and vegetables was banned by the U.S. Food and Drug Administration in 1986. The review of Laurila et al. (1998) shows that the inhibition of browning in minimally processed fruits and the search for practical and functional alternatives to sulfite agents have received a great deal of attention but with limited success. Apple slices treated with combinations of some reducing agents, enzymatic inhibitors, and calcium-containing compounds derived from natural products resulted in the extension of storage life and better maintenance of quality (Buta et al., 1999).

One of the important stages of pretreatment process is draining. After the peeling, slicing, and disinfection washing, the high water content on the surface serves as a good environment for the propagation of microorganisms. Thus, it is necessary to reduce water on the surface. The most effective draining is centrifuging and removing about 2–3% water content from the surface, depending on the type and texture of the fruit.

The final stage of processing is packaging. The pretreated product is placed into the pack and sealed. Sometimes, vacuum is used to reduce the oxygen content of the fruit tissue. A mixed gas with increased CO_2 and reduced O_2 is introduced into the pack to provide a good environment to reduce respiration and to prolong the shelf life. It usually takes place in the packaging room, the most critical zone in the processing chain. This room has to be appropriate for the temperature of the product and the hygienic requirement of the processing.

Handling and Distribution

The quality of modified atmosphere packaged and minimally processed fruits or fruits products in the distribution chain is affected by many factors. The quality maintenance is aided by the following procedures in the distribution chains (Faith, 1994):

- minimizing handling frequency
- providing continued control of temperature, RH, CA conditions during storage and transport
- transferring product from truck to refrigerated storage immediately
- rotating products on a first-in/first-out basis
- stacking individual cases no more than five cases high.

The most important factor is the temperature. High temperature or its fluctuation increases the human risk of the product and decreases the quality and shelf life.

Temperatures to be used during processing and distribution and their relationship to the sensitivity of products are described in Table 7.2.

MICROBIOLOGICAL SAFETY OF MINIMALLY PROCESSED FOODS

Although the presence of pathogenic bacteria on food products have historically been associated mainly with food of animal origin, recently vegetable products have come under increased interest as sources of food-borne illnesses (Beuchat, 1996).

Minimal processing, such as MAP, can retard microorganisms causing quick spoilage. However, the extended shelf life of chilled foods may present potential microbial hazards due to the possible growth of psychrotrophic pathogenic bacteria, without spoilage symptoms, particularly because there is always a risk of temperature abuse (Beuchat, 1996). At seriously abusive temperatures, even some mesophilic pathogens can proliferate (Nguyen and Carlin, 1994). Studies show that CO_2 has little impact on some food-borne pathogens, such as *Clostridium botulinum, Yersinia enterocolitica, Salmonella typhimurium, E. coli, Staphylococcus aureus, L. monocytogenes,* and *Campylobacter* (Hintlian and Hotchkiss, 1986).

Fruits and Fruit Products

Prepackaged fresh-cut fruits that have been peeled or cut in a raw state, ready-to-eat, and unpasteurized fresh juices are one category of minimally processed foods. Among problems that limit the shelf life and commercial development of fresh-cut fruit are browning, softening, flaccidity (loss of water), microbial decay, and safety (King and Bolin, 1989; Rajkowski and Baldwin, 2003).

The pH of many fruits is lower than 4.0. This low pH, combined with the presence of organic acids, generally prevents the growth of pathogenic bacteria. Therefore, incidents of outbreaks and cases of fruit-borne microbiological diseases are low as compared to those of other foods (Beuchat, 1996). However, changes are taking place in agricultural practices: greater use of animal waste and municipal waste on land; an increasing use of fruits grown in and transported from all parts of the world. All these changes including development of novel types

Table 7.2. Suggested Parameters for MAP of Fruits Regarding Their Temperature Sensitivity (Gorris, 2000)

Type of Product	Temperature (°C)	RH (%)	O_2 (%)	CO_2 (%)
Fruit (cool ripened)				
Apple	1–4	90–95	1–3	0–6
Pear	0–1	90–95	2–3	0–2
Fruit (cool ripened)				
Apricot	0–1	90	2–5	0–2
Blackberry	0–2	90		
Cherry	0–2	85–90		
Currant red	0–1	90		5–10
Grape	1–2	90		
Kiwi	0–2	85–90		3–5
Peach	0–2	85–90	1–2	5
Plum	0–1	85–90	2	2–5
Raspberry	0–1	85–90		
Strawberry	0–1	90		
Fruit (warm ripened)				
Avocado	12–13	90	2–3	4–7
Banana	12–14	85–95	2–3	8
Fruit (warm ripened)				
Mango	8–12	90		2–5
Pineapple	11–13	85–90	2–4	5–10
Fruit (cool-ripened citrus product)				
Grapefruits	10–16	90	5–10	0–1
Lemon	3–5	85–90	5–10	0–1
Mandarin	1–4	85–90	5–10	0–1
Orange	1–6	85–90	5–10	0–1

of product may give rise to new problems (Lund and Snowdon, 2000; Beuchat, 2002; Rajkowski and Baldwin, 2003). Conditions during the growth of fruits and their handling influence contamination and affect the microbiological safety of fruits and fruit products. Major sources of microbiological contamination are animal feces, biosolids, or contaminated water used (Beuchat, 1996; Ait and Hassani, 1999). Fruits should not be collected from orchard grounds used for grazing. Following production, the processes of harvesting, washing, cutting, slicing, packaging, and shipping can create additional conditions where contamination can occur.

The formation of a mycotoxin, patulin, by the mold *Penicillium expansum* in rots occurring in apples is a major problem in the production of apple juice (Stratford et al., 2000).

If acidic fruits and fruit products are contaminated with certain pathogenic bacteria, then they may survive for a sufficient time to cause disease (Miller and Kasper, 1994; Parish, 1997). Unpasteurized juices from apple and citrus have been commercially available for many years and have a record of safety problems (Rajkowski and Baldwin, 2003). Contamination of apples with manure from grazing cattle was probably the cause of outbreaks of infection with *Salmonella, Escherichia coli* O157, and *Cryptosporidium parvum,* a parasitic protozoon, associated with unpasteurized apple juice and apple cider (Besser et al., 1993; McLellan and Splittstoesser, 1996; Centers for Disease Control and Prevention, 1997). Similarly, oranges that have fallen to the ground and contaminated with *Salmonella* and have been processed without adequate washing were suspected as the cause of infection associated with unpasteurized orange juice. Consumers, particularly the young, elderly, or immunocompromised for any reason, are now warned by government advisories that drinking unpasteurized fruit drinks can make them ill (Tauxe et al., 1997).

Escherichia coli O157:H7 appears to have acquired a *Shigella*-like toxin gene and is capable of causing illness from a very low infective dose. Infections can cause death from hemorrhagic colitis and hemolytic uremic syndrome. The bacterium has been associated with animal feces from many sources, including cattle. *Escherichia coli* O157:H7 is acid tolerant, and its viability may remain for weeks in apple cider (Zhao et al., 1993). The bacterium is heat sensitive; its D value in apple juice proved to be 18 min at 52°C (Splittstoesser et al., 1995).

Some less acidic fruit products may even permit the growth of certain pathogens and may be thereby a potential source of microbiological disease of the consumer (Zhuang et al., 2003). The pH of cantaloupe and honeydew melons is between 6.2 and 6.7 and that of watermelons and papaya is between 5.8 and 6.0 and between pH 4.5 and 6.0, respectively (Splittstoesser, 1996). Because melons are grown on the ground, it is difficult to prevent contamination with microorganisms. If melons with a contaminated rind are cut, then the edible part may be contaminated by the cutting knife and maintenance of the cut melon at ambient temperature can also result in the growth of pathogenic bacteria. *Salmonella* spp. and *Shigella* spp. have been shown to multiply on the cut surface of these produce at 23°C (Escartin et al., 1989; Golden et al., 1993). *Escherichia coli* O157:H7 multiplied on cubes of cantaloupe and watermelon at 25°C and high humidity (Del Rosario and Beuchat, 1995). Even *Campylobacter jejuni* was reported to survive on sliced watermelon and papaya for sufficient time to present a risk to the consumer (Castillo and Escartin, 1994). When fresh-cut melons were inoculated with *C. botulinum*, then treated with UV light to inactivate vegetative microorganisms, and packaged using passive MAP, storage resulted in marginal spoilage and botulinal toxin formation (Larson and Johnson, 1999).

The growth of *C. botulinum* was also demonstrated in fresh tomatoes with metabolic association of fungi *Fusarium, Alternaria*, and *Rhizoctonia* (Draughon et al., 1988). Transfer of *Salmonella Montevideo* into the inner tissue of tomato by cutting has also been demonstrated (Lin and Wei, 1997). *Salmonella* spp. have been reported to multiply at 22–25°C on the cut surface of tomatoes with a pH between 3.99 and 4.37 (Asplund and Nurmi, 1991; Wei et al., 1995). *L. monocytogenes* was able to maintain its original population density numbers in chopped tomatoes for up to 2 weeks of storage at 10 or 21°C (Beuchat and Brackett, 1991). All these observations also reflect the importance of cold temperature chain management.

The possible use of certain plant volatiles to prevent microbial growth, thereby improving the shelf life and safety of minimally processed fruits, have been widely investigated recently in response to consumer pressure to eliminate chemically synthesized additives. Literature data (Lanciotti et al., 2004) indicate that aroma compounds and their combination with other hurdles such as CO_2, mild heat treatment, pH reduction, etc., can represent useful tools to increase shelf life and safety of specific minimally processed fruits. The *in vitro* efficacy of a number of plant volatiles has been demonstrated, e.g., Utama et al. (2002); however, their application potential may also be limited by their eventual phytotoxicity and human toxicity.

Using certain lactic acid bacteria to improve the microbial safety of minimally processed refrigerated fruits and vegetables, e.g., fruit-based salads through competitive inhibition of pathogenic bacteria, also in combination with other hurdles, seems to be a promising research field (Breidt and Fleming, 1997).

Due to the above-mentioned problems, prevention of microbial contamination of fresh fruits is important to control microbiological safety of them or of their products. To minimize these hazards, growers, packers, and shippers should use good agricultural and management practice (FDA, 1998). In 1998, the FDA issued guides, which describe good agricultural and manufacturing practices for fresh fruits and vegetables covering water quality, manure management, worker training, field and facility sanitation, and transportation (FDA, 2001). Kvenberg et al. (2000) developed a generic hazard analysis critical control points (HACCP) plan mandated for the production of fruit and vegetable juices.

Antimicrobial Agents in Wash Water

Washing with water can reduce the number of microorganisms on the produce. However, when antimicrobial agents such as chlorine are used in wash water in usual concentrations (e.g., 100 mg free chlorine per liter at pH 6.5–7.0 for 20 min), the reduction may only be of the order of 10- to 100-fold (Beuchat, 1998; Beuchat et al., 1998; FDA, 1998). The antibacterial activity of chlorine solutions is mainly due to hypochlorous acid and it is influenced strongly by the pH of the solution. A pH of 6.5–7.0 is suitable. Below a pH of 6.0, the hypochlorite solutions become too unstable for use. In addition to chlorine, treatments with chlorine dioxide, trisodium phosphate, organic acids, hydrogen peroxide, and ozone have also been investigated as alternative antimicrobial chemicals; however, none of them appears to be more effective than chlorine for decontamination of fruits (Lund and Snowdon, 2000; Sapers, 2003). This role of chlorine is mainly to prevent the spread of bacteria in the wash water rather than to kill them on the surface of the fruit. Golden et al. (1993) showed that chlorinated water reduced but did not eliminate *Salmonella* contamination once it was on the rind of melon. They concluded that chlorine was only a risk reduction factor, and other preventive measures were needed to further reduce the risk of *Salmonella* on melon rind. Novel means of applying sanitizing agents such as vacuum infiltration, vapor-phase disinfection, or surface pasteurization with hot water washing show promise for more powerful antimicrobial effects; however, they are not yet fully developed (Sapers, 2003).

Considering the relatively poor efficacy of chemical sanitizers, it is a challenge for further research to understand the competitive inhibition of pathogens by naturally occurring microorganisms on produce in fresh-cut packages during storage.

NEW METHODS OF MINIMAL PROCESSING WITH MICROBICIDAL EFFECT

Recently, a new type of mild heat processing, i.e., cooking food in vacuum packaging (so-called "sous-vide" treatment) is increasingly used, especially for the catering sector. However, this technology is basically for other types of commodities, not for fruit products consumed raw, because, except juices, they cannot be heat pasteurized without the loss of quality preferred by consumers. Several nonthermal physical treatments, however, such as ionizing irradiation, high hydrostatic pressure (HHP), or PEF (for liquid products) are emerging and show utility to improve the microbiological safety of minimally processed foods including fruit products (Parish, 1997; Tewari, 2003). For optimizing new technologies, an item-by-item approach is required to design processing conditions of the food materials, and it is essential to know the tolerance level of different microorganisms in specific situations. A comprehensive review on the kinetics of microbial inactivation for alternative food processing technologies has been made available by the U.S. FDA Center for Food Safety and Applied Nutrition (USFDA/CFSAN, 2003) on the basis of a task-force work of the Institute of Food Technologists.

IRRADIATION

In addition to the microbial decontamination of food of animal origin by irradiation, a combination of ionizing radiation and other treatments has been investigated for the reduction of microbial load of fruits and fruit products. Limited studies using apple slices in the dose range of 1–5 kGy showed microbiological efficacy (Hanotel et al., 1990). However, the limiting factors in irradiation of horticultural products are sensorial changes, particularly softening of fruit and vegetable tissues (Yu et al., 1996). Therefore, doses higher than 1–2 kGy are not feasible for these practical reasons. Low-dose irradiation may be implemented as a terminal, postpackaging treatment for fruit juices, fruit salads, etc. Irradiation of fresh and minimally processed fruits, vegetables, and juices has been discussed recently in detail by Niemira (2003). The FAO/IAEA Joint Division runs presently a coordinated research program on the use of irradiation to ensure hygienic quality of fresh, precut fruits and vegetables, and other minimally processed food of plant origin (FAO/IAEA, 2001).

The present regulatory limit in the United States for fresh fruits and vegetables is 1.0 kGy. Furthermore, irradiation is limited to specific uses, basically for disinfestation and delay of overripening. In 1999, a coalition of U.S. food processors petitioned to the U.S. Food and Drug Administration to amend the regulation to allow doses up to 4.5 kGy for a wide variety of refrigerated foods including juices and up to 10 kGy for frozen foods, including juices too. Elimination of human pathogens is the primary goal of these requested dose limits; the potential for extension of shelf life is regarded as a secondary goal (NFPA, 2000).

In recent studies in Hungary, the low-dose irradiation of 1 kGy reduced the viable cell number of *L. monocytogenes* 4ab No.10 strain by two log cycles and *E. coli* O157:H7 ATCC 4388 strain by more than 5 log cycles, when they were inoculated on sliced tomatoes and precut cantaloupe and water melon samples. (Mohácsi-Farkas et al., 2001).

Regarding fruit juices, the majority of research efforts in the past were related to irradiation targeted at spoilage organisms such as yeasts and molds rather than bacterial pathogens (Monk et al., 1994). Recent studies, however, provide information on matters of microbiological safety. A radiation dose of 1.0 kGy effectively eliminated three isolates of *Escherichia coli* O157:H7 tested in inoculated commercial apple juices showing D_{10} values of 0.12, 0.16, and

0.21 kGy, respectively (Buchanan et al., 1998). (D_{10} value is the required radiation dose to reduce the number of viable microorganisms by a factor of 10 or one log cycle.) The same authors showed that the radiation sensitivity of the three strains was reduced from 54 to 67% by previous growth of their cultures in acid environments. A D_{10} value of 0.35 kGy has been reported for *Salmonella enteritidis* irradiated in reconstituted orange juice (Niemira, 2001). Four *Salmonella* serotypes irradiated in orange juice also varied in their radiation resistance with D_{10} values of 0.71 kGy (*S. anatum*), 0.48 kGy (*S. newport*), 0.38 kGy (*S. stanley*), and 0.35 kGy (*S. infantis*) (Niemira et al., 2001), providing a 5-log reduction even for the most resistant isolate (*S. anatum*) at 3.5 kGy dose level.

Irradiation is known to oxidize a portion of total ascorbic acid (vitamin C) to its dehydro form (Romani et al., 1963). However, both these forms of the vitamins are biologically active, suggesting minimal nutritional impact (Thayer, 1994).

Sensorial properties of irradiated juices have been the subject of several studies. Dose limits for detectable flavor degradation and browning may vary greatly, as a function of the differences in composition, variety, and maturity of the source fruit (Niemira, 2003).

High Hydrostatic Pressure (HHP) Treatment

HHP processing of foods has been explored extensively during the past two decades and several HHP-processed commercial fruit products are already available in countries such as Japan, the United States, and some Western-European countries (Tewari, 2003). Actually, it is easier to adapt this process to acid foods than to low-acid products.

HHP in the range of 200–900 MPa for several minutes inactivates the vegetative cells of microorganisms, compared to heat pasteurization (Patterson et al., 1995), without damaging the low molecular weight food components, well maintaining thereby the nutritional and sensory characteristics of high-moisture foods in flexible packaging. The isostatic principle of HHP treatment implies that the transmittance of pressure is uniform and instantaneous (independent of size and geometry of food). The extent of microbial inactivation of HHP is not only species dependent but also influenced by the physicochemical environment such as water activity and pH. Inactivation of bacterial spores requires combined processes: pressure with elevated temperatures or other inimical agents. Besides shelf-life extension, already widely studied, more systematic efforts are needed to develop databases to predict extents of inactivation of specific pathogenic microorganisms under specific conditions to ensure the reliability of HHP as an alternative to traditional preservation processes for contaminants of target foods. For example, isostatic HHP reduced the counts of *Escherichia coli* O157: H7 and various serovars of *Salmonella* in fruit juices. The pathogens were found to be most sensitive in grapefruit juice and least sensitive in apple juice (Rajkowski and Baldwin, 2003).

High-Voltage Pulsed Electric Field (PEF)

This nonthermal technology, which is aimed for liquid food pasteurization, was recently comprehensively reviewed, e.g., by Ho and Mittal (2000). The PEF system consists of a set of electrodes introduced into the fluid food in the treatment chamber, a pulse generator, a capacitor, and a switch. The pulse generator charges the capacitor. When it is discharged, the resulting high-energy field pulse creates electrical potential difference across the cell membrane of the suspended cells. When the electrical potential exceeds a certain critical value by a large amount, the change in the cell membrane becomes irreversible, leading to cell death. The critical electrical potential for vegetative bacterial cells is approximately 15 kV/cm (Mertens and Knorr, 1992; Knorr et al., 1994).

The microbicidal effects in aqueous systems depend not only on the voltage and electric field strength, pulse period, and number of pulses applied but also on the fluid properties (electrical conductivity, density and rheological properties). The process parameters used for batch processing are of a wide range: 2.5–43 kV DC voltage, 0.6–100 kV/cm electric field strength, 1 μs–10 ms pulse width, 0.2–50 Hz pulse frequency, and 1–120 pulse number applied (Tewari, 2003). Studies on real food products such as fresh fruit juices are still limited and the majority of researchers have been using small-size, batch mode equipments. Even fewer studies report yet the experimental work on continuous-flow and pilot-scale systems.

CONCLUSIONS

Implementation of good agricultural, manufacturing, and distribution practices is needed to prevent contamination of minimally processed fruits and fruit products. New washing and decontamination technologies are emerging and will eventually become part of HACCP programs. Strict temperature control of minimally processed commodities is of eminent importance. However, if sufficient temperature control during the chill distribution chain cannot be achieved, additional hurdles are necessary to ensure product safety. Safety criteria and recommendations for their harmonization have been summarized by an EU FAIR Concerted Action (Martens, 1999). In addition, inactivation of several enzymes is necessary to ensure stability of fruit juices by new technologies alternative to thermal technologies while minimally affecting the quality and organoleptic characteristics (Barrett and Anthon, 2003). Selection of optimal process conditions requires an item-by-item approach. Regarding regulatory issues associated with novel processing technologies, the safety assessment of process-specific effects needs to be carried out on a case-by-case basis.

REFERENCES

Ait Melloul, A. and Hassani, L. 1999. *Salmonella* infection in children from wastewater-spreading zone of Marrakesh city (Morocco). J Appl Microbiol 87:536–539.

Asplund, K. and Nurmi, E. 1991. The growth of salmonellae in tomatoes. Int J Food Microbiol 13:177–182.

Barrett, D.M. and Anthon, G. 2003. Inactivation of Apple Juice using Thermal and Non-thermal Processing Methods. Proceedings. Workshop on Nonthermal Food Preservation, 7–10 September, Wageningen, The Netherlands.

Besser, R.E., Lett, S.M., Weber, J.T., Doyle, M.P., Barret, T.J., Wells, J.G. and Griffin, P.M. 1993. An outbreak of diarrhea and hemolytic uremic syndrome from *Escherichia coli* O157:H7 in fresh-pressed apple cider. J Am Med Assoc 269:2217–2220.

Beuchat, L.R. 1996. Pathogenic microorganisms associated with fresh produce. J Food Sci 59(2):204–216.

Beuchat, L.R. 1998. Surface Decontamination of Fruits and Vegetables Eaten Raw. A Review. WHO/FSF/FOS/98.2. Food Safety Unit, World Health Organization, Geneva.

Beuchat, L.R. 2002. Ecological factors influencing survival and growth of human pathogens on raw fruits and vegetables. Microb Infect 4: 413–423.

Beuchat, L.R. and Brackett, R.E. 1991. Growth of *Listeria monocytogenes* on tomatoes as influenced by shredding, chlorine treatment, modified atmosphere packaging, temperature and time. Appl Environ Microbiol 57:1367–1371.

Beuchat, L.R., Nail, B.V., Adler, B.B. and Clavero, M.R.S. 1998. Efficacy of spray application of chlorinated water in killing pathogenic bacteria on raw apples, tomatoes and lettuce. J Food Sci 61:1305–1311.

Breidt, F. and Fleming, H.P. 1997. Using lactic acid bacteria to improve the safety of minimally processed fruits and vegetables. Food Technol 51:44–51.

Buchanan, R.I., Edelson, S.G., Snipes, K., Boyd, G., et al. 1998. Inactivation of *Escherichia coli* O157:H7 in apple juice by irradiation. Appl Environ Microbiol 64(1):4533–4535.

Burg, S.P. and Burg, E.A. 1967. Molecular requirements for the biological activity of ethylene. Plant Physiol 42:144–141.

Buta, J.G., Moline, H.E., Spaulding, D.W. and Wang, C.Y. 1999. Extending shelf-life of fresh-cut apples using natural products and their derivatives. J Agric Food Chem 47:1–6.

Castillo, A. and Escartin, E.F. 1994. Survival of *Campylobacter jejuni* on sliced watermelon and papaya. J Food Prot 57:166–168.

Centers for Disease Control and Prevention. 1997. Outbreaks of *Escherichia coli* O157:H7 infection and cryptosporidiosis associated with drinking unpasteurized apple cider—Connecticut and New York, October 1996. Morb Mortal Wkly Rep 46(01):4–8.

Dégen, Gy. 2002. Gyorsfagyasztott termékek csomagolása. In Hütöipari Kézikönyv Technológiák 2, 2nd edition. Ed. Beke,Gy. pp. 373–416. Mezögazda Kiadó, Budapest, Hungary.

Del Rosario, A. and Beuchat, L.R. 1995. Survival and growth of enterohemorrhagic *Escherichia coli* O157:H7 in cantaloupe and watermelon. J Food Prot 58:105–107.

Dixon, N.M. and Kell, D.B. 1989. The inhibition by CO_2 of the growth and metabolism of micro-organisms. J Appl Bacteriol 67:109–136.

Draughon, F.A., Chen, S. and Mundt, J.O. 1988. Metabolic association of *Fusarium, Alternaria,* and *Rhizoctonia* with *Clostridium botulinum* in fresh tomatoes. J Food Sci 53:120–123.

Escartin, E.F., Ayala, A.C. and Lozano, J.S. 1989. Survival and growth of *Salmonella* and *Shigella* on sliced fruits. J Food Prot 52:471–472.

Faith, Y. 1994. Initial preparation, handling, and distribution of minimally processed refrigeration of minimally processed refrigerated fruits and vegetables. In Minimally Processed Refrigerated Fruits and Vegetables, Ed. Willey, R.C. 1994. pp. 15–66. Chapman and Hall, New York, London.

Fellows, P. 2000. Food Processing Technology: Principles and Practice. Referred by: Ohlsson, T., Bengtsson, N. 2002. Minimal processing technologies in the food industry, Woodhead Publishing Limited Cambridge, England.

Fidler, J.C., Wilkinson, B.G., Edney, K.L. and Sjharples, R.O. 1973. The Biology of Apple and Pear Storage. Research Review. No. 3. Commonwealth Bureau of Horticulture and Plant crops. East Malling, Maidstone, Kent, U.K. Referred by Solomos, T. 1994. Some biological and physical principles underlying modified atmosphere packaging. In Minimally Processed Refrigerated Fruits and Vegetables, Ed. Willey, R.C. 1994. pp. 183–225. Chapman and Hall, New York, London.

Food and Agriculture Organization and International Atomic Energy Agency (FAO/IAEA). 2001. The Co-ordinated Research Programme on Use of Irradiation to Ensure Hygienic Quality of Fresh, Pre-cut Fruits and Vegetables and Other Minimally Processed Food of Plant Origin. Report of First FAO/IAEA Research Co-ordination Meeting, 5–9 November, Rio de Janeiro, Brasil. D6-RC-844; 319-D6-10-22. Vienna, International Atomic Energy Agency.

Food and Drug Administration (FDA). 1986. Sulfiting agents: revocation of GRAS status for use on fruits and vegetables intended to be served or sold raw to consumers. Fed Reg 51:25021–25026.

Food and Drug Administration (FDA). 2001. Guidance for Industry-Guide to Minimize Microbial Food Safety Hazards for Fresh Fruits and Vegetables. http://www.cfsan.fda.gov/~dms/prodsur9.html.

Food and Drug Administration and Center for Food Safety and Applied Nutrition (FDA/CFSAN). 1998. Guide to Minimize Microbial Food Safety Hazards for Fresh Fruits and Vegetables, 33 p. Food and Drug Administration, Washington, D.C.

Geeson, J.D., Browne, K.M., Maddison, K., Shepperd, J. and Guaraldi, F. 1985. Modified atmosphere packaging to extend the shelf life of tomatoes. J Food Technol 20:339–349.

Golden, D.A., Rhodehamel, E.J. and Kautter, D.A. 1993. Growth of *Salmonella* spp. in cantaloupe, watermelon and honeydew melons. J Food Prot 56:194–196.

Gorris, L.G.M. 2000. The principle of modified atmosphere packaging of prepared produce. In: Kíméletes feldolgozás az élelmiszeriparban, - Hungarian Scientific Society for Food Industry (MÉTE), Budapest, Hungary, pp. 1–32.

Hanotel, L., Libert, M.F. and Boisseau, P. 1990. Effet de l'ionisation sur le brunnisement enzymatique des végetaux frais prédécoupés. Cas des pomme Golden Delicious, p. 269. Proc. 9th Colloque Recherches Fruitiéres (Colloquium on Fruit Research). Institut National de la Recherche Agronomique, Centre Technique Interprofessionnel des Fruits et des Légumes, Paris.

Hintlian, C.B. and Hotchkiss, J.H. 1986. The safety of modified atmosphere packaging: a review. Food Technol 40(12):70–76.

Ho, S.Y. and Mittal, G.S. 2000. High voltage pulsed electric field for liquid food pasteurization. Food Rev Int 16(4):395–434.

Hong, J.H. and Gross, K.C. 2001. Maintaining quality of fresh-cut tomato slices through modified atmosphere packaging and low temperature storage. J Food Sci 66(7):960–965.

Huis in't Veld, J.H.J. 1996. Minimal Processing of Foods: Potential, Changes and Problems. A paper presented to the EFFoST Conference on the Minimal Processing of Food, Cologne, 6–9 November.

Kader, A.A., Zagory, D. and Kerbel, E.L. 1989. Modified atmosphere packaging of fruits and vegetables. Crit Rev Food Sci Technol 28:1–30.

King, A.D. and Bolin, H.R. 1989. Physiological and microbiological storage stability of minimally processed fruits and vegetables. Food Technol 43:132–135, 139.

Knorr, D., Geulen, M., Grahl, T. and Sitzmann, W. 1994. Food applications of high electric field pulse. Trends Food Sci Technol 5:71–75.

Kuai, K. and Dilley, D.R. 1992. Extraction, partial purification and characterisation of 1-aminocyclopropane-1-carboxilic acid oxidase from apple fruit. Postharvest Bio Tech 1:203–211.

Kvenberg, J., Stolfa, P., Stringfellow, D., Garret, E.S. et al. 2000. HACCP development and regulatory assessment in the United States of America. Food Control 11:387–401.

Lamikanra, O. and Richard, O. 2002. Effect of storage on some volatile aroma compounds in fresh-cut cantaloupe melon J Agric Food Chem 50:4043–4047.

Lanciotti, R., Gianotti, A., Patrignani, F., Belletti, N., Guerzoni, M.E. and Gardini, F. 2004. Use of natural aroma compounds to improve shelf-life and safety of minimally processed fruits. Trends Food Sci Technol 15:201–208.

Larson, A.E. and Johnson, E.A. 1999. Evaluation of botulinal toxin production in packaged fresh-cut cantaloupe and honeydew melons. J Food Prot 62:948–952.

Laties, G.G. 1978. The development and control of respiratory pathways in slices of plant storage organs. In Biochemistry of Wounded Plant Tissues, Ed. Kahl, G. pp. 421–466. Walter de Gruyter, Berlin.

Laurila, E., Kerviven, R. and Ahvenainen, R. 1998. The inhibition of enzymatic browning in minimally processed vegetables and fruits—Review article. Postharvest News Inf 47:53N–66N.

Lin, C.M. and Wei, C.I. 1997. Transfer of *Salmonella montevideo* onto the interior surfaces of tomatoes by cutting. J Food Prot 60:858–863.

Lund, B.M. and Snowdon, A.L. 2000. Fresh and processed fruits. In The Microbiological Safety and Quality of Food, vol. 1, Ed. Lund, B.M., Baird-Parker, T.C. and Gould, G.W. pp. 738–758. Aspen Publishers, Inc., Gaithersburg, Maryland.

Martens, T. (ed.) 1999. Harmonization of Safety Criteria for Minimally Processed Foods. Rational and Harmonization Report. FAIR Concerted Action FAIR CT96-1020. Alma University Restaurants, Leuven, Belgium.

McLellan, M.R. and Splittstoesser, P.F. 1996. Reducing the risk of *E. coli* in apple cider. Food Technol 50:174.

Mertens, B. and Knorr, D. 1992. Developments of non-thermal processes for food preservation. Food Technol 46(5):124–126, 132.

Miller, L.G. and Kasper, C.W. 1994. *Escherichia coli* O157:H7 acid tolerance and survival in apple cider. J Food Prot 57:460–464.

Mohácsi-Farkas, C., Farkas, J., Andrássy, É., Polyák-Fehér, K. and Brückner, A. 2001. Improving the Microbiological Safety of Some Fresh-Cut and Prepackaged Chilled Produce by Low-Dose Gamma Irradiation. Progress Report to the IAEA, Contract No. 1119. Faculty of Food Science, Szent István University, Budapest, Hungary.

Molin, G. 2000. Modified atmospheres. In The Microbiological Safety and Quality of Food, vol. 1, Ed. Lund, B.M., Baird-Parker, T.C. and Gould, G.W. pp. 204–234. Aspen Publishers, Inc., Gaithersburg, Maryland.

Monk, J.D., Beuchat, L.R. and Doyle, M.P. 1994. Irradiation inactivation of food-borne microorganisms. J Food Prot 58(2):197–208.

National Food Processors Association (NFPA). 2000. Petition to Amend 21CFR179 (Irradiation in the Production, Processing and Handling of Food). http:/www.nfpa-food.org/petition/petition.pdf.

Nguyen-the, C. and Carlin, F. 1994. The microbiology of minimally processed fresh fruits and vegetables. Crit Rev Food Sci Nutr 34:371–401.

Niemira, B.A. 2001. Citrus juice composition does not influence radiation sensitivity of *Salmonella enteritidis*. J Food Prot 64(6):869–872.

Niemira, B.A. 2003. Irradiation of fresh and minimally processed fruits, vegetables and juices. In Microbial Safety of Minimally Processed Foods, Ed. Novak, J.S., Sapers, G.M. and Juneja, V.K. pp. 279–299. CRC Press, Boca Raton, London, New York, Washington, D.C.

Niemira, B.A., Sommers, C.H. and Boyd, G. 2001. Irradiation inactivation of four *Salmonella* species in orange juice with varying turbidity. J Food Prot 64(5):614–617.

Ohlsson, T. and Bengtsson, N. 2002. Minimal Processing Technologies in the Food Industry. Woodhead Publishing Limited, Cambridge, England.

Parish, M.E. 1997. Public health and non-pasteurized fruit juices. Crit Rev Microbiol 23:109–119.

Patterson, M.F., Quinn, M., Simpson, R. and Gilmore, A. 1995. Sensitivity of vegetative pathogens to high hydrostatic pressure treatment in phosphate-buffer saline and foods. J Food Prot 58:524–529.

Rajkowski, K.T. and Baldwin, E.A. 2003. Concerns with minimal processing in apple, citrus, and vegetable products. In Microbial Safety of Minimally Processed Foods, Ed. Novak, J.S., Sapers, G.M. and Juneja, V.K. pp. 35–52. CRC Press, Boca Raton, London, New York, Washington, D.C.

Romani, R.J., Van Kooy, Lim, L., Bowers, B. 1963. Radiation physiology of fruit–ascorbic acid, sulfhydryl and soluble nitrogen content of irradiated citrus. Rad Bot 3:58.

Sapers, G.M. 2003. Washing and sanitizing raw materials for minimimally processed fruit and vegetable products. In Microbial Safety of Minimally Processed Foods, Ed. Novak, J.S., Sapers, G.M. and Juneja V.K. pp. 221–253. CRC Press, Boca Raton, London, New York, Washington, D.C.

Seymour, I.J., Burfoot, D., Smith, R.L. et al. 2002. Ultrasound decontamination of minimally processed fruits and vegetables. Int J Food Sci Tech 37(5):547–557.

Solomos, T. 1994. Some biological and physical principals underlying modified atmosphere packaging. In Minimally Processed Refrigerated Fruits and Vegetables, Ed. Willey, R.C. pp. 183–225. Chapman and Hall, New York, London.

Splittstoesser, D.F. 1996. Microbiology of fruit products. In Processing Fruits: Science and Technology, Biology, Principles and Applications, Ed. Somogyi, L.P., Ramaswamy, H.S. and Hui, Y.H. pp. 261–292. Technomic Publishing Co, Lancaster, Pennsylvania.

Splittstoesser, D.F., McLellan, M.R. and Churey, I.J. 1995. Heat resistance of *Escherichia coli* O157:H7 in apple juice. J Food Prot 59:226–229.

Stiles, M.E. 1991. Scientific principle of controlled/modified atmosphere packaging. In Modified Atmosphere Packaging of Food, Ed.

Ooraikul, B. and Stiles, M.E. pp. 18–25. Elis Horwood, Chichester, England.

Stratford, M., Hofman, P.D. and Cole, M.B. 2000. Fruit juices, fruit drinks, and soft drinks. In The Microbiological Safety and Quality of Food, vol. 1, Ed. Lund, B.M., Baird-Parker, T.C. and Gould, G.W. pp. 836–869. Aspen Publishers, Inc., Gaithersburg, Maryland.

Tauxe, R., Kruse, H., Hedberg, C., Potter, M., Madden, J. and Wachsmuth, K. 1997. Microbial hazards and emerging issues associated with produce: a preliminary report to the National Advisory Committee on microbiological criteria for food. J Food Prot 60(11):1400–1408.

Tewari, G. 2003. Microbiological safety during nonthermal preservation of foods. In Microbial Safety of Minimally Processed Foods, Ed. Novak, J.S., Sapers, G.M. and Juneja, V.K., eds. pp. 185–204. CRC Press, Boca Raton, London, New York, Washington, D.C.

Thayer, D.W. 1994. Wholesomeness of irradiated foods. Food Technol 48(5):132–136.

U.S. FDA Center for Food Safety and Applied Nutrition (USFDA/CFSAN). 2003. Kinetics of Microbial Inactivation for Alternative Food Processing Technologies. International Atomic Energy Agency, Vienna. http://vm.cfsan.fda.gov/~comm/ift-pref.html.

Utama, I.M.S., Willis, R.B.H., Ben-Yehoshua, S. and Kuek, C. 2002. In *vitro* efficacy of plant volatiles for inhibiting the growth of fruit and vegetable decay microorganisms. J Agric Food Chem 50:6371–6377.

Wei, C.I., Huang, T.S. and Kim, J.M. et al. 1995. Growth and survival of *Salmonella Montevideo* on tomatoes and disinfection with chlorinated water. J Food Prot 58:829–836.

Wiley, R.C. 1994. Preservation methods for minimally processed refrigerated fruits and vegetables. In Minimally Processed Refrigerated Fruits and Vegetables, Ed. Willey, R.C. 1994. pp. 66–127. Chapman and Hall, New York, London.

Yu, L., Reitmeier, C.A. and Lore, M.H. 1996. Strawberry texture and pectin content as affected by electron beam irradiation. J Food Sci 61(4):844–846.

Yuan, I.T.C. 2003. Modified atmosphere packaging for shelf life extension. In Microbial Safety of Minimally Processed Foods, Ed. Novak, I.S., Sapesrs G.M. and Uneja, U.K. pp. 205–219. CRC Press, Boca Raton, London, New York, Washington, D.C.

Zhao, T., Doyle, M.P. and Besser, R.E. 1993. Fate of enterohemorrhagic *Escherichia coli* O157:H7 in apple cider with and without preservatives. Appl Environ Microbiol 59:2526–2530.

Zhuang, H., Barth, M. and Hankinson, T.R. 2003. Microbial safety, quality and sensory aspects of fresh-cut fruits and vegetables. In Microbial Safety of Minimally Processed Foods, Ed. Novak, J.S., Sapers, G.M. and Juneja, V.K. pp. 255–278. CRC Press, Boca Raton, London, New York, Washington, D.C.

8
Fresh-Cut Fruits

Olga Martín-Belloso, Robert Soliva-Fortuny, and Gemma Oms-Oliu

INTRODUCTION

Fresh-cut fruits appeared in the market as a response of a consumer trend toward fresh-like high-quality products as well as an increase in popularity of ready-to-eat products. Therefore, fresh-cut fruit produce requires new preservation techniques capable of keeping the safety and quality of commodities long enough to make distribution feasible and achievable.

Fresh-cut products were first introduced in the market by western countries such as the United States, which has an efficient commercial distribution.

Fresh-cut fruits and vegetables are also called lightly processed, ready-to-eat, and minimally processed fruits and vegetables. These products are usually defined as those processed by appropriate unit operations such as washing, peeling, slicing, and packaging, including chemical treatments which may have a synergistic effect by using a combination of them (Wiley, 1994). The processing of fresh-cut produce includes packaging in modified atmosphere (MA) as well as storage at 2–4°C, during the whole shelf life of about 7–10 days (Fig. 8.1).

For processors, achieving microbiologically safe products with sensorial and nutritional fresh-like values is still a challenge despite the amount of research already spent on this topic (Bett et al., 2001). As a result of peeling, cutting, and preparation of

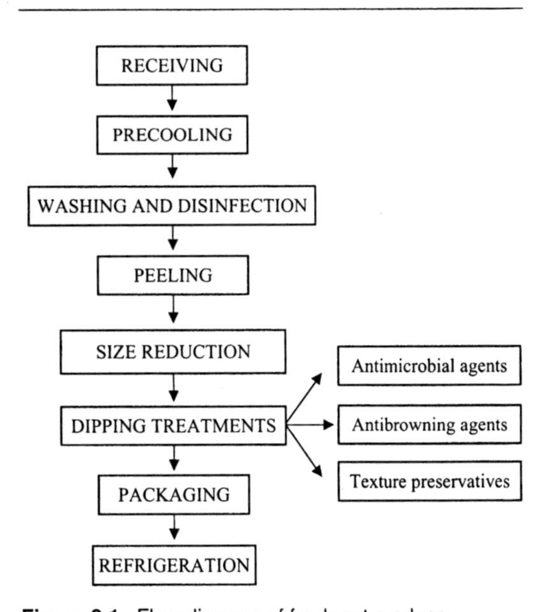

Figure 8.1. Flow diagram of fresh-cut produce.

ready-to-eat fruits, a large number of physiological phenomena such as biochemical changes and microbial spoilage takes place and may result in degradation of color, texture, and flavor. As the acceptance of consumers for fresh-cut fruits is mainly based on appearance as well as flavor and texture, these attributes will determine the commercial durability of fresh-cut produce.

Processing factors are not only related to deteriorating processes but also to a great number of physiological factors such as cultivar, growing and harvesting practices, postharvest treatments, and/or maturity at processing. These could determine the feasibility if fruits can be processed as a fresh-cut product (Rocha et al., 1998; Soliva-Fortuny et al., 2002a).

This chapter aims to introduce the main factors of minimal processing, which affect the handling of raw material, processing, packaging, and distribution of fruits. Thus, the key requirements for extending microbiological, sensory, and nutritional shelf life of fresh-cut fruits are analyzed in each step of the production chain.

HYGIENIC REQUIREMENTS FOR FRESH-CUT FRUIT PROCESSING FACILITIES

Because the natural barrier of skin is removed and damage is inflicted to the fruit tissue, fresh-cut fruits need to be processed under strict sanitary conditions. Unlike whole fruits, fresh-cut products can be spoiled by hazardous human pathogens (FDA, 2001). Microbiological risks must be reduced and controlled from the orchard to the consumer.

The appropriate handling of whole fruits during harvest and postharvest can significantly reduce the risk of contamination prior to processing. If fruits were contaminated with pathogenic microorganisms during these steps, sanitation with chlorine or other disinfectants would not assure the product safety. Cleaning operations of field bins, drenchers, storage chambers and other possible sources of cross contamination must be carefully controlled to ensure proper sanitation.

The risk of product contamination does not stop at this point. Adherence to good manufacturing practices and the implementation of hazard analysis critical control points (HACCP) are strongly advised in a processing plant for such commodities. Within the processing plant, hygienic conditions should be maintained. All workers, maintenance personnel, and visitors should be required to wear gloves, caps and appropriate smocks, and footwear. Hygiene training should insist on the importance of hand washing using bactericidal soap and water before entering the plant. Gloves should also be renewed several times during a working shift. General advice should be followed regarding the conditions during processing. Proper sanitation of the processing facilities must be conducted to ensure the exclusion of harmful bacteria. In addition, low temperatures are helpful to inhibit microbial growth and to minimize the respiration of the cut fruit in response to mechanical bruises (Toivonene and DeEll, 2002). Refrigeration temperatures below 8–10°C are advised. Also, cooling of dipping solutions or transport water is an effective way of controlling microbial spoilage, although some microorganisms, even pathogenic ones such as *Listeria monocytogenes*, can grow under those conditions (Zagory, 1999).

Do not use field bins. Use only plastic washable boxes or pallet boxes that are more appropriate for maintaining the sanitation. Compartmentalization of each step of the process, especially the first, may help to prevent cross contamination of the product throughout processing. Before the cutting operations, a rinse in chlorinated water may reduce the initial loads of naturally occurring microorganisms. Design of hygienic equipment for these processes is required. Most equipment available in the market has been adapted from the existing machinery used by fruit preserve producers. Mechanical peelers that remove the skin, seed, and debris, with isolation of the parts in contact with the product from other components of the machine, are under development. It is also important to assure that the cutting blades are kept clean and sharp. An alternative to blade peeling technologies could be the use of water jet cutting systems, which would substantially reduce the risk of cross contamination during this step. However, the high cost of water jet systems can limit their use in the fruit industry.

The next processing steps are generally conducted to extend the fresh-cut fruit shelf life, although little can be done to reduce much of the contamination. Dips to stabilize the cut surfaces are often blends of antioxidant and/or antisoftening agents (Rosen and Kader, 1989; Gorny et al., 1998a). Since it is not possible to replace these solutions after a single batch, because of the high cost of the products used, they represent a high risk of cross contamination through the plant process. New methods are being developed to sanitize these solutions, given that chlorine, ozone, ultraviolet (UV) light, and other possible methods of decontamination, such as thermal treatments, are not compatible with most antioxidant treatments (FDA, 2001).

QUALITY OF RAW MATERIAL

Fruits, once detached from the plant, undergo a physiological response consisting of processes such as transpiration, respiration, and ripening. Transpiration leads to loss of water and the consumption of substrates during respiration that converts the stored energy into usable energy to sustain life. Thus, the higher the transpiration and respiration, the shorter the shelf life. Such processes are mainly responsible for wilting, shrinking, and loss of firmness among other phenomena, which adversely affect the sensorial quality of produce (Amiot et al., 1995; Kader, 2002; Soliva-Fortuny et al., 2004).

The quality of fresh-cut fruits depends directly on the quality of the raw material and other factors related to processing, storage, and distribution (Gorny et al., 1998b). These factors include the condition of the raw materials such as firmness, size, variety, and ripeness at processing. These significantly affect the shelf life and quality of produce.

The ripening process is induced by the gaseous plant hormone ethylene and related factors characterized by physical, chemical, physiological, and metabolic changes. Ripening seems to have an important influence on sensorial features related to the flavor and texture of fruit. Ethylene may be responsible for the synthesis of enzymes that lead to ripening and subsequent senescence and degradation (Perera and Baldwin, 2001). In fact, these adverse effects of ethylene on quality tend to be enhanced in climacteric fruits, since ripening implies a larger evolution of ethylene and an abrupt rise in respiration rates when ripeness is about to take place. Generally, climacteric immature fruits are more likely to shrivel and undergo physiological alterations during processing. Overripe fruits may produce less ethylene but are more sensitive to mechanical damage and fungal development. Softness, mealiness, and insipid flavor are some sensorial characteristics related to excessive maturity at processing (Kader, 2002). Ethylene production observed during the first weeks of storage under passive packaging conditions can double that of ripe fruits, compared to fresh-cut apple and pear slices processed in a partially-ripe state (Fig. 8.2). Therefore, the ripeness stage of fruit at processing was clearly determinant for shelf life of fresh-cut apple slices (Soliva-Fortuny et al., 2002a).

The quality of fresh-cut produce is related to an appropriate selection of cultivars. In pears, the shelf life based on flesh firmness and cut surface discoloration can be affected by cultivar. Barlett pears showed the longest shelf life among cultivars such as Bosc,

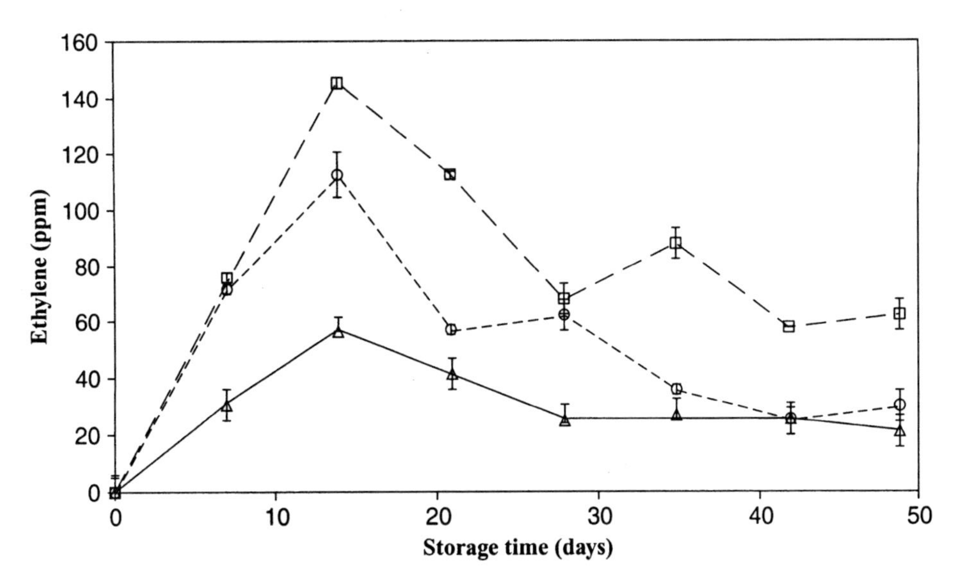

Figure 8.2. Ethylene concentrations in the package headspace of fresh-cut pears processed at different maturity states: circles, mature-green pears; squares, partially-ripe pears; and triangles, ripe pears.

Anjou, and Red Anjou pears (Gorny et al., 2000). Additionally, the most important factors that determine the shelf life of fresh-cut peach and nectarine slices (Gorny et al., 1999) are as follows:

- selection of cultivars
- maturity at harvest, and
- proper storage: temperature (0°C) and relative humidity (RH) (90–95%)

PROCESSING

WASHING AND SANITIZING RAW MATERIALS FOR FRESH-CUT FRUIT PRODUCTS

Washing and sanitizing of raw fruits is required to remove pesticide residues, plant debris, and other possible contamination as well as microorganisms responsible for quality loss and decay. Fruit products undergo fermentative spoilage by lactic acid bacteria or yeasts resulting in the production of acids, alcohol, and CO_2. Fermentative species of yeasts such as *Kloeckera* and *Hanseniaspora* occur naturally on the surfaces of fruits and are capable of causing fermentative spoilage (Barnett et al., 2000)

Raw material is generally immersed in tap water, whereas sanitizing agents are added to process water to effectively reduce the microbial loads on the fruit surface. The use of chlorine at a concentration no greater than 200 ppm has been widely reported as an effective sanitation treatment of both whole and fresh-cut fruits (Lanciotti et al., 1999; Gorny et al., 2000; Dong et al., 2000, Bett et al., 2001; Soliva-Fortuny et al., 2002b). In melon and watermelon, the sanitation of the whole fruit is usually achieved by using dips ranging from 50 to 1000 ppm of sodium hypochlorite (NaOCl) (Qi et al., 1999; Portela and Cantwell, 2001). The effectiveness of NaOCl on microbicidal activity is related to the concentration of the sanitizer as well as pH and temperature. On the other hand, chlorine efficacy may be influenced by the type of produce and diversity of microorganisms that fruits contain (Beuchat, 2000).

New sanitizing agents have been introduced in the past few years because of concern about the products obtained when chlorine is decomposed by organic matter, resulting in the formation of potentially harmful substances.

Other sanitizers such as hydrogen peroxide (H_2O_2), chlorine dioxide, peroxyacetic acid, and organic acids have been used for washing produce. Hydrogen peroxide (H_2O_2) demonstrates a broad-spectrum efficacy against virus, bacteria, yeasts, and bacterial spores, although it is less active against fungi than bacteria (Block, 1991). Its bacteriocidal effect is based on the production of hydroxyl free radicals (OH), which attack essential cell components, including lipids, proteins, and DNA (McDonnell and Russell, 1999). Thus, this treatment was recommended for fresh-cut melon and analogous fruits because it extended the shelf life by 4–5 days in comparison with the chlorine treatment (Sapers and Simmons, 1998). Sapers (1996) also showed that hydrogen peroxide vapor treatments were highly effective in reducing loads of microorganisms on whole prunes and table grapes. Hydrogen peroxide solutions used alone or in combination with commercial sanitizing agents achieved more effectiveness in decontaminating apples, which contained non-pathogenic strains of *Escherichia coli*, than by using chlorine or other commercial sanitizing agents for fruits or vegetables (Sapers et al., 1999). However, exposure to H_2O_2 vapor caused bleaching of anthocyanins in strawberries and raspberries. This treatment could also be unfit for pome fruits due to the presence of residual contents in the product (Sapers and Simmons, 1998).

Several published studies have assessed the efficacy of different sanitizers against *E. coli* O157:H7 on inoculated apples. Apples washed with 80 ppm of peroxyacetic acid reduced the microbial loads by about 2 logs and a 5% acetic acid wash reduced the load by about 3 logs when compared to water wash (Wright et al., 2000). On the other hand, 80 μg/ml of chlorine dioxide, 16 times the recommended concentration, was needed to reduce the population of *E. coli* O157:H7 by 2.5 logs (Wisniewsky et al., 2000).

Ozone and UV light could be other alternatives to traditional sanitizing agents as these sanitizing processes are not only effective in destroying microorganisms but they could also improve the safety of fruits because of the lack of residues on produce. Fungal deterioration of blackberries and grapes was decreased by ozonation of the fruits (Beuchat, 1992). Recent studies supported this work, ozone exposure at 0.3 ppm inhibited the normal aerial growth of the mycelia and prevented sporulation on peach wounds inoculated with *Monilinia fructicola*, *Botrytis cinerea*, *Mucor piriformis*, and *Penicillium expansum* and stored for 4 weeks at 5°C and 90% RH. Under 0.3 ppm ozone, gray mold, caused by *B. cinerea*, spread from the decayed fruit to adjacent healthy fruit among table grapes was also completely inhibited, when fruits were stored for 7 weeks at 5°C (Palou et al., 2002). In citrus fruit, the exposure to ozone did not reduce the final incidence of postharvest green mold, caused by *Penicillium digitatum*, and postharvest blue mold, caused by *Penicillium italicum Wehmer*, although infections developed more slowly on fruits stored in an ozonated atmosphere than on fruits stored in an ambient air atmosphere (Palou et al., 2001).

UV light could also be effective as a minimal processing alternative for extending the shelf life of fresh-cut fruits. The effect of UV light (UVC, $\lambda = 254$ nm) may be based on its direct effect on pathogens because of DNA damage as well as its ability to simulate biological stress in plants and consequently, by inducing resistance mechanisms in different fruits against pathogens. Actually, the exposure of melon slices to UV light decreased the concentrations of most of the aliphatic esters by over 60% of the amounts present in fresh-cut fruit and resulted in the production of terpenoid compounds in response to biological stress, particularly β-ionone, which is capable of inhibiting the microbial growth in the fruit tissue (Lamikanra et al., 2002). UV light at a wavelength of 253.7 nm (UVC) was applied to apples inoculated with *E. coli* O157:H7, achieving a log reduction of approximately 3.3 logs at 24 mW/cm^2 (Yaun et al., 2004).

Mechanical Operations

Mechanical operations during minimal processing damages fruit tissues, which in turn limits the shelf life of products. Operations including peeling, coring, cutting, and/or slicing are responsible for such phenomena as microbial spoilage, desiccation, discoloration or browning, textural changes, and development of off-flavor or off-odor. During the preparatory steps of minimal processing, the natural protection of fruit (the peel) is generally removed and hence, they become highly susceptible to microbial spoilage. During processing, the leakage of juices and sugars from damaged tissues allow the growth and fermentation of some species of yeasts such as *Saccharomyces cerevisiae* and *Saccharomyces exiguous* (Heard, 2002).

Damage on plant tissues may make them more susceptible to attack by pathogenic microorganisms and contamination with human pathogens. Cross contamination may occur during cutting and shredding operations because sanitation in raw fruits may not have been carried out properly (Garg et al., 1990). The whole fresh fruits with bacterial soft rot and fungal rot were shown to have a high incidence of contamination with *Salmonella* spp. (Wells and Butterfield, 1997, 1999).

Although safety is the most important attribute to be taken care in food, color, texture, flavor, and nutritional value are also the primary limiting factors in determining the product's acceptability by the consumer. Therefore, the influence of cutting operations on quality should be taken into account. It is clear that turgor pressure has a great effect on the textural response, as it has been reported for minimally processed melon by Rojas et al. (2001). In bananas, less ethylene production and lowest respiration rates were observed when a 1-cm thick transverse cutting section was chosen (Abe et al., 1998). In apples or pears, the core and adjacent tissues should be removed during cutting operations because these parts are susceptible to browning (Soliva-Fortuny et al., 2001).

Enzymatic browning is regarded as one of the most important problems related to color deterioration in fresh-cut fruit produce. Such phenomenon is caused by the discoloration of fruit by the action of a group of enzymes called polyphenol oxidases (PPOs). This enzymatic reaction consists of the oxidation of phenolic substrates, found naturally in many fruits, to o-quinones, which is highly reactive and will react with (Whitaker and Lee, 1995):

- other quinone molecules
- other phenolic compounds
- the amino group of proteins, peptides, and amino acids
- aromatic amines, thiol compounds, ascorbic acid (AA), etc

Browning phenomena are caused when, after mechanical operations during processing, enzymes, which are liberated from the tissues, come in contact with phenolic compounds. However, several factors may contribute to the development of brown pigments due to enzymatic browning. The tendency toward browning may be influenced by high concentration or types of phenolic compounds in fruits as well as high PPO activity (Garcia and Barrett, 2002), ripeness stage, activity of oxidative enzymes, oxygen availability, and compartmentalization of enzymes and substrates (Nicoli et al., 1994; Rocha, 1998). According to Soliva-Fortuny (2002b), in mature apples, the chloroplast begins to disintegrate, causing a solubilization of PPOs, which would increase the oversensitivity of browning. In pears, browning is related to phenolic and PPO compositions, whose contents may vary according to cultivar, stage of maturity, and postharvest storage conditions (Amiot et al., 1992). It was found that, in pear fruits of different varieties, the susceptibility to browning and the phenolic content were not greatly different, although a significant decrease in the phenolic content occurred with delayed harvest times (Amiot et al., 1995). Reduced rates of enzymatic browning in pears may be related to low levels of PPO (Soliva-Fortuny et al., 2002b).

It has been shown that the pectinolytic and proteolytic enzymes may be responsible for softening when they are exuding from bruised cells during slicing operations. Not only do these enzymatic mechanisms play a significant role in the softening process but also affect the morphology, cell wall–middle lamella structure, cell turgor, water content, and biochemical components (Harker et al., 1997). Peeling and cutting also result in high rates of moisture loss from cut surfaces as was reported in pears by Gorny et al. (2000). Increased rates of water loss lead to wilting and/or shriveling, limiting factors of quality in fresh-cut produce (Toivonene and DeEll, 2002).

Low temperatures minimize the effects of mechanical injuries because they are able to reduce enzymatic activity, metabolic reactions, and microbial growth. Processing is performed at around 10–15°C and washing water is generally refrigerated (Ahvenainen, 1996). Rinsing the peeled and/or cut product in cold water is suggested to keep products in a suitable range of temperature or for removing cellular exudates released during mechanical operations.

Dipping Treatments

Dipping treatments after peeling and/or cutting reduce microbial loads and rinse tissue fluids, thus reducing enzymatic oxidation during storage and the growth of microorganisms.

Due to the low pH values of most fruits, the main typical flora consists of moulds and yeasts. Both fungi and yeasts are responsible for the production of a wide range of enzymes. Among these, pectic enzymes should be taken into account because of their role in the degradation process of plant polymers. *B. cinerea and Aspergillus niger* were found to be important fungi on fruits as well as yeasts such as *Canidia*, *Cryptococcus*, *Fabospora*, *Kluyveromyces*, *Pichia*, *Saccharomyces*, and *Zygosaccharomyces* (Chen, 2002). Also, the ability of lactic acid bacteria to alter the food flavor might contribute to the relatively rapid flavor loss in fresh-cut fruits. In fact, the deterioration of fresh-cut cantaloupe stored at 20°C was related to Gram-positive bacteria and an increased production of lactic acid (Lamikanra et al., 2000). During the spoilage of fruits, Gram-negative bacteria such as pseudomonads are believed to degrade the fruit tissues by the production of pectic enzymes.

Consumption of fresh-cut fruits is associated with foodborne disease due to some pathogenic bacteria such as *Cyclospora cayetanensis* in raspberries, *Salmonella* spp. in precut watermelons, and *Shigella* spp. in fruit salad, among others (Heard, 2002). In general, pathogens may often be able to grow on some fruit surfaces such as melon, watermelon, papaya, or avocado because of the high pH value of the fruits. For example, *Shigella* species can survive on sliced fruits, including watermelon and raw papaya (Escartín et al., 1989). A recent study suggests that, after contamination, *Campylobacter jejuni*, a common cause of foodborne bacterial gastroenteritis in developed countries worldwide, may continue to survive on cantaloupe pieces and strawberries (Kärenlampi and Hänninen, 2004). *Escherichia coli* O157:H7 can grow within damaged or wounded apple tissues (Dingman, 2000). The ability of *E. coli* O157:H7 to grow in the moderate pH of a bruise will likely predispose the bacterium for survival in a fresh-cut fruit. Therefore, the use of damaged fruits will increase the risk for contamination of fresh-cut products in comparison to surface contamination of whole apples.

Citric acid has been widely used as an effective preservative because it is able to reduce the pH of cut fruits such as orange (Pao and Petracek, 1997), apple (Rocha et al., 1998), peach, apricot, kiwifruit (Senesi and Pastine, 1996), avocado (Dorantes et al., 1998), and bananas (Moline et al., 1999). However, there is a consumer demand for more natural food where the use of chemical additives is reduced or eliminated. Hence, the use of antimicrobial agents from plants and plant products can represent a natural alternative to food additives. These substances, generally regarded as safe (GRAS), are able to inhibit microorganisms and determine flavor and quality because of the presence of some volatile compounds (Utama et al., 2002). Some natural constituents, such as hexanal, hexanol, 2-(E)-hexenal, and 3-(Z)-hexenol, responsible for the aroma of some vegetables and fruits, provide protective action against microbial proliferation in wounded areas (Gardini et al., 2002). The effectiveness of hexanal in improving the quality of minimally processed apples is based on its antimicrobial activity, its ability to delay color deterioration of slices, and its interconversion to volatile compounds giving an enhancement of aromatic properties. The formation of volatile compounds such as hexanol and hexyl acetate may be beneficial as they are regarded as inhibitors of the polyphenol oxidase (Valero et al., 1990). Hexanal totally inhibited mesophilic bacteria at 4°C and considerably prolonged the lag phase of psychrotrophic bacteria. Its presence also significantly inhibited, at abuse temperatures, the growth of moulds, yeasts, mesophilic, and psychrotrophic bacteria (Lanciotti et al., 1999). Hexanal, 2-(E)-hexenal, as well as hexyl acetate are also capable of inhibiting some pathogenic bacteria. In fresh apple slices, their addition at levels of 150, 150, and 20 ppm for hexanal, hexyl acetate, and 2-(E)-hexenal, respectively, may have a bactericidal effect on *L. monocytogenes*, and caused a significant extension of lag phase of *E. coli* and *S. enteritidis* inoculated at levels of 10^4–10^5 cfu/g (Lanciotti et al., 2003) In addition, the antimicrobial activities of hexanal, 2-(E)-hexenal, and hexyl acetate are positively affected by a rise in temperature, since their action is dependent on vapor pressure (Lanciotti et al., 1999). The antimicrobial action of essential oil constituents seems to be related to their solubility in the microbial membrane (Karatzas et al., 2000), their partition in the cytoplasmatic microbial membranes (Juven et al., 1994) or the perturbation of membrane permeability (Tassou et al., 2000). Fruit essential oils may either reduce the growth of *S. cerevisiae* inoculated at levels of 10^2 cfu/ml or increase the death rate of *E. coli* inoculated at levels of 10^6cfu/ml, under temperature abuse conditions. Citrus essential oils may be compatible with the organoleptic characteristics of minimally processed fruit. Thus, some research carried out by Lanciotti et al. (1999) suggested the addition of citrus, mandarin, cider, lemon, and lime essential oils to fresh sliced fruit mixtures (apple, pear, grape, peach, and kiwifruit) to inhibit the proliferation of naturally occurring microbial populations. In fact, citrus oxygenated monoterpenes have been reported as molecules with the highest antifungal activity, and citral as the most active compound against *P. digitatum* and *P. italicum* (Caccioni et al., 1995).

The addition of chemical agents is the most common way to control browning phenomenon. They can either affect the enzyme or their substrates. AA has been generally used as an antibrowning agent. This reducing agent indirectly inactivates the PPO enzyme by degrading the free radical of the histidine molecule at the active site and by reducing the cofactor Cu^{2+} to Cu^{+}, thereby causing the cuprous ion to dissociate more readily from the enzyme (Osuga and Whitaker, 1995). AA is able to prevent the browning caused by PPO reducing quinones back to phenolic compounds before they undergo further reaction to form brown-colored pigments. The antibrowning effects of AA have been widely demonstrated in several fresh-cut fruits under a wide range of conditions (Agar et al., 1999; Gorny et al., 1999; Rocha et al., 1998; Dorantes et al., 1998; Soliva-Fortuny et al., 2002b; Senesi et al., 1999; Buta et al., 1999; Soliva-Fortuny et al., 2001). Cysteine as a reducing agent is also capable of inhibiting enzymatic browning, although the amount required is often incompatible with product taste (Richard-Forget et al., 1992). However, among a wide variety of antibrowning compounds, Dorantes et al. (1998) chose cysteine as the best inhibitor of browning in minimally processed avocado slices.

Acidulants, such as citric acid, are effective in preventing the fresh-cut produce from browning due to its dual effect on PPO enzyme by chelating copper and its action as an acidulant (Sapers, 1993). Optimum PPO activity is observed at pH 6.0–6.5, while little activity is detected below pH 4.5 (Whitaker, 1994). Sodium chloride, as potassium chloride, is known to control browning when they are used at pH < 3.5 (Rouet-Mayer and Philippon, 1986). Calcium chloride may confer undesirable bitterness to the product when it is used at concentrations in excess of 1% (Perera and Baldwin, 2001). However, acidulants are not often used alone because it is difficult to achieve efficient browning inhibition. Nevertheless, the acid combination with a chemical reductant may show a major effect. According to Pizzocaro et al. (1993), above 90% inhibition of PPO in apples cubes was reported by using a mixture of 1% AA + 0.2% citric acid or 1% AA + 0.5% sodium chloride. Effects of citric acid and/or AA dips were not effective in controlling the browning of pear slices but an improvement in color was shown by adding 1% $CaCl_2$ and storage under 2.5°C for 1 week, rather than water-treated control slices (Rosen and Kader, 1989). Sapers and Miller (1992) suggested that PPO inhibition could be due to the firming action of calcium, which reduces the leakage of PPO and its substrates at the exposed cut surfaces. Gorny et al. (1998a) also reported the effectiveness of a dip for 1 min in 1.0% $CaCl_2$ + 2% AA, in reducing pear slice surface browning.

4-Hexylresorcinol (4-HR) inhibitory action is based on its interaction with PPO, which compromises the ability of the enzyme to catalyze the reaction. Luo and Barbosa-Cánovas (1996) showed a synergistic effect in browning inhibition using 0.01% 4-HR and 0.5% AA in combination. Thus, not only an improvement in the inhibition of browning in apple slices was provided but also the residual level of 4-HR in apple slices decreased in comparison with the use of 4-HR alone.

Some blends of additives have been proved to extend the storage life of fresh-cut produce. A mixture of 0.001 M 4-HR + 0.5 M isoascorbic acid + 0.05 M calcium propionate + 0.025 M homocysteine maintained the freshness of apple slices for 4 weeks at 5°C (Buta et al., 1999). Other combinations have been suggested as effective in preventing fresh-cut pears from browning. For instance, 4-HR in combination with sodium erythorbate had a significant effect on maintaining the color of fresh-cut Anjou pears (Sapers and Miller, 1998). A dip in 0.01 4-HR + 0.5% AA + 1–0% calcium lactate solutions also provided color stability during 30 days refrigerated storage (Dong et al., 2000). Recently, Gorny et al. (2002) reported minor changes in the surface color of Barlett pear slices treated with 2% AA + 1% calcium lactate + 0.5% cysteine provided that the latter chemical was added at pH 7 or otherwise the appearance of pinkish-red colored compounds may occur (Richard-Forget, 1992). In banana slices, among the antioxidants tested by Moline et al. (1999), it was concluded that citric acid + N-acetylcysteine provided the best results in browning inhibition during 1 week storage.

Residual core tissues on fresh-cut apple and pear slices are more susceptible to browning than parenchyma tissue, and the product shelf life is usually limited (Sapers and Miller, 1998). Therefore, the addition of polyphosphate to an acidic browning-inhibitor treatment such as a pH 2.9 dip containing AA + citric acid suppressed core browning for at least 3 weeks at 4°C, whereas formulations without hexametaphosphate failed within 1 week (Pilizota and Sapers, 2004).

Softening is a major factor limiting the shelf life of fresh-cut produce. It is generally regarded that pectinase enzymes such as pectin methylesterase and polygalacturonase are responsible for texture losses in plant tissues. Polygalacturonase hydrolyzes the α-1,4-glycosidic bond between the anhydrogalacturonic acid units resulting in the degradation of texture because of hydrolysis of the pectin polymers. On the other hand, pectin methylesterase hydrolyzes the methyl ester bonds of pectin to give pectic acid and methanol. Stability and integrity of plant tissues depend on the maintenance of cellular structures, which may undergo a progressive degradation due to processing and storage. Calcium salts as chloride and lactate may be added to maintain such a structural integrity. This benefit is related to the high affinity of pectin acid to form calcium bridges. Hydrolysis of pectin to pectic acid in the presence of Ca^{2+} results in texture firming due to the formation of crossbridges between Ca^{2+} and carboxyl groups of the pectic acids (Perera and Baldwin, 2001). The effect of calcium salts is focused on its ability to increase the number of calcium-binding sites. The effect of $CaCl_2$ dips at concentrations ranging from 0.1% to 1% (Bett et al., 2001; Rosen and Kader, 1989; Sapers and Miller, 1998; Soliva-Fortuny et al., 2002c, 2003) has been widely reported in texture preservation. A concentration of 2.5% of $CaCl_2$ has been regarded as optimal to preserve the texture of minimally processed melon (Luna-Guzmán et al., 1999). In kiwifruit slices, differences between 1% and 2% $CaCl_2$ treatments were not found. In addition, a retention of AA was observed in the slices (Agar et al., 1999). However, the use of calcium chloride at concentrations above 0.5% has been reported as responsible for off-flavors in cantaloupe slices, whereas this effect was not observed in samples treated with calcium lactate (Luna-Guzmán and Barret, 2000).

Drainage

Prior to packaging, the cut fruit should be submitted to a drainage step. The excess water or juice is undesirable because it may be an excellent medium for the growth of microorganisms. Moreover, some enzymatic reactions can be accelerated leading to a rapid degradation of the fruit flavor and/or appearance. A good drainage is also important between steps of the same process to avoid cross contamination throughout the processing line. When the mechanical integrity of the fruit allows it, spinning could be an alternative to drainage but at much lower speeds than those currently used for leaf vegetables (Rosen and Kader, 1989; Bett et al., 2001; Gorny et al., 2002).

Packaging

Modified atmosphere packaging (MAP) has allowed fresh-cut produce to take a leap in the market. This technology consists of the use of low oxygen (O_2) and/or high carbon dioxide (CO_2) atmospheres to slow the degradation processes that occur throughout storage. It also provides to the product a moisture barrier that keeps high values of RH in the environment, thus avoiding dehydration of the cut surfaces.

Because tissue respiration is substantially increased after processing, it can often account for the atmosphere modification that is needed to reach a desired equilibrium of gas concentrations. With this aim, plastic materials with high enough O_2 transmission are under development to prevent excessive modification of the package headspace atmosphere. Too low O_2 levels and/or excessive amounts of CO_2 in the package headspace are often detrimental to the fruit shelf life because anaerobic respiration, leading to fermentative processes, and the subsequent production of undesirable metabolites and physiological disorders is induced (Zagory and Kader, 1988; Soliva-Fortuny et al., 2002a). Sometimes, the modification is not achieved soon enough to avoid browning and other undesirable reactions of quality loss, and the packaged needs to be initially flushed with an appropriate gas mixture. The package area and the ratio of product/gas are also important to reach an adequate O_2/CO_2 balance.

However, the recommended values differ between fruits and even between cultivars and physiological state. Neither are threshold values to avoid such undesirable quality degradation processes known in detail.

Low O_2 atmospheres have been extensively used to extend the shelf life of cut fruits. Recommended concentrations for each fruit are displayed in Table 8.1. The benefits of reducing the amounts of O_2 surrounding media are well reported in literature. Depleted O_2 levels help to reduce respiration of the cut fruits and thus limiting the consumption of sugar, starch, and other energy storage products that are responsible for texture and flavor changes. Also, a large number of enzymatic processes in which O_2 is involved can be limited, especially those related to browning. Together with elevated CO_2 concentrations, appropriate amounts of O_2 may also control ethylene production by injured tissues, probably because O_2 is necessary for the conversion of 1-aminocyclopropane-1-carboxylic acid to ethylene (Yang, 1981).

Rosen and Kader (1989) reported a substantial decrease in browning of sliced pears and in texture loss of strawberry slices throughout MAP storage. Sensesi et al. (1999) reported that fresh-cut pears were preserved for 15 days under passive packaging conditions. These results are in agreement with those of Soliva-Fortuny et al. (2002a, 2004) in apples and pears, although oxygen concentrations facilitated a dramatic increase in ethylene evolution. Gorny et al. (1999) suggested the use of 0.25 kPa O_2 + 10 kPa CO_2 atmospheres to extend the shelf life of fresh-cut peach and nectarine slices to control ethylene production and respiration rates. Higher CO_2 concentrations (20 kPa) were discarded because off-flavor formation was detected in the product. Atmospheres of 2 kPa O_2 + 10 kPa CO_2 have been shown to be suitable for improving the quality of fresh-cut processed mangoes. Qi et al. (1999) tested similar gas compositions (2 kPa O_2 + 10 kPa CO_2) with honeydew melons, achieving a good visual appearance during 6 days storage, but results by Ayhan et al. (1998), who extended the shelf life of the product for 15 days with 5 kPa O_2 packaging atmospheres, may indicate that CO_2 can be even more detrimental than beneficial to the fruit product. The results achieved in this field

Table 8.1. Recommended Modified Atmosphere Concentrations for Different Fresh-Cut Fruits

Commodity	Atmosphere	Bibliographic Source
Apple	<1 kPa O_2	Gil et al. (1998)
		Soliva-Fortuny et al. (2002b)
Pear	0.5 kPa O_2	Rosen and Kader (1989)
		Gorny et al. (2000)
	2 kPa O_2	Soliva-Fortuny et al. (2004)
Peach	2 kPa O_2 + 12 kPa CO_2	Palmer-Wright and Kader (1997)
	0.25 kPa O_2 + 10 kPa CO_2	Gorny et al. (1999)
Kiwifruit	2 kPa O_2 + 5 kPa CO_2	Agar et al. (1999)
Cantaloupe melon	4 kPa O_2 + 10 kPa CO_2	Bai et al. (2001)
Honeydew melon	2 kPa O_2 + 10 kPa CO_2 5 kPa O_2	Qi et al. (1999) Ayhan et al. (1998)
Watermelon	3 kPa O_2 + 15 kPa CO_2	Cartaxo et al. (1997)
Mango	2 kPa O_2 + 10 kPa CO_2	Nithiya et al. (2001)
Persimmon	2 kPa O_2 + 12 kPa CO_2	Palmer-Wright and Kader (1997)
Strawberries	1–2 kPa O_2 + 10 kPa CO_2	Watada et al. (1996)
Citrus	Air	Palma et al. (2003)

are sometimes contradictory and rely on numerous factors but it is agreed that MAP alone is not enough to prevent fresh-cut fruit from senescence.

Atmosphere modification is also very important to control microbial spoilage of fresh-cut fruits. The proliferation of aerobic microorganisms can be substantially delayed with reduced O_2 levels. Gram-negative aerobes such as Pseudomonas are especially inhibited in front of Gram-positive, microaerophilic species such as Lactobacillus. CO_2 inhibits most aerobic microorganisms, specifically Gram-negative bacteria and moulds (Al-Ati and Hotchkiss, 2002). Anaerobic conditions inhibit the growth of aerobic spoilage microorganisms, which usually warn consumers of spoilage. On the contrary, the growth of anaerobic pathogenic microorganisms may be favored by these conditions. Maintenance of refrigeration conditions is also crucial. In fruits with high pH, such as melon, presence of *Clostridium botulinum* toxin has been reported after 9 days storage at 15°C (Larson and Johnson, 1999).

During the past several years, the use of enriched O_2 atmospheres has been suggested in many studies. This MAP technique has been shown to be particularly effective in inhibiting enzymatic browning, preventing anaerobic fermentation reactions, and inhibiting both aerobic and anaerobic microbial growth in fresh produce (Kader and Ben-Yehoshua, 2000).

Edible Coatings

Edible coatings may be a good alternative or complementary to MAP packaging. Appropriate formulations may be a way of controlling surface quality losses due to damaged tissues. They may reduce the gas exchange rates and especially the water vapor rates between the fruit product and its environment and also represent an excellent way of incorporating additives to control reactions that are detrimental to the quality (Baldwin et al., 1995). Reduction of surface water activity can be achieved by infiltration of fruit pieces with juices, sucrose syrups, or glycerol with suitable water-soluble polymers (Wong et al., 1994).

Edible coatings can be composed of one or more ingredients of proteic, lipidic, or polysaccharide nature. Alone, they are unlikely to be effective for fresh-cut fruits. Polysaccharides and proteins are normally hydrophilic and do not behave well as moisture barriers (Nisperos-Carriedo, 1994). Lipid coatings, on the contrary, have good barrier properties for water vapor but may be incompatible for fresh-cut fruits from the point of view of flavor (Hernández, 1994). Combined or emulsified, some coatings may be an alternative to improve the quality of some precut fruits. Casein-lipid emulsions and cellulose polymers have already been reported to reduce surface desiccation of peeled carrots (McHugh and Krochta, 1994; Howard and Dewi, 1994; Avena-Bustillos et al., 1994).

Browning of fresh-cut apples has been reported to be reduced when appropriate antioxidants are combined with different coatings such as cellulose (Baldwin et al., 1996), chitosan/lauric acid (Pennisi, 1992), alginic acid/casein/lipid (Wong et al., 1994), or whey protein concentrate (Sonti et al., 2003). Bilayer coatings made of polysaccharide and lipid also reduced water loss and respiration processes in fresh-cut apples (Wong et al., 1994). Chitosan, a polysaccharide coating, has shown to extend the shelf life of fresh fruits (El-Ghaouth et al., 1991; Kittur et al., 2001) and has a great potential to be used in fresh-cut products because of its natural preservative effect against fungi (Baldwin, 1999).

A great amount of materials, especially polysaccharides or emulsions, require in-depth studies. Nevertheless, it is difficult for edible coatings to become a preeminent technology for fresh-cut fruit processing, but using them as a way of improving the effect of other treatments and reducing surface dehydration is more feasible. Therefore, accurate nonempirical knowledge about the processes that undergo in coated products is still necessary to further develop applications for fresh-cut produce (Cisneros-Zevallos and Krochta, 2002).

Distribution and Commercial Storage of Fresh-Cut Fruit Commodities

The commercial shelf life of fresh-cut fruits is mostly determined by storage temperature. Because fresh-cut products continue to respire and are highly susceptible to be spoiled by microorganisms, the chill chain (temperatures ranging from 0°C to 4°C) must be kept throughout every stage to achieve optimum freshness, quality, and safety. Consumers should also be aware of the storage requirements of this produce and information about handling at home should be provided by producers and retailers. Special attention must be dedicated to the temperature of display cases. Overfilled shelves, blocked return airflow, or even the position of the product in the shelf may impact the product temperature significantly. In addition, optimal temperatures are rarely achieved. Refrigeration may account for more than 50% of the annual electric energy costs of a supermarket (RTTC, 2004). Therefore, small temperature changes are commercially significant because supermarkets operate on a narrow profit margin and increased energy costs impact their competitiveness.

There are few research works that take into account temperature oscillations throughout storage and the subsequent limitation of the product shelf life. Gorny et al. (1998b) reported that the shelf life of fresh-cut "Flavorcrest" peach and "Zee Grand" nectarine processed at optimal maturity and stored at 0°C under a continuous flow of humidified air was reduced by one half, when the storage temperature was increased to 10°C. This degradation was attributed to an increase in respiration rates and ethylene production at high temperatures.

REGULATIONS FOR FRESH-CUT PRODUCE

As pointed out above, concerns about safety issues need to be strengthened for fresh-cut produce. This is being translated into higher requirements for processors and distributors but information about storage and handling should also reach consumers. Nevertheless, the development of HACCP specific plans for the fresh-cut product industry is not yet mandatory. A deficient manipulation of these produce, especially during the primary stages, may entail a high risk of contamination with human pathogenic microorganisms and the possibility of foodborne illnesses.

Regulatory initiatives specific for fresh-cut products are still under development. However, the United States and most European countries already have regulations relative to fresh-cut produce. Most of them limit the counts of aerobic microorganisms to 10^6 cfu/g at expiration date. Pathogenic microorganisms are not allowed (*Salmonella*) or greatly restricted (*E. coli, L. monocytogenes*) in ready-to-eat meals prepared from raw vegetable products.

On September 2001, United States Food and Drug Administration (FDA), together with the Institute of Food Technologists (IFT) issued a document entitled "Analysis and Evaluation of Preventive Control Measures for the Control and Reduction/Elimination of Microbial Hazards on Fresh and Fresh-Cut Produce" (FDA, 2001). This document presents a comprehensive guide about potential microbial risks in fresh-cut produce and how to handle them throughout production, processing, and distribution.

The Codex Alimentarius Commission is currently developing a code of hygienic practice for precut fruits and vegetables to ensure consumer health protection and adequate trade practices in this field. The draft of this code, CX/FH 00/5 (FAO, 2001), was introduced by France and has been submitted to allegations prior to a definitive proposal. The Code will represent a valuable tool for the standardization of this market all over the world.

Another important issue is regarding the safety of the chemical substances involved in fresh-cut fruit processing. GRAS-listed additives (generally recognized as safe by the United States Food and Drug Administration) should be used. The list of additives and their specifications of the Codex Alimentarius Commission is also a world reference in this subject.

FINAL REMARKS

The fresh-cut fruit industry is expected to continue growing during the forthcoming years. The International Fresh Produce Association forecasts sales of $15 billion in 2005 in the United States up from the current $10–12 billion sales.

However, there are still many research goals to be achieved to allow the continued growth of the fresh-cut fruit market.

Future challenges include the development of procedures that ensure high quality and safety standards for fresh-cut fruit produce. Special attention should be dedicated to investigate new treatments to improve quality and to extend the microbiological shelf life to prevent the growth of pathogenic microorganisms. The effect of nontraditional sanitizers and new compounds from natural sources that appear to be healthier for consumers also needs to be assessed. Furthermore, it would be important to study the influence of traditional and new packaging systems on safety, specifically with a proper regard for the growth of human foodborne pathogens. Last but not least, it is necessary to develop better technology for the entire processing chain, from the growing fields to consumers. This will permit processors to achieve more stable products and to meet the highest hygienic standards.

REFERENCES

Abe, K.; Tanase, M.; Chachin, K. 1998. Studies on physiological and chemical changes of fresh-cut bananas. I. Deterioration in fresh-cut green tip bananas. Journal of the Japanese Society for Horticultural Science. 67: 123–129.

Agar, I.T.; Massantini, R.; Hess-Pierce, B.; Kader, A.A. 1999. Postharvest CO_2 and ethylene production and quality maintenance of fresh-cut kiwifruit slices. Journal of Food Science. 64: 433–440.

Ahvenainen, R. 1996. New approaches in improving the shelf-life of minimally processed fruit and vegetables. Trends in Food Science and Technology. 7: 179–187.

Al-Ati, T.; Hotchkiss, J.H. 2002. Application of packaging and modified atmosphere to fresh-cut fruits. In O. Lamikanra, ed. Fresh-Cut Fruits and Vegetables: Science, Technology, and Market, pp. 305–338. CRC Press, FL, USA.

Amiot, M.J.; Tacchini, M.; Aubert, S.; Nicolas, J. 1992. Phenolic composition and browning susceptibility of various apple cultivars at maturity. Journal of Food Science. 57(4): 958–962.

Amiot, M.J.; Tacchini, M.; Aubert, S.Y.; Leszek, W. 1995. Influence of cultivar, maturity stage, and storage conditions on phenolic composition and enzymatic browning of pear fruits. Journal of Agricultural and Food Chemistry. 43(5): 1132–1137.

Avena-Bustillos, R.J.; Cisneros-Zevallos, L.A.; Krochta, J.M.; Saltveit M.E. Jr. 1994. Application of casein-lipid edible film emulsions to reduce white blush on minimally processed carrots. Postharvest Biology and Technology. 4: 319–329.

Ayhan, Z.; Chism, G.W.; Richter, E.R. 1998. The shelf life of minimally processed fresh-cut melons. Journal of Food Quality. 21: 29–40.

Bai, J.H.; Saftner, R.A.; Watada, A.E; Lee, Y.S. 2001. Modified atmosphere maintains quality of fresh-cut cantaloupe (Cucumis melo L.). Journal of Food Science. 66: 1207–1211.

Baldwin E.A. 1999. Surface treatments and edible coatings in food preservation. In M. Shafiur Rahman, ed. Handbook of Food Preservation. Marcel Dekker, Inc., New York, NY, USA.

Baldwin, E.A.; Nisperos-Carriedo, M.O.; Baker, R.A. 1995. Edible coatings for lightly processed fruits and vegetables. Hortscience. 30: 35–38.

Baldwin, E.A.; Nisperos, M.O.; Chen, X.; Hagenmaier, R.D. 1996. Improving storage life of cut apple and potato with edible coating. Postharvest Biology and Technology. 9: 151–163.

Barnett, J.A.; Payne, R.W.; Yarrow, D. 2000. Yeasts: Characteristics and Identification, 3rd edition, pp. 395–401. Cambridge University Press, Cambridge, UK.

Bett, K.L.; Ingram, D.A.; Grimm, C.C.; Lloyd, S.W.; Spanier, A.M.; Miller, J.M.; Gross, K.C.; Baldwin, E.A.; Vinyard, B.T. 2001. Flavor of fresh-cut gala apples in barrier film packaging as affected by storage time. Journal of Food Quality. 24: 141–156.

Beuchat, L.R. 1992. Surface disinfection of raw produce. Dairy, Food and Environmental Sanitation. 12(1): 6–9.

Beuchat L.R. 2000. Use of Sanitizers in Raw Fruit and Vegetable Processing. In Minimally Processed Fruits and Vegetables. Fundamental Aspects and Applications, edited by S.M. Alzamora, M.S. Tapia amd A. López-Malo, pp. 63-78. An Aspen Publication Gaithersburg, MD. Block, S.S. 1991. Peroxygen compounds. In S.S. Block, ed. Disinfection, Sterilization, and Preservation, 4th edition, pp. 182–190. Lea & Febiger, Philadelphia.

Buta, J.G.; Moline, H.E.; Spaulding, D.W.; Wang, C.Y. 1999. Extending storage life of fresh-cut apples using natural products and their derivatives. Journal of Agricultural and Food Chemistry. 47: 1–6.

Caccioni, D.R.L.; Deans, S.G.; Ruberto, G. 1995. Inhibitory effect of citrus fruits essential oil components on *Penicillium italicum* and *Penicillium digitatum*. Petria. 5: 177–182.

Cartaxo, C.B.C.; Sargent, S.A.; Huber, D.J.; Chia, M.L. 1997. Controlled atmosphere storage supresses microbial growth on fresh-cut watermelon. Proceedings of the Florida State Horticultural Society. 110: 252–257.

Chen, J. 2002. Microbial enzymes associated with fresh-cut produce. In O. Lamikanra, ed. Fresh-Cut Fruits and Vegetables: Science, technology and Market, pp. 249–266. CRC Press, Boca Raton, FL, USA.

Cisneros-Zevallos, L.; Krochta, J.M. 2002. Internal modified atmospheres of coated fresh fruits and vegetables: Understanding relative humidity effects. Journal of Food Science. 6: 1990–1995.

Dingman, D.W. 2000. Growth of *Escherichia coli* O157:H7 in Bruised Apple (Malus domestica). Tissue as influenced by cultivar, date of harvest, and source. Applied and Environmental Microbiology. 66(3): 1077–1083.

Dong, X.; Wrolstad, R.E.; Sugar, D. 2000. Extending shelf life of fresh-cut pears. Journal of Food Science. 65: 181–186.

Dorantes, L.; Parada, L.; Ortiz, A.; Santiago, T.; Chiralt, A.; Barbosa-Cánovas, G. 1998. Effect of anti-browning compounds on the quality of minimally processed avocados. Food Science and Technology International. 4: 107–113.

El-Ghaouth, A.; Arul, J.; Ponnampalam, R.; Boulet, M. 1991. Chitosan coating effect on storability and quality of fresh strawberries. Journal of Food Science. 56(6): 1618–1620.

Escartín, E.F.; Castillo, A.A.; Lozano, J.S. 1989. Survival and growth of *Salmonella* and *Shigella* on sliced fresh melon. Journal of Food Protection. 52: 471–472, 483.

FAO. 2001. Proposed draft code of hygienic practice for pre-cut vegetable products ready for human consumption. Retrieved August 10, 2004, from http://www.fao.org/docrep/meeting/005/X4296E/x4296e0c.htm#bm12

FDA. 2001. Analysis and evaluation of preventive control measures for the control and reduction/elimination of microbial hazards on fresh and fresh-cut produce. Retrieved August 10, 2004, from http://vm.cfsan.fda.gov/~comm/ift3-toc.html

Garcia, E.; Barrett, D.M. 2002. Preservative treatments for fresh-cut fruits and vegetables. In O. Lamikanra, ed. Fresh-Cut Fruits and Vegetables: Science, Technology and Market, pp: 267–303. CRC Press, Boca Raton, FL, USA.

Gardini, F.; Lanciotti, R.; Belletti, N.; Guerzoni, M.E. 2002. Use of natural aroma compounds to control microbial growth in foods. In R. Mohan, ed. Research Advances in Food Science, vol. 3, pp. 63–78. Global Research Network, Kerala.

Garg, N.; Churey, J.J.; Splittstoesser, D.F. 1990. Effect of processing conditions on the microflora of fresh-cut vegetables. Journal of Food Protection. 53: 701–703.

Gil, M.I.; Gorny, J.R.; Kader, A.A. 1998. Responses of "Fuji" apple slices to ascorbic acid treatments and low-oxygen atmospheres. Hortscience 33: 305–309.

Gorny, J.R.; Cifuentes, R.A.; Hess-Pierce, B.; Kader, A.A. 2000. Quality changes in fresh-cut pear slices as affected by cultivar, ripeness stage, fruit size, and storage regime. International Journal of Food Science and Technology. 65: 541–544.

Gorny, J.R.; Gil, M.I.; Kader, A.A. 1998a. Postharvest physiology and quality maintenance of fresh-cut pears. Acta Horticulturae. 464: 231–236.

Gorny, J.R.; Hess-Pierce, B.; Cifuentes, R.A; Kader, A.A. 2002. Quality changes in fresh-cut pear slices as affected by controlled atmospheres and chemical preservatives. Postharvest Biology and Technology. 24: 271–278.

Gorny, J.R.; Hess-Pierce, B.; Kader, A.A. 1998b. Effects of fruit ripeness and storage temperature on the deterioration rate of fresh-cut peach and nectarine slices. Hortscience. 33: 110–113.

Gorny, J.R.; Hess-Pierce, B.; Kader, A.A. 1999. Quality changes in fresh-cut peach and nectarine slices as affected by cultivar, storage atmosphere and chemical treatments. Journal of Food Science. 64: 429–432.

Harker, F.R.; Redgwell, R.J.; Hallett, I.C.; Murray, S.H. 1997. Texture of fresh fruit. Horticultural Review. 20: 121–224.

Heard, G.M. 2002. Microbiology of fresh-cut produce. In O. Lamikanra, ed. Fresh-Cut Fruits and Vegetables. Science, Technology, and Market. CRC Press, Boca Raton, FL, USA.

Hernández, E. 1994. Edible coatings from lipids and resins. In Edible coatings and films to improve food quality, edited by J.M. Krochta, E.A. Baldwin, and M.O. Nisperos-Carriedo. pp. 279–303. Technomic Publishing Co.

Howard, L.R.; Dewi, T. 1994. Sensory, microbiological and chemical quality of mini-peeled carrots as affected by edible coating treatment. Journal of Food Science. 60: 142.

Juven, B.J.; Kanner, J.; Sched, F.; Weisslowicz, H. 1994. Factors that interact with the antibacterial of thyme essential oil and its active constituents. Journal of Applied Bacteriology. 76: 626–631.

Kader, A.A. 2002. Quality parameters of fresh-cut fruit and vegetable products. In O. Lamikanra, ed. Fresh-Cut Fruits and Vegetables: Science, Technology and Market, pp. 11–19. CRC Press, Boca Raton, FL, USA.

Kader, A.A.; Ben-Yehoshua, S. 2000. Effects of superatmospheric oxygen levels on postharvest physiology and quality of fresh fruits and vegetables. Postharvest Biology and Technology. 20: 1–13.

Karatzas, A.K.; Bennik, M.H.J.; Smid, E.J.; Kets, E.P.W. 2000. Combined action of S-carvone and mild heat treatment on *Listeria monocytogenes* Scott A. Journal of Applied Bacteriology. 89: 296–301.

Kärenlampi, R.; Hänninen, M.L. 2004. Survival of *Campylobacter jejuni* on various fresh produce. International Journal of Food Microbiology. 97: 187–195.

Kittur, F.S.; Saroja, N.; Habibunnisa; Tharanthan, R.N. 2001. Polysaccharide-based comoposite coating formulations for shelf-life extension of fresh banana and mango. European Food Research and Technology. 213: 306–311.

Lamikanra, O.; Chen, J.C.; Banks, D.; Hunter, P.A. 2000. Biochemical and microbial changes during the storage of minimally processed cantaloupe. Journal of Agricultural and Food Chemistry. 48(12): 5955–5961.

Lamikanra, O.; Richard, O.; Parker A. 2002. Ultraviolet induced stress response in fresh cut cantaloupe. Phytochemistry. 60: 27–32.

Lanciotti, R.; Belletti, N.; Patrignani, F.; Gianotti, A.; Gardini, F.; Guerzoni, M.E. 2003. Application of hexanal, 2-(E)-hexenal and hexyl acetate to improve the safety of fresh sliced apples. Journal of Agricultural and Food Chemistry. 47: 4769–4776.

Lanciotti, R.; Corbo, M.R.; Gardini, F.; Sinigaglia, M.; Guerzoni, M.E. 1999. Effect of hexanal on the shelf-life of fresh apple slices. Journal of Agricultural and Food Chemistry. 47: 4769–4775.

Larson, A.E.; Johnson, E.A. 1999. Evaluation of botulinal toxin production in packaged fresh-cut cantaloupe and honeydew melons. Journal of Food Protection. 62: 948–952.

Luna-Guzmán, I.; Barrett, D.M. 2000. Comparison of calcium chloride and calcium lactate effectiveness in maintaining shelf stability and quality of fresh-cut cantaloupes. Postharvest Biology and Technology. 19(1): 61–72.

Luna-Guzmán, I.; Cantwell, M.; Barrett, D.M. 1999. Postharvest CO_2 and ethylene production and quality maintenance of fresh-cut kiwifruit slices. Journal of Food Science. 64: 433–440.

Luo, Y.; Barbosa-Cánovas, G.V. 1996. Preservation of apple slices using ascorbic acid and 4-heylresorcinol. Food Science and Technology International. 2: 315–321.

McDonnell, G.; Russell, A. 1999. Antiseptic and disinfectant: Activity action, and resistance. Clinical Microbiology Reviews. 12:147–179.

McHugh, T.H.; Krochta, J.M. 1994. Milk-protein-based edible films and coatings. Food Technology. 48: 97.

Moline, H.E.; Buta, J.G.; Newman, I.M. 1999. Prevention of browning of banana slices using natural products and their derivatives. Journal of Food Quality. 22: 499–511.

Nicoli, M.C.; Anise, M.; Severinc, C. 1994. Combined effects in preventing enzymatic browning reactions in minimally processed fruit. Journal of Food Quality. 17: 221–229.

Nisperos-Carriedo, M.O. 1994. Edible coatings and films based on polysaccharides. In J.M. Krochta, E.A. Baldwin and M.O. Nisperos-Carriedo. Edible Coatings and Films to Improve Food Quality, pp. 305–335. Technomic Publishing Co., Lancaster, PA, USA.

Nithiya, R.; Yuen, L.; Tianxia, W.; Watada, A.E. 2001. Quality and microbial changes of fresh-cut mango cubes held in controlled atmosphere. Hortscience. 36: 1091–1095.

Osuga, D.T.; Whitaker, J.R. 1995. Mechanisms of some reducing compounds that inactivate polyphenol oxidases. In Y. Chang, C.Y. Lee and J.R. Whitaker. Enzymatic Browning and Its Prevention. Oxford University Press, Oxford, UK.

Palma, A.; D'Aquino, S.; Agabbio, M. 2003. Physiological response of minimally processed citrus fruit to modified atmosphere. Acta Horticulturae (ISHS). 604: 789–794.

Palmer-Wright, K.; Kader, A.A. 1997. Effect of controlled-atmosphere storage on the quality and carotenoid content of sliced persimmons and peaches. Postharvest Biology and Technology. 10: 89–97.

Palou, L.; Crisosto, C.H.; Smilanick, J.L.; Adaskaveg, J.E.; Zoffoli, J.P. 2002. Effects of continuous 0.3 ppm ozone exposure on decay development and physiological responses of peaches and table grapes in cold storage. Postharvest Biology and Technology. 24: 39–48.

Palou, L.; Smilanick, J.L.; Crisosto, C.H.; Mansour, M. 2001. Effects of gaseous ozone exposure on the development of green ang blue molds on cold stored citrus fruit. Plant Disease. 85: 632–638.

Pao, S.; Petracek, P.D. 1997. Shelf life extension of peeled oranges by citric acid treatment. Food Microbiology. 14: 485–491.

Pennisi, E. 1992. Sealed in edible film. Science News. 141: 12.

Perera, C.O.; Baldwin, E.A. 2001. Biochemistry of fruits and its implications on processing. In: D. Arthey and R.P. Ashurst. Fruit Processing. Nutrition, Products, and Quality Management, 2nd edition. Aspen Publishers, Inc., New York, NY, USA.

Pilizota, V.; Sapers, G.M. 2004. Novel browning inhibitor formulation for fresh-cut apples. Journal of Food Science. 69(4): 140–143.

Pizzocaro, F.; Torreggiani, D.; Gilardi, G. 1993. Inhibion of apple polyphenoloxidase (PPO) by ascorbic acid, citric acid and sodium chloride. Journal of Food Processing and Preservation. 17: 21–30.

Portela, S.I.; Cantwell, M.I. 2001. Cutting blade sharpness affects appearance and other quality attributes of fresh-cut cantaloupe melon. Journal of Food Science. 66: 1265–1270.

Qi, L.; Wu, T.; Watada, A.E. 1999. Quality changes of fresh-cut honeydew melons during controlled atmosphere storage. Journal of Food Quality. 22: 513–521.

Richard-Forget, F.C.; Goupy, P.M.; Nicolas, J.J. 1992. Cysteine as an inhibitor of enzymatic browning. 2. Kinetic studies. Journal of Agricultural and Food Chemistry. 40: 2108–2113.

Rocha, A.M.C.N.; Brochado, C.M.; Morais, A.M.M.B. 1998. Influence of chemical treatment on quality of cut apple (cv. Joangored). Journal of Food Quality. 21: 13–28.

Rocha, A.M.C.N.; Brochado, C.M.; Morais, A.M.M.B. 1998. Influence of chemical treatment on quality of

cut apple (cv. Joangored). Journal of Food Quality. 21: 13–28.
Rojas, A.M.; Castro, M.A., Alzamora, S.M.; Gerschenson, L.N. 2001. Turgor pressure effects on textural behaviour of honeydew melon. Journal of Food Science. 66: 111–117.
Rosen, J.C.; Kader, A.A. 1989. Postharvest physiology and quality maintenance of sliced pear and strawberry fruits. Journal of Food Science. 54(3): 656–659.
Rouet-Mayer, M.A.; Philippon, J. 1986. Inhibiton of catechol oxidases from apples by sodium chloride. Phytochemistry. 25(12): 2717–2719.
RTTC, 2004. Southern California Edison. Retrieved August 10, 2004, from http://www.sce.com/sc3/002_save_energy/002e_show_engy_eff/002e1_commercial/002e1c_testing_perf.htm
Sapers, G.M. 1993. Scientific status summary. Browning of foods: Control by sulfites, antioxidants, and other means. Food Technology. 47(10): 75–84.
Sapers, G.M. 1996. Hydrogen peroxide as an alternative to chlorine. Abstract 59-4. IFT Ann. Mtg: Book of Abstracts, p. 140. Institute of Food technology, Chicago.
Sapers, G.M.; Miller, R.L. 1992. Enzymatic browning control in potato with ascorbic acid-2-phosphates. Journal of Food Science, 57: 1132–1135.
Sapers, G.M. and Miller, R.L. 1998. Browning inhibition in fresh-cut pears. Journal Food Science. 63: 342–346.
Sapers, G.M.; Miller, R.L.; Mattrazzo, A.M. 1999. Effectiveness of sanitizing agents in inactivating *Escherichia coli* in Golden Delicious apples. Journal of Food Science. 65: 529–532.
Sapers, G.M.; Simmons, G.F. 1998. Hydrogen peroxide disinfection of minimally processed fruits and vegetables. Food Technology. 52(2): 48–52.
Senesi, E.; Galvis, A.; Fumagalli, G. 1999. Quality indexes and internal atmosphere of packaged fresh-cut pears (abate fetel and kaiser varieties). Italian Journal of Food Science. 2(11): 111–120.
Senesi, E.; Pastine, R. 1996. Pre-treatments of ready-to-use fresh-cut fruits. Industrial Alimentary (Italy) 35: 1161–1166.
Soliva-Fortuny, R.C.; Grigelmo-Miguel, N.; Odriozola-Serrano, I.; Gorinstein, S.; Martín-Belloso, O. 2001. Browning evaluation of ready-to-eat apples as affected by modified atmosphere packaging. Journal of Agricultural and Food Chemistry. 49: 3685–3690.
Soliva-Fortuny R.C.; Oms-Oliu, G.; Martín-Belloso, O. 2002a. Effects of ripeness stages on the storage atmosphere, color, and textural properties of minimally processed apple slices. Journal of Food Science. 67: 1958–1963.
Soliva-Fortuny, R.C.; Biosca-Biosca, M.; Grigelmo-Miguel, N.; Martín-Belloso, O. 2002b. Browning, polyphenol oxidase activity and headspace gas composition during storage of minimally processed pears using modified atmosphere packaging. Journal of the Science of Food and Agriculture. 82: 1490–1496.
Soliva-Fortuny, R.C.; Grigelmo-Miguel, N.; Hernando, I.; Lluch, M.A.; Martín-Belloso, O. 2002c. Effect of minimal processing on the textural and structural properties of fresh-cut pears. Journal of the Science of Food and Agriculture. 82: 1682– 1688.
Soliva-Fortuny, R.C.; Lluch, M.A.; Quiles, A.; Grigelmo-Miguel, N.; Martín-Belloso, O. 2003. Evaluation of textural properties and microstructure during storage of minimally processed apples. Journal of Food Science. 68: 312–317.
Soliva-Fortuny, R.C.; Alòs-Saiz, N.; Espachs-Barroso, A.; Martín-Belloso, O. 2004. Influence of maturity at processing on quality attributes of fresh-cut Conference pears. Journal of Food Science 69(7): 290–294.
Sonti, S.; Prinyawiwatkul, W.; No, H.K.; Janes, M.E. 2003. Maintaining quality of fresh-cut apples with edible coating during 13-days refrigerated storage. IFT Annual Meeting. Book of Abstracts, Session 45F. Chicago.
Tassou, C.C.; Koutsoumanis, K.; Nychas, G.-J.E. 2000. Inhibition of *Salmonella enteritidis* and *Staphylococcus aureus* in nutrient broth by mint essential oil. Food Research International. 33: 273–280.
Toivonen, P.M.A.; DeEll, J.R. 2002. Physiology of Fresh-cut Fruits and Vegetables. In Fresh-cut Fruits and Vegetables: Science, Technology and Market, edited by Olusola Lamikanra, pp. 91–123, CRC Press, Boca Ratón, FL, USA.
Utama, I.M.S.; Willis, R.B.H.; Ben-Yehoshua, S.; Kuek, C. 2002. In vitro efficacy of plant volatiles for inhibiting the growth of fruit and vegetable decay microorganisms. Journal of Agricultural and Food Chemistry. 50: 6371–6377.
Valero, E.; Varon, R.; Garcia-Carmona, F. 1990. Inhibition of grape polyphenol oxidase by several aliphatic alcohols. Journal of Agricultural and Food Chemistry. 38: 1097–1103.
Yaun, B.R.; Sumner, S.S.; Eifert, J.D.; Marcy, J.E. 2004. Inhibition of pathogens on fresh produce by

ultraviolet energy. International Journal of Food Microbiology. 90: 1–8.

Watada, A.E.; Ko, N.P.; Minott, D.A. 1996. Factors affecting quality of fresh-cut horticultural products. Postharvest Biology and Technology. 9: 115–125.

Wells, J.M.; Butterfield, J.E. 1997. *Salmonella* contamination associated with bacterial soft rot of fresh fruits and vegetables in the marketplace. Plant Disease. 81: 867–872.

Wells, J.M.; Butterfield, J.E. 1999. Incidence of *Salmonella* on fresh fruit and vegetables affected by fungal rots or physical injury. Plant Disease. 83: 722–726.

Whitaker, J.R. 1994. Principles of Enzymology for the Food Sciences, 2nd edition, pp. 431–530. Marcel Dekker, New York.

Whitaker, J.R.; Lee, C.Y. 1995. Recent advances in chemistry of enzymatic browning: An overview. In C.Y. Lee and J.R. Whitaker, ed. Enzymatic Browning and its Prevention, pp. 2–7. ACS Symp. Ser. 600, Washington, D.C.

Wiley, R.C. 1994. Introduction to minimally processed refrigerated fruits and vegetables. In R.C. Wiley, ed. Minimally Processed Refrigerated Fruits and Vegetables, pp. 1–14. Chapman & Hall, New York.

Wisniewsky, M.A.; Glatz, B.A.; Gleason, M.L.; Reitmeier, C.A. 2000. Reduction of *Escherichia coli* O157:H7 counts on whole fresh apples by treatment with sanitizers. Journal of Food Protection. 63: 703–708.

Wong, D.W.S.; Camarind, M.W.; Pavlath, A.E. 1994. Development of edible coatings for minimally processed fruits and vegetables. In J.M. Krochta, ed. Edible Coatings to Improve Food Quality, p. 65. Technomic Publishing Co., Lancaster, PA.

Wright, J.R.; Sumner, S.S.; Hackney, C.R.; Pierson, M.D.; Zoecklein, B.W. 2000. Reduction of *Escherichia coli* O157:H7 on apples using wash and chemical sanitizer treatments. Dairy, Food and Environmental Sanitation. 20: 120–126.

Yang, S.F. 1981. Biosynthesis of ethylene and its regulation. In Recent advances in the biochemistry of fruit and vegetables, edited by J. Friend and M.J.C. Rhodes, pp. 89–106. Academic Press, London, UK.

Zagory, D. 1999. Effects of post-processing handling and packaging on microbial populations. Postharvest Biology and Technology. 15: 313–321.

Zagory, D.; Kader, A.A. 1988. Modified atmosphere packaging of fresh produce. Food Technology. 42: 70–77.

9
Food Additives in Fruit Processing

P. S. Raju and A. S. Bawa

INTRODUCTION

Food processing as such had its onset as a practice when mankind stepped into culinary operations using agricultural and livestock products. The kitchen-based operations include cleaning, cutting, and conditioning for ultimate consumption. The conditioning of food as such includes addition of taste-promoting ingredients. The preservative aspects followed immediately and the classical examples included fruits and vegetable preserves like preserves and pickles. The earliest oriental foods included fermented products made from rice, soya, and milk. Similarly the wine making included earliest practices of fermentation and product flavoring. All these operations needed extraneous substances to enhance the product quality (Giese, 1993). These substances included ingredients such as sugar, salt, spices, oils, etc., of natural origin. The modern concept of food processing has given a new dimension, and conventional household food processing has undergone metamorphosis to grow into a full-fledged food technology. The concept of food additives keeping in pace with the technological advancement enlarged their horizon encompassing a number of functional additives to improve the quality, shelf life, nutritional status, economies, as well as aesthetics of the products.

At present, we have more than 2500 additives for an intentional use during production, processing, packaging, or storage of foods. Therefore, the general definition acceptable for any additive reads as "any substance added to food in restricted quantities other than the original food components during production, processing, packaging, or storage." The general version of defining food additives needs further elaboration to explain the nature of direct and indirect food additives:

- *Direct food additives*: Any substance added to the food intentionally in smaller quantities for functional purpose during food processing.
- *Indirect food additives*: Substances entering into the food in small quantities during processing or packaging.

Direct additives are classified further depending on the functionality (Branen and Haggerty, 2002). However, flavors occupy an important and dominant position among them, due to their widespread requirement in different types of food products. Food additives have a key role to play in bringing about many advantages, i.e., safety and improved nutrition; diversity in product profiles; increased sensory value of the products through optimization of unit operations in processing; and finally offering affordably priced food products.

Safe and Nutritious Foods

The safety and nutritional aspects are the most important aspects of direct food additives. Safety includes all the preservatives and constitutes an important aspect of food processing. As such food safety goes far beyond food preservation and includes the usage of food additives viz. antioxidants, which restrict the formation of toxic substances besides maintaining the vitamin status. Vitamin and mineral supplementation caters to the nutritional requirements and prevents possible deficiencies of the same.

Product Diversity

In the case of processed fruits, we have a variety of products and to name a few major categories, e.g., beverages (fruit-flavored carbonated beverages and noncarbonated beverages); thermally processed, dehydrated, and frozen products; structured, glazed, and candied fruits; salads, desserts, pies, bars, jams, jellies, and marmalades.

The use of additives is involved at one step or the other in all the technologies involved. Various types of functionalities are derived from these products for optimizing sensory properties such as color, aroma, taste, and texture besides being instrumental in ensuring the preservative aspect directly or in synergism with physical conditioning such as water activity regulation or thermal processing.

Low-calorie products are the watchwords of today, and artificial sweeteners and fat substitutes play an important role in their developments. The use of saccharin and cyclamates opened the market for various food products with reduced calories. Aspartame is the latest sweetener, which is showing immense potential in low-calorie fruit beverages. Similarly, emulsifiers and stabilizers besides fat substitutes such as sucrose polyesters have significantly reduced the use of fat in food formulations. In the next decade, it is likely that functional additives and neutraceuticals will dominate the global market (Sloan, 2000). The fruit-based products may contribute substantially in this section as reservoirs/carriers of many health-promoting substances such as dietary fiber, vitamins, minerals, and natural antioxidants. In fact, the modern biotechnological methods have further enhanced the functional ingredients in the fresh produce itself and a separate class of additives with genomic origin has emerged.

Economic Benefits

The economic benefits derived from the use of food additives are plenty. The use of enzymes in juice clarification results in significant rise in the yield of fruit juices. Similarly, the nonnutritional sweeteners could provide economic benefits through judicious use of their sweetness potency vis-à-vis conventional sweeteners (DuBois, 1992). The other direct use of additives is food preservation itself as the raw materials, i.e., fresh fruits, in the case of fruit products, are procured during the glut seasons and their preservation in the form of processed pulp/products, directly contributes to the value addition.

Optimization of Unit Operations

Several food additives are used as aids in food processing. Clarification enzymes such as pectin methylesterase and polygalacturonase optimize the juice extraction operations. Similarly, the osmophilic coatings facilitate better osmoregulation of solid contents during osmotic dehydration of fruits. It is a well-known concept that additives such as acidulants and ascorbic acid when charged in dip solutions can impart better color profiles to the products on size reduction. In this case, additive charged solutions can allow size reduction operations for enough time, without discoloration of the product, for further processing (Sapers and Miller, 1995).

Sensory and Convenience Optimization

Sensory value is also one of the important aspects of value addition during food processing, and there are a number of food additives, which enhance sensory perception. A host of nonpreservative additives has the functionality to improve the sensory value of food products (Ollikainen et al., 1984). Color, aroma, taste, and texture constitute the sensory value, and each one of them is enhanced by a range of food additives.

Convenience is another important aspect of food products and there are a number of food additives, which are potential promoters of convenience. Quick cooking nature of the processed foodstuffs is an important aspect, and it is achieved by additives such as bicarbonates and polyphosphates in processed products such as legumes (Uebersax and Occena, 1993).

All the abovementioned factors illustrate the usefulness of food additives and their significant role in the establishment of food processing as a full-fledged technology. However, along with the increase in the use of food additives a number of doubts have also developed with regards to the safety aspects and the associated health risks, warranting the necessity to have a balanced code of regulation to minimize the risks involved.

CLASSIFICATION

Food additives are classified primarily as direct and indirect food additives as per the definitions mentioned earlier. The direct food additives encompass all the intentionally added additives. They are classified based on their chemical nature as well as functionality (Branen and Haggerty, 2002) (Table 9.1).

Table 9.1. Classification of Food Additives

	Food Additives		
	Direct Food Additives		
No.	Chemical Classes	Functional Classes	Indirect Food Additives
1.	Inorganics	**Preservatives**	Catalysts
	Phosphates	*Antimicrobials*	Lubricants
	Sulfites	*Antioxidants*	Propellants
	Salt, etc.	*Antibrowning agents*	
2.	Synthetic chemicals	**Nutritional additives**	
	Dyes	*Vitamins*	
	Silicones	*Amino acids*	
	Benzoates	*Fiber*	
	Vitamin A, etc.	*Minerals*	
		Fat substitutes	
3.	Extraction products	**Coloring agents**	
	Gums	*Natural colorants*	
	Essential oils	*Synthetic colorants*	
	Vitamin E, etc.		
4.	Fermentation derived products	**Flavoring additives**	
	Enzymes	*Sweeteners*	
	Yeasts	*Natural and synthetic flavors*	
	Citric acid, etc.	*Flavor enhancers*	
		Texturizing agents	
		Emulsifiers	
		Stabilizers	
		Miscellaneous additives	
		Chelating agents	
		Enzymes	
		Antifoaming agents	

The classification of food additives includes multifunctional food additives. There are many additives, which qualify as multifunctional. Sulfur and its compounds have different functionalities such as antimicrobial, antienzymatic, and anti-nonenzymatic (Josyn and Braverman, 1954). Acidulants such as citric acid too have multiple functions, i.e., acidification, metal chelation, antimicrobial, and antioxidative. The classification is flexible enough for the multifunctional additives to be included in different functional categories.

GOVERNMENTAL REGULATIONS

Fruit and vegetables and their products inclusive of food additives and pesticides/biocides are regulated both by international and national regulations by means of food laws. Food additives need to be approved by the regulating bodies both in terms of usage as well as dosage, as they are basically extraneous substances added to food products to bring about a variety of benefits such as sensory quality, nutritional value, and storage stability of the products besides being the processing aids (Sumner and Eifert, 2002). The regulation of food additives is an absolute necessity as misuse can cause far-reaching health implications in children as well as adults. Food safety as such demands stringent regulatory measures to ensure total food safety to the consumers. At the same time, the food laws need to be flexible enough to render techno–commercial feasibility, optimal sensory quality, and shelf stability. Therefore, in order to maintain a dynamic balance between the legislation and the product development, the specifications are periodically subjected to revision by national and international bodies concerned with coining of standards and monitoring of the same.

Different countries have their own regulatory norms encompassing the list of approved additives and their enforcement procedures. The foremost among the governmental regulations are Food and Drug Act (FDA), European Union standards, and Codex Alimentarius, which constitutes the FAO/WHO joint regulatory body (Somogyi, 1996). These bodies as such dominate the global food sector as they encompass an umbrella of developed and developing countries involved in world trade. Certain salient features of FDA and Codex are described below.

Food and Drug Act (FDA)

The FDA had come to the fore with the first law tabled during 1906. The Delany Committee report in 1952 gave a comprehensive recommendation to include newer additives in quick succession to fill the vacuum as the food industries felt the increasing demand for food additives in their product developmental activities (FAO, 1996). The amendment in 1958 gave rise to three distinct classes of food additives: (a) substances approved by FDA prior to 1958, (b) substances that are generally recognized as safe (GRAS), and (c) substances without prior sanction or GRAS status and defined as food additives.

GRAS Substances

The GRAS feature is a novelty brought in by the FDA and it has paved way for the filing of many affirmation petitions seeking the status (FDA, 1995). The seekers of GRAS status find the GRAS clause an ideal way to make the extraneous substances of nonfood/food origin in their process, compatible with the law.

The major features considered for affirmation of GRAS status are as follows:

1. General recognition of safety may be based only on the views of experts qualified by scientific training.
2. General recognition of safety based on scientific procedures shall require the same quantity and quality of scientific evidence as is required to obtain the approval of a food additive regulation of the ingredient.
3. General recognition of safety through experience based on common use in food prior to January 1, 1958, may be determined without the quantity or quality of scientific procedures required for the approval of the food additive regulation.
4. Any food substance of biological origin with known use record prior to January 1, 1958, for which no known safety hazard exists, shall qualify to be GRAS.
5. Distillates, isolates, extracts, and concentrates of extracts of GRAS substances.
6. Reaction products of GRAS substances.
7. Substances not of a natural biological origin, including those for which evidence is offered that they are identical to a GRAS counterpart.
8. Substances of natural biological origin intended for consumption for other than their nutrient properties.

The GRAS status compliance makes a substance safe subject to limits prescribed for products with standards of identity. The GRAS list is constantly reviewed with addition and deletions. Natural acids and synthetic color additives are subject to the color additive act of 1960 and are not included in the food additive regulations. Some of the successfully petitioned substances include aspartame, acesulfame-K, gellan gum, and polydextrose. The safety issues shroud additives such as butylated hydroxy anisole (BHA) and saccharin.

Indirect Food Additives

Indirect additives are also important as many extraneous substances may come in contact with food and contaminate it during processing, i.e., lubricating oils and parts of machinery (Somogyi, 1996). Under the general provisions of indirect food additives (Congress for Federal Regulation, CFR 21.174.5), the following aspects need to be taken into notice:

1. Any substance used as a component of articles that contact food shall be of pure quality suitable for its intended use.
2. The quantity of any food additive substance that may be added to food as a result of use in articles that contact food shall not exceed, where no limits are specified, that which results from use of the substance in an amount not more than reasonably required to accomplish the intended physical or technical effects.
3. Substances that under conditions of good manufacturing practice (GMP) may be safely used as components of articles that contact food include the following, subject to any prescribed limitations.
   (a) GRAS in or on food
   (b) GRAS for intended use in food packaging
   (c) Substances of prior sanction or approval.

Codex Alimentarius

The Codex Alimentarius Commission was established to implement the joint FAO/WHO standards program, the purpose of which is, as set in the statutes of the commission, to protect the health of consumers and to ensure fair practices in the food trade, to coordinate all work on food standards in different countries, to determine priorities in the coining of standards, and to finalize the standards after acceptance by governments (Anon, 1992). Basically, the Codex Alimentarius is a collection of internationally adopted food standards presented in a uniform manner to ensure consumer health and fair practices in trade.

The general principles for the use of food additives specify that food additives shall conform to an approved specification, i.e., the specifications of identity and purity recommended by the commission. A list of 450 additives has been indexed since 1997 as a compendium of all specifications prepared by the FAO/WHO joint expert committee on food additives (JECFA). The other important bodies/committees associated with the Codex include joint meeting on pesticide residues (JMPR) (Anon, 1998) with FAO and WHO jointly constituting the body.

The status of GMP is a novelty brought in by the Codex to include a host of substances added during fruit processing. GMP encompasses a number of substances, which allows restricted usage of additives to a quantity not more than what is required to achieve the desired technological effect and in accordance with the Codex general principles for the use of food additives, with emphasis on the allowed daily intake (ADI) of specific substances (Anon, 1992). GMP necessarily involves the following aspects:

- The quantity of the additive added to food does not exceed the amount reasonably required to accomplish its intended physical, nutritional, and other technical effect in food.
- The quantity of the additive that becomes a component of food as a result of its use in the manufacturing, processing, or packaging of a food and which is not intended to accomplish any physical or other technological effect in the food itself is reduced to the extent reasonably possible.
- The additive is of appropriate food-grade quality and is prepared and handled in the same way as a food ingredient. Food-grade quality is achieved in compliance with the specifications as a whole and not merely with individual criteria in terms of safety.

Codex Alimentarius has adopted an international numbering system for all approved food additives and the food categories have also been specified under food category numbers. Some of the individual numbers for food additives have been specified as follows:

- glacial acetic acid, 280
- BHA, 320
- ascorbic acid, 300
- calcium alginate, 404
- aspartame, 951
- carbon dioxide, 290
- beet red, 162
- chlorophyll, copper complexes, 1411
- benzoic acid, 210
- cyclamates, 952.

The fruit product categories are as follows:

4	Fruits and vegetables (including mushrooms and fungi, roots and tubers, pulses and legumes) and nuts and seeds
4.1	Fruit
4.1.1	Fresh fruit
4.1.1.1	Untreated fruit
4.1.1.2	Surface treated fruit
4.1.1.3	Peeled or cut fruit
4.1.2	Processed fruit
4.1.2.1	Frozen fruit
4.1.2.2	Dried fruit
4.1.2.3	Fruit in vinegar, oil, or brine
4.1.2.4	Canned or bottled (pasteurized) fruit
4.1.2.5	Jams, jellies, and marmalades
4.1.2.6	Fruit-based spreads other than 4.1.2.5 (e.g., chutney)
4.1.2.7	Candied fruit
4.1.2.8	Fruit preparations including pulp and fruit toppings
4.1.2.9	Fruit-based desserts, including fruit-flavored, water-based desserts
4.1.2.10	Fermented fruit products
4.1.2.11	Fruit fillings for pastries
4.1.2.12	Cooked or fried fruit

STATUS OF FOOD ADDITIVE INDUSTRY

The food additive market is estimated at $20 billion according to Leatherhead Food International (Anon, 2002). The largest market segment within the food additives is flavors at 30% followed by hydrocolloids at 17%, acidulants at 13%, flavor enhancers at 12%, sweeteners at 6%, colors at 5%, emulsifiers at 5%, vitamins and minerals at 5%, enzymes at 4%, chemical preservatives at 2%, and antioxidants at 1%. Annual growth is projected at 2–3% over the next 5 years, with the market expected to reach $22 billion in 2005.

The strongest growth section will include vitamins and minerals. There is an increasing demand for natural varieties of flavors and colors at the expense of synthetics. Similarly, the demand for natural antioxidants over synthetic ones is also growing. It is also believed that more emphasis will be laid on health. The use of hydrocolloids as fat substitutes along with enhanced growth in artificial sweeteners may dominate the proceedings in the food industry based on food additives. As far as the preservatives are concerned, the sector may further grow from its current position of 2%, as Latin American, Asian, and Eastern European countries may continue their quest toward foods with longer shelf life. There can also be a greater demand for ethnic food flavorings, particularly as ingredients in preprepared dressings for salads to restrict microbiological contaminants otherwise originating from spice ingredients (Mannikes, 1992).

The demand for food additives in various sectors varies considerably depending on consumer requirements of different countries. The level of utilization also differs within various food manufacturing sectors depending on the functional requirements of such applications viz, solubility, thermal and light stability, and compatibility with human metabolism (Robach, 1980).

The demand for food additives is primarily distributed among the three sectors, i.e., commodity processing sector, pharmaceuticals and drugs sector, and the food manufacturing sector (Fig. 9.1). The blenders and bulk suppliers play a major role in coordinating the overall blending, repackaging, and distribution to the utility sectors. The commodity suppliers are the sources of primary and secondary processed, value-added commodities such as purees and juice concentrates for the manufacturing sector as such, which may be termed as tertiary processing during which fruit ingredients are used as per the standards of identity (21 CFR 150.160).

Of the two phases of fruit product manufacturing, i.e., commodity processing and ultimate fruit product manufacturing, the use of additives is the maximum in the second phase accounting for nearly 60–70% of total food additives used in the fruit-based industries. Among unit operations carried out in commodity processing, operations such as extraction/refining, clarification, and concentration require the bulk of the additives, while formulating, extrusion, freezing, dehydration, and fermentation involve their intensive use for the product manufacture (Somogyi, 1996).

The use of additives in fruit processing is expected to grow especially in the areas such as artificial sweeteners, flavors, and texturizing agents. However, the regulations are expected to be more stringent, necessitating the need for more emphasis on the use of natural ingredients, such as natural colors, flavors, and preservatives, which qualify to be in the GRAS list (Robach, 1980). The nutritional additives may gain an increased demand, as fruit-based functional foods are likely to constitute a major food category.

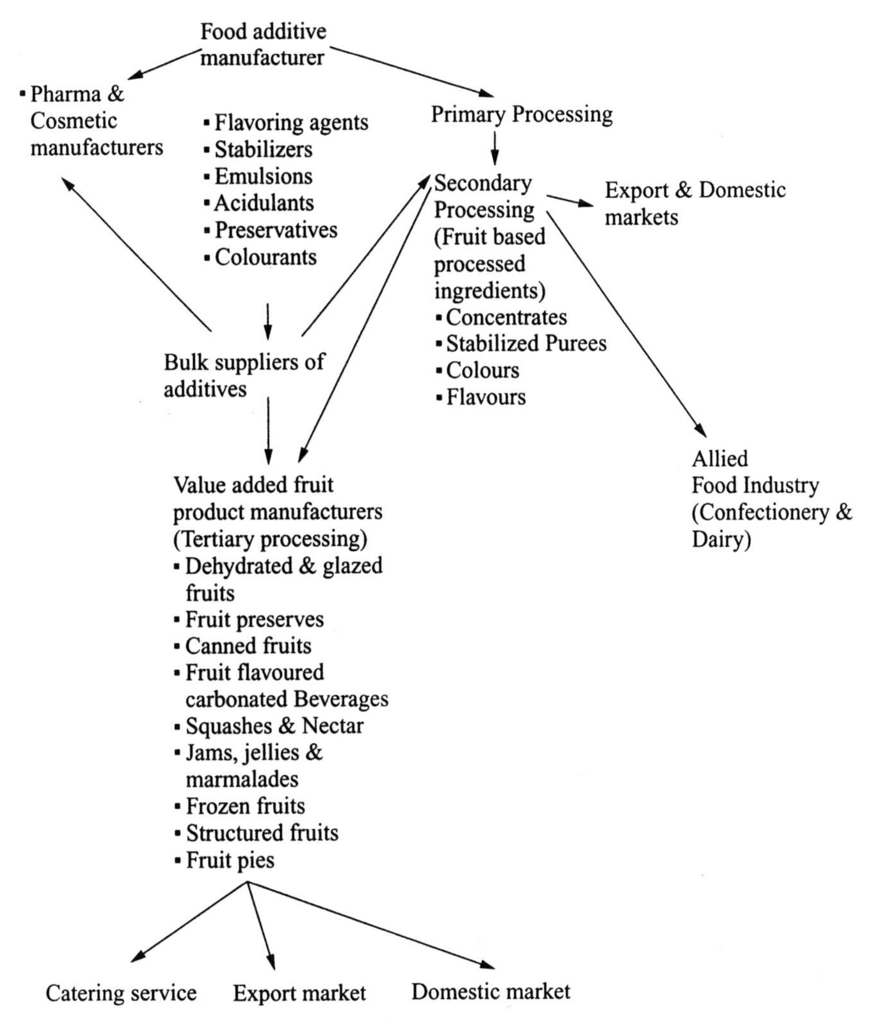

Figure 9.1. Utility profile of additives in fruit processing.

ACIDULANTS

Fruit processing as such involves physical, biochemical, and microbial stabilization of the processed products. Acidification has been in practice and its origin dates back to the onset of human civilization. Oriental preserves, such as fermented vegetable products and pickles, clearly indicate the identification of the preservation potential of acidulants. The dual functionality of acidulants in imparting tartness to the product besides the preservative function makes them congenial to widespread food applications. Fruit and vegetable products are the major ones to use acidification, as the sensory value of the product is not impaired and at times the taste is improved as a result of flavor enhancement (Gardner, 1972). In fruit products, acidification is usally accompanied by adjustment of the total soluble solids so that the brix/acid value is comparable to that of fresh fruits. An adjustment in the Brix–acid ratio has been found to take care of the sensory perception, enabling a variety of fruit products to undergo acidification and have the advantages such as preservative and antioxidative effects. The oriental vegetable products such as pickles require a characteristic sourness/tartness to attain an optimal sensory value. The advantages of acidification can be best utilized by clear understanding of product profile and an appropriate acidulant can be selected to obtain the optimal sensory perception and shelf life of the finished product (Dziezak, 1990).

Selection of Acidulants

The following aspects help in selecting the suitable acidulants for specific food products.

Regulatory Considerations

1. The GRAS nature of the acidulant needs to be ascertained. The GRAS substances may be used in foods not covered by standards of identity and which do not have restrictions on their usage levels, provided GMP are followed.
2. If the acidulant is covered as a food additive, as in the case of fumaric acid, it has to be regulated by the food additive regulation.
3. For foods covered by standards of identity, the maximum levels prescribed need to be followed.
4. If any local regulations exist, they should be considered while selecting an acidulant.

Functional Considerations

1. Effect of acidulant on the overall product profile.
2. Matching of solubility characteristics with the process conditions and acidulant concentrations required.
3. Hygroscopic characteristic requirements of dry mixes.
4. Suitability of the acidulant to impart an optimal level of tartness at the functional pH.
5. Physical form and particle size for application in dry mixes.
6. Screening of several acidulants based on feedback from the market/user for optimum product applications.

Functions

Acids and their salts serve a variety of functions in foods that include the following, with the dominant ones being antimicrobial (Levine and Fellers, 1940) and flavoring (Hartwig and McDaniel, 1995):

1. Flavoring to provide a desired taste and intensity, which enhances, blends, or modifies the overall flavor of the product.
2. Reduction in pH to prevent or retard the growth of microorganisms as well as germination of spores and to increase the lethality of the process.
3. Maintenance or establishment of pH through buffering action. Usually a combination of free acids and salts are used.
4. Chelation of metal ions to assist in minimizing lipid oxidation (Cu and Fe), reducing color changes, and controlling texture in some fruits and vegetables.
5. Alteration of the structure of foods including gels made from gums (pectin and carrageenan) and proteins.
6. Modification of sugar crystallization in hard candy manufacture.

Diversity in Different Acidulants

Different acidulants offer a host of functional diversity for various food applications warranting closer scrutiny for optimal utility: (1) flavor, (2) acidity, (3) metal-chelating activity, (4) antimicrobial activity, (5) solubility, (6) hygroscopicity, and (7) cost.

Mechanism of Action

Acidification is one of the major functions of acidulants apart from preservation and flavor enhancement. At equal concentrations, the acidulants vary in their ability to depress pH and the degree/intensity of the tartness produced. The percentage required to replace anhydrous citric acid varies from 55% to 60% in case of phosphoric acid; 67–72% for fumaric acid; 80–85% for tartaric acid; 78–83% for malic acid; and 110–115% for adipic acid. The use of glucono-delta-lactone (GDL) is gaining an increasing popularity due to the novel features such as slow hydrolysis and mild acid flavor. In fruit processing, acidulants are used extensively and the Brix–acid ratio is maintained at appropriate levels to optimize the sensory perceptions. The products include nectars, squashes, jams, and jellies (Seiferi, 1992).

The other major function of acidulants is preservation and safety assurance. In 1970s, considerable research took place on the action of organic acids on the microbial cell at the molecular level. It is well known that the mode of action of an acid was related to the undissociated portion of the molecule (Huntur and Segel, 1973). This action is deemed more important than any other external change in pH brought about by the addition of acids. The undissociated moiety of a weak acid penetrates rapidly to the interior of the cell because of its lipid solubility and discharges the gradient diffusion through the plasma membrane and dissociates internally. The dissociated forms of weak acids, on the other hand, could not be absorbed by microorganisms to any great extent.

Murdock (1950) found that citric acid inhibited the flat-sour organisms isolated from tomato juice, and this inhibition appeared to be related to the inherent pH of the product. Citric acid, rather than acetic or lactic acids, also inhibited thermophilic bacteria (Fabian and Graham, 1953). In addition to the pH-lowering effects of citric acid, a secondary inhibitory effect is attributed to the chelation of essential minerals. It is believed that inhibition may have been attributable to chelation of essential metal ions by citrate rather than inherent acid inhibition. Chelation has also been believed to be an influencing factor in inhibiting the growth of *Staphylococcus aureus* (Rammell, 1962). There has been a number of studies that have attributed the antimicrobial activity of citric acid to the chelation of metal ions, which are essential for microbial growth (Beuchat and Golden, 1989).

Citric acid is also used conventionally as a synergist for antioxidants and to retard browning reaction. Usage levels for citric acid have been found to be 0.1–0.3% with an antioxidant level of 100–200 ppm (Dziezak, 1986). The specialized uses include the function as a gelling agent. The structured fruits can be taken as an example to illustrate the specific aspect. Alginates are gelled with calcium salts under acidic pH (Kaletunc et al., 1990). Pectinacious fruits are largely subjected to structuring using a variety of acidulants especially GDL.

Range of Products

A wide variety of fruit-based products possess acidulants as a direct food additive:

(1) Beverages
  (a) fruit-flavored carbonated beverages
  (b) fruit-flavored noncarbonated beverages
  (c) dry beverage powders
  (d) low-calorie beverages (diet beverages).

Among the abovementioned products fruit-flavored carbonated beverages are the recent ones. Usually these beverages are made with 10% natural juices for health-conscious consumers. Citric and malic acids are used in fruit-flavored carbonated beverages. Tartaric acid is generally used in lime-flavored beverages. The noncarbonated beverages include fruit drinks, nectars, and isotonic beverages as "thirst quenchers." Fruit-flavored dry beverages make use of acidulants such as citric, malic, and fumaric acids for imparting tartness along with the release of carbon dioxide from carbonate salts of sodium and calcium. Fruit-flavored diet beverages yielding 50% fewer calories than comparable products make use of acidulants to control pH of the beverage so that desired sweetness characteristics can be achieved (Dziezak, 1990).

(2) *Candies, structured fruits, and fruit gums*: A wide variety of acidulants are used in these fruit-based products. Fruit-flavored candies are popular products and several patents exist in this area (Dwivedi, 2003). The candied fruit products are extruded in a ribbon/belt format, capable of being extruded and rolled on to itself to form a candy roll. The product as such includes sweeteners, fruit flavors, binders, water, stabilizers, and acidulants. Fassin and Bachmueller (2000) described the manufacture of fruit gum confectionary, based on the preparation of fruit gum mass by heating water, sweetener, gelling agent, acidulants, flavorings, and fruit/vegetable extracts.

In the case of structured fruits, GDL has been used as an acidulant to prepare structured hydrocolloid gels with fruit pulp and sugar as the major ingredients (Nussinovitch et al., 1991). Apple sauce and grape juice concentrates have been texturized to obtain gelled products with excellent consistency using GDL as an acidulant and gelling agent to liberate calcium from calcium hydrogen phosphate, which ultimately caused the gelation of the product (Kaletunc et al., 1990).

(3) *Thermal processed fruits*: Acidulants are also widely used in thermally processed fruit and vegetable products. In the canning process, citric, lactic, malic, and other organic acids are used to lower pH to 4.6 or below. This addition allows lower temperatures and shorter processing times to be used for the inhibition of sporulation and growth of microorganisms without any loss of flavor, color, and texture in the finished product (Rajashekhara et al., 2000).

Commonly Used Acidulants

The most commonly used acidulants are acetic acid, citric acid, malic acid, phosphoric acid, fumaric acid, and tartaric acid. The physical and functional characteristics of the commonly used acidulants are described in Table 9.2. Except fumaric acid, the other acidulants are generally recognized as safe (GRAS).

Table 9.2. Physical and Functional Characteristics of Major Acidulants Used in Fruit Processing

No.		CFR	Physical Form	PKa	Range of Appln in Fruit Industry	Taste	Type of Products
1.	Citric acid	184.1033 (GRAS)	Crystalline powder	3.14 4.77 6.39	Very high	A burst of tartness	Carbonated and noncarbonated beverages Wines, jams, and jellies Desserts and fruit squashes Canned and frozen products
2.	Fumaric acid	172.350 (Food additive)	White granules or crystalline powder	3.03 4.44	Low	Tart	Frozen concentrates Cider and apple drinks
3.	Malic acid	184.1069 (GRAS)	Crystalline powder	3.4 5.11	Medium	Smooth tartness	Fruit-flavored sodas
4.	Phosphoric acid	182.1073 (GRAS)	Liquid	2.12 7.21 12.67	Low	Acrid	Buffering agent in jams and jellies
5.	GDL	184.1318 (GRAS)	White crystalline powder	3.7	Medium	Neutral taste acidic paste upon hydrolysis	Salad dressings
6.	Tartaric acid	182.1073 (GRAS)	Crystalline powder	2.98 4.34	Medium	Extremely tart	Cranberry and grape-flavored fruits Jams and jellies
7	Acetic acid	184.1009 (GRAS)	Clear colorless liquid	4.75	High	Tart and sour	Pickled fruits

Citric, malic, and acetic acids are widely used in a variety of fruit products inclusive of canned and frozen products, fruit-flavored carbonated and noncarbonated beverages, jams, jellies, and pickled fruits. Sudden burst of tartness is a major characteristic of acidulants such as citric acid, and the use of GDL is gaining increasing popularity to overcome the problem.

Glucono-Delta-Lactone (GDL)

It is an inner and neutral ester of gluconic acid and gets hydrolyzed in aqueous solutions to form gluconic acid (Fig. 9.2). Gluconic acid is a natural constituent of juices and honey and an intermediate in glucose oxidation. The functionality involves slow hydrolysis under moist conditions, resulting in a gradual and continuous decrease in pH. At the end of hydrolysis equilibrium mixture exists, consisting of gluconic acid as well as delta- and gamma-lactones. The rate of acid formation increases with temperature and the intensity of acidification depend on the concentration of GDL and the temperature.

The slow rates of acidification by GDL and its mild taste set it apart from other acidulants. The use of GDL is gaining increasing popularity. In case of fruit

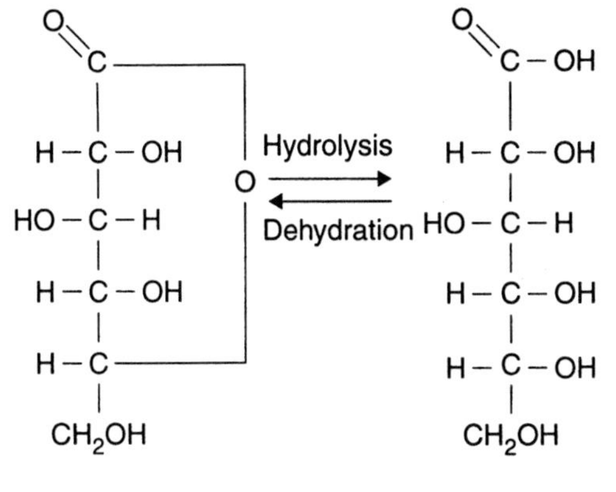

Glucono-deltalactone ⇌ Gluconic acid (Hydrolysis / Dehydration)

Figure 9.2. Hydrolysis of glucono-delta-latone.

products apart from the juices it is widely used in jellies and structured fruits. The relatively higher cost of GDL is a drawback for its extensive use in lieu of acidulants such as citric and malic acids (Montinez et al., 1997).

PRESERVATIVES

The preservative category of food additives is a large group encompassing a number of functionalities. Though the basic understanding of preservatives is about their antimicrobial activity, it encompasses other functions such as antioxidation and antibrowning activities. Chemical preservatives are defined by FDA (1979) as "any chemical that when added to food tends to prevent or retard deterioration but does not include common salt, sugars, vinegars, spices and oils extracted from spices, substances added to food by direct exposure thereof to wood smoke, or chemicals applied for their respective insecticidal or herbicidal properties." The antioxidants and antibrowning agents constitute the preservatives responsible for restricting the chemical deterioration of the products, whereas the antimicrobials are responsible for the protection from biological hazards. Often the synergistic effects of antioxidants with antimicrobials could give the best result in terms of shelf life (Branen et al., 1980).

During fruit processing a number of antimicrobials, antioxidants, and antibrowning agents are used as additives. The number of antimicrobials approved for use in food is remarkably limited. The primary food additives cited in FDA are (1) sodium benzoate, (2) calcium and sodium propionates, (3) sorbic acid and potassium sorbate, and (4) parabens. These preservatives are often used in combination with other methods of preservation such as refrigeration, freezing, and dehydration to obtain better control of deleterious organisms. Certain important preservatives such as organic acids and sulfites are multifunctional and their functionalities include preservation as well as antioxidation and antibrowning activities.

ANTIMICROBIALS

Selection of Antimicrobials

The selection of antimicrobials has to be carried out appropriately to obtain the best possible preservative function. The following criteria need to be followed for the selection of approximately 30 compounds, which can legally be used as antimicrobials in food products (Fulton, 1981):

1. Antimicrobial and chemical properties of the compound.
2. Composition of the target food.
3. The type of preservation technique adopted for the product.
4. The type and quantum of microbiological load.
5. The safety and regulatory aspects of the antimicrobial for use in the specific product.
6. Cost effectiveness of the antimicrobials.

Mode of Action

1. The mode of action of antimicrobials can be either bactericidal or bacteriostatic and generally falls into one of the following categories (Davidson et al., 2002).
2. Reaction with the cell membrane causing increased permeability and loss of cellular constituents.
3. Inactivation of essential enzymes.
4. Destruction of functional activity.

ANTIMICROBIALS IN PROCESSED FRUIT PRODUCTS

In processed fruits, the widely used antimicrobials are benzoates, sorbates, propionates, and parabens, apart from acidulants and sulfites. The selection of appropriate antimicrobial for use in fruit products takes notice of the low pH as well as the target organisms, mostly yeasts and fungi due to the low pH.

Benzoates

Benzoates have a typical aromatic ring structure. Benzoic acid and sodium benzoate are widely used in a number of fruit products with an effective functional pH range of 2.5–4.0. These compounds are used primarily as antimycotic agents, and most yeasts and fungi are inhibited by 0.05–0.1% of the undissociated acid. Food-poisoning and spore-forming bacteria are generally inhibited by 0.01–0.02% of undissociated acid, but many spoilage bacteria are much more resistant. Therefore, benzoic acid cannot be relied upon completely for effective preservation of foods capable of supporting bacterial growth (Baird-Parker, 1980). The antimicrobial properties have been attributed to undissociated benzoic acid according to the results of a study involving the uptake of benzoates by *Saccharomyces cereviceae* (Macris, 1975). Benzoates have an advantage of low cost compared to other antimicrobial additives. The lower pH makes benzoates suitable for use in maraschino cherries, fruit pie fillings, fruit-based carbonated and noncarbonated beverages, pickles, sauces, fruit preserves, and minimally processed acidified vegetables (Raju et al., 2000). The benzoate compounds are most effective in the low-pH acid foods such as apple cider, soft drinks, tomato sauce/ketchup, and the like and not as effective in low-acid vegetables such as peas, beans, lettuce, etc. The pKa value of benzoate is 4.2 making a pH range of 4.0–5.0 as the functional pH, and most of the fruit products fall within this range. At a pH of 6.0, which is normal for many vegetables, only 1.5% of the benzoate is undissociated (Jay, 1986). Care should be taken with the addition of benzoates to acid foods because they can deliver a "peppery" or a burning taste sensation at levels of about 0.1%.

As far as regulatory status is concerned, benzoates are considered as GRAS substances with a maximum limit of 0.1% set apart (21 CFR 184.1021 and 21 CFR 189.1733). In most of the countries, the maximum permissible-use concentration is 0.15–0.25%. Sodium benzoate is used as an antimicrobial in carbonated and still beverages (0.03–0.05%); syrups (0.1%); cider (0.05–0.1%); jams, jellies, and preserves (0.1%); and pie pastry fillings and salads (0.1%).

Parabens

Alkyl esters of p-hydroxy benzoic acid are collectively known as parabens and are permitted in the United States with reference to methyl, ethyl, and heptyl parabens. Parabens are most effective against molds and yeasts and as such less effective against bacteria especially the Gram-negative bacteria. The pKa of these compounds is around pH 8.47. Their antimicrobial activity tends to increase with the length of the alkyl chain and extends up to pH 7.0 (Dziezak, 1986). It is also known that the parabens are effective inhibitors of growth and toxin production of *Clostridium botulinum* (Robach and Pierson, 1978).

Parabens are used in fruit juices, salads, and artificially sweetened jams and jellies. The methyl and propyl esters of parabens are considered GRAS with a maximum total content of 0.1% (21 CFR 184.1490 and 21 CFR 184.1670).

Sorbates

Sorbic acid is a monocarboxylic fatty acid that is used to preserve foods. Sorbic acid is slightly soluble in water, whereas the potassium salt is highly water soluble up to 58.2 g/100 ml at 20°C (Chichester and Tanner, 1975). The optimum pH range for effectiveness extends up to 6.5, higher than the upper range of benzoates and propionates but below that of parabens. A number of reports exist regarding the antimicrobial activity of sorbates. Spoilage-causing and heat-resistant fungi such as *Neosortorya fischeri* along with the ascospores were subjected to studies on thermal death rate, and the sorbate was found to control the organism without impeding the sensory value of preserved fruit juices such as grape and mango (Rajashekhara et al., 1998). Effective inhibition of fungi was also observed in low-sugar preserves through the synergistic effects of sorbates and water activity regulation. Sorbates are currently used to preserve dehydrated prunes, figs, and many beverages such as orange juice, lemonade, and apple juice (Robach, 1980).

Sorbates are considered GRAS substances (21 CFR 182.3089), and they have been used in more than 90 food products having standards of identity. Sorbic acid is considered as one of least harmful antimicrobial preservatives, even at levels exceeding those normally used in foods (Sofos and Busta, 1981).

Acidulants

The preservative role of acidulants is discussed earlier under the category of acidulants. The undissociated moiety is responsible for the bactericidal property of acidulants. They act to reduce the pH, minimizing microbial growth and often enhancing the effect of weak acid preservatives. The mechanism of action leading to preservative function may be attributed to lowering of pH as well as metal ion chelation. Citric acid has been primarily used in many fruit-based products, representing more than 60% of all food acids used. Malic acid and GDL are relatively newer and emerging acidulant preservatives, with the potential to impart excellent sensory property to the fruit products. The GRAS status of food acidulants along with the cost effectiveness favor the widespread use in a variety of fruit products including fruit-flavored carbonated and noncarbonated beverages.

Sulfites

Elemental sulfur and sulfur compounds are known to show antimicrobial activity and sulfur obtained from volcanic lava as well as hot spring water containing sulfur are used extensively for various dermal infections and wounds. The pKa values for sulfur dioxide are 1.76 and 7.2, indicating a rather weak dibasic acid. It is useful to have sulfur dioxide in a salt form. The dry salts are easier to store and less of a problem to handle than the gaseous or liquid forms (Ough, 1983). In water solutions, sulfur dioxide shows the following reaction equilibriums:

$$SO_2 + H_2O \leftrightarrows [H_2SO_3]$$
$$[H_2SO_3] \leftrightarrows HSO_3^- + H^+$$
$$HSO_3^- \leftrightarrows SO_3^{-2} + H^+$$

The growth inhibiting or lethal effects of sulfurous acid are most intense when the acid is in the unionized form (Hailer, 1911). It was also noted that bacteria were much more sensitive to sulfur dioxide than were yeasts and molds. It is also known that the bisulfites had lower activity than sulfur dioxide against yeasts and the sulfites had none. The three main groups of microbes of interest in the high acidic beverages and fruits are (1) acetic acid-producing and malolactic bacteria, (2) fermentation and spoilage yeasts, and (3) fruit molds.

As far as food applications are concerned, sulfites are used in a number of fruit products, i.e., dried fruits, frozen fruits, fruit-based beverages, glazed fruits, jams, and jellies within the purview of GMP and GRAS. However, any product with sulfur dioxide levels above the detectable limits needs to specify on the labels the nature of sulfitation and the residue levels (Taylor et al., 1986).

The FDA considers sulfur dioxide and several sulfite salts as GRAS (21 CFR 182). However, sulfites cannot be used in fruits and vegetables intended to be served, presented, or sold raw/fresh to the consumers. They are allowed in fruit juices and concentrates, dehydrated fruits, vegetables and wine. The maximum level of sulfur dioxide allowed in wine has been set at 350 mg/l by the regulating body for the U.S. alcoholic beverage industry. Sulfur elicits allergenic responses in certain individuals, especially steroid dependent and therefore, the usage levels in ready-to-eat fruit and vegetable products have been under stringent scrutiny, leading ultimately to its ban in such products (Anon, 1990).

BIOPRESERVATIVES

The use of biopreservatives is gaining an increasing popularity. These are basically of biological origin and therefore can easily be considered GRAS as compared to the chemical additives. The biopreservative "nisin" is the foremost among them as the use of nisin is gaining momentum for a range of food applications. The compound is a peptide produced by the lactic bacteria *Lactococcus lactis* sp. *lactis*. The structure and amino acid content of nisin was determined by Gross and Morell (1971). The solubility is 56 mg/ml at pH 2.2, while at pH 5.0 the solubility is 3 mg/ml.

Nisin by itself has a narrow spectrum affecting only Gram-positive bacteria including *Bacillus*, *Clostridium*, *Enterococcus*, *Lactobacillus*, *Listeria*, *Pediococcus*, and *Staphylococcus*. The spectrum of activity of nisin can be expanded to include Gram-negative bacteria by combining it with chelating agents, such as ethylene diamine tetra acetic acid (EDTA) (Carneiro et al., 1998).

Nisin as a food additive has been approved in many countries. The FDA has approved nisin as a preparation (21 CFR 184.1538) with a content of not less than 900 IU/mg. It is approved to inhibit the growth of *C. botulinum* spores in pasteurized cheese spreads with fruits and vegetables. It has been accorded a GRAS status with a maximum use level of 250 ppm. As such it is used to reduce the thermal process levels in different canned products and the future for nisin appears bright as a need is being felt to decrease the thermal processing levels to protect finished product flavor (Davidson et al., 2002).

Antibrowning Agents

Antibrowning agents are of special significance in fruit products as a majority of them are susceptible to browning reaction during processing and storage. The fruit products are highly susceptible toward the browning reactions of both the nonenzymatic and enzymatic nature. The major reasons for higher rate of browning in fruit products can be cited as, abundance of sugars with particular reference to reducing sugars.

Nonenzymatic browning is a maillard reaction between carbonyl and amino groups with a host of intermediates, finally resulting in the formation of nitrogenous polymers and copolymers known as melanoidins. Nonenzymatic browning reactions can also result in the loss of vital nutrients such as ascorbic acid, which gets oxidized to dehydroascorbic acid, which further undergoes aldol condensation or reaction with amino groups to form brown pigments (Loescher et al., 1991).

Browning due to ascrobic acid is very important in processed fruit juices enriched with vitamin C. Nonenzymatic browning can also take place due to sugar degradation, iron complexing of polyphenols (Smith, 1987), and oxidation of polyphenols by hypochlorites (Choi and Sapers, 1994).

Nonenzymatic browning reaction in fruits and vegetables depends on a number of factors such as (1) product composition, (2) moisture content of the product, and (3) storage temperature and exposure to oxygen. The compositional factors include maillard precursors or ascorbic acid (Kennedy et al., 1990). Nonenzymatic browning in fruits and vegetables can be inhibited by refrigeration and through the control of water activity in dehydrated foods (Labuza and Saltmarch, 1981). Other methods of control include use of glucose oxidase for reduction of glucose levels, reduction of amino nitrogen content in juices by ion exchange, packaging with oxygen scavengers, and use of sulfites (Bolin and Steele, 1987). Sulfhydryl-containing amino acids have been found to be nearly as effective as bisulfite in inhibiting nonenzymatic browning in a model system (Friedman and Molnar, 1990). However, cysteine treatment was ineffective in dried apple (Bolin and Steele, 1987).

Additives to control nonenzymatic browning are used in a number of fruit products such as dehydrated fruits, glazed fruits, beverages, fruit bars, texturized fruit products, and fruit candies. The sulfite treatment levels vary in foods widely depending on the application. Residual levels usually do not exceed several hundred parts per million but may approach 1000 ppm in certain fruit and vegetable products (Taylor et al., 1986). FDA has proposed that maximum residual sulfur dioxide levels of 300, 500, and 2000 ppm be permitted in fruit juices, dehydrated potatoes, and dried fruits, respectively (FDA, 1988).

ANTIOXIDANTS

Antioxidants as food additives have a highly significant role in preserving many oxidation-susceptible fruit products. However, fruit products are different compared to cereal and meat products in terms of lipid contents and the use of antioxidants is primarily aimed at restricting the discoloration due to enzymatic browning, besides the loss of carotenoid contents.

The FDA has defined the antioxidants as substances used to preserve food by retarding deterioration, rancidity, or discoloration due to oxidation (CFR 21.170.303). These reactions cause browning, discoloration of endogenous pigments, loss of nutritional value from destruction of vitamin A, C, D, or E, and essential fatty acids such as linolenic acid. Special problems arise in fruits such as avocado due to higher lipid content leading to the possible development of rancid off-flavors and toxic oxidation products (Dziezak, 1986).

The basic antioxidant effect arises out of four types of antioxidants, i.e., free radical scavengers such as BHA and butylated hydroxyl toluene (BHT); reducing agents such as ascorbic acid; chelating agents like EDTA; and secondary antioxidants like dilauryl acid. In case of fruit products, reducing and chelating agents are of prime importance due to their functionality in the prevention of enzymatic browning.

Enzymatic Browning

Enzymatic browning is the discoloration that results in the oxidation of monophenolic compounds of plants or shellfish in the presence of atmospheric oxygen and polyphenoloxidase (PPO). The monophenolic compounds are hydroxylated to o-diphenols and the latter are oxidized to o-quinones (Fig. 9.3) (Mayer and Hanel, 1979). PPO (EC 1.14.18.1) is also known by other names such as tyroninase, polyphenoloxidase, and catechol oxidase, etc. The quinones condense and react nonenzymatically with other phenolic compounds, amino acids, etc., to produce pigments of indeterminate structure. A variety of phenolic compounds is oxidized by PPO (Fig. 9.4), the most important substrates being catechins, cinnamic

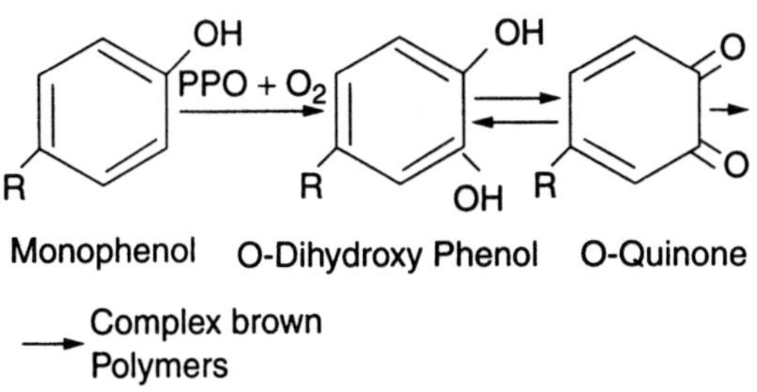

Figure 9.3. Enzymatic browning reaction.

acid esters, 3,4-dihydroxy phenylalanine, and tyrosine. The optimum pH for PPO activity is between 5 and 7. The enzyme is relatively heat labile and can be inhibited by acids, halides, phenolic acids, sulfites, chelating agents, reducing agents such as ascorbic acid, and quinone couplers such as cysteine and various substrate-binding compounds.

Enzymatic browning is a significant problem in a number of important commodities, specifically fruits such as apples, pears, peaches, bananas, and grapes. The discoloration limits the shelf life of many minimally processed foods (Huxsoll et al., 1989) and is also a problem in the production of dehydrated and frozen fruits and vegetables (Shewfelt, 1986). Enzymatic browning can be controlled in some fruit and vegetable products by blanching to inactivate PPO (Hall, 1989). However, blanching cannot be used for certain products with delicate flavors as the heat can be detrimental to some, necessitating the need for the use of chemical additives, apart from the physical measures such as moisture regulation, molecular oxygen exclusion, and lowering of storage temperatures.

Sulfites are effective inhibitors of PPO and hence widely used for the regulation of enzymatic browning in a variety of fruit products. Sulfiting agents such as sulfur dioxide, sodium sulfite, and sodium and potassium bisulfites and metabisulfites are used extensively to prevent enzymatic browning apart from their effective role in restricting the nonenzymatic browning process. These additives can also act as bleaching agents, antioxidants, and reducing agents as well as to check the microbial growth (Taylor et al., 1986). However, the increasing awareness toward the health hazards of sulfur compounds as food additives has necessitated the search for potential substitutes.

Sulfite Substitutes

Sulfite substitutes for the prevention of enzymatic browning include a host of additives viz. ascorbic acid, metal chelators, sulfhydrl amino acids, and acidulants. Such substitutes must be cost effective and approved for food use by FDA (Sapers, 1993). The search for sulfite alternatives has largely remained inconclusive, as sulfites are multifunctional, whereas the alternatives are effective substitutes only for one or two of the functionalities obtained with sulfites. It is unlikely that a multifunctional sulfite substitute can be developed easily. Rather, a

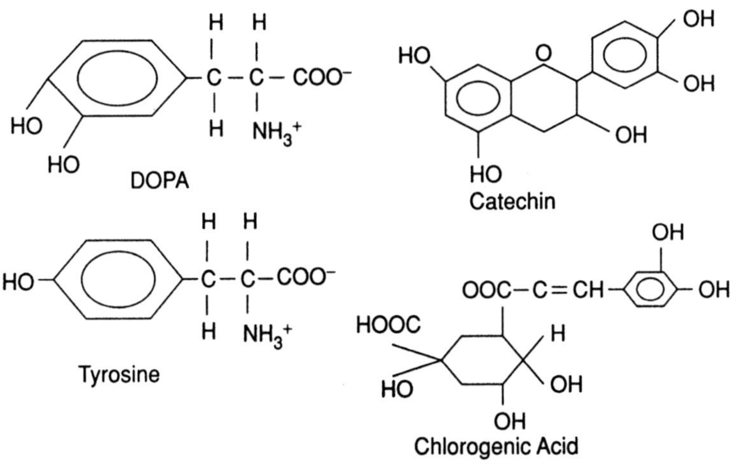

Figure 9.4. Common substrates for polyphenol oxidase.

combination of several active ingredients formulated to meet the needs of specific commodities and product types could be developed.

The best known alternative to sulfites is ascorbic acid, which is a highly effective inhibitor of enzymatic browning, primarily because of its ability to induce quinones, generated by PPO-catalyzed oxidation of polyphenols, back to phenolic compounds before they can undergo further reaction to form brown pigments. However, ascorbic acid once added is completely oxidized to dehydro ascorbic acid (DHAA) by this reaction, quinones can accumulate and undergo browning. In addition, DHAA can itself lead to nonenzymatic browning. At high concentrations, ascorbic acid can directly inhibit PPO (Vamos-Vigyazo, 1981).

Ascorbic acid and its isomer erythorbic acid have been used for a long time as inhibitors of enzymatic browning (Sapers and Miller, 1995). This compound is added to syrups or applied by dipping the fruit in solutions containing the browning inhibitor. Penetration can be enhanced by vacuum infusion (Guadagri, 1949).

Most of the ascorbic acid formulations contain ascorbic acid or its sodium salts; an acidulant such as citric acid, a calcium salt, a phosphate salt, sodium chloride, and cysteine; and preservatives such as benzoates or sorbates. A list of potential sulfite substitutes is given in Table 9.3. The research and development activities in this line are gaining increasing popularity resulting in a spurt in the development of formulations as substitutes for sulfites. However, the FDA approval is likely to be obtained for these formulations with proven use record of ingredients and GRAS substances as such.

FOOD COLORS AND FLAVORS

Colors occupy an important place among the food additives and play an important role in the sensory perception of the products. The functions to be considered for understanding human reactions to foods can be listed as follows:

- Perception – Food selection
- Motivation – Increase or decrease in appetite
- Emotion – Attractive foods for pleasure
- Learning – Prediction of food properties based on color and
- Thinking – Understanding of unusual food properties based on color.

Both artificial and natural colors are used extensively for several food applications including fruit products. Prior to deciding the type of color to be used for a specific food, the following considerations need to be made (Meggos, 1994):

- target shade
- physicochemical attributes of food
- nontechnical and marketing requirements
- target countries
- food processes involved
- packaging type
- storage conditions.

The FDA defines any color additive as "dye, pigment or any other substance made by a process of synthesis or similar artiface, or extracted, isolated or otherwise derived with or without intermediate or final change of identity, from a vegetable, animal, mineral, or other source and that, when added or applied to a food, drug or cosmetic or to the human body or any part thereof, is capable of (alone or through reac-

Table 9.3. Some of the Potential Sulfite Substitutes and Their Functionalities

No.	Additive	Functionality
1.	Ascorbic acid	PPO inhibition, O_2 exclusion
2.	Sodium chloride (inorganic halide)	PPO inactivation
3.	Carrageenan (sulfated polysaccharide)	O_2 exclusion
4.	Xanthan gum	O_2 exclusion
5.	Protease enzymes	PPO inactivation
6.	4, O-Hexyl resorcinol	PPO inhibition, O_2 exclusion
7.	Kojic acid	PPO inhibition, reduction of O-quinones to O-diphenols
8.	EDTA	PPO inhibition by chelation
9.	Polyvinyl pyrollidone	Polyphenol binding
10.	Cyclodextrins	Complex formation with PPO
11.	Cysteine and N-acetyl cysteine	Complexing with PPO
12.	Sodium pyrophosphate	PPO inhibition by chelation

tion with another substance) imparting color thereto." (FDA, 1986)

As per the federal regulations, food colors are termed as certified and uncertified colors. The permanent list of dyes consists of the following and the specific lakes are provisionally listed as follows (Von Elbe and Schwartz, 1996).

No.	Dyes	Lakes
1.	Blue no. 1 (brilliant blue FCF)	Provisional
2.	Blue no. 2 (indigo)	Provisional
3.	Red no. 3 (erythrosine)	–
4.	Red no. 40 (allura red AC)	Provisional
5.	Yellow no. 5 (tartrazine)	Provisional
6.	Yellow no. 6 (sunset yellow FCF)	Provisional
7.	Green no. 3 (fast green FCF)	Provisional

The approved dyes and lakes are used depending on the nature of the food product. Dyes are water-soluble compounds, which manifest their color by being dissolved. The dyes are used to color a variety of fruit products inclusive of beverages and jellies with the maximum GMP limit of 300 ppm. Lakes are extensions of food, drugs, and cosmetics (FD&C) and are water-soluble dyes on a substrate of alumina hydrate. Lakes are used in oily products or products with less moisture. Lakes are also used in products requiring a distinct separation of color.

Uncertified Colors

The permanent list of uncertified colors include a number of natural or natural-identical colorants inclusive of pigments derived from different fruits, vegetables, spices, algae, flowers, food grains, and oil seeds such as corn, cotton seed, etc. Apart from the derivatives of sugar-like caramel, the list also includes compounds such as ferrous gluconate. As such the permanent list of uncertified colors consists of (Von Elbe and Schwartz, 1996) (a) annatto extract, (b) canthaxanthin, (c) β-carotene, (d) β-apo-8-carotenal, (e) beet powder, (f) ferrous gluconate, (g) grape color extract, (h) grape skin extract, (i) saffron, (j) fruit and vegetable juices, (k) paprika, (l) oleoresin, (m) riboflavin, (n) carrot oil, (o) caramel, (p) turmeric, (q) turmeric oleoresin, (r) cochineal extract, (s) defatted and cooked cotton seed flour, (t) Synthetic iron oxide, (u) dried algae, (v) aztec marigold extract, (w) corn endosperm oil, and (x) titanium dioxide.

Among the uncertified colors, three synthetic carotenoids appropriately termed as natural-identical compounds are permanently listed by FDA as food color additives that are exempt from certification. They are β-carotene, β-apo-8-carotenal (apocarotenal), and canthaxanthin (FDA, 1986). Carotenoid crystals are sensitive to light and oxidation and require storage under vacuum or inert gas (Emodi, 1978). Carotenoids can be converted into products with both water and fat dispersibilities. Carotenoids are generally used within a range of 1–25 ppm of food and are among the colorants with the highest tinctorial potency (Dziezak, 1987). β-carotene, apocarotenal, and canthaxanthin were approved by FDA in 1956, 1963, and 1969, respectively. The maximum usage levels are guided by GMP and standards of identity.

A variety of natural plant extracts and fruit/vegetable juices chiefly constitute natural colors of biological origin among the uncertified colors. The stabilized forms of natural colorants are gaining increasing popularity and soon may dominate the food colorant industry as such.

The uncertified colors are covered under Title 21 section 401. However, wherever the standards of identity are promulgated the added color needs to be authorized by such standards. The maximum use levels of some specific colorants are canthaxanthin at 66 mg/kg solid or pint of liquid; similarly, apocarotenal is slated at a maximum level of 33 mg/kg food. Titanium dioxide should not exceed 1% by weight of the food. As such the regulations favor the widespread use of the uncertified colorants.

Flavors

Flavors are aroma-rendering additives in the form of extracts or concentrates. They are complex ingredients that play a key role in food acceptance. When flavors are perceived as the food is eaten, their sensation immediately evokes feelings about the degree of pleasure in the immediate moment and at the same time strongly influences the intentions about the consumption of that type of food at a later date.

The flavors are usually classified using a definitive scheme based on both origin as well as usage (Hall and Merwin, 1981), as shown in Table 9.4.

Table 9.4. Some Important Natural Flavors

Types	Examples
Condiments	Mustard, catsup, and vinegar
Spices	Black pepper, ginger, celery, basil, and caraway
Fruit juice concentrates	Concentrated lemon, orange, cherry, and apple flavorings and juices
Processed flavors	Starter distillate and hydrolyzed vegetable protein
Oleoresins	Oleoresins of cinnamon, celery, ginger, and black pepper
Essential oils	Nutmeg, celery, and cinnamon
Aromatic chemicals	Vanillin, anethol, menthol, citral, and isobutyl methoxy pyrazine

Table 9.5. Forms of Commercial Flavor Products

a.	Solid form	Encapsulated flavors, crystals, spray-dried powders, dried extracts, plated powders, and freeze-dried powders.
	Advantage	Highly volatile compounds such as dimethyl sulfide, methyl mercaptan can be encapsulated to provide flavors in solid form
b.	Semisolid paste	Oleoresins and fruit concentrates
	Advantage	Easy dispersion and uniform flavoring
c.	Liquid flavors	Flavors in emulsified form with a compatible solvent base
	Advantage	Suitability for beverages and other liquid products

Commercial Forms

Flavors are available in different commercial forms and their usage depends on the nature of the flavor as well as the type of product, as shown in Table 9.5.

Functionalities of Flavors

1. To render new taste to the product or to enhance the already existing flavor.
2. To substitute for flavor losses during processing.
3. To replace the components missing from the overall flavor of the product.
4. To mask the undesirable flavors.

As far as product applications are concerned, fruit flavors, both natural and synthetic, represent around 48% of the total sales value. Fruit beverages constitute the largest user of fruit flavors followed by desserts, jams, and jellies. Citrus flavors are the most popular flavors among the fruit flavors.

As far as the regulatory status is concerned, FDA has been working closely with Flavor and Extract Manufacturers Association (FEMA) to obtain information on the identity and safety of flavors (Sethi and Sethi, 2004). Some of the GRAS flavors are mentioned below (Table 9.6).

SWEETENERS

Sweeteners are widely used in fruit processing and apart from sensory properties a number of functions are also attributed to the sweeteners. Sweeteners are one of the earliest food additives subjected to extensive use. The preservative function was the major one put to practice for natural sweeteners such as sucrose. Ever since, extensive research has led to the development of several sweeteners of natural or synthetic origin (Fig. 9.7).

Characteristics of Sweeteners

1. Sweeteners are of nutritive as well as of nonnutritive nature. In other words, they are either metabolized for obtaining energy or not metabolized.
2. Sweeteners can be of natural or synthetic origin. Products arising out of modification of natural sweeteners are also considered as natural.
3. Low-calorie sweeteners are also known as dietetic sweeteners. The high-intensity sweeteners pave

Table 9.6. Some of the GRAS flavoring substances

FEMA No.	Substance Primary Name
3909	Cyclo hexanone
3912	9-Decenal
3923	3-Hexenal
3941	Maltol propionate
3958	Phenyl acetate
3961	2-Propyl pyragine

way for low-calorie inputs due to lesser quantity of sweetener used.

4. Fruit products such as juice concentrates and dehydrated juice powders can also be used as natural sweeteners.
5. Sweeteners such as sugars have high degree of humectant function and are, therefore, used successfully as preservatives to regulate water activity in the food products, which ultimately leads to the protection against microbial spoilage.
6. Sweeteners are successfully used as effective tools in fruit product design, paving way for the development of a variety of beverages and concentrates.

The list of high-intensity sweeteners extensively used for food applications is given in Table 9.7.

POLYOLS

Polyols (sugar alcohols) are reduced carbohydrates and are used widely for food applications due to their special sensory and dietary functions, e.g., sorbitol, xylitol, mannitol, and lactitol (Giese, 1993). They are used in glazed/dried and texturized fruits. The FDA had accorded an "interim approval" for sorbitol, xylitol, mannitol, and lacitol within an overall range of 1.8–2.4 Kcal/g. The additional functionalities of polyols include high viscosity, hygroscopicity, cool taste, sequestering ability, retardation of crystallization, and bulking ability. Polyols such as sorbitol are widely used in dietetic foods, as it can be metabolized without insulin and is noncariogenic. However, excessive intake of polyols has laxative effect.

Although honey, sugar, and other traditional sweeteners have been exploited for their taste, caloric value, and functional abilities in foods for hundreds of years, the discovery and use of most alternate sweeteners date back to the past century. Three alternate sweeteners (Fig. 9.7) are currently approved for use in the United States, i.e., acesulfame-K, aspartame, and saccharin. Cyclamates were considered GRAS at one time but are now banned from use. Apart from steviocide, thaumatin is a natural alternate sweetener obtained from the fruit proteins of the West African plant *Thaumatococcus danielli*. Under the trade name "Talin," thaumatin has undergone numerous safety tests and results indicate that it is safe at the levels that it would be consumed (Higginbotham, 1986). Talin has an unusual taste profile, with a lingering sweet taste 2000 times sweeter than that of sucrose. Use of this compound as a flavor enhancer is permitted in chewing gum in the United States, where it is in the FEMA's GRAS list. Thaumatin is likely to

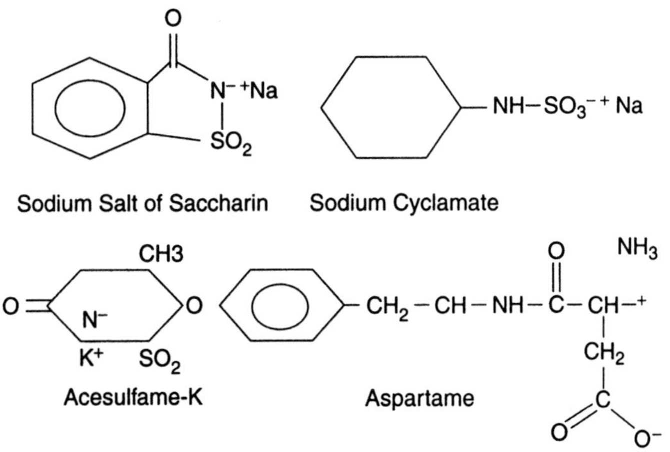

Sodium Salt of Saccharin

Sodium Cyclamate

Acesulfame-K

Aspartame

Figure 9.5. Chemical structures of widely used alternate sweeteners.

Table 9.7. Features of Some Important High-Intensity Sweeteners

Sweeteners	Relative Sweetness	Application	Regulatory Status
Acesulfame-K	200	Canned fruits, low-sugar jams and jellies, and dry beverage mixes	FDA approved
Aspartame	200	Tabletop sweeteners, and dry beverage mixes	FDA approved
Cyclamates	30	Fruit flavor enhancer, making of tartness in citrus products	Approval pending
Saccharin	200–700	Stewed fruits, canned fruits, jams, jellies, and diet drinks	FDA approved
Sucralose	600	Jams, jellies, and canned fruits	Approval pending
Steviocide	300	Less use in fruit products	Approval pending

attract further attention in the area of natural alternate sweeteners.

HYDROCOLLOIDS: THICKENERS, STABILIZERS, AND GUMS

Thickeners, stabilizers, and gums are the basic texturizing additives and are widely used in fruit processing. In fact among the food additive industry, apart from flavor industries, the ones concerned with the texturizers form the core of the industry in terms of net turn over of the finished products. Hydrocolloids are long-chain polymers that function as thickeners and stabilizers.

FUNCTIONS OF HYDROCOLLOIDS

1. Suspension of particulate matter in food products along with the regulation of crystallization.
2. Optimization of rheological properties of solid as well as liquid food products. The major parameters include flow and mouth-feel properties.
3. Stabilizers for oil and water emulsion systems.
4. Binding of dry and semidry products.
5. Optimized gelation to give both hard and soft gels.
6. Foam stabilization and flavor fixation.

The major sources of hydrocolloids are plant products viz. gum exudates, seeds, seaweeds. Products obtained by fermentation and chemically modified polysaccharides also contribute to the development of hydrocolloids for different food applications. A variety of terms are often used to describe hydrocolloids. The terms are based either on the origin of the product or on the function for which it is being used. Three of the most common terms employed are gums, stabilizers, and hydrocolloids. The term "gum" describes a wide variety of water-soluble thickening and gelling polysaccharides (Carr, 1993).

"Stabilizer" is another term used in the food industry to describe products that prevent separation of multicomponent food systems during storage. Another term that addresses both the behavior and physical characteristics of food gums is "hydrocolloids." This term is a contraction of two terms "hydrophilic colloid" and describes the water loving nature and colloidal characteristics of this class of compounds (Table 9.8).

The use of hydrocolloids in fruit processing is on the rise. Apart from the general use of pectin in fruit bars for texturization and in jams for the purpose of gelling, newer fruit products such as structured fruits involve novel applications of state-of-art gelling agents. Ample information is available on the mechanical properties of gelling agents such as alginates with or without other polysaccharides (Pelaez and Karel, 1981). Nevertheless, for texturizing pulp or concentrated fruit juice, optimal gelling conditions need to be introduced, i.e., the pH of the pulp/juice concentrate. Calcium alginate gels have been found to be superior causing continuity in gel formation (Wood, 1975). Fruit pulps such as mango pulp have been texturized using hydrocolloids such as alginates (Mouquet et al., 1992).

As far as the regulatory aspects are concerned, the above-listed hydrocolloids are approved by FDA and are classified as either food additives or GRAS

Table 9.8. Classification of Hydrocolloids

No.	Type of Hydrocolloid	Functions	Sources	Food Applications
1.	Unmodified starch	Thickener Gelling, adhesive, and film former	Potato, cereals tapioca, and arrow root	Pie fillings Jams and jellies
2.	Modified starch	Bodying and gelling Improvement of viscosity and other rheological parameters Thermal resistance against higher as well as lower temperatures Modification in gelling To improve solubility and gelling in cold water	Products of chemical modification of natural polysaccharides	Pie fillings Canned fruits Fruit-based desserts
3.	Casein	Film former with cohesive and adhesive properties	Milk	Edible films for packaging of precut fruits and vegetables
4.	Guar gum	Stabilizer	Guar beans	Beverages Water-based frozen fruit desserts
5.	Gum Arabic	Stabilizer Emulsifier	Exudate from genus *Acacia*	Beverages Emulsifier for citrus oils and flavors
6.	Carrageenan	Gelation	Irish moss	Frozen desserts Structured fruits
7.	Pectin	Gelation Texturization as binder Stabilizer	Fruit wastes	Jams and jellies Beverages Fruit bars Fruit snacks and desserts
8.	Xanthan gum	Stabilizer Gelation	Fermentation product from *Xanthomonas comprestris*	Beverages Puddings Dry mixes of beverages
9.	Carboxy methyl cellulose	Thickener Stabilizer	Modified cellulose	Dry drink mixes Fruit-flavored syrups

substances (CDR 121.172.580-CDR121.172.874). Their safety has been promulgated by JECFA and EEC.

EMULSIFIERS

Emulsions are necessary for obtaining homogeneity in liquid foods having a tendency toward phase separation during processing/storage. Emulsifiers have lipophilic as well as hydrophilic groups enabling the products to bring together the water and oil moieties without phase separation.

Apart from phase separation, the other functions of food emulsifiers are to enhance stability in flavors and fats and oils by restricting the onset of rancidity. Emulsions are also used for better crumb texture in baked products due to optimal starch complexing property (Thompson and Buddemeyer, 1954).

Usually except lecithin most of the emulsifiers are used in combinations. The important emulsifiers used widely are as follows:

- Mono- and diglycerides
- Acetylated monoglycerides
- Sucrose fatty acid esters
- Stearoyl-2-lactylates
- Propylene glycol esters
- Sorbitan esters
- Diacetyl tartaric acid esters of monoglycerides (DATEM)

FDA permits lecithin, mono- and diglycerides, and DATEM as GRAS. The other emulsifiers are approved under the standards of identity at specific levels. In fruit products, emulsifiers find applications in flavor emulsions, beverages, pie fillings, fruit desserts, and salad dressings (Mahungu and Artz, 2002).

ENZYMES

Enzymes are biological catalysts and proteinaceous in nature. The use of enzymes in food processing in general and in fruit industries in particular is an offshoot of advances in fermentation/biotechnology. Enzymes, which exist within the fruits, perform a number of functions such as softening, flavor development, ethylene biosynthesis, etc. Vegetables and fruits are also rich in oxidases, a class of enzymes, such as PPO and peroxidase, which cause browning reaction and off-flavor development, which is detrimental to the fruit quality (Whitaker, 1996).

In case of fruit processing, enzymes are used mainly as processing aids aimed at specific functions, such as:

- improvement in juice extraction yields
- increment in solids recovery
- improvement in filtration
- removal of nonnutritional factors
- flavor enhancement
- viscosity modifications
- anticlouding operations
- antifouling operations for membrane concentrations.

Enzymes of commercial importance used in fruit processing are (a) pectinases, (b) cellulases, (c) amylases, and (d) glucose oxidase. Pectic enzymes, i.e., pectin methylesterase and polygalacturonase are used often in combination with amylases and cellulases in fruit and vegetable juice clarifications to obtain higher juice yields and clarity. The turbidity/cloudiness of fresh fruit juices can be decreased with pectinase treatment due to the removal of negatively charged pectin deposits on particulate matter, which ultimately results in the coagulation of turbidity causing materials (Yamasaki et al., 1967). Enzymatic treatment of soft fruit pulp facilitates pressing and improves juices and anthocyanin pigment yields (Neubeck, 1975). Pectin degrading enzymes are also used to degrade highly esterified apple pectin and increase juice yields (Devos and Pilnik, 1973). Amylases are often used along with pectinases to clarify juices such as banana to obtain optimal juice yields. Similarly, application of cellulases also facilitates higher recovery of juices due to the degradation of cellulosic matrix. Cellulases are also used for waste treatments from fruit processing units for the development of value-added products. Glucose oxidase in combination with catalase is used to protect citrus juices from off-flavor development (Scott, 1975) and in the prevention of enzymatic browning in frozen fruits (Somogyi, 1996).

As far as the regulatory aspects are concerned, enzymes are considered as direct food additive as per FDA. The source organisms play an important role in the affirmation of GRAS status. Pectinases as well as glucose oxidase derived from *Aspergillus niger* are considered as GRAS. Same holds good with α-amylase and cellulase derived from *A. niger*.

VITAMINS

Vitamins occupy an important place in nutrition and participate in a variety of biological processes. Fruits are rich sources of both water-soluble as well as fat-soluble vitamins. Vitamin A precursors in the form of carotenoids are available in significant proportions in several fruits. On the other hand, fruits are also rich sources of vitamin C. Certain fruits, i.e., seabuckthorn is known to be a good source of vitamin E. Apart from this, vitamins are usually used as supplements. The supplements may be directed toward compensating the processing losses. The processing loss of water-soluble vitamins is significant and supplementation is required to maintain the nutritional balance. Thermally processed fruit extracts need such supplementations due to heavier loss of vitamins during heat processing as well as subsequent storage of the product (Kanner et al., 1982).

Apart from supplementation, vitamins also have certain functional properties. Vitamin C is widely used as ascorbic acid in dextro- or levo-rotatory forms as a natural antioxidant. Vitamin E (tocopherols) is also known as a potential antioxidant. Ascorbic acid on oxidation can inhibit PPO activity by restricting the availability of molecular oxygen required for the reaction. Fruit processing unit operations such as blanching, thermal processing, and freezing considerably reduce vitamin C content. The storage losses add to the processing losses, demanding necessary supplementation (Kacem et al., 1987). The other functionalities include use of riboflavin-5-phosphate as a colorant.

The FDA considers vitamins as GRAS and nutritional supplements. Pantothenate is allowed as a food additive and requires label-mediated expression with regards to the concentration used. Some of the commercial forms are marketed as follows:

- vitamin A as vitamin A acetate;
- thiamine as thiamine hydrochloride;
- pantothenic acid as calcium pantothenate;
- pyridoxin as pyridoxine hydrochloride;
- ascorbic acid as ascorbic acid, calcium ascorbate, sodium ascorbate; and
- vitamin E as tocopherol acetate, DL-α-tocopherol.

SAFETY AND HEALTH IMPLICATIONS OF FOOD ADDITIVES

The safety and health implications of food additives are given utmost importance in framing the standards and regulatory measures. The FDA as well as Codex Alimentarius and other national and international bodies stress on the need of ADI as a premier aspect to restrict cytotoxicity-induced health hazards. ADI limits are usually expressed in terms of chemical exposure of body on unit body weight basis and are based on authentic risk assessment (Winter and Francis, 1997).

The concept of ADI renders more flexibility to maximum limits prescribed for various food categories with regards to standards of identity. Lowest observed effect level (LOEL) and no observed effect level (NOEL) form the bases of risk assessment in framing the ADI levels. ADI is the NOEL value divided by 100 when the NOEL is derived from animal studies or the NOEL value divided by 10 when the NOEL relates to human data (Renwick, 1996).

The risks involved in overuse or underuse of additives can give rise to several types of risks, i.e., (a) microbial hazards, (b) nutritional hazards, (c) color additive hazards, (d) environmental hazards, and (e) immunological/physiological hazards. The typical symptoms of health-based hazards include hypersensitivity and associated allergic reactions. The allergic reactions may cause respiratory problems such as asthma. The cross reactions of sulfites and steroid dependency in asthma patients have given rise to novel debates and amendments in regulations, which ultimately resulted in the ban of sulfites in ready-to-eat fruit- and vegetable-based products (Anon, 1990).

Toxicological research continues to throw light on implications of chemical toxicity hitherto unknown. Colorants, antimicrobial compounds, antioxidants, antibrowning agents, and sweeteners continue to receive regulatory restrictions. The advent of functional foods, dietary supplements, and transgenic products offer plethora of challenges to ensure safety to the consumers. Food sector is basically a buyers market, and it is important to keep the "psyche" of the consumer in a satisfaction mode. Mere toxicological certification may not necessarily satisfy the consumer. The regulatory authorities, therefore, are striking a fine balance between consumer safety and the manufacturer's product requirements to keep up optimal quality and commercial feasibility (Branen and Haggerty, 2002).

Labeling is being optimally used to overcome many problems and at the same time keeps the consumer aware of the compositions involved along with the statutory warnings, if any. The usefulness of advanced labeling can come a long way in bridging the gap among regulator, processor, and consumer.

FUTURE TRENDS

The future trends in research and development as well as in commercial application involving direct food additives may be outlined as follows:

1. Increased stress on natural preservatives, sweeteners, colorants, and antioxidants.
2. Minimal or no use of chemical additives with emphasis on physical conditioning of the products.
3. Advanced research on gene products for transgenics and comprehensive studies on the health implications thereof.
4. Restriction in additive use by adopting hurdle-based processing.
5. Labeling strategies to counter growing toxicological concerns.
6. Consumer awareness and technical strategies to prevent misuse of food additives.

REFERENCES

Anon. 1990. Sulfiting agents: revocation of GRAS status for use on fresh potatoes preserved or sold unpackaged and unlabeled to consumers. Federal Register. 55(51): 9826–9833.

Anon. 1992. In: General Requirements. Codex Alimentarius Brochure. Vol. I, 2nd ed. Issued by secretariat of the Joint FAO/WHO food standards programme, FAO, Rome, pp. 1–7.

Anon. 1998. In: Pesticide Residues in Foods. Codex Alimentarius Brochure. Vol. 20, 2nd ed. Issued by secretariat of the joint FAO/WHO food standards programme, FAO, Rome. pp. 1–14.

Anon. 2002. Leatherhead International Report on The Food Additive Market—Global Trends and Developments, 2nd ed. May (www.Just.food.Com/store/products_detail asp?art).

Baird-Parkar, A. 1980. Organic acids. In: Microbial Ecology of Foods. Vol. 1. Siliker, J.H., ed. Academic Press, New York, pp. 126–148.

Beuchat, L.R., Golden, D.A. 1989. Antimicrobials occurring naturally in foods. Food Technology. 43(1): 134–142.

Bolin, H.R., Steele, R.J. 1987. Nonenzymatic browning in dried apples during storage. Journal of Food Science. 52: 1654–1657.

Branen, A.L., Davidson, P.M., Katz, B. 1980. Antimicrobial properties of phenolic antioxidants and lipids. Food Technology. 34(5): 42–53.

Branen, A.L., Haggerty, R.J. 2002. Introduction to food additives. In: Food Additives, 2nd edn. Branen, A.L., Davidson, P.M., Salminen, S., Thorngate, J.H., III., eds. Marcel and Dekker Inc., New York-Basel, pp. 1–9.

Carneiro de melo, A.M.S., Cassar, C.A., Miles, R.J. 1998. Trisodium phosphate increases sensitivity of gram-negative bacteria to lysozyme and nisin. Journal of Food Protection. 61: 839–844.

Carr, J.M. 1993. Hydrocolloids and stabilizers. Food Technology. (47): 10, 100.

Chichester, D.F., Tanner, F.W. 1975. Antimicrobial food additives. In: Handbook of Food Additives. Furia T.E., ed. CRC Press, Cleveland, pp. 137–207.

Choi, S.W., Sapers, G.M. 1994. Effects of washing on polyphenols and polyphenol oxidase in commercial mushrooms (*Agaricus bisporus*). Journal of Agricultural and Food Chemistry. 42(10): 2866–2290.

Davidson, P.M., Juneja, V.K., Branen, A.L. 2002. Antimicrobial agent. In: Food Additives, 2nd ed. Branen, A.L., Davidson, P.M., Salmeinen, S., Thorngate, J.H., III., eds. Marcel Dekker Inc., New York-Basel, pp. 563–620.

Devos, L., Pilnik, W. 1973. Proteolytic enzymes in apple juice extraction. Process Biochemistry. 8:18–19.

DuBois, G.E. 1992. Sweeteners: non nutritive. In: Encyclopedia of Food Science and Technology. Hui, Y.H., ed. John Wiley and Sons Inc., New York, pp. 2470–2487.

Dwivedi, B.K. 2003. Extrudable candy fruit flavored food product. U.S. Patent No. US 6 548 090 B2.

Dziezak, J.D. 1986. Preservative systems in foods, antioxidants and antimicrobial agents. Food Technology. 40(9): 94–136.

Dziezak, J.D. 1987. Application of food colorants. Food Technology. 41(4): 78–88.

Dziezak, J.D. 1990. Acidulants: ingredients that do more than meet the acid test. Food Technology. 45(1): 76–83.

Emodi, A. 1978. Carotenoids-properties and applications. Food Technology. 32(5): 38–42.

Fabian, F.W., Graham, H.T. 1953. Viability of thermophilic bacteria in the presence of varying concentration of acids, sodium chloride and sugars. Food Technology. 7: 212–217.

FAO. 1996. Food fortification technology and quality control FAO technical meeting Rome. FAO Food and Nutrition paper 60.

Fassin, K., Bachmueller, J. 2000. A process for manufacture of fruit gum confectionary. European Patent No. EP 1002 4 65 A1.

FDA. 1979. Specific Food labeling requirements. Food and Drug Administration. Code of Federal Regulations. Title 21, paragraph 101.22(a). Washington, D.C.

FDA. 1986. Colour additives. Food and Drug Administration, Code of Federal Regulations. Title 21, 70.3(f), Washington, D.C.

FDA. 1988. Sulfiting agents; Affirmation of GRAS status. Food and Drug Administration. Federal Reg. 53; 51065–51084. Washington, D.C.

FDA. 1995. The FDA food additive review process: backlog and failure to observe statutory dead line: HR Rep No. 104-436, 104th Cong., 1st session, Washington, D.C.

Friedman, M., Molnar, P.I. 1990. Inhibition of browning by sulfur amino acids. 1. Heated amino acid glucose systems. Journal of Agricultural and Food Chemistry. 38: 1642–1647.

Fulton, K.R. 1981. Surveys of industry on the use of food additives. Food Technology. 35(12): 80–83.

Gardner, W.H. 1972. Acidulants in food processing. In: Handbook of Food Additives, 2nd ed., vol. 1. Furia T.E., ed. CRC Press, Cleveland, Ohio, pp. 225–270.

Giese, J.H. 1993. Alternative sweetners and bulking agents. Food Technology. 57(1): 114–126.

Gross, E., Morell, J.L. 1971. The structure of nisin. Journal of American Chemical Society. 93:4634–4635.

Guadagri, D.G. 1949. Syrup treatment of apple slices for freezing preservation. Food Technology. 3:404–408.

Hailer, E. 1911. Experiments on the properties of free sulfurous acid of sulfites and a few complex compounds of sulfurous acid in killing germs and rehandling their development. Arlo Keis Gesundh. 36.297 [Chem. Abst. 5:1805 (1911)].

Hall, G.C. 1989. Refrigerated, frozen and dehydrofrozen apples. In: Processed Apple Products. Downing, D.D., ed. Avi-Van Nostrand Reinhold, New York, pp. 239–256.

Hall, R.L., Merwin, E.J. 1981. The role of flavors in food processing. Food Technology. 35(6): 46–52.

Hartwig, P., McDaniel, M.R. 1995. Flavor characteristics of lactic, malic, citric and acetic acids of various pH levels. Journal of Food Science. 60:384–388.

Higginbotham, J.D. 1986. Talin protein (thaumatin). In: Alternate Sweetners. Nabors, L.O., Gelardi, R.C., eds. Marcel Dekker Inc., New York, pp. 103–134.

Huntur, D., Segel, I.H. 1973. Effect of weak acids on amino acid transport by pencillium chrysogenum: evidence for a proton or charge gradient as the driving force. Journal of Bacteriology. 113:1184–1192.

Huxsoll, C.C., Bolin, H.R., King, A.D., Jr. 1989. Physicochemical changes and treatments for highly processed fruits and vegetables. In: Quality Factors of Fruits and Vegetables. Chemistry and Technology. Jen, J.J., ed. ACS Symp. Series. 405, 203–215, American Chemical Society, Washington, D.C.

Jay, J.M. 1986. Food preservation with chemicals. In: Modern Food Microbiology. Van Nostrand Reinhold, New York, pp. 259–296.

Josyn, M.A, Braverman, J.B.S. 1954. The chemistry and technology of the pretreatment and preservation of fruit and vegetable products with SO_2 and sulphites. Advanced Food Research. 5:97–160.

Kacem, B., Cornnell, J.A., Marshall, M.R., Shireman, R.B., Mathews, R.F. 1987. Nonenzymatic browning in aseptically packaged orange drinks: Effect of ascorbic acid, amino acids and oxygen. Journal of Food Science. 52(6):1668–1672.

Kaletunc, G., Nussinovitch A., Peleg, M. 1990. Alginate texturisation of highly acid fruit pulps and juices. Journal of Food Science. 55(6):1759–1761.

Kanner, J., Fishbein, J., Shalom, P., Harel, S., Ben-Gera, I. 1982. Storage stability of orange juice concentrate packaged aseptically. Journal of Food Science. 47:429–435.

Kennedy, J.F., Rivers, Z.S., Lloyd L.L., Warne F.P., Jumel, K. 1990. Studies on nonenzymatic browning in orange juice using a mode/system based on freshly squeezed orange juice. Journal of Science Food and Agriculture. 52:85–95.

Labuza, T. P., Saltmarch, M. 1981. In Water Activity: Influence of Food Quality. Rockland, L.B., Stewart, G.F., eds. Academic Press, New York, pp. 605–650.

Levine, A.S., Fellers, C.R. 1940. Action of acetic acid on food spoilage microorganisms. Journal of Bacteriology. 39:499–514.

Loescher, J., Kroh, L., Westpal, G., Vogel, J. (1991). L ascorbic acid a carbonyl component of nonenzymatic browning reactions 2, amino carbonyl reactions of L-ascorbic acid Zeieschreft Fuer Lebensm. UntersForch. 192:323–327.

Macris, B.J. 1975. Machanism of benzoic acid uptake by Saccharomyces cerevisiae. Applied Microbiology. 30:503–510.

Mahungu, S.M., Artz, W.E. 2002. Emulsifiers. In: Food Additives, 2nd ed. Branen, A.L., Davidson, P.M., Salminen, S., Thorngate, J.H., III ed. Marcel and Dekker Inc., New York-Basel, pp. 1–9.

Mannikes, A. 1992. Mayonnaices and salad dressings Dragoco Report. Flavoring Information Service. 37(4):139–146.

Mayer, A.M., Hanel, E. 1979. Polyphenol oxidases in plants. Phytochemistry. 18:193–215.

Meggos, H.N. 1994. Effective utilization of food colours. Food Technology. 1:112.

Montinez, A., Fernanez, I.S., Rodvigo, E., Rodvigo, M.C. 1997. Methods of minimal process. European Food and Drink Review. 39:41–42.

Mouquet, C., Dumas, J.C., Guilbert, S. 1992. Texturization of sweetened mango pulp. Optimization using response surface methodology. Journal of Food Science. 57(6):1395–1400.

Murdock, D.I. 1950. Inhibitory action of citric acid on tomato juice flatsour organism. Food Research. 15:107–113.

Neubeck, C.E. 1975. Fruits, fruit products and wine. In: Enzymes in Food Processing, Reed, G., ed. Academic Press, New York, pp. 397–442.

Nussinovitch, A., Kopelman, J., Mizrahi, S. 1991. Mechanical properties of composite fruit products based on hydrocolloid gel, fruit pulp and sugar. Lebensmittel-Wissenchaft und Technologie. 24:214–217.

Ollikainen, H.T., Kultanen, S.M., Kurkela, R. 1984. Relative importance of colour, fruity flavor and sweetner in the overall liking of soft drinks. Journal of Food Science. 49:1598–1600, 1603.

Ough, C.S. 1983. Sulfur dioxide and sulfites. In: Antimicrobials in Foods. Branen, A.L., Davidson, P.M., eds. Marcel and Dekker, Inc., New York-Basel, pp. 177–203.

Pelaez, C., Karel, M. 1981. Improved method for preparation of fruit stimulating alginate gels. Journal of Food Processing and Preservation. 5:63–81.

Rajashekhara, E., Suresh, E.R., Ethiraj, S. 1998. Thermal death rate of ascospores of Neasortorya fischeri ATCC 200957 in the presence of organic acids and preservatives in fruit juices. Journal of Food Protection. 61:1358–1362.

Rajashekhara, E, Suresh, E.R., Ethiraj, S. 2000. Modulation of thermal resistance of ascospores of Neosaortorya fischeri by acidulants and preservatives in mango and grape juice. Food Microbiology. 17(3):269–275.

Raju, P.S., Ashok, N., Mallesha, Das Gupta, D.K. 2000. Physiological and quality changes during minimal processing and storage of shredded cabbage. Indian Food Packer. 4:51–58.

Rammell, C.G. 1962. Inhibition of citrate of the growth of coagulase positive Staphylococci. Journal of Bacteriology. 84:1123–1124.

Renwick, A.G. 1996. Needs and methods for priority setting for estimating the intake of food additives. Food Additives and Contaminants. 13(4):467–475.

Robach, M.C. 1980. Use of preservatives to control microorganisms in foods. Food Technology. 10:81–84.

Robach, M.C., Pierson, M.D. 1978. Influence of parahydroxy benzoic acid esters on the growth and toxin production of *Clostridium botulinum* 10755A. Journal of Food Science. 43:787–789, 792.

Sapers, G.M. 1993. Browning of foods: Control by sulfites, antioxidants and other means. Food Technology. 10:75–84.

Sapers, G.M., Miller, R.L. 1995. Heated ascorbic/citric acid solution as browning inhibitor for prepeeled potatoes. Journal of Food Science. 60:762–766, 776.

Scott, D. 1975. Applications of glucose oxidase. In: Enzymes in Food Processing. Reed, G., ed. Academic Press, New York, pp. 519–549.

Seiferi, D. 1992. Functionality of food acidulants. International Journal of Food Ingredients. 3:4–7.

Sethi, V., Sethi S. 2004. Importance of food additives in food industry. Beverages Food World. 1:27–30.

Shewfelt, R.L. 1986. Flavor and color of fruits affected by processing. In: Commercial Fruit Processing. Woodroof J.G., Luh, B.S., ed. AVI Publication Co., Westport Conn., pp. 481–529.

Sloan, A.E. 2000. The top ten functional food trends. Food Technology. 54(4):33–62.

Smith, O. 1987. Transport and storage of potatoes. In: Potato Processing, 4th ed. Talburt, W.F., Smith, O., ed. Avi-Van Nostrand Reinhold, New York, pp. 203–285.

Sofos, J.N., Busta, F.F. 1981. Antimicrobial activity of sorbates. Journal of Food Processing and Preservation. 44:614–622.

Somogyi, L.P. 1996. Direct food additives in fruit processing. In: Processing Fruits Science and Technology. Somogyi, L.P., Ramaswamy, H.S., Hui, Y.H., ed. Technomic Publ. Co. Inc., Lancaster, Basel. pp. 293–361.

Sumner, S.S., Eifert, J.D. 2002. Risks and benefits of food additives. In: Food Additives, 2nd ed. Branen, A.L., Davidson, P.M., Salminen, S., Thorngate, J.H., III., eds. Marcel and Dekker, New York-Basel, pp. 27–42.

Taylor, S.L., Higley, N.A., Bush, R.K. 1986. Sulfites in foods: uses, analytical methods, residues, fate, exposure assessment, metabolism, toxicity and hyper sensitivity. Advanced Food Research. 30(1):1–76.

Thompson, J.E., Buddemeyer, B.D. 1954. Improvement in flour mixing characteristics by a steryl lactylic acid salt. Cereal Chemistry. 31:296–302.

Uebersax, M.A., Occena, L.G. 1993. Legumes in the diet. In: Encyclopaedia of Food Science and Technology, vol. IV. Macrae, R., Robinson, R.K., Sadler, M.J., eds. Academic Press, New York, pp. 2718–2725.

Vamos-Vigyazo, L. 1981. Polyphenol oxidase and peroxidase in fruits and vegetables. CRC Critical Review in Food Science and Nutrition. 15:49–127.

Von Elbe, J.H., Schwartz, S.J. 1996. Colourants. In: Food Chemistry, 3rd edn. Fennama, O.R., ed. Marcel Dekker Inc., New York-Basel-Hongkong, pp. 651–722.

Whitaker, J.R. 1996. Enzymes. In: Food Chemistry, 3rd ed. Fennamma, O.R., ed. Marcel Dekker Inc., New York-Basel-Hong Kong, pp. 431–530.

Winter, C.K., Francis, F.J. 1997. Assessing, managing and communicating chemical food risks. Food Technology. 51(5):85–92.

Wood, F.W. 1975. Artificial fruit and process thereof. United States Patent No.3, 892, 870.

Yamasaki, M., Kato, A., Chu, S.Y., Arima, K. 1967. Pectic enzymes in the clarification of apple juice. Part II. The mechanism of clarification. Agricultural and Biological Chemistry. 31:552–560.

10
Fruit Processing Waste Management

Judit Monspart-Sényi

INTRODUCTION

Since the middle of the 20th century environmental pollution has increased, as a result of technological development and an explosion in the population. Consequently, international, states, and civil organizations joined forces to limit economic development that contribute to environmental pollution—water, air, soil, and landscape—that endangers human life (Anon, 2003).

The concept of "sustainable development" triggers the interest of an increasing number of people who pay attention to the environment not only in their workplace but also in their homes.

National leaders often refer to cooperation between the economic sector and the state as an essential requirement to success. Thus environmental protection and economic growth are closely related. Economic growth enables the implementation of environmental protection. However, within the framework of a market economy, economic growth cannot be sustained without protection of the environment and the living world. Industry should operate in a more efficient way, producing more products with less input and waste. Moreover, customer preferences should also be more sustainable (Vermes, 1998).

THE IMPACT OF FOOD PRODUCTION ON THE ENVIRONMENT

Production and acquisition of food, which is necessary to sustain life, is, at the same time, continuously changing the environment. Traditional rural production was environmentally friendly; closed systems were developed since farmers utilized the by-products and wastes of food processing. There was a dynamic balance between nature and humans: plants, animals, and humans formed an ecological system.

However, since the beginning of the industrial revolution, the emphasis of agricultural and industrial development has been on short-term economic gains instead of ecological equilibrium (Anon, 1996a).

Today, with improved technology there is more efficient production with greater yields, consequently, there is less attention given to the preservation of natural resources and utilization of wastes and by-products (Henningsson et al., 2004).

FOOD INDUSTRY AS AN ENVIRONMENTAL POLLUTER

Waste management, placement and disposal, are critical aspects of food processing. The percentage of by-products and wastes in food processing is 30% (Fehr et al., 2002; Anon, 2001; Tuncel et al., 1995). The majority of food industry wastes and by-products (of plant origin) is used as animal feed (Adebowale, 1985). Only 2–3% is marketed for human use. Some 30% of all production waste are so-called dangerous waste (mainly of animal origin), which requires special treatment, for example, the disposal of sludge resulting from sewage treatment (Pándi, 2001). The food industry is not a heavy environmental polluting industry. Production and packaging wastes are significant, however, because of the size of the food processing industry. The role of food industry is to convert agricultural raw materials of plant and animal origin to foodstuffs suitable for human consumption. This process unavoidably results in wastes and by-products unfit for consumption. Food industry wastes can be further classified as technological waste (some of them are "dangerous") and sludge and packaging wastes. This chapter will deal only with fruit production waste, since further processing can decrease the burden to the environment.

The amount and ratio of waste depend on the given branch of the food industry (Di Blasi et al., 1997; Huebner and Kienzle, 2001; Poonam-Nigam et al., 2001).

However, because of differences in ownership and size of individual food industry companies (e.g., increasing role of small and middle scale enterprises), certain branches have different impacts on the environment (Woods, 1998).

Some technological waste is of animal origin (80%), which will not be included in this chapter (Williams, 1995; Apellaniz et al., 1996; Anon, 2003). Technological waste of plant origin is usually not considered dangerous. Wastes may be burned, used for animal feed, or used for human applications. However, there are some wastes of plant origin that are "dangerous" (Békési and Pándi, 2001)

- Asbestos and silica filters used by the beverage industry
- "Blue precipitate" that is the result of "blue clarification" in wine making (it can be used for tartaric acid production, but this leads to sewage containing cyanide compounds and disposal is still a problem)

BIOLOGICAL AND ECONOMICAL ENVIRONMENT, JUSTIFYING WASTE MANAGEMENT

Agricultural production and further processing can become one of the most serious sources of pollution (Di Blasi et al., 1997). Because of cheap energy and raw materials following World War II, little attention was given to utilizing by-products. An increase in the number of industrial plants in turn increased the volume of by-products. Further treatment and environmentally friendly disposal and distribution were not the main interest of the industry. This paradox could be handled by directing wastes back into production (Howitt, 1997). This is not a new solution, since classic agricultural technology also utilized by-products, e.g., the fodder industry used waste from vegetable oil production (Bulla, 1994).

The process is also influenced by economical conditions since the expense of converting waste products into useable products will happen only if the expenses do not exceed the final selling price. Such decisions can be supported by operations research to determine inaccurate economic value of by-products (Cheeseborough, 2000; Henningsson et al., 2004).

OPPORTUNITIES FOR UTILIZING FOOD INDUSTRY WASTES AND BY-PRODUCTS

Food industry wastes and by-products are geographically scattered, of large volume and low nutritional value. Consequently, their collection, transportation, and processing costs of by-products can exceed the selling price. The real difference between the main product and by-products is the level of profit. If we could produce valuable products from food industry waste through new scientific and technological methods, environmentally polluting by-products could be converted into products with a higher economic value than the main product.

The utility of raw materials—to what extent the valuable components can be converted to the end product—depends on the following factors:

- Nature and quality of raw material
- Type, characteristics, and quality criteria of the finished product
- Technological process
- Technical level and condition of the machinery
- Human factors

The difference between the useful components of the raw material and the end product is called the production loss. The role of waste utilization is important in terms of environmental protection, too. Food industry wastes are usually organic substances and can pollute our environment if not disposed prudently. All in all, food industry wastes and by-products are substances that originate from production and can be further utilized in other ways. For example, fruit wastes from cleaning and seeding are used for feed without further treatment.

Solid and liquid food industry wastes and by-products possess long traditions. Although large-scale industry methods are not always efficient, the technical development of food companies conform more and more to the BAT principle, applied in Europe (Anon, 1996b). The term "BAT" (best available techniques) is defined in Article 2(11) of the EC Directive (Integrated Pollution Prevention and Control) as "the most effective and advanced stage in the development of activities and their methods of operation which indicate the practical suitability of particular techniques for providing in principle the basis for emission limit values designed to prevent and, where that is not practicable, generally to reduce emissions and the impact on the environment as a whole."

In Europe, the BAT group is working to develop and distribute information on how to avoid pollution when using production methods for various commercial products. The aim is to further the utilization of the BAT within selected fields (for example in the field of fruit processing). The object is to reduce the burden on the environment as much as possible. Due to the diversity of the fruit industry, examples of techniques and methods proven successful in some cases are not necessarily applicable directly to other cases. Each technique should be assessed case by case, taking into consideration the age, the scale, and the geographic location of the processing plant, as well as the product portfolio and the type of process. The best strategy is to establish a working environment, which means defining policies, monitoring environmental impact of specific activity, and setting up goals to enhance performance.

The application of this principle contributes to decreased quantity of waste and water consumption and has several environmental advantages. Nevertheless, the amount of food processing wastes and by-products—that are the most important issues of food processing environmental strategies—is significant even if this principle is applied properly (Perédi-Vásárhelyi, 2004).

UTILIZATION OF BY-PRODUCTS FOR FEEDING

Feed production has become a new industry. More and more plants are processing by-products as raw materials, many of which were not used in traditional feeding technologies. The finished products (feed) thus produced can be full (value) mixtures or concentrates to be mixed at local fodder-plants. Sometimes these products are patented and producers enclose directions for use, which is usually controlled. Factories preparing, producing, and using the feed are interdependent and operate in a complex ecological system that can build on agricultural and industrial production plants to utilize waste (Adebowale, 1985).

There is no sharp distinction between the two alternatives as plants (e.g., forage mixers) processing for only one company can be involved in production or paid work. On the other hand, vertical integration can also establish complex systems. Disposition can be the top priority in waste management in addition to profitability (Subburamu et al., 1992).

The Role of Biomass in Waste Utilization

Biomass production and utilization are of great importance both in the short and the long run. Biomass refers to all organisms (microorganisms, plants, and animals), both living and recently dead; products of biotechnology-related industries; all biological products, wastes and by-products (human, animal, processing industry, etc.). Food industry production, where the agricultural raw materials become foodstuffs, is a primary source of biomass transformation (Hammond et al., 1996; Stabnikova et al., in press).

Evaluating the potential of biomass production and utilization and developing a system that includes relevant recommendations for practical applications has been discussed in scientific literature (Mahadevaswamy and Venkataraman, 1990; Viswanath et al., 1992; Bouallagui et al., 2004). Animal feeding is such a potential utilization. Mass feed production takes substantial amounts of land; however, these areas could be used for plant production for export, meanwhile by-products could be partially substituted for feed. The profitability of ruminant farming requires a cost decrease without a decline in production.

Evaluation of Waste Utilization Techniques

The above enumeration expresses the value of utilization procedures. The highest new values can be obtained by the introduction of by-products in human nutrition; meanwhile burning (if its energy is not utilized) is primarily "waste elimination." Primary conditions for the introduction and implementation of technologies selected for utilization are as follows:

- Preserving and storing by-products of different origins geographically. This means that they are decentralized. To gather them, distant transport is required.
- Reasonable solutions for transport. This means to achieve cost effectiveness. That is, the best traffic logistics for the minimum cost and the most effective transportation technology.
- Ensuring the space, building, machinery, equipment, energy, and personnel requirements of processing.

Before making a decision to utilize by-products, an evaluation of the by-products should be made on the following issues:

- Concentrated or widely scattered origin?
- Seasonal or continuous availability?
- Quantity (large or small)?
- Concentration of valuable substances (high or low)?

The ideal utilization choice is usually determined by complex evaluation and short-run economical analysis.

According to international experiences, by-product utilization and waste-free technologies are most sophisticated in developed countries, where food production is also on a high level. In less developed countries the use of such technologies commensurate with the financial resources (Polpraser, 1996), thus

- the easiest use for by-products of plant origin is plant fertilizer
- biogas production and burning combined with energy recovery are still not widespread
- the use of food industry by-products (bran, coarse or oleaginous seeds, and feed yeast) as a feed supplement is a cheap and simple procedure
- by-products with high water content need to be dried or concentrated prior to further processing
- primarily by-products, rich in protein, are used for feeding, and
- human application is the most expensive, due to high requirements for equipment and energy. Fruit processing (e.g., pectin production) is a good example to illustrate this kind of utilization

WASTES AND BY-PRODUCTS OF FRUIT PROCESSING

The aim of fruit processing is to transform fresh fruits into preserved products. Therefore, the selection and elimination of components unsuitable for human consumption, lead to by-products and wastes. There is no sharp limit between the two categories as industries processing substances of biological origin are in close cooperation; the by-product of one industry can be a valuable secondary raw material for another.

Apple processing is a good example for the aforementioned principle. Although apples (canned apples, apple juice, etc.) are the finished product, and the by-product, apple pomace, is the secondary raw material of pectin production. However, if this by-product goes to the refuse, it becomes waste and pollutes the environment. Generally, in Hungary as well as the rest of the world, the majority (60–65%) of food industry by-products become waste and a burden to the environment (Szenes, 1995).

In order to meet stringent environmental requirements, modern fruit processing should minimize the amount of by-products and waste, decrease energy utilization, produce high-quality foodstuffs without polluting the soil, air, and spas (Barta et al., 1997).

Fruit processing wastes differ from other wastes

- they are organic and therefore, decompose. Most go back into the soil, due to natural biomass circulation, or decompose without pollution;
- they are large in volume with high water content. In spite of the high volume their origin is scattered, making gathering and utilization difficult and expensive; and
- they tend to deteriorate, thus limiting the storage period, even under appropriate circumstances, which include low temperatures, controlled humidity, and storage in dark and dry places.

Apart from the main finished product, unused substances are considered as waste and by-products. These can be utilized in different ways depending on their texture and content.

Certain by-products can be valuable resources for human nutrition if special technologies are used. They include

- precooling techniques (Brosnan and Sun, 2001)
- solid-state production (Zheng and Shetty, 2000)
- Gas chromatic evaluation of residues (Pugliese et al., 2004)
- the beneficial effects of grinding, soaking, and cooking on the degradation of dangerous matters in fruit waste (Tuncel et al., 1995)
- Others

Some examples are

- pectin from apple pomace
- aromas and coloring agents from fruit waste
- oils from seeds
- tartaric acid from wine lees

Also, further processing of by-products can transfer their valuable compounds into new products:

- Distillery wastes can be added to feed after appropriate treatment.
- Household and gardening wastes are utilized for soil improvements.
- All organic by-products can be utilized in a profitable way, if used for biogas production (methane production) or burning (especially if combined with recovery) (Prema Viswanath et al., 1992; Mahadevaswamy and Venkataraman, 1990).

Pollution Prevention and Control in Fruit Processing

Reductions in wastewater volumes up to 95% have been reported through implementation of good practices. Where possible, adopt the following measures:

- Use clean raw fruit, reducing the concentration of dirt and organics (including pesticides) in the effluent.
- Use dry methods such as vibration or air jets to clean raw fruit. Dry peeling methods reduce the effluent volume (up to 35%) and pollutant concentration (organic load reduced up to 25%).
- Separate and recirculate processed wastewaters.
- Use countercurrent systems where washing is necessary.
- Use steam instead of hot water to reduce the quantity of wastewater to be treated.
- Remove solid wastes without the use of water.
- Reuse concentrated wastewaters and solid wastes for production of by-products.

As an example, recirculation of processed water from fruit preparation reduces the organic load by 75% and water consumption by 95%. Similarly, the liquid waste load (in terms of biochemical oxygen demand, BOD) from apple juice processing can be reduced by 80%.

Good water management should be adopted, where feasible, to achieve the levels of consumption presented in Table 10.1 (Anon, 1998).

Target Pollution Loads

Implementation of cleaner production processes and pollution prevention measures can yield both

Table 10.1. Water Usage in the Fruit Processing Industry

Product Category	Water Use (cubic meters per metric ton of product)
Canned fruit	2.5–4.0
Frozen fruit	5.0–6.0
Fruit juices	6.5
Jams	6.0
Baby food	6.0–9.0

Source: Anon, 1998.

Table 10.2. Target Loads Per Unit of Production, Fruit Processing Industry

	Fruit		
Product	Waste Volume (m^3/U)	BOD (kg/U)	Solid Waste (kg/t product)
Apricots	29.0	15.0	
Apples			90
All products	3.7	5.0	
All except juice	5.4	6.4	
Juice	2.9	2.0	
Cranberries	5.8	2.8	10
Citrus	10.0	3.2	
Sweet cherries	7.8	9.6	
Sour cherries	12.0	17.0	
Bing cherries	20.0	22.0	
Dried fruit	13.0	12.0	
Grapefruit			
Canned	72.0	11.0	
Pressed	1.6	1.9	
Olives	38.0	44.0	20
Peaches			180
Canned	13.0	14.0	
Frozen	5.4	12.0	200
Pears	12.0	21.0	
Pineapples	13.0	10.0	
Plums	5.0	4.1	
Raisins	2.8	6.0	
Strawberries	13.0	5.3	60

Source: Anon, 1998.

economic and environmental benefits. The target loads per unit of production are shown in Table 10.2 (Anon, 1998). The data refer to the waste loads arising from the production processes before the implementation of pollution control measures. These levels are derived from the average loads recorded in a major study of the industry and should be used as maximum levels of unit pollution in the design of new plants.

By-product Treatment and Utilization in the Case of Certain Fruit-based Products

The processing of fruits produces two types of waste: solid waste [e.g., peel/skin (Larrauri et al., 1997; Negi et al., 2003; Fernández et al., 2004), seeds (Noguchi and Tanaka, 2004), and stones (Lussier et al., 1994)] and liquid waste [juice (Gil et al., 2000) and wash water]. A serious waste disposal problem can attract flies and rats in the processing room, if not corrected properly. If there is no plan to use the waste products, they should be buried or fed to animals in distant locations.

Solid Fruit Wastes

There are possible ways to use some solid fruit wastes, which are discussed below. However, it is stressed that a full financial evaluation should be done before the implementation of any of the suggestions.

One major goal in using fruit wastes is to ensure a reasonable microbiological quality in them. This means that one should process waste products on the same day that they become available. It is not advisable to store wastes until the end of the week's production before processing them. Even with this precaution, the wastes being used will most likely contain moldy fruit (discarded during processing), insects, leaves, stems, soils, etc. This will contaminate any products derived from such wastes.

Therefore, some preliminary separation is needed during processing, such as

- peel and waste pulp in one bin
- moldy parts, leaves, soil, etc in a second bin, which may be discarded
- stones, seeds, etc., in a third bin

Possible Products

The six main products from wastes include

- candied peel
- oils
- pectin
- re-formed fruit pieces
- enzymes
- wine/vinegar

Each is discussed below.

Candied Peel. Peel from citrus fruits (orange, lemon, and grapefruit) can be candied for use in, for example, baked goods and snack food. In addition, shreds of peel are used in marmalades, similar to the process of candying. That is, boil the slices or shreds of peel in a 20%-sugar syrup for 15–21 min and progressively increase the sugar concentration in the syrup to 65–70 Brix (percentage of sugar monitored by a refractometer) during soaking of the food for 4–5 days. It is then removed, rinsed, and given a final drying in the sun or in the hot air drier. This can serve as a secondary product for a fruit juice or jam processor. This assumes that a large food company is interested in buying the candied peel as an ingredient for their products. In one application, candied melon skin has been used to substitute for sultanas in baked goods and, in another, candied root vegetables have found a similar market.

Oils. The stones of some fruits (e.g., mango, apricot, and peach) contain appreciable quantities of oil or fat, some of which have specialized markets for culinary or perfumery/toiletry applications. Palm kernel oil is well known as a cooking and industrial oil. In addition, some seeds (e.g., grape, papaya, and passion fruit) contain oil with a specialized market. Of course, for any commercial product in any country, the goal is to identify the import/export agents interested in such products. After that, the processor's responsibility is to produce the oil to satisfy the customer in terms of sufficient quantity and stringent quality standards. Obviously, the manufacturer has to secure proper equipment to produce the oils at a reasonable cost.

The process involves grinding the seeds and nuts to release the oil without a significant rise in temperature, which would spoil their delicate flavors, with the exception of palm kernel oil. Generally, a powered hammer mill is needed for nut and kernels. A press is needed to extract the oil. Since the existing manual presses have not been tried for this application, a certain amount of experimentation is needed to establish oil yields and suitability of the equipment. Solvent extraction is not recommended for small-scale applications. However, steam distillation of citrus peel oils is well established for small-scale operations.

The crude oil may be sold to be refined elsewhere, but it is likely that the producer is responsible for the initial refining.

Pectin. This is a gelling agent used in jams and some sweets and occurs in most fruits, ranging from a low to a high level. Commercially, pectin is extracted from citrus peel and apple pomace, the residue left after apple juice has been removed. Other tropical fruits may contain high levels of pectin, passion fruit being a notable example. The utilization of the "shells" remaining after pulp removal may permit pectin extraction.

In most developing countries, pectin is imported from Europe or United States. This may look like a good market or opportunity for processors in these countries to provide pectin locally to substitute for imports. However, there are major problems:

- In countries where this has been tried, it has not been possible to produce pectin at a cost lower than that for imported products.
- It is difficult to produce pectin powder on a small scale, although liquid pectin is possible.
- There are many types of pectin, each with specific properties suitable for a particular application. For example, pectin for jams as a preserve differs from that used in jam as an ingredient in baked goods.

Re-formed Fruit Pieces. Fruit pulp can be recovered and formed into fruit pieces. Although the process is relatively simple, the demand for this product is low. Therefore, a thorough evaluation of the potential market is recommended before investing in the enterprise (Kilham, 1997).

The process involves preparing a concentrate by boiling the fruit pulp, followed by sterilization. Sugar may also be added. A gelling agent, sodium alginate, is then combined with the cooled pulp and then mixed with a strong solution of calcium chloride. All ingredients are safe for human consumption, being legal food additives in most countries. The calcium and the alginate combine to form a solid gel structure and the pulp can therefore be re-formed into fruit pieces. The most common way is to pour the mixture into fruit-shaped moulds and allow it to set.

It is also possible to allow drops of the fruit/alginate mixture to fall into a bath of calcium chloride solution where they form small grains of re-formed fruit, which can be used in baked goods. Commercially, the most common product of this type is glaced cherries.

Enzymes. Commercially, the three most important enzymes from fruit are papain (from papaya), bromelain (from pineapple), and ficin (from figs). Each is a protein-degrading enzyme used in such applications as meat tenderizers, and washing powders and is also used in leather tanning and beer brewing. However, it is unlikely to be economical to harvest these enzymes from fruit processing waste. Currently, even the more efficient process of collecting enzymes from fresh whole fruit is no longer economical. Changes in both large-scale production with higher quality standards and use of biotechnology to produce "synthetic" enzymes mean that small-scale producers will be unlikely to compete effectively. In addition, there are proposals to phase out the use of these enzymes in food products in Europe and United States. Their market is therefore declining. Consequently, it is not cost effective to harvest enzymes from fruits processing waste.

Wine/Vinegar. Although products such as wine or vinegar should be produced from fresh, high-quality fruit juices in order to obtain high-quality products, it is technically feasible to produce them from both solid and liquid fruit wastes. Solid wastes should be shredded and then boiled for 20–30 min to extract the sugars from the fruit and to sterilize the liquid. Several batches of waste may be boiled in the same liquid to increase the sugar concentration. This is then filtered through boiled cloth to remove the solids and cooled in preparation for inoculation with yeast. Liquid wastes should be separated during production to ensure that fruit juice is kept separate from wash water. For example, the juice could be drained from a peeling/slicing table into a separate drum. The juice is then boiled for 10–15 min and treated as above.

The liquid is then inoculated with "wine" yeast and not bread or beer yeast and fermented in the normal way for wine production. This can then undergo the standard second fermentation to produce fruit vinegar.

In summary, each of the above uses of fruit waste requires

- a good knowledge of the potential market for the products and the quality standards required
- an assessment of the economics of production
- a basic familiarity with the production technique
- a reasonable capital investment in equipment
- a fairly large amount of waste available to make utilization or harvesting worthwhile

For small-scale operations, where reducing pollution or increasing waste disposal is more important than process economics, the most likely solution is to use wastes as animal feeds.

Some Example of Research Areas for the Utilization Opportunities of Fruit Processing By-products

Several valuable substances—fibers, coloring agents, gelling agents—can be extracted from the wastes of fruit-based products. The way and goal of utilization is determined by the economic efficiency of extraction and the market potential. It is not easy to collect data about the quantity of fruit waste and widespread treatment techniques.

The following literatures review the types of by-products, modern treatment technologies, and approaches of utilization. It does not deal with traditional methods, which can be found in standard literature:

- Fruit stones constitute a significant waste disposal problem for the fruit processing industry. High-quality activated carbon can be produced from waste cherry stones (Lussier et al., 1994).
- Fruit processing wastes including apple, cranberry, and strawberry pomace were used as substrates for polygalacturonase production by *Lentinus edodes* through solid-state fermentation (Zheng and Shetty, 2000).
- Watermelon peel constitutes 44% of the whole fruit weight. In the study of Madhuri and Kamini-Devi (2003), the potential to produce preserved products such as pickles, tutti-fruiti, vadiyams, and cheese using the white portion of watermelon rind was investigated.
- Fruit wastes (pineapple, mixed fruit, and maosmi) were investigated as possible substrates for citric acid production by solid-state fermentation using *Aspergillus niger* (Kumar et al., 2003).
- In 1996, a laboratory study was conducted by Hammond et al., to assess ethanol production potential from banana waste.
- *Citrus junos* is one of the important citrus fruits in Japan. The fruit juice is an ingredient used in sauces and salad dressings for its special flavor.

After juice extraction, the fruit pulp is usually dumped as waste at a large cost. The manipulation of food processing wastes is now becoming a very serious environmental issue. The peel of *C. junos* fruit was found to possess potent allelopathic activity and a methanol extract of the peel inhibited the growth of several weed species (Fujihara and Shimizu, 2003).
- A method is described by Drunen and Hranisavljevic (2003) for the enrichment of fruit products with beneficial substances (e.g., antioxidants) extracted from processing waste, e.g., fruit peel.
- Progress is described in an ongoing project in European Union (EU) (QLKI-1999-00124) on anthocyanin bioactivities. The investigation covers the functional properties and the effects of anthocyanins and anthocyanin-rich food ingredients on heart disease. This study aims to use such compounds as colorants and in the development of new anthocyanin-rich functional foods (Anon, 2002).

Successful Waste Utilization in Wine Making

Among food industry by-products of fruit or vegetable origin the products of wine industry are outstanding. Grape marc and wine lees are the basic by-products of wine making. Depending on the variety, ripeness, vintage, and harvesting time, 100 kg of grapes produces 15–20 kg of grape marc.

The majority of grape marc goes to refuse; a small percentage is burned or used as compost. Unfortunately, it is also used for wine adulteration. Grape marc contains seeds that possess valuable substances such as oil, proteins, and tannins. Grape seed oil is the most valuable because of its health-protecting and cholesterol-lowering effects. This oil can be obtained by pressing or extraction. It has the main advantage of being pesticide free. Due to its valuable fatty acid content it can be used as a foodstuff, lubricant, and raw material for cosmetics as well. In the world's vine-producing lands, the quantity of marc has been registered since 2001. In Hungary, after the EU accession, like in other wine-producing countries, the by-products need be distilled or further utilized. EU requirements, for instance, are that grape marc be withdrawn from circulation within the framework of a controlled process, and then used for feed, organic manure, oil, distilled spirits, and tartaric acid production (Szenes, 1995).

BY-PRODUCT UTILIZATION IN APPLE PROCESSING

In the temperate zone, apples are the most significant fruit economically. Apple production achieves approximately 10% of the world's fruit production, with one fourth produced in Europe.

There are opportunities for the utilization of apple press cakes

- drying
- feeding
- composting
- storing

However—both environmentally and economically—the best technique is drying for pectin extraction.

What is Pectin?

Pectin for use in food is defined as a polymer containing galacturonic acid units (at least 65%). The acid groups may be free, combined as a methyl ester, or as sodium, potassium, calcium, or ammonium salts, and in some pectins amide groups may also be present.

Commercial Production

Process details vary between different companies, but the general process is as follows:

The pectin factory receives apple residues [pomace (Carson et al., 1994)] or citrus–orange peels from a number of juice producers (El-Nawawi and Heikal, 1996). In most cases this material has been washed and dried, so it can be transported and stored without spoilage.

If the raw material is dry, it can be assessed and selected from storage when the need arises. If wet citrus peel is needed, it has to be used immediately on receipt because of rapid deterioration (Kim et al., 2004).

The raw material is added to hot water containing a processing aid, usually a mineral acid, although others such as enzymes could be used (Schieber et al., 2003). Water alone will extract only a very limited amount of pectin.

After pectin is extracted, the remaining solids are separated, and the solution clarified and concentrated by removing some of the water. The solids can be separated by filter, centrifuge, or other means. The solution is then filtered again for further clarification if necessary.

Either immediately or after a holding period to modify the pectin, the concentrated liquid is mixed with an alcohol to precipitate the pectin. The pectin can be partly de-esterified at this stage, or earlier or later in the process.

The precipitate is separated, washed with more alcohol to remove impurities, and dried. The alcohol wash may contain salts or alkalis to convert the pectin to a partial salt form (sodium, potassium, calcium, and ammonium).

The alcohol (usually isopropanol) is recovered very efficiently and reused to precipitate further pectin.

Before or after drying, the pectin may be treated with ammonia to produce an amidated pectin if required (Braddock, 1999). Amidated pectins are preferred for some applications.

The dry solid is ground to a powder, tested, and blended with sugar or dextrose to a standard gelling power or a product with other functional property such as viscosity or stabilizing effect.

Pectins are also blended with other approved food additives for use in commercial applications.

The various raw materials yield different amounts of extractable pectin: Pomace, 10–15%; Sugar beet chips, 10–20%; Sunflower-infructescence, 15–25%; and Citrus peels, 20–35%.

Application of Pectins

Pectin is one of the most versatile stabilizers available. Its gelling, thickening, and stabilizing properties make it an essential additive in the production of many food products.

Traditionally, pectin was primarily used in the production of jams and fruit jellies—industrially as well as domestically and in low- as well as high-sugar products. It produces the desired texture, limits the creation of water/juice on top of the surface as well as an even distribution of fruit in the product. With the change in lifestyle, pectin is primarily sold for industrial use. In some European markets it is still sold to the consumers as an integrated component in gelling sugar, though.

Product and application development by the major pectin producers has over the years resulted in a large expansion of the opportunities and applicability of pectin. Pectin is a key stabilizer and is used in many food products as

- fruit applications in jams, jellies, and desserts;
- bakery fillings and toppings in fruit preparations for dairy applications
- dairy applications in acidified milk and protein drinks, yogurts (thickening)
- confectionery in fruit jellies, neutral jellies;
- beverages
- nutritional and health products
- pharmaceutical and medical applications

This wide range of applications explains the need for many different types of commercial pectin, which are sold according to their application, for example

- rapid set pectin traditionally used for jams and marmalades
- Slow set pectin used for jellies and some jams and preserves, especially for vacuum cooking at lower temperatures. It is also important for higher sugar products like bakery and biscuit, jams, sugar confectionery, etc.
- stabilizing pectin used for stabilizing acidic protein products such as yogurts, whey, and soya drinks during thermal processing
- Low methyl ester aminated pectin used in a wide range of low-sugar products, reduced sugar preserves, fruit preparations for yogurts, dessert gels and toppings, and savory applications such as sauces and marinades. It can also be used in low-acid and high-sugar products such as preserves containing low-acid fruits (figs and bananas) and confectionery

By-products: Coloring and Noncoloring Sweetener

After distilling the alcohol used for the precipitation of pectin and fruit extracts, such as sugar and fruit acids, the natural flavors will remain. For example, apple extract obtained from pomace will be used as sweetening agents for the preservation of freshness and/or coloring of food. A further possibility is to ferment it to form apple ethanol.

At a further processing stage, special technologies are used to remove dark natural coloring agents, mineral substances, and fruit acids from these fruit extracts. The resulting products will only contain the sugars of the respective raw material that has been processed. They will be used by the food industry as sweetening agents (Khachatourians et al., 2001)

By-product: Fodder

After pectin is extracted, the various residues of the original raw material are dried and pressed into pellets. Due to their high energy content and nutritive value, these products are in demand as fodder. The residual moisture and the fodder value of these products are checked continuously so as to ensure that products of uniform quality are obtained (Bennett and Bendigo, 2002).

Other Ways for Apple Pomace Processing (Fiber Utilization)

After adding wine yeast to the apple pomace, remaining from fruit juice and apple pulp production, the marc is fermented at 30°C in solid phase, resulting in a liquid with a 4–5% ethyl alcohol content. Then it is concentrated to 10% by means of vacuum distillation. With further fermentation high-quality apple vinegar can be obtained.

If we add "*A. niger*" mold and methyl alcohol to the apple pomace, its sugar content will decrease by 81% in 5 days. Meanwhile, from 1 kg of apple marc we can extract 90 g of citric acid or a yield of 88%, if expressed in sugar. If apple marc is treated with a thin alkali solution we get two fractions: fibers comprising of alpha-cellulose pentosanes (26%) and pectin (10–18%). Both fractions can be used for apple products as a thickener and a calorie-free texture modifier.

The Importance of Dietary Fibers and Fiber Sources

In civilized societies, there is a preference for refined and cleaned foodstuffs. However, consumers deprive themselves of many substances that are considered healthy. A lack of fiber, for example, would result in diseases and abnormalities, which are unknown in uncivilized societies. Nutrition scientists are researching the degree to which refined food will trigger health problems (Barta, 1993). There are ongoing efforts to add back important substances, such as dietary fibers (Larrauri, 1999; Miguel and Belloso, 1999), coloring matters, aromas, volatile compounds, vitamins, etc. These substances have been removed or cleared during operations to purify the food for a convenient "end product." They may also be the result of a negative effect from an essential processing. However, some of such "removed" substances are very important for a healthy human life. Examples include fibers or vitamins (Ramadan and Mörsel, 2003), which are added back, for health reasons or legal requirements, after removal during processing. Today, fiber products are very important dietary supplements. The indigestible parts of plant cell wall, such as cellulose and lignin, were considered as unnecessary parts of foodstuffs that decrease the energy, compositional, and sometimes even the sensory values. After gaining further information, this approach changed and scientists ascertained that plant-based fiber helps to prevent several diseases. They found a relationship, for example, between the fiber content of the food and serum cholesterol of consumers. Significant intestinal diseases can be cured and prevented with an increase in food fiber.

The nutritional effect of dietary fiber components is due to their physical and chemical properties. The human body does not have enzymes to digest fibers. Fibers are resistant to digestion by gastric juices, only some bacterium can decompose a certain quantity. Consequently, fibers possess slight nutritive value, but they play an important role in digestion (Barta et al., 1989).

ENVIRONMENTAL GUIDELINES FOR FRUIT PROCESSING WASTE MANAGEMENT

Process Description

Fruits can be processed in many different ways depending on raw materials and end products. The techniques most frequently used are canning or bottling accompanied by heat treatment, refrigeration or freezing, fermentation, drying, pickling, and chemical preservation. In most cases the aim is to lengthen the shelf life (to reduce the perishability) of the product, but there are often secondary objectives related to consumer acceptance, for example, convenience (preparation and recipes), appeals (packaging and presentation), eating quality, novelty, or new products (juices, purees, jams, or wine).

The manufacturing steps include some or all of the following: receipt and weighing of raw materials, storage, washing, grading, peeling, cutting, crushing, filtrations, heating, cooling, preservation, pickling, drying, concentration, fermentation, packaging (cans, jars, vacuum packs, tetra-paks, etc.) and storage. This includes grading and packing of fresh fruit for market. Common examples of processed fruit products include fruit juices (apple, orange, etc.), canned peaches and pears, dried fruits (apricots, prunes, dates, raisins, etc.), and wine and fruit purees for commercial applications.

Fruit Waste Characteristics

The fruit industry typically generates large volumes of effluents and solid waste. The effluents contain high organic loads, cleansing and blanching agents, salt, and suspended solids such as fibers and soil particles. They may also contain pesticide residues from the washing of raw materials or discarded fruits. Odor problems can occur with poor management of solid wastes and effluents.

By-products are Produced During Many Steps of the Fruit Production Chain

Such by-products may come from

- overstock merchandize
- screenings, tops, stems, pulps, pomace, skins, hulls, peels, meals, seeds, fines, green chop, pressed cake, dried fruit, and fresh fruit waste
- unsold merchandize because of passed "sell by" date
- rinse water or cooking materials
- below standard (size, color, and texture) merchandize
- trimmings
- fruit harvest

The by-products can result from the processing of almonds, apples, apricots, bananas, cantaloupe, coconuts, dates, figs, grapes, grapefruits, lemons, melons, nectarines, pecans, oranges, peaches, pears, pineapples, plums, prunes, pumpkins, raisins, tangerines, walnuts, and many others.

Key Environmental Risks/Liability Factors

The risk/liability factors associated with water supply and wastewater management are as follows:

- Sources of processing and potable water are municipal, abstraction wells, and boreholes. The quality of water is important; pretreatment may be required depending on the initial quality and intended use.
- Large volumes of water are needed for washing raw materials, factory cleaning, and transport of materials within the factory.
- Pollutants in wastewaters include detergents, pesticides, suspended solids, dissolves solids, nutrients, and microbes.
- Seasonality of production can place heavy demands on treatment/disposal systems during the peak season.

Other issues to consider include

- Permits and fees for the usage of water.
- Availability of acceptable water during the peak processing season(s).
- Potential for minimizing water use and/or recycling water for washing.
- Integrity of the drainage system to prevent contamination of surface or ground water. Check drains, pipes, screens, interceptors, etc.
- Potential for the use of wastewater treatment plant (facility/municipal); check type, effectiveness, monitoring final effluent disposal to sewers or lagoons.
- Regulatory compliance like discharge permits, impact statement, enforcement, costs, etc.

Solid Waste Disposal

The following considerations are important:

- Production of large volumes of bulky perishable solid waste such as peels, stems, shells, rinds, pulps, seeds, pods, rejected raw material, etc.
- The need to separate solid from liquid waste by screening, sedimentation, flotation, etc.
- The cost of transporting waste to approved disposal sites.
- Microbial action in stored solid waste can produce odors.
- Vermin (e.g., rats) and insects may be attracted to solid waste storage areas (Cholos and Cheremisinoff, 2003).

Issues related to the above considerations are

- Permits and charges are usually required for solid waste disposal.
- Approval or licensing of solid waste disposal contractors.
- Security of access to solid waste disposal sites.
- Potential for process adjustment to minimize solid waste production.
- Plans to utilize solid waste for fuel, fertilizer, or animal feed.
- Potential treatment of solid waste (e.g., drying) to facilitate disposal.
- Compliance with solid waste disposal regulations and sanitary regulations in importing countries may be expensive or difficult.

Refrigerants in Fruit Processing

Fruit processing plants will usually have large cold storage facilities. The refrigerants used may be ozone-depleting chemicals, such as CFCs, the production of which is being phased out under the Montreal Protocol. Ammonia, which has no such restriction, is also used as a refrigerant. The release of ammonia into atmosphere, due to leaks from cooling equipment, is a primary health and safety concern.

MEANS OF MODERN WASTE MANAGEMENT

Feasibility Strategic Principles

Sustainable development means a sustainable use of the environment to improve the quality of human life without exceeding the availability of natural resources. Prevention is of strategic importance. It is the main direction of regulatory, research, and development activities (Hollingdele, 2000).

Issues of Environmental policy, Planning, and Regulation

Modern environmental policy requires a prudent, target-oriented, integrated approach and universal planning involving every region. Environmental regulation needs to be efficient regarding the cost compared to environmental protection achievements. Regulation should be based on general legal principles and enforcement by the "polluter pays" principle. Environmental fines should be allocated for solving problems with appropriate interventions (Sherwood et al., 1995).

Research and Technical Development

Research and technical development is of strategic importance, providing the knowledge, procedures, methods, and technologies necessary for the implementation of the given tasks.

Environmental Information System

Environmental data collection and processing are just partly done in many countries. Therefore, it is necessary to

- develop an environmental information system involving central systems and professional systems as well
- establish an environmental database within the given countries and their commercial partners, too
- work out recommendations concerning the index numbers and the content of environmental reports (Woodard et al., in Press)

Institutional System

The institutional system of environment protection should be decentralized by financing and strengthening the organization of regional institutions (Read et al., 1997). The system of horizontal relationships, the activity of professional organizations, environment protection units, and the branches involved also need to be strengthened.

Social Participation

The implementation of environmental projects requires social participation. The importance of education, training, and information spreading is considerable, in order to make people aware of the proper solutions' and their decisions' environmental consequences. Furthermore, the conditions of environmentally conscious, healthy lifestyle need to be ensured and communicated (Fehr et al., 2002).

International Cooperation

The compliance of countries to future environmental standards is of crucial importance. On the one hand, it should be accomplished by maintaining the existing environmental and regulatory advantages. On the other hand, it needs a reasonable adaptation of international (e.g., EU) requirements, tailored to conditions unique to each country (Di Blasi et al., 1997).

Studying the situation of fruit processing wastes and by-products, it can be stated that processing wastes should be further utilized as secondary raw materials (for feeding or even for human applications). As far as packaging materials are concerned, the amount needs to be decreased, moreover selective collection and reuse needs to be organized, with special regard to the principles of a "closed-loop economy" approach.

CONCLUSIONS

The latest research and development have resulted in new methods of modern waste management solutions around the world. Biotechnological (bioconversion) procedures (e.g., aerobe composting, biogas production) are outstanding in this respect and can successfully be applied. However, in the future more extensive applications will become necessary in the field of, e.g., animal feed supplementation, soil conditioning products, and energetic ingredient production.

REFERENCES

Adebowale, E.A. 1985. Organic waste ash as possible source of alkali for animal feed treatment, *Animal Feed Science and Technology*, Vol. 13, Issues 3–4, pp. 237–248.

Anon. 1996a. Processing fruits, *Science and Technology*, vol.1, Biology, Principles and Applications, *Technomic Publishing Co. Inc.*, Lanchester, PA, pp. 461–499, 49 ref.

Anon. 1996b. *IICP 96 /61/ EC*, Integrated Pollution Prevention and Control, EC Directive.

Anon. 1998. Pollution Prevention and Abatement, Handbook World Bank Group, Environment Department, Washington, D.C.

Anon. 2001. East Sussex Country Council. Waste management statistics (http://www.eastsussexcc.gov.uk/env/waste/statsitics.htm).

Anon. 2002. Healthy colours from berries, *Flair-Flow EuropeReports*. FFE 546/02/CG54, 1p.

Anon. 2003. *Environmentally-Friendly Food Processing*, *Woodhead Publishing Ltd.*, Cambridge CB1 6AH, U.K., pp. 29–53, 46 ref., pp. 54–69. 23 ref., pp. 218–240, 46. ref. www.woodhead-publishing.com.

Apellaniz, I., Elorriaga, J.L., Casis, O., Apellaniz, A. 1996. Handling, processing and storage of toxic wastes in the university of the basque country, *Toxicology Letters*, Vol. 88, p. 81.

Barta, J. 1993. Jerusalem artichoke as a multipurpose raw material for food products of high fructose or inulin content, In: A. Fuch (ed.) *Inulin and Inulin-Containing Crops Studies in Plant Science*, Vol. 3, *Elsevier Science Publishers*, Amsterdam, The Netherlands, pp. 323–339.

Barta, J., Förster, H., Porcsa, I., Rák, I., Sósné, M., Vukov, K. 1989. Method for fiber-rich fruit drinks processing to promote lead and heavy metal detoxication (In Hungarian: Eljárás az emberi szervezetbe kerülő ólom, és egyéb nehéz fémek detoxikálásátő elősegítő gyümölcstermékek, főként rostos italok előállítására.), *Patent 203 960.* Hungary.

Barta, J., Pátkai, G.Y., Gion, B., Körmendy, I. 1997. Presentation of an alternative, waste-free processing technology illustrated by the example on inulin containing crops. *Acta Alimentaria*, Vol. 26, Issue 3, pp. 88–89.

Békési, Z. Pándi, F. 2001. Position and tasks of the environmental protection in the domain fruit distilling industry (In Hungarian: A környezetvédelem helyzete és feladatai a hazai gyümölcsszesziparban.), *Élelmezési Ipar*, Vol. 55, Issue 3, pp. 76–79.

Bennett, B. 2002. *Feeding Crop Waste to Livestock and the Risk of Chemical Residues, Notes Information Series, Department of Primary Industries,* Victoria, Australia.

Bouallagui, H., Torrijos, M., Godon, J.J., Moletta, R., Cheikh, R.B., Touhami, Y., Delgenes, P., Hamdi, M. 2004. Two-phases anaerobic digestion of fruit and vegetable wastes: Bioreactors performance, *Biochemical Engineering Journal*, Vol. 21, Issue 2, pp. 193–197.

Braddock, R.I. 1999. *Handbook of Citrus By-product and Processing Technology*, John Wiley and Sons, Canada, pp. 39–149.

Brosnan, T., Sun, D.W. 2001. Pre cooling techniques and applications for horticultural products—a review, *International Journal of Refrigeration*, Vol. 24, Issue 2, pp. 154–170.

Bulla, M. 1994. Inspection of the environment (Környezetvizsgálat) *SZIF*, Győr, pp. 55–70.

Carson, K.J., Collins, J.L., Penfield, M.P. 1994. Unrefined, dried apple pomace as a potential food ingredient, *Journal of Food Science*, Vol. 59, Issue 6, pp. 1213–1215.

Cheeseborough, M. 2000. Waste reduction and minimisation. In Conference: *Waste Reduction for the Third Millennium.* EMBRU, Environmental Management and Business Research Unit, EcoTech Centre, Swaffham (UK).

Cheremisinoff, N.P. 2003. *Handbook of Solid Waste Management and Waste Minimization Technologies*, Elsevier Science, USA, pp. 39–66.

Di Blasi, C., Tanzi, V., Lanzetta, M. 1997. A study on the production of agricultural residues in Italy, *Biomass and Bioenergy*, Vol. 12, Issue 5, pp. 321–331.

Drunen, J. van, Hranisavljevic, I. 2003. Process for enriching foods and beverages. *Patent US 6 572 915 B1 (US6572915B1).*

El-Nawawi, S.A., Heikal, Y.A. 1996. Production of pectin pomace and recovery of leach liquids from orange peel, *Journal of Food Engineering*, Vol. 28, Issues 3–4, pp. 341–347.

Fehr, M., Calcado, M.D.R., Romano, D.C. 2002. The basis of a policy for minimizing and recycling food waste, *Environmental Science and & Policy*, Vol. 5, Issue 3, pp. 247–253.

Fernández-López, J., Fernández-Ginés, J.M., Carbonell, L.A., Sendra, E., Sayas-Barberá, E., Pérez-Alvarez, J.A. 2004. Application of functional citrus by-products to maet products, *Trends in Food Science and & Technology*, Vol. 15, Issues 3–4, pp. 176–185.

Fujihara, S., Shimizu, T. 2003. Growth inhibitory effect of peel extract from *Citrus junos*, *Plant Growth Regulation*, Vol. 39, Issue 3, pp. 223–233.

Gil, M.I., Tomas-Barberán, F.A., Pierce, B.H., Holcroft, D.M., Kader, A.a 2000. Antioxidant activity of pomegranate juice and its relationship with phenolic composition and processing, *Journal of Agricultural and Food Chemistry*, Vol. 48, Issue 10, pp. 4581–4589.

Hammond, J.B., Egg, Diggins, D., Coble, C.G. 1996. Alcohol from bananas, *Bioresource Technology*, Vol. 56, Issue 1, pp. 125–130.

Henningsson, S., Hyde, K., Smith, A., and Campbell, M. 2004. The value of resource efficiency in the food industry: A waste minimisation project in East Anglia, UK, *Journal of Cleaner Production*, Vol. 12, Issue 5, pp. 505–512.

Hollingdele, R.J. 2000. *The Waste Books, The New York Review of Book*, pp. 231–235.

Howitt, S. (ed.) 1997. *Waste Management. Keynote Market Report*, 3rd ed., Key Note Ltd., Hampton, UK.

Huebner, M., Kienzle, M. 2001. Retentate-waste or a valuable product? New solutions, *Food Processing*, Vol. 12, Issue 9, pp. 358–363.

Khachatourians, G.G., Arora D.K. 2001. Applied mycology and biotechnology, Vol. I. *Agriculture and Food Production*, Elsevier Science, The Netherland, pp. 353–387.

Kilham, C. 1997. *The Whole Food Bible: How to Select and Prepare Safe*, Healthful Foods, Book, Healing Art Press, Rochester, Vermont, pp. 41–52.

Kim, W.C., Lee, D.Y., Lee, C.H., Kim, C.W. 2004. Optimization of narirutin extraction during washing step of the pectin production from citrus peels, *Journal of Food Engineering*, Vol. 63, Issue 2, pp. 191–197.

Kumar, D., Jain, V.K., Shanker, G., Srivastava, A. 2003. Utilization of fruits waste for citric acid production by solid state fermentation, *Process Biochemistry*, Vol. 38, Issue 12, pp. 1725–1729.

Larrauri, J.A. 1999. New approaches in the preparation of high dietary fibre powders from fruit by-products, *Trends in Food Science and Technology*, Vol. 10, Issue 1, pp. 3–8.

Larrauri, J.A., Rupérez, P., Saura-Calixto, F. 1997. Effect of drying temperature on the stability of polyphenols and antioxidant activity of red grape pomace peels, *Journal of Agricultural and Food Chemistry*, Vol. 45, Issue 4, pp. 1390–1393.

Lussier, M.G., Shuff, J.C., Miller, D.J. 1994. Activated carbon from cherry stones, *Carbon*, Vol. 32, Issue 8, pp. 1493–1498.

Madhuri, P., Kamini-Devi, 2003. Value addition to watermelon fruit waste, *Journal of Food Science and Technology*, Vol. 40, Issue 2, pp. 222–224.

Mahadevaswamy, M., Venkataraman, L.V. 1990. Integrated utilization of fruit-processing wastes for biogas and fish production, *Biological Wastes*, Vol. 32, Issue 4, pp. 243–251.

Miguel, G.N., Belloso, M.O. 1999. Comparison of dietary fibre from by-products of processing fruits and greens and from cereals, *Lebensmittel-Wissenschaft und Technologie*, Vol. 32, Issue 8, pp. 513–508.

Negi, P., Jayaprakasha, G.K., Jena, B.S. 2003. Antioxidant and antimutagenic activities of pomegranate peel extracts, *Food Chemistry*, Vol. 80, Issue 3, pp. 393–397.

Noguchi, H.K., Tanaka, Y. 2004. Allelopathic potential of *Citrus junos* fruit waste from food processing industry, *Bioresource Technolgy*, Vol. 94, Issue 2, pp. 211–214.

Pándi, F. 2001. The position, aims and the tasks of the environmental management in the food industry (In Hungarian: A környezetgazdálkodás helyzete, célkitűzései az élelmiszeriparban és az abból adódó feladatok.), *Élelmezési Ipar*, Vol. 55, Issue 1, pp. 21–25.

Perédi-Vásárhelyi, K. 2004. In utilization of the waste materials of the plant origin raw materials/fruit and vegetables pruduce (In Hungarian: Növényi nyersanyag/gyümölcs, zöldség/feldolgozási hulladékainak hasznosítása c. 4/005/2001 *NKFP pr.*).

Polpraser, C. 1996. *Organic Waste Recycling: Technology and Management*, John Wiley, *Chichester.*

Poonam-Nigam, Dadel-Singh, Ashok-Pandey, 2001. Utilization of agricultural and food waste and by-products by biotechnology. *Agro Food Industry hi tech*, Vol. 12, Issue 3, pp. 26–29.

Prema-Viswanath, S., et al. 1992. Anaerobic digestion of fruit and vegetable processing wastes for biogas production, *Bioresource Technology*, Vol. 40, Issue 1, pp. 43–48.

Pugliese, P., Moltó, J.C., Damiani, P., Marín, R., Cossignani, L., Manes, J. 2004. Gas chromatographic evaluation of pesticide residue contents in nectarines after non-toxic washing treatments, *Journal of Chromatography A*, Vol. 1050, Issue 2, pp. 185–191.

Ramadan, M.F., Mörsel, J.T. 2003. Recovered lipids from prickly pear/*Opuntia ficus-indica* (L.) Mill/peel: a good source of polyunsaturated fatty acids, natural antioxidant vitamins and sterols, *Food Chemistry*, Vol. 83, Issue 3, pp. 447–456.

Read, A.D., Phillips, P., Robinson, G. 1997. Landfill as a future waste management option in England: the view of landfill operators, *Resources, Conservation and Recycling*, Vol. 20, Issue 3, pp. 183–205.

Schieber, A., Hilt, P., Streker, P., Endres, H-U, Rentschler, C., Carle, R. 2003. A new process for the combined recovery of pectin and phenolic compounds from apple pomace, *Innovative Food Science* and *Emerging Technologies*, Vol. 4, Issue 1, pp. 99–107.

Sherwood, C., Crites, R.W., Middlebrooks, E.J. 1995. *Waste Management and Treatment*, 2nd ed., Mc Graw-Hill Special Reprint Edition, Mc Graw-Hill, New York, pp. 43–60.

Stabnikova, O., Wang, J.Y., Ding, H.B., Tay, J.H. 2005. Biotransformation of vegetable and fruit -processing wastes into yeast biomass enriched with selenium, *Bioresource Technology*, Vol. 96, Issue 6, pp. 747–751.

Subburamu, K., Singaravelu, M., Nazar, A., Irulappan, I. 1992. A study on the utilization of jack fruit waste, *Bioresource Technology*, Vol. 40, Issue 1, pp. 85–86.

Szenes, Ené. 1995. Environmental protection in food industry (In Hungarian: Környezetvédelem az élelmiszer-ipari kis- és középüzemekben.), *IntegraProjekt Kft.*, Budapest, pp. 10–21, pp. 119–122.

Tuncel, M., Nout, M.J.R., Brimer, L. 1995. The effects of grinding, soaking and cooking on the degradation of amygdalin of bitter apricot seeds, *Food Chemistry*, Vol. 53, Issue 4, pp. 447–451.

Vermes, L. 1998. Waste management and utilization (In Hungarian: Hulladékgazdálkodás, hulladékhasznosítás.), *Mezőgazda Kiadó*, Budapest, pp. 14–15.

Viswanath, P., Devi, S.S., Nand, K. 1992. Anaerobic digestion of fruit and vegetable processing wastes for bio-gas production, *Bioresource Technology*, Vol. 40, Issue 1, pp. 43–48.

Williams, P.E.V. 1995. Animal production and European pollution problems, *Animal Feed Science and Technology*, Vol. 53, Issue 2, pp. 135–144.

Woods, C. 1998. Waste minimization: where is it going?, *Waste Management*, Vol. 16, Issue 1, p. 37.

Woodard, R., Bench, M., Harder, Stantzos, N. 2004. The optimisation of household waste recycling centres for increased recycling—a case study in Sussex, UK, *Resources, Conservation and Recycling*, Vol. 43, Issue 1, pp. 75–93.

Zheng, Z., Shetty, K. 2000. Solid state production of polygalacturonase by Lentinus edodes using fruit processing wastes, *Process Biochemistry*, Vol. 35, Issue 8, pp. 825–830.

Part II
Products Manufacturing

11
Manufacturing Jams and Jellies

H. S. Vibhakara and A. S. Bawa

INTRODUCTION

Fruits have been part of man's food since times immemorial. Among the preserved fruits, jams and jellies constitute important products and afford means of utilizing a large amount of sound cull fruits unsuitable for other purposes. It is an older method than canning and freezing (Peckham, 1964; Thakur et al., 1997). Historically, jams and jellies may have originated as an early effort to preserve fruits for consumption in the off-season. As sugar manufacture became more affordable, the popularity and availability of these fruit products increased (Anon, 1983). Jams in their various forms are probably the easiest by-products made of citrus fruits. The earliest published record of jelly making appeared in the later part of the 18th century.

The preparation of jams and jellies was developed as an art by the housewife, and served as a means of preserving fruit, corresponding to the time, when the fruits were being harvested. Science as applied to jam manufacture seems often to be regarded by the general public as synonymous with sophistication. Factory-made jam must conform to certain specifications and standards, not essential for a home-made jam, for instance; it must be of a consistency firm enough to meet the demands of confectioneries and to withstand handling during transport.

The consistency depends upon the presence of pectin. Hence, scientific jam manufacture is largely based on the correct application of knowledge about pectin and the laws governing the formation of the pectin–sugar–acid gel. Jam and jelly products were prepared with a high concentration of dissolved solids so that fermentation could not occur. However, only pectin and sugar are not sufficient for the formation of the products. Equally important is the acidity of the fruit, resulting in a definite equilibrium in the "pectin–acid–sugar" system (Breverman, 1963).

Pectin is the most essential thing in the formation of jam and jelly. Lack of knowledge about the requirements necessary for the pectin gel formation, frequently contributes to products of undesirable consistency. The preservation industry awaited the accumulation of sufficient information on the chemistry of pectin and its gels to control the consistency under commercial conditions. As the knowledge of pectin chemistry increased, production grew and to some extent replaced home-prepared products. Consequently, the jam and jelly industry assumed considerable magnitude.

INGREDIENTS FOR JAMS AND JELLIES

Definition and standards of identity for various fruit preserves and jellies have been issued by the Food and Drug Administration under the Food, Drug and Cosmetic Act under which ingredients for jams and jellies have been discussed. A brief outline of these standards has been presented here. The U.S. standards for Grades were established by the production and marketing administration (Anon, 1974, 1975).

Jellies are viscous or semi-solid foods made from a mixture of not less than 45 parts by weight of saccharine ingredient. The mixture is concentrated by heat to such a point that the soluble solids content of the finished jelly is not less than 65%. Spices, sodium citrate, sodium potassium tartrate, sodium benzoate, benzoic acid, mint flavor, and harmless artificial green coloring may be optional ingredients. Optimal saccharine ingredients are corn sugar, invert sugar syrup, sucrose, honey, or combinations of these. Pectin and designated organic acids may be added to compensate for deficiencies of these substances in fruit juice. Inducement for adding pectin or acid in quantities greater than required to supply the natural deficiency of the fruit juice is eliminated by fixing the minimum fruit juice content. The name of the fruit or fruits present as well as spices, chemical preservatives, honey, or corn syrup used must be indicated on the label.

Standards for jams and preserves are similar to those for jelly except that fruits are used rather than fruit juice ingredients, and mint flavor and green coloring are not optional ingredients. The fruit mixture is concentrated by heat to such an extent that the total soluble solids content is not less than 65% for certain specified fruits, and 68% for others.

It is extremely difficult to account for the behavior of gels formation. Gels are a form of matter intermediate between a solid and a liquid. They consist of polymeric molecules cross-linked to form a tangled, interconnected molecular network (Fig. 11.1) immersed in a liquid medium (Oakenfull, 1991). The water, as a solvent, influences the nature and magnitude of the intermolecular forces that maintain the integrity of the polymer network; the polymer network holds the water, preventing it from flowing away in acid medium; pectin with sugar affects the pectin water equilibrium and forms a network of fibers throughout the jelly (Mitchell, 1979; Rees, 1969; Thakur et al., 1997). This structure is capable of supporting liquids. Figure 11.2 shows the factors affecting the strength of the network.

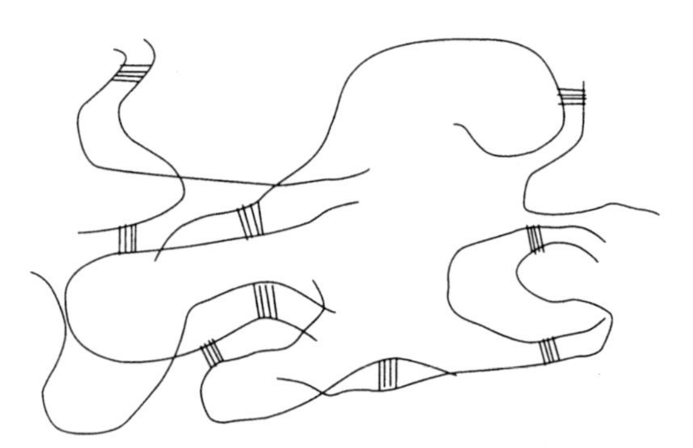

Figure 11.1. Gel network in jams and jellies (the hatched areas represents junction zones).

TYPES AND VARIETIES OF FRUITS

A jam manufacturer can choose a fruit from among the following five categories:

1. Fresh fruit
2. Frozen, chilled, or cold stored fruit
3. Fruit or fruit pulp preserved by heat

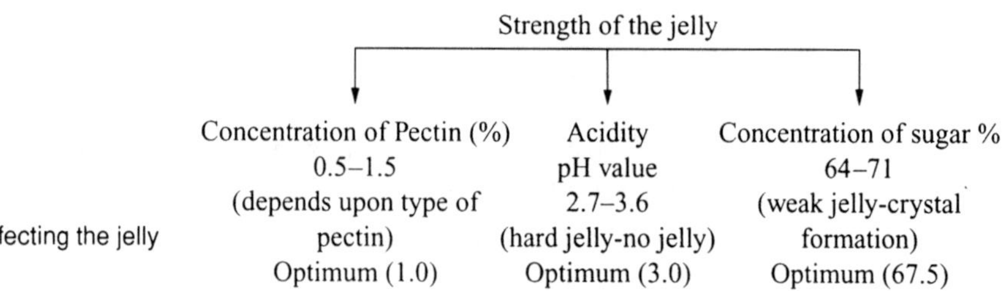

Figure 11.2. Factors affecting the jelly strength.

4. Sulfited fruit or fruit pulp, i.e., fruit preserved with sulfur dioxide
5. Dried dehydrated fruits

Fresh fruits generally give the best jams. As pectin is the main ingredient in the fruit that gives a set to the jam, it is preferable to use a slightly underripe fruit that is rich in pectin along with the ripe fruit to secure the desirable gelling effect in the jam. Apple, pear, sapota, apricot, loquat, peach, papaya, karonda, plum, strawberry, mango, tomato, grapes, and muskmelon have been used for preparation of jams. It can be prepared from a single fruit or a mixture of two or more.

In jelly making, pectin is the most essential constituent. Although there is difference of opinion about the exact nature of pectin, it is generally accepted that pectin forms jellies, when mixed and boiled with proper amounts of sugar, acid, and water. All these constituents must be present in a particular proportion for making a good jelly (Kratz, 1993, 1995).

Guava, sour apple, plum, grapes, karonda, wood apple, loquat, papaya, and gooseberry are generally used for preparation of jelly. Apricot, pineapple, strawberry, raspberry, etc. can be used but only after addition of pectin powder, because these fruits have low pectin content. Fruits can be divided into four groups on the basis of their pectin and acid contents (NIIR Board, 2002).

1. Rich in pectin and acid: sour and crab apple, grape, sour guava, lemon, orange (sour), plum (sour), jamun.
2. Rich in pectin but low in acid: apple (low acid varieties), unripe banana, sour cherry, fig (unripe), pear, ripe guava, peel or orange, and grapefruit.
3. Low in pectin but rich in acid: apricot (sour), sweet cherry, sour peach, pineapple, and strawberry.
4. Low in pectin and acid: ripe apricot, peach (ripe), pomegranate, raspberry, and any other over-ripe fruit.

Types and varieties of fruits selected should be used without undue delay for making jam and jelly because, if kept for a long time, degradation of pectin proceeds rapidly.

PEELS

Apart from the sorting and removal of leaves, stalks, and undesirable portions of the fruit, which can only be done by hand, most fruits require treatment of some kind before they enter the boiling pan. For example, strawberries require sometime light crushing, plums require heating with minimum of water until soft. When it is required, cherries are similarly treated. Currants are passed through machines that remove the stalks, gooseberries are whirled in machine (Morries, 1951).

Sour and bitter oranges are utilized by hand or machine and the peel cut of any desirable size or shape by special machines. Sometimes admixture of certain proportion of grapefruit or lemons is used, peel of which constitutes only a quarter to a half if utilized in proportion to the total weight of fruit. These slices need to be softened either by prolonged boiling, or more rapidly by heating. The slices may be covered with water and cooked until tender, the water being changed at least twice during cooking. It can be done in autoclaves under pressure of 1 atm. More rapid softening is attained by boiling the peel in a solution of carbonate of soda or ammonium hydroxide, which removes from the cell wall certain substances that otherwise render the peel tough after boiling with sugar. Ammonia is less drastic in its action than that of soda and is preferable for use since less danger of the peel breaking up into small fragments, which is due to the fact that soda dissolves both hemicelluloses and pectic subtances, whereas the former are said to be insoluble in ammonia. It would appear that some more research is required as to the precise causes of the softening.

The time required to soften peel in an autoclave depends on the size of the slices and on the pressure and temperature employed. Some discoloration may occur if too high pressure is used. To avoid handling the peel after it has been softened, Morris (1935) suggested placing the peel intended for softening in perforated baskets of non-corrosible metal, each containing sufficient peel for one batch. After having been cooked in ammonium hydroxide or soda, the peels may be re-cooked for a short time with a weak solution of citric acid to remove any traces of the hydroxide.

GELLING AGENTS

Gelling agents are used in the food industry in a wide range of products both traditional and novel, and this use is increasing rapidly with the increase of convenience foods. An ideal gelling agent should not interfere with the odor, flavor, or taste of the product to which it is added (Fishman and Jen, 1986). Improvements to existing and development of new ones require basic understanding of the processes of gelatin and the properties of gels at the molecular level (Doublier and Thibault, 1984; May and Stainsby, 1986).

Gels are a form of matter intermediate between a solid and a liquid. They consist of polymeric molecules cross-linked to form a tangled, interconnected molecular network immersed in a liquid medium (Flory, 1953). The polymer network holds the water, preventing it from flowing away (Oakenfull, 1987; Meyer, 1960). In gels, the molecules are held together by a combination of weak intermolecular forces such as hydrogen bonds, electrostatic forces, Vanderwaals forces, and hydrophobic interactions. The cross-linkages are not point interactions but involve extensive segments from two or more polymer molecules, usually in well-defined structures called junction zones (Rees, 1969). The gelation process is essentially the formation of these junction zones (Fig. 11.1).

The physical characterizations of gel are the consequence of the formation of a continuous three-dimensional network of cross-linked polymer molecules on a molecular level; an aqueous gel consists of three elements (Jarvis, 1984):

1. Junction zones where polymer molecules are joined together.
2. Interjunction segments of polymers those are relatively mobile.
3. Water entrapped in the polymer network.

Gels are always formed in an aqueous environment. Thus, the interactions of protein and polysaccharides with water are by themselves important factors in the gelation process. Both types of polymers are strongly hydrated in aqueous solution, so that some water molecules are so tightly bound that they fail to freeze even at temperatures as low as −60°C (Eagland, 1975). Although the formation of a stable intermolecular junction is a critical requirement for gelation, some limitation on the interchain association is also necessary to give a hydrated network rather than an insoluble precipitate (Axelos and Thibault, 1991).

It is important to know the condition for the onset of gelation in technological processes involving gelling food products. Several methods are used to characterize this change in consistency (Doesburg and Grevers, 1960; Walter and Sherman, 1981; Beveridge and Timber, 1989; Dhame, 1992; Rao, 1992; Rao and Cooley, 1993). Physically, the critical stage of gelation may be monitored from the loss of fluidity or from the rise of the elastic property of the growing network (Shomer, 1991). Table 11.1 gives different types of jelling agents used in the manufacture of jellies.

Gelatin is a water-soluble protein formed by initial degradation of collagen from animal skin and bones. Gelatin jellies have a rather soft or rubbery texture. For these, it is normal to use an additional gelling agent such as thin boiling starch. This involves the texture incidentally. Gelatin gels forms reversibly on cooling a gelatin solution. It is now well established that the protein molecules are cross-linked to form a network by junction zones, where the protein chain

Table 11.1. Different Types of Jelling Agents Used in the Manufacture of Jellies

Type of Gelling Agent	Origin	Use
Gelatin	A protein of animal origin extracted from bones and purified	Generally, must not be boiled. To be added to warm syrup for setting on cooling
Agar/Alginates	Extracted from various sea weeds	Various products such as neutral jellies, weakened by boiling in acid solution
Gum Arabic/Acacia	Exudates from trees	Used to produce hard gums, and as an extender and thickener in products, e.g., Marshmallow
Starch/Modified starch	Seeds and various roots	These have been completely and partly replaced by other jelling agents in gums—Turkish delight Glazer
Pectin	Fruit residues particularly citrus and apple pomace	Used largely in acid fruit jellies but with low melting point is used in neutral jellies

have partly refolded in the collagen triple helix structure (Veis, 1964).

Agar/alginates are the major structural polysaccharides of algae. Agar jellies have a very soft texture. Straight agar jellies have a characteristic "shortness" that may be modified by the addition of gelatin, gum Arabic, pectin, starch, etc. Alginates with a high ratio of poly-β-D-mannuronic acid (M) and poly-α-L-guluronic acid (G) form weak, forbid gels, whereas low M/G alginates give transparent, stiff, brittle gels, and the gel strength depends on the nature of the divalent cation (Smidsred, 1974).

Gum Arabic is the most water-soluble of the natural gums (up to 50%) and their solutions are of relatively low viscosity. Other advantages of gum Arabic are its absence of odor, color, and taste. Hard jellies can be produced with gum Arabic.

Unmodified starches, produced by wet milling of field corn, supply the major amount of thickening material. Modified starch is starch that has undergone one of the varieties of treatment to alter its physical property and/or functionality (Mauro et al., 1991; Furcsik and Mauro, 1991). They are used to extend the bodying or gelling effect of normal starches, to modify gelling tendencies, and to improve texture. Starch is an essentially linear polymer of α-$(1 \rightarrow 4)$ linked D-glucose (Wolfram and El Khadem, 1965). Starch gels consequently have a composite structure of open, porous amylopectin molecules threaded by an amylose matrix. Thus, actual gel-forming polymer in starch is amylose. The molecular weight distribution of amylose depends on the plant source and molecular weights of several millions with broad distribution have been reported (Rao et al., 1993).

Pectin is the most frequently used hydrocolloids in processed fruits. Jams and jellies are the major food type using larger amounts of pectin. Pectin is a class of complex hetero polysaccharides found in the cell walls of higher plants, where they function as a hydrating agent and cementing material for the cellulosic network (Muralikrishna and Taranathan, 1994).

When pectin-rich plant materials are heated with acidified water, the protopectin is liberated and is hydrolyzed into pectin that is readily soluble in water. It happens in plant tissues during ripening of fruits with the aid of an enzyme protopectinase. As the ripening of fruit proceeds, more and more of the insoluble protopectin is converted into soluble pectin (Woodmansee et al., 1959). Their composition varies with the source and conditions of extraction, location, and other environment factors (Chang et al., 1994). Pectic substances in the primary cell wall have a relatively higher proportion of oligosaccharides chain on their backbone, and the side chains are much longer than those of the pectin of the middle lamella (Sakai et al., 1993).

Pectins are primarily a polymer of D-galacturonic acid (homopolymer of $[1 \rightarrow 4]$ α-D-galacto pyranosyluronic acid units with varying degrees of carboxyl groups methylated estrified) and rhamnogalacturonan (hetero polymer of repeating $[1 \rightarrow 2]$ α-L-rhamnosyl [1-L] α-D-galactosyluronic acid disaccharide units), making it an α-D-galacturonan (Lau et al., 1985). The molecule is formed by L-1,4-glycosidic linkages between the pyranose rings of D-galacturonic acid units. As both hydroxyl groups of D-galacturonic acid at carbon atoms 1 and 4 are on the axial position, the polymer formed is 1,4-polysaccharide (Sakai et al., 1993; Oakenfull, 1991).

The chemical structure of galacturonic acid and its methyl ester are shown in Figure 11.3 and the linkages between different galacturonic acids and their methyl esters in pectic and pectinic acid are shown in Figure 11.4 (Swaminathan, 1987).

Pectic acid is composed mostly of colloidal polygalacturonic acid molecules, and is essentially free from methyl ester groups. The salts of pectic acids are either normal or acid pectates, whereas pectinic acid ones are colloidal polygalacturonic acids containing more than a negligible portion of methyl ester groups. Pectinic acid under suitable conditions is capable of forming gels with sugar and acid or, if suitably low in methoxyl content, with certain metallic ions. The salts of pectinic acids are either normal or acid pectinates.

Studies on esterified residue in pectin claimed that they are randomly distributed (Garnier et al., 1993; De vries et al., 1986) or non-randomly distributed (Mort et al., 1993a,b). However, ion exchange chromatography showed a random distribution of change in citrus pectin that had undergone acid catalyzed deesterification (Garnier et al., 1993). Such disparate findings may, impart, be due to the length of galacturonate residues being examined or due to differences in pectin source (Baker et al., 1996).

Polygalactoronic acid could be considered as a rod in solution, whereas pectins are segmented rods with flexibility at the rhamnose tees (Fry, 1986). The size, charge density, charge distribution, and degree of substitution of pectin molecules can be changed biologically or chemically (Kerstez, 1951; Kratz, 1993).

The most unique and outstanding property of pectin is their ability to form gel in the presence of Ca^{2+} ions in sugar and acid solution (Gordon et al., 2000; Halliday and Bailey, 1954). Degree of

Galacturonic acid and its methyl ester

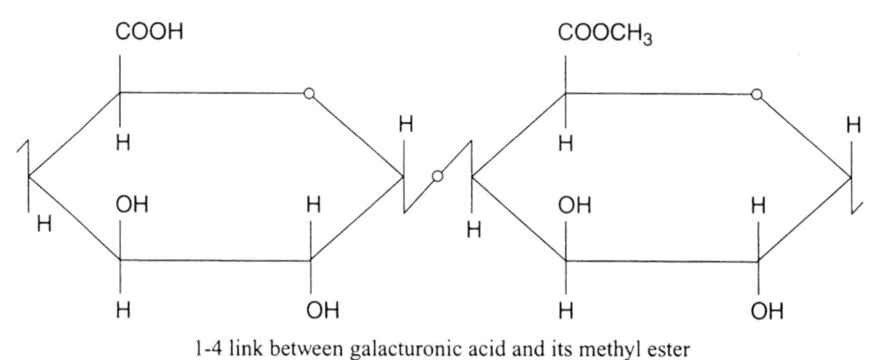

1-4 link between galacturonic acid and its methyl ester

Figure 11.3. Chemical structure of galacturonic acid and its methyl ester.

esterification, attached chain of neutral sugars, acetylation, and cross-linking of pectin molecules also affect the texture of pectin gels (Guichard et al., 1991). Pectin chains carry negative charge and the change density is higher at higher pH and lower degree of methoxylation (DM) (Gross et al., 1980). Conformation of the pectin molecules is not affected by the branching, but side branching in pectin can result in significant entanglement in concentrated solutions (Hawang and Kokini, 1992; Ring and Oxford, 1985).

Depending on the DM, pectins are classified into (1) low methoxy (LM) (25–50%) and (2) high methoxy (HM) (50–80%) pectins, and form gels of two types with occasional intermediates. They are called acid and calcium gels and formed from HM and LM pectins, respectively. In HM pectins, the effect of sugars depends specially upon molecular geometry of the sugar and the interactions with neighboring water molecules (Oakenfull and Scott, 1984). Non-covalent forces (i.e., hydrogen bonding and hydrophobic interactions) are believed to be responsible for gel formation in HM pectins (DaSilva et al., 1992; Walkinshaw and Arnott, 1981). In LM pectins, gel is formed in the presence of Ca^{2+}, which acts as a bridge between pairs of carboxyl groups of pectin molecules. The two kinds of pectins are relatively stable at the low pH levels existing in jams and jellies (Pilgrim et al., 1991). HM pectin are used to form gels in acid media of high sugar content and LM pectins are used in products of lower sugar content (Axelos and Thibault, 1991). Pectins can be further divided into rapid-set, medium-set, and low-set pectin, depending upon the time the gel takes to set.

The functionality of pectin molecules is determined by a number of factors, including DM and molecular size (Doublier et al., 1992). Pectins of 100–500 grades are available in the market. Their application as a food hydrocolloid is mainly based on their gelling properties (Voragen et al., 1986). Selection of pectin for a particular food depends on many factors, including the texture required, pH, processing temperature, presence of ions, proteins, and the expected shelf life of the product (Hoefler, 1991).

Combinations of gelling agents are often used because a combination gives a desirable texture. For this reason, mixed systems are of great technological importance (Christensen and Trudsoe, 1980; Hughes

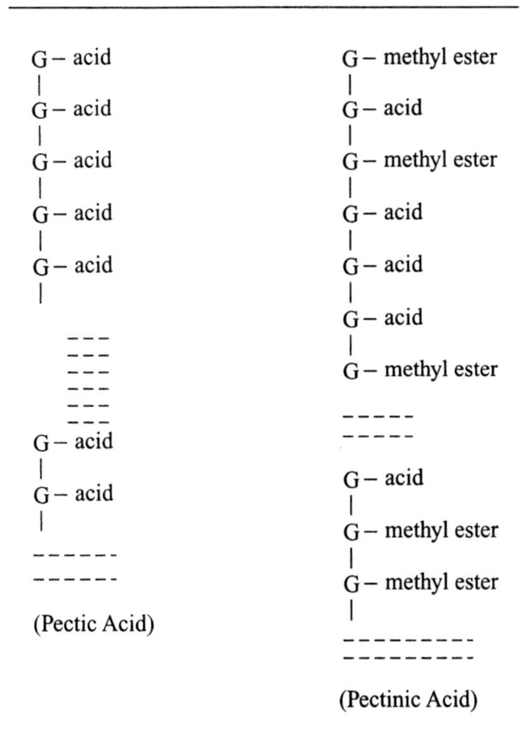

Figure 11.4. Pectic acid and Pectinic acid (G—represents galacturonic acid).

et al., 1980). The physical chemistry of mixed system is obviously very complex. However, there has recently been significant progress in developing a suitable theoretical framework, at least for two component systems (Morris, 1986).

SWEETENING AGENTS

Sweeteners are used in fruit processing for many functional reasons as well as to impart sweetness. They add flavor, body, bulk, and control viscosity that contribute to texture and prevent spoilage. It binds moisture in fruits that is required by detrimental microorganism. Too little sugar prevents gelling and may allow yeast and mold growth. Sugar serves as a preserving agent and aids in jelling.

Sucrose, commonly known as table sugar (or refined sugar), is the standard against which all sweeteners are measured in terms of quality, taste, and taste profile. However, glucose syrups have been widely used as a part of sugar source in recent years. An invert sugar component is necessary to prevent sucrose crystallization in high solids jellies and jams during storage. Such crystallization is rare in products containing less than 68% solids. If the concentration of sugar is high, the jelly retains less water resulting in a stiff jelly, probably because of dehydration (Giridhari Lal et al., 1986).

The optimum percentage of invert sugar is between 35% and 40% of the total sugar in the jam. During the process of inversion, molecular water is taken into the sugars; that is the reason why 95 parts of sucrose yield 100 parts of invert sugar. The rate of inversion is influenced by three factors:

1. Hydrogen ion concentration (pH) of the product
2. Boiling temperature
3. Boiling time

During the process of boiling, sucrose undergoes a chemical change. Cane and beet sugars are nonreducing sugars. However, when boiled with acid or treated with some enzymes, sucrose is converted into two reducing sugars, namely dextrose and levulose in equal parts. Sucrose has a molecular weight of 342, invert sugar 360, the difference of 18 being the molecular weight of water molecule added during inversion (Rauch, 1965).

Invert sugar syrup can be produced from sucrose by the action of an acid. After the process is completed, a suitable amount of sodium bicarbonate is added in order to neutralize the acid, as it is frequently undesirable to use strong acid in a food containing sugar. Methods using hydrochloric, tartaric, or citric acid are satisfactory when pure white sugar is used. Lower grade of sugar may require larger amount of acid. A chemist should be consulted for the quantities of acid needed.

The addition of sugar in LM pectin gels can increase gel strength and reduce syneresis (Axelos and Thibault, 1991). For HM pectin gels, a number of other sugars, alcohols, and polyols will permit gelation. From a practical point of view, it may be advantageous to substitute other sugars for sucrose, either because of cost, or to reduce the likelihood of crystallization, or for flavor modification (Ahmed, 1981). Partial replacement of sucrose with other sugars such as maltose, glucose, syrups, or high fructose corn syrup altered the setting times and certain rheological properties of model gels (May and Stainsby, 1986). For example, addition of maltose reduced the setting time and extended the pH range of gelation, while fructose delayed setting time. Partial or complete replacement of sucrose with other sugars alters the water activity of the system and can modify the hydrophobic interactions contributing to gelation.

HM pectin gel forms only under acidic conditions and when the sugar content is at least 55% (Oakenfull and Scott, 1984). Low pH suppresses dissociation of free carboxylic acid groups, reducing their electrostatic repulsion (Watase and Nishinari, 1993), while sugars stabilize hydrophobic interactions between the methyl ester groups (Oakenfull and Scott, 1984; Brosio et al., 1993; Rao et al., 1993). Junction zone size and the standard free energy of gelation increase as the degree of esterification increases, being proportional to the square of the degree of esterification. As pectin ester content is lowered to 50%, jelly strength increases, but only at progressively lower pH values (Smit and Bryant, 1968).

Esterification levels of pectins can also have an effect on the flavor perception of the jellies. For a substance to be tasted, it must contact the taste buds. Therefore, if a gel delays diffusion of the substance to the taste bud surface perception may be controlled by diffusion and not by the taste reaction. Guichard et al. (1991) found, at similar gel consistencies, high methoxyl pectin reduces taste intensity more than low methoxyl ones.

ACIDULANTS

Acidulants are acids that either occur naturally in fruits and vegetables or are used as additives in food processing. Acidulants serve many functions in fruit processing. They perform a variety of functions in fruit processing, the primary being acidifier, pH regulator, preservative and curing agent, flavoring agent, chelating agent, buffering agent, gelling/coagulating agent, and antioxidant synergist.

The choice of the particular acid for any specific food application is dependent upon a number of factors. Since each acid has its own unique combination of physical and chemical properties, the choice should be made on the basis of the qualities required (Watase and Nishinari, 1993). The simplest possible use of a food acid involves decreasing the pH only. As an example, it might be desired simply to acidify a solution where pH adjustments are all that are required. In many cases, the selection is based upon the ability of the acid to bring out and enhance the food flavor. For such purposes, compatibility and blending are important factors in the choice of acid. Each acid behaves somewhat independently of pH and not necessarily preferred for another. Thus, citric acid is preferred in most fruits, tartaric acid for grape, and malic acid for apple. Since taste stability cannot always be predicted, the product should be examined both before and after storage, when blending or substituting the acids used.

Apples differ in their organic acid profiles. Frauzen and Helwert (1923) found oxalic 0.001, succinic 4.22, malic 69.57, citric 24.95, and lactic 1.18% in apples. Nelson (1925), on the other hand, found a much less complicated mixture in American apples, the acid being entirely malic. The total acidity may be about 1.2% (as malic acid) in apples. It shows that the same kinds of fruits may vary, not only in the amount but also in the kind of acid they contain. The H ion concentration is affected very considerably by buffering substances such as organic bases and inorganic and organic salts and is of the utmost importance in connection with fruit preservation and a determining factor in the formation of the pectin–sugar–acid gel.

It has been found that gel formation occurs only within a certain range of hydrogen ion concentration, the optimum acidity figure for ions and jellies being around pH 3.0. The gel strength fouls slowly on decreasing and rapidly on increasing the pH value. Beyond pH 4, no jelly formation occurs in the usual soluble solid range. The pH value is also critical in determining the temperature at which jellies set. Insufficient acidity is one of the common causes of jelly failure. The pH value of jelly should be taken when the jelly is concentrated sufficiently to pour. Different juices will require different amounts of additional acid depending upon the original acidity of the juice and buffering capacity of the juice. The pH may be adjusted to attain optimum flavor to control or modify the rate of setting and to modify the degree of sugar inversion (Smit and Bryant, 1968).

Control of pH is critical to successful gel formation with pectin, particularly in the case of HM pectins. Low pH increases the percentage of unionized carboxyl groups, thus reducing electrostatic repulsion between adjacent pectin chains. Rapid-set pectins with high degree esterification will gel at higher pHs than the one with lower degree esterifications, slow-set pectin; however, this difference is slight, with the optimum pH for slow-set pectins being about 3.1 and for rapid-set pectins being 3.4 (Crandall and Wicker, 1986). Substitution of other sugars for sucrose, by modifying hydrophobic interactions between chains, allows gels to be formed at somewhat higher pHs (May and Stainsby, 1986). Since they rely on calcium bonding to effect gelation, low methoxyl pectin can form gels at higher pHs than high methoxyl pectins. Gels can be made at pH values near neutrality (Chang and Miyamoto, 1992; Garnier et al., 1993), which is advantageous in producing dairy-based products (Baker et al., 1996).

COLORING AND FLAVORING AGENTS

The color of jam is a factor of considerable importance (Abers and Wrolstad, 1979). A good jam should appeal to the eye as well as to the palate. No coloring is required for jams produced from fresh fruit, provided the boiling time is short and the heat not excessive. The natural color of the fruit is, however, always affected by presentation with SO_2 and in some cases by the process of boiling, necessitating the addition of artificial color. The aim should be to restore the original natural appearance. Only permitted edible food colors should be used.

Coal tar colors are most frequently used to a lesser extent. It is essential that the colors should be intensive, readily soluble, and stand up to high concentration in solution. As a rule, acid colors are of higher stability than basic ones. Color should be acid proof; many are affected by acids and particularly by sulfur dioxide present in glucose and fruit pulp. They must also be light proof and have certain stability to heat. All colors suffer from prolonged and excessive heat and should therefore be added at the last stage of the boiling process (Rauch, 1965).

Traditionally, coal tar dyes have been used to color jams and marmalades, particularly those prepared from sulfited fruit. Color testification of jams with natural colors, notably with anthocyanins from grape skins, is increasing as artificial additives become less acceptable to consumers (Encyclopedia Food Science, 1993). Colors available in powder form should be prepared just before the time of addition, since many colors have a tendency to precipitate on standing. When dissolving, the color is mixed to a paste with a little cold water then adding the required amount of boiling water and stirring well.

Ordinarily jams do not require the addition of flavors. If desired, they may be added when jam boiling is nearing completion. Citrus oils and fruit volatile compounds (recovered in the concentration of fruit juices) can be added toward the end of the process to improve product aroma.

PRODUCT TYPE AND RECIPES

According to the U.S. federal standard, a jam should have a minimum of 45 parts of prepared fruit and 55 parts of sugar, concentrated to 65% or higher solids, resulting in a semi-solid product. Jellies are similar to jams with 45 parts of clarified fruit juice and 55 parts of sugar, resulting in a minimum of 65% solids. Both categories can utilize a maximum of 25% corn syrup as sweetener as well as pectin and acid to achieve the gelling texture required (Baker et al., 1996).

On the basis of specification and the allowance for soluble solids such as acids and pectin, it will be seen that the finished jam or jelly, no matter what the grade, will contain approximately the quantity of sugar necessary to give the maximum strength to the pectin–sugar–acid gel. It may be assured that about 3–5% of the total weight of the jam is represented by sugar derived from the fruit; hence, about 65% is added sugar. This figure will therefore be based on the nature of the jam or jelly that the manufacturer wishes to produce. If the fruit that is being used is deficient in pectin or acid or both, or if the quantity of fruit in the recipe is less, it may be necessary to add pectin or acid, so that a gel with sufficient firmness can be obtained.

Some typical recipes found useful in large-scale manufacture of jams mentioned by Giridhari Lal et al. (1986) are given below.

To prepare jam, the requirement of pulp (fresh or canned) is 75 kg, with sugar 75 kg, citric acid 35 g, pectin 150 grade 565 g, and pineapple essence 75 ml.

To prepare orange jam, 50 kg of lye peeled segments require 50 kg of sugar, citric acid 250 g, pectin 150 grade 375 g, and sweet orange essence 50 ml.

To prepare mango jam, 40 kg of mango pulp requires sugar 40 kg, pectin 150 grade 500 g, citric acid400 g, and mango essence 70 ml.

To prepare apple jam, 40 kg of apple pulp requires sugar 44 kg, pectin 150 grade 400 g, citric acid 500 g, and apple essence 60 ml.

To prepare mixed fruit jam, blends of mango, pineapple, orange, apricots, papaya, guava, etc., and equal weight of sugar to that of blended pulp taken, and citric acid to the extent of about 0.75–1.0% by weight of blended pulp containing pectin to the extent of 0.5–1.0% by weight of blended pulp are required depending upon the fruits used. A blend of predominantly red food grade colors may be added along with an appropriate essence to the desired extent.

Jams from cherry, mulberry, strawberry, muskmelon, jack fruit, cashew apple, etc. also can be made in the usual way. It may, however, be necessary to vary slightly the fruit sugar ratio and the percentage of acid added. In some cases, it may be useful to supplement the flavor of the jam by adding extra fruit flavor.

To prepare jelly from guava, ripe fruits are washed well and cut into thin slices and covered with an equal weight of water containing 2 g of citric acid for every kg of guava taken. The mixture is heated gently for about 30 min to extract pectin from the slices. The extract is drained and to the residue, water equal to one-fourth of the weight of slices is added and a second extract is taken. The two extracts are combined and strained through a piece of thick cloth to remove the coarse particles. The strained extract is allowed to settle overnight in a tall vessel. The clear supernatant liquid is tested for its pectin content. The required quantity of sugar (generally an equal quantity) is added to the extract and the mixture boiled down to the desired consistency. The finished jelly is packed in glass jars or in cans. By adopting the procedures described for guava jelly, jellies can be prepared from fruits such as apple, grape, jamun, jack fruit, etc.

Uncooked jam and jelly recipes are a newer method of preserving fruit produce to keep fresh from the wine flavor. These are also called freezer jams, since they need to be stored in the freezer to preserve them. In these recipes, the uncooked fruit is just mixed with sugar, lemon juice pectin, and stirred. Powdered pectin is usually dissolved in water first and then added to the fruit. Care is taken to stir well to dissolve as much sugar as possible, so that it would not have a sugary or gritty jam or jelly. The jam is then usually poured into covered plastic freezer container or can or freezer jars and allowed to set at room temperature. It may take 24 h for the product to jell properly. Recipes for uncooked freezer jams and jellies come with all the pectin products available and the methods of preparation vary slightly for each (Peckham, 1964).

METHOD OF MANUFACTURE

Jam and jelly making is essentially a process that lends itself to be controlled by a chemist. The basic important steps in selection of fruit, boiling for setting, filling of the finished product, and packaging will be discussed below. Manufacture of jams and jellies may be considered rather simple; however, unless scientific approaches are not adhered to, the finished product will not be perfect.

Fruit Preparation

Fruits for jam making should be fully mature, possess a rich flavor and be of the most desirable texture. Fruits are washed thoroughly with water to remove any adhering dirt. If the fruit has been sprayed with lead or arsenical sprays, it should be washed in a warm solution of 1% hydrochloric acid and then rinsed in water (Giridhari Lal et al., 1986).

All berries must be carefully sorted and washed. Strawberries must be stemmed; peaches, pears, apples, and other fruits with heavy skin must be peeled, while apricots, plums, and fresh prunes can be pitted by machine. Stone fruits such as plum and apricots require a very heavy pulping screen because of abrasive action of the pits. Berries should not be softened by boiling before the addition of sugar, but need only to be crushed (Cruess, 1948).

The proportion of sugar to fruit varies with the variety of the fruit, its ripeness, and the effect desired, although the most common ratio of sugar to fruit is 1:1. This is usually a suitable ratio for berries, currants, plums, apricots, pineapple, and other fruits. Sweet fruits with low acidity such as ripe peaches, sweet prunes, and sweet varieties of grapes normally require less than an equal weight of sugar.

Jams may be made from practically all varieties of fruits. Various combinations of different varieties of fruits can often be made to advantage, pineapple being one of the best for blending purposes because of its pronounced flavor and acidity.

Fruits for Jelly

Fruits required for jelly making should contain sufficient acid and pectin to yield good jelly without the addition of these substances, although often in commercial practice, this ideal condition is not attained. Some fruits contain enough of both pectin and acid for the purpose, while others are deficient in either one or both the substances (Smith, 1972).

In the case of fruits deficient in pectin but rich in acid, fruits rich in either pectin or commercial pectin should be added. Both these methods have their own merits. Combining of fruits rich in acids with those rich in pectin is expensive than using acid or commercial pectin to supplement the deficiency, but the drawback is that the flavor of the jelly is affected. Special care is, therefore, necessary to ensure that the fruits are mixed in proper and judicious proportions; on the other hand, commercial pectin does not have any deleterious effect on the final flavor of the jelly made from any particular fruit.

Of the fruits rich in pectin and acid, table varieties of apples, sour blackberries, currants, lemons, limes, grapefruits, sour oranges, sour guavas, and sour varieties of grapes, plums, and cherries are good examples. Of the fruits low in acid but rich in pectin are sweet varieties of cherries, unripe figs, ripe melon, and unripe bananas. Fruits that are rich in acid but low in pectin are apricots and most varieties of strawberries. Fruits that may be classified as containing a moderate concentration of both acid and pectin are ripe varieties of grapes, ripe blackberries, ripe apples, and loquats. Fruits low in both acid and pectin are represented by pomegranates, ripe peaches, ripe figs, and ripe pears (Cruess, 1948).

It is customary to blend fruits deficient in acid or pectin, or both, with fruits that have an abundance of the required constituents.

Boiling

Boiling is one of the most important steps in the jelly making process, as it dissolves the sugar and causes union of the sugar, acid, and pectin to form a jelly. It usually causes a coagulation of certain organic compounds that can be skimmed from the surface during boiling, and their removal renders the jelly clearer. The principal purpose of boiling is to increase the concentration of the sugar to the point where jelling occurs (Ashish Kumar, 1988).

The boiling operation, while normally being a necessary step in jelly making, should be as short as possible. Prolonged boiling results in loss of flavor, change in color, and hydrolysis of the pectin; consequently, it is a frequent cause of jelly failure (Lesschaeve et al., 1991). In jam making, the fruits resistant to boiling are desirable to concentrate the product by evaporation of excess moisture. Boiling in commercial practice is usually conducted in open steam jacketed stainless steel kettles. During boiling, the juice or pulp should be skimmed, if necessary, to remove coagulated material and should be stirred to cause thorough mixing. The boiling is continued until on cooling, the product will form a jam or jelly of the desired consistency. The concentration of the mixture when this point is reached will depend upon several factors viz., the concentration of pectin, the concentration of acid, the ratio of sugar to pectin and acid, and the texture desired.

The most common method of determining the end point is by allowing the liquid to sheet from a wooden paddle or a large spoon. If it drips from the instrument as thin syrup, the process is not complete. If it partly solidifies and breaks from the spoon in sheets, or forms jelly-like sheets on the side of the spoon, the boiling is considered to be complete.

Boiling of the sugar solution in the presence of the fruit acid results in inversion of some of the cane sugar. Hence, a jelly that is boiled for a long period is less liable to develop crystals of sucrose than a jelly boiled for a short time. Prolonged boiling may result in loss of flavor through evaporation, hydrolysis, or other forms of decomposition.

Sometimes jam is boiled in a vacuum pan at a lower temperature in the range of 65–76°C. The vacuum boiling or cooking minimizes the undesirable changes in color and prevents loss of vitamin C. However, the jam mixture has to be boiled for a long time to soften the fruit pieces resulting in some loss of flavor which can, however, be restored by recovering the volatile esters and putting them back into the jam (Giridharilal et al., 1986). Proper control of boiling is necessary to avoid over concentration of soluble solids, over inversion of sugar, and hydrolysis of pectin. Manufacture of jam can be considered rather a simple process, but unless controlled properly may lead to undesirable results. Figure 11.5 gives a scientific approach to different factors involved in jam production.

Filling

The finishing of jams comprises three main steps: (1) pre-cooling prior to filling, (2) filling, and (3) cooling after filling.

Prolonged heating affects the appearance as well as the keeping quality of the finished product. As the inversion of sugar is greatly influenced by the temperature, it is obvious that an efficient cooling system

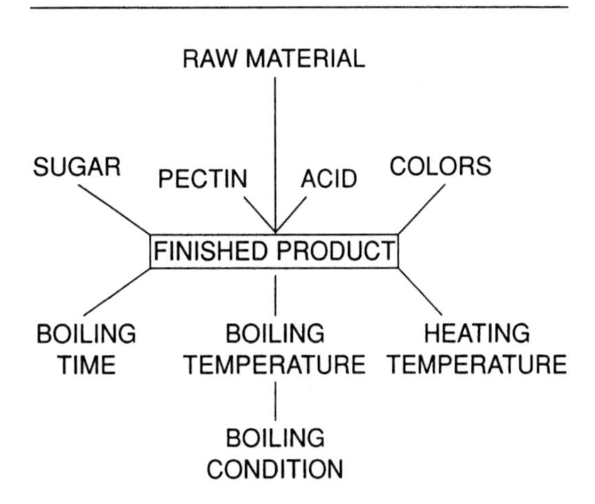

Figure 11.5. Factors controlling process of jam manufacture.

is necessary to control and check the process. Difficulties are also experienced in filling, as some fruit varieties show a tendency to float, the most susceptible being strawberry, cherry, black currant, and stone fruit jams.These jams should be cooled until they are near setting point, but great care must be taken not to exceed the limit, otherwise the set will break and the jam curdles, more so in the case of jellies (Rauch, 1965).

The efficient method in trade is filling on a roller conveyor. The empty jars are packed into trays, each one holding a certain number. The trays are then moved on roller conveyer to the filling operators. The trays containing filled jars are moved to cooling room after the preliminary setting has taken place. This prevents discoloration due to caramalization of the sugar.

After being filled, jam in jars must not be cooled too quickly. As far as canned jam is concerned, the procedure is simple enough, the cans being passed through a water bath. Glass jars and large containers have to be cooled by air. This can be done by passing them slowly through a tunnel fitted with an air blast or keeping them in a cooling room constructed on the same principle until the jam is well set. Automatic filling machines that measure a definite volume of jelly into each container are in general used in large factories. They greatly reduce the cost of filling as compared to that of hand filling and give more uniform net contents.

Packaging Materials

The final stage of the production process is packaging. A wide variety of sizes and shapes of containers are used for jellies. Glass is the usual material, although enamel-lined tin cans and special containers are also used.

Jelly should be hermetically sealed in glass containers. A paraffin seal is not adequate to prevent spoilage of the product. Container filled scalding hot (in excess of 83°C) need not be pasteurized, as the hot-filled jelly itself will sterilize the container. The jar should be filled to at least 90% full leaving not more than 1.25 cm head space. The scalded lids should be loosely placed on the containers immediately following filling, and then tightened firmly within 2–3 min. This allows time for exhausting of air from the head space. The steam in the head space condenses when the jelly cools, creating a vacuum seal. Capping with superheated steam injection is often used to attain a hermetic seal. Where the product is not filled sufficiently hot to ensure head space sterilization, or where superheated steam injection is not used, a post-capping sterilization treatment is employed.

Some jellies form during boiling and filling a layer of bubbles on the surface of the jar of hot jelly. The jelly should be quickly skimmed while in the kettle just prior to pouring. If the jelly can be drawn from the bottom of the kettle, clear jelly can be filled into the jars.

If jelly is to be poured into glass jars, the sides of the jar should be smooth so that the jelly can be turned out without breaking its shape or structure. Before pouring the jelly, glass containers should be warmed to prevent breakage. After filling, the jars should be cooled rapidly to about 21°C. Pectin jellies set more quickly at this temperature than at lower temperatures. If the jelly fails to set or is weak, it is placed in a drier to evaporate the excess of water in it and promote setting (Giridhari Lal et al., 1986).

Deterioration in storage is now largely prevented by hermetically sealing the jars, while still hot, in sterile manner using metal caps fitted with rubber gaskets. Many patented caps of this kind have been devised which can be placed in position, sterilized by steam jets in a steam box and, finally, held firmly by creating a partial vacuum in the head space of the jars, either in a specially constructed vacuumizing chamber or merely by screwing them up tightly while still hot. Jars sterilized and sealed in this way form an ideally hygienic package and are to a very large extent independent of storage conditions (Morries, 1951).

Two types of vacuum-sealed jars are in common use. In one of these, the seal between the glass and jar cover is made with a rubber composition ring attached to the cap. This composition melts during pasteurization, and after cooling of the jar and contents, it solidifies to form an air-tight seal. The lid must be held in place with a clamp during pasteurization. The second type of jar is sealed with a rubber gasket similar to a fruit jar rubber, but this rubber is pressed against the side of the jar rather than the top. It is held in place by friction and the cap is rolled in much the same manner as an ordinary sanitary can top. The cap needs no clamp to hold it in place during pasteurization. A similar lid is pressed into position but is not rolled (Cruess, 1948).

Small enamel-linked jam cans and gallon cans are sometimes used for jelly. Wooden tubes or buckets are often used for cheap jellies for baker's use, the product usually being preserved with sodium benzoate. The inside of cans used for jellies made from fruits of red color should be heavily lacquered to prevent bleaching of the color by tin salts. Some of the jams are sold in cans. Glass containers are almost universally used for jams and jellies. For jams also, the processes of filling and sealing are done by automatic machinery as described for jellies.

FUTURE RESEARCH NEEDS

Jam and jelly production relies on the native pectins of incorporated fruit for gel formation. Modern manufacturing requirements of uniform gel strength and appearance preclude reliance on fruit maturity and variety. In spite of the current availability of other gelling agents, pectin remains the universal choice for jams and jellies, in part because of its presence as a natural fruit ingredient, and also because of the characteristic consistency that pectin imparts to a gel. The joint FAO/WHO committee on food additives recommended pectin as a safe additive with no limit on acceptable daily intake, except as dictated by good manufacturing practice.

In recent years, pectin has been used as a fat or sugar replacer in low calorie foods. It is estimated that 80–90% of commercial pectin production which totals 6–7 million kg is used in the production of jellies and jams (Crandall and Wicker, 1986). In spite of its availability in a large number of plant species, commercial sources of pectin are very limited. There is therefore a need to explore other sources of pectin or modify the existing sources to obtain pectin of desired quality attributes. Modern food sciences such as genetic engineering can be used to modify pectin *in vivo*. Current knowledge of the molecular basis of gelation in pectin has helped us to understand some aspects of this complex phenomenon. There are still some areas where our knowledge is scanty. So a systematic study of these observations will help in understanding the reaction processes in pectin gel formation, resulting in better control of processes and products.

REFERENCES

Abers, J.E., Wrolstad, R.E. 1979. Causative factors of colour deterioration in strawberry preserves during processing and storage. Journal of Food Science. 44, 75–78.

Ahmed, G.E. 1981. High methoxyl pectins and their uses in jam manufacture—a literature survey. The British Manufacturing Industries Research Association. Scientific and Technical Surveys No. 127, 16.

Anon. 1974. Fruit jelly and preserves revised standards. Federal Register. Title 21, part 29, 31304–31309, Aug 28.

Anon. 1975. Effective dates for new food labeling regulation. Federal Register Title 21, part 29, 2798, Jan 16.

Anon. 1983. Jams and jellies. Institutional Distribution. 19(12), 218, 222–224.

Ashish Kumar Pal. 1988. The effect of ingredient on the quality of confectionery jellies—dissertation report. CFTRI, Mysore. pp. 1–22.

Axelos, M.A.V., Thibault, J.F. 1991. The chemistry of low methoxy pectin gelation. In: The Chemistry and Technology of Pectin. Walter, R.H. (ed.) Academic Press. New York. pp. 109–118.

Baker, R.A., Berry, N., Hup, Y.H. 1996. Fruit preserves and jams in processing fruits—Science and Technology, Vol 1. Biology, Principles and Application. Technomic Publishing Co., Inc. Switzerland. pp. 117–133.

Beveridge, T., Timber, G.E. 1989. Small amplitude oscillatory testing (SAOT): Application of pectin gelation. Journal of Texture Studies. 20, 317–324.

Breverman, J.B.S. 1963. Pectic Substances in Introduction to Biochemistry of Foods. Elsevier Publishing Company Inc. New York. 94–107.

Brosio, E., Delfini, M., Dinola, A., Dubaldo, A., Lintas, C. 1993. H and Na NMR relaxation times study of pectin solutions and gels. Cellular and Molecular Biology. 39, 583–588.

Chang, K.C., Dhurandhar, N., You, X., Miyamoto, A. 1994. Cultivar/location and processing methods affect yield and quality of sunflower pectin. Journal of Food Science. 59, 602–605.

Chang, K.C., Miyamoto, A. 1992. Gelling characteristics of pectin from sunflower head residues. Journal of Food Science. 57, 1435–1438.

Christensen, O., Trudsoe, J. 1980. Effect of other hydrocolloids on the texture of Kappa Carrageenan gels. Journal of Texture Studies. 11, 137–147.

Crandall, P.G., Wicker, L. 1986. Pectin internal gel strength: Theory, measurement and methodology. In: Chemistry and Function of Pectin. Fishman, M.L. and Jen, J.J. (eds.) Amercian Chemical Society, Washington, DC. ACS Symposium series. pp. 88–102.

Cruess, W.V. 1948. Commercial Fruit and Vegetable Products, III Edn. MCGraw Hill Book Company. New York. pp. 377–426.

Dhame, A. 1992. Gelpoint measurement on high methoxy pectin gels by different techniques. Journal of Texture Studies. 23, 1–11.

De vries, J.A., Voragen, A.G.J., Rombouts, F.M., Pilnik, W. 1986. Structural studies of apple pectin with pectolitic enzyme. ACS Symposium series 310, 38–48, American Chemical Society. Washington, DC.

DaSilva, J.A.L., Goncalves, M.P., Rao, M.A. 1992. Rheological properties of high methoxy pectin and low cut bean gum solutions in steady shear. Journal of Food Science. 57, 443–448.

Doesburg, J.J., Grevers, G. 1960. Setting time and setting temperature of pectin jellies. Food Research. 25, 634–645.

Doublier, J., Thibault, J. 1984. Les agents epaississants de fabrication dans les industries agroalimentories. In: Tec et Doc. Multon, J.C. (ed.) Apria, Lavoisier. Paris. p. 305.

Doublier, J.L., Launay, B., Cavelier, G. 1992. Visco elastic properties of food gels. In: Visco Elastic Properties of Food. Rao, M.A. and Steffe, J.F. (eds.) Elsevier Applied Science. New York. pp. 371–434.

Eagland, D. 1975. Nucleic acid peptides and proteins. In: Water: A Comprehensive Treatise, Vol 4. Franks, F. (ed.) Plenum Press. New York. p. 306.

Encyclopedia Food Science. Food Technology and Nutrition. 1993. Jam and Preserves, Vol IV, Academic Press, Harcourt Brace Jovanovich Publishers. London. pp. 2612–2621.

Fishman, L., Jen, J.J. 1986. Chemistry and functions of pectins. ACS Symposium series 310, Washington, DC.

Flory, P.J. 1953. New Approaches to Investigation of Fruit Gels in Gums and Stabilizers for the Food Industry. Phillips, G.O., Wedlock, D.J., Williams, P.A. (eds.) Elsevier Applied Science Publihsers, London, 432 p.

Frauzen, E.H., Helwert, F. 1923. Ubesr die chemischen Bestandteile gruner and the concord grape. Journal of American Chemical Society. 127, 14.

Fry, S.C. 1986. Cross linking of matrix polymers in the growing walls of angio sperms. Annual Review of Plant Physiology. 37, 165.

Furcsik, S.L., Mauro, D.J. 1991. Starch jelly candy. United States Patent. 540360.

Garnier, C., Axelos, M.A.V., Thibault, J.F. 1993. Dynamic visco elasticity and thermal behaviour of pectin-calcium gels food hydrocolloids. 5(1/2), 105–108.

Giridhari Lal, Siddappa, G.S., Tandon, G.L. 1986. Preservation of fruits and vegetables. Indian Council of Agricultural Research. New Delhi.

Gordon, D.L., Schwenn, K.S., Ryan A.L., Roy S. 2000. Gel products fortified with calcium and method of preparation. United States Patent. 6077557.

Gross, M.O., Rao, V.N.M., Smit, C.J.B. 1980. Rheological characterization of low methoxyl pectin gel by normal creep and relaxation. Journal of Texture Studies. 11, 271–290.

Guichard, E., Issanchou, S., Descourveres, A., Etievant, P. 1991. Pectin concentration, molecular weight and degree of esterification; influence on volatile composition and sensory characteristics of strawberry jam. Journal of Food Science, 56(6), 1621–1627.

Halliday E. and Bailey J. 1954. Effect of calcium chloride on acid sugar pectin gels. Industrial Engineering Chemistry, 16, 595.

Hawang, J., Kokini, J.L. 1992. Contribution of the side chain to rheological properties of pectins. Carbohydrate Polymer. 19, 41.

Hoefler, A.L. 1991. Other pectin food products. In: The Chemistry and Technology of Pectin. Walter, R.H. (ed.) Academic Press. San diego. p. 51.

Hughes, L., Ledward, D.A., Mitchell, J.R., Summerlin, C. 1980. The effect of some meat protein on the rheological properties of pectate and alginate gels. Journal of Texture Studies. 11, 247–256.

Jarvis, M.C. 1984. Structure and properties of pectin gels in plant cell wall. Plant Cell and Environment. 7, 153.

Kerstez, Z.I. 1951. Preparation and purification of pectic substances in the laboratory. In: The Pectic Substances. Interscience publishers Inc. New York. pp. 94–129.

Kratz, R. 1995. Recent developments: in pectin technology. Instant Pectins. Food-Technology-International, Europe 35–36.

Kratz. 1993. Jam, jellies, marmalade II. The phenomenon of syneresis and method of manufacturing. Food Marketing & Technology 17–21.

Lau, J.M., McNeil, M., Darvill, A.G., Alan, G., Darvill, Albersheim, P. 1985. Structure of backbone of rhamono galacturonan I, A pectic polysaccharide in the primary cell walls of plants. Carbohydrate Research. 137, 111–125.

Lesschaeve, I., Langlois, D., Etievant, P. 1991. Effect of short term exposure to low O_2 and high CO_2 atmosphere on quality attributes strawberries. Journal of Food Science. 56(1), 50–54.

May, C.D., Stainsby, G. 1986. Factors Affecting Pectin Gelation in Wedlock. Williams, P.A. (eds.) Elseveir Applied Science Publishers. New York. pp. 515–523.

Meyer, L.H. 1960. Food Chemistry. Reinhold Publishing Corporation. New York. 125.

Mitchell, J.R. 1979. On the nature of the relationship between the structure and rheology of food gels. In: Food Texture and Rheology. Sherman, P. (ed.) Academic Press. London. p. 425
Morris, T.N. 1935. Softening orange peel. Food Manufacture. 10, 167.
Morries, T.N. 1951. Principle of Fruit Preservation. Chapman and Hall Ltd. London.
Morris, V.J. 1986. Analysis, structure and properties of biopolymer mixtures, In: Gums and Stabilizers for the Food Industry. Vol 3. Williams, P.A., Wedlock, D.J. (eds.) Pergammon Press. Oxford. p. 87.
Mort, A.J., Maness, N.O., Qiu, F., Otiko, G., Anj Westpant Komalavilas, P. 1993a. Extraction of defined fragments of pectin by selective cleavage of its backbone allow new structural features to be discovered. Journal of Cellular Biochemistry. Issue 517a, 9pp.
Mort, A.J., Qiu, F., Maness, N.O. 1993b. A determination of the pattern of methyl esterification in pectin distribution of contagious non esterified residues. Carbohydrate Research. 247, 21–35.
Mauro, D.J., Furcsik, S.L., Kvansnica, W.P. 1991. Starch jelly candy. United States Patent. 540367.
Muralikrishna, G., Taranathan, R.N. 1994. Characterisation of pectic polysaccharides from pulse husks. Food Chemistry. 50, 87–89.
Nelson, G.K. 1925a. The nonvolatile acids of the straweberry the pineapple and the concord grape. Journal of American Chemical Society. 47, 1177.
Nelson, G.K. 1925b. The non volative acids of the blackberry. Journal of American Chemical Society. 47, 568.
NIIR Board. (2002). Jam, jelly and Marmalade. In: Hand Book on "Fruits, Vegetables and Food Processing with Canning and Preservation". Asia Pacific Business Press Inc. Delhi. pp. 238–252.
Oakenfull, D., Scott, A. 1984. Hydropholic interaction in the gelation of high methoxy pectins. Journal of Food Science. 49(2), 1093–1098.
Oakenfull, D.G. 1987. Gelling agents. CRC Chemical Reviews in Food & Nutrition 26(1), 1–25.
Oakenfull, D.G. 1991. The Chemistry of High Methoxyl Pectins in the Chemistry and Technology of Pectin. Walter, R.H. (ed.) Academic Press. New York. pp. 87–108.
Peckham, G.C. 1964. Jams, jellies and conserves. In: Foundation of Food Preparation. The Macmillan Company. New York. 443–448.
Pilgrim, G.W., Walter, R.H., Oakenfull, D.G. 1991. Jam, jellies and preserves. In: The Chemistry and Technology of Pectins. Walter, R.H. (ed.) Academic Press. San Diego. p. 23.
Rao, M.A. 1992. Measurement of visco elastic properties of fluid and semi solid foods. In: Visco Elastic Properties of Foods. Rao, M.A., Steffy, J.F. (eds.) Elseveir Applied Science Publishers. New York. pp. 207–231.
Rao, M.A., Cooley, H.J. 1993. Dynamic rheological measurement of structure and development in high methoxy pectin/fructose gels. Journal of Food Science. 58, 876–879.
Rao, M.A., Van Buren, J.P., Cooley, H.J. 1993. Rheological changes during gelation of high methoxy pectin/fructose dispersions; effect of temperature and ageing. Journal of Food Science. 58, 173–176.
Rauch, G.H. 1965. Jam Manufacture. Leonard Hill Books. London.
Rees, D. 1969. Structure conformation and mechanism in the formation of polysaccharide gels and networks. Advances in Carbohydrate Chemistry-Biochemistry. 24, 267–332.
Ring, S.G., Oxford, P.D. 1985. Recent observations on the retrogradation of amino pectin in gums and stabilizers for the food industry. Glyn, O.P., David, J.W., Peter, A.W. (eds.) Elsevier Applied Science Publishers. London. pp. 159–165.
Sakai, T., Sakamoto, T., Hallaert, J., Vandamme, E.J. 1993. Pectin, pectinase and protopectinase: Production, properties and application. Advances in Applied Microbiology. 39, 213.
Shomer, J. 1991. Protein coagulation cloud in citrus fruit extract I formation of coagulates and their bound pectin and neutral sugars. Journal of Agricultural and Food Chemistry. 39, 2263–2266.
Smidsred, O. 1974. Molecular basis for some properties in the gel state. Faraday Discuss. Chemical Society. 57, 263.
Smit, C.J.B., Bryant, E.F. 1968. Ester content and jelly pH influences on the grade of pectin. Journal of Food Science. 33, 262–264.
Smith, P.S. 1972. July, Jelly gum-manufacturing and the problems. The manufacturing confectioner. 40–41.
Swaminathan, M. 1987. Fruits, pectic substances and fruit products. in: Food Science Chemistry and Experimental Foods. Bappco, Bangalore, India, pp. 175–195.
Thakur, B.R., Singh, R.K., Handa, A.K. 1997. Chemistry and uses of pectin—a review. Critical Review in Food Science and Nutrition. 37(1), 47–73.
Veis, A. 1964. Macromolecular Chemistry of Gelation, Academic Press. New York.

Voragen, A.G.J., Schols, H.A., Pilnik, W. 1986. Determination of the degree of methylation and acetylation of pectin by HPLC. Food Hydrocolloids. 1, 65.

Walkinshaw, M.D., Arnott, S. 1981. Conformation and interactions of pectin II, Models for junction zones in pectinic acid and calcium pectate gels. Journal Molecular Biology. 153, 1075.

Walter, R.H., Sherman, R. 1981. Apparent activation energy of viscous flow in pectin jellies. Journal of Food Science. 46(2), 1223–1225.

Watase, M., Nishinari, K. 1993. Effect of pH and DMSO content on the thermal and rheological properties of high methoxyl pectin-water gels. Carbohydrate Polymers. 20(3), 175–181.

Wolfram, M.L., El Khadem, H. 1965. Chemical evidence for the starch. In: Starch-Chemistry and Technology. Whistler, R.L., Paschall, E.F. (eds.) Academic Press. New York. p. 25.

Woodmansee, C.W., Mc clendon, J.H., Somers, G.F. 1959. Chemical changes associated with the ripening of apples and tomatoes. Food Research. 24, 503–514.

12
Manufacturing Fruit Beverages

Emőke Horváth-Kerkai

Fruits have always played an important role in human nutrition. However, before the 20th century, drinking squeezed fruit juices was the privilege of a few.

Welch was the first to preserve grape juice with heat treatment in America in 1869, followed by Müller-Thurgan in Switzerland in 1896 (Kardos, 1962). Thus began the production of preserved fruit juices, which was followed by a huge development in the 20th century. The role of vitamins and minerals in the human body was discovered at that time, which triggered substantial changes in eating habits. Consequently, fruit consumption has become an everyday need.

Due to the revolutionary development of technical equipment, the appearance of chemicals, and biological substances (enzymes, clarifying and flavoring agents), and the application of new technological procedures, especially the aseptic technique—which enabled the production of fruit juices without preservatives—of fruit juice production became widespread.

MAIN FRUIT DRINK CATEGORIES

Today, there are countless fruit juice products in the world's markets. They may differ substantially in terms of raw material, composition, quality, nutrient content, sensory traits, and packaging. In some cases, the biggest difference is the brand name.

Generally, fruit juice-based drinks are classified according to their fruit content. Thus, we can declare three categories:

- juices and fruit musts
- fruit nectars, and
- soft drinks with fruit content

Juices and fruit musts are obtained by mechanical procedures. They possess the color, taste, and aroma of the original fruit and their composition is identical as well. In the case of certain products sugar addition and vitamin enrichment are allowed, but these have to be declared on the labelling. Juices and fruit musts are not allowed to contain food industry additives (preservatives, aromas, and coloring agents). Therefore, it is consumed in fresh form soon after production or it is preserved by heat treatment. The permitted ingredients of different fruit beverages are shown in Table 12.1.

According to the type of fruit, these products can be divided into two subcategories. They can be filtered to be transparent (e.g., grape, apple juice) or they can be cloudy juices containing colloids like all citrus based juices. Products belonging to this latter group may contain fruit fibers.

Fruit nectars are made of sieved juices or from fruit juices that are diluted with sugar syrup. They usually contain only one fruit—such as orange,

Table 12.1. Permitted Ingredients of Different Fruit Beverages in Europe

Components	Juice or Fruit Must	Nectar	Soft Drinks
Fruit, min. (%)	100	25–50[a]	free
Sugar, max. (%)	1, 5	20	free
Acid, max. (g/l)	3[b]	3[b]	free
Preservatives (mg/l)	–	–	++
Food additives	–	–	++

Note: ++ indicates according to the legislation of additives.

[a] Depending on fruits.

[b] With natural lemon juice.

apple, or peach—but they can also be made from blends of more than one fruit juices or pulps (Szenes, 1991). In Europe, the preparation of blends and minimum fruit content are regulated by government standards, industrial specifications, and other voluntary and mandatory requirements. In order to ensure wide international trade, these standards conform to the recommendations of the Codex Alimentarius of the FAO/WHO Food Standard Program (Várkonyi, 2000). However, many western countries impose their own restrictions on imports and exports.

FRUIT DRINK RAW MATERIALS

The most important raw materials of fruit drinks, that are available in international trade are citrus, pomaceous fruits, stone fruits, grape, and berries. However, all cultivated and wild grown fruits are used for drink production. Some of the raw materials used are suitable for juice production (e.g., apple, orange), since their juices are enjoyable themselves. Meanwhile, the juice of other fruits (e.g., sea blackthorn, currant sorts) is delightful only if blended with sugar syrup.

The quality of fruit drinks, made without additives, is basically determined by the quality of raw materials. Generally, acidulous, juicy fruits with high-sugar content and distinctive aromas are suitable for drink production. The ripeness of raw materials is of critical importance because optimally ripe fruits possess the ideal sugar/acid ratio and the most advantageous flavor and aroma components. Prior to the optimal ripeness, the fruit contains much less aromas and sugar. On the other hand, overripe fruits may lose their acids (e.g., vitamin C), coloring agents, and consumption value (Stéger-Máté et al., 2002).

Due to the development of fruit drink consumption, raw material production and juice consumption were separated both geographically and in time, then versus now. Consequently, fruit pulps and concentrates—that are easier to store and transport—came to the front in production. Fruit drinks, made of these preserved semifinished products, can only be competitive if their composition and sensory traits are close to those made of fresh fruits. According to different surveys, 70% of fruit drink quality complaints are rooted in the raw materials. These facts made experts, involved in production, quality control and sales, to set up uniform quality requirements for the clarity and origin of fruit drinks. Recommendations of this RSK (Richtwerte und Schwankungsbreiten bestimmter Kennzahlen) system—worked out in Germany—for compositional features are generally accepted in the European commercial practice (Bielig et. al., 1987). However, fruit variety, origin, climate, production, and processing technology can often change the composition, resulting in quality problems. The further development of this system led to the publication of the European Economic Commission's Association for Juice and Nectar Production (AIJN), called "Code of Practice." In this publication, RSK values are supplemented with other analytical features—that are generally accepted—of apple, grapefruit, orange, and grape juices. Besides consumption value, quality, genuineness, raw material, and technological deficiencies, these criteria comprise factors concerning the environmental pollution of the production land, such as arsenic and heavy metal level. Technological deficiencies usually result in high concentration of biogenic acids, HMF (hydroxymethylfurfural), ethylalcohol, and patulin; thus, their maximum levels are under regulation. The values determined by the RSK system and the AIJN

Code of Practice are primarily used in Europe but significant deviations may endanger the competitiveness of products in the world market.

PRODUCTION OF FILTERED AND CLOUDY FRUIT DRINKS

Filtered and cloudy fruit drinks are made of mechanically pressed and cleaned juice directly or from the dilution of concentrated semifinished products. As it can be seen in the flow chart, production technology comprises five main operations (Fig. 12.1).

Juice extraction—the elimination of the juice from fibrous, solid particles—is a basic technological step of fruit juice production. The fruit has to be prepared prior to juice extraction, which is then followed by juice clarification and drink completion. Subsequently, the finished drink is packed and preserved.

PREPARATION STEPS

Raw Material Reception

Only those raw materials are allowed for fruit drink production that meet the following criteria: appropriate ripeness and flavor, no signs of deterioration, and free from foreign ingredients, pathogenic organisms, and their effect. Moreover, raw materials have to conform to the regulations and standards in force. During reception, huge attention has to be paid to the cleanliness of berries in which washing may cause substantial damages (Szenes, 1991). The conformity of each batch has to correspond to the methods and examinations of the relevant descriptions. Then, it has to be labelled for further identification and traceability. In addition, conformity to the production technology requirements also has to be checked (crop spraying records).

Washing

The aim of this step is to remove every contamination from the surface of the fruit, i.e., to increase physical, chemical, and microbiological cleanliness (Fellows, 2000).

The surface of raw materials is strongly contaminated by microorganisms; it can attain 10^5–10^9 microorganisms per gram. Even with effective washing, it can be decreased only by 3–5 orders of magnitude. Therefore, washing efficiency has a significant impact on the heat treatment necessary for preservation.

Physical and chemical surface contaminations are eliminated by water soaking, since these substances are water soluble or their adhesion properties decrease in aqueous solution. The efficacy of the dissolving process can be increased with higher water flow. The latter can be achieved by streaming, air injection, and by mechanical means. Due to the water flow, close contacts between fruit particles increase washing efficiency but potentially leading to damages to the fruits. Thus, the texture of the raw material always has to be taken into consideration when choosing washing equipment. Washing usually consists of three main steps. The first phase is soaking, which breaks up surface contamination and eliminates soil particles. In the case of fruits covered with wax layer or oleaginous skin, warm water (50–60°C) is applied.

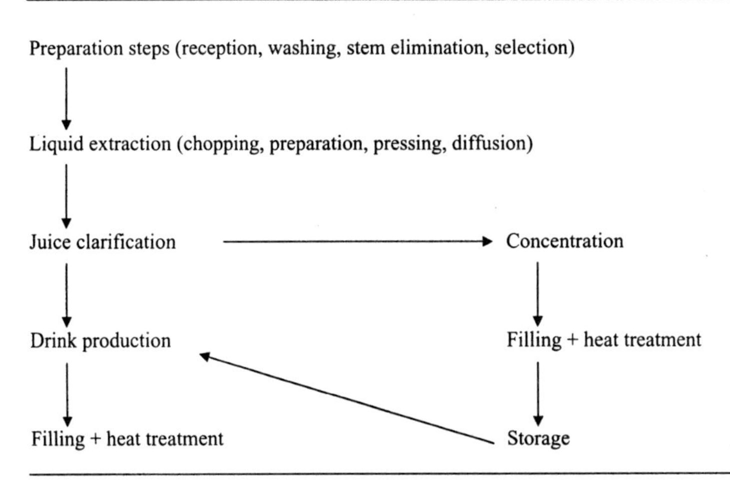

Figure 12.1. The flow chart of fruit juice production.

Warm water soaking or a long soaking period may result in substantial loss of valuable fruit components (Barta and Körmendy, 1990).

The active phase of washing is intended to remove every contamination, and it is always followed by a clean water rinse. Washing means water flows all around the fruit, meanwhile rinsing means water spraying in order to remove washing water residues from the fruit's surface.

Stem Elimination

To prepare fruits for juice extraction, in the case of certain fruit species (e.g., cherry, sour cherry, and plum), long green peduncle parts have to be removed. Otherwise, they will spoil the color and other quality traits of the juice. Mechanized stem elimination can only be carried out in raw materials of homogenous size that do not tend to damage and burst. The most frequently used equipment is the belt-based solution, but in Eastern Europe roller-based machines are still widely used.

Selection

This step, which usually follows washing, separates everything from the raw material that is unsuitable for processing. These can be foreign substances, stem and leaf particles, or mouldy, deteriorated fruits. This activity is performed manually and requires close attention. Therefore, the necessary job environments (e.g., proper lighting and reasonably positioned waste containers) must be provided for the workers. The selection table on which this operation is done should be able to roll the fruits, enabling the workers to observe the entire fruit surface. Both the roller-based and belt-based machines comply with this requirement. In order to achieve better efficacy, proper adjustments must be made for single layer fruit flow and optimal belt speed (Parker, 2003).

Juice Extraction

This operation can be divided into two steps: fruit chopping and preparation and the separation from solid fruit particles.

Chopping

The aim of this step is to smash, cut the fruit, increase its surface, and launch cell-fluid elimination. However, this can lead to enzymatic reactions damaging valuable components. Therefore, the fruit has to be processed immediately after chopping.

If this step is done appropriately, the fruit is not pulpy but consists of homogenous, irregular-shaped, few-millimeter-sized particles, which tend to form channels to drain the liquid when pressed.

Chopped Fruit Preparation

Procedures designed to prepare chopped fruits are to increase juice yield and prevent undesirable changes (chemical, biological, mechanical, etc.) to achieve better aroma, flavor, and color properties. The type of preparation will depends largely on the type of fruit and production technology.

There are several methods for this operation, such as mechanical, freezing, enzymatic, vibration, ultrasonic, electro-plasmolytic, ion-radiation procedures, and heat treatment (Szenes, 1991). In practice, mechanical operations, heat treatments, and enzymatic solutions are widely used.

Mechanical preparation is used to chop fruit flesh, smash the tissues, and increase the surface.

Stiff raw materials (e.g., apple) are usually crushed; meanwhile, soft ones (e.g., red currant) are only cracked. Crushing opens up the tissues, damaging some of the cells as well, and the draining of cell-fluid begins.

The degree of chopping is determined by the method of juice extraction. If pressing is applied, the chopped fruit releases the juice under a relatively small amount of pressure. Appropriately prepared chopped fruits contain particles of nearly identical size, enabling channels to form for the liquid to drain.

If the fruit is chopped into very fine pieces, it spreads easily, expands under pressure and does not tend to form channels to drain the juice.

Diffusion-based liquid extraction requires chopping to minimize the thickness of the slices and strips. In addition the size of these pieces should form channels to ensure the flow of the extraction liquid. There are different devices for the crushing of fruits. These can be specialized for a given fruit (e.g., apple crusher) or generally used as hammer- and roller-based machines. Their common feature is the rotating system and the pressing, shearing, pulling, and striking forces applied.

Preparation with heat treatment is mainly used prior to the pressing of berries, since it can increase juice yield by 5–10%. Furthermore, this procedure contributes to a better color. Within the framework of this procedure, the crushed/cracked fruit is rapidly heated to 80–85°C and then quickly cooled back. This short heat treatment enables different physical, chemical, and microbiological processes to take place. The denaturation of proteins and the hydrolysis of the protopectin lead to the inactivation of enzymes, making the cell walls permeable; thus, accelerating the diffusion of water-soluble substances. In the case of technological failures—too long heat treatment—tissues become too soft and damaged, then fruit will be difficult to press, and juice taste changes as well (Szenes, 1991).

There are different heat-exchanger devices for this procedure.

Enzymatic treatments are also frequently used before pressing, to make the process easier and to increase the yield.

Fruit raw materials possess different amounts and types of pectin, depending on the species and the variety. Pectin can be found between the cell wall layers connecting the solid shells that contain cellulose and hemicellulose. Pectin can also be found in dissolved form in the tissues, increasing their density and sticking properties. High pectin content negatively influences the following juice producing steps. Therefore, the level and the composition of pectin have to be decreased or modified according to the quality criteria of the finished product or the production technology. The most general solution to this issue is the enzymatic treatment of the cracked/crushed fruit with pectin decomposing enzymes such as pectin transeliminase and polygalacturonase (Aehle, 2004). The use of these enzymes leads to the decomposition of glycoside bounds, rapidly decreasing the viscosity of the mash (Reising, 1990).

The enzyme products added also contain cellulase and hemicellulase enzymes in order to decompose the cell wall and improve the permeability. Enzyme treatment can also be carried out under cold and warm circumstances.

Cold treatment at 20–25°C takes more hours, which endangers the juice quality (Schmitt, 1990).

Meanwhile, warm treatment takes place in 0.5–1 hour, at 50–55°C. As enzymes are protein-based molecules, these are heat sensitive and are only active at certain pH values. If the temperature and pH conditions of the mash are not optimal, successful pectin decomposition requires longer time or higher enzyme concentration (Dietrich, 1998).

The pressing waste of high-pectin fruits (e.g., apple) is usually used for pectin production. In these cases, enzyme treatment should not be applied.

Juice Extraction

In this process, the liquid phase of fruits is detached from solid particles. There are different methods for this separation: pressing, diffusion, centrifugal procedures, and reverse-osmosis (Fellows, 2000). The type of equipment applied depends on the fruit species, production line, and economical background. The most widely used solution is pressing.

Pressing separates a food system into two phases. In this case, fruit tissues mean the solid phase, while the liquid between the particles is the liquid phase.

Pressing needs outside forces to create tension in the system, drain liquid, resulting in shape modification. The equipment hinders the disposal of the solid phase and the liquid gathers in a vessel. The remaining material, with low liquid content, is called marc. The most important parameter of pressing is the liquid yield, which means the percentage of juice extracted, compared to the raw material at the beginning of the process. Juice yield is basically determined by the type of the pressing device, and the quality and preparation of the raw material (Lengyel, 1995). Fruit processing industry applies continuous—such as belt- and screw-based—and intermittent—like the package and basket type—pressing machines. In addition, decanters are based on centrifugal forces (Nagel, 1992).

The juice of fruits can also be detached with extraction. It means that semipermeable cell walls are made permeable following a heat treatment and the cell fluid is then dissolved with water.

This process is featured by the degree of extraction, expressing the amount of extracted valuable substances, compared to the total valuable matter content of the fruit.

The amount of substances diffused is in direct proportion with the diffusion coefficient, the active surface, and the concentration gradient (Pátkai and Beszedics, 1987).

In order to increase the diffusion coefficient and the permeability of the cell walls, diffusion fluid extraction is performed at 50–70°C. Active surface can be increased by proper chopping. The concentration gradient is determined by the stream conditions and the solvent–cell fluid ratio. However, the amount of solvent applied is limited by the concentration decrease of the liquid extracted. Diffusion juice extraction is usually carried out in double-screw extractor devices.

Juice Clarification

Extracted fruit juices are usually turbid, due to the plant particles that are water insoluble (fibers, cellulose, hemicellulose, protopectin, starch, and lipids) and colloid macromolecules: pectin, proteins, soluble-starch fractions, certain polyphenols, and their oxidized or condensed derivatives. Depending on the finished product, these substances must be partially or entirely eliminated to avoid further turbidity and precipitation and to improve sensory attributes (taste, smell, and color). Juice clarification can be performed by physical–chemical methods, mechanical procedures, and their combinations.

A physical–chemical clarification is applied when eliminating all substances causing turbidity. Clarifying agents and enzymes are added during this procedure. The effect of mineral clarifying agents is based on their surface activity and electric charge. For the clarification of fruit juices, bentonite and solid silicic acid are used. Bentonite is of volcanic origin belonging to the group of montmorillonites. It possesses big surface and good thickening properties, its negatively charged particles strongly adsorb positively charged proteins. Solid silicic acid is a negatively charged colloid solution. It is usually combined with other clarifying agents or with enzyme treatment. Its clarifying effect is good, with ashort clarification period. In the case of enzymatic pectin decomposition, solid silicic acid is added to the juice together with the enzyme. Gelatine is a protein-based clarifying agent that precipitates negatively charged particles (polyphenols, decomposed pectin). It is often completed with tannin, which reacts with protein molecules. Polyvinylpolypyrrolidon is a water-insoluble powder that primarily adsorbs and precipitates polyphenols. During juice clarification, the so-called protecting colloids (mainly pectin molecules) need to be decomposed, since they hinder aggregate formation and the settling of floating substances. This process can be accomplished with the use of enzymes. Pectin decomposition is usually completed with starch and protein decomposition, thus enzyme products marketed contain other enzyme components as well (Dietrich, 1998; Grassin, 1990).

Mechanical clarification targets the elimination of suspended fibers and precipitation. This process is usually carried out in centrifuges and filtration devices. In this practice, the first filtration phase is performed in settling centrifuges, meanwhile decanters are used to eliminate fibers from cloudy juices (Welter, 1991; Nagel, 1992).

Filtration is an important step of fruit juice production. Traditional filtration is performed in slurry layer–based devices. First filtration additives (silica, perlite) are added to the liquid to be filtered (Szenes, 1991). The equipment can be frame, column, and vacuum based. For the clarification of filtered juices, membrane and ultrafiltration techniques are widely applied (Szabó, 1995; Galambos, 2003; Capannelli et al., 1994). These latter devices enable clarification and filtration in one step. To increase the active period of the membrane, enzyme treatment is usually performed prior to the filtration (Kinna, 1990).

Clarified, filtered, or cloudy juices are ready for consumption. These can be further processed to preserved products in two ways: they are packaged right after production or concentrated to semifinished products.

Concentrate Production

The aim of concentration is to increase the dry matter content and decrease the water content of juices, in order to extend shelf life and to improve transportation and storage properties. This operation has to be implemented with minimal loss of valuable ingredients and minimal damage to sensory traits (Braddock, 1999).

Evaporation

This is the most frequently used concentration solution. From a physical point of view, it means water evaporation by means of boiling. This operation is carried out in evaporators, and steam ensures the energy necessary for boiling. Some of the solution's water content evaporates during boiling. The vapour thus formed is then driven out from the device and condensed. As the valuable juice components are heat sensitive, short time, low-temperature condensation is desired. In order to ensure low boiling point, the process is performed under vacuum. Usually more evaporators are applied in sequence to minimize the energy costs. Such systems of three–four elements are commonly used (Fábry, 1995; Fellows, 2000).

Chemical, rheological, and thermal juice properties play an important role in the condensation process. As these features depend on the raw material, operation parameters may vary with the use of different fruit species. Evaporators should be chosen according to the juice properties. The most widely used devices are the film, pipe, plate, and centrifugal-based ones (Szenes, 1991).

Evaporator systems are usually combined with aroma-recovery units. These are generally connected to the first part of the evaporator, and condense the most volatile aroma compounds. Aromas thus condensed are often remixed into the concentrate to improve its smell and flavor. Otherwise, these can be concentrated and applied as natural aroma extracts for other fruit products.

Concentration by Freezing

This method is used for the concentration of valuable, heat-sensitive fruit juices. During this process, the water content of the juice is frozen with ice-crystal formation. These crystals contain clean water, thus solvent loss occurs in the solution. As the procedure goes on, the fluid gets more and more concentrated and contains more and more crystals. Then the two phases can be separated mechanically (Fellows, 2000).

This type of concentration is a very gentle process, as there is no aroma, color, and vitamin loss due to the low temperatures. Concentrates thus prepared contain almost every valuable ingredients of the original juice. Its disadvantages are high energy consumption and lower concentration efficiency compared with the heat treatment procedure (Várszegi, 2002).

Reverse Osmosis

This membrane-based fruit concentrate producing technology is becoming more important. This means that some of the water is filtered out of the solution. Due to the rapid increase of osmotic pressure, concentrates up to 30 Brix can be produced. It is usually used as a preconcentration step prior to the aforementioned freezing devices in order to increase capacity (Beaudry and Lampi, 1990; Hribar and Sulc, 1990).

Fruit Concentrate Storage

The method of storage largely depends on the properties of the raw material and the characteristics of the concentrate (Stéger-Máté and Horváth, 1997).

From the filtered, clarified juice of less valuable fruit (apple, grape) concentrates containing 70% water-soluble dry-matter can be prepared. These are microbiologically stable enough to be stored in cooled stainless steel containers until further use.

However, colored fruits and berries can only be concentrated up to 45–55 Brix percent depending on the fruit species. Moreover, there are concentrates with special composition (chandy) and ones made by freeze concentration that hardly achieve 40–45 Brix percent. These can only be stored in frozen form or with aseptic technology.

Drink Production

Clarified, filtered, and cloudy juices are usually packed in their original composition. Meanwhile, there are juices to which sugars or vitamins are added.

Juices are frequently made of concentrates by dilution. In this process the finished drink composition should be as similar to that of the original juice as possible. The water applied for dilution is condensed or softened; flavors and aromas are adjusted with the aroma compounds condensed during concentration. As the sensory value of the product thus made is not as good as the one made with the direct procedure, labelling must contain the information that it was diluted from the concentrates.

Fruit nectars can be made of juices and concentrates as well. These are made of the appropriate amount of juice or concentrate and heated sugar syrup. Their acid content can be adjusted with lemon juice. As far as their raw material is concerned they can be made of one or more fruits. Their filling and preservation is identical to that for the juices.

Packaging and Preservation of Fruit Juices and Nectars

Prepared fruit drinks are filled into glass or plastic bottles of different shape and size, but they can also be packed into combined boxes. Fruit juices are preserved with heat treatment. According to the traditional method, the juice is heated up to 82–85°C, then filled at that temperature, closed, and pasteurized afterwards. Pasteurization is carried out at 84–88°C for 15–45 min depending on the size of the packaging. Following heat treatment, products are cooled back to room temperature. However,the aseptic procedure preserves juice quality much better. This means that the juice is pasteurized when flowing in a closed system, then cooled under conditions where no infection may occur, and finally filled into previously sterilized containers (Buchner, 1990). This process applies heat treatment of 100–110°C for 0.5–1.5 min.

PRODUCTION OF FRUIT NECTARS WITH FRUIT FLESH CONTENT

These products are made of fruit pulps. They are obtained by passing the raw materials through sieves, thus fruit flesh can be separated from seed and skin particles. Before performing this operation, fruits have to undergo preparation steps. Fruit nectars containing fruit flesh can be made of fruit pulp directly after production or later from the concentrated pulp (Fig. 12.2).

PREPARATION STEPS

This above operation includes steps such as raw material reception, washing, stem elimination, and selection, which are identical with the previously introduced technology. However, the last steps are different: coarse chopping and preheating.

Coarse Chopping

Coarse chopping is the cracking and crushing of pomaceous fruits, which results in coarse fruit texture. This is necessary to ensure the efficiency and smoothness of preheating.

Preheating

Prior to sieving most of the raw materials are preheated and is done so for two reasons. Heat treatment makes fruit texture soft and lax, improving the efficacy of sieving. It also inactivates the enzymes found in the fruit. It is particularly important in the case of oxidative and pectin-decomposing enzymes, since these can damage the color and texture of the pulp later (Binder, 2002; Fellows, 2000).

To avoid dilution during preheating, closed systems are applied. The time and temperature values of the process depend on the type of equipment, properties of raw material, and the degree of chopping.

SIEVING

Sieving is a separation procedure from which pulp, which consists of small fruit particles, is obtained.

The equipment applies centrifugal forces to make tender fruit parts pass through the sieve, while hard skin and seed particles remain on the surface. The perforation of the sieve determines the size of particles detached. In order to increase sieving efficiency, this separation is performed in more steps with decreasing perforation sieves in line. The perforation of sieves

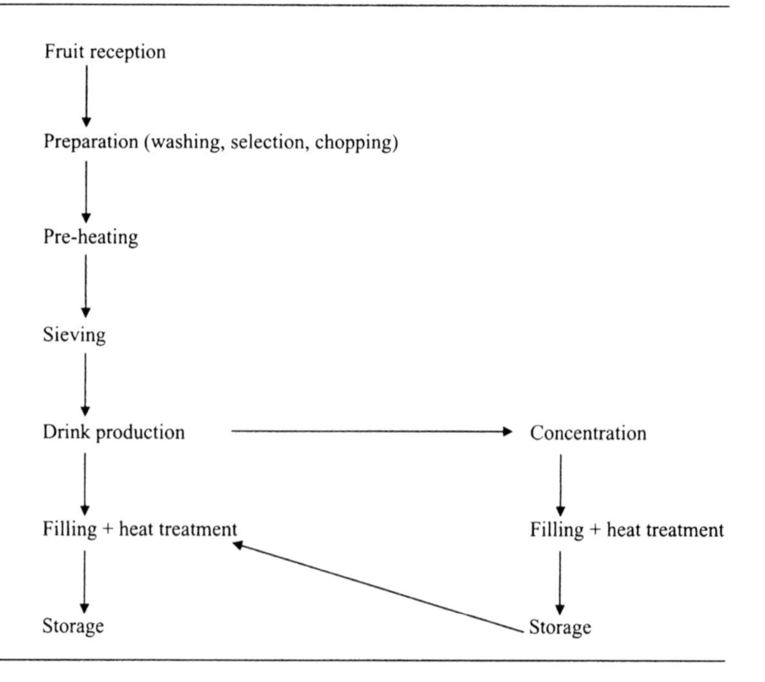

Figure 12.2. The flow chart of fruit drink production.

used for fruits is between 0.4–6.0 mm (Hidegkuti and Körmendy, 1990).

In the case of fruits containing seeds, an additional sieve has to be built into the process to detach seeds. This device must be more massive and possess bigger perforations.

Sieving can be carried out under both warm and cold conditions. Soft and vulnerable fruits (raspberry, strawberry) are sieved cold and the pulp is heated afterwards, in order to inactivate enzymes. Meanwhile, in the case of more solid, heat-resistant fruits (apple, pear, quince) warm conditions are used to obtain better juice yield and less sieving waste.

Fruit Pulp Processing and Storing

Fruit pulp is a valuable raw material for drink production but possesses no refreshing effect itself. Therefore, it is used for fruit nectar production. Fruit nectar can be made right after pulp production or later at another point in time. Fruit pulp thus prepared is ready to be stored and transported. For storage it is often concentrated to 26–32 Brix percent.

Fruit pulps can safely be stored under aseptic circumstances, i.e., heat treatment and storage in sterile containers. Aseptic systems are mechanically closed under slight pressure to exclude microbiological infections (David et al., 1996).

Heat treatment in aseptic technology is usually performed above 100°C (90–120°C) for a relatively short period (1–2 min). After pasteurization, the pulp is cooled to about 30°C to stop further chemical reactions (Ott, 1990).

Filling takes place in special aseptic machines, which also sterilizes the packaging material. Fruit pulps are generally stored in large containers that can range from "bag in a barrel" (120 l) to stainless steel tanks of more than a hundred cubic meters. The main advantage of "bag in a barrel" packaging is the easy transportation and the compatibility with international carriage system (Szenes, 1991).

Preparation of Fruit Nectars Containing Fruit Flesh

These nectars contain not only the juice but also the fruit flesh in refined and homogenized form. Besides the necessary fruit content, which is regulated, sugar syrup is also added to nectars. These products can be flavored with lemon juice or concentrate and enriched with vitamins. Carbohydrates or sweeteners may also be added. Nectars can be made of one or more fruits. As far as the raw materials are concerned, fruit pulps and concentrated pulps can be used, combined with filtered juices or concentrates.

First the water and sugar are blended, then the syrup is boiled, and the fruit pulp is added. Subsequently, it is supplemented with lemon juice and vitamin solutions if necessary.

Refining is an operation that determines the nectar quality. It is performed in colloid mills, which are based on hydrodynamic shearing forces. Homogenization chops fibers to fine particles and creates a dispersal system. The fluid thus prepared is stable, no settling will occur in it, even if stored. This procedure substantially decreases visible viscosity and results in a tender smooth texture and taste (Hidegkuti and Körmendy, 1990).

The packaging of nectars is identical with that for juices. General references relating to manufacturing fruit beverages: Ashurst, 1998; Knee, 2001; Shewfelt, 2000; Barrett et al., 2004; Jongen (ed.), 2002; Nagy et al., 1993; Arthey and Ashurst, 1996.

REFERENCES

Aehle, W. 2004. Enzymes in Industry, 2nd ed. Wiley-VCH Verlag GmbH & Co., KGaA, Weinheim, The Netherlands, pp. 113–123.

Barta, J. and Körmendy, I. 1990. Washing of vegetable raw materials (in Hungarian), para 3.4. In: Körmendy, I. and Török, S. Z. (eds.). Konzervtechnológia növényi eredetü nyersanyagok feldolgozásához (Canning Technology for Processing Vegetable Raw Materials). University of Horticulture and Food Industry, Faculty of Food Science Press, Budapest, Hungary, pp. 165–175.

Beaudry, E.G. and Lampi, K.A. 1990. Osmotic concentration of fruit juice (in German). *Flüssiges Obst.* 57(10):652–656.

Bielig, H.J., Faethe, W., Fuchs, G., Koch, J., Wallrauch, S. and Wucherpfennig, K. 1987. RSK-Values. The Complete Manual. Verband der deutschen Fruchtsaftindustrie e.V. Bonn.

Binder, I. 2002. Pre-heating (in Hungarian). In: Beke, Gy. (ed.). Hütöipari Kézikönyv (Handbook of Industrial Refrigeration), 2nd ed. Mezögazda Kiadó, Budapest, Hungary, pp. 316–320.

Braddock, R.J. 1999. Handbook of Citrus By-Product and Processing Technology. John Wiley & Sons, Inc., Canada, pp. 43–60.

Buchner, N. 1990. Aseptic filling of glass and plastic containers (in German). *Verpacken. Transportiere.* Lagern. 41(5):295–300.

Capannelli, G., Bottin, A. and Munari, S. 1994. The use of membrane processes in the clarification of orange and Lemon juices. *J. Food Eng.* 21(9):473–483.

Dietrich, H. 1998. Enzymes in fruit juice processing. *Fruit Processing.* 8(3):105–107.

Fábry, Gy. 1995. Evaporation (in Hungarian), para 16. In: Fábry, Gy. (ed.). 1995. Élelmiszeripari eljárások és berendezések (Processes and Apparatuses of Food Industry). Mezögazda Kiadó, Budapest, Hungary, pp. 392–428.

Fellows, P.J. 2000. Food Processing Technology, 2nd ed. Woodhead Publishing, Cambridge, England. Part II–IV.

Galambos, I. 2003. The basis of membrane operations and the opportunities of their application in the drink-industry (in Hungarian). *Ásványvíz, üdítöital, gyümölcslé.* 4(2):36–41.

Grassin, C. 1990. Improvement of juice clarification with enzymes (in German). *Flüssiges Obst.* 57(8):501–506.

Hidegkuti, Gy. and Körmendy, I. 1990. Chopping, homogenization, mixing (in Hungarian), para 3.7. In: Körmendy, I. and Török, S. Z. (eds.). Konzervtechnológia növényi eredetü nyersanyagok feldolgozásához (Canning Technology for Processing Vegetable Raw Materials). University of Horticulture and Food Industry, Faculty of Food Science Press, Budapest, Hungary, pp. 201–240.

Hribar, J. and Sulc, D. 1990. Preconcentration of apple juice with reverse osmosis (in German). *Flüssiges Obst.* 57(9):590–592.

David, J.R.D., Graves, R.H. and Carlson, V.R., 1996. Aseptic Processing and Packaging of Food: A Food Industry Perspective. CRC Press, Boca Raton, Florida, pp. 41–49.

Kardos, E. 1962. Gyümölcs- és zöldséglevek, üdítöitalok (Fruit and Vegetable Juices Soft-Drinks) (in Hungarian). Müszaki Könyvkiadó, Budapest, Hungary, pp. 17–18.

Kinna, J. 1990. Ultra- and mikrofiltration in the food industry (in German). *Flüssiges Obst.* 57(9):593–605.

Lengyel, A. 1995. Pressing (in Hungarian), para 10. In: Fábry, Gy. (ed.). 1995. (Élelmiszeripari eljárások és berendezések). Processes and Machineries of Food Industry. Mezögazda Kiadó, Budapest, Hungary, pp. 252–259.

Nagel, B. 1992. Continuous production of high quality cloudy apple juices. *Fruit Processing.* 2(1):3–5.

Ott, J. 1990. Aseptic technology (in Hungarian), para 4.2. In: Körmendy, I. and Török, S.Z. (eds.) 1990. Konzervtechnológia növényi eredetü nyersanyagok feldolgozásához (Canning Technology for Processing Vegetable Raw Materials). University of Horticulture and Food Industry, Faculty of Food Science Press, Budapest, Hungary, pp. 395–435.

Parker, R. 2003. Introduction to Food Science. Thomson Learning, Delmar, Albany, USA. pp. 111–117.

Pátkai, Gy. and Beszedics, Gy. 1987. Apple juice production by diffusion extraction (in Hungarian). *Élelmezési Ipar.* 41(1):19–22.

Reising, K. 1990. Successful dejuicing enzymes (in German). *Flüssiges Obst.* 57(8):495–499.

Schmitt, R. 1990. Use of enzymes for the production of stable pulps, purees and cloudy juices (in German). *Flüssiges Obst.* 57(8):508–512.

Stéger-Máté, M. and Horváth, E. 1997. Relationship Between the C-Vitamin Content and the Storability of Apple Juices. *Horticult. Sci.* 29(3–4):61–66.

Stéger-Máté, M., Horváth, E. and Sipos, B.Z. 2002. The examination of the composition and processing possibilities of the different currant varieties II. (in Hungarian) *Ásványvíz, üdítöital, gyümölcslé.* 3(2):27–32

Szabó, G. 1995. Membrane filtration (in Hungarian), para 22. In: Fábry, G.Y. (ed.). 1995. Élelmiszripari eljárások és berendezések (Processes and Apparatuses of Food Industry). Mezögazda Kiadó, Budapest, Hungary, 616–631.

Szenes, E. 1991. Fruit juices (in Hungarian), para. 7.1.3.5. In: Szenes, E. and Oláh, M. (eds.). Konzervipari Kézikönyv. (Handbook of Canning). Integra-Projekt Kft., Budapest, Hungary, pp. 253–255.

Várkonyi, G. 2000. Conference on the activity of the FAO/WHO Codex Alimentarius Com and on the inland effect of his activity (in Hungarian). *Ásványvíz, üdítöital, gyümölcslé.* 1(2):55–57.

Várszegi, T. 2002. Crioconcentration (in Hungarian). In: Beke, Gy. (ed). Hütöipari Kézikönyv (Handbook of Industrial Refrigeration), 2nd ed. Mezögazda Kiadó, Budapest, Hungary, pp. 404–406.

Welter, C.C., Hartmann, E. and Frei, M. 1991. Production of very light coloured cloudy juices (in German). *Flüssiges Obst.* 58(5):230–233.

GENERAL REFERENCES

Arthey, D. and Ashurst, P.R. 1996. Fruit processing. Blackie Academic and Professional, England.

Ashurst, P.R. 1998. The Chemistry and Technology of Soft Drinks and Fruit Juices. Sheffield Academic Press, England.

Barrett, D.M., Somogyi, L. and Ramaswamy, H. 2004. Processing Fruits: Science and Technology, 2nd ed. CRC Press, Florida, USA.

Jongen, W.M.F. and Jongen, W. 2002. Fruit and Vegetable Processing: Improving Quality. Woodhead Publishing, Abington Cambridge, England.

Knee, M. 2001. Fruit Quality and its Biological Basis. CRC Press, Florida, USA.

Nagy, S., Chen, C.S. and Shaw, P.E. 1993. Fruit Juice Processing Technology. Florida Science Source, Florida, USA.

Shewfelt, R.L. 2000. Fruit and Vegetable Quality: An Integrated View. CRC Press, Florida, USA.

13
Fruit as an Ingredient in a Fruit Product

Györgyi Pátkai

Humans consume food of both plant and animal origin, because we need mono- and polysaccharides, organic acids, oils of plant origin, minerals, and vitamins, which are found primarily in plants, and also need proteins of full value, essential amino acids, fats, and fat-soluble vitamins, which are found in raw materials of animal origin. Most food products consist of several different components of plant and animal origin.

Food products based on raw materials from plant and animal origin are found to be among the products of the dairy, bakery, confectionery, canning, cold-storage, frozen food, and distilling industries as well. Fruits are important value-adding food components not only in fresh, unbroken, or chopped form, but also as preserved fruit preparations, jams, marmalades, jellies, dried or candied products, extracts, juices, concentrates, etc. They are harvested seasonally and deteriorate quickly. This is the reason why the production of food containing fruit components is profitable, only if the fruit component needed for the product has been provisionally preserved until further utilization.

The present chapter deals with production, characteristics, and quality requirements of preserved fruit preparations and semi-prepared products that are being applied in different sectors of the food industry. Some characteristic groups of food products containing fruit components have also been described.

FRUIT PRODUCTS AND SEMI-PREPARED PRODUCTS TO BE APPLIED AS FOOD COMPONENTS

FRUIT PUREES

Fruit puree is the edible part of a fruit, pulped, homogenized, or manufactured by a similar process (Codex Alimentarius Hungaricus, 1996). It is a semi-prepared product, it contains the fruit in a sieved, homogenous form, and contains no recognizable fruit pieces. Fruit puree is not suitable for direct consumption for most people except infants, elderly individuals with less than adequate amount of original teeth, and some individuals under medical care. However, fruit puree is the raw material for production of jams, marmalades, fruit juices, and bakery fillings, and it is used as a component of many food products. Industrially processed fruit purees can be preserved by preservatives, by cooling or freezing, by heat, by concentration, and by addition of sugar. The best way to preserve fruit puree is with aseptic technology. Each healthy and ripened fruit is suitable for puree production. Regarding maturity, exceptions can be made in the case of fruits which contain a high concentration of pectin in the half-mature state (e.g., apple, gooseberry, etc.).

Preparatory operations of fruit puree processing are washing and selection. Blanching is usually necessary after selection, prior to pulping with a view to decrease losses, energy requirement, and mechanical stress of the sieve. Crushing of the blanched fruit is performed gradually by passing it through a crusher and a coarse sieve, finally through a fine sieve and possibly through a homogenizer as well.

The homogenized fruit puree can be preserved chemically, by adding preservatives—usually SO_2 (max. 2 g/kg), benzoic acid (max. 1.5 g/kg), sorbic acid (max. 1.0 g/kg)—or by traditional pasteurization at 94–96°C after having filled the puree in 5 l jars and hermetically closed. In the modern fruit processing industry, fruit purees are preserved by *aseptic technology*. In this case, a continuous, so-called "scratched-wall" heat-exchanger (type α-Laval, Crepaco, Manzini, APV, etc.) is used for killing yeast and mold cells which endanger the durability of the fruit puree. The *sterile product* is filled under *aseptic conditions* into *sterile packaging* (laminated foil-bags, bag-in-box, barrels, containers, etc.), hermetically closed, and stored until further usage (Körmendy and Török, 1990; Potter and Hotchkiss, 2001).

Single-strength apple puree can be concentrated to an intermediate extract content of 20–25°Brix in a multi-stage vacuum-evaporator or a film-evaporator with scraper. These evaporators are suitable also for the production of a concentrate of 40°Brix. Concentrated fruit purees not only enrich the product with valuable extracts, characteristic taste, flavor, and color, but also contribute to its functional properties because of their high pectin and fiber content (Hegenbart, 1994).

Most demanded raw materials for fruit preparations are berries and cherries because of their high vitamin and mineral content, and attractive taste and color. The vitamin content of fruits can be fairly protected by use of mild processing parameters. In this case, the processing of fruits with high vitamin content (elderberries, rosehip, sea-buckthorn, etc.) results in fruit preparations with valuable composition; and when used as additives, increase the nutritive value of the combined product (Barta et al., 2002; Lampe, 1999).

The above-mentioned berries and red grape-vine are processed to give anthocyanin extracts and concentrates as well, which are used as natural food colorings. The stability and intensity of elderberry anthocyanins are increased by the fact that the high anthocyanin content is coupled with high vitamin C content of the fruit (Stéger-Máté et al., 2001). A similar synergic effect between anthocyanins and vitamin C was proved in case of blackberries as well (Stéger-Máté et al., 2002). The anti-oxidative effect of the anthocyanins helps to protect the vitamin C content (Gardner et al., 2000).

Fruit Pulps

Pulp is the edible part of the fruit, possibly without the skin, shell, stones, etc., sliced or crushed, but not sieved (Codex Alimentarius Hungaricus, 1996). It is a semi-prepared product containing crushed fruit-flesh with recognizable pieces, processed from washed and selected raw material. It is not suitable for direct consumption. The fruit processing industry is producing jams from fruit pulp and fresh fruit as well.

Fruit pulps differ from the above-mentioned fruit purees in the technology of crushing. "Pulping" means to crush the washed and selected fruit-flesh after separation of the peel, shell, stones, and stalks, as needed, by a crusher, fruit-milling machine, or hammer-grinder without crushing it by a sieve. These machines crush the fruit-flesh to coarse ground, recognizable pieces. The provisory preservation of the cleaned and crushed fruit-flesh can be performed by preservatives, by heat, by cooling, or freezing, respectively.

In case of chemical preservation, the pulp has to be mixed with the given preservative (0.03% SO_2, 0.15% benzoic acid, 0.1% sorbic acid, or with their Na or K salts, respectively). The chemical preservatives are effective below pH 3.5. If the pH value is higher, a combination of acid and preservatives is used. This step of the technology requires careful mixing of the pulp by a mixer of moderate revolution per minute, continuous injection, or filling up layer-by-layer. Attention must be given not to crush the fruit pieces; they must be protected in the course of the manufacturing process.

The advantage of the heat-preserved fruit preparation is that it does not require preservatives, whereas the considerable demand on package materials is a disadvantage, as the pasteurization of the pulp is often performed in 5 l jars. Great disadvantage of cooling in a large container is slowness, resulting in deterioration of color, flavor, and nutrients. In case of pulps, it is problematic to use the aseptic technology because of the lumpy character of the product. The cleaned and crushed fruit-flesh will be blanched in water before pasteurization in a duplicator-kettle or continuous blancher. (Added water cannot be more than the water quantity evaporated during blanching.) Acid content of the blanched product can be adjusted to 1% by adding citric acid, and the hot pulp has to be glassed-in and pasteurized without delay. The temperature of the thermal center must attain 86–90°C.

Desiccated Fruits

Dehydration of food is one of the most ancient ways of preservation, and it is based on the realization that the lowest limit of water-activity for microbial growth is $a_w = 0.62$. Under this a_w value, neither microbes nor the spores of the xerophyle molds are able to grow.

Only healthy and ripened fruits are suitable for drying. Quality requirements for fruits as raw material for drying include: solid fruit-flesh, easy separation of pits and skin, and small pits. Washed and cleaned fruit-flesh is treated with SO_2, if necessary, to prevent enzymatic browning. This treatment gives some protection also against microbial deterioration, pests, and non-enzymatic browning (Maillard-reaction).

Withdrawal of water from the prepared fruit-flesh can be performed in two different ways.

- Under tropical climate, people take advantage of the sunshine, and dry fruits in thin layer by slow, *natural drying* (figs, bananas, peaches, and apricots).
- *Dehydration* means treatment in tunnel- or band-dryers by blowing through warm air of 60–70°C (apples, plums).

Residual water-content of dehydrated fruits is about 12–16%, and when adequately packaged they are long-lasting without any damage because of their high sugar content and low ($a_w = 0.72 - 0.75$) water activity.

Quality requirements related to the dried product are the following.

- Solid, plastic, and creamy, but not sticky consistency and leather-like surface of dried prunes and apricots
- Bright color of dried apple-rings, light color of peaches and other fruits.
- High sugar content, harmonic sugar/acid ratio, and easy disconnection of the stones
- Bigger size, thin skin, and a pleasant, pliable consistency (dried figs)

Dried grapes are taken in trade as "raisins," "sultanas," and "chorinths." Raisins have a brown color, and do not contain any pits; sultanas have a light color without pits; and chorinths are smaller than raisins and sultanas, and are dark colored (Herrmann, 1966).

Dehydrated cherries and dehydrated apple products are widely used in foods such as pastry, confectionery products, ice-cream, frozen desserts, sweets, fruit salads, cheese, and yogurt (Somogyi et al., 1996).

Sweetened and Candied Fruits

Sweetened fruits, soughs, candied fruits, and dehydrated-sweetened fruits belong to the above-mentioned product-family. The sugar content of these products is increased to a degree, which prevents the growth of microorganisms. The important point here for preservation is to achieve a low water activity with the addition of sugars. The significant increase of sugar content of the fruit preparation can be performed by soaking in sucrose or glucose solutions of gradually increasing concentration, under atmospheric pressure or in vacuum. The concentration difference between the sugar syrup and fruit cells will be equalized in consequence of the difference in osmotic pressure. A saccharose content of 70–80% of the fruit-flesh can be achieved during soaking in the last, most concentrated sugar syrup. Its water content is decreased to 12–21% by careful drying. Cherries and sour cherries, pineapples, pears, quinces, peaches, apricots, plums, figs, green almonds and nuts, chestnuts, gooseberries, strawberries, and raspberries can be processed to candied fruits of good quality. The surface of sweetened fruits will be coated with sugar-glazing or with a thin layer of crystallized sugar for confectionery and sweets industrial use. The first mentioned procedure is called glazing, and the second one is candying. A special group of this type of fruit preparations is produced by treating the blanched peel of citrus fruits with sugar syrup, followed by glazing and drying it carefully (Herrmann, 1966). Preparatory steps of the technology are similar in all types of sweetened products, but they are adapted to the characteristics of the raw material (e.g., washing, cleaning, pitting, peeling, separation of the stems if needed, crushing, and blanching). Ready-made fruit preparations (sweetened fruits, soughs, candied fruits, and dehydrated-sweetened fruits) differ from each other in finishing and final appearance. The raw material of all those products is the so-called "egute"—the cleaned, blanched fruit, soaked in syrup of starch or saccharose or in high-fructose corn syrup.

In case of fruits with a high fructose content (pears, quinces, melons, etc.), producing fruit preparations of a low caloric content can be realized. In this case, a concentrated syrup of Jerusalem artichoke juice with high fructose content (75–80%) can be used for the sweetening treatment (Barta, 1993; Bray et al., 2004).

The aim of soaking in syrup is to saturate the cells, which have been opened during blanching, with sugar. Form and consistency of the fruits will be stabilized as a consequence of this treatment, and their volume remains constant. Following the above-mentioned preparative operations, the fruit pieces should be fed into a flat vessel and filled with hot sugar syrup of 28–30°Brix. The product filled with the syrup stays in the syrup for 2–3 days, after that the syrup will be poured off, heated, and sweetened with sugar to a concentration value of 35°Brix. This procedure will be repeated 5–6 times, gradually increasing the concentration of the syrup up to 60°Brix. After having achieved this value, the hot, sweetened fruit pieces should be dripped, filled into 5 l jars, and filled up with high-fructose corn syrup or starch at a concentration of 68–70°Brix and hermetically closed. This semi-prepared product (*egute*) can be stored and used continuously, without posing any potential hazard.

Sweetened fruit is made from the *egute* after being dripped, dried in a warm current of air, rolled in crystalline sugar, and packed in cellophane. Sweetened berry preparations are manufactured also by blending whole, sliced, or crushed fruits with sugar in ratios such as 4:1, 3:1, or 7:1 (berries:sugar). A usual procedure is to "cap" or sprinkle sugar onto the surface of the berries, after they have been filled into pails or drums. In the United States, the quality of the berries is USDA Grade A or B. These products are typically used in ice-cream, yogurt, bakery preparations, or fillings (Somogyi et al., 1996).

The sough (glazed egute covered with a shining fondant layer) is manufactured by soaking the semi-prepared product in a hot, over-saturated sugar solution, finally put on a grating and dried in a hot room. A bright fondant layer crystallizes on the surface of the fruit pieces, which makes the product attractive and protects its fresh, plastic consistency. This procedure is called "glazing."

Dripped, dried egute is also used to manufacture *candied fruit* (fruit pieces, coated with a thin layer of crystallized sugar). The semi-prepared product will be soaked in cool sugar syrup of 59–60°Brix. The surface of the fruit pieces is covered with a thin layer of granulated sugar. After the sweetening treatment, the syrup will be racked from the vessel, which is equipped with an outlet and a grating at the bottom. The fruit layer remains on the grating, and can drip and dry. The sugar layer on the surface of the fruit pieces, consisting of closely united crystals, protects the product from drying out quickly. An attractive packaging improves the protecting effect of the sugar layer and also the marketability of the product.

To manufacture *sweetened-dehydrated fruits*, the raw material is prepared in a similar way to the procedure used in the technology of canned or bottled fruit processing (washing, cleaning, peeling, pitting, de-stemming, halving, or slicing, and treatment of the surface in the case of light colored fruits with a diluted solution of citric acid or H_2SO_3). The prepared raw material should be soaked 10–20 min under vacuum in a sugar solution of 15–20°Brix at 60°C. Duration of the treatment depends on fruit ripeness. After the sweetening treatment, the fruits become transparent. Because of the vacuum-effect, the inter-cellular capillaries will become de-aerated, and the sugar solution will penetrate into the air space. The less aroma the fruit contains, the lower sugar concentration has to be adjusted (apples, pears: 30°Brix; apricots and peaches: 35°Brix; quinces: 40°Brix).

Sweetened fruits should be carefully placed on a sieve, slowly dried to a water content of 15–17%, at the beginning at a temperature of 70–75°C, and at the end at 65°C. To decrease hygroscopicity, the sweetened-dried fruits should be rolled in sugar powder. The sugar surplus can be selected by sieving, and the product should be packed in cellophane or in ornamental boxes (Szenes, 1995; Mathlouthi, 2003).

Fruit Jams, Sweetened Fruit Purees, Jellies and Marmalades

These products are, according to Codex Alimentarius Hungaricus (Magyar Élelmiszerkönyv, 1996), fruit preparations of jelly-like consistency, produced of fruit purees, pulps, juices, or watery extracts of one or more kinds of fruits and by adding sugar.

Sweetened fruit purees are fruit preparations, manufactured from fruit purees (pulped and homogenized fruit-flesh, fresh or preserved) or of fruit juices by adding sugar, citric acid, pectin, and food colorings possibly. They have a spreadable, bounded, jelly-like consistency. They are generally made of one sort of fruit, except "mixed sweetened fruit puree," which contains several fruits. It has a harder consistency, and it is sliceable with a knife. Concentrated plum puree of good quality can also be made without sugar added. Soluble solid content of the sweetened fruit purees achieves 56°Brix; 10–12% of this value originates from the fruit, the rest is added in the form of beet-sugar. The Department of Canning Technology of the Corvinus University of Budapest successfully used high-fructose Jerusalem artichoke concentrate for sweetening fruit preparations (Barta, 1993). In limited amounts, fructose tolerance by diabetics is accepted in Europe, but those products are sweetened, in addition to fructose as a natural sweetener, by the simultaneous addition of artificial sweeteners and gellying additives, because in the absence of saccharose, pectin has no gellying effect (Pátkai and Barta, 2000; Bassoli, 2003).

In the course of putting together a batch of sweetened fruit puree, the pectin has to be added in the form of a colloidal solution, which is prepared by mixing the pectin with cold or warm water. The sweetened fruit preparation has to be cooked in a vacuum evaporator. First, the fruit puree and one-third of the needed sugar quantity should be fed into the machine. The superfluous water quantity should be evaporated during 25–40 min under maximal vacuum value and at the boiling temperature corresponding with the vacuum. Some minutes before the end of evaporation, the missing part of the sugar quantity and the colloidal pectin solution must be added to the puree. After having added the rest of sugar and the pectin solution, the evaporation is continued for about 5 min under vacuum; finally, the evacuation is stopped, and the product is heated to 80–90°C. The pH value of the sweetened, evaporated puree is adjusted to a value of 2.8–3.2 by adding citric acid. After having controlled the soluble solid content of the product, it can be filled into specified packages; the filled packages are then closed and pasteurized. If the volume of the package is smaller than 5 l, a processing period of 10 min at 94–96°C is adequate for pasteurization.

Jams are solid gels made of fruit pulp or juice, sugar, and added pectin. They can be processed from one single kind of fruit or from a combination of fruits. The fruit content should be at least 40%. In mixed fruit jams, the amount of the first-named fruit should achieve at least 50% of the total fruit quantity (based on UK legislation) (Anon, 2004b).

The semi-prepared product (pulp), preserved by adding SO_2 should be fed into the vacuum cooking kettle, and at first SO_2 has to be removed by boiling under vacuum. If the semi-prepared product has been preserved by addition of sorbic acid or by heat treatment, the above-mentioned pre-treatment is not necessary. A measured quantity of pulp can be fed directly into the evaporator and the batch can be prepared. The solid content of the jam is composed of the solid content of fruit, sugar, pectin, and citric acid. Its recommended value is 68°Brix. "Fruit rate" (rate of the solid content originating from the pulp) should be between 20% and 50%, and it is always an actual quality prescription. Added pectin and citric acid quantities should be established by cooking-test, on the basis of quality prescriptions as well. The pectin quantity has to guarantee the jelly-like, suitably bounded consistency. The pH value of the jam should be adjusted also to a value of 2.8–3.2 by adding citric acid. The quantity of pectin and citric acid to be added is generally about 0.3–0.5%, depending on the original pectin- and acid content of the fruit. The sugar quantity needed can be calculated on the basis of those data. Both pectin and citric acid are always used in the form of solutions; and similar to the case of sweetened fruit purees, they are added only at the end of cooking. Cooking time for jams is 5–15 min, depending on the fruit type.

Cooking apparatus used are open, double-walled, steam-heated, tiltable stainless steel kettles with stirrer, or closed, pressure-tight vacuum-kettles.

The cooked jam is cooled to 90°C or 60–70°C, respectively, depending on the pectin quality and depending on the type of packaging material, filled into packages with a screw- or piston loader. Jams filled into jars should be heated at 95–100°C for a short time. In the case of wooden or plastic packages, the product must be preserved by preservatives (0.15% benzoic acid, 0.1% sorbic acid, or their Na or K salts, respectively) (Körmendy and Török, 1990; Smith, 2003).

Fruit jellies and puddings are manufactured from filtered, clear fruit juice with high pectin content (apple-, quince-, strawberry-, gooseberry, or citrus juice) or of strained fruit-flesh by adding sugar, pectin, and food acids. Fruits for jelly processing should not be fully ripe (Szenes, 1995). The intact, healthy, and cleaned fruit is cooked in little water, the juice should be filtered, sweetened, concentrated, packaged, and chilled without moving it. The package should be closed only if the surface is already dry (Körmendy and Török, 1990; Smith, 2003).

Fruit pudding is a special fruit preparation, which is made of homogenized puree of one single kind of fruit. It can be flavored and decorated with shelled nut varieties (Szenes, 1995; Paltrinieri et al., 1997) (Tables 13.1 and 13.2).

Fruit Juice Concentrates

Fruit juice concentrates are produced by evaporation of the superfluous water quantity of fruit juices. Withdrawal of water can be performed by

1. Heating and evaporation of water at boiling temperature, generally under vacuum at low temperature in a multi-stage evaporator. Preliminary separation of the aroma content by distillation and the re-usage of it is important.
2. Freezing of the juice and separation of crystallized ice by centrifuging (cryoconcentration or freezeconcentration) or sublimation of ice crystals under vacuum (lyophilization or freeze-drying).
3. Reverse osmosis (separation of water by a semipermeable membrane).

Natural fruit juice should be cleaned of fiber content and colloids, by clarification and sieving, but

Table 13.1. Chemical Composition of Jams, Marmalades, and Jellies, Frequently Used as Food Components

Name	Product	Soluble Solids (%)	Soluble Carbohydrates (%)	Red Sugars (%)	Saccharose (%)	Pectin (%)	Acids (Total) (%)	Sugar/Acid Ratio
Apples	Jelly	65.0	64.87	–	–	–	–	–
Apricots	Jam	66.2	62.0	33.1	28.2	0.50	0.71	86.34
Raspberries	Jam	67.8	61.3	35.7	24.7	0.37	0.88	68.64
Raspberries	Jelly	65.0	64.78	–	–	–	–	–
Cherries	Jam	66.9	62.2	32.55	29.1	0.42	0.55	112.1
Red-currants	Jam	66.1	58.65	47.6	10.1	0.43	0.95	60.73
Red-currants	Jelly	66.3	66.3	–	–	–	–	–
Oranges	Jam	68.0	60.4	43.9	16.0	–	0.52	115.2

Source: Souci, Fachmann, and Kraut (1989).

Table 13.2. Mineral and Vitamin C Content of Jams, Marmalades, and Jellies, Used as Food Components

Name	Product	Minerals (mg/100 g)	Vitamin C (mg/100g)
Apples	Jelly	130	–
Apricots	Jam	360	–
Raspberries	Jam	–	2.7
Raspberries	Jelly	220	–
Cherries	Jam	380	1.2
Red-currants	Jam	340	20.6
Red-currants	Jelly	Na +K +Ca = 90.0	–
Oranges	Jam	140	4.0

Source: Souci, Fachmann, and Kraut (1989).

it is possible to concentrate the opal juice as well. Soluble solid content of the concentrate should attain 62–65°Brix to get a long-life product. The sugar ratio of the solids is variable depending on the sort and ripeness of the fruit. The sweetening effect of the concentrate may also vary in case of an equal sugar concentration because of sugar interactions. The fruit-juice concentrates contain nearly all valuable components of the fresh juice: sugars, organic acids, minerals, colorings, and natural antimicrobials. Therefore, they are popular additives in the bakery-, confectionery- and dairy industry as natural sources of colorants and sweeteners.

Fruit-based sweeteners tend to be more expensive than high-fructose corn syrup and sugar, but they provide several advantages. They add soluble and insoluble dietary fiber, color, a unique flavor profile, several vitamins, minerals, and "label appeal." Sometimes, the technologist has to use a juice-concentrate as a source of natural sugar with reduced color, flavor, and acidity in order to design products without imparting a characteristic taste or color of the fruit.

Natural color extracts currently permitted by the US Food and Drug Administration (US FDA) for industrial use are red cabbage, beet juice or powder, carmine, grape skin extract, and color extractives from grapes. These natural colorants are defined by the FDA as exempt from certification and are listed in US 21 CFR 73. More than 50% of the total reducing sugars are present as fructose which, being naturally hygroscopic, helps to prolong the shelf life of breads and bakery products (Cantor, 2004).

Frozen Fruits

The Hungarian cooling and refrigeration industry is processing and preserving by freezing all characteristic fruits cultivated in Hungary: raspberries, elderberries, apples, quinces, strawberries, gooseberries, red- and black-currents, cherries, sour-cherries, apricots, peaches, and chestnuts. Fruits are frozen after having been harvested, receipted, selected, washed, cleaned, crushed, or sliced. Freezing, repeated selection, and separation of debris are followed by packaging and storage of the frozen product. Chemical composition of frozen fruits is nearly equivalent to that of the fresh ones, therefore it can be used similarly to fresh fruits in the bakery-, confectionery-, and dairy industry, for ice-creams, desserts, sauces, and ready-made meals. In case of expert processing, freezing, and storage of the frozen product, undesirable changes are insignificant; and by a suitable technology, these changes can be decreased further. Undesirable changes of color are controlled by the following factors:

- blanching
- other color-fixing methods
- packaging in concentrated sugar (fructose) syrup
- vacuum-packaging of the frozen product (e.g., vacuum-packaging of sliced apples soaked in 0.5% solution of $CaCl_2$ or NaCl)
- ascorbic or erythorbic acids

The most perishable quality parameter of frozen fruits is their consistency, which must be taken into consideration in the case of sensible fruits (raspberries, strawberries, sour cherries, etc.). Deterioration of the consistency causes serious troubles in the case, if the user needs recognizable whole fruits or pieces of well-defined form for his product (e.g., bakery fillings, desserts, ice-cream, yogurts containing whole fruits, etc.).

Color changes (darkening) of frozen, sliced apricots, peaches, and apples during storage indicate the presence of polyphenoloxidase, the activity of which can be prevented by blanching, color-fixing, precluding the possibility of oxidation, and by the selection of suitable sorts. Discoloring of sour cherries is caused by oxidation of kera-cianin or of polyphenols. The former causes discoloration, the latter induces browning (Pai, 2003).

Time of harvest and the period between harvesting and processing (freezing) play a significant role in the development of taste and fragrant flavor. Expertly processed and packaged frozen fruits—stored under minimal fluctuation of temperature—do not undergo significant loss of taste and fragrant flavor. They have a quality nearly equivalent to fresh fruits as ingredients in bakery, confectionery, and dairy products (Beke, 2002; James, 2003).

Kernel of Nut Varieties, Used as Bakery and Confectionery Additives

Generally known as "shell-fruits" in Europe, China, and in the United States are nuts (*Juglans regia L.*), almonds (*Amygdalus communis*), and hazelnuts (*Corylus avelana L.*). Kernels of the shell-fruits are frequently used as valuable additives of bakery and confectionery products, desserts, muesli, and ice-cream. They can be used unbroken, coarse broken or as grist, sweetened cream (marzipan: fondant, flavored with almonds), and syrup. These additives have an intensive, pleasant and characteristic

Table 13.3. Chemical Composition of Soluble Solids of Shelled Nut Varieties, Used as Food Components

Name	Nutritional Value (kJ/100 g)	Protein (g/100 g)	Fat (g/100 g)	Carbohydrate (g/100 g)	Calium (mg/100 g)	Phosphorus (mg/100 g)	Iron (mg/100 g)
Nut (*Juglans regia*)	2601	14.4	62.5 ± 6.5	12.14	544	409	2.5
Almond (*Amygdalus communis*)	2318	18.72	54.1 ± 0.9	9.08	835	454	4.1
Hazel-nut (*Corylus avelana*)	2521	11.96	61.6 ± 1.5	11.36	636	333	3.8

Source: Souci, Fachmann, and Kraut (1989).

flavoring and decorating effect, and increase significantly the nutritional value of the product because of their high content of digestible solids. The composition of soluble solids of shelled nut varieties is shown in Table 13.3.

DAIRY PRODUCTS CONTAINING FRUIT COMPONENTS

In the food market, the most important dairy products containing fruit are fruit yogurts, milky desserts, and functional drinks.

Soluble solid content of these products generally does not exceed 20–30%. Sweetened berries (strawberries, raspberries) are very popular as dairy additives. These fruit preparations are usually either hot filled or aseptically packaged and consist of fruit, sweeteners, starch or other stabilizers, and flavorings (Somogyi et al., 1996).

FRUIT YOGURTS

Fruit yogurts are milk products fermented by using special cultures of lacto-bacteria. Their consistency may be jelly-like or fluid. They contain the fruit additives either homogenously distributed or in layers. Fruit preparations in dairy products call for pectin. Pectin provides the required rheological properties and assures the following:

- good dosing
- a regular fruit distribution in the container due to their yield point
- a homogenous mixing with the fermented milk product and a good shelf life of the finished product

In layered products, the special pectin has a stabilizing effect and keeps the fruit preparation separated from the yogurt without a jellying effect (Szakály, 2001; Anon, 2004f). In fruit yogurts, pectin provides the fruit preparations with a smooth and creamy structure, fruit specific flavor, and prevention of syneresis. Low-methoxy pectin or amid-pectin additives offer a solution to the problem (Imeson, 1998; Anon, 2004g). The same result can be achieved by adding high nutritive and healthy whey-proteins to the preparation (Anon, 2004a).

Average chemical composition of fruit yogurts can be seen on Table 13.4.

MILK-FRUIT DESSERTS

Milk-fruit desserts are semi-finished products, made of sugar, buffer substance, fruit, and water mixed with an equal amount of cold milk. The result is a gel which forms within minutes after mixing. Stable, jelly-like consistency, prevention of syneresis, and a characteristic fruit aroma are requirements for the fruit additives of milk-fruit desserts (Anon, 2004a).

Individually quick-frozen (IQF) berries and cherries are appropriate for premium dairy products such as refrigerated desserts. Some processors offer IQF fruits infused with a sugar solution to prevent them from freezing solid in frozen applications (Berry, 2001).

CHEESES WITH FRUIT ADDITIVES

Dried fruits are typically not used in dairy applications. However, finely chopped or diced versions can add a great deal of flavor to more unique applications such as cream cheese spreads, Cheddar cheese, and butter. Cheeses contain either finely pulped fruit

Table 13.4. Average Chemical Composition of Fruit Yogurts

Contents	Fruit Yogurts	Liquid Fruit Yogurts	Fermented Milk with Fruit Juice
Protein (g/100 g)	2.9–3.0	3.1	2.5
Carbohydrate (g/100 g)	12.1–13.8	11.6	11.0
Fat (g/100 g)	2.6–4.6	1.1	1.0
Energy (kJ/100 g)	353–454	290	266

Source: Anon (2004c).

additives uniformly dispersed or layers of fruit pieces, similar to fruit yogurts. Berry and cherry preparations are most suitable as cheese additives (Berry, 2001). Cherries have been used successfully in the production of spreads and cottage cheese sauce (Hendrick et al., 1969).

Raisins endow some types of cheeses with a characteristic taste and flavor (Anon, 1990).

Alcohol-Free Milky Beverages

Fruit components are generally added to milky beverages in the form of fruit juice concentrates. Carrageen or other hydrocolloid additives ensure homogeneity of the product. Milk beverages flavored with small quantities of banana puree have been developed mainly for children, as a product of very high nutritional value.

CONFECTIONERY PRODUCTS AND FROZEN SWEETS CONTAINING FRUIT COMPONENTS

The confectionery industry is using fresh and preserved fruits as filling for chocolates and candies and to decorate and enrich frozen desserts, fruit creams, and parfaits. Most frequently used fruit additives of sweets are—just like those of bakery products—cherries, sour cherries, and berries; but the confectionery industry is using other fruits as well.

Enzyme hydrolyzed *Jerusalem artichoke* juice concentrate can be successfully used as sweetener in dietary and low-energy fruit preparations, because 75–80% of its sugar content is fructose. Limited amount of fructose can be metabolized by humans in the absence of insulin—20–80 g daily are tolerated by diabetics. Its sweetening effect is 20–50% higher than that of sucrose, so it can guarantee the same sweet taste even at lower concentrations (Barta, 2000; Bray et al., 2004).

Chocolate Covered Fruits, Chocolates with Fruit Fillings

Cherries, strawberries, blueberries, apricots, whole apples dipped in caramel, diced and sweetened orange, or grapefruit peel can be processed to chocolate-coated products. Cherries, apricots should be pitted, apples have to be peeled; the cleaned fruits have to be slightly dried, dipped in caramel and/or milk-chocolate, covered with colored candy glaze, and packed in presentation tins (Anon, 2004d).

Very valuable and popular special products of the Hungarian confectionery industry are "sour cherries in rum," pitted cherries soaked in rum and coated in chocolate.

Confections contain high levels of sugar and have low value of water activity, many undesirable reactions are either slowed or stopped. Consequently, fruit juice pigment systems tend to be fairly stable in confections (Hegenbart, 1994).

DISTILLERY PRODUCTS OR OTHER ALCOHOLIC DRINKS CONTAINING FRUIT COMPONENTS

Prepared, cleaned, pitted, and chopped or whole fruits are valuable components of distillery products (liquors, fruit punches, etc.) as well.

Brine cherries are popular components of maraschino cocktails and desserts. The procedure of cherry brining is described by Somogyi et al. (1996). After brining, maraschino cherries may be placed in NaCl solution, to further remove skin discoloration (Wagenknecht and Van Buren, 1965; Anon, 1968).

Fruit cocktail requires bleaching, firming, and dyeing of the fruit. Pitted cherries are firmed by soaking in hot 0.5% $CaCl_2$; bleached, neutralized, and colored by treating with erythrosine dye; rinsed with water; and acidified by soaking in citric acid solution to prevent bleeding of the dye (Chandler, 1965; Woodroof and Luh, 1975). After the prescribed treatment, maraschino cherries are sugared in a 48% sugar syrup. Once sugared, the fruit is drained, and packed in a 45% sugar syrup with flavoring and enough citric acid to produce a final pH of 3.6. The maraschino pack is vacuum-sealed and pasteurized (Woodroof and Luh, 1975).

Attempts have been made to use cherry in the production of flavored beers (Peill, 1976). Production of this speciality necessitates the neutralization of the inherent malt liquor flavor. The liquor is then sweetened, colored, and flavored with cherry juice as necessary.

FROZEN DESSERTS, PARFAITS, AND ICE-CREAMS CONTAINING FRUIT COMPONENTS

In addition to yogurt, most dairy fruit preparations are used in frozen desserts. This category of preparation can be classified into three basic, discreet types:

1. Straight, no-particulate flavor systems added directly to the mix tank.
2. Variegates that do not contain particulates and are run through the variegating pump on the ice-cream freezing system.
3. Fruit feeder systems that have a higher percentage of particulates and are actually pumped into the ice-cream stream. Each of these categories has specific formulation requirements. Fruit feeder preparations must be thicker so that the juice does not drain out of the feeder into the ice-cream stream. They must be injected cleanly, without any excessive liquid run-off. Variegates can be either thick or thin, and they usually do not contain much particulate fruit. The variegate has to be evenly distributed, but there is no concern with fruit destruction, so the feeder construction is more simple.

A fruit preparation for a frozen application must have its freezing characteristics controlled so that it will not freeze solid in the ice-cream or frozen yogurt product. The fruit content, solid content, and the stabilization system all play a role here. The stabilizer system affects the size and rate of ice crystal growth. Finding the correct solid content to control the freezing point depression requires looking at the actual Brix of the fruit used. The interplay between the total fruit content and the product's final solid content must be balanced for an equilibrium of solids in the finished preparation. In a fruit preparation, you have your fruit and your matrix. These will come to an equilibrium of sugar concentration and that is important in controlling whether the fruit becomes icy or if it maintains its softness.

An emerging trend in the realm of frozen fruit preparations is the use of aseptic preserved fruits in place of traditional whole or individually quick-frozen fruit used in the fruit feeder. The aseptic fruits help to reduce concerns about microbial contamination of the mix by the fruit, and ensuring a matrix that firms to maintain texture at freezer temperatures (Hegenbart, 1994).

BAKED PRODUCTS CONTAINING FRUIT COMPONENTS

Baked products combined with fruit preparations are very popular, due to their fresh-fruity character.

Bakery fillings include a large variety of products. They include simple pectin-based fillings with little or no bake stability; high fruit low-solids pie fillings; homogenous, creamy preparations processed from fruit purees; to high-solids cookie fillings that must endure a severe heating. Pie filling contains about 30% solids, while a cookie filling has minimally 80% solid content. Fruit components for breakfast products (croissant, muesli) have a lower a_w value than pie fillings. Bakery fruit preparations come in two general types: those that are designed thermally stable and those designed to be cold-filled into the prebaked product. They must be easy to process, their organoleptic and technological quality should be protected during processing. Preparations, being baked together with the dough, must endure high temperature without quality losses. Special pectin additives are playing a significant role in ensuring quality requirements. Besides pectin, other stabilizers such as starch concentrates and micro-crystalline cellulose are also used (Kuntz, 1997).

The technologist must know the baking parameters, interactions between the filling and the cake or dough, and if the product is going to be open ended or totally encrusted (Hegenbart, 1994).

Requirements on Baking Stable Fruit Preparations

Fruit preparations which are baked together with the dough are produced as bucket, drum, or container goods based on individual requirements. The properties of the fruit preparations are influenced by processing technology, recipe parameters (Ca^{++}-ion concentration, type of sugar, pH value, and type of fruit), the used pectin, and/or the combination of these factors.

Requirements before Baking

The fruit preparation for bakery products is expected to be well processable and which does not change its texture after mechanical stressing. In order to obtain a product which is easily pumpable and dispensable after filling, the fruit preparations are stressed mechanically during cooling and after they are filled cold. In this way, the forming of an elastic gel is avoided and a non-gelled, creamy product with the required firmness results.

Requirements during Baking

Both the dough and the fruit preparation are exposed to a defined heat for a certain time during the baking process in the oven (e.g., breakfast cookies, "croissants" filled with jam or marmalade). Baking stability of the preparation means that it does not start boiling or melting. The fruit filling must have an excellent shape stability, but a limited melting on the surface will result in a nice gloss, giving the cake an attractive surface after cooling and firming. This is called "limited baking stability".

Requirements after Baking

After baking, the products are usually packaged and stored. As baking stable fruit preparations are mainly used for baked products with a long shelf life, it is especially important that the cakes keep their optimal quality over a longer period of time. Therefore, the fruit preparation is expected to be stable also after baking, and it may not release water or show any tendency to syneresis. Ideally a_w value of the fruit preparation complies additionally with that of the baked product. In contrast to jams and marmalades, a pre-gelled texture is desired in the production process of baking stable fruit preparations with low methyl-ester pectin. Pregelling by using special pectin or gellan gum additives can prevent precipitation of Ca-pectinate and syneresis and very good processing properties can be achieved.

Application of Different Fruits in Baking Stable Fruit Preparations

Due to their components, the different fruit pulps influence texture and baking stability of the finished products. Also, the following factors can influence the quality of the final product:

- soluble solids content
- the content of fibers (and/or dietary fibers)
- pH
- total acid and calcium content of the fruit

In order to produce products with constant properties, it may be necessary to consider the different kinds of fruit used in the recipe. A mixture of sugar syrup and berries, cherries, apricot, or apples is most frequently used as fruit filling of pies (Anon, 2004e).

Among dried fruits, raisins play a significant role as fruit components for high quality bakery products, milk loafs, and cookies (Anon, 1990).

Cherry pie filling is a typical extension of the canned, pitted cherry pack. Formulations for this pack vary, depending on the source starch (Somogyi et al., 1996).

Alternative and novel combinations have been developed, e.g., a reduced sugar cherry-apple filling (Wittstock et al., 1984).

HEAT TREATED AND FROZEN READY-MADE MEAL PRODUCTS CONTAINING FRUIT COMPONENTS

Sterilized or frozen ready-meals generally contain besides the meat component, potatoes, rice, vegetables, or pastry as trimmings, but the increasing demand for healthy nutrition makes fruit trimmings and sauces more and more popular. Rice trimming with fruit components or other fruit trimmings which can be preserved by heat or freezing are considered to characterize the early nutrition habits of the Asian people.

Sauces, Dressings, and Ready Meals Sterilized by Heat

Fruits have become popular ingredients in sauces and dressings. Good tasting products may be prepared from cherries, sour cherries, and particularly from apples of desirable varieties such as Gravenstein, Pippin, and Golden Delicious. Nearly all apple cultivars can be used for processing applesauce, but only a few are considered ideal. Quality attributes in raw apples that produce a high-quality finished product include high-sugar solids, high-acid content, aromatic, bright golden or white flesh, variable grain or texture, and sufficient water-holding capacity (LaBelle, 1971). The finished product can be flavored with spices or combined with other fruits.

The interaction between the stabilizer system and the fruit preparation during and after the heat treatment can cause severe problems. Starch may hydrolyze under low pH value, or may not cook out properly. The heat and high pH of the system can harm the fruit pigments, but acidity combined with heat can harm the starch. To fix the problem, the due course of cooking the sauce and of adding starch, acids, and fruit preparation must be settled exactly by pilot production (Hegenbart, 1994).

Frozen Ready-Made Meals and Fruit-Filled Pastas

Individual components of multi-component food are generally cooked or baked separately under parameters considering the aspects of food safety and storage. All components must be equivalent and delightful, having been prepared according to the "consuming proposal."

Some components are used fresh, others half-cooked or totally cooked. If the product contains fruit components as well, it must be taken into consideration that the cooking time of fruits is generally much shorter than that of vegetables and meats. Packaging and closing (practically vacuum-closing) of ready-made meals should protect organoleptic and nutritional values of the food. Cooling of the packed and hermetically closed product must begin instantly and the product must have been cooled under 10°C as soon as possible. The storage life of a food with multi-components is determined by the component with the fastest quality changes. However, normally, the fruit is not the most critical component (Beke, 2002; James, 2003).

Industrial production of frozen pastas and fruit-filled frozen pasta-meals have been developed in Hungary recently. Preferred products of this type are fruit-filled dumplings and jam pockets. The dough containing water-soluble additives (sugar), egg-yolk, or starch composed of amylo-pectin (rye-flour, potatoes), congeal and store well for some months at a temperature of -18°C or lower. Requirements related to the fruit filling are determined by the user on technological and economical basis. The prescriptions for the maturity of the fruit used for fruit fillings have technological causes, and those for fruit measurement an economical one.

The deep-frozen "plum-dumplings" processed from potato-dough are filled with whole plums or plum jam. The maturity of the fruit may be max. 70–80 %, the greatest diameter of the fruits is max. 26 mm. The true variety is required. Deep-frozen "apricot-dumplings" differ only in the filling from "plum-dumplings." The longitudinal diameter of the fruits with sweet kernel may be max. 30 mm. The maturity of the fruits should not exceed 90%. Only healthy apricots, being true to variety can be used for processing pasta-fillings (Almási, 1977).

CONCLUSIONS

Fruit preparations are important components of many products in nearly all sectors of the food industry, but as additives they play significantly the greatest role in the dairy and bakery industry.

Most important quality requirements of fruit additives from the viewpoint of the users are consistency, ease of processing, characteristic color, aroma, taste, and stability.

For this reason, the most popular raw materials for fruit preparations used as components in combined foods are cherries, sour cherries, and berries—alone or mixed with apple preparations.

REFERENCES

Almási, E. 1977. Food Preservation by Freezing (in Hungarian). Mezögazdasági Kiadó, Budapest, pp.154–155.

Anon. 1968. Oregon state develops cherry process. *Pacific fruit news*, 146(4196): 6.

Anon. 2004a. Pectins by H&F and Their Use in the Dairy Industry. http://www.herbstreith-fox.de.

Anon. 2004b. Jams, Jellies and Marmelade. Technical Brief. www.itdg.org.

Anon. 2004c. Se pa Se Yoghurt. www.lj-mlek.sl.

Anon. 2004d. Chocolate Covered Fruit. www.piglette.com/chocolatecoveredfruit.html.

Anon. 2004e. Fruit Preparations for Baked Products. http://www.herbstreith-fox.de.

Anon. 2004f. Dairying. The Columbia Encyclopedia. South Edition.

Anon. 2004g. Danisco Textural Ingredients. http://www.danisco.com/texturalingredients/pectin/lowester.asp.

Barta, J. 1993. Jerusalem artichoke as a multipurpose raw material for food products of high fructose or inulin content. In: Fuchs, A. (ed.). Inulin and Inulin-containing Crops. Studies in Plant Science. Elsevier Science Publishers, Amsterdam.

Barta, J. 2000. Composition, storage and processing of Jerusalem artichoke tubers (in Hungarian). In: Angeli, I., Barta, J., and Molnár L. (eds.). A Gyógyító Csicsóka. Mezögazda Kiadó, Budapest, pp. 77–147.

Barta, J., Horváth-Kerkai, E., Stéger-Máté, M., and Tátraházi R. 2002. Manufacture of fruit products containing biologically active substances (in Hungarian). *Konzervújság* 50(3): 68–70.

Bassoli, A. 2003. Sweeteners: Intensive. In: Caballero, B. (ed.). Encyclopedia of Food Sciences and Nutrition. Elsevier/Academic Press, Amsterdam, pp. 5688–5695.

Beke, Gy. (ed.). 2002. Handbook of the Cold-Storage Industry. 2. Technologies (in Hungarian). Mezögazda Kiadó, Budapest, pp. 451, 460.

Berry, D. 2001. The Latest and Greatest on Cherries and Berries. The use of fruit in dairy products. (Nov. 2001) Dairy Foods. www.findarticles.com.

Bray, G.A., Nielsen, S.J., and Popkin, B.M. 2004. The American Journal of Clinical Nutrition, 79(4): 537–543.

Cantor, S. (ed.). 2004. Juicing-Up Products With Fruit Based Ingredients. www.foodproductdesigne.com/archive/1996/1296AP.html.

Chandler, B.V. 1965. Fruit salad cherries. *Food Presv., Quart.*, 25(1): 16–18.

Codex Alimentarius Hungaricus (Magyar Élelmiszerkönyv). 1996. MÉ 1–3 79 /693.

Gardner, P.T., White, T.A.C., Mc Phail, D.E., Duthie, G.G. 2000. The relative contributions of Vitamin C, carotenoids and phenolics to the antioxidant potential of fruit juices. *Food chemistry* (68): 471–474.

Hegenbart, S. (ed.). 1994. Harvesting the benefits of fruit-containing ingredients. www.foodproduct designe.com/archive/1994/1294CS.html.

Hendrick, T., Markakasis, P., and Wagnitz, S. 1969. Cherries in spreads and cottage cheese sauce. *Journal of Dairy Science*, 52(12): 2057–2059.

Herrmann, K. 1966. Fruits, Preserved and Processed Fruit Products (in German). Verlag Paul Parey, Berlin-Hamburg, pp. 136, 140.

Imeson, A. 1998. Thickening and Gelling Agents for Food. Culinary Hospitality Industry. C.H.I.P.S. Publication Services, Texas, USA.

James, Ch. 2003. Freezing: Structural and flavor changes. In: Caballero, B. (ed.). Encyclopedia of Food Sciences and Nutrition. Elsevier/Academic Press, Amsterdam.

James, S. 2003. Chilled storage: Chemical and physical conditions. In: Caballero, B. (ed.). Encyclopedia of Food Sciences and Nutrition. Elsevier/Academic Press, Amsterdam.

Körmendy, I. and Török, Sz. 1990. Canning Technology of Raw Materials of Plant Origin (in Hungarian). University Press, University of Horticulture and Food Industry, Budapest, pp. 600–604, 607, 610.

Kuntz, L.A. 1997. Fruitfull Designs for Fillings and Preps. www.foodproductdesign.com/archive/1997/0197DE.html.

LaBelle, R.I. 1971. Heat and Calcium treatment for firming red and tart cherries in a hot-fill process. *Journal of Food Science*, 36(2): 323–326.

Lampe, J.W. 1999. Healt effects of vegetables and fruit assessing mechanism of action in human experimental studies. *The American Journal of Clinical Nutrition*, 70(9): 475–490.

Anon. 1990. The Performance of California Raisins in Selected Product Types. British Food Manufacturing Industry Research Organisation. Leatherhead Food RA. Leatherhead. U.K.

Mathlouthi, M. 2003. Packaging: Packaging of solids. In: Caballero, B. (ed.). Encyclopedia of Food Sciences and Nutrition. Elsevier/Academic Press, Amsterdam.

Pai, J.S. 2003. Freezing: Nutritional value of frzen food. In: Caballero, B. (ed.). Encyclopedia of Food Sciences and Nutrition. Elsevier/Academic Press, Amsterdam.

Paltrinieri, G., Figuerola, F., and Rojas, L. 1997. Technical Manual on Small Scale Processing of Fruits and Vegetables. FAO Regional Office for Latin America and the Caribian, Santiago, Chile.

Pátkai, Gy and Barta, J. 2000. Development of sweet taste in fruit processing (in Hungarian). *Konzervújság*, 48(2): 51–54.

Peill, A.J.C. 1976. Flavored beer: Will it succed in the U.S.? *Food Engineering International*, 1(5): 46–47. Science Publishers. Amsterdam. pp. 323–339.

Potter, N.N. and Hotchkiss, J.H. 2001. Food Science. Culinary Hospitality Industry. C.H.I.P.S. Publication Services, Texas, USA.

Smith, D.A. (2003) Jams and preserves: Methods of manufacture. In: Caballero, B. (ed.). Encyclopedia of Food Sciences and Nutrition. Elsevier/Academic Press, Amsterdam.

Somogyi, L.P., Barrett, D.M., and Hui, Y.H. (ed.). 1996. Processing Fruits: Science and Technology. Major Processed Products. Pa: Technomic

Publishing Co., Lancaster, Basel, pp. 83–86, 87, 89, 137, 149.

Souci, S.W., Fachmann, W., and Kraut, H. 1989. Food Composition and Nutrition Tables. Wissenschaftliche Verlagsgesellschaft mbH, Stuttgart, pp. 876, 893, 891, 921–923, 928, 930–932.

Stéger-Máté, M., Horváth-Kerkai, E., and Barta J. 2001. Evaluation of elder (Sambucus nigra) varieties and candidates for the canning industry. Results of the composition studies. *Journal of Horticultural Science*, 7(1): 102–107.

Stéger-Máté, M., Horváth-Kerkai, E., and Sipos, B. 2002. The examination of the composition and processing possibilities of the different currant varieties I. (in Hungarian). *Ásványvíz, üdítőital, gyümölcslé*, 3(1): 3–9.

Szakály, S. (ed.). 2001. Dairying (in Hungarian). Dinasztia Kiadó, Budapest. p. 211.

Szenes, M. (ed.). 1995. Fruit Preservation in Small-Scale Plant and Household (in Hungarian). Integra-Projekt Kft, Budapest, pp. 59, 106, 117–118.

Wagenknecht, A.G. and Van Buren, J.P. 1965. Preliminary observations on secondary oxydative bleaching of sulfited cherries. *Food technology*, 19(4): 658–661.

Wittstock, E., Neukirch, I., and Nobis, L. 1984. Fruit preparations with reduced sugar content—new filling for cakes and patisserie products. *Bäcker und Konditor*, 32(3): 75–76.

Woodroof, J.G. and Luh, B.S. 1975 Commercial Fruit Processing. The AVI Pub. Co. Inc., Westport.

14
Fruit Processing Plant

József Barta

This chapter discusses the development of the fruit processing plant that includes the technological procedures of the food processing and the steps of planning of the plant.

GENERAL CONDITIONS FOR ESTABLISHING A FRUIT PROCESSING PLANT

Before establishing and operating a fruit processing plant, it is necessary to evaluate all legal, engineering, and economic factors. *Legal conditions* are: laws related to the establishment and operation of a plant. *Engineering conditions* are: the architectural, environmental, electrical, technological, sanitary engineering and hygienic, work safety, and other features of the plant. These make it possible to manufacture quality products. *Economic condition* is: evaluating whether there is a demand for the products. The assessment and decisions based on these conditions are the functions of the marketing plan.

COURSE OF PLANNING

Planning, establishing, and developing a plant are complex engineering and economic tasks demanding the cooperation and coordination of various professionals. The planning scheme is summarized in Figure 14.1.

A technological planner with knowledge of processing a given product and conditions for establishing a plant is important. Pre-planning starts with an examination of demand for the product, whether demands will be long-lasting. A feasibility study should be done before a financial investment is made. It is necessary to examine the technological, manufacturing, financial, and environmental conditions required for the economic production of the given product followed by details of the conditions needed for building the plant. The basis of the technical planning is the food processing being applied. The long-lasting and safety management of the plant must be ensured by the work of the professional designers. Dimensions of the plant are based on the requirements for food processing. The engineer applies the principles of engineering to food processing. Professionals of the given branches make planning tasks of the settlement—starting with the establishment of the building through the construction of the sanitary

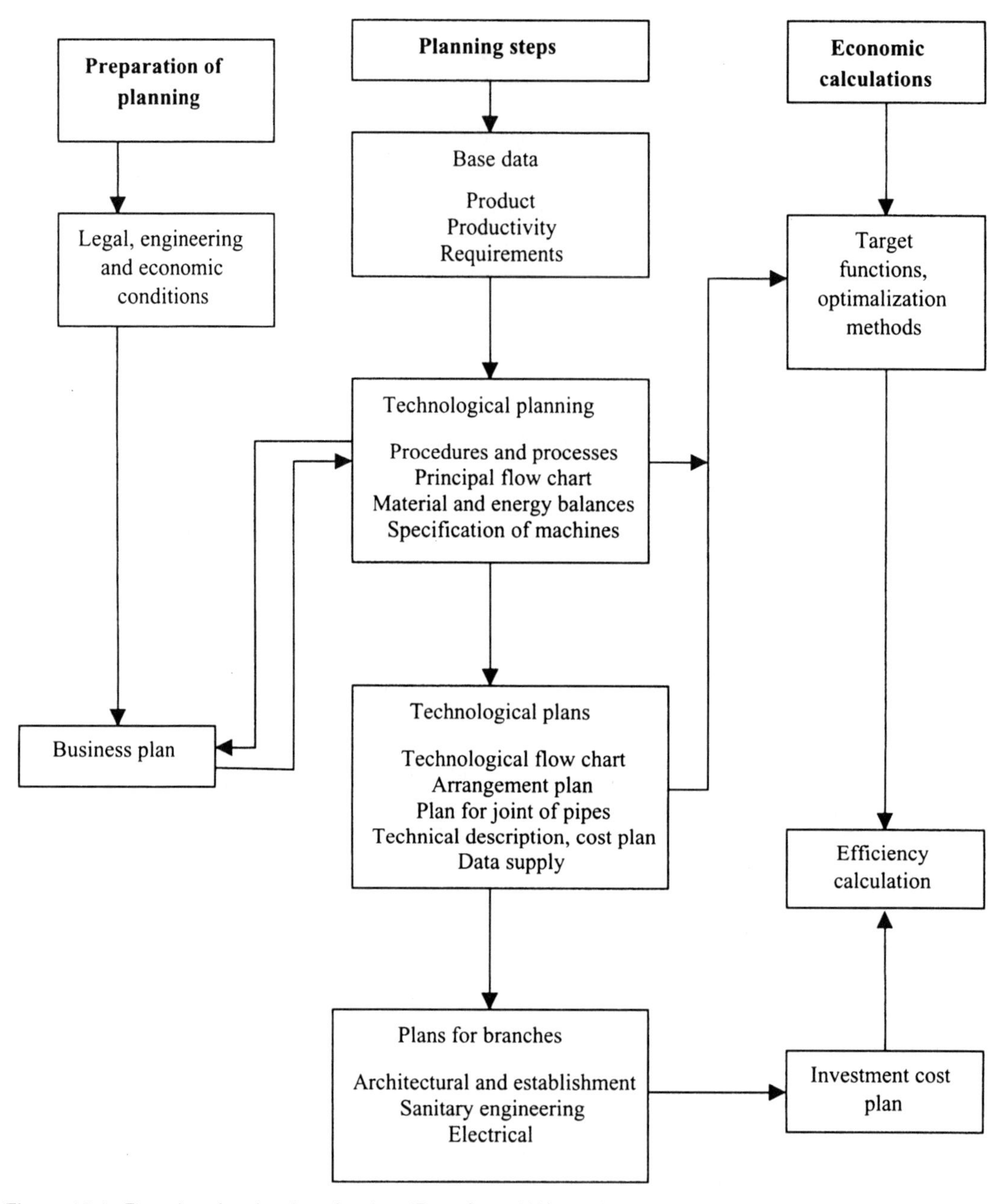

Figure 14.1. Flow chart for planning of a plant (Berszán and Várszegi, 2000).

engineering and electrical circuits through the transporting the final product. The work and tasks are determined by the technological plan. The most important document is the technological flow chart, showing detailed architectural, sanitary engineering, and electrical plans. The basis of technical planning is the steps in processing the final food product while assuring the safety of the employees and the customer. Investment as well as planning can be made within the scope of various professionals. The investor requires assurance that the plant is ready to use, while the task of the designer is to ensure that the plant meets all of the technological requirements (Bastian, 1996).

FEATURES OF THE TECHNOLOGICAL PLANNING OF FRUIT PROCESSING

A fruit processing plant transforms the raw materials—base, additives, and auxiliaries—into preserved finished products that are then packaged and stored. The products are transported to the customer through the trade. Some steps of this complex process are shown in Figure 14.2.

The planning of a fruit processing plant starts with answering the following questions:

- For the choice of products: what do we want to produce?
- For the quantity of the products: what amount of product do we want to produce?
- For the technological procedure: how do we want produce?

The product is the carrier of the following information and basic data for the plant calculations:

- The formula of the product: gives the amounts of raw materials needed for the given product. This is the basic data for calculation of the material balance.
- The shape and dimension of the product: influence the type of processing machines and dimensions of material handling and storing equipment and spaces.
- Quality of the product: the palatability or quality-preserving period, the composition, the organoleptic properties, the packaging and functional properties affect the technological process. Determination of these properties is done by evaluating customer demands and preferences (marketing plan).
- Product volume: determines the requirement for the productivity of the machinery, storage space, and the economics of production (Rouweler, 1991).

Processing technology consists of all the methods, procedures, and processes used to produce the finished product from raw materials. The processing technology is realized by a definite sequence of various procedures (e.g., sorting, classifying, washing, chopping, etc.) and of processes (e.g., heat treatment, parboiling, pre-cooling).

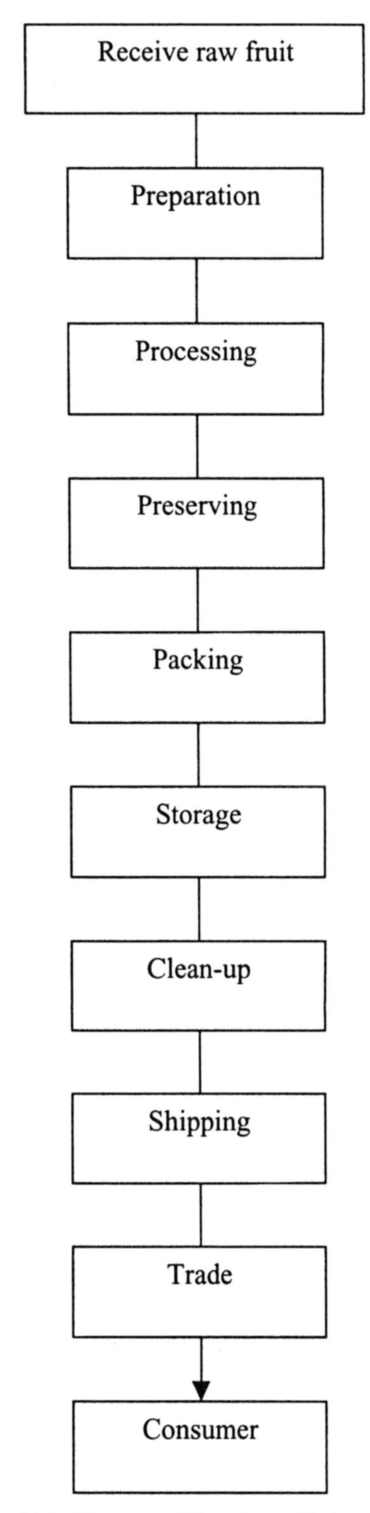

Figure 14.2. The general flow chart of fruit processing.

Fruit Processing Technology, Manufacturing Process

Planning of a plant based on technological design means the technical realization of manufacturing can be varied even with the same formula and the same processing sequence. Generally, the raw material is seasonal. The type of raw material, the harvesting time, and the quality determine the production time and the continuous or campaign feature of the manufacturing process. Fruits degrade quickly, therefore, it is necessary to process soon after harvesting. Processing is generally done in two steps. The first step is the so-called postharvest treatment that helps to save the freshness of the raw materials. Another choice can be the primary processing of the fruits, resulting in semi-finished products. At the second step, or secondary processing, extension of shelf life and producing end products are also possible (De Raaijl, 1991; Kushwaha et al., 1995; Shewfelt and Prussia, 1993). The form of primary processing is the production of paste, purée, as well as pulp and dried products. Secondary processing of, e.g., juices, syrups, beverages, jelly, jams, dairy- and confectionery fruit products, muesli, tea may be performed at a later period, depending on market demand. The theoretical scheme of fruit processing is shown in Figure 14.3.

In order to perform the same tasks, various procedures and processes can be used. For example, the concentration, i.e., the production of a paste, can be performed with the following processes: thermal evaporation, concentration by freezing, separation by a membrane (ultrafiltering or reversed osmosis), or application of thickening agents to improve the consistency of the product. Thus, the raw materials as well as the final food product can be preserved. In the following, we summarize how to evaluate each process and choose the most appropriate for manufacturing. In the course of process analysis, the following are examined:

- Operation alternatives
- The effect of the procedure on the material
- Productivity: incoming and outgoing materials, losses
- The realization mode of a procedure: manual or mechanized
- Machines used for the procedures
- Character of the human labor: numbers of the workers, craftsmanship
- Linkability of the procedures

Table 14.1 summarizes the incoming and outgoing data of the procedure analysis.

As a result of procedure analysis, features affecting quality and quantity characteristics of product manufacturing are determined. Features affecting the quality of the product are:

- Types of procedures and processes
- Working quality of the machines
- Labor requirements
- Monitoring system, degree of instrumentation
- Material handling
- Continuity, off-times

Features affecting the quantity of the product and productivity are:

- Dimension of the material
- Yield, residual products (off quality raw material and process waste)
- Productivity (capacity) demand
- Degree of mechanization, automation
- Continuity, off-times

Procedure alternatives: differences among the quality or quantity of products made with various procedures for the same task, marketing expectations, and preferences. All of these affect the economy of the production. The working quality of the machines means the effect on the product, for example, whether the given machine could meet quality demands. As an example, determination of the working quality of a chopper is done by analyzing the size and shape of the material (e.g., cube or grits). Analysis shows the sizes of the particles present in the product and the percentage distribution of the various size fractions. On the basis of this study, it can be decided which type and how much of the product can be produced by a given machine. At chopping, the productivity of the machine is also an important criterion. The task is that the peeled product needs to be chopped as soon as possible. The demand for workers and the number of them can be an obstacle as well as an advantageous feature. Obstacles can be, for example, at some dangerous processes or at ones that are critical from a hygienic point of view (packaging for consumers), or in procedures that need highly skilled workers and control of the operations (heat treatment, fermentation). For small factories, technologies with high-labor requirements are characteristic. Human labor might have an advantage over machines if labor is less expensive, and there is no obstacle as mentioned above (craftsmanship, hygiene, and worker safety).

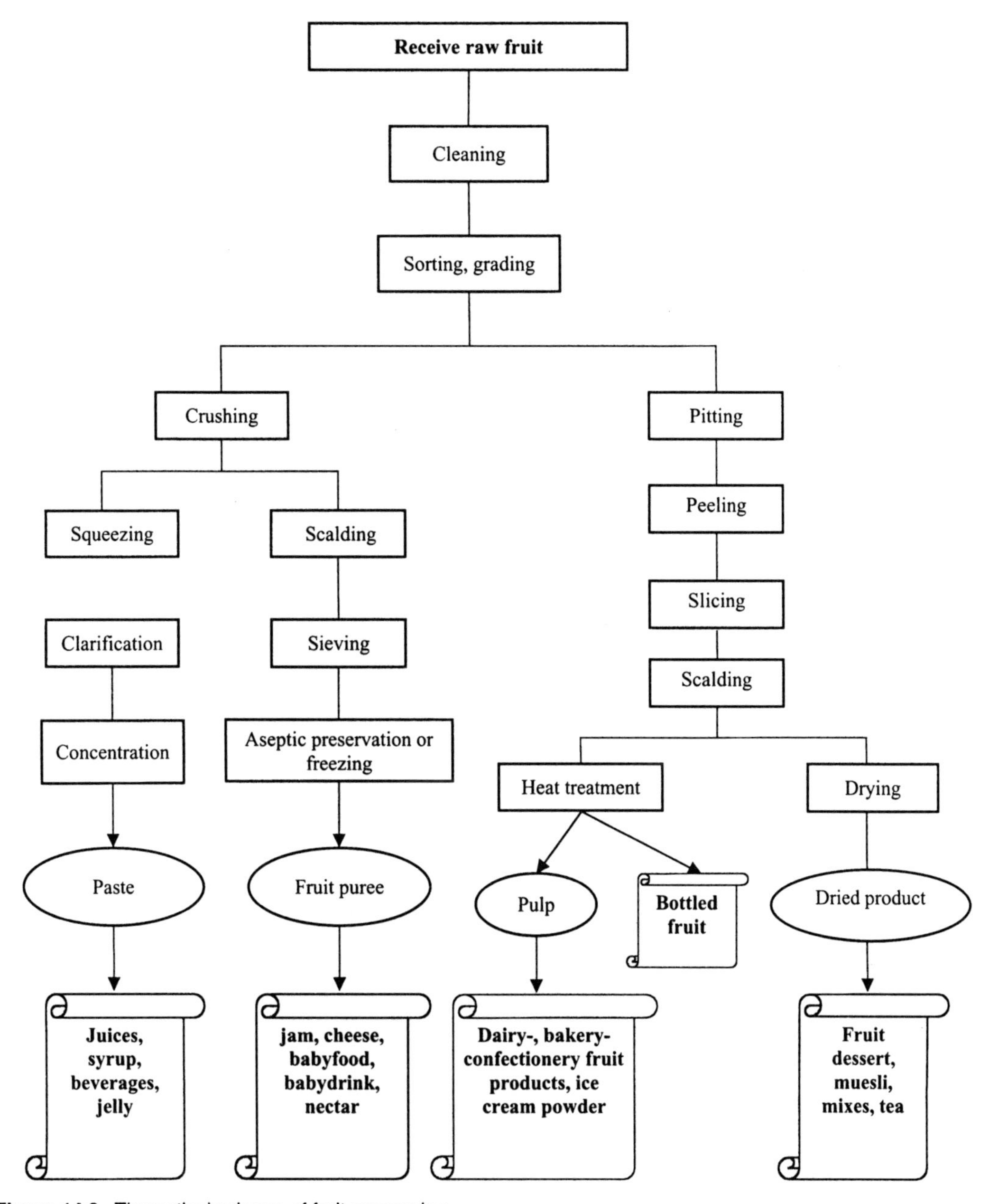

Figure 14.3. Theoretical scheme of fruit processing.

Product safety can be improved by automation of the machines and production. The quality of the product, especially the organoleptic properties, needs the craftsmanship and care of the boiler master in addition to the quality of the raw materials (Floros, 1992; Locin and Merson, 1979; Okos and Balint, 1990).

Material Transport of Fruit Processing

We summarize the calculation methods that are connected to the transport of raw materials used at processing of the product. These calculations make a

Table 14.1. Procedure Analysis

Incoming Data	Operation Analysis	Outgoing Data
Raw material: base, additives, and auxiliaries	Material balance, capacity calculation	Processed product (finished product), residual product, waste matter
Machines, tools: electric energy, water, steam, compressed air, vacuum, cleaning matters and disinfectants	Specification of machines, energy balance, hygiene, work safety	Vapor, drain, technological sewage, noise, pollution, odor
Workers	Demand for numbers of workers, craftsmanship, hygiene	Municipal sewage

basis for choosing the machines and determining the labor requirements.

The amount of materials yielded from the raw materials can be calculated on the basis of some normative. The yielding norm gives the types of materials yielded from the raw materials as well as their percentage ratios. Normative can be found in some technical books (Burits and Berki, 1974; Szenes and Oláh, 1991). The yield is a function of the quality of the raw material, the type of the product as well as the applied technology. In the following, some data are given for the usual or expected yields (Barta et al., 1990). For drying some fruits, the following yields can be taken into account for calculations:

- Dried fruit made of peeled apple: 8–10%
- Dried fruit made of unpeeled apple: 14–16%
- Dried plums: 30–35%
- Rose hips: 40–50%

The demand for raw material can be calculated similar to calculating yield data. The formula for the product, i.e., the composition gives the amount of material needed for manufacturing a given product. The compositional normative, generally, is the quantity of materials needed to produce 100 kg of a finished product including material losses (Baert, 1995). Yield is a phrase used to explain disintegration of raw materials during processing. The yield relates to how much finished product can be gained from given amount of base matter (raw materials), additives, and auxiliaries. While the requirement for additives is included in the normative, the demand for packaging material must be calculated for each type of packaging material. The quantities of flasks, metal, and cartons (e.g., daily, weekly demands) can be calculated on the basis of the product and the volume of the given packaging materials. Generally, it is necessary to account for some losses, e.g., glass breakage (Szenes and Oláh, 1991).

The material balance is the sum of incoming materials and outgoing products from the technological process, that is, the balance of the materials. Material balance can be determined by calculation. The simplified equation for material balance to the yield is

$$
\begin{aligned}
m_{\text{in}} &= \sum_{i=1}^{n} m_{i,\text{out}}, \\
m_{\text{in}} &= \sum_{i=1}^{n} \left(m_{\text{in}} \times x_{i,\text{out}}\right),
\end{aligned}
\tag{14.1}
$$

where

m_{in}—the mass or mass flow of the starting material (kg or kg/s)

$m_{\text{i,out}}$—the mass or mass flow of the ith product, auxiliary or residue (kg or kg/s)

$x_{\text{i,out}}$—yield ratio of the ith product, auxiliary or residue (kg/kg).

Material balance can be illustrated in three different formats:

- Narrative description of the calculation
- Table format
- Graphic presentation (Shankey diagram) (Berszán and Várszegi, 2000)

If a plant produces several types of products, first it is necessary to calculate the material balance for each type of product. Then, these can be summarized and the material balance for the plant can be calculated.

Capacity Calculation

Capacity means efficiency and cubic capacity. The capacity of the production equipment, tools, or plant determines the possibility of the economical performance (product volume) during production.

Therefore, capacity is the possibility. The ratio of product volume to capacity is the yield of capacity. The yield is less than 100% when expressed as a percentage. The capacity can be expressed as the product of the unit capacity (capacity norm) and of the production time (time base):

$$K = k_n I,$$

where k_n is the capacity norm (kg/h, m^3/h or piece/h) and I is the time base (h).

Capacity norm is related to production units (machine, production line, "cross section"). Time basis can be day, week, month, year, as well as season, depending on the character of production.

Capacity Norm

The capacity norm of a batch type machine can be determined as the ratio of the single cubic capacity to the total processing time (useful, i.e., running time + service (attendance) time):

$$k_{nl} = \frac{60 \times m_a}{t_c}, \qquad (14.2)$$

where m_a is the single cubic capacity of the machine, i.e., the mass of one charge (kg) and t_c is the processing time (cyclic time) (min).

Capacity norm of a continuously running machine can be calculated based on the technical parameters of the machine. The performance (capacity norm) of a chopper with pulley transport is given as

$$k_{n2} = a \times 60\frac{\pi}{4}\left(D^2 - d^2\right) n \times s \times p, \qquad (14.3)$$

where
a—the charging, filling parameter
n—the number of revolutions of the pulley (l/min)
D—outer diameter of the pulley (m)
s—pitch of the pulley (m)
d—core diameter of the pulley (m)
ρ—density of the material (kg/m^3).

The capacity norm of a rotary filling machine is determined by using a single-filling volume, the mass of transported material per cycle, and the number of revolutions. For liquids, the basis of the calculation is the performance of the pump carrying the liquid. Capacity norms are determined for the total production as well as production at a specific point in time (cross section). Cross section is a machine or tool by which one process of the total production is performed. The main cross-section process (e.g., heat treatment, juice production) is the process having primary importance for the total production. Extension in production of the main cross section is very expensive. A bottleneck in production is a machine or tool, with the lowest capacity. Often, the main cross section and the bottleneck are the same (Berszán and Várszegi, 2000).

Mechanization of Fruit Processing: Specifications of Machines

Production can be performed manually, by using hand-driven machines, or by automatic equipment. Material-handling devices linking technological tools are considered as machines. In the planning stage, technological tools that are well constructed and reliable are chosen for all steps in the production of a product. The modern technological arrangement (realization of the technological plan) is characterized by large number of various machines and devices. An important part of the machine specifications is the mobile material handling and auxiliary production tools. These are vehicles for transport, and containers, tanks, etc., used for transporting the materials. Specifications are necessary for the types, parameters, and quantities of machines and auxiliary devices needed for production of a specific product. While calculating the quantity of devices, the following must be take into account:

- The device is a part of the production.
- The device being used must be cleaned before the next run.
- A part of the device must be in maintenance.

All the features of a machine that affect production must be put into the specifications; the effect of a machine on production is as follows:

1. Effect on the quality of the product, such as the effect of a blade of a chopper on the quality of the material; the effect of heat treatment on the taste of the product, as well as the effect of automation on quality control.
2. Effect on hygiene, which is also a quality feature, e.g., how quickly can the equipment be disassembled, cleaned, sanitized, and reassembled.
3. Effect on the choice of products. How can the features of a machine be changed for varying product type?

The effect of a machine on the technological plan and on the production run is the following:

- Capacity
- Energy demand
- Emission effects on the environment

- Dimension and mass data
- Safety techniques
- Maintenance demand

The following data must be added to the specification of a machine in the technological plans (based on the above evaluations):

- Denomination of the machine (what is meant by denomination?)
- Type of machine (instrument, device)
- Number of pieces of the machine
- Engineering properties of the machine: capacity, dimensions
- Notes related to the machine: line production, individual machine, mounting instructions

Specifications of a machine are included in the following technological documents:

- Technological flow chart
- Technological arrangement plan
- Engineering arrangement and mounting plan
- Technological plan for joining of pipes (Ábrahám, 1980; Berszán and Várszegi, 2000)

PACKING MATERIALS FOR FRUIT PRODUCTS

By using various packaging materials and auxiliary products (tags, adhesives, clips, caps, and taps), several packages (bag, box, flask, pocket, keg, etc.) and parcels (customer, collecting and transferring packs) can be made. The function of packaging materials is to ensure the safety and quality of the product from production through transportation to the customer including storage, transporting, and selling the product and informing the customer (Fellows, 1988; Floros and Gnanasekharan, 1994). Table 14.2 summarizes the requirements for packing. Selection of the packaging material must be made using the knowledge of product characteristics and requirements for food safety and quality control.

Various packing materials are used for packaging processed fruits. However, plastics are applied in the largest quantity. Combined packaging materials were developed for better product safety and various forms of the product. Processing of foil combinations of paper, metal (Al), and plastics are used most frequently. For frozen fruits, the best quality is given by the polyethylene (PE)/paper, Al/PE combinations. For dehydrated (dried) fruits, the paper/Al/PE or paper/Al/ionomer is best, while for packing under vacuum or some protective gases, combinations of polyamide (PA)/PE or PETP (polyester)/Al/PA/PE combinations are best. Besides plastics, the following packaging materials are frequently applied: paper, carton, glass, and metal (Monspart-Sényi, 2000; Szenes and Oláh, 1991). Table 14.3 summarizes packaging materials and tools.

Table 14.2. Some Requirements for Packing for Saving the Quality of the Product

Environmental Effect	Form of Deterioration	Packing Requirements
Oxygen	Fatty acid oxidation, vitamin degradation, protein loss, coloring, material oxidation	Sealing against oxygen
Moisture	Loss of nutrients, organoleptic changes, lipid oxidation	Sealing against moisture
Light	Oxidation, rancidity, vitamin degradation, protein and amino acid transformation, coloring agent, oxide	Light tightness
Micro- and macro-organisms	Formation of faulty product, loss in nutrients and in quality, potential risk to infection	Sealed packing
Mechanical burdening (dropping down, pressure, vibration, attrition, coarse handling)	Organoleptic change, deterioration and other changes in the quality, susceptibility for deterioration due to changes of the sealing, formation of micro holes	Mechanical strength, tightness
Customer handling, faulty usage	Product loss, quality deterioration, loss in nutrients, organoleptic degradation	Mechanical strength, clear information

Source: Berszán and Várszegi (2000).

Table 14.3. Features of Some Packing Tools

Packing Tools	Features
Tools with rigid walls	
Metal boxes	Stiffness, acknowledged by customers, safety
Metal kegs, vessels	High capacity, low costs
Glasses	Strength, stiffness, fragility
Paper-based tools	
Paper–plastic–plastic cartoon	Environmental friendly
Pre-boarded cartoon	Easily formable, etc.
Semi-rigid plastic tools	
Heat-formed boxes, trays	High productivity, low-space requirement, easily formable, variability
Pre-mould boxes, trays	
Tools with elastic walls	
Blown plastic flasks	Sterile and prefabricated, no need for space for the packing tools
Bags, pockets, bag-in-box	
	Low costs, simple, mechanization, minimum space requirement for storage

Besides requirements for saving the quality of the product, the effects of the packing on the environment are also very important.

ENERGY REQUIREMENTS OF FRUIT PROCESSING

The basis of planning for energy needs for a given technological procedure including the energy needed by individual machines is the energy (heat) demand of the technology as well as the recovery possibilities of the heating energy.

ENERGY BALANCE OF FRUIT PROCESSING

The energy balance is the application of the law of conservation of energy. The mathematical formula can be described as follows:

$$\Sigma Q_{\mathrm{m\,in}} + Q = \Sigma Q_{\mathrm{m\,out}} + Q_{\mathrm{l}} \pm Q_{\mathrm{r}}$$

where
ΣQ_{min}—heat content of the incoming materials
$\Sigma Q_{\mathrm{m\,out}}$—heat content of the outgoing materials
Q_{l}—energy loss
Q_{r}—heat demand or heat production of the chemical reactions
Q—energy demand of the procedure.

The unit of all parameters is kJ or kW h. On the left side of the equation is energy coming into the procedure, while on the right side is energy outgoing from the procedure. The heat content of incoming and outgoing materials can be calculated as follows:

$$\Sigma Q_{\mathrm{m}} = m_{\mathrm{m}} c_{\mathrm{m}} T_{\mathrm{m}},$$

where
m_{m}—the mass of the material or its components or phases processed during the given period (kg)
c_{m}—specific heat of the material, or its components or phases (kJ/kg °C)
T_{m}—temperature of the material, or its components or phases (°C)

Losses are determined empirically or by calculation. In the simplified calculations, only those components that are the main part of the total energy demand are taken into account, and processes without losses are estimated. For example, at boiling temperatures, the heat energy needed for evaporation is taken into account and the energy demand required for warming to the boiling point is eliminated since the energy required for this step is of several orders of magnitude less than the energy needed for evaporation. The energy demand of the process is determined by summing the energy demands of all procedures and taking into account heat recovery and coincidental losses (Barta et al., 1991).

Energy Optimization of the Procedures

Optimization of operations, procedures, and processes can be done from various viewpoints; one is minimalization of the cost of applied energy. Generally, the energy cost (E_c) can be calculated as follows:

$$E_c = (Q/\eta)e_c,$$

where
Q—energy demand on the basis of the energy balance (kJ or kW h)
η—efficiency of the system
e_c—energy costs ($/kJ or $/kW h).
Calculation of the units: 1 kWh = 3600 kJ.

It can be seen from this expression that minimalization can be achieved by decreasing the amount of energy used or by using cheaper energy. An evaluation of the technical and engineering processes must be done together with a complex economical analysis of the product for a given quality requirement (Körmendy and Török, 1990; Locin and Merson, 1979; Okos and Balint, 1990).

Energy Consumption of Fruit Processing by Energy Carriers

In food processing technology, electrical energy is required by electrical engines of machines and tools. Planning this system is a responsibility of an electrical engineer, who must consider the energy demand on the machines and coincidentals.

Energy demands of machines and tools used for heating or cooling are given by brochures and technical books. Often, heat exchange occurs during the pre-heating or pre-cooling stages of fruit production. Heating energy (warm water or steam) used in heat exchangers is ensured from the supply system. For calculating the average energy demand of heat exchangers, the following relationships are used. For a batch process,

$$Q = cm\Delta T,$$

where
c—specific heat of the material (kJ/kg K)
m—mass of the charge (kg)
ΔT—temperature difference between the starting and ending temperatures of the product (°C or K).

For a continuous process,

$$Q = cm\Delta T t$$

where
m—mass flow of the material (kg/s)
t—process time (s).

Temperature and time are technological data obtained from the technical literature. The actual warming or cooling of a product to the required temperature (in a device) depends on circumstances for transmittance of heat. The calculation is as follows. For a batch type device, the energy uptake of pre-heating or pre-cooling is

$$Q = cm\Delta T = kA\Delta T_m t.$$

For a continuous device,

$$Q = cm\Delta T t = kA\Delta T_m t.$$

In both cases,
k—parameter of heat transmittance of a heat exchanger (kJ/s m^{2}K)
A—heat transferring surface of a heat exchanger (m^2)
ΔT_m—logarithmic average temperature of heat exchange or the difference between the average temperatures of the two media (heat carrier and the product) (°C)
t—process time (s).

By determining the various energy demands, it is possible to determine the quantities of the heat carriers as well as the temperatures and pressures. Planning requires determining the quantities of the heat carriers instead of the heat energy. Quantities of heat carriers (warm water, steam) can be calculated as follows. The mass flow of the warm water is

$$m_w = Q/(c_w\Delta T_w).$$

The mass flow of the steam is

$$m_g = Q/r,$$

where
c_w—specific heat of the water: c_w= 4.2 kJ/kg K
ΔT_w—temperature decrease of the heating water: $\Delta T_w \leq 20$°C
r—evaporation heat of the water vapor: $r \approx 2400$ kJ/kg (mean value due to the pressure dependence of it).

Energy Consumption of Fruit Processing

The users of energy in fruit processing are:

- Processing equipment
- Equipment for storage, transportation, and material handling

- Devices used by the informatics system of the enterprise
- Lighting, heating, and climatization of work place

Users mentioned above require various sources and carriers of energy. Energy sources are:

- Electrical energy
- Heat energy, fuel sources: natural gas, fuel oil, wood, coal
- Renewable energies:
    - Biological materials: wood, wood waste, sawdust, cuttings, biogas
    - Solar energy, energy of some thermal sources
    - Wind energy

In practice, the first two types of energy are important; however, from a regional point of view, some other sources can have importance, too. The quantities of the natural sources are not constant, there is uncertainty in their presence, and therefore sole application is not generally proposed. Energy carriers are:

- Heat energy: warm water, water vapor, dew, and steam
- Cool energy: recirculation (from a cooling tower) cooling water, natural cooling water, icy water, brine, glycol, liquid nitrogen, ice water, and dry ice
- Compressed air and vacuum.

Production of the above energy carriers can be done in the plant or obtained from public utilities.

DOCUMENTATION OF THE TECHNOLOGICAL PLANNING OF FRUIT PROCESSING

The technological calculations, the analysis of the material, machine, and labor demands and that of the information system, i.e., determination of the base data, make it possible to outline the plant, construct the arrangement drawings, and the technological plan. The flow chart shows the materials for production, the production procedures, and sometimes the machines and their connections. The principal connections of the systems, materials, and tools are shown without the accurate data of the spatial arrangement; however, the main directions of the technological process are indicated. The technological arrangement drawings show the connections of equipment and tools in the space and their arrangement. By means of the architectural and sectional drawings, the technological arrangement plan shows all the tools needed in the manufacturing of the product and those in direct contact with the raw material and product. The knowledge of the construction and dimensions of the plant are required for making the base drawings and cross sections. The basis for establishing the architectural plan is the technological plan. The apparent contradiction can be solved by the following:

1. The designer of the technological plan has architectural knowledge (the technological designer is generally a mechanical engineer or a food technologist).
2. The technological designer adjusts the preliminary plan with the architect.
3. In the course of the technological planning, further adjustments are required. During this time, compromises are made to achieve an optimum plan that has a criterion for product quality and operational efficiency (Burits and Berki, 1974; Mikus and Barta, 1998).

The general principle is that the architectural and other planning must meet the requirements of the technology. The technological plan consists of the following documentation:

- Plans for the main technological procedures and manufacturing of the food product
- Plans of the technological auxiliary procedures or sub-systems: processing of the subsidiary products, waste disposal, cleaning of sewage water, and a neutralizing system to prevent air pollution
- Plans of the service sub-systems:
    - Analytical, organoleptic, and microbiological laboratories
    - Personnel and tools of the hygiene department
    - Personnel facilities for plant employees including changing rooms, bathrooms, WC, break room, and dining room
    - Auxiliary (sub)-factories connected to the technology, e.g., box-making factory
    - Establishments of the vehicle stock carrying the raw material and the finished product
    - Maintenance shops.
- Forms of documentation are:
    - Technological flow chart
    - Arrangement plan for machines, technological arrangement plan

- Base drawings
- Sectional drawings
- List of the items (list of the machines or machine specifications)
- Plan for routing material transport and personnel traffic
- Plan for tubes, technical description
- Financial plan

According to the targets (permission plan, data supply), flow charts can be made in several variations. Designing of the arrangement plan for machines is more or less uniform. Flow charts can be done:

- For the authority that issues the necessary permits
- Flow chart made by the designer or producer of the technological process to present its offer
- For informing the technological designer and the employer
- For construction of the quality, safety, and process control systems

Technological flow charts are of three types depending on the depth of content:

- Principal schematics
- Block schematics
- Material transport schematic and Shankey diagram
- (Heat) Energy transport schematics and Shankey diagram
- Flow charts
- Drawings for joining pipes (Ábrahám, 1980; Berszán and Várszegi, 2000; Szenes and Oláh, 1991)

General references relating to fruit processing plant: Alzamora (2000), Arthey and Ashurst (2000), Dris (2003), and Thompson (2003).

FOOD SAFETY AND QUALITY

Food products must fit the dual requirements of food safety and food quality. Besides, there is a third requirement: a continuous supply of the required products in the quantity needed by the customers. Food safety means that the food is free from microbes and/or toxins causing foodborne illness and free from foreign matter dangerous to human health. In other words, the product must meet public health requirements for microbiological quality (Schlotke et al., 2000). The concept of food quality means that the product meets expectations for nutrient content and organoleptic characteristics (appearance, taste, and smell); that labeling accurately describes the product; and that packaging maintains

Table 14.4. Food Safety and Quality Features

Quality Features	Features of Quality Changes and Their Phenomena
Physical quality features	*Features of the physical quality changes*
Appearance: color, shine, smoothness	Fainting, browning, taint, wrinkling
Texture: fragile, fibrous, tough, crackling, crumbling	Softening, desiccation, hardening
	Melting, recrystallization, phase separation
Phase state: liquid/solid	Physical impurities: metal, glass, powder
Shortage of physical impurities	
Chemical quality features	*Features of changing of the chemical quality*
Protein and amino acid composition	Browning reactions
Sugar content, types of starch	Vitamin deterioration, coagulation of mineral salts
Nutrients, vitamins, mineral salts	Solvent residues, alien matters
Permanent composition	
Free from chemical impurities	
Enzymolitic quality features	*Features of changing of the enzymolitic quality*
Shortage of the rotting enzymes	Enzymolitic browning
Presence of the anti-rotting enzymes	Oxidation
Presence of starch-hydrolyzing enzymes	
Microbiological quality features	*Features of changes of the microbiological quality*
Shortage of toxin producers	Microbial growth and toxin production
Shortage of degradation causing	Degradation: smell bad and bad taste, degradation of nutrients

Source: Berszán and Várszegi (2000).

quality while meeting customer expectations. The quality of similar products can vary due to the variations in above-mentioned features (Balla and Binder, 2002; De Wit, 1991; Jen, 1989; Stauffer, 1994; Surak, 1992; Thorner and Manning, 1983). Food safety and quality are determined by several parameters. These quality features are classified into four groups in Table 14.4. This table contains also the features and the phenomena of quality changes. Some can be controlled organoleptically and provide information on the quality of the product.

SPECIFIC REFERENCES

T. Ábrahám. 1980. A betakarítástól a csomagolásig. A konzervgyártás műveleteinek gépei (From Harvesting to Packing. Machines of Procedures for Manufacturing Canned Products). Budapest, Hungary, Mezőgazdasági Kiadó (Agricultural Press).

L. Baert. 1995. Advanced Food Technology. Course Notebook. Belgium, Gent University.

Cs. Balla, I. Binder. 2002. Fagyasztott élelmiszerek tárolása (Storing of frozen foods). In: Beke, Gy, ed. Hűtőipari Kézikönyv 2. Technológiák (Handbook for Cooling Industry 2. Technologies). Budapest, Hungary, Mezőgazda Kiadó (Agronomist Press). pp. 417–476.

J. Barta, K. Vukov, B. Gion. 1990. Vízelvonás szárítással (Dehydration with drying). In: Körmendy, I, ed. Konzervtechnológia. Növényi Eredetű Nyersanyagok feldolgozásához (Canning Technology for Processing Vegetable Raw Materials). Budapest, Hungary, Budapest Univ. of Economic Sciences and Public Administration, Faculty of Food Science. pp. 482–521.

J. Barta, J. Farkas, K. Vukov, E. Zukál. 1991. A tartósító és állatitermékfeldolgozó iparágakban közös technológiák (Common Technologies in Conserving and Animal Product Processing Industries). Budapest, Hungary, Budapest Univ. of Economic Sciences and Public Administration, Faculty of Food Science. pp. 61–115.

E. D. Bastian. 1996. Technology of Food Processing. Course Outline. University of Minnesota.

G. Berszán, T. Várszegi. 2000. Agrárgazdasági élelmiszerelőállító üzem (Agricultural Food Processing Plant). Budapest, Hungary, Agroinform Kiadó (Agricultural Information Press) pp. 11–35.

O. Burits, F. Berki. 1974. Zöldség- és gyümölcsszárítás (Drying of Vegetables and Fruits). Budapest, Hungary, Mezőgazdasági Kiadó (Agricultural Press).

I. De Raaijl. 1991. Post-harvest Technology and Control. Course Notebook. Wageningen, The Netherlands, IAC.

J. C. De Wit. 1991. Introduction to Food Hygiene. Course Notebook. Wageningen, The Netherlands, IAC.

P. Fellows. 1988. Packaging. Food Processing Technology. Chichester, England, Ellis Norwood Ltd. pp. 421–447.

J. D. Floros. 1992. Optimization methods in food processing and engineering. In: Hui, Y. H., Wiley, R. C., eds. Reprinted from Encyclopedia of Food Science and Technology. New York, Wiley. pp. 1952–1965.

J. D. Floros, V. Gnanasekharan. 1994. Principles, technology and applications of destructive and nondestructive package integrity testing. Reprinted from Advances in Aseptic Processing Technology. London. Elsevier Sci. Publ.

J. J. Jen. 1989. Quality Factors of Fruits and Vegetables. Washington, DC, USA American Chemical Society.

I. Körmendy, Sz. Török. 1990. Konzervtechnológia. Növényi eredetű nyersanyagok feldolgozásához (Canning Technology for Processing Vegetable Raw Materials). Budapest, Hungary, Budapest Univ. of Economic Sciences and Public Administration, Faculty of Food Science. pp. 482–521.

L. Kushwaha, R. Serwatowski, R. Brook. 1995. Harvest and Postharvest Technologies for Fresh Fruits and Vegetables. Proceedings of the International Conference, Guanajuato, Mexico. St. Joseph, Michigan, A.S.A.E.

M. Locin, L. R. Merson. 1979. Food Engineering. Principles and Selected Applications. New York, USA Academic Press.

I. Mikus, J. Barta. 1998. Az Európai Unió agrárrendszere a gyakorlatban (Agrarian system of the United Europe). Budapest, Hungary, Szent István University, Faculty of Food Science. pp. 94–103.

J. Monspart-Sényi. 2000. 4.9. Packing medium, 4.10. Container and wrapping, 4.11. Food contact surface. In: A. Moeller, J. Ireland, ed. LANGUAL 2000 The Langual thesaurus, Luxemburg, Belgium (COST European cooperation in the field of scientific and technical research), pp. 302–309.

M. R. Okos, A. Balint. 1990. Simulation of Multiproduct Food Processing Operations Using Batches. Presented at ASAE/EPEI Food Processing Automation Conference. Lexinglott, KY, May 6–8.

J. Rouweler. 1991. Main Unit Operations in Food Processing. Course Notebook. Wageningen, The Netherlands, IGC.

F. Schlotke, W. Becker, J. Ireland, A. Moeller, M. L. Ovaskainen, J. Monspart-Sényi, I. Unwin. 2000. Eurofoods Recommendations for Food Composition Database Management and Data Interchange. Luxemburg, Belgium (COST European cooperation in the field of scientific and technical research). pp. 1–79.

L. R. Shewfelt, E. S. Prussia. 1993. Postharvest Handling: A Systems Approach. New York, USA Academic Press Inc.

J. E. Stauffer. 1994. Quality Assurance of Food. Trumbull, USA Food & Nutrition Press Inc.

J. G. Surak. 1992. The ISO 9000 standards. Establishing a foundation for quality. Food Technol. (November):74–80.

E. Szenes, M. Oláh. 1991. Konzervipari kézikönyv (Handbook for Canning Industry). Budapest, Hungary, Integra-Projekt Kft. pp. 363–369.

M. E. Thorner, P. B. Manning. 1983. Quality Control in Food Service. Werstport, CT, Avi Publishing Company Inc. pp. 58–232.

GENERAL REFERENCES

S. M. Alzamora. 2000. Minimally Processed Fruits and Vegetables: Fundamental Aspects and Applications. Aspen Publishers, New York, USA, 360 pp.

D. Arthey, P. R. Ashurst. 2000. Fruit Processing: Nutrition, Products, and Quality Management. Kluwer Academic Pub., ISBN No.: 0834217333.

R. Dris. 2003. Crop Management and Postharvest Handling of Horticultural Products: Fruits and Vegetables. Science Pub Inc., New Hampshire, USA, 422 pp.

A. K. Thompson. 2003. Fruit and Vegetables: Harvesting, Handling, and Storage. Blackwell Pub Professional, Oxford, UK, 496 pp.

15
Fruits: Sanitation and Safety

Sameer Al-Zenki and Husam Al-Omariah

INTRODUCTION

Fruits are raw agricultural products given a high level of priority by the government with respect to food safety and security. Recently, safety concerns have been raised on the potential contamination of fresh whole, cut, and minimally processed fruits and fruit juices with foodborne pathogens (FDA, 2001a). These reasons were justified in view of the fresh nature of the product, together with the various sources of contamination and the lack of an effective treatment to eliminate/control the growth of these pathogens once present on the fruit.

The key means to improve the safety of perishable products such as fruits is to minimize their initial microbiological contamination with foodborne pathogens at the preharvest stage. Fruits are produced in an environment that supports contamination. Bacterial contamination from the soil, irrigation and wash water, farm workers, rodents, and other animals are all vectors for the spread of pathogenic bacteria. Under temperature-abuse conditions, these microorganisms will multiply rapidly, resulting in the deterioration of the fruit and, in some cases, the transmission of foodborne diseases. Potential cross-contamination could also occur during postharvesting, handling, in-plant processing, and during transport and distribution (Brackett, 1992). Therefore, preventive measures must be adopted at the primary production stage, the processing plant, and transport and distribution chain to limit the level of contamination and growth of pathogenic microorganisms. This chapter deals with the characteristics of microbial pathogens associated with fruits, Hazard Analysis of Critical Control Points (HACCP), and sanitation measures necessary for lowering the incidences of food-poisoning outbreaks involving fresh fruits.

FOODBORNE OUTBREAKS ASSOCIATED WITH FRUITS

Traditionally, fruits have been perceived as wholesome, nutritious foods that pose no significant health risk. However, in recent years, there has been an increase in the incidence of foodborne diseases associated with the consumption of fruits. About 41% of foodborne illness cases in the United States were associated with the consumption of produce (fruits, vegetables, and juices). In addition, outbreaks related to contaminated produce (fruits, vegetables, and juices) in the United States doubled between 1973–1987 and 1988–1992 (Bean and Griffin, 1990; Griffiths, 2000). The reasons for this were attributed to the increase in consumption of fruits in response to the consumer desire for "fresh" foods, their nutritional value, and assumed beneficial health effects as well as the increase in importation of fruits and improved surveillance.

Food-poisoning outbreaks are defined as "the occurrence of two or more cases of a disease transmitted by a single food." There are two exceptions, botulism and chemical poisoning in which one case constitutes an outbreak (CDC, 1990). There is little published information on the natural occurrence of human pathogens in raw and fresh-cut fruits and fruit juices. This is because of the large number of variables to be considered that influence the contamination of fruits (procedure for sampling, location of source, number/size of samples, area or portion to be tested, etc.) and the lack of optimized methods for pathogen detection and isolation from fruits (FDA, 2001a). However, the survival of foodborne pathogens on fresh fruits has been clearly demonstrated. Pathogens will survive but not multiply on the uninjured fruit due to the protective natural barrier on the fruit that provides unsuitable conditions for pathogen multiplication. The survival/growth of foodborne pathogens is enhanced once the protective layer of the fruit is removed by physical damage (bruising, puncture), by plant pathogen degradation (bacteria, fungi), or by mechanical operation (peeling, cutting, slicing). Water, insects, or birds are the most common mode of transmission of pathogens resulting in the contamination of damaged or decayed sites on the rind and subsequently may infiltrate the fruits through these damaged sites. In addition, fruit can become contaminated if the warm intact fruit is submerged into cold, contaminated water. This pressure differential will draw the contaminated water into the fruit through the stem scar, calyx, or other openings in the fruit. Processing equipment and personnel also have been shown to cross-contaminate fruits during processing.

Fresh whole, cut, and minimally processed fruits and juices are frequently implicated in outbreaks of bacterial food poisoning. *Clostridium botulinum, Staphylococcus aureus, Campylobacter jejuni, Listeria monocytogenes, Bacillus cereus, Shigella, Salmonella,* and *Escherichia coli* have all been identified as a food safety concern for fruits. Some of these microorganisms have also been responsible for fruit-associated foodborne outbreaks (Beuchat, 1998).

C. botulinum is a Gram-positive spore-forming anaerobe that grows well in the absence of oxygen (Farber, 1989). The strains of *C. botulinum* can be classified according to the toxin they produce into seven groups designated A through G. *C. botulinum* types A, B, and E are responsible for botulism in humans. Botulism results from the consumption of a neurotoxin that is produced by this microorganism while growing on the food. However, the rate of botulism associated with fruits is very low compared to the foods of animal origin. This is due to the fact that most fruits are sold fresh and under aerobic conditions so that the development of anaerobic conditions suitable for outgrowth of clostridial spores and toxin production is not possible. More recently, safety concerns have been raised about the potential of *C. botulinum* to grow and produce toxin in fresh-cut packaged produce (Austin et al., 1998). The high rate of respiration of both the produce and the microorganisms could create anaerobic conditions for spore outgrowth and toxin production. In fact, Larson and Johnson (1999) have reported the presence of *C. botulinum* toxin in artificially inoculated fresh-cut cantaloupe and honey dew melons after 9 days of storage at 12°C, packaged under modified atmosphere packaging. Thus, it is necessary to package produce in a high oxygen/carbon dioxide permeable film and store the produce if possible at temperatures not less than 3°C.

St. aureus is another major food-poisoning microorganism frequently involved in fruit-associated foodborne illnesses. It is a Gram-positive, facultative anaerobic coccus with a temperature range from 7°C to 45°C (Farber, 1989). Food products become contaminated with *St. aureus* from noses, skin, or infected lesions of field workers, handlers, and processing personnel (Bryan, 1980). The microorganism is also capable of producing different heat-stable enterotoxins. These enterotoxins are high molecular weight proteins and are produced in the lag phase of the bacterial growth (Genigeorgis, 1989). High levels of *St. aureus* in fruits indicate poor hygiene and improper storage conditions. This pathogen is a problem in the postharvest processing of fruits because of the widespread direct hand contact involved in the process of sorting, packing, and repacking of fruits. Nevertheless, this microorganism is a poor competitor and does not grow well in the presence of other microorganisms normally present on fruits. Mukhopadhyay et al. (2002) reported the presence of coagulase-positive *St. aureus* in 17% of the sliced papaya samples examined by them. The authors also reported that despite the risk associated with the presence of this pathogen, the microbiological counts were not high enough for toxin production (80–100 CFU/g).

Campylobacter jejuni has been recognized as the most common cause of bacterial diarrheal disease in humans (Griffiths and Park, 1990). It is a Gram-negative, spiral-shaped, microaerophilic bacterium that belongs to the Spirillaceae family (Farber, 1989). Foods of animal origin are commonly the implicated sources of *Campylobacter* foodborne illness. Although there are no reported outbreaks associated with *Campylobacter* in fruits, this pathogen has been isolated from a variety of vegetables and has been shown to survive on sliced watermelon and papaya (Beuchat, 1998; Castillo and Escartin, 1994). Since the infectious dose of *Camplyobacter* is low (800 cells), intervention must be put in place to destroy any pathogenic bacteria that may be present on the raw fruit (Black et al., 1988).

L. monocytogenes is a Gram-positive, rod-shaped, motile bacterium isolated from various foods including cheese, milk, and fresh and processed meats. In accordance with other psychrotrophic bacteria, *L. monocytogenes* exhibits a wide temperature range from 1°C to 45°C with an optimum of 35–37°C (Farber, 1989). Listerosis is the disease contracted by the ingestion of contaminated food. Although the minimum infectious dose is still unknown, a high number of viable cells ($>10^6$ CFU/g) are required to cause illness in healthy adults (Farber, 1989). High-risk groups including pregnant women and their fetuses, the elderly, and immuno-compromised individuals will show clinical symptoms of listerosis at around 10^3–10^4 CFU/g (Farber, 1989). The importance of *Listeria* as a causative foodborne agent in fruit stems from the following: (i) the ubiquity of *Listeria* in the environment, (ii) the ability of the microorganism to grow in extended shelf-life refrigerated products at temperatures as low as 1°C, and (iii) high mortality rates as high as 30% (Farber and Losos, 1988). Great concerns have been expressed about fruits as major vehicles of transmission of listerosis to humans. Several reports have shown that low temperature is not a hurdle to *Listeria* growth. Conway et al. (2000) reported the growth of *L. monocytogenes* in apples incubated at 5°C. Furthermore, while the pH of most fruits and fruit juices (pH < 4) is in the range unsuitable for *Listeria* growth, some fruits such as watermelon (pH 5.2–5.7) and cantaloupe (pH 6.2–6.9) may serve as good substrates for *Listeria* growth. In fact, Penteado and Leitao (2004) have reported that low-acid fruits (melon, watermelon, and papaya) were good substrates for the growth of *L. monocytogenes* at various incubation temperatures of 10°C, 20°C, and 30°C.

B. cereus are rod-shaped, Gram-positive spore formers that are commonly present in soil, on the surface of plant material, and in many raw and processed foods. Two distinct types of illness have been attributed to the fruits contaminated with *B. cereus*, a diarrheal and an emetic toxin. Fruits incriminated in past outbreaks include orange juice (FDA, 2000).

Shigella spp. are rod-shaped, Gram-negative, facultative anaerobic bacteria that have been epidemiologically associated with foods or water contaminated with human feces. Shigellosis is the collective term used for the dysentery disease resulting from the infection by a species of *Shigella*. There are four species of *Shigella* known to cause bacillary dysentery. These include *S. dysenteriae*, *S. flexneri*, *S. boydii*, and *S. sonnei*. Shigellosis infections accounted for 3.1% of the total foodborne outbreaks reported from 1988 to 1992 in the United States with most of these outbreaks being attributed to the consumption of contaminated vegetables including parsley, lettuce, and vegetable salads (Bean et al., 1997; Davis et al., 1988; Dunn et al., 1995). Recently, the Food and Drug Administration (FDA) conducted a survey on 151 cantaloupe samples exported to the United States from nine countries. They reported an incidence of *Shigella* in these products to be around 2% (FDA, 2001b). This pathogen is of great concern to the fruit industry because of its low infectious dose (10 cells) (Wu et al., 2000).

Non-typhoid *Salmonella* species continue to be the most reported foodborne disease with incidence rates varying between 17.4 and 187 cases per 100,000 population and an estimated number of 2 million cases per year world wide (D'Aoust, 1991; Siliker, 1982). *Salmonella* is a genus of the family Enterobacteriaceae that includes other genera such as *E. coli*, *Shigella*, and *Proteus*. *Salmonella* species are high-temperature mesophiles that grow at temperatures from 5.2°C to 45°C (Farber, 1989). These microorganisms are ubiquitous in nature and have been isolated from several sources including irrigation water, sewage, rodents, and dust (Mackenzie and Bains, 1976; Oosterom, 1991). Recently, the EPA scientific advisory board panel specifically identified *Salmonella*, *L. monocytogenes*, and *E. coli* O157:H7 as pathogens of concern for fresh produce.

Furthermore, the panel recommended testing five outbreak strains in a cocktail for each pathogen when conducting challenge studies (EPA, 1997). Most outbreaks of human Salmonellosis have been recently linked to fresh-cut melons (Tauxe et al., 1997). These fruits are considered as potentially hazardous foods in the FDA Food Code because of their low acidity (pH 5.2–6.7) and high water activity (0.97–0.99). Thus, according to the FDA time/temperature requirements for potentially hazardous foods, melons should be prepared under sanitary conditions and cut melons should be kept at or below 7°C and displayed no longer than 4 h if they are not refrigerated (Golden et al., 1993). *Salmonella* have also been shown to adapt to reduced pH, and numerous outbreaks related to unpasteurized juices have also been reported (WHO, 1998).

E. coli are rod-shaped, Gram-negative, facultative anaerobic bacteria that are part of the microflora of the intestinal tract of humans and animals. Pathogenic *E. coli* are classified into five major groups based on their virulence properties, mechanism of pathogenicity, clinical symptoms, and antigenic characteristics (Beuchat, 1998). These five groups include the Enterotoxigenic *E. coli* (ETEC), Enteroinvasive *E. coli* (EIEC), Enteropathogenic *E. coli* (EPEC), Enterohemorrhagic *E. coli* (EHEC), and Enteroadherent *E. coli* (EAEC). The serotypes of major concern to fruits are the ETEC and EHEC. Sources and the mechanism of contamination are similar to those described for *Shigella* and *Salmonella*. *E. coli* O157:H7 have been commonly associated with outbreaks of unpasteurized apple juice (Besser et al., 1993). Although apple juice and cider are regarded as high acidic foods (pH 3–4), *E. coli* O157:H7 have shown to be acid-tolerant and survive in apple juice and cider for weeks in storage (Zhao et al., 1993). *E. coli* O157:H7 have also been shown to survive in stored apple juice for up to 24 days at refrigeration temperatures (Miller and Kasper, 1994). The major factor related to unpasteurized fruit juice safety is that these products receive no heat treatment. These numerous outbreaks have led the FDA to take action to improve the safety of fresh and processed juice by requiring processors to use an inactivation step that would result in a 5-log reduction of the target pathogen or else place a warning label on their products (FDA, 1998a).

Raw fruits may also harbor many non-bacterial pathogens of public health concerns including viruses and parasites. These pathogens are usually not part of the natural microflora of fresh produce and are primarily introduced by symptomatic food handlers via fecal-oral route. Secondary modes of parasite and viral transmission to produce include sewage, water-containing untreated sewage, and sludge from primary or secondary municipal water treatment facilities (Beuchat, 1998).

An increasing proportion of fruit-associated foodborne outbreaks have been recently traced to viruses. Hepatitis A virus and the small round structured viruses (SRSVs) are the most common viral contaminants of fresh produce. Historically, Hepatitis A virus has been recognized as the major cause of viral foodborne outbreaks. However, more recently, SRSVs have emerged as potential pathogen in fresh produce. Indeed, O'Brien et al. (2000) reported that SRSVs represented 20% of the total produce (vegetable, fruits, salads) associated outbreaks during 1992–1999 in England and Wales. Hepatitis A and SRSVs have been linked to outbreaks of strawberries and raspberries (Niu et al., 1992; CDC, 1996a).

Parasites including *Giardia*, *Cryptosporidium parvum*, and *Cyclospora cayetanensis* have been historically associated with waterborne outbreaks (Rose and Slifko, 1999). However, in recent years, these pathogens have also emerged as potential risk for foodborne illness and a concern to the fresh produce industry. The CDC reported that these parasites contributed to 2% of the foodborne outbreaks between 1988 and 1992 in the United States (CDC, 1996b). However, these numbers could be misleading due to the limitations of improved methods for recovery and detection of parasites in foods. Contamination of fruits with parasites occurs when crop irrigation water becomes contaminated with sewage or untreated wastewater or by runoff from non-point sources. Thurston-Enriquez et al. (2002) studied the occurrence of *Giardia* and *Cryptosporidium* in irrigation water in the United States and several Central American countries. They reported that 60% and 36% of irrigation water sampled tested positive for *Giardia* and *Cryptosporidium*, respectively. In a recent survey conducted in Norway, out of 475 fruit and vegetable samples examined, 6% were found to be positive for *Giardia* and *Cryptosporidium*. No *Cyclospora* were detected in any of the samples. Furthermore, water samples were also positive for those two parasites, clearly indicating that irrigation and wash water are the sources of parasite transmission to fruits (Robertson and Gjerde, 2001). Outbreaks linked to these protozoan parasites include fruit salad, apple cider, and raspberries (Anon, 2000; Tauxe et al., 1997; CDC, 1997).

FRUIT PROCESSING

Raw and minimally processed fruits could be produced in many different ways depending on the type of fruit and the end product. For such products, the techniques most commonly used include whole fruits, fresh-cut (peeled/cut/sliced into pieces), and in the form of unpasteurized juices. From a microbiological safety point of view, all critical processing factors should be controlled to minimize risks associated with the contamination of such products. In a fruit processing plant, the major sources of contamination are the plant's environment and the other fruits that transmit both spoilage and pathogenic bacteria. Thus, it is important to ensure the safety of all steps of the fruit processing operations at the plant, in addition to the cleanliness and hygiene of personnel working in the plant.

CRITICAL CONTROL POINTS OF FRUIT PROCESSING

Critical control points are defined as the location, practice, processing step, or procedure where control must be exercised to prevent one or more of the identified hazards (Baird Parker, 1987). This concept underlies the HACCP system. This system provides a structure for anticipating foodborne, microbiological, chemical, and physical hazards depending on their associated risk and on effective measures to prevent these hazards from occurring (Notermans et al., 1994). Since HACCP is regarded as an internationally accepted standard for food safety, for many years, many regulatory agencies in several countries have recognized the importance of implementing HACCP system in their operations. Various countries have already implemented mandatory HACCP in federally regulated food sectors such as meat and the seafood industry and now require that other products such as fruits be produced under HACCP system. Thus, the industry with the aid of government agencies has taken numerous steps to develop and implement HACCP programs for fruit processing by developing guides to improve the preharvest and postharvest operations. In fact, in 1998, the FDA in collaboration with the food industry has issued guidance for the industry "*Guide to Minimize Food Safety Hazards for Fresh Fruits and Vegetables*" (FDA, 1998b). This guide covers areas such as good agricultural practices (GAP), good manufacturing practices (GMP), water quality, manure management, workers training in field and facility, sanitation, and transport. The U.S. industry has also provided farm and processing personnel with codes of practices to improve the washing, handling, distribution, and storage of fruits such as the melon quality assurance program, the strawberry assurance program (US Department of Health, 1993; California Strawberry Commission, 1997), and the "*Food Safety Guidelines for the Fresh-Cut Produce Industry*" (IFPA, 2001). Internationally, the World Health Organization issued a report on surface decontamination of fruits and vegetables that can be eaten raw (WHO, 1998).

The first step toward developing a HACCP program for a process is to properly identify the risks associated with each operation in the process. Because microbial hazards are of major concern in the fruit processing industry, the identification of hazards involves a list of pathogenic bacteria associated with the particular product; the possible sources of contamination; and the growth, survival, and death of these microorganisms during processing (Pierson and Corlett, 1992). These CCP must be known before the implementation of additional hygienic practices in the production process. Because raw and fresh-cut fruits usually receive no or a small degree of processing (i.e., no harsh treatment), contamination of fruits still occurs at the different stages of processing. The most important stages to monitor include: (i) primary production and harvesting of the fruit, (ii) washing stage, (iii) mechanical operation, and (iv) storage, transport, and distribution.

The primary production and harvesting stage is a crucial stage for limiting the contamination of fruits in the processing plant. Fruits are frequently in contact with the soil that contains a considerable amount of microorganisms, dirt, dust, and feces. Foodborne pathogens such as *C. botulinum*, *B. cereus*, and *L. monocytogenes* are all normal inhabitants of the soil, whereas *Salmonella*, *Shigella*, *E. coli*, and *Campylobacter jejuni* could contaminate fruits through contact with the feces, sewage, or irrigation water. Insects, birds, and dust can act as vectors for human pathogens, especially on bruised or injured fruits (Beuchat, 1996). Houseflies have shown to harbor different pathogens (Olsen, 1998). Janisiewicz et al. (1999) have also demonstrated that fruit flies carrying a fluorescent-tagged non-pathogenic strain of *E. coli* O157:H7 transmitted the microorganism to apple wounds after 48-h exposure of the apples to the flies. Thus, ensuring the safety of the fruits at the preharvest stage is a very crucial stage for limiting the presence of pathogenic bacteria on fruits. Once present, these pathogenic microorganisms can be established in the plant and serve as a source of cross-contamination for incoming products that enter the processing line.

SORTING AND FIRST WASH INSPECTION

Following harvesting, the preparation of fruits for processing involves reducing and/or eliminating external contamination by visual inspection and sorting of the incoming fruits (color, size, maturity), the removal of decayed and/or injured fruits, and packing into containers. Direct hand contact is a common practice during inspection and sorting. In a 1999 survey of production in the United States, 93% of farms that grew fruits harvested fruits by hands (USDA, 2001). Furthermore, about 50% of fruit packers only require their employees to wear gloves (USDA, 2001). Hand contact of fruits is a problem due to the potential risk of cross-contamination from infected food handlers. Thus, it is necessary to ensure the cleanliness of workers' hands by implementing proper hand-washing techniques or the wearing of gloves.

Containers used to pack whole fruits from the field to the packinghouse and/or processing plant are another sources of microbial contamination. Fruits are usually packed using one container (bags, bins, wooden boxes, buckets, etc.) or exposed to a variety of containers during the postharvest operation. Most of these containers are stored for a long period of time prior to use and are often reused and difficult to clean. Thus, it is necessary to ensure that containers are washed and disinfected regularly to remove gross foreign residue, bird droppings, and rodent nests.

Initial washing of fruits with clean potable water is the next step to remove plant debris, pesticide residues, decomposed products, other extraneous matter, and bacteria. Washing in water will also remove the field heat, thereby maintaining the quality of the fruit and its shelf life. Washing of fruits such as strawberries and grapes will impair their quality; therefore, other treatments such as dry cleaning, brushing air blowers, and vacuum have been proposed for their decontamination. For most fruits, the selection of wash equipment depends on the characteristic of the fruit commodity. Soft fruits are washed on conveyer belts using water sprayers. Solid fruits are washed in flume water, while root crops are washed using oscillating brushes. In all cases, the water used for washing should be replaced daily to prevent the build-up of organic material and prevent cross-contamination to other fruits.

The washing stage presents the earliest opportunities for cross-contamination. This is typically the case for fresh-cut and unpasteurized fruit juices (apple cider, orange juice) because external contamination of the skin would be spread to the edible part of the fruit during processing or to other processing equipment. Various reports have demonstrated wash water as the source of pathogen contamination in minimally processed fruits and fruit juices. An outbreak of mangoes was associated with an outbreak containing *Salmonella newport* from wash water during postharvest handling (CDC, 2000). Similarly, orange juice was linked to *Salmonella typhimurium* originated from wash water that contaminated the peel (Bates, 1999). Furthermore, the temperature of the wash water plays an important role in the internalization of pathogens. A common practice of the fruit industry is to wash fruits with cold water to reduce the respiration rate of the fruit and maintain its quality. Merker et al. (1999) reported that potential internalization of pathogens from contaminated water in intact oranges and grapefruit most often occurred when the warm fruits were placed into cold water, so that the resulting pressure differential favors the uptake and internalization of the pathogen into the fruit. Similar results were observed by Buchanan et al. (1999) who reported the internalization of *E. coli* O157:H7 when warm apples were dipped into cold peptone water. Thus, it is recommended that prewashing be conducted at temperatures 10°C above the temperature of the fruit. One method to do that is to use an initial air-cooling step prior to washing to minimize the temperature differential between the fruit and the wash water. Subsequent washing should then be conducted in cold water to remove field heat and maintain quality (FSAI, 2001). Therefore, the use of sanitary potable pathogen-free water for washing of the rind or surface of the fruit before processing is a critical point in juice processing.

SANITIZERS IN FRUIT WASHING

Various forms of washing techniques have been adopted by the industry; these include the use of sprays to wash off the microorganisms off the fruit and prevent the risk of recontamination. Another approach involves the use of a double wash technique that involves a preliminary wash treatment to remove field heat, excess soil, and dirt followed by another more efficient washing/dipping that includes water containing an approved sanitizer for dipping/rinsing application. These sanitizers will only prevent the wash water from being a cross-contamination step in fruit processing. In fact, several researchers have reported that sanitizer washing may enhance the growth of pathogenic microorganisms on fruits, particularly when these products are temperature abused or mishandled after disinfection (Bennik et al., 1996). This is mainly because pathogens hide in cracks and crevices of the fruit and are often protected while the competing epiphytic bacteria are more accessible to sanitizers.

Chlorine-based sanitizers, especially sodium hypochlorite, are the most commonly used chemical sanitizers allowed for the disinfection of fruits (Beuchat et al., 1998). Sodium hypochlorite is allowed at a maximum concentration of 0.2% (2000 ppm) in wash water. However, water containing 50–2000 ppm of chlorine is widely used to sanitize fruits. For example, a concentration of 50–200 ppm of available chlorine is usually used for pome fruits, while a higher concentration of 50–1000 ppm is used for melons and watermelons (Soliva-Fortuny and Martin-Belloso, 2003). Residual chlorine on the fruit surface should then be rinsed by potable water after the chlorination treatment and only a residual concentration of 2–7 ppm of chlorine is accepted in the washed fruits.

The effectiveness of most chlorine-based sanitizers is influenced by several factors such as pH, temperature, exposure time, type of pathogen, and the surface morphology of the fruit (FDA, 2001a). Hypochlorous acid (HOCl) is formed when these chlorine compounds are mixed with water. The antimicrobial action of chlorine-based sanitizers is due to the formation of hypochlorous acid (HOCl) in solution, which is a measure of the available chlorine responsible for the inactivation of microorganisms. At pH 5, nearly all of chlorine is in the form of HOCl; however, at pH 7, 75% of the chlorine is in the form of HOCl. Therefore, the wash water should always be kept at pH 6.0–7.5 for optimum chlorine effect. Furthermore, chlorine-based sanitizers are more effective at low temperatures. Maximum solubility of chlorine in water is achieved at low temperature (~4°C). However, to prevent microbial infiltration due to temperature-generated pressure differential, the process water is always kept 10°C higher than that of the fruit. The exposure time is mainly dependent on the product. For some products, long exposure time will have a dramatic impact on the organoleptic quality of the fruit product. Hypochlorites also have different effects on different microorganisms. For example, *L. monocytogenes* are more resistant to chlorine than *E. coli*, *Shigella*, and *Salmonella* (Beuchat, 1998). Furthermore, *L. monocytogenes* will grow better on post-chlorine disinfected produce compared to non-disinfected water-rinsed produce (Bennik et al., 1996). The surface of the fruit also plays a major role on the effectiveness of the chlorine-based sanitizers. The presence of crevices, creases, pockets, and natural openings in the skin will ultimately reduce the accessibility of chlorine to reach the target pathogens. Pao et al. (1999) reported that immersing inoculated orange in hot water using 200 ppm chlorine was effective at reducing *E. coli* population by more than 2 log CFU/cm^2 on the surface of the fruit. However, only 1-log reduction was achieved on the stem scar. Du et al. (2002) reported that treatment with 4 ppm ClO_2 gas resulted in a 5.5-log reduction in *L. monocytogenes* on the skin surface of apples and a 3-log reduction on the stem and calyx cavities. Furthermore, the natural waxy cuticles, which are a barrier to microbial presence on the surface, removed after brushing of hardy fruits are usually replaced by commercial waxes after washing. Kenney et al. (2001) have reported that microorganisms can become enmeshed in the waxy cuticle making their removal more difficult and thus reducing the effectiveness of chlorine.

Sanitation of fruits with chlorine has shown to have a limited effect at reducing microbial loads. Produce may contain up to 10^6 CFU/g and only 1–2-log reduction is achieved by washing with chlorine (Farber et al., 1998). Furthermore, parasites and viruses generally exhibit higher resistant to chlorine than bacteria. Several decontamination procedures have been introduced including hot water sanitization, rinsing with organic acids, hydrogen peroxide, irradiation, and ultraviolet radiation (Beuchat, 1998; Yuan et al., 2004).

Hot water sanitization has been investigated for various fruits including apples, cherries, lemon, mango, melons, and pears to control insect and plant pathogens (FDA, 2001a, b). This treatment has been mainly proposed for fresh-cut fruits and juices where the outer surface of the fruit is removed due to the unacceptable sensory impact of hot water on intact fruits (Pao and Davis, 1999). Disadvantages of this treatment also include the FDA rule 21 CFR part 101.95 (CFR, 2000a) that considers thermally treated produce not "Fresh".

Organic acids have been one of the most effective treatment methods for the inhibition of pathogens.

These acids have been used successfully in spray washing, rinsing, and dipping application for the decontamination of meat/poultry carcasses. These acids include acetic acid, malic acid, sorbic acid, lactic acid, and citric acid. These acids are "generally regarded as safe" (GRAS) food additives, and are present naturally in various fruits. These acids act by lowering the pH of the food. In their undissociated form, the acids penetrate the bacterial cell, releasing hydrogen ions, and thus eliminating the proton gradient across the cell membrane. Therefore, they are more effective at acidic pH levels. Several reports have shown the inhibitory effect of organic acids on various pathogens present on fruits. Treatment of citric acid reduced populations of *S. typhi* inoculated onto cubes of papaya and jacana (Fernandez Escartin et al., 1989). Furthermore, the use of acetic acid (5% for 2 min) resulted in more than 3-log reduction of *E. coli* O157:H7 inoculated onto the surface of apples (Wright et al., 2000). Similarly, Park and Beuchat (1999) found that a concentration of 40–80 ppm of peracetic acid significantly reduced *Salmonella* and *E. coli* O157:H7 inoculated onto cantaloupe and honeydew melons.

Hydrogen peroxide has also been recommended as a replacement to chlorine (Sapers and Simmons, 1998). The use of hydrogen peroxide has been recommended for sanitation of fresh-cut melon and cantaloupe fruits. Residual hydrogen peroxide is then removed by the action of endogenous catalase or by rinsing with potable water (Soliva Fortuny and Martin-Belloso, 2003). Unfortunately, the application of hydrogen peroxide have shown to have negative effect on strawberries, raspberries, and black berries by causing bleaching of the anthocyanin pigments.

Treatment of fruits with low doses of ionized radiations has been very effective in reducing the number of foodborne pathogens and in extending its shelf life while maintaining the products' nutritive and sensory qualities (Thayer and Reykowski, 1999). Food irradiation is currently approved for certain plant and animal products in over 30 countries. In the United States, as an example, products cleared for irradiation include wheat and wheat flour, potatoes, spices and dry vegetable seasonings, dry or dehydrated enzyme preparations, shell eggs, meat, poultry, pork carcasses, and fresh fruits (Morehouse, 2001). This process involves exposing the food to an energy source in the form of gamma rays, X-rays, or a beam of high-energy electrons. These rays penetrate the product, damaging the genetic material of all living cells including bacteria, so they cannot survive or multiply (Morehouse, 2001).

Irradiation at doses up to one kilo gray (kGy) has been shown to be sufficient to eliminate most foodborne pathogens found on fruits (Howard and Gonzalez, 2001). However, higher doses (> 1 kGy) required to inactivate spores, viruses, and parasites caused sensory defects and/or accelerated senescence (Kader, 1986). Only strawberries have shown to withstand higher doses of irradiation and maintain their quality attributes. However, fruits treated at doses higher than 1 kGy cannot be termed as "fresh" according to the Code of Federal Regulation (CFR, 2000a). Although this technology produces a microbiologically safe product, consumers' objection to irradiated foods may limit the commercial application of this technology and sale of irradiated fruits.

In addition to the aforementioned antimicrobial agents, novel treatments have been tested and in some instances approved for dipping application. These include electrolyzed water, ozone, acidified sodium chlorite (ASC), ultraviolet energy, and natural aroma compounds.

Recently, electrolyzed oxidizing (EO) water has emerged as a potential process proposed as an alternative to chlorination to eliminate or substantially reduce bacterial population. EO water is prepared by electrolysis of a dilute salt solution (NaCl) in an electrolysis chamber to produce an acidic fraction and a basic fraction. The acidic fraction of the EO water has shown to be effective in inactivating various foodborne pathogens including *Salmonella*, *L. monocytogenes*, and *E. coli* O157:H7 (Venkitanarayanan et al., 1999; Kim et al., 2000). The antimicrobial effectiveness of acidic EO water is due to its high oxidative-reduction potential (1100 mV), its low pH (pH 2.7), and its production of hypochlorous acid. Information on the utilization of EO as a disinfectant for fruits is limited. EO water has several advantages over other processes such as (i) it requires only pure water and sodium chloride to produce EO water, no adverse hazardous chemicals required, (ii) is produced onsite and on demand with no dilution from concentrated chemicals, and (iii) has less potential hazard to workers due to lack of need to handle concentrated chemicals for microbial inactivation (Park et al., 2002).

Ozone is a bluish water-soluble gas that has long been used to disinfect, detoxify, and deodorize water in water-treatment plants. Ozone is initially generated by passing air or oxygen through a small chamber charged with high-voltage electricity that converts oxygen (O_2) into ozone (O_3). The water-soluble gas is then bubbled into water, pressurized, and sprayed. Ozone has shown to be an excellent alternative to chlorinated water for materials, equipments, and facilitate cleaning in various process operations because of its superior oxidative potential and low toxicity. Furthermore, since ozone is produced onsite, it requires no storage area (Beuchat, 1998).

The application of ozone for fruit processing is yet another process that is being reevaluated. Recently, an expert panel in the United States affirmed the GRAS status of ozone to be used as a disinfectant or sanitizer in food processing plants (Kim et al., 1999). Bacteria, mold, yeast, and viruses all have been shown to be sensitive to ozone. The biocidal effect of ozone is caused by its high oxidative potential. The antimicrobial effect of ozone is a result of a number of factors including changes in cellular morphology, genetic mechanism, and biochemical reactions (Kim et al., 1999). Several researchers have examined the use of ozone to inactivate microorganisms and to extend the shelf life of various food products. Ozonated water at a concentration of 20 ppm was effective against *Salmonella typhimurium*, *St. aureus*, *L. monocytogenes*, and *Yersinia enterocolitica* (Restaino et al., 1995). Ozone treatment has been effective for shelf-life extension of oranges, grapes, raspberries, apples, and pears (Beuchat, 1998). Dipping apples and strawberries into ozonated water solution (3 ppm) decreased the number of viable *E. coli* O157:H7 and *L. monocytogenes* cells by about 5.6 $\log_{10}$ CFU/g without adversely affecting the sensory qualities of the inoculated apples and strawberries (Rodgers et al., 2004). Disadvantages of ozone use in fruit-processing plants include its corrosiveness and concerns regarding workers safety (Howard and Gonzalez, 2001).

ASC has long been used as a sterilant for non-porous surfaces and devices in the medical and pharmaceutical industries, as a disinfectant in automobiles air conditioning systems and as a skin antiseptic in the dairy industry (Kemp et al., 2000). ASC is prepared by mixing an aqueous solution of sodium chlorite with a GRAS acid, usually citric or phosphoric acid. ASC has been approved by the FDA as a pretreatment for raw agricultural commodities or after mechanical operation. The FDA has also approved its use as a dip or spray at a concentration between 500 and 1200 ppm (CFR, 2000c). The antimicrobial action of ASC is due to sodium chlorite acidification which when applied to organic matter will produce oxycholorous antimicrobial intermediates (Gordon et al., 1972). These intermediates cause a disruption of oxidative bonds on the cell membrane surface of bacteria (Kross, 1984). Park and Beuchat (1999) studied the effect of sodium chlorite on cantaloupe, honeydew melon, and asparagus spears artificially inoculated with *Salmonella*. They reported approximately 3-log reduction following a 5-s exposure to 1200 ppm of ACS.

Ultraviolet energy at a wavelength of 200–280 nm (UVC) is also a promising sanitizer for intact and processed fruits, although it has not been approved for use by the FDA. The antimicrobial action of UVC is due to the penetration of the cell membrane and damaging the DNA and preventing cell replication (Shafiur, 1999). Yuan et al. (2004) reported a 3.3-log reduction in *E. coli* O157:H7 on apples after UVC exposure at 24 mW/cm^2. Furthermore, Yuan et al. (2004) have also reported that UVC has great potential in fruit-processing applications because it leaves no residues and is non-corrosive to equipment.

Another promising decontamination treatment of fruits relies on the use of natural essential oils and aroma compounds present in fruits. The primary target of these compounds is the cytoplasmic membrane of sensitive cells whereby they will disrupt the proton motif force of the sensitive cell and inhibit all energy-dependent cellular processes leading to cell death. Although most of these compounds are GRAS, their use has been often limited because of their detrimental effect on the organoleptic quality of the fruits (Lanciotti et al., 2004).

MECHANICAL OPERATIONS

Several studies have shown that mechanical operations such as cutting, slicing, and peeling play an important role in cross-contamination of the finished product (Garg et al., 1990; Brackett, 1987). Fresh-cut fruits are a highly perishable product due to their limited shelf life (5–7 days), high a_w ($a_w > 0.85$), and nutrient content (Soliva Fortuny and Martin-Belloso, 2003). Mechanical operations for minimally processed fruits will provide a source of nutrients for the pathogens as a result of the removal of the protective layer of the fruits or damage to its natural structure (punctures, wounds, cuts).

STORAGE, TRANSPORT, AND DISTRIBUTION

Storage and transport are the last stage where the risk of microbial contamination may occur. Once processed, fruits are usually kept for at least a short period of time prior to transportation. Temperature abuse and cross-contamination as a result of mishandling are fairly common at all stages of storage and transport, and pose a major threat to the safety of these products (Brackett, 1992). This is because the safety of these products relies solely or primarily on refrigeration, in addition to the long-term effect of the wash water sanitizer. Optimum temperature for storage and transportation of fruits depends greatly on the respiration rate of the fruit and resistance to chill and freezing injury (FDA, 2001a, b). In most cases, a temperature between 1°C and 3°C is regarded as a hurdle to pathogen growth. Minimally processed fruits are usually stored at lower temperatures, preferably 0°C for quality retention (FSAI, 2001; FDA, 2001a).

Vehicles for transporting fruits should be clean and properly refrigerated. Workers involved in loading and unloading of fruits should be trained on proper hand-washing techniques. Containers and storage facilities should be cleaned, sanitized regularly, and secured from rodents and insects. Thus, truck sanitation and temperature management are critical points during storage, transport, and distribution of fresh fruits (FDA, 2001a).

PLANT DESIGN AND SAFETY

The processing plant is an important place for the safe production of fruits. Processing plants with a poor hygienic design and practices may result in the contamination of final product. Thus, most food processors are now adhering to GMP in their processing operation. In the United States, the FDA code of federal regulation 21 CFR part 110 (CFR, 2000b) covers all aspects of a safe and sanitary processing environment. These GMP include the overall design of the processing plant, construction material selection, processing facilities, water, air, and traffic flow, sanitary facilities, water quality program, storage, and distribution.

LOCATION OF PROCESSING PLANT

The location of the fruit processing plant has a dramatic impact on plant sanitation (CFIA, 2000). Fruit processing plants should be located in areas away from physical, chemical, or microbial hazards such as industrial activities or environmentally polluted areas (waste disposal areas or areas prone to pest infestation).

DESIGN OF PROCESSING PLANT

Fruit processing plants should be designed and constructed to minimize contamination, ease of cleaning, and sanitation. A well-designed fruit processing plant should be constructed so that the product flow is "linear" by segregating raw product areas from processing areas and finished product areas to avoid cross-contamination (FDA, 2001a). This is a recommended practice especially for minimally processed fruits. Personnel working in such areas are segregated within their area and excluded from entering other areas. In a similar manner to the overall design of the processing facilities, the process water, waste streams, and airflow should be countercurrent to the product flow (FDA, 2001a).

Exterior surfaces such as floors, walls, and ceilings should be constructed of approved food grade materials, which are smooth, impervious, and easily cleaned. These surfaces should be constructed in a way to prevent or attract soils, pests, insects, and microorganisms. All floors should be constructed with a slight slope to prevent water accumulation and allow adequate drainage and cleaning. Surfaces that are in direct contact with fruits should be relatively inert to the fruits and non-absorbent (FDA, 2001a; Schmidt, 1997a).

The processing facilities should also be designed and constructed to deter the entrance of airborne pathogens and pests (rodents/birds). Ventilation systems should be adequate in a sense that a continuous positive air pressure should be maintained in the production area. In addition, adequate filtration of air entering the processing area using close fitting screens or filters to reduce airborne contamination should always be considered (Schmidt, 1997a).

All equipment that comes in direct contact with the fruit should be of stainless steel or plastic because these materials are easily maintained, cleaned, and sanitized. In particular, this equipment should be designed to have a minimum number of pockets, crevices to prevent bacterial attachment, and biofilm formation, and should not have any loose bolts/knobs or movable parts that may accidentally fall off and contaminate the final product (Marriott, 1999a).

Separate storage facilities should be designed for incoming fruits, finished products, and non-food chemicals (cleaners, sanitizers, etc.) to prevent the risk of cross-contamination. A separate facility should also be considered for the sanitary storage of waste and inedible material prior to their removal from the plant or surroundings (Marriott, 1999a).

The sanitary facilities should be designed to be separate from the processing and finished product area. Such facilities should be provided with self-closed doors to prevent the risk of cross-contamination (Marriott, 1999a).

The hand-washing facilities should be equipped with a sufficient number of hand-washing sinks, with hot and cold potable water, soap, sanitary hand-drying supplies or devices. It is also imperative that hand-washing sinks should be located adjacent to the toilets to encourage personnel to wash their hands after going to the toilet and should be separate from sinks used for equipment cleaning and other operations (Marriott, 1999a).

A potable pathogen-free water supply is crucial for sanitary fruit processing since water has been demonstrated to be a vector for pathogen transmission to raw/processed fruits. Compliance with appropriate regulations and standards must be verified through testing programs. Water treatments (chlorination, ozonation, filtration) should be routinely monitored by continuous confirmation of the chlorine/ozone concentrations and by the periodic analysis of water to ensure that the water source meets the recognized microbiological criteria for potable water (CFIA, 2000).

EFFECTIVENESS OF SANITATION PROGRAM

In general, the facilities and various product and non-product contact surfaces and equipment must be evaluated to assess the effectiveness of the sanitation program in place. This is usually carried out by visual inspection and microbiological testing. Visual inspection is usually conducted with the aid of a sanitation evaluation sheet whereby the inspector/sanitarian will use a numerical rating system for each area in the processing plant. Microbiological testing will include swabs and contact plates from processing equipment surfaces and personnel to assess the overall sanitation in the process environment and then the results of such testing are measured against microbiological criteria set by the industry itself, enforcement agencies, or international committees (FSAI, 2001).

SANITATION RULES AND REGULATIONS

The term *sanitize* for food contact surface is defined as "to adequately treat food contact surfaces by a process that is effective in destroying vegetative cells of microorganisms of public health significance" (CFR, 2000b).

REGULATORY CONSIDERATIONS FOR SANITIZER USE

The major regulatory issues concerning chemical sanitizers are their antimicrobial efficacy, the safety of their residues on food contact surfaces, and environmental safety (Schmidt, 1997b). In the United States, the registration of chemical sanitizers and antimicrobial agents for use on food and food contact surfaces, and on non-food contact surfaces comes under the jurisdiction of the U.S. Environmental Protection Agency (EPA). These chemical sanitizers are recognized by the EPA as pesticides (Marriott, 1999b). The Food and Drug Administration (FDA) responsibility lies in evaluating residues from sanitizer use that may enter the food supply. Thus, any antimicrobial agent and its maximum allowable use concentrations for direct use on food or on food contact surfaces must be approved by the FDA (Schmidt, 1997b).

CLEANING AND SANITATION OF PROCESSING PLANTS

In a fruit processing plant, processors need to process fruits under hygienic conditions. Fruit processors should always ensure that an appropriate cleaning and sanitation program is in place. A detailed cleaning and sanitation procedures must be developed for all food contact surfaces as well as for non-food contact surfaces (Marriott, 1999a).

The objective of cleaning and sanitizing food contact surfaces is to remove food soils so that microorganisms will not have a suitable environment for survival and multiplication. The correct order of events for cleaning/sanitizing of food contact surface in a processing plant is the removal of gross debris, rinsing, presoaking in cleaning agents, rinsing, and sanitizing. Cleaners, also known as detergents, are used first to remove food soils. The choice of detergent to be used depends on the type of food soil present. Food soils can be classified as those soluble in water (sugars, starches, salts); soluble in acid (limestone and most mineral deposits); soluble in alkali (protein, fat emulsions); soluble in water, alkali, or acid (Schmidt, 1997b).

After cleaning and rinsing the food contact surface with water, a sanitizer is usually applied to the surface. The sanitizer is used to destroy the presence of bacteria. The selection of a sanitizer varies with the type of equipment to be sanitized, the hardness of the water, the application equipment available, the effectiveness of the sanitizer under site conditions, and the cost. Various sanitizers are used in the food industry. These sanitizers include but are not limited to steam, hot water, and chemical sanitizers. Steam and hot water sanitization are usually carried out through immersion, spraying, or circulating systems. Steam and hot water sanitization are easy to apply and readily available, generally effective over a broad range of microorganisms and relatively non-corrosive. However, they have some disadvantages including film formation, high-energy costs, and safety concerns for employees (Schmidt, 1997b).

Chlorine

Chemical sanitizers are one of the most effective treatment methods for the reduction of foodborne pathogens on food and non-food contact surfaces. Some of these sanitizers have been approved by the FDA for use in process water, processing equipment, and facilities, while others have been approved to be used in direct contact washes of fruits such as spray washing, rinsing, and dipping application in the fruit processing plants. For example, chlorine and chlorine dioxide have been approved by the FDA to be used to reduce the number of pathogenic bacteria on the surface of fruits and on food contact and non-food contact surfaces.

Chlorine-based sanitizers are the most commonly used sanitizer in food processing and handling applications. Commonly used chlorine compounds include liquid chlorine, hypochlorites, inorganic chloramines, and organic chloramines. Recently, safety concerns have been raised about the use of chlorine as a sanitizer and antimicrobial agent for food processing. This concern is based on the reaction of chlorine with the organic matter to produce potential carcinogens such as trihalomethanes (THMs), chloroform, and chlorophenols, thus other treatments are being considered as a replacement of chlorine (Hurst, 1995).

Chlorine Dioxide

Chlorine dioxide (ClO_2) is currently being considered as a replacement for chlorine, since it appears to be more environmentally friendly. Stabilized ClO_2 has FDA approval for most applications in sanitizing equipment or for use as a foam for environmental and non-food contact surfaces. Approval has also been granted for use in flume waters for uncut fruit operations. ClO_2 has 2.5 times the oxidizing power of chlorine and is less affected by pH and organic matter than chlorine (Dychdala, 1991). A maximum concentration of 200 ppm is permitted for processing equipment, while 1–5 ppm has been approved for whole fresh fruits (Cherry, 1999).

Iodine

Like chlorine compounds, iodine exists in various forms as mixtures of elemental iodine with a non-ionic surfactant as a carrier and is termed as iodophors. Iodophors have a very broad spectrum, being active against bacteria, viruses, yeasts, molds, fungi, and protozoa. However, iodophors have had limited use in the fruit industry as sanitizers because of their corrosiveness, reduced efficacy at low temperature, staining effect on plastic surfaces, and their reaction with starch-containing food commodities to form a bluish purple-colored complex, thus restricting their use to non-starch fruit commodities. Iodophors are also not approved for direct food contact. The general recommended usage for iodophors for food contact surfaces and equipment is 10–100 ppm (Cherry, 1999).

Quaternary Ammonium Compounds

Quaternary ammonium compounds (QACs) are mainly used as sanitizers of walls, ceilings, food contact surfaces, and equipment used for fruit processing. Since QACs are positively charged cationic surface-active sanitizers, their mode of action is related to the disruption of protein membrane function of bacterial cells. Unlike chlorine, QACs are non-corrosive to metals, are active and stable over a broad temperature range, and are less affected by organic matter. However, QACs are generally more active against Gram-positive than Gram-negative bacteria and their effectiveness could be improved by the incorporation of chelating agents such as EDTA. Disadvantages of QACs also include limited activity under alkaline conditions, excessive foaming, especially for Clean-in-Place (CIP) applications and incompatibility with certain detergents after cleaning operation (Schmidt, 1997b).

MICROBIAL BIOFILMS

Microbial biofilms are defined as "functional consortium of microorganisms attached to a surface and embedded in the extra cellular polymeric substance (EPS) produced by microorganisms" (Costerton et al., 1987). EPS will help the proper colonization of microorganisms by trapping nutrients and protecting the pathogen from the action of sanitizers. The resistance of biofilms to sanitizers has caused severe problems such as fouling of water transport systems, and contamination of dental caries and medical implant devices (Carpenter and Cerf, 1993).

Microbial biofilms also constitute a major problem in the fruit processing industry. Pathogenic and spoilage microorganisms can attach firmly to food contact surfaces, and thus become a source of post-processing contamination. LeChevallier et al. (1988) have reported that biofilms once formed are 150–3000 times more resistant to the chlorine effect. Sanitation programs have had limited success in the removal of biofilms from processing lines and equipments. Efforts to remove biofilms include identifying the high-risk production areas and the critical sites of production in a plant where biofilms may occur and then expose these critical sites to frequent sanitation and special treatments. Special treatments such as scrubbing or brushing have shown to be effective in removing bacterial biofilms (Gibson et al., 1999).

FUTURE RESEARCH NEEDS

Increased consumer demands toward eating fresh and minimally processed fruits (fresh-cut fruits and unpasteurized fruit juices) have imposed a great responsibility on the food industry to provide them with a safe and wholesome food product. Future needs for the purpose of assuring the safety of these products include the continued and increased surveillance studies on the presence of these pathogens in imported and domestic fruits and the development of optimized methods for the isolation and detection of these pathogens in fruits. Further safety issues that need to be addressed include better understanding of the interaction between plant pathogens and human pathogens; investigation on factors affecting pathogen infiltration into fruits and their significance to fruit safety; development of new treatments/technologies to reduce/eliminate these pathogens from fruits and studies on the behavior of these pathogenic bacteria in the presence/absence of the natural microflora on fruits due to these proposed treatments/technologies.

REFERENCES

Anon. 2000. Pathogens and produce: Food borne outbreaks/incidents. European chilled food federation. Presented to the EC Scientific Committee for food. March 2000. International Fresh-cut Produce Association. 2001. Food Safety Guidelines for the fresh-cut produce Industry.

Austin, J.W., K.L. Dodds, B. Blanchfield, and J.M. Farber. 1998. Growth and toxin production by *Clostridium botulinum* on inoculated fresh-cut packaged vegetables. *J. Food Prot.* 61:324–328.

Baird Parker, A.C. 1987. The application of preventive quality assurance. In: Elimination of Pathogenic Organisms from Meat and Poultry (Ed. J.M. Smulder). Elsevier Sci. Publ., London, pp. 149–179.

Bates, J.L. 1999. Nippy's *Salmonella* outbreak. *Food Aust.* 51:272.

Bean, N.H., J.S. Goulding, M.T. Daniels, and F.J. Angulo. 1997. Surveillance for food borne disease outbreaks-United States, 1998–1992. *J. Food Prot.* 60:1265–1286.

Bean, N.H. and P.M. Griffin. 1990. Food borne disease outbreaks in the United States, 1973–1987: Pathogens, Vehicles and Trends. *J. Food Prot.* 53:804–817.

Bennik, M.H.J., H.W. Peppelenbos, C. Nguyen-the, F. Carlin, E.J. Smid, and L.G.M. Gorris. 1996. Microbiology of minimally processed, modified atmosphere packaged chicory endive. *Postharvest Biol. Technol.* 9:209–221.

Besser R.E., S.M. Lett, J.T. Weber, M.P. Doyle, T.J. Barett, J.G. Wells, and P.M. Griffin. 1993. An outbreak of diarrhea and hemolytic uremic syndrome from *Escherichia coli* O157:H7 in fresh prepared apple cider. *J. Am. Med. Assoc.* 17:2217–2220.

Beuchat, L.R. 1996. Pathogenic microorganisms associated with fresh produce. *J. Food Prot.* 59: 204–216.

Beuchat, L.R. 1998. Surface decontamination of fruits and vegetables: A review. World Health Organization (FSF/FOS/98.2.).

Beuchat, L.R., B.V. Nail, B.B. Adler, and M.R.S. Clavero. 1998. Efficacy of spray application of chlorinated water in killing pathogenic bacteria on raw apples, tomatoes and lettuce. *J. Food Prot.* 61:1305–1311.

Black, R.E., M.M. Levine, M.L. Clements, T.P. Hughes, and M.J. Blaser. 1988. Experimental

Campylobacter jejuni infections in humans. *J. Infect. Dis.* 157:472–479.
Brackett, R.E. 1987. Microbiological consequences of minimally processed fruits and vegetables. *J. Food Qual.* 10:195–206.
Brackett, R.E. 1992. Shelf stability and safety of fresh produce as influenced by sanitation and disinfection. *J. Food Prot.* 55:808–814.
Bryan, F.L. 1980. Food borne diseases in the United States associated with meat and poultry. *J. Food Prot.* 43:140–150.
Buchanan, R.L., S.G. Edelson, R.L. Miller, and G.M. Sapers. 1999. Contamination of intact apples after immersion in an aqueous environment containing *Escherichia coli* O157:H7. *J. Food Prot.* 62: 444–450.
California Strawberry Commission. 1997. Quality Assurance Program: Growers and Shippers Checklist, Watsonville, California.
Canadian Food Inspection Agency. 2000. Code of practice for minimally processed ready to eat vegetables. Food of Plant Origin Division. http://www.inspection.gc.ca/english/plaveg/fresh/read-eat_e.shtml.
Carpenter, B. and O. Cerf. 1993. Biofilms and their consequences, with particular reference to hygiene in the food industry. *J. Appl. Bacteriol.* 75:499–511.
Castillo, A. and E.F. Escartin. 1994. Survival of *Campylobacter jejuni* on sliced watermelon and papaya. *J. Food Prot.* 57:166–168.
Center for Disease Control. 1990. Food borne disease outbreaks, 5 year summary (1983–1987). *CDC Surveill. Summ. Morb. Mort. Wkly. Rep.* 39:15–57.
Center for Disease Control. 1996a. Surveillance for food borne-disease outbreaks-United States (1988–1992). *Morb. Mort. Wkly. Rep.* 45:1–65.
Center for Disease Control. 1996b. Update: Outbreaks of *Cyclospora cayetanensis* infection-United States and Canada. *Morb. Mort. Wkly. Rep.* 45:611–612.
Center for Disease Control. 1997. Update: Outbreaks of Cyclosporiasis. *Morb. Mort. Wkly. Rep.* 46:521.
Center for Disease Control. 2000. *CDC surveill. summ.* 49:1–51.
Cherry, J.P. 1999. Improving the safety of fresh produce using antimicrobials. *Food Technol.* 53:54–59.
Code of Federal Regulations. 2000a. Title 21, Part101.95. Food Labeling: "fresh" "freshly frozen," "fresh frozen" "frozen fresh". Available from:http://www.access.gpo.gov/nara/cfr/index .html>.
Code of Federal Regulations. 2000b. Title 21, Part110.3. Current good manufacturing practices in Manufacturing, Packing, or holding human food: Definitions Available from:http://www.access.gpo.gov/nara/cfr/index.html>.
Code of Federal Regulations. 2000c. Title 21, Part173.325. Secondary direct food additives permitted in food for human consumption: acidified sodium chlorite solutions. Available from: http:// www.access.gpo.gov/nara/cfr/index .html>.
Conway, W.S., B. Leverentz, and R.A. Saftner. 2000. survival and growth of *Listeria monocytogenes* on fresh cut apple slices and its interaction with *Glomerella cingulata* and *Penicillium expansum*. *Plant Dis.* 84:177–181.
Costerton, J.W., K.J. Chang, G.G. Geesey, T.I. Ladd, J.C. Nickel, M. Dasgupta, and T.J. Marrie. 1987. Bacterial biofilms in nature and disease. *Annu. Rev. Microbiol.* 41:435–464.
D'Aoust, J.Y. 1991. Psychrotrophy and food borne *Salmonella*. *Int. J. Food Microbiol.* 13:207–216.
Davis, H.J., P. Taylor, J.N. Perdue, G.N. Stelma, J.M. Humphreys, R. Rowntree, and K.D. Greene. 1988. A shigellosis outbreak traced to commercially distributed lettuce. *Am. J. Epidemiol.* 128:1312–1321.
Du, J., Y. Han, and R.H. Linton. 2002. Inactivation by chlorine dioxide gas (ClO_2) of *Listeria monocytogenes* spotted onto different apple surfaces. *Food Microbiol.* 19:481–490.
Dunn, R.A., W.N. Hall, J.V. Altamirano, S.E. Dietrich, B. Robinson-Dunn, and D.R. Johnson. 1995. Outbreak of *Shigella flexneri* linked to salad prepared at a central commissary in Michigan. *Public Health Rep.* 110:580–586.
Dychdala, G.R. 1991. Chlorine and chlorine compounds. In: Disinfection, Sterilization and Preservation (Eds. S. Block). Lea and Febiger Press, Philadelphia. pp. 131–155.
Environmental Protection Agency. 1997. Final report. A set of scientific issues being considered by the agency in connection with efficacy testing issues concerning public health antimicrobial pesticides. http://www.epa.gov/oscpmont/sap/1997/September/finalsep.htm#3.
Farber, J.M. 1989. Food borne pathogenic microorganisms: Characteristics of the organisms and their associated diseases. I. Bacteria. *Can. Inst. Food Sci. Technol. J.* 22:311–321.
Farber, J.M. and J.Z. Losos. 1988. *Listeria monocytogenes*: A food borne pathogen. *Can. Med. Assoc. J.* 138:413–418.
Farber, J.M., S.L. Wang, Y. Cai and S. Zhang. 1998. Changes in populations of *Listeria monocytogenes* inoculated on packaged fresh-cut vegetables. *J. Food Prot.* 61:192–195.

Fernandez Escartin, E.F., A. Castillo Ayala, and J. Saldana Lozano. 1989. Survival and growth of *Salmonella* and *Shigella* on sliced fresh fruits. *J. Food Prot.* 52:471–472.

Food and Drug Administration. 1998a. Hazard analysis and critical control points (HACCP); procedures for the safe and sanitary processing and importing of juice; food labeling: warning and notice statements; labeling of juice products; final rules. *Fed. Regist.* 63:37029–37056.

Food and Drug Administration. 1998b. Guidance for industry-Guide to minimize microbial food safety hazards for fresh fruits and vegetables. http://vw.cfsan.fda.gov.~dms/prodguid.html.

Food and Drug Administration. 2000. Experience with microbial hazards in fresh produce. Lee Anne Jackson, PhD, Center for Food Safety and Applied Nutrition. Food and Drug Administration. Presented to the EC Scientific Committee for food. March 2000.

Food and Drug Administration. 2001a. Center for Food Safety and Applied Nutrition. Analysis and evaluation of preventive control measures for the control and reduction/elimination of microbial hazards on fresh and fresh-cut produce. http://www.cfsan.fda.gov/~comm/ift3-4a html.

Food and Drug Administration. 2001b. Center for Food Safety and Applied Nutrition. FDA survey of imported fresh produce. FY 1999 field assignment. http://www.cfsan.fda.gov/~dms/prodsur6.html.

Food Safety Authority of Ireland. 2001. Code of Practice for food safety in the fresh produce supply chain in Ireland. Code of practice No. 4.

Garg, N., J.J. Churey, and D.F. Splittstoesser. 1990. Effect of processing conditions on the microflora of fresh-cut vegetables. *J. Food Prot.* 53:701–703.

Genigeorgis, C. 1989. Present state of knowledge on *Staphylococcus aureus* intoxication. *Int. J. Food Microbiol.* 9:327–360.

Gibson, H., J.H. Taylor, K.E. Hall, and J.T. Holah. 1999. Effectiveness of cleaning techniques used in the food industry in terms of the removal of bacterial biofilms. *J. Appl. Microbiol.* 87:41–48.

Golden, D.A., E.J. Rhodehamel, and D.A. Kautter. 1993. Growth of *Salmonella spp.* in cantaloupe, watermelon and honeydew melons. *J. Food Prot.* 56:194–196.

Gordon, G., G. Kieffer, and D. Rosenblatt. 1972. The chemistry of chlorine dioxide. In: Progress in Organic Chemistry (Ed. S. Lippard). Wiley Interscience, NY, New York, pp. 201–286.

Griffiths, M. 2000. The new face of food borne illness. *Can. Meat Sci. Assoc.* 1:6–9.

Griffiths, P.L. and R.W. Park. 1990. *Campylobacter* associated with human diarrhoeal disease. *J. Appl. Bacteriol.* 69:281–301.

Howard, L.R. and A.R. Gonzalez. 2001. Food safety and produce operations: What is the future? *Hort. Sci.* 36:33–39.

Hurst, W.C. 1995. Disinfection methods: A comparison of chlorine dioxide/ozone and ultra violet alternatives. Cutting edge, Fall issue. International Fresh-Cut Produce Association. Alexandria, Va. pp. 4–5.

International Fresh Produce Association. 2001. Food safety guidelines for the fresh-cut produce. 4th Edition.

Janisiewicz, W.J., W.S. Conway, M.W. Brown, G.M. Sapers, and P. Fratamico. 1999. Fate of *Escherichia coli* O157:H7 on fresh-cut apple tissue and its potential for transmission by fruit flies. *Appl. Environ. Microbiol.* 65:1–5.

Kader, A.A. 1986. Potential applications of ionizing radiation in postharvest handling of fresh fruits and vegetables. *Food Technol.* 40:117–121.

Kemp, G., M.L. Aldrich, and A.L. Waldroup. 2000. Acidified sodium chlorite antimicrobial treatment of broiler carcass. *J. Food Prot.* 63:1087–1092.

Kenney, S.J., S.L. Burnett, and L.R. Beuchat. 2001. Location of *Escherichia coli* O157:H7 on and in apples as affected by bruising, washing and rubbing. *J. Food Prot.* 64:1328–1333.

Kim, C., Y. Hung, and R.E. Brackett. 2000. Role of oxidation-reduction potential in electrolyzed oxidizing and chemically modified water for the inactivation of food-related pathogens. *J. Food Prot.* 63:19–24.

Kim, J., A.E. Yousef, and G.W. Chism. 1999. Use of ozone to inactivate microorganisms on lettuce. *J. Food Saf.* 19:17–34.

Kross, R. 1984. An innovate demand-release microbicide. In: Second Annual Conference on Progress in Chemical Disinfestations. Suny, Binghamton, New York.

Lanciotti, R., A. Gianotti, N. Belletti, M.E. Guerzoni, and F. Gardini. 2004. Use of natural aroma compounds to improve shelf-life and safety of minimally processed fruits. *Trends Food Sci. Technol.* 15:201–208.

Larson, A.E. and E.A. Johnson. 1999. Evaluation of botulinum toxin production in packaged fresh-cut cantaloupe and honey dew melons. *J. Food Prot.* 62:948–952.

LeChevallier, M.W., C.D. Cawthon, and R.G. Lee. 1988. Inactivation of biofilm bacteria. *Appl. Environ. Microbiol.* 54:2492–2499.

Mackenzie, M.A. and B.S. Bains. 1976. Dissemination of *Salmonella* serotypes from raw food ingredients to chicken carcasses. *Poult. Sci.* 55:957–960.

Marriott, N.G. 1999a. Fruit and vegetable processing plant sanitation. In: Principles of Fruit Sanitation, 4th Edition. Chapman & Hall Food Science Book, Aspen publishers Inc, Maryland, pp. 291–302.

Marriott, N.G. 1999b. Sanitation in the food industry. In: Principles of Fruit Sanitation, 4th Edition. Chapman & Hall Food Science Book, Aspen publishers Inc, Maryland, pp. 1–10.

Merker, R., S. Edelson-Mamel, V. Davis, and R.L. Buchanan. 1999. Preliminary experiments on the effect of temperature differences on dye uptake by oranges and grapefruit. U.S. Food and Drug Administration, Center for Food Safety and Applied Nutrition, 200 C Street SW, Washington, D.C. 20204.

Miller, L.G. and C.W. Kasper. 1994. *Escherichia coli* O157:H7 acid tolerance and survival in apple cider. *J. Food Prot.* 57:460–464.

Morehouse, K.M. 2001. Food irradiation-US regulatory consideration. *Radiat. Phys. Chem.* 63:281–284.

Mukhopadhyay, R., A. Mitra, R. Roy, and A.K. Guha. 2002. An evaluation of street vended sliced papaya (Caica papaya) for bacteria and indicator microorganisms of public health significance. *Food Microbiol.* 19:663–667.

Niu, M.T., L.B. Polish, B.H. Robertson, B.K. Khanna, B.A. Woodruff, C.N. Shapiro, M.A. Miller, J.D. Smith, J.K. Gedrose, M.J. Alter, and H.S. Margolis. 1992. Multistate outbreak of Hepatitis A associated with frozen strawberries. *J. Infect. Dis.* 166:518–524.

Notermans, S.H., M.H. Zwietering, and G.C. Mead. 1994. The HACCP concept: Identification of potentially hazardous microorganisms. *Food Microbiol.* 11:203–214.

O'Brien, S., R. Mitchell, I. Gillespic, and G. Adak. 2000. PHLS Communicable Disease Surveillance Centre. Microbiological status of ready to eat fruits and vegetables. Advisory Committee on the Microbiological Safety of Food (ACMSF). ACM 476. pp. 1–34.

Olsen, A.R. 1998. Regulatory action criteria for filth and other extraneous materials. III. Review of flies and food borne enteric disease. *Regul. Toxicol. Pharmacol.* 28:199–211.

Oosterom, J. 1991. Epidemiological studies and proposed preventive measures in the fight against human salmonellosis. *Int. J. Food Microbiol.* 12:41–52.

Pao, S. and C.L. Davis. 1999. Enhancing microbiological safety of fresh orange juice by fruit immersion in hot water and chemical sanitizers. *J. Food Prot.* 62:756–760.

Pao,S., C.L. Davis, D.F. Kelsey, and P.D. Petracek. 1999. Sanitizing effects of fruit waxes at high pH and temperature on orange surfaces inoculated with *Escherichia coli. J. Food Sci.* 64:359–362.

Park, C. and L. Beuchat. 1999. Evaluation of sanitizers for killing *Escherichia coli* O157:H7, *Salmonella* and naturally occurring microorganisms on cantaloupe, honeydew melon and asparagus. *Dairy Food Environ. Sanit.* 19:842–847.

Park, H., Y. Hung, and R.E. Brackett. 2002. Antimicrobial effect of electrolyzed water for inactivating *Campylobacter jejuni* during poultry washing. *Int. J. Food Microbiol.* 72:77–83.

Penteado, A.L. and M.F. Leitao. 2004. Growth of *Listeria monocytogenes* in melon, watermelon and papaya pulps. *Int. J. Food Microbiol.* 92:89–94.

Pierson, M.D. and D.A. Corlett. 1992. HACCP: Principles and Applications: Van Nostrand Reinhold, New York.

Restaino, L., E.W. Frampton, J.B. Hempfill, and P. Palnikar. 1995. Efficacy of ozonated water against various food-related microorganisms. *Appl. Environ. Microbiol.* 61:3471–3475.

Robertson, L.J. and B. Gjerde. 2001. Occurrence of parasites on fruits and vegetables in Norway. *J. Food Prot.* 64:1793–1798.

Rodgers, S.L., J.N. Cash, M. Siddiq, and E. Ryser. 2004. A comparison of different chemical sanitizers for inactivating *Escherichia coli* O157:H7 and *Listeria monocytogenes* in solution and on apples, lettuce, strawberries and cantaloupe. *J. Food Prot.* 67:721–731.

Rose, J.B. and T.R. Slifko. 1999 . *Giardia, Cryptosporidium* and *Cyclospora* and their impact on foods: A review. *J. Food Prot.* 62:1059–1070.

Sapers, G.M. and G.F. Simmons. 1998. Hydrogen peroxide disinfection of minimally processed fruits and vegetables. *Food Technol.* 52:48–52.

Schmidt, R.H. 1997a. Basic elements of a sanitation program for food processing and handling. Series of Food Science and Human Nutrition. Department of Florida Cooperative extension Services, Institute of Food and Agricultural Sciences. University of Florida. Fact sheet FS15. http://edis.ifasufl.edu /FS076.

Schmidt, R.H. 1997b. Basic elements of equipment cleaning and sanitizing in food processing and handling operations. Series of Food Science and Human Nutrition. Department of Florida Cooperative extension Services, Institute of Food and Agricultural Sciences. University of Florida. Fact sheet FS14. http://edis.ifasufl.edu/FS077.

Shafiur, R.M. 1999. Light and Sound in Food Preservation (Ed. M.R. Shafiur). *Handbook of food preservation* 22, pp. 669–686.

Siliker, J.H. 1982. The Salmonella problem: Current status and future direction. *J. Food Prot.* 45: 661–666.

Soliva-Fortuny and O. Martin-Belloso. 2003. New advances in extending the shelf-life of fresh-cut fruits: A review. *Trends Food Sci. Technol.* 14:341–353.

Tauxe, R., H. Kruse, C. Hedberg, M. Potter, J. Madden, and K. Wachsmuth. 1997. Microbial hazards and emerging issues associated with produce: A preliminary report to the National Advisory Committee on Microbiological Criteria for Foods. *J. Food Prot.* 60:1400–1408.

Thayer, D.W. and K.T. Reykowski. 1999. Developments in irradiation of fruits and vegetables. *Food Technol.* 52:62–65.

Thurston-Enriquez, J., P. Watt, S.E. Dowd, R. Enrrquez, I.L. Pepper, and C.P. Gerba. 2002. Detection of protozoan parasites and microsporidia in irrigation waters used for crop production. *J. Food Prot.* 65:378–382.

U.S. Department of Agriculture, National Agricultural Statistics Service. 2001 June. Fruit and Vegetable Agricultural Practices–1999. USDA. http://usda.gov/nass/pubs/rpts106.htm.

US Department of Health and Human Services. 1993. Food Code.

Venkitanarayanan, K., T. Zhao, and M.P. Doyle. 1999. Inactivation of *Escherichia coli* O157:H7 by combination of GRAS chemicals and temperatures. *Food Microbiol.* 16:75–82.

World Health Organization. 1998. Surface decontamination of fruits and vegetables eaten raw: A review. WHO/FSF/FOS 98.2.

Wright, J.R., S.S. Sumner, C.R. Hackney, M.D. Pierson, B.W. Zoecklein. 2000. Reduction of *Escherichia coli* O157:H7 on apples using wash and chemical sanitizer treatments. *Dairy Food Environ. Sanit.* 120:6.

Wu, F.M., M.P. Doyle, L.R. Beuchat, J.G. Wells, E.D. Mintz, and B. Swaminathan. 2000. Fate of *Shigella sonnei* on parsley and methods of disinfection. *J. Food Prot.* 63:568–572.

Yuan, B.R., S.S. Sumner, J.D. Effert, and J.E. Marcy. 2004. Inhibition of pathogens on fresh produce by ultraviolet energy. *Int. J. Food Microbiol.* 90:1–8.

Zhao, T., M.P. Doyle, and R.E. Besser. 1993. Fate of enterohemorrhagic *Escherichia coli* O157:H7 in apple cider with and without preservatives. *Appl. Environ. Microbiol.* 59:2526–2530.

Part III
Commodity Processing

16
Apples

Nirmal Sinha

INTRODUCTION

Apple (*Mauls domestica*), a climacteric fruit, is commercially grown in the temperate regions of the world in the latitudes between 30° and 60° north and south. The apple tree belongs to Rosaceae (rose) family (Anon, 2004). It adapts well in climates where the average winter temperature is near freezing for at least 2 months. The fruit production phase begins in the late spring when trees bear white blossoms (flowers) that look like tiny roses. Blossoms produce pollen and nectar to attract bees and other insects that pollinate the blossoms, which grow into ripened fruits in about 140–170 days. A new apple tree would produce fruits in about 6–8 years.

PRODUCTION AND CONSUMPTION

Apples are one of the leading fruits produced in the world (Table 16.1). According to the Food and Agricultural Organization (FAO), world production of apple in 2002 was estimated at about 57 million metric tons, with China and the United States as the leading producers with approximately, 36% and 7% of world production, respectively (Table 16.2; FAO, 2003). The acreage devoted to apple production in China in 2002 was 15 times greater than that in the United States (Table 16.3). Yields per acre vary significantly in the apple producing regions of the world. New Zealand, Italy, France, Chile, and Brazil have shown higher apple yield than India, the Russian Federation, and China, where yields in 2002 were lower than the world average of 10 t/ha (Table 16.4). Yields in the United States have typically been more than twice the world average, but that appears to be leveling off (Table 16.4).

Advances in production, storage, processing, product development, and marketing efficiencies have created viable apple industries in many parts of the world. Next to banana, apples are the most consumed fruit in the United States, with a per capita fresh consumption of approximately 18.0 lb (Table 16.5). When apple juice and cider are factored into the equation, apples consumption far exceeds banana (average of 38.6 lb apple vs 28.2 lb banana; Tables 16.5 and 16.6). About 39% of apples produced in the United States are utilized as fresh, while the remaining 61% are processed. Of the processed apples, juice and cider comprised 46%, canned apples (mostly applesauce) about 10%, frozen and dried 2% each, and

Table 16.1. Leading Fruits and Leading Producing Countries—2002

Leading Fruit	World Production	Leading Country
	(Metric ton, Mt): 1 Mt = 1000 kg = 2200 lb	
1. Banana	69,510,944	India (16,000,000)
2. Oranges	64,712,847	Brazil (18,694,412)
3. Grapes	62,389,467	Italy (8,500,000)
4. Apples	57,982,587	China (20,507,763)
5. Mangoes	25,760,848	India (11,500,000)

Source: FAO (2003).

Table 16.2. Apple Production (Mt) in Leading Countries and World

Country	1997	1998	1999	2000	2001	2002
1. China	17,227,752	19,490,501	20,809,846	20,437,065	20,022,763	20,434,763
2. U.S.A.	4,682,000	5,282,509	4,822,078	4,681,980	4,277,000	3,857,000
3. Turkey	2,550,000	2,450,000	2,500,000	2,400,000	2,450,000	2,500,000
4. France	2,473,000	2,209,900	2,165,800	2,156,900	2,397,000	2,477,790
5. Poland	2,098,297	1,687,226	1,604,221	1,450,376	2,433,940	2,168,856
6. Iran	1,998,107	1,943,627	2,137,037	2,141,655	2,353,359	2,353,359
7. Italy	1,966,470	2,143,300	2,343,800	2,232,100	2,340,677	2,222,191
8. Russia	1,500,000	1,330,000	1,060,000	1,832,000	1,682,000	1,800,000
9. Germany	1,602,100	2,296,200	2,268,400	3,136,800	1,928,800	1,600,000
10. India	1,308,390	1,320,590	1,380,000	1,040,000	1,230,000	1,420,000
11. Argentina	1,117,690	1,033,520	1,116,000	833,322	1,428,802	1,000,000
12. Chile	845,000	795,000	1,175,000	805,000	1,135,000	1,050,000
13. Japan	993,300	879,100	927,700	799,600	930,000	911,900
14. Brazil	793,585	791,437	937,715	1,153,269	716,030	857,824
15. Spain	983,700	736,000	988,400	838,246	962,000	652,500
16. New Zealand	567,000	523,000	545,000	620,000	485,000	536,999
17. World (Total)	57,532,889	56,808,052	58,045,081	59,155,512	58,146,188	57,094,939

Source: FAO (2003).

Table 16.3. Apple Fruit—Area Harvested in Leading Countries and World

	(Area harvested, Hectare, Ha)					
	1997	1998	1999	2000	2001	2002
1. China	2,839,388	2,622,497	2,439,868	2,254,759	2,066,922	2,500,772
2. U.S.A.	189,390	189,230	186,486	173,900	168,573	163,878
3. Turkey	107,080	106,460	106,826	107,600	108,392	108,600
4. France	78,000	78,000	78,000	78,000	78,000	78,000
5. Poland	165,000	157,800	144,266	147,337	148,666	148,666
6. Iran	141,045	157,868	144,266	147,337	148,666	148,666
7. Italy	64,984	64,233	63,599	62,527	62,961	61,175
8. Russia	420,000	415,000	420,000	425,000	430,000	435,000
9. Germany	65,200	90,000	90,000	70,000	70,000	70,000
10. India	222,700	227,680	231,000	230,000	240,000	250,000
11. Argentina	44,302	45,327	45,000	46,000	48,000	53,000
12. Chile	39,900	38,400	37,400	35,790	35,090	35,000
13. Japan	46,600	45,500	44,600	43,900	43,200	42,400
14. Brazil	26,418	26,318	28,555	30,043	30,938	31,070
15. Spain	49,619	49,305	46,400	46,500	46,500	47,000
16. New Zealand	15,800	15,000	14,500	14,114	14,114	14,114
17. World (Total)	6,186,012	5,855,054	5,677,185	5,452,192	5,250,686	5,675,372

Source: FAO (2003).

Table 16.4. Apple Fruit—Yield per Hectare in Leading Countries and World

	(Yield: ton/hectare- Mt/Ha)					
	1997	1998	1999	2000	2001	2002
1. China	6.07	7.43	8.53	9.06	9.69	8.17
2. U.S.A.	24.72	27.92	25.86	26.92	25.37	23.54
3. Turkey	23.81	23.01	23.40	22.30	22.60	23.02
4. France	31.71	28.33	27.77	27.65	30.73	31.77
5. Poland	12.72	10.69	9.71	8.78	14.63	12.77
6. Iran	14.17	12.31	14.81	14.54	15.83	15.83
7. Italy	30.26	33.37	36.85	35.70	37.18	36.33
8. Russia	3.57	3.20	2.52	4.31	3.91	4.14
9. Germany	24.57	25.51	25.20	44.81	27.55	22.86
10. India	5.88	5.80	5.97	4.52	5.13	5.68
11. Argentina	25.23	22.80	24.80	18.12	29.77	18.87
12. Chile	21.18	25.39	31.42	22.49	32.35	30.00
13. Japan	21.32	19.32	20.80	18.21	21.53	21.51
14. Brazil	30.04	30.07	32.84	38.39	23.14	27.61
15. Spain	19.83	14.93	21.30	18.03	20.69	13.88
16. New Zealand	35.89	34.87	37.59	43.93	34.36	38.05
17. World (Total)	9.30	9.70	10.22	10.82	11.07	10.06

Source: FAO (2003).

Table 16.5. Per Capita Consumption of Fresh Fruit in the United States (lb)

Fruit	1997	1998	1999	2000	2001
1. Apple	18.1	19.0	18.5	17.4	15.8
2. Banana	27.2	28.0	30.7	28.4	26.5
3. Orange	13.9	14.6	8.4	11.7	12.3
4. Grapes	7.9	7.1	8.0	7.3	7.6
5. Peach	5.5	4.8	5.4	5.5	5.3
6. Pears	3.4	3.3	3.3	3.2	3.1
7. Strawberries	5.1	5.1	5.7	6.4	5.7
8. Pineapples	2.3	2.8	3.0	3.2	3.2

Source: USDA (2003).

Table 16.6. Per Capita Consumption of Apple and Apple Products in United States (lb, fresh weight equivalent)

	Fresh	Canned	Juice/Cider	Frozen	Dried	Others	All
1997	18.1	5.6	18.5	1.3	0.9	0.7	45.1
1998	19.0	4.4	21.6	1.0	1.2	0.3	47.5
1999	18.5	4.8	21.4	1.0	1.0	0.4	47.1
2000	17.4	4.4	21.6	0.7	0.8	0.3	45.2
2001	15.8	4.7	21.1	0.9	0.9	0.3	43.7
Average	17.8	4.8	20.8	1.0	1.0	0.4	45.7
% Share	38.85	10.45	45.58	2.14	2.10	0.87	

Source: USDA (2003).

other items (apple jam, apple butter, etc.) about 1% (Table 16.6).

HARVEST, POSTHARVEST, AND STORAGE

Harvest

Apple harvesting takes into account intended use and storage requirements because premature harvesting affects flavor, color, size, and storability (i.e., susceptibility to bitter pit and storage scald). Similarly, harvesting too late can give softer fruits and a shorter storage life. After harvest, apples can be stored in controlled atmosphere (CA) for up to 12 months.

Although early varieties may be ready for harvest during August or early September, most apples in the United States are harvested later in September through October. Producers generally use "Days after full bloom (DAFB)" as a guideline for maturity. For Gala and Fuji apples, DAFB could be 110–120 days and 170–185 days, respectively. Objective measurements (Anon, 2003) to determine harvesting date include pressure tests (for measuring firmness) using an Effigi tester or a Magness–Taylor pressure tester. Apples for storage and processing generally require pressures above 15 lb, while pressures in the 13 lb range indicate fruit that is ripe for immediate consumption. Other maturity indicators include fruit size and appearance, soluble solids (measured as Brix with ranges from about 9.0 to 15.0 depending on variety, and growing location), acidity (measured as %malic acid ranging 0.3–1.0%), starch content, because starch is converted to sugars during ripening process (a starch reading of 1–2 on a scale of 1–8 indicates that the fruit is immature, reading of 5–6 indicates that the fruit is suitable for fresh consumption; starch is estimated by applying iodine solution to cut apple surfaces, a high starch content gives a complete blue–black reaction with a reading of 1), and internal ethylene concentration (the onset of maturity is preceded by production of ethylene in the tissues, which induces ripening).

A preharvest foliar spray of calcium can maintain apple fruit firmness and decrease the incidences of physiological disorders such as water core, bitter pit, and internal breakdown. Application of bioregulators such as ethephon may also improve fruit quality and storability (Drake et al., 2002).

Apples are picked by hand to avoid bruising. Fresh market apples should be free of bruises, blemishes, and other defects. Bruising during harvest, transport, grading, storage, and packing is always a concern.

Table 16.7. Season–Average Grower Price of Apple in the United States

Price	1997	1998	1999	2000	2001
1. Fresh (cent/lb)	22.10	17.3	21.3	17.8	22.9
2. Processed (Dollar/ton)	130.0	94.0	128.0	102.0	106.0

Source: USDA (2003).

Historically, retail prices for fresh apples are higher than prices paid for processing apples (Table 16.7).

Postharvest Handling and Storage

Postharvest handling of apples includes hydrocooling, washing, culling, waxing, sorting, and packing. Shellac- and wax-based fruit coatings having high permeability to carbon dioxide (CO_2) and oxygen (O_2), and relatively low permeability for water vapor and fruit volatiles have been reported to improve the storage and shelf life of some apples (Saftner, 1999).

Apples are stored at 0° to −1°C with 92–95% humidity in regular cold storage for a short duration. Some apples such as McIntosh are susceptible to internal disorders at lower storage temperatures and are often stored at 0°C. Prompt cooling prevents softness in stored apples. Apples stored in this manner are meant to be utilized within 2–3 months. Controlled atmosphere storage has greatly extended the marketing season of apples. This type of storage involves holding apples at approximately 0°C in a facility whose atmosphere contains 1–3% O_2 and 1–3% CO_2 to slow down the respiration of the fruit. The relative humidity in storage is maintained at 92–95%. Under CA conditions, apples can be stored for about 1 year without any appreciable loss of quality. The CA storage requires airtight refrigerated rooms that are sealed after apples are stored inside (and must not be opened for at least 90 days after the seal is affixed). The O_2 content is reduced from atmospheric 21% to 1–3% and CO_2 content is increased from atmospheric 0.25% to 1–3%. Storage disorders (such as "corky" flesh browning, bitter pit, firmness loss, chilling injury, superficial scald, water core, etc.) are not uncommon, and result from inconsistent storage protocols and questionable fruit quality going into long-term storage (Anon, 2003). Lavilla et al. (1999) showed that low oxygen (1.8–2% O_2 /1.8–2% CO_2) and ultra-low oxygen (0.8–1% O_2 /0.8–1% CO_2) CA had better effect on sensory qualities of stored apples than a standard low oxygen (2.8–3.0% O_2 /2.8–3% CO_2) storage.

APPLE BIOCHEMISTRY (FLAVOR, COLOR, TEXTURE, PLANT FLAVONOIDS, ANTIOXIDANT CAPACITY) AND NUTRITION

The good quality that consumers prefer is based on a variety of simple attributes, such as the balance of sweetness (Brix) and tartness (acidity), fresh aroma, firm juicy texture, color, and appearance of apple flesh (Table 16.8). Table 16.9 gives typical characteristics of two popular commercial apple varieties, Golden Delicious and Granny Smith at harvest for two growing seasons. These two varieties are similar in Brix, but the acidity of Granny Smith is almost three times more than Golden Delicious. A Brix to acid ratio is indicative of sweetness and tartness. For example, Golden Delicious and Granny Smith's

Table 16.8. Selected Commercial Apple Varieties

Variety	Main Characteristics	Use
1. Braeburn	Originated in New Zealand; harvest, mid to late October; appearance, pale pink over a yellow-green background; yellow flesh; sweet taste with moderate tartness; texture is crisp and juicy	An all-purpose apple for fresh eating and processing as sauce, pie and baked, juice and cider
2. Empire	A cross between McIntosh and Red Delicious; harvest, late September to early October; dark red appearance, size small to medium (2 3/8″ to 2 $^3/_3$″); sweet with mild tartness; crisp; flesh is creamy white	A good apple for fresh eating and for salad but can be used in other applications as well
3. Fuji	A cross between Ralls Janet and Red Delicious; bred in Japan; popular apple in Japan and China; introduced in United States in 1980s; harvest, September to late October (in some regions late October to mid-November); appearance, golden hued to red; round shape; large to extra large in size (1 $^1/_2$″ to 4″); sweet and aromatic; crisp texture; flesh is whitish yellow; good storage apple	For multiuse. Holds texture after processing
4. Gala	Originated in New Zealand from 'Kidd's Orange Red × Golden Delicious'; harvest, mid to late August; appearance, orange-red strips over creamy yellow; mild sweet flavor; flesh is yellow, firm, and juicy. Good aroma apple	For fresh eating and salad
5. Golden Delicious	Genetically not related to Red Delicious; a famous apple from West Virginia; parentage is believed to be Golden Reinette and Grimes Golden; harvest, mid September to early October; small to medium size (2 3/8″ to 3″) pale greenish-yellow appearance; sweet apple with excellent flavor; juicy yellow flesh; aromatic apple	This is an all-purpose apple used in fresh apple cider; Produces cream colored applesauce
6. Granny Smith	Raised from a chance seed thrown out by an Australian grand mother Maria Ann Smith; a long season fruit, may not ripen before frost, harvest, October to early November; appearance, signature green; size, large (2 $^1/_2$″ to 3 $^1/_2$″); a slightly tart apple with a crisp texture and white flesh	This apple has good eating and cooking qualities, excellent for applesauce, salad and apple juice

(Continued)

Table 16.8. *(Continued)*

Variety	Main Characteristics	Use
7. Ida Red	This variety is developed from Jonathan and Wagener; harvest, mid to late October; large apple (2 1/2″ to 3 1/4″); appearance, bright red; flavor, tangy and tart, flesh is white, firm and juicy	Used as fresh, frozen, canned, sauces and pies
8. Jonagold	A cross between Jonathan and Golden Delicious; harvest, mid-October; size, medium large (2 1/2″ to 3″); appearance is orange red; well-flavored greenish white flesh; fine taste; crisp and juicy	Excellent fresh eating and cooking (pie and baked) quality apple
9. Jonathan	Discovered as a chance seedling in 1820s on a farm in Woodstock, New York; named after the man who promoted it; harvest, mid-September to mid-October; crimson color with touches of green; flavorful with small to medium slices; flesh is off-white, sweet and juicy; blends well with other varieties in sauces and cider; stays firm during cooking	Dual-purpose variety suitable both for fresh eating and for processing; makes good textured applesauce
10. McIntosh	An important commercial variety; harvest, early to mid-September. Appearance, red striped; white juicy flesh, tender skin, medium large fruit (2 1/2″ to 3″); good aroma apple	Mainstay of fresh cider, for eating out of hand and sauce making
11. Mutsu (Crispin)	Origin, Japan; a cross between Golden delicious and Japanese variety Indo; one of the later variety apple with mid-October harvest; Light green to yellow; Very large fruit (1 1/2″ to 3 3/3″); moderately sweet; creamy flesh	Preferred use, fresh, pies and baked
12. Northern Spy	Harvest, mid-October; large size (2 1/2″ to 3 1/4″); yellow-green skin and creamy white flesh; crisp and juicy; spicy, aromatic flavor	Can be used in various processing, including, slices, sauces and pie
13. Red Delicious	Believed to be the most popular variety in the world and in the United States; discovered as a chance seedling on a farm of a non-apple region – central Iowa, glossy red with a distinctive "typey" five-pointed elongated shape. Sweet and flavorful; harvested late August; can generally be found in market in Fall and early winter; good aroma apple	Fresh and salad
14. Rome	Discovered in Rome township, Ohio; harvest, mid to late October; size, large (2 1/2″ to 3 1/4″) bright red skin; sweet, juicy, white flesh; somewhat neutral in flavor	Good processing apple for pies, sauce and as baked

Sources:

- www.michiganapple.com
- www.applejournal.com
- www.paapples.org/varieties.htm
- www.raa.nsw.gov.au; Peterson Farm Inc. Michigan

Table 16.9. Characteristic of Golden Delicious and Granny Smith Apples

Characteristics	Golden Delicious 1993–1994	Golden Delicious 1994–1995	Granny Smith 1993–1994	Granny Smith 1994–1995
Brix (soluble solids)	12.2 ± 0.2	11.8 ± 0.2	11.8 ± 0.2	10.3 ± 0.3
Acidity (% malic acid)	0.39 ± 0.1	0.45 ± 0.7	0.93 ± 0.7	1.0 ± 0.13
Texture (kg/cm^2)	6.7 ± 0.1	6.5 ± 0.3	7.3 ± 0.4	7.8 ± 0.3

Source: Lopez et al. (1998).

Brix/acid values from Table 16.9 are 26–31 and 11.8–12.7, respectively. Thus, Golden Delicious would be perceived twice as sweeter than Granny Smith.

Typically, apples consist of approximately 85% water, 12–14% carbohydrate, about 0.3% protein, insignificant amount of lipids (<0.10%), minerals, and vitamins (Table 16.10, USDA National Nutrient Database, 2003). Variation in these components can be expected because of differences in growing location, variety, harvest maturity, agronomical and environmental conditions.

Table 16.10. Proximate Composition of Apples (raw, with peel)

Component	Value/100 g	Value/serving (154 g)
1. Water	85.56 g	131.76 g
2. Protein	0.26 g	0.40 g
3. Total lipids	0.17 g	0.26 g
5. Ash	0.19 g	0.29 g
4. Total Carbohydrate (by difference)	13.81 g	21.27 g
5. Sugars (total)	10.39 g	16.00 g
6. Sucrose	2.07 g	3.19 g
7. Glucose	2.43 g	3.74 g
8. Fructose	5.90 g	9.09 g
9. Starch	0.05 g	0.08 g
10. Dietary Fiber	2.40 g	3.70 g
11. Calcium	6.00 mg	9.00 mg
12. Iron	0.12 mg	0.18 mg
13. Magnesium	5.0 mg	8.00 mg
14. Phosphorus	11.0 mg	17.00 mg
15. Potassium	107 mg	165 mg
16. Sodium	1.0 mg	2.0 mg
17. Zinc	0.04 mg	0.06 mg
18. Vitamin C	4.60 mg	7.10 mg
19. Vitamin A	54.0 IU	83.0 IU
20. Cholesterol	0.0 mg	0.00 mg
21. Calorie	52.0 Kcal	80.0 Kcal

Source: USDA National Nutrient Database (http://www.nal.usda.gov/fnic/foodcomp/).

Approximately 80% of carbohydrates present are the soluble sugars, sucrose (about 2%), glucose (2.4%), and fructose (6.0%). The total fiber content is about 2%, and 0.2% sorbitol has also been shown to be present in apple juice (Gorsel et al., 1992). Malic acid is the primary organic acid (0.3–1.0%) in apples and the quantity present can vary considerably due to variety, maturity, and environmental conditions during growth and storage (Ackermann et al., 1992). Lopez et al. (2000) reported changes in firmness, acidity, soluble solids, and skin color of Golden Delicious apples during storage with fruits in normal cold storage maturing more quickly than fruits stored under CA.

Flavor

Flavor and aroma of apple are a function of many variables including variety, maturity, environmental conditions of growth, biochemical and metabolic processes regulating ripening, and storage conditions. At harvest, the fruit is at an early preclimacteric stage of ethylene (0.7 μl/kg/h) production and esters comprise more than 185 μg/kg of total volatile compounds. Ethyl acetate, ethyl propionate, and propyl acetate accounted for 73 μg of the total volatile fractions and 2-methylbutyl acetate, 8.0 μg/kg (Lopez et al., 2000). Cunningham et al. (1986) showed beta-damascenone (sweet, perfumy, and fruity odor), butyl isomyl and hexyl hexanoates, along with ethyl, propyl, and hexyl butanoates, are important to the flavor of most apple cultivars. Ester compounds such as ethyl propionate and butyl acetate give the characteristic "apple" flavor; hexyl acetate, "sweet-fruity;" and 1-butanol, "sweetish sensation." Other compounds in volatile fractions of apple are propyl acetate, butyl butyrate, *t*-butyl propionate, 2-methylpropyl acetate, butyl acetate, ethyl butyrate, ethyl 3-methylbutyrate, and hexyl butyrate. Compounds responsible for undesirable flavors such as acetaldehyde (piquancy), *trans*-2-hexanal and butyl

Table 16.11. Concentration of Flavor Volatiles (μg/kg)[a]

Compound	Concentration (μg/kg)
1. Methyl acetate	178.5 ± 78.5
2. Hexyl acetate	70.5 ± 18.9
3. 2-methylbutyl acetate	81 ± 2.8
4. Ethyl propionate	120.0 ± 69.3
5. Ethyl butyrate	147.5 ± 48.8
6. Hexyl butyrate	25.5 ± 0.7
7. Ethyl 2-methyl butyrate	167.0 ± 19.8
8. Hexyl 3-methyl butyrate	35.0 ± 5.7
9. Ethyl hexanoate	169.0 ± 67.9
10. 1-propanol	56.5 ± 7.8

Source: Lavilla et al. (1999).

[a]In Granny Smith apples after 1 day at 20°C followed by 3 months under CA (2% O_2/2% CO_2) storage.

propionate (bitter), 3-methylbutylbutyrate and butyl 3-methylbutarate (rotten) were not found in apple varieties analyzed by Lopez et al. (1998).

The flavor volatiles responsible for characteristic apple flavor are maintained even after 3 months of CA storage (Table 16.11).

Plant Flavonoids

Apples are an important source of plant flavonoids, which are secondary plant metabolites with antioxidant properties, and they have been shown to assist in the amelioration of free radicals that may contribute to aging, development of cancer, and coronary heart disease. Apple and apple products contain flavonols, quercetin glucosides, catechins, anthocyanidins, and hydroxycinnamic acids (Table 16.12–16.14). Flavonols are present more in peel than in flesh, while hydroxycinnamics such as chlorogenic acid is present more in flesh than in peel (Table 16.13; Tsao et al., 2003). Apple varieties differ in their flavonoid concentration. Jonagold was shown to have relatively higher concentration than Golden Delicious, Elstar, and Cox's orange (Table 16.12; Sluis et al., 2001). Brands of apple juice differ in their phenolic content and fresh apple juice has greater phenolics than stored juice (Table 16.14). Recent investigation (Wu et al., 2004) of total antioxidant capacity (TAC) of selected apple varieties indicated a range of 3578–5900 μmol of TE/serving of apple (Table 16.15). Apples with peel had higher TAC values than without peel. Among the varieties analyzed, Red Delicious had the highest TAC (5900 μmol) followed by Granny Smith (5381 μmol), Gala (3903 μmol), Golden Delicious (3685 μmol), and Fuji (3578 μmol).

Nutritional Quality

Apart from various plant flavonoids present in apple, it is endowed with natural sugars and dietary fiber (80% of which are soluble fibers), various minerals, and vitamins (Table 16.10). Fresh eating apples are considered as a natural dessert with little fat and cholesterol, and less than 100 calorie per serving size. Apples are also a good source of potassium, which have beneficial role in blood pressure regulation. Nutrient profile of various apple products is given in Table 16.16.

PROCESSED APPLE PRODUCTS

Table 16.6 shows utilization of apples in various forms. Apples are used as ingredients in bakery, dairy products, cereals, snack bars, confection, etc. In this section, selected processed apple products are discussed.

Apple Juice and Cider

Apple juice is low in calories and is a natural source of sugars (Table 16.16). Apple juice (and cider) making

Table 16.12. Concentration of Flavonoids in Selected Apple Varieties[a] (μg/gram of fresh weight)

Compound	Jonagold	Golden Delicious	Cox's Orange	Elstar
1. Total Q-glycosides[b]	95 ± 11	67 ± 11	64 ± 12	63 ± 12
2. Total catechins[c]	145 ± 37	121 ± 29	106 ± 47	152 ± 42
3. All compounds[d]	467 ± 86	385 ± 108	265 ± 98	326 ± 100

Source: Sluis et al. (2001).

[a]Mean (± SD)

[b]Total quercetin glycosides

[c]Catechin and epicatechin

[d]Sum of all phenolic compounds analyzed

Table 16.13. Average Concentration (μg /gram fresh weight) of Phenolic Compounds in Peel and Flesh of Apple[a]

Compound	Peel	Flesh
1. Total hydroxycinnamics[b]	148.5	193.0
2. Total procyanidins[c]	958.2	267.7
3. Total flavonols[d]	288.2	1.3
4. Total dihydrochalcones[e]	123.7	19.3
5. Total polyphenolics (HPLC)[f]	1604.4	481.3
6. Total phenolic content (F-C)[g]	1323.6	429.6

Source: Tsao et al. (2003).

[a]Based on eight apple varieties

[b]Chlorogenic and p-coumaroylquinic acid

[c]Catechin, epicatechin and other procyanidins

[d]Quercetin -3-galactoside, glucoside, xyloside, arabinocide, rhamnoside

[e]Phloridzin and ploreitin

[f]Calculated on the basis of total phenolics calculated by HPLC

[g]Phenolic content measured by Folin–Ciocalten method

requires separation (extraction/expression) of liquid (juice) from solids. Thus, these products are extracted liquid containing soluble solids. Cider is cloudy because of pulp (solids), amber golden in appearance, and is often unfiltered and unpasteurized (which can pose safety concerns). However, it is not a fermented product like a "hard cider" with more than 0.15% alcohol by volume. The characteristic cider flavor varies from region to region and is based on types of apples used in the manufacturing process.

Due to its acid tolerance and low infectious dose (10–2000 cfu/g), *Escherichia coli* 0157:H7 has become an organism of concern in unpasteurized juices. In 2001, the FDA promulgated rules that required juice processors to use science-based HACCP principles and sanitary standard operating procedures (SSOPs) for juice-making because of the risk of foodborne illness from consumption of natural and unpasteurized juice products. Chikthimmah et al. (2003) reported a 5-log cycle reduction in *E. coli* 0157:H7 in cider by adding fumaric acid (0.15% w/v) and sodium benzoate (0.05% w/v) to the product. The use of these preservatives lowered the natural pH of cider (i.e., 3.40–3.87) to pH 3.19–3.41 and inactivated any *E. coli* organisms in the product.

Patulin, a mycotoxin produced by certain species of Penicillium, Aspergillus, and Byssochylamys molds that may grow on harvested apple is a regulatory concern. The regulations typically limit patulin content in apple juice to no more than 50 μg/kg (Kryger, 2001).

Generally, apple juice sold in the United States is pasteurized. However, a large part of apple juice is sold as frozen concentrate as well. The processing steps and mechanism for making apple juice may vary between processors, but the basic steps are summarized below.

Fruit selection and preparation (air cleaning, washing, inspection, grading, quality checks, etc.) –> Milling/slicing (a critical step to crush/mash/slice the apples using bars/knives) –> Pressing/Extraction –> Clarification /Filtration –> Pasteurization –> Packaging (hermetically sealed in cans or bottles) or asceptically processed and packaged. Usually a blend of several apple varieties are used to provide flavorful apple juice and cider. A batch hydraulic press (can be a rack and frame press where milled apple pieces are placed in thick fabric separated by wooden racks, and hydraulic pressure is applied from the top of the rack to release the juice) and various types of mechanical pressing systems can be used. The latter type of integrated system is often the equipment of choice in commercial operations because of better efficiencies in yield, quality, and process controls. It also enables low residual moisture in the press (filter cake or pomace), which provides lower disposal costs. Apple juice yield in a commercial operation ranges from 70% to 80%.

Press designs such as belt press (where milled apple pulp is held between belts and juice is released through pressure from rollers) and screw press (made up of tapered screws to convey and squeeze juice against a close fitting screen) are also used.

Table 16.14. Concentration of Phenolic Acids, Flavonoids and Total Polyphenols (mg/L) in Fresh and Stored Apple Juice

Component	Fresh Juice	Stored Juice
1. Phenolic acids	43.54 ± 0.45 to 93.07 ± 0.39	34.44 ± 0.17 to 84.46 ± 0.09
2. Flavonoids[a]	20.30 ± 0.31 to 92.11 ± 3.19	17.55 ± 0.17 to 74.77 ± 0.39
3. Total polyphenols	63.84 ± 0.71 to 163.35 ± 3.62	51.99 ± 0.18 to 139.43 ± 0.36

Source: Gliszczynska-Swiglo and Tyrakowska (2003).

[a]Quercetin glucosides, phloridin and kaempferol

Table 16.15. Lipophilic (L-ORAC$_{FL}$), Hydrophilic (H-ORAC$_{FL}$), Total Antioxidant Capacity and Total Phenolics of Selected Apple Varieties

Apple	% Moisture	L-ORAC$_{FL}$ (μmole of TE/g)	H-ORAC$_{FL}$ (μmole of TE/g)	TAC[a] (μmole of TE/g)	TP[b] (mg GAE per/g)	Serving size (g)	TAC/s[c] (μmole of TE)
1. Fuji	84.2	0.21 ± 0.11	25.72 ± 6.96	25.93	2.11 ± 0.32	138 g (1 fruit)	3578
2. Gala	85.8	0.35 ± 0.08	27.93 ± 1.42	28.28	2.62 ± 0.29	138 g (1 fruit)	3909
3. Golden Delicious (with peel)	86.1	0.26 ± 0.06	26.44 ± 1.61	26.70	2.48 ± 0.18	138 g (1 fruit)	3685
4. Golden Delicious (No peel)	86.9	0.05	22.05	22.10	2.17	128 g (1 fruit)	2829
5. Red Delicious (with peel)	85.5	0.41 ± 0.02	42.34 ± 4.08	42.75	3.47 ± 0.38	138 g (1 fruit)	5900
6. Red Delicious (No peel)	86.7	0.07	29.29	29.36	2.32	128 g (1 fruit)	3758
7. Granny Smith	85.7	0.39 ± 0.11	38.60 ± 4.69	38.99	3.41 ± 0.38	138 g (1 fruit)	5381

Source: Wu et al. (2004).

[a]TAC = Total antioxidant capacity (L-ORAC$_{FL}$ + H-ORAC$_{FL}$) as Trolox equivalent/g.

[b]TP = Total phenolics as mg of gallic acid equivalent/g.

[c]TAC/s = Total antioxidant capacity per serving size.

Table 16.16. Nutrient Profile of Apple Products

Nutrients/100 g	Apple Juice[a]	Apple Juice Concentrate[b]	Apple Sliced, Canned[c]	Applesauce[d]	Dehydrated Apple[e]	Infused Dried Apple[f]
Calories (Kcal)	47.0	166.0	67	43	346.0	336.0
Total fat (g)	0.11	0.37	0.43	0.05	0.58	1.93
Saturated fat (g)	0.019	0.06	0.07	0.008	0.095	0.20
Polyunsaturated fat (g)	0.033	0.108	0.126	0.014	0.171	0.30
Monounsaturated fat (g)	0.005	0.015	0.017	0.002	0.024	1.4
Cholesterol (mg)	0.00	0.0	0.0	0.0	0.0	0.0
Sodium (mg)	3.0	25	3.0	2.0	124.0	31.0
Potassium (mg)	119.0	448	70.0	75.0	640.0	64.0
Total carbohydrate (g)	11.68	41.0	16.84	11.29	93.53	81.7
Total fiber (g)	0.1	0.4	2.0	1.2	12.4	6.3
Total sugar (g)	10.90	38.83	14.84	10.09	81.13	75.5
Sucrose (g)	1.70	NA	NA	NA	NA	NA
Glucose (g)	2.50	NA	NA	NA	NA	NA
Fructose (g)	5.60	NA	NA	NA	NA	NA
Protein (g)	0.06	0.51	0.18	0.17	1.32	2.26
Calcium (mg)	7.0	20.0	4.0	3.0	19.0	65.0
Iron (mg)	0.37	0.91	0.24	0.12	2.00	0.42
Vitamin C (mg)	0.9	2.1	0.2	1.2	2.2	372.0
Vitamin A (IU)	1.0	0.0	56.0	29.0	81.0	20.0
Water (g)	87.93	57.0	82.28	88.35	3.0	12.7

Source: http://www.nal.usda.gov/fnic/foodcomposition/cgi-bin/list_nut_edit.pl

NA: Not available.

[a]Canned or bottled, unsweetened, without added ascorbic acid

[b]Frozen concentrate without added ascorbic acid

[c]Canned, sweetened sliced apple: Drained and heated

[d]Canned unsweetened apple sauce without added ascorbic acid

[e]Low moisture dehydrated, sulfured apples

[f]Data of commercial product, courtesy Graceland Fruit Inc, Frankfort, MI, USA: Infused with sugar prior to drying, contains ascorbic acid and high oleic sunflower oil.

Juice extraction can also be done using a centrifuge. In counter current extraction, juice from apple pieces is extracted using liquid such as water to strip juice solids. Reverse osmosis and ultra filtration are then used to concentrate the juice.

Presence of colloidal pectic substances, starch, cellulose, hemicelluloses, etc. cause haze, cloudiness, and sedimentation in apple juice. Therefore, after extraction, apple juice is treated with enzymes to remove these suspended particles, and subsequently, the juice is clarified. Use of a combination of ascorbic acid and citric acid can be used in place of sulfites to prevent enzymatic browning.

Three types of enzymes hydrolyzing pectic substances termed pectinases are: (1) Pectin esterase, or pectin methyl esterase, or pectase, or pectin demethoxylase (EC 3.1.111), this enzyme is naturally present in fruits, it hydrolyzes methyl ester on number 6 carbon of the methylated galactose unit in pectin, producing pectic acid. This enzyme enables firm texture and has little application in apple juice production.

(2) Polygalacturonases, which act on glycosidic linkage, are responsible for fruit softening. This enzyme is also found in fruits. It has an endo form (EC 3.2.1.15) and two exoforms (EC 3.2.1.67 and EC 3.2.1.82). Use of these enzymes is not so much as a clarification aid but as a press aid to help in juice extraction. (3) Pectin lyase (EC 4.2.2.10): is found in microorganisms. This enzyme breaks the pectin chain by acting on glycosidic linkage through a β-elimination process (where unlike hydrolytic action water is not required). A combination of endo- and exo-lyases can be used to degrade pectin as a clarification aid. Commercial enzyme preparation may contain other enzymes (amylases and cellulases). The required enzyme doses, time, and temperature would depend upon specific enzymes and the clarification

time is inversely proportional to the enzyme concentration used. Use of diatomaceous earth, gelatin (preparation containing 0.005–0.01% gelatin, works by agglomeration due to electrostatic attraction), centrifugation, and filtration are practiced to remove haze-forming particles (Kilara and VanBuren, 1989). Following the enzymatic treatment and removal of suspended plant materials, the juice is filtered and then pasteurized (88°C/1 min).

Apple Slices

Fresh pack and frozen apple slices are in demand for various applications and as snacks. A number of different techniques are used to maintain freshness, color, flavor, texture, and firmness of apple slices. The softening of apple tissue due to loss of cell fluids and enzymatic browning are the subject of many investigations including modified atmosphere packaging, low-temperature storage, and addition of preservatives. The polyphenol oxidase enzyme catalyzes oxidation of phenolic compounds into *o*-quinines, which polymerize to form dark brown pigments. Sulfites are very effective in preventing this type of browning but have been linked to allergic reaction, especially in asthmatic individuals; consequently, the FDA regulates its use. The alternatives to sulfites are based on ascorbic acid, citric acid, and calcium salts.

Santerre et al. (1988) reported sulfite alternatives, ascorbic acid, or D-araboascorbic acid/citric acid, and calcium chloride combinations are effective in preserving color of frozen apple slices. The role of oxygen in enzymatic browning and cellular damage was evident from lower firmness values and greater amount of loss of cell fluid in apples packed in a 2.5% O_2 and 7% CO_2 atmosphere package than in 100% N_2 atmosphere package (Soliva-Fortuny et al., 2001, 2003). Calcium ascorbate or calcium erythorbate and ascorbic acid applications ,and storage temperatures of -7–20°C have been claimed to extend the shelf life of fresh-cut fruits including apples (Chen et al., 1999; U.S. patent 5,939,117). Powrie and Hui (1999) (U.S. patent 5,922,382) described a method of preserving fresh apples by immersing apple pieces in an acid solution containing 5–15% ascorbic acid and erythorbic acid (pH 2.2–2.7) for up to 3 min, followed by removal of excess solution from fruit surfaces, quick chilling, and storing at 0°–10°C.

Applesauce

Besides apples, applesauce can contain sugar and other sweeteners, honey, acidulants, salt, flavorings, spices, preservatives, etc. The desired characteristics of applesauce are golden, creamy color, a balance of sweetness and tartness, glossy texture, which is not soft and mushy. The Brix of unsweetened applesauce is about 9.0. However, the Brix of sweetened applesauce is about 16.0. The applesauce has a pH of approximately 3.4–4.0. A summary of processing steps involved are:

Dice/quarter or chop washed apples (a blend of several apple cultivars, in peeled or unpeeled forms can be used) –> process/cook apple pieces, which is similar to blanching of apples using steam (this step is critical for consistency and texture of the finished product and is closely controlled) –> Pass through a pulper/finisher (screen size, 0.16–0.32 cm; for coarse grainy sauce, 0.25–0.32 cm) to remove seeds, peels, etc. –> Fill and seal in containers (the fill weight is 90% of the container's capacity and allow for a head space of about 0.6 cm upon cooling) –> Heat process (if the fill temp is $\leq$88°C, heat containers in boiling water for 10–15 min in an open container to insure microbiological safety; a high-pressure processing ensures product safety and longer shelf life by destroying spoilage organisms) –> Invert containers and cool.

Processed Frozen Apples

Apple can be infused and pasteurized for use in frozen desserts such as ice cream, frozen yogurt, etc., or for addition into baked good. The water content of apple is stabilized by infusion with sweeteners or by incorporating stabilizer such that fruit cells are not hard or icy when frozen. The products are pasteurized and are ready to use as ingredients in frozen desserts which cannot use IQF or frozen fruits. The process maintains fruit piece identity, natural color, and flavor (Sinha, 1998).

Dried Apples

Drying as a preservation method for foods has been practiced since the earliest times of recorded history. A number of drying techniques are available (i.e., sun drying, atmospheric drying using heated air, vacuum, and freeze drying) but they all must take into consideration the economics and efficiencies of each type of operation. Prior to the advent of modern techniques of preservation, drying was one of the few methods available to conserve surplus quantities of products. Since drying was more or less a salvage operation, very little emphasis was placed on utilizing important parameters (i.e., variety, maturity of product, postharvest handling, uniform size, and shape, etc.) that are required for obtaining good quality dried products. This is not true in the current market where consumers demand quality and are willing to pay for it. Thus, the aim of drying is to preserve as much quality as possible, while providing shelf stability of the product.

For a given type of dryer, the drying rate at a given temperature depends on pre-treatments (blanching, etc.), type of product (fresh or frozen) with or without skin, composition, size and geometry (sliced, diced, etc.), and drying load or feed rate. Generally, the drying profile can be divided into: (i) a short period of quick water evaporation per unit time. In this phase, the fruit contains relatively high amounts of free moisture not tightly bound to the constituents of the fruit cell and this free moisture can be evaporated fairly quickly; (ii) followed by a steady or constant drying rate; and (iii) a falling or declining rate of water evaporation toward the end of cycle, because the remaining water is tightly held by fruit components and is difficult to dislodge.

Percent moisture is commonly referred to in dried fruit; however, completion of drying should be determined by monitoring water activity (a_w). The cutoff a_w of shelf stable dried fruit should be below 0.65; above this a_w mold can grow on the dried fruits (Beuchat, 1981) if it does not contain antimycotic agents.

The color of dried apple is often of concern. About 20% of the non-enzymatic browning reaction occurring during storage have been attributed to non-oxidative, or Maillard reaction and about 70% to oxidation. Cysteine, which is helpful in preventing non-enzymatic browning, did not reduce browning during storage (Bolin and Steele, 1987). Pineapple juice was an effective browning inhibitor in both fresh and dried apples. In this case, the inhibitor(s) was not a high molecular weight protein such as bromelain, but a neutral compound of low molecular weight (Lozano-de-Gonzalez et al., 1993).

Most dried foods with reduced moisture are partly or completely amorphous. There is a glass transition (T_g) theory, which is based on the concept that a glass (amorphous material) is changed into a supercooled melt or liquid during heating, or to reverse transformation during cooling. The glass transition phenomena have implications for textural transformation in sugar-containing dried fruits. Mobility of water is high in glassy food systems; a heterogeneous distribution of water is often desired to provide soft dried texture. In foods, plasticization (T_g decreases) is mainly due to water, but other solutes such as glycerin may also act as plasticizers to enable soft textured dried products (Le Meste et al., 2002).

The T_g of freeze-dried apples has been reported to be 33.8°C ± 3.4°C. The T_g is shown to increase with the temperature of drying. The glass transition represents the thermal limit between a state where the molecular mobility is rather low (no reaction) and a state of increased molecular mobility, which favors diffusion and other reactions. Thus, a freeze-dried strawberry with a T_g of 45.4°C ± 2.2°C would be less susceptible to color changes than freeze-dried apples (T_g of 33.8°C ± 3.4°C) or pear (T_g of 24°C ± 2.6°C). However, other factors (besides T_g) such as internal fruit structure, porosity, thermal conductivity, matrix elasticity, temperature within the product, and rate of heat transfers during drying can affect fruit quality (Khalloufi and Ratti, 2003).

The hygroscopic properties of dried apple chips are mainly due to their composition of sugars and pectins, in combination with their porous structure. At a_w below 0.12, apple chips demonstrated excellent crispness (Konopacka et al., 2002). During air-drying of apple rings, non-uniform moisture and/or temperature distribution can cause varying degree of shrinkage. Apples dried at 60–65°C showed higher cellular collapse than those dried at 40–45°C and 20–25°C (Bai et al., 2002). Infused dried apples retain their shape and texture much better than those that are traditionally dried, so apples thus dried have become value-added ingredients for uses in various foods and as snack items. Unlike other forms of dried apples, infused dried apples can be diced into pieces for incorporation into foods. With infusion-drying, use of sulfites is not warranted (Sinha, 1998). The infused dried apples are also not high in calorie (Table 16.16).

REFERENCES

Ackermann J, Fischer M, Amado R. 1992. Changes in sugars, acids, and amino acids during ripening and storage of apples (Cv. Glockenapfel). J Agric Food Chem 40:131–134.

Anon. 2004. Apple—*Malus domestica* Borkh. http://www.uga.edu/fruit/apple.htm

Anon. 2003. Harvest and Post Harvest Handling. http://tfpg.cas.psu.edu

Bai Y, Rahman S, Perera CO, Smith B, Melton LD. 2002. Structural changes in apple rings during convection air-drying with controlled temperature and humidity. J. Agric Food Chem 50:3179–3185.

Beuchat LR. 1981. Microbial stability as affected by water activity. Cereal Food World 25 (7):345–349.
Bolin HR, Steele RJ. 1987. Nonenzymatic browning in dried apples during storage. J. Food Sci 52 (6): 1654–1657.
Chen C, Trezza TA, Wong DWS, Camirand WM, Pavlath AE. 1999. U.S. Patent 5,939,117. Methods for preserving fresh fruit and product thereof.
Chikthimmah N, Laborde LF, Beelman RB. 2003. Critical factors affecting the destruction of *E. coli* 0157:H7 in apple cider treated with fumaric acid and sodium benzoate. J. Food Sci 68 (4):1438–1442.
Cunningham DG, Acree TE, Bernard J, Butts RM, Braell PA. 1986. Charm analysis of apple volatiles. Food Chem 19: 137–147.
Drake MA, Drake SR, Eisele TA. 2002. Influence of Bioregulators on Apple Fruit Quality. http://ift.confex.com/ift2002/techprogram/paper_11812.htm
FAO. 2003. FAO Crop Database. Food and Agriculture Organization. http://www.faostat.org
Gliszczynska-Swiglo A, Tyrakowska B. 2003. Quality of commercial apple juices evaluated on the basis the polyphenol content and the TEAC antioxidant activity. J Food Sci 68 (5):1844–1849.
Gorsel H, Li Chingying, Eduardo L, Kerbel, Smits, Mirjam, Kader AA. 1992. Compositional characterization of prune juice. J Food Agric Chem 40: 784–789.
Khalloufi S, Ratti C. 2003. Quality deterioration of freeze-dried foods as explained by their glass transition temperature and internal structure. J Food Sci 68 (3): 892–903.
Kilara A, VanBuren JP. 1989. Clarification of apple juice. In Processed Apple Products. Edited by DW Downing. Van Nostrandt Reinhold, New York, p. 83–96.
Konopacka D, Plocharski W, Beveridge T. 2002. Water sorption and crispness of fat-free apple chips. J Food Sci 67 (1):87–92.
Kryger RA. 2001. Volatility of patulin in apple juice. J Agric Food Chem 49:4141–4143.
Lavilla T, Puy J, Lopez ML, Recasens I, Vendrell M. 1999. Relationships between volatile production, fruit quality, and sensory evaluation in Granny Smith apples stored in different controlled atmosphere treatments by means of multivariate analysis. J Agric Food Chem 47: 3791–3803.
Le Meste M, Champion D, Roudaut G, Blound G, Simatos D. 2002. Glass transition and food technology. J Food Sci 67 (7): 2444–2458.
Lopez ML, Lavilla MT, Riba M, Vendrell M. 1998. Comparison of volatile compounds in two seasons in apples: Golden delicious and Granny Smith. J Food Qual 21: 155–156.
Lopez ML, Lavilla MT, Recasens I, Graell J, Vendrell M. 2000. Changes in aroma quality of Golden Delicious apples after storage at different oxygen and carbon dioxide concentrations. J Sci Food Agric 80: 311–324.
Lozano-de-Gonzalez PG, Barrett DM, Wrolstad RE, Drust RW. 1993. Enzymatic browning inhibited in fresh and dried apple rings by pineapple juice. J Food Sci 58(2): 399–404.
Powrie WD, Hui WC. 1999. U.S. Patent 5,922,382. Preparation and preservation of fresh, vitaminized, flavored and unflavored cut apple pieces.
Saftner RA. 1999. The Potential of Fruit Coating and Film Treatments for Improving the Storage and Shelf-life Quantiles of "Gala" and "Golden Delicious" Apples. J. Amer. Soc. Hort. Sci. 124 (6): 682–689.
Santerre CR, Cash JN, Vannorman DJ. 1988. Ascorbic acid/Citric acid combinations in the processing of frozen apple slices. J Food Sci 53 (6): 1713–1716, 1736.
Sinha NK. 1998. Infused-dried and processed frozen fruits as food ingredients. Cereal Food World 43 (9): 699–701.
Sluis AA, Dekker M, Jager A, Jongen WM, 2001. Activity and concentration of polyphenolic antioxidants in apple: Effect of cultivar, harvest year, and storage conditions. J Agric Food Chem 49: 3606–3613.
Soliva-Fortuny RC, Lluch MA, Quiles A, Miguel N, Belloso O. 2003. Evaluation of textural properties and microstructure during storage of minimally processed apples. J Food Sci (1): 312–317.
Tsao R, Yang R, Young C, Zhu Honghui. 2003. Polyphenolic profile in eight apple cultivars using high-performance liquid chromatography. J Agric Food Chem 51: 6347–6353.
USDA. 2003. Fruit and Tree Nuts Outlook. United States Department of Agriculture, Economic Research Service. http://www.ers.isda.gov/publication/FTS/ Yearbook03/fts2003.pdf
USDA 2003. National Nutrient Database. http://www.nal.usda.gov/fnic/foodcomp/
Wu X, Beecher GR, Holden JM, Haytowitz DB, Gebhardt SE, Prior RL. 2004. Lipophilic and hydrophilic antioxidant capacities of common foods in the United States. J Agric Food Chem 52: 4026–4037.

17
Apricots

Muhammad Siddiq

INTRODUCTION

Apricot (from Latin meaning early ripe) tree belongs to genus *Prunus* in the Rosaceae (rose) family. Apricot (*Prunus armeniaca*) is categorized under "stone fruits," along with peaches, plums, almonds, and some cherries, due to its seed being enclosed in a hard, "stone" like endocarp. This fruit is native to temperate Asia, first discovered growing wild on the mountain slopes of China, and long cultivated in Armenia later. Early Spanish explorers are credited with introducing apricots to California. The apricot tree is of medium size, usually held under 18 ft by pruning. The fruit is generally globose to slightly oblong, 1.25–2.5 in. in diameter; the fruit flesh is yellow and the skin is yellow or blushed red. The early apricot bloom is highly susceptible to injury by frost. Fruits are subject to cracking in humid climates; in the U.S. commercial production is primarily limited to regions west of the Rocky Mountains (Anon, 2004a; Magness et al., 1971).

World production of apricots from 1997 to 2003 is shown in Figure 17.1. Turkey is the leading apricot producer next to Iran; these countries accounted for about 30% share of the total world production in 2003 (Table 17.1). Turkey and Iran also lead in area under apricot cultivation with about one-fourth of the world total (Table 17.2). However, on yield per hectare basis, Greece is the leading country with 16 mt of apricots per hectare, which is about 2.5 times the world average; in comparison, China's yield per hectare is only 2.9 mt (Table 17.3). The United States, which is ninth in world production and eleventh in area harvested, is ranked second on yield per hectare basis, mainly due to the use of good agricultural practices. The U.S. average yield of 12.2 mt/ha is almost twice the world average.

The first major production of apricots in the United States was recorded in California in 1792. California continues to be the leading apricot producing state in the United States (see Box 17.1 for more details). Over 90% of the apricots in the United States are grown in California, with much smaller production in Washington and Utah. The per capita consumption of fresh apricots and their processed products in the United States, is shown in Table 17.4. More than half of per capita consumption of apricots is in the dried form.

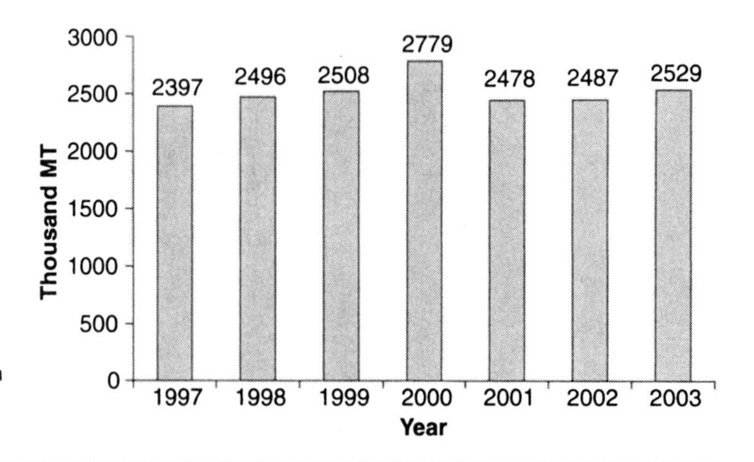

Figure 17.1. World apricot production (1997–2003). (Adapted from FAO (2004) data.)

Table 17.1. Apricot Production in Leading Countries (1997–2003)

Country	Production (1000 mt)						
	1997	1998	1999	2000	2001	2002	2003
Turkey	306.0	540.0	378.0	579.0	470.0	315.0	440.0
Iran	225.0	243.3	240.7	262.4	282.9	284.0	284.0
Spain	141.9	149.8	148.9	142.5	134.8	119.2	142.3
Pakistan	189.0	190.9	120.5	125.9	124.7	125.0	125.0
France	158.1	80.3	180.9	138.9	103.2	171.8	111.0
Italy	102.9	135.6	212.2	201.4	193.8	200.1	108.5
Syria	34.7	67.2	62.9	78.9	66.0	100.9	100.9
Morocco	103.6	116.8	106.4	119.6	104.3	86.2	98.0
United States of America	126.1	107.5	82.1	79.7	74.8	81.7	88.8
Ukraine	101.4	59.6	57.2	102.1	44.0	85.0	80.0
Greece	47.2	38.9	85.3	82.0	68.5	74.7	75.0
Russian Federation	57.0	38.0	50.0	67.0	50.0	80.0	75.0

Source: FAO (2004).

Table 17.2. Apricot: Area Harvested in Leading Countries and World Total (1997–2003)

Country	Area Harvested (1000 ha)						
	1997	1998	1999	2000	2001	2002	2003
Turkey	61.2	61.0	62.3	63.7	64.1	63.5	63.5
Iran	27.3	27.7	28.0	29.0	30.6	32.0	32.0
Spain	24.5	25.0	24.8	23.5	20.8	23.3	22.6
Russian Federation	16.7	16.8	17.0	18.0	19.0	20.0	20.0
Italy	14.8	15.0	15.2	15.3	15.3	15.3	17.2
France	16.1	16.0	15.9	15.9	15.7	15.0	15.0
Pakistan	12.1	12.2	12.5	12.9	13.0	13.0	13.0
Syria	9.0	12.4	12.4	12.4	12.5	12.6	12.6
Morocco	13.8	13.6	13.9	13.9	13.7	12.5	12.6
Ukraine	19.3	12.0	12.0	11.6	11.0	10.1	10.0
United States of America	8.7	8.7	8.2	8.2	7.9	7.5	7.3
Greece	4.7	4.7	4.7	4.7	4.7	4.7	4.7
World	397.1	378.0	383.0	390.6	390.8	401.0	398.9

Source: FAO (2004).

Table 17.3. Apricot: Yield Per Hectare, Leading Countries and World Average (1997–2003)

Country	Yield (mt/ha)						
	1997	1998	1999	2000	2001	2002	2003
Greece	10.0	8.3	18.1	17.5	14.6	15.9	16.0
United States of America	14.6	12.4	10.0	9.7	9.5	10.9	12.2
Pakistan	15.6	15.7	9.6	9.8	9.6	9.6	9.6
Iran	8.2	8.8	8.6	9.0	9.2	8.9	8.9
Syria	3.9	5.4	5.1	6.4	5.3	8.0	8.0
Ukraine	5.3	5.0	4.8	8.8	4.0	8.4	8.0
Morocco	7.5	8.6	7.6	8.6	7.6	6.9	7.8
France	9.8	5.0	11.4	8.7	6.6	11.5	7.4
Turkey	5.0	8.9	6.1	9.1	7.3	5.0	6.9
Spain	5.8	6.0	6.0	6.1	6.5	5.1	6.3
Italy	7.0	9.1	14.0	13.1	12.6	13.1	6.3
Russian Federation	3.4	2.3	2.9	3.7	2.6	4.0	3.8
World (Average)	6.0	6.6	6.5	7.1	6.3	6.2	6.3

Source: FAO (2004).

Box 17.1. California Apricot Industry Facts

- California produces over 90% of the apricots grown in the United States
- There are over 200 plus growers producing apricots from orchards covering 17,000 acres in the San Joaquin Valley, with the leading production area being Stanislaus County
- A consistent producer, the Patterson variety accounts for the majority of all California apricots
- The harvest period typically begins in May and ends in early July
- In 2003, California harvested 77,300 t of apricots

Source: Anon (2004b).

Table 17.4. U.S. Per Capita Consumption of Apricots, in lb (1998–2002)

			Processed			
Year	Total	Fresh	Total	Canned	Frozen	Dried
1998	1.2	0.1	1.0	0.3	0.1	0.7
1999	1.0	0.1	0.8	0.2	0.1	0.5
2000	1.2	0.2	1.1	0.2	0.1	0.8
2001	1.1	0.1	1.1	0.2	0.1	0.7
2002	0.9	0.1	0.8	0.2	0.1	0.5

Source: Adapted from USDA-ERS (2004).

PRODUCTION AND POST-HARVEST PHYSIOLOGY

PLANT BREEDING

Apricot cultivation is most successful in mild, Mediterranean-type climates, where spring frost is minimal. Deep, fertile, and well-drained soils are best suited to produce tall and healthy trees. Apricot trees exhibit moderate tolerance to high pH soils and salinity but are intolerant to waterlogging.

Most American cultivars are self-pollinating, exceptions being "*Riland*" and "*Perfection*." Apricots are pruned fairly heavily as they bear too many fruits. Generally, most new growth and interfering wood are removed each year, exposing the spurs to maximum sunlight (Rieger, 2004; Westwood 1993). Perez-Gonzales (1992) reported that the important morphological traits of the apricot trees that correlated with fruit weight were tree growth habit, apical and basal diameter of fruiting spurs, and bud and leaf size. Timing and type of fertilizer application are reported to improve external and internal fruit quality and storage ability, reduce production costs, maintain soil fertility, avoid nutrient deficiency or excess, and control tree vigor (Tagliavini and Marangoni, 2002).

VARIETIES

A list of apricot varieties grown in California along with their maturity dates, size, and flavor characteristics is shown in Table 17.5. *Patterson* continues to be the most widely grown apricot variety in California. It has firm texture, good flavor, and an excellent shelf-life. Some other varieties include: *Perfection*—a mid to late season variety that is oval, oblong in shape with excellent flavor, its skin is clear yellow/orange with a blush, and its flesh is firm and yellow/orange in color; *Rival*—a mid-season variety that is oval in shape, and has sweet/tart flavor with firm texture; the skin is light orange with a red blush; *Moorpark*—large fruit with red blush, yellow flesh, sweet, and juicy, with excellent flavor, and is good for canning and eating fresh; *Moongold*—soft golden colored fruits are of medium size, attractive orange yellow flesh, firm and sweet with excellent flavor, good for eating fresh or making preserves; *Sungold*—medium sized, brightly colored clear gold with attractive orange blush, nearly round, and tender skinned, good for eating fresh or making preserves (Anon, 2004c, d).

MATURITY AND FRUIT QUALITY

Changes in the skin ground color from green to yellow are used to determine maturity and harvest date in California. The yellowish-green color is cultivar-dependent. Since apricots are susceptible to high bruising when fully ripe and soft, picking is recommended when the fruit is still firm. Important quality criteria include fruit size, shape, freedom from defects (including gel breakdown and pit burn), and decay. Soluble solid contents of over 10% and a moderate titratable acidity of 0.7–1.0% are critical for high consumer acceptance. Fruit with flesh firmness of 2–3 lb force (8.9–13.3 N) is ready for fresh consumption (Crisosto and Kader, 2004). Femenia et al. (1998) reported that the decrease in the pectic galactans and inhibition of cross-linking within the pectic backbone was related to the softening during apricot ripening. During post-harvest storage of apricots, increase in β-galactosidase activity results in lower total pectin and hence the loss of fruit firmness (Kovacs and Nemeth-Szerdahelyi, 2002).

HARVEST AND STORAGE

Apricots for fresh consumption and processing are mostly picked by hand and usually picked over 2–3 times each when fruits are firm. There is a little flavor development in apricots ripened off the trees (Gomez and Ledbetter, 1997). Apricots are generally handled in half bins, tray-packed in single/double layers, and shipped in shallow containers to prevent crushing and bruising. Apricots should be uniform in size, and by count, not more than 5% apricots in each container may vary more than 6 mm when measured at the widest part of the cross-section.

Apricots can be kept well for 1–2 weeks (or even 3–4 weeks for some cultivars) at −0.5 to 0°C with RH of 90–95%. Susceptibility to freezing injury depends on soluble solid contents, which vary from 10% to 14%. Cultivars that are sensitive to chilling develop and express chilling injury symptoms (gel breakdown, flesh browning), and loss of flavor, more rapidly at 5°C than at 0°C. In order to minimize chilling injury on susceptible cultivars, storage at 0°C is recommended. Controlled atmosphere (CA) storage conditions of 2–3% O_2 + 2–3% CO_2 are suggested to retain fruit firmness and ground color. The development of off-flavors may hasten at O_2 exposure of <1% and CO_2 exposure in excess of 5% for over 2 weeks. Pre-storage treatment with 20%

Table 17.5. Maturity Dates and Quality Characteristics of Apricot Varieties Grown in California

Apricot Variety	Average Maturity Date	General Size Trend	Profile/Flavor
Poppy	7 May	Medium/large	Solid orange/good flavor
Earlicot	11 May	Large	Firm/high color
Lorna	12 May	Large	Deep orange/juicy
Robada	17 May	Medium	Red blush/high sugar
Ambercot	25 May	Medium	Firm/sweet
Jordanne	25 May	Large	High color/flavor
Castlebrite	28 May	Medium	Firm/full flavor
Katy	5 June	Large	Firm/good flavor
Helena	8 June	Large	Firm/juicy
Tri Gem	11 June	Large	Firm/good flavor
Goldbar	15 June	Medium	Firm/deep red blush
Patterson	15 June	Medium	Firm/good flavor
Tomcot	15 June	Large	Orange/sweet
Blenheim	19 June	Medium	Intense flavor
Tilton	25 June	Large	Firm/tart flavor

Source: Anon (2004d).

CO_2 for 2 days may reduce incidence of decay during subsequent transport and/or storage in CA or air. As compared to 0°C, respiration rate is more than double at 10°C. Ethylene production rate increases significantly with temperature from <0.1 μl/kg/h at 0°C to 4–6 μl/kg/h at 20°C for firm-ripe apricots and even higher for soft-ripe apricots. The greatest hazard in handling and shipping apricots is decay, mainly brown rot and rhizopus rot, and accelerated ethylene production can hasten the development of such decay. In order to retard ripening, softening, and decay, quick cooling to temperatures of 4°C or lower and storing at 0°C is recommended (Crisosto and Kader, 2004).

A variety of methods and treatments have been investigated to extend the shelf-life and preserve flavor, firmness, and other quality attributes of apricots. In recent years, the use of ethylene inhibitor 1-methylcyclopropene (1-MCP), to delay ripening and prolong storage life of apricots, has shown good results for ethylene inhibition, decline in pectin methylesterase activity, and preserving fruit firmness and flavor (Fan et al., 2000; Botondi et al., 2003). DeMartino et al. (2002) examined interactions between temperature, ethylene generation, and impact injury symptoms in apricots. Martinez-Romero et al. (2002) studied the effects of post-harvest putrescine treatment on the physicochemical properties and physiological response to mechanical damage of apricots (cv. Mauricio) during storage. Apricots harvested at the commercial ripening stage were treated with 1 mM putrescine by pressure infiltration, then mechanically damaged with a 25 N force and stored at 10°C for 6 days. Putrescine treatment increased fruit firmness and reduced the bruising due to mechanical damage. Color changes, weight loss, ethylene emission, and respiration rates were reduced in putrescine-treated fruits.

Physiological Disorders

Two main physiological disorders in apricots are "chilling injury" also termed as gel breakdown, and "pit burn." Chilling injury, characterized by the formation of water-soaked pockets that subsequently turn brown, is developed usually at 2.2–7.6°C during extended cold storage (Crisosto and Kader, 2004). Breakdown of tissue is sometimes accompanied by sponginess and gel formation. In addition to a short market life, fruit stored at elevated temperatures also loses flavor. Pit burn occurs as a result of softening of the flesh that turns brown around the stone when apricots are exposed to temperatures above 38°C before harvest. Tagliavini and Marangoni (2002) recommended early diagnosis of bitter pit for guiding applications of calcium sprays. Bussi et al. (2003) evaluated the effects of N and K fertilization on incidence of pit burn defect (PD), yields, and fruit quality of apricot (cv. *Bergeron*). They reported that fruit N increased with N fertilization level; high fruit N was shown to be a predisposing factor for PD, whereas higher Ca contents tended to reduce PD during storage.

PROCESSED PRODUCTS

About 15–20% of apricots produced are consumed fresh and the rest are processed as canned, dried, frozen, jam, juice, and puree. The processing information in this section relates to apricots produced and processed in the United States, unless noted otherwise. Table 17.6 shows utilization and market value of apricots produced in California. Canned apricots continue to lead the processed apricot products.

The use of apricot pit/kernel has been explored to extract oil or as an alternative source of proteins (Hallabo et al., 1975; Rahma et al., 1994; Sharma and Gupta, 2004). The oil from apricot kernel is very rich in essential fatty acids, especially linoleic acid.

CANNED APRICOTS

Canned apricots are convenient and easy to use in recipes such as baked goods, salads, and sauces. Canned apricots can be added to plain oatmeal, cottage cheese, yogurt, or ice cream to extend flavor and nutritional profile.

Processing

Apricots for canning are selected at the peak of ripeness, which helps to retain quality and nutrients. The optimum harvesting period is short; if the fruit is green, the flavor is astringent and if it is over ripe, it is too soft to handle during processing. Apricots are canned within 24 h of delivery to the processing plant, which ensures that the processed fruit maintains nutritional value and flavor (Anon, 2004d; Rieger, 2004). A brief description of processing steps for canning apricots is as below (Lopez, 1981a):

(1) *Grading and washing*: Since the fruit is received at the cannery with wide range of sizes, the first operation is to grade for size by running them over screens of 40/32, 48/32, 56/32, 64/32, and 68/32 in. After grading, apricots are spray washed and bad smutty spots are trimmed.

(2) *Pitting and cutting*: Apricots are split into halves (or other styles) and pitted using mechanical pitters and cutters; splitting is done on the natural dividing line in order to make symmetrical pieces and to have the pit come out easily. The pitters make a preliminary sorting by placing the prime fruit in one receptacle, the hard and irregular in another, and the soft in the third one.

(3) *Filling*: The filling of cut apricots is done by hand or by hand pack fillers. Standard of Identity requires a minimum cut-out Brix. Therefore, in order to estimate the strength of the syrup to use, it is important to know soluble solids of fruit, and also how much fruit goes into each can. The syrup is added hot by means of rotary and straight-line syrupers or by pre-vacuumizing syrupers.

(4) *Exhausting*: Apricots contain considerable trapped air, which presents a danger of can pin holing if not driven out. Cans must be either thoroughly exhausted prior to closing, or closed in a steam-flow machine. Depending on the texture of fruit, exhaust procedure of up to 10 min at 180–190°F (82.2–87.8°C) is used. If necessary, additional syrup is added before the cans are closed.

(5) *Processing and cooling*: To attain a commercially sterile product, canned apricots should be processed long enough for the center temperature of the product to reach a minimum of 190°F (87.8°C) or 195°F (90.5°C) for air- and water-cooling, respectively. Most canned apricots are processed in continuous reel cookers at 212°F (100°C) for 17–30 min, depending on the size of the cans and the texture of the fruit. The cans are generally partially cooled in water and then air-cooled on trays.

Table 17.6. Utilization and Price of California Processed Apricots (2001–2003)

Apricot Product	Quantity (t)			Price/ton (U.S. $)		
	2001	2002	2003	2001	2002	2003
Canned	31,000	30,500	30,000	292	286	298
Frozen	9,000	10,500	11,000	308	299	313
Dried (fresh basis)	6,000	8,000	6,800	268	223	233
Juice	9,000	7,000	9,200	102	158	210

Source: USDA-NASS (2004).

An important quality attribute of canned apricots is its firm texture. If the texture is too soft, the product acceptability is low despite its acceptable color and flavor. Textural quality of canned apricots has been researched extensively (Luh et al., 1978; Brecht et al., 1982; Sharma et al., 1992; Mallidis and Katsaboxakis, 2002). Chitarra et al. (1989) studied fruit softening and cell wall pectin solubilization in canned "*Patterson*" apricots and reported that canning of slightly immature fruit led to immediate softening and an unacceptable texture. The softening occurred in conjunction with solubilization of polysaccharides (presumably pectic), rich in uronic acid, arabinose, and galactose, into the canning liquid. Since this solubilization was not accompanied by pectin depolymerization, it was suggested that some process interactions other than acid-catalyzed polymer hydrolysis were responsible for fruit softening during processing.

Generally, canning apricots are not peeled. However, lye peeling method, the same as for peaches, is employed for those that are peeled before canning. Toker and Bayindirli (2003) investigated enzymatic peeling of apricots at 20°C, 35°C, and 50°C. They concluded that enzymatic peeling could be an alternative to chemical or mechanical peeling of stone fruits due to (1) better quality, as fruit retained their structural integrity and fresh fruit properties, (2) reduced heat treatment, and (3) less industrial waste.

U.S. Standards of Quality and Styles of Canned Apricots

The Standard of Quality for canned apricots has requirements regarding minimum size, uniformity of size, trimming, blemishes, and texture, which must be met. The U.S. Standards for Grade require uniformity of color, a minimum drained weight, and certain numbers of units in the usual sizes of cans. The standard for "Fill of Container" must always be reached or exceeded. The amount of fruit in each can is also highly important with respect to the strength of syrup that must be used to justify the label statement. Altogether, the filling of the cans is a critical part of the canning of apricots. The canned apricots may be seasoned with one or more of the optional ingredients permitted in the Food and Drug Standards of Identity (Lopez, 1981a). Complete detail on quality standards of canned apricots can be found on the web site of Agricultural Marketing Service of the USDA (USDA-AMS, 1976).

Styles of canned apricot are: (a) *Halves*—pitted apricots cut approximately in half along the suture from stem to apex; (b) *Slices*—pitted apricots cut into thin sectors or strips; (c) *Whole*—is unpitted apricots with stems removed; (d) *Pieces or irregular pieces*—cut apricot units that are predominantly irregular in size and shape which do not conform to a single style, or which are a mixture of two or more of such styles; (e) When the apricots are unpeeled, the name of the style is preceded or followed by the word "unpeeled;" (f) When the apricots are peeled, the name of the style is preceded or followed by the word "peeled" (USDA-AMS, 1976).

Cut-out requirements for liquid media in canned apricots are not incorporated in the grades of the finished product since syrup or any other liquid medium, as such, is not a factor of quality for the purposes of these grades. The designations of liquid packing media and the Brix measurements are shown in Table 17.7.

Frozen Apricots

Apricots to be processed as frozen should have even ripening, good color and flavor, low browning tendency, a tender and smooth skin, and firm texture.

Table 17.7. Cut-Out Brix Levels of Syrups Used in Canned Apricots

Designations	Brix Measurements
"Extra heavy syrup" or "Extra heavily sweetened fruit juice(s) and water" or "Extra heavily sweetened fruit juice(s)"	25° or more but less than 40°
"Heavy syrup" or "Heavily sweetened fruit juice(s) and water" or "Heavily sweetened fruit juice(s)"	21° or more but less than 25°
"Light syrup" or "Lightly sweetened fruit juice(s) and water" or "Lightly sweetened fruit juice(s)"	16° or more but less than 21°
"Slightly sweetened water" or "Extra light syrup"; or "Slightly sweetened fruit juice(s) and water" or "Slightly sweetened fruit juice(s)"	10° or more but less than 16°

Source: USDA-AMS (1976).

In California, the principal variety is *Patterson*, followed by *Blenheim, Tilton*, and *Modesto*, which are processed, on a smaller scale. Apricot halves are preferred to sliced fruit and can be processed with or without peeling. Halves and slices are sold to bakers, ice cream makers, and frozen desserts makers, whereas multiple-scored apricots (called machine-pitted) are primarily used for jam, jelly, and preserve making (Boyle et al., 1977; Scorza and Hui, 1996). Most of the frozen apricots are bulk-packed for use in food applications listed above. Retail marketing of frozen apricots has not gained popularity on a scale similar to some other frozen fruits.

Processing

Apricot freezing method (Boyle et al., 1977; Scorza and Hui, 1996) is described briefly here. Upon arrival at the processing plant, apricots are graded and inspected on a conveyor belt and then passed through a halving and pitting machine. Apricots to be processed as "machine-pitted" (for jam, jelly, and preserve) are pitted using Elliott pitters. After separation of the pits, halves, slices, or dices are inspected and washed to remove small fruit pieces or skin. They are then treated to prevent browning before they are syruped, packed, and frozen. One of the following two methods are used for browning control: (1) the fruit may be blanched in hot water or steam or (2) treated with an anti-browning agent, like ascorbic acid. Blanching is done in a single layer on a mesh belt for 3–4 min in steam to give satisfactory results for firm fruit. In the case of softer fruit batch, it is best to treat it with ascorbic acid or other anti-browning agents at a level of 0.05% or more into the syrup used for packing. The packing medium can vary from 15°Brix syrup to dry sugar (fruit to sugar ratio is usually 3:1), which may be sprinkled on the fruit as it is filled into the containers or can be distributed evenly by light mixing.

Another process for frozen apricots, "Osmodehydrofreezing" (osmotic dehydration followed by air drying and freezing) has been proposed for the production of intermediate moisture apricot ingredients without SO_2 application, with natural and agreeable color (Forni et al., 1997). Ten millimeter apricot cubes were osmodeyhydrated in sucrose, maltose, or sorbitol syrups at 56% and 3:1 fruit/syrup ratio. Ascorbic acid (1%) and sodium chloride (0.1%) were added as antioxidants. Osmodehydration was carried out at 25°C for 15 min under vacuum (700 mm Hg), and for 45 and 120 min at atmospheric pressure. Drying was conducted in an upward air-circulated dryer, at the dry-bulb temperature of 65°C and an air speed of 1.5 m/s, to achieve a soluble solids content corresponding to a_w of 0.86. Osmo-air-dehydrated apricot cubes were then frozen in an air-blast tunnel at −40°C and an air speed of 4 m/s. Results showed that incorporation of different sugars into the apricot cubes affected their low-temperature phase transitions and the percentage distribution of the sugars. Ascorbic acid retention during air-drying was affected by the composition of the syrup. Maltose exhibited the greatest protective effect on color stability during subsequent 8-months frozen storage. Giangiacomo et al. (1994) and Erba et al. (1994) also studied the feasibility of osmodehydrofreezing apricots.

U.S. Standards of Quality and Styles of Frozen Apricots

Details on quality standards of frozen apricots can be found on the web site of Agricultural Marketing Service of the USDA (USDA-AMS, 1963). Styles of frozen apricots are: (a) *Halves*—cut approximately in half along the suture from stem to apex and from which the pit has been removed; (b) *Quarters*—apricot halves cut into two approximately equal parts; (c) *Slices*—apricot halves cut into sectors smaller than quarters; (d) *Diced*—are apricots cut into approximate cubes; (e) *Cuts*—apricots that are cut in such a manner as to change the original conformation and do not meet any of the foregoing styles; (f) *Machine-pitted*—mechanically pitted in such a manner as to substantially destroy the conformation of the fruit in removing the pit.

Dried Apricots

Product Description

Dried apricots are the halved and pitted fruit of the apricot tree (*P. armeniaca*) from which the greater portion of moisture has been removed. Before packing, the dried fruit is processed to cleanse the fruit and may be treated with SO_2 to retain a characteristic color. Federal inspection certificates shall indicate the moisture content of the finished product, which should not be more than 26% by weight for sizes No. 1, No. 2, and No. 3, and for slabs, and not more than 25% by weight for other sizes (USDA-AMS, 1967).

Processing

Turkey produces almost half of the world's total dried apricots, where apricots are traditionally sun-dried after pretreatment with SO_2 (obtained by burning sulfur in specially constructed rooms), to moisture content of 23–28%. Sun-drying produces a product with a rich orange color, translucent appearance, and desirable texture. However, the long drying time, dependence on weather, and manual labor requirements are some of its disadvantages (Abdelhaq and Labuza, 1987; Mahmutoglu et al., 1996).

In the United States, almost all of dried apricots are produced in California. Cultivars that retain their color and flavor during processing and storage are best suited for drying. Ledbetter et al. (2002) studied the effects of fruit maturity at harvest on dried *Patterson* apricot quality with regard to color changes during 8 months of storage. While dried fruits of the immature class were of low quality at the end of storage period, both medium and most mature dry fruits were of sufficient quality to warrant marketing even after 8 months of cold storage.

Fruit is harvested at fully ripe stage and treated with SO_2 to preserve the color of the finished product. Drying is either natural, in the sun, or in large dehydrators used for prunes (Rieger, 2004; see Chapter 29 for more details). California apricots are sun-dried in halves while Turkish (Mediterranean) dried apricots are whole with the pit squeezed out (Anon, 2004b). For drying apricots in a single layer, the use of a solar energized rotary dryer was reported by Akpinar et al. (2004).

Sulfites are the most widely used chemicals in the production of dried apricots, mostly to preserve the color quality during storage. Traditionally, sulfited apricots can have SO_2 contents from 1000 to 6000 ppm. However, maximum legal limit allowed in most countries is 2000 ppm. Various methods of desulfiting apricots, with the objective to meet legal limits for SO_2, have been reported by Ozkan and Cemeroglu (2002a, b). Nisar-Alizai and Ahmad (1997) reported that apricot products treated with polyphenol oxidase (PPO) inhibitors prior to dehydration were of better color and textural quality than those prepared with sulfite treatment alone. However, shelf-life after drying was restricted, suggesting that PPO inhibitors usually do not provide the same extent of protection as sulfiting does. In a storage study of 17 different dried fruits, Winus (2000) showed that sulfite content of the dried fruit decreased during 3-month storage, the majority decrease (~80%) was in dried salted Japanese apricot while the smallest decrease (~11%) occurred in dried apples.

U.S. Standards of Quality and Sizes of Dried Apricots

A complete detail on quality standards of dried apricots can be found on the web site of Agricultural Marketing Service of the USDA (USDA-AMS, 1967). The various sizes of dried apricots, except for slabs are as follows:

- No. 1 (Jumbo Size): 1-3/8 in., or larger in diameter.
- No. 2 (Extra Fancy): 1-1/4 to 1-3/8 in. in diameter.
- No. 3 (Fancy Size): 1-1/8 to 1-1/4 in. in diameter.
- No. 4 (Extra Choice Size): 1 to 1-1/8 in. in diameter.
- No. 5 (Choice Size): 13/16 to 1 in. in diameter.
- No. 6 (Standard Size): less than 13/16 in. in diameter.

Other Products—Jam, Juice, and Puree

Apricot jam, juice/nectar, and puree are processed on a much smaller scale than canned, dried, and frozen apricots. A typical formulation for making apricot jam or preserve is shown in Table 17.8. Benamara et al. (1999) prepared a "light" apricot jam and studied the effects of partial substitution of aspartame for sucrose on the quality of jam. Substitution of aspartame for 90% or 100% of the sugar did not give satisfactory results; 80% level was the maximum substitution possible. Changes in color of apricot jam during storage increased with increasing concentration of added aspartame; this was probably due to gradual breakdown of the aspartame to release free aspartic acid and phenylalanine, which underwent a browning reaction with sugars.

Apricot juice and puree are used mainly in baby foods. Apricot juice production method is the same as for peaches (see Chapter 27). The use of pectinolytic enzymes (pectinesterase and polygalacturonase) aids in the liquefaction of apricot pulp and juice extraction (Chauhan et al., 2001). Juice thus obtained had higher total soluble solids, sugars, and acidity, but lower crude fiber and vitamin C. The use of apricot puree is gaining popularity as a new substitute for oil in many high-calorie, high-fat recipes. Unlike prunes, which can darken the color of some baked goods, or applesauce, which may cause recipes to be watered down, apricot puree reduces the fat content and adds a touch of flavor without any negative effect (Anon, 2004d). McHugh et al. (1996) investigated the use of apricot puree as an edible film and

Table 17.8. Formulation for Apricot Jam and Preserve

Ingredient	Unit	Higher Quality (50/50)	Standard Quality (45/55)
Water	lb	20	20
Fruit	lb	100	82
Pectin, rapid set	Ounces	5.50–6.75	5.50–6.75
Sugar	lb	100	100
Acid solution[a]	Fl. Ounces	Approx. 16	Approx. 14
Cooking temperature to finish[b]	°F	221	221
Soluble solids	%	65	65
Yield	lb	160	157
pH		3.3	3.3

Source: Lopez (1981b).

[a]The acid solution is prepared by mixing 1 lb of citric acid with 1 pint of hot water. Sufficient quantity is added to adjust to the indicated pH.

[b]The finishing temperature applies to cooks at or near sea level. At higher elevation, the temperatures are corrected for difference between 212°F and the boiling point of water.

Table 17.9. Composition of Apricots and Their Processed Products (per 100 g Edible Portion)

Nutrient	Unit	Raw Apricots	Canned Apricots[a]	Frozen Apricots[b]	Dried Apricots	Apricot Nectar
Proximate						
Water	g	86.35	82.56	73.30	30.89	84.87
Energy	kcal	48	63	98	241	56
Protein	g	1.40	0.53	0.70	3.39	0.37
Total lipid (fat)	g	0.39	0.05	0.10	0.51	0.09
Fatty acids, total saturated	g	0.027	0.003	0.007	0.017	0.006
Carbohydrate, by difference	g	11.12	16.49	25.10	62.64	14.39
Fiber, total	g	2	1.6	2.2	7.3	0.6
Sugars, total	g	9.24	14.89	—[c]	53.44	13.79
Vitamins						
Vitamin A, IU	IU	1926	1322	1680	3604	1316
Vitamin C, total ascorbic acid	mg	10	2.7	9.0	1	0.6
Thiamin	mg	0.03	0.016	0.02	0.015	0.009
Riboflavin	mg	0.04	0.02	0.04	0.074	0.014
Niacin	mg	0.6	0.304	0.80	2.589	0.26
Pantothenic acid	mg	0.24	0.092	0.20	0.516	0.096
Vitamin B-6	mg	0.054	0.054	0.06	0.143	0.022
Folate, total	mcg	9	2	2	10	1
Vitamin E (*alpha*-tocopherol)	mg	0.89	0.60	—[c]	4.33	0.31
Vitamin K	mcg	3.30	2.20	—[c]	0.05	1.20
Minerals						
Calcium	mg	13	11	10	55	7
Iron	mg	0.39	0.39	0.90	2.66	0.38
Magnesium	mg	10	8	9	32	5
Phosphorus	mg	23	13	19	71	9
Potassium	mg	259	138	229	1162	114
Sodium	mg	1	4	4	10	3
Zinc	mg	0.20	0.11	0.10	0.39	0.09
Copper	mg	0.078	0.079	0.064	0.343	0.073
Manganese	mg	0.077	0.052	0.050	0.235	0.032

Source: USDA (2004).

[a]In light syrup (solids and liquids).

[b]Sweetened.

[c]No values given.

concluded that fruit puree edible barriers may be used in food system, not only for favorable sensory characteristics, but also to control mass transfer, improve product quality, and extend shelf-life. Fruit barrier films that are good oxygen barriers, can be used for low to intermediate moisture food systems such as nuts, confections, and baked goods.

CHEMICAL COMPOSITION, NUTRIENT PROFILE, AND DIETARY BENEFITS

Apricots and their processed products are low in fat, specially saturated ones, and are rich source of some important nutrients. Chemical and nutritional composition of fresh apricots and their processed products is shown in Table 17.9. Varietal differences can contribute to variations in the composition of raw and finished products. The values shown here are for the fruit grown and processed in the United States, therefore, some differences can be anticipated in composition of apricots and their processed products in other regions of the world due to different climatic and soil conditions, agricultural practices, post-harvest handling, and processing techniques.

Apricots are rich in β-carotenes. The β-carotenes play a critical role in fighting disease and infections by maintaining strong immunity, protect the eyes, help keep skin, hair, gums, and various glands healthy, and help build bones and teeth. Apricots are an excellent source of potassium, iron, and magnesium. In addition to many other food uses, fresh or processed apricots could be an excellent source of a healthy and nutritious breakfast. Kartashov et al. (2003) reported that people who eat breakfast are significantly less likely to be obese and diabetic than those who usually do not. Eating just half cup of preserved or three fresh apricots provide 35–45% of the daily recommended intake for vitamin A and one serving of fruit (Anon, 2004d).

Apricots, especially unpeeled, are a good source of fiber, which is important to a healthy diet and can help control weight and lower cholesterol levels. Dried apricots are a concentrated source of fiber and one of the highly nutrient-dense dried fruits (Rieger, 2004). The American Heart Association continues to recommend an overall healthy dietary pattern that is rich in fruits, vegetables, whole grains, low-fat dairy products, lean meats, poultry, and fish. Diets rich in fruits, vegetables, whole grains, and fish have been associated, in many studies, with a lower risk of cardiovascular disease and stroke (AHA, 2003). Apricots, available throughout the year, in fresh, frozen, canned, or dried form, are a flavorful source of nutrients and are a convenient way to accomplish the "five-a-day" requirement of five servings of fruits and vegetables daily.

Apricots, in comparison to other major fruits, contain significantly higher amounts of flavonoids, especially catechin and epicatechin at 4.40 and 20.20 mg/100 g fruit, respectively. A variety of dietary flavonoids have been found to inhibit tumor development (Shahidi and Naczk, 1995). Apricot seeds were used to treat tumors as early as 502 A.D., and apricot oil was used against tumors and ulcers in England in the 1600s (Rieger, 2004).

REFERENCES

Abdelhaq EH, Labuza TP. 1987. Air drying characteristics of apricots. J Food Sci 52:342–5.

AHA (American Heart Association). 2003. Efficiency and safety of low-carbohydrate diets. Media Advisory (http://www.americanheart.org).

Akpinar EK, Sarsilmaz C, Yildiz C. 2004. Mathematical modeling of a thin layer drying of apricots in a solar energized rotary dryer. Int J Food Energy Res 28:739–52.

Anon. 2004a. Apricot. The Columbia Electronic Encyclopedia, 6th ed. Columbia University Press, New York (http://www.columbia.edu).

Anon. 2004b. California Apricots. Apricot Producers of California (http://www.apricotproducers.com).

Anon. 2004c. Apricot Varieties (http://www.ewbrandt.com/ewb/varieties/apricots.html).

Anon. 2004d. Apricot Varieties (http://www.cayugalandscape.com).

Benamara S, Messoudi Z, Bouanane A, Chibane H. 1999. Formulation and analysis of color of a light apricot jam. Ind Aliment Agricoles 116:27–33.

Botondi R, DeSantis D, Bellincontro A, Vizovitis K, Mencarelli F. 2003. Influence of ethylene inhibition by 1-methylcyclopropene on apricot quality, volatile production, and glycosidase activity of low- and high-aroma varieties of apricots. J Agric Food Chem 51:1189–200.

Boyle FP, Feinberg B, Ponting JD, Wolford ER. 1977. Freezing fruits. In: Desrosier ND, Tressler DK, editors. Fundamentals of Food Freezing. Westport, AVI Publ. Co. pp. 151–2.

Brecht JK, Kader AA, Heintz CM, Norona RC. 1982. Controlled atmosphere and ethylene effects on quality of California canning apricots and clingstone peaches. J Food Sci 47:432–6.

Bussi C, Besset J, Girard T. 2003. Effects of fertilizer rates and dates of application on apricot (cv. Bergeron) cropping and pitburn. Sci Hort 98:139–47.

Chauhan SK, Tyagi SM, Singh D. 2001. Pectinolytic liquefaction of apricot, plum and mango pulps for juice extraction. Int J Food Prop 4:103–9.

Chitarra AB, Labavitch JM, Kader AA. 1989. Canning-induced fruit softening and cell wall pectin solubilization in the 'Patterson' apricot. J Food Sci 54:990–2, 1046.

Crisosto CH, Kader AA. 2004. Apricot, peach, nectarine. In: Gross, K.C., Wang, C.Y., and Saltveit, M. Agriculture Handbook Number 66—The Commercial Storage of Fruits, Vegetables, and Florist and Nursery Stocks. Washington, DC, Agricultural Research Service of the United States Department of Agriculture.

DeMartino G, Massantini R, Botondi R, Mencarelli F. 2002. Temperature affects impact injury on apricot fruit. Postharvest Biol Technol 25:145–9.

Erba ML, Forni E, Colonello A, Giangiacomo R. 1994. Influence of sugar composition and air dehydration levels on the chemical-physical characteristics of osmodehydrofrozen fruit. Food Chem 50:69–73.

Fan X, Argenta L, Mattheis JP. 2000. Inhibition of ethylene action by 1-methylcyclopropene prolongs storage life of apricots. Postharvest Biol Technol 20:135–42.

FAO. 2004. World primary crops data. Food and Agriculture Organization of the United Nations (http://www.fao.org).

Femenia A, Sanches ES, Simal S, Rosello C. 1998. Developmental and ripening related effects on the cell wall of apricot (*Prunus armeniaca*) fruit. J Sci Food Agric 77:483–93.

Forni E, Sormani A, Scalise S, Torreggiani D. 1997. The influence of sugar composition on the color stability of osmodehydrofrozen intermediate moisture apricots. Food Res Int 30:87–94.

Giangiacomo R, Torreggiani D, Erba ML, Messina G. 1994. Use of osmodehydrofrozen fruit cubes in yogurt. Ital J Food Sci 6:345–50.

Gomez E, Ledbetter CA. 1997. Development of volatile compounds during fruit maturation: Characterization of apricot and plum x apricot hybrids. J Sci Food Agric 74:541–6.

Hallabo SAS, El-Wakeil FA, Morsi MKS. 1975. Chemical and physical properties of apricot kernel, apricot kernel oil and almond kernel oil. Egypt J Food Sci 3:1–6.

Kartashov AI, Van Horn L, Slattery M, Jacobs DR, Ludwig DS. 2003. Eating breakfast may reduce risk of obesity, diabetes, and heart disease. The American Heart Association's 43rd Annual Conference on Cardiovascular Disease Epidemiology and Prevention, Miami, FL.

Kovacs E, Nemeth-Szerdahelyi E. 2002. β-galactosidase activity and cell wall breakdown in apricots. J Food Sci 67:2004–8.

Ledbetter CA, Aung LH, Palmquist DE. 2002. The effect of fruit maturity on quality and color shift of dried 'Patterson' apricot during eight months of cold storage. J Hort Sci Biotechnol 77:526–33.

Lopez A. 1981a. Canning of fruits—apricots. A Complete Course in Canning, Book II: Processing Procedures for Canned Food Products, 11th ed. Baltimore, The Canning Trade. pp. 137–9.

Lopez A. 1981b. Jams, jellies, and related products. A Complete Course in Canning, Book II: Processing Procedures for Canned Food Products, 11th ed. Baltimore, The Canning Trade. p. 359.

Luh BS, Ozbilgin S, Liu YK. 1978. Textural changes in canned apricots in the presence of mold polygalacturonase. J Food Sci 43:713–6.

Magness JR, Markle GM, Compton CC. 1971. Food and Feed Crops of the United States. New Jersey, Agricultural Experiment Station, Bulletin 828.

Mahmutoglu T, Saygi YB, Borcakli M, Ozay G. 1996. Effects of pretreatment-drying method combinations on the drying rates, quality and storage stability of apricots. Lebens Wiss Technol 29:418–24.

Mallidis CG, Katsaboxakis C. 2002. Effect of thermal processing on the texture of canned apricots. Int J Food Sci Technol 37:569–72.

Martinez-Romero D, Serrano M, Carbonell A, Burgos L, Riquelme F, Valero D. 2002. Effects of post-harvest putrescine treatment on extending shelf life and reducing mechanical damage in apricot. J Food Sci 67:1706–12.

McHugh TH, Huxsoll CC, Krochta JM. 1996. Permeability properties of fruit puree edible films. J Food Sci 61:88–91.

Nisar-Alizai M, Ahmad Z. 1997. Comparative investigation of sun-drying of apricots and their products produced in N.W.F.P. and northern areas of Pakistan. Sarhad J Agric 13:501–9.

Ozkan M, Cemeroglu B. 2002a. Desulfiting dried apricots by exposure to hot air flow. J Sci Food Agric 82:1823–8.

Ozkan M, Cemeroglu B. 2002b. Desulfiting dried apricots by hydrogen peroxide. J Food Sci 82:1631–5.

Perez-Gonzales S. 1992. Associations among morphological and phenological characters representing apricot germplasm in central Mexico. J Am Soc Hort Sci 117:486–90.

Rahma EH, El-Adawy TA, Lasztity R, Gomaa MA, El-Badawey AA, Gaugecz J. 1994. Biochemical studies of some non-conventional sources of proteins. VI. Physicochemical properties of apricot kernel proteins and their changes during detoxification. Nahrung 38:3–11.

Rieger M. 2004. Mark's Fruit Crops Homepage, University of Georgia (http://www.uga.edu/fruit).

Scorza R, Hui YH. 1996. Apricots and peaches. In: Somogyi LP, Barret DM, Hui YH, editors. Processing Fruits: Science and Technology: Vol. 2—Major Processed Products. Lancaster, Technomic Publ. Co., Inc. pp. 37–76.

Shahidi F, Naczk, M. 1995. Food Phenolics—Sources, Chemistry, Effects and Applications. Lancaster, Technomic Publishing Co. pp. 1–5, 95.

Sharma A, Gupta MN. 2004. Oil extraction from almond, apricot and rice bran by three-phase partitioning after ultrasonication. Eur J Lipid Sci Technol 106:183–6.

Sharma TR, Sekhon KS, Saini SPS. 1992. Studies on canning of apricot. Ind J Food Sci Technol 29: 22–5.

Tagliavini M, Marangoni B. 2002. Major nutritional issues in deciduous fruit orchards of northern Italy. Hort Technol 12:26–31.

Toker I, Bayindirli A. 2003. Enzymatic peeling of apricots, nectarines and peaches. Lebens Wiss Technol 36:215–21.

USDA. 2004. Nutrient Database (http://www.nal.usda.gov).

USDA-AMS. 1963. United States Standards for Grades of Frozen Apricots (http://www.ams.usda.gov/standards/fzapricots.pdf).

USDA-AMS. 1967. United States Standards for Grades of Dried Apricots (http://www.ams.usda.gov/standards/driaprco.pdf).

USDA-AMS. 1976. United States Standards for Grades of Canned Apricots and Canned Solid-Pack Apricots (http://www.ams.usda.gov/standards/aprict.pdf).

USDA-ERS. 2004. US per capita consumption data. USDA-Economic Research Service (http://www.ers.usda.gov/).

USDA-NASS. 2004. National Agricultural Statistics Service (http://www.usda.gov/nass/).

Westwood MN. 1993. Fruit growth and thinning. In: Temperate-Zone Pomology: Physiology and Culture. Portland, Timber Press, Inc. pp. 254–74.

Winus P. 2000. The study of relationship of total sulfite quantity and the other properties of dehydrated fruits with storage times. Food 30:283–91.

18
Horticultural and Quality Aspects of Citrus Fruits

María-Jesús Rodrigo and Lorenzo Zacarías

INTRODUCTION

Fruit is an indispensable part of the human diet and is an important source of fiber, vitamins, minerals, and many other compounds essential for health. Moreover, fruit crops have an agricultural and commercial impact worldwide and are a considerable source of income for both the developed and the developing countries. World export trade for four of the major fruit crops (citrus, bananas, grapes, and apples) was over 15 billion U.S. dollars in 2002, a 11% increase over the export trade in 2000, indicating the growing economical importance of fruit crops. Table 18.1 summarizes the top 10 fruit crops produced in the world in 2003; citrus fruits are first not only in total production, but also in economic value (FAO, 2004). In the citrus fruit market, there are clearly two different sectors: for fresh fruit consumption, oranges, tangerines, and mandarins dominate; and for processed citrus products, orange juice ranks first. Other chapters of this book cover specific aspects of fruit processing and juice technology and, therefore, this chapter is mainly focused on fresh citrus fruits, with special emphasis on more relevant horticultural, botanical, and quality aspects of this important commodity.

Table 18.1. Production (Mt) of Top 10 Fresh Fruit Crops in the World in 2003

Fruit	Mt
Citrus	103,821,013
Watermelons	91,790,226
Bananas	69,286,046
Grapes	60,883,454
Apples	57,967,289
Coconuts	52,940,408
Plantains	32,974,339
Mangoes	25,563,469
Pears	17,191,205
Peaches and nectarines	14,787,539

Source: FAO (2004).

WORLD CITRUS PRODUCTION

World production of citrus fruits has experienced a continuous growth in recent decades, and total production in 2003 was over 103 million tons. Even though the total production of oranges is decreasing slightly, it still represents the bulk of citrus production, accounting for about 58% of the total. Mandarins and tangerines are second with 20% of the total, followed by grapefruit and pummelos that account for 12%. Citrus fruits are produced around the world and more than 130 countries are citrus producers, although the total production in countries of the Northern Hemisphere is much higher than in those of the Southern Hemisphere. The main citrus fruits producing countries are listed in Table 18.2. Brazil, United States, and China represent slightly less than 50% of the world production. The states of Sao Paolo (Brazil) and Florida (USA) produce 85% of the citrus juice processed in the world. A difference between these two countries is that 99% of the processed juice produced in Brazil is dedicated to exportation, whereas 90% of that produced in the United States is for local consumption. Mediterranean countries are the main suppliers of fruit for fresh consumption (about 60%), with Spain being the world leader in exportation. A detailed analysis of data in different developed countries reveals that the consumption of fresh oranges has gradually decreased over the last few years. The reasons are higher consumption of juice and the new fresh citrus fruit, the seedless easy peeler mandarins that are becoming a favorite, especially for children. Moreover, the improvement in transportation and storage conditions favor the availability of other fruits in the Northern Hemisphere. The search for new cultivars that can be harvested either earlier or later than the traditional season, and thus extend the marketing period of fresh citrus fruits, is a major objective for citrus breeders.

ORIGIN, TAXONOMY, AND MAJOR CULTIVARS

Although the origin of the genus *Citrus* is not yet clear, it is believed to originate in a geographical area from Southeastern Asia, including China, Thailand, Malaysia, and Indonesia. Recent evidence suggests that South-central China may be the area of origin (Gmitter and Hu, 1990). The first written report of citrus is dated 5 centuries BC in China, but there is archaeological evidence of the uses of citrus fruit several thousands of years BC. Dissemination of citrus throughout the world is the result of migration of many cultures and trade routes. It is thought that citrus were grown in Europe from several centuries BC and introduced in America in late 1400. Citrus culture proliferated in Florida in late 1700, but much later in California and other western states. Currently, the citrus culture is located between the 40°N and 41°S parallel (Davies and Albrigo, 1998).

Table 18.2. Distribution of Citrus World Production in 2003

Country	Oranges	Tangerines and Mandarins	Lemons and Limes	Grapefruits and Pomelos	Other Citrus Fruits	Total
World	60,046,286	20,950,238	12,451,680	4,696,707	5,676,102	103,821,013
Brazil	16,935,512	1,263,000	950,000	67,000		19,215,512
United States	10,473,450	486,250	939,000	1,871,520	900	13,771,120
China	1,647,681	9,000,000	600,568	356,224	940,222	12,544,695
México	3,969,810	360,000	1,824,890	257,711	63,000	6,475,411
Spain	3,091,400	2,081,600	1,065,700	30,453	15,000	6,284,153
India	2,980,000		1,370,000	137,000	93,000	4,580,000
Iran	1,850,000	710,000	1,040,000	37,000	66,000	3,703,000
Italy	1,962,000	562,368	548,808	4500	26,000	3,103,676
Egypt	1,725,000	500,000	296,776	3100	2400	2,527,276
Argentina	700,000	400,000	1,200,000	170,000		2,470,000
Turkey	1,215,000	525,000	500,000	130,000	3000	2,373,000
Pakistan	1,400,000	496,000	99,000			1,995,000
South Africa	1,082,330	100,172	149,647	380,000		1,712,149
Greece	1,200,000	110,000	160,000	8000	4500	1,482,500
Japan	108,000	1,147,000			200,000	1,455,000

Source: FAO (2004).

The genus *Citrus* belongs to the Rutaceae family, subfamily Aurantoideae. This family contains many edible species and the tribe Citrinae includes the three commercially important genera: *Fortunella*, *Poncirus*, and *Citrus*.

- *Fortunella* spp. (kumquat). Species of evergreen shrubs or small trees that can be grown in subtropical areas and are used for decoration. Produces small fruit with an edible peel that is usually sweeter than the pulp.
- *Poncirus trifoliate*. Tree of deciduous habit with trifoliate leaves, very tolerant to freezing when acclimated. Very important as rootstock.
- *Citrus*. The number of species is still a matter of debate, but it is thought that 16 species of the genus originated from three "basic" species:
  - *Citrus medica*. The citron is used only for the peel and may be the progenitor of all acid fruits (lemons and limes). It is believed to be the only citrus fruit in Roman and Greek times.
  - *Citrus reticulata*. Mandarin and may be the ancestor of all oranges and tangerines currently cultivated.
  - *Citrus grandis* or *Citrus maxima*. Pummelo or shaddock and may be the progenitor of pummelo and grapefruit.

In the genus *Citrus*, the tree is an evergreen and variable in size, according to the species. Most of the fruits are edible, variable in size and shape, and characterized by a juicy pulp composed of vesicles within segments surrounded by a leathered colored peel. Although there are a large number of cultivated citrus species, some vital characteristics of the most economically important are given in the following sections.

Citrus Sinensis

The widely accepted name for this species is sweet oranges, and the different cultivars can be placed in four groups.

Navel Oranges

Navel oranges are the most important group for fresh fruit. The main feature of navel oranges is the small secondary fruit embedded in the stylar end of the primary fruit, giving an external appearance of a navel, also seedless. "Washington navel" is the major cultivar within this group, and there are many others, namely "Navelina" (the main cultivar of this group in Spain), "Navelate," "Thompson," and "Lane late."

Common Oranges

They are also known as blond oranges. Oranges of this group belong to the Blanca group and are mostly used for juice and processing. The fruit does not have a navel, is seedless, and is usually less acidic than oranges from other groups. "Valencia" is the most important cultivar within this group in the world. Other oranges are "Shamouti" (of great importance in Israel), "Pera," "Hamlin," "Pineapple," "Parson Brown," and "Salustiana."

Blood Oranges

They are distinctive for the red color of the peel and pulp because of anthocyanins accumulation. Main cultivars of this group are "Tarocco," "Moro," and "Sanguinelli." The total production in the world is low, but of importance in Italy.

Acidless Oranges

A less common group with very low acid content (about 0.2%), rendering a bland sweet flavor.

Citrus Reticulata

Cultivars of this species probably originated in China and are recognized as mandarins, although the name tangerine is synonymously used in the United States. This is the most varied group of citrus. The production of mandarins is of high importance in Japan and in Spain, and has increased in recent years. The Clementine mandarin was originally detected in Algeria and is becoming one of the most important mandarins in the world. There are more than a dozen cultivated varieties with wide consumer acceptance as a fresh fruit because of the seedless and easily peeled qualities. Clementine mandarins are very prone to spontaneous bud mutations. Satsuma is the major group of mandarins in Japan, and many different mutants have been detected. The Satsuma is of lesser quality than Clementines and easily develops peel puffing. Satsuma mandarins are more parthenocarpic than Clementine mandarins. Another mandarin is "Ponkan," of significance in India, China, and Brazil. There is a large group of mandarin hybrids that may be of particular importance in different countries, as "Fortune," "Wilkings," "Michal," "Encore," or "Fremont." Other hybrids from crosses between mandarins and other citrus species should not be botanically considered mandarins, for example, hybridization with grapefruit, called tangelo ("Orlando" and "Minneola," important in Florida), or with sweet oranges, called tangors ("Ortanique," "Ellendale," "Temple," and "Murcott") or in the case of "Nova" a cross between a mandarin and a tangelo (Saunt, 2000).

Citrus paradisi

The origin and cultivation of grapefruit is more recent than other citrus species and may originate from cross-hybridization between a pummelo and a sweet orange. Grapefruit is divided into two groups: white and pigmented cultivars. Both groups contain seed and seedless cultivars, and in general have a long maturation period, between 8 and 12 months. The main white cultivar is the seedless "Marsh" grapefruit that gradually replaced the seeded "Duncan." Pigmented grapefruit varieties are characterized by a reddish tint of the peel and a more intense red color of the pulp due to lycopene. Red grapefruit originated in Texas and Florida (Redblush) as mutations of white grapefruit, and currently, the main cultivars are "Star Ruby," "Ruby Red," "Rio Red," and "Flame" (Gmitter, 1995).

Citrus limon

The origin of lemons is unknown, but may be a hybrid related to citron that originated in India. The areas of production are limited to semiarid and arid regions, since, with the exception of limes, the tree is more sensitive to low temperature than other citrus species. The tree is very vigorous, having several flushes of growth and producing fruits throughout the year. Fruit shape may be spherical or spheroid as in the most common cultivars, with a nipple at the stylar end. There are several groups of lemons. "Femminelo" is the Italian lemon cultivar, which is also grown in South Africa and in the United States. Fruits of this cultivar are moderately acidic and can be harvested throughout the year. "Eureka" lemons are the most important cultivar. The tree is less densely foliated, and fruits, of ovate shape and excellent quality, are less frost-hardy than other cultivars and can be harvested throughout the year. "Verna" is the major cultivar in Spain, producing an elongated fruit that has two differentiated harvest seasons. "Fino" and "Villafranca" are important cultivars grown in the Mediterranean area and "Lisbon" in California, Australia, and Argentina.

Citrus aurantifolia

Limes probably originated from a tropical area because of the extreme sensitivity to freezing. Its culture is limited to tropical and subtropical warm and humid regions. There are acid and sweet cultivars, although only the former have commercial interest. The main cultivars of acid limes are the "Tahiti" or "Persian" lime and the "Mexican" lime. Fruit are green at maturity and can be harvested throughout the year.

Citrus grandis

Pummelo is also known as shaddock and is native to southern China. The size of the fruit is variable but in general much larger with a thicker peel than grapefruit; it has very low acid content and poor flavor. Production is limited to very few countries but is popular in China. Hybrids between pummelos and grapefruit, as "Oroblanco", are of importance in some countries.

Citrus auratium

Referred to as sour oranges, was widely used in the past as rootstock, but currently discarded because of the sensitivity to the tristeza virus. Mature fruit contains neohesperidin, responsible for the bitter flavor, and is only used for marmalade production. Some trees are also used as ornamentals.

ANATOMY AND PHYSIOLOGY OF CITRUS FRUITS

The fruit of the citrus species is a unique type of modified berry named hesperidium. The fruit arises from the growth and development of the ovary that is divided into several segments or carpels (more than eight in the genus *Citrus*) that are joined to a central axis and separated by thin septa (Fig. 18.1). The fruit is composed of two morphologically distinct parts: the pericarp, usually referred to as peel or rind, and the endocarp or pulp, which is the edible portion of the fruit. The pulp is composed of a variable number of segments or locules enclosed in a locular membrane (also named septa). These locules contain the juice vesicles that are enlarged cavities linked by a small stalk to the epidermal surface of the endocarp (Spiegel-Roy and Goldschmidt, 1996).

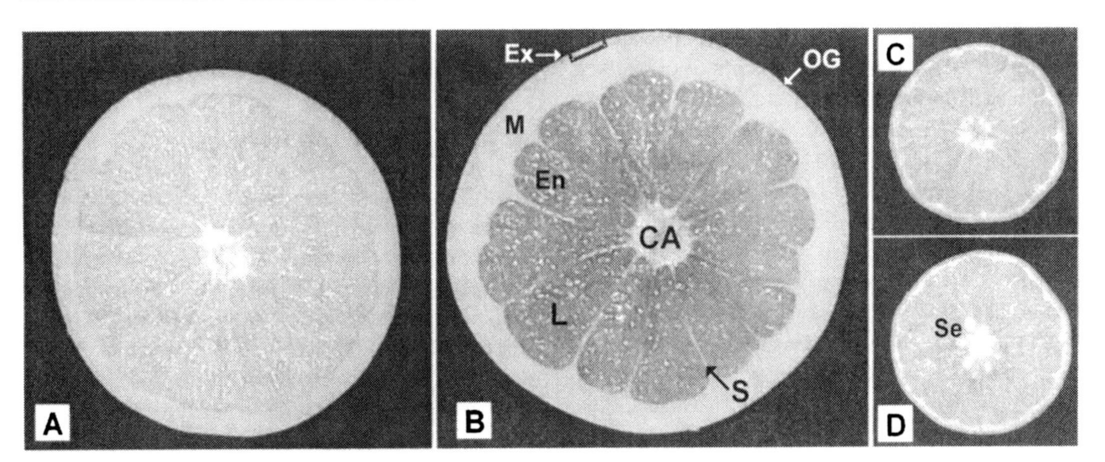

Figure 18.1. Equatorial section of "Navelate" orange (a), "Star Ruby" grapefruit (b), seedless "Clementine" mandarin (c) and a cross-pollinated Fortune mandarin (d). Abbreviations of the different parts of the fruit are: Ex, exocarp or flavedo; M, mesocarp or albedo; En, endocarp or pulp; L, locule; CA, central axis, S, septa; OG, oil gland, Se, seed.

The external layer of the peel (exocarp, Fig. 18.1) is usually named as flavedo. The cuticle that is rich in epicuticular waxes covers the epidermal cells of the flavedo. A small layer of compacted parenchyma cells containing chloroplast or chromoplast that provides the typical coloration of the fruit composes the flavedo. Embedded in the flavedo are the multicellular structures of oil glands, which contain terpenes and essential oils distinguishable in the ovary of the flower. The mesocarp or albedo is the white portion of the peel. This layer is of variable thickness in the fruit of different species. Whereas in grapefruit, it may be several centimeters wide (Fig. 18.1b), and in mandarin fruit, it is almost completely degraded to a small portion of few millimeters with vascular bundles (Fig. 18.1c and d). This layer is composed of typical eight-armed cells connected to each other to form a spongy mesh with large intercellular air spaces. This network of cells is connected to the vascular bundles of the segments to which the juice vesicles are attached (Albrigo and Carter, 1977).

The endocarp is the edible portion of the fruit and is composed of juice vesicles or juice sacs (Fig. 18.1b). These structures originate from the epidermal cells of the endocarp, and mature vesicles are large cavities of epidermal cells containing juice that may cross the locule. The juice cells are highly vacuolated with a narrow cytoplasm and are covered by a fine layer of waxes. The stalk of the juice vesicles is not connected to the vascular bundles, and therefore their water and assimilates supply should be through postphloem cell-to-cell transport (Koch et al., 1986). Under different postharvest conditions or long storage periods, juice vesicles may accumulate calcium oxalate, hesperidin, or naringinin, where naringinin contributing to bitterness of the juice (Horowitz and Gentill, 1977).

Citrus fruits have a long growth period; flowering to maturation takes 6–12 months, depending on the climate. Growth and development of citrus fruit can be divided into three stages: cell division, cell enlargement, and fruit maturation (Bain, 1958). The cell division period (stage I) commences at fruit set, immediately after anthesis, and the rapid increase in fruit size is mainly due to cell division of different tissues. During this period, a massive abscission of flowers and young ovaries occurs, which has a strong influence in the final number of fruits that will be harvested. At the end of this period, the peel reaches its maximum width. The content of gibberellins appears to play a major role in ovary growth, and some cultivars (Clementine mandarin) require application of gibberellins to set parthenocarpic fruits. This period lasts for about 8 weeks (Agusti, 2000). The phase of cell enlargement (stage II) is characterized by an important increase in fruit size due to a large augmentation of pulp segments and juice. The cells of the peel also elongate, and at the end of this period (3–6 months), the fruit reaches its final size. The transition from cell division to cell enlargement is separated by the abscission of developing fruits, also referred to as "June drop." Auxins appear to be implicated in cell enlargement. Application of synthetic auxins after June drop has been demonstrated to increase the final fruit size, mainly by the effect on growth of the pulp, which has important commercial relevance (Agusti, 2000). When fruit has nearly reached its final size, stage III of development or fruit maturation begins; recognized by an initial signal of chlorophyll degradation and changes in the external fruit color. This stage also encompasses a decline in total acidity (TA) and an increase in sugar content, usually expressed as total soluble solids (TSS). Thus, the ratio TSS/TA increases throughout maturation and is used as a maturity index to determine the harvest time for each cultivar (Spiegel-Roy and Goldschmidt, 1996). Maturation in citrus is different than other fruits. An example illustrating this behavior is that the typical fruit softening by cell wall degradation in many edible fruits does not happen in citrus, rather a loss in turgor is responsible for softening in citrus. Other transformations of relevant nutritional components during maturation of citrus fruit are discussed below.

Maturation of citrus fruit is classified as nonclimacteric, as the typical changes in fruit quality parameters are not associated with an increase in respiration rate and ethylene production (Watkins, 2002). The respiration rate continuously declines during fruit ontogenity and remains nearly constant during natural maturation or in detached mature fruits (Eaks, 1970). Even though citrus fruit has very low ethylene production during maturation, application of ethylene is a common postharvest practice for degreening in many citrus cultivars. The peel of fruits of early harvested varieties or from warm climates remains green after the pulp has reached an optimum maturity index. Treatment with ethylene induces chlorophyll degradation and accumulation of carotenoids, and these effects are widely used in the packinghouses to induce color development and uniform color between shipments (Grierson et al., 1986). Ethylene treatment only affects external fruit coloration and does not significantly affect internal fruit quality. On the contrary, preharvest application of gibberellic acid is a common treatment to delay fruit coloration. This hormone has a well-recognized anti-senescent activity, and its application delays many of the external maturation-related processes, including fruit coloration and softening (Coggins, 1981). Besides hormone signaling, nutritional factors play an essential role in the regulation of fruit maturation. Nitrogen strongly inhibits fruit coloration, while sugars promote the process (Huff, 1983, 1984). Thus, a complex interplay between the nutritional status of the fruit and hormonal signals drives the maturation of citrus fruits (Alferez and Zacarías, 1999; Iglesias et al., 2001)

POSTHARVEST PHYSIOLOGY

Citrus fruits have a considerable storage potential that can be very different among cultivars. Unlike other fruits, mature citrus fruits can be left unharvested on the tree for relatively long periods. Valencia oranges, grapefruit, and lemons may be left unharvested on the tree for a few months, whereas Clementine mandarins and Navel oranges should be picked relatively shortly after reaching commercial maturity. Abscission of mature fruit is the main problem associated with an excessive delay in the harvesting period. Application of synthetic auxins increased fruit retention force and with gibberellins is a practical treatment to extend the harvesting period on the tree (Agusti, 2000). A delay in harvesting may induce in some Clementine mandarins, a senescence-related disorder manifested as cracking in the peripeduncular and equatorial areas of the peel (Fig. 18.2) that is aggravated in humid winters.

Fruit color is not a good indicator of maturity; therefore, the ratio of soluble solids/acidity (TSS/TA) is currently used as a maturation index. A ratio from 7 to 10 is generally accepted as a measure of minimum maturity and from 10 to 16 is considered as an acceptable quality, although these values may vary among the different markets. In lemon, the percentage of juice content is the main parameter used for maturity (Baldwin, 1993).

Citrus fruits are very sensitive to physiological disorders in the peel during storage, and comprehensive reviews have considered these problems (Grierson, 1986; Martínez-Javega and Cuquerella,

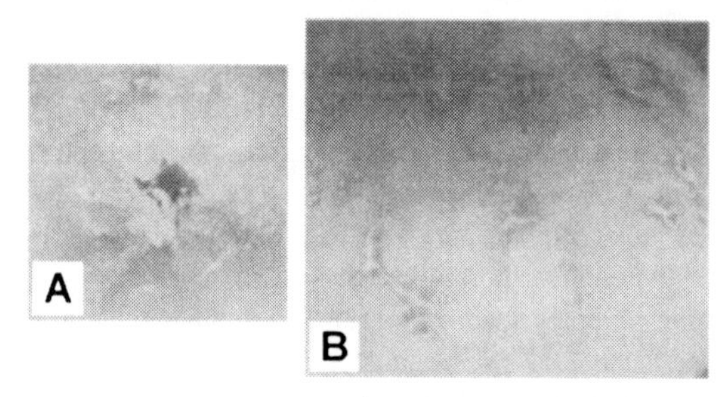

Figure 18.2. Peel cracking in Clemenules mandarin. Initial symptoms around the peduncular area (a) and advanced symptoms on the equatorial surface (b).

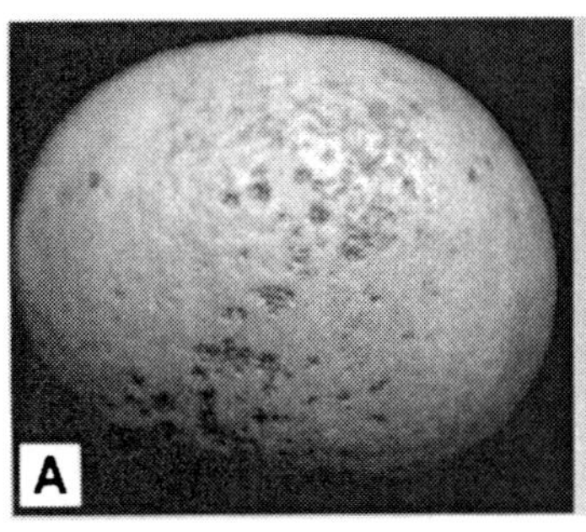

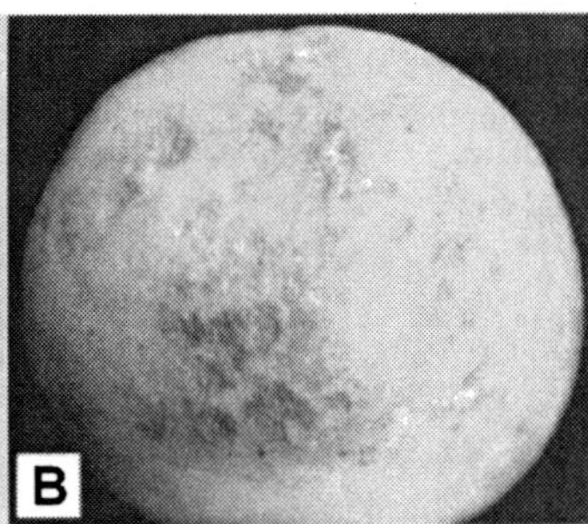

Figure 18.3. Chilling injury symptoms in fruit of the hybrid "Fortune" mandarin and in white "Marsh" grapefruits stored at 1°C.

1995; Agusti, 2000). Storage at low temperature is the best practice to extend the postharvest life and to maintain fruit quality. However, many citrus species and cultivars (grapefruit, lime, lemon, and mandarin hybrids) are sensitive to chilling injury when stored at temperatures below 5°C. Chilling injury is manifested as small pitting areas that arc progressively extended over the fruit surface as collapsed brown clusters (Fig. 18.3a and b). Although internal quality is not affected, the disorder affects external appearance and fruit marketability. Chilling injury is thus a serious postharvest problem for sensitive cultivars, as exposure to low temperature is required in the quarantine treatment for disinfestations of citrus fruits from the Mediterranean fruit fly (*Ceratitis capitata*). Postharvest treatment to induce chilling tolerance in citrus fruit has been described and, among them, heat treatments show promising results (Schirra and Ben-Yehoshua, 1999).

It is interesting to mention that the development of different physiological and pathological disorders are associated with a substantial increase in ethylene production (Cooper et al., 1969; Riov and Yang, 1982; Achilea et al., 1985; Zacarías et al., 2003). Different experimental evidence indicates that this stress-ethylene is a part of the protective mechanism developed by citrus fruit to cope with these postharvest stress conditions. This assumption is based on the fact that treatment of citrus fruits with an inhibitor of ethylene action 1-MCP (Blankenship and Dole, 2003) enhanced natural and artificial *Penicillium* infection in Shamouti and Navel oranges (Porat et al., 1999; Marcos et al., unpublished results), accelerated chilling injury in Fortune mandarins (Lafuente et al., 2001) and induced stem-end rind breakdown in Valencia oranges (Zacarías et al., unpublished results). Thus, low ethylene levels are important during postharvest (Porat et al., 1999) and special care should be taken in postharvest practices aimed to reduce ethylene action.

Application of waxes to citrus fruit is a common postharvest practice worldwide with extensive efforts to develop new components and wax formulations (Petracek et al., 1999). The challenge is to provide a coating that reduces water loss but does not modify internal gas concentration that would induce anaerobic respiration. Reduction of internal oxygen and increasing CO_2 concentrations are responsible for an elevation in ethanol and acetaldehyde, and off-flavors (Baldwin, 1993). Postharvest wax treatment of citrus fruit should then provide the fruit shine requirement for each particular country and to reduce water loss during the postharvest period with minimum effect on internal gas concentrations. Formulas using carnauba, shellac, and polyethylene-based waxes are currently used in the postharvest industry of citrus in different countries (Petracek et al., 1999).

NUTRITIONAL AND QUALITY CONSTITUENTS OF CITRUS FRUITS

Citrus fruit contains a large number of different constituents, but the nutritional importance of citrus is due to their particular composition. Table 18.3 summarizes the nutrient value of fresh citrus fruit. In the following section, some of the more relevant components of citrus fruits are described and discussed, highlighting their nutritional properties, biosynthesis, and evolution during fruit maturation, economic importance, and differences among citrus species.

Carotenoids

The external and internal color of citrus fruit is one of the main attributes of quality and a major

Table 18.3. Nutrients in Different Fresh Citrus Fruits. Values are per 100 g of Edible Portion

Nutrient	Orange	Tangerine and Mandarin	Pink and Red Grapefruit	White Grapefruit	Lemon
Calories (kcal)	44	44	39	32	29
Water (g)	86	87	90	91	89
Protein (g)	0.91	0.63	0.50	0.63	1.10
Fat (g)	0.15	0.19	0.10	0.10	0.30
Carbohydrates (g)	12.5	11.2	9.2	8.19	9.32
Fiber (g)	2.2	2.3	2.5	2.6	2.8
Calcium (mg)	43	14	9	15	26
Phosphorous (mg)	23	10	15	7	16
Iron (mg)	0.13	0.10	0.20	0.05	0.60
Sodium (mg)	1	1	1	0	2
Potassium (mg)	166	157	162	150	138
Vitamin C (mg)	59	31	38	37	53
Vitamin A (IU)	200	675	440	10	32
% RDI	4%	13.5%	8.8%	0.2%	0.4%
Vitamin E (mg)	0.15	0.15	n.r.	n.r.	0.15
Thiamine (mg)	0.068	0.105	0.040	0.040	0.040
Riboflavin (mg)	0.051	0.022	0.020	0.020	0.020
Niacin (mg)	0.420	0.160	0.200	0.200	0.100

% RDI, Percentage of the Recommended Daily Intake; n.r., not reported.
Source: USDA National Nutrient Database, 2004.

parameter for consumer acceptance. The development of citrus color is the result of coordinated changes in carotenoid content and composition, and chlorophyll degradation (Gross, 1987). Carotenoids are a large family of C_{40} isoprenoids, which accumulate in the chromoplasts of ripened fruit (Gross, 1987). Carotenoids also play an important role in human health as precursors of vitamin A and for antioxidant (Olson, 1989) and anticancer properties (Tsushima et al., 1995). In the past, content and carotenoid composition in fruits of different citrus cultivars were extensively studied, showing that the flavedo of mature fruit is one of the richest and more complex sources of carotenoids in plants (Stewart and Leuenberger, 1976; Oberholster et al., 2001). As occurs in other fruits, the rate of carotenoid biosynthesis is higher in the flavedo than in the juice vesicles, and flavedo contains approximately 70% of the carotenoid content (Baldwin, 1993).

The peel of immature citrus fruit shows a carotenoid profile characteristic of chloroplastic-containing tissue, with lutein being the main carotenoid (Gross, 1987; Rodrigo et al., 2003; Kato et al., 2004). A noticeable decrease in carotenes and lutein occurs in the peel at the onset of fruit coloration, with a parallel accumulation of specific xanthophylls (oxygenated carotenoids) (Gross, 1987). In the pulp of citrus, a continuous increase of carotenoid content occurs during ripening. The predominant carotenoids in the flavedo and pulp of mature sweet orange and mandarins are the xanthophylls violaxanthin and β-cryptoxanthin, which are found mainly in esterified form (Gross, 1987; Oberholster et al., 2001). β-Cryptoxanthin, which has provitamin A activity, is especially abundant in the peel and pulp of mandarins. The characteristic reddish color of the peel of some colored varieties (mandarins and oranges) is provided not solely by xanthophylls, but also by citrus-specific C_{30}-apocarotenoids (Gross, 1987; Oberholster et al., 2001). Grapefruit (red, pink, and white cultivars) and lemons contain lower levels of carotenoids than oranges and mandarins (Table 18.4).

The light yellow color of white grapefruit and lemon is due to the low carotenoid content. In addition, white grapefruit also contain colorless carotenes, phytoene, and phytofluene, which account for 70–80% of the total content in the peel and pulp (Yokoyama and White, 1967; Romojaro et al., 1979). The typical pale yellow color of lemons is due to the accumulation of β-cryptoxanthin (30%), colorless carotenoids (25%), β-carotene (14%), and ζ-carotene (only in the peel) (17%) (Yokoyama and Vandercook, 1967) (Table 18.4). Red and pink

Table 18.4. Distribution of Carotenoid Content in Peel and Pulp of Several *Citrus* Species at Maturity

	Total Carotenoid Content μg/gFW	
	Peel	Pulp
Sweet Orange *Citrus sinensis*		
Valencia	120	12
Navel	67	10
Mandarins		
Dancy *C. reticulata*	195	n.r.
Clemenules *C. clementina*	102	7
Satsuma *C. unshiu*	250	25
Hybrids		
Michal Dancy × clementine	174	13
Grapefruit *C. paradisi*		
Marsh white	3	0.8
Star ruby red	12	10
Lemon *C. limon*	1.4	0.6

n.r., not reported.
Source: Gross, 1987; Kato et al., 2004; Rodrigo, unpublished.

grapefruits, which originated as mutants of white grapefruit, have a carotenoid content in the pulp similar to that of orange and mandarin, but much lower in the peel (Table 18.4). The main carotenoids identified in pink and red grapefruits are lycopene and β-carotene (with provitamin A activity), and minor amounts of xanthopylls (violaxanthin and neoxanthin). Interestingly, red and pink grapefruits are the only citrus fruits that have similar levels of total carotenoids in both peel and pulp.

Recently, the relationship between carotenoid accumulation and the expression of carotenoid biosynthetic genes in different citrus varieties has been reported. Carotenoid accumulation in *Citrus* is highly regulated by the coordinated expression of the different carotenoid biosynthetic genes (Ikoma et al., 2001; Kato et al., 2004; Rodrigo et al., 2004).

Vitamin C

Vitamin C is one of the most important nutritional benefits of citrus fruit. Ascorbic acid (vitamin C) is the only vitamin present in citrus fruit in amounts of major nutritional significance; one orange has 50 mg of vitamin C, which is nearly the double of the recommended daily intake. The characteristic levels of vitamin C in juice of different citrus species are summarized in Table 18.3. The concentration of ascorbic acid has been reported to decrease with maturity or remain constant until late in the season and then decline (Baldwin, 1993). Only 25% of ascorbic acid in the fruit is in the juice, the remainder is found in the peel, especially in the flavedo (Kefford, 1959).

Organic Acids

Maturity standards for sweet citrus are often based on chemical indicators of fruit flavor, including sweetness and acidity, and the ratio between these components. Therefore, acidity levels have a major impact on internal fruit quality and consequently affect the time when the fruit reaches minimum market standards (Marsh et al., 2000). Organic acids significantly contribute to juice acidity, citric acid being primary organic acid (70–90% of total). Interestingly, organic acid composition differs in the different fruit tissues; citric acid is the major acid in juice, malate predominates in the flavedo, and oxalate is the most abundant in albedo (Clements, 1964). Other minor organic acids found in the juice of citrus fruits are acetate, pyruvate, glutarate, formate, succinate, and α-ketoglutarate (Yamaki, 1989).

Citric acid content in juice of different citrus cultivars at harvest varies from 0.1% to 6% (Clements, 1964). The pulp of orange contains about 0.8–1% citric acid, grapefruit 1.5–2.5% of acid (sour), and lemons and limes 5–6% (notably sour). Consumer acceptance of mandarins, tangelos, and tangors is also influenced by acidity, and only mandarins with acidity of 0.8% are fully accepted by tasters (Kefford and Chandler, 1970). In "sweet" citrus fruits, the level of citric acid peaks around 4% at early stages of fruit development and then declines as juice sacs expand and the fruit matures (Erickson, 1968; Marsh et al., 1999). Many factors influence acidity of citrus fruits, the scion-rootstock combination is one of the most important. For instance, juice acidity declines more rapidly in Navel than in Valencia oranges and, as such, decline is temperature dependent. Navel oranges are less suited to warm environments (Reuther, 1988). Fruit position and cultural practices (nutrition and irrigation) may also be used to manipulate fruit acidity levels (Marsh et al., 2000).

Citric acid is synthesized in the mitochondria of the juice sac cells as a part of the TCA cycle and is translocated and accumulated into the vacuoles, where it is metabolized via the glycolitic pathways (Tucker, 1993). The expression of genes related to citric acid synthesis, transport to the vacuole and later breakdown regulate citric acid content in the fruit. Many of the genes have been cloned, allowing the candidate gene to be used in genetic studies in regulation of citric acid content (Canel et al., 1996; Roose, 2000; Echeverria et al., 2000; Sadka et al., 2000; Sadka et al., 2001).

Flavonoids

Flavonoids are important secondary plant metabolites and are present in plant tissues at relatively high concentrations as sugar conjugates. The basic flavonoid structure is the flavan nucleus, which consists of 15 carbon atoms arranged in three rings (C6-C3-C6). Flavonols, flavanols, anthocyanidins, flavones, and flavanones are flavonoids with chemical variants and substitutions of the C15 flavonoid molecule. Differences in structure and substitutions influence the stability of phenoxyl radical and thereby the antioxidant property. Flavonoids may act as antioxidants or as part of other mechanisms contributing to anticarcinogenic or cardioprotective action (Castelluccio et al., 1995; Rice-Evans et al., 1996). Citrus fruits are particularly abundant in flavonoids, which may account for up to 75% of the total solids (Kefford, 1959; Rouseff et al., 1987) and over 60 have already been characterized (Vandercook and Stevenson, 1966; Albach and Redman, 1969; Ting et al., 1979; Park et al., 1983).

Previous studies in *C. sinensis* revealed the presence of the polymethoxyflavones sinensetin (5, 6,7,3′,4′-pentamethoxyflavone), nobiletin (5,6, 7,8,3′,4′-hexamethoxyflavone), tangeretin (5,6,7, 8,4′-pentamethoxyflavone), and 3,5,6,7,8,3′,4′-heptamethoxyflavone, and the flavanones hesperidin and isonaringin (Del Río et al., 1998a, b). The highest levels of these secondary compounds were found in young developing fruits (Castillo et al., 1992; Del Río et al., 1998a, b; Ortuño et al., 1997), although it has been suggested that some polymethoxyflavones might be also related to maturation in other *Citrus* species (Ortuño et al., 1997). These compounds are mainly located in the peel, polymethoxyflavones in the flavedo and flavanones in the albedo (Kanes et al., 1992; Ortuño et al., 1997), and may function as a first and second defense barrier, respectively, from pathogenic attack (Ben-Aziz, 1967; Arcas et al., 2000; Ortuño et al., 2002). Polymethoxyflavones are biologically more active than flavanones, although they occur in lower concentration (Arcas et al., 2000; Del Río et al., 2000). The flavanone hesperidin, which is tasteless, is the predominant flavonoid in most citrus fruits. The phenolic profile in the fruit shows quantitative differences as a function of cultivars, environmental growing conditions and maturity stage (Mouly et al., 1997; Rapisarda et al., 1998), and the antioxidant capacity might vary accordingly.

Anthocyanins are an important subgroup of flavonoids, responsible of the typical red color of the peel and pulp of blood orange varieties (Maccarone et al., 1983, 1985). The main cultivars are Moro, Tarocco, and Sanguinello, which cover 70% of the orange production in Italy. Anthocyanins have considerable pharmacological properties related to their ability to scavenger free radicals and to their vasoprotective and antiplatelet activities (Proteggente et al., 2003). Anthocyanin content is an important quality index for fresh and processed products of blood orange and appears to be one of the more determinant factors of their antioxidant activity (Rapisarda et al., 1999). Cyanidin-3-glucoside and cyanidin-3-(6 malonyl)-glycoside are predominant anthocyanins in blood oranges. Anthocyanin biosynthesis in blood oranges is still unclear but certainly genetic traits, environmental factors, as low temperature or light, and fungal elicitors greatly affect the content. Recently, two gene fragments of chalcone and anthocyanidin synthase have been cloned. The expression of chalcone synthase is much higher at any maturation stage in Moro than in sweet oranges, suggesting that this enzyme may be a key element for anthocyanin biosynthesis (Recupero et al., 2000).

Limonoids

Limonoids are a group of highly oxygenated triterpenes found in the families of Rutaceae and Meliaceae. Citrus limonoids occur as aglycones and glycosides. There are at least 38 different limonoids in citrus, and limonin and nomilin are the more relevant. A number of studies have suggested that limonoids might have health-promoting properties and could inhibit the development of carcinogen-induced cancers in a variety of different animal models (Lam et al., 1994, 2000). Most of the early research on citrus limonoids concentrated on the fact that some of these chemicals, primarily limonin and nomilin, are extremely bitter (Rouseff, 1982; Rouseff and Matthews, 1984). The threshold of bitterness for limonin and nomilin in citrus juices is about 3–6 ppm, and concentrations exceeding 6 ppm can lead to significant consumer rejection. To reduce and eventually solve this bitterness problem, considerable research has been conducted to understand the transport and metabolism of these compounds (Hasegawa, 2000). Nomilin is most likely the precursor of all known limonoids in citrus, synthesized from acetate, mevalonate, and/or farnesyl pyrophosphate in the stem phloem. This precursor then migrates to leaves, fruit, and seeds where other limonoids are biosynthesized. The highest limonin concentration in fruits and leaves is found at the earliest stages of growth, whereas seed accumulate limonin during fruit growth and maturation. In leaves and fruit, total limonoid content increases during growth and maturation, and decreases after maturation. In contrast, limonoid concentration does not decrease in seeds after fruit maturity, indicating that seeds act as storage tissues for these compounds. Limonoid aglycones are converted to limonoid glycosides during maturation, and this natural debittering process is catalyzed by the enzyme limonoid glucosyltransferase. This enzyme appears to be responsible for the glucosidation of all limonoid aglycones to their respective glucosides. Limonoid glucosides are major secondary metabolites and accumulate in fruit tissues and seeds in significant quantities. Limonoid glucosyltransferase has been identified and tagged for genetic manipulation to generate transgenic citrus fruit free from limonin bitterness (Moriguchi et al., 2003).

Soluble Sugars

Carbohydrates are the main soluble components of nonacid citrus fruit, mainly in the form of sucrose, glucose, and fructose, as well as other trace sugars. These make up to 75–80% of the TSS in orange juice. Sugar levels in citrus range from 1% to 2.3% glucose, 1% to 2.8% fructose, and 2% to 6% sucrose for oranges and mandarins, 2% to 5% reducing sugars and 2% to 3% sucrose in grapefruit, and around 0.8% glucose and fructose, and 0.2% to 0.3% sucrose for lemons and limes (Ting and Attaway, 1971). Small quantities of mannose and galactose have been also found in citrus juices (Davies and Albrigo, 1998). As mandarins, oranges, and grapefruits ripen, sugars, especially sucrose, increase (Koch et al., 1986), and this disaccharide is the primary nonreducing sugar and the major translocable carbohydrate. Starch levels are less abundant than soluble carbohydrates and increase continuously with maturation. In Fortune mandarins, it has been shown that the changes in reducing sugars during maturation correlated with changes in sucrose synthase, acid and alkaline invertase activities, but acid invertase was less active than the other sucrose-metabolizing enzymes and that the evolution of sugars appears not to be related to the chilling tolerance of theses fruits during cold storage (Holland et al., 1999). As with many other constituents of citrus fruits, sugar accumulation is affected by temperature and light intensity, and fruits exposed to more sunlight accumulate more sugars than unexposed fruits (Kimball, 1984).

Pectin

All citrus contain pectin and the richest sources are limes, lemons, oranges, and grapefruit, in decreasing importance. The albedo is the main source of pectin. With the generic term of pectin are recognized all the esterified polygalacturonic acids at different degree of neutralization. Molecular weight, viscosity, and behavior in aqueous solution are variable. In the presence of saccharine and small quantities of organic acids (usually citric acid), pectins gelatinized, and this property is exploited by the agrochemistry and pharmaceutical industries for pectin isolation. In 1825, Braconnot found that pectin was a very abundant component in many fruits and protopectina, pectin linked to cellulose that can be separated by acid hydrolysis, is the major form. In oranges, pectin accounted for one-third of their weight and is the most important commercial element of the peel of citrus fruit. During maturation, pectin content in the peel and pulp decreases as insoluble pectins are converted into water-soluble pectins and pectinates (Nagy and Attaway, 1980). The fresh or desiccated pulps of some citrus fruit, like lemon or grapefruit, are an excellent raw material for pectin extraction.

Volatiles

The volatile components of citrus fruit are responsible for much of the aroma and flavor perceived by consumers of fresh fruits or their derivates. In addition, some of these volatiles are industrially valuable for the production of fragrances, pharmaceuticals, and agrochemicals for aphid pheromones (Tavera-Loza, 1999).

Citrus volatile compounds consist mostly of mono- (C10) and sesquiterpenes (C15), which are the major components of citrus essential oils, including their derivatives, such as alcohols, esters, and acetates. These compound accumulate in the oil glands of the flavedo and in oil bodies of the juice sacs. Although monoterpene limonene normally accounts for over 90% of essential oils (Weiss, 1997), several unique and less-abundant sesquiterpenes and monoterpenes have a profound effect on aroma and flavor, like valencene, α- and β-sinensal (Maccarone et al., 1998; Vora et al., 1983). Nootkatone, a putative derivative of valencene, is a small fraction of the essential oils with a dominant role in the flavor and aroma of grapefruit (Shaw and Wilson, 1981).

In spite of the importance of aromatic compounds in citrus fruit quality, much is still unknown about the physiological, biochemical, and genetic regulation of their production. The isolation and characterization of a key enzyme, valencene synthase, in aroma production of citrus fruits has been recently reported (Sharon-Asa et al., 2003). Studies on the accumulation of valencene and the pattern of valencene synthase gene expression showed that valencene production is tightly regulated at the transcriptional level. In addition, four monoterpene synthase genes have been cloned from *C. limon*: two limonene synthase, one β-pinene synthase, and one γ-terpinene synthase, and these synthases produced 10 monoterpenoids, including minor by-products (Lücker et al., 2002). In Satuma mandarin, the expression of monoterpene synthase genes is well correlated with monoterpene production, which occurred mainly in peel at early developmental stages (Shimada et al., 2004). Because of the large composition of monoterpenoids in citrus and the differences between fruits and leaves, more monoterpene synthase genes may be expected (Vekiari et al., 2002).

REFERENCES

Achilea O, Chalutz E, Fuchs Y, Rot I. 1985. Ethylene biosynthesis and related physiological changes in *Penicillium digitatum*-infected grapefruit *Citrus paradisi*. *Physiol. Plant Pathol.* 26:125–134.

Agusti M. 2000. *Citricultura*. Ediciones Mundi Prensa, Madrid, Spain.

Albach RF, Redman GH. 1969. Composition and inheritance of flavanones in citrus fruits. *Phytochem.* 8:127–143.

Albrigo LG, Carter RD. 1977. Structure of citrus fruits in relation to processing. In *Citrus Science and Technology*, Vol. 1. S. Nagy, P.E. Shaw, and M.K. Veldhuis, Eds. AVI Publishing Co. Inc., Westport, CT, pp. 38–73.

Alferez F, Zacarías L. 1999. Interaction between ethylene and abscisic acid in the regulation of Citrus fruits maturation. In *Biology and Biothecnology of the Plant Hormone Ethylene II*. A.K. Kanellis, C. Chang, H. Klee, A.B. Bleecker, J.C. Pech, and D. Grierson, Eds. Kluwer Academic Press, Dordrecht, pp. 183–184.

Arcas MC, Botía JM, Ortuño A, Del Río JA. 2000. UV irradiation alters the levels of flavonoids involved in the defence mechanism of *Citrus aurantium* fruits against *Penicillium digitatum*. *Eur. J. Plant Pathol.* 106:617–622.

Bain JM. 1958. Morphological, anatomical and physiological changes in the developing fruit of the Valencia orange, *Citrus sinensis* L. Osbeck. *Aust. J. Bot.* 6:1–28.

Baldwin EA. 1993. Citrus fruit. In *Biochemistry of Fruit Ripening*. G.B. Seymour, J.E. Taylor, and A. Tucker, Eds. Chapman and Hall, London, pp. 107–149.

Ben-Aziz A. 1967. Nobiletin is main fungistant in tangerines resistant to Mal Secco. *Science* 155:1026–1027.

Blankenship SM, Dole JM. 2003. 1-Methylcyclopropene: A review. 2003. *Postharvest Biol. Technol.* 28:1–25.

Canel C, Bailey-Serres JN, Roose ML. 1996. Molecular characterization of the mitochondrial citrate synthase gene of an acidless pummelo *Citrus maxima*. *Plant Mol. Biol.* 31:143–147.

Castelluccio C, Paganga G, Melikian N, Bolwell GP, Pridham J, Sampson J, Rice-Evans C. 1995. Antioxidant potential of intermediates in phenylpropanoid metabolism in higher plants. *FEBS Lett.* 368:188–192.

Castillo J, Benavente-García O, Del Río JA. 1992. Naringin and neohesperidin levels during development of leaves, flower, buds, and fruits of *Citrus aurantium*. *Plant Physiol.* 99:67–73.

Clements RL. 1964. Organic acids in citrus fruits. I. Varietal differences. *J. Food Sci.* 29:276–280.

Coggins CW. 1981. The influence of exogenous growth regulators on rind quality and internal quality of citrus fruits. *Proc. Am. Soc. Hort. Sci.* 76:199–207.

Cooper WC, Rasmusen GK, Waldon ES. 1969. Ethylene evolution stimulated by chilling in *Citrus* and *Persea* sp. *Plant Physiol.* 44:1194–1196.

Davies FS, Albrigo LG. 1998. *Citrus*. CAB International, Oxford.

Del Río JA, Arcas MC, Benavente O, Sabater F, Ortuño A. 1998a. Changes of polymethoxylated flavones levels during development of *Citrus aurantium* cv. Sevillano fruits. *Planta Med.* 64:575–576.

Del Río JA, Arcas MC, Benavente-García O, Ortuño A. 1998b. *Citrus* polymethoxylated flavones can confer resistance against *Phytophthora citrophthora*, *Penicillium digitatum*, and *Geotrichum* species. *J. Agric. Food Chem.* 46:4423–4428.

Del Río JA, Arcas MC, Botía JM, Báidez A, Fuster MD, Ortuño A. 2000. Involvement of phenolic compounds in the antifungal defense mechanisms of *Olea europaea* L. and *Citrus* sp. In *Recent Research Developments in Agricultural and Food Chemistry*, Vol. 4. S.G. Pandalai, Ed. Research Signpost, Trivandrum, pp. 331–341.

Eaks II. 1970. Respiratory responses, ethylene production and responses to ethylene of citrus fruit during ontogenity. *Plant Physiol.* 45:224–338.

Echeverria E, Brune A, Gonzalez P. 2000. Citrate uptake into tonoplasts vesicles from acid lime juice cells. *Proc. Intl. Soc. Citricult. IX Congress*, pp. 649–653.

Erickson LC. 1968. The general physiology of citrus. In *Citrus Industry: Anatomy, Physiology, Genetics and Reproduction,* Vol II. W. Reuther, L.D. Batchelor, and H.J. Webber, Eds. University of California Press, Riverside, CA, pp. 86–126.

FAO. 2004. http://faostat.fao.org/faostat/

Gmitter JG Jr. 1995. Origin, evolution and breeding of the grapefruit. *Plant Breed. Rev.* 13:345–363.

Gmitter FG, Hu X. 1990. The possible role of Yunnan, China, in the origin of contemporary Citrus species Rutaceae. *Econ. Bot.* 44:267–277.

Grierson W. 1986. Physiological disorders. In *Fresh Citrus Fruits*. W.F. Wardowski, S. Nagy, and W. Grierson, Eds. AVI Publishing Co., Inc., Westport, CT, pp. 361–378.

Grierson W, Cohen E, Kitagawa H. 1986. Degreening. In *Fresh Citrus Fruits*. W.F. Wardowski, S. Nagy, and W. Grierson, Eds. AVI Publishing Co., Inc., Westport, CT, pp. 253–274.

Gross J. 1987. *Pigments in Fruits*. Academic Press, London, UK.

Hasegawa S. 2000. Biochemistry of limonoids in citrus. In *Citrus Limonoids: Functional Chemicals in Agriculture and Food*. M. Berhow, S. Hasegawa, and G. Manners, Eds. American Chemical Society, Washington, DC, pp. 9–30.

Holland N, Sala JM, Menezes HC, Lafuente MT. 1999. Carbohydrate content and metabolism as related to maturity and chilling sensitivity of cv. Fortune mandarins. *J. Agric. Food Chem.* 47:2513—2518.

Horowitz RM, Gentill B. 1977. Flavonoid constituents of Citrus. In *Citrus Science and Technology*, Vol. 1. S. Nagy, P.E. Shaw, and M.K. Veldhuis, Eds. AVI Publishing Co., Inc., Westport, CT, pp. 397–426.

Huff A. 1983. Nutritional control of regreening and degreening in citrus peel segments. *Plant Physiol.* 73:243–249.

Huff A. 1984. Sugar regulation of plastic interconversions in epicarp of citrus fruits. *Plant Physiol.* 73:307–312.

Iglesias D, Tadeo FR, Legaz F, Primo-Millo E, Talon M. 2001. *In vivo* sucrose stimulation of colour change in citrus fruit epicarps: interactions between nutritional and hormonal signals. *Physiol. Plant* 112:244–250.

Ikoma Y, Komatsu A, Kita M, Ogawa K, Omura M, Yano M, Moriguchi T. 2001. Expression of a phytoene synthase gene and characteristic carotenoid accumulation during citrus fruit development. *Physiol. Plant* 111:232–238.

Kanes K, Tisserat B, Berhow M, Vandercook C. 1992. Phenolic composition of various tissues of rutaceae species. *Phytochemistry* 32:967–974.

Kato M, Ikoma Y, Matsumoto H, Sugiura M, Hyodo H, Yano M. 2004. Accumulation of carotenoids and expression of carotenoid biosynthetic genes during maturation in Citrus Fruit. *Plant Physiol.* 134:824–837.

Kefford JF. 1959. The chemical constituents of citrus fruits. *Adv. Food Res.* 9:285–372

Kefford JF, Chandler BV. 1970. General composition of citrus fruits. In *The Chemical Constituents of Citrus Fruits*. C.O. Chichester, E.M. Mrak, and G.F. Stewart, Eds. Academic Press, New York, pp. 5–22.

Kimball DA. 1984. Factors affecting the rate of maturation of citrus fruits. *Proc. Fol. State Hort. Soc.* 97:40–44.

Koch KE, Lowell CA, Avigne WT. 1986. Assimilate transfer through citrus juice vesicles stalks: A nonvascular portion of the transport path. In *Phloem Transport*. Alan R., Ed., A.R. Liss Inc., New York, pp. 247–258.

Lafuente MT, Zacarías L, Martínez-Tellez MA, Sanchez-Ballesta MT, Dupille E. 2001. Phenylalanine ammonia-lyase as related to ethylene in the development of chilling symptoms during cold storage of Fortune mandarin Citrus fruits. *J. Agric. Food Chem.* 49:6020–6025.

Lam LKT, Hasegawa S, Bergstrom C, Lam SH, Kenney P. 2000. Limonin and nomilin inhibitory effects on chemical-induced tumorigenesis. In *Citrus Limonoids Functional Chemicals in Agriculture and Foods*. M. Berhow, S. Hasegawa, and G. Manners, Eds. American Chemical Society, Washington, DC, pp. 185–200.

Lam LKT, Zhang J, Hasegawa S. 1994. Citrus limonoid reduction of chemically induced tumorigenesis. *Food Technol.* 48:104–108.

Lücker J, El Tamer MK, Schwab W, Verstappen FW, van der Plas LH, Bouwmeester HJ, Verhoeven HA. 2002. Monoterpene biosynthesis in lemon *Citrus limon*. cDNA isolation and functional analysis of four monoterpene synthases. *Eur. J. Biochem.* 269:3160–3171.

Maccarone E, Campisi S, Fallico B, Rapisarda P, Sgarlata R. 1998. Flavor components of Italian orange juices. *J. Agric. Food Chem.* 46:2293–2298.

Maccarone E, Maccarrone A, Perrini G, Rapisarda P. 1983. Anthocyanins of the Moro orange juice. *Ann. Chim.* 73:533–539.

Maccarone E, Maccarrone A, Rapisarda P. 1985. Acylated anthocyanins from oranges. *Ann. Chim.* 75:79–86.

Marsh KB, Richardson AC, Erner Y. 2000. Effect of environmental conditions and horticultural practices on citric acid content. *Proc. Intl. Soc. Citricult. IX Congress*, pp. 640–643.

Marsh KB, Richardson AC, MacRae EA. 1999. Early and mid-season temperature effects on the growth and composition of Satsuma mandarins. *J. Hort. Sci. Biotech.* 74:443–451.

Martínez-Javega JM, Cuquerella J. 1995. Alteraciones fisiológicas en la postrecolección de frutos cítricos 2ª parte. *Fruticultura Prof.* 66:57–67.

Moriguchi T, Kita M, Hasegawa S, Omura M. 2003. Molecular approach to citrus flavonoid and limonoid biosynthesis. *Food Agric. Environ.* 1:22–25.

Mouly PP, Gaydou EM, Faure R, Estienne JM. 1997. Blood orange juice authentication using cinnamic acid derivatives. Variety differentiations associated with flavanone glycoside content. *J. Agric. Food Chem.* 45:373–377.

Nagy S, Attaway JA. 1980. *Citrus Nutrition and Quality*. American Chemical Society, Washington, DC.

Oberholster R, Cowan K, Molnar P, Toth G. 2001. Biochemical basis of color as an aesthetic quality in *Citrus sinensis*. *J. Agric. Food Chem.* 49:303–307.

Olson JA. 1989. Provitamin-A function of carotenoids: the conversion of β-carotene into vitamin-A. *J. Nutr.* 119:105–108.

Ortuño A, Arcas MC, Botía JM, Fuster MD, Del Río JA. 2002. Increasing resistance against *Phytophthora citrophthora* in tangelo Nova fruits by modulating polymethoxyflavones levels. *J. Agric. Food Chem.* 50:2836–2839.

Ortuño A, Reynaldo I, Fuster MD, Botía JM, García-Puig D, Sabater F, García-Lidón A, Porras I, Del Río JA. 1997. *Citrus* cultivars with high flavonoid contents in the fruits. *Sci. Hortic.* 68:233–236.

Park GL, Avery SM, Byers JL, Nelson DB. 1983. Identification of bioflavonoids from citrus. *Food Technol.* 37:98–105.

Petracek PD, Hagenmaier RD, Dou H. 1999. Waxing effects on citrus fruit physiology. In *Advances in Postharvest Diseases and Disorders Control of Citrus Fruit*. M. Schirra, Ed. Res. Signpost, Trivandrum, pp. 71–92.

Porat R, Weiss B, Cohen L, Daus A, Goren R, Droby S. 1999. Effects of ethylene and 1-methylcyclopropene on the postharvest qualities of 'Shamouti' oranges. *Postharvest Biol. Technol.* 15:155–163.

Proteggente AR, Saija A, De Pasquale A, Rice-Evans CA. 2003. The compositional characterisation and antioxidant activity of fresh juices from Sicilian sweet orange *Citrus sinensis* L. Osbeck varieties. *Free Radic. Res.* 37:681–687.

Rapisarda P, Carollo G, Fallico B, Tomaselli F, Maccarone E. 1998. Hydroxycinnamic acids as markers of Italian blood orange juices. *J. Agric. Food Chem.* 46:464–470.

Rapisarda P, Tomaino A, Lo Cascio R, Bonina F, De Pasquale A, Saija A. 1999. Antioxidant effectiveness as influenced by phenolic content of fresh orange juices. *J. Agric. Food Chem.* 11:4718–4723.

Recupero G, Russo MP, Rapisarda M, la Rosa M, Guardo M, Lo Piero AR, Petrone G. 2000. Anthocyanin biosynthesis in blood oranges. *Proc. Int. Soc. Citricult. IX Congress*, pp. 681–682.

Reuther W. 1988. Climate and fruit quality. In *Factors Affecting Fruit Quality*. J.J. Ferguson and W.F. Wardowski, Eds. Uni. Fla. Citrus Short Course Proc., Gainesville, FL.

Rice-Evans CA, Miller NJ, Paganga G. 1996. Structure-antioxidant activity relationships of flavonoids and phenolic acids. *Free Radic. Biol. Med.* 20:933–956.

Riov J, Yang SF. 1982. Autoinhibition of ethylene production in citrus peel disks- Suppression of 1-aminocyclopropane-1-carboxylic acid synthesis. *Plant Physiol.* 69:687–692.

Rodrigo MJ, Marcos JF, Alferez F, Mallent MD, Zacarías L. 2003. Characterization of Pinalate, a novel *Citrus sinensis* mutant with a fruit-specific alteration that results in yellow pigmentation and decreased ABA content. *J. Exp. Bot.* 54:727–738.

Rodrigo MJ, Marcos JF, Zacarías L. 2004. Biochemical and molecular analysis of carotenoid biosynthesis in flavedo of orange *Citrus sinensis* L. during fruit development and maturation. *J. Agric. Food Chem.* 52, 6724–6731.

Romojaro F, Banet E, Llorente E. 1979. Carotenoides en flavedo y pulpa de pomelo Marsh. *Rev. Agroquimica Tecn Aliment.* 19:385–392.

Roose ML. 2000. Citric acid content in citrus fruit: inheritance and genetic manipulation. *Proc. Intl. Soc. Citricult. IX Congress*, pp. 647–648.

Rouseff RL. 1982. Nomilin, a new bitter component in grapefruit juice. *J. Agric. Food Chem.* 30:504–507.

Rouseff RL, Martin SF, Youtsey CO. 1987. Quantitative survey of narirutin, naringin, hesperidin, and neohesperidin in *Citrus*. *J. Agric. Food Chem.* 35:1027–1030.

Rouseff RL, Matthews RF. 1984. Nomilin, taste threshold and relative bitterness. *J. Food Sci.* 49:777–779.

Sadka A, Dahan E Or E, Cohen L. 2000. NADP+-isocitrate dehydrogenase gene expression and isozyme activity during citrus fruit development. *Plant Sci.* 158:173–181.

Sadka A, Dahan Eor E, Roose ML, Marsh KB, Cohen L. 2001. Comparative analysis of mitochondrial citrate synthase gene structure, transcript level and enzymatic activity in acidless and acid-containing *Citrus* varieties. *Aust. J. Plant Physiol.* 28:383–390.

Saunt J. 2000. *Citrus Varieties of the World*. Sinclair International, Norwich, UK.

Schirra M, Ben-Yehoshua S. 1999. Heat treatments: A possible new technology in citrus handling. Challenges and prospects. In *Advances in Postharvest Diseases and Disorders Control of Citrus Fruit*. M. Schirra, Ed. Res. Signpost, Trivandrum, pp. 133–145.

Sharon-Asa L, Shalit M, Frydman A, Bar E, Holland D, Or E, Lavi U, Lewinsohn E, Eyal Y. 2003. Citrus fruit flavor and aroma biosynthesis: isolation, functional characterization, and developmental regulation of Cstps1, a key gene in the production of the sesquiterpene aroma compound valenceno. *Plant J.* 36:664–674.

Shaw PE, Wilson CW. 1981. Importance of nootkatone to the aroma of grapefruit oil and the flavor of grapefruit juice. *J. Agric. Food Chem.* 29:677–679.

Shimada T, Endo T, Fujii H, Hara M, Ueda T, Kita M, Omura M. 2004. Molecular cloning and functional characterization of four monoterpene synthase genes from Citrus unshiu Marc. *Plant Sci.* 166:49–58.

Spiegel-Roy P, Goldschmidt EE. 1996. *Biology of Citrus*. Cambridge University Press, Cambridge.

Stewart I, Leuenberger U. 1976. Citrus color. *Alimenta* 15:33–36.

Tavera-Loza H. 1999. Monoterpenes in essential oils; biosynthesis and properties. In *Chemicals via Higher Plant Bioengineering*. E. Shahidi, Ed. Kluwer Academic Publishers and Plenum Press, New York.

Ting SV, Attaway JA. 1971. Citrus fruits. In *The Biochemistry of Fruits and their Products*, Vol. 2. A.C. Hulme, Ed. Academic Press, London, pp. 107–169.

Ting SV, Rouseff RL, Dougherty MH, Attaway JA. 1979. Determination of some methoxylated flavones in citrus juices by high-performance liquid-chromatography. *J. Food Sci.* 44, 69–71.

Tsushima M, Maoka T, Katsuyama M, Kozuka M, Matsuno T, Tokuda H, Nishino H, Iwashima A. 1995. Inhibitory effect of natural carotenoids on Epstein–Barr virus activation activity of a tumor promoter in Raji cells. *Biol. Pharm. Bull.* 18:227–233.

Tucker GA. 1993. Introduction. In *Biochemistry of Fruit Ripening*. G.B. Seymour, J.E. Taylor, and A. Tucker, Eds. Chapman and Hall, London, pp. 3–43.

Vandercook CE, Stevenson RG. 1966. Lemon juice composition. Identification of the major phenolic compounds and estimation by paper chromatography. *J. Agric. Food Chem.* 14, 450–454.

Vekiari SA, Protopapadakis EE, Papadopoulou P, Papanicolaou D, Panou C, Vamvakias M. 2002. Composition and seasonal variation of the essential oil from leaves and peel of a Cretan lemon variety. *J. Agric. Food Chem.* 50:147–153.

Vora JD, Matthews RF, Crandall PG, Cook R. 1983. Preparation and chemical composition of orange oil concentrates. *J. Food Sci.* 48: 1197–1199.

Watkins CB. 2002. Ethylene synthesis mode of actions and consequences and control. In *Fruit Quality and its Physiological Basis*. M. Knee, Ed. Sheffield Academic Press Ltd, Sheffield, pp. 180–224.

Weiss EA. 1997. *Essential Oils Crops*. Wallingford: CAB International.

Yamaki YT. 1989. Organic acids in the juice of citrus fruits. *J. Jpn. Soc. Hort. Sci.* 58:587–594.

Yokoyama H, Vandercook CE. 1967. Citrus carotenoids. I. Comparison of carotenoids of mature green and yellow lemons. *J. Food Sci.* 32:42–48.

Yokoyama H, White MJ. 1967. Carotenoids in the flavedo of Marsh seedless grapefruit. *J. Agric. Food Chem.* 15:693–696.

Zacarías L, Lafuente MT, Marcos JF, Saladie M, Dupille E. 2003. Regulation of ethylene biosynthesis during cold storage of the chilling-sensitive Fortune mandarin fruit, In *Biology and Biotechnology of the Plant Hormone Ethylene*. M. Vendrell, H. Klee, J.C. Pech, and F. Romojaro, Eds. IOS Press, Oxford,, pp. 112–117.

19
Oranges and Citrus Juices

Kulwant S. Sandhu and Kuldip S. Minhas

INTRODUCTION

The oranges include the bitter (sour) orange (*Citrus quarantium*), the sweet orange (*C. sinensis*), and the mandarin orange (*C. reticulata*). The most important of all these is the sweet orange. It is widely grown in all regions of the world adapted to citrus, although each region usually has its own characteristic varieties. Distribution of over 100 varieties of oranges of variable commercial importance is described by Nagy et al. (1977).

The world production of oranges is 60.04 million tons in 2003 according to FAO Database (2004). Brazil ranks first with the production of 16.93 million tons followed by USA with a total production of 10.47 million tons. Mexico, Spain, India, Italy, and China are the other major producers of oranges (Table 19.1). Brazil, the world's largest producer of oranges and orange juice, is a major exporter of orange juice to the United States. It accounts for 80% of the frozen, concentrated orange juice exported to the world. Now Brazil earns about $1.6 billion in foreign exchange through exports of frozen concentrated orange juice (FCOJ) alone. The United States is the world's second largest orange juice exporter. Other orange juice-producing countries are: Greece, Israel, Italy, Mexico, Morocco, Spain, Turkey, Argentina, Australia, and South Africa utilizing mainly the fruit that fails to meet the quality requirements for the fresh market.

The major markets for orange juice in the world are the United States, Europe, Canada, and Japan. In recent years, U.S. market alone has accounted for more than half of the orange juice consumption in the world. In the future, the growth rates for orange juice demand in the United States and Canada are expected to be 1.5% or less, whereas the orange juice demands growth rates in Europe and rest of the world are expected to be about 4%.

The main varieties of sweet oranges grown around the world are classified into four groups, namely the common orange, the acidless orange, the pigmented orange, and the navel orange (Hodgson, 1967). The

Table 19.1. Estimated Orange, Grapefruit, Pomelos, Lemons and Limes Production (Million Tons) of Countries in the World 2003

	Production (Million Tons)		
Country	Orange	Grapefruit and Pomelos	Lemons and Limes
World	60.04	4.696	12.45
Argentina	0.700	0.170	1.200
Australia	0.437	0.014	0.031
Brazil	16.935	0.067	0.950
China	1.647	0.356	0.600
Cuba	0.492	0.227	0.026
Greece	1.200	0.008	0.160
India	2.980	0.137	1.370
Iran	1.850	0.037	1.040
Israel	0.200	0.270	0.027
Italy	1.962	0.004	0.548
Mexico	3.969	0.257	1.824
Morocco	0.821	0.002	0.012
Pakistan	1.400	–	0.099
South Africa	1.082	0.380	0.149
Spain	3.091	0.030	1.065
Turkey	1.215	0.130	0.500
USA	10.473	1.871	0.939

Source: FAO database 2004 (www.fao.org).

acidless orange is of minor commercial importance. Among the common oranges are Valencia, Pineapple, Hamlin, Parson Brown, Jaffa, and Shamouti, which are generally grown in various parts of the world. Pera, Corriente, and Bahianinha are some of this type grown especially in Latin America.

The pigmented oranges are widely grown in the Mediterranean area. The principal cultivars include: Doblefina, Eutrifina, Moro, Tarocco, Ovale, Sanguinello Commun, Ruby Blood, and Maltese Blood. Navel oranges are mostly grown for fresh market because of their tendency to develop a bitter taste in processed products. The Washington Navel is probably the most important cultivar of this group. The Frost Washington and Dream are some of the other promising cultivars. Also, many fine-quality sweet orange varieties, including Tien cheng, Lui cheng, Sekkan, Yinkan, and Hwang Kuo, are known to exist in China. One reason that the Chinese sweet oranges have not received much attention from the Western world is that Asian people generally prefer the mandarin type of citrus.

Mandarins are a group of loose-skinned oranges of primary importance to the Far East, but they are also becoming increasingly popular in the United States. The Satsumas are an important citrus crop in Japan; Dancy tangerine is widely grown in the United States; and Clementine is an important cultivar of the Mediterranean areas, especially Algeria. Among other commercially grown varieties are King, Robinson, Page, Ponkan, and Murcott. The cultivar Murcott, according to many horticulturists could be a tangelo that is hybrid of mandarin and orange. Temple, which is also believed to be a tangelo, has excellent flavor and is widely planted in the United States, especially in Florida. The most important cultivars of the tangelo are Orlando and Minneola, both of which are the crosses between Dancy tangerine and Duncan grapefruit.

VARIETIES SUITABLE FOR JUICE PRODUCTION

The sweet orange varieties, the blood, navel, and the common white-fleshed (non-blood) oranges, yield more juice and soluble solids when grown in conditions other than a dry Mediterranean climate. Most of the blood varieties are only weakly pigmented when grown in Florida. Pera, Hamlin, Parson Brown, and Natal are the main varieties of oranges grown in Brazil. In Florida, they are Temple, Parson Brown, Pineapple, Hamlin, and Valencia (similar to the Pera). The Mediterranean region is a secondary center of diversity in the sweet orange species. Here, many new and improved sweet orange varieties have developed through bud mutations and chance seedlings. These include the high-quality blood oranges (Sanguinelli of Spain) and the high-quality standard sweet orange varieties. The Spanish Sanguinelli, Moro, and Tarocco varieties have very distinctive flavors and are more likely to show blood coloration than other blood orange varieties. A comprehensive account of world production of citrus varieties for processing has been given by Nagy et al. (1977).

The entire common white-fleshed, round orange varieties are suitable for processing, but some are preferred to others. The Valencia has excellent processing quality and is the most widely grown variety for processing. The early-maturing varieties, such as Hamlin, Parson Brown, Marrs, and Salustiana, tend to produce poorer juice color, lower yields of soluble solids, and more after-processing bitterness (if harvested early) than most mid-season and late varieties. Good fertilization practice may improve the processing quality of the early-maturing varieties, but will not convert them to high-quality varieties. The Hamlin, Parson Brown, Pineapple, and Valencia sweet orange varieties are used for the production of processed juice in Florida.

The main deterrents in the use of Washington navel oranges in processing are the high levels of after-processing bitterness in the juice and low fruit yields. The new Lane's Late navel variety from Australia is a distinct improvement over the Washington navel. It produces higher yields of soluble solids and juice than fruit grown in Mildura. Victoria is free of after-processing bitterness. Juice from Belladonna variety has an appreciably higher limonene concentration than that from Valencia variety (Di Giacomo et al., 1976). Baladi and Succari varieties of sweet orange are also recommended for the production of juice (Noaman and Husssein, 1973).

Some seedy varieties have good processing quality, but the seeds are an objectionable feature in the juice extraction process. The various common round sweet orange varieties of the world are classified according to earliness of maturity and seediness. Parson Brown and Mars Early are early seedy, while Hamlin, Diller, and Salustiana are early seedless varieties. By virtue of their mid-season-to-late maturity, seedlessness, and good flavor, the Mediterranean varieties Verna, Cadenera, Ovale Calabrese, and Belladonna all possess good processing quality. The Calderon, Tajamur, Natal, and Pera have been used successfully for processing in South America. Juice of the Shamouti, reported to develop after-bitterness, has been processed extensively in Israel.

A selection from the cultivar "Valencia" described by Stewart et al. (1975), has much deeper orange colored juice than the normal. The improved color was related to the increased levels of cryptoxanthin (152 μg/ml) in juice samples taken in early April, compared to 90 μg/ml in normal "Valencia" juice. The juice tends to have a lower acidity, and juice volume ratio is also higher than normal. In spite of differences between samples from the various regions, both the "blood" oranges and those with pigmented flesh are suitable for processing (Giannone and Matliano, 1977). Only the juice from blood oranges occasionally had a not wholly satisfactory natural color.

In Australia, Valencia and navel oranges have always dominated the citrus crop (Chandler and Nicol, 1983). Valencia is the preferred variety for juice production (Anon, 1993a). Australian juice processors have had supply problems with Valencias because of their alternate bearing, and the possibility of bitterness development in juice processed from navel oranges. Trials carried out on other varieties show that Hamlins performed well on Rough lemon with good juice yield, high total soluble solids (TSS) content, high TSS:acid ratio, low acidity, negligible bitterness, and good processing quality when harvested after June. Silettas on Rough lemon also performed well, but not as sweet as Hamlins and had lower TSS:acid ratios. Of the other cultivars, Mediterraneans and St. Michaels showed promise. The former gave a very acceptable juice in September, whereas the latter does so if the fruit is allowed to hang for longer time on the tree. For the maximization of profitability from processing of oranges, the juice yields of 38.0%, 30.6%, 43.2%, and 42.5% for Washington, Thompson, Chilena, and Valencia cultivars, respectively, are essential (Erazo et al., 1984).

Four commercial varieties of Malta oranges, i.e., Blood Red, Pineapple, Jaffa, and Valencia Late were analyzed for physico-chemical composition and suitability for processing and waste utilization. Recovery of juice, peel, and pomace in Malta variety ranged from 50.8% to 55.4%, 23.2% to 30.9% and 13.6% to 22.1%, respectively. New citrus hybrid, Ambersweet (1/2 orange, 3/8 tangerine, and 1/8 grapefruit) yields orange juice with good color and flavor, and the fruit is easy to peel and matures early in the season (Wade, 1995). Some technological characteristics of five orange cultivars (Hamlin, Magnum Bonum, Dortyol Native, Kozan Native, and Valencia) used in the juice industry have been evaluated by Altan (1995). Based on his results, cultivar Kozan Native was found superior for production of orange juice. The processing properties of three early cultivars of oranges (NT-18, NT-34, and NT-36) in two successive seasons were studied by Pinera et al. (1995). They concluded that the weak aroma, flavor, and pale yellow color of juice from these early cultivars of Cuban oranges would limit their use for the manufacture of fresh and concentrated orange juice, although ascorbic acid content was adequate and yields were high.

MANDARINS

The Kinnow (King × Willowleaf) was released by H.B. Frost at the University of California Citrus Research Centre at Riverside in 1935. It is commercially grown in Arizona and California in USA as well as in the Punjab states of India and Pakistan. Because of its higher juice content, a lot of efforts have been made to explore its utilization for processing into juice and juice-based products in India and Pakistan.

Kinnow orange has been analyzed for physico-chemical composition, and suitability for processing and waste utilization. Recoveries of juice, peel, and pomace in Kinnow were 55.8%, 26.8%, and 17.4%, respectively (Pruthi et al., 1984). The composition of Kinnow juice is reported as: total solids 11.15%, TSS 10.0%, pH 3.6, titratable acidity 1.15%, reducing sugars 3.2%, total sugars 6.74%, pectin 0.29%, ascorbic acid 12.20 mg%, and β-carotene 1.35 mg% (Sandhu and Bhatia, 1985). Sandhu and Singh (1999) studied the factors affecting the physico-chemical and organoleptic properties of Kinnow juice. Incorporation of additives in the juice considerably improved the organoleptic quality of Kinnow juice.

Several high-quality mandarins (Ponkan, Tankan, and Dancy types) have come out of China. Dancy, Robinson, Orlando tangelo, Nova, Minneola tangelo, Page, and Kinnow juice are suitable for blending with orange juice to improve the color of the finished product. Orlando tangelo juice does not develop post-processing bitterness, and Minneola tangelo contains no limonin and is suitable for processing. Page and Nova have good processing quality (Scott and Hearn, 1966). Dancy, Robinson, Orlando tangelo, and Nova juices do not develop off-flavors and are suitable for blending to the juice of Temple mandarin.

MATURITY CHARACTERISTICS AND JUICE QUALITY

The maturity of fruit is assessed from the color, juice content, TSS, and acidity of the juice. The juice content and composition of juice vary widely for a variety grown at different places. The different varieties of oranges have been reported to produce 26.3–59% juice, 8.8–14.8% TSS, and 0.64–1.77% acid content (Gaetano, 1975). Before harvest, fruit for processing must meet certain minimum maturity requirements established by the regulatory agencies. These requirements may vary from one citrus-producing area to another. These requirements are usually based on: (1) color break, (2) minimum juice content, (3) minimum acid content, (4) minimum percentage of TSS, and (5) °Brix:acid ratio. The single-strength orange juice as per USDA, shall have specific gravity (at 20°C) of 1.0473, i.e., 11.75°Brix. There is a steady rise in specific gravity at 20°C, soluble solids, reducing and total sugars, °Brix, ratio of soluble solids to acidity, ascorbic acid, formol index, and pH, but a decrease in acidity and ash contents of juice obtained from Sanguinella oranges harvested between 9 January and 26 March.

Color scores of Hamlin orange (*C. sinensis* L. Osbeck) juice increase with late harvesting (Wutscher and Bistline, 1988). The effect of time of harvest on alternate cropping yields and fruit quality of Valencia orange trees, are subject to the alternate cropping phenomenon. There is a tendency for soluble solids, acid, and juice concentration to decrease, and for fruit weight and peel thickness to increase with increasing lateness of harvest (Gallasch, 1978). Data are given for the weight, juice yield, peel thickness, and TSS, and acid contents of oranges grown in the 3 years and harvested early, mid-season, or late.

The stage of maturity, variety, and processing, affect the chloramine-*T* values and total amino acid content of orange juices (Maraulja and Dougherty, 1975). The total amino acid content of the juices increases from the beginning of the "Hamlin" season to the "Pineapple" maturity stage, and declines slightly during the "Valencia" season. All three varieties show higher chloramine-*T* values as maturity increases. Both sets of test results are slightly higher for hard squeeze juices than for soft squeeze juices. The stage of maturity, variety, and processing also affects the color, cloud, pectin, and water-insoluble solids of orange juice (Huggart et al., 1975). "Valencia" juice had the best color, and average color values were higher for soft squeeze-soft finish than for hard squeeze-hard finish juices. "Hamlin" juice showed less cloud than the other cultivars. Total pectin and water-insoluble solids increased in juices with increased extractor–finisher pressures. It has been observed that the concentration of some amino acids in orange juice vary with the harvesting date, while that of others remain practically constant in the samples from Brazil and various Mediterranean countries (Wallrauch, 1980a).

Juice produced at Piana di Rosarno from Biondo Comune oranges between December and May when analyzed (Di Giacomo et al., 1975a,b), produced the following values: degree of concentration 5.75–6.25; acidity (as citric acid in 10°Brix) 0.90 (May)–2.15 (December); soluble solids:acidity ratio 5.26:1 (December)–12.32:1 (May); formol index (in 11°Brix) 1.53–2.17; ascorbic acid 5.44–7.73 mg/unit Brix; reducing sugars 50.56–74.12% of total sugars; β-carotene 3.0–6.11%; carotene esters 18.58–22.67% of total carotenoids; alkalinity of ash 47.10–57.60%, with 34.80–53.80% K, 0.12–0.37% Na, 1.74–2.91% Ca, 1.95–2.65% Mg, 2.63–4.04% P, and 0.020–0.113% Fe; Na:K ratio 1:114–1:363; flavanoids 1420–1940 ppm. Acidity, total sugars, ash, K, and Cl measured in samples of Israeli orange and grapefruit juices over growing seasons were measured and the total sugars and chlorides were found to be strongly affected by the growing season (Cohen, 1982).

The early, mid-season, and late sweet orange varieties are available for processing throughout most of the periods of the year. Specific types are classified according to the time required between blossoming and harvesting, and the time of year when fruit normally gains maturity. Throughout the United States, citrus trees normally blossom at about the same time, between late February and mid-April. Early-season Parson Brown and Hamlin usually mature from October through December in Florida and Texas. Mid-season Pineapple orange matures during the first quarter of the year (January–March). Late-season cultivars (Valencia) mature from around mid-March through June and in some seasons are harvested even up to July depending upon the particular weather and climatic factors (Harding et al., 1940).

Washington Navel and Valencia are predominant cultivars in California and Arizona. Washington Navel matures from November to May, and Valencia matures from March through October. Because of climatic differences in these two states, times between bloom and harvest are longer there than in Florida and Texas. Specific details of differences and characteristics of these cultivars and other similar cultivars are covered (Harding et al., 1940). In California and Arizona, most of the fruits are marketed fresh, but excess fruits or fruits unsuitable for fresh market are processed. In Florida and Texas, most of the mid- and late season fruit are processed. Although considerable early-season fruits are marketed fresh, more than half of the Hamlins and Parson Browns are processed.

FACTORS AFFECTING THE QUALITY OF JUICE

The juice quality characteristics vary with variety, rootstock, scion, fertilization, frequency of irrigation, date of harvesting, age of tree, tree spacing, position of fruit on the tree, climatic conditions, and place of growing.

ROOTSTOCK

Isaacs (1980) summarized the effects of rootstock on the quality of the juice from mandarins. Bitterness development in the fruit was found to be greatly affected by rootstock; mandarins budded on Rough lemon rootstock yield juices, which develop a considerable degree of bitterness soon after extraction. Tree spacing and rootstock affect the growth, yield, fruit quality, and freeze damage of young "Hamlin" and "Valencia" orange trees (Wheaton et al., 1986). The influence of nine rootstocks on the composition of juice from "Valencia Late" and "Moro" cultivar oranges was studied (Di Giacomo et al., 1977), with special reference to limonene contents. The results showed that rootstock could be classified into three groups, the first determining low limonene contents (less than 10 ppm) in pasteurized orange juice, the second determining intermediate contents, and the third determining high contents (greater than 20 ppm). With respect to juice yield, Brix:acid ratio, and limonene content, best results were obtained with the rootstock Mandarino Cleopatra.

IRRIGATION

Juice yield, soluble solids, citric acid, suspended solids, and pH of Marrs and Valencia oranges, are affected by irrigation plus rainfall (Wiegand et al., 1982). Vitamin C, juice yield, pH, and suspended solids are only occasionally affected by irrigation treatment where rainfall contributes about half the annual water requirement (Cruse et al., 1982). Citric acid content of "Valencia" orange juice was consistently higher in less frequently irrigated treatments, regardless of rainfall. Partial fertigation treatments lowered the concentration of the soluble solids of the juice. Irrigation increased fruit production by 39–64% over the no irrigation control (Koo and Smajstrla, 1985). Partial fertigation treatments had minimal influence on fruit production.

The climate has effect on the quality of Corsican clementines with respect to soluble extract, acidity, and maturity index and these parameters are influenced by rainfall, with low quality associated with high rainfall prior to harvest and vice versa (Sanchez et al., 1978). Fruit quality was not appreciably affected by temperature greater than 12.8°C.

FERTILIZATION

Variations in Brix:acid ratio, formol number, total acidity (TA), citric, malic and isocitric acid contents, citric:isocitric ratio, K, PO_4, Mg and Ca, and glucose:fructose ratio are affected by the harvesting time (Wallrauch, 1980b). Higher levels of nitrogen fertilization of Satsuma mandarin trees increased fruit weight, peel ratio, °Brix of juice, Kjeldahl and amino-N, alkalinity of juice, and levels of the main amino acids in the juice (Kodama et al., 1977). Phosphorus contents were decreased, but no effect was discernible on ash and ascorbic acid contents.

AGE OF TREE

Age (years) of tree and cultivar influence the juice content, TSS, acidity, and ripeness index of oranges (Frometa and Echazabal, 1988). The number of years necessary to obtain reliable results on juice characteristics was determined as four for early cultivars of "Hamlin," "Salustiana," and "Victoria."

POSITION AND FRUIT LOAD ON TREE

The color of the fruit and juice is influenced by the position of the fruit on the tree (Stewart, 1975). The juice color was consistently brighter and deeper from orange fruits growing on the north side of the trees. Peel colors showed similar trends to those of juice. In the Seto Inland Sea area, Satsuma mandarin (*C. unshiu* Marc. cv. Sugiyama) fruit quality is expressed in °Brix, and titratable acidity is little affected by variations in microclimate depending on the fruit locations within the tree canopy (Daito et al., 1981). Fruit load and fruit thinning influenced the fruit character, shoot growth, and flower bud formation in the following season in young Satsuma mandarin ("Miyagawa Wase") trees (Morioka, 1987).

PRETREATMENTS

Peel oil composition and processed juice quality of Hamlin oranges (*C. sinensis* L. Osbeck), degreened with ethylene gas showed no significant differences between control and experimental samples in flavor tests, vitamin C levels, Brix:acid ratios, color scores, or volatile peel oil constituents (Moshonas and Shaw, 1977). Forty-three peel oil compounds were identified, 29 of which are reported for the first time specifically as constituents of Hamlin orange peel oil.

Ethephon, gibberellic acid (GA), and light exclusion, affect rind pigments, plastid ultrastructure, and juice quality of Valencia oranges in green, colored, and re-greened fruits (El-Zeftawi and Garrett, 1978). In the early phase of re-greening, an increase occurs in the plastid lamellae in the two outer layers of rind cells, corresponding to decreases in carotenoids and starch. Plastids containing lamellae without grana stacks are most frequent in re-greened fruits. GA inhibits carotenoids and increases chlorophylls at the re-greening stage only, but hastens the loss of starch at the colored stage, whereas ethephon decreases chlorophylls and increases starch contents in the plastids.

Light exclusion slightly decreases flavonoids and polyphenols in juice of green and colored fruits, and carotenoids in colored fruit, however ethephon increases flavonoids and polyphenols markedly at the green and re-greening stages and carotenoids at the colored and re-greened stages, but carotenoids are decreased at the green stage. GA decreases flavonoids and polyphenols in the juice of green and colored fruits but decreases carotenoids only at the green stage. Six growth regulators: (i) gibberellic acid (10–20 ppm), (ii) Planofix (200–300 ppm), (iii) 2, 4, 5-T (50–100 ppm), (iv) 2, 4, 5-TP (100–200 ppm), (v) Ethrel (200–300 ppm), and (vi) lead acetate (250–500 ppm) were sprayed 2–3 times on mandarins (Kaula) and sweet oranges (Kinnow, Pineapple, and Dancy) as preharvest sprays, to determine their effect on granulation and bitterness of juice. Treatments (i) and (ii) effectively reduced the bitterness of juice; bitterness was not observed up to 12 h after juice extraction in treatment (ii).

ORANGE JUICE: TYPES AND THEIR CHARACTERISTICS

Juice is the cell sap that is present in the cell vacuoles and expressed from sound fruits by squeezing. Orange juice is consumed in a natural cloudy state. The clarification would impair the appearance and flavor of the juice. Different types of orange juices are available in the market. The chilled single-strength orange juice has limited shelf life and requires installation of expensive refrigerated tanks. The conventional pasteurized single-strength orange juice in cans is widely used, but the FCOJ is now a commodity, which is traded worldwide. Concentrated juices are distributed in large containers as a base for the manufacture of a variety of soft drinks. The same is reconstituted to single-strength juice for direct consumption. Comminuted orange products are prepared for use in beverages. Dehydrated juices in powder form are also available in the market.

FRESH JUICE

Freshly squeezed, un-pasteurized orange juice is very desirable for the consumer because of its fresh aroma and flavor, but the shelf life is less than 20 days at 1°C, as it is highly susceptible to microbial spoilage. The manufacturing operations from fruit washing to packaging must be exceptionally clean to minimize product spoilage. Pectin esterase activity in un-pasteurized juice results in loss of cloudiness (Wicker et al., 2003). Due to this reason, product has to be maintained near freezing point throughout its distribution, however, cloud separation, flavor changes due to reactions with oxygen, and color instability still occur, although at slower rate. After several days of packaging, flavor from diacetyl, fused oils, and other microbiologically generated off-flavors, make the product inferior to good quality pasteurized juice. There is a risk of food-borne illness from consumption of un-pasteurized packaged fruit juice. This includes serious incidence of salmonellosis from the consumption of contaminated fresh orange juice. FDA has proposed juice regulations to mandate the use of Hazard Analysis and Critical Control Point (HACCP) by most juice-producing companies and procedures for implementing HACCP have been published (Schmidt et al., 1997).

The distribution of volatile compounds in pulp, serum, and cloud of freshly squeezed orange juice, has no relationship between the retention of aroma compounds in pulp or cloud and their lipid content

Table 19.2. Total Solids, Total Sugars, Acidity, and Pectin Contents of Oranges

Fruit	Total Solids, %	Total Sugars, %	Acidity No. cc 0.1N Per 100 g	Pectin, % as Calcium Pectate
Orange (bitter)				
Edible portion	13.59	5.49	3.30	0.86
Peel and pith	27.27	5.86	0.46	0.89
Juice	10.72	5.74	3.77	
Orange (sweet)				
Edible portion	12.98	7.88	0.79	0.59
Peel and pith	25.52	6.81	0.27	
Juice	11.09	8.47	1.17	0.13

Source: Money and Christian (1950).

or composition (Brat et al., 2003). Juice monoterpene and sesquiterpene hydrocarbons are primarily present in the pulp (74.0% and 87.2%, respectively) and cloud (7.3% and 14.9%, respectively). Esters and monoterpene alcohols are mainly found in the serum (90.4% and 84.1%, respectively). Long chain aliphatic aldehydes tend to concentrate in the pulp. The relative proportions of individual volatile compounds are similar in the pulp and cloud. Half of the alcohol insoluble residues in pulp and cloud are made of non-cell wall proteins and the rest are made of cell wall materials. Pulp and cloud total and neutral lipids have similar fatty acid distribution, although cloud is much richer in total lipids than the pulp. The composition of orange juice in terms of total solids, sugars, acidity, and pectin is given by Money and Christian (1950) (Table 19.2). Table 19.3 shows the vitamin (USDA, 1957) and mineral contents (McCance and Widdowson, 2002).

Pasteurized Juice

The consumer preference is increasing towards single-strength chilled juice. The necessity for food safety and quality requires pasteurization of juice before packaging and distribution. Many important nutrients in citrus juices including sugar, acid, vitamins, minerals, some flavonoids, and other components are quite heat stable under the conditions of pasteurization. Pasteurization process is designed to inactivate the thermally stable isoenzyme of pectin esterase. The temperature necessary for enzyme inactivation is higher than that required for killing the microbes. At a lower pH, the enzyme inactivation is achieved in a shorter time, thus producing a better quality juice. Juice treatment with carbon dioxide at above supercritical conditions, has the advantage of enzyme inactivation without heat, thus preserving the natural flavor. The juice maintains color and cloud stability throughout its shelf life (Lotong et al., 2003).

Table 19.3. Vitamin and Mineral Contents of Orange Juice

Vitamins[a]	Content	Minerals[b]	Content, mg/100 g
Vitamin A (β-carotene) (IU/100 ml)	190–400	Na	1.7
Vitamin C (mg/100 g)	50	K	179.0
Thiamin (μg/100 ml)	60–145	Ca	11.5
Niacin (μg/100 ml)	200–300	Mg	11.5
Riboflavin (μg/100 ml)	11–90	Fe	0.30
Pantothenic acid (μg/100 ml)	130–210	Cu	0.05
Biotin (μg/100 ml)	0.1–2.0	P	21.7
Folic acid (μg/100 ml)	1.2–2.3	S	4.6
Inositol (mg/100 ml)	98–210	Cl	1.2
Tocopherols (mg/100 ml)	88–121		

[a]USDA (1957).
[b]McCance and Widdowson (2002).

Aseptic Single-Strength Juice

Now the technology is available on a large scale to extract, process, and store single-strength juice in bulk aseptic refrigerated tanks, minimizing microbial spoilage and product quality deterioration. This technology enables provision of blended juices to consumers on a year-round basis, when the fruit is not in season. Depending on the processing capacity of the plant, number of tanks of capacity 950–3800 m^3 each, are installed in refrigerated rooms or insulated with refrigeration. With proper nitrogen blanketing and mixing, the juice quality may be maintained for a year or more (Wilke, 2002).

Single-Strength Juice from Concentrate

A significant amount of orange juice is packaged from reconstituted concentrate as chilled juice. Because of the economics of storing large bulk quantities of concentrated citrus juice and the consumer preference for a ready-to-serve product, the volume of this product is large now. Pasteurized juice is packaged in cartons or glass containers and is microbiologically stable. The flavor of juice from reconstituted concentrate is not comparable with single-strength juice because of the two thermal treatments and the loss of volatiles during the concentration process. Addition of aromas and essences can improve the quality of the finished product (Ranganna et al., 1983).

Frozen Concentrated Juices

Concentrated orange juice with soluble solids content of 65°Brix is now largely produced in the world. The primary water removal technology is high-temperature short-time evaporation, although freeze concentration and membrane processes are also used. The concentration process is accompanied by aroma recovery. The concentrate is blended with a small amount (less than 0.01%, v/v) of cold-pressed oil to mask the off-flavors that develop during storage. The small quantity of fresh juice can also be added back to concentrate to make up the losses of flavor during concentration process. The concentrate is chilled to −9°C by passing through heat exchanger and pumped to large stainless steel tanks maintained at desirable temperature in cold rooms. This concentrate is blanketed with nitrogen and carefully monitored for quality characteristics, so that the juice with different characteristics may be accurately blended to produce a uniform-quality finished product. Under these conditions, the concentrate can be stored for over a year with little loss in quality (Ranganna et al., 1983).

JUICE PRODUCTION

Harvesting

Fruits of suitable quality may be harvested manually using clippers or mechanically, depending upon the facilities available. Manual harvesting may be preferred in the countries where cheap labor is available and comparatively small acreage of orchards is managed. In developed countries, mechanical harvesting is practiced and a number of abscission chemicals are applied to facilitate detachment of fruits from the tree. Care should be taken to avoid any damage to fruit during handling. The fruits are packed in bags or bins and transported to the processing factory. A detailed description of types of harvester used and important factors related to mechanical harvesting of citrus fruits is given by Whitney (1995).

Receiving

After reaching the processing plant, the fruit goes through inspection lines for removal of bruised or damaged fruits. The sorted fruits are conveyed to storage bins and sufficient quantity is accumulated for continuous operation of the processing plant. The laboratory draws a small portion of fruit at this stage for testing the titratable acidity, °Brix, and juice yield. The tests for fruits have been discussed by Miller and Hendrix (1996) and Kimball (1991). The testing record of individual lots is maintained to determine which bins are to be blended for uniform product quality.

Washing

Fruits from the bins are conveyed to a washer. The fruits are first soaked briefly in water containing a detergent, scrubbed by revolving brushes, rinsed with clean water, and inspected again to remove the damaged ones. Sanitizing is essential for control of spoilage microbes, which may contaminate conveying equipment and juice extractors, and affect the juice quality. Applied citrus plant sanitation requirements have been published (Winniczuk, 1994). A HACCP plan should be followed for complete sanitization of fruits during washing (Schmidt et al.,

1997). The fruits are then separated automatically depending on their sizes and allowed to enter into the juice extractors.

EXTRACTION

The development of automatic orange juice extractors has been a major breakthrough in the progress of the fruit juice industry. Various types of extractors and finishers including Rotary Juice Press, FMC In-Line Extractor, and various Brown Model extractors have been discussed by different workers (Sigbjoern, 1975; Woodroof and Luh, 1975; Sutherland, 1977; Nagy et al., 1977). The juice extractor and finisher are both important to the nature, yield, quality, and characteristics of the orange juice and concentrate and can be adjusted to control the amounts of pulp, oil, etc., in the final products. According to Florida state regulation, the orange juice should not contain suspended pulp more than 12% for USDA Grade A (Braddock, 1999). The finishing process removes the excess of pulp, bits of peel, rag, and seeds. The yield is important to the grower who wants the highest return of his fruit, and to the processor who is responsible for the quality of the finished product.

Guintrand (1982) patented an appliance for extraction of orange juice. This appliance drops the orange into hinged hands, cuts the orange into two halves using a circular saw, and presses the fruit interior using a double press. Subsequently, the pulp and pits are separated from the juice in a centrifuge, and the juice is collected in a dispenser. The main advantage of this system is that all these operations are integrated without any human intervention, other than to introduce oranges into the appliance. The appliance can be used both in the public sector (e.g., airports, stations, etc.) and in more specialized sectors, including fruit-juice factories.

Ohta et al. (1982) compared the juice prepared from Satsuma mandarin by two methods viz. (i) an FMC In-Line Extractor that extracts the whole fruit, or (ii) by a new screw press extractor that extracts peeled fruit. Concentration of 20 volatile components in the two juices are given; the (i) juice had much higher concentration of hydrocarbons (such as D-limonene, γ-terpinene, myrcene, and α-pinene) and linalool than did (ii) juice, but (ii) juice had more favorable linalool:D-limonene and citronellal:D-limonene ratios, i.e., for (i) 0.0021 and 0.0003, and for (ii) 0.0048 and 0.0028, respectively.

The performance of a new screw press type of juice extractor has been evaluated by Watanabe et al. (1982) with peeled fruits of natsudaidai (*C. natsudaidai*) and satsuma (*C. unshiu*) oranges. Soluble solids content, amino-N, and turbidity increased with pressure of extraction, and acidity and soluble sugars decreased in the same order. Concentration of naringin, limonin, and monilin were higher in last stage extraction juice from natsudaidai than in first stage juice. The new extractor gave juice yields a few percent higher than an in-line extractor. At each stage, the extractor expressed (% of total juice): first 63%, second 24%, and final 13%. A slit screen was more effective than a punched screen in terms of susceptibility to plugging.

Sunkist Growers' juice processing plant at Tipton, CA, USA uses up to 1800 t of oranges/day to produce frozen concentrate, single-strength juice, oil, and cattle feed (Mans, 1983). Every step in the plant operation, from the conveyors bringing fruit out of the storage bins to the evaporator feed tanks, is controlled by a computer and operated from a graphic panel. Three computer-controlled FMC extractor lines of orange and one of grapefruit in a FMC Corp/Citrus World-built plant (Lake Wales, FL) handle 4300 t of oranges and 1375 t of grapefruits daily (Anon, 1984). Six programmable control computers constantly receive information from numerous sensors on the lines. Information may be fed into any stage or reports requested at any moment in the process by a single operator.

A continuous process system for orange juice, as installed by Farmland Dairies Inc., incorporating an in-line refractometer for °Brix measurements, is explained by Polizzi et al. (1987). Processing steps are briefly outlined, with special reference to the in-line refractometer and an automatic bench-top refractometer to confirm accuracy of the in-line unit. An increase of orange juice production by 30% is achieved with the new system.

A machine for extracting juice from citrus fruits, particularly oranges, is described by Antonio (1992). It includes an inclined chute conveying the fruit to be squeezed against a step, a spoon for raising the fruit resting on the step, and two squeezing plates below the step. The front plate is pivoted at the top and is pulled towards the back plate by a spring. The back plate is joined to a connecting rod and crank, which is driven by a speed reducer used to slide the back plate either towards or away from the front plate. As the two plates converge, fruit between them is squeezed. Extracted juice is collected, as it drips from the fruit, in an underlying hopper from which it is collected in a container. After squeezing, as the plates separate, the fruit residue falls down an inclined grill over the hopper and is collected in a drawer.

Di Giacomo et al. (1989) determined the color and anthocyanin contents in industrially produced orange juices (1989 harvest) from Calabria (31 samples) and Sicily (24 samples); the extraction lines used were in general usage (FMC juice extractor, Polycitrus machine). Absorption spectra for a model anthocyanin solution and of an orange juice extract in the 400–600 nm range are shown: both the curves are of similar shape, with a maximum at 538 nm. Tabulated data for Calabrian juices obtained between January and April showed range of anthocyanin contents of 91.6–497.0 mg/l (early April), with mean ± SD of 199.7 ± 106.6, and for Sicilian juices obtained in March and April a range of 497–1038.7 mg/l, with mean ± SD of 721.0 ± 153.3 (early April).

Physical and chemical characteristics and volatile constituents of orange juice, as affected by method of extraction have been studied by Mohsen et al. (1986a, b). They observed slight differences in orange juice characteristics (TSS, ascorbic acid, sugars, free amino-nitrogen (FAN), pulp, and 5-hydroxymethyl furfural content) extracted by pressing or by rotary extractor. Rotary extraction increased content of pulp, pectin esterase, and carotenoids. Extraction by pressing increased contents of limonene and oxygenated terpenes, while rotary extraction increased amounts of esters and water-soluble volatiles in juice. Taste and color were more acceptable in juice extracted by rotary method, but color of pressed juice was preferred. Limonene contents of orange juice are influenced during industrial processing by the Polycitrus juice extractor (Di Giacomo et al., 1976). The variations in limonin content of the Italian Sanguinello cultivar juices due to extraction technique (fixed or revolving reamers, FMC In-Line Extractors, rotary extractors), screening and centrifuging treatments, and juice recovery from peel and rag residues by pressing, are reported by Trifiro et al. (1983).

The choice of machinery depends upon the capacity, yield, and quality of the product required by the processor. For small-scale work, halving and burring machines, plunger type press, continuous screw expeller press, superfine pulper, and cup-type extractor can be used. For large-scale commercial production, automatic plants are being used.

Blending

Processors are aware of variations in the color of juice from different varieties and different seasons of fruit. The color of juice obtained from the fruits harvested in early season is poor. The poor color of early season juice can be improved by blending juice or concentrate from the oranges rich in color. Attention is given to the blending of different lots to achieve a balance of solids, acidity, color, and flavor. After finishing, the juice flows to large stainless steel tanks where it is checked for acidity and soluble solids; and sugar may be added to increase sweetness, if needed (Hyoung and Coates, 2003).

Deoiling

Previously, the oil level in juices was controlled only by adjusting the extractor setting or by choice of the type of extractor. The oil content could be controlled by softening the peel by immersing fruits for 1–2 min in hot water, but the oil in the juice varied from lot to lot and the control became difficult. Deoilers have been developed to control the peel oil level in citrus juices. Currently, deoiling in commercial operations is done by using small vacuum evaporators where the juice is heated to about 51.4°C and about 3–6% of the juice is evaporated. After the vapors are condensed, the oil is separated by centrifugation or decantation, and the aqueous layer is returned to the juice. With this treatment, about 75% of the volatile peel oil can be removed. The Standards for U.S. Grade A orange juice, permit not more than 0.035 vol.% of peel oil (USDA, 1959). For most commercial orange juices produced in the United States, oil content is maintained at 0.015–0.025 vol.%. Use of hermetic separators for deoiling (removal of essential oils) of single-strength orange juice is discussed by Puglia and Harper (1996), with special reference to: developments in separator design; types of separators for the citrus industry; design of hermetic separators; benefits of hermetic over centrifugal type separators for juice deoiling; and deoiling of single-strength juice using hermetic separators. A process for commercial debittering of navel orange juice by reducing the limonin content is described by Kimball (1990). The system includes a commercial deoiler to reduce the essential oil concentration in freshly extracted juice from 0.180 to less than 0.015 vol.%.

Deaeration

The single-strength juices are deaerated because dissolved oxygen lowers the vitamin C levels and causes flavor deterioration. The current tendency is to recommend that oxygen levels be kept low in all processed citrus juices. Dissolved oxygen disappears rapidly in canned juices, particularly at high temperatures. A definite benefit from deaeration has been a decrease of frothing in the filler bowl. Vacuum deoilers simultaneously deaerate juice and hence modern juice canneries do not have separate deaerators. Deaeration methods are known to affect the quality attributes of orange juice with respect to browning, vitamin C, sensory and Hunter Lab color values (Mannheim and Passy, 1979). Hot-filling and storage at less than 15°C gives bottled citrus juices a shelf life of almost 1 year.

Pasteurization

The pasteurization is aimed at inactivating the spoilage organisms and enzyme pectin methylesterase (PME) (pectin esterase) responsible for loss of cloud stability and discoloration in juice. Citrus juices are sensitive to heat. Their vitamin content and delicate fresh aroma and flavor may be lost or damaged by undue exposure to heat, so they are usually pasteurized as rapidly as possible. pH plays an important role in pasteurization of juice. Optimization of microbial destruction, enzyme inactivation, and vitamin C retention during pasteurization of pH-adjusted orange juice, is reviewed by Uelgen and Oezilgen (1993). The pH–temperature optimum determined by response surface methodology in the ranges 65–75°C and pH 2.5–4.0 has shown that no pectin esterase activity below pH 3.5 is observed. *Leuconostoc mesenteroides* had its maximum and minimum thermal resistances at pH 3.5 and 2.7, respectively. For an ideal theoretical process requiring four-log cycles of microbial reduction, the optimum pasteurization conditions are 12 min at 75°C and pH 2.7. The natural pH of juices varies with the variety of oranges. With the aim of optimizing pasteurization temperature for orange juice, thermal death characteristics of *Aspergillus niger* spores, *Saccharomyces cerevisiae*, and *Lactobacillus fermentum* have been studied by Hasselbeck et al. (1992). Thermal inactivation of all investigated microorganisms occurred at about 75°C. *D* values at 75°C were 0.004 s for *S. cerevisiae* and 0.53 s for *L. fermentum* in orange juice. Chemical and sensory tests showed that thermal treatment in the investigated time–temperature regime (65–95°C, 3–30 s) did not lower the orange juice quality. Time–temperature relationships are also important for heat inactivation of enzyme pectin esterase in orange juice under different conditions (Lee et al., 2003).

Commercially, the juice is rapidly heated to about 92°C and the exact temperature depends on the type of equipment used and on rate of juice flow. Juice may be in the pasteurizer from a fraction of a second to about 40 s. Recent trends are toward the use of high-temperature short-time pasteurization with either tubular or plate-type heat exchangers that are heated either by steam or hot water. Heating usually takes about 30 s or less and the juice is heated rapidly without local overheating. Modern heat exchangers and automatic controls are usually designed so that scorching or under-heating of portions of the juice is prevented.

Recently, a few other techniques have been employed for pasteurization of juice, and the effect of pasteurization on the composition and quality of citrus juices has been widely studied. The chemical, physical, organoleptic properties, and volatile components of juice are affected by pasteurization to varying extents depending upon the technique of pasteurization used (Min et al., 2003; Gil et al., 2002).

Effect of Pasteurization on Physico-chemical Properties

Pasteurization of orange juice produces sub-taste-threshold levels of p-vinylguaiacol (PVG) and induces ascorbic acid degradation but has almost no effect on browning (Naim et al., 1997). Fortification with glutathione, L-cysteine, or *N*-acetyl-L-cysteine at concentrations below 4.0 mM has no effect on PVG formation and browning, but inhibits ascorbic acid degradation during pasteurization and improves juice acceptance. The ascorbic acid concentration, density, cloud, and fructose levels of the juices are significantly influenced by the processing method when unpasteurized orange juice is bottled and frozen at −18°C; pasteurized juice is either bottled and frozen at −18°C or stored in plastic bins at 1°C; all stored for 9 months (Farnworth et al., 2001).

Pigment loss and potential visual color changes are associated with thermal processing of Valencia orange juice, which is quantitatively the richest in carotenoids amongst sweet oranges (Hyoung and Coates, 2003). Orange juice was pasteurized at 90°C for 30 s and then rapidly chilled. Color values were determined spectrophotometrically, and carotenoids were analyzed by RP-HPLC. Fresh unpasteurized juices contained (mg/l) 0.574 violaxanthin, 0.387 luteoxanthin (one of two HPLC peaks), 0.810 *cis*-violaxanthin, 0.819 antheraxanthin, 0.67 lutein, 0.582 isolutein, 0.662 zeaxanthin, and 0.337 β-cryptoxanthin, as well as several other carotenoids at lower concentration. Pasteurization significantly reduced concentration of violaxanthin, luteoxanthin, *cis*-violaxanthin, and antheraxanthin to 0.308, 0.351,

0.655 and 0.616 mg/l, respectively, whereas concentration of lutein was increased to 0.76 mg/l. Pasteurization also significantly reduced the concentration of neoxanthin and an unidentified peak, and increased concentration of luteoxanthin (the other peak) and mutatoxanthin. Isomerization of the major 5, 6-epoxide carotenoids to 5, 8-epoxides probably caused the perceptible color changes on pasteurization, which involved juices becoming lighter and more color-saturated.

The effect of pasteurization on volatile components has been most widely studied. Pasteurization of orange juice extracted by the two methods caused a slight decrease in limonene and oxygenated terpenes, a noticeable decrease in esters and water-soluble volatiles, and a slight increase in carbonyl compounds (Mohsen et al., 1986a, b). Major volatile components were δ-limonene, linalool, α-terpineol, and terpinen-4-ol. Taste and color were more acceptable in juice extracted by rotary method, but color of pressed juice was preferred. Pasteurization decreased acceptability of taste and odor of juice prepared by the rotary method. Yield of flavor compounds from fresh juice compared to heated juice from Satsuma mandarin were 4.0 and 5.2 mg/kg, respectively (Araki and Sakakibara, 1991). The comparison of hydrocarbons and oxygenated compounds in fresh and heated juice, has shown that β-terpineol and β-damascenone were found only in heated juice. β-Damascenone had an odor similar to overripe fruit and seemed to contribute to the characteristic odor of the heated juice.

Volatile flavor compounds present in fresh and heated juices of Satsuma mandarin and three cultivars of sweet oranges (Hamlin, Pineapple, and Valencia), were studied by Araki et al. (1992). δ-Limonene, linalool, octanal, and ethyl butyrate decreased and α-terpineol increased after heating the juices. Heat-related changes in these compounds were smaller in sweet orange juices than in Satsuma mandarin juice. Moshonas and Shaw (2000) evaluated flavor quality of pasteurized orange juice during a normal shelf-life period by analysis of sensory quality and qualitative and quantitative changes in 46 volatile flavor compounds. Farnworth et al. (2001) found that while un-pasteurized juice contained the highest levels of acetaldehyde, ethyl acetate, α-pinene, β-myrcene, limonene, α-terpineol, 1-hexanol, 3-hexen-1-ol, and sabinene, and the lowest concentration of valencene, compared to pasteurized juice. They suggested that un-pasteurized orange juice should be frozen rapidly to be more acceptable to consumers than the pasteurized juices. Jordan et al. (2003) reported that generally the pasteurization did not further modify the aromatic profile of deaerated orange juice with the exception that the concentration of δ-3-carene decreased significantly after pasteurization ($P < 0.05$) and neryl acetate, the concentration of which decreased significantly after a combination of deaeration and pasteurization.

ADVANCES IN PASTEURIZATION TECHNOLOGY

PULSED ELECTRICAL FIELD

It is the latest development in the field of food processing that has found a number of applications in the recent past (Min et al., 2003). This technology relies on the lethal effect of strong electric fields for inactivation of microorganisms. By using this technique, the undesirable changes, such as protein denaturation and vitamin losses during processing are avoided. The technique is effective against only vegetative cells while spores are not affected. A pilot plant scale pulsed electric field continuous processing system was integrated with an aseptic packaging machine to demonstrate the efficacy of pulsed electric field technology as a non-thermal pasteurization method for fresh orange juice (Qiu et al., 1998). Major system components include a 40,000 V/17 MWp high-voltage pulse generator, a set of multiple stage co-field flow pulsed electrical field (PEF) treatment chambers, a fluid handling system for flow control and a Benco Aspack/2 aseptic thermal form packaging machine. Orange juice is processed at a 75–125 l/h flow rate and packed aseptically into 200 ml plastic cup containers. Microbiologically enhanced orange juice tests validated the microbial inactivation effect of this integrated pulsed electric field system. The PEF treated and aseptically packaged fresh orange juice demonstrated the feasibility of pulsed electric field technology to extend the product shelf life with very little loss of flavor, vitamin C, and color.

Pasteurization of fresh orange juice using low-energy PEF was conducted by Hodgins et al. (2002). Low energy PEF of orange juice was optimized with respect to effects of temperature, pH and number of pulses on microbial contamination. In addition, the effect of PEF combined with addition of nisin, lysozyme, or a combination of two antimicrobial agents on the decontamination of orange juice was also investigated. Optimal conditions were determined as 20 pulses of an electric field of 80 kV/cm, at pH 3.5 and 44°C with 100 U nisin/ml. Using these conditions, a greater than six-log cycle reduction in the microbial population occurs. Following treatment, 97.5% retention of vitamin C, along with a 92.7% reduction in pectin methyl-esterase (pectin esterase) activity is obtained. The microbial shelf life of the orange juice is also improved, and is determined to be greater than or equal to 28 days when stored at 4°C without aseptic packaging. GC revealed no significant differences in aroma compounds before and after PEF treatment.

PEF treatments can be applied to un-pasteurized Valencia orange juice using a bench-top PEF system to study effects of PEF on the activity of pectin methyl-esterase (Yeom et al., 2002). Electric field strengths up to 35 kV/cm were applied to orange juice at a constant water bath temperature of 30°C. An increase in the electric field strength caused a significant inactivation of pectin methyl-esterase with an increase in orange juice temperature. An electric system using a high AC electric field can inactivate *Bacillus subtilis* spores in orange juice (Uemura and Isobe, 2003). Higher degree of inactivation of *B. subtilis* spores can be achieved with electric field at 121°C than at 100°C with reduced loss of ascorbic acid and more acceptable product aroma. Electric field-treated samples of Satsuma mandarin juice has more acceptable flavor than those subjected to conventional heat treatment.

High-Pressure Treatment

High pressures in the range of 500–10,000 bar at the ambient temperatures have been known to have lethal effect on microorganisms for a long time. The technology is not totally effective in destroying bacterial spores and many enzymes. The combination of high pressure with moderate temperature may have to be used to have a desirable effect. Combination of high pressure with chilled storage and distribution may also be desirable if quality deterioration by enzymes is to be prevented. Addition of CO_2 during high-pressure processing (HPP) of fresh orange juice increases the rate of pectin methyl-esterase inactivation beyond that achieved by pressure alone (Truong et al., 2002). This technology is also compatible with existing range of semi-rigid and flexible packaging materials (Parish, 1998; Sellahewa, 2002). High-pressure treatment may provide an adequate alternative to thermal pasteurization for minimally processed citrus juice products. Bioactive compounds present in freshly squeezed orange (*C. sinensis* L.) juice are more stable in high-pressure treatment than the conventional heat processing (Sanchez Moreno et al., 2003). Application of 350–450 MPa of high-pressure treatment of orange juice not only increases the extraction of flavanones but also enhances potential health-promoting attributes of the juice that are retained during cold storage.

Microwave Heating

Microwave heating could be used for continuous pasteurization of citrus juice (Nikdel and MacKellar, 1993). Orange juice is pasteurized by pumping it through a coil of Teflon tubing in an oven heated with microwave energy (at 2450 MHz). Juice temperature (95°C) is controlled by varying the flow rate at 100% microwave power. Over 90% of the 570 W microwave power is absorbed by the juice. Pectin methyl-esterase activity is reduced by more than 99.9% by pasteurizing for 15–25 s at 95°C. Bacteria are also rapidly inactivated at either 70°C or 90°C. Microwave pasteurization of juice does not cause any flavor change compared to fresh un-pasteurized juice.

Gamma Radiation

Use of γ-irradiation for the preservation of a number of food commodities, has been extensively investigated. While safety of food processed by using γ-irradiation up to 10 kGray has been well established, yet feasibility of gamma radiation for pasteurization of orange juice has been ruled out due to its adverse effect on quality of juice (Spoto et al., 1993). The adverse changes in flavor of irradiated samples are described as "oil," "cooked," and "medicine" notes. While irradiation is effective in destroying pathogens such as *Listeria monocytogenes* and *Salmonella* spp. in fresh orange juice, off-flavors are generated precluding its use as an alternative processing technology (Foley et al., 2002).

BITTERNESS IN JUICE

The post-processing development of bitterness is a common problem in orange juice, which is the main deterrent in the utilization of many orange varieties for juice production. Roughly 15–20% of navel oranges harvested in a given season are not used directly and are processed into a juice that is bitter and requires processing or back-blending with other sweeter juice. The bitter fraction of the orange juice arises from a tetracyclic triterpenoid called limonin. The limonin is produced over time from limonic acid or limonin monolactone, which is found in the seeds and membranes of most citrus fruits. Because of low solubility of limonin, heating or prolonged storage increases its concentration in juice. A concentration of 2 ppm of limonin imparts a definite bitterness to the juice (Kefford, 1959).

Limonoids

The non-bitter precursor in citrus fruits is identified as limonin monolactone which after acid-catalyzed conversion forms limonin (Maier and Beverly, 1968). Limonin monolactone is stable in the tissues of intact fruit (which is not bitter) because it is apparently not in direct contact with the acidic juice. It slowly converts into limonin (and the juice becomes bitter) when fruit tissues come in contact with the acid in juice, after the juice is expressed from the fruit. The limonin content of the fresh mature Valencia oranges fruit is 2–3 ppm after extraction, rising to 11–13 ppm after pasteurization, but undergoes little subsequent change during storage (Tariq et al., 1974). The majority of the limonin is formed from its non-bitter precursor limonate A-ring lactone. N and K fertilization have significant effects on limonin content of Washington Navels, the limonin contents falling at higher N and K doses (Rodrigo et al., 1978). Average limonin and nomilin concentrations in orange juice have been reported to be 0.75 and 0.03 ppm, respectively (Rouseff and Fisher, 1980). The early Marrs and Hamlin oranges from five locations in the South Texas citrus have less limonin (6.2 ppm) by mid-November which reaches a minimum of 1.8 ppm in January (Albach et al., 1981). Linear correlations between limonin and time of harvest were much better than between limonin and °Brix, percentage of acid, or the °Brix:acid ratio. Rodrigo et al., (1985) reported strong correlation between limonin and acid content in juice of Washington navel oranges.

Limonin content varies with the Navel orange cultivars and could range from 15.37 to 25.02 ppm (Navarro et al., 1983). A lower limit of detection of 1 ppm for limonin on a 2.1 mm internal diameter micropore column and 2 ppm on a 4.6 mm internal diameter analytical column by high performance liquid chromatography can be achieved (Shaw and Wilson, 1984). Limonin concentration (5–6 ppm) in the juices from Italian Tarocco and Sanguinello oranges is not affected by maturity and is lower in juices from first than from second pressing; it could be reduced by removing the peel immediately after juice extraction and by low temperature processing (Trifiro et al., 1984). The addition of hydrocolloid to Kinnow juice has been reported to decrease the limonin content (Aggarwal and Sandhu, 2004a). Nomilin found in orange has been shown to have anticarcinogenic properties and to induce detoxifying enzymes in animals (Herman et al., 1990).

Flavonoids

The principle flavonoids of citrus are the anthocyanins, flavones, flavonols, or flavanones. Certain flavonoids are used to detect the adulteration of orange juice with other citrus juices. The presence of hesperidin is associated with the fouling of heating surfaces. The composition of juices in relation to flavonoids has been studied widely. The major flavonoids present in orange are hesperidin, rutinoside of naringenin, isosakuranetin, 4-β-D-glucoside of naringin, 4-β-D-glucoside of naringenin rutinoside and neohesperidin. The bitter principle of bitter (sour) oranges is naringin in place of hesperidin of sweet oranges.

Amount and distribution of polymethoxylated flavones in juice, peel and pulp wash vary considerably and may be used to detect adulteration. Flavone content of range peel is about 100× higher than that of juice (Heimhuber et al., 1988). The polymethoxylated flavones including: 5,6,7,8,4′-pentamethoxyflavone; 5,6,7,4′-tetramethoxyflavone; 3,5,6,7,8,3′,4′-heptamethoxyflavone; 5,6,7,8,3′,4′-hexamethoxyflavone; 5,6,7,3′,4′-pentamethoxyflavone; 5,7,8,3′,4′-pentamethoxyflavone; and 3,5,6,7,3′,4′-hexamethoxyflavone have been identified in Valencia orange juice and peel oil (Hadj Mahammed and Meklati, 1987) while another compound, 6,7,8,4′-tetramethoxy-5-hydroxyflavone, is found only in peel oil. Fully methoxylated flavones such as nobiletin, tangeretin, 3,5,6,7,8,3,4-heptamethoxyflavone, tetra-*O*-methylscutellarein, and sinensetin (5,6,7,3,4-pentamethoxyflavone) in FCOJ range from 4.2 to 7 ppm and these substances are not important contributors to the flavor of orange juice (Veldhuis et al., 1970).

The bioflavonoids contents (mg/100 ml) of orange juice determined by GLC method are: rutin 0.73, hesperidin 11.17, and naringin 3.72 (Drawert et al., 1980). Hesperidin solubility during production and storage of clarified Satsuma mandarin juice is important to avoid the formation of haze or sediment. In turbid juice (raw material for clarified juice), soluble hesperidin, which comprised about 20% of total hesperidin immediately after extraction and concentration by evaporation, remains soluble during storage at −20°C, but at higher storage temperature its solubility rapidly decreases. With clarification by ultrafiltration (UF) (batch operation), higher operating temperature gives higher hesperidin content in the permeate. These findings may be useful in determining whether clarified juice will develop haze or precipitate during storage. The hesperidin glycosides solubilize hesperidin in canned Satsuma mandarins and in orange juice, and inhibit crystallization of hesperidin for more than 6 months during storage (Nishimura et al., 1998).

Flavonoid glycoside detection has been used to assess the authenticity of orange juice (Pupin et al., 1998). Concentration of narirutin and hesperidin in hand squeezed juices may vary from 16 to 142 and 104 to 537 mg/l, respectively. Varietal differences in the ratio of narirutin to hesperidin are observed; the ratio is highest for Pera (mean 8.4) and lowest for Baia (mean 3.6). Authentic FCOJ has higher flavonoid glycoside levels than those in hand squeezed juices (narirutin 62–84 mg/l and hesperidin 531–690 mg/l, after dilution to 12°Brix) while frozen concentrated pulp wash has levels ranging from 155 to 239 mg/l for narirutin and 1089 to 1200 mg/l for hesperidin.

A method based on flavone glycosides and polymethoxyflavone patterns for differentiating sweet orange (*C. sinensis*) juices from other citrus juices by their flavonoid content is proposed (Ooghe and Detavernier, 1999). Analysis of a number of fruit juices such as lemon, grapefruit, and pommerans showed that flavone glycosides fingerprints could be used to differentiate *C. sinensis* from other juices such as grapefruit, mandarin, sour/bitter oranges, and bergamot.

Naringin and neohesperidin are flavonoids found only in certain citrus fruits; sweet orange cultivars do not contain these compounds and their presence in orange juice indicates adulteration with juice from certain other citrus fruits such as grapefruit. A method adopted by the First Action AOAC International demonstrated the reliability of the method in detecting the presence of grapefruit juice in orange juice as indicated by a finding of 10 ppm naringin and 2 ppm neohesperidin (Widmer, 2000).

DEBITTERING PROCESSES

Efforts have been made to reduce the accumulation of bitter principles during the development and maturing of orange fruit through chemical sprays, agronomic practices, and post-harvest treatment of fruits. Many debittering technologies have been developed based on chemical, physical, and microbial processes.

Tree Treatment

Treating Washington Navel orange trees after bloom with 200 or 300 ppm of 2-(3,4-dimethyl phenoxy) triethylamine or 200 ppm 2-(4-chlorophenyl thio) triethylamine is reported to reduce the concentration of limonin in the juice of oranges at the beginning of maturity, compared with untreated control, with a concomitant small decrease in soluble solids and acidity (Casas et al., 1979). However, these differences in limonin contents decreased with advancing maturity.

Microbial Debittering

The study conducted by Brewster et al. (1976) revealed that use of 200 m-units of the limonoate dehydrogenase of Pseudomonas sp. 321–18 per milliliter of navel orange juice reduced the eventual limonin content of 21 ppm to 3 ppm, a level below the general bitterness threshold, whereas comparable activity levels of the enzyme from *Arthrobacter globiformis* caused substantially smaller decreases in eventual limonin content. This wide difference in activity at low pH is explained by the instability of the limonoate dehydrogenase of *A. globiformis* at pH 3.5 and the relative stability of this enzyme from Pseudomonas spp. Naringinase produced by solid-state fermentation of soy meal with *A. niger* ZG86 (2%) at 32°C for 4 days has been used for reduction of bitterness from orange juice. Orange juice extracted from whole orange fruits or peeled fruits with or without centrifugation was mixed with naringinase at 0–32 μm/100 ppm naringin at 50°C for 1–2 h. Naringin concentration for the enzyme treatment (16 μm/100 ppm) was reduced to 30 ppm compared with the initial concentration of 320 ppm in the orange juice. Bitterness was decreased by 90% by this naringinase treatment.

Adsorption

A process for debittering citrus juices by removing limonin using a selective absorption process has been reported. Cellulose esters in the form of gel beads (2–4 mm in diameter) selectively remove limonin from Navel orange juice without significantly affecting the color and flavor constituents or the ascorbic acid content (Anon, 1976). Juice is passed upward through the column of beads. It should be pasteurized for optimum results, but it may also be a freshly extracted juice. About 10 bed volume of juice may be passed through the column to achieve the removal of about 40–60% of the limonin before the column requires re-activation.

Chandler (1977) reviewed the processes for overcoming this problem of bitterness with special reference to the development of gel bead sorption process for separation of the bitter principle (limonin) from orange juices. Cellulose acetate as flake or powder is an efficient and selective sorbent for removing the bitter principle limonin from orange juice. Treatment of orange juice serum with cellulose acetate powder (10 g/l) removed 44–70% of the limonin content in less than 1 h, at the same time removing relatively negligible amounts of hesperidin and ascorbic acid (Chandler and Johnson, 1977). Bitter orange juice was successfully debittered by holding for several days in containers with linings of the cellulose esters in gel form, by batchwise agitation with gel cubes or beads, and by passage through columns packed with these materials (Chandler and Johnson, 1979). These processes provide the basis for methods suitable for in-storage, batchwise or in-line debittering treatments on an industrial scale. Cellulose acetate gel beads were used to debitter eight successive batches of juice, the sorptive capacity of the beads being regenerated by washing with water between the debittering treatments. Johnson and Chandler (1978) described the removal of limonin from navel orange juice by passage through a column of cellulose acetate gel beads. The beads may be easily fabricated to a size permitting the passage of whole juice, and the column can be reactivated by simple water washes.

Cellulose acetate gel beads, when used commercially for removing the bitter principle, limonin, from orange juice, may eventually become so loaded with organic matter that simple water washes are no longer adequate to reactivate them. Such exhausted beads may be readily and economically restored to their original activity by washing with a small volume of warm water, which is recycled through a small active carbon bed (Johnson, 1981). β-Cyclodextrin (0.3% w/w) can also be used to overcome bitterness of citrus juices due to limonin and naringin (Konno et al., 1981; Shaw and Wilson, 1983; Shaw et al., 1984). Other components such as naringenin 7-β-rutinoside, coumarins, and flavonoids were removed, but the TSS (°Brix), TA, and ascorbic acid content of the juice were unchanged. The polymer was regenerated by treatment with dilute aqueous alkali or ethanol.

There are significant reductions in limonin, citrus oils, and pulp in California navel orange juice during commercial debittering using a hydrophilic absorbent (Kimball and Norman, 1990a, b). Citrus oils and pulp can be replenished without violation of federal standards of identity. Bitter components, such as limonin, may be effectively removed from citrus juices, particularly navel orange juices, by contacting the juices with an adsorbent resin (Norman et al., 1990). Couture and Rouseff (1992) used neutral XAD-16 and weak base anionic exchange IRA-93 resins for increasing palatability of juice from four cultivars of sour orange (*C. aurantium*), namely Seville, Bigaradier, Sour, and Bittersweet. Bitter compounds were removed by both resin treatments. Average naringin concentration was reduced by 50–66% in high acid juice and 89% in low acid juice using IRA-93.

With the aim of improving naringin removal from orange juice, the naringin absorption capacity of five resins (A–E) produced in Tianjin, China, was tested (Wu et al., 1997). The selective removal of limonin and naringin from orange juice by batch adsorption to various materials to reduce bitterness, was investigated (Ribeiro et al., 2002). Since reducing sugars, pigments, and vitamin C may be removed simultaneously with flavonoids, adsorption of these compounds was also investigated. The highest adsorption efficiency for the bitter compounds occurred with Amberlite XAD-7. Adsorption of sugars and pigments was low and no adsorption of vitamin C was detected for any adsorbent. The bitterness of navel orange juice is reduced with the addition of 100 ppm of neodiosmin (Guadagni et al., 1977). As this treatment suppresses the bitterness, a level of 50–150 ppm is recommended to improve the acceptability of juice.

Enzymatic Debittering

Study conducted on Shamouti orange juice products revealed that the non-bitter limonin precursor is located mainly in the non-soluble parts of the fruits, i.e., albedo, membranes of segments and of juice sacs, and rags (Levi et al., 1974). An enzymic activity reducing the limonin monolactone concentration by 50% at the natural pH of 5.4 at optimum temperature (40°C) was observed in aqueous extracts of albedo and segment membranes. Industrial trials showed that 40–50% reduction in limonin content and hence significant reduction in bitterness was achieved by using only 0.04 in. strainers with only minimum production loss (about 2%), thus allowing production of high-quality orange juice earlier in the season.

Immobilized Enzyme

Limonin debittering of navel orange juice serum was successfully demonstrated with *A. globiformis* cells immobilized in acrylamide gel treatment of 30 ml serum (10–27 ppm limonin), for instance, on a 1.5 cm diameter column packed with 1.6 g immobilized cells (16-ml bed volume) reduced limonin content by greater than 70% (Hasegawa et al., 1982). This column was used 17 times without losing its effectiveness. The treatment did not affect juice composition as measured by total acids, TSS, pH and sugars. Removal of limonin by free and immobilized cells of *Rhodococcus fascians* in a model buffer system and in a real system (orange juice), has been studied using chitin as immobilization support because of its high yield, simplicity, and cheapness (Bianchi et al., 1995). It is concluded that biodegradation of limonin by *R. facians* depends largely on limonin conversion to its acid forms and therefore on the pH value.

Studies were conducted on metabolism of nomilin (a bitter limonoid constituent of citrus juices) in orange serum by immobilized cells of *Corynebacterium fascians* (Hasegawa et al., 1984). The main metabolite of nomilin was identified as obacunone by TLC and NMR. The immobilized *C. fascians* system was highly active: nomilin was completely converted when orange serum containing 20–25 ppm nomilin was passed through the column only once; the column was used 15 times without any loss of effectiveness. The enzyme responsible is probably nomilin acetyl lyase. Cell-free extracts of *C. fascians* also catalyzed the conversion of nomilin. These results are of significance in relation to debittering of citrus juices.

ACID REDUCTION

Resins have been used to process orange juice in two major applications, (i) for the reduction of acid content and (ii) to remove bitter components (Norman, 1990). Some orange juice consumers prefer a reduced acid version of the juice. Fresh juice, stabilized concentrate, or concentrates reconstituted back to 15°Brix can be treated this way. First, the pulp is removed from the juice via centrifugation and the pulp free juice can then be passed through a weak base anion column, which reduces the citric acid content of the juice. The resin will also retain the ascorbic acid (vitamin C) and folic acid portions of the juice. The resin is more selective for the citric acid so if enough juice is passed, the citric acid portion of the feed juice will displace the ascorbic acid and folic acid held by the resin. This can be monitored by following the pH of the juice effluent coming off the column. Below a pH of 4.6, the ascorbic and folic acid portions are washed through the resin. The processed deacidfied juice effluent can then be blended back to unprocessed juice to achieve the right balance of reduced acid and fresh juice tastes. The pulp can also be added back, or the juice can be sold as a reduced pulp juice. Juices from four cultivars of sour orange (*C. aurantium*) i.e., Seville, Bigaradier, Sour, and Bittersweet, were treated with neutral (XAD-16) and weak base anionic exchange (IRA-93) resins for increasing palatability of juice (Couture and Rouseff, 1992). Average acidity was reduced 57–87% using IRA-93 before depletion, and sensory acceptability of the juice increased.

CLOUD STABILIZATION

Cloud loss is a major quality defect in orange juice. Cloud stability decreases with increasing amounts of pulp. Juices and concentrates containing 3% pulp have been found to be more stable than ones with higher pulp content (PC) of about 9%. Combination of homogenization and pasteurization gives better stability than either of these treatments alone. The combination of pasteurization and addition of stabilizers such as pectin and gum acacia give good stability to both juices and concentrates (Ahmad and Bhatti, 1971). TIC Gums, Inc. has produced a gum system TICALOID No. 1004, which suspends fruit pulp without increasing viscosity (Anon, 1983). In processed orange juices with original viscosities between 10 and 25 cP, the addition of 0.15% TICALOID No. 1004 provided complete suspension of the fruit pulp and the final viscosity was between 55 and 60 cP. The juice manufacturer has two options: to maintain the original mouthfeel of the juice, and to provide complete suspension of the pulp; or to develop products with a pulpier consistency.

Methods of preserving cloud without the extreme temperature use in commercial pasteurization (techniques) are desirable. HPP is used as a means of preserving cloud in freshly squeezed orange juice (Goodner et al., 1999). Pressures from 500 to 900 MPa have been investigated at dwell times of 1 s, 1, and 10 min. Higher pressures and longer processing times are more effective at preserving cloud; all treatments yield a microbiologically stable product. A 90-day shelf life under refrigeration conditions could be achieved using pressures of 700 MPa combined with treatment times of 1 min.

PACKAGING AND STORAGE OF ORANGE JUICE

Different types of packaging including cans, bottles, cartons, drums, and barrels made up of glass, metal, plastic, or laminates are used for the packaging of orange juice and concentrates. The packaging and storage of orange juice have been studied extensively. Packaging of orange juice in metal cans is becoming obsolete. The latest trends are towards aseptic packaging in flexible plastic films and laminates.

CANNING

Plain tin cans are used for single-strength orange juice, because they prevent discoloration of juice upon storage and are least expensive. Enamel-lined cans or lids have been used, but appear to be unnecessary. The cans varying in sizes from about 200 ml to over a liter are used for packing.

Hot juice from the pasteurizer is pumped to the large stainless steel filler bowl and filled directly into the cans. The juice is kept in the filler bowls for a minimum time to prevent damage of flavor by the heat. The cans are filled automatically by opening the valve as they pass around the turntable beneath the filler bowls. It is desirable to minimize the amount of oxygen in the final container. Much of the air in the juice is removed by deoiling or deaeration process. Live steam injected into the headspace as the can is closed replaces the air and helps to create a vacuum during closure. They are closed automatically as they leave the filling machine. The cans are inverted for about 20 s to sterilize the lid by the heat of the juice, then, while spinning in a roller conveyer, the cans are rapidly cooled to 37.8°C by cold water spray to facilitate drying and prevent subsequent rusting of the outside of the can. High speed filling and closing machines handle up to 500 cans/min (Kefford et al., 1959).

STORAGE

The orange juice undergoes various physical and chemical changes, depending on the type of packaging and storage conditions. When orange juices are compared, immediately after canning, with samples of the original juice, changes in flavor and other quality factors during the actual canning procedure are minimal. Changes during storage of canned juice, however, are much more profound. The storage temperature is the major determinant influencing the flavor and vitamin content of the juice. Kefford (1973) summarized studies of ascorbic acid retention and flavor stability in canned citrus juices during storage at different temperatures, and stated that from the point of view of practical nutritionists, canned citrus juices should be stored at the coolest possible temperature.

Some workers reported, in different studies, that over 90% of ascorbic acid was retained and flavor deteriorated little in canned citrus juices stored at 21.1°C for 1 year or longer (Freed et al., 1949; Riester et al., 1945). Other workers indicated that ascorbic acid retention decreased and flavor deteriorated at higher temperatures (Martin et al., 1995; Petersen et al., 1998). Khan and Khan (1971) found that canned orange juice had better retention of color, better flavor, and higher retention of vitamin C than bottled orange juice. Hashimoto et al. (1995) reported that exclusion of dissolved oxygen before heat treatment and filling into epoxy resin coated cans effectively preserved fresh flavor during storage at 10°C or 23°C for up to 12 weeks.

Some studies have indicated that tin content may reach 150–200 ppm in canned orange juices stored at temperatures approaching 30°C for 6 months or longer (Bielig, 1973). The Codex Alimentarius proposed a maximum of about 150 ppm tin in orange juice for infants. Omori et al. (1973) reported that high concentration of tin in orange juice is a major cause of toxicity. They reported that tin in excess of 300 ppm, can cause undesirable physiological reactions in large animals and human beings.

Changes in dissolved oxygen concentration, during storage of packaged orange juice were studied by Manso et al. (1996). Single-strength Valencia orange juice aseptically packaged and stored up to 5 months at 4–50°C was analyzed for dissolved oxygen. Dissolved oxygen concentration reached equilibrium in a few days from an initial level of approximately 2 ppm; equilibrium concentration was independent of temperature of storage.

Sorption of food components, particularly volatile compounds, by polymeric packaging materials is an unsolved problem for the food industry. The food itself develops an unbalanced flavor profile (termed flavor scalping), and the pack if recycled (e.g., PET bottles), can transfer the adsorbed aroma compound to the next product. Sensory properties of orange juices are highly related to their levels of D-limonene. The effect of packaging and storage conditions on the quality of orange juice has been summarized in Table 19.4. Decreases in sensory quality (overall scores

Table 19.4. Interaction of Different Packaging Materials with Orange Juice During Storage

Packaging Material	Nutrient/Component	Changes	Conditions	Storage Period	Reference
Glass bottles	Ash, Na, Mg, and P	No change	5–30°C	6 months	Martin et al. (1995)
	Tiratable acidity and Sucrose	Decreased			
	Vitamin C	Decreased			
	Hydroxymethyl furfural	Increased			
	Sensory quality	Deterioration			
Accelerated storage	α-Terpineol and β-terpineol	Increased	–	–	Petersen et al. (1998)
	Sensory quality	Oxidized flavor appeared			
	Sourness	No change			
Polyethylene/ barrier material laminated cartons	D-Limonene	50% reduction	4°C	24 weeks	Pieper et al. (1992)
Laminated multi-layered aseptic package	α-Terpineol and ethyl acetate	Increased	21–26°C	5 weeks	Moshonas and Shaw (1989b)
	2-(2-Butoxy-ethoxy)ethanol	Appeared as new compound			
	D-Limonene	40% reduction			
	Vitamin C	26.09% reduction			
	Oil	Reduction from 0.01% to 0.0066%			
Polyethylene-lined cartons	D-Limonene	40% reduction	20°C	6 days	Duerr and Schobinger (1981)
	α-Terpineol	Increased		90 days	
	Neral, geranial, and octanal	Decreased			
Aluminum-lined cartons	D-Limonene	Decreased	20°C	3 months	Mottar (1989)

Aseptic package	1-pentene-3-one, hexanal, ethyl butyrate, octanal, neral, and geranial,	Decreased	21–26°C	8 months	Moshonas and Shaw (1989a)
	Oil content	Decreased			
	Vitamin C	Decreased			
	α-Terpineol and furfural	Increased			
Paperboard sandwiched between two polymer layers	α-Pinene, octanal, and D-limonene	Decreased	4°C	4 weeks	Paik and Venables (1991)
	Microbial growth	No growth			
Polypropylene and polyethylene	Flavor compounds	Scalping (absorption) by package	–	–	Risch and Hotchkiss (1991)
LDPE, HDPE, Polypropylene, and Surlyn	Terpene, sesquiterpene, and aldehydes	Scalping (absorption) by package	–	4 days	Charara et al. (1992)
Aluminum foil/tie resin/LDPE laminate	Free fatty acid content at 1.1–1.4% level	Caused delamination	–	Long-term storage	Pieper and Petersen (1995)
Silica coated PE/PET cartons versus coextruded ethylenevinyle alcohol (EVOH)	Ascorbic acid and aroma	Better retained in EVOH	2°F (−16.6°C)	11 weeks	Lee and Barros (1996)
LDPE and EVOH versus PVDC	α-Pinene and D-limonene	Less absorption in PVDC	–	–	Sheung et al. (1996)
LDPE, PET, PVDC, and EVOH	Ethyl butyrate and octanal	Equal absorption in all four			
PET bottles	Octanol, decanol, and linalool	Decreased	–	–	Will et al. (2000)
	Furfural and hydroxymethyl furfural	Increased	20–37°C	3 months	
	Ascorbic acid	Decreased			
	*L*a*b** color	Deteriorated			
	α-Terpineol	Formed			

Continued

Table 19.4. *Continued*

Packaging Material	Nutrient/Component	Changes	Conditions	Storage Period	Reference
Hexamethylene	Lactic acid bacteria	Inhibited	Frozen	–	Devlieghere et al. (2000)
tetramine (HMT)	Yeast growth	Not retarded			
containing LDPE film	HMT	Dissolved from film into juice			
Polylactate, HDPE, and polystyrene	Quality retention	All equally effective	–	–	Haugaard et al. (2002)
LDPE	Valence, decanal, hexyl acetate, octanal, nonanone, and D-limonene	Absorbed	20°C	29 days	Willige et al. (2003)
Polycarbonate	Valence	Absorbed			
PET	Decanal, D-limonene, myrecene	Absorbed			
LDPE	Myrecene, α-pinene	Absorbed			
Laminated brick style cartons	D-Limonene	Maximum absorption			Hirose et al. (1988)
LDPE					
Syrlyn 1601					
Syrlyn 1652 with antioxidant and antibacterial agents					

for color, appearance, aroma, and flavor) found during storage in glass bottles, are greater at higher storage temperature and with exposure to light. Significant deterioration in sensory quality occurred after 3 months at ambient temperature, and after 1–2 months at 30°C. Changes in bitterness are similar to those of oxidized flavor, but less pronounced, while no significant differences were found for sourness. Sorption of D-limonene by plastic packaging is affected by the external factors such as temperature, relative humidity, and other storage conditions. A gradual decrease in several flavor components, 1-penten-3-one, hexanal, ethyl butyrate, octanal, neral, and geranial, is observed during storage. Contents of some volatile components (α-terpineol and ethyl acetate) increase during storage (Moshonas and Shaw, 1989a).

Microbial growth and ascorbic acid concentration are sensitive to variations in D-limonene concentration, within the range of values typically observed in commercial orange juice. Consequently, packaging materials that absorb D-limonene, potentially influence microbial stability and ascorbic acid content in single-strength orange juices. However, scalping of flavor volatiles by LDPE, PET, polyamide, and ethylene (co)vinyl alcohol does not result in significant differences in flavor of high oil, typical oil, low oil, and thermally abused orange juice samples made from high-quality orange juice concentrate (Sadler et al., 1977).

Off-Flavors

Bielig and Askar (1974) reported the changes occurring during (i) bottling (80°C/10 min) and (ii) storage of orange juice: (i) increase in α-terpineol, *n*-octanal, *n*-hexanal and (ii) increase in *n*-hexanal, *n*-hexenal, and *n*-octanal during 2 months then decrease; increase in *n*-decanal during 6 months; decrease in fatty acids possibly by autoxidation or enzymic oxidation (causing "cardboard off-flavor"), and increase in α-terpineol at the expense of α-limonene and linalool. α-Terpineol is proposed as an indicator for predicting storage life. Correlation coefficient between flavor scores and furfural from "Hamlin" orange, "Valencia" orange, and grapefruit juices were highly significant, as reported by Maraulja et al. (1973). Rate of flavor deterioration was dependent on both storage temperature and storage time. However, temperature was the more significant factor. Retention of initial flavor quality was much better when the canned products were stored at less than 60°F (15.5°C).

Orange juice (reconstituted from 65°Brix concentrate) with added peel oil (100 or 200 ppm) was packed in polyethylene-lined cartons (Tetra Brik) and stored at 4°C, 20°C, and 32°C, and in glass bottles at 20°C in the dark to study the effect of packaging on aroma quality (Duerr et al., 1981). Samples were analyzed before and after filling, and periodically until 90 days after filling. Limonene was reduced by 40% after 6 days in the soft packages (by absorption into the polyethylene lining), versus 10% after 90 days in glass. There was a linear rise in α-terpineol, formed from limonene, which was much more dependent on storage temperature than on initial limonene concentration. The rise was greater in glass bottles than in soft packages, at the same storage temperature. The juice was described as stale and musty after 13 days at 32°C, or 90 days at 20°C (62 days in glass), but was still acceptable after 3 months at 4°C. The loss of limonene into the polyethylene lining was considered as an advantage, while desirable volatiles were scarcely absorbed.

Oxidation of D-limonene (constituting 90–95% of orange oil) is a major commercial problem, e.g., in orange drinks. Nearly all the limonene disappears from oil in air after storage for 9 weeks at room temperature. Autoxidation products are responsible for poor odor and flavor. A major hydroperoxide fraction possessed the typical strong odor of oxidized orange oil, and a 2% addition to fresh orange oil spoiled the taste and odor of orange drinks. The hydroperoxide was highly unstable and decomposed to carbonyl and hydroxy products (El-Zeany, 1977).

Walsh et al. (1997) found two major off-flavors in orange juice, PVG, and 2,5-dimethyl-4-hydroxy-3 (2H)-furanone (DHMF or Furaneol) by modified chromatographic procedure. In a study conducted by Fallico et al. (1996), hydroxycinnamic acids and their corresponding decarboxylation products, PVG and p-vinylphenol (PVP), were determined in two-processed blood orange juices (freshly squeezed or made from concentrate) produced in Italy; samples were analyzed during 4 months of storage at 4°C and 25°C. Phenols (free and total ferulic and *p*-coumaric acids) were determined in the juices by HPLC. Results showed that, there was a 35% reduction in concentration of total *p*-coumaric and ferulic acids after 4 months of storage. Juices did not initially contain vinylphenols, but they were observed after 3 months of storage. PVP and PVG concentration were higher at 25°C. The flavor threshold of authentic vinylphenols in aqueous solution and in the blood orange juice was evaluated by sensory analysis. Concentration of these off-flavors in the stored juices abundantly exceeded the threshold values, especially in the juice from concentrate. It is concluded that the content of vinylphenols might provide a reliable index of blood orange juice quality.

A gradual decrease in several flavor components, 1-penten-3-one, hexanal, ethyl butyrate, octanal, neral, and geranial, and an increase in two undesirable components, furfural and α-terpineol, were observed, in addition to other changes in aseptically packaged orange juice during 8 months of storage at 21°C and 26°C (Moshonas and Shaw, 1989b). After 2 months of storage, an experienced taste panel found a significant difference in stored juices compared with a control sample, and significantly preferred the starting control juice. Continuous reductions in oil content and ascorbic acid were noted during the 8-month storage period.

ORANGE JUICE CONCENTRATE

Concentration of fruit juices permits economic advantages in packaging, storage, and distribution. It also helps in the economic utilization of perishable fruits during the peak harvest periods, thus stabilizing the market prices of fresh produce. Orange juice contains about 85–90% water; during the concentration step, their bulk is considerably reduced by removing most of this water. Orange juice concentrate is prepared either from freshly extracted and pasteurized single-strength juice or from a stored and pasteurized single-strength juice. The major equipment used in the concentration process is the evaporator, which warrants a detailed discussion about its design and other relevant aspects as elaborated below.

EVAPORATORS

Orange juice concentrate is the major item of trade. Most of the juice is converted into concentrate in major orange-producing areas of Brazil and Florida. With improved designs of evaporators, it has now become possible to produce high-quality product. The first commercial low temperature orange juice evaporator was built at Plymouth, Florida by Vacuum Foods Company, which was designed as a result of research carried out at the Winter Haven laboratories. This company later became Minute Maid Corporation, Minute Maid Division of Coca Cola Corporation and was designated as the Foods Divisions of the Coca Cola Corporation. The designed evaporator was a falling film type and consisted of a series of vertical stainless steel cylinders with spray device, water heating, and vapor removal system. This could produce concentrate of 42°Brix. Later on, Skimmer Company, Majonnier Brothers, Buflo-vak Division of Blaw-Knox, Kelly and Gulf Machinery Company, developed falling film evaporators for low temperature concentration of juice. The efficiency was improved by developing the multi-stage and multi-effect evaporators. These combinations however required long residence time for the product in the evaporator. Sutherland (1977) discussed the evaporators for concentration of orange juice along with the other equipments required for packaging and storage of juices, and their relative advantages.

Modeling of multiple-effect falling-film evaporators was studied by Angeletti and Moresi (1983). The classic mathematical model of multi-effect evaporators, operating in forward flow was combined with an accurate estimation of the overall heat transfer coefficient (HTC) in falling-film evaporators. For this, a correlation developed by Narayanamurthy and Sarma (1977) was used. It was possible also to estimate the order of magnitude of the fouling factor (Rd). Finally, design of an orange-juice double-effect falling-film evaporator system was carried out according to two different strategies.

Concentration of orange juice using an agitated thin film evaporator (ATFE) system, was studied by Nongluk et al. (2001). Experiments were performed on a laboratory-scale ATFE system and a pilot plant scale system capable of handling larger flow rates. The effect of PC on the rates of evaporation and the maximum obtainable concentration was studied for orange juice of the Pera variety (Stolf et al., 1973/1974). The PC was between 2% and 10%; the evaporator was a Centri-Therm, Model CT-1B pilot plant, centrifugal manufactured by Alfa-Laval, Sweden.

Thermally Accelerated Short Time Evaporator (TASTE) has been designed to use high temperature at first stage, which would simultaneously evaporate, pasteurize and enzyme-stabilize the juice. TASTEs have relatively low initial cost and ease of cleaning. The principal disadvantage is their relative inflexibility. It is not convenient to blend add-back concentrate with the feed juice, as might be done with other types of evaporators. The high temperatures cause more rapid fouling of heat exchanger tubes and necessitate more frequent cleaning (Chen et al., 1979). Changes in main constituents of orange juice were evaluated by Pino et al. (1987) during concentration on a TASTE. Mean retention of ascorbic acid was

93.5% and concentration caused slight browning of juice, but sugar content was not affected. Viscosity reduction by homogenization of orange juice concentrate in a pilot TASTE was studied by Crandall et al. (1988). Controlling viscosity is critical for the efficient evaporation and pumping of citrus concentrates. Viscosity of an orange (cv. Pineapple) juice concentrate was reduced by installation of a commercial homogenizer between the third and fourth stages of a pilot plant "TASTE".

Possibilities for manufacture of orange juice concentrates by using evaporator equipment (TASTE and a five-stage falling film evaporator from GEA Wiegand); aroma recovery; concentrate manufacture by freeze concentration; manufacture by UF and reverse osmosis; and storage of orange juice concentrate, are discussed by Loeffler (1996). Losses of volatile compounds in orange juice during UF and subsequent evaporation by a TASTE, were investigated by Johnson et al. (1996). Physico-chemical changes during preparation of Kinnow mandarin juice concentrate, was studied by Sandhu and Bhatia (1985). Juice was extracted using a superfine pulper and concentrated at 50–55°C under vacuum in rotary glass evaporator. The desirable concentration of juice was 40–42°Brix under the conditions of experiment. The composition of 40°Brix Kinnow concentrate is reported as: total solids 44.50%, TSS 40.00%, titratable acidity 4.57%, reducing sugars 14.80%, total sugars 26.90%, ascorbic acid 44.88 mg%, β-carotene 5.36 mg%, pectin 1.16%, and pH 3.60. The ready to serve drinks prepared from concentrate of 40°Brix after reconstitution were comparable with those prepared from freshly extracted juice.

The economics of using mechanical vapor recompression (MVR) evaporators to concentrate orange juice, has been discussed by Kelso et al. (1980). Evaporators using MVR have been operating successfully in various food industries for several years. Energy consumption in a concentrated orange juice plant, was determined by Filho et al. (1984). The consumption of electricity and thermal energy for FCOJ and citrus pulp pellets (CPP) was determined. Thermal energy accounted for 90% of total energy consumption in the plant and its consumption for CPP exceeded that for FCOJ. The kilocalories of thermal energy/kg of water evaporated increased as the feed rate of single-strength juice was decreased. At the design evaporation capacity, the steam efficiency of two tubular evaporators and two plate evaporators was found to be 0.85 N and 0.82 N , respectively, N being the number of effects of the evaporator. Fouling of the waste heat evaporator was a major reason for the high-energy consumption in the CPP unit.

Essence Recovery

The orange essence is a volatile fraction recovered from fresh juice. The flavor of orange products is enhanced by addition of essence. The quality of essence varies considerably and depends upon the variety, seasonal variation, degree of maturity, condition of fruit, and other factors. Morgan et al. (1953) investigated the recovery of essence from orange juice under vacuum at 110°–115°F (43.3–46.1°C). The volatile oil content of orange juice has been reported to vary from 0.016% to 0.075%. The details of orange aroma compounds are given in Table 19.5 (Nursten and Williams, 1967). In many citrus processing plants, recovery of essence during concentration of juice has been incorporated, as a part of the process for FCOJ. A method for essence recovery under vacuum, has been developed by Wolford and Attaway (1967) and Wolford et al. (1968). In this method, the essence is recovered by evaporating about 15% of juice just before it is fed to the evaporator, and the volatile fraction is concentrated in a series of condensers until the volume is reduced to about 1/100 that of the original juice. The product has been reported to be very fragrant and is used to impart fresh flavor and aroma to FCOJ.

A system in which the vapors are recovered by a pump with a liquid seal has been described by Bomben et al. (1966) and Brent et al. (1968). Currently, the most widely used system is recovery of essence from the first effect of a TASTE. Essence is stripped off the aqueous phase during the first stage of' concentration by evaporating 15–25% of the juice. By distillation through a packed column, the fragrant vapor is allowed to pass over to a refrigerated condenser while the heavier fractions and waste are condensed separately. This method can be extended to aroma recovery from peel and juice waste streams (Veldhuis et al., 1972). To economize the process, the essence recovery unit should be incorporated as part of a multi-effect evaporator. The blending of large quantities of essence concentrates is recommended to maintain the uniformity of product quality. The essence is extremely sensitive to oxidation; it must be stored in filled containers with no air headspace and often is packed under nitrogen or carbon dioxide. Essence is usually kept at low temperature (about 2°C).

Table 19.5. Aroma Compounds of Orange Juice

Hydrocarbons: Methane, 3-methylpentane, *n*-hexane, cyclohexane, benzene, toluene, *o*-xylene, *m*-xylene, *p*-xylene, limonene, *p*-cymene, α-thujene, myrcene, α-phellandrene, α-terpinene, β-terpinene, γ-terpinene, terpinolene, sabinene, α-pinene, β-pinene, car-3(4)-ene, camphene, *p*-methyl-*i*-propenyl benzene, 2,4-*p*-menthadiene, farnesene, ylangene, β-elemene, α-copaene, β-copaene, caryophyllene, α-humulene, α-humulene valencene, and cadinene

Alcohols: Methanol, ethanol, *n*-propanol, *n*-butanol, 3-methylbutan-1-ol, 2-methylbut-3-en-2-ol, *i*-butanol, *n*-pentanol, *i*-pentanol, *n*-hexanol, *i*-hexanol, *n*-hex-x-en-1-ol, *n*-hex-2-en-1-ol, *n*-hex-3-en-1-ol, *n*-hept-3-en-1-ol, *n*-octanol, methylheptenol, *n*-nonanol, 2-nonanol, decanol, linalool, α-terpineol, *i*-pulegol, borneol, *trans*-carveol, geraniol, nerol, farnesol, terpinen-4-ol, and phenylethanol

Carbonyls: Acetaldehyde, acetone, 2-butanone, pentanal, *n*-hexanal, hexenal (2 isomers), 2-hexenal, 4-methyl-2-pentanone, *n*-octanal, *n*-nonanal, *n*-decanal, 2-decanone, 2-decenal, *n*-undecanal, 2-dodecenal, geranial, citral, neral, carvone, citronellal, furfural, and nootkatone

Acids: Formic, acetic, *n*-propionic, *n*-butyric, *n*-pentanoic, *i*-pentanoic, *n*-hexanoic, *i*-hexanoic, *n*-octanoic, and *n*-decanoic

Esters: Ethyl formate, geranyl formate, ethylacetate, *n*-octyl acetate, *n*-decyl acetate, citronellyl acetate, terpinyl acetate, linalyl acetate, bornyl acetate, geranyl acetate, ethyl butyrate, 3-octyl butyrate, geranyl butyrate, methyl *i*-pentanoate, ethyl *i*-pentanoate, *n*-octyl *i*-pentanoate, ethyl hexanoate, methyl 3-hydroxyhexanoate, ethyl 3-hydroxyhexanoate, ethyl octanoate, methyl 2-ethylhexanoate, methyl anthranilate, methyl *N*-methylanthranilate, and ethyl benzoate

Miscellaneous: Ethyl 2-butyl ether and 1,1-diethoxyethane

Source: Nursten and Williams (1967).

Freezing and Freeze Concentration

Orange juice concentrated in one of the evaporators as described above and carried through the process is prechilled by passing it through a slush freezer (e.g., Votator). This model has three cylinders, 15 cm × 122 cm, while newer models have two cylinders, 15 cm × 183 cm. Similar units have been studied for concentration of citrus juice by freezing ice crystals and separating these crystals from the concentrate by centrifuging. A principal advantage of freeze concentration is that it tends to concentrate certain soluble fragrant components in the liquid product. Another advantage is that, theoretically, it should take less energy to separate waste by freezing than by evaporation. However, multi-effect evaporators tend to nullify this advantage because optimum engineering design has increased their efficiency. A disadvantage is that freezing water is a relatively slow process in comparison to evaporation, and concepts for speeding it up are not readily apparent. Another principal disadvantage is that suspended matter and soluble solids tend to separate with the ice. Such losses, in most trial runs, became large enough to require partial recovery. This increases the time to reach a desired concentration point as well as the cost. One citrus processing plant used a freeze concentration process to recover partially concentrated juice for blending in the final product to improve fresh flavor. However, the development of equipment to recover high-quality citrus essences made freeze concentration less attractive. Although the freeze concentration can produce a frozen concentrate of superior quality, no commercial production has been carried out due to economical reasons (Muller, 1967).

Storage of Concentrates

The most desirable temperature for storage of frozen concentrate is −18°C. Cans or bulk storage stainless steel tanks are used commercially. Storage studies have been conducted in different types of packages under different conditions. Tocchini et al. (1979) compared the effect of pasteurization methods on the quality of canned concentration from "Pera" cultivar orange juice in 0.5 l cans. Three pasteurization regimes were tested: (i) cans sealed under vacuum, followed by spin-cooking and cooling; (ii) as (i), but cans sealed under N_2; and (iii) hot-filling of the cans, followed by spin-cooling. All samples were pasteurized at 74 ± 2°C for 40s. Samples (i)–(iii) were then stored for less than 180 days at 5°C. Frozen samples stored at −15°C were used as a control. The results show that (i) canned samples had the best keeping quality; its organoleptic properties did not differ significantly from those of frozen samples after storage for 90 days.

Distinct differences in Brix:acid ratio, bottom pulp, and apparent viscosity of the concentrates from the control and freeze-damaged samples have been observed (Hendrix and Ghegan, 1980). In the freeze-damaged samples, the Brix:acid ratio was initially higher and continued to increase with storage. Bottom PC increased with storage in the freeze-damaged samples, as did the apparent viscosity.

Kenawi et al. (1994) investigated the effects of storage and packaging material on properties of calcium fortified orange juices. Orange juice concentrate was fortified with calcium phosphate to provide the consumer with 20% of the recommended daily intake. Fortified and unfortified juices were aseptically packaged in cans, laminated pouches, or LDPE bags. Quality of orange juices was evaluated in these packaging materials during storage for 10 weeks at room temperature. Vitamin C content of juice concentrate decreased during storage in fortified and unfortified samples. The decreasing trend was similar for all packaging materials. Titratable acidity increased and pH decreased during storage. Color of fortified and unfortified concentrates changed little during storage, Fe contents varied little, but a slight loss in Ca content occurred in all samples (4–4.5%) during the 10 weeks of storage. Acceptability of fortified juice was higher than for unfortified juice; sensory evaluation was affected by packaging material and storage time. Samples in LDPE bags rated poorer than those in laminated pouches or cans.

FREEZE DRYING OF JUICE

In the freeze-drying process, the moisture is removed from the product by sublimation, and ice is directly converted into water vapor, thus minimizing the quality deterioration during drying. Pereira de Almeida (1974) gave a general account of the production of freeze-dried juice, covering: extraction from the fresh oranges and elimination of oil to give a maximum residual content of 0.020–0.025%; concentration by freezing in an inert atmosphere to preferably 18–20% solids, separation of free ice and centrifugation of juice concentrate from pulp solids; vacuum treatment to remove oxygen; homogenization (preferably at 300 kg/cm^2) to disperse fine pulp particles; pasteurization at 95°C for 10 s and rapid cooling; freezing and granulation (−50°C); freeze-drying at maximum temperature of 35–40°C to give 1.0–1.5% residual moisture content; and packaging with minimum exposure to oxidation. While it is possible to obtain a high-quality product, doubts are expressed on the economic possibilities due to demands on basic technical resources and potential marketing difficulties.

The most important parameters for correct operation of the freeze–drying cycle were evaluated in relation to the original concentration of the juice (Mowzini et al., 1974). Special attention was paid to the relation between the internal temperature of frozen granules and pressure in the freeze-drying chamber, in view of the difficulties of industrial processing caused by melting, puffing, and collapse. To avoid puffing, limiting temperature were −34°C for juice concentration to 40°Brix, −30°C for 30°Brix, −28°C for 20°Brix, and −26°C for natural juice. Maximum tolerable temperature for frozen granule surface was 35°C, for heating plates 90–120°C, and temperature at the condenser surface 8–10°C below that of the heating chamber. The results were used to devise optimal industrial freeze–drying cycles for processing orange juices at different concentrations. To ensure product quality, 30–35°Brix was suggested as a safety limit.

BEVERAGES MADE FROM ORANGE JUICE

Ready to serve drinks, cordial, squash, crush, and orange syrup are prepared from orange juice or concentrate, squash being the most commonly prepared product. Both oranges and mandarins are used for preparation of squash. The juice is extracted in the usual way and pasteurized. Sugar syrup is prepared according to recipe, cooled, and filtered. Food grade citric acid is used for acidification of the product. Peel oil or good quality orange essence is used for flavoring. Two recipes based on 25% and 33.33% juice are used for preparation of squash. Squash contains moderate quantity of pulp. Minimum TSS 40% and acidity 1.0% are maintained in squash as per food laws (FPO, 1955). Potassium metabisulfite equivalent to 350 ppm sulfur dioxide is permitted in the product. Orange color and essence are added to increase the aesthetic appeal of the product. The product is filled into bottles leaving about an inch headspace. It does not need pasteurization and is stable under ambient conditions. Aggarwal and Sandhu (2004b) studied the effect of hydrocolloids on the quality of Kinnow squash. The addition of carboxymethylecellulose, sodium alginate, and gum acacia individually at a level of 0.6%, improves the physico-chemical and organoleptic characteristics of Kinnow squash. Among these hydrocolloids, carboxymethylecellulose have the most desirable effect in improving the flavor, appearance, and cloud stability of the product.

Crush and fruit syrup contains minimum 55 and 65% TSS, respectively, with at least 25% of orange juice content. Cordial is prepared from clarified juice. It is sparkling clear beverage containing 25% juice and minimum 30% TSS. Permitted preservatives and flavor are added (FPO, 1955).

Ready to serve beverages based on 10% juice are prepared which contain sugar and citric acid to suit the individual taste, generally in the ratio 12:0.3 (Brix:acid ratio 40:1). Permissible amount of potassium metabisulfite (sulfur dioxide 70 ppm) or sodium benzoate (benzoic acid 120 ppm), color, and flavor can be added. The bottled product is processed to increase its shelf life at room temperature. The time and temperature of processing depends upon the size of bottles. The glass bottles of 200 ml capacity require 20 min processing in boiling water for sterilization. The aseptic packaging is ideally suited for ready to serve beverages, which permits use of heat labile plastic containers. This product has good consumer demand in the Indian market where there are hot climatic conditions for most of the part of the year. Ready to serve beverages may also be carbonated (Khurdia, 1990).

Garbagnati (1978) described a machine for producing beverages in granulated form of the Vomm-Chemipharma production line incorporating a Turbo-Granulator. Applications include the production of cold orange and lemon beverages. Influence of different packages and incandescent light, on the quality of an HTST-pasteurized single-strength orange drink during ambient storage (25–30°C) was investigated by Sattar et al. (1989). Both in terms of physicochemical and sensory characteristics of the drink, amber colored packages were found to be superior to the green glass bottles and co-extruded wax laminated paper (Tetra Pak).

Ascorbic acid retention in juice and drinks packaged aseptically into two flexible films (retort pouch and polyethylene), is affected by processing, concentration of added amino acids, and the type of packaging material. Non-enzymic browning of single-strength orange juice and synthetic orange drinks containing 10% (v/v) orange juice is linearly related to added amino acids concentration, and is more pronounced in the presence of high levels of ascorbic acid (Kacem et al., 1987). Deaeration and anaerobic storage resulted in increased retention of ascorbic acid. There was a very little change in the flavor score, browning, color, or amino acid content in aseptically packaged beverages in 250 ml Tetra Brik packs during storage at 24°C for 16 weeks.

Herrera et al. (1979) manufactured a beverage-clouding agent from orange pectin pomace leach water prepared from Valencia and Hamlin oranges. A commercial pectinase was used to hydrolyze pectin in the leach water. Clouding agent solids recovered (as percentage of peel solids) were 29% for Valencia orange peel and 27% for Hamlin orange peel. The stability of the prepared cloud was evaluated at a solids level of 1%. Several food proteins, together with some naturally occurring water-soluble gums (viscosity builders) and lecithin, were evaluated as clouding agents for formulated orange drinks (Garti et al., 1991). Use of these three types of natural products, represents a significant advantage over the formulations currently in use consisting of synthetic emulsifiers and wetting agents.

Orange carbonated beverages prepared from orange concentrate were treated with the stabilizers such as sodium alginate (0.2%, 0.4%, and 0.6%, w/v), guar gum (0.25%, 0.35%, and 0.45%, w/v), and gum Arabic (2.3%, 2.6%, and 2.9%, w/v), and their cloud stability was monitored spectrophotometrically during storage at 4.4°C for 45 days (Ibrahim et al., 1990). Best inhibition of cloud separation was obtained with 0.35% guar gum (10 days), 2.9% gum Arabic (8 days), or 0.2% sodium alginate (28 days).

El-Wakeil et al. (1974) reported that the contents of nitrogenous compounds, total ash, Na, K and P, and the K/Na ratio can be used to differentiate orange beverages prepared from synthetic and natural concentrates. The formol value, while being a useful statistical quality index, cannot be applied as a sole parameter for individual products or for the determination of the orange juice content of beverages (Fogli, 1975). Determination of the natural fruit content in orange juices and beverages on the basis of the free amino acid content (FAA), is described by Habegger and Sulser (1974). Empirical formulae are given for determining the approximate percentage content of natural fruit in an orange product, viz. (total FAA $\times$ 100)/3, or total $N \times 100$, SD of $\pm 20\%$ is given for the former. Benk (1969) described a method for detection of adulteration in orange beverages with soybean extract that contains raffinose and stachyose. Since raffinose and stachyose are absent in orange juice and orange peel, it is possible to detect the use of soybean extracts in orange beverages using circular paper chromatography. It was found that glucose syrup interferes with the detection.

BY-PRODUCTS UTILIZATION

The large amount of fruit being processed dictates evolution of a significant by-products industry to utilize the peel residue, essential oils, and other components. There are many by-products from the manufacture of orange juice. These include peels, pulps, and seeds, which are pressed and dried for cattle feed. The press liquid yields citrus molasses, which is also used for cattle feed as well as terpene oils for paints and plastics. Citric acid, pectin, and essential oils are also recovered from fruit processing waste.

CITRIC ACID

Citric acid can be extracted from culled sour oranges or pomace left after extraction of juice. Juice may be fermented first to remove the gums, pectin, and sugars to facilitate the filtration, mixed with filter aids like Kieselguhr at 60°–66°C and then filtered. Calcium citrate is precipitated with the addition of hydrated lime and calcium carbonate, and precipitate is separated and dried quickly to avoid discoloration. For the conversion of citrate into citric acid, the wet paste is treated with calculated amount of concentrated sulfuric acid. The precipitated calcium sulfate is removed and the liquor is concentrated to crystallize the citric acid (Cruess, 1997). The composition of orange juice and peel with respect to major organic acids given by Clements (1964) is presented in Table 19.6.

ESSENTIAL OILS

Orange oil finds uses in the food, beverage, pharmaceutical, perfumes, and soap industries. Mandarin essential oil expressed from the peel is also employed commercially in flavoring a number of food products. Mandarin essential oil paste is a standard flavoring for carbonated beverages. The essential oils are obtained from peel by means of scraping the whole fruit and pressing the scrapings in vertical or continuous presses, then separating the essential oil and water by centrifugation. In some factories, the essential oil is extracted at the time of juice extraction in the squeezing machines (FMC In–Line). Essential oils are also obtained in very small quantities by the stiletto systems, and by S-fumatric type presses, which use squeezed half fruits (Lal et al., 1986). The essential oils from green or mature bitter oranges are obtained by a system of scraping, followed by pressing the scrapings, and are considered as the products of excellent quality available on the market.

The highest quality natural oils known in the trade are cold-pressed peel oils. The fresh orange peels yield about 0.54% of oil by the cold press method. Kesterson et al. (1971) described seven different types of equipment used to express citrus peel oils. The general commercial methods in the production of cold-pressed oils include oil recovery from peel after juice extraction, simultaneous extraction of juice and emulsion from whole fruit, and recovery of oil from peel flavedo following removal from the whole fruit by abrasion or shaving. Natural or phenolic antioxidants may be added to oils to increase the stability under adverse storage conditions. Concentrated or folded oils are prepared by vacuum distillation of cold-pressed oils.

Developments in the analytical techniques have lead to the identification of hundreds of volatile compounds in the orange fruit depending upon the variety. Essential oils further undergo changes during packing and storage. The most important volatile

Table 19.6. Organic Acids of Juice and Peel of Oranges and Tangerines

	Juice (g/100 ml)		Peel (meq/g Dry Weight)			
Fruit	Malic	Citric	Malic	Citric	Oxalic	Malonic
Orange						
Washington Navel I	0.06	0.56	0.02	0.01	0.11	0.02
Washington Navel II	0.20	0.93	0.02	0.01	0.10	0.03
Valencia	0.16	0.98	0.02	Trace	0.13	0.03
Tangerine						
Dancy I	0.18	1.22	0.06	0.02	0.15	0.01
Dancy II	0.21	0.86	0.09	0.02	0.20	0.02

Source: Clements (1964).

materials of citrus fruit are those associated with flavor and aroma. These include terpene hydrocarbons, carbonyl compounds, alcohols, esters, and volatile organic acids. They are generally present in peel oil in flavedo but are also found in oil sacs embedded in the juice vesicles. Variable amounts are present in different parts of fruit and their concentration is affected by processing and storage conditions. One compound D-limonene alone accounts for 80–96 wt.% of all citrus oils. α-Pinene, sabinene, β-myrcene, and D-limonene play an important role in orange flavor. Linalool and myrcene make a positive contribution and 2-hexanol and α-terpineol have a negative contribution to orange aroma (Pino, 1982).

Chemical composition and aroma of orange were investigated by Rodopulo (1988). The results of analysis showed 12 aldehydes, 12 alcohols (usually as esters with organic acids), five terpenes, and three esters. Referring to these results, recipes were formulated for the manufacture of aromatic essences enriched with vitamins and trace elements, and sugar. The following composition was recommended (mg/l): limonene, 125; citral, 0.25; citronellal, 0.26; linalool, 1.2; α-pinene, 0.9; hexanal, 0.08; citric acid, up to pH of 4.5. Ethanol (less than 1%) can be replaced with glycerol or 2,3-butylene glycol, 3 g/l each. After addition of sugar, 75 mg/l of ascorbic acid and trace elements (B, Mo, Mn, and Zn) are incorporated up to maximum quantities of 0.3–0.5 mg/l. The aroma components of orange as reported by Nursten and Williams (1967) are given in Table 19.5.

PECTIN

After the extraction of orange juice, about 50% of the fruit ends up as peel and pomace. Citrus peel is the major raw material for citrus pectin manufacture. The maximum pectin grade obtainable under optimum conditions may be about 150–250 for orange peel, which is quite low as compared to that of lemon and grapefruit peels. Peel for citrus pectin is used after extraction of peel oil. Pectic enzymes of peel have to be inactivated to prevent the decomposition of pectin during processing and storage. For commercial production of pectin, peel passes through several steps including removal of peel oil, shredding or comminuting the peel to facilitate the washing and extraction, screening or reeling the peel in rotating cages to remove seeds, and rag and leaching the peel with water to remove soluble sugars and other undesirable materials. If the peel is not to be utilized immediately for pectin extraction, it has to be heated for 10 min at 95°–98°C to inactivate the pectic enzymes. The extraction conditions such as pH, time and temperature have to be controlled carefully to get maximum extraction of pectin. Both organic and inorganic acids can be used for hydrolyzing the protopectin into soluble pectin. Cheaper mineral acids like hydrochloric and sulfuric acids are preferred for extraction. Wooden extraction tanks or vats are commonly used. Raw shredded or dry stored peel may be used for extraction of pectin.

Owens et al. (1949) described a continuous process using a countercurrent extractor for the manufacture of pectin from orange peel. Process conditions involved are a water to peel ratio of 3:1, pH of 1.3–1.4 and temperature of 90–100°C with heating for 1 h. After separation, pectin is given several washings in 50–70% solvent and final dehydration/washing in 80–90% solvent. Finally, the pectin is dried to approximately 6–10% moisture content with warm air or in a heated drum. Low methoxyl pectin can be prepared by controlled acid, alkali, or enzymatic deesterification. The pectin is recovered from comminuted orange peels pretreated with (i) fungal crude enzyme (prepared from *Aspergillus terreus*), (ii) yeast crude enzyme (prepared from *Kluyveromyces lactis*), and (iii) Irgazyme M10 (a commercial preparation from cultures of *A. niger*).

In order to establish optimum processing parameters, extraction of pectin from dried orange peel was studied by Gentschev et al. (1991). Mechanical composition, water uptake, and swelling properties of orange pomace were studied during initial rinsing; weight and volume of pomace increased by 6× and 2×, respectively, in the first 20–30 min of rinsing. Optimum washing conditions were achieved using a hydromodule of 24:1, and three separate washings with a time–temperature combination of 20 min at 20°C, 15 min at 30°C, and 10 min at 40°C. Effect of extraction variables on yield and quality of pectin extracted from dry orange waste with nitric acid was studied by Aravantinos-Zafiris and Oreopoulou (1992). A factorial experimental design was used and effects of temperature, time, and pH of extraction on pectin yield and "jelly units," i.e., pectin yield × viscosity of pectin solution, were determined. Optimum conditions for pectin extraction are also discussed. According to ash and methoxyl content determinations, the product can be classified as low ash and high methoxyl pectin. Its purity expressed as anhydrogalacturonic acid content varied from 68.5% to 75.0%.

HESPERIDIN

Hesperidin can be recovered from the wastewater of orange juice processing (yellow water) by concentration of diluted extracts on styrene-divinylbenzene resin (Di Mauro et al., 2000). Turbid raw material flowing out from centrifuges of essential oil separation contains considerable amount of hesperidin (approximately 1 g/l), mainly associated with solid particles. Yellow water is treated with calcium hydroxide until pH 12 is reached, to solubilize hesperidin before being filtered, neutralized at pH 6, and loaded on resin up to a saturation point. Desorption with 10% ethanol aqueous solutions at different NaOH concentrations (0.23–0.92 M) is carried out to ensure high concentration of hesperidin in selected fractions (10–78 g/l), from which it is precipitated in high yield and purity immediately after acidification to pH 5. Best results are obtained using 0.46 M NaOH as eluent: 71.5% of the adsorbed hesperidin is desorbed in 300 ml, with an overall 64% yield of isolated product at 95.4% purity (HPLC).

COLORING MATTER

The orange peel is a good source of pigments and yields about 350 mg/kg orange peel (Wilson et al., 1975). The crude extracts and yields from 20-fold concentration of cold-pressed oils of mid-season orange, Valencia orange, and tangerine were 116, 95, and 116 mg/kg peel, respectively. A dispersible powder for coloring and flavoring orange drinks was produced from the essential oil and carotenoid pigments of orange peel (Rovesti, 1978). Recycling orange essential oil on flavedo from whole oranges produces an extract with 336 ppm of total carotenoids (87.7% xanthophylls). To obtain a highly colored water-dispersible product, color extract is dispersed in an aqueous emulsion containing natural and modified polysaccharides (a mixture of unmodified hydrolyzed starches, enzyme-treated water extract of orange peel, and 3% gum acacia or sodium caseinate) and is spray dried at less than 75°C in a 75% N_2, 21% CO_2, 4% O_2 gas mixture at a water soluble:fat soluble substance ratio of 1.8 and a water:total DM ratio greater than 10, atomizer rotation rate 7000 rev/min. The product is a deep yellow powder of medium density, easily dispersible in water at 40°C (or in cold water with stirring). About 7% of the carotenoids in the extract are lost in spray drying. No significant changes in carotenoid contents or composition are observed after 120 days storage at 35°C away from direct light. Color changes in prepared drink are prevented by addition of 100 ppm of ascorbic acid. The pigments from orange peels have also been produced by using microbial enzymes (Elias et al., 1984).

GRAPEFRUIT *(C. grandis; C. aurantium,* var. *grandis; C. decumana)*

The world production of grapefruit and pomelos is 4.69 million tons. The cultivars of grapefruit used for juice production are Marsh seedless and Duncan. Red Blush and Foster are recommended for cultivation in India. Grapefruit juice prepared from pigmented rather than white grapefruit often yields a product with poor color (Lee, 1997). It is suggested that a grapefruit juice of good color could be produced through careful selection of highly pigmented grapefruits, controlled blending, and further color enhancement. "Star Ruby," a new cultivar of red grapefruit, was found to have excellent color both in the flesh and in the juice, even in late season (Ting et al., 1980). The color is sensitive to heat, but the juice after pasteurization and concentration retains sufficient color. The juice of the Star Ruby is higher in soluble solids and acid than the Ruby variety (Hensz, 1971). Thompson is another pigmented cultivar. Grapefruit is processed both for single-strength juice and frozen concentrate.

Fruit in the processing plant is graded, sampled for analysis, and stored in bins after removal of split and decay fruits. Fruit is given another grading inspection before washing, to remove additional inferior fruit. The fruit is then dumped in a soaking tank or hot pit, which has multiple purposes of soaking, washing, and oil control.

JUICE EXTRACTION AND FINISHING

The grapefruit juice can be extracted using the same machinery as used for orange juice with adjustment of extractors depending upon the size of fruit. Specially designed extractors are also available for different size fruits. Modern extractors can handle over 2 t of fruit/min. Some of the latest types of juice extractors produce juice of low oil content due to their better adjustment for fruit size, thus eliminating the need for hot-water treatment in hot pits. Bireley Citromat, Brown and the Food Machinery In-Line Extractors are used for juice production. The juice yield varies from about 334 l/t of fruit at beginning, to a maximum of about 542 l/t at mid-season, dropping considerably towards the end of season. The average yield is reported about 480 l/t of fruit. Mohsen et al. (1986a) reported that rotary extraction increased PC and pectin esterase (PE) in the juice. Pressing extraction increased contents of limonene, oxygenated terpenes, and carbonyl compounds. Rotary extraction increased contents of esters and water-soluble volatiles. Effect of method of extraction and pasteurization on grapefruit juice properties and its volatile components was studied. Taste and color were not affected by the treatment, but odor of juice extracted by pressing was more acceptable than that obtained by rotary extraction.

Mechanically extracted juice contains seeds, pips, and segment membranes, which must be removed quickly to avoid leaching of naringin and limonin from suspended particles into the juice. The juice is finished in the same type of finisher that is used for orange juice. In the case of grapefruit juice, the pressure applied during finishing has marked effect on yield and flavor of juice. The naringin and limonin contents of Florida grapefruit juice have been reported to range from 218 to 340 ppm and 2 to 10 ppm, respectively, throughout a commercial packing season (Tatum et al., 1972; Barros, 1992).

Coloring Pigments

Cruse et al. (1979) reported that juice of the Star Ruby grapefruit (*C. paradisi* Macfad) extracted by commercial machinery and canned during the regular harvest season contained from 0.34 to 0.56 mg/100g β-carotene and 0.57 to 0.72 mg/100 g lycopene, as compared to the corresponding respective value ranges of 0.09–0.14 mg/100 g β-carotene and 0.04–0.08 mg/100 g lycopene for the Texas Ruby Red variety. Blending with Star Ruby juice at about a 30% level in the early processing season and 40% in the late processing season is proposed as a means of maintaining a desired level of pink color in canned Texas grapefruit juice. In Texas, the processing of pink grapefruit juice is emphasized, as contrasted to white grapefruit juice in Florida and California. The carotenoid composition of the juice of red-fleshed grapefruit is reviewed by Benk and Bergmann (1973). For carotenoids, the ranges reported are: total carotenoids, 1.4–3.5 mg/l; β-carotene, 0.6–1.6 mg/l (39–53% of total); lycopene, 0.4–0.9 mg/l (23–38%); and cryptoxanthin, 0.03 mg/l (1.6%).

Deaeration and Deoiling

Deaeration and Deoiling are important features in canning of grapefruit juice. The freshly extracted juice contains 2–4 vol.% gases as reported by Pulley and von Loesecke (1939). The oxygen has adverse effect on color, vitamin C, and citrus oils. It must be removed immediately. The acid-catalyzed dehydration of the D-limonene produces turpentine-like taste even in the absence of molecular oxygen (Blair et al., 1952). It is a serious problem with canned and chilled grapefruit juice because of its high acidity. Shearon and Burdick (1948) and Scott (1941) discussed technical and engineering phases of deaeration and deoiling of juice. For optimum flavor and storage life, the juice should contain 0.003–0.005% recoverable oil.

Debittering

Debittering of grapefruit juice has been achieved with β-cyclodextrin polymers or XAD resins in a fluidized bed process (Wilson et al., 1989), polystyrene divinylbenzene adsorbents (Manlan et al., 1990), polyvinylpyrrolidone (Nisperos and Robertson, 1982), naringinase entrapped in cellulose triacetate fibers (Tsen and Yu, 1991), naringinase immobilized in packaging films (Soares and Hotchkiss, 1998b), active packaging (Soares and Hotchkiss, 1998a), enzymes (Prakash et al., 2002) and Amberlite IR 120 and Amberlite IR 400 and alginate entrapped naringinase enzyme (Mishra and Kar, 2003).

Deacidification

Optimum conditions for use of chitosan in deacidification of grapefruit juice and effects of chitosan treatment on the composition and sensory properties of the product have been studied (Rwan and Wu, 1996). Chitosan with a particle size of 40–60 mesh and a degree of deacetylation of 90% showed superior deacidification properties. Deacidification carried out by adding chitosan into juice at 0.015 g/ml with stirring at room temperature for 30–60 min gave deacidified juice with a Brix:acid ratio approximately 13.4; total acid content was reduced by approximately 52.6%, contents of citric, tartaric, L-malic, oxalic, and ascorbic acids being reduced by approximately 56.6%, 41.2%, 38.8%, 36.8%, and 6.5%, respectively. Sugar, amino acid, mineral, and naringin contents and Hunter *L*, *a* and *b* color values were not affected. Deacidified juices had higher scores for total acceptability than the original juice. Chitosan could be successfully regenerated and reused. The economic feasibility of adsorptive deacidification and debittering of Australian citrus juices was studied by Johnson and Chandler (1985). Aspects covered are: basis for costing (capital outlay, loan repayment and depreciation, maintenance, and labor cost); deacidification of citrus juice (regeneration of deacidifying resin, replacement of deacidifying resin, evaporation of diluent water, and cost of citric acid removal); debittering of citrus juice (regeneration and replacement of debittering resin, evaporation of diluent water, and cost of debittered navel orange juice); return by saving in sucrose addition; and return by increased value of debittered juice. It is concluded that the operation of both processes for 3 months of the year each should repay capital in less than 1 year.

Sweetening

The relationship between the ratio of °Brix to percent acid and sensory flavor in grapefruit juice was studied by Fellers (1991). A significant correlation between ratio and flavor was found; the higher the ratio the better the flavor. Because of high acidity and tartness, most grapefruit juice is sweetened. For sweetening, calculated amount of sugar or highly concentrated sugar syrup (65°Brix) is added to the tanks containing the juice, so that °Brix is increased to desired level.

Pasteurization and Packing

In the case of grapefruit juice, higher temperature is required to deactivate pectin esterase compared to orange juice. The effect of time and temperature of pasteurization on cloud stability of canned grapefruit juice was studied by Kew et al. (1957). The cloud was stabilized in a high-speed pasteurization at 98.9°C in 1.75 s, at 90°C in about 13 s, and at 85°C in about 43 s. Effects of thermal pasteurization on color of red grapefruit juice have been studied (Lee and Coates, 1999). Juices were pasteurized at 91°C using a plate-type heat exchanger. Thermal pasteurization affected all three-color parameters (CIE L^*, a^*, b^*) within the juice, causing a slight shift in color (lighter and brighter). Thermal pasteurization especially affected CIE b^* values and chroma within the juice. Reflectance spectra within the visible region (400–700 nm) clearly showed changes in spectral distribution of light reflected from juice after pasteurization. There were no changes ($P > 0.05$) in major carotenoid pigments (β-carotene and lycopene) of the juices after pasteurization.

Two new "cold" pasteurization techniques available to the food industry, ultrahigh pressurization pasteurization, and electric impulse poration pasteurization are described (Anon, 1993b). With ultrahigh pressurization pasteurization, the pressure treatment destroys the cellular integrity of microorganisms without detrimentally affecting heat labile flavor compounds, vitamins, and other essential components. With electric impulse poration pasteurization, liquid foods are passed through very narrow orifices across which are passed high-voltage electric pulses. The electric pulses inactivate microorganisms by electroporation.

Pasteurization caused an inactivation of about 94% in pectin esterase and decreased contents of limonene, oxygenated terpenes, esters, and water-soluble volatiles. Carbonyl compounds showed a slight increase after pasteurization. The major volatile components of grapefruit juice, identified by GC, were: limonene; terpinen-4-ol; α-terpineol; linalool; methanol; ethyl butyrate; and octanol (Mohsen et al., 1986b). Pasteurized grapefruit juice is filled into cans, closed, and cooled by the same procedures described for canned single-strength orange juice.

Concentrated Grapefruit Juice

Praschan (1951) has discussed the different types of evaporators and different methods of freezing from engineering and technical angle. After pasteurization, juice is concentrated under high vacuum at temperature below 26.7°C to about fivefold, diluted or cut-back with fresh unconcentrated, deaerated, heat-treated juice having a higher pulp and oil content to about fourfold. Often cold-pressed grapefruit oil is added to replace that lost during the concentration. Product is filled into cans of appropriate size.

Aroma and total volatile compounds composition of a single unpasteurized Marsh grapefruit juice and its 65°Brix concentrate (obtained from juice using a thermally accelerated short time evaporator) reconstituted with water to 10°Brix, were examined using GC-olfactometry and GC-FID (Lin et al., 2002). Total volatile compounds (measured by FID) in the reconstituted concentrate were reduced to <5% of initial values, but 57% of total aroma compounds (as judged by GC-olfactometry) remained. Forty-one aroma-active compounds were detected in unpasteurized single-strength juice, whereas 27 components were found in the unflavored reconstituted concentrate. Aroma-active compounds were classified into grapefruit/sulfury, sweet/fruity, fresh/citrusy, green/fatty/metallic, and cooked/meaty groups. Five of the 6 components in the sweet/fruity, and 14 of 18 green/fatty/metallic components survived thermal concentration. However, only 4-mercapto-4-methyl-2-pentanone in the grapefruit/sulfury group, and linalool and nootkatone from the fresh/citrusy group, were found in the reconstituted concentrate. Methional was the only aroma compound in the cooked/meaty category found in both juice types. β-Damascenone and 1-*p*-menthen-8-thiol were found only in the reconstituted concentrate. 4-Mercapto-4-methyl-2-pentanol was detected for the first time in grapefruit juice.

Storage

Grapefruit juice is probably the most stable of the citrus group, though the changes in flavor are probably masked to some extend by high acidity and bitterness due to naringin that has not been found in significant amount in other citrus juices. Under normal storage temperature and conditions, properly processed juice should retain its normal flavor and appearance for about 9 months; it should still possess a good flavor for about 15 months, after which definite off-flavor and off-colors develop. The most significant changes in stored canned juice are a decrease in limonene and an increase in linalool monoxide, α-terpineol, and furfural. Canned or bottled grapefruit juice will retain its normal flavor almost indefinitely when stored at 0°C. Changes in aromatic volatile constituents of grapefruit juice were studied during 8 months storage of concentrated juice at −9°C and −18°C (Pino, 1986). Results obtained by GLC indicated decreases in terpene hydrocarbon and esters, while other compounds such as α-terpineol, epoxydihydrolinalool, furfural, and aldehydes increased during storage at −9°C. Changes were less evident during storage at −18°C. Alterations in composition corresponded with results of sensory analysis.

Changes in color and pigment composition (β-carotene and lycopene) of red grapefruit juice concentrates during storage at −23°C for 12 months have been studied (Lee and Coates, 2002). Concentrate (38°Brix) was packed in both plastic (16 oz) and metal (6 oz) cans. Decrease in red intensity (CIE a^*) in juice color and slight increases in CIE L^*, b^*, and hue values from analysis of reconstituted juices were the characteristic color changes in concentrate during frozen storage. With respect to fresh concentrate, juice color in stored concentrate shifted towards the direction between negative DELTAC* and positive DELTAL*, indicating that the color became slightly paler. A color difference seemed to exist between the two containers, especially for the magnitude of DELTAL*; color changes were more pronounced in concentrates packed in plastic. There were significant changes ($P < 0.05$) in major carotenoid pigments (β-carotene and lycopene) in the concentrates. More than 20% loss of lycopene and approximately 7% loss of β-carotene occurred with plastic containers after 12 months. Regression analysis showed that the rate of decline was approximately 0.291 ppm/month ($r = 0.990$) for lycopene compared to 0.045 ppm ($r = 0.817$) for β-carotene in concentrate stored in plastic. In the metal can, the same trends were observed but pigment losses were slightly smaller than those with plastic. An estimated shelf life for lycopene was 26.1 months in the metal can compared to 18 months in plastic. Shelf life for β-carotene was > 39 months, which was more than twice that of lycopene in the plastic container.

Roig et al. (1994) studied the possible additives for extension of shelf life of single-strength reconstituted citrus juice, aseptically packaged in laminated cartons. The results show the importance of eliminating oxygen in systems where L-ascorbic acid is present for its nutritional value. Of the additives studied for the purpose of extending the shelf life of the juices, sodium metabisulfite was the most effective.

LEMON (*C. limonia*)

Eurekas, Lisbons, and Villa Francas are the commercial varieties. Predominant cultivar in Mediterranean countries is Femminello. The production of lemons and limes is reported to be 12.45 million tons in the world. Lemons are picked by hand according to size rather than color when they meet maturity standards. Lemons are considered mature when they contain 30 vol.% or more juice. Very immature fruit that has not gone through the curing process may impart a green-fruit taste, whereas fruit stored for very long time may impart an over-mature or stale taste. The blending of fruit may sometimes be necessary to obtain satisfactory, uniform flavor, and acidity.

Lemons used for juice production in commercial operation are graded and brush-and-spray washed with warm detergent solution and rinsed before going to juice extractors (Kieser and Havighorst, 1952). Biesel (1951) described in detail the methods for controlling the microorganisms in citrus processing plants. The citric acid is the most important single constituent of the lemon and consequently concentrated juices are standardized on the basis of acidity. To counter the variations in acidity, the juice from different lots is blended and stored in concentrated form. The concentrated juice can be manufactured to any desired acid content by adjustment of the ratio between volume and concentration.

JUICE EXTRACTION

Extraction of lemon juice can be accomplished with different types of machines. Most of the juice is extracted with automatic machines such as Brown Extractor and Food Machinery and Chemical Corporation machine. Both these machines produce juice free of peel extractives. The amount of peel oil, pulp, and pectic enzyme incorporated into juice during extraction are the deciding factors for selecting the type of extractor for specific products. FMC In-line machine is used for simultaneous recovery of both juice and oil. The Citrus Equipment Corporation unit, the Citromat, and the Elpico machines are chiefly used for peel oil recovery. Juice yield from 36% to 57% is obtained by different methods of extraction (Dupaigne, 1971).

The juice should be screened as soon as possible to remove the insoluble solids, which contain leachable substances that may impair the flavor, color, and cloud stability of the juice. For better juice yields, screening in paddle finishers or screw presses is done to remove coarse pulp. Ray et al. (2003) used an ultrafiltration unit for clarification of lemon juice and reported that the membrane with a molecular weight cut-off of 300 kDa was best suited for clarification of lemon juice.The juice should be chilled to 10°C to avoid undesirable changes at higher temperature due to atmospheric oxidation.

The lemon peel oil content is very important from the standpoint of both flavor and stability. The amount of cold-pressed peel oil in lemon juice is limited to 0.025% because of oil burn as noted by throat feel at higher level. It is necessary to lower the oil content of lemon juice for specified uses. Deoiling may be accomplished by injecting preheated juice into a vacuum chamber, such as Majonnier deoiler.

A process for production of a shelf stable lemon juice without use of sulfite preservatives is described (McKenna et al., 1991). Lemon concentrate and/or lemon oil, sodium benzoate, and water are mixed; ascorbic acid, sodium acid pyrophosphate, glucose oxidase, or sodium hexametaphosphate are added with stirring; and an inert gas, e.g., CO_2, He, or N_2, is bubbled through the mixture, to produce a juice with a shelf life of greater than 9 (preferably greater than 12) months.

Donsi et al. (1998) carried out high-pressure stabilization of lemon juice. All the benefits of HPP were achieved at the 3000 bar. It gives high-quality lemon juice with a satisfactory shelf life.

PASTEURIZATION

The current practice is to heat the juice to a temperature for sufficient time, to assure practical sterility as well as cloud stability by inactivating the natural juice enzymes (Rothschild et al., 1975). A temperature of 77°C for 30 s is used in commercial operations. The juice is cooled immediately after pasteurization by passing through the heat exchanger.

CONCENTRATION OF JUICE

The concentration of heat-stabilized juices is carried out in vacuum evaporators. TASTE and AVP rising and falling film plate evaporators are used for concentration. A centrifugal evaporator developed by Alfa-Laval is suitable for concentration of lemon juice of 47.7% dry matter at 38°C, with no adverse effects on flavor or vitamin contents (Thormann, 1972).

A triple-effect plate evaporator for the production of lemon juice concentrate features a novel density-control system for single-pass, non-recirculatory flow operation, which regulates steam pressure and condenser vacuum (Dinnage, 1970). The evaporator has three effects arranged in four stages of evaporation and additional use of steam jet recompression gives overall evaporative efficiency of approximately equal to 4:1 while concentrating lemon juice to 50°Brix. The feed is regeneratively preheated to 54.5°C, with concurrent cooling of the concentrate. Liquid holding time is minimized by use of extraction pumps at each evaporator stage; total holding time is 135 s. A feed input of 35,700 lb lemon juice/h (8% solids, 4.5°C) results in 5700 lb final concentrate/h (50% TS, 7.2°C). The final product is standardized on the basis of grams of anhydrous citric acid per liter, usually about 325 g/l of concentrate.

PACKAGING AND STORAGE

Plastic or steel drums, glass containers, and cans of different sizes can be used for packing depending upon the storage time and temperature, as well as on the basis of economical considerations. All types of lemon juice products should be stored at lowest possible temperature. Frozen single-strength juice chilled to −1°C is filled into cans and quick-frozen, and stored at −23°C. Frozen concentrated juice after standardization of acid is also frozen at −23°C in drums for storage.

Vandercook (1970) reported that when lemon juice was stored under adverse conditions, the long wavelength peak (323–335 nm) in the UV spectra of the juice shifted to shorter wavelengths. With continued juice deterioration, the peak became a shoulder. The deterioration and spectral changes were proportional to the storage temperature and greatly increased by air.

Effects of adding SO_2, Sn^{2+}(tin) or cysteine to concentrated lemon juice on its color during storage have been investigated (Nunez et al., 1989). The results show that at 45°C, browning was inhibited by greater than 125 ppm SO_2, and the degradation of ascorbic acid, formation of furfural and hydroxymethyl furfural by greater than 250 ppm SO_2. Browning rate was reduced by Sn^{2+}(tin) depending on the concentration used: at 1000 mg Sn^{2+}(tin)/kg juice, browning was reduced to about one-third of the initial rate. Cysteine inhibited color formation only slightly at high concentration and affected the aroma of the juice at concentration greater than 500 mg/kg. Use of Sn^{2+}(tin) was promising because of its low toxicity and high legal tolerance levels, especially as its concentration would be reduced when the juice was diluted for use.

Studies were conducted on effects of cysteine on browning of concentrated lemon juice during storage (Tateo and Bianco, 1984). The results show that addition of cysteine before concentration reduces browning during subsequent storage; it also reduced browning under UV or IR irradiation. Addition of cysteine after concentration did not control browning.

LIME (*C. acida; C. aurantifolia*)

The only Key (Maxican or West Indian) or common lime and the large-fruited Persian (Tahitian) seedless limes have gained commercial importance. The Key is the most important cultivar. Lime resembles the lemon in structure and composition but it is usually smaller in size. The composition of the lime varies considerably with variety of fruit and the location where the fruit is grown.

Juice Extraction

Anon (1993c) outlined small-scale manufacture of lime juice while Seelig (1993) discussed in detail the various aspects of lime juice production. Because of difference in size of two principal types of limes, the method of juice extraction varies. For large Persian lime, the standard juicing equipment is used while because of small size of Key lime, the use of conventional extractors becomes impractical. From Key lime, the juice is generally obtained by crushing the whole fruit in a screw press.

The fruit is carefully graded, washed, and sometimes sterilized before passing to the high-speed juice presses or rollers. The juice is screened, analyzed, blended, deaerated, and deoiled before final canning. The lime juice is handled and pasteurized in the same way as lemon juice. The juice from Persian lime is passed through a finisher after extraction with standard juicing equipment. During finishing, it is important to keep pressure comparatively light to avoid extraction of bitter constituents from peel. Pruthi and Lal (1951) observed that deaeration combined with flash pasteurization at 90–96°C gives the best retention of vitamin C and overall juice quality. Study conducted on the factors affecting the keeping quality of lime juice (El-Shiaty et al., 1972), has shown that pasteurization improved cloud stability, had no effect on palatability, percentage of TA or pH; but resulted in a slight decrease in ascorbic acid content. The acid content of juice varies from 4.94 to 8.32 g/100 g of juice for different cultivars grown in different places. In the juice obtained by crushing the whole fruit in a screw press, the reduction of peel oil content by vacuum deoiling or centrifuging is desirable. The lime juice may be clarified by passing through a suitable plate-and-frame or other type of filter presses after mixing with filter aid. The juice may also be clarified by storing in wooden tanks with addition of either sulfur dioxide or sodium benzoate. The clear juice is siphoned off. Shaila-Bhatawadekar (1981) studied the clarification of lime (*C. aurantifolia*) juice by cellulase from *Pencillium funiculosum* at pH 2.2–5.6, using an enzyme concentration of 0.82–3.2 filter paper units (FPU), with subsequent incubation of the juice at 30°C, 40°C, and 50°C for 2–32 h. Maximum turbidity reduction (greater than 90%) was obtained at pH 4.0–4.5, temperature 30–40°C, an incubation period of 18–24 h, and addition of 1.64 FPU enzyme. For freezing, the juice is precooled before filling into cans.

Edwards and Marr (1990) analyzed the volatile components of lime juice oil and lime peel oil by capillary GC and MS. α-Santalene and β-santalene (0.06% and 0.11%, respectively) were found in lime juice oil, while β-santalene and β-santalol were found in lime peel oil (0.03–0.04% and 0.5%, respectively). These were not previously reported in lime.

CONCENTRATION

Askar et al. (1981) described the production of lime juice concentrates using the serum–pulp method. Lime juice (*C. aurantifolia* Swingle) was concentrated using four methods: (i) freeze concentration of the juice, (ii) vacuum evaporation of the "cut-back" juice with pasteurized juice, centrifugal separation into pulp and serum followed by (iii) freeze concentration, or (iv) evaporation of serum and recombination with the pulp. The chemical composition of fresh, pasteurized, and concentrated lime juice has shown that the best quality concentrates were obtained by method (iii), followed by methods (i), (iv), and (ii). Freeze-concentration does not markedly affect palatability or cloud stability but results in a decrease in percentage of TA and ascorbic acid (El-Shiaty et al., 1972).

STORAGE

Under unfavorable conditions, the lime oil in contact with high acid lime juice develops characteristic off-odors and flavors. Ikeda et al. (1961) reported that γ-terpinene is readily oxidized to off-flavored product, *p*-cymene. Pasteurized juice can be stored at 2°C for 15 months without appreciable change in flavor. In untreated samples, changes occur and storage life is limited to about 4.5 months at 27°C.

Sensory and physico-chemical stability of frozen Tahiti lime juice, natural and sweetened have been studied (Pedrao et al., 1999). Samples of natural Tahiti lime (*C. latifolia*) juice and Tahiti lime juice sweetened with 60% sucrose were frozen and stored for up to 60 days. At intervals during storage, ascorbic acid and citric acid concentration, optical density at 420 nm, pH, Brix value, and sensory properties were determined. Optical density at 420 nm increased during frozen storage, indicating darkening; this change was greater for the sweetened than for the non-sweetened samples. This darkening was not detected in sensory tests. The other physico-chemical and sensory properties studied were little affected by freezing or frozen storage, and most properties were similar for sweetened and natural juices.

El-Shiaty et al. (1972) reported that the storage temperature, time, and type of packaging material affected the organoleptic characteristics of the lime juice. Plain tin cans were superior to glass bottles and plastics or polyethylene packages. If plastics or polyethylene containers are to be used for packaging of lime juice concentrate, the storage temperature should not exceed 0°C.

The shelf life of lime juice, made from *C. aurantifolia* picked at maturity, was investigated under various pretreatment, packaging, and storage conditions (El-Ashwah et al., 1981). Lime concentrates were pasteurized or pasteurized and treated with 200 or 300 ppm SO_2before being packed in polyethylene bags. The remaining samples were pasteurized or pasteurized and treated with 200 ppm SO_2 or 200 ppm SO_2 + 10% NaCl and filled into glass bottles. Analyses for TSS, TA, pH, total SO_2, ascorbic acid (AA), FAN, PC, color index (CI), pectin methylesterase activity, and sensory quality (taste, odor, and color) were conducted. TSS, TA, and pH were 8.8%, 7.7%, and 1.9, respectively, for all juices and storage did not affect these values. SO_2 levels decreased during storage, with a reduction of 8% after 1 month and 5% at the end of storage. NaCl did not affect SO_2 retention, which was higher in glass bottles than in plastic bags. Ascorbic acid decreased from 30 to 26.7 mg/100 ml although addition of SO_2 enhanced AA retention. AA retention in glass bottles was greater than that in plastic bags. FAN increased from 18.7 to 21.3 mg/100 ml during pasteurization, yet storage did not affect this value. PC decreased from initial 6% to 4% after pasteurization and it increased in all samples during storage. CI decreased after pasteurization and during storage, although in unpasteurized control samples addition of SO_2 slowed down this decrease especially in plastic bag samples. NaCl accelerated discoloration in a linear manner. PME activity was completely inhibited in all samples after 4 months storage. Organoleptic evaluation showed that pasteurization reduces juice quality slightly, but addition of SO_2 maintained it at acceptable levels for up to 6 months. Juice in glass bottles was unacceptable after 4 months, but addition of NaCl and SO_2 extended its shelf life to 6 months.

El-Ashwah et al. (1974) studied effect of storage on frozen lime juice. Juice was extracted from mature limes with a hand-reamer and strained through a thin cloth. One batch was pasteurized at 76°C for 1 min, and another was not pasteurized. Three preservatives were applied to sub-batches before bottling: (i) 100 ppm sorbic acid, (ii) 1000 ppm sodium benzoate, and (iii) a combination of 500 ppm sodium benzoate and 250 ppm potassium metabisulfite. Storage was carried out at −12.2°C for less than 12 months. Unpasteurized juice retained better flavor for a greater time than pasteurized juice. Partial cloud separation commenced at 6 months for all, except treatment (iii), which remained stable until 12 months. (ii) and (iii) had a better effect on pasteurized juices than (i). Total soluble solids and total titratable acidity remained essentially constant throughout, but ascorbic acid was gradually destroyed. Pasteurization and preservative (iii) helped ascorbic acid retention. Amino-nitrogen levels were inconsistent during storage, and volatile oils gradually declined, irrespective of treatment.

A study was made about the effects of different storage conditions on the keeping quality of concentrated lime juice (Heikal et al., 1972). Juice was mechanically extracted from Baladi limes and concentrated under vacuum to 42% soluble solids. The concentrated lime juice was either (i) bottled without any preservative, or (ii) treated with 0.15% calcium carbonate plus 0.3% potassium metabisulfite, and then bottled. Bottles were stored at room temperature, 0°C or −10°C for less than 30 weeks. No significant change was observed in pH or TA of the concentrated lime juice during processing or storage. Prolonged storage resulted in increased loss of ascorbic acid, the loss being greater at higher temperature. The samples subjected to (ii) and stored at −10°C retained the highest concentration of ascorbic acid; discoloration and off-flavor development were also satisfactorily retarded by this treatment.

FUTURE RESEARCH NEEDS

It would be desirable to propagate the orange trees with as diverse a genetic makeup as possible of both scion and rootstock, to minimize the universal problem of post-processing bitterness and to improve the other quality characteristics. Extraction machinery has greater scope for improvement. Instantaneous extraction processes should be developed that would extract the juice under vacuum preferably by centrifugal forces with minimal incorporation of pulp particles but still keeping a highly acceptable quality of juice. The methods to improve organoleptic acceptability of grapefruit, lemon, and lime juices need to be developed.

Consumer demand for orange juice being as natural as possible and without extensive processing is increasing. Problems related to production of freshly squeezed (non-pasteurized) orange juice are mainly microbial, and its stability depends on strict sanitation practices during production and on storage at low temperature during distribution and retailing. Since shelf life and sensory properties of fresh orange juice depend almost entirely on initial bacterial counts and on storage temperature, importance is given to hygiene and sterilization in the manufacturing plant. Good production practices of raw material and product manufacturing technology should be developed throughout the world to meet the quality requirements of unprocessed juice. Other safety concerns regarding pesticides and mycotoxins in citrus fruit juices are of utmost importance. Pesticides are likely to enter the food chain from environmental or direct exposure in the fruit groves or from pest control applications. Rapid methods for screening and estimation of pesticide residues in citrus juice should be developed.

Inert packaging material with low cost and desirable quality characteristics specific to fresh and frozen juice is the need of the hour. The efforts should be done to commercialize the technologies such as HPP, pulsed electric field, and the processes using carbon dioxide gas under high pressure. Economically viable technologies should be developed to recover the newer and valuable by-products from the wastes of citrus processing industries. All efforts should be concentrated to promote the natural citrus juices as far as possible, vis-à-vis ensuring the safety of product for human consumption.

REFERENCES

Aggarwal, P. and Sandhu, K.S. 2004a. Effect of hydrocolloids on the limonin content of Kinnow juice. Journal of Food Agriculture & Environment 2(1):44–48.

Aggarwal, P. and Sandhu, K.S. 2004b. Effect of hydrocolloids on the quality of Kinnow squash. Journal of Food Science and Technology 41(2):64–75.

Ahmad, N. and Bhatti, M.B. 1971. Studies on the stabilization of cloud in orange juices and concentrates. Agriculture, Pakistan 22(1):41–47.

Albach, R.F., Redman, G.H. and Lime, B.J. 1981. Limonin content of juice from Marrs and Hamlin oranges [*Citrus sinensis* (L.) Osbeck]. Journal of Agricultural and Food Chemistry 29(2):313–315.

Altan, A. 1995. Determination of some technological characteristics of five cultivars of oranges grown in

the Cukurova region for the juice industry. Gida 20(4):215–225.
Angeletti, S. and Moresi, M. 1983. Modelling of multiple-effect falling-film evaporators. Journal of Food Technology 18(5):539–563.
Anon. 1976. Debittering navel orange juice. Food Technology in Australia 28(9):357.
Anon. 1983. New gum suspends fruit pulp without increasing viscosity. Processed Prepared Food 152(3):154.
Anon. 1984. Computers help citrus plant handle 4,300 tons of oranges daily. Food Engineering 56(4):135.
Anon. 1993a. Australian orange juice industry. Food Australia 45(4):163.
Anon. 1993b. How to make lime juice. Food Chain No. 10:15.
Anon. 1993c. Pasteurization revisited. Prepared Foods 162(2):49–50.
Antonio, C. 1992. Machine for extracting juice from citrus fruit, particularly oranges. United States Patent US 5 097 757.
Araki, C., Ito, O. and Sakakibara, H. 1992. Changes of volatile flavor compounds in sweet orange juices by heating. Journal of Japanese Society of Food Science and Technology [Nippon Shokuhin Kogyo Gakkaishi] 39(6):477–482.
Araki, C. and Sakakibara, H. 1991. Changes in the volatile flavor compounds by heating satsuma mandarin (*Citrus unshiu* Marcov.) juice. Agricultural and Biological Chemistry 55(5):1421–1423.
Aravantinos-Zafiris, G. and Oreopoulou, V. 1992. The effect of nitric acid extraction variables on orange pectin. Journal of the Science of Food and Agriculture 60(1):127–129.
Askar, A., El-Samahy, S.K., Abd-El-Baki, M.M., Ibrahim, S.S. and Abd-El-Fadeel, M.G. 1981. Production of lime juice concentrates using the serum–pulp method. Alimenta 20(5):121–128.
Barros, S.M. 1992. Limonin content of Florida packed grapefruit juice. Proceedings of the Florida State Horticultural Society 105:105–108.
Benk, E. 1969. Soya extract as a beverage base and possible adulterant of concentrated orange juice. Sojaextrakte als Getraenkegrundstoffe und moegliche Faelschungsmittel fuer Orangensaftkonzentrate. Brauereitechniker 21(3):18–20.
Benk, E. and Bergmann, R. 1973. The red-fleshed grapefruit and its juice. Industrielle Obst und Gemueseverwertung 58(15):437–439.
Bianchi, G., Setti, L., Pifferi, P.G. and Spagna, G. 1995. Limonin removal by free and immobilized cells. Cerevisia 20(2):41–46.
Bielig, H.J. 1973. Tin uptake in canned orange juices. Chemie Mikrobiologie Technologie der Lebensmittel 2:129–136. (German).
Bielig, H.J. and Askar, A. 1974. Aroma deterioration during the manufacture and the storage of orange juice in bottles. IV-International Congress of Food Science and Technology 1b:33–34.
Biesel, C.G. 1951. Working out the fruit bug. Food Engineering 23(11):82–84, 204, 205, 207.
Blair, J.S., Godar, E.M., Masters, J.E. and Riester, D.W. 1952. Flavour deterioration of stored canned orange juice. Food Research 17:235–260.
Bomben, J.L., Kitson, J.A. and Morgan, A.J. Jr. 1966. Vacuum stripping of aroma. Food Technology 20:1219–1222.
Braddock, R.J. 1999. Juice processing operations. In Handbook of Citrus By-products and Processing Technology. John Wiley & Sons Inc., p. 46.
Brat, P., Rega, B., Alter, P., Reynes, M. and Brillouet, J.M. 2003. Distribution of volatile compounds in the pulp, cloud and serum of freshly squeezed orange juice. Journal of Agricultural and Food Chemistry 51(11):3442–3447.
Brent, J.A., Dubois, C.W. and Huffman, C.F. 1968. Essence recovery. United States Patent 3 248 233.
Brewster, L.C., Hasegawa, S. and Maier, V.P. 1976. Bitterness prevention in citrus juices. Comparative activities and stabilities of the limonoate dehydrogenases from Pseudomonas and Arthtobacter. Journal of Agricultural and Food Chemistry 24(1):21–24.
Casas, A., Rodrigo, M.I. and Mallest, D. 1979. Prevention of limonin precursor accumulation in Washington navel oranges by treating the trees with triethylamine derivatives. Revista de Agroquimica y Tecnologia de Alimentos 19(4):513–519.
Chandler, B.V. 1977. One of the '101 most interesting problems in food science'—bitterness in orange juice, a case history. Food Technology in Australia 29(8) 303–305, 307–311.
Chandler, B.V. and Johnson, R.L. 1977. Cellulose acetate as a selective sorbent for limonin in orange juice. Journal of the Science of Food and Agriculture 28(10):875–884.
Chandler, B.V. and Johnson, R.L. 1979. New sorbent gel forms of cellulose esters for debittering citrus juices. Journal of the Science of Food and Agriculture 30(8):825–832.
Chandler, B.V. and Nicol, K.J. 1983. Alternative cultivars for orange juice production. CSIRO Food Research Quarterly 42(2):29–36.
Charara, Z.N., Williams, J.W., Schmidt, R.H. and Marshall, M.R. 1992. Orange flavor absorption into

various polymeric packaging materials. Journal of Food Science 57(4):963–966, 972.

Chen, C.S., Carter, R.D. and Buslig, B.S. 1979. Energy requirements for the TASTE citrus juice evaporator. (In 'Changing energy use futures' [see FSTA (1981) 13 3A125].) Lecture pp. 1841–1848.

Clements, R.L. 1964. Organic acids in citrus fruits. 1 Varietal differences. Journal of Food Science 29:276, 281.

Cohen, E. 1982. Seasonal variability of citrus juice attributes and its effect on the quality control of citrus juice. Zeitschrift fuer Lebensmittel Untersuchung und Forschung 175(4):258–261.

Couture, R. and Rouseff, R. 1992. Debittering and deacidifying sour orange (*Citrus aurantium*) juice using neutral and anion exchange resins. Journal of Food Science 57(2):380–384.

Crandall, P.G., Davis, K.C., Carter, R.D. and Sadler, G.D. 1988. Viscosity reduction by homogenization of orange juice concentrate in a pilot TASTE evaporator. Journal of Food Science 53(5):1477–1481.

Cruess, W.V. 1997. Commercial Fruit and Vegetable Products. Allied Scientific Publishers, Bikaner, India, pp. 767–773.

Cruse, R.R., Lime, B.J. and Hensz, R.A. 1979. Pigmentation and color comparison of Ruby Red and Star Ruby grapefruit juice. Journal of Agricultural and Food Chemistry 27(3):641–642.

Cruse, R.R., Wiegand, C.L. and Swanson, W.A. 1982. The effects of rainfall and irrigation management on citrus juice quality in Texas. Journal of the American Society for Horticultural Science 107(5):767–770.

Daito, H., Tominaga, S., Ono, S. and Morinaga, K. 1981. Yield and fruit quality at various locations within canopies of differently trained satsuma mandarin trees. Journal of the Japanese Society for Horticultural Science 50(2):131–142.

Devlieghere, F., Vermeiren, L., Jacobs, M. and Debevere, J. 2000. The effectiveness of hexamethylenetetramine-incorporated plastic for the active packaging of foods. Packaging Technology & Science 13(3):117–121.

Dinnage, D.F. 1970. Multi-stage evaporation gives 4:1 steam efficiency. Food Engineering 42(4):62–65.

Donsi, G., Ferrari, G., Matteo, M.-di. and Bruno, M.C. 1998. High-pressure stabilization of lemon juice. Italian Food & Beverage Technology 14:14–16.

Drawert, F., Leupold, G. and Pivernetz, H. 1980. Quantitative determination of rutin, hesperidin and naringin in orange juice by gas liquid chromatography. Quantitative gaschromatographische Bestimmung von Rutin, Hesperidin and Naringin in Orangensaft. Chemie Mikrobiologie Technologie der Lebensmittel 6(6):189–191.

Duerr, P. and Schobinger, U. 1981. The contribution of some volatiles to the sensory quality of apple and orange juice odour. (In 'Flavour '81' G [see FSTA (1983) 15 G3T130].) pp. 179–193.

Duerr, P., Schobinger, U. and Waldvogel, R. 1981. Aroma quality of orange juice after filling and storage in soft packages and glass bottles. Alimenta 20(4):91–93.

Dupaigne, P. 1971. The determination of percentage of juice in fresh fruits. Fruits 26(4):305–308.

Edwards, D.J. and Marr, I.M. 1990. Previously unreported sesquiterpenes of lime oil (*Citrus latifolia* Tanaka). Journal of Essential Oil Research 2(3):137–138.

El-Ashwah, E.T., Tawfik, M.A., El-Hashimy, F.S., Raouf, M.S. and Sarhan, M.A.I. 1981. Chemical and physical studies on preserved Benzahir lime juice. Sudan Journal of Food Science and Technology 13:64–68.

El-Ashwah, F.A., El-Manatawy, H.K., Habashy, H.N. and El-Shiaty, M.A. 1974. Effect of storage on fruit juices: frozen lime juice. Agricultural Research Review 52(9):79–85.

Elias, A.N., Foda, M.S. and Attia, L. 1984. Production of pectin and pigments from orange peels by using microbial enzymes. Egyptian Journal of Food Science 12(1/2):159–162.

El-Shiaty, M.A., El-Ashwah, F.A. and Habashy, H.N. 1972. Effect of storage on fruit juices. I. Study of some factors affecting lime juice storage. Agricultural Research Review 50(5):215–229.

El-Wakeil, F.A., Hamed, H.G.E., Heikal, H.A. and Foda, I.O. 1974. Detection of accepted natural juices in carbonated beverages. II. Studies with 'Baladi' orange. Egyptian Journal of Food Science 2(1):59–69.

El-Zeany, B.A. 1977. Isolation of the fraction responsible for the bad odour and flavour of oxidised orange oil. Egyptian Journal of Food Science 3(1/2):73–79.

El-Zeftawi, B.M. and Garrett, R.G. 1978. Effects of ethephon, GA and light exclusion on rind pigments, plastid ultrastructure and juice quality of Valencia oranges. Journal of Horticultural Science 53(3):215–223.

Erazo, G.S., Beuchemin, C.L.F. and Abbot, C.F.J. 1984. Preliminary study on processing of oranges of the cv. Washington, Thompson, Chilena and Valencia. Alimentos 9(2):9–16.

Fallico, B., Lanza, M.C., Maccarone, E., Asmundo, C.N. and Rapisarda, P. 1996. Role of hydroxycinnamic acids and vinylphenols in the

flavor alteration of blood orange juices. Journal of Agricultural and Food Chemistry 44(9):2654–2657.
Farnworth, E.R., Lagace, M., Couture, R., Yaylayan, V. and Stewart, B. 2001. Thermal processing, storage conditions, and the composition and physical properties of orange juice. Food Research International 34(1):25–30.
Fellers, P.J. 1991. The relationship between the ratio of degrees Brix to percent acid and sensory flavor in grapefruit juice. Food Technology 45(7):68, 70, 72–75.
Filho, J.G., Vitali, A.A., Viegas, F.C.P. and Rao, M.A. 1984. Energy consumption in a concentrated orange juice plant. Journal of Food Process Engineering 7(2):77–89.
Fogli, A. 1975. Determination of the juice content of beverages by the formol number. Essenze Derivati Agrumari 45(3/4):308–314.
Foley, D.M., Pickett, K., Varon, J., Lee, J., Min, D.B., Caporaso F. and Prakash, A. 2002. Pasteurization of fresh orange juice using gamma irradiation: microbiological, flavor and sensory analyses. Journal of Food Science 67(4):1495–1501.
FPO. 1955. Fruit Products Order. Government of India, New Delhi.
Freed, M., Brenner, S. and Wodicka, V.O. 1949. Prediction of thiamine and ascorbic acid stability in stored canned foods. Food Technology 3:148–151.
Frometa, E. and Echazabal, J. 1988. Influence of age and cultivar on the juice characteristics of early oranges. Agrotecnia de Cuba 20(1):71–75.
Gaetano, O. 1975. Characteristics of the juice of Sanguinello oranges from Ragusa province. Essenze Derivati Agrumari 45(1):34–37.
Gallasch, P.T. 1978. Effect of time of harvest on alternate cropping yields and fruit quality of Valencia orange trees. Australian Journal of Experimental Agriculture and Animal Husbandry 18(92):461–464.
Garbagnati, G. 1978. Granulated beverages. Industrie delle Bevande 7(4):258–260.
Garti, N., Aserin, A. and Azaria, D. 1991. A clouding agent based on modified soy protein. International Journal of Food Science & Technology 26(3):259–270.
Gentschev, L., Vladimirov, G. and Grantschev, D. 1991. Modelling and optimum conditions for the extraction of citrus pectin. Modellierung and Optimierung des Extraktionsvorganges von Citruspektin. Fluessiges Obst 58(2):65–67.
Di Giacomo, A., Bovalo, F. and Postorno, E. 1975a. Industrially produced orange juice from Piana di Rosarno fruit (1972–1973). IV. Essenze Derivati Agrumari 45(1):42–57.
Di Giacomo, A., Calvarano, M., Calvarano, I. and Bovalo, F. 1976. Limonene content of orange juice. III. Role of processing technology. Essenze-Derivati-Agrumari 46(3):247–263.
Di Giacomo, A., Calvarano, M., Calvarano, I., Giacomo, Gdi. and Belmusto, G. 1989. Juice of Italian coloured oranges. Essenze Derivati Agrumari 59(3):273–289.
Di Giacomo, A., Calvarano, M. and Tribulato, E. 1977. Limonene content of orange juice. IV. The effect of the rootstock on 'Valencia Late' and 'Moro' cultivars. Essenze Derivati Agrumari 47(2):156–166.
Di Giacomo, A., Postorino, E. and Bovalo, F. 1975b. Industrially produced orange juice from Piana di Rosarno (1973–1974). V. Essenze Derivati Agrumari 45(3/4):315–335.
Giannone, L. and Matliano, V. 1977. Suitability of some orange cultivars for production of deep frozen juice. Industria Conserve 52(2):100–104.
Gil, I.A., Gil, M.I. and Ferreres, F. 2002. Effect of processing techniques at industrial scale on orange juice antioxidant and beneficial health compounds. Journal of Agricultural and Food Chemistry 50(18):5107–5114.
Goodner, J.K., Braddock, R.J., Parish, M.E. and Sims, C.A. 1999. Cloud stabilization of orange juice by high pressure processing. Journal of Food Science 64(4):699–700.
Guadagni, D.G., Horowitz, R.M., Gentili, B. and Maier, V.P. 1977. Method for reducing bitterness in citrus juices. United States Patent 4 031 265.
Guintrand, P. 1982. Automatic presser and distributor of juice from fresh natural fruit. French Patent Application FR 2 498 056 A1.
Habegger, M. and Sulser, H. 1974. Determination of the natural fruit content in orange juices and beverages on the basis of the free amino acid content. Bestimmung des natuerlichen Fruchtanteiles in Orangensaeften und getraenken anhand der freien Aminosaeuren. Lebensmittel Wissenschaft ù Technologie 7(3):182–185.
Hadj Mahammed, M. and Meklati, B.Y. 1987. Qualitative determination of polymethoxylated flavones in Valencia orange peels oil and juice by LC-UV/VIS and LC-MS techniques. Lebensmittel Wissenschaft und Technologie 20(3):111–114.
Harding, P.L., Winston, J.R. and Fisher, D.F. 1940. Seasonal changes in Florida oranges. United States Department of Agriculture, Technical Bulletin 753.
Hasegawa, S., Dillberger, A.M. and Choi, G.Y. 1984. Metabolism of limonoids: conversion of nomilin to obacunone in *Corynebacterium fascians*. Journal of Agricultural and Food Chemistry 32(3):457–459.

Hasegawa, S., Patel, M.N. and Snyder, R.C. 1982. Reduction of limonin bitterness in navel orange juice serum with bacterial cells immobilized in acrylamide gel. Journal of Agricultural and Food Chemistry 30(3):509–511.

Hashimoto, K., Matsunaga, M., Oikawa, H., Suzuki, T. and Watanabe, E. 1995. Effects of dissolved oxygen and containers on aseptic orange juice. IFT Annual Meeting 1995, p. 42.

Hasselbeck, U., Ruholl, T., Popper, L. and Knorr, D. 1992. Fruit juice pasteurization under reduced thermal load. Fruchtsaftpasteurisation mit reduzierter thermischer Belastung. Fluessiges Obst 59(10):592–593.

Haugaard, V.K., Weber, C.J., Danielsen, B. and Bertelsen, G. 2002. Quality changes in orange juice packed in materials based on polylactate. European Food Research and Technology 214(5):423–428.

Heikal, H.A., El-Manawaty, H., Shaker, G. and Gamali, L. 1972. Concentration of citrus juices. I. Factors affecting the quality and stability of concentrated lime juices by the vacuum method. Agricultural Research Review 50(4):139–147.

Heimhuber, B., Galensa, R. and Herrmann, K. 1988. High-performance liquid chromatographic determination of polymethoxylated flavones in orange juice after solid-phase extraction. Journal of Chromatography 439(2):481–483.

Hendrix, D.L. and Ghegan, R.C. 1980. Quality changes in bulk stored citrus concentrate made from freeze-damaged fruit. Journal of Food Science 45(6):1570–1572.

Hensz, R.A. 1971. Star Ruby, a new deep-red-fleshed grapefruit variety with distinct tree characteristics. Journal of the Rio Grande Valley Horticultural Society 25:54–58.

Herman, Z., Fong, C.H., Ou, P. and Hasegawa, S. 1990. Limonoid glucosides in orange juices by HPLC. Journal of Agricultural and Food Chemistry 38(9):1860–1861.

Herrera, M.V., Matthews, R.F. and Crandall, P.G. 1979. Evaluation of a beverage clouding agent from orange pectin pomace leach water. Proceedings of the Florida State Horticultural Society 92:151–153.

Hirose, K., Harte, B.R., Giacin, J.R., Miltz, J. and Stine, C. 1988. Sorption of D-limonene by sealant films and effect on mechanical properties. (In 'Food and packaging interactions' [see FSTA (1989) 21 2F3].) ACS Symposium Series 365, 28–41, 11.

Hodgins, A.M., Mittal, G.S. and Griffiths, M.W. 2002. Pasteurization of fresh orange juice using low-energy pulsed electrical field. Journal of Food Science 67(6):2294–2299.

Hodgson, R.W. 1967. The Citrus Industry, Rev. Ed., Vol. 1. Reuther, W., Webber H.J. and Batchelor, L.D. Eds., University of California Press, Berkeley, pp. 431–592.

Huggart, R.L., Rouse, A.H. and Moore, E.L. 1975. Effect of maturity, variety and processing on color, cloud, pectin and water-insoluble solids of orange juice. Proceedings of the Florida State Horticultural Society 88:342–345.

Hyoung, S.L. and Coates, G.A. 2003. Effect of thermal pasteurization on Valencia orange juice color and pigments. Lebensmittel Wissenschaft und Technologie 36(1):153–156.

Ibrahim, M.M., Badei, A.Z.M. and El-Wakeil, F.A. 1990. Stabilization of the colloidal state of citrus carbonated beverages by application of some stabilizers. Egyptian Journal of Food Science 16(1/2):1–7.

Ikeda, R.M., Stanlley, W.L., Vannier, S.H. and Rolle, L.A. 1961. Deterioration of lemon oil. Formation of *p*-cymene from γ terpinene. Food Technology 15:379–380.

Isaacs, A.R. 1980. Citrus processing research in Queensland. Australian Citrus News 56 (September):10–11.

Johnson, J.R., Braddock, R.J. and Chen, C.S. 1996. Flavor losses in orange juice during ultrafiltration and subsequent evaporation. Journal of Food Science 61(3):540–543.

Johnson, R.L. 1981. The reactivation of 'exhausted' cellulose acetate gel beads used commercially for debittering orange juice. Journal of the Science of Food and Agriculture 32(6):608–612.

Johnson, R.L. and Chandler, B.V. 1978. Removal of limonin from bitter navel orange juice. Proceedings of the International Society of Citriculture, pp. 43–44.

Johnson, R.L. and Chandler, B.V. 1985. Economic feasibility of adsorptive de-acidification and debittering of Australian citrus juices. CSIRO Food Research Quarterly 45(2):25–32.

Jordan, M.J., Goodner, K.L. and Laencina, J. 2003. Deaeration and pasteurization effects on the orange juice aromatic fraction. Lebensmittel Wissenschaft und Technologie 36(4):391–396.

Kacem, B., Cornell, J.A., Marshall, M.R., Shireman, R.B. and Matthews, R.F. 1987. Nonenzymatic browning in aseptically packaged orange drinks: effect of ascorbic acid, amino acids and oxygen. Journal of Food Science 52(6):1668–1672.

Kefford, J.F. 1959. Chemical constituents of citrus juices. Advances in Food Research 9:351.

Kefford, J.F. 1973. Citrus fruits and processed citrus products in human nutrition. World Review of Nutrition and Dietetics 13:60–120.

Kefford, J.F., McKenzie, H.A. and Thompson, P.C. 1959. Effects of oxygen on quality and ascorbic acid retention in canned and frozen orange juices. Journal of Science of Food and Agriculture 10:51–63.
Kelso, L.R., Rowan, C.M. Jr. and Holladay, K.L. 1980. The economics of using mechanical vapor recompression evaporators to concentrate orange juice. Activities Report 32(2):108–123.
Kenawi, M.A., Shekib, L.A. and El-Shimi, N.M. 1994. The storage effects of calcium-fortified orange juice concentrate in different packaging materials. Plant Foods for Human Nutrition 45(3):265–275.
Kesterson, J.W., Hendrickson, R. and Braddock, R.J. 1971. Florida citrus oils. University of Florida, Agricultural Experimental Station Bulletin 749.
Kew, T.J., Veldhuis, M.K., Bissett, O.W. and Patrick, R. 1957. The effect of time and temperature of pasteurization on the quality of canned citrus juices. United States Department of Agriculture ARS- 72-6.
Khan, S.A. and Khan, R. 1971. Canning and bottling of different grades of sweet orange juices. Science and Industry, Pakistan 8(2):210–213.
Khurdiya, D.S. 1990. Orange concentrate based carbonated beverage. Journal of Food Science and Technology, India 27(6):394–396.
Kieser, A.H. and Havighorst, C.R. 1952. They use every part of fruit in full product line. Food Engineering 24(9):114–116, 156–159.
Kimball, D.A. 1990. The industrial solution of citrus juice bitterness. Perfumer & Flavorist 15(2):41–44.
Kimball, D.A. 1991. Citrus Processing Quality Control and Technology. Van Nostrand Reinhold, New York, pp. 7–135.
Kimball, D.A. and Norman, S.I. 1990a. Changes in California navel orange juice during commercial debittering. Journal of Food Science 55(1):273–274.
Kimball, D.A. and Norman, S.I. 1990b. Processing effects during commercial debittering of California navel orange juice. Journal of Agricultural and Food Chemistry 38(6):1396–1400.
Kodama, M., Akamatsu, S., Bessho, Y., Owada, A. and Kubo, S. 1977. Effect of nitrogen fertilizing on the composition of Satsuma mandarin juice. Journal of Japanese Society of Food Science and Technology [Nippon Shokuhin Kogyo Gakkaishi] 24(8):398–403.
Konno, A., Miyawaki, M., Misaki, M. and Yasumatsu, K. 1981. Bitterness reduction of citrus fruits by beta-cyclodextrin. Agricultural and Biological Chemistry 45(10):2341–2342.
Koo, R.C.J. and Smajstrla, A.G. 1985. Effects of trickle irrigation on fruit production and juice quality of 'Valencia' orange. Citrus Industry 66(1):14–15, 17, 19.
Lal, G, Siddappa, G.S. and Tandon, G.L. 1986. Preservation of Fruits and Vegetables. Indian Council of Agricultural Research, New Delhi, pp. 313.
Lee, H.S. 1997. Issue of color in pigmented grapefruit juice. Fruit Processing 7(4):132–135.
Lee, H.S. and Barros, S.M. 1996. Evaluation of silica-coated packing materials for refrigerated storage of orange juice. Fruit Processing 6(9):363–365.
Lee, H.S. and Coates, G.A. 1999. Thermal pasteurization effects on color of red grapefruit juices. Journal of Food Science 64(4):663–666.
Lee, H.S. and Coates, G.A. 2002. Characterization of color fade during frozen storage of red grapefruit juice concentrates. Journal of Agricultural and Food Chemistry 50(14):3988–3991.
Lee, J.-Y., Lin, Y.-S., Chang, H.-M., Chen, W. and Wu, M.-C. 2003. Temperature–time relationships for thermal inactivation of pectinesterases in orange juice. Journal of the Science of Food and Agriculture 83(7):681–684.
Levi, A., Flavian, S., Harel, S., Ben-Gera, I., Stern, F. and Berkovitz, S. 1974. The Bitter Principle and the Prevention of Bitterness in Shamouti Orange Juice Products. Special Publication, Volcani Center, Israel No. 29, 25 pp.
Lin, J., Rouseff, R.L., Barros, S. and Naim, M. 2002. Aroma composition changes in early season grapefruit juice produced from thermal concentration. Journal of Agricultural and Food Chemistry 50(4):813–819.
Loeffler, C. 1996. Possibilities for manufacture of orange juice concentrate. Moeglichkeiten zur Herstellung von Orangensaftkonzentrat. Fluessiges Obst 63(12):695, 698–701.
Lotong, V., Chambers, E. and Chambers, D.H. 2003. Categorization of commercial orange juice based on flavor characteristics. Journal of Food Science 68(2):722–725.
Maier, V.P. and Beverly, G.D. 1968. Limonin monolactone, the nonbitter precursor responsible for delayed bitterness in certain citrus juices. Journal of Food Science 33(5):488–492.
Manlan, M., Matthews, R.F., Rouseff, R.L., Littell, R.C., Marshall, M.R., Moye, H.A. and Teixeira, A.A. 1990. Evaluation of the properties of polystyrene divinylbenzene adsorbents for debittering grapefruit juice. Journal of Food Science 55(2):440–445, 449.
Mannheim, C.H. and Passy, N. 1979. The effect of deaeration methods on quality attributes of bottled

orange juice and grapefruit juice. Confructa 24(5/6):175–187.

Mans, J. 1983. New Sunkist plant capitalizes on latest equipment and computer controls. Processed Prepared Food 152(5):87–89.

Manso, M.C., Ahrne, L.M., Oste, R.E. and Oliveira, F.A.R. 1996. Dissolved oxygen concentration changes during storage of packaged orange juice. United States of America, Institute of Food Technologists 1996 Annual Meeting. 1996 IFT annual meeting:book of abstracts, p. 128 ISSN 1082–1236.

Maraulja, M.D., Blair, J.S., Olsen, R.W. and Wenzel, F.W. 1973, publ. 1974. Furfural as an indicator of flavour deterioration in canned citrus juices. Proceedings of the Florida State Horticultural Society 86:270–275.

Maraulja, M.D. and Dougherty, M.H. 1975. Effect of maturity, variety, and processing on chloramine-*T* values and total amino acid content of orange juices. Proceedings of the Florida State Horticultural Society 88:346–349.

Martin, J.J., Solanes, E., Bota, E. and Sancho, J. 1995. Chemical and organoleptic changes in pasteurized orange juice. Alimentaria No. 261:59–63, 31.

Di Mauro, A., Fallico, B., Passerini, A. and Maccarone, E. 2000. Waste water from citrus processing as a source of hesperidin by concentration on styrene-divinylbenzene resin. Journal of Agricultural and Food Chemistry 48(6):2291–2295.

McCance, R.A. and Widdowson, E.M. 2002. Chemical Composition of Foods. Agribios, Jodhpur, India. p. 126.

McKenna, R.J., Keller, D.J. and Bibeau, L.S. 1991. Process for preserving lemon juice utilizing a non-sulfite preservative. United States Patent US 5 021 251.

Miller, W.M. and Hendrix, C.M. 1996. Fruit quality inspection, handling, sampling and evaluation. In Quality Control Manual for Citrus Processing Plants, Vol. 3, Chap. 7. Redd, J.B., Shaw, P.E., Hendrix, C.M. and Hendrix, D.L. Eds., AgScience, Auburndale, FL, pp. 233–251.

Min, S., Jin, Z.T., Min, S.K., Yeom, H. and Zhang, Q.H. 2003. Commercial-scale pulsed electric field processing of orange juice. Journal of Food Science 68(4):1265–1271.

Mishra, P. and Kar, R. 2003. Treatment of grapefruit juice for bitterness removal by Amberlite IR 120 and Amberlite IR 400 and alginate entrapped naringinase enzyme. Journal of Food Science 68(4):1229–1233.

Mohsen, S.M., El-Hashimy, F.S.A. and El-Ashmawy, A.G. 1986a. Effect of method of extraction and pasteurization on orange juice properties and its volatile components. Egyptian Journal of Food Science 14(2):301–312.

Mohsen, S.M., El-Hashimy, F.S.A. and El-Ashmawy, A.G. 1986b. Effect of method of extraction and pasteurization on grapefruit juice properties and its volatile components. Egyptian Journal of Food Science 14(2):397–407.

Money, R N. and Christian, W.A. 1950. Analytical data of some common fruit. Journal of Science of Food and Agriculture 1:8–12.

Morgan, D.A., Veldhuis, M.K., Eskew, R.K. and Phillips, G.W.M. 1953. Studies on the recovery of essence from orange juice. Food Technology 7:332.

Morioka, S. 1987. Influences of fruit load and fruit thinning treatment on fruit character, shoot growth and flower bud formation in the following season in young Satsuma mandarin trees. Journal of the Japanese Society for Horticultural Science 56(1):1–8.

Moshonas, M.G. and Shaw, P.E. 1977. Evaluation of juice flavour and peel oil composition of ethylene-treated (degreened) Hamlin oranges. International Flavours and Food Additives 8(4):147, 152.

Moshonas, M.G. and Shaw, P.E. 1989a. Changes in composition of volatile components in aseptically packaged orange juice during storage. Journal of Agricultural and Food Chemistry 37(1):157–161.

Moshonas, M.G. and Shaw, P.E. 1989b. Flavor evaluation and volatile flavor constituents of stored aseptically packaged orange juice. Journal of Food Science 54(1):82–85.

Moshonas, M.G. and Shaw, P.E. 2000. Changes in volatile flavour constituents in pasteurized orange juice during storage. Journal of Food Quality 23(1):61–71.

Mottar, J. 1989. The usefulness of polypropylene for the aseptic packaging of orange juices. Zeitschrift fuer Lebensmittel Untersuchung und Forschung 189(2):119–122.

Mowzini, A., Maltini, E. and Bertolo, G. 1974. Optimal processing conditions for freeze-drying of concentrated orange juices. Scienza e Tecnologia degli Alimenti 4(6):335–340.

Muller, J.G. 1967. Freeze concentration of food liquids: theory practice and economics. Food Technology 21(1):49–52, 54–56, 58, 60–61.

Nagy, S., Shaw, P.E. and Veldhuis, M.K. 1977. Citrus Science and Technology, Vol. 2. Fruit Production, Processing Practices, Derived Products and Personnel Management. The AVI Publishing Company, Inc., Westport, CT, pp. 1–127, 188–199.

Naim, M., Schutz, O., Zehavi, U., Rouseff, R.L. and Haleva T.E. 1997. Effects of orange juice fortification with thiols on p-vinylguaiacol formation, ascorbic-acid degradation, browning, and acceptance during pasteurization and storage under moderate conditions. Journal of Agricultural and Food Chemistry 45(5):1861–1867.

Narayanamurthy, V. and Sarma, P.K. 1977. Falling film evaporators—A design equation for heat transfer rate. Canadian Journal of Chemical Engineering 55:732–735.

Navarro, J.L., Diaz, L.S. and Gasque, F. 1983. Determination of limonin in orange juice by HPLC. Revista de Agroquimica y Tecnologia de Alimentos 23(2):276–280.

Nikdel, S. and MacKellar, D.G. 1993. Continuous pasteurization of citrus juice with microwave heating. Fluessiges Obst 60(12) (Fruit Processing 3(12):433–435).

Nishimura, T., Kometani, T., Okada, S., Kobayashi, Y. and Fukumoto, S. 1998. Inhibitory effects of hesperidin glycosides on precipitation of hesperidin. Journal of Japanese Society of Food Science and Technology (Nippon Shokuhin Kagaku Kogaku Kaishi) 45(3):186–191.

Nisperos, M.O. and Robertson, G.L. 1982. Removal of naringin and limonin from grapefruit juice using polyvinylpyrrolidone. Philippine Agriculturist 65(3):275–282.

Noaman, M.A. and Husssein, M.A. 1973. Some physical characters and chemical composition of important commercial varieties of sweet orange. Research Bulletin, Faculty of Agriculture (K.E.S.), Tanta University No. 3, 13 pp.

Nongluk, C., Supaporn, C., Douglas, P. and Wilai, L. 2001. Simulation of an agitated thin film evaporator for concentrating orange juice using AspenPlusRegistered. Journal of Food Engineering 47(4):247–253.

Norman, S.I. 1990. Juice enhancement by ion exchange and adsorbent technologies. In Production and Packaging of Non-Carbonated Fruit Juices and Fruit Beverages. Hicks, D. Ed., Van Nostrand Reinhold, New York, NY, pp. 259–260.

Norman, S.I., Stringfield, R.T. and Gopsill, C.C. 1990. Removal of bitterness from citrus juices using a post-crosslinked adsorbent resin. United States Patent 4 965 083.

Nunez, J.M., Laencina, J. and Saura, D. 1989. Effect of adding chemical reducing agents for storing concentrated lemon juice. Essenze Derivati Agrumari 59(4):386–387.

Nursten, H.E. and Williams, A.A. 1967. Fruit aromas: a survey of compounds identified. Chemistry and Industry 486–497.

Ohta, H., Tonohara, K., Watanabe, A., Iino, K. and Kimura, S. 1982. Flavor specificities of Satsuma mandarin juice extracted by a new-type screw press extraction system. Agricultural and Biological Chemistry 46(5):1385–1386.

Omori, Y., Takanaka, A., Akeda, Y. and Furuya, T. 1973. Experimental studies on toxicity of tin in canned orange juice. Journal Food Hygienic Society 14:69–74.

Ooghe, W. and Detavernier, C. 1999. Flavonoids as authenticity markers for *Citrus sinensis* juice. Fruit Processing 9(8):308–313.

Owens, H.W., McCready, R.M. and Maclay, W.D. 1949. Gelation characteristics of acid–precipitated pectinates. Food Technology 3:77–82.

Paik, J.S. and Venables, A.C. 1991. Analysis of packaged orange juice volatiles using headspace gas chromatography. Journal of Chromatography 540(1/2):456–463.

Parish, M.E. 1998. Orange juice quality after treatment by thermal pasteurization or isostatic high pressure. Lebensmittel Wissenschaft und Technologie 31(5):439–442.

Pedrao, M.R., Beleia, A., Modesta, R.C.D. and Prudencio Ferreira, S.H. 1999. Sensory and physicochemical stability of frozen Tahiti lime juice, natural and sweetened. Ciencia e Tecnologia de Alimentos 19(2):282–286.

Pereira de Almeida, R. 1974. Freeze-drying of orange juice. Reordenamento No. 32:39–43.

Petersen, M.A., Tonder, D. and Poll, L. 1998. Comparison of normal and accelerated storage of commercial orange juice—changes in flavour and content of volatile compounds. Food Quality and Preference 9(1/2):43–51.

Pieper, G., Borgudd, L., Ackermann, P. and Fellers, P. 1992. Absorption of aroma volatiles of orange juice into laminated carton packages did not affect sensory quality. Journal of Food Science 57(6):1408–1411.

Pieper, G. and Petersen, K. 1995. Free fatty acids from orange juice absorption into laminated cartons and their effects on adhesion. Journal of Food Science 60(5):1088–1091.

Pinera, R., Ferandez, M., Pino, J.A., Garcia, A.L. and Nunez, M. 1995. Evaluation of three early Cuban orange crops obtained by selection. Alimentaria No. 259:21–23.

Pino, J. 1982. Correlation between sensory and gas-chromatographic measurements on orange volatiles. Acta Alimentaria 11(1):1–9.

Pino, J. 1986. Changes caused by storage temperature on volatile constituents of concentrated grapefruit juice. Tecnologia Quimica 7(2):67–72, 87.

Pino, J., Ramos, M., Sanchez, S. and Torricella, R. 1987. Changes in orange juice during production of frozen concentrate and ways to increase its quality. Tecnologia Quimica 8(2):6–13, 81.

Polizzi, F., Gormley. T., Kavolius, L. and LaBell, F. 1987. Switch from batch to continuous increases juice production 30%. Food Processing, USA 48(6):151–152.

Prakash, S., Singhal, R.S. and Kulkarni, P.R. 2002. Enzymic debittering of Indian grapefruit (*Citrus paradisi*) juice. Journal of the Science of Food and Agriculture 82(4):394–397.

Praschan, V.C. 1951. Chemical engineering in the frozen food industry. Chemical Engineering Progress 47:325–330.

Pruthi, J.S. and Lal, G. 1951. Preservation of citrus fruit juices. Journal of Science and Industrial Research 10B:36–41.

Pruthi, J.S., Manan, J.K., Teotia, M.S., Radhakrishna Setty, G., Eipeson, W.E., Saroja, S. and Chikkappaji, K.C. 1984. Studies on the utilization of Kinnow and Malta oranges. Journal of Food Science and Technology, India 21(3):123–127.

Puglia, J.A. and Harper, D.P. 1996. Deoiling single-strength orange juice. Transactions of the Citrus Engineering Conference No:42, 27–44.

Pulley, G.N. and von Loesecke, H.W. 1939. Gases in the commercial handling of citrus juices. Industrial and Engineering Chemistry 31:1275–1278.

Pupin, A.M., Dennis, M.J. and Toledo, M.C.F. 1998. Flavanone glycosides in Brazilian orange juice. Food Chemistry 61(3):275–280.

Qiu, X., Sharma, S., Tuhela, L., Jia, M. and Zhang, Q.H. 1998. An integrated PEF pilot plant for continuous nonthermal pasteurization of fresh orange juice. Transactions of the ASAE 41(4):1069–1074.

Ranganna, S., Gobindarajan, V.S. and Ramanna, K.V. 1983. Citrus fruits. II. Chemistry, technology and quality evaluation. B. Technology. CRC Critical Reviews in Food Science and Nutrition 19:1–98.

Ray, K., Raychowdhury, U. and Chakraborty, R. 2003. Physico-chemical characteristics of lemon juice clarified through ultrafiltration membrane. Journal of Food Science and Technology 40(2):194–196.

Riester, D.W., Braun, O.G. and Pearce, W.E. 1945. Why canned citrus juices deteriorate in storage. Food Industry 17:742–744, 850, 852, 854, 856, 858.

Ribeiro, M.H.L., Silveira, D. and Ferreira Dias, S. 2002. Selective adsorption of limonin and naringin from orange juice to natural and synthetic adsorbents. European Food Research and Technology 215(6):462–471.

Risch, S.J. and Hotchkiss, J.H. 1991. Food and packaging interactions. II. ACS-Symposium Series No. 473, xv + 262pp. ISBN 0-8412-2122-7.

Rodopulo, A.K. 1988. Aromatic compounds in orange juice. Pishchevaya Promyshlennost' No. 12:24–25.

Rodrigo, M.I., Casas, A. and Mallent, D. 1978. Factors affecting the limonin precursor content in Washington navels. II. Influence of nitrogen, phosphorus and potassium fertilization. Revista de Agroquimica y Tecnologia de Alimentos 18(2):193–198.

Rodrigo, M.I., Mallent, D. and Casas, A. 1985. Relationship between the acid and limonin content of Washington navel orange juices. Journal of the Science of Food and Agriculture 36(11):1125–1129.

Roig, M.G., Bello, J.F., Rivera, Z.S. and Kennedy, J.F. 1994. Possible additives for extension of shelf-life of single-strength reconstituted citrus juice aseptically packaged in laminated cartons. International Journal of Food Sciences and Nutrition 45(1):15–28.

Rothschild, G., Vliet, C. and Karsenty, A. 1975. Pasteurization conditions for juices and comminuted products of Israeli citrus fruits. Journal of Food Technology 10(1):29–38.

Rouseff, R.L. and Fisher, J.F. 1980. Determination of limonin and related limonoids in citrus juices by high performance liquid chromatography. Analytical Chemistry 52(8):1228–1233.

Rovesti, G. 1978. Hydrodispersible natural colour extracted by means of orange oil from citrus waste materials. Rivista Italiana Essenze, Profumi, Piante Officinali, Aromi, Saponi, Cosmetici, Aerosol 60(2):66–68.

Rwan, J.-H. and Wu, J.-I. 1996. Deacidification of grapefruit juice with chitosan. Food Science, Taiwan 23(4):509–519.

Sadler, G., Parish, M., Clief, D. and Davis, J. 1997. The effect of volatile absorption by packaging polymers on flavor, microorganisms and ascorbic acid in reconstituted orange juice. Lebensmittel Wissenschaft und Technologie 30(7):686–690.

Sanchez, C.D., Blondel, L. and Cassin, J. 1978. Effect of climate on the quality of Corsican clementines. Fruits 33(12):811–813.

Sanchez Moreno, C., Plaza, L., de Ancos, B. and Cano, M.P. 2003. Vitamin C, provitamin A carotenoids, and other carotenoids in high-pressurized orange juice during refrigerated storage. Journal of Agricultural and Food Chemistry 51(3):647–653.

Sandhu, K.S. and Bhatia, B.S. 1985. Physico-chemical changes during preparation of fruit juice concentrate. Journal of Food Science and Technology, India 22(3):202–206.

Sandhu, K.S. and Singh, N. 1999. Studies on the factors affecting the physico-chemical and organoleptic properties of Kinnow juice . Journal of Food Science and Technology, India 38(3):266–269.

Sattar, A., Durrani, M.J., Khan, R.N. and Hussain, B. 1989. Effect of different packages and incandescent light on HTST-pasteurized single strength orange drink. Chemie Mikrobiologie Technologie der Lebensmittel 12(2):41–45.

Schmidt, R.H., Sim, C.A., Parish, M.E., Pao, S. and Ismail, M.A. 1997. A model HACCP plan for small-scale, fresh-squeezed (non-pasteurized) citrus juice operations. University of Florida Cooperative Extension Service Circular No. 1179. Gainesville, Florida, 20 pp.

Scott, W.C. 1941. Pretreatment of grapefruit for juice canning. Canner 93 No. 18:11.

Scott, W.C. and Hearn, C.J. 1966. Processing qualities of new citrus hybrids. Proceedings Florida State Horticultural Society 79:304–306.

Seelig, W. 1993. Processing of limes in Mexico. Fluessiges Obst 60(5):236, 238.

Sellahewa, J. 2002. Shelf life extension of orange juice using high pressure processing. Fruit Processing 12(8):344–350.

Shaila-Bhatawadekar, P. 1981. Clarification of lime juice by cellulase of *Pencillium funiculosum*. Journal of Food Science and Technology, India 18(5):207–208.

Shaw, P.E., Tatum, J.H. and Wilson, C.W. III. 1984. Improved flavor of navel orange and grapefruit juices by removal of bitter components with beta-cyclodextrin polymer. Journal of Agricultural and Food Chemistry 32(4):832–836.

Shaw, P.E. and Wilson, C.W. III. 1983. Debittering citrus juices with beta-cyclodextrin polymer. Journal of Food Science 48(2):646–647.

Shaw, P.E. and Wilson, C.W. III. 1984. A rapid method for determination of limonin in citrus juices by high performance liquid chromatography. Journal of Food Science 49(4):1216–1218.

Shearon, W.H. Jr. and Burdick, E.M. 1948. Citrus fruit processing. Industrial and Engineering Chemistry 40:370–378.

Sheung, K.S., Min, D.B. and Sastry, S.K. 1996. Flavor sorption interaction between polymeric packaging materials and orange juice flavor compounds. United States of America, Institute of Food Technologists 1996 Annual Meeting. 1996 IFT annual meeting: book of abstracts, p. 151 ISSN 1082–1236.

Sigbjoern, B. 1975. Manufacture of orange juice concentrate: extraction (pressing) and finishing. Nordisk Mejeriindustri 2(11):460–463, 467.

Soares, N.F.F. and Hotchkiss, J.H. 1998a. Bitterness reduction in grapefruit juice through active packaging. Packaging Technology & Science 11(1):9–18.

Soares, N.F.F. and Hotchkiss, J.H. 1998b. Naringinase immobilization in packaging films for reducing naringin concentration in grapefruit juice. Journal of Food Science 63(1):61–65.

Spoto, M.H.F., Domarco, R.E., Walder, J.M.M., Hoekstra, R.M.S. and Andrade, D.F. 1993. Preservation of concentrated orange juice by gamma radiation. II. Sensorial characteristics. Boletim da Sociedade Brasileira de Ciencia e Tecnologia de Alimentos 27(2):96–104.

Stewart, I. 1975. Influence of tree position of citrus fruit on their peel and juice color. Proceedings of the Florida State Horticultural Society 88:312–314.

Stewart, I., Bridges, G.D., Pieringer, A.P. and Wheaton, T.A. 1975. Rohde Red Valencia, an orange selection with improved juice color. Proceedings of the Florida State Horticultural Society 88:17–19.

Stolf, S.R., Siozawa, Y., Miya, E.E. and Silva, Sdda. 1973/1974. Influence of pulp content on the concentration of orange juice. Coletanea do Instituto de Tecnologia de Alimentos 5:145–170.

Sutherland, C.R. 1977. Orange juice processing: storage and packing in Florida. Proceedings of the International Society of Citriculture 748–750.

Tariq, A.M., Chaudry, M.S. and Qureshi, M.J. 1974. Effect of processing and storage on the development of bitterness in the orange juice. Pakistan Journal of Scientific and Industrial Research 17(1):27–28.

Tateo, F. and Bianco, M.G. 1984. Use of L-cysteine in concentration of lemon juice and production of derived preparations: studies on anti-browning action. Rivista della Societa Italiana di Scienza dell'Alimentazione 13(6):471–478.

Tatum, J.H., Lastinger, J.C. Jr. and Berry, R.E. 1972. Naringin isomers and limonin in canned Florida grapefruit juice. Proceedings of Florida State Horticultural Society 85:210–213.

Thormann, H.U. 1972. The Centri-Therm-Evaporator in the fruit juice industry. Gordian 72(1):7–8.

Ting, S.V., Huggart, R.L. and Ismail, M.A. 1980. Colour and processing characteristics of 'Star Ruby' grapefruit. Proceedings of the Florida State Horticultural Society 93:293–295.

Tocchini, R.P., Ferreira, V.L.P. and Shirose, I. 1979. Factors influencing the quality of pasteurized concentrated juice of oranges of the cv. Pera. Boletim do Instituto de Tecnologia de Alimentos, Brazil 16(3):325–335.

Trifiro, A., Gherardi, S., Bigliardi, D. and Bazzarini, R. 1983. Limonin in orange juices from the Italian variety Sanguinello. Industria Conserve 58(1):19–22.

Trifiro, A., Gherardi, S., Bigliardi, D., Bazzarini, R. and Castaldo, D. 1984. Limonin, L-malic acid and D-isocitric acid contents of Italian Tarocco and Sanguinello oranges. Industria Conserve 59(1):12–17, 23.

Truong, T.T., Boff, J.M., Min, D.B. and Shellhammer, T.H. 2002. Effect of carbon dioxide in high-pressure processing on pectinmethylesterase in single-strength orange juice. Journal of Food Science 67(8):3058–3062.

Tsen, H.Y. and Yu, G.K. 1991. Limonin and naringin removal from grapefruit juice with naringinase entrapped in cellulose triacetate fibers. Journal of Food Science 56(1):31–34.

Uelgen, N. and Oezilgen, M. 1993. Determination of optimum pH and temperature for pasteurization of citrus juices by response surface methodology. Zeitschrift fuer Lebensmittel Untersuchung und Forschung 196(1):45–48.

Uemura, K. and Isobe, S. 2003. Developing a new apparatus for inactivating *Bacillus subtilis* spores in orange juice with a high electric field AC under pressurized conditions. Journal of Food Engineering 56(4):325–329.

USDA. 1957. Handbook No. 98, 99pp. Agricultural Research Service, United States Department of Agriculture.

USDA. 1959. Standards for grades of chilled orange juice. Agricultural Marketing Service, United States Department of Agriculture. Washington, DC.

Vandercook, C.E. 1970. Changes in ultraviolet spectral properties of lemon juice under adverse storage conditions. Journal of Food Science 35(4):517–518.

Veldhuis, M.K., Berry, R.E., Wagner, C.J. Jr., Lund, E.D. and Bryan, W L. 1972. Oil and water-soluble aromatics distilled from citrus fruit and processing waste. Journal of Food Science 37:108–112.

Veldhuis, M.K., Swift, L.J. and Scott, W.C. 1970. Fully-methoxylated flavones in Florida orange juices. Journal of Agricultural and Food Chemistry 18(4):590–592.

Wade, R.L. 1995. Use of citrus hybrids in orange juice production. Fruit Processing 5(11):358–360.

Wallrauch, S. 1980a. Natural amino acid content of orange juice and effect of havesting date. Der natuerliche Aminosaeuregehalt von Orangensaeften und seine Abhaengigkeit vom Erntetermin der Fruechte. Fluessiges Obst 47(2):47–52, 57.

Wallrauch, S. 1980b. Composition of Brazilian orange juices and effects of harvesting date. Beitrag ueber die Zusammensetzung brasilianischer Orangensaefte und deren Abhaengigkeit vom Erntetermin der Fruechte. Fluessiges Obst 47(7):306–311.

Walsh, M., Rouseff, R. and Naim, M. 1997. Determination of furaneol and p-vinylguaiacol in orange juice employing differential UV wavelength and fluorescence detection with a unified solid phase extraction. Journal of Agricultural and Food Chemistry 45(4):1320–1324.

Watanabe, A., Iino, K., Ohta, H., Ohtani, T. and Kimura, S. 1982. Development of a juice extractor for Satsuma mandarin. I. Performance of the new type of juice extractor for Satsuma mandarin. Journal of Japanese Society of Food Science and Technology [Nippon Shokuhin Kogyo Gakkaishi] 29(5):277–282.

Wheaton, T.A., Whitney, J.D., Castle, W.S. and Tucker, D.P.H. 1986. Tree spacing and rootstock affect growth yield, fruit quality, and freeze damage of young 'Hamlin' and 'Valencia' orange trees. Proceedings of the Florida State Horticultural Society 99:29–32.

Whitney, J.D. 1995. A review of citrus harvesting in Florida. Transaction of Citrus Engineers Conference American Society of Mechanical Engineers 41:33–59.

Wicker, L., Ackerley, J.L. and Hunter, J.L. 2003. Modification of pectin by pectinmethylesterase and the role in stability of juice beverages. Food Hydrocolloids 17(6):809–814.

Widmer, W. 2000. Determination of naringin and neohesperidin in orange juice by liquid chromatography with UV detection to detect the presence of grapefruit juice: collaborative study. Journal of AOAC International 83(5):1155–1165.

Wiegand, C.L., Swanson, W.A. and Cruse, R.R. 1982. Marrs, Valencia and Ruby Red juice quality as affected by irrigation plus rainfall. Journal of the Rio Grande Valley Horticultural Society 35:109–120.

Wilke, B. 2002. Aspetic packaging of fruit juices in thermoformed containers. Fruit Processing 13(1):13–16.

Will, F., Schoepplein, E., Ludwig, M., Steil, A., Turner, A. and Dietrich, H. 2000. Analytical and sensorial alterations in orange juice after hot bottling in PET. Deutsche Lebensmittel Rundschau 96(8):279–284.

Willige, R.W.G., Linssen, J.P.H., Legger Huysman, A. and Voragen, A.G.J. 2003. Influence of flavour absorption by food-packaging materials (low-density polyethylene, polycarbonate and polyethylene terephthalate) on taste perception of a model solution and orange juice. Food Additives and Contaminants 20(1):84–91.

Wilson, C.W., Shaw, P.E. and Kirkland, C.L. 1975. Improved method for purifying crude citrus pigments. Proceedings of the Florida State Horticultural Society 88:314–318.

Wilson, C.W. III, Wagner, C.J. Jr. and Shaw, P.E. 1989. Reduction of bitter components in grapefruit and navel orange juices with beta-cyclodextrin polymers or XAD resins in a fluidized bed process. Journal of Agricultural and Food Chemistry 37(1):14–18.

Winniczuk, P.P. 1994. Effects of sanitizing compounds on the microflora of orange fruit surfaces and orange juice. Thesis, Citrus Research & Education Center, University of Florida, Lake Alfred, FL.

Wolford, R.W., Atkins, C.D., Dougherty, M.H. and MacDowell, L G. 1968. Recovered volatiles from citrus juices. Florida Section of American Society of Mechanical Engineers 14, 64–81.

Wolford, R W. and Attaway, J A. 1967. Analysis of recovered natural orange flavor enhancement materials using gas chromatography. Journal of Agricultural and Food Chemistry 15:369–377.

Woodroof, J.G. and Luh, B.S. 1975. Commercial Fruit Processing. The AVI Publishing Company, Inc., Westport, CT, pp. 293–297.

Wu, H.J., Jiao, B.L., Wang, H., Sun, Z.G., Wang, X.H., Tang, Z.H., Jiang, D.B. and Yu, E.H. 1997. Absorption of naringin by resins. Food & Fermentation Industries 23(4):37–39, 57.

Wutscher, H.K. and Bistline, F.W. 1988. Rootstock influences juice color of Hamlin orange. HortScience 23(4):724–725.

Yeom, H.W., Zhang, Q.H. and Chism, G.W. 2002. Inactivation of pectin methyl esterase in orange juice by pulsed electric fields. Journal of Food Science 67(6):2154–2159.

20
Sweet Cherries

Jesús Alonso and Rafael Alique

INTRODUCTION

Cherries are stone fruits belonging to the *Rosaceae* family, subfamily *Prunoideae*. Within the *Prunus* genus there are two main species: the sweet cherry (*Prunus avium* L.) and the sour or tart cherry (*Prunus cerasus* L.). The fruit is an oblong or heart-shaped drupe. The edible part is formed by the development of the external layers of the wall of the ovary, the pulp (mesocarp), and the skin (exocarp). The pit or stone (endocarp) encloses the seed.

The cherry seems to have originated in the countries on the shores of the Caspian and Black seas, and it is currently grown in most of the countries with a mild and Mediterranean climate. The annual world production of sweet cherries during the 3-year period (2001–2003) was more than 1.8 million metric tons (MT) over an area of 370,600 ha (FAO). The main producing countries were Turkey (238,000 MT), the Islamic Republic of Iran (220,000 MT), the United States (196,000 MT), Germany (128,000 MT), Italy (115,000 MT), and Spain (97,000 MT). The Turkish cherry is gradually increasing its penetration into the European market with the incorporation of new processing lines, packaging in modified atmospheres, and a careful sorting and handling of the product; its star variety is the Ziraat 0900. The United States continues to increase the growing area devoted to the cherry tree, and the states on the Pacific coast (Washington, Oregon, and California) are the main producers. The Bing variety predominates production followed by Rainier. The main export destinations for the American cherry are Taiwan, Canada, the United Kingdom, Japan, and Hong Kong. The European Union has gradually increased its production in the last 10 years from 139,500 ha in 1993 to 148,000 ha in 2003, with an average production of 606,000 MT a year. The main producer countries are Germany, Italy, Spain, and France, primarily for the European consumer market. They grow a wide range of local varieties and have also introduced new varieties mainly from Summerland (Canada). Spain markets, with the designation *picota*, four traditional late varieties of sweet cherry harvested without a stem, which are grown mainly on terraces at altitudes between 700 and 1100 m. In South America, production is increasing in Argentina and Chile, and during the last 10 years the area for growing cherry trees has doubled (8500 ha), and most of the 42,000 MT produced are exported.

FRUIT PHYSIOLOGY

Sweet cherries (*P. avium* L.) are highly perishable non-climacteric fruits. With a high respiration rate, their shelf life is very short and can be seen in the browning and drying of the stems, the darkening of the fruit color, shriveling, and development of decay.

Cherries are classified as non-climacteric fruits as a basis of their production of ethylene, and the

Table 20.1. Respiratory Intensity of Bing Cherries According to the Storage Temperature

Temperature	mg CO_2/kg/h
0°C	5–12
5°C	10–15
10°C	22–32
20°C	30–50

Table 20.2. Range of Respiratory Intensities of Some Sweet Cherry Cultivars at 20°C at the Optimum Stage of Maturity

Cultivar	mg CO_2/kg/h
Burlat	45–50
Navalinda	35–40
Sunburst	40–45
Bing[a]	40
Van	35–40
Lapin	30–35
Ambrunes	20–25

[a]Crisosto et al. (1993).

deterioration process of the fruits starts when they are harvested. Their production of ethylene is lower than 1 μl/kg/h at 20°C, and their response to the application of exogenous ethylene is minimal (Palou et al., 2003).

The respiratory intensity of the cherry will depend on the variety and agroclimatic conditions during the development of the fruit. Respiration rates increase with increasing temperature (Crisosto et al., 1993), following the Arrhenius law (Jaime et al., 2001). At 20°C, the respiratory intensity in different varieties ranges from 30 to 50 mg CO_2/kg/h (Table 20.1). A relation has been established between respiratory intensity and the date of harvesting, and the RI decreases in cherries harvested late such as "Ambrunes" and is higher in early cherries such as "Burlat" (Table 20.2).

HARVEST INDICATOR AND FRUIT QUALITY

The freshness, size, and color of the fruits are the main attributes that are valued by the cherry consumer. The fruits must be uniform in color and size, and fruits with a diameter larger than 26–28 mm are especially prized. The prime indicator of the freshness of the fruit is the stem, which must have a fresh, green, and turgent appearance, easily identifiable in the recently picked fruit with no browning and discoloring.

The color of the skin is the main indicator of maturity used for harvesting, and it is due to the accumulation of anthocyans during the development of the fruit. During senescence, the darkening of the fruit is due to deterioration of the anthocyans. The fruits must have a bright and shiny appearance with a range of color, depending on the variety, extending from dark red—mahogany color (Bing, Ambrunes), red (Summit, Sweetheart) to yellowish with reddish blushes (Pico Colorado, Rainier). The sugar/acid balance is responsible for the fruit's flavor. The main sugars are glucose and fructose, and the predominant acid is malic acid. The total soluble solid (TSS) content in harvested fruits must be higher than 14–16% and is more tolerant in the early varieties of rapid development such as Burlat. The initial titulable acid (TA) of the fruits is 8–10 mg of malic acid per gram of fresh weight, decreasing during refrigerated storage of the fruits. The TSS/TA quotient is the main indicator of the quality of the fruits. Another suggested indicator of quality is also the firmness of the fruits ($QI = TSS + 10TA + 10F$) (Alique et al., 2003). The firmness of the fruits (F) depends on the variety. The softness of varieties such as Burlat increases during their development, and firmness in varieties such as Ambrunes increases at a more advanced stage of maturity. The concentration of aromatic compounds in cherry is low and irrelevant from the point of view of the fruit's quality.

The main nutritional component of the cherry is carbohydrates, and their proportion varies according to the variety, physiological state of the fruit, and the agroclimatic conditions (Table 20.3). Cherries are moderately rich in vitamin C and have considerable concentrations of pantothenic acid, vitamin B-6, folate, vitamin B-12, and vitamin A. The main minerals are calcium and by far potassium. The concentration of phytosterols is around 12 mg/100 g of fresh weight.

PHYSIOLOGICAL DISORDERS

During the post-harvest period, pitting and bruising are the most frequent physical damage in cherries (Facteau and Rowe, 1979). They are produced from breaking and collapse of the mesocarp cells, damaged by impact and compression of the fruits during their development, harvesting, transport, and

Table 20.3. Nutritional Composition of Cherry (Value × per 100 g of Fresh Weight)

Components		Minerals		Vitamins	
Water	80.76 g	Potassium	224 mg	Vitamin C	7 mg
Energy	72 kcal	Calcium	15 mg	Pantothenic acid	0.127 mg
Protein	1.2 g	Phosphorus	19 mg	Riboflavin	0.06 mg
Total lipid	0.96 g	Magnesium	11 mg	Thiamin	0.05 mg
Carbohydrate	16.55 g	Iron	0.39 mg	Vitamin B-6	0.036 mg
Fiber total	2.3 g	Copper	0.095 mg	Vitamin A	214 IU

USDA Nutrient Database for Standard Reference.

handling. The damage appears after a number of days at atmospheric temperature or in the long term at low temperatures. Most bruising occurs during harvesting, and the processing lines are responsible for approximately 39% of the pitting and 10% of the bruising to the fruits. The main damage on the processing lines is from the cluster cutters and the shower-type hydrocoolers (Thomson et al., 1997). To reduce the damage to the fruits, it is best for the cherries not to bounce too much and also to drop near the processing line, to increase the protection in the flumes and on the belts, to decrease the speed of the cluster cutters, and have the hydrocooler water droplets fall near the cherry (Mitchell et al., 1980).

The physical injuries increase the respiration of the fruits, the production of ethylene, and their susceptibility to rot (Mitchell et al., 1980), thereby decreasing their post-harvest life. The sensitivity of cherries to pitting has been inversely related to their soluble solid content and the fruit weight of the "Lambert" variety; it has not been associated with other post-harvest factors including the firmness of the fruits (Facteau and Rowe, 1979). The fruit at low temperature is more susceptible to injury from impact and therefore to pitting (Griggs, 1995), and a processing temperature of 7°C is recommended to reduce the damage (Olmstead, 1994).

Susceptibility to pitting has also been related to the firmness of the fruits. Thus, pre-harvest treatments with gibberellic acid are recommended to increase the size and firmness of the fruits, making them less susceptible to pitting (Facteau and Rowe, 1979). Pre-harvest treatments with calcium have also been recommended to achieve greater firmness of the fruit and to reduce pitting (Patten et al., 1983). Lidster et al. (1979) found that post-harvest treatments with $CaCl_2$ at 4% reduced the effect of pitting in "Van" cherries. However, Facteau et al. (1987) did not find any reduction in pitting or bruising from numerous applications of calcium before harvesting the fruit. Treatments with controlled atmospheres were not effective in reducing pitting or bruising.

POST-HARVEST DISEASES

The main post-harvest diseases of cherries are produced by the development of rot of fungal origin. Among the pathogen fungi responsible are *Monilinia fructigena* (brown rot), *Botrytis cinerea* (gray mold), *Penicillium expansum* (blue mold), *Alternaria* spp., and *Rhizopus stolonifer*.

Infection from brown rot appears particularly in wet years during the flowering and development of the fruit, and rapidly infects the fruits in the presence of humidity and temperatures of 20–26°C. Rot appears as the development of a dark brown spot that attacks all the fruits and then becomes a brown powdery growth (spores) on the surface of the fruit (Fig. 20.1a). Pre-harvest fungicide applications with captan and tebuconazole, elimination of field heat using hydrocoolers and chlorinated water to avoid the germination of the spores, and the infestation of the fruit, are recommended for controlling the rot.

Infection by gray mold (*B. cinerea* Pers. ex Fr.) is likely to occur in senescent fruits and fruits stored for a long time. Rot begins with a light-brown spot on the skin followed by browning and liquefaction of the pulp. In relative low-humidity conditions, gray spores are produced abundantly, whereas in humid conditions an abundant whitish mycelium develops (Fig. 20.1b). Its development, although slow, can occur at 0°C. The application of pre-harvest treatments with tebuconzole, low storage temperature, and careful handling to prevent its post-harvest development, are recommended.

Infection by blue mold (*P. expansum* LK. ex. Thom.) occurs on damaged or senescent fruit. The infected fruit has a circular, flat area that later develops

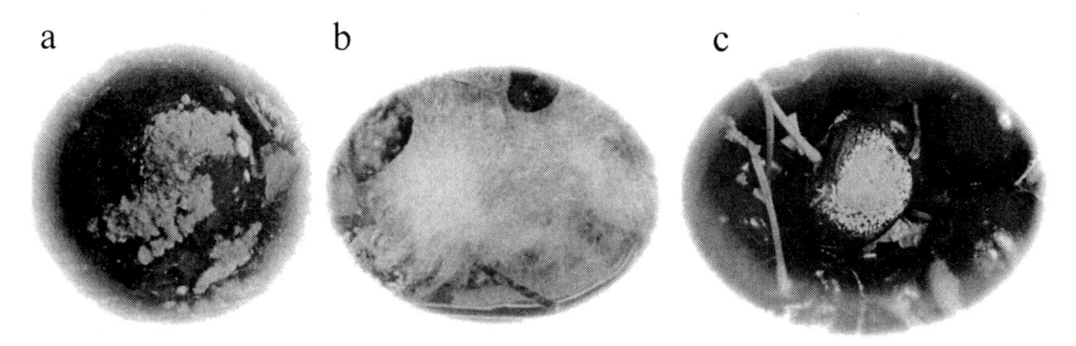

Figure 20.1. Major post-harvest decays of sweet cherry produced by *Monilia fructigena* **(a)**, *Botrytis cinerea* **(b)**, and *Penicillium expansum* **(c)**.

into a white mold with countless tiny bluish-green spores that easily spread (Fig. 20.1c). Early varieties of cherries such as Burlat or Navalinda are more sensitive to *P. expansum* than the late varieties. It is best not to store cracked fruit and handle the fruit carefully. Low storage temperatures retard the development of the disease.

Infection by Alternaria (*Alternaria* spp.) begins by penetrating the spores into the fruit wounds produced by cracking and harvesting. The rot is a dark color covered in olive green spores and whitish mold. Treatments with Iprodione are effective for controlling the disease. It is advisable to store the fruit at 0°C, and if possible in CO_2-rich (10–20%) atmospheres to prevent its development.

Infections by Rhizopus (*R. stolonifer* (Ehrenberg: Fries) Vuillemin) occur during the post-harvest of the fruits. It develops in fruits with skin injuries caused by cracking and handling, and leads to softening and liquefaction of the fruits. It is characterized by the formation of an abundant air mycelium covered by sporangiophores, and in the sporangiophores endings there are tiny dark sporangias that give it a grayish appearance. It rapidly propagates and infects via the spores. The fungus develops at temperatures higher than 4°C, and it is recommended that the fruits are stored at 0°C to avoid its development.

Post-harvest diseases are the result of adverse meteorological conditions during the development of the fruit, senescence of the fruits, or poor pre-harvest or post-harvest handling. The presence of disease limits the capacity for storage and commercial distribution of the cherries. Therefore, the control of diseases requires combined practices ranging from the application of fungicide treatments during pre- and post-harvest, careful harvesting at an optimum state of maturity, elimination of field heat, careful handling, low storage temperature, packaging in atmospheres and adequate cleaning of the handling equipment and disinfection of water.

CHERRY PROCESSING FOR THE FRESH MARKET

The variety of cherry crop, its nutrition, climatological conditions during its development, and the application of fitosanitary treatments to control plagues and diseases, will be the main factors for obtaining a quality fruit and a more or less prolonged shelf life. Of special relevance is the application of pre-harvest treatments with fungicides for controlling post-harvest diseases, especially produced by *Monilinia* spp. and *B. cinerea* infections. In the United States, sterol-inhibitor fungicides such as tebuconazole, propiconazole, and fenbuconazole are registered in IPM programs for their pre-harvest use (Adaskaveg and Forster, 2003).

A pre-harvest treatment that is widely used is the application of gibberellic acid (GA_3). Applications with gibberellic acid improve the quality of the fruit and facilitate the handling and storage of the fruits. The applications are done 3 weeks before harvesting at concentrations of 10–20 ppm. The main effects on cherries treated with gibberellic acid are that they are firmer and larger, fruit coloring is delayed and hence the date of harvesting, they are also less likely to develop pitting and disease during the post-harvest period (Looney and Lidster, 1980).

Harvesting of cherries for the fresh market is done by hand using wicker baskets or plastic or aluminum buckets. Harvesting with machinery is not viable due to the severe damage it causes to the fruit. The fruit should not drop more than 15 cm and the bottom of the baskets should be cushioned to avoid damage from bumps. Harvesting of cherries must be done with care to avoid bumps and unnecessary grazings. After harvesting, the fruit stems and then the fruits rapidly dehydrate; the stems shrivel and turn brown; and the fruits wrinkle. The transpiration of the stem is much higher than that of the fruit and results in weight losses of 4% and 1%, respectively, in conditions of relative low humidity. In high temperatures, harvesting is recommended in the early morning to reduce the excessive transpiration. The harvested fruits must be placed in the shade or in packaging protected with reflective tarps to avoid transpiration until the field heat is eliminated (Schick and Toivonen, 2002). In the United States, fruit is transported in 200-kg bins, whereas in Europe plastic boxes 40 cm × 60 cm with 8–10 kg of fruit are used. Early variety cherries such as Burlat, are more resistant to handling and storage if they are harvested at an early stage of maturity, although their level of soluble solid content is very low. By contrast, other varieties such as the Ambrunes require a developed maturity, which increases the firmness of the fruit and its resistance to mechanical damage during its handling.

The principal means of reducing transpiration and respiration of sweet cherries and prolonging their shelf life, is the rapid elimination of field heat within 4 h after harvesting (Petracek et al., 2002). Transport to the packing house must be done in closed, preferably refrigerated lorries with an insulated cover. If the packing house is far away, the field heat must be eliminated in the orchard using portable hydrocoolers. Although it is accepted that the optimum temperature of the fruit must be near 0°C during the processing, storage, and transport of the fruits, it has been observed that the damage to the fruit is higher during handling at low temperatures (Crisosto et al., 1993).

Room cooling, forced-air tunnel, and hydrocooling are the pre-cooling methods used in the cherry industry for the fresh market. Room cooling is a refrigeration system that is too slow and inefficient for the processing of cherries and it produces non-uniform cooling in the palletized boxes. Forced-air tunnel is quicker and more efficient. It is based on generating a pressure gradient, forcing cold air to circulate through air passages between the stacked boxes, thereby facilitating exchange of heat in the product. A forced-air cooling system with humidification is recommended to avoid dehydration in the fruits during their cooling. Forced-air tunnels are very common in European packing houses with water-sensitive varieties. Hydrocooling is the fastest, most uniform, and effective way of cooling cherries, and showering is the system that is preferred by the industry (Looney et al., 1996; Bahar and Dundar, 2001). Packing houses use different systems during the processing of the cherry. The field heat is eliminated with a hydrocooler of the rain-type for pallets or bins and the temperature of the fruit falls below 10°C. Within the processing lines, most of the packing houses have a hydrocooler of the rain-type, where the fruit is carried on a belt under a constant flow of cold water. A less efficient but softer system is the immersion hydrocooler where the fruit is carried on a belt immersed in water.

The preparation of the fruit for the fresh market consists of a series of processes in the cherry packing houses, namely, sorting, sizing, and packaging of the fruit. The different processes are usually developed in just one stage; the fruit is carried by a flow of water or statically on belts. Proper cleaning of the equipment used in the packing houses is essential as is also a strict chlorination of the water. The effectiveness of the chlorination will depend on the pH of the water, chlorine concentration, temperature of the water, organic matter, and exposure time. A right chlorination requires constant monitoring of the solution and a thorough knowledge of the factors involved. Monitoring every hour of the chlorination of water of the hydrocoolers is recommended, using a surfactant to decrease the surface tension and increase the effectiveness of the chlorination. To achieve a maximum proportion of hypochlorous acid and increase the effectiveness of the chloride, the pH of the water must be controlled in a range of 6.5–7.5, especially if we use sodium hyperchlorite, which increases the pH. The concentrations of chloride must be within a range of 40–70 ppm, depending on the temperature and the organic contamination. The maximum concentration of chloride must not exceed 100 ppm, and the fruit must then be rinsed with tap water.

The application of post-harvest fungicides is banned in the European Union where only the chlorination of water is used. In 1996, the use of Iprodiona for the post-harvest of cherries in the United States was banned because the producing company voluntarily decided to stop manufacturing it, due to the possible carcinogenic risk of the product from high exposure. Nonetheless, some producing countries still use this product on cherries during the post-harvest period. The United States has currently allowed the use of fludioxonil (Scholar 50WP) and a biological fungicide based on a strain of *Pseudomonas*

Figure 20.2. Cherries travel down the sizer of diverging rollers lubricated with cold water.

syringae (BIO-SAVE) for controlling post-harvest diseases.

A typical cherry processing line for the fresh market consists of the following elements: tub reception, leaf eliminator, cluster cutter, first fruit sizing (fruit industry eliminator), sorting tables, hydrocooler, fruit sizing, and box filling (Adaskevag and Foster, 2003).

The first stage of the process consists of pouring the cherries out onto a bed of water in the reception tub. In the United States, the bins are slowly tilted to empty the fruits into the tub, causing compression processes that damage the fruit. In Europe, small field boxes (10 kg) are used enabling the fruit to be carefully poured out onto the bed of water and minimizing the risk of damage to the fruits.

A belt elevator takes the cherries out of the water and carries them to the leaf eliminator and cluster cutter. The cherry lines usually have a leaf eliminator, which via an air aspiration system removes the leaves from the processing line. New leaf eliminators have been developed using a two-tub reception, which by using jets of water remove the leaves before they pass to the second tub.

Cherries picked with stems, all except the *picotas*, are usually picked in clusters. The clusters have to be eliminated using a cluster cutter consisting of a fruit conveyor belt that crosses several series of alternate revolving saws with protectors that prevent the fruit from being cut. The cluster cutter is the part of the line where the fruit suffers the greatest damage. New cluster cutters have been developed using a water flume system to carry the fruit and reduce the damage.

Each fruit with its stem is carried to the first fruit sizing (fruit industry eliminator). Generally, the lines have two sizings consisting of pairs of divergent rollers coated with hard rubber or chrome and lubricated with fresh water that move outwards and are situated on a sloping plane to facilitate the flow of the fruits (Fig. 20.2). In the first fruit sizing, called the eliminator, the fruits less than the commercial size pass through the rollers and this fruit is used for industrial processing.

Fruits with commercial sizes for the fresh market pass along the rollers and are carried along the flume toward the sorting lines. The sorting is done by hand, eliminating the cherries that do not meet the quality norms: double cherries, cracked, soft, too ripe, etc. The fruit is carried in one layer on belts with lighting at 4.6 lux (Kupferman, 1991) at a variable speed, depending on the percentage of faulty fruit. The sorting belts usually set the speed of the processing line according to the discard of the fruit.

After the sorting, some lines have the second fruit sizing and then a segmented hydrocooler for processing together the different sizes obtained. Other lines have a common hydrocooler first for all the fruits and then a fruit sizing. The hydrocoolers used are usually the rain-type because they are more effective than the immersion hydrocoolers. The water droplets must fall less than 30 cm to avoid damage to the surface of the fruit. The temperature of the fruits decreases in a few minutes to 0°C and then they are raised and carried to the fruit sizing.

In the second fruit sizing, we obtain the different commercial sizes according to the size of the fruit. The water lubricated fruits flow along sloping divergent rollers that check their lower equator diameter. The cherries drop onto a bed of water and are carried along flumes to the different packagers. The shape of the cherries differs according to the variety, and because of this, the rollers must be adjusted each time that we work with a new variety. New electronic sizing is being developed for cherries, especially for those varieties sensitive to hydropathies or with a variable range of color that makes it difficult to put fruits with the same color in boxes. These sizings are able to classify the fruits by size and color. Their main disadvantage with respect to the divergent roller is the low yield of the manufactured product.

The commercial size of fruits in Europe is determined by the largest diameter of the equator section. Although there are regulations that demand minimum sizes of the fruit, the market differentiates fruits with the following sizes: 18–21 mm, +22 mm, +24 mm, +26 mm, and +28 mm. There are three different categories depending on the percentage of defects: Extra Category—admits up to a maximum of 5% of fruits that do not meet the characteristics of the category and a minimum fruit size of 20 mm; Category I—admits up to 10% of fruits that do not correspond to the characteristics of the category; and Category II—admits up to 15% of the fruits. In the United States, the Pacific coast cherries are sold based on their size or "row size." "Row size" is the number of fruits of the same size that fit in a packed row across a container with a 10.5 in. (26.7 cm) width. The most common commercial sizes are in descending order 9 Row (29.8 mm), 10 Row (26.6 mm), 11 Row (24.2 mm), 12 Row (21.4 mm), and 13 Row (20.6).

The packaging of the fruit differs according to the market. The United States packages most of its product in boxes of 18–20 lb using polyethylene liners and a damp-absorbent cellulose pad. In Europe, the formats are smaller: 5 and 2 kg of fruits packaged in cardboard boxes for the wholesaler market and small formats of 400 g–1 kg packaged in punnets or clam-shell plastic containers for supermarkets and hypermarkets. The 400–500 g punnets are film wrapped using a horizontal flow-pack system with macroperforated or microperforated films (Fig. 20.3). The punnet, bags, and small 2 kg boxes are filled automatically using horizontal multi-head weigher that weigh the exact amount of cherries for the boxes.

STORAGE

Cherries are highly tolerant to low temperatures, the storage conditions recommended for keeping the stems green and fruits fresher are the temperatures between −0.50°C and 0°C and high relative humidity (90–95%). Quality cherries can be stored in these conditions for 2–4 weeks. The conditioning temperature of the cherry must also depend on the immediacy of its sale and consumption and also how far it is being dispatched. Thus, for fruits for immediate consumption (1–2 days) conditioning temperatures between 8°C and 12°C are recommended, those fruits requiring 4–6 days must be conditioned at temperatures not higher than 4–8°C, whereas fruits that are to be sold after more than 6 days must be stored at temperatures of 0°C (Looney et al., 1996). Although the cherry is a fruit that is resistant to low temperatures, the application of low temperatures induces a small respiratory stress, which increases the respiratory intensity of the fruits when they are exposed to atmospheric temperature, thereby accelerating their senescence (Alique et al., 2003).

MODIFIED ATMOSPHERE PACKAGING

Modified atmosphere packaging (MAP) is a common practice in the commercialization of sweet cherries. MAP effectively retards deterioration of certain cherry quality parameters (Meheriuk et al., 1995; Artés et al., 2001; Wang and Vestrheim, 2002) and decay caused by fungal growth (Tian et al., 2001). In cv. Bing sweet cherries, MAP with CO_2/O_2 concentrations of 8%/5% and 10%/5% effectively reduced rotting, browning of stems, darkening of fruit color, and loss of firmness and acidity as compared to fruit packed in macroperforated box liners (Crisosto et al., 2002). In cv. Burlat, MAP with 9–12% CO_2 and 1–3% O_2 effectively prolonged shelf life, especially of fruit harvested at the red color stage (Remón et al., 2000). The range of optimum concentrations of atmosphere components for post-harvest preservation of cherries, has been established as 3–10% for O_2 and 10–15% for CO_2 (Kader, 2003).

The use of modified atmospheres for cherry preservation requires strict temperature control during storage, distribution, and sale. High temperatures increase respiration; this leads to hypoxia levels low enough to stimulate anaerobiosis and induce fermentation, producing off-flavors. Concentrations of $O_2 \leq 1\%$ have been reported as crucial for the

Figure 20.3. Horizontal flow-pack wrapping of sweet cherry punnets.

onset of pitting and off-flavors in "Bing" cherries, although anaerobiosis is considered to exist at O_2 concentrations lower than 3%. CO_2 levels in MAP of cherries do not appear to influence the rate of respiration or the production of ethanol or acetaldehyde (Jaime et al., 2001). However, CO_2 concentrations greater than 30% have been associated with brown skin discoloration and off-flavor (Zoffoli et al., 1988).

REFERENCES

Adaskaveg, J.E. and Forster, H. 2003. Strategies for postharvest decay management of sweet cherries in 2003. Central Valley Postharvest Newsletter 12(1):1–5.

Alique, R., Martínez, M.A. and Alonso, J. 2003. Influence of the modified atmosphere packaging on shelf life and quality of Navalinda sweet cherry. European Food Research and Technology 217:416–420.

Artés, F., Tudela, J.A. and Artés-Hdez, F. 2001. High carbon dioxide effects on keeping quality of sweet cherry. ISHS Acta Horticulturae 553:663–664.

Bahar, A. and Dundar, O. 2001. The effects of hydrocooling and modified atmosphere packaging system on storage period and quality criteria of sweet cherry cv. Aksehir Napolyonu. ISHS Acta Horticulturae 553:615–616.

Crisosto, C.H., Garner, D., Doyle, J. and Day, K.R. 1993. Relationship between respiration, bruising susceptability and temperature in sweet cherries. HortScience 28(2):132–135.

Crisosto, C.H., Zoffoli, J.P. and Garner, D. 2002. Evaluation of different box liners for the California 'Bing' cherry industry. Central Valley Postharvest Newsletter 11(2):3–7.

Facteau, T.J. and Rowe, K.E. 1979. Factors associated with surface pitting of sweet cherry. Journal of the American Society for Horticultural Science 10:706–710.

Facteau, T.J., Rowe, K.E. and Chestnut, N.E. 1987. Response of 'Bing' and 'Lambert' sweet cherry fruit to preharvest calcium chloride applications. HortScience 22(2):271–273.

Griggs, D. 1995. Packers can reduce cherry damage. Good Fruit Grower 46(10):10–11.

Jaime, P., Salvador, M.L. and Oria, R. 2001. Respiration rate of sweet cherries: 'Burlat', 'Sunburst' and 'Sweetheart' Cultivars. JFS: Food Chemistry and Toxicology 66:43–47.

Kader, A. 2003. A summary of CA requirements and recommendations for fruits other than apples and pears. ISHS Acta Horticulturae 600:737–740.

Kupferman, E.M. 1991. Cherry sorting table lighting. American Society of Agricultural Engineers Paper No. 913552, 9 pp.

Lidster, P.D., Porritt, S.W. and Tung, M.A. 1979. Effect of a delay in storage and calcium chloride dip on surface disorder incidence in 'Van' cherry. Journal of the American Society for Horticultural Science 104(3):298–300.

Looney, N.E. and Lidster, P.D. 1980. Some growth regulator effects on fruit quality, mesocarp composition, and susceptibility to postharvest surface marking of sweet cherries. Journal of the American Society for Horticultural Science 105:130–134.

Looney, N.E., Webster, A.D. and Kupferman, E.M. 1996. "Harvest and handling sweet cherries for the fresh market." In Cherries: Crop Physiology, Plant Materials, Husbandry and Product Utilization, edited by Webster, A.D. and Looney, N.E., pp. 411–441. Ames: Washington State University Press CAB.

Meheriuk, M., Girard, B., Moyls, L., Beveridge, H.J.T., McKenzie, D.L., Harrison, J., Weintraub, S. and Hocking, R. 1995. Modified atmosphere packaging of 'Lapins' sweet cherry. Food Research International 28(3):239–244.

Mitchell, F.G., Mayer, G. and Kader, A.A. 1980. Injuries cause deterioration of sweet cherries. California Agriculture 34(3):14–15.

Olmstead, B.S.O. 1994. Cherry growers strive to extend shelf-life. Good Fruit Grower 45(10):11–13.

Palou, L., Crisosto, C.H., Garner, D. and Basinal, L.M. 2003. Effect of continuous exposure to exogenous ethylene during cold storage on postharvest decay development and quality attributes of stone fruits and table grapes. Postharvest Biology and Technology 27:243–254.

Patten, K.D., Patterson, M.E. and Kupferman, E. 1983. Reduction of surface pitting in sweet cherries. Post Harvest Pomology Newsletter 1(2):15–19.

Petracek, P.D., Joles, D.W., Shirazi, A. and Cameron, A.C. 2002. Modified atmosphere packaging of sweet cherry (*Prunus avium* L., cv. 'Sams') fruit: metabolic responses to oxygen, carbon dioxide, and temperature. Postharvest Biology and Technology 24(3):259–270.

Remón, S., Ferrer, A., Marquina, P., Burgos, J. and Oria, R. 2000. Use of modified atmospheres to prolong the postharvest life of Burlat cherries at two different degrees of ripeness. Journal of the Science of Food and Agriculture 80:1545–1552.

Schick, J.L. and Toivonen, P.M.A. 2002. Reflective tarps at harvest reduce stem browning and improve fruit quality of cherries during subsequent storage. Postharvest Biology and Technology 25(1):117–121.

Thomson, J.F., Grant, J.A., Kupferman, E.M. and Knutson, J. 1997. Reducing sweet cherry damage in postharvest operations. Hort Technology 7(2):134–138.

Tian, S., Fan, Q., Xu, Y., Wang, Y. and Jiang, A. 2001. Evaluation of the use of high CO_2 concentrations and cold storage to control of *Monilinia fructicola* on sweet cherries. Postharvest Biology and Technology 22(1):53–60.

Wang, L. and Vestrheim, S. 2002. Controlled atmosphere storage of sweet cherries (*Prunus avium* L.). Acta Agriculturae Scandinavica B: Soil and Plant Science 52(4):136–142.

Zoffoli, J.P., Lavanderos, J.C. and Zarate, M.M. 1988. Posibles alternativas de embalaje para la exportación de cerezas. Revista Frutícola 1:13–15.

21 Cranberry, Blueberry, Currant, and Gooseberry

Kristen K. Girard and Nirmal Sinha

Section 1: Cranberry

INTRODUCTION

Cranberries are a healthy fruit that contribute color, flavor, nutritional value, and functionality. Because of their versatility and blendability, applications using cranberries are extensive. This chapter presents information on cranberry history, cultivation, physicochemical characteristics, nutritional and health considerations, and process for cranberry products.

HISTORICAL

Cranberries are one of the only three Native American fruits (the other two being the blueberry and the Concord grape), widely utilized by Native Americans long before the Pilgrims arrived in 1620. They combined crushed cranberries with deer meat and melted fat to make pemmican—a food that was kept for an extended time without refrigeration. Native American women also used cranberry juice as a dye in making rugs and blankets. Legend also has it that cranberries were served at the first Thanksgiving in Plymouth, Massachusetts.

To various Native American tribes, the cranberry was known by many different names. The Cape Cod Pequot and New Jersey Leni-Lanape tribes called the red berry "ibimi" or bitter berry. To the Pilgrims, the shape of the cranberry blossom resembled the head of a crane; therefore, the berry was named "crane berry," and later shortened to "cranberry".

The first recorded harvest of cranberry was documented in Dennis, Massachusetts, in 1816. Soon after, cranberries became a shipboard staple during trans-Atlantic voyages to prevent scurvy, caused by vitamin C deficiency. During World War II, American troops consumed about 1 million pounds of dehydrated cranberries a year.

The modern cranberry industry took shape soon after the turn of the 20th century. At that time, fresh fruit marketing was the focus for use in sauces and relishes. Since fresh cranberries are harvested during September through November, their usages peaked during the Thanksgiving and Christmas holidays. The development of canned cranberry sauce made cranberry consumption possible on a year-round basis. Today, cranberry products ranging from cranberry juice cocktail, cranberry juice blends to sweetened dried cranberries (SDCs) are available around the world.

Classification

The North American cranberry *(Vaccinium macrocarpon)* is recognized by the US Department of Agriculture (USDA) as the standard for fresh cranberries and cranberry juice cocktail. The European variety, grown in parts of central Europe, Finland, and Germany, is known as *V. oxycoccus*. This is a smaller fruit with slightly different anthocyanin and acid profiles similar to that of the North American variety. In Europe, this fruit is commonly known as lingonberry or English mossberry.

Cultivation

Cranberries are a unique fruit, which grow and survive only under a special condition of acid peat soil, an adequate fresh water supply, and a growing season that extends from April to November. Principal cranberry growing areas in North America all exist in the northern latitudes of 40–50°.

Cranberries grow on hearty vines in beds layered with sand, peat, and gravel. These beds, commonly known as bogs or marshes, were originally formed as a result of glacial deposits. Normally, growers do not replant each year since an undamaged cranberry vine will survive indefinitely. Some vines on Cape Cod are more than 150 years old and are still bearing fruit. As a matter of fact, the majority of cranberry growers are multi-generational families, some fifth and sixth generation, with two to three generations working and living together on their cranberry farms.

Harvesting

Cranberries are typically harvested from mid-September to early-November. Originally, cranberries were harvested by hand. Changes in techniques have been developed such that two types of harvesting procedures can be used.

Dry harvest—Until the 1950s, dry harvested cranberries were scooped by hand off the vines. In the early mid-1960s, mechanical pickers were developed and their use became common practice. These mechanical pickers comb the berries off the vine using moving metal teeth. Dry harvested cranberries are used to supply the fresh fruit market. These cranberries are most often used for cooking and baking.

Wet harvest—Cranberries have pockets of air inside the fruit. Because of this, cranberries float in water, and thus, the bogs can be flooded to aid in removal of the fruit from the vines. Water reels, nicknamed "egg-beaters," are used to stir up the water in the bogs (Fig. 21.1). By this action, cranberries are dislodged from the vines and float to the surface of the water. "Booms" are used to corral the berries into one area (Fig. 21.2) so they can be vacuumed into trucks and shipped to the receiving station for cleaning, sorting, and grading. The quality-graded fruit is sent to freezer storage. Frozen fruits can be kept for over 1 year, allowing production of cranberry juices, sauces, and ingredients to occur year-round. Wet harvested cranberries are used for juices, sauces, and as an ingredient in processed foods.

Berry color is one of the key factors in determining when a bog is ready to be harvested. Some important factors that affect color are growing regions, variety, weather, and time of harvest. The fruit with the redder color comes from the Northwest region. Growers here are able to delay harvest longer because of reduced risk of frost, thereby enhancing color development in the cranberry.

Figure 21.1. Water reels dislodging fruit from vines.

Figure 21.2. Cranberries "boomed" and waiting for receiving trucks.

WORLD PRODUCTION AND CONSUMPTION

The majority of the world's cranberries are grown in five U.S. states and two Canadian provinces. On the basis of volume, the proportions of the cranberry crop grown in each of these locations are Wisconsin, 36%; Massachusetts, 28%; New Jersey, 12%; Oregon, 7%; Washington, 4%; and Quebec and British Columbia, 13%.

The total U.S. commercial crop in 2003 was 617 million pounds (6.2 million barrels—1 barrel is equal to 100 pounds). The total global commercial crop for cranberries in 2003 was 763 million pounds or 7.6 million barrels. This crop is then processed into four major categories as: Fresh cranberries 10%, sauce products 15%, shelf-stable SDCs 10%, and juice/juice drink products 60%. The overall retail value of these consumer products in 2003 was approximately 1.2 billion dollars. Americans consume over 400 million pounds of cranberries a year, 20% of this during Thanksgiving week alone.

Industry wide, the estimated area of cranberry production is approximately 40,000 acres. On average, every acre of cranberry bog is supported by about 4–10 acres of wetlands, uplands, and woodlands. These areas provide refuge to a variety of wildlife.

PHYSICOCHEMICAL AND NUTRITIONAL QUALITY

Determination of Quality

Cranberries bounce! This is one way the quality of cranberries is judged. As mentioned earlier, cranberries have pockets of air inside which enables them to bounce.

A damaged or spoiled cranberry does not bounce, an observation that led to the development of the first cranberry bounce board separator in the late 1800s. This method is still used today to select quality fruit.

Chemical Composition

Proximate composition of raw cranberries are listed in Table 21.1. These numbers may tend to vary slightly from one crop year to another, but are generally representative of average values.

The cranberry has a unique chemical composition that sets it apart from other North American fruits. The combination of high-acid content (~2.0%) and low Brix level (~7.5°Brix) gives pure cranberry juice a Brix/acid ratio of about 3.75, which makes it extremely tart and unpalatable in a single-strength form. In contrast, apple juice and orange juice have Brix/acid ratios of above 10.

Another unique characteristic of the cranberry is astringency created by significant quantities of tannins. The cranberry also contains an unusual mixture of organic acids. Citric and malic are the predominant acids followed by quinic (~1.0%) and benzoic (~0.01%), which is uncommon in most popular fruits. The chemical composition of cranberry juice is shown in Table 21.2.

Table 21.1. Composition of Raw Cranberries

Component	Percent
Water	86.5
Protein	0.4
Ash	0.2
Fat (lipids)	0.2
Dietary fiber	4.2
Available carbohydrates	8.5

USDA (1999)

Table 21.2. Chemical Composition of Cranberry Juice

Component	Percent
Acids	
Citric	1.0
Quinic	1.0
Malic	0.7
Benzoic	0.01
Sugars	
Glucose	2.9
Fructose	1.0
Tannins	0.3
Pectin	0.1
Anthocyanins (mg/100 ml)	40.0
Flavonols (mg/100 ml)	17.75

Source: Courtesy Ocean Spray Cranberries, Inc.

Anthocyanins, Total Phenolics, and Antioxidant Capacity

The characteristic red color of cranberries, juice and beverage, and products is due to anthocyanin pigments. Zheng and Wang (2003) reported concentration of major anthocyanins in cranberry (cv. Ben Lear) as: peonidin 3-galactoside (213.6 μg/g fresh weight), peonidin 3-arabinoside (99.7 μg/g), cyanidin 3-galactoside (88.9 μg/g), cyanidin 3-arabinoside (48.0 μg/g), and peonidin 3-glucoside (40.4 μg/g). Other phenolic compounds analyzed were: vanillic acid (49.3 μg/g), caffeic acid (42.5 μg/g), quercetin 3-galactoside (70.4 μg/g), and quercetin 3-arabinoside (34.4 μg/g). Wang and Stretch (2001) reported average anthocyanin and total phenolic concentrations in 10 cranberry cultivars as: 34.8 mg of cyanidin 3-galactoside per 100 g, and 141.2 mg of gallic acid equivalent per 100 g of fresh weight, respectively. The average antioxidant capacity of these cranberry cultivars expressed as μmol of Trolox equivalents (TE) per gram of fresh weight was 10.4. Recently, Wu et al. (2004) reported total (sum of lipophilic and hydrophilic) antioxidant capacity (μmol TE) of 8983 in 1 whole cup (95 g) serving of cranberry. Among the fruits analyzed by these researchers, cranberries showed high (2.00 μmol TE/g) lipophilic antioxidant capacity (L-ORAC$_{FL}$), second to avocado (5.52 μmol TE/g).

NUTRITIONAL QUALITY AND HEALTH CONSIDERATIONS

Nutritional Quality

Besides, phytochemicals, cranberry products contain dietary fiber and certain vitamins and minerals. Cranberries contain nutritionally significant quantities of vitamin C and potassium (Table 21.3). In an unsweetened form, cranberries are low in calories, sodium, and free from cholesterol and saturated fats.

Health Considerations

Historically, the health-promoting properties of cranberries have been based on folklore remedies, which have existed for centuries. Native American Indians and early settlers recognized the health-giving properties of this fruit.

Cranberries have been used in many different forms as a folk remedy for the treatment of urinary tract infections (UTIs). These infections cause frequent and painful urination. The first reported use of cranberries by conventional medical practitioners was in 1923 (Blatherwick and Long, 1923). It was suggested that cranberries acidify urine, killing the bacteria causing UTIs. A study conducted at the Harvard Medical School (Avorn et al., 1994) determined that regular consumption of cranberry juice reduced the amount of bacteria in the urinary tracts of elderly women. Rather than acidification of the urine, however, these researchers concluded that something specific to the cranberry actually prevented bacteria from adhering to the lining of the bladder. Later, Howell et al. (1998) identified condensed tannins or proanthocyanidins from the cranberry fruit as the component that prevented *Escherichia coli,* the primary bacteria responsible for UTIs, from attaching to cells in the urinary tract. These organisms are flushed out from the urinary tract rather than being allowed to adhere, grow, and lead to infection.

The anti-adhesion mechanism may work beyond the bladder in fighting certain bacteria in other parts

Table 21.3. Nutritional Values of Cranberries (Per 100 g Product)

	Cranberry Product				
Nutrient	Frozen[a]	Concentrate	Sweetened/Dried[b]	Flavored Pieces[c]	Powder
Calories	48	198	298–367	337–342	360
Calories from fat (%)	0	0	11–12	5	2
Total fat (g)	0.5	0	1.2–1.4	0.5	0.2
Saturated fat (g)	0	0	0	0	0
Cholesterol (mg)	0	0	0	0	0
Sodium (mg)	3	14	3–4	2–3	29
Potassium (mg)	73	500	40–90	11	734
Total carbohydrate (g)	10	49	82–88	83–84	89
Dietary fiber (g)	4	<0.5	6–9	5–6	6
Sugars (g)	4	22	64–69	67–68	37
Protein (g)	0.6	<0.5	<0.5	<0.5	<0.5
Vitamin A (IU[d])	0	0	70[e]	16,200[f]	0
Vitamin C (mg)	18	58	0	1	5
Calcium (mg)	10	39	10–18	4	184
Iron (mg)	0.6	1.7	0.5	0	4

Source: Courtesy Ocean Spray Cranberries, Inc.

[a]Whole or sliced.

[b]Regular, soft and moist, and glycerated forms.

[c]Orange, blueberry, cherry, strawberry, or raspberry flavored fruit pieces.

[d]As provitamin A.

[e]Value for glycerated forms of sweetened dried cranberries. Regular and soft and moist forms contain 0 IU.

[f]Value for orange flavored fruit pieces. Other flavors contain 0 IU.

Table 21.4. Cranberry and Processed Products

Product	Product Description	General Industrial Packaging	Storage/ Shelf Life	Comments	Applications
Fresh	Whole, fresh cranberries	12 ounce consumer bag or 40 lb box	32–34°F (0–1°C), 3 months	Available September-November	Bakery products, sauces
Frozen					
Whole	Whole cranberries	40 lb box	0 ± 15°F (−18 ± 9°C), 18 months	Available year round	Bakery products, sauces, condiments, dairy products
Sliced	3/8 in. (10 mm) thick	20 lb box	0 ± 15°F (−18 ± 9°C), 18 months	Available year round; individually quick frozen (IQF)	Bakery products, sauces, condiments, dairy products
Liquid					
Single-strength juice	7.5°Brix	44 gallon drum or tanker	0 ± 15°F (−18 ± 9°C), 2 years	Direct expressed juice	Beverages, natural colorant
Concentrate	50°Brix, 14 ± 1.5% titrable acidity	5 gallon pail, 50 gallon drum or tanker	0 ± 15°F (−18 ± 9°C), 2 years	Highly colored, pure cranberry concentrate	Beverages, natural colorant, condiments, dairy products, confections
Puree	5.4 or 6.1°Brix	33 lb pail or 50	0 ± 15°F (18 ± 9°C), 18 months	Well-colored, high pectin content	Sauces, beverages, bakery products

Shelf-stable products					
Sweetened dried cranberries	Sugar-infused, dried fruit	25 lb box	12 months at <65°F (18°C) or 18 months at <45°F (7°C); shelf stable in cool, dry conditions	No artificial color, flavor, or preservative; excellent color retention	Bakery products, cereals, trail mix, snack foods, dairy products, confections
Flavored fruit pieces	Sugar-infused cranberry-based, dried fruit with natural flavor topically coated	25 lb box	12 months at <65°F (18°C) or 18 months at <45°F (7°C); shelf stable in cool, dry conditions	Firm yet tender fruit texture; versatile and cost effective; available in raspberry, cherry, strawberry, blueberry, orange flavors	Bakery products, cereals, trail mix, snack foods, dairy products, confections
Cranberry powder	Spray-dried cranberry concentrate	100 lb drum, 50 lb drum	2 years in dry ambient storage	Soluble, hygroscopic fruit	Nutracueticals, confections, beverages, colorant, teas

Variations of sweetened dried cranberries:

1. Regular: Distinct cranberry flavor, tart, 11–14% moisture.
2. Soft and moist: Eat out of hand, less tart, softer, 13–16% moisture.
3. Glycerated: Remains soft in low-moisture system ($a_w = 0.45–0.53$).

Source: Ocean Spray Cranberry.

of the body including the oral cavity (periodontal gum disease) and stomach (ulcers). For example, Weiss et al. (2002) suggested that compounds in the cranberry prevent certain bacteria found in the mouth from adhering to teeth and gums. Burger et al. (2000) suggested that the same anti-adhesion mechanism fights *Heliocobacter pylori*, the bacteria that cause stomach ulcers. This study suggests that the cranberry's anti-adhesion effect prevents the bacteria from attaching to the stomach lining and causing an ulcer.

Howell and Foxman (2002) suggested that regular consumption of cranberry juice cocktail may decrease the need for certain antibiotics.

Maher et al. (2000) reported potential benefits of cranberry juice in protecting against cholesterol oxidation. In their study, cranberry juice was tested for its ability to inhibit oxidation of LDL cholesterol and proved to be an effective antioxidant.

There is strong epidemiological evidence that diets high in vegetables and fruits contribute to an overall anti-cancer effect. Preliminary evidence suggests that powerful cancer-fighting antioxidants are found in the cranberry seeds. The cranberry seeds have been found to contain a high level of tocotrienols. Cranberry seed oil contains significant amounts of these potent forms of Vitamin E. Rich in flavonoids as well, the cranberry may have potential anti-cancer activity.

CRANBERRY PRODUCTS

Table 21.4 lists various cranberry products along with information regarding storage conditions, shelf life, packaging, and various applications. As indicated before in terms of processed products, 60% of cranberries in the United States are utilized as juice and drinks, about 15% as sauce, and 10% as SDCs. Figure 21.3 shows steps involved in making various cranberry products. In this section, aspects of cranberry juice processing and SDCs are discussed.

CRANBERRY JUICE PROCESSING

After cranberries are harvested, they are cleaned and sorted for color and then frozen in 1000 pound bins. During the rest of the year, they are then pulled out of frozen storage as needed to be extracted, filtered, and then concentrated into a 50°Brix juice concentrate. This concentrate is utilized for processing into cranberry juice drinks or sold, as is, in the market.

There are several different methods for extracting juice from the fruit: pressing, mash depectinization, and counter current extraction. Prior to pressing or mash depectinization, the fruit is milled to a specific size. Pressing uses a larger piece size to reduce the amount of insoluble solids in the pressed juice. When depectinization is followed, the piece size is reduced to increase the rate of depectinization and thus minimize the amount of pectinase required.

- *Pressing*—This method uses a giant press (e.g., Reitz-Willmes) (Fig. 21.4) and mechanical action to express the juice. This cold process produces a high-quality juice with excellent flavor attributes. The juice color is also very stable since no heat has been used to degrade the anthocyanin pigments. The yield is approximately 75%.
- *Mash depectinization*—This process involves blending size reduced cranberries with enzymes to digest the fruit into a mash. The entire process can take 4–12 h at approximately 52°C. This mash is then extracted via either a Bucher Press or by centrifugation. This method generates extremely high yields—sometimes over 100%. The negatives include length of processing time, and heat degradation, which affects color, flavor, and shelf life of beverage.
- *Counter current extraction*—Counter current technology is a very gentle process involving a large screw (Fig. 21.5). Sliced fruit is deposited in one end and water into the opposite end. Through counter current flow of the two material streams, juice extraction takes place. The benefits are that this cold process produces very high-quality juice with yields exceeding 90%. The other benefit is that the co-product produced is an intact fruit piece that can be further processed into SDCs by adding a sweetener (sucrose syrup, etc.) and then drying to a specific moisture content.

Once the cranberry juice is extracted by one of the methods previously mentioned, it is then filtered. Filtration can take on many different forms:

Types of Filters

❑ Screen	Plastic or stainless steel
❑ Bag	Woven fiber
❑ Media	Diatomaceous earth, perlite
❑ Membrane	Polymer, ceramic, stainless steel

Pore Size

❑ Media	$< 5 \mu m$
❑ Microfilter	0.1–1.5 μm
❑ Ultrafilter	0.005–0.1 μm
❑ Nanofilter	0.001–0.008 μm

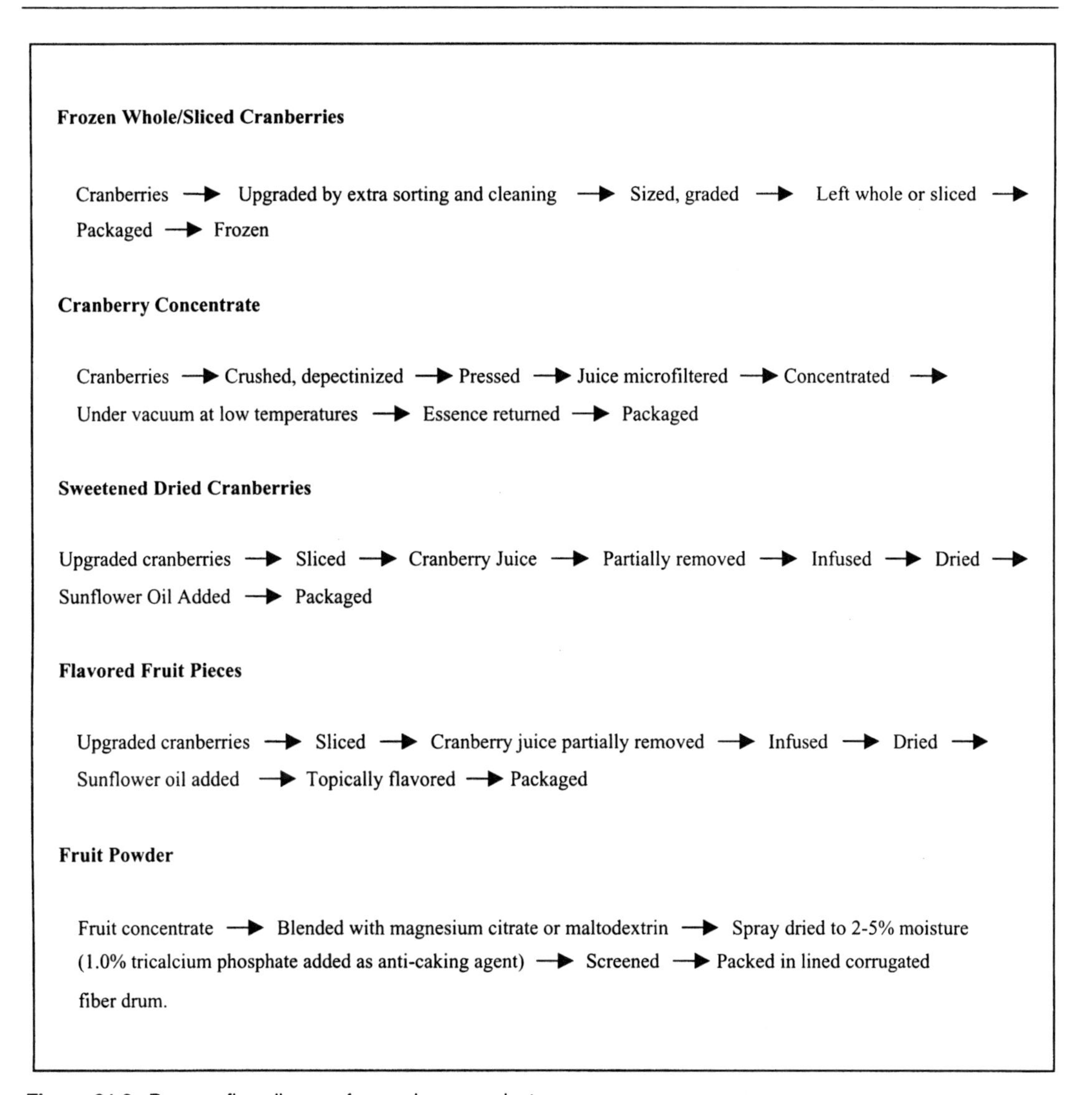

Figure 21.3. Process flow diagram for cranberry products.

Once filtered, the juice is concentrated. Concentration improves microbial stability and also reduces storage and shipping volumes. Single-strength cranberry juice is about 7.5°Brix, whereas standard concentrate is 50°Brix. Concentration to 50°Brix involves reverse osmosis and evaporation. Figure 21.6 shows how a typical RO system operates.

Evaporation can occur using single effect, triple effect, or triple effect with regeneration as shown in Figures 21.7–21.9. Single pass, multiple effect evaporators having no re-circulation are considered the preferred method since the juice sees minimal time at elevated temperatures.

The concentrate is then filled into drums and stored at −18°C for industrial sale. Refrigerated bulk storage can also be used for short-term periods prior to use.

Sweetened Dried Cranberries

Because cranberry is highly acidic, infusing this fruit with sweeteners prior to drying reduces its acidity and makes it more acceptable in the dried form. The infused-dried cranberries also known as SDCs are used in many ways (Table 21.1). U.S. Patent 5,320,861 (Mantius and Peterson, 1994) described a combined process of extracting cranberry juice and

Figure 21.4. Reitz-Willmes press.

Figure 21.5. Counter current technology.

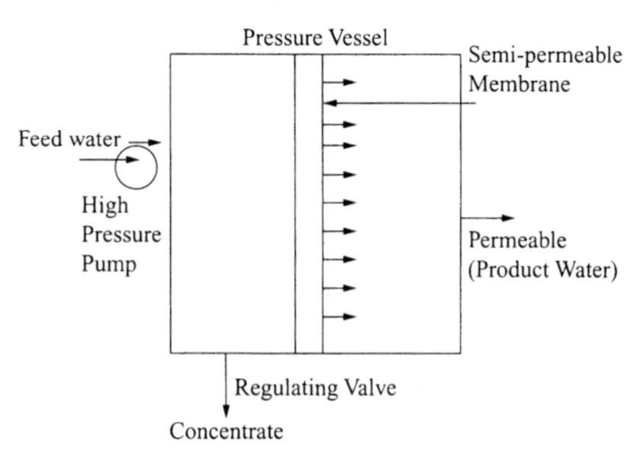

Figure 21.6. Reverse osmosis operation.

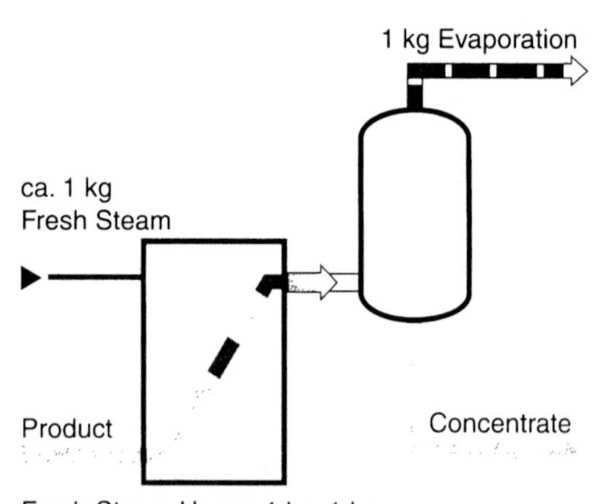

Figure 21.7. Single-effect evaporation.

infusing and drying extracted cranberries to make SDCs. This sequential two-step process of extraction and infusion based on counter current principle using flights in a screw conveyer (Fig. 21.5) with intermittent forward and reverse motion works very well for a firm fruit such as cranberries. In this process, the raw fruit is frozen prior to extraction. The residence time and temperature for juice extraction is about 120–150 min and 24°C, respectively. The residence time and temperature for infusing cranberries prior to drying is 120–300 min and 38–54°C,

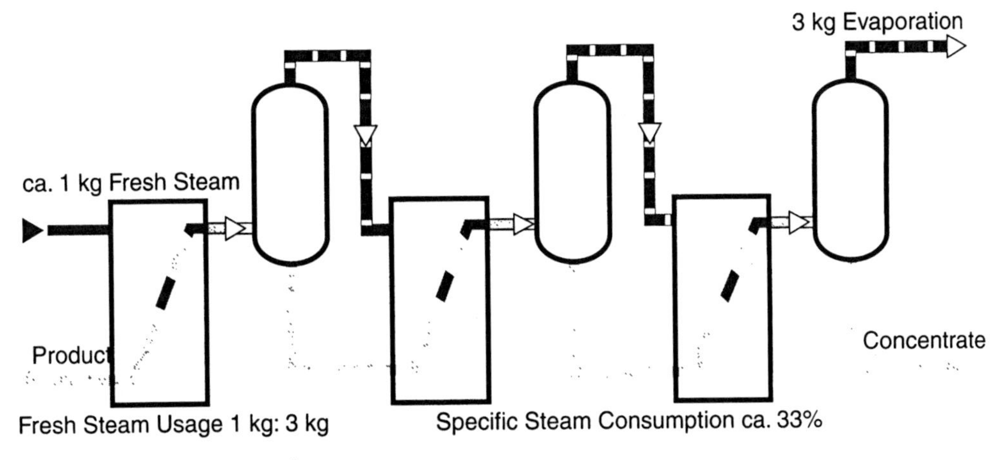

Figure 21.8. Triple-effect evaporation.

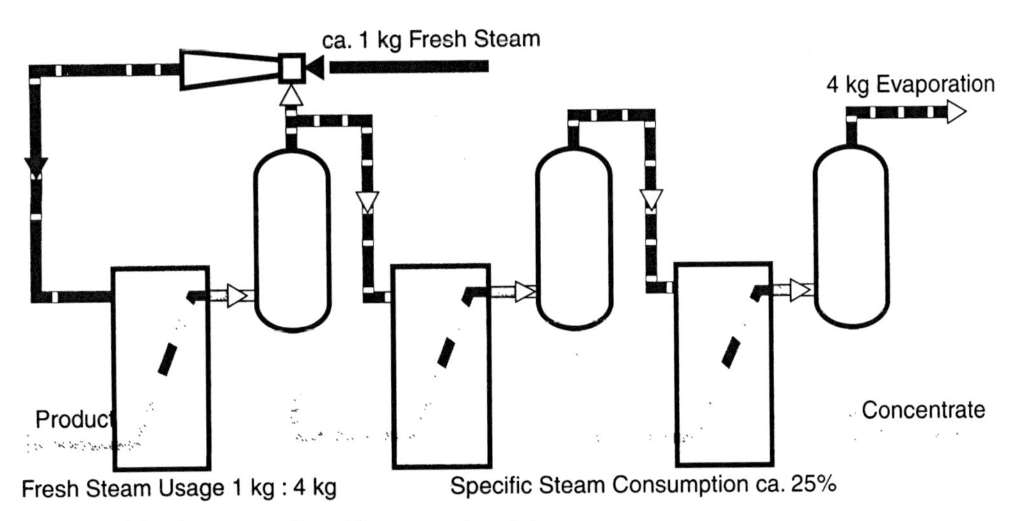

Figure 21.9. Triple-effect evaporation with regeneration of steam.

respectively. The extracted cranberries are infused to 40–55°Brix and dried to a water activity of 0.50–0.55. As mentioned before, this product finds application in many products including ready-to-eat cereals and trail mixes.

SUMMARY

Cranberries are an extremely versatile fruit. When incorporated into other food products, they provide refreshing flavor as well as a characteristic red color. Used in combination with other fruits, cranberries can accentuate and enhance the flavors of these fruits. Because of their health benefits, cranberries are experiencing an expansion into the food and beverage industry. Available year-round and in a variety of forms, cranberries can be used to enhance numerous products and applications in the food and beverage industry.

REFERENCES

Avorn J, Monane M, Gurwitz JH, Glynn RJ, Choodnovsky I, Lipsitz LA. 1994. Reduction of bacteriuria and pyruia after ingestion of cranberry juice. J Am Med Assoc. 271:751.

Blatherwick NR, Long ML. 1923. Studies on urinary acidity: The increased acidity produced by eating prunes and cranberries. J Biol Chem. 57:815.
Burger O, Ofeck I, Tabak M, Weiss EI, Sharon N, Neeman I. 2000. A high molecular mass constituent of cranberry juice inhibits *Heliocobacter pylori* adhesion to human gastric mucus. FEMS Immunol Med Microbiol. 29:295–301.
Howell AB, Foxman B. 2002. Cranberry juice and adhesion of antibiotic-resistant uropathogens. J Am Med Assoc. 287(23):3082–3083.
Howell AB, Vorsa N, der Marderosian A, Foo LY. 1998. New research identifies substance that prevents urinary tract infections. N Engl J Med. 339:1085.
Maher M, Mataczynski H, Stefaniak H, Wilson T. 2000. Cranberry juice induces nitric oxide-dependent vasodilation *In Vitro* and its infusion transiently reduces blood pressure in anesthetized rats. J Med Food. 3:141–147.
Mantius HL, Peterson PR. 1994. U. S. Patent 5,320,861. Fruit Extraction and Infusion.
USDA. 1999. Nutrient Database for Standard Reference, release 13. http://www.nal.usda.gov/fnic/foodcomp/Data/index.html.
Wang SY, Stretch AW. 2001. Antioxidant capacity of cranberry is influenced by cultivar and storage temperature. J Agric Food Chem. 49:969–974.
Weiss EI, Lev-Dor R, Sharon N, Ofeck I. 2002. Inhibitory effect of a high-molecular-weight constituent of cranberry on adhesion of oral bacteria. Crit Rev Food Sci Nutr. 42:285–292.
Wu X, Beecher GR, Holden JM, Haytowitz DB, Gebhardt SE, Prior RL. 2004. Lipophilic and hydrophilic antioxidant capacities of common foods in the United States. J Agric Food Chem. 52:4026–4037.
Zheng W, Wang SY. 2003. Oxygen radical absorbing capacity of phenolics in blueberries, cranberries, chokeberries, and lingonberries. J Agric Food Chem. 51:502–509.

Section 2: Blueberry

INTRODUCTION

Blueberries (family, Ericaceae; genus, *Vaccinium*) are a soft fruit native to North America. The Northeast Native American tribes revered blueberries. Parts of blueberry plant were used as medicine and blueberry juice was used to treat coughs. Dried blueberries were added to stews, soups, and meats. In the 1880s, a blueberry canning industry began in the Northeast United States (Anon, 2004a).

Blueberry is a perennial crop that can produce for more than 30 years. The ideal conditions for cultivating blueberries require sandy soil, high in organic matter, pH of 4.5–5.0, and a water table 2–3 ft deep to provide moisture during the growing season. Blueberry planting takes about 2–3 year to establish and harvesting can begin in third or fourth growing season. However, the plants require 6–8 years to reach full production potential. As the berries do not mature uniformly, harvesting is done two to five times during the season. Table 21.5 lists blueberry production in leading countries of the world.

CLASSIFICATION

Three classes of blueberries (highbush or cultivated, rabbit eye, lowbush or wild blueberry) are commercially grown in the United States. Table 21.6

Table 21.5. Blueberry Production in Leading Countries (Mt)

Country	1999	2000	2001	2002	2003
1. United States	110,859	134,446	121,745	115,480	122,380
2. Canada	63,794	59,035	67,708	64,861	78,608
3. Poland	20,500	21,500	30,000	16,400	16,500
4. Ukraine	0	5,500	3,000	5,000	5,000
5. The Netherlands	3,800	3,800	3,800	4,000	4,000
6. Romania	5,000	4,000	3,000	2,500	2,500
7. Lithuania	3,000	3,000	2,000	1,600	2,000
8. Italy	1,647	1,896	1,880	1,811	1,800
9. New Zealand	1,100	1,740	1,750	1,750	1,750
10. France	1,000	1,000	1,000	1,000	1,000
World (Total)	212,920	238,537	238,703	217,272	238,358

Source: FAO (2004).

Table 21.6. Characteristics of Selected Blueberry Varieties

Variety	Fruit Characteristics
I. Highbush (cultivated blueberries)	
1. Bluecrop	A leading commercial variety released in 1952; high-dependable yields; height to 4–6 ft; mid-season; small scar; light blue color; firm; good sugar–acid balance; medium to large size; good for fresh pack and for processing
2. Duke	Released in 1987; medium size; light blue; firm berry with a small scar; good for IQF and fresh shipping
3. Elliott	Released in 1973; high-yielding berry with firm fruit and strong to mildly acid flavor; light blue color; medium size; late season fruit
4. Jersey	Released in 1928; mid to late season fruit; light blue color; medium size, firm texture; mild flavor; machine harvests well
5. Legacy	Released in 1995; insufficient winter hardiness; late season fruit; high fruit quality; stores well
6. Rubel	Released in 1911; machine harvests well; mid-season; small-medium size; medium blue color; firm texture; mild flavor
7. Sierra	Released in 1988; large, firm berries, high fruit quality; suitable for machine harvest
II. Southern highbush	
1. O'Neal	Low winter chill requirements (200–300 h); large size fruit with sweet flavor
2. Reveille	Medium size fruit, medium blue color, firm texture; excellent flavor; fruit can crack during extended periods of rain
III. Rabbiteye	
1. Brightwell	Large fruit, gritty texture; tough skin
2. Premier	Large fruit with excellent color and taste; one of the best eating rabbiteye blueberries
3. Powderblue	Small-medium fruit with light blue color; the berries have slight white powdery coat
4. Tifblue	Released in 1955. Regarded as a benchmark for all rabbiteye varieties; medium-large light blue fruit

Note: Early season varieties ripen in late June; late-season varieties continue through August.
Sources: Anon (2004b); Anon (2004c); Anon (2004d).

lists fruit characteristics of selected blueberry varieties.

1. *Highbush* (*Vaccinium corymbosum* L.; also called cultivated) blueberries: The first highbush varieties were transplanted from wild. There are at least 40 improved highbush varieties. Northern highbush and southern highbush blueberries have been developed through artificial selections from a limited germplasm. About 95% of cultivated blueberries consist of the northern highbush varieties grown in cooler climates of northern temperate zone. These berries generally need 160 frost-free days. Severe winter temperatures (about −20°C or below) will injure most highbush varieties. However "half-high" varieties, which are hybrids of highbush and lowbush blueberries (plants are 2–4 ft tall), can tolerate severe winter conditions (Anon, 2004b). The highbush plants are woody shrubs and may grow higher than 10 ft. The fruits are bigger than lowbush berries. Kalt et al. (2001) reported that the berry weight in highbush and southern highbush varied from <1 to 4.0 g/berry, although most highbush fruits were between 1 and 3 g/berry. In contrast, berries from lowbush clones were smaller and had a much narrower range of fruit weight; all clones had fruit no greater than 0.5 g. The fruit size distribution was symmetrical for the fruits of the lowbush genotypes but not for highbush genotypes analyzed. It should be noted that the berry size is also governed by genetic and environmental factors.
2. *Southern highbush* varieties with low chill tolerance have been developed to boost blueberry production in the southern United States. Besides being adaptable to growing conditions of this region,

Table 21.7. Characteristics of Selected Lowbush (Wild Blueberries) Clones at Various Maturity Stages

Maturity	Blomidon	Cumberland	Fundy
I. *Unripe*			
1. Berry fresh wt (g/berry)	0.217	0.220	0.305
2. Soluble solids (%)	7.52	8.71	7.70
3. Dry matter (%)	13.80	14.62	12.63
4. Firmness (N)	97.52	72.28	95.68
5. Titratable acidity (mEq/g dry wt)	0.730	0.928	1.227
6. Anthocyanins (mg/g dry wt)	4.38	6.77	7.08
II. *Ripe*			
1. Berry fresh wt (g/berry)	0.292	0.344	0.357
2. Soluble solids (%)	10.72	12.18	10.55
3. Dry matter (%)	15.95	16.23	14.03
4. Firmness (N)	79.17	49.00	72.76
5. Titratable acidity (mEq/g dry wt)	0.466	0.461	0.845
6. Anthocyanins	9.51	10.14	13.46
III. *Overripe*			
1. Berry fresh wt (g/berry)	0.631	0.600	0.761
2. Soluble solids (%)	13.13	14.22	11.92
3. Dry matter (%)	16.81	17.32	14.89
4. Firmness (N)	58.45	49.94	58.74
5. Titratable acidity (mEq/g dry wt)	0.354	0.332	0.652
6. Anthocyanins	8.88	8.91	12.86

Source: Kalt and McDonald (1996).

these new varieties inherit some characteristics of the northern highbush, such as a late bloom date and a shorter ripening period. With late bloom date, these varieties tend to face a lower risk of frost damage during the flowering stage, a critical period of fruit development. With the shorter ripening period, blueberries from Southern highbush varieties can be harvested around mid-April through late May, earlier than most rabbiteye blueberries. This extends marketing season and allows growers to take advantage of the premium prices typically available early in the season. The southern highbush berries tend to be not as soft as northern highbush berries and lend to machine harvested well.

3. *Lowbush* (*V. angustifolium* Ait; also called wild) berries are not commercially planted, but natural stands are pruned, sprayed, and harvested in the northeastern United States (primarily Maine) and eastern provinces of Canada. The strands of lowbush blueberries are made up of numerous wild clones; as a result the commercial lowbush berries are more heterogeneous than the commercial highbush or rabbiteye blueberries. The wild blueberry plants are about 1 ft high. The fruits are typically smaller in size than highbush or rabbiteye blueberries. Table 21.7 gives maturity-related differences for selected wild blueberry clones in fresh berry weight, dry matter, soluble solids, acidity, firmness, and anthocyanins. When fully ripened, the variety Cumberland had higher percentage of dry matter and soluble solids than the other varieties analyzed. However, this blueberry variety was also softer than others.
4. *Rabbit eye* (*V. ashei* Reade) blueberries grow in the relatively warmer climates of the southeastern United States. They are not winter-hardy, but drought tolerant. The plants are comparable to highbush blueberries; the berries have slightly fibrous mouth feel. They are sold for fresh consumption.

PRODUCTION AND CONSUMPTION

More than 80% of blueberries are commercially grown in United States and Canada. Next to strawberry, blueberries are the second most important berries in the United States (USDA, 2003). Outside the United States and Canada, Poland is the largest producer with about 10% of the world production (Table 21.5). In the United States, Michigan leads in cultivated blueberry production, followed by New Jersey, Oregon, North Carolina, Georgia, and Washington. The state of Maine is the largest producer of wild blueberries.

Most of the wild blueberries produced are utilized as processed. In contrast, a little over 40% of the cultivated blueberries are consumed as fresh and the remaining as processed. The states of New Jersey and North Carolina specialize in growing blueberries for the fresh market. In these states, more than 70% of blueberries produced are sold for fresh consumption.

Frozen berries are bulk frozen or individually quick frozen (IQF), a process that ensures freshness while preserving nutrients. Processed blueberries are used as an ingredient in manufacture of bakery, dairy, and convenience foods. Dried berries are used as ingredients in ready-to-eat cereals and many snack food products. Blueberries are also processed into jam/jellies, syrup, juice/concentrates, and baby food. Demand for blueberries in the United States has increased since the 1980s. The per capita fresh blueberry consumption, which averaged 0.30 pound annually during the 1990s, has increased to 0.34 pound during 2000–2003. The per capita consumption of frozen blueberries is also about 0.35 pound (USDA, 2003).

PHYSICOCHEMICAL AND NUTRITIONAL QUALITIES

Brix, Acidity, and Sugars

Unlike some fruits, this soft and small fruit does not require peeling or cutting before use. The berries are mildly sweet and not tart like cranberries, strawberries, and raspberries. In our laboratory, typical cultivated blueberry Brix (soluble solids) and titratable acidity were analyzed as about 12.0 and 0.80%, respectively. However, commercially available cultivated blueberries from Pacific Northwest can have higher Brix (17.5) and titratable acidity (1% as citric acid). Conner et al. (2002) reported Brix and acidity of cultivated blueberries in the range of 11–12.6° and 0.90–2.46%, respectively. On an average, the wild blueberries have lower acidity (0.4–0.7% as citric) than cultivated blueberries. Their Brix ranges from 7° to 14° depending on harvest time and year of production. Both cultivated and wild blueberries contain mostly glucose and fructose.

Flavor

Blueberries have fruity and floral notes due to esters, ethyl acetate, and 3-isopropyl-butyrate. Other flavor compounds are: branched aldehydes 3-methyl-butyraldehyde and 2-methyl-butyraldehyde (fruity notes); benzaldehyde (almond flavor); hexanol, heptanal, and nonanal; terpenes, 1,8-cineole, and linalool with traces of cymene, limonene, 1-terpineol, and 3-carene. Citral, which has a strong lemon note, was identified in Ranococas and Bluecrop varieties. Alcohols, 1-butanol, 1-pentanol, 1-hexanol, 2-ethyl-1-hexanol, 1-heptanol, 1-octanol, and 1-nonanol, and several furan derivates have also been found (Anon, 2004e).

Anthocyanins, Phenolics, and Antioxidant Capacity

The typical blue color of blueberries is due to the presence of anthocyanin pigments. Zheng and Wang (2003) reported four major anthocyanins, malvidins (purple color), petunidins (blue-purple), delphinidins (blue-violet), and cyanidin (red) in highbush blueberries (cv. Sierra). The total anthocyanin pigment concentration measured was 1.1 mg/g, of which malvidins were 34.5%; petunidins, 31.1%; delphinidins, 18.7%; and cyanidins, 15.8%. Individually, delphinidin 3-galactoside was the lead pigment (17.0%) followed by malvidin 3-galactoside (13.6%), petunidin 3-galactoside (13.0%), and cyanidin 3-galactoside (13.0%). Kalt et al. (2001) reported anthocyanins and phenolics content in highbush blueberries as 1.18 mg/g (cyanidin 3-glucoside equivalent) and 1.91 mg/g (gallic acid equivalent) fresh weight, respectively. They reported anthocyanin and phenolic contents in lowbush blueberries as 1.63 mg/g (cyanidin 3-glucoside equivalent) and 3.76 mg/g (gallic acid equivalent), respectively. The average antioxidant capacities (oxygen radical absorbency capacity, or ORAC) of highbush and lowbush berries were 45.2 and 69.8 μmol of Trolox equivalent (TE) per gram fresh weight (Kalt et al., 2001). In highbush blueberry (cv. Sierra), anthocyanins and chlorogenic acid accounted for 56.3% and 20.9% of antioxidant activity, respectively (Zheng and Wang, 2003).

It is suggested that unlike fruits such as grapes, for which significantly higher antioxidant levels are found in seeds due to tannins, in blueberries, majority of the antioxidants are concentrated in skins. Further, blueberry leaf tissues had greater phenolics and

Table 21.8. Antioxidant Activity, Anthocyanins and Phenolic Concentrations of Blueberries

Blueberries	ORAC (μmol TE/g)	Anthocyanins (mg/100 g)	Phenolics (mg/100 g)
I. *Northern highbush*			
1. Bluecrop	17.0 ± 1.2	93.1 ± 1.6	189.8 ± 10.9
2. Duke	18.1 ± 0.2	101.2 ± 1.5	181.1 ± 10.4
3. Jersey	20.8 ± 0.6	100.1 ± 2.3	206.2 ± 4.1
4. Rubel	37.1 ± 0.5	235.4 ± 6.1	390.5 ± 6.5
II. *Southern highbush*			
1. O'Neal	16.8 ± 1.9	92.6 ± 4.6	227.3 ± 6.9
III. *Rabbiteye*			
1. Tifblue	23.0 ± 2.6	87.4 ± 5.6	361.1 ± 16.6
2. Climax	13.9 ± 4.1	90.8 ± 5.2	230.8 ± 7.3
3. Brightwell	15.3 ± 2.8	61.8 ± 1.8	271.4 ± 12.7
IV. *Lowbush*			
1. Cumberland	27.8 ± 2.6	103.6 ± 0.9	295.0 ± 13.2
2. Blomidin	28.8 ± 2.1	91.1 ± 0.7	313.0 ± 6.4
3. Fundy	42.0 ± 2.0	191.5 ± 2.5	433.0 ± 45.5

Source: Prior et al. (1998).

antioxidant capacities than the fruit tissues (Ehlenfeldt and Prior, 2001). Wu et al. (2004) reported that based on a serving size (one cup containing 145 g blueberries), the wild (lowbush) blueberry had the highest total antioxidant capacity (13,247 μmol TE) among fruits analyzed followed by cultivated (highbush) blueberry (9019 μmol TE) (Table 21.8).

Nutritional Qualities

Table 21.9 provides the nutritional values of blueberry and its products. As is seen, raw blueberries have low calories and can be a source of fiber, natural sugars, vitamins, and minerals. The dried blueberries are high in dietary fiber and potassium.

Health Considerations

In addition to potential effects on cancer and heart disease, phytochemicals present in antioxidant-rich fruits such as blueberries are suggested to reverse the age-related declines in balance and coordination, muscle strength, and brain functions (learning and memory) (Joseph et al., 1999).

Similar to bilberry, intake of blueberry is believed to improve eye health and reduce eyestrain. This beneficial effect of blueberries on eyes is believed to be one of the reasons for increased demand of blueberries in Japan.

BLUEBERRY PRODUCTS

As indicated previously, most of the wild blueberries and about 60% of cultivated blueberries are processed into various products. IQF and frozen blueberries can be used directly in some applications. The IQF process consists of various steps beginning with passing the berries through a multi-layer shaker dock that removes, sticks, leaves, etc., which are too large to fall through the holes in the dock. Good berries are led into successive pans that contain water to separate heavy solids, such as stones, etc. Subsequently, the berries are sanitized with chlorine water spray (20–35 ppm chlorine). For the IQF process, the berries pass through a freezer tunnel with a high-velocity cold air of −40°F for about 10 min. Then they are led through a de-stemming reel to remove stems before being laser sorted and inspected for defects. The culled berries are generally made into puree.

Infused Frozen Blueberries

Infused frozen blueberries, which are infused with sweeteners to about 25–35°Brix and pasteurized, have been finding increasing use in bakery industry where direct use of frozen products can create problems because of excess and freely available moisture. Further, the unpasteurized berries can be a source of enzymes such as amylase, which affect pie fillings (lower viscosity) if added directly.

Table 21.9. Nutrient Values of Blueberries

Nutrients/100 g	Blueberries, Raw[a]	Canned Blueberry in Syrup[a]	Infused-Dried Cultivated Blueberries[b]	Infused-Dried Wild Blueberries[b]	Infused-Dried Organic Wild Blueberries[b]	Dehydrated Blueberries[c]
Calories (kcal)	57.0	88.0	290	305	280.0	353.0
Calories from fat (kcal)	3.0	3.0	20.0	19.0	11.0	21.5
Total fat (g)	0.33	0.33	2.19	2.06	1.17	2.39
Saturated fat (g)	0.028	0.027	0.3	0.3	0.1	NA
Polyunsaturated fat (g)	0.146	0.144	0.4	0.8	0.8	NA
Monounsaturated fat (g)	0.047	0.047	1.4	1.0	0.3	NA
Cholesterol (mg)	0.00	0.0	0.0	0.0	0.0	0.0
Sodium (mg)	1.0	3.0	18.0	15.0	22.0	38.0
Potassium (mg)	77.0	40.0	252.0	166.0	144.0	561.0
Total carbohydrate (g)	14.49	22.06	77.9	80.3	78.6	89.0
Total fiber (g)	2.4	1.6	16.6	15.4	15.1	8.19
Total sugar (g)	9.96	20.46	61.2	64.9	60.5	80.80
Sucrose (g)	0.11	NA	NA	NA	NA	NA
Glucose (g)	4.88	NA	NA	NA	NA	NA
Fructose (g)	4.97	NA	NA	NA	NA	NA
Protein (g)	0.74	0.65	2.03	2.43	0.84	4.22
Calcium (mg)	6.0	5.0	255.0	380.0	49.0	38.00
Vitamin C (mg)	9.7	1.1	<0.10	76.0	<0.10	81.90
Vitamin A (IU)	54.0	36.0	14.0	33.0	4.0	630.0
Water (g)	84.21	76.78	16.8	13.8	NA	3.00

NA: Not Available.

[a]Data from USDA: http://www.nal.usda.gov/fnic/foodcomposition/cgi-bin/list_nut_edit.pl.

[b]Data from Graceland Fruit Inc, Frankfort, MI, USA (www.gracelandfruit.com).

[c]Data from Esha Nutritional Database.

U.S. Patents 6,254,919 (Phillips 2001) and 4,713,252 (Ismail 1987) described methods to prepare sugar-added blueberry products having moisture in the range of 30–50% and 10–40%, respectively.

Blueberry Juice

Blueberry juice can be made from both fresh and frozen berries. With this soft fruit, a screw juice expresser with appropriate screens to separate the juice from other parts of the fruit can be used. The juice can be further clarified with bentonite and gelatin, filtered, and pasteurized at 85°C for 90 s (Chung et al., 1994). A single-strength blueberry juice is about 8–12°Brix. Frozen blueberry juice concentrates of 45–65°Brix can be made by vacuum concentrating single-strength blueberry juice and freezing to less than −18°C.

Often blueberry juice is blended with other juice products.

Blueberry Puree

Single-strength blueberry puree and puree concentrates can be made by crushing berries, passing through a pulper/finisher, pasteurizing, cold filling (for single-strength puree and vacuum concentrating to make puree concentrates) and freezing as above. These products are suitable for use in sauce and fillings.

Canned Blueberries

Canned blueberries can be light or heavy syrup packed or water packed. For this, fresh or frozen blueberries are placed in cans, light syrup or water is added to cover the head space, and cans are sealed and heat processed at about 93–95°C (200–203°F) with 25–30-min holding time.

Blueberry bakery fillings can be made from fresh/frozen blueberries and/or blueberry puree or concentrates by adding sweeteners and stabilizers according to requirements, heat processing, and packaging.

INFUSED-DRIED BLUEBERRIES

Infused-dried and freeze-dried blueberries are commercially available for use in ready-to-eat cereals and as snacks. Infused-dried blueberries, which are produced by infusing either wild or cultivated blueberries with a sweetener prior to drying, typically have moisture content of 10–14% and a water activity of 0.45–0.60. These products have fresh-like texture, color, and flavor, and are shelf stable. They are utilized as ingredients in snacks, breakfast bars, energy bars, and ready-to-eat cereals.

FREEZE-DRIED BLUEBERRIES

Freeze-dried blueberries are light, crispy, of low bulk density (about 0.10), and their water activity is slightly above 0.20. These products have found increasing use in ready-to-eat cereals because of their light texture, low bulk density, and water activity. The freeze-drying is based on a direct conversion of ice to vapor (sublimation) to dry a product. Freeze-dried blueberries can hold their shape and not look shrunk or shriveled. Freeze drying process utilizes IQF blueberries. The frozen berries are placed in a freeze-dryer heating plate having same temperature as the frozen berries and vacuum (e.g., less than 6 mbar at which the boiling point of water is less than 0°C) is applied. Subsequently, the plates are heated to a desired temperature and the product dried to a desired moisture level. Generally, freeze-drying is a lengthy process and the cost of the freeze-dried products is about 3–5 times higher than products dried by other methods.

REFERENCES

Anon. 2004a. Blueberries. http://www.ushbc.org/blueberry.htm.

Anon. 2004b. Blueberry Information from Michigan State University Extension: http://www.msue.msu.edu/fruit/bbvarbul.htm.

Anon. 2004c. Blueberries Varieties Evaluated: http://mtvrnon.wsu.edu/frt_hort/blueberry.htm.

Anon. 2004d. Blueberry Information from University of California Cooperative Extension. http://www.mastergardeners.org/picks/bluvar.html.

Anon. 2004e. Volatile Organic Composition of Blueberries. http://www.sisweb.com/conference/applenote/app-43.htm.

Conner AM, Luby JJ, Hancock JF, Berkheimer S, Hanson EJ. 2002. Changes in Fruit Antioxidant Activity Among Blueberry Cultivars During Cold-Temperature Storage. J Agric Food Chem 50, 893–898.

Chung T-S, Siddiq M, Sinha NK, Cash JN. 1994. Plum Juice Quality Affected by Enzyme Treatment and Fining. J Food Sci 59(5), 1065–1069.

Ehlenfeldt MK, Prior RL. 2001. Oxygen Radical Absorbance Capacity (ORAC) and Phenolic and Anthocyanin Concentrations in Fruit and Leaf Tissues of Highbush Blueberry. J Agric Food Chem 49, 2222–2227.

FAO. 2004. FAOSTAT—Agricultural Data. http://faostat.fao.org/faostat/collections?subset=agriculture.

Ismail AA. 1987. U.S. Patent 4,713,252. Process for Producing a Semi-moist Fruit Product and the Products Therefrom.

Joseph JA, Shukitt-Hale B, Denisova NA, Bielinski D, Martin A, McEwen JJ, Bickford PC. 1999. Reversal of Age-Related Declines in Neuronal Signal Transduction, Cognitive, and Motor Behavioral Deficits with Blueberry, Spinach, or Strawberry Dietary Supplementation. J Neurosci 19(18), 8114–8121.

Kalt W, McDonald JE. 1996. Chemical Composition of Lowbush Blueberry Cultivars. J Am Soc Hort Sci 121(1), 142–146.

Kalt W, Ryan DAJ, Duy JC, Prior RL, Ehlenfeldt MK, Kloet SPV. 2001. Interspecific Variation in Anthocyanins, Phenolics, and Antioxidant Capacity Among Genotypes of Highbush and Lowbush Blueberries (Vaccinium Section cyanococcus spp.). J Agric Food Chem 49, 4761–4767.

Phillips RM. 2001. U.S. Patent # 6,254,919. Preparation of Shelf stable Blueberries and moist shelf stable product.

Prior RL, Cao G, Martin A, Sofic E, McEwen J, O'Brien C, Lischner N, Ehlenfeldt M, Kalt W, Krewer G, Mainland MC. 1998. Antioxidant Capacity as Influenced by Total Phenolic and Anthocyanin Content, Maturity, and Variety of Vaccinium Species. J Agric Food Chem 46, 2686–2693.

USDA. 2003. Fruit and Tree Nuts Outlook/FTS-305/ July 30, 2003.
Wu X, Beecher GR, Holden JM, Haytowitz DB, Gebhardt SE, Prior RL. 2004. Lipophilic and Hydrophilic Antioxidant Capacities of Common Foods in the United States. J Agric Food Chem 52, 4026–4037.
Zheng W, Wang SY. 2003. Oxygen Radical Absorbing Capacity of Phenolics in Blueberries, Cranberries, Chokeberries, and Lingoberries. J Agric Food Chem 51, 502–509.

Section 3: Currant and Gooseberry

INTRODUCTION

Currant (*Ribes sativum*) and gooseberry (*Ribes grossularia*) are small fruits that by and large serve a niche market. The plants are shrubby bush (small tree) that grow well in cool climates. Unlike gooseberries, which have soft thorns, currants are thornless. Currants grow in grape-like clusters and can be red, black, pink, or white in color. Gooseberries are generally bright red fruit; however, they can also be pale green. A hybrid between gooseberry and black currants produces jostaberries (*R. nidigrolaria*). These fruits are nearly black in appearance and 2–3 times the size of a typical currant.

PRODUCTION

Table 21.10 gives data on production of currant in top 10 leading countries and the world aggregate. The Russian Federation leads world currant production and is closely followed by Poland and Germany. These three countries are also significant producers of gooseberry (Table 21.11).

Table 21.10. Currants Production in Leading Countries (Mt)

Country	1999	2000	2001	2002	2003
1. Russian Federation	206,000	220,000	203,000	228,000	222,000
2. Poland	153,353	146,780	175,300	157,434	192,475
3. Germany	155,400	158,300	148,100	148,000	148,000
4. Austria	19,537	22,861	19,072	19,581	18,342
5. United Kingdom	8,400	12,300	15,400	12,900	18,800
6. Ukraine	14,594	19,887	19,372	16,500	17,000
7. Czech Republic	22,792	18,089	16,597	13,487	13,472
8. Hungry	12,304	11,848	12,194	8,138	12,000
9. France	8,364	8,382	8,604	9,191	9,145
10. Denmark	5,000	4,000	3,500	3,500	3,500
World (Total)	633,809	650,900	650,543	645,543	687,806

Source: FAO (2004).

Table 21.11. Gooseberry Production in Leading Countries (Mt)

Country	1999	2000	2001	2002	2003
1. Germany	84,800	88,200	90,300	90,100	90,100
2. Russian Federation	38,000	35,000	30,000	50,000	42,000
3. Poland	31,463	28,514	29,622	21,738	20,345
4. Ukraine	4,500	8,900	8,000	7,500	7,000
5. Czech Republic	8,350	6,824	4,816	4,091	4,046
6. Hungary	4,641	4,649	3,683	4,000	4,000
7. Austria	1,668	1,815	1,800	1,800	1,800
8. United Kingdom	1,700	1,500	1,380	1,580	1,610
World (Total)	177,031	177,629	171,622	182,557	172,553

Source: FAO (2004).

CURRANT AND GOOSEBERRY VARIETIES

The following is a list of selected currant and gooseberry varieties.

SELECTED CURRANT VARIETIES

1. *Titania*: This is a black currant variety, which produces good quality large fruits that can be machine harvested. The plants are upright and vigorous and can reach a height of 2 m. Normally, the plants reach full maturity in three seasons as opposed to four to five seasons for most popular varieties.
2. *Ben Alder*: This is a late season black currant. The plants are compact and suitable for machine harvesting. The fruits are small but high in anthocyanins and vitamin C. Because of color stability, it is liked for juice production.
3. *Red Lake*: This is a mid-season ripening red currant variety. Fruits are large and of good quality. This is a good red currant for juice production.
4. *Jonkheer Van Tets*: This is a popular early season red currant. The plants are susceptible to frost damage.The fruits are medium size and of good flavor.

SELECTED GOOSEBERRY VARIETIES

1. *Invitica*: This variety produces a pale green, large fruit with good flavor suitable for fresh market and processing. The plants of this variety are vigorous with spreading bush.
2. *Xenia*: This variety produces dark red fruit with oval shape. The fruit ripens early- to mid-season.
3. *Hinnonmaki Red*: This variety produces dark red fruit of medium size. The flavor is tangy-sweet. It adapts well to different growing conditions and can be machine harvested.

PHYSICOCHEMICAL AND NUTRITIONAL QUALITIES

BRIX AND FLAVOR

Black currant has higher Brix (16.2) than red (9.7) and white counterparts (13.0). Similarly, the Brix of red gooseberries is slightly higher (12.0) than pale green/yellow varieties (9.3) (Maatta-Riihinen et al., 2004). The aroma of black currants emanates from naturally present ester compounds (methyl acetate, ethyl acetate, methyl butanoate, ethyl butanoate, etc.), terpenes (3-Caene, terpinolene and *cis*- and *trans*-ocimene are major terpenes), terpenoids, alcohols, aldehydes, and ketones (Mikkelsen and Paul, 2002). Generally, esters are linked with fruity, sweet notes; terpenes with nutty; dry, camphor with chemical type of sensation; and aldehydes and ketones with old, earthy notes.

ANTHOCYANINS, PHENOLICS, AND ANTIOXIDANT CAPACITY

Fifteen anthocyanins have been identified in black currant extracts. Of these, delphinidin 3-glucoside, delphinidin 3-rutinoside, cyanidin 3-glucoside, and cyanidin 3-rutinoside accounted for >97% of the total anthocyanin contents. Further, the amounts of anthocyanin 3-O-rutinosides were higher than 3-O-glucosides (Slimestad and Solhelm, 2002). The notable anthocyanin in red currants was cyanidin 3-O-sambubioside (72 mg/kg fresh weight), and as expected red currants did not contain any delphinidins (blue-violet pigments) (Maatta et al., 2003). Similarly, red gooseberries contain only cyanidins (240 mg/kg fresh weight) (Maatta-Riihinen et al., 2004). Table 21.12 shows concentration of anthocyanin, flavonol, and total phenolics in black currant, red currant, and gooseberry (Kahkonen et al., 2001). As can be seen black currants contain a very high concentration of these bioactive compounds.

Table 21.12. Anthocyanins, Flavonol, and Total Phenolic Content (mg/100 g) of Black Currants, Red Currants, and Gooseberries

Fruit	Anthocyanins[a]	Flavonol[b]	Total Phenolics[c]
1. Black currant	756–1297	72–87	2230–2790
2. Red currant	113	9.5	1400
3. Gooseberry	83	51	1320

Source: Kahkonen et al. (2001).
[a]As cyanidin 3-glucoside.
[b]As rutin.
[c]As gallic acid.

Table 21.13. Nutritional Values of Selected Currants and Gooseberries

Nutrients/100 g	Black Currants (European), Raw	Currants, Red and White, Raw	Gooseberries Raw	Gooseberries, Canned, Light Syrup Pack
Calories (kcal)	63.0	56.0	44.0	73.0
Total fat (g)	0.41	0.20	0.58	0.20
Saturated fat (g)	0.034	0.017	0.038	0.013
Polyunsaturated fat (g)	0.179	0.088	0.317	0.110
Monounsaturated fat (g)	0.058	0.028	0.051	0.018
Cholesterol (mg)	0.00	0.00	0.0	0.00
Sodium (mg)	2.0	1.0	1.0	2.0
Potassium (mg)	322.0	275	198.0	77.0
Carbohydrate (g)	15.38	13.80	10.18	18.75
Total fiber (g)	NA	4.3	4.3	2.4
Sugars (g)	NA	7.37	NA	NA
Protein (g)	1.40	1.40	0.88	0.65
Calcium (mg)	55.0	33.0	25.0	16.0
Vitamin C (mg)	181.0	41.0	27.7	10.0
Vitamin A (IU)	230.0	42.0	290.0	138.0
Water (g)	81.96	83.95	87.87	80.10

NA: Not available.
Source: USDA: http://www.nal.usda.gov/fnic/foodcomposition/cgi-bin/list_nut_edit.pl.

Nutritional Quality

Table 21.13 shows typical nutritional data of currants and gooseberry. Black currants have high vitamin C content (Hakkinen et al., 2000). Black currants also have higher vitamin A and potassium content.

PROCESSED PRODUCTS

Various processes can extend availability of these seasonal fruits throughout the year. Like other fruits, currants and gooseberries can be frozen as IQF or block frozen, and stored for processing.

These fruits find increasing use in pie fillings, jelly, jam, juice, etc. Black currant juice is liked in many parts of Europe.

Black Currant Juice

A process for making black currant juice (Mikkelsen and Poll, 2002) is as below:

Thaw frozen fruit → Mill/Crush → Heat (75°C/2 min) → Cool to 50°C → Add enzyme (Pectinase, Grindamyl LB, Danisco @ 0.4 ml per kg of black currant) → Hold at 50°C for 2 hr → Press (using High pressure Tincture press HP-5 M) → Pasteurize at 98°C for 1 min → Clarify with gelatin (0.03 g galatin per liter juice and stir) → Add kieselsol (0.25 g per liter juice, without stirring, to further aid in clarification) → Filter at room temperature → Pasteurize at 98°C for 30 sec → Fill in package → Chill → Store

In this process, approximately 75% of anthocyanins remain in the juice. Important aroma compounds in this juice were terpenes, 3-carene, α-terpinolene, and *cis*-ocimene, and esters, ethyl acetate, methyl butanoate, and ethyl butanoate. During juice processing, the relative concentration of methyl butanoate, ethyl butanoate, and ethyl hexanoate decreased to less than 10%.

Black Currant Nectar

Iversen (1999) made black currant nectar of 16°Brix (standardized with honey and sugar) and a pH of 3.0 from frozen berries of (cv. Ben Lomond). The nectar was pasteurized at 80°C for 27 s in an APV Pasilac plate heat exchanger, model 1090, at flow rate of 180 l/h. Pasteurization caused about 3–18% and 8–9% loss of ascorbic acid and anthocyanins, respectively. The choice of pectinex enzyme used in juice extraction mainly affected the recovery of ascorbic acid and not anthocyanins.

Black Currant and Gooseberry Juice Concentrate

Black currants juice can be concentrated to 65°Brix. This concentrate usually has 16–18% acidity as citric acid. This concentrate can be used to make single-strength juice of 11°Brix. Red currant and gooseberry concentrates can also be used to make single-strength juice of 10.5°Brix and 8.3°Brix, respectively.

Jelly and Jam

Currants and gooseberries can be made into jellies (45 parts fruit components and 55 parts sweetener solids; the finished soluble solids content is 65°Brix) and jams and preserves (47 parts by weight of the fruit component to 55 parts of sugar; the finished product should of minimum 65°Brix).

REFERENCES

FAO. 2004. FAOSTAT-Agricultural Data http://faostat.fao.org/faostat/collections?subset=agriculture.

Hakkinen SH, Karenlampi SO, Mykkanen HM, Torronen AR. 2000. Influence of Domestic Processing and Storage on Flavonol Contents in Berries. J Agric Food Chem 48, 2960–2965.

Iversen CK. 1999. Black Currant Nectar: Effect of Processing and Storage on Anthocyanin and Ascorbic Acid Content. J Food Sci 64, 37–41.

Kahkonen MP, Hopia AI, Heinonen M. 2001. Berry Phenolics and Their Antioxidant Activity. J Agric Food Chem 49, 4076–4082.

Maatta KR, Kamal-Eldin A, Torronen AR. 2003. High-Performance Liquid Chromatography (HPLC) Analysis of Phenolic Compounds in Berries with Diode Array and Electrospray Ionization Mass Spectrometric (MS) Detection: Ribes Species. J Agric Food Chem 51, 6736–6744.

Maatta-Riihinen KR, Kamal-Eldin A, Mattila PH, Gonzalez-Paramas AM, Torronen AR. 2004. Distribution and Contents of Phenolic Compounds in Eighteen Scandinavian Berry Species. J Agric Food Chem 52, 4477–4486.

Mikkelsen BB, Poll L. 2002. Decomposition and Transformation of Aroma Compounds and Anthocyanins during Black Currant (Ribes nigrum L.) Juice Processing. J Food Sci 67(9), 3447–3455.

Slimestad R, Solhelm H. 2002. Anthocyanins from Black Currants (Ribes nigrum L.). J Agric Food Chem 50, 3228–3231.

22
Date Fruits Production and Processing

Jiwan S. Sidhu

INTRODUCTION

Date fruit (*Phoenix dactylifera* L.) has always occupied an important place in the diets of people in the Arab world. In spite of the enormous socioeconomic transformation that has taken place in this part of the world, date fruit continues to form an essential component of the daily diet. Unlike many other fruits, dates can be consumed or used for human consumption in every stage of fruit development. Date fruit is usually classified into four main maturity stages, i.e., *kimri, khalal, rutab,* and *tamer* (Hussein, 1970). *Kimri* stage fruit is young, green in color with hard texture, thus can be used for the preparation of pickles and chutney. Depending on the cultivar, the date fruit changes its green color during the next stage. The *khalal (or bisr)* stage date fruit attains maximum size and weight; develops a typical yellow, purplish-pink, red, or yellow-scarlet color but retains a firm texture and is largely consumed raw as fresh fruit or can be used for jam, butter, or dates-in-syrup (Al-Hooti et al., 1995b). During the *rutab* stage, half of the fruit becomes soft, less astringent, and sweeter but darkens in color and can be used for jam, butter, date bars, and date paste. *Rutab* stage fruits of a few cultivars are being popularly consumed as fresh. During the final or ripe stage, *tamer*, the whole fruit attains maximum total solids, highest sweetness and lowest astringency with a dark brown color, a soft texture, and a wrinkled appearance. The physical appearance of a few date cultivars at various stages of maturity is shown in Figures 22.1a–g. For prolonged shelf life and storage, *tamer* stage fruits can also be sun dried. A major portion of the date produce enters world trade as *tamer* fruit for human consumption (Mikki et al., 1986).

In the Arabian countries, the value and importance of the date palm tree and its fruits have immense importance. Every household feels proud to grow at least one tree in their backyard for fruit production. Date tree is also mentioned and honored in the Holy Quran, and recommended by the Prophet—"Peace Be Upon Him." Date fruit is known to be a rich source of carbohydrates (mainly glucose and fructose sugars) and certain vitamins, minerals, and dietary fiber (Al-Hooti et al., 1997f). Date fruit is cherished for

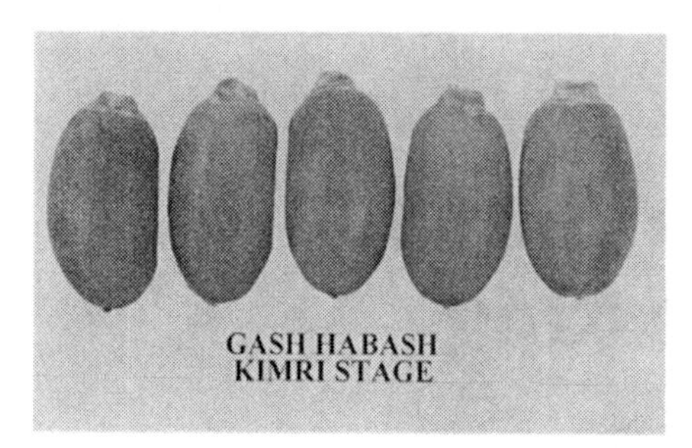

Figure 22.1a. *Gash Habash* cultivar—*kimri* stage of maturity.

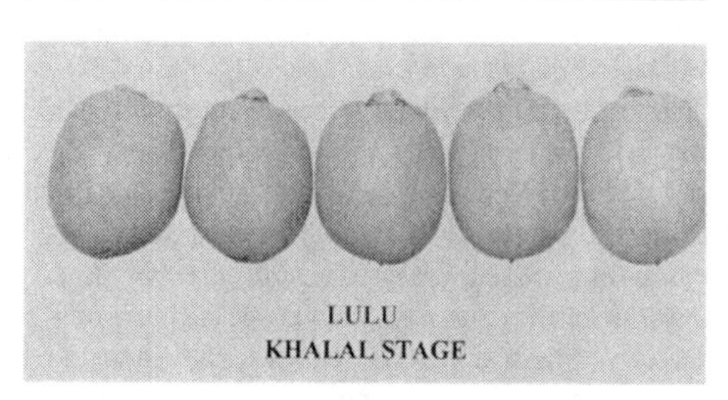

Figure 22.1b. *Lulu* cultivar—*khalal* stage of maturity.

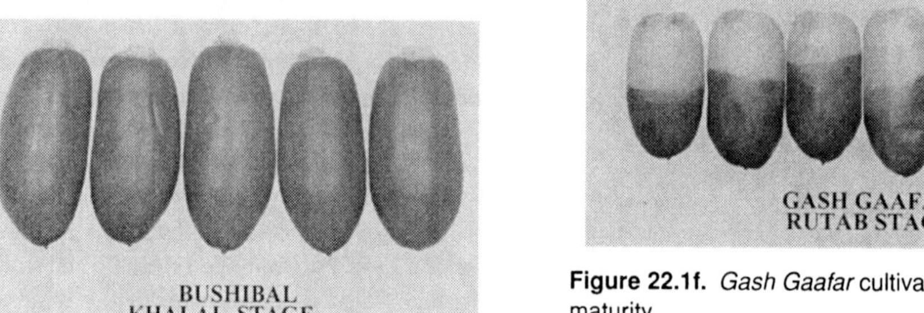

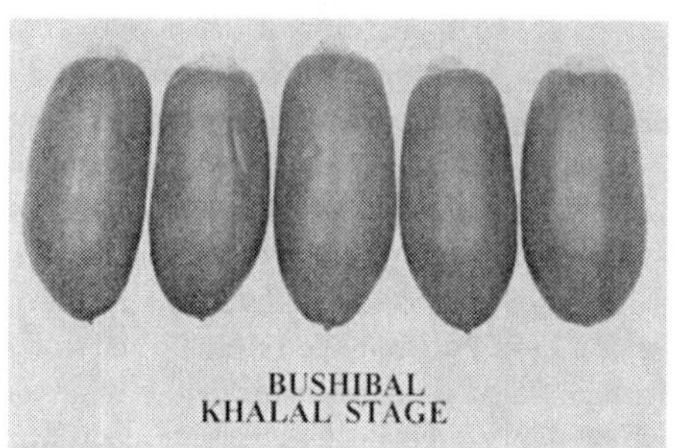

Figure 22.1c. *Bushibal* cultivar—*khalal* stage of maturity.

its flavor and nutritive value all over the Arab world, as this plant is well suited to grow in arid regions where there are hot and dry climates with limited rainfall. Date trees are usually propagated vegetatively by offshoots. The trees grown from seeds show high genetic heterozygosity that result in not true-to-type male and female plants (Toutain, 1967). To overcome

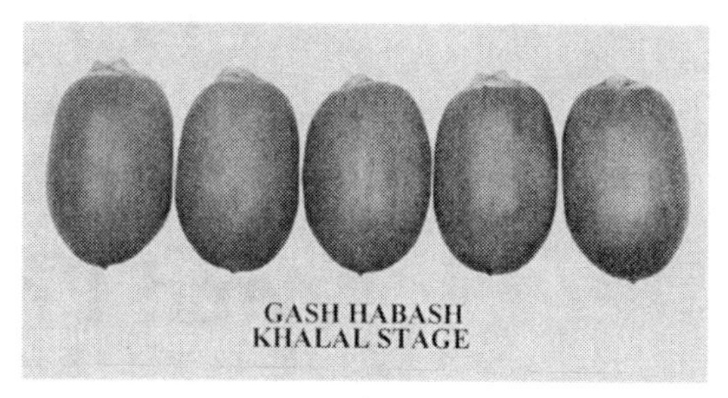

Figure 22.1d. *Gash Habash* cultivar—*khalal* stage of maturity.

Figure 22.1e. *Shahla cultivar—khalal* stage of maturity.

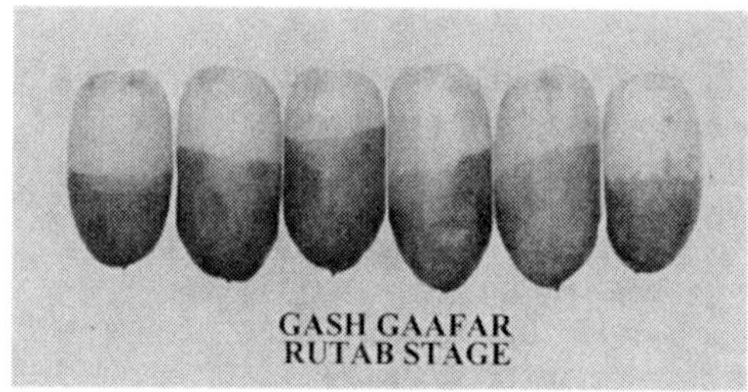

Figure 22.1f. *Gash Gaafar* cultivar—*rutab* stage of maturity.

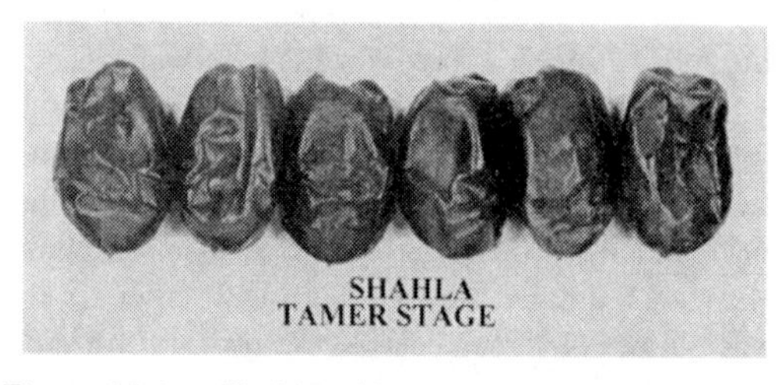

Figure 22.1g. *Shahla* cultivar—*tamer* stage of maturity.

these drawbacks of longer generation times and genetic heterozygosity, newly emerging tissue culture techniques are now being employed for the propagation of date palm trees (Zaid, 1986; Sudhersan et al., 1993a, b).

Recent research is producing a lot of evidence to strengthen the evidence of a vital link between the food we consume and our ultimate health. Our knowledge about the beneficial role of various food constituents (such as date fruit nutrients) in the prevention and treatment of various disease conditions is growing rapidly. New terms, such as functional foods, medical foods, nutritional foods, hypernutritional foods, designer foods, therapeutic foods, super foods, prescriptive foods, nutraceuticals, and pharmafoods, are being commonly used in one context or another by food scientists as well as by consumers. Generally speaking, functional food can be defined as any food that provides benefits to one's health, physical appearance, endurance, or state of mind (mood) in addition to its usual nutritive value. Japan has been a leader in stimulating the growth of the functional food market, but this trend is picking up worldwide, especially in the past 10 years. The functional food areas relating nutrition and health will definitely revolutionize the food industry in the future. It is predicted that these foods will be the fastest growing segment of the food industry during the coming decades (Kevin, 1995).

Keeping in view the importance of date palm trees and date fruits to this Arabian region, the history of date cultivation, cultivars, production, marketing, physicochemical characteristics, preharvest treatments, postharvest handling of fresh fruits, storage problems, nutritive value and health benefits, date-fruit-based value-added products, standards, regulations, and future research needs are covered in this chapter.

HISTORY OF DATE CULTIVATION

The date palm (*P. dactylifera* L.) belongs to the Arecaceae (or Palmae) family and has been cultivated for a long time in the semiarid and desert areas of the Middle East, Pakistan, and India; in California, USA; in the Canary Islands; and in the northern African countries for fuel, shade, fiber, food, and as building material (Nixon, 1951). In addition to date palm, this family also includes other kinds of palm trees such as oil palms, coconut palms, and Washington palms. Although it is not known exactly where the date palms originated, it is suggested that they first originated in Babel, Iraq or in Dareen or Hofuf, Saudi Arabia, or Harqan, an island in Bahrain, from where it spread to other places (Marei, 1971). Date palms were first introduced to Andalus by the Arabs, during the 7th and 8th centuries and later spread throughout the deserts of the Middle East and North Africa by the Bedouin tribes of the Arab countries. It is also believed that date palm trees were introduced to India after the victory over India by Alexander, the Great around 327 B.C. In around 1769, date seeds were introduced to the arid areas in the United States, namely the states of New Mexico, California, and Arizona (Al-Tayeb, 1982). Evidently, the Middle East and North African countries are the major date-fruit producing countries in the world, although the United States also produces sizable quantities of date fruits in North America, and Spain in the European Union.

DATE PALM CULTIVATION

For millennia, date fruits have been an important part of the dietary patterns of the people in the Middle Eastern countries. Being a tree of the desert, it is well established in this region. Date trees grow to about 23 m tall; the stems are strongly marked with the pruned stubs of old leaves. The plant has a crown of graceful, shining, pinnate leaves, measuring about 5 m long. Depending on the cultivar and other agricultural practices, 1000 dates may appear in a single bunch and weigh 8 kg or more. Date trees start bearing fruit in 4–5 years and reach their full potential in 10–15 years, producing 40–80 kg of fruit per tree. Although date trees can live as long as 150 years, due to declining fruit production, commercial cultivators replace them at an earlier age. Propagation of date palm trees through seeds is not successful because of the high genetic variation (not true-to-type) in male and female seedlings. Thus, date palms are mainly propagated by offshoots produced by younger palm trees, and these offshoots maintain true-to-type make up of the cultivars (Nixon and Farr, 1965). With the recent developments in tissue culture techniques, date palm breeding has been given a boost. Now, this technique is being used extensively to clone a wide range of economically important palms such as coconut palms (Eeuwens and Blake, 1977), oil palms (Rabechault and Martin, 1976), and date palms (Tisserat, 1979).

Date cultivation is not mechanized in the Middle Eastern and North African countries; yet most of the operations such as pollination, pruning, and harvesting are carried out manually. Attempts are now being made to mechanize date production due to the rising cost of production and shortage of skilled labor (Shabana and Mohammad, 1982). The number of palm trees per hectare is an important parameter that determines the yield and quality of date fruits. The number of trees per hectare varies from 100 to 250 depending on the cultivars. For most cultivars, palm trees are usually planted in square patterns with about 9 m gap between the rows. For the initial few years, intercropping of fodder or vegetable crops can be carried out. In the United States, most operations, such as pollination, bunch thinning, pruning, bagging, and harvesting, are carried out with the help of machines (Perkins and Brown, 1966; Brown, 1982). Harvesting is the most expensive operation and consists of handpicking individual, mature fruits from each bunch. Pollination is the next most expensive operation. To ensure good yields of quality fruits, pollination must be performed according to the opening of the flowers. One part of pollen is usually mixed with 6–10 parts of wheat flour a day or so before the application to the female flowers (Brown et al., 1969). Thinning is carried out at that time, bunches are tied down so that optimal amounts of fruits are retained in these bunches to attain an acceptable fruit size.

With regard to the irrigation requirements of date trees, an old Arab saying, "The date palm, queen of trees, must have her feet in running water and her head in the burning sky," has been quoted by Hussein and Hussein (1982a), who recommended moderate irrigation of 12 applications per year with 300 m^3/*Feddah* each time at intervals of about 4 weeks. A few reports on the irrigation requirements of date palm trees in California, USA, are available. Prolonged periods of severe water restriction during the growing period have been reported to affect the growth of leaves, and the size, grade, and yield of fruits (Aldrich, 1942; Furr, 1956). Unsuitable irrigation practices result in small-sized early ripening fruit of poor quality (Hilgman et al., 1957). Deep irrigation at long intervals is more beneficial in maintaining higher moisture levels in the root zones compared with shallow irrigation at short intervals (Hilal, 1986). Although the date tree can tolerate the hot, dry conditions of desert climates, sufficient quantities of irrigation water must be provided during the fruiting season to obtain the maximum yield of higher quality fruits.

The amount of nitrogen fertilizers used affects the vegetative growth and yield of fruit (Furr and Armstrong, 1959). The addition of nitrogen to palm trees increases the fruit yield but the dry matter, total soluble solids, total sugars, and sucrose remain lower (Hussein and Hussein, 1982b). When the fruits have higher moisture contents, the fruit maturity is delayed. About 750 g of nitrogen per palm tree is recommended to obtain the highest fruit yield and quality. The Saudi date palm growers are generally satisfied with the use of organic manure only. Bacha and Abo-Hassan (1982) investigated the effect of seven fertilizer treatments comprised of nitrogen, phosphate, potassium, and organic manure on the yield and fruit quality of the *Khudari* date variety grown in Saudi Arabia, and showed that chemical fertilizers increased the fruit yield and size but did not affect the mineral contents of the fruits. In addition to organic manure, they recommended 1500 g of nitrogen per tree.

Pruning is one of the important cultural practices in palm cultivation, and it affects the fruit-bearing capacity of the date tree. The number of green leaves present on a date tree depends on the cultivar and determines its photosynthetic potential and ultimately its fruit-bearing capacity (Nixon, 1943). Too many leaves may affect the fruit quality adversely by shading and may increase disease incidence. The presence of an adequate number of leaves on palm trees is important; the most suitable leaf–bunch ratio has been suggested as being 5.4–9.0 by a number of researchers (Miremadi, 1970; Miremadi, 1971; Nixon, 1957; Nixon and Wedding 1956). Abdulla et al. (1982) reported that leaf–bunch ratio affected the fruit yield and other properties except the titratable acidity, tannins, and crude fiber of *Hayany* dates grown in the Kalubia Province of Egypt. In Saudi Arabia, 12 leaves are usually left in each bunch, to obtain the best yield and fruit quality (Hussein et al., 1977), but farmers in the western province cut off only dry leaves after harvest (Khalifa et al., 1983). During pruning, the spikes (thorns) are also removed to facilitate pollination and fruit picking during the harvesting operation.

Date trees bear flowers that can be bisexual or unisexual and monoecious or dioecious. The occurrence of hermaphroditism in male date palm has been reported (Sudhersan and El-Nil, 1999). Although the date tree normally produces only unisexual flowers, the occurrence of bisexual flowers has also been reported in this species (De Mason and Tisserat, 1980). Because of the unisexual nature of this plant, the female flowers, within a few days of opening, are pollinated manually (a tedious and expensive operation) by sprinkling pollen from male flowers on them. The production, storage, and handling of pollen, therefore, are very important operations for the success of date cultivation. In the United States, pollination is done mechanically from the ground, thus minimizing the drudgery (Nixon and Carpenter, 1978). To economize on the use of pollen, it is mixed with diluents such as talcum powder or wheat flour. The time of pollination, type of male flowers, and storage history of the pollen are known to affect the fruit set as well as its quality. Use of 20–40% pollen in talcum powder gives satisfactory yield and fruit quality for *Zegloul* date trees. Use of 20% pollen results in lower bunch weight but the fruit quality is improved (El-Kassas and Mahmoud, 1986). Better fruit set and quality are obtained if pollination is carried out just before sunset rather than in the morning (Moustafa et al., 1986). Pollen stored either at room temperature or at chilling temperature in a refrigerator produces higher fruit set compared with the pollen stored in a deep freezer (Shaheen et al., 1986). Male palms having greater (spathe) weight, greater length and number of strands, and more than 15 g of pollen per spathe should be selected for pollination to obtain the best fruit set and quality (Nasr et al., 1986). Mixing one part pollen with nine parts wheat flour or in 10% sucrose solution has been reported to be more effective than the use of pure pollen in smaller amounts (Khalil and Al-Shawaan, 1986). Wheat flour and sugar solution media are good carriers of date palm pollen grains.

Due to overbearing of fruits, alternate bearing is common among many cultivars of date palm, resulting in smaller fruits of inferior quality. Fruit and bunch thinning usually overcomes this problem and improves fruit quality (El-Hamady et al., 1982). Commercially, thinning operations could be accomplished mechanically by reducing the number of fruits per bunch or the number of bunches per tree. To reduce alternate bearing and to improve fruit quality, certain chemicals such as 2, 4-D (2,4-dichlorophenoxyacetic acid), 2, 4, 5-T (2,4,5-trichlorophenoxyacetic acid), and ethephon are used as thinner for many fruit crops including the date palm (Ketchie, 1968; El-Zeftawi, 1976; Weinbaum and Muraoka, 1978). Use of 400 ppm of ethephon significantly reduces the biennial bearing behavior in date trees (El-Hamady et al., 1982). Compared with controls, significantly higher fruit weight and total soluble solids but lower tannin contents were obtained in fruits treated with ethephon. Thinning treatments improve the fruit quality but result in an overall lower yield of fruit per tree. To expose the date fruits to sunlight and to prevent fruit-bearing stalks from breaking, bunches are pulled downward and tied to a nearby leaf stalk at the *kimri* stage. This practice is particularly important for those cultivars having long fruit stalks.

DATE PRODUCTION AND MARKETING

The total annual world production of date fruits was reported to be 5190 thousand tons during the 2000 crop year, and is expected to increase further due to the efforts taken by various countries to encourage and popularize date palm cultivation through modern tissue culture techniques (FAO, 2000). Iran leads the world in date fruits with a total production of 930 thousand tons in 2000 (Table 22.1). The other

Table 22.1. Major Date Fruit Producing Countries of the World (1000 tons)

Country	Production	Country	Production
World	5190	Iraq	400
Algeria	430	Israel	10
Egypt	890	Kuwait	9
Libya	133	Qatar	17
Sudan	176	Oman	135
Tunisia	103	Spain	7
United States	20	Saudi Arabia	712
Bahrain	17	Morocco	74
China	125	UAE	318
Iran	930	Yemen	29
Mauritania	22	Pakistan	580
Chad	18	Turkey	9

Source: FAO, 2000.

major producers of date fruits are Egypt, Saudi Arabia, Pakistan, Algeria, Iraq, and United Arab Emirates (UAE). The recent advances in tissue culture techniques and the improved agricultural practices will definitely boost the date fruit production in this part of the world in the coming years. For marketing purposes, most of the fresh dates at the *khalal* and *rutab* stages of maturity are packaged in wooden or plastic crates (2–3 kg/crate) and offered for sale in the fruit and vegetable markets in these countries. However, dates at the *tamer* stage of maturity are packaged in tin cans, plastic bags, or straw baskets, either in a pressed or in an unpressed form. These date fruits are bought by the consumers from the date markets usually located in a central place in the town (Sabbri et al., 1982). Although the major portion of dates are consumed at the *tamer* stage, a significant amount of *khalal* date fruit is also being consumed. Thus, the marketing prospects for fresh dates at the *khalal* stage have opened up interesting opportunities to extend their shelf life and consumer acceptability (Al-Hooti et al., 1995a). Some of the pertinent problems in date marketing such as the lack of a proper layout of markets, improper handling of produce, an unclean environment, inferior packaging and wrapping materials, and unhygienic storage methods need to be tackled by the food technologists. Asia is the largest importer and consumer of date fruits in the world. Within Asia, China and India are the major importers of date fruits. In the European Union, France is the leading importer followed by Germany. In North America, the United States is the main importer of date fruits, followed by Canada.

DATE CULTIVARS

A large number of date cultivars do exist but their exact number is not known. Hussain and El-Zeid (1975) have reported the existence of 400 cultivars, but Nixon (Nixon, 1954) has indicated the probability of only about 250 named varieties. But, of all these cultivars, only a few dozen have attained economic and commercial importance. Of the four imported cultivars grown commercially in the United States, three fourths of the cultivated area is under *Deglet Noor* cultivation and 10% under the remaining three cultivars, *Zahidi, Khadrawy,* and *Halawy*. Knight (1980) has reviewed extensively the most important cultivars that are being grown commercially all over the world. The cultivar *Deglet Noor* (which literally means the date of the light or translucent seedling) is popular because of its large size, light color, delicate flavor, and outstanding shelf life during storage. This cultivar is also known for prolific bearing, and ripens in October–November. Like most other date cultivars, it also requires high temperature and low humidity for proper ripening. Another high-yielding *Yahidi* cultivar is being commercially grown in Iraq. This cultivar is hardy and drought resistant, and exhibits vigorous growth. *Hallawi* (which means sweet) cultivar bears a light-colored fruit and is one of the leading cultivars grown in Iraq and exported from Basra. Another cultivar *Sayer* (meaning widespread), which is not only known for its very sweet fruit but is also a hardy plant capable of tolerating adverse climatic conditions. It has shown the highest production and is quite important in commercial trade. *Medjool* is an important commercial cultivar from Morocco known for its sweetness. Asif et al. (1986) have described the characteristics of 15 commercially important cultivars, i.e., *Burhi, Gur, Hilali, Khalas, Khasab, Majnaz, Ruzeiz, Sahal, Shashi, Tanjeeb, Tayyar, Um Rahim,* and *Zamil* that are being grown in the Hassa area of Saudi Arabia. Some of the cultivars like *Khudari, Nabbut-Al-Seif, Sullaj, Sukai, Maktumi, Sultana, Shagra, Nabtat Ali, Shbibi, Barni, Rabiaa, Safawi, Shalabi*, and *Sifri* are the major cultivars of commercial importance in Saudi Arabia (Anon, 1984). Some of the other important cultivars being grown commercially are *Zaghloul, Duwiki,* and *Hayani* in Egypt; *Kabkabe* and *Khustawai* in Iran and *Barhee, Maktoom, Shalabi, Sukkari,* and *Khustawai* in Iraq (Sawaya, 1986; El-Kassas, 1986; Popenoe, 1973).

PHYSICOCHEMICAL CHARACTERISTICS

MORPHOLOGY OF DATE FRUITS

Depending on the cultivars as well as the stage of maturity, date fruit shows the greatest extent of variation in its shape, size, and color. Apart from changes in physical characteristics, date fruit varies greatly in chemical composition during its various stages of maturity. As the date fruit matures, it undergoes major morphological and chemical changes that subsequently determine the overall quality and acceptability of this fruit. A lot of information on the physicochemical changes occurring in date fruits during different stages of maturity is now available from all the major date fruit growing countries of the world (Al-Hooti et al., 1997f; Mohammed et al., 1983; Shabana et al., 1981; Sawaya et al., 1982; Salem and Hegazi, 1971).

Date fruits may have round, oval, oblong, or cylindrical shapes depending on the cultivar. Al-Hooti et al. (1997f) reported the physical measurements and colors of five major date fruit cultivars grown in the UAE. All the cultivars were green in color at the *kimri* stage but at the *khalal* stage, the color varied among the cultivars. At the *khalal* stage, *Shahla* and *Bushibal* were red, *Gash Gafaar* and *Lulu* were yellow, whereas *Gash Habash* fruits were yellow-scarlet (Figs 22.1a–g). At the last stage of maturity (i.e., *tamer*), the fruits of all the cultivars turned dark brown and shriveled considerably. The fruit weight, pulp–seed ratio, and physical measurements at the various stages of maturity of these five cultivars have been described in detail elsewhere (Sidhu and Al-Hooti, 2004). At the *kimri* stage, the *Bushibal* and *Gash Gafaar* fruits are cylindrical in shape, whereas the *Lulu* fruits are nearly round. These shapes are more or less retained by all cultivars throughout the various stages of fruit development. The lengths of *Bushibal, Gash Gafaar, Gash Habash,* and *Shahla* are 26.5, 24.8, 23.7, and 27.7 mm, respectively, with the corresponding values for width being 15.4, 15.5, 17.7, and 18.6 mm. In contrast, the *Lulu* cultivar has a length and width of 19.3 and 18.6 mm, respectively, indicating its nearly round appearance. Fruit weights are generally the highest at the *khalal* stage (6.1–10.0 g) and decrease subsequently toward the *tamer* stage (4.9 g). The pulp percentage varies from 83% to 90%, and the seed percentage varies from 10% to 17% among these five cultivars.

The physicochemical characteristics of 55 Saudi cultivars (Sawaya et al., 1986a) at *khalal* and *tamer* stages of maturity are quite similar to those of the UAE cultivars reported by Al-Hooti et al. (1997f). Some of Saudi cultivars had a bigger fruit size (25.6–26.8 g) at the *khalal* stage, but it was reduced to 13.7–14.1 g at the *tamer* stage. The fruit weights ranged from 5.8 to 26.8 g (with an average of 13.5 g) at the *khalal* stage and from 4.8 to 18.3 g (with an average of 9.8 g) at the *tamer* stage. The weight of seeds ranged from 0.7 to 1.8 g at *khalal* stage and 0.6–1.3 g at the *tamer* stage of maturity of these cultivars. The pulp percentage of these cultivars was reported to be in the range of 86–96% and was slightly higher than that reported by Al-Hooti et al. (1997f), although some of the smaller fruit cultivars studied by them had similar fruit size and pulp percentages. Sourial et al. (1986) evaluated four local cultivars, namely, *Sofr-Eldomain, Kabooshy, Sergy,* and *Homr-Baker,* in relation to their control cultivar, *Hayany*, grown in Egypt for their physicochemical characteristics. The fruit lengths for these five cultivars ranged from 4.9 to 5.6 cm and fruit weights ranged from 15.79 to 25.30 g, seed weights ranged from 1.87 to 2.38 g, and pulp percentage ranged from 88.15% to 90.70%. These values were quite comparable to those reported earlier in the literature (Sawaya, 1986). In another study, Nour et al. (1986) reported the physical characteristics of nine dry palm cultivars (i.e., *Balady, Bartamuda, Degna, Garguda, Gondalia, Kolma, Malkabi, Sakkoti,* and *Shamia*) grown in Aswan, Egypt. The fruit weights, fruit lengths, fruit widths, and seed weights ranged from 6.5 to 16.9 g, from 3.89 to 5.40 cm, from 1.75 to 3.32 cm, and from 1.0 to 1.63 g, respectively. The cultivar *Malkabi* had the highest fruit weight (16.9 g), *Shamia* was the longest (5.4 cm), and *Bartmuda* had the smallest seed (1.0 g only).

Carbohydrates

Like most other fruits, dates are a rich source of carbohydrates, mainly fructose and glucose, but smaller quantities of vitamins, minerals, and other minor constituents are also present. The major chemical constituents of date fruit are carbohydrates, mainly reducing sugars, such as glucose and fructose, and also a nonreducing sugar, sucrose. Carbohydrates (i.e., sugars) are, therefore, the most widely studied constituents of date fruits. The chemical composition of five major cultivars grown in the UAE at various stages of maturity has been described by Al-Hooti et al. (1997f). In a majority of the cultivars, the sucrose content increased rapidly as the date fruit matured, reaching the highest level at the *khalal* stage (42.58%) but subsequently decreased to a nondetectable level at the *tamer* stage of maturity. As the date fruits matured, the glucose and fructose sugars increased rapidly to reach a level of 38.47–40.04%. When the date fruit matured from the *kimri* to the *tamer* stage, the fructose content increased approximately threefold, which accounts for the characteristic sweet taste of *tamer* date fruits. The total sugar contents, which were 32.99–38.20% at the *kimri* stage, reached nearly 80% by the *tamer* stage of maturity. The presence of equal amounts of glucose and fructose in soft-type cultivars is responsible for their enhanced levels of sweetness. On the other hand, some of the semidry and dry cultivars are reported to retain higher levels of glucose than fructose or the unhydrolyzed sucrose. One of the earliest studies also reported that the total sugars and invert sugars increased with ripening, reaching a maximum level by the later stages of development (Vinson, 1924).

Soft dates with almost no sucrose, semidry cultivars like *Deglet Noor* with higher levels of sucrose, and dry dates with equimolar concentrations of sucrose and reducing sugars are also available (Sawaya et al., 1982).

The total reducing sugar contents are related to the cultivar as well as to the stage of maturity. In the semidry varieties of Egyptian *tamer* dates, both the sucrose and reducing sugars are about 35–40% each. The total sugar concentration at this stage reaches between 80 and 90% of the dry weight. During the curing stage, the sucrose content of soft varieties disappears completely (Ragab et al., 1956). The total sugar contents of 39 Saudi Arabian cultivars falls between 61 and 80%, while the total sugar contents of three cultivars amounted to 80–90% (Hussein et al., 1976). The sucrose content is usually the highest (10–30%) at the *khalal* stage in most of the cultivars, but it declines to low levels of 0–2% at the *tamer* stage. In contrast, the reducing sugars generally increase with fruit development, reaching 29–85% at the *tamer* stage of maturity. Reducing sugars are mainly the predominant sugars at *tamer* stage in most of the cultivars, with the exception of two cultivars, namely, *Sukkari* and *Sukkarat Al-Shark,* which contain more sucrose at the *tamer* than at the *khalal* stage of maturity. In a given cultivar, the sucrose and reducing sugar contents are related to the quality and texture of the date fruit (Coggins and Knapp, 1969). Therefore, the majority of these Saudi Arabian cultivars (Hussein et al., 1976) having higher concentrations of reducing sugars at the *tamer* stage, are of the soft-type date fruits. The five cultivars from the UAE reported by Al-Hooti et al. (1997f) with nondetectable sucrose contents at the *tamer* stage also belonged to the soft-type date fruits. The declining moisture content coupled with the rapid increase in glucose and fructose contents render the *tamer* date fruits extreme resistance to fungal spoilage during storage.

As the texture and color of dates are the important attributes affecting fruit quality and acceptability, most of the biochemical and enzyme studies have been limited to these aspects of date fruit physiology. The higher activity of the sucrose-hydrolyzing enzyme invertase present in soft-type date fruit cultivars is the most important enzyme influencing the date fruit quality and is considered to be mainly responsible for the highest levels of reducing sugars present at the *tamer* stage of maturity (Vinson, 1911; Vandercook et al., 1980). The changes in invertase activity in *Deglet Noor* date fruits during maturation and ripening have been studied by Hasegawa and Smolensky (1970). Soluble invertase increases dramatically when the date fruit matures from the green stage to the early red stage. The insoluble form of this enzyme is present in substantial amounts during the green stage, when it decreases to 50% of its original activity and then remains fairly constant later on. Both the insoluble and soluble invertase hydrolyzes sucrose, raffinose, and melezitose in a similar manner. Among the four grades of dates evaluated in their study, soft, good quality dates had a higher activity of this enzyme than the tougher dates of inferior quality. Invertase can be used to improve the quality and market value of date cultivars, which have crystalline sucrose present in their tissues (Smolensky et al., 1975). The enzyme concentration, temperature, and the time of treatment are important to bring the ratio of sucrose–reducing sugars to a level low enough to prevent sucrose crystallization later on during storage. Soluble invertase and insoluble invertase have been isolated from date fruits (*Zehdi* var.), and for both enzymes 45°C is an optimum temperature for activity (Marouf and Zeki, 1982). The optimum pH ranges for soluble and insoluble invertase are 3.6–4.8 and 3.6–4.2, with K_m values of $3.12 \times 10^{-3}$and 4.35×10^{-3}mM, respectively. The specific activity of soluble invertase is 40.2 μmol/mg protein/min, while the specific activity of insoluble invertase is 1.1 μmol/mg protein/min. Sodium Dodecyl Sulfate inhibits both the enzymes.

Proteins

In addition to the major constituent carbohydrates, date fruits also contain significant amounts of protein, crude fiber, pectin, tannins, minerals, and vitamins. Al-Hooti et al. (1997f) have analyzed five important date cultivars from the UAE at different stages of maturity. The crude protein content in these cultivars is highest at the *kimri* stage (5.5–6.4%) and gradually decreases to 2.0–2.5% as the fruit reaches the *tamer* stage of maturity. Although the protein content is not high in *tamer* date fruits, the essential amino acid content of these proteins is quite good. Similar protein content in other cultivars have been reported by various other workers (Hussein et al., 1976; Sawaya et al., 1986b).

Fat

The date fruit is quite low in crude fat, which usually ranges from 0.5% at the *kimri* stage to 0.1% at the *tamer* stage of maturity (Al-Hooti et al., 1997f). Evidently, date fruit, like most other fruits, cannot be considered as an important source of fat or fatty acids in our diet. These values for crude fat content are similar to those of some of the cultivars reported by others (Sawaya et al., 1982, 1986b; Ragab et al., 1956).

Crude Fiber

The crude fiber content of date fruits at the *kimri* stage is substantially higher (6.2–13.2%) than that at the *tamer* stage (2.1–3.0%) of maturity (El-Kassas, 1986). The crude fiber content of date fruits is not a good indication of their dietary fiber content (Yousif et al., 1982). The total dietary fiber content (comprised of pectin, hemicellulose, cellulose, gums, mucilages, resistant starch, and lignin) depends on the stage of maturity of the date fruits (El-Zoghbi, 1994). Xylan has been identified as one of the components of date fruit fiber (Hag and Gomes, 1977). Alcohol-extractable material from date fruit, when further treated with water, dilute acid, and aqueous alkali yields polysaccharide, which contains varying proportions of D-galactose, D-glucose, L-arabinose, D-galacturonic acid, and L-rhamnose. Glucomannan is another polysaccharide found in date fruits. The structure of glucomannan isolated from the seeds of Libyan dates has been elucidated (Ishrud et al., 2001). This polysaccharide is extracted with 80% hot ethanol (Fraction-I) and 0.1 M phosphate solution (Fraction-II), fractionated and purified by ion exchange and gel filtration chromatography. According to methylation and hydrolysis analysis, the main chains of FI and FII consist of 1–4, linked glucomannan with only traces of branched sugar residues.

The total fiber decreases as the date fruits lose their firm texture and become soft at the *tamer* stage. A large variation in the total dietary fiber content of date fruits comes from the type of method employed in its determination. The Southgate method does not determine resistant starch, whereas the Fibertec and Englyst methods do (Kirk and Sawyer, 1991). So, to obtain comparable results, internationally accepted methods of analysis for dietary fiber must be employed. The total dietary fiber content measured by the enzymatic method has been reported to be 9.2%, with 6.9% as insoluble and 2.3% as soluble fiber (Lund et al., 1983). The total fiber content in some of the dates from Saudi Arabian, Egyptian, Iraqi, and Irani cultivars, determined by the Fibertec system, ranged from 8.1% to 12.7% (Al-Shahib and Marshall, 2002). Research conducted during the past three decades has shown that an adequate intake of dietary fiber (20–25 g daily) lowers the incidence of colon cancer, heart diseases, diabetes, and other diseases. Obviously, the consumption of 100 g of date fruit (six to seven dates) would provide us with about 50% of the recommended daily amount of dietary fiber. The total dietary fiber of dates decreases from 13.7% at the *kimri* stage to 3.6% at the *tamer* stage of maturity (Ishrud et al., 2001). The decrease in the pectin, hemicellulose, cellulose, and lignin contents during date-fruit ripening range from 1.6 to 0.5, from 5.3 to 1.3, from 3.4 to 1.4, and from 3.5 to 0.3%, respectively. This shows that maximum benefit can be obtained by consuming fresh dates (i.e., those at the *kimri, khalal,* and *rutab* stages) rather than by consuming the fully mature *tamer* fruits. The presence of resistant starch in the fresh dates will provide an additional advantage as it may be prebiotic, promoting conducive conditions for the growth of desirable bifidobacteria in the lower gastrointestinal tract (Topping and Clifton, 2001).

Pectins

The pectic substances (considered a part of the soluble dietary fiber) are a complex mixture of polysaccharides that are important constituents of plant cell wall structures. The pectin contributes to the adhesion between cells and also plays an important role in some of the processed fruit products such as jams, jellies, and preserves (Jarvis, 1984). As the date fruit ripens, protopectin is converted into water-soluble pectin through the combined action of two pectolytic enzymes, i.e., polygalacturonase and pectin methyl esterase (Al-Jasim and Al-Delaimy, 1972; Coggins et al., 1968). There is a close relationship between polygalacturonase activity and fruit softening during ripening (Hasegawa et al., 1969). Like invertase, polygalacturonase activity is relatively higher in soft dates than in the tough dates. Unlike cellulase and polygalacturonase, the activity of pectin esterase increases as the fruit grows reaching a maximum during the *khalal* and *rutab* stages of maturity (Al-Jasim and Al-Delaimy, 1972). Cellulase activity, which is absent in *kimri-* stage fruit, increases as the fruit develops, reaching its peak at the late *rutab* stage and remains constant during the *tamer* stage (Hasegawa and Smolensky, 1971). As the date fruit ripens, the pectin content increases, reaching a maximum level at the *khalal* stage (7.0–14.3%), and then decreases at the *tamer* stage (1.3–1.9%) of maturity. Graces-Medina (1968) estimated the pectin contents of banana, mango, pineapple, and mountain apple to be 0.62%, 0.38%, 0.13%, and 0.47%, respectively. The date fruit, therefore, can be a better source of soluble dietary fiber in our diet than some of the more common fruits such as banana, mango, pineapple, and apple.

Minerals

Dates are known to be a reasonably good source of many minerals. The mineral composition of date fruit is largely affected by the level of soil fertility as well as by the amount of chemical fertilizers and manures applied to the trees (Hussein et al., 1976). The ash content of date fruits decreases from the *kimri* stage (3.5–3.9%) to the *tamer* stage (1.3–1.8%) of maturity (Al-Hooti et al., 1997f). The ash content of some common fruits like grapes, apples, plums, and oranges is also similar (1.9–3.1%) to that of date fruits (FAO, 1982). Obviously, date fruits are an equally good source of important minerals. Although the mineral content decreases at the *tamer* stage of maturity, the change is small when compared with changes in other constituents such as sugars. The composition of date fruits from five major cultivars being grown in the UAE, in terms of important mineral contents, has been reported in detail by Al-Hooti et al. (1997f). These date cultivars are rich in most of the macroelements but are poor in microelements. Like most other fruits, these cultivars are low in sodium (1.5–9.4 mg%) but high in potassium (402.8–1668.6 mg%). Minerals, particularly potassium, accumulate in date fruits during ripening (Ragab et al., 1956). This low sodium–potassium ratio obviously makes the date fruit a desirable food for persons suffering from hypertension.

Date fruits are considered a rich source of iron (Anwar-Shinwari, 1987). Different varieties generally contain significantly varied amounts of iron per unit weight or per fruit, the variation is attributed to genetic differences. Boron is another important mineral present in date fruits. The formation of a red complex between boron and the quinalizarin reagent can be used to determine the boron content in date fruits through a simple and sensitive spectrophotometric method. At 620 nm, the absorbance is linear ($r = 0.999$) over the 0.25–2.5 μg/ml concentration range (Al-Warthan et al., 1993). This method can detect the prevailing wide variation in the mineral contents of Saudi Arabian date cultivars.

Tannins

The tannins in date fruit play an important role not only in flavor perception but also in the development of color in the date fruits during ripening and storage. The color of dates is primarily due to the pigments produced by browning reactions during ripening, processing, and storage. These brown pigments could be produced by three possible mechanisms: browning of sugars, enzymatic oxidation of polyphenols, and oxidative browning of tannins (Coggins and Knapp, 1969). Generally, the oxidative browning reactions occur more rapidly at elevated temperatures than at low refrigerated temperatures. Even at room temperature, the enzymatic browning of polyphenols and tannins is much faster than sugar browning reactions. However, at temperatures above 38°C, sugar browning predominates. As the date fruit matures, tannins decrease rapidly from the highest value of 1.8–2.5% at the *kimri* stage to about 0.4% at the *tamer* stage of maturity (Al-Hooti et al., 1997f). This trend of reduction in tannins with ripening was also observed earlier in other date cultivars (Hussein et al., 1976).

Kimri fruit is quite astringent and unpalatable but this decreases drastically at *tamer* stage when the level of tannins reaches a very low level, indicating some probable contribution of tannins in the flavor of date fruits. The *Lulu* cultivar had the lowest tannin content at the *khalal, rutab,* and *tamer* stages of maturity, when compared with other cultivars. The major enzyme involved in the metabolism of tannins in date fruits, polyphenol oxidase (PPO), has been studied at different stages of ripening (Benjamin et al., 1979). Using catechol as a substrate, the optimum pH and temperature for its activity are 6.4 and 37°C, respectively. This enzyme has no monophenol oxidase activity and varies in specific activity toward several diphenols. Its heat inactivation follows first-order reaction conditions. During the ripening of date fruit, the PPO activity is the highest at the *kimri*, followed by *khalal* and *tamer* stages. The PPO activity can be completely inhibited by 0.01 M of ascorbic acid, cysteine, sodium sulfite, and sodium dithiocarbamate.

Vitamins

Compared with other fruits, the *tamer* date fruit is not considered a good source of vitamins, but *khalal*-stage fruits contain appreciable amounts of ascorbic acid and β-carotene (Watt and Merrill, 1963). *Khalal* dates are reported to contain 1.8–14.3 mg and *tamer* dates 1.1–6.1 mg of ascorbic acid per 100 g of fresh fruit pulp. Similarly, the β-carotene content (Pro-vitamin A), expressed as IU of vitamin A, ranged between 20 and 1416 IU in *khalal* and 0–259 IU per 100 g fresh fruit pulp in *tamer* date fruits

(Hussein et al., 1976). Date fruit is also a reasonably good source of thiamin, riboflavin, and niacin (Watt and Merrill, 1963).

PREHARVEST TREATMENTS FOR DATE FRUITS

The use of growth regulators is not uncommon in fruit cultivation. Plant regulators are used to increase the fruit size, maturity rate, and quality characteristics. Preharvest application of naphthalene acetic acid (NAA) at 10, 20, 30, 40, and 60 ppm to immature date fruits of the *Zahdi* date palm 15–16 weeks after pollination (i.e., late in the *kimri* stage) influences fruit size, fruit weight and volume, pulp–seed ratio, and moisture content (Mohammed and Shabana, 1980). The highest increase in fruit weight (39%) can be obtained with the 60-ppm application of NAA. Total solids are not changed with the 40- to 60-ppm application of NAA but fruit ripening is delayed by about a month. Ethrel significantly increases the length, diameter, fresh weight, volume, and pulp–seed ratio of 2,4,5-trichlorophenoxy-propionic-acid-treated fruits but fails to show such effects in NAA-treated fruits (Mohammed et al., 1980). Ethrel increases the moisture content of NAA-treated date fruits and reduces the total soluble solid contents of both types of auxin-treated fruits. Gibberellic acid (GA) can be applied to unpollinated date flowers to produce seedless fruits, but the fruit yield decreases and fruit ripening is retarded. Additional application of 200 ppm of ethephon improves the fruit quality in terms of color, total soluble solids, and titratable acidity (Maximos et al., 1980). However, the higher dosages of ethephon (0, 125, 250, 500, 1000, or 2000 ppm) applied to *Shahani* cultivar date fruits, from Iran, during harvest show minor increases in the dry weight percentage of the pulp–seed ratio, titratable acidity, soluble solids, and respiration rates; and a minor decrease in pH, firmness, and astringency (Rouhani and Bassiri, 1977).

The effect of GA treatment of date fruits on date palm trees depends on the level and rate of application. Mohammed et al. (1986) sprayed the date bunches of three *Zahdi* and three *Sayer* cultivars with 0, 50, 100, or 150 ppm GA during the slow period of growth (i.e., 12–14 weeks after full bloom and pollination). About 18 weeks after the treatment, fruit bunches were harvested and the fruits were analyzed for physical parameters and chemical composition. GA had no pronounced effect on total soluble solids, except at the highest level applied to *Zahdi* cultivar, in which the sucrose content was higher. Total and reducing sugars increased with GA application but varied with the cultivar and the rate of GA application. The natural plant hormone, indole-3-acetic acid (IAA) is suggested as being involved in the control of date fruit development (Abbas et al., 2000). The female flowers of the date tree are very rich in IAA, and if not pollinated, this leads to the setting of parthenocarpic fruit (i.e., fruit which fails to ripen fully, due to the absence of a climacteric rise in respiration). The concentration of IAA declines 2 weeks after pollination (at fruit set) probably because it is used in cell division. The IAA concentration rises again in 2 weeks, possibly due to the embryo development but remains high up to 8 weeks after pollination, before it declines to a minimum in fully ripe fruits, i.e., around 18 weeks after pollination. The fall in IAA is accompanied by a marked increase in gibberellins in date plants.

The application of plant growth regulators, alone or in combinations, is known to produce varied results in date fruit characteristics and in the productivity of the date palm tree (Al-Juburi et al., 2001). Application of 150 ppm of GA or 1000 ppm of ethephon on flower clusters of the *Barhee* date palm tree showed no consistent effect on fruit characteristics or productivity. However, 100 ppm of 2-(1-naphthyl) acetic acid (NAPA) or the above growth regulator mixtures reduced fruit dry matter and fruit ripening percentage, but increased the fruit weight per bunch per tree. NAPA, when applied to *Barhee* date palm flowers 20 days after pollination, resulted in the increase of the fruit flesh percentage and the date palm yield.

POSTHARVEST HANDLING OF DATE FRUITS

At the *tamer* stage of maturity, date fruit has good storage stability mainly because of low moisture and high sugar contents. Due to its good shelf life during storage, date fruit is also known as a self-preserving fruit. As discussed earlier, most of date fruits are consumed at the *tamer* stage of maturity but because of higher nutritional value, substantial amounts are also consumed in the perishable, immature *khalal* and *rutab* stages. Date fruits at these immature stages are rich in dietary fiber, ascorbic acid, and β-carotene; hence they are traditionally quite popular in date-growing regions (Al-Hooti et al., 1997a). Date fruits of some cultivars at the *khalal* and *rutab* stages of maturity are preferred by consumers (Al-Mulhim and Osman, 1986). *Khalas*, *Shahl*, and *Khenaizi* in Saudi Arabia have been the preferred fresh dates among consumers there. Over 84% of the consumers of fresh dates are reported to prefer *Khalas* dates and 73% of the consumers purchase fresh dates packaged in baskets. But unfortunately, date fruits at these immature stages of maturity not only have higher moisture contents and are susceptible to microbial spoilage but are also available only for a very limited period during the season.

Among the potential pathogenic bacteria, *Escherichia coli, Staphylococcus aureus,* and *Bacillus cereus* have been identified in date fruits together with lactic acid bacteria, yeasts, *Aspergillus flavus*, and *A. parasiticus.* Date fruits at the *khalal* and *rutab* stages being high in moisture are the most heavily contaminated (Aidoo et al., 1996). Development of any postharvest techniques to extend the shelf life of date fruits at the *khalal* and *rutab* stages would enhance the commercial and economic value of this crop, on the one hand, and would also provide consumers with a greater choice of delicious products with desirable nutritional contributions to their diet over an extended period of time.

One approach to achieving this objective of shelf life extension is the use of antifungal agents such as potassium sorbate (Al-Hooti et al., 1997a). Compared with the control, the shelf life of *khalal* stage fruits of *Bushibal* and *Lulu* cultivars treated with 0.05% potassium sorbate when stored at 4°C, remained acceptable for an additional 8 and 10 weeks, respectively. The microbial load on these date fruits stayed within the acceptable limits. When these treated fruits are stored at −2°C and −20°C, no coliforms or enterobacteriaceae were detected, while the aerobes and mold counts stayed within the acceptable limits. These subzero temperatures retard the growth of naturally occurring microflora present on these fruits. The fruits of both the cultivars matured from the *khalal* stage to the *rutab* stage during the storage period, thus making the fruits more acceptable to consumers.

The frozen storage of date fruits at the *rutab* stage also leads to changes in various physicochemical characteristics. During 6 months of storage at -18 ± 2°C, *rutab* date fruits increased in moisture content, reducing sugars, and pH but decreased in tannins (Mikki and Al-Taisan, 1993). The fruits developed an acceptable sweet taste with the disappearance of astringency. At the end of storage, the thawed product became soft in texture and darker in color. Similar findings on the chemical composition of Egyptian dates during frozen storage have also been reported (Goneum et al., 1993). Some of the Saudi cultivars at the *rutab* stage of maturity are suitable for preservation by refrigeration and frozen storage (Yousif and Abou-Ali, 1993; Al-Mashadi et al., 1993). The *Haynai* var. of date fruits, when picked at the red stage, can be stored in frozen conditions at −18°C. On thawing, it has a relatively short shelf life when stored at varying temperatures (6°C, 15°C, 20°C, or 26 ± 0.2°C). Predominantly, the presence of yeasts and lactobacilli is responsible for the relatively short shelf life, and storage temperature is the major influencing factor in the rate of spoilage. Considering the initial microbial load, and time–temperature conditions in storage, the organoleptic shelf life of soft dates can be predicted during transport, handling, and retailing.

Date cultivation poses another unique problem in countries like India and Pakistan, where the crop ripening period coincides with the rainy season (Chatha et al., 1986), which leads to extensive spoilage of the date crop. However, to circumvent these circumstances, date fruits are harvested at the *khalal* stage and then cured using various techniques such as the use of sodium chloride, sodium hydroxide, acetic acid, 2,4-dichlorophenoxy acetic acid, etc. Even using these methods, the shelf life of *khalal*-stage fruits cannot be extended beyond 2 months, after which the fruits turn black in color.

Antimicrobial agents and refrigeration are used for the preservation of high-moisture dates (43–56%) (Hassan et al., 1979). Date fruits can be immersed for 1 min in 0.25%, 0.5%, 1%, 2%, 5%, and 10% solutions of calcium propionate, sodium benzoate, potassium metabisulfite ($K_2S_2O_5$), potassium sorbate, dehydroacetic acid, and combinations of sodium benzoate with calcium propionate or with SO_2, inoculated with molds and yeast spores, and stored at 18–25°C or 2–5°C. Treatment with 10% calcium propionate or 5% potassium metabisulfite did not provide adequate protection during 13 weeks of storage at any of these temperatures. Potassium sorbate provided 2 weeks protection at 0.5%, 7 weeks at 2%, and 14 weeks at 5% at low temperatures. At low temperatures, 5% benzoate-treated dates stayed acceptable for 4 weeks. Use of 1% sodium benzoate with 2% potassium metabisulfite or calcium propionate protected these date fruits against microbial spoilage for 9 months with refrigerated storage.

The *khalal* stage fruit can also be used for the preparation of *Chhuhara (khalal matbukh*) by cooking for 10, 15, 20, or 25 min and then drying in a solar drier (Gupta and Siddiqui, 1986). *Chhuhara* (dried dates) prepared from *Khadrawi* cultivar yields better quality in terms of pulp content, sugars, proteins, and sensory characteristics, than those prepared from *Shamran* cv. The quality of the *chhuhara* from *Khadrawi* dates improved with prolonged boiling, whereas the quality deteriorated with prolonged boiling for *Shamran* cultivar. The ripening of date fruits shows a small peak in ethylene production initially, which increases as the fruit matures, thus suggesting that dates could be considered a climacteric fruit, and the plant hormone ethylene is responsible for changes in its color, fruit texture, soluble solids, and acidity. Fruit firmness decreases during various stages of ripening. The greatest loss of fruit firmness correlates with the greatest increases in both the polygalacturonase and the β-galactosidase activity (Serrano et al., 2001).

OTHER STORAGE PROBLEMS

All varieties of date fruits are commonly attacked by a number of insect pests and fungi during storage. Most of the insects belong to the Lepidoptera and Coleoptera orders. Some species belonging to the order Acarina can also cause considerable damage to date fruits. The commonly reported insects found in stored date fruits are *Oryzaephilus surinamensis, O. mercator, Tribolium confusum, Plodia interpunctella, Cadra cautella, C. calidella*, and *C. figulilella* (Abdelmonem et al., 1986b; Carpenter and Elmer, 1978). The commonly isolated fungi from date fruits are *Aspergillus* sp., *Alternaria* sp., *Fusrieum* sp., *Penicillium* sp., *Rhizopus* sp., and *Saccharomyces* sp. (Carpenter and Klotz, 1966; Chohan, 1972). Most important is the fact that any insect (at any stage of their growth) and/or insect fragments, sharply reduce the market value of such fruits. Moreover, the presence of these insects in date fruit products create serious quarantine problems in the international trade (Abdelmonem et al., 1986a).

Apart from insects, mycotoxins (low molecular weight chemicals) can be produced by many common filamentous fungi found on various food and feed products. Aflatoxins are among the most potent carcinogenic, mutagenic, and teratogenic chemicals that can be produced by *Aspergillus flavus* and *A. parasiticus*, under conditions suitable for growth. Four major aflatoxins, namely, B_1, B_2, G_1, and G_2, are commonly found in foods in tropical and subtropical climates. Date fruit under conditions of high humidity and moderate temperature can be contaminated with aflatoxins. Some varieties of date fruit stored under simulated conditions of 98% relative humidity and 30°C were found to contain significant levels of aflatoxin B_1 or B_2 ranging from 35 to 11,610 μg/kg (Shenasi et al., 2002).

The presence of *A. flavus* and *A. parasiticus* on date fruits has been reported from a number of countries (Abu-Zinada and Ali, 1982; Emam et al., 1994; Ahmed et al., 1997; Ahmed and Robinson, 1999; Ahmed and Robinson, 1997; Salik et al., 1979). Two of the isolates from the date fruits treated with methyl bromide (MB) and stored in polyethylene bags for 8 months at 60–75% relative humidity and 20–25°C, produced aflatoxins B_1, G_1, and G_2 in synthetic medium and on date fruits (Emam et al., 1994). The presence of *A. flavus* and *A. parasticus* in some of the date fruits imported to the United Kingdom has been observed (Ahmed et al., 1997). The mold *A. parasiticus* is able to penetrate the intact date fruit tissue and can produce aflatoxin in 10 days at 28°C in all stages of maturity except at the *tamer* stage, which do not support mold growth (Ahmed and Robinson, 1999), but the extracts from all four stages of date fruit were able to support growth of this mold for aflatoxin production (Ahmed and Robinson, 1997), and the amounts of aflatoxin produced increased with ripeness. Generally, the pattern of aflatoxin production appears to be broadly in line with the changes in the sugar content and chemical composition of maturing date fruits. Therefore, maximum care should be taken during the processing and handling of date fruits to avoid aflatoxigenesis.

Detection of aflatoxins in date fruits requires an accurate method for extraction and derivatization. The Romer minicolumn method is able to detect aflatoxins in contaminated date fruits. However, using high-performance liquid chromatography and post-column derivatization, the contaminants branch (CB) method gives average recoveries of 75.7% and 83.5% for the *Lulu* and *Naghal* date varieties, respectively (Ahmed and Robinson, 1998). The recovery of total aflatoxins by the best food (extraction and purification procedure) is about 35% less than that with the CB method. The available Association of Official Analytical Chemists methods, with slight modifications, give better recoveries of aflatoxins from date fruits.

Date fruits are a vital component of the diet in most of the Arabian countries. Besides insect infestation, it is not known whether or not the dates available in the market are also contaminated with aflatoxins. Consequently, a number of control measures have been suggested to overcome these problems of insect infestation and the growth of undesirable mycotoxin-producing microorganisms on date fruits during marketing and storage. The commercially packed *Zahdi* cultivar treated with ionizing radiation or with MB fumigation can provide an insect-free product for about 25 days (Ahmed et al., 1982) but with longer storage, re-infestation takes place. The major fungi, *Botrytis cinerea* and *Penicillium expansum*, causing postharvest decay of soft-type date fruits of the *Zagloul* var. can be drastically reduced without causing any shrinkage of the fruits when irradiated with up to 200 krad gamma ray dosages (El-Sayed, 1978). Additionally, the reduction of the tannin content with irradiation leads to reduced astringency and improved sensory quality. Fortunately, irradiation does not produce any changes in the amino acids, sugars, and protein contents of these treated date fruits during their extended storage life.

Irradiation of date fruits can be used to control the growth of undesirable microorganisms without adversely affecting the sensory quality of the fruit. To reach a decimal reduction value (D_{10} value), a dosage of 1.4 kGy is sufficient for total plate counts, 1.2 kGy for yeasts, and 0.9 kGy for bacterial spores present in some of the date fruit varieties being grown in Saudi Arabia. A dosage of 4–5 kGy is required to reduce the microorganisms to an undetectable level without adversely affecting the sensory quality of date fruit (Grecz et al., 1986). Emam et al. (1994) found that irradiation (1.5 and 3.0 kGy) is more effective in preventing insect infestation than MB fumigation of Egyptian semidried date fruits of the *El-Seidi* cultivar during storage for 8 months at room temperature. Neither irradiation nor MB caused any significant changes in moisture content, pH, or titratable acidity, but both produced significant changes in browning, total, reducing, and nonreducing sugars; and in the sugar–acid ratio. An irradiation dosage of 3.0 kGy was more effective than MB fumigation in inhibiting fungal growth and aflatoxin production, which is recommended for maintaining date fruit quality during long-term storage. Fumigation and aeration of mature, dry date fruits of the *Zahdi* cultivar with phosphine gas at various levels from 0 to 56.7 mg/l for 72 h did not produce significant differences in the sugar and protein contents, but the threonine and methionine contents decreased significantly (Al-Hakkak et al., 1986).

A method exists to determine if a date fruit has been irradiated or not. An unirradiated date stone contains a radical with a single line $g = 2.0045$, feature A. Irradiation up to a dosage of 2.0 kGy (the recommended dosage for irradiation of fruits in the United Kingdom) induces the formation of additional radicals with signals $g = 1.9895$ and 2.0159, feature C. The single line having $g = 2.0045$ decays both in the irradiated and control date fruit samples, whereas the additional signals $g = 1.9895$ and 2.0159 remain almost unaltered for 15 months of storage at room temperature and at 4°C (Ghelawi et al., 1996).

Three fumigants, namely, phostoxin, MB, and hydrogen cyanide; dry heat treatment (55 ± 2°C); and cold storage (at −15°C and 5°C) have been tested against the most common insects present in stored date fruits (Abdelmonem et al., 1986a). The three fumigants caused 100% mortality of all insect species (*Cadera cautella*, *Oryzaephilus surinamensis*, and *Tribolium confusum*) after 24 h of exposure. Dry heat treatment also produced similar results after 1.5 h of exposure against all insects. Chilling temperature could achieve only up to a maximum of 84% mortality in 30 h, whereas −15°C, treatment produced 100% mortality in just 6 h. These fumigants are also effective in reducing the fungi commonly present on date fruits, with MB being the most effective (Hegazi et al., 1986a). The use of a 100% carbon dioxide environment to control the insects and fungi commonly present in stored date fruits has been tried, but the success rate is variable against these organisms (Mohammed et al., 1980). Carbon dioxide gas in saturated chambers produced 100% mortality of all insects after 48 h but the fungi and bacteria survived in the stored date fruits.

VALUE-ADDED FOODS FROM DATE FRUITS

A number of epidemiological studies that have been undertaken during the past few decades have provided support for an inverse relationship between the intake of fruits and vegetables and the risk of cancer and other chronic diseases in humans. The reasonably high intake of fruits and vegetables (at least 400 g daily) lowers the risk of these chronic diseases. It is the presence of a number of phytochemicals in fruits and vegetables that may, in part, explain their beneficial effects (Gerber et al., 2002). Studies using animal models and cell cultures have generated a lot of information about the possible mechanisms by which a diet rich in fruits and vegetables may reduce the risk of these chronic diseases. Although date fruits are known to be rich in some of these phytochemicals, the scientific information generated on this fruit so far is scanty.

To date, most of the date fruits at *khalal, rutab*, and *tamer* stages of maturity are being consumed directly with little or no processing, but the quantity of processed date products is growing rapidly in this region. A number of value-added date products are now available in the local market throughout the year. For more detailed description of value-added functional foods prepared from date fruits, the reader is referred to a recently published book chapter (Sidhu and Al-Hooti, 2004). A brief discussion about some of the important value-added food products prepared from date fruits will be discussed here.

Commercially Packed Dates

Apart from sizeable quantities of dates being consumed at perishable immature stages (*khalal* and *rutab*), the majority of date fruits are consumed in the dry *tamer* stage with moisture content of less than 20% (Sabbri et al., 1982). These *tamer* date fruits are bulk packed in bags, metal tins, or baskets without fumigation or even without normal washing and are offered for marketing. To maintain the high quality expected by the consumers, the date producing and exporting countries have established a number of bulk-packing houses with modern facilities. The Government of Saudi Arabia has initiated an ambitious project to modernize the date processing industry and to establish modern date-packing plants. To conform to the Saudi date standards for quality, the dates are fumigated with MB to kill all insects, if present (Anon, 1983). MB is used at a concentration of 1.0 lb/1000 ft^3 for an exposure time of 24 h. Chloropicrin may also be added to the MB at a rate of 2%. After fumigation, the dates are transferred to a shaker for preliminary washing with water sprayers to remove dust or other coarse foreign materials. The washed dates are then graded and sorted to remove the defective and inferior dates. The dates are finally washed using fresh water containing a food-grade detergent superchlore (sodium salt of dodecylbenzene sulfonic acid) and dubois 317 (sodium mono- and dimethyl naphthalene sulfonate) as a disinfectant. Excess water is removed from the dates by blowing hot air before packing in 20-kg corrugated cartons, which are then pressed, sealed, and wrapped (Mikki et al., 1986). If they are not sent for immediate shipping and marketing, the packed dates are transferred to cold storage ($5 \pm 2°C$) to maintain better shelf life up to 6 months with minimal changes in their original texture and flavor (Hegazi et al., 1986b).

Preserved Products

A number of preserved products such as pickles, chutney, jam, date butter, dates-in-syrup, paste, candy, and confectionery items have been prepared from date fruits (USDA, 1973). For preparing pickles and chutney, date fruits at the *kimri* and *khalal* stages of maturity are suitable. Pickles-in-oil and chutney prepared from *kimri* date fruits (Al-Hooti et al., 1997b, c) can substitute the extremely popular products being commercially prepared from raw mango fruit (Das-Thakur et al., 1976). Typical hard texture and the ample amounts of sugars present at *kimri* stage are conducive for producing good-quality pickles and chutney. The shape, size, and green color of *kimri* stage date fruits make them look similar to olives. Except for their lower acidity values, the sweetness and textural characteristics of *kimri* -stage date fruits are similar to raw mango fruit, thus suitable for preparing pickles-in-oil and sweet chutney for local consumption and export purposes. Brine and salt-stock pickles are other popular products that could be prepared from *kimri* date fruits (Hamad and Yousif, 1986). These pickles are microbiologically safe as coliforms were absent, and the products had acceptable sensory quality even after 3 months of storage. The duration of the pickling process varies from prolonged fermentation for brine pickles to very limited fermentation for fresh-pack pickles or no fermentation as for mango and other fruit pickles (Das-Thakur et al., 1976). Detailed information on the most important factors for pickling, such as brine concentration, use of antimold additives like sorbic acid and acetic acid, thermal processing requirements, and shelf life is available (Yousif et al., 1985; Al-Ogaidi et al., 1982; Khatchadourian et al., 1986). Some of the processed products prepared from date fruits are presented in Figure 22.2.

Traditionally, jam is defined as a self-preserved, cooked mixture of fruit and sugar (honey is often qualified as a sugar), with a total soluble solid content of 68.5% or higher (Al-Hooti et al., 1997d). For preparing a good jam, 65% of sugar, 1% of pectin, and a pH of about 3.0–3.2 are required. If the fruit is low in acidity, citric acid is often added. The basis of jam preservation is related to the water activity of the product. Mainly the sugar and pectin present in jam are responsible for attaining the desired water activity. Usually a sugar–date pulp ratio of 55:45 is used for jam making. Date fruits, having high sugar contents, are suitable for jam manufacture (Khatchadourian et al., 1986). The *rutab* stage date fruits have a reasonable quantity of sugar as well as the pectin required for jam preparation. Certain date fruit cultivars, such as *Khalas*, *Sukkary*, and *Ruzeiz*, have been shown to possess the desirable sugar and pectin contents and are highly suitable for jam making (Yousif et al., 1993a). For making date butter

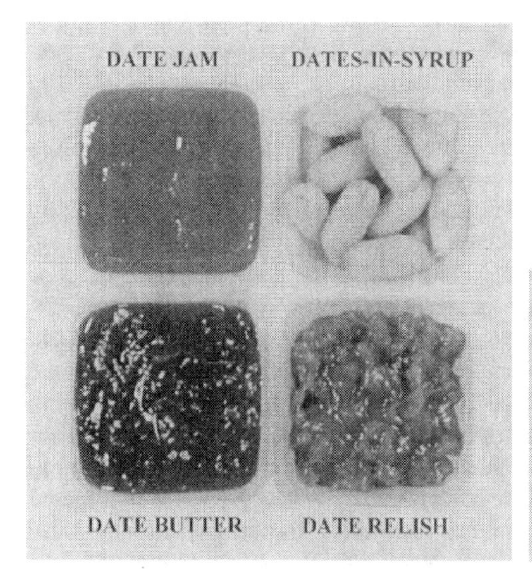

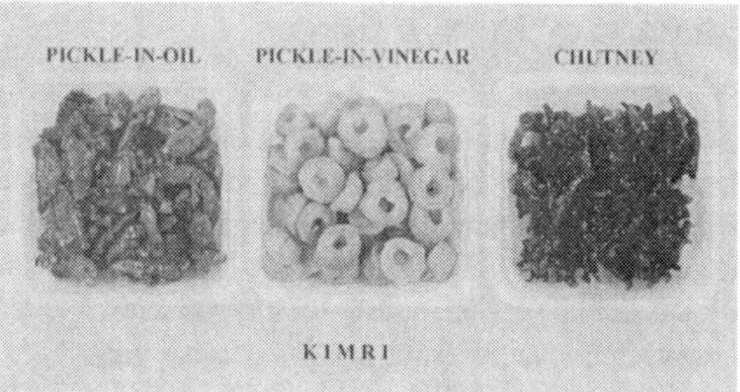

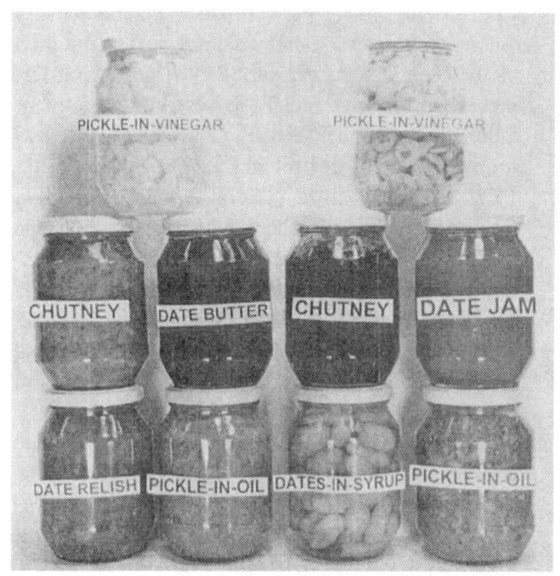

Figure 22.2. Some of the processed date fruit products.

(a product similar to peanut butter in usage), *tamer* fruits having the highest sugar content are used. All the steps are similar to jam making, except the pH of the pulp and sugar mixture is adjusted to 4.5–4.7, and the total soluble solid content at finishing stage is 74–75 Brix. Usually a sugar–date pulp ratio of 40:60 is used in date butter making. For the preparation of dates-in-syrup, peeled, pitted whole date fruits at the *khalal* stage of maturity are used (Al-Hooti et al., 1997e). After adjusting the pH of sugar syrup (50 Brix) to 2.8–3.0, it is boiled to reach a concentration of about 75–80 Brix. The hot syrup is poured into glass jars containing peeled, pitted date fruits, and the jars are capped immediately. The minimum drained weight of processed fruit should be kept at 55%. To achieve microbial sterility, the capped jars are processed in hot water (95°C) for 30 min, then cooled to room temperature and labeled.

Date syrup (dibs) is another useful product that can be prepared from *tamer* stage fruits. Recently,

Al-Hooti et al. (2002) made use of pectinase and cellulase enzymes to obtain almost double the recovery of soluble solids than were obtained with the conventional hot water and autoclaving extraction methods. The date syrup extracted with pectinase and cellulase can be used as a good substitute for sucrose in bakery products (Sidhu et al., 2002). Compared with the traditional heating methods, the use of microwave heating is another alternative to obtain better uniformity in product temperature, in a comparatively shorter time period that leads to better quality and yield of syrup (Ali et al., 1993). Date syrup produced by these methods is used in a variety of food products, such as cakes (El-Samahi et al., 1993), carbonated beverages (Hamad and Al-Beshr, 1993), soft frozen yogurt (Hamad et al., 1993), milk-based drinks (Alhamdan, 2002; Yousif et al., 1986a, 1996), nutritious creamy foods (Alemzadeh et al., 1997), and ready-to-serve date juice beverages (Godara and Pareek, 1985; Yousif et al., 1993b). Based on date syrup, butter, hazelnuts, dried skim milk, cocoa, starch, lecithin, and baking powder, a food can be formulated to have 6.13% protein, 19.86% fat, 47.8% total sugars, and a good amount of minerals (Alemzadeh et al., 1997). The hot weather prevailing in this part of the world for most of the year offers a very good potential for the commercial production of these date juice or syrup-based drinks.

Traditionally, a number of fruits, such as apple, apricot, mango, raisin, and strawberry, are converted into paste on a commercial scale for use in baby foods, baked goods, and confectionery (Ziemke, 1977; Anon, 1981), but so far date fruit has not been exploited to its full potential. Date fruit is not only the richest source of sugars but also contains various vitamins, minerals, and phytochemicals. The production of date paste is, therefore, of particular interest to the food industry as it also results in reduced transportation and storage costs, since the stones (10–20% of the whole fruit weight) are removed in the process. This will also ensure the availability of date fruit paste for the food industry throughout the year. For the preparation of date paste, pitted *tamer* date fruits are either soaked in hot water at 95°C for 5–15 s or steamed at 10 psig for about 3 min. To maintain the desirable color and good shelf life, citric acid or ascorbic acid (0.2% on a fruit basis) is added to lower the pH of date paste. The water activity (a_w) and pH of date paste prepared by this method are kept within the safe limits of 0.57 and 5.4, respectively (Yousif et al., 1986b, c). Date paste offers an opportunity to convert even the lower grade date fruits into an intermediate value-added product by the date processing industry (Mikki et al., 1983). Date paste and date fruit chunks can also be added to a number of food products such as baked goods and ice cream. Up to 50% of the sucrose in ice cream can easily be replaced with date paste without adversely affecting its quality (Hamad et al., 1986). Addition of date pieces (10%) to ice cream reduces the overrun slightly. Use of 4–8% date paste in bread formulation results in marked improvements in the dough rheological properties, delays gelatinization, improves gas production and retention, prolongs the shelf life, retards staling, and improves the crumb and crust characteristics (Yousif et al., 1991).

Date fruits serve mainly as a source of calories as these are rich in carbohydrates (about 78%) but low in proteins (2–3%) and fat (1%). To convert date fruits into nearly a complete food would, therefore, require supplementation with proteins, dietary fiber, and fats. Recently, the trend is shifting toward the use of blends of vegetable and dairy proteins to formulate a variety of candies, energy bars, and confectionery, which are becoming popular among children and adolescents. One similar product made with *tamer* date pulp, sesame seeds, almonds, and oat flakes has been found to be quite acceptable to consumers (Al-Hooti et al., 1997e). The average ash, fat, and protein contents of 1.78%, 6.09%, and 7.83% in the control date bars (containing date paste and almonds) changed to 2.60%, 3.90%, and 9.56% in these date bars fortified with sesame seeds, almonds, skim milk powder, and rolled oats, respectively. In another type of date bars fortified with soy protein isolate, single-cell proteins, almonds, and skim milk powder, the protein content was increased from 4.9–5.3% in the control to about 10.7–12.1% (dry basis) in samples containing the high-protein ingredients. Such formulated bars not only supply calories but can also provide a reasonable amount of fat, fiber, and minerals. These supplemented date bars not only have increased protein content but also possess significantly higher chemical scores of essential amino acids (Khalil, 1986).

A variety of candied or glace fruits are being prepared from a number of fruits for use in new food product development by the dairy and bakery industries. To enhance the penetration of sugars, the fruit is pierced and is also dipped in dilute calcium chloride solution to toughen the texture. Use of citric acid and ascorbic acid is also commonly used in the preparation of invert sugar syrup (about 30–45 Brix) required for cooking such fruit. During the preparation of candied fruit, the cooking of fruit with sugar syrup is repeated for short intervals over a period of many days until the soluble solids content of the cooked fruit reaches 70 Brix or higher. This higher sugar content enhances the shelf life of candied fruit even when stored for many months at room temperature.

Khalal stage fruits can be used for the preparation of glace dates (Sawaya et al., 1986b). *Khalal* fruits from two varieties, *Hallaw* (red) and *Khuwaildi* (yellow), were washed, air-dried, and pricked to facilitate sugar penetration. Sugar syrup of 35 Brix was prepared from 4.25 kg each of sucrose and glucose in 20 l of water. To the syrup, 10 g each of calcium chloride and potassium sorbate are added, and the pH is adjusted to 2.8 with a solution made from a 4:1 mixture of citric acid and ascorbic acid. The fruit and syrup are cooked slowly over a period of time (with intermittent overnight rests) till the fruit gets to 75 Brix. The glace fruit can be flavored or coated with milk chocolate for improved acceptability.

BY-PRODUCTS FROM DATE PROCESSING

Waste date pulp, low-grade rejected date fruits, and the date seeds (pits, stones) are the three major by-products coming out of the date fruit processing plants. As the waste date pulp and rejected date fruits are rich in many components such as soluble sugars, proteins, vitamins, and minerals, these can be used as a fermentation substrate for oxytetracycline production by some suitable mutants of *Streptomyces rimosus* (Baeshin and Abou-zeid, 1993). At the end of 96 h of fermentation period, the cell biomass is harvested and the antibiotic is recovered. The presence of naturally occurring glucose, fructose, sucrose, proteins, amino acids, certain B-complex vitamins, and minerals in date fruit pulp is conducive for the synthesis of oxytetracycline by the chosen strain, *S. rimosus.* In addition to waste date fruit pulp, low-grade date fruits, immature fruits, and other wastes coming out of date fruit processing plants can also be used for the production of many pharmaceuticals, citric acid, and ethanol.

Date pit is another major date by-product, constituting about 10% by weight of the whole fruit. Although a lot of information is now available about the chemical composition and utilization of date fruit pulp, the information on date seed (pits, stone) composition and utilization has started accumulating. As the date seeds are rich in oil, proteins, minerals, and fiber, these could serve as valuable raw materials for animal feeds. After extracting edible oil for humans, date seed meal can be used for animal feeds (Rygg, 1977). The reported average values for date seeds are 6–7% protein, 9–10% fat, 1–2% minerals, and 20–24% fiber (Al-Hooti et al., 1998; Sawaya and Khalil, 1986). The fats from date seeds are rich in many unsaturated fatty acids such as oleic acid (58.8%) and linoleic acid (12.8%). Considering the nutritional importance of the fatty acid profile, date seeds have the potential for the production of edible oil for human consumption. The major minerals present in date seeds are potassium, phosphorus, calcium, and magnesium. The essential amino acid profile of date seed proteins is quite comparable with those of other comparatively expensive oilseeds such as soybean, groundnut, cottonseed, and sesame. To improve their nutritional value as animal feeds, the higher content of cell wall materials present in date seeds needs to be solubilized by treating them with 4.8–9.6% solution of sodium hydroxide (Al-Yousef et al., 1986). The ligninase enzyme has been shown to be useful in solubilizing date seeds and thus improving their nutritional value for use in animal feeds. For the production of this enzyme, date seeds rich in carbohydrates, proteins, lipids, and minerals can be utilized as a substrate to grow *Phanerochaete chrysosporium.* A slurry of 10% crushed seeds in water is inoculated with *P. chrysosporium* and incubated at 30°C for 7 days. The pH of the substrate is adjusted to 4.0 for achieving the highest yield of ligninase enzyme (El-Nawawy et al., 1993). After increasing the supply of carbon and nitrogen in the fermentation medium, the date seeds can also be used for the production of oxytetracycline by *Streptomyces rimosus* as the date seeds, as such, cannot produce appreciable amounts of this antibiotic (Abou-zeid and Baeshin, 1993).

NUTRITIVE VALUE AND HEALTH BENEFITS

Long before recorded history, date fruits were so important in Arabian diets that they were called the "Fruit of Life." Date crops in the desert regions of the Middle Eastern countries have been the bases for survival for its residents. During the Holy Month of Ramadan, at the time of breaking fast in the evening, a few date fruits are eaten. Dates are consumed in many forms such as fresh, preserved, candied, and as a constituent of the processed food products. The largest amount of date fruit is eaten as a fully ripe fruit (*tamer*). Many advances have recently taken place in date processing. Now a number of companies are producing cleaned, washed, and pressed *tamer* dates in a variety of attractive packaging. Small amounts are also pitted, stuffed with nuts (almonds) and/or mixed with anise or fennel seeds and offered for retail sale (El-Shaarawy, 1986).

Date fruits are mainly considered a source of readily available carbohydrates. At the *tamer* stage of maturity, dates are a rich source of readily available carbohydrates in our diet, as dates (without pits) contain more than 80% of total monosugars (Myhara et al., 1999). The sucrose present at the *khalal* and *rutab* stages of maturity in most of the cultivars is almost completely converted to glucose and fructose through the action of the invertase enzyme. *Tamer* dates are also a rich source of total dietary fiber (about 12.97–13.32%, on a dry basis), both water-soluble and water-insoluble fractions, both having proven health benefits. Dietary fiber from date fruits when fed to white albino rats for 8 weeks has been shown to significantly lower the total cholesterol, triglycerides, and phospholipids in rat livers (Jwanny et al., 1996). Apart from lowering the total serum lipids and low-density lipoprotein cholesterol by 32–48%, even the serum triglycerides and total cholesterol are decreased by 23–35%. The practice of eating date fruits at *khalal* and *rutab* stages of maturity would also supply reasonable amounts of soluble dietary fiber such as pectins (4–5% dry basis), thus making significant nutritional contributions to the dietary fiber intake among humans.

An average person in Saudi Arabia consumes daily about 100 g of dates (El-Shaarawy, 1986), which would meet 13% of their daily requirement for total energy, more than 11% of their daily requirement for iron, about 7% of their daily requirement for ascorbic acid, and 6% of their daily requirement for proteins. Some "date lovers," are known to consume even higher amounts of dates at the *tamer, Khalal,* and *rutab* stages; thus the intake of these valuable nutrients will be much higher in their diets. Traditionally, the date fruits are consumed with milk, especially during the fasting month of Ramadan, which has a strong scientific logic. In addition to providing calories, the vitamin C and fructose present in date fruits are known to enhance the absorption of iron. This makes the date fruit and milk a good nutritional combination in terms of iron, vitamin C, and proteins. There are a number of chemical constituents that are known to enhance the absorption of iron. Ascorbic acid is one such enhancer that improves the absorption of the iron present in the date fruits (Fleming et al., 1998).

Fortunately dates, at all stages of maturity, are extremely low in fat (about 1%), but at the same time quite rich in minerals, certain B-complex vitamins, and polyphenolic compounds. Among the minerals, dates are especially rich in potassium, but at the same time, are low in sodium, thus serve as an excellent food source for persons suffering from hypertension. Among the other minerals, dates are reasonably good sources of iron, copper, sulfur, and manganese; and fair sources of calcium, chloride, and magnesium. Dates also contain moderate amounts of a few B-complex vitamins in relation to the calories they contain, especially thiamin, riboflavin, and folic acid. In terms of recommended levels of thiamin, riboflavin, and nicotinic acid of 0.4, 0.6, and 6.6 mg/1000 calories, the date fruits are known to provide 0.32, 0.35, and 8.0 mg/1000 calories, respectively (Vandercook et al., 1980).

It is our folly to consider date fruit only as a source of carbohydrates. Experimental evidence is accumulating about the contributions of date fruits in terms of their health-promoting properties due to the presence of various phytochemicals. Dates are now known to be a rich source of many phytochemicals such as phenolic compounds. The astringency of *kimri* stage date fruit is due to the presence of phenolic substances, generally known as tannins. Many types of tannins are found in date fruits but two main groups, phenolic acids and condensed tannins, are thought to be important mainly in producing astringency sensations. These phenolic acids are comprised of cinnamic acid derivatives originating from the amino acid, phenylalanine, while condensed tannins, or proanthocyanidins, are polyphenolics (Myhara et al., 2000). *Kimri* stage date fruit is known to contain maximum amount of tannins, which decrease as the fruit matures to the *tamer* stage (Al-Hooti et al., 1997f). These phenolics are known to possess strong antioxidative properties and prevent oxidative damage to DNA, lipids, proteins, and cell membranes, which may play a role in preventing chronic diseases such as cancer and cardiovascular disease (Hollman, 2001). Similarly, the *tamer* date fruits are not only good sources of sugars and dietary fiber, but they also supply reasonable amounts of potassium, phosphorus, calcium, iron, thiamin, riboflavin, and nicotinic acid in our diet. However, to achieve the maximum nutritional benefits, consumption of *khalal* and *rutab* fruits needs to be encouraged as these are much better sources of some of the nutrients, especially dietary fiber, ascorbic acid, β-carotene, and many phytochemicals. This viewpoint is now well accepted that the plant-based diets (especially those rich in fruits, vegetables, and whole grains) have contributed to greater longevity and a lower risk of coronary artery disease among the people in the Mediterranean and Asian countries (Fung and Hu, 2003). Diets based on the consumption of *khalal* and *rutab* fruits fit very well into this health-promoting strategy by the nutritionists.

STANDARDS AND REGULATIONS

The quality of a date-fruit-based food material depends on various factors such as cultivars, maturity stage, level of microorganisms, and methods of processing, storage, and handling. For the successful marketing of any food material, appropriate food standards and regulations are necessary. Except for packed dates, food grades or standards for all the processed date fruit products need to be developed by these countries. Food standards for the packed dates have been developed by the State of Kuwait, and similar standards do exist for the other Gulf Cooperation Council countries (Anon, 1998). These standards are only for the packed dates and cover labeling, contamination with sand or grit, good manufacturing practices to be followed, expected shelf life, limits for microorganisms, and the suggested methods of analysis required for testing of these products. Some of the packaging requirements for *tamer* date fruits are described in Tables 22.2 and 22.3.

Table 22.2. Kuwaiti Standards for Quantity Requirements for Packed *Tamer* Date Fruits

	Number of Date Fruits/500 g	
Fruit Size	Without Seeds	With Seeds
Small fruits	>110	>90
Medium fruits	90–110	80–90
Large fruits	<90	<80

Source: Anon, 1998.

Apart from meeting the weight and number of fruit requirements as per the above Kuwaiti standard, the packed dates should also conform to the identity of the declared cultivar on the package as well as its stage of maturity, free from insects, their fragments or excreta, should possess the characteristic flavor of the cultivar, fruits should be of same color and size. The packed date fruits (without seeds) should not contain more than two whole seeds or four parts of broken seeds/100 date fruits. However, other nuts (such as almonds, coconut) and condiments (e.g., fennel) of suitable quality (fit for human consumption) may also be added to these packed dates. The other specifications for the packed dates are <7% blemished dates; <6% damaged, unripe, and unpollinated dates; <7% dirty, insect infested dates; and <1% souring, moldy, and decayed dates. The packaging material used should be clean, dry, impermeable to moisture, and capable of preventing contamination of dates from dirt, etc. The packed dates should be stored in a cool dry place, free from humidity, away from direct sunlight, and rodents.

FUTURE RESEARCH NEEDS

The date palm tree (*P. dactylifera* L.) is a very hardy plant capable of growing under extremely hot, dry, and arid climates prevailing in the desert regions of the world. The value and importance of the date palm tree lie in providing timber, shade, and food to humans. We must agree that there is a lot more to the date fruit than just its sweet taste. At present, our knowledge about the detailed chemical composition, especially about the micronutrients and various phytochemicals that may provide immense nutritional benefits of eating date fruits, is extremely limited (Vayalil, 2002).

The recent advances in biotechnology and tissue culture techniques need to be further exploited

Table 22.3. Kuwaiti Standards for Microbiological Quality of Packed *Tamer* Date Fruits

Type of Microorganism	No. of Positive Samples Observed	Maximum No. of Samples to Exceed Permissible Limits	Maximum Permissible Microbial Load CFU/g	Maximum Microbial Load in Any of the Sample Tested CFU/g
Yeasts	5	2	10	10^2
Molds	5	2	10^2	10^3
E. coli	5	2	0	10

Source: Anon, 1998.

Figure 22.3a. Fruit bearing date tree.

not only to increase the fruit size, bearing capacity, uniformity of fruit maturity, and total yield per hectare but also to improve the processing and nutritional qualities, especially concerning the amount of phytochemicals and other micronutrients present in the date fruit (Figs. 22.3a,b). We need to identify clearly the chemical constituents responsible for the distinct flavors of date fruits from different cultivars. Recent advances in the analytical techniques such as gas chromatography coupled with mass spectrophotometry would be of great help in achieving such goals. Moreover, if we are interested in developing the date processing industry on scientific lines for exploiting economic gains through global marketing, the development of suitable food standards and regulations for such processed date-fruit products are essential. Currently, no other food standard is available except that for the packed *tamer* date fruits. In addition, we also need to identify, characterize, and estimate the various micronutrients and phytochemicals present in date fruits, and their bioavailability and metabolism in humans. One of the most challenging and beneficial areas of research should be focused on the determination of various potential antioxidants and their stability during date fruit processing, storage, and distribution. Like most other fruits (Liu, 2003), date fruit at all stages of maturity is also expected (and probably is) to be a storehouse of these valuable micronutrients and phytochemicals, which can provide tremendous health benefits to humans.

Figure 22.3b. Date fruit trees in an orchard.

REFERENCES

Abbas, M.F., M.J. Abbas, and O.I. Abdel-Basit. 2000. Indole-3-acetic acid concentration during fruit development in date palm (*Phoenix dactylifera* L. cv *Hillawi*). Fruits 55(2):115–118.

Abdelmonem, A.E., S.H. Fouad, and E.M. Hegazi. 1986a. Fumigation and thermal treatments on stored date insects. Proceedings of the Second Symposium on the Date Palm in Saudi Arabia, Vol. II, Al-Hassa, pp. 441–451.

Abdelmonem, A.E., A.A. Rokaibah, and S.H. Fouad. 1986b. Effect of CO_2 fumigation on the fauna and flora of stored dates in Qassim, Saudi Arabia. In Proceedings of the Second Symposium on the Date Palm in Saudi Arabia, Vol. II, Al-Hassa, pp. 469–479.

Abdulla, K.M., M.A. Meligi, and S.Y. Rysk. 1982. Influence of crop load and leaf/bunch ration on yield and fruit properties of *Hayany* dates. In Proceedings of the First Symposium on the Date Palm in Saudi Arabia, Al-Hassa, pp. 223–232.

Abou-zeid, A.A. and N.A. Baeshin. 1993. Utilization of Saudi date seeds in formation of oxytetracycline. Program and Abstracts of the Third Symposium on the Date Palm in Saudi Arabia, Al-Hassa, Abstract No. I-8, pp. 158–159.

Abu-Zinada, A.H. and M.I. Ali. 1982. Fungi associated with dates in Saudi Arabia. J Food Prot 45(9):842–844.

Ahmed, I.A., A. Ahmed, and R.K. Robinson. 1997. Susceptibility of date fruits (*Phoenix dactylifera*) to aflatoxin production. J Sci Food Agric 74(1):64–68.

Ahmed, M.S.H., Z.S. Al-Hakkak, S.R. Ali, A.A. Kadhum, I.A. Hassan, S.K. Al-Maliky, and A.A. Hameed. 1982. Disinfection of commercially packed dates, *Zahdi* variety, by ionizing radiation. Date Palm J 1(2):249–259.

Ahmed, A.I. and R.K. Robinson. 1997. Incidence of *Aspergillus flavus* and *Aspergillus parasiticus* on date fruits. Agric Eng Int 49:136–139.

Ahmed, I.A. and R.K. Robinson. 1998. Selection of a suitable method for analysis of aflatoxins in date fruits. J Agric Food Chem 46(2):580–584.

Ahmed, A.I. and R.K. Robinson. 1999. The ability of date extracts to support the production of aflatoxins. Food Chem 66(3):307–312.

Aidoo, K.E., R.F. Tester, J.E. Morrison, and D. MacFarlane. 1996. The composition and microbial quality of pre-packed dates purchased in greater Glassgow. Int J Food Sci Technol 31(5):433–438.

Aldrich, W.W. 1942. Some effects of soil moisture deficiency upon *Deglet Noor* fruit. Date Growers' Inst Rep 19:7–10.

Alemzadeh, I., M. Vossoughti, A. Keshavarz, and V. Maghsoudi. 1997. Use of date honey in the formulation of nutritious creamy food. Iran Agric Res 16(2):111–117.

Al-Hakkak, Z.S., H. Auda, and J.S. Al-Hakkak. 1986. Effect of high dosages of phosphine gas on the amino acid, protein and sugar composition of Iraqi dates. Date Palm J 4(2):235–246.

Alhamdan, A.M. 2002. Rheological properties of a new nutritious dairy drink from milk and date extract concentrate (*dibs*). Int J Food Properties 5(1):113–126.

Al-Hooti, S.N., J.S. Sidhu, J. Al-Otaibi, and H. Qabazard. 1995a. Extension of shelf life of date fruits at the *khalal* stage of maturity. Indian J Hortic 52(4):244–249.

Al-Hooti, S.N., J.S. Sidhu, and H. Qabazard. 1995b. Studies on the physicochemical characteristics of date fruits of five UAE cultivars at different stages of maturity. Arab Gulf J Sci Res 13(3):553–569.

Al-Hooti, S.N., J.S. Sidhu, H. Al-Amiri, J. Al-Otaibi, and H. Qabazard. 1997a. Extension of shelf life of two UAE date fruit varieties at *Khalal* and *Rutab* stages of maturity. Arab Gulf J Sci Res 15(1):99–110.

Al-Hooti, S.N., J.S. Sidhu, J. Al-Otaibi, H. Al-Amiri, and H. Qabazard. 1997b. Utilization of date fruits at different maturity stages for variety pickles. Adv Food Sci 19(1/2):1–7.

Al-Hooti, S.N., J.S. Sidhu, J. Al-Otaibi, H. Al-Amiri, and H. Qabazard. 1997c. Processing of some important date cultivars grown in United Arab Emirates into chutney and date relish. J Food Process Preserv 21:55–68.

Al-Hooti, S.N., J.S. Sidhu, J.M. Al-Saqer, H. Al-Amiri, and H. Qabazard. 1997d. Processing quality of important date cultivars grown in the United Arab Emirates for jam, butter and dates-in-syrup. Adv Food Sci 19(1/2): 35–40.

Al-Hooti, S.N., J.S. Sidhu, J. Al-Otaibi, H. Al-Amiri, and H. Qabazard. 1997e. Date bars fortified with almonds, sesame seeds, oat flakes and skim milk powder. Plant Foods Hum Nutr 51:125–135.

Al-Hooti, S.N., J.S. Sidhu, and H. Qabazard. 1997f. Physicochemical characteristics of five date fruit cultivars grown in the United Arab Emirates. Plant Foods Hum Nutr 50:101–113.

Al-Hooti, S.N., J.S. Sidhu, J.M. Al-Saqer, and A. Al-Othman. 2002. Chemical composition and quality of date syrup as affected by pectinase/cellulase treatment. Food Chem 79(2):215–220.

Al-Hooti, S.N., J.S. Sidhu, and H. Qabazard. 1998. Chemical composition of seeds of date fruit cultivars of Unites Arab Emirates. J Food Sci Technol 35(1):44–46.

Ali, A.I., A.I. Mustafa, E.A. Elgasim, H.A. Alhashem, and S.E. Ahmed. 1993. The use of microwave heating in the production of date syrup (*Dibs*). Program and Abstracts of the Third Symposium on the Date Palm in Saudi Arabia, Al-Hassa, Abstract No. I-23, p. 167.

Al-Jasim, H.A. and K.S. Al-Delaimy. 1972. Pectinesterase activity of some Iraqi dates at different stages of maturity. J Sci Food Agric 23:915–917.

Al-Juburi, H.J., H.H. Al-Masry, and S.A. Al-Muhanna. 2001. Effect of some growth regulators on some fruit characteristics and productivity of the *Barhee* date palm tree cultivar (*Phoenix dactylifera* L.). Fruits 56(5):325–332.

Al-Mashadi, A., A. Al-Shalhat, and A.K. Fawal. 1993. Storage and preservation of date in *rutab* stage. Program and Abstracts of the Third Symposium on the Date Palm in Saudi Arabia, King Faisal University, Al-Hassa, Saudi Arabia, Abstract No. I-16, p. 163.

Al-Mulhim, F.N. and G.E. Osman. 1986. Household date consumption patterns in Al-Hassa: A cross-sectional analysis. Proceedings of the Second Symposium on the Date Palm in Saudi Arabia, Vol. I, Al-Hassa, pp. 513–522.

Al-Ogaidi, H.K., J. Al-Baradei, and M. Abdel-Maseih. 1982. Possibility of pickling *Zahdi* dates at *al-kimri* stage. J Res Agric Water Resour (Iraq) 1(1):51–55.

Al-Shahib, W. and R.J. Marshall. 2002. Dietary fiber content of dates from 13 varieties of date palm *Phoenix dactylifera*, L. Int J Food Sci Technol 37:719–721.

Al-Tayeb, A. 1982. History of date cultivation in the world. Al-Yawin Newspaper, Hofuf, Kingdom of Saudi Arabia. March 19, p. 18.

Al-Warthan, A.A., S.S. Al-Showiman, S.A. Al-Tamrah, and A.A. BaOsman. 1993. Spectrophotometric determination of boron in dates of some cultivars grown in Saudi Arabia. J AOAC Int 76(3):601–603.

Al-Yousef, Y., R.L. Belyea, and J.M. Vandepopuliere. 1986. Sodium hydroxide treatment of date pits. Proceedings of the Second Symposium on the Date Palm in Saudi Arabia, Vol. II, Al-Hassa, pp. 197–206.

Anon. 1981. Fruit pastes enrich the summer assortment. CCB Rev Chocolate Confect Bakery 6(2):24–26.

Anon. 1983. Packed Dates: Saudi Specifications. Saudi Arabian Standards Organization. Saudi Government Press, Riyadh, Saudi Arabia, pp. 1–8.

Anon. 1984. Study on the Development of Date Cultivation, Production, Processing and Marketing in the Kingdom of Saudi Arabia. Arab Organization for Agricultural Development, Arab League Countries, Khartoum, pp. 23–27.

Anon. 1998. Kuwaiti Standards for Packed Dates. Ministry of Commerce and Industry, State of Kuwait, Standards No. 98/894, pp. 1–10.

Anwar-Shinwari, M. 1987. Iron contents of date fruits. J Coll Sci King Saudi Univ 18(1):5–13.

Asif, M.I., A.S. Al-Ghamdi, O.A. Al-Tahir, and R.A.A. Latif. 1986. Studies on the date palm cultivars of Al-Hassa oasis. Proceedings of the Second Symposium on the Date Palm in Saudi Arabia, Vol. I, Al-Hassa, pp. 405–413.

Bacha, M.A. and A.A. Abo-Hassan. 1982. Effects of soil fertilization on yield, fruit quality and mineral content of *Khudari* date palm variety. Proceedings of the First Symposium on the Date Palm in Saudi Arabia, Al-Hassa, pp. 174–180.

Baeshin, N.A. and A.A. Abou-zeid. 1993. Saudi dates as fermentation media for oxytetracycline production by some mutants of *Streptomyces rimosus*. Program and Abstracts of the Third Symposium on the Date Palm in Saudi Arabia, Al-Hassa, Abstract No. I-7, pp. 157–158.

Benjamin, N.D., K.C. Tonelli-Peres, N.M. Ali, and N.A. Al-Drobi. 1979. Date polyphenol oxidase: Partial purification and characterization. Tech Bull Palm Dates Res Cent No. 9/79:1–25.

Brown, G.K. 1982. Date production in the USA. Proceedings of the First Symposium on the Date Palm in Saudi Arabia, Al-Hassa, pp. 2–13.

Brown, G.K., R.M. Perkins, and E.G. Vis. 1969. Mechanical pollination experiments with the *Deglet Noor* date palm in 1969. Date Growers' Inst Rep 47:19–24.

Carpenter, J.B. and H.S. Elmer. 1978. Pests and diseases of the date palm. Agricultural Handbook No. 227, USDA, USA, pp. 46–59.

Carpenter, J.B. and L.J. Klotz. 1966. Diseases of the date palm. Date Growers' Inst Rep 43:15–21.

Chatha, G.A., A.H. Gilani, and M. Bashir. 1986. Effect of curing agents on the chemical composition and keeping quality of date fruit. Proceedings of the Second Symposium on the Date Palm in Saudi Arabia, Vol. II, Al-Hassa, pp. 27–33.

Chohan, J.S. 1972. Diseases of date palm (*Phoenix dactylifera* L.) and their control. Punjab Hort J 12:25–32.

Coggins Jr, C.W. and J.C.F. Knapp. 1969. Growth, development and softening of *Deglet Noor* date fruit. Date Growers' Inst Rep 46:11–14.

Coggins Jr, C.W., J.C.F. Knapp, and A.L. Ricker. 1968. Postharvest softening of *Deglet Noor* date fruit: Physical, chemical and histological changes. Date Growers' Inst Rep 45:3–5.

Das-Thakur, S.P., D.R. Chaudhuria, S.N. Mitra, and A.N. Bose. 1976. Quality aspects of processed mango products. Indian Food Packer 30(5):45–50.

De Mason, D. and B. Tisserat. 1980. The occurrence and structure of apparently bisexual flowers in the date palm, *Pheonix dactylifera* L. (Arecaceae). Bot J Linnean Soc 81:283–292.

Eeuwens, C.J. and J. Blake. 1977. Culture of coconut and date palm tissues with a view to vegetative propagation. Acta Hortculturae 78:277–286.

El-Hamady, M.M., A.S. Khalifa, and A.M. El-Hamady. 1982. Fruit thinning in date palm with ethephon. Proceedings of the First Symposium on the Date Palm in Saudi Arabia, Al-Hassa, pp. 284–295.

El-Kassas, S.E. 1986. Effect of some growth regulators on the yield fruit quality of *Zaghloul* date palm. Proceedings of the Second Symposium on the Date Palm in Saudi Arabia, Vol. I, Al-Hassa, pp. 179–186.

El-Kassas, S.E. and H.M. Mahmoud. 1986. The possibility of pollinating date palm by diluted pollen. Proceedings of the Second Symposium on the Date Palm in Saudi Arabia, Vol. I, Al-Hassa, pp. 317–322.

El-Nawawy, M.A., A.K. Abdel-Latif, and M.S. Al-Jassir. 1993. Ligninase production from micromycetes. Program and Abstracts of the Third

Symposium on the Date Palm in Saudi Arabia, Al-Hassa, Abstract No. I-6, p. 157.

El-Samahi, S.K., S.I. Goneim, S.S. Ibrahim, M.G.A. El-Fadeel, and S.M. Mohamed. 1993. Date syrup (*Dibs*) and its utilization in cake making. Program and Abstracts of the Third Symposium on the Date Palm in Saudi Arabia, Al-Hassa, Abstract No. I-22, p. 166.

El-Sayed, S.A. 1978. Control of post-harvest storage decay of soft-type fruits with special reference to the effect of gamma irradiation. Egypt J Hort 5(2):175–182.

El-Shaarawy, M.I. 1986. Dates in Saudi Diet. Proceedings of the Second Symposium on the Date Palm in Saudi Arabia, Vol. II, Al-Hassa, pp. 35–47.

El-Zeftawi, B.M. 1976. Effects of ethephon and 2,4,5-T on fruit size, rind pigments and alternate bearing of Imperial mandarin. Sci Horticulturae 5:315–320.

El-Zoghbi, M. 1994. Biochemical changes in some tropical fruits during ripening. Food Chem 49:33–37.

Emam, O.A., S.E.A. Farag, and A.I. Hammad. 1994. Comparative studies between fumigation and irradiation of semi-dry date fruits. Nahrung/Food 38(6):612–620.

FAO. 1982. Food Composition Tables for the Near East, Publication No. 26. FAO/UNO, Rome, p. 64.

FAO. 2000. FAO Production Year Book: FAO Statistics Series, Vol. 54. FAO, Rome, pp. 168–169.

Fleming, D.J., P.F. Jacques, G.E. Dallal, K.L. Tucker, P.W.F. Wilson, and R.J. Wood. 1998. Dietary determinants of iron stores in a free-living elderly population: The Framingham Heart Study. Am J Clin Nutr 67(4):722–733.

Fung, T.T. and F.B. Hu. 2003. Plant-based diets: What should be on the plate? Am J Clin Nutr 78:357.

Furr, J.R. 1956. The seasonal use of water by *Khadrawy* date palms. Date Growers' Inst Rep 33:5–7.

Furr, J.R. and W.W. Armstrong. 1959. The relation of growth, yield and fruit quality of *Deglet Noor* dates to variations in water and nitrogen supply and to salt accumulation in the soil. Date Growers' Inst Rep 36:16–18.

Gerber, M., M.C. Boutron, S. Hercberg, E. Riboli, A. Scalbert, and M.H. Siess. 2002. Food and cancer: State of the art about the protective effect of fruits and vegetables. Bull Cancer 89(3):293–312.

Ghelawi, M.A., J.S. Moore, and N.J. Dodd. 1996. Use of ESR for the detection of irradiated dates (*Phoenix dactylifera* L.). Appl Radiat Isot 47(11/12):1641–1645.

Godara, R.K. and O.P. Pareek. 1985. Effect of temperature on storage life of ready-to-serve date-juice beverage. Indian J Agr Sci 55(5):347–349.

Goneum, S.I., S.K. El-Samahy, S.S. Ibrahim, M.G.A. El-Fadeel, and S.M. Mohammed. 1993. Compositional changes in the date fruits during ripening by freezing. Program and Abstracts of the Third Symposium on the Date Palm in Saudi Arabia, King Faisal University, Al-Hassa, Saudi Arabia, Abstract No. I-14, p. 162.

Graces-Medina. M. 1968. Pectin, pectin esterase, and ascorbic acid in tropical fruit pulps. Arch Latinoam Nutr 18:410–412.

Grecz, N., R. Al-Harithy, R. Jaw, M.A. El-Mojaddidi, and S. Rahma. 1986. Radiation inactivation of microorganisms on dates from Riyadh and Al-Hassa area. Proceedings of the Second Symposium on the Date Palm in Saudi Arabia, Vol. II, Al-Hassa, pp. 155–164.

Gupta, O.P. and S. Siddiqui. 1986. Effect of time of cooking for the preparation of *Chhuhara* from date fruits. Date Palm J 4(2):185–190.

Hag, Q.N. and J. Gomes. 1977. Studies on xylan from date fruits (*Phoenix dactylifera* L.). Bangladesh J Sci Ind Res 12(1/2):76–80.

Hamad, A.M. and A.A. Al-Beshr. 1993. Possibility of utilizing date in the production of carbonated beverage. Program and Abstracts of the Third Symposium on the Date Palm in Saudi Arabia, Al-Hassa, Abstract No. I-25, p. 168.

Hamad, A.M., H.A. Al-Kanhal, and I. Al-Shaieb. 1986. Possibility of utilizing date puree and date pieces in the production of milk frozen desserts. Proceedings of the Second Symposium on the Date Palm in Saudi Arabia, Vol. II, Al-Hassa, pp. 181–187.

Hamad, A.M., A.I. Mustafa, and S.S. Al-Sheikh. 1993. Date syrup as a potential sweetener for the preparation of soft frozen yogurt (ice cream). Program and Abstracts of the Third Symposium on the Date Palm in Saudi Arabia, Al-Hassa, Abstract No. I-28, p. 169.

Hamad, A.M. and A.K. Yousif. 1986. Evaluation of brine and salt-stock pickling of two date varieties in the *kimri* stage. Proceedings of the Second Symposium on the Date Palm in Saudi Arabia, Vol. II, Al-Hassa, pp. 245–257.

Hasegawa, S., V.P. Maier, H.P. Kaszycki, and J.K. Crawford. 1969. Polygalacturonase content of dates and its relation to maturity and softness. J Food Sci 34:527–529.

Hasegawa, S. and D.C. Smolensky. 1970. Date invertase: Properties and activity associated with maturation. J Agric Food Chem 18(5):902–904.

Hasegawa, S. and D.C. Smolensky. 1971. Cellulase in dates and its role in fruit softening. J Food Sci 36(6):966–967.

Hassan, H.K., M.S. Mikki, M.A. Al-Doori, and T.S. Jaffar. 1979. Preservation of high-moisture dates (*Rutab*) by antimicrobial agents. Technical Bulletin Palm & Dates Res Center No. 2/79, pp. 1–18.

Hegazi, E.M., A.E. Abdelmonem, and A.A. Rokaibah. 1986a. Efficacy of three fumigants on the microflora associated with stored dates infested by insects in Qassim region. Proceedings of the Second Symposium on the Date Palm in Saudi Arabia, Vol. II, Al-Hassa, pp. 453–467.

Hegazi, A.M., M.S. Mikki, A.A. Abdel-Aziz, and S.M. Al-Taisan. 1986b. Effect of storage temperature on the keeping quality of commercially packed Saudi dates cultivars. Proceedings of the Second Symposium on the Date Palm in Saudi Arabia, Vol. II, Al-Hassa, pp. 61–71.

Hilal, M.H. 1986. Studies on irrigation and fertilization on date palm. Proceedings of the Second Symposium on the Date Palm in Saudi Arabia, Vol. I, Al-Hassa, pp. 286–302.

Hilgman, R.H., G.C. Sharples, and L.H. Howland. 1957. Effect of irrigation and leaf-bunch ratio on shrivel and rain damage of the *Maktoom* date. Date Growers' Inst Rep 34:2–345.

Hollman, P.C.H. 2001. Evidence for health benefits of plant phenols: Local or systemic effects? J Sci Food Agric 81:842–852.

Hussain, F. and A. El-Zeid. 1975. Studies on Physical and Chemical Characteristics of Date Varieties of Saudi Arabia. Ministry of Agriculture and Water, Kingdom of Saudi Arabia, pp. 1–60. (Arabic).

Hussein, F. 1970. Date Cultivars in Saudi Arabia. Ministry of Agriculture and Water, Dept. of Research and Development, Saudi Arabia, pp. 33–34.

Hussein, F. and M.A. Hussein. 1982a. Effect of irrigation on growth, yield and fruit quality of dry dates grown at Asswan. Proceedings of the First Symposium on the Date Palm in Saudi Arabia, Al-Hassa, pp. 168–173.

Hussein, F. and M.A. Hussein. 1982b. Effect of nitrogen fertilization on growth, yield and fruit quality of *Sakkoti* dates grown at Asswan. Proceedings of the First Symposium on the Date Palm in Saudi Arabia, Al-Hassa, pp. 182–189.

Hussein, F., S. Moustafa, and M. El-Kahtani. 1977. Effect of leaves/bunch ratio on quality, yield and ripening of *Barhi* dates grown in Saudi Arabia. First Agricultural Conference of Muslim Scientists, Riyadh, Saudi Arabia, pp. 18–25.

Hussein, F., S. Mostafa, F. El-Samirafa, and A. Al-Zaid. 1976. Studies on physical and chemical characteristics of eighteen date cultivars grown in Saudi Arabia. Indian J Hort 33:107–113.

Ishrud, O., MM Zahid, V.U. Ahmad, and Y. Pan. 2001. Isolation and structure analysis of a glucomannan from the seeds of Libyan dates. J Agric Food Chem 49(8):3772–3774.

Jarvis, M.C. 1984. Structure and properties of pectin gels in plant cell walls. Plant cell Environ 7:153–164.

Jwanny, E.W., M.M. Rashad, S.A. Moharib, and N.M. El-Beih. 1996. Studies on date waste dietary fiber as hypolipidemic agent in rats. Z Ernahrungswiss 35(1):39–44.

Ketchie, D.O. 1968. Chemical tests for thinning *Medjool* dates. Date Growers' Inst Rep 45:19–20.

Kevin, K. 1995. Functional foods: Designed for good health. Food Technol 56(4):72, 74, 77.

Khalifa, T., Z.M. Zuana, and M.I. Al-Salem. 1983. Date palm and Dates in the Kingdom of Saudi Arabia. Agriculture Research Department, Ministry of Agriculture and Water, Riyadh, Saudi Arabia, pp. 1–7.

Khalil, J.K. 1986. Date bars fortified with soy or yeast proteins and dry skim milk. In: W.N. Sawaya (ed.). Dates of Saudi Arabia. Safir Press, Riyadh, pp. 143–165.

Khalil, A.R. and A.M. Al-Shawaan. 1986. Wheat flour and sugar solution media as carriers for date palm pollen grains. Proceedings of the Second Symposium on the Date Palm in Saudi Arabia, Vol. I, Al-Hassa, pp. 68–71.

Khatchadourian, H.A., W.N. Sawaya, W.J. Safi, and A.F. Al-Shalhat. 1986. Date products. In: W.N. Sawaya, (ed.). Dates of Saudi Arabia. Safir Press, Riyadh, pp. 95–142.

Kirk, R.S. and R. Sawyer. 1991. Pearson's Composition and Analysis of Foods, 9th ed. Longman Scientific and Technical, London, pp. 28–31.

Knight Jr, R.J. 1980. Origin and world importance of tropical fruit crops. In: S. Nagy and P.E. Shaw, (eds.). Tropical and Subtropical Fruits. AVI Publishing Co, Westport, Connecticut, pp. 1–45.

Liu, R.H. 2003. Health benefits of fruits and vegetables are from additive and synergistic combinations of phytochemicals. Am J Clin Nutr 78(suppl):517S–520S.

Lund, E.D., J. Smoot, and N.T. Hall. 1983. Dietary fiber content of 11 tropical fruits and vegetables. J Agric Food Chem 31:1013–1016.

Marei, H.M.1971. Date Palm Processing and Packing in the Kingdom of Saudi Arabia. Ministry of Agriculture and Water, Riyadh, pp. 5–7 (in Arabic).

Marouf, B.A. and L. Zeki. 1982. Invertase from date fruits. J Agric Food Chem 30(5):990–993.

Maximos, S.E., A.B. Abou-Aziz, I.M. Desouky, N.S. Antoun, and N.R.E. Samra. 1980. Effect of GA3 and ethephon on the yield and quality of *Sewy* date fruits (*Phoenix dactylifera* L.). Annals Agric Sci Moshtohor 12:251–263.

Mikki, M.S. and S.M. Al-Taisan. 1993. Physico-chemical changes associated with freezing storage of date cultivars at their *rutab* stage of maturity. Program and Abstracts of the Third Symposium on the Date Palm in Saudi Arabia, King Faisal University, Al-Hassa, Saudi Arabia, Abstract No. I-11, p. 160.

Mikki, M.S., W.F. Al-Tai, and Z.S. Hamodi. 1983. Canning of date pulp and *khalal* dates. Proceedings of the First Symposium on the Date Palm in Saudi Arabia, Al-Hassa, pp. 520–532.

Mikki, M.S., A.H. Hegazi, A.A. Abdel-Aziz, and S.M. Al-Taisan. 1986. Suitability of major Saudi date cultivars for commercial handling and packing. Proceedings of the Second Symposium on the Date Palm in Saudi Arabia, Vol. II, Al-Hassa, pp. 9–26.

Miremadi, A. 1970. Fruit counting and thinning in six date varieties of Iran. Date Grower's Inst Rep 47:15–18.

Miremadi, A. 1971. Principles of date pruning in relation to fruit thinning. Date Grower's Inst Rep 48:9–11.

Mohammed, S. and H.R. Shabana. 1980. Effect of naphthalene acetic acid on fruit size, quality and ripening of '*Zahdi*' date palm. HortScience 15(6): 724–725.

Mohammed, S., H.R. Shabana, and N.D. Benjamin. 1986. Response of date fruit to gibberellic acid application during slow period of fruit development. Trop Agr 63(3):198–200.

Mohammed, S., H.R. Shabana, and E.A. Mawlod. 1983. Evaluation and identification of Iraqi date cultivars: Fruit characteristics of 50 cultivars. Date Palm J 2(1):27–55.

Mohammed, S., H.R. Shabana, and H.A. Najim. 1980. Effect of ethrel on quality and ripening of auxin-treated date palm fruits. Technical Bulletin Palm & Dates Res Center No. 3/80:1–9.

Moustafa, A.A., H.M. El-Hennawy, and S.A. El-Shazly. 1986. Effect of time of pollination on fruit set and fruit quality of *Seewy* date palm grown in El-Fayoum. Proceedings of the Second Symposium on the Date Palm in Saudi Arabia, Vol. I, Al-Hassa, pp. 323–329.

Myhara, R.M., A. Al-Alawi, J. Karkalas, and M.S. Taylor. 2000. Sensory and textural changes in maturing Omani dates. J Sci Food Agric 80:2181–2185.

Myhara, R.M., J. Karkalas, and M.S. Taylor. 1999. The composition of maturing Omani dates. J Sci Food Agric 79:1345–1350.

Nasr, T.A., M.A. Shaheen, and M.A. Bacha. 1986. Evaluation of date palm males used in pollination in the central region, Saudi Arabia. Proceedings of the Second Symposium on the Date Palm in Saudi Arabia, Vol. I, Al-Hassa, pp. 337–346.

Nixon, R.W. 1943. Flower and fruit production of the date palm in relation to the retention of older leaves. Date Grower's Inst Rep 20:7–8.

Nixon, R.W. 1951. The date palm-tree life tree in the subtropical deserts. Econ Bot 31:15–20.

Nixon, R.W. 1954. Date culture in Saudi Arabia. Date Growers' Inst Rep 31:15–20.

Nixon, R.W. 1957. Effect of age and number of leaves on fruit production of the date palm. Date Grower's Inst Rep 34:21–24.

Nixon, R.W. and J.B. Carpenter. 1978. Growing Dates in the United States. Department of Agriculture, Washington, DC, United States, pp. 1–62.

Nixon, R.W. and J.R. Farr. 1965. Problems and progress in date breeding. Ann Rep Date Growers Inst 42:2–5.

Nixon, R.W. and R.T. Wedding. 1956. Age of date leaves in relation to efficiency of photosynthesis. Proc Am Soc Hort Sci 67:265–269.

Nour, G.M., A.S. Khalifa, A.A.M. Hussein, and A.A. Moustafa. 1986. Studies on the evaluation of fruit characteristics on nine dry date palm cultivars grown at Aswan. In: Proceedings of the Second Symposium on the Date Palm in Saudi Arabia, Vol. I, Al-Hassa, pp. 163–171.

Perkins, R.M. and G.K. Brown. 1966. Date harvest mechanization. Calif Agr 20(2):8–10.

Popenoe, P. 1973. The Date Palm. Field Research Projects. Coconut Grove, Miami, Florida, pp. 1–9.

Rabechault, H. and J.P. Martin. 1976. Multiplication vegetative du palmier a huile (*Elaeis guineensis* Jacq.) a l'aide de cultures de tissue foliaires. C R Acad Sci Paris 283:1735–1737 (in French).

Ragab, M.H.H., A.M. El-Tabey Shehata, and A. Sedky. 1956. Studies on Egyptian dates. II. Chemical changes during development and ripening of six varieties. Food Technol 10:407–411.

Rouhani, I. and A. Bassiri. 1977. Effect of ethephon on ripening and physiology of date fruits at different stages of maturity. J Hort Sci 52(2):289–297.

Rygg, G.L. 1977. Date Development, Handling and Packing in the United States, Agricultural Handbook No. 482. USDA, Washington, DC, pp. 25–41.

Sabbri, M.M., Y.M. Makki, and A.H. Salehuddin. 1982. Study on dates consumers preference in different regions of the Kingdom of Saudi Arabia. Proceedings of the First Symposium on the Date Palm in Saudi Arabia, Al-Hassa, pp. 23–25.

Salem, S.A. and S.M. Hegazi. 1971. Chemical composition of Egyptian dry dates. J Sci Food Agric 22:632–633.

Salik, H., B. Rosen, and I.J. Kopelman. 1979. Microbial aspect and the deterioration process of soft dates. Lebensm Wiss Technol 12(2):85–87.

Sawaya, W.N. 1986. Overview. In: W.N. Sawaya (ed.). Dates of Saudi Arabia. Safir Press, Riyadh, pp. 1–44.

Sawaya, W.N. and J.K. Khalil. 1986. Date seeds: Chemical composition and nutritional quality. In: W.N. Sawaya (ed.). Dates of Saudi Arabia. Safir Press, Riyadh, pp. 167–176.

Sawaya, W.N., J.K. Khalil, A.F. Al-Shalhat and A.A. Ismail. 1986b. Processing of glace dates. Proceedings of the Second Symposium on the Date Palm in Saudi Arabia, Vol. II, Al-Hassa, pp. 113–119.

Sawaya, W.N., H.A. Khatchadourian, J.K. Khalil, W.M. Safi, and A. Al-Shalhat. 1982. Growth and compositional changes during various development stages of some Saudi Arabian date cultivars. J Food Sci 47:1489–1493.

Sawaya, W.N., A.M.A. Miski, and A.S. Al-Mashhadi. 1986a. Physical and chemical characteristics of the major Saudi Arabian date cultivars. In: W.N. Sawaya (ed.). Dates of Saudi Arabia. Safir Press, Riyadh, pp. 45–73.

Serrano, M., M.T. Pretel, M.A. Botella, and A. Amoros. 2001. Physico-chemical changes during ripening related to ethylene production. Food Sci Technol Int 7(1):31–36.

Shabana, H.R., N.D. Benhamin, and S. Mohammed. 1981. Pattern of growth and development in date palm fruit. Date Palm J 1(1):31–42.

Shabana, H.R. and S. Mohammad. 1982. Mechanization of date production. Proceedings of the First Symposium on the Date Palm in Saudi Arabia, Al-Hassa, pp. 714–723.

Shaheen, M.A., T.A. Nasr, and M.A. Bacha. 1986. Date palm pollen viability in relation to storage conditions. Proceedings of the Second Symposium on the Date Palm in Saudi Arabia, Vol. I, Al-Hassa, pp. 331–336.

Shenasi, M., A.A.G. Candish, and K.E. Aidoo. 2002. The production of aflatoxins in fresh date fruits and under simulated storage conditions. J Sci Food Agric 82(8):848–853.

Sidhu, J.S., Al-Hooti, S.N., Al-Saqer, J.M., and Al-Othman, A. 2002. Chemical composition and quality of date syrup as affected by pectinase/ cellulase treatment. Food Chem. 79(2):215–220.

Sidhu, J.S. and S.N. Al-Hooti. 2005. Functional Foods from Date Fruits. In: J. Shi, C.T. Ho, and F. Shahidi, (eds.). Asian Functional Foods. Marcel and Dekker Inc., New York, pp. 491–524.

Sidhu, J.S., J.M. Al-Saqer, S.N. Al-Hooti, and A. Al-Othman. 2004. Quality of pan bread as affected by replacing sucrose with date syrup produced by pectinase/cellulase enzymes. Plant Foods Hum Nutr (in press).

Smolensky, D.C., W.R. Raymond, S. Hasegawa, and V.P. Maier. 1975. Enzymic improvement of date quality: Use of invertase to improve texture and appearance of sugar wall dates. J Sci Food Agric 26(10):1523–1528.

Sourial, G.F., A.S. Khalifa, S.I. Gafaar, A.A. Tewfik, and I.A. Mousa. 1986. Evaluation of some selected date cultivars grown at Sharkiya Province, Egypt. I. Physical characters. In: Proceedings of the Second Symposium on the Date Palm in Saudi Arabia, Vol. I, Al-Hassa, pp. 127–140.

Sudhersan, C. and M. Abo El-Nil. 1999. Occurrence of hermaphroditism in the male date palm. Palms 43(1):18,19,48–50.

Sudhersan, C., M.M. Abo El-Nil, and A. Al-Baiz. 1993a. Occurrence of direct somatic embryogenesis on the sword leaf of *in vitro* plantlets of *Phoenix dactylifera* L. cultivar *barhee.* Curr Sci 65(11): 887–888.

Sudhersan, C., M.M. Abo El-Nil, and A. Al-Baiz. 1993b. Direct somatic embryogenesis and plantlet formation from the leaf explants of *Phoenix dactylifera* L. cultivar *barhe* e. J Swamy Bot Cl, 10(1&2):37–43.

Tisserat, B.H. 1979. Tissue culture of the date palm. J Hered 70:221–222.

Topping, D.L. and P.M. Clifton. 2001. Short-chain fatty acids and human colonic function: Roles of resistant starch and nonstarch polysaccharides. Physiol Rev 81:1031–1064.

Toutain, G. 1967. Le Palmier dattier: Culture et Production. Al Awamia 25:81–151 (in Arabic).

USDA. 1973. Complete Guide to Home Canning, Preserving and Freezing. Dover Publications Inc., New York, pp. 9–15.

Vandercook, C.E., S. Hasegawa, and V.P. Maier. 1980. Dates. In: S. Nagy, P.E. Shaw (eds.). Tropical and Subtropical Fruits. AVI Publishing Co, Westport, Connecticut, pp. 506–541.

Vayalil, P.K. 2002. Antioxidant and antimutagenic properties of aqueous extract of date fruit (*Phoenix dactylifera* L. Arecaceae). J Agric Food Chem 50(3):610–617.

Vinson, A.E. 1911. Chemistry and ripening of the date. Ariz Agric Exp Sta Bull 66:403–435.

Vinson, A.E. 1924. Chemistry of the date. Date Growers' Inst Rep 1:11–12.

Watt, B.K. and A.L. Merrill. 1963. Composition of Foods. USDA Handbook No. 8, United States Department of Agriculture, Washington DC, pp. 25–67.

Weinbaum, S.A. and T.T. Muraoka. 1978. Chemical thinning of prune: Relation of assimilate deprivation to ethyl-mediated fruit abscission. Hortscience 13:159–160.

Yousif, A.K. and M. Abou-Ali. 1993. Suitability of fresh Saudi dates (*rutab*) for refrigeration and freezing storage. Program and Abstracts of the Third Symposium on the Date Palm in Saudi Arabia, King Faisal University, Al-Hassa, Saudi Arabia, Abstract No. I-15, p. 162.

Yousif, A.K., M. Abou-Ali, and A. Abou-Idrees. 1993a. Processing, evaluation and storability of date *katter.* Program and Abstracts of the Third Symposium on the Date Palm in Saudi Arabia, Al-Hassa, Abstract No. I-27, p. 169.

Yousif, A.K., M. Abou-Ali, and A. Bou-Idress. 1993b. Processing evaluation and storability of date jam. Program and Abstracts of the Third Symposium on the Date Palm in Saudi Arabia, Al-Hassa, Abstract No. I-20, p. 165.

Yousif, A.K., S.S. Ahmad, and W.A. Mirandilla. 1986a. Developing of a nutritious beverage from concentrated date syrup and powdered milk. Proceedings of the Second Symposium on the Date Palm in Saudi Arabia, Vol. II, Al-Hassa, pp. 121–131.

Yousif, A.K., A.S. Alghamdi, A. Hamad, and A.I. Mustafa. 1996. Processing and evaluation of a date juice-milk drink. Egyp J Dairy Sci 24(2): 277–288.

Yousif, A.K., N.D. Benjamin, A. Kado, S.M. Alddin, and S.M. Ali. 1982. Chemical composition of four Iraqi date cultivars. Date Palm J 1:285–294.

Yousif, A.K., A.M. Hamad, and W.A. Mirandella. 1985. Pickling of dates at the early *khalal* stage. J Food Tech 20:697–701.

Yousif, A.K., I.D. Morton, and A.I. Mustafa. 1986b. Studies on date paste. I. Evaluation and standardization. Proceedings of the Second Symposium on the Date Palm in Saudi Arabia, Vol. II, Al-Hassa, pp. 85–92.

Yousif, A.K., I.D. Morton, and A.I. Mustafa. 1986c. Studies on date paste. II. Storage stability. Proceedings of the Second Symposium on the Date Palm in Saudi Arabia, Vol. II, Al-Hassa, pp. 93–112.

Yousif, A.K., I.D. Morton, and A.I. Mustafa. 1991. Functionality of date paste in bread making. Cereal Chem 68(1):43–47.

Zaid, A. 1986. Review of date palm (*Phoenix dactylifera* L.) tissue culture. Proceedings of the Second Symposium on the Date Palm in Saudi Arabia, Vol. I, Al-Hassa, pp. 67–75.

Ziemke, W.H. 1977. Raisins and raisin products for the baking industry. The Bakers' Dig 51(4):26–29.

23
Grape Juice

Olga Martín-Belloso and Angel Robert Marsellés-Fontanet

INTRODUCTION

GRAPE HARVESTING AND CULTIVARS

Mankind has a close relationship with grapes. There are signs of its cultivation around the Mediterranean Sea during the Bronze Age. It is well accepted that *Vitis vinifera* had its origin near the Black and Caspian seas. It has also been reported that grapes were carried around the world as civilization spread. Today, there are grapevines in all the temperate regions of the world and there have been some trials done to introduce them in tropical regions (Bombardelli and Morazzoni, 1995; Hidalgo-Togores, 2002).

Despite the fact that *V. vinifera* and its hybrids are the most cultivated varieties, each country also has its own wild or harvested grapevines. Thus, more than 8000 grape types have been described. The family *Vitaceae* includes 11 genera and nearly 600 species but only genus *Vitis* has edible ones and comprises two subgenera (Table 23.1). *Euvitis* , whose species are known as bunch grape, have 38 somatic chromosomes. The number of species belonging to this subgenus depends on taxonomical criteria and they are usually classified by geographical origin. *Muscadinia* grapevines have 40 somatic chromosomes and contain only three species, which are endemic to north and central America (Heinonen and Meyer, 2002; Hidalgo-Togores, 2002).

Only a few of these species are harvested. As mentioned before, the varieties of *V. vinifera* are most widely used in grape production especially in Europe and California. However, on the U.S. coast, American species such as *Vitis labrusca* or their hybrids such as *Vitis labruscana* are currently harvested because they are more resistant to cold winters and some diseases. Even varieties belonging to the subgenus *Muscadinia* are used in the food juice industry as they give better color and flavor to juice (Patil et al., 1995).

GRAPE CONSUMPTION

Sources about grape history have reported that grapes were harvested for consumption as fresh fruit or processed as raisins and in wine production. Today, grapes are still cultivated to obtain the same commodities; thus, approximately 65% of the world's grape crop is used in wine and juice manufacturing, 20% is consumed as fresh grapes, and 10% is used in the production of raisins. Juice consumption has grown in developed countries in recent years and forecasts show the same growing trends (Baron, 2002). One of the reasons is that people wish to live more healthy lives and they know that food can contribute to their well-being. So, thanks to the fame of the French paradox or the Mediterranean diet,

Table 23.1. Taxonomical Chart of Grapevine

Family	Genus	Subgenus	Species (Origin)
Vitaceae	*Vitis*	*Euvitis*	18–28 (America)
			10–15 (Asia)
			vinifera (Europe)
		Muscadinia	*munsoniana* (America)
			rotundifolia (America)
			popenoeii (America)

Source: Gray and Meredith (1992).

drinking beverages produced from fruits and vegetables is recognized as a healthy habit.

CHEMISTRY AND CHARACTERISTICS OF GRAPE JUICE

NUTRITIONAL COMPOSITION

Grape juice is an attractive and healthy commodity extracted from the edible part of berries. Its sensorial properties and nutritive value are determined by both chemical composition and particle size, which are highly dependent on grape variety, berry ripeness, and manufacturing process.

Grapes are composed of flesh, which accounts for 75–85% of berry weight, skin, which is 15–20%, and the remaining parts are seeds. Flesh cells have large vacuoles filled with juice, which is a cloudy slightly yellow colored liquid, with a density range within 1.065–1.110 g/l. A typical grape juice chemical composition is shown in Table 23.2.

After water, carbohydrates are the most abundant component in grape juice. Both glucose and fructose account for the major part of the total carbohydrate content with a mean glucose/fructose ratio of 0.92–0.95. Saccharose level is low, in general about 1–3 g/l, because it is quickly hydrolyzed to glucose and fructose by enzymatic actions (e.g., invertase). Arabinose, ramnose, galactose, xylose, rafinose, or galacturonic acid have been reported at trace levels because they are monomeric constituents of polymeric carbohydrates. Sugar moieties with low polymerized degree or oligosaccharids are usually present as a part of glycoproteins and flavoring or colorant precursors (Belitz and Grosch, 1999; Hidalgo-Togores, 2002; Pellerin and Cabanis, 2000).

Table 23.2. Proximate Grape Juice Composition

Compound	Range Content (g/l)
Water	700–850[a]
Carbohydrates	120[a]–250[b]
Organic acids[c]	3.6–11.7[a]
Volatile acids	0.08–0.25[a]
Phenolic compounds	0.1–1[a]
Nitrogen compounds	4–7[b]
Minerals	0.8[b]–3.2[d]
Vitamins	0.25–0.8[b]

[a]Belitz and Grosch (1999).
[b]Cabanis et al. (2000).
[c]Calculated as grams of tartaric acid.
[d]Ribéreau-Gayon et al. (1989).

Grapes have several polymeric carbohydrates such as cellulose, hemicelluloses, and pectin that come from cell walls. The well-established enzymatic cleavage of pectin during berry ripening tends to dissolve some fragments but the remainders stay insoluble (Crouzet et al., 2000; Tucker and Seymour, 2001; Ribéreau-Gayon et al., 1989). Pectin is a polymeric chain built chiefly by galacturonic acid although it includes different contents of ramnose, galactose, or arabinose among others. This fact results in pectin with three principal structures called homogalacturonane, rhamnogalacturonane I, and rhamnogalacturonane II, and two less frequent structures xilogalacturonanes and apiogalacturonanes. In addition, rhamnogalacturonane I fragments contain three other arrangements known as arabane, arabinogalactane I, and arabinogalactane II (Crouzet et al., 2000; Hidalgo-Togores, 2002; Schols and Voragen, 2001). A typical pectin structure is shown in Figure 23.1. Homogalacturonane areas are larger in proportion and are called smooth regions. Naturally occurring pectin enzymes break the pectin chain by the cleavage of smooth regions. The other types of arrangements are known as hairy regions, and grapes have no enzymes to break down them. There is no evidence of cellulase and hemicellulase enzymatic activity in grapes, so cellulose and hemicellulose are insoluble due to their huge size (Crouzet et al., 2000).

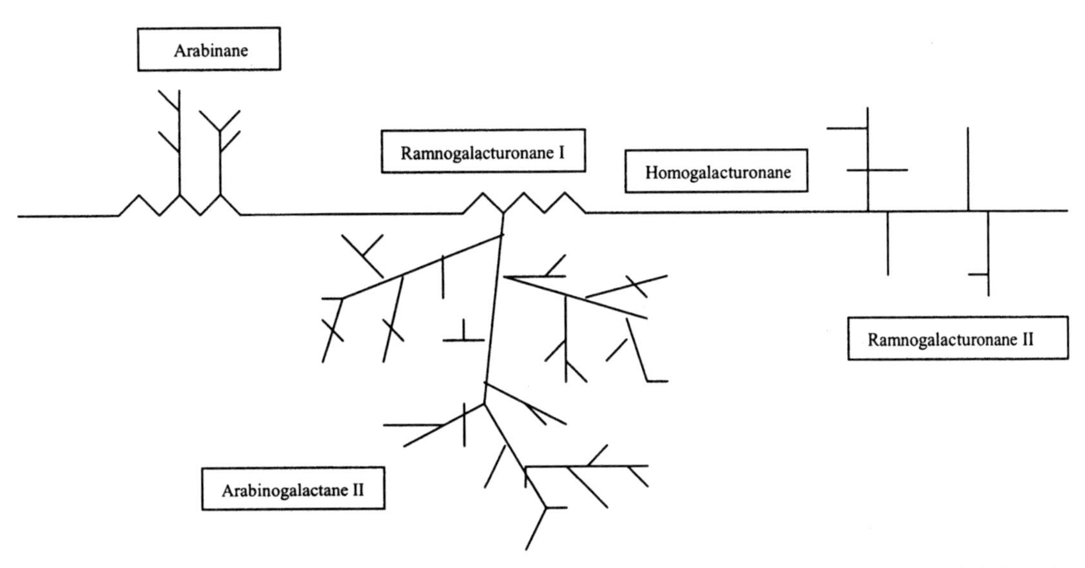

Figure 23.1. A schematic representation of a pectin chain structure. *Source:* Hidalgo-Togores (2002); Schols and Voragen (2001).

Grape juice pH varies from 3.3 to 3.8 due to organic acids. Principal acids in grape juice are tartaric acid and malic acid, which constitute more than 90% of the total acidity, although the values vary in a wide range depending on grape variety (Hidalgo-Togores, 2002; Ribéreau-Gayon et al., 1989). There are many other organic acids in smaller quantities including citric acid, galacturonic acid, which is the major monomer of pectin chains, phenolic and fatty acids, which are originated by hydrolytic cleavage of juice esters (Belitz and Grosch, 1999; Patil et al., 1995). Despite this acidity level, which decreases the pathogen risk, spoilage microbes can grow in grape juice.

Nitrogen content of grape juice accounts for 20–30% of total grape berry content. It is distributed in inorganic forms (25%), which are chiefly ammonium salts, amino acids (70%), peptides (3%) that have mass below 10 kDa, and proteins (2%)(Feulliat et al., 2000; Hidalgo-Togores, 2002; Ribéreau-Gayon et al., 1989). Glycoproteins and enzymes are the main proteins, their size ranging from 10 to 90 kDa. Grape juice contains oxidases such as polyphenol oxidases and peroxidases pectin enzymes like pectin methylesterase (PME) and polygalacturonases (PG) and also proteases (Cheynier et al., 2000; Crouzet et al., 2000; Shahidi and Naczk, 1995).

Mineral compounds are found in grape juice as salts. Principal cationic species are those with structural functions as potassium, calcium, magnesium, and sodium. In addition, many other cationic species, such as iron and copper, occur in grape juice at lower quantities, but they have catalytic roles. On the other hand, major anionic species present are phosphates, sulfates, and chlorides (Hidalgo-Togores, 2002; Ribéreau-Gayon et al., 1989).

Grape juice is a poor source of lipids since it has only 1–2% of the grape lipid content. The most abundant types of lipids are phospholipids (65–70%), neutral lipids (15–25%), and glycolipids (10–15%), which have a high content in polyunsaturated fatty acids (Cabanis, 2000; Hidalgo-Togores, 2002; Pueyo and Polo, 1992).

Vitamin content of grape juice is shown in Table 23.3. Grape juice has greater content of water-soluble vitamins than fat-soluble vitamins and the most important is ascorbic acid. Among the fat-soluble vitamins, grape juice contains only small quantities of carotenoids like lutein and β-carotene, which is a precursor of vitamin A (Cabanis, 2000; Hidalgo-Togores, 2002; Pueyo and Polo, 1992).

All these components have a huge influence over the nutritional value and the typical taste of grape juice. Thus, sugars give sweetness and acids confer acidity, whereas the ratio of sugar content–acid content indicates the palatability of grape juice. However, the antioxidant activity or appetizing attributes

Table 23.3. Proximate Vitamin Content of Grape Juice

Name	Range Content (μg/l)
Thyamine (B_1)	160–450[a]
Riboflavin (B_2)	3–60[a]
Nicotinamide (B_3)	0.68–2.6[a]
Pantothenic acid (B_5)	0.5–1.4[a]
Pyridoxine (B_6)	0.16–0.5[a]
Folic acid	0–1.8[a]
Biotin (B_8)	1.5–4.2[a]
Cianocobalamine (B_{12})	0–0.2[a]
Ascorbic acid (C)[c]	4[b]
Vitamin A[d]	100[b]

[a]Ribéreau-Gayon et al. (1989).
[b]Salunkhe and Kadam (1995).
[c]Expressed as mg/l.
[d]Expressed as International Units (IU).

like flavor and color of grape juice are due to other naturally occurring substances.

Grape Juice Flavor

Many compounds are responsible for typical grape flavor including those already discussed, such as polyunsaturated fatty acids and carotenoids. They participate in aromatic development of grape juice since it is converted to smaller molecules by enzymes and in industrial manufacturing. Those compounds belong to alcohol and carbonyl derivatives of hexane and norisoprenoids, respectively (Bayonove et al., 2000; Belitz and Grosch, 1999; Hidalgo-Togores, 2002).

Nevertheless, the most common flavoring agents in grape juice are terpenes, although its content depends on varieties, which range from 500 to 1700 μg/l. The major terpenes are hydrocarbons, alcohols, aldehydes, and oxides of monoterpenes (C10) and sesquiterpenes (C15). They are usually bound to oligosaccharides in odorless glycosidic forms and require glycosidase enzymes to liberate them (Bayonove et al., 2000; Hidalgo-Togores, 2002).

Other odor components are habitual in concrete varieties; thus, there are metoxipyrazines and some mercaptans in Cabernet Sauvignon. Several American species have distinctive flavors; thanks to different compounds like methyl anthranilate that confers the peculiar foxy smell of *V. labrusca* and *V. rotundifolia* . They also contain some thiols and even furanones (Hidalgo-Togores, 2002; McLellan and Race, 1995).

Grape Juice Color

The most significant contributors to grape juice color are flavonoid compounds whose content varies from 500 to 3000 mg/kg. Flavonoids are products with the same structural core of 2-phenylbenzopirone and are also found in several families. Principal flavonoid compounds present in grapes are summarized in Figure 23.2 (Belitz and Grosch, 1999; Cheynier et al., 2000; Hidalgo-Togores, 2002).

Anthocyanidins (Fig. 23.2a) are skin components responsible for red, blue, or violet colors that depend on the molecule type, pH, and the number of attached groups (e.g., hydroxylic or methoxylic groups). The anthocyanidin pattern is specific for each specie and variety and is usually linked to monosaccharides by glycosidic bonds. Therefore, the major component of *Vitis* species is malvidin glucoside whereas *Muscadinia* have higher quantities of other anthocyanidins. In addition, only *V. vinifera* has 3-monoglucosides with hydroxylic and methoxylic substitutions, whereas other varieties (*V. labrusca* or *V. rupestris*) have 3,5-diglucosides (Belitz and Grosch, 1999; Cheynier et al., 2000).

Flavonols (Fig. 23.2b) and flavononols (Fig. 23.2c) are faintly yellow compounds that give pale colors to white grapes although they are also present in dark varieties. They are located only in the skin in the form of 3-glycosides or 3-glucuronides (Cheynier et al., 2000; Shahidi and Naczk, 1995).

Flavan-3-ols (Fig. 23.2d) are colorless compounds present as monomers and polymers in great quantities both in seeds and skin. They do not affect the color of juice but are responsible for its unwanted astringent taste. However, polymeric forms are precursors of anthocyanidins, which is why they are called proanthocyanidins. Both forms are known as tannins because of their capacity to precipitate proteins in aqueous solution (Belitz and Grosch, 1999; Cheynier et al., 2000; Shahidi and Naczk, 1995).

Grape juice has another family of compounds related to flavonoids because both have phenol groups although they have other structures. They are tartarate esters of benzoic and cinnamic acids, and stilbene derivatives. These substances give taste and flavoring agents to juice once transformed by enzymes (Gray and Meredith, 1992). Moreover, many grape polyphenols are under research because they have interesting pharmacological activities such as antioxidant, antiproteasic, antiglycosidasic, and antimutagenic effects (Bombardelli and Morazzoni, 1995).

(a)

Name	R_1	R_2
Cyanidin	OH	H
Peonidin	OCH_3	H
Delphinidin	OH	OH
Petunidin	OCH_3	OH
Malvidin	OCH_3	OCH_3

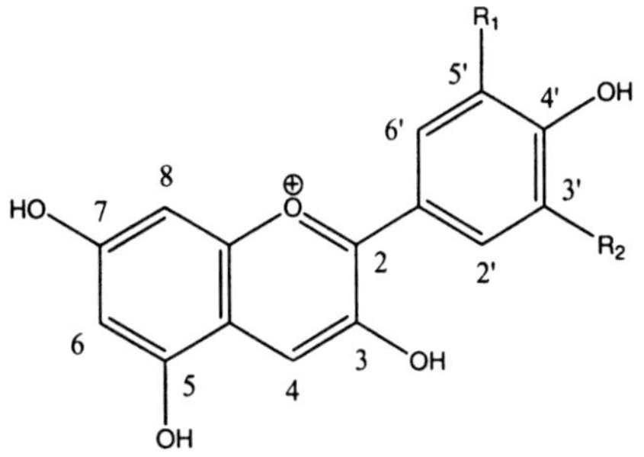

(b)

Name	R_1	R_2
Kaempferol	H	H
Quercetin	OH	H
Myricetin	OH	OH
Isorhamnetin	OCH_3	H

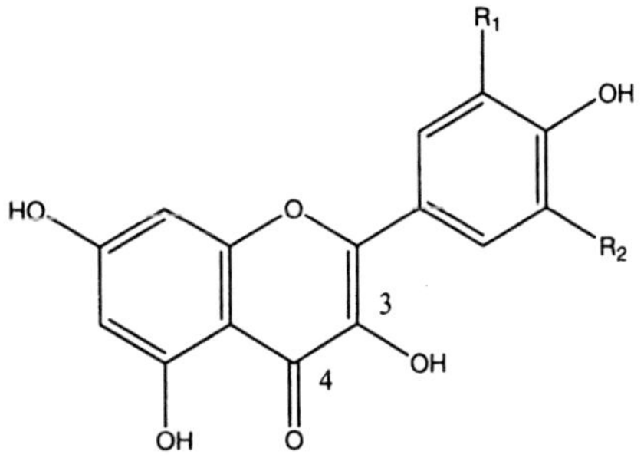

(c)

Name	R_1
Engeletin	H
Astilbin	OH

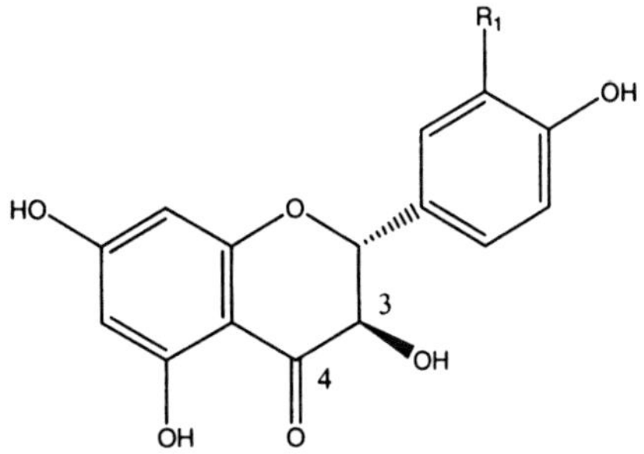

(d)

Name	R_1	R_2	R_3
Catechin	H	OH	H
Gallocatechin	H	OH	OH
Epicatechin	OH	H	H
Epigallocatechin	OH	H	OH

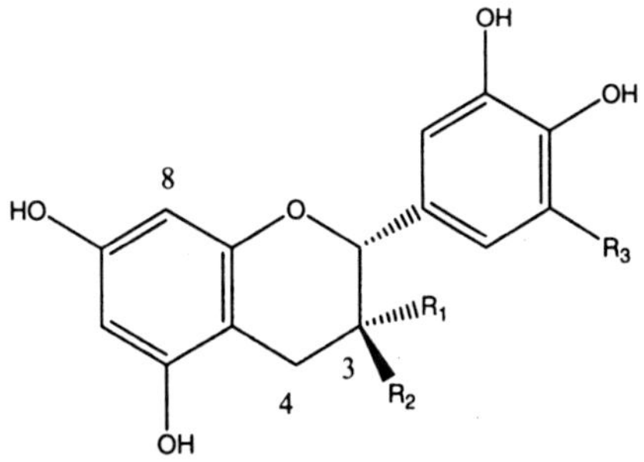

Figure 23.2. Structures of flavonoid grape compounds. (a) Anthocyanidins, (b) Flavonols, (c) Flavanonols, and (d) Flavan-3-ols. *Source:* Belitz and Grosch (1999); Bombardelli and Morazzoni (1995); Cheynier et al. (2000).

All these flavoring and coloring compounds or their precursors are usually located in solid parts of berries; therefore, the manufacturing process will play a weighty role in modifying their content in juice, which will have an effect on quality.

GRAPE JUICE MANUFACTURING

Quality Considerations for Raw Material

Factors influencing the quality of grape juice manufacturing can be combined into preharvest and postharvest factors (Kader and Baret, 1996; Morris and Striegler, 1996; Morris, 1998). The most important preharvest factors are classified according to their nature and summarized in Table 23.4.

Permanent and modifiable preharvest factors are selected to maximize the sensorial characteristics and productivity. In addition, selected factors (e.g., variety, fertilization, or irrigation) should reduce the negative effects on juice quality of remaining parameters (e.g., diseases or climatic disasters).

Several varieties of grapes around the world are currently used to obtain juice, and many countries of the European Union (EU) and Asia have a long tradition in wine making. *Vitis vinifera* grapes are usually processed to make grape juice. There, the production of white juice is greater than that of the dark and some of the most important varieties are Gorda, Thompson Seedless, and Muscat (Urlaub, 2002).

Table 23.4. Quality Preharvest Factors that Influence Grape Juice Quality

Permanents	Climate
	Soil
	Variety
	Grapevine density
	Conduction system
Variables	Temperature
	Sunlight
	Relative humidity
	Grapevine age
Accidentals	Pests
	Diseases
	Climatic disasters
Modifiables	Prunning
	Fertilization
	Irrigation
	Other harvesting practises

Source: Hidalgo-Togores (2002); Ribéreau-Gayon et al. (1989).

In the United States, current bunch grape varieties to obtain dark juice are Fredonia, Van Buren, Sheridan, and Ives among others but Concord grapes are the widest cultivar. The main reasons are that consumers are used to its typical color and foxy aroma and secondly these attributes are quite resistant throughout manufacturing (Morris and Striegler, 1996; Patil et al., 1995). White varieties used are Niagara, which is the most important, or Thompson Seedless (Morris and Striegler, 1996). Even *Muscadinia* species are used in blended grape juices since they modify its color and give the mix a refreshing taste (Sistrunk and Morris, 1985).

Postharvest factors include harvesting, storage, and transport conditions to the factory (Hidalgo-Togores, 2002; Morris, 1998; McLellan and Race, 1995; Morris and Striegler, 1996). It has been established that at temperatures beyond 29°C, grape crushing or long periods of waiting time for processing increase microbial spoilage and favor the beginning of fermentative processes. These facts lead to grape juice with higher content in alcohol and acetic acid that reduce grape juice quality (Sistrunk and Morris, 1985). Thus, berries must be picked and delivered either manually or mechanically but this must be done quickly, to avoid berry damaging and at temperatures as low as possible. Therefore, harvesting at night or the addition of sulphur dioxide to grapes are some strategies that have been developed to prevent fermentation and quality losses in juice (Morris and Striegler, 1996; Morris, 1999; Morris, 1998).

Grape ripeness and harvest time are worthy of attention since grapes do not show any desired development after harvest. In addition, picking the fruits when they are industrially ripe produce highest quality raw materials and yield. Many efforts have been made to predict the optimum moment to harvest the grapes, and several indicators of grape ripeness have been developed. Table 23.5 shows several of these indicators.

Chemical indicators are the most widely used because they are easily measurable, give reproducible results and are linked to grape sensory properties. Among these, sugar content and acidity of grape juice are usually measured to define grape ripeness. Typical values of sugar content from ripe grapes measured refractometrically vary within 14–20 Brix depending on the variety (Morris and Striegler, 1996). The sugar/acid ratio gives information about the palatability of juice and a value of approximately 10 confers a

Table 23.5. Some Grape Ripeness Indicators Sorted by Type of Measurement

General	Physical	Chemical	Physiological
Berry visual check	Berry color	Sugar and acid evolution	Cluster respiration rate
Seeds visual check	Bunch or berry weight	Glucose–fructose ratio	Ethylene production rate
Berry tactile check	Flesh and skin firmness	Density–acidity ratio	Chlorophyll a content
Flesh taste	Yield	Sugar content–acidity ratio	Polyphenol content
Skin taste	Density		Flavor analysis
Seeds taste			

Source: Carbonneau et al. (2000); Hidalgo-Togores (2002); Ribéreau-Gayon et al. (1989).

good refreshing taste to juice (Espiard, 2002). Some sensorial properties such as grape softening and color or taste and flavor are used as complementary measures because they are easy to measure but they are more subjective.

Quality of Grape Juice

Defining grape juice quality is a difficult task because it is associated with the whole of characteristics. The simplest solution consists of using several parameters whose values can be correlated to high-quality grape juice. These attributes are usually easy chemical measurements of sugar, acid, ethanol, and volatile acids content. Recently, the evolution of juice analyses methods allows the measure of some complementary physiological indicators to avoid grape juice manipulation such as anthocyanidin or color content, cloudless or sugar, and acid pattern. Until recently there was a minor grape juice trade. Some countries have their own quality regulations for grape juice. These regulations usually define the quality attributes of grape juice and their values that reflect those typical of grapes in their regions.

There is no international trade regulation for quality attributes of grape juice, only worldwide standards from Codex Alimentarius about grape juice (Codex Stan 82-1981), concentrated grape juice (Codex Stan 83-1981), and concentrated grape juice from *V. labrusca* grapes (Codex Stan 84-1981). These standards provide for soluble solids, ethanol and volatile acids content, and some general sensory properties. Any other quality parameter and analytical method are usually defined by manufacturers before reaching an international commercial agreement. There are also frequent inquiries about pesticide or toxin contents to avoid improper manufacturing process. In some countries, the presence of genetically modified organisms is not allowed, so this aspect must be taken into account by grape juice manufacturers.

Manufacturing Process

It is important to note that the final product will define the manufacturing process. Here, it is described in a logical order. However, the typical processes for manufacturing grape juice and its concentrate (Fig. 23.3) and other grape juice derivatives require additional details since each process is unique. Also, some minor derivatives such as grape powder or deionized grape juice concentrate are usually used as ingredients in other food products. The processes for their manufacture are also different and unique.

Critical Process Parameters

Cleaning is the first step in fruit juice manufacturing because fermentation is better controlled and it prevents equipment damage. Cleaning usually consists of several separation steps. Unwanted inorganic materials (soil, stones, or metallic pieces) are eliminated by putting the grapes in a water bath or by manual selection and removal of rotten bunches. Organic materials (barks, canes, leaves, petioles, and staples) collected during harvesting of grapes are mechanically separated by a piece of equipment called a destemmer. The drum spin is set at a predefined speed and the berries are moved through holes while the unwanted material remains inside and is collected for disposal. The last step is often a chlorinated water rinse to reduce the microbial load on the berries.

After cleaning, grapes are crushed to facilitate juice extraction during the pressing step. The crusher consists of several ruffled rotary rollers that break the berries while they are passing through. The distance between rollers must be correctly adjusted because if seeds are crushed, they will release large amounts of tannins increasing the astringency of the juice.

Prepressing enzymatic treatment is commonly used to improve juice yields. In the past, limited use of prepressing enzymatic treatment was preferred.

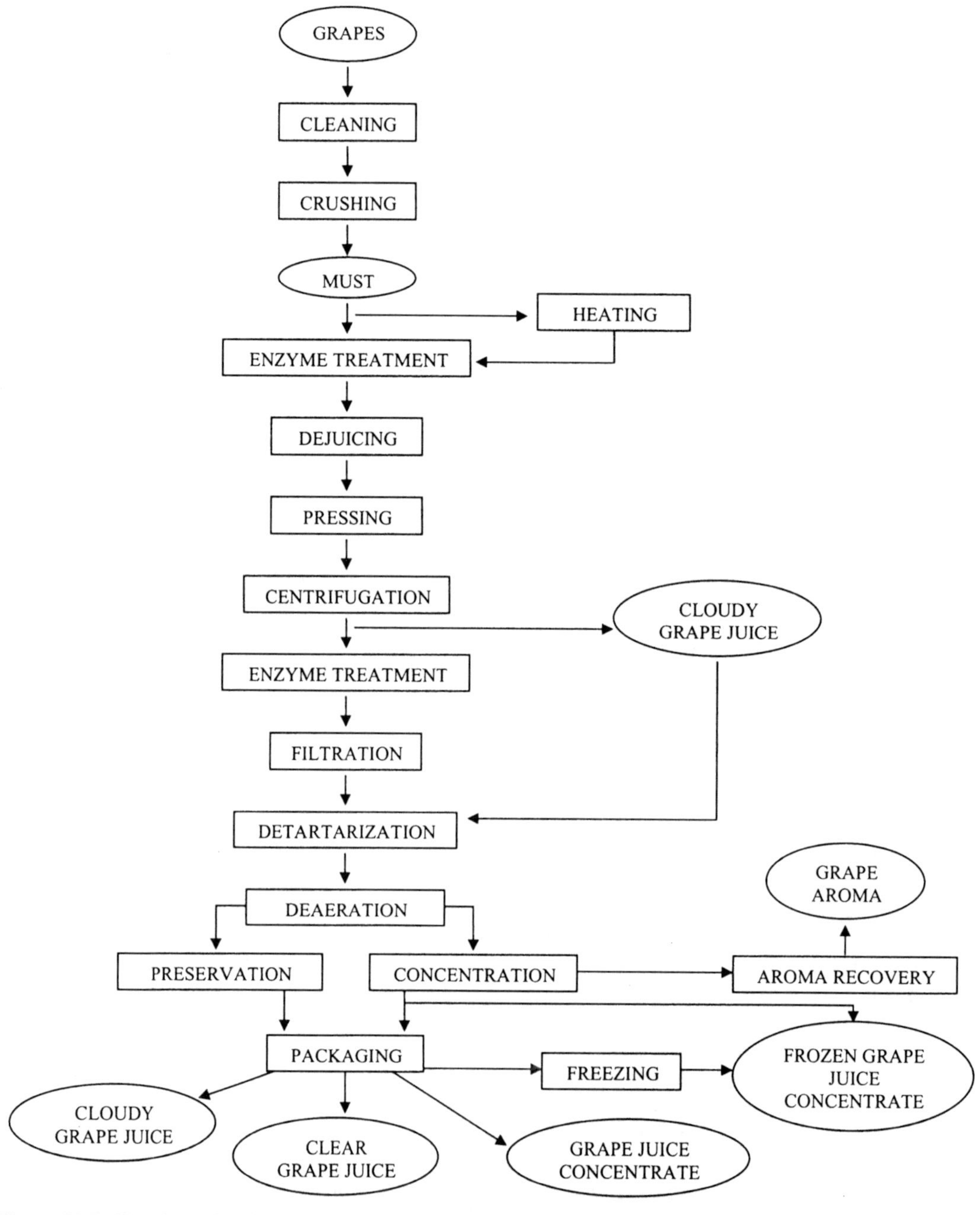

Figure 23.3. Flowchart of typical grape juice manufactuing. *Source:* Enachescu-Dauthy (1995); Espiard (2002).

The viscosity of smashed berries is affected by soluble pectin that is released as a part of the process. In addition to naturally occurring pectic enzymes in grapes, liquefying endo-PG and endo-pectin lyases (PL) are used to break the chains of polymeric backbone into small fractions resulting in a lower juice viscosity (Pilnik and Voragen, 1989; Pilnik and Voragen, 1991; Urlaub 2002). Depending on the press process, rice hulls or paper pulp can be used as a pressing aid as they give the pressurized cake a drainage system that facilitates the removal of the juice (McLellan and Race, 1995).

Enzymatic treatments are applied either cold or hot depending on the color, flavor, and yield desired in the final product. In cold break process, the enzyme mixture is added to temperate crushed berries that are kept at 15–20°C in stirred holding tanks for 2–4 h. Although press yield with cold break process improves with regard to directly pressed juice, it is poorer than the values obtained with hot break process. Cold process is suitable to obtain white grape juice or juice with few losses of temperature sensitive compounds as the lowest possible temperatures are employed (McLellan and Race, 1995; Urlaub, 2002).

Hot break process is similar except for the temperature used. In this case, crushed grapes are heated with tube heat exchangers up to 60°C or 65°C and after the addition of enzymes, the crushed grapes are placed into stirred tanks, keeping the temperature within 60–63°C for 30–60 min. This combination of moderate temperature and enzymatic action improves the extraction of components from grape solid parts leading to juice with higher total solids, polyphenols, and pigments than other treatments (McLellan and Race, 1995; Morris, 1998). Alternative methods include holding the mash a few minutes at 88–90°C to achieve the highest extraction of pigments. This is followed by the actions of enzymes for about 1 h over mashed berries held at 58–60°C (Urlaub, 2002). Conversely, excessive heating or extended holding times to obtain higher yields or intense color are harmful to juice quality because fresh flavor disappears and tannin content raises augmenting its astringent taste. In addition, an excessive dissolution of lipids and waxes occurs, which produces an increase in the final juice turbidity.

Dejuicing is a prepressing step commonly used in juice production that consists of removing the free run juice from the mash. This high-quality juice accounts for 30–60% of the total juice depending on the enzyme treatment, thus duplicating the press capacity. There are special devices to achieve dejuicing including one named dejuicer, although it can be done in the presses as well (Hidalgo-Togores, 2002). Unpressed juice has high sensory quality and is usually blended with pressed juice that is poorer in quality. Sometimes, it can be manufactured separately to obtain products with greater added value.

Pressing is the final step to obtain raw grape juice. There are several kinds of presses where juice is squeezed out of mashed grapes as shown in Table 23.6 (Hidalgo-Togores, 2002; McLellan and Race, 1995; McLellan, 1996). Batch presses undergo successive pressing steps where each cycle reaches a final pressure higher than that of the previous one, whereas continuous presses gradually increase the pressure over the grapes until the maximum is reached. The choice of a press requires bearing in mind some of its

Table 23.6. Advantages and Disadvantages of Currently Used Presses

Operation Mode	Press Type	Advantages	Disadvantages
Batch	Vertical press	Few grape maceration and low turbidity.	High pressure, several pressing steps or further presses, manual cake discharge, high time consuming, and requires press aid.
	Horizontal press	Lower pressure, automatized cake discharge, and good productivity and yield.	High aeration and high solid content.
	Membrane press	Lower pressure, automatized cake discharge, good productivity and yield, and good quality juice.	High solid content and membrane care.
Continuous	Screw press	The highest productivity and yield and poor quality juice.	High berry breakage and the highest solid content.
	Belt press	Low pressure, high-quality juice, and fast juice extraction.	High solid content and pomace requires further pressing to drain juice.

Source: Hidalgo–Togores (2002); McLellan and Race (1995).

characteristics parameters such as pressing time, flow rate (which is related with productivity), energy requirements, juice yield, and also juice quality. (Table 23.6).

Supplying the consumers with high quality (e.g., uniform, microbiologically safe) grape juice throughout the year requires physical, chemical, and microbiological stabilization of industrial grape juice. Other industrial purposes such as diminishing grape juice volume or manufacturing new products based on grape juice need further processing steps.

Physical stabilization of grape juice means keeping the juice appearance constant from manufacturing until consumption because several changes in grape juice can take place leading to an unhomogenized product presentation. First of these changes is the drop of insoluble particles due to gravitational force. A separation of these insoluble solids by grape juice decantation, filtering, or centrifugation avoids this problem. Centrifuges and decanters are the most used devices at present since they allow the manufacturers to work in continuous mode with reasonable yields without quality losses (Hidalgo-Togores, 2002; McLellan and Race, 1995; McLellan, 1996).

The remaining insoluble particles present in juice after centrifugation are stable against precipitation due to the presence of pectin. Pectin works at two levels, serving as a protective coating of positively charged protein cores and increasing grape juice viscosity. Thus, electrostatic repulsion, a small particle size, and viscosity prevent precipitation leading to a grape juice with stable clouding (Endo, 1965; Yamasaki et al., 1964; Yamasaki et al., 1967). However, remaining PME activity will split the methyl esters of the pectin, resulting in a spontaneous clarification of juice. The reason is that unprotected carboxyl groups of pectin react with calcium ions present in juice and form jammy pectin aggregates. Several alternatives are used to try avoiding this trouble and the most effective technique is to split the soluble pectin into smaller chains of less than 10 units, which are not calcium sensitive (Patil et al., 1995; Pilnik and Voragen, 1991). This effect can be achieved by using enzymatic mixtures with high-PG activity in prepressing enzymatic treatment or after centrifugation.

Clarified grape juice is obtained by removing all juice solids to achieve a sparkling juice. The method currently in use is called depectinization, which consists of processing the juice using an enzymatic treatment for 1–2 h with temperatures ranging from 15°C to 30°C (Urlaub, 2002). Again, pectin enzymes promote the most rapid pectin hydrolysis, enabling the subsequent solid elimination by clarification or filtration. In this stage, the enzymatic mixture (e.g., arabanases) used should mainly affect side chains to break up the hairy regions of pectin (Crouzet et al., 2000; Urlaub, 2002). Hemicelullase and proteinase activities are also needed because grape juice does not show these enzymatic activities.

Filtration is the technology currently used to achieve grape juice clarification after depectinization. Until recent years, finishing agents such as jam, bentonite, tannin, or silica gel were used to remove turbidity prior to final filtration. Today, the improvement of membrane filtration technology has replaced traditional finishing agents because it produces a comparable quality product with several advantages (Cheryan, 1998; McLellan, 1996), which are as follows:

- It allows continuous processing.
- It allows the use of automatic machinery that saves labor costs and time.
- It improves the juice yield because of a higher recovery of juice.
- It does not need clarification and finishing agents.
- It reduces the tank space requirements.

Figure 23.4 shows how crossflow membrane filtration works and the reasons for these advantages. Despite the manufacture of a large volume, this filtration method decreases only slightly its ability to discriminate compounds by molecular weight because the filtering surface is continuously rinsed by new juice.

Typical arrangement of a filtration module in juice manufacturing to obtain an optimum efficiency is displayed in Figure 23.5; it consists of a modified batch stage where retentate is recycled and permeate is the required product. Several modules could be used in parallel to improve juice productivity (Cheryan, 1998).

In grape juice production, microfiltration membranes, whose average pore size range from 0.1 to 10 μm, eliminating high molecular weight polysaccharides and glucoproteins without modifying the remaining components (Escudier et al., 2000). Ultrafiltration membranes, whose pore sizes vary within 0.001–0.1 μm, are reserved only for specific circumstances because they cause huge reductions in colorant content (Peri et al., 1988). They achieve complete protein elimination or even sterilization of grape juice (Cheryan, 1998; Daufin et al., 1995; Hsu et al., 1987).

The last step to guarantee physical stabilization of both cloudy and clear grape juice is the elimination

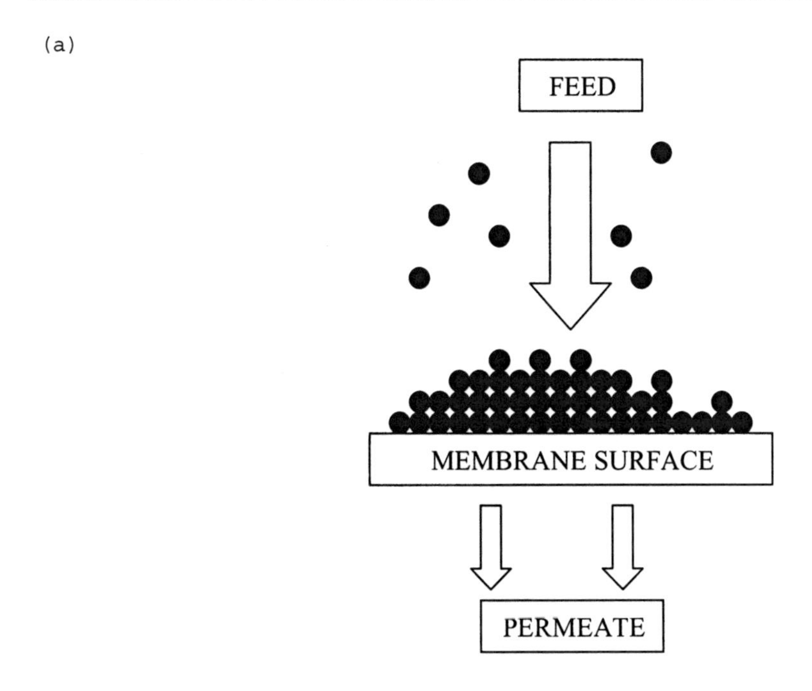

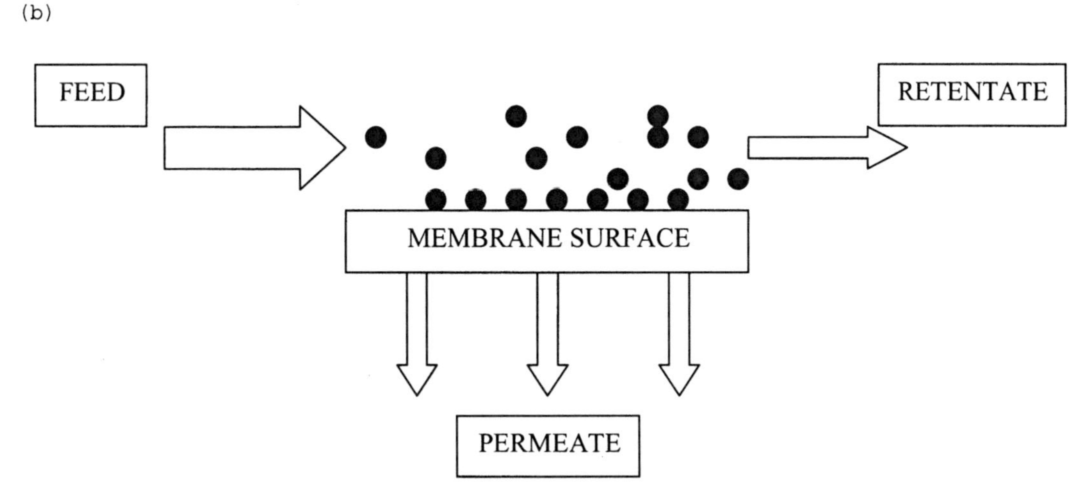

Figure 23.4. Comparison chart between depth (a) and crossflow filtration (b).

of excessive content in potassium and calcium tartarate salts, which could precipitate as soon as some factors modify their unstable equilibrium. Some of these factors are internal, such as pH or the presence of inhibitory compounds, but others are external like temperature change or light exposure. A preventive precipitation of salt excess is usually the method chosen by manufacturers and the most ancient method used is cooling juice down to –2.0 to 0.0°C. At this temperature, the solubility of tartarate salts is reduced and the salt excess precipitates falling to the bottom of the tanks. Nevertheless, the natural process can be accelerated by sowing the juice with crystals of calcium tartarate that act as a crystal nucleus or with continuous crystallizers. Reverse osmosis (RO) technology is used to remove part of the juice water, which makes the precipitation of tartarate salt easier. RO membranes have the lowest pore size range

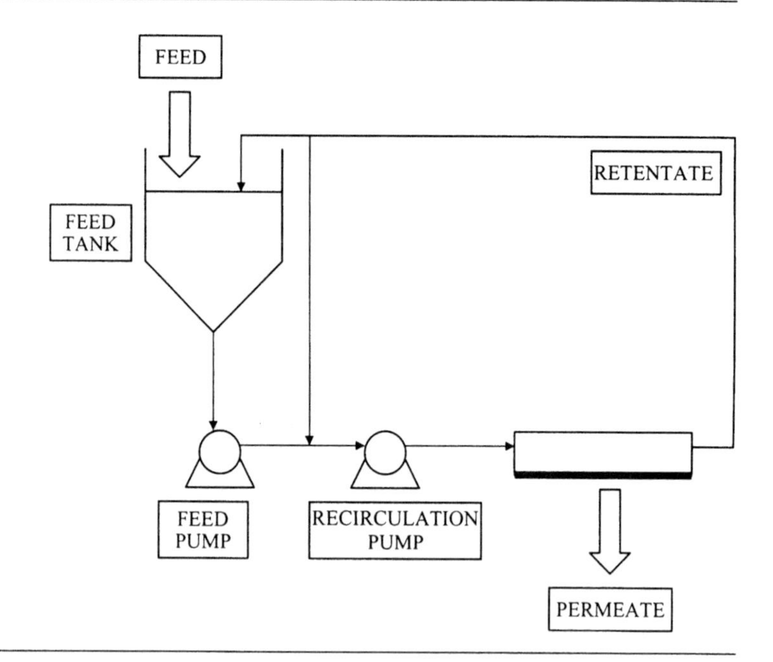

Figure 23.5. A Schematic representation of typical batch arrangement for juice filtration after depectinization. *Source:* Cheryan (1998); Hidalgo-Togores (2002).

(0.0001–0.001 μm). Reducing naturally occurring potassium and calcium in the juice content is another solution to avoid tartarate precipitation. It requires ionic exchange resins or electrodialysis equipment that permits the exchange of cations by sodium (Escudier et al., 2000; Moutounet et al., 1998).

The chemical stabilization of grape juice is the next step. It consists of eliminating oxygen from grape juice. Oxygen is the most important chemical destabilizer in juices because it participates in all spoilage reactions of pigments and vitamins once juice has been packaged. This step is done in chambers with a mild vacuum where juice is sprayed in at room temperature or lower (Ramesh, 1999) prior to any preservative treatment.

The last stage in manufacturing grape juice is microbiological stabilization. Common microorganisms present in grape juice are yeasts of the genera *Kloeckera* and *Saccharomyces*. Lactic and acidic bacteria of *Leuconostoc*, *Lactobacillus*, or *Gluconobacter* genera are also typical. All these spoilage microbes are resistant to low grape juice pH that avoids the growing of pathogen cells. Thermal preservation technologies are the most commonly used and their application to grape juice depends on later processes such as the packaging system (Table 23.7).

Pasteurization treatments achieve the elimination of vegetative microbial cells and the denaturalization of nonthermal resistant enzymes with slight collateral effects on sensorial and nutritive juice properties (Ramesh, 1999). To diminish even more the effect of heat on sensory and nutritive properties, new preservation technologies are being developed. Ohmic heating seems to be a good option for liquid foods as rapid heating is achieved by Joule effect when an alternating current is applied across the product (Ruan et al., 2001). Other promising technology in the fruit juice field is pulsed electric field treatments although they were used as a milk pasteurization method in the United States during the 1930s (Espachs-Barroso et al., 2003a, b; Moses, 1938). High hydrostatic pressure is pointing the way to commercial expansion as there are several fruit industries, which are manufacturing juice by means of this technology (Ludikhuyze et al., 2002; Palou et al., 1999).

Concentration of grape juice is a manufacturing step to reduce its volume, so it also reduces packaging, storage, and transportation costs of the final product. It also has a preservative effect due to the low water activity of the product. This slows down spoilage and microbial growth rate, which is reduced further when the concentrate is frozen. In addition, concentrated grape juice might be used in other foodstuff such as fillings in bakery products or as sweetener for other juices. Evaporation is the most common

Table 23.7. Thermal Preservative Technologies in Juice Industry

Preservation Type	Typical Packaging System	Usual Devices		Typical parameters		Observations
		Clear Juice	Cloudy Juice and Concentrate	Temperature (°C)	Time	
Bottle pasteurization	Prior treatment	Retort	Retort	65[a]/68[b]	30 min[a]	It requires heat-resistant packages
High temperature short time	Bulk storage	Plate heat exchanger	Tube heat exchanger	77[a]/±80[b]	1 min[a], 10–60 s[b]	It requires sterile tanks and refrigeration
	Hot filling	Plate heat exchanger	Tube heat exchanger	77[a]/±80[b]	1 min[a], 10–60 s[b]	It requires heat-resistant packages
	Cold filling	Plate heat exchanger	Tube heat exchanger	77[a]/±80[b]	1 min[a], 10–60 s[b]	It requires refrigerated conditions for storage
	Aseptic	Plate heat exchanger	Tube heat exchanger	88[a]	15 s[a]	It requires sterilized packages
Sterilization	Aseptic	Plate heat exchanger	Tube heat exchanger	93[a]	30 s[a]	It requires sterilized packages

[a] Ramesh (1999).
[b] Enachescu-Dauthy (1995).

Table 23.8. Packaging Systems and Package Materials for Grape Juice

Product	Packaging System	Main Purpose	Package Materials	Observations
Grape juice	Cold filling	Direct consumption	Glass, cans, plastic	Refrigeration
		Transportation	Tanks, drums	
	Hot filling	Direct consumption	Glass, cans	
	Aseptic filling	Direct consumption	Laminated paperboard	Sterilization prior to filling
		Transportation Bulk storage	Tanks, drums	
Grape juice concentrate	Cold filling	Transportation	Tanks, drums	Refrigeration
	Aseptic filling	Direct consumption	Cans, laminated paperboard	Sterilization prior to filling
		Transportation Long-term storage	Tanks, drums	Refrigeration, freezing

way to achieve concentrate in the grape juice industry although freeze concentration and reverse osmosis are good alternatives when sensory and nutritive factors are important. The latter two technologies are typically used as preconcentrating treatments to reduce evaporation costs and sensorial losses (Rao and Vitali, 1999). The only way to reduce sensorial losses when juice is evaporated is by means of aroma recovery using a distillation system, which separates the volatile compounds dissolved in steam outflow.

PACKAGING

Packaging is the final step of grape juice manufacturing, and it must guarantee juice stability and preservation from external contamination until consumption. Packaging is also used as a good communication tool with consumers.

There are many packaging systems and package types (Table 23.8). At present, aseptic filling of sterile packages with grape juice is preferred because they apply the advantage of using high-temperature short-time preservation systems. In addition, aseptic juice has the same or longer shell life than juice packed using hot and cold filling systems, which are still used. Although glass bottles and cans are still used, the tendency is to apply the new development of heat-sealed paperboard laminates associated with aseptic packaging systems. Concentrated grape juice has a very important market in the United States, where it is usually distributed as canned frozen concentrate (Castberg et al., 1995; Driscoll and Paterson, 1999).

FINAL REMARKS

Grape juice and derivatives are very important commodities in international juice trade because they are used not only as a single beverage but also as an ingredient for others foodstuffs. There are considerable efforts to enhance sensory and nutritive properties of grape juice because today's consumers demand juice without preservatives but with high taste, color, and nutritional qualities. Therefore, grape juice producers must improve and diversify their products constantly to adapt them to the changing juice market (Brown et al., 1993; Schols and Voragen, 2001).

Some innovations could come from areas like technology, biotechnology, or engineering. Engineers and technologists are working to improve actual processing devices and to develop new technologies so that the grape juice is elaborated with a minimum loss of quality. Biotechnology also allows breeders to select or clone the best grapevines to improve its industrial aspects or quality attributes (Gray and Meredith, 1992; Romig and Orton, 1989).

REFERENCES

Baron A. 2002. Jus de fruits. In: Albagnac G, Varoquaux P, and Montigaud JC, coordonateurs, Technologies de Transformation des Fruits. Collection Sciences & Techniques Agroalimentaires. Paris: Editions Tech & Doc, pp. 287–344.

Bayonove C, Baumes R, Crouzet J, and Günata Z. 2000. Aromas. In: Flanzy C, coordonateur.

Enologia: Fundamentos Científicos y Tecnológicos. Madrid: Ediciones Mundi-Prensa & A. Madrid Vicente, pp. 137–176.

Belitz HD, Grosch W. 1999. Food Chemistry, 2nd ed. Würzburg (Germany): Springer.

Bombardelli E, Morazzoni P. 1995. Vitis vinifera L. Fitoterapia 66(4): 291–317.

Brown M, Kilmer RL, Bedigian K. 1993. Overview and trends in the fruit juice processing industry. In: Nagy S, Chen CS, Shaw PE, editors. Fruit Juice Processing Technology. Auburndale, FL (USA): Agscience, pp. 1–22.

Cabanis JC. 2000. Ácidos orgánicos, sustancias minerales, vitaminas y lípidos. In: Flanzy C, coordonateur. Enologia: Fundamentos Científicos y Tecnológicos. Madrid: Ediciones Mundi-Prensa & A. Madrid Vicente, pp. 43–65.

Cabanis JC, Cabanis MT, Cheyner V, Teissedre PL. 2000. Tablas de composición. In: Flanzy C, coordonateur. Enologia: Fundamentos Científicos y Tecnológicos. Madrid: Ediciones Mundi-Prensa & A. Madrid Vicente, pp. 218–231.

Carbonneau A, Champagnol F, Deloire A, Sevila F. 2000. Vendimia y calidad de la uva. In: Flanzy C, coordonateur. Enologia: Fundamentos Científicos y Tecnológicos. Madrid: Ediciones Mundi-Prensa & A. Madrid Vicente, pp. 406–417.

Castberg HB, Osmundsen JI, Solberg P. 1995. Packaging systems for fruit juices and non-carbonated fruit beverages. In: Ashurt PR, editor. Production and Packaging of Non-Carbonated Fruit Juices and Fruit Beverages, 2nd ed. London: Blackie Academic & Professional, pp. 290–309.

Cheryan M. 1998. Ultrafiltration and Microfiltration Handbook, 2nd ed. Lancaster, PA (USA): Technomic Publishing.

Cheynier V, Moutounet M, Sarni-Manchado P. 2000. Los compuestos fenólicos. In: Flanzy C, coordonateur. Enologia: Fundamentos Científicos y Tecnológicos. Madrid: Ediciones Mundi-Prensa & A. Madrid Vicente, pp. 114–136.

Crouzet J, Flancy C, Günata Z, Pellerin P. 2000. Las enzimas enologia. In: Flanzy C, coordonateur. Enologia: Fundamentos Científicos y Tecnológicos. Madrid: Ediciones Mundi-Prensa & A. Madrid Vicente, pp. 245–273.

Daufin G, René F, Aimar P. 1995. Les Séparations Par Membrane Dans Les Procédés de l'Industrie Alimentaire. Paris:Tec & Doc.

Driscoll RH, Paterson JL. 1999. Packaging and food preservation. In: Rahman, MS, editor. Handbook of Food Preservation. New York, Basel (USA): Marcel Dekker, pp. 687–734.

Enachescu-Dauthy M. 1995. Fruit and Vegetable Processing. Rome: FAO Agricultural services bulletin, p. 119.

Endo A. 1965. Studies on pectolytic enzymes of molds. Part XIII: Clarification of apple juice by join action of purified pectolytic enzymes. Agric Biol Chem 29: 129–136.

Escudier JL, Moutounet M, Batlle JL, Boulet JC, Brugirard A, Dubernet M, Saint-Pierre B, Vernhet A. 2000. Clarificación, estabilización de los vinos. In: Flanzy C, coordonateur. Enologia: Fundamentos Científicos y Tecnológicos. Madrid: Ediciones Mundi-Prensa & A. Madrid Vicente, pp. 558–607.

Espachs-Barroso A, Barbosa-Cánovas GV, Martín-Belloso O. 2003a. Microbial and enzymatic changes in fruit juice induced by high intensity pulsed electric fields. Food Rev Int 19(3): 253–273.

Espachs-Barroso A, Silvia B, Martín-Belloso O. 2003b. Aplicación de pulsos electricos de alta intensidad de campo en zumos de frutas. Alimentaria 345(3): 75–84.

Espiard E. 2002. Introduction à la Transformation Industrielle Des Fruits. Paris: Tec & Doc.

Feuillat M, Charpentier C, Mauhean A. 2000. Los compuestos nitrogenados. In: Flanzy C, coordonateur. Enologia: Fundamentos Científicos y Tecnológicos. Madrid: Ediciones Mundi-Prensa & A. Madrid Vicente, pp. 87–113.

Gray DJ, Meredith CP. 1992. Grape. In: Hammerschlag FA, Litz RE, editors. Biotechnology of Perennial Fruit Crops. Oxon (England): CAB International, pp. 229–262.

Heinonen IM, Meyer AS. 2002. Antioxidants in fruits, berries and vegetables. In: Jongen W, editor. Fruit and Vegetable Processing. Woodhead publishing in food science and technology collection. Boca Raton, FL (USA): Woodhead Publishing & CRC Press, pp. 23–51.

Hidalgo-Togores J. 2002. Tratado de Enología. Madrid: Ediciones Mundi-Prensa.

Hsu JC, Heatherbell DA, Flores JH, Watson BT. 1987. Heat-unstable proteins in grape juice and wine II. Characterization and removal by ultrafiltration. Am J Enol Viticult 38: 17–22.

Kader AA, Baret DM. 1996. Classification, composition of fruits and postharvest maintenance of quality. In: Somogyi LP, Ramaswamy LP, Huy YH, editors. Processing Fruits: Science and Technology, vol. 1, Biology, Principles and Applications. Lancaster (USA): Technomic Publishing, pp. 1–24.

Ludikhuyze L, Van Loey A, Indrawati I, Hendrickx M. 2002. High pressure processing of fruit and vegetables. In: Jongen W, editor. Fruit and Vegetable

Processing. Woodhead publishing in food science and technolgy collection. Boca Raton, FL (USA): Woodhead Publishing & CRC Press, pp. 346–362.

McLellan MR. 1996. Juice processing. In: Somogyi LP, Ramaswamy LP, Huy YH, editors. Processing Fruits: Science and Technology, vol 1, Biology, Principles and Applications. Lancaster (USA): Technomic Publishing, pp. 67–94.

McLellan MR, Race EJ. 1995. Grape juice processing. In: Ashurt PR, editor. Production and Packaging of Non-Carbonated Fruit Juices and Fruit Beverages, 2nd ed. London: Blackie Academic and Professional, pp. 89–105.

Morris JR. 1998. Factors influencing grape juice. Horttechnology 8(4): 471–478.

Morris JR. 1999. Developing mechanized systems for producing, harvesting, and handling brambles, strawberries, and grapes. Horttechnology 9(1): 22–31.

Morris JR, Striegler K. 1996. Grape juice: Factors that influence quality, processing technology, and economics. In: Somogyi LP, Ramaswamy LP, Huy YH, editors. Processing Fruits: Science and Technology, vol 1, Biology, Principles and Applications. Lancaster (USA): Technomic Publishing, pp. 197–234.

Moses D. 1938. Electric pasteurization of milk. Agric Eng 19: 525–526.

Moutounet M, Escudier JL, Vernhet A, Cadot Y, Saint-Pierre B, Mikolajczack M. 1998. Les séparacions par membrane dans les procédés de l'industrie alimentaire. In: Daufin G, René F, Aimar P, coordonateurs. Collection Sciences & Techniques Agroalimentaires. Paris: Editions Tech & Doc, pp. 444–473.

Palou E, Lopez-Malo A, Barbosa-Cánovas GV, Swanson BG. 1999. High pressure treatment in food processing. In: Rahman MS, editor. Handbook of Food Preservation. New York, Basel (USA): Marcel Dekker, pp. 533–576.

Patil VK, Chakrawar VR, Narwadkar PR, Shinde GS. 1995. Grape. In: Salunkhe DK, Kadam SS, editors. Handbook of Fruit Science and Technology: Production, Composition, Storage and Processing. New York: Marcel Dekker, pp. 7–38.

Pellerin P, Cabanis JC. 2000. Los glucidos. In: Flanzy C, coordonateur. Enologia: Fundamentos Científicos y Tecnológicos. Madrid: Ediciones Mundi-Prensa & A. Madrid Vicente, pp. 66–96.

Peri C, Riva M, Decio P. 1988. Cross-flow membrane filtration of wines: Comparison of performance of ultrafiltration, microfiltration and intermediate out-off membranes. Am J Enol Viticult 39: 162–168.

Pilnik W, Voragen GJ. 1989. Effect of enzyme treatment on the quality of processed fruit and vegetables. In: Jen JJ, editor. Quality Factors of Fruits and Vegetables: Chemistry and Technology. ACS Symposium Series 405. ACS, pp. 251–269.

Pilnik W, Voragen GJ. 1991. The significance of endogenous and exogenous pectic enzymes in fruit and vegetable processing. In: Fox PF, editor. Food Enzymology. vol 1. Belfast: Elsevier Science Publishers, pp. 303–336.

Pueyo E, Polo MC. 1992. Composición lipídica de las uvas y el vino. Alimentación, equipos y tecnologia 2: 77–81.

Ramesh MN. 1999. Food preservation by heat treatment. In: Rahman MS, editor. Handbook of Food Preservation. New York, Basel (USA): Marcel Dekker, pp. 95–172.

Rao MA, Vitali AA. 1999. Fruit juice concentration and preservation. In: Rahman MS, editor. Handbook of Food Preservation. New York, Basel (USA): Marcel Dekker, pp. 217–258.

Ribéreau-Gayon J, Peynaud E, Ribéreau-Gayon P, Sudraud P. 1989. Tratado de enología. Ciencias y técnicas del vino. volume II. Caracteres de los vinos. Maduración de la uva. Levaduras y bacterias. Spanish translation. Argentina: Hemisferio Sur.

Romig WR, Orton TJ. 1989. Applications of biotechnology to the improvement of quality of fruits and vegetables. In: Jen JJ, editor. Quality Factors of Fruits and Vegetables: Chemistry and Technology. ACS Symposium Series 405. ACS, pp. 381–393.

Ruan R, Ye X, Chen P, Doona CJ, Taub I. 2001. Ohmic heating. In: Richardson P, editor. Thermal technologies in food processing. Boca Raton, FL (USA): Woodhead Publishing & CRC Press, pp. 241–265.

Salunkhe DK, Kadam SS. 1995. Introduction. In: Salunkhe DK, Kadam SS, editors. Handbook of Fruit Science and Technology: Production, Composition, Storage and Processing. New York: Marcel Dekker, pp. 1–6.

Schols HA, Voragen AGJ. 2001. The chemical structure of pectins. In: Seymour GB, Knox JP, editors. Pectins and their Manipulation. Great Britain: Blackwell Publishing & CRC Press, pp. 1–29.

Shahidi F, Naczk M. 1995. Food Phenolics. Sources, Chemistry, Effects, Applications. Lancaster, PA (USA): Technomic Publishing.

Sistrunk WA, Morris JR. 1985. Quality acceptance of juices of two cultivars of muscadine grapes mixed with other juices. J Am Soc Hort Sci 110: 328–332.

Tucker GA, Seymour GB. 2001. Modification and degradation of pectins. In: Seymour GB, Knox JP, editors Pectins and their Manipulation. Great Britain: Blackwell Publishing & CRC Press, pp. 150–173.

Urlaub R. 2002. Enzymes in fruit and vegetable juice extraction. In: Whitehurst RJ, Law BA, editors. Enzymes in Food Technology. Sheffield, England: Sheffield Academic Press, pp. 144–183.

Yamasaki M, Kato A, Chu SY, Arima K. 1967. Pectic enzymes in the clarification of apple juice. Part II: The mechanism of clarification. Agric Biol Chem 31: 552–560.

Yamasaki M, Yasui T, Arima K. 1964. Pectic enzymes in the clarification of apple juice. Part I: Study on the clarification reaction in a simplified model. Agric Biol Chem 28: 779–787.

24
Grapes and Raisins

N. R. Bhat, B. B. Desai, and M. K. Suleiman

INTRODUCTION

Grapes and grape products are among the world's most important horticultural products and consequently are of major commercial interest. They are served as a fresh fruit, dried into raisins, preserved or canned in jellies and jams, and crushed for making juice or wine. Grapes accounted for nearly 13% of total fruit, melon, and nut output during 1995–1997 (Moulton and Possingham, 1998). During this period, grapes and grape products exports accounted for approximately one-quarter of all horticultural exports. Grapes probably originated in Asia Minor, in the region between and to the south of the Black and Caspian seas; from there, they have spread to six continents. They are now grown everywhere with a reasonably favorable environment. North America is the native habitat for more than 70% of the world's grape species. Commercially, grapes are grouped into four major classes and one minor group: (i) table grapes, (ii) raisin grapes, (iii) wine grapes, (iv) sweet juice grapes, and (iv) canning grapes (the minor group).

Raisins are the second most important product of the grape vine, wine being the first. The word raisin comes from the French "raisin sec," meaning dry grape. The term has become limited to the dried grapes of mainly three varieties, viz., Thompson seedless, Black Corinth and Muscat of Alexandria. Preservation of grapes in the form of raisins is a major profit-making business in several countries where grapes are grown. The world raisin production in 1995 was estimated to be 1,072,000 t (FAO, 1995). United States and Turkey are the major producers and the exporters of raisins (USDA, 2004). The U.S. raisin production during 2002–2003 was 386,279 t, majority of which was produced from Thompson seedless. During the same period, Turkey produced 220,000 t of raisins. The other raisin-producing countries are Greece, Mexico, Chile, Australia, South Africa, Iran, and India.

TYPES OF RAISINS

The trade names of raisins may signify, besides grape cultivar, the method of drying (natural, golden-bleached, sulfur-bleached, and lexia), the principal place of origin (Vostizza, Patras, Pyrgos, Smyrne, Malaga, and Valencia), the conditions in which the product is offered for sale (layers, loose, and seeded), the size grades (4 crown, 3 crown, 2 crown, and so on), the U.S. maturity grades (B or Better, C, and substandard), and the quality grades (extra standard, standard, substandard, extra fancy, fancy, choice, etc.). Raisins are also classified as currents, golden raisins, monukka raisins, Muscat raisins, dark raisins, and sultanas.

VARIETIES SUITABLE FOR RAISIN PRODUCTION

Raisins are produced mainly from four varieties: Thompson seedless, Muscat, Sultana, and Black Corinth. Historically, Thompson seedless constitutes 90% of the total raisin supply in the United States.

DRYING CHARACTERISTICS OF GRAPE BERRIES

FRUIT PROPERTIES

Maturity, Harvesting, and Handling of Grapes for Processing

Both physical (berry size, berry shape, berry color, the nature of waxy cuticle) and chemical fruit properties (moisture content, sugar content, and acidity) at harvest affect raisin quality. These properties are influenced by several factors, some of which cannot be manipulated by the grower (variety and root stock, the age of the vine, soil and climatic conditions) and others such as soil improvement, irrigation management, nitrogen and potassium nutrition, growth regulator application, pruning, and crop load, which can be altered by the grower. Certain factors like exposure to pathological conditions and diseases also affect the maturation and processing quality of grapes.

Gibberellins (GA_3) are commonly used during the bloom to loosen clusters and improve the quality of raisins. Peacock and Beede (2004) evaluated the effects of GA_3 application at different stages of berry development on dry down ratio, yield, and quality of raisins. They showed that bloom time application of GA_3 improved raisin quality, whereas the raisin and fresh weight yield were greatest when GA_3 was applied at the rate of 20 g/ha 7 days after the bloom (berry set stage).

Grape ripening is normally associated with an increase in sugars (Brix), decrease in acidity, and the development of characteristic color, texture, and flavor. These changes continue as long as the grapes remain on the vine, and stop after they are harvested. Such changes result in a gradual improvement until the optimum stage is reached, followed by a steady deterioration. The optimum harvest maturity of grapes for processing represents a compromise of various factors, although the consumer satisfaction should usually be the deciding factor (Winkler et al., 1974). Firm-ripe grapes ship and store better than either underripe or overripe ones.

The appearance of the berries, associated with the brilliance of their color and taste, help in the selection of clusters to be picked. The red and black grapes become intensely colored as ripening advances. The green color of white cultivars becomes more nearly white or yellow. The cluster stems also mature along with the berries obtaining a wood, straw, or yellow color. Owing to the absence of reserve material such as starch, which might be hydrolyzed to sugars after picking, grapes do not ripen after harvest. Thus, if the grape is not at its best in respect of maturity and quality at the time of harvest, it will never be because all changes after harvest are deteriorative. Appropriate postharvest handling and storage techniques, however, can slow the rate of deterioration. While stem browning and excessive water loss can deteriorate quality, trimming and careful handling improve the fruit quality.

Criteria for harvesting grapes include sugar (Brix), acidity, pH, and sugar:acid ratio, the most important criterion being the sugar content of grape. It is often determined as the total soluble solids by Brix or some other hydrometer. Abbe's refractometer or saccharometer may also be used. While hand refractometer is an efficient tool to judge harvest maturity of individual clusters of table grapes, it is not recommended for use to determine the harvest maturity of wine grapes. Portable near-infrared spectroscopy can be used for nondestructive, accurate measurements of several chemical compounds, including sugars and acidity in grapes (Temma et al., 2002; Saranwong et al., 2003; Chauchard et al., 2004). The Brix:acid ratio is a better criterion than either sugar or acid alone to decide the correct time of harvest. Depending upon cultivar and climate, the Brix:acid ratios may fluctuate from 20:1 to 35:1 (Winkler et al., 1974). With an advancement in grape maturity, the acid content may drop significantly, especially in warm tropical regions as a result of prolonged hot periods during ripening, thus, lowering the fruit quality. Winkler (1954) reported a similar effect resulting from overcropping, which may delay maturation.

Weather conditions and date of harvest have significant effects on drying time, with later harvests and shorter days slowing down drying by a few days under Californian conditions (Peacock and Christensen, 1998).

In the United States, grade- and standard-based maturity and berry characteristics, such as free from decay, visual mold, immature berries, sunburn, freezing, insect injury, and foreign material, are available for grapes used in processing and freezing. According to these standards, grapes to be graded as U.S. No. 1 should be mature (soluble solids concentrations not $<15.5\%$ as determined by an approved refractometer), have similar varietal characteristics (skin and pulp color), and be free from defects due to decay, visible mold, immature berries, sunburn, freezing, attached insects or insect injury, and foreign materials, or any cause within the established tolerance limits (USDA, 1997). However, grapes with lower maturity level or fairly well matured (soluble solids concentration of not $<14.5\%$) with defects within the tolerance limits are graded as U.S. No. 2. However, no tolerance is provided grapes in both standards which fail to meet maturity requirement.

Grapes are harvested manually for table purpose or by mechanical harvesters for raisin and juice production. Mechanical grape harvesters are large machines that pass over the vines and vibrate the grapes from their stems into collecting troughs, which are then emptied into boxes to be carried by trucks to a processing plant.

DRYING CHARACTERISTICS

Drying and Dehydration

Drying is probably the oldest and one of the most cost-effective method for preserving fruits. However, with the development of new preservation methods and ease of obtaining fresh fruits in the market, drying is considered as one more method for diversifying products range for consumer's convenience. Nevertheless, preservation of grapes by drying is a major industry in several parts of the world where grapes are grown. Drying of grapes on the vine or by open sun drying, shade drying, or mechanical drying produces raisins.

The grape with an outer waxy cuticle and a pulpy material inside is a complex product for dehydration. The outer waxy cuticle controls the moisture diffusion rates during drying. A chemical or physical treatment is generally applied to decrease skin resistance and hence, improving moisture diffusion through waxy cuticle (Ponting and McBean, 1970). Loss of moisture from berries during air-drying is accompanied by changes in fruit structure and texture, such as fruit softening or loss of firmness, which are related to their microstructure (Bolin and Huxsoll, 1987; Aguilera and Stanley, 1999). Texture is an important quality criterion of raisin when they are consumed after rehydration in breakfast cereals and dairy and bakery products.

Sorption Equilibrium

The quality of the dehydrated products like raisin largely depends on the water activity of the product, which in turn depends on the moisture content and temperature (Singh and Singh, 1996). The adsorption and desorption isotherms demonstrate a concurrent increase in equilibrium moisture content with increasing equilibrium relative humidity. This relationship is manifested in the form of sigmoid-shaped curves. The desorption curve identifies the type of water present in the product and thus provides preliminary information on the terms driving the mass transport. The state of equilibrium resulting from multiple interactions on a microscopic scale is described by a relationship between the water content at equilibrium of the product to be dried and the relative humidity of the atmosphere which surrounds it at a constant temperature. Desorption isotherms of grapes can be determined by placing the product in an atmosphere whose relative humidity is controlled by solutions of sulfuric acid (Azzouz et al., 2002). The controlled humidity environments can also be created using saturated solutions of various inorganic salts (LiCl, MgCl, $MgNO_3$, $NaCl_2$, and KCl), and the equilibrium moisture contents for desorption isotherms can be determined by gravimetric method (Suthar and Das, 1997; Greenspan, 1977).

A number of models have been suggested to describe the relationship between equilibrium moisture content and equilibrium relative humidity (Mujumdar, 1995). These have been adopted as standard equations by the American Society of Agricultural Engineers for describing the moisture sorption of biological materials. Of these, the modified Henderson, modified Oswin, modified Chung-Pfost, and Gaggenheim, Anderson, and de Boer (GAB) equations take into account the effects of temperature. According to the Henderson equation, the relationship can be written as:

$$a_w = \text{Experimental}\left(-BX_{eq}^{C}\right), \qquad (24.1)$$

where a_w is the water activity, X_{eq} the equilibrium moisture content on dry basis (kg/kg), and B and C the empirical constants. For Sultanin (Turkish variety) grapes, Azzouz et al. (2002) estimated the following values for B and C coefficients:

$$B = -0.892T - 314,$$
$$C = -0.086T + 31,$$

where T is the absolute temperature (K).

This relationship can be fitted to the experimental curves.

Product Shrinkage

Shrinkage of tissues is a major physical change during drying of grape berries. It is now recognized as an important consequence of fruit drying that has to be accounted for, since it modifies the dimension of the product, which in turn affects the mass transport phenomena (Wang and Brennan, 1995; Ramos et al., 2003). Cellular shrinkage causes modifications in the global structure of grape berries and is directly related to the loss of water from cells. According to Hills and Remigereau (1997), drying results in loss of water from vacuolar compartment, with minor changes in the water content of cytoplasm or cell wall compartments in parenchyma apple tissue. Loss of water during drying leads to loss of turgor pressure and cellular integrity (Jewell, 1979).

Shrinkage of fruits also results in "case hardening," a phenomenon that occurs when the drying rate is rapid. As the fruit surface dries faster than its pulp, internal stresses develop and fruit interior becomes cracked and porous. Nonvolatile compounds migrate with diffusing water, precipitate on the product's surface, and form a crust that keeps fruit dimension thereafter. Consequently, the overall shrinkage is lesser when the drying velocity is higher.

Shrinkage and other changes in geometric features of cells can be quantified by image analysis (Bolin and Huxsoll, 1987) by stereomicroscope (Ramos et al., 2003). These changes follow a smooth exponential decrease with time and a first-order kinetic model easily fits the data (Ramos et al., 2003). Cellular shrinkage, which follows an Arrhenius-type behavior, increases with temperature from 20°C to 60°C. Drying conditions play an important role in determining the textural properties in raisins. Slow drying achieved by low temperature, low air velocity, and high relative humidity produces uniform, dense products (Brennan, 1994). On the other hand, fast drying rates result in less dense, tougher product with crust on the surface and with higher dehydration rate and rehydrated products with soft texture.

Azzouz et al. (2002) developed the following model to estimate shrinkage in Chasselas and Sultanin varieties:

$$\text{Shrinkage} = 0.79 \times \frac{\text{Average local moisture content}}{\text{Initial moisture content}} + 0.22. \qquad (24.2a)$$

Chemical Changes During Drying of Grapes

Drying either on the vine or on the ground provokes changes in the physical, chemical, and biological properties and modifies characteristics of grape berries. The moisture loss in dry-on-vine raisins occurs in graduated stepwise manner, with rapid decline from an initial 86 ± 2% to 60 ± 5% after 108 days from first bloom, a slower loss, and a final accelerated loss to 25 ± 4% after 151 days from first flowering (Aung et al., 2004). Although the pattern of dry-matter accumulation was the same in large, medium, and small berries, the large berries contained higher dry matter than the medium and small berries. Sucrose exhibited two maxima, on the 96th day and 123rd day from the first bloom. Rise in sucrose levels preceded the rise in sucrose, fructose, and sorbitol (Aung et al., 2004). Since sorbitol was not detected in mature berries, but increased during drying process, the authors proposed sorbitol or its biosynthetic enzyme as a useful indicator for determining the raisin harvest. The moisture loss in untreated and pretreated grape berries is far more rapid in solar or mechanical drying process compared to the dry-on-vine method, but the pattern of moisture loss is the same (Fig. 24.1; Di Matteo et al., 2000).

Air-Drying Kinetics

In the dehydration process of grape berries by means of warm air, simultaneous heat and water (liquid and vapor) transport in the pulp, in the peel (if present), and in the gaseous film surrounding the grapes takes place. Since the duration of the thermal transient is generally far less than the duration of the drying process, mass transport occurs under the isothermal conditions. In other words, the whole drying process is controlled by mass transport only (Bird et al., 1960; Peri and Riva, 1984). Under the assumptions that pulp and peels (if present) are uniform and isotropic, and the grape berries are spherical, the mathematical model of grape dehydration can be reduced to that of mass diffusion from a spherical body (Bird et al., 1960; Carslaw and Jaeger, 1980). Any mathematical model that describes the diffusion of water through whole grape berries must account for its diffusion both in the pulp and in the peel (Di Matteo et al., 2000). Therefore, both processes are described by the model:

$$\frac{\partial C_i}{\partial t} = D_i\left(\frac{\partial^2 C_i}{\partial r^2} + \frac{2\partial C_i}{r\partial r}\right), \qquad (24.2b)$$

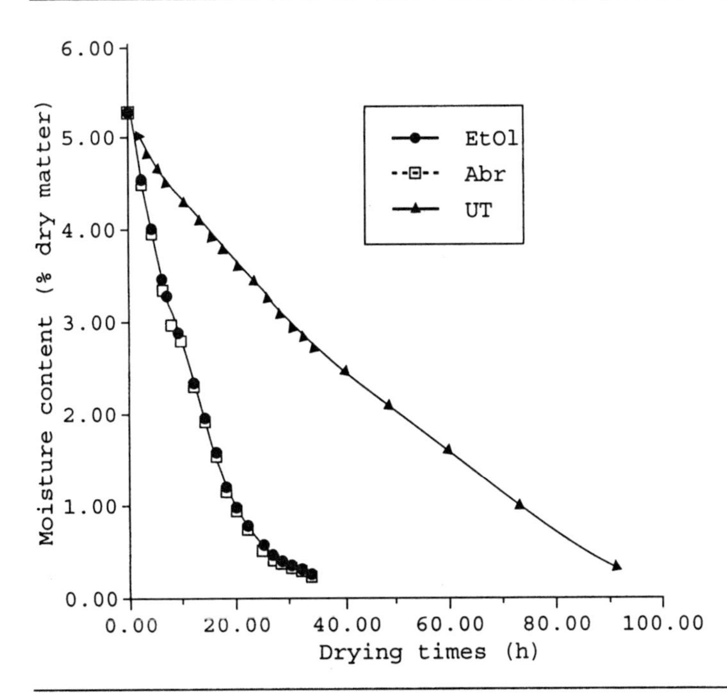

Figure 24.1. Moisture content (% dry matter) in untreated (UT) and pretreated (abrasion of peel, Abr, and pretreatment with 2% ethyl oleate + 2.5% K_2CO_3, EtOl) grape berries during drying process. [Reprinted from Di Matteo et al. (2000), with permission from Elsevier.]

where the index $i = 1$ refers to the pulp [i.e., for $r \in (0, R_1)$] and $i = 2$ to the peel [i.e., $r \in (0, R_1, R_2)$], D_1 is the water diffusivity in the grape pulp, which is much higher than that in the grape peel (D_2), C_1 the water concentration in grape pulp, C_2 the water concentration in the peel, r the distance from the grape center, and δ is the peel thickness.

The drying curves for various values of drying conditions obtained for single layer of grapes do not clearly indicate the existence of a first phase of drying. Since the beginning of the process, diffusion of water from the interior to the surface where it evaporates has been limiting. The diffusion becomes more difficult because of shrinkage of grapes and because of the formation of dry layers at the surface (Diamante and Murno, 1991).

Knowledge of the drying rate constant of grapes and its dependence on drying air conditions is necessary to design and optimize the grape dryers. Several studies have been conducted to determine the drying constant for grapes under different drying conditions: at a single air temperature and two airflow rates for berries of different weights with various pretreatments (Martin and Stott, 1957), effects of velocity, temperature of ambient air, and pretreatment of grapes on their drying time (Eissen et al., 1985), under open sun drying and forced convective drying (Riva and Peri, 1986), and microwave drying of grapes at different temperatures, but one air velocity (Tulasidas et al., 1993).

Pangavhane et al. (2000) determined the drying rate constant for Thompson seedless grapes using a commercial pretreatment (cold dip) under a wide range of controlled drying air conditions (temperature, velocity, and relative humidity of drying air) normally used for commercial drying and obtained the values of constants for empirical relations of Arrhenius type. He described the drying behavior of grapes with time for the given conditions of the drying air by the following equation:

$$MR = \frac{M - M_e}{M_0 - M_e} = c \exp(-kt), \qquad (24.2c)$$

where MR is the moisture ratio, M the moisture content on % dry basis at time t, M_e the equilibrium moisture content on % dry basis, M_0 the initial moisture content on % dry basis, c the constant of equation, k the drying rate constant (h^{-1}), and t the time (h).

The equilibrium moisture content, M_e, for the raisin at different air temperature and humidity conditions can be computed using well-known GAB equation (Singh and Singh, 1996). However, this model was found to be inadequate for loss of moisture from composite materials. To describe drying kinetics of such products, the Page's equation was shown to be better (Diamante and Murno, 1991; Tulasidas et al., 1993)

$$MR = \exp(-kt^N),$$

where N is the Page's number.

The dependence of the drying constant, k, and N of Page's equation on the drying air condition variables are modeled in the form of Arrhenius-type model:

$$k \text{ or } N = \alpha_0 V^{\alpha 1} H^{\alpha 2} \text{Experimental} \frac{-\alpha_3}{T_{abs}}, \quad (24.2d)$$

where α_0, α_1, α_2, and α_3 are the empirical constants of Arrhenius-type equation, V is the airflow velocity, H the absolute humidity, T the air temperature, abs the absolute values of temperature (K), and exp is experimental value.

According to these authors, Page's equation adequately describes the drying kinetics of a thin layer of Thompson seedless grapes, and drying constant is influenced more by temperature compared to velocity or relative humidity of drying air. The experimental data fitted better into the Arrhenius model compared to the power model.

Pretreatment of Grape Before Drying

In grape drying, the rate of moisture diffusion through the berries is controlled by the waxy cuticle of grapes. A number of authors have reported the effects of pretreatments on the drying rates and quality parameters (Aguilera et al., 1987; Bolin et al., 1975; Guadagni et al., 1975; Mahmutoglu et al., 1996; Pala et al., 1993; Raouzeos and Saravacos, 1986; Saravacos et al., 1988; Tulasidas et al., 1996). Dipping in hot water and the use of chemicals such as sulfur, caustic, and ethyl oleate (EO) or methyl oleate emulsions are some of the pretreatments widely used in grape drying. These chemicals facilitate the drying process by altering the structure of waxy layer, thus reducing the resistance to water diffusion (Doymaz 1998; Petrucci et al., 1973; Di Matteo et al., 2000; Ponting and McBean, 1970; Riva and Peri, 1986). EO acts on grape skin by dissolving the waxy components, which offer high resistance to moisture transfer, yet higher alkali concentrations and longer dipping times can cause adverse changes in dried grapes (Saravacos et al., 1988).

Golden-bleach process is commonly used to produce light-colored (bleached) raisins in California. In this process, SO_2 is employed as a bleaching agent, which is also necessary during drying and storage to preserve the yellow color of the raisins. In the "Greek process," Thompson seedless grapes are immersed in an aqueous solution containing 4.5% K_2CO_3, 0.5% Na_2CO_3, and 1% completely emulsified 0.4% olive oil for about 5 min. The dipped grapes are drained, then spread on trays under direct sunlight for drying. In "soda-dip-process," grapes are dipped for 2–3 s in a solution of 0.2–0.3% NaOH (caustic soda) at 93.3–100°C. Faint checks develop in the skins after the grapes are cooled by rinsing with cold water. Weaker sodas (Na_2CO_3 and $NaHCO_3$) are preferred to avoid the danger of overdipping. A small quantity of olive oil is often used in "soda-dip-process."

Raisins obtained from 2% dipping oil pretreatment were lighter in color with better skin integrity and scored the highest points for quality (41.1 of a total of 50 points) compared to 0.5% hot NaOH (35.7), 2.0% EO + 2.5% K_2CO_3 (39.9) and 0.4% olive oil + 7.0% K_2CO_3 (38.7), and control (36.4). Pretreatment with 0.5% NaOH at 93.0 ± 1.0°C for 5 s produced raisins with gummy or sticky surface caused by oozing of syrup through microcracks in the skin and pretreatment with 0.4% olive oil + 7.0% K_2CO_3 produced raisins having oily surface (Pangavhane et al., 1999b). NaOH pretreatment adversely affect the uniformity of color, whereas 0.4% olive oil + 7.0% K_2CO_3 treatment imparts reddish color to raisins. Untreated berries produce brownish raisins.

Doymaz and Pala (2002) dipped grapes in alkaline emulsion of EO (AEEO) and potassium carbonate (PC) solutions for 1 min prior to drying and evaluated the effects on drying rate and color of raisins. Dipping time in AEEO and PC solutions was around 1 min at the ambient conditions. Grapes were dried as a single layer in a batch at 50–70°C with air velocity of 1.2 m/s. Grapes dipped in AEEO and PC solutions exhibited shorter drying times than untreated grapes. AEEO enhanced drying rate to a greater extent than PC and resulted in raisins with higher color values. Highest raisin quality was obtained with grapes pretreated with AEEO and dried at 60°C. Mixed fatty esters were prepared in the laboratory from five vegetable oils (groundnut, safflower, sunflower, cotton seed, and rice bran oil) by the CFTRI processes and used as dipping oil for grapes before drying. Drying rates for grapes treated with mixed fatty acid esters were higher than those obtained with the commercial dipping oil (Giridhar et al., 2000).

Doreyappa Gowda et al. (1997) evaluated 17 new grape hybrids consisting of seven yellow seedless, five black seedless or soft seeded, and five seeded grapes and compared with Thompson seedless and Arkavati to test their suitability for dehydration. Sensory evaluation of the dry product revealed that hybrids "E-29/5," "E-12/3," and "E-12/7" produced raisins superior to Thompson seedless and were comparable to Arkavati with a distinctly lower acidity.

Vazquez et al. (1997) described a pilot scale drying plant comprising a closed circuit, hot-air convection chamber with a heat pump for drying grapes that were variously pretreated.

A novel physical treatment process to enhance the drying rate of seedless grapes has been used in the production of raisins, as an alternative to the conventional EO dipping method. This process consists of preliminary abrasion of the grape peel using an inert abrasive material. Grapes subjected to either abrasion drying or EO dipping pretreatments were dried in a convection oven at 50°C (air speed 0.5 m/s) until average moisture content was reduced to about 20% (w/w). Assessing drying rate, drying time, and microstructure of pretreated grapes and color of the dried samples compared the effectiveness of the two processes. The physical abrasion method was found to be as effective as the conventional dipping process for removing the waxy outer layer from grapes prior to drying. Although a darker product is obtained with the physical method, it makes no use of chemical additives, thus allowing production of safer raisins (Di Matteo et al., 2000).

Factors Affecting Drying of Grapes

Weather conditions and date of harvest have greatest influence on sun drying grapes, but roll type, tray type, and tray filling also have considerable influence on drying time (Peacock and Christensen, 1998). They compared five tray types: (1) standard wet strength (60 X 90 cm); (2). standard wet strength extra wide (65 X 85 cm); (3). Polycoated; (4). Polycoated with venting; (5). Extra wide with surface sizing for drying rates over a 16 day period. Although there were minor differences in drying rates among the five tray types, raisins in polyvented and extra wide trays with surface sizing were slightly drier during the last four days of drying as compared to those in the standard wet strength and standard wet strength extra wide trays. Grape berries in flop and cigarette roll lost about 1.2% moisture per day compared to 0.8% in those on biscuit rolls. Similarly, filling of 8.2 kg or less in standard wet strength trays (60 × 90 cm) reduces the drying time.

DRYING TECHNOLOGY

Drying practices for grapes are largely traditional and vary with variety and geographical locations. Certain grapes such as Currants (Black Corinth raisins and Zante currants) are usually dried without any preliminary treatments as in "natural sun-dried" Thompson seedless raisins. "Dry-on-vine" method may be followed to dry the clusters of Black Corinth. The vine-dried currants are more attractive than those dried on trays. The California raisins are called "loose Muscat" if the seeds are removed. In Spain and Australia, raisins are sold as either Valencia or Lexia, the latter being a rack-dried product. Sultana is a light-colored, tender raisin made from Thompson seedless, prepared by various processes other than natural sun drying in California.

In the traditional (open sun) drying method, grape bunches are spread over either the ground or on a platform in a thin layer directly exposed to the sun. This method is most cost-effective and takes approximately 8–10 days to produce dried product (Pangavhane and Sawhney, 2002). There is, however, a considerable risk of deterioration of quality due to dust, insect infection, undesirable changes in color, and the presence of foreign matter. To overcome these problems, grapes are covered with transparent sheet to reduce the effects of uncertain weather and protect from dust or insect infection. This practice might reduce the drying time by a day to two. Grapes are dried until the moisture content is reduced to about 14%.

Another improved traditional drying method is natural rack drying, which is used in Australia and India. In this method, pretreated bunches of grapes are laid on a long narrow wire screen racks under the iron roof which is slightly wider than the stack of racks. The sides may also be covered with jute cloth, plastic, or tarpaulin sheets to protect the berries from water droplets carried by air. However, it is not desirable to use side curtains when the relative humidity is high. Solar radiation in the early morning and late afternoon hours usually supplies most of the heat required for drying (Lof George, 1962). The solar energy absorbed by berries is subsequently utilized for water vaporization during the rest of the day when bunches remain shaded. Depending on the weather conditions, the drying time in this method is usually 2–3 weeks. After taking the dried grapes out of these racks, they are laid on the ground in shade to remove any moisture left on the berries.

The above methods are not commercially used because of mass losses and low quality of the finished products. For these reasons, artificial or mechanical drying process, which is rapid and controllable, is more suitable for commercial operations. Both the initial investment and the operating costs, particularly the energy cost, are high in these methods. Because of increasing fuel prices, use of renewable energy sources (solar radiation) for drying process is gaining popularity. Gee (1980) pointed out that low temperatures are desired for quality dehydrations, which can be easily obtained from indirect solar or waste heat sources.

Various types of solar dryers are now available for drying the grapes. The rack-type solar dryers (shade drying) are used extensively for drying sultana grapes in Australia and Thompson seedless grapes in India. This is the simplest and most effective type of solar dryer. In this structure, untreated sultana grapes get dried to moisture content of about 14% (wet basis) within 10–14 days. In Afghanistan, greenish colored (amber) raisins with smooth surface and good texture are produced in special drying houses called "Soyagi-Hana." According to Grnacarevic (1969), pretreated grapes get shriveled to certain extent, retaining their greenish colors, compared to the untreated grapes within 4 days. Several types of solar dryers have been tested for grape drying, including solar cabinet dryer (Sharma et al., 1986, 1992), glass roof solar dryer (Nair and Bongirwar, 1994), solar dryers with natural ventilation (Sharma et al., 1986), indirect-type solar drier (Sharma et al., 1993), solar multiple layer batch dryer (Eissen et al., 1985), solar dryer with greenhouse as a collector (Fohr and Arnaud, 1992), solar tunnel dryer integral collector (Eissen et al., 1985; El-Shiatry et al., 1991; Lutz et al., 1987), hybrid solar dryer (Tsamparlis, 1990), and multipurpose natural convection solar dryer (Pangavhane, 2000; Pangavhane et al., 1999a, b). Since the drying time and inside temperatures vary considerably in these dryers, the quality of the raisins produced is also different. Solar dryers dehydrate grapes (the moisture content in berries was reduced from 34.9% to 17% on dry weight basis) within 4 days compared to 15 days in shade drying and 7 days in open sun drying (Pangavhane and Sawhney, 2001).

Freeze and microwave vacuum dehydration have also been tried to minimize the compositional changes in grape berries during drying process (King, 1973; Clary and Ostrom, 1995; Tulasidas et al., 1993, 1997). Tulasidas et al. (1993) produced Thompson seedless raisins by drying grapes under combined convective and microwave drying in a single mode cavity type of applicator at 2450 mHz with varying operating parameters: air temperature, microwave power, density, and superficial air velocity. A novel physical treatment process to enhance the drying rate of seedless grapes has been used in the production of raisins, as an alternative to the conventional EO dipping method. This process consists of preliminary abrasion of the grape peel using an inert abrasive material. Grapes subjected to either the abrasion drying or the EO dipping pretreatments were dried in a convection oven at 50°C (air speed 0.5 m/s) until average moisture content was reduced to about 20% (w/w). Assessing drying rate, drying time, and microstructure of pretreated grapes and color of the dried samples compared the effectiveness of the two processes. The physical abrasion method was found to be as effective as the conventional dipping process for removing the waxy outer layer from grapes prior to drying. Although a darker product is obtained with the physical method, it makes no use of chemical additives, thus allowing production of safer raisins (Di Matteo et al., 2000).

Since commercial sun drying of raisins requires a long time (2–4 weeks) and chemical pretreatments are required to enhance drying rate, Kostaropoulos and Saravacos (1995) assessed the feasibility of improving the sun drying process of grapes by microwaves. Sultana seedless grapes dipped in alkali solution (2.5% K_2CO_3 + 0.5% olive oil to increase water permeability of the grape skin) were pretreated in a domestic microwave oven and dried by direct solar radiation. Microwave pretreatment reduced the moisture by 10–20% and these grapes dried nearly two times faster than the controls. Blanching in boiling water had the same effect on the drying rate as microwaves. Color and appearance of treated grapes were comparable to those of commercial products. This study concluded that microwave-pretreated grape had improved color after drying, due to reduction in enzymic browning as enzymes are partially inactivated by the absorption of microwave energy.

McLellan et al. (1995) studied the effects of honey as an antibrowning agent to produce light-colored golden raisins without the addition of SO_2 and concluded that raisins produced using honey without addition of SO_2 were ranked highest by each of the 18 panelists. The results were based on the Friedman rank analysis.

QUALITY OF RAISINS

Winkler et al. (1974) developed the criteria for determining the quality of raisins, which include color and appearance, texture, flavor, and taste. The color, taste, and texture are the main attributes for estimating the quality and consumer acceptance of raisins. These are judged by sensory evaluation on a 9-point hedonic scale (Ranganna, 1986; Guadagni et al., 1978).

Several factors affect the quality of raisins, including size of the berries; hue, uniformity, and brilliance of the color; condition of the berry surfaces; texture of the skin and pulp; moisture content; chemical composition; presence of decay (rot), mould, yeast, and of foreign matter; insect infestation; and drying time and conditions during the drying process.

CHEMICAL COMPOSITION

Grapes and raisins are essentially the same fruits, as raisins are the dehydrated grapes. Dehydration process induces certain compositional changes at a molecular level, especially in total proteins, type of carbohydrates, and enzymes such as polyphenoloxidase (Krueger et al., 2003). Raisins were found to contain hexose furanose, whereas grapes have pentose furanose. Raisins also contain higher levels of total proteins compared to grapes. The enzyme, polyphenoloxidase, which is present in grapes, does not occur in raisins. One hundred grams of raisins contains 15.4 g water, 3.3 g proteins, 0.4 g lipids, 79.2 g carbohydrates including fiber (5.3 g), 1.8 g ash, 49.00 mg calcium, 0.31 mg copper, 2.08 mg iron, 33.00 mg magnesium, 0.31 mg manganese, 97.00 mg phosphorus, 751.00 mg potassium, 12.00 mg sodium, 0.27 mg zinc, 3.3 mg ascorbic acid, 0.16 mg thiamin, 0.09 mg riboflavin, 0.82 niacin, 0.25 mg vitamin B_6, 3.3 μg folacin, 8.0 IU vitamin A, 0.70 mg vitamin E, 4.00 μg biotin, and 0.05 mg pentathenic acid (Californian Raisin Marketing Board, 2004).

NUTRITIONAL CONTRIBUTION

Based on a 2000 calorie daily diet, 40 g (1/4 cup) of raisins provides 125 kilo calories of energy and contributes 9% of daily requirement of potassium, 10% carbohydrate, 8% dietary fiber, 2% calcium, 6% iron, and <2% vitamin A. Addition of raisins to beef jerky lowered the amount of fat in the jerky, and increased antioxidants and fibers in the product (Mercola, 2003).

Grapes and grape products are rich in phenolic compounds, which have demonstrated a wide range of biochemical and pharmacological effects, including anticarcinogenic, antiatherogenic, anti-inflammatory, antimicrobial, and antioxidant activities. Based on total phenols per serving, Karakaya et al. (2004) categorized solid foods in the order of red grapes > raisins > tarhana > dried black plum > dried apricot > grape > fresh paprika > fresh black plum > *Utrica* sp. > cherry > fresh apricot > paprika pickle > paprika paste. They also found a high degree of correlation between total phenols and total antioxidant activity ($r^2 = 0.95$). Therefore, regular consumption of grape products would have a long-term health benefits. Specifically, raisins have been found to prevent heart disease, colon cancer, and intestinal tumors. Consumption of sun-dried raisin was found to increase fecal weight and short-chain fatty acid excretion, and decrease fecal bile acids in healthy adult humans (Gene et al., 2004) in a way that has health benefits.

PACKAGING AND STORAGE OF RAISINS

Raisins may be stored at room temperature without noticeable loss of color or flavor for a few months. The stability at room temperature will depend on the moisture levels in the raisins and the relative humidity of the atmosphere. If the temperature exceeds 10°C, the relative humidity should be kept below 55%. In addition, precautions against insect damage are necessary. The optimum relative humidity for storage of raisins is between 45% and 55%. However, raisins can be stored indefinitely in the refrigerator. It is important to provide natural ventilation of air around raisins which is necessary to avoid condensation, especially when they are stored at higher temperature and relative humidity.

Methyl bromide is widely used as a fumigant for dried fruits, including sultanas and raisins (Hilton and Banks, 1997). Johnson et al. (2003) proposed a treatment strategy combining initial disinfestation treatment using low oxygen (0.4%) with protective treatments (storing at 10°C, storage in controlled atmosphere with 5% oxygen or application of Indian meal moth granulosis virus) as an alternative to chemical fumigation of raisins for controlling postharvest insect populations.

Femenia et al. (1999) studied the effects of temperature on cell walls of raisin tissues during modified (controlled) atmospheric storage. Raisins obtained from seedless grapes (var. Flame) were kept under modified atmosphere composed of 60% CO_2 and 40% N_2, and stored at 10°C, 20°C, 30°C, and 40°C. Based on the changes in color and cell wall components, it was concluded that the combined use of relatively low temperature and a modified atmosphere helped to preserve raisin color and maintain levels of their cell wall material at values similar to initial concentration.

Aspergillus niger and *Zygosaccharomyces rouxii* are the dominating contaminants of Moroccan raisins. El-Halouat et al. (1998) examined the effects of modified atmosphere and certain antifungal preservatives on the growth of *A. niger* and *Z. rouxii* on high-moisture raisins. A combination of modified atmosphere (40% or 80% CO_2) and either 383/321 ppm sodium benzoate or 417/343 ppm potassium sorbate was found to be sufficient to inhibit microbial growth and extend shelf life of prunes and raisins for 6 months, respectively.

Cabras and Angioni (2000) reviewed pesticide residues in grapes and their behavior during grape processing, focusing on research carried out in the 1990s. Individual aspects considered included residues on grapes (fungicides and insecticides), residues on raisins, pesticide residues and fermentative microflora, effect of winemaking on residues, wine clarification, and pesticide residues in other alcoholic beverages produced from wine and wine by-products.

FOOD APPLICATION OF RAISIN

A new product termed "Baking Raisins" (BR), which does not require conditioning at the bakery, has been described (Bruno, 1996). These raisins are more uniform in moisture content, with approximately 8.3% more sugars than in conventional raisins. BR were packed in airtight bags with reduced O_2 levels to avoid fermentative changes during storage. BR could be stored in this way for up to 1 year under the ambient conditions. Baked products containing BR exhibited longer shelf life and improved texture.

Fruit anthocyanins provide color and health benefits, and addition of these pigments to breakfast cereals can improve their functional values. The stability and acceptability of anthocyanins from blueberry and Concord grape juice concentrates were tested in a model corn-based breakfast cereal. The overall acceptability was higher for the syrup and grape cereals, sweetness and flavor acceptability being correlated with overall acceptability (Camire et al., 2002).

Raisins have been used as a valuable ingredient in bread and other bakery products. Toops (2002) described these aspects with reference to the healthy image of raisins; consumer demand for healthier bakery products containing natural ingredients; the high levels of propionic acid natural preservatives in raisins; role of raisins in promoting flavor, visual appeal, and shelf life of bakery products; manufacture of raisin bread; incorporation of raisin products (e.g., juice concentrate, syrup, pastes) into other breads; and role of raisin conditioning in the production of high-quality bakery products.

From part of the crop, processors make raisin juice and raisin paste. Raisins are leached with water to produce a pure extract, which is then evaporated under vacuum to produce a self-preserving concentrate. Raisin juice concentrate contains 70% natural fruit soluble solids. It is added to a number of foods including dairy, confectionery, and bakery items. It is a natural substitute for preservatives and it sweetens and colors natural baked goods. It is also a sugar substitute for confectionery items, and filling for hard candies and molded chocolates. In cookies and crackers, raisin juice helps control breakage and maintains moisture. It is also used as binding agent for cereal bars. It enhances color and flavor in ice-cream and chocolate milk.

Raisin paste is made from 100% raisins, by extruding them through fine mesh screens. It is added for visual appeal and flavor in sundae-style yogurt and cottage cheese as wells as in frozen novelties and soft-centered candies. It is a stable ingredient that sweetens naturally.

Raisin may soon become an alternative to sodium nitrite, a preservative commonly used in processed meat, such as beef-jerky. Addition of raisins to beef-jerky lowers the amount of fat while providing additional benefits like antioxidant properties, fiber supplementation, lower sodium, reduced off-flavor, and check microbial growth.

FUTURE RESEARCH NEEDS

The heated air used to dehydrate grapes induces compositional changes including polyphenoloxidase activity, which changes color from light green of fresh grapes to dark brown to purple black color of dehydrated product. Sulfur dioxide is used to preserve the color. Since the use of excess sulfur dioxide causes allergy reactions, alternative to sulfur dioxide fumigation is being attempted. Researchers are evaluating various drying techniques such as liquid media dehydration and treating raisin with honey to prevent color change.

More research is necessary to further refine and optimize the drying process and prevention of undesirable changes in color, flavor, and composition of raisins. Like in other crops, organically produced grapes and raisins have become a leading commodity. There is need to develop production practices and certification standards for organic raisins.

REFERENCES

Aguilera, J. M., K. Oppermann, and F. Sanchez. 1987. Kinetics of browning of sultana grapes. *Journal of Food Science* 52:990–993.

Aguilera, J. M., and D. W. Stanley. 1999. *Microstructural principles of food processing and engineering*. Gaithersburg: Aspen Publishers.

Aung, L. H., D. W. Ramming, and R. Tarailo. 2004. Changes in moisture, dry matter and soluble sugars of dry-on-the-vine raisins with special reference to sorbitol. *The Journal of Horticultural Science and Biotechnology* 77(1):100–105.

Azzouz, S., A. Guizani, W. Jomaa, and A. Belghith. 2002. Moisture diffusivity and drying kinetic equation of convective drying of grapes. *Journal of Food Engineering* 55:323–330.

Bird, B. R., W. E. Stewart, and E. N. Lightfoot. 1960. *Transport phenomena*. New York: Wiley.

Bolin, H. R., and C. C. Huxsoll. 1987. Scanning electron microscope/image analyzer determination of dimensional postharvest changes in fruit cells. *Journal of Food Science* 6:1649–1650.

Bolin, H. R., V. Petrucci, and G. Fuller. 1975. Characteristics of mechanically harvested raisins produced by dehydration and by field drying. *Journal of Food Science* 40:1036–1038.

Brennan, J. G. 1994. *Food dehydration: a dictionary and guide*. Oxford: Butterworth-Heinemann.

Bruno, R. C. 1996. Thermally processed raisins: what are the benefits? Paper read at 1996 IFT Annual Meeting: Book of Abstracts, p. 69 (ISSN 1082–1236).

Cabras, P., and A. Angioni. 2000. Pesticide residues in grapes, wine and their processing products. *Journal of Agricultural and Food Chemistry* 48(4):967–973.

Californian Raisin Marketing Board. 2004. Raisin research. Available from World Wide Web: http://www.calraisins.org./nutrition/news.html.

Camire, M. E., A. Chaovanalikit, M. P. Dougherty, and J. Briggs. 2002. Blueberry and grape anthocyanins as breakfast cereal colorants. *Journal of Food Science* 67(1):438–441.

Carslaw, H. S., and J. C. Jaeger. 1980. *Conduction of heat in solids*. Oxford: Claredon Press.

Chauchard, F., R. Cogdill, S. Roussel, J. M. Roger, and V. Bellon-Maurel. 2004. Application of LS-SVM to non-lenear phenomena in NIR spectroscopy: development of robust and portable sensor for acidity prediction in grapes. *Chemometrics and Intelligent Laboratory Systems* 71:141–150.

Clary, C. D., and G. A. Sawyer Ostrom. 1995. Use of microwave vacuum for dehydration of Thompson seedless grapes. *Research Bulletin*, University of California, CATI Publication No. 950405, p. 5.

Diamante, L. M., and P. A. Murno. 1991. Mathematical modeling of hot air drying of sweet potato slices. *International Journal of Food Science and Technology* 26:99–109.

Di Matteo, M., L. Cinquanta, G. Galiero, and S. Crescitelli. 2000. Effect of a novel physical pretreatment process on the drying kinetics of seedless grapes. *Journal of Food Engineering* 46(2):83–89.

Doreyappa Gowda, I. N., R. Singh, and B. N. S. Murthy. 1997. Evaluation of new grape hybrids for dehydration. *Journal of Food Science and Technology (India)* 34(4):286–290.

Doymaz, I. 1998. Investigation of drying characteristics of grape and Kahramanmaras pepper. Ph.D. Dissertation. Science Institute, Yildiz Technology University, Istanbul, Turkey.

Doymaz, I., and M. Pala. 2002. The effects of dipping pretreatments on air-drying rates of the seedless grapes. *Journal of Food Engineering* 52(4):413–417.

Eissen, W., W. Muhlbauer, and H. D. Kutzbach. 1985. Solar drying of grapes. *Drying Technology* 3:63–74.

El-Halouat, A., H. Gourama, M. Uyttendaele, and J. M. Debevere. 1998. Effects of modified atmosphere packaging and preservatives on the shelf life of high moisture prunes and raisins. *International Journal of Food Microbiology* 41(3):177–184.

El-Shiatry, M. A., J. Muller, and W. Muhlbauer. 1991. Drying fruits and vegetables with solar energy in Egypt. *Agricultural Mechanization in Asia, Africa and Latin America* 22:61–64.

FAO. 1995. *FAO yearbook of production*. QBS 8, no. 3/4.

Femenia, A., E. S. Sanchez, S. Simal, and C. Rossello. 1999. Effect of temperature on the cell wall composition of raisins during storage under a modified atmosphere. *European Food Research and Technology* 209(3/4):272–276.

Fohr, J. P., and G. Arnaud. 1992. Grape drying: from sample behavior to the drier project. *Drying Technology* 10(2):445–465.

Gee, M. 1980. Some flavor and color changes during low temperature dehydration of grapes. *Journal of Food Science* 45:146–147.

Gene, A. S., J. A. Story, T. A. Lodics, M. Pollack, S. Monyan, G. Butterfield, and M. Spiller. 2004. Effects of sun-dried raisins on bile acid excretion, intestinal transit time, and fecal weight: a dose-response study. *Journal of Medicinal Food* 6:87–91.

Giridhar, N., A. Satyanarayana, K. Balaswamy, and D. G. Rao. 2000. Effect of mixed fatty acid esters prepared from different vegetable oils on the drying rate of Thompson seedless grapes. *Journal of Food Science and Technology (India)* 37(5):472–476.

Greenspan, L. 1977. Humidity of fixed points of binary saturated aqueous solutions. *Journal of Research of Natural Bureau of Standards—A. Physical and chemistry* 81A(1):89–96.

Grnacarevic, M. 1969. Drying and processing grapes in Afghanistan. *American Journal of Enology and Viticulture* 20:1198–2003.

Guadagni, D. G., A. E. Stafford, and G. Fuller. 1975. Taste thresholds of fatty acid esters in raisins and raisin paste. *Journal of Food Science* 40:780–783.

Guadagni, D. G., C. L. Storey, and E. L. Soderstrom. 1978. Effects of control atmosphere on flavor stability in raisins. *Journal of Food Science* 43:1726–1728.

Hills, B. P., and B. Remigereau. 1997. NMR studies of changes in subcellular water compartmentation in parenchyma apple tissue during drying and freezing. *International Journal of Food Science and Technology* 32:51–61.

Hilton, S. J., and H. J. Banks. 1997. Methyl bromide sorption and residues on sultanas and raisins. *Journal of Stored Products Research* 33:231–249.

Jewell, G. G. 1979. Fruits and vegetables. In: *Food microscope*. J. G. Vaughan (Ed.). London: Academic Press.

Johnson, J. A., P. V. Vail, D. G. Brandl, J. S. Tebbets, and K. A. Valero. 2003. Integration of nonchemical treatments for control postharvest pyralid moths (Lepidoptera: Pyralidae) in almonds and raisins. *Journal of Economic Entomology* 95:190–195.

Karakaya, S., S. N. El, and A. A. Tas. 2004. Antioxidant activity of some foods containing phenolic compounds. *International Journal of Food Science and Nutrition* 52:501–508.

King, C. J. 1973. Freeze drying. In: *Food dehydration*. W. B. Van Arsdel, M. J. Copley, and A. I. Morgan (Eds). West Port, CN: AVI Publishing Co., pp. 161–200.

Kostaropoulos, A. E., and G. D. Saravacos. 1995. Microwave pre-treatment for sun-dried raisins. *Journal Food Science* 60:344–347.

Krueger, J., M. Kuelbs, L. May, and J. Wolcott. 2003. Carbohydrate, paper chromatography and enzyme tests distinguish chemical differences between grapes and raisins. Available from World Wide Web: http://www.msu.edu/course/lbs/145/luckie/inquiriesf 2003/cellular4.html.

Lof George, O. G. 1962. Recent investigations in the use of solar energy for the drying of solids. *Solar Energy* 6:122–128.

Lutz, K., W. Muhlbauer, J. Muller, and G. Reisinger. 1987. Development of multipurpose solar crop dryer for arid zones. *Solar and Wind Technology* 4:417–424.

Mahmutoglu, T., F. Emir, and Y. B. Saygi. 1996. Sun/solar drying of differently treated grapes and storage stability of dried grapes. *Journal of Food Engineering* 29:289–300.

Martin, R. J. L., and G. L. Stott. 1957. The physical factors involved in the drying of Sultana grapes. *Australian Journal of Agricultural Research* 8:444–459.

McLellan, M. R., R. W. Kime, C. Y. Lee, and T. M. Long. 1995. Effect of honey as an antibrowning agent in light raisin processing. *Journal of Food Engineering and Preservation* 19:1–8.

Mercola, J. 2003. Raisins may be alternative to nitrites. Available from World Wide Web: http://eurekalert.org.

Moulton, K., and J. Possingham. 1998. Research needs and expectations in the grape and grape products sector. Paper read at World Conference on Horticultural Research International Society for Horticultural Science, 17–20 June 1998.

Mujumdar, A. S. 1995. *Handbook of industrial drying*. New York: Marcel Dekker, Inc.

Nair, K. K. V., and D. R. Bongirwar. 1994. Solar dryer for agricultural products. A do it yourself solar dryer. *Indian Chemical Engineering* 36(3):103–105.

Pala, M., Y. B. Saygi, and H. Sadikoglu. 1993. A study on the drying of Sultana grapes by different techniques and effective parameters. In: *Food flavors, ingredients and composition*. G. Charalambous (Ed.). Elsevier: Amsterdam, pp. 437–444.

Pangavhane, D. R. 2000. Analytical and experimental studies on grape drying. Ph.D. Dissertation, Devi Ahilya Vishwavidyalaya, Indore, India.

Pangavhane, D. R., and R. L. Sawhney. 2001. Review of research and development work on solar dryers for grape drying. *Energy Conservation and Management* 43:45–61.

Pangavhane, D. R., and R. L. Sawhney. 2002. Design, development and performance testing of a new natural convection solar dryer. *Energy* 27:579–590.

Pangavhane, D. R., R. L. Sawhney, and P. N. Sarsawadia. 1999a. Comparative studies on drying time of different methods for grape dehydration: renewable energies and energy efficiency for sustainable development. Proceedings of 23rd National Renewable Energy Convention Devi Ahilya Vishwavidyalaya, Indore, India, pp. 260–263.

Pangavhane, D. R., R. L. Sawhney, and P. N. Sarsawadia. 1999b. Effect of various dipping pretreatment on drying kinetics of Thompson seedless grapes. *Journal of Food Engineering* 39:211–216.

Pangavhane, D. R., R. L. Sawhney, and P. N. Sarsawadia. 2000. Drying kinetic studies on single-layer Thompson seedless grapes under controlled heated air conditions. *Journal of Food Processing and Preservation* 24:335–352.

Peacock, W., and B. Beede. 2004. Improving maturity of Thompson seedless for raisin production. *Grape News* 1:1–5.

Peacock, W., and P. Christensen. 1998. *Science of sun dried raisins*. Publication No. RG 4-96. Tulare County, CA: University of California Cooperative Extension. pp. 1–6.

Peri, C., and M. Riva. 1984. Etude du sechage des raisin 2: effet des traitments de modification de la surface sur la qualite du produit. *Science des Alimentes* 4:273–286.

Petrucci, V., N. Canata, H. R. Bolin, G. Fuller, and A. E. Stafford. 1973. Use of oleic acid derivatives to accelerate drying of Thompson seedless grapes. *Journal of the American Oil Chemists' Society* 51:77–80.

Ponting, J. D., and D. M. McBean. 1970. Temperature and dipping treatment effects on drying rates and drying times of grapes, prunes and other waxy fruits. *Food Technology* 24:1403–1406.

Ramos, I. N., C. L. M. Silva, A. M. Sereno, and J. M. Aguilera. 2003. Quantification of microstructural changes during first stage air drying of grape tissue. *Journal of Food Engineering* 62(2): 159–164.

Ranganna, S. 1986. *Handbook of analysis and quality control for fruits and vegetable products*. New Delhi, India: Tata McGraw Hill.

Raouzeos, G. S., and G. D. Saravacos. 1986. Solar drying of raisins. *Drying Technology* 4:633–649.

Riva, M., and C. Peri. 1986. Kinetics of sun drying of different varieties of seedless grapes. *Journal of Food Technology* 21:199–208.

Saranwong, S., J. Sournrivichai, and S. Kawano. 2003. Performance of a portable near infrared instrument for Brix value determination of intact mango fruit. *Journal of Near Infrared Spectroscopy* 11:1750–1810.

Saravacos, G. D., S. N. Marousas, and G. S. Raouzeos. 1988. Effect of ethyl oleate on the rate of air-drying of foods. *Journal of Food Engineering* 7:263–270.

Sharma, P. C., K. D. Sharma, and R. S. Parashar. 1992. Prospects of raisin production in tribal areas of Himachal Pradesh. *Indian Food Packer* 16:9.

Sharma, V. K., A. Colangelo, and G. Spagna. 1993. Experimental performance of an indirect type solar fruit and vegetable dryer. *Energy Conservation and Management* 34:293–308.

Sharma, V. K., S. Sharma, R. A. Ray, and H. P. Garg. 1986. Design and performance of a dryer suitable for rural applications. *Energy Conversion and Management* 26(1):111–119.

Singh, P. C., and R. K. Singh. 1996. Application of GAB model for water sorption isotherm of food products. *Journal Food Processing and Preservation* 20:203–220.

Suthar, S. H., and S. K. Das. 1997. Moisture sorption isotherms for Karingda [*Citrullus lanatus* (Thumb) Mansf] seed, kernel and hull. *Journal of Food Process Engineering* 20:349–366.

Temma, T., K. Hanamatsu, and S. Kawano. 2002. Performance of a portable near infrared sugar measuring instrument. *Journal of Near Infrared Spectroscopy* 10:77–83.

Toops, D. 2002. Raisins' rising popularity. *Food Processing (USA)* 63(9):73–75.

Tsamparlis, M. 1990. Solar drying for real applications. *Drying Technology* 8(2):261–285.

Tulasidas, T. N., C. Ratti, and G. S. V. Raghavan. 1997. Modeling of microwave drying of grapes. *Canadian Agriculture Engineering* 39(1):57–67.

Tulasidas, T. N., G. S. V. Raghavan, and E. R. Noris. 1993. Microwave and convective drying of grapes. *Transactions of American Society of Agricultural Engineers* 36:1861–1865.

Tulasidas, T. N., G. S. V. Raghavan, and E. R. Noris. 1996. Effects of washing pre-treatments on microwave drying of grapes. *Journal of Food Process Engineering* 19:15–25.

USDA. 1997. United States Standards for Grades of Grapes for Processing and Freezing.

USDA. 2004. FAS Quarterly Reference Guide to World Horticultural Trade: Production, Supply and Distribution of Key Commodities. United States Department of Agriculture, Foreign Agricultural Service Circular Series 1-04 January 2004.

Vazquez, G., F. Chenlo, R. Moreira, and E. Cruz. 1997. Grape drying in a pilot plant with a heat pump. *Drying Technology* 15(3/4):899–920.

Wang, N., and J. G. Brennan. 1995. Changes in structure and density and porosity of potato during dehydration. *Journal of Food Engineering* 24:61–76.

Winkler, A. J. 1954. Effects of over cropping. *American Journal of Enology and Viticulture*. 5: 4–8.

Winkler A. J., J. A. Cook, J. A. Kliewer, and L. A. Lider. 1974. *General viticulture*. California: University of California Press.

25
Grape and Wine Biotechnology: Setting New Goals for the Design of Improved Grapevines, Wine Yeast, and Malolactic Bacteria

Isak S. Pretorius

INTRODUCTION

HOW THE DEMAND ECONOMY IS DRIVING GRAPEGROWING AND WINEMAKING

The wine industry is driven by the consumer—what the market wants is what the winemaker must provide and those who do not provide the right package of taste, quality, price, and value will lose a sale to those who do. Despite the abundance of, and affection for, traditional methods of grapegrowing and winemaking, the industry in its totality will follow the path of technological progress because that is how the market's future needs will be met. The challenge for grape and wine biotechnology in the 21st century is to maximize the potential of technological innovation to enable the industry to deliver the best product to the greatest number of consumers at prices they will pay while retaining the traditional milieu of wine (Pretorius and Høj, 2005).

However, the entire process of technological innovation is often as gradual as the maturation of a good Shiraz and this is positive because, while it might be true that big events erupt noisily to dominate the headlines and thrust themselves onto the world, changes that have more impact tend to take place much more slowly and quietly. This is the approach of grape and wine biotechnology, and many in the discipline agree with those futurists who say that the world at the start of the third millennium is on the verge of an era of massive and unprecedented change so dynamic and far-reaching that many facets of society will be almost unrecognizable in the next few decades. If they are right, and we are already feeling the tremors of massive change, then it is time to remind ourselves of the adage that "the future belongs to those who prepare for it" (Pretorius and Høj, 2005).

The need for preparation to lead the industry into a future that will meet the expectations of consumers has caused some major wine-producing countries to undertake in-depth analyses to anticipate global business trends and position their industries to be ready for the most probable scenarios. If the threat of growing wine surpluses proves to be more than just a threat, they will need vision and long-term strategies to help their industries to work through the clutter of technological innovation and adjust to, and succeed in the face of, changes in consumer preference while recognizing the consequences both for their industry and their countries of failure to adjust.

This is not a new challenge. The wine industry has never ceased changing in the seven millennia of recorded history that has brought wine, which was initially seen as a storable (preserved), dietary beverage, to a position where it is now viewed by millions as a desirable luxury. This metamorphosis has catapulted wine into a situation where it is caught between the forces of "market pull" and "technology push." It needs to maintain that traditional appeal, which gives it the luster of luxury, while the industry, to meet the quantities consumers will buy, must embrace methodologies that enhance quality, purity, safety, and consistency. For example, a winemaker might be producing limited bottles of traditional Pinot Noir from selected blocks of vines. The wine is well-known and popular but enough cannot be produced to meet demand. The options for the winemaker are therefore to restrict production and try to raise the price—or to adapt from tradition to technology to make greater quantities of the same wine using supplementary grape sources that will still consistently meet the expectations of its market. This requires tradition and innovation to become partners in meeting the needs of producers and the preferences of consumers because the continued sustainable existence of the producer depends on the ability of the industry as a whole to meet the expectations of a wine consumer who is health- and environmentally conscious and looking for an affordable quality product. It is clear that the demand side—the discerning consumer—is driving the supply side—the industry—and this places the producer under pressure.

While the consumer looks for quality, the wine industry globally has achieved quantity with about 29 billion liters produced annually from some 8 million hectares of vines, which results in about 5 billion liters—or between 15% and 20% of total production—without a ready market. There are two main factors driving this disparity. First is the 40–50% decline in per capita consumption in the traditional European wine-drinking countries, where it appears that the changing nature of society is having its effect (Høj and Pretorius, 2004). Second, non-European production of wine has increased strongly while market growth for this new product, added to the surplus European product, has not kept pace. New World countries (Argentina, Australia, Chile, Canada, New Zealand, South Africa, USA, etc.), which now produce about 20% of the world's wine, gained significant market share of the world export market, moving from 2% to 15%, in the past few decades because they were quick to respond to market forces and shifting perceptions of wine quality (Pretorius and Bauer, 2002). Old World countries (France, Italy, Spain, Portugal, Germany, etc.) now produce slightly more than half the world's wine with their own per capita consumption falling by 40–50% (Pretorius and Høj, 2005). This represents a fundamental realignment of the global wine league table. However, the supply and demand problems facing some countries cannot be related only to the recent export successes of newer producing countries such as Australia, and a deeper analysis with a more multifaceted response is required to understand this mismatch between supply and demand.

Once again it appears that the consumer is the driving force. The global excess is concentrated in the bulk wine category, which would probably indicate that there has been a move from "basic" wines to premium, super-premium, ultra-premium, and even icon wines. Globalization and the wider availability of information and therefore knowledge about wine have combined to make modern consumers more sophisticated in their understanding of quality and value, and they have higher expectations in all price categories as a result. Another indication of this spread of sophistication in consumer choice is the growing demand for artisan wines, currently between 2% and 3% of the market (Bisson et al., 2002).

Rather than the bulk wine buyer of yesteryear, the consumer today is image-conscious and price-sensitive, and this has led to a change in the rules of the marketplace to a degree where quality is defined as "sustainable customer and consumer satisfaction" (Pretorius and Bauer, 2002). Today's consumer looks for strong brand names and expects them to deliver consistently on style, purity, quality, uniqueness, and diversity, and this has led to the need for a winemaker, if he or she is to be successful and enterprising, to be responsive to the new breed of consumer by delivering products of outstanding quality at every price point. This new breed vote relentlessly with their wallets, which means, for the wine producer, that innovation at all stages of the value chain from vineyard to palate is no longer an option but a necessity for an industry that wants to remain profitable and socially responsible.

The oversupply situation, which is probably more a case of lack of demand, invites a range of responses (Høj and Pretorius, 2004). One way would be to allow market forces to eliminate unprofitable producers through bruising competition, but this does nothing to prevent a further decline in global consumption. Another way would be to use those market forces to grow, responsibly, the total demand for wine by creating new markets and by increasing or reinvigorating the demand in existing markets where per capita consumption is presently in decline. This will almost certainly mean that the wine industries are transforming themselves from production-driven enterprises into ventures that are driven by the demands of consumers. In other words, rather than continuing to make what they have always made, and hoping to be paid for it one way or another, they become ventures that make what the various consumers actually want to drink (Høj and Pretorius, 2004).

The New World and the Old World can benefit from working together on this change of direction because the old, conservative approach, which has entailed a divisive stand-off between different wine-producing countries, will scare away potential new consumers in the emerging economies that will become the dominant powers of this new century (Høj and Pretorius, 2004). To achieve this, it will be necessary for those traditional wine producers who continue to depict innovative winemaking as "*industrial*" and conventional winemaking as "*agricultural*" to change their approach because such divisive and poisonous language will be counterproductive in convincing potential consumers of the merits of wine as a pleasurable product. This approach is very disappointing because it will hurt all producers, both in the Old World and in the New World, and can only be negative for wine while being positive for the myriad of other sectors vying for the consumer dollar. It will ultimately diminish the total global market for wine at a time when all wine producers should have the same interest in ensuring that existing consumers continue to drink wine and potential new consumers choose wine—bottled poetry—over other alcoholic beverages (Høj and Pretorius, 2004).

The challenge, therefore, for all producers is to produce wines that are competitive on quality and price and meet consumers' expectations for an enjoyable beverage that is healthier, produced in an environmentally sensitive manner and still measures up to the traditional mystique of wine while making a profit along the production chain (Bisson et al., 2002; Pretorius and Bauer, 2002). Success in this "demand-chain" environment (as opposed to the outdated "supply-chain" approach) means that the wine industry must keep abreast of the application of new technologies if it is to meet the challenges presented by consumers and innovative competitors. These pacesetting grapegrowers and winemakers see it as a minimum requirement that they accurately measure and meet consumer preferences for style and quality and maintain the smart application of efficient practices all along the supply-chain while at the same time leaving a minimal environmental footprint. This expectation is a challenge for wine science and has led to a fundamental change in the focus of research and development (R&D) in leading wine-producing countries.

The Changing Focus of Grape and Wine Research

Grape and wine research has recently shifted attention from a focus on risk factors in grape growing and wine production to a focus on achieving that poetic quality for the consumer. It has played a key role in past decades in identifying and providing solutions or ways of managing problems in the vineyard and the winery. It has played a key role in improving product quality by reducing known risks such as disease in the vineyard, microbially induced faults in the product, and suboptimal bottle sealing (Høj et al., 2003). At the same time, research has introduced improved mechanization to both vineyards and wineries that has enabled labor cost reduction, in turn resulting in quality wine being available at competitive prices (Pretorius and Høj, 2005). There are still targets to achieve in these practical areas, such as the more efficient use of limited quantities of irrigation water, further reduction in the risk of unintended microbial activity, and better ways to seal a bottle while allowing the wine inside to mature and retain its quality.

While these targets of eliminating faults are still valid and are still being tackled, research in more recent years has started to look at ways of enhancing the attributes and value of the product in the bottle. A holy grail for the wine industry is to develop objective measures for grape and wine quality but, as with the grail of legend, this is as elusive as ever despite considerable progress (Pretorius and Høj, 2005). The challenge for grape and wine researchers is even more difficult because consumer and market preferences are changeable and therefore keep altering the type of measurement perceived to be important (Bisson et al., 2002; Menashe et al., 2003; Pretorius et al., 2004b). Individual taste compounds this problem because individual wine consumers rank quality attributes differently. Thus, grape and wine scientists need to extract clear goals from variable messages so that they can provide grapegrowers and winemakers with meaningful and tested options.

It is always the case that worthwhile outcomes from R&D rely on basic strategic research to create new knowledge, test it under applied conditions, and translate it into a commercial advantage. Now, with an exacting and more specific consumer, research is increasingly more targeted and synergistic partnerships between practitioner and scientist are becoming even more important (Høj et al., 2003). These partnerships have changed the nature of the relationship from one similar to a relay race, in which the baton is passed from researcher to practitioner, to a team approach in which all are players in the same football team passing the ball back and forth, moving the full distance to the goal line as a unit (Pretorius and Høj, 2005). It is in this way that basic and applied wine research continually enlighten one another and are no longer separate activities. Science must now be responsive to the needs of producers and consumers both when problems are selected for research and when experiments are designed for the outcomes to be what industry wants to meet consumers' expectations (Høj et al., 2003; Pretorius, 2003; Pretorius and Høj, 2005). And both researcher and practitioner realize that once the mind has been stretched by a new idea, it will never again return to its original size.

The Nature of Consumer Demands and Possibilities of Technology

It is as true for the wine industry as for all science that one of the biggest challenges is to show value and benefit in what can be achieved when innovators look over the horizon and see a future that awaits the general public, and in particular the consumer. We are reminded constantly in our daily lives of the contributions of science and technology that were barely foreseen when the research was being done decades earlier. Many household goods we enjoy today are the fruits of technological research planted decades ago. Yet it is as popular now as it was then among many self-proclaimed experts without scientific or technological backgrounds to malign technology as fundamentally bad.

The same tendency can be detected in some sections of the wine media, by whom wines with a high technical input are portrayed as uniform in flavor and lacking in character (Day, 1998). Ironically, this is in direct contradiction with reality because the consumer seeks diversity, yet expects consistency with each bottle of the same label. It is technology that enables grapegrowers and winemakers to respond to those market needs. Technology ensures that the wonderful diversity nature itself provides within grapes grown under different conditions is not obscured by ubiquitous faults (e.g., volatile acidity, spoilage by *Brettanomyces/Dekkera*, etc.) (Pretorius and Høj, 2005). It might be true that "the world hates change, yet it is the only thing that has brought progress," and it might also be true to say that if the methods used by the Romans made better wines, we would still be using them today (Day, 1998). Consistent with the adage that 'the Stone Age did not end because man ran out of stones', wine quality will continue to improve as man keeps stretching the boundaries of technological progress.

While today's market-driven wine producers support the credo "spare the innovation and ruin the industry," it is only in the past six to seven decades that researchers have come to understand the biochemical pathway for fermenting sugar to alcohol, after thousands of years imbibing (Pretorius and Høj, 2005). Now researchers are working out how to identify, measure, and manage compounds that can be detected on the palate or the nose of a wine even when present in levels as low as a few nanograms per liter. One example of a compound that shows itself at these low threshold concentrations is the passionfruit character in some white wines that emanates from

3-mercaptohexyl acetate (3MHA) (Darriet et al., 1995; Tominaga et al., 1998a, b, 2000; Dubourdieu et al., 2000). One gram of 3MHA would be enough to give passionfruit character to the equivalent of 100 Olympic-sized swimming pools filled with wine (Pretorius and Høj, 2005). Just as potent are compounds such as isobutyl methoxy-pyrazines (IBMP), which give rise to the "green" and herbaceous capsicum character in white wines (particularly important to Sauvignon blanc). Grape and wine researchers are looking at how to accurately control such potent impact compounds as 3MHA and IBMP in the composition of wine and it is likely that the work might need to start in the vineyard and finish with consumer education, as is supported by Australia's Strategy 2025 credo of "a total commitment to innovation and style from vine to palate" (Pretorius and Høj, 2005).

With such "grape-to-glass" R&D, the researchers need to start with some hypotheses or testable questions. For 3MHA, one could assume white wines with a strong hint of tropical fruit characters (including passionfruit) have been identified as desirable through market research. The challenge facing the R&D team is to find out how 3MHA is formed, how it behaves, and how it can be dealt with to deliver the wine the consumer is looking for. If the grapegrower can provide precise management of vines on a particular soil type, pick the crop when a precursor of 3MHA is known to be at its peak, and use appropriate juice extraction techniques, the winemaker will have a head start. The correct choice of yeast might then be made to ensure that 3MHA is formed to provide full passionfruit expression (Dubourdieu et al., 2000). Finally, the wine producer must provide the right bottle and closure, drawing on recent R&D (Godden et al., 2001; Capone et al., 2003), before the wine is placed before the consumer.

This approach, from "grape to glass," involves many cross-cutting, multidisciplinary research groups; extensive field and laboratory work; close interaction between scientists, grapegrowers, and winemakers; scientifically sound project design; and some old-fashioned luck. It might take millions of dollars of investment and several vintages to progress from the grape to the glass. For such a task, the challenges would require a substantial and sustained effort from researchers in many disciplines, including senso-chemistry, molecular biology, bioinformatics, and biotechnology (Pretorius, 2003; Pretorius and Høj, 2005). The rewards for success, however, are correspondingly large and science should not be afraid to think outside the box when developing an appropriate arsenal of research tools, even if they include those that are today considered by some as "brutally radical."

Without radical approaches, the wine industry might not be what it is today. Consider how radical it was in the 19th century to graft Europe's noble varieties onto American rootstocks to control phylloxera and consider whether it is any less radical in the 21st century to attempt to develop transgenic grapevines resistant to potentially equally devastating diseases, such as Pierce's disease (Pretorius and Høj, 2005). If long-term wine industry research is guided by negativity toward new technologies (e.g., gene technology and nanotechnology), consider whether such negativity should be allowed to be fueled by emotive arguments that are not underpinned by scientific evidence. Or should wine innovators recall the old saying "one experiment might be more valuable than 100 so-called expert opinions" (Høj and Hayes, 1998), arm themselves with a trans-disciplinary approach aimed at benefiting consumers, producers, and the environment, and look over the horizon.

PERSPECTIVES CONCERNING BIOTECHNOLOGICAL OUTCOMES IN THE WINE SECTOR

An Optimistic Viewpoint

The digital code of DNA has generated an explosion in the biological sciences in the half-century since the discovery of the double helix, which perhaps can be said to be the birth of modern biotechnology. It generated the tools that cast light on the molecular genetics of living cells and enabled the cloning of individual genes and the sequencing of whole genomes. The addition of modern computer knowledge and capabilities enabled biotechnology to expand further because its application allowed science to transform the nature and interactions of molecules into an "information and technology" science (Pretorius and Høj, 2005). Thus, the digital code of DNA generated a new view of biology as an information science, now called bioinformatics and recognized as a knowledge watershed in biology.

Wine production has benefited in all this because yeast has been a pacesetter among organisms for which a large body of experimental databases have been created by the development of high-throughput analytical techniques for the analysis of the gen*ome*, transcript*ome*, prote*ome*, and metabol*ome* (Winzeler et al., 1999; Østergaard et al., 2000; Allen et al., 2003; Dollinski et al., 2003). The new discipline, Bioinformatics, is continuously upgrading the content in these databases while mathematical models are currently being developed to integrate the information from the different "omes." This in itself has given rise to a further sub-discipline, referred to as "Systems Biology" (Delneri et al., 2001; Bro et al., 2003), because integration requires analysis of a complete system rather than individual components or pathways.

Biotechnologists are only now appreciating the power of gen*omics*, transcript*omics*, prote*omics*, interact*omics*, flux*omics*, and metabol*omics* (Hauser et al., 2001; Fiehn, 2002; Perez-Ortin et al., 2002a; Bro et al., 2003; Huh et al., 2003; Zhang and Gernstein, 2003). There are many who believe "omics" have the credentials to propel R&D activities from the laboratory to the marketplace. In this scenario, genetic engineering is lauded as a spectacular achievement that will soon translate all these frontier-shifting tools into a positive "economic" outcome for the wine industry (Pretorius and Høj, 2005).

A Clouded Viewpoint

It is important for grape and wine research bodies always to look at the potential negative impacts and risks accompanying the introduction of new technology, and to do this in an objective way. Unfortunately, the debate about genetic engineering has been muddied over the past two decades by anti-genetic engineering campaigners who have branded DNA science with emotive fear-mongering, bloodying the nose of the biotechnology community on numerous occasions (Ferber, 1999; Pretorius, 2000; Vivier and Pretorius, 2002; Arntzen et al., 2003; Arvanitoyannis, 2003; Pretorius and Høj, 2005). Such modern-day Luddites, while masquerading as genuinely concerned, have portrayed shaky scientific argument as truth to prey on vague public misgivings and stir up a storm of protest about scientists playing God, poisonous "Frankenfood," and the alleged global havoc that will be caused by genetically modified organisms (GMOs). These scenarios have been eagerly swallowed and regurgitated by others protesting against globalization. The result of this continuing misinformation has been to create a situation where a significant proportion of people suspect that GM food will prove unhealthy in the long term and that organisms with transplanted genes will somehow escape into the environment causing degradation of the earth's biodiversity (Pretorius and Høj, 2005). Even worse, many are deeply suspicious that knowledge about, and the rights to, these organisms are "quarantined" in the hands of large multinational companies, giving them unfair economic power to wield over the individual, thus prompting restrictive regulation and legislation in many jurisdictions around the world, notably in Europe (Pretorius, 2000).

Although legislation and regulations have been brought in to govern the use of GMOs and the planting of GM food crops, they have been along the same lines but break down into one of the two main categories (Arvanitoyannis, 2003), an "avant-gardist" approach and a "traditionalistic" approach (Pretorius and Høj, 2005). The American approach follows the reasoning that there is no *a priori* rationale to assume that there is a risk from GM any more than there is a risk from conventional methods of introducing new genes into organisms and products, as has been undertaken for centuries. In this approach, each case of a new GMO is considered on its merits in more of an innocent-until-proven-guilty approach where proponents must use arguments of "substantial equivalence" and provide assurances such as no additional threat to human health or the environment, usually with satisfactory evidence from highly controlled field trials, before statutory approval is granted (Pretorius and Høj, 2005).

The European Union's regulations exemplify the second approach. This takes a guilty-until-proven-otherwise stance and assumes that all and any GM material has an inherent level of risk higher than anything that might apply from a non-GM product. By framing regulations around this proposition, the European Union places demands on the product of any GMO-based material that are substantially more rigorous than those placed on food material bred by conventional genetic means. This can be illustrated in the recent introduction by the European Union of Labelling Legislation under which any food product with more than 0.9% GM content must be labeled as a GM product (Reiss, 2002). All this has led to a chasm of difference between the world's biggest trading blocs into which has been thrown an arsenal of argument to justify trade bans and technical barriers to free trade.

Wine is not without its own genetic modification debate and its own issues (Vivier and Pretorius, 2002), but it will always be true to say that natural diversity in the characteristics of wine will be enhanced by the application of technology while failure to apply new knowledge will diminish this diversity. Despite the populist anti-GMO stampede, there are obvious benefits biotechnology can offer the wine sector. It is already an industry bound by restrictions as well as tradition and regional culture. The traditionalists, who view wine as a natural beverage produced by time-honored procedures and cloaked in a mantle of romanticism, are "refuseniks" when it comes to change in much the same way as the anti-GM lobby. As a food product, it is one of the most natural and long-lived, and contains only a tiny amount of added substances when compared with most packaged food. It is already hidebound by restrictive laws in most wine-producing countries with a list of additives that are permitted and a ban on anything not in the list. The market itself provides its own form of restriction through expectations of label integrity and product identity. This has led to concerns that obligatory labeling of wines with any level of GM input, however great the improvements in quality and purity, would be detrimental to sales and would lead to GM wines that exhibit novel characteristics being sidelined by the trade, disallowed the use of varietal names, and characterized as "standardized" or "McDonaldized" (Pretorius and Høj, 2005).

However, the practical application of transgenic grapevines and recombinant microbial starter strains is remote. Given the misinformation of the anti-GM lobbyists that has created concerns among consumers, the set-in-stone opposition of the traditionalists, and a business environment in which product-liability litigation is becoming more prevalent, the general view is that it would be commercial suicide to market the first GM wine (Pretorius and Høj, 2005).

A Realistic Viewpoint

The question therefore arises as to how grape and wine research can derive benefits from DNA technology without directly using GM organisms, because there remain major advantages in the scientific learning to be gleaned for wine producers seeking to meet future market needs. To realize these benefits, the wine industry must be prepared to work within the current political and social environment and implement different strategies for the immediate, middle, and distant future (Pretorius and Høj, 2005).

To look at a short-term option first, DNA technology can be used as a tool for research that will authenticate plant varieties and microbial strains, and provide more data on the fundamentals of the three main organisms involved in wine—the grapevine, wine yeast, and malolactic bacteria. This would involve the use of existing, non-GM varieties of the three organisms to produce non-GM wine while using DNA technology to achieve a result (de Barros Lopes et al., 2003). The approach would enable the use of molecular tools to create GM prototype plants and microbes, from which fundamental and complex aspects could be unraveled and hypotheses tested while the original applied goal of the research is reached through an alternative non-GM strategy. Although this approach is clearly cumbersome, it is viable and acceptable within the current climate.

A further approach, also fairly unwieldy but which could be a medium-term solution, involves the use of a product produced from a genetically modified organism, such as a GM enzyme, instead of the live GM organism itself. This would effectively be an "arm's-length" approach because neither the GMO nor its DNA would be in the wine on the shelf or even on the premises of the grapegrower or the winemaker (Pretorius and Høj, 2005). Scientists could use products derived from a GMO to replace problematic processing agents such as crude enzyme preparations, proteinaceous fining aids, and bentonite.

This could give rise to solutions for some of the undesirable side effects of some of these agents—side effects that are not related to their primary functions but which nevertheless cause trouble for the winemaker. There is no doubt that pectinase preparations degrade grape pectin; proteinaceous fining agents modulate the phenolic content of wine to make it more palatable; and bentonite can make wine protein stable and prevent haze formation (Pretorius and Høj, 2005), but while doing good with one hand they can do harm with the other. Some commercial enzyme preparations for pectin, for example, can introduce changes to the composition of wine that are detrimental to the health and sensory values of the wine because they can be accompanied by actions (e.g., pectin methyl esterase, cinnamyl esterase) that might lead to the formation of methanol (which can be oxidized in the body to the toxic compounds, formaldehyde, and formic acid), 4-vinylphenol, and 4-vinylguaicol (which can give rise to a medicinal flavor that will mar taste), and to oxidation artifacts during the hydrolysis of authentic norisoprenoid glycosides (which complicates interpretation of analytical data) (Van Rensburg and Pretorius, 2000; Smit et al., 2003). Similarly, the use of bentonite for clarification and fining, although essential for the wine, causes sediment that has to be disposed of and can alter the flavor of the wine itself.

The technological possibilities of this arm's-length approach have potential not just for the wine industry but also for food and other beverages. It is possible, through generating GM products as processing aids, that the entire food industry could benefit in the not-so-distant future (Pretorius and Høj, 2005). The speed with which this might happen is unknown, although it is likely that some degree of acceptance is possible in the initial stage with the use of a protein (e.g., a pectinase) or a metabolite (e.g., a vitamin) indistinguishable from its non-GM counterpart and produced in a process undertaken away from its place of use. This means that it would be possible and probably acceptable for researchers, instead of using a recombinant yeast producing highly specific pectinases or glycosidases, to use a preparation of ultra-pure enzymes produced from a GMO. This would be prepared at a remote and secure location before being added to the grape must. Such use of DNA technology would be entirely possible if the GMO-derived enzymes were indistinguishable from that which is found in the traditional but crude preparations now in use.

The approach for the distant future, which could only be undertaken with eventual consumer acceptance, would be to generate and introduce GM grape varieties and microbial starter strains into grapegrowing and winemaking; and contrary to the belief of many, this would serve only to increase the diversity of wine. Genetic design work is already under way to create superior grapevines, wine yeast, and malolactic bacteria for cost-competitive production of grapes and wine of high quality that will require relatively low resource inputs and have a low environmental impact (Pretorius and Bauer, 2002; Vivier and Pretorius, 2002; de Barros Lopes et al., 2005; Pretorius et al., 2005; Pretorius and Høj, 2005). This is not to imply that the wine industry is on the verge of realizing the potential benefits of biotechnology because there is a whole host of complex and intertwined scientific, technical, economic, marketing, safety, legal, social, and ethical issues to be worked through that will delay and possibly exclude the commercialization and benefits that can be derived from genetically engineered grape varieties and microbial starter strains. However, it would be equally unwise for the wine community to buckle under these issues and discard the vast potential benefits the biotechnology has to offer.

Having said that, it is clear that grape and wine research must take a highly moral and ethical position and guarantee that at all costs, it will avoid conducting obviously risky experiments; suppressing "inconvenient" scientific data or simply lying about food and environmental safety; slanting scientific data and exploiting consumer confusion to justify political and economic protectionism; and "force-feeding" GM products and GMOs for profit when there is no clear advantage to the wine consumer (Pretorius, 2000). Researchers must also ensure that wine's most attractive talisman, its diversity of style, must not be threatened by made-to-order grape varieties and microbial starter strains; instead they should look at biotechnology to expand the diversity that makes wine so interesting. Similarly, researchers must use DNA technology judiciously, systematically, and with high regard for wine's unique personality if producers and consumers are to benefit (Pretorius, 2003, 2004a; Pretorius and Høj, 2005).

SETTING NEW GOALS FOR THE DESIGN OF SUPERIOR GRAPEVINES, WINE YEAST, AND MALOLACTIC BACTERIA—CHALLENGES, OPPORTUNITIES, AND POTENTIAL BENEFITS

Main Species of Grapevine, Yeast, and Bacteria Involved in Wine Production

Three basic organisms produce wine: grapevines, wine yeast, and malolactic bacteria. From the vine comes the juice of the grapes and the must, which are the raw materials and the principal contributors to the flavors and aromas of the bottled product. The yeast brings about the alcoholic fermentation during which grape sugars (mainly glucose and fructose) are converted into ethanol, carbon dioxide, and other minor, but important, metabolites. The final stage in winemaking after alcoholic fermentation, although it is not restricted to this stage, is malolactic fermentation. In this process, malolactic bacteria convert L-malic acid to the "softer" L-lactic acid, thus playing an important role in removing acidity, microbial stability, and flavor modification of red wines and some white wines.

Grapevines are members of the genus *Vitis* (*V.*), consisting of two sub-genera, *Euvitis* and *Muscadinia* (Jackson, 1994; Vivier and Pretorius, 2000, 2002; Pretorius, 2004b; Pretorius and Høj, 2005). The preferred species for wine grapes, *Vitis vinifera*, originated in Europe and consists of about 5000 cultivars (Jackson, 1994). However, the industry uses a limited number of these, including such well-known varieties as Chardonnay, Sauvignon blanc, Riesling, Cabernet sauvignon, Merlot, Shiraz, and Pinot Noir, for producing popular styles with established varietal names.

Across the centuries, the yeasts that occur naturally on grape skins have been responsible for the conversion of grape juice into wine. In this natural and spontaneous fermentation, there is an ordered growth pattern of indigenous yeasts (*Kloeckera/Hanseniaspora, Candida, Metschnikowia, Pichia, etc.*) with the alcohol-tolerant species of the genus *Saccharomyces* (*S.*) invariably dominating the final stages (Fleet and Heard, 1993; Pretorius et al., 1999). Although other members of the *Saccharomyces sensu stricto* group (e.g., *S. bayanus* and *S. paradoxus*) are sometimes used, *S. cerevisiae* is widely preferred for triggering fermentation and is known as the "wine yeast" (Henschke, 1997; Pretorius, 2000). The selection of starter strains of *S. cerevisiae* on the basis of their fermentation performance has led to significant improvements in the control of fermentation and therefore the quality of wine. It is now common practice to achieve a predetermined style of wine by inoculating grape juice with a specifically selected active dried starter strain of *S. cerevisiae*.

The bacteria associated with malolactic fermentation belong to the family of lactic acid bacteria. They are encompassed in four genera, *Lactobacillus (Lb.)*, *Leuconostoc (Lc.)*, *Oenococcus (O.)*, and *Pediococcus (P.)* from within the larger group of lactic acid bacteria. Those most commonly associated with wine are *Lb. brevis*, *Lb. plantarum*, *Lb. hilgardii*, *Lc. mesenteroides*, *O. oeni*, *P. damnosus*, and *P. pentosaceus* (Wibowo et al., 1985; Axelsson, 1993). *O. oeni* (formerly known as *Leuconostoc oenos*) is particularly well adapted to the harsh environment within wine (low pH, high ethanol content, low nutrients) and is the preferred species for initiating malolactic fermentation.

Genetic Features of Vine, Yeast, and Bacteria

Knowledge of the genetic characteristics of the three basic organisms needed to produce wine varies widely, with comparatively little known about malolactic bacteria, a great deal about yeast and much work to be done further to understand *V. vinifera*.

International groups of researchers are working on several initiatives with *V. vinifera* to establish its molecular markers and sequence the entire genome to make the species more amenable to genetic study and direction. When the full sequence is available, it is likely to open up several new possibilities for grapevine development. What is currently known is that the genome of *V. vinifera* is relatively large and complex. Only 4% of the 483 megabase (mb) genome (distributed along 38–40 chromosomes) is transcribed (Vivier and Pretorius, 2000, 2002; Pretorius and Høj, 2005).

Much more is understood about the genetic characteristics of yeast. It is known that the yeast strains are predominantly diploid or aneuploid, and occasionally polyploid; *S. cerevisiae* has a relatively small genome, a large number of chromosomes, little repetitive DNA, and few introns. Haploid strains contain 12–13 mb of nuclear DNA, distributed along 16 linear chromosomes (Snow, 1983; Pretorius, 1999, 2000), with each chromosome being a single DNA molecule approximately 200–2200 kilobases (kb) long. The genome of a laboratory strain of *S. cerevisiae* was sequenced as long ago, in research terms, as 1996, and the genomes of *S. bayanus*, *S. paradoxus*, and *S. mikatae* have all been sequenced recently (Goffeau et al., 1996; Oliver, 1996; Kellis et al., 2003; Rokas et al., 2003). These three are species that are evolutionarily close to *S. cerevisiae* and the first two are sometimes used by winemakers. There has also been comparative gene analysis that has shown that the genome of *S. cerevisiae* contains 5538 genes encoding peptides/proteins consisting of $\geq$100 amino acids, while 5, 8, and 19 genes were found to be unique to *S. paradoxus*, *S. mikatae*, and *S. bayanus*, respectively (Kellis et al., 2003; Rokas et al., 2003). R&D consortia are relishing the prospect of identifying key genes and regulatory elements in wine yeast, as well as the functional analysis of its genome, transcriptome, proteome, fluxome, interactome, and metabolome, as a result of comparative genome sequencing and the availability of yeast deletion libraries (different mating types and ploidy that possess a set of *S. cerevisiae* strains, each with a different single gene deleted).

While yeast is well explored, knowledge of the genetics of malolactic bacteria is incomplete. The *O. oeni* genome, consisting of one "chromosome", has almost been sequenced in its entirety and that information will greatly expand the limited number of genes that have been characterized to date. This in itself will allow improved targeted development of malolactic strains of *O. oeni* (Ze-Ze et al., 1998, 2000).

Techniques to Improve the Grapevine, Wine Yeast, and Malolactic Bacteria

Tissue culture and transformation biotechnology have widened the potential of grapevine development programmes largely replacing the recently preferred methods of clonal selection and hybridization. To take a step further back in the long tradition of growing grapes and making wine, improvements to rootstocks and cultivar grafts were achieved when grapegrowers selected natural mutations they believed would enhance cultivation, some aspect of the fruit, or wine quality. These classical techniques were replaced in the main by more targeted clonal selection schemes but, while both clonal selection and classical breeding have had significant impacts on improving rootstock varieties through such areas as increasing resistance to soil-borne pests and pathogens, only a few of the scion varieties have become commercially successful. However, in 1989, after tissue culture systems and embryonic cell lines became available that could be regenerated and withstand *Agrobacterium* and biolistic transformation and subsequent selection regimes (Icco et al., 2001; Vivier and Pretorius, 2000, 2002; Pretorius and Høj, 2005), a grapevine was successfully transformed with naked DNA, leading to the breakthrough that enabled a great leap forward in grapevine development.

As far as wine yeast is concerned, several genetic methods are available for its improvement. Some alter limited regions of the genome, while others can be used to recombine or rearrange the entire genome. Among the available methods are clonal selection of variants, mutation and selection, hybridization, rare-mating, and spheroplast fusion (Pretorius and van der Westhuizen, 1991; Barre et al., 1993; Querol and Ramón, 1996; Grossmann and Pretorius, 1999). These genetic methods have been used successfully to develop new strains but lack the precision and capability of gene cloning (Masneuf et al., 1998; Eglinton et al., 2003). This limitation was lifted by the first successful transformation of *S. cerevisiae* in 1978, which led to a wide variety of integrating, autonomously replicating, and other specialized vectors being developed by which specific genes can be introduced, deleted, or altered in yeast spheroplasts or intact cells (Pretorius, 2003) in transformation methods that are based on chemical, electrical, or biolistic procedures. In addition, the combined use of tetrad analysis (micromanipulation), replicaplating, mutagenesis, hybridization, and recombinant DNA methods have expanded the genetic diversity that can be introduced into yeast cells (Pretorius et al., 2003).

While the grapevine and yeast forge are ahead in this area, there is still no effective method to transform *O. oeni* with cloned genes despite the work that has been done with other lactic acid bacteria, particularly from the dairy industry (de Barros Lopes et al., 2005; Pretorius et al., 2005; Pretorius and Høj, 2005). There has been a substantial body of work on the genetics of *Lactobacillus* species and, to a lesser extent, of *Pediococcus* species, and numerous vectors have been constructed to introduce modified genes into *Lactobacillus* species. These are designed to enable easy movement between Gram-positive and Gram-negative bacteria, utilizing dual replicons (origin of replication) and antibiotic markers. Features sensitive to temperature are also available on a small number of vectors while some also function in *Leuconostoc*, which is the closest genus to *Oenococcus*, and many might be suitable for *O. oeni*.

There is a further approach to improving industrially important *O. oeni* strains, which is a new technique called genome shuffling (Stemmer, 1994; Zhang et al., 2002). It involves using a classical method to improve a strain to generate populations with subtle improvements that are shuffled by recursive pool-wise protoplast fusions. It remains to be seen whether genome-shuffling methods can enable improvement of *O. oeni* starter strains for malolactic fermentation but they have been successful in improving acid tolerance in a poorly characterized industrial *Lactobacillus* strain, and they appear to be useful for the rapid development of tolerance and other complex phenotypes in industrial organisms (Patnaik et al., 2002).

Targets for Improvement of the Three Organisms

Disease-resistant and stress-tolerant varieties that will facilitate improvements in productivity, efficiency, sustainability, and environmental friendliness, especially in relation to improved pest and disease control, water-use efficiency, and grape quality, are being focused on by grape biotechnologists researching *V. vinifera* (Pretorius, 2004b; Pretorius and Høj, 2005). In wine biotechnology, the search is on to develop strains of *S. cerevisiae* and *O. oeni* that provide winemakers with improvements in fermentation, processing, and biopreservation as well as the power to enhance the wholesomeness and sensory wonders of wine (Pretorius and Høj, 2005).

Improving Grapevine Cultivation and Crop Yield

Research to improve cultivation and crop yield aspires to greater understanding of how plants tolerate stress, how they cope with adverse climatic conditions, such as colder weather, and key aspects of their growth and development (Vivier and Pretorius, 2000, 2002; de Barros Lopes et al., 2005; Pretorius, 2004b; Pretorius et al., 2005; Pretorius and Høj, 2005). Current research into the development of transgenic vines that offer improved cultivation and higher yields is looking at processes such as carbon-partitioning, how sugar is translocated, water transport, and the role of aquaporins, and the regulation of these processes. Some important limitations to cultivation, such as drought and salt stress, photo-damage, and freezing tolerance (Tsvetkov et al., 2000) are also the focus in current research.

An example of this already reported is a transgenic grapevine that expresses antifreeze genes developed from Antarctic fish to be used as a mechanism to provide tolerance to cold weather (Tsvetkov et al., 2000). Plant stress responses are intricate and involve pathways of interacting proteins driven by a range of signals attenuated and augmented by equally intricate processes. However, it is no simple matter to manipulate this biological interaction with single or even multiple gene additions and it might be easier to reach the desired outcomes with knowledge of the control mechanisms and alterations. Two examples that have been studied in grapevines are the accumulation of proline and polyamines; the regulation of key genes in these biosynthetic pathways is unraveling their involvement in abiotic stress (Paschalidis et al., 2001; Van Heeswijck et al., 2001). In addition, it has been shown that several stress-related genes are activated during the grape ripening (Davis and Robinson, 1996; Pratelli et al., 2002).

Because one of the main challenges for grapegrowers is managing the quantity and quality of their crop to meet market demand while coping with seasonal and environmental variations, research into crop production has become important (Vivier and Pretorius, 2000, 2002; Pretorius, 2004b). Recently, a study of how related genes in grapevines are organized and interact was made possible by the identification of the genes involved in the flowering and fruitfulness in *Arabidopsis thaliana*. An example that can be understood from this is how the chimera that results in Pinot Meunier can be separated into a conventional Pinot and a mutant Pinot form in which fruitfulness is spectacularly increased even in juvenile plants. This dramatic increase is the result of a single DNA base change in a single gene affecting vegetative and floral development of the vine (Boss and Thomas, 2002).

Improving Disease and Pest Resistance in Vines

The entire industry of grapegrowing and winemaking depends on the ability to deliver an affordable and consistent product to consumers. This means, for an industry that depends almost entirely on a single grapevine species, chemical pesticides and fertilizers must be used to guarantee the crop. Because of this basic requirement, the potential to glean agricultural, economic, and environmental benefits from science has impelled research to investigate the ability of plants to become accustomed to biotic and abiotic stresses. The disease pathways in plants are being mapped and much emphasis is currently being placed on describing and understanding the trigger systems of defense and the subsequent signal transduction processes leading to the various forms of defense (Bezier et al., 2002).

Grapevines are no different from other plants in that they use both passive and active defense mechanisms against invading pathogens. Passive defense mechanisms already formed on the plant include structural barriers such as a waxy cuticle and strategically positioned antimicrobial compounds to prevent tissue colonization. Active defenses the plant can generate include hypersensitivity responses (HR), such as cell and tissue death at the site of infection to limit the spread of an infection, which are characteristic features of incompatible plant–pathogen interactions associated with disease resistance (Colova-Tsolova et al., 2001). Hypersensitivity responses are often the trigger for an array of non-specific, defense-related processes throughout the plant that are collectively known as systemic acquired resistance. There are 11 classes of pathogenesis-related (PR) proteins included among the products of these processes (Tattersall et al., 2001). These include PR-1 (antifungal), PR-2 (acidic and basic β-1,3-glucanases), PR-3 (chitinase), PR-4 (antifungal), PR-5 (thaumatin-like protein), and PR-8 (acidic and basic class III chitinases), as well as defensins, cyclophilin-like proteins, ribosome-inactivating proteins, etc., which are capable of providing resistance to a wide range of pathogens for several days. Whether a specific pathogen attack will lead to disease in the plant or trigger one of its defense response mechanisms is determined by the recognition process at the initial contact between host and pathogen, the genetic basis of which is known as the gene-for-gene hypothesis. It is also known that single genes can give plants resistance to disease, which means the transfer of individual genes into plant genomes is probably the best approach to control insect, fungal, bacterial, and viral pathogens.

The natural interactions between host plant and pathogen are the focus of practically all research approaches aimed at enhancing disease tolerance (Colova-Tsolova et al., 2001) and for grapevines, there are currently two major routes of approach. The first is through the introduction of a gene product with known antipathogen activity, at high copies or in an inducible manner, into the host to optimize the plant's innate defense responses. The other approach relies on pathogen-derived resistance (PDR) by which a pathogen-derived gene is expressed inappropriately in time, form, or amount during the infection cycle, preventing the pathogen from maintaining the infection.

PDR is the preferred approach to combat all 49 viruses currently known to infect grapevines. The three most important complexes of these are degeneration and decline (caused by members of the *Comoviridae)*, rugose wood (caused by the trichoviruses), and leafroll (caused by members of the *Closteroviridae)*. Transgenic vines have been generated that provide viral resistance; in most cases, expressing the virus coat protein achieved PDR (Colova-Tsolova et al., 2001). Subsequently, resistance to infection was achieved by expressing anti-sense to virus movement proteins and replicase proteins (Martinelli et al., 2002).

Bacterial pathogens cause the three most common diseases in grapevines, which are Pierce's disease (*Xylella fastidiosa*), crown gall (*Agrobacterium vitis*), and bacterial blight (*Xanthomonas ampelina*). The first of these, Pierce's disease, is sweeping through the Californian wine industry with such devastation that grapegrowers and winemakers around the world are being reminded of the 19th century phylloxera outbreak. The *X. fastidiosa* pathogen is transmitted by the glassy-winged sharpshooter leafhopper *Homalodisca coagulata*. The pathogen kills the plant because it invades the grapevine's xylem leading to the release of gums and the production of tyloses in the vessels, both of which disrupt water flow (Jackson, 1994) placing the vine under stress. *X. fastidiosa* also produces several phytotoxins that disrupt cellular function. The humble silkworm has provided science with one strategy to combat Pierce's disease because resistant *V. vinifera* varieties can be developed by engineering the cecropin-encoding gene from the silkworm into the grapevine. Cecropin disrupts the cell wall and cell membrane of bacteria, killing pathogens such as *X. fastidiosa*. Among other approaches to combat bacterial diseases are expression of the lytic peptide, Shiva-1, and induced expression of several bacterial defense genes that have been detected (Robert et al., 2001, 2002). In general, however, the search for transgenic plants capable of resisting pathogenic bacteria has been less successful than the battle against viral and fungal pathogens.

The disease range of fungal pathogens is wide and encompasses fruit, leaf, wood, and root damage, but the most common and best-known are powdery mildew (*Uncinula necator*), downy mildew (*Plasmopara viticola*), grey rot (alias botrytis bunch rot, *Botrytis cinerea*), and dieback (*Eutypa lata*). There have been a number of studies on fungal infections and their effects on grapes, including examining the expression of grapevine chitinases, β-1,3-glucanases, thaumatin-like proteins, stilbene synthetic enzymes (resveratrol production), and polygalacturonase-inhibiting proteins (PGIPs). Antifungal activity has been found in several of these enzymes and inhibiting proteins, and it has been demonstrated that in some cases, transgenic expression of genes encoding the proteins leads to increased resistance (Giannakis et al., 1998; Salzman et al., 1998; Coutos-Thevenot et al., 2001; Kikkert et al., 2001; Shouten et al., 2002). One example is how the existence of elicitors of general plant defense responses, such as long-chain oligogalacturonides, can be prolonged by PGIPs, which are cell wall bound proteins that inhibit the action of cell wall macerating polygalacturonases produced by fungal pathogens. Expression of PGIPs is regulated by the stage of the vine in its season, peaking at véraison, and it can be induced in a tissue-independent manner by wounding, osmotic stress, indole acetic acid, salicylic acid, and *B. cinerea* infection. Ways to shield grapevines from phytopathogens that rely on polygalacturonase-induced cell wall damage for penetration into the plant are being sought through studies seeking to enhance the expression levels of the grapevine (*Vvpgip1*) and heterologous PGIP-encoding genes. Research has already shown that transgenic expression of a detoxifying enzyme enhances resistance to *E. lata* (Legrand et al., 2003), while a gene from the American wild grape species *Muscadinia rotundifolia* has been introgressed into a succession of pseudo backcrosses of *V. vinifera*. Using this strategy, resistance to *Oidium tuckerii* has continued to segregate as a single gene. That gene has now emerged as an important candidate for the development of a transgenic method to make *V. vinifera* resistant to powdery mildew.

Field-testing is already under way on some transgenic grapevine prototypes expressing antipathogenic genes, but the technology will certainly develop further making multiple gene transfers and long-term and stable expression of antipathogenic genes along with the use of highly specific inducible regulatory sequences more likely. This work is clearing the way for research to develop grapevine varieties that are resistant to disease and pests—with the additional, ultimate benefit of a decrease in the use of fungicides and pesticides. This promises a new horizon of tantalizing prospects that include reduced input costs for grapegrowers and therefore cheaper prices for consumers, reduced exposure of vineyard employees and the environment to agricultural chemicals, and the elimination of any chemical residues from wine. The last prospect, however, is already a minimal issue in Australia where residue is not only very low but well below the limits set by government. There is already an example of what can be achieved from such an approach in a non-food crop; insect-resistant cotton has drastically reduced pesticide use enabling an increase in populations of beneficial insects (Arvanitoyannis, 2003).

Improving Grape Quality

Researchers are studying the basic process of how grape berries ripen and what signals they release to indicate their progress through the sequence because the indicators of sugars, acids, and phenolics are the measures the winemaker uses to achieve the premium quality that relies on small grapes of specific color and ripeness. Because the grapevine is a fruit that does not have a decisive period, it must be studied for the biochemical, hormonal, and environmental signals that influence ripening processes, which include pigment production, sugar accumulation, and the formation of components that will give the wine its aroma. Among these, the process for the accumulation of sugar—glucose/fructose—is being revealed through the identification and analysis of grape berry genes encoding invertases and sugar transporters (Fillion et al., 1999), while the process of color development—expression of flavonoid biosynthetic genes—is known to coincide with the accumulation of hexose in the berries (Sparvoli et al., 1994; Boss et al., 1996a; Ford et al., 1998). Work to compare the gene expression of red and white grapes has brought to light the importance of UDP glucose flavonoid-3-glucosyl transferase (UFGT) in the control of berry color, and recent results indicate that Myb genes are involved in regulating UFGT (Boss et al., 1996b; Kobayashi et al., 2001, 2002). Some of the genes that influence berry softening have also been isolated (Davis and Robinson, 2000; Barnavon et al., 2001; Nunan et al., 2001). Two of the natural hormones (auxin and abscisic acid) that regulate growth and physiology are known to affect the expression of genes involved in ripening and are thought to influence the ripening process (Davis et al., 1997). The aim of these approaches is to meet the quality parameters of the grapes being required by winemakers through the formation of the most desirable metabolites by changing the metabolic flux through the important biosynthetic pathways active in the ripening berry.

These are the goals of grape quality research by biotechnologists but a lot more knowledge is needed of the underpinning processes and more improvements in transformation technology need to be discovered. Technologies that make targeted gene insertion and deletion possible will make these and other innovative prospects more feasible.

Improving Fermentation Performance

One of the added costs that affects the affordability of the end product results from the major problems that happen during fermentation, which are known to winemakers as "runaway," "sluggish," "stuck," and "incomplete" (Pretorius, 2000, 2001; de Barros Lopes et al., 2005; Pretorius et al., 2005; Pretorius and Høj, 2005). Alcoholic fermentations that happen too quickly result in reduced fermentor space, because of foaming, and loss of the desired aroma compounds, because of entrainment with evolving carbon dioxide. On the other hand, inefficient utilization of fermentor space and wine spoilage caused by the low rate of protective carbon dioxide evolution and high residual sugar content are often blamed for financial losses that are really due to alcoholic fermentations that cease prematurely or proceed too slowly (Pretorius, 2000). Winemakers are also exasperated by malolactic fermentations that are sluggish or slow to happen because they also contribute to spoilage, either as a result of prolonged exposure to oxygen or by permitting the access of other microorganisms (de Barros Lopes et al., 2005; Pretorius et al., 2005; Pretorius and Høj, 2005).

Predictability of fermentation and wine quality depend directly on two major factors (Henschke, 1997;

de Barros Lopes et al., 2005). The first of these is the set of wine yeast attributes that assist in the rapid establishment of numerical and metabolic dominance in the early phase of alcoholic fermentation and that determine the evenness and efficiency of the fermentation with the desired concentrations of ethanol and residual sugar. The second is the set of malolactic bacterial attributes that support the rapid conversion of malic acid to lactic acid.

While there are many factors that affect how yeasts and malolactic bacteria perform, a series of general targets for improvement have been identified. These include increased resilience and resistance to stress by active dried starter cultures; improvements to the way nutrients are taken up and assimilated; enhancements to their ability to resist ethanol and other inhibitory metabolites, proteins, and peptides; their resistance to sulfite and other antimicrobial compounds; and their tolerance to other environmental stress factors (Pretorius, 2000, 2003; Pretorius et al., 2003; Pretorius, 2004a; de Barros Lopes et al., 2005; Pretorius et al., 2005; Pretorius and Høj, 2005).

During winemaking, wine yeast and malolactic bacteria enter a very stressful arena where they must overcome harsh environmental conditions to enable the winemaker to produce a product of high quality (Ivorra et al., 1999). As living organisms, they have the means to respond and their general environmental stress reaction contributes to a wide range of cellular functions. These functions include cell wall modification, fatty acid metabolism, protein chaperone expression, DNA repair, detoxification of reactive oxygen species, and energy generation and storage (Gasch et al., 2000). In addition to these general responses, microbial starter cultures need specific abilities to combat the discrete stresses they undergo during winemaking.

During the production of active wine yeast, the material undergoes a desiccation process that challenges it to a combination of stresses that, although not clearly defined, appear to have similarities to freezing and osmotic stress (de Barros Lopes et al., 2005). There is a strong incentive to develop wine yeast strains with greater ability to accumulate beneficial compounds, such as the proteins (e.g., aquaporins) and metabolites (e.g., sterols, trehalose, glycogen), that increase the survival of *S. cerevisiae* cells exposed to physical or chemical stresses and that have been implicated in general stress tolerance, resilience, fitness, and vigor of reactivated dried wine yeast starter cultures.

It has been shown (Kim et al., 1996; Shima et al., 1999) that overexpression of the genes encoding trehalose-6-phosphate synthase (*TPS1*) and trehalose -6-phosphate phosphatase (*TPS2*), together with the blocking of the trehalose hydrolysis pathway in yeast, thereby increasing cellular trehalose concentrations by deleting the *NTH1* or *ATH1* (neutral and acidic trehalase) genes, increases freezing and dehydration tolerance in a laboratory and commercial baker's yeast. The build-up of a second carbohydrate storage molecule, glycogen, is thought to provide dried yeast with a ready-made source of energy when it is reactivated. It has also been reported that a wine yeast strain overexpressing the glycogen synthase genes (*GSY1* and *GSY2*) accumulates glycogen and has enhanced viability under glucose limitation conditions (Perez-Torrado et al., 2002) and that overexpression of the *S. cerevisiae AQY1* and *AQY2* (aquaporin) genes confers freeze tolerance on both laboratory and commercial baking strains (Tanghe et al., 2002).

However, while it has not yet been shown that these changes will prove to be an advantage to active dried yeast, it has been shown that proline and charged amino acids endow tolerance to desiccation and freezing. Two research results provide evidence for this—deletion mutations in the *PUT1* gene, encoding proline oxidase, and a dominant mutation in the proline biosynthetic *PRO1* gene (γ-glutamyl kinase) increase intracellular proline and show higher desiccation, freeze, and/or osmotolerance (Takagi et al., 2000; Morita et al., 2003). Secondly, a deletion of the arginase gene (*CAR1*) of *S. cerevisiae* prevents the degradation of arginine, increases intracellular arginine and glutamic acid concentrations, and provides freeze tolerance (Shima et al., 1999).

Work has also been carried out on how malolactic bacteria respond to environmental pressures in winemaking and it has been shown that they respond to stresses such as high alcohol, acid, and sulfur dioxide (SO_2) concentrations by producing heat shock and stress proteins (*hsp18*, *clpX*, and *trxA*) (Guzzo et al., 1997). The protease (encoded by *fstH*) associated with the ATPase complex has also been implicated in stress tolerance associated with tannins and fatty acids (Bourdineaud et al., 2003). It is possible that modifying the expression levels of *hsp18*, *clpX*, *trxA*, and *fstH* might result in improving the tolerance of *O. oeni* to wines where the conditions are close to upper limits. Additionally, elevated expression of these heat and stress proteins that better prepares the bacterial cells for the harsh wine environment might aid commercial production of *O. oeni* for direct inoculation into wine.

Wine yeast and malolactic bacteria also have to cope with the stress conditions during fermentation as well as those associated with the desiccation process during their preparation as active starter cultures (Bauer and Pretorius, 2000; Erasmus et al., 2003; Trabalzini et al., 2003). Among these fermentation stresses is the high concentration of sugar in grape juice, which imposes an osmotic stress on yeast. Glycerol is the main osmolyte in yeast and, in response to an osmotic stress, a MAPK (mitogen-activated protein kinase) cascade is activated and the two key genes in glycerol biosynthesis—*GPD1* (glycerol-3-phosphate dehydrogenase) and *GPD2* (glycerol-3-phosphatase)—are rapidly and transiently induced more than 50-fold (Rep et al., 2000). Mutations in these genes result in osmosensitivity, while mutations that increase their expression enhance osmotolerance (Hohmann, 2002). Furthermore, it has been shown that commercial wine yeast strains overexpressing either *GPD1* or *GPD2* have a slight growth advantage at the beginning of fermentation (Michnick et al., 1997; de Barros Lopes et al., 2000).

Winemakers and researchers know that the main cause of most sluggish and stuck fermentations is a high concentration of ethanol, and that this can be augmented by several inherent and environmental factors. The physiological basis of ethanol toxicity is complex and not fully understood (Alexandre et al., 2001), but cell membranes, in which ethanol increases fluidity and permeability of the membrane and impairs the transport of sugars and amino acids, appear to be a major target of the toxicity. In yeast, the modification of the *SUT1*, *SUT2*, *PMA1*, and *PMA2* genes results in increased sterol accumulation and ATPase activity, respectively, thereby increasing resistance to ethanol (Henschke, 1997; Walker, 1998). Combined expression of the *S. cerevisiae ERG1* (squalene epoxidase) and *ERG11* (14α-demethylase) genes also leads to the accumulation of sterols (Veen et al., 2003).

The presence of wild yeast can also contribute to sluggish or stuck fermentation because it might have zymocidal capabilities that inhibit the growth of the inoculated yeast strain (Perez et al., 2001); however, there is rapid progress in understanding cultured wine yeasts' sensitivity to killer toxins (zymocins), which is encouraging the development of broad-spectrum zymocidal resistance in yeast (Lowes et al., 2000; Breinig et al., 2002; Fichtner et al., 2002; Page et al., 2003). Similarly, feral bacteria can produce bactericidal compounds that might inhibit malolactic bacteria. Resistance of *O. oeni* against these and other toxic metabolites could be improved by introducing "immunity" genes into selected starter strains.

Other challenges inhibiting the fermentation performance of wine yeasts can emanate from chemical preservatives (e.g., sulfite) and agrochemicals containing heavy metals (e.g., copper), and therefore another way to improve fermentation is to increase their resistance to these by, for instance, an increase in the copy number of the *CUP1* copper chelatin gene that enables the yeast to tolerate higher levels of copper residues in the grape must (Fogel et al., 1983; Henderson et al., 1985).

A further factor that affects the efficacy of microbial starter cultures in fermentation is the inefficient use of nutrients. An ideal example of this is found by looking at how sugar uptake and assimilation by wine yeast cells appears to limit complete sugar utilization during vinification and how it is strongly influenced by conditions such as ethanol concentration and the availability of nitrogen. Firstly, the efficient use of glucose and fructose in the grape by *S. cerevisiae* and a rapid rate of glycolytic flux relies on the presence of functional alleles of genes encoding the key glycolytic enzymes, hexokinase, glucokinase, phosphoglucose isomerase, phosphofructokinase aldolase, triosephosphate isomerase, glyceraldehyde-3-phosphate dehydrogenase, phosphoglycerate kinase, phosphoglycerate mutase, enolase, pyruvate kinase, pyruvate decarboxylase, and alcohol dehydrogenase (Pretorius, 2000). Secondly, it has been shown that the low affinity hexose transporter, Hxt3p, and the high affinity transporters, Hxt6p and Hxt7p, play particularly important roles in fermentation (Luyten et al., 2002). From this it can be surmised that, as wine yeast strains are glucophilic, it is possible that increased expression of these *S. cerevisiae* transporters together with fructose-specific transporters from fructophilic yeasts (e.g., *S. pasteurianus* and *Zygosaccharomyces bailii*) will decrease the occurrence of those "stuck" fermentations that result in excessive residual fructose levels. Furthermore, two hexose transporter homologues, Snf3p and Rgt2p, are required for glucose sensing. Snf3p is specifically required for inducing a number of *HXT* genes under low glucose conditions. Dominant *SNF3* mutants constitutively express hexose transporters and are resistant to translational inhibition upon glucose withdrawal, which provides a potential mechanism for improving yeast (Ozcan et al., 1996; Ashe et al., 2000). It can be seen in these "stuck" fermentations that the preference of wine yeasts for glucose rather than fructose can lead to excessive residual fructose levels that compromise the quality of the wine.

Sugars aside, the nutrient believed to be the main cause of failed fermentation is nitrogen, because lack of it can cause fermentation to stop as well as create off-flavors such as hydrogen sulfide. It has been shown that a strain of yeast carrying a mutation of the *URE2* gene, which represses expression of genes involved in proline transport and metabolism, increases the availability of proline and arginine as nitrogen sources (Salmon and Barre, 1998; Martin et al., 2003). However, *ure2*Δ strains possess pleotropic phenotypes. To improve nitrogen presence during fermentation, specifically targeting the proline permease and utilization proteins (the *PUT1*-encoded proline oxidase and *PUT2*-encoded pyrroline-5-carboxylate dehydrogenase) might prove to be a more direct method (Poole, 2002). There is a clear need for further study of wine yeast growing under differing nitrogen conditions if other genes that can improve fermentation under limiting nitrogen conditions are to be identified (Backhus et al., 2001; Marks et al., 2003). Because it is known that autolysis of the yeasts themselves is an alternative source of nitrogen in vinification, it is possible that a subset of yeast could be targeted to lyse towards the end of fermentation by adjusting the regulation of the cell integrity pathway (Martinez-Rodriguez and Polo, 2000; Lagorce et al., 2003).

Most of these strategies are concerned with improving the performance of wine yeast during fermentation. However, the malate catabolism in *O. oeni* could also be targeted to improve the efficiency of malolactic fermentation. Three genes for malate metabolism have been cloned from *O. oeni,* and identified as *mleR* (regulator), *mleA* (enzyme), and *mle* P (permease) (Volschenk et al., 1997). If biotechnologists are to increase the conversion efficiency of L-malate to L-lactate, they will need to understand the regulation of *mleA* by *mleR* and determine the rate-limiting step for malate metabolism, which is especially important at low cell density. Another mechanism that might also ultimately enhance the catabolism of L-malate is the rapid adaptation of the *O. oeni* cell to its harsh wine environment.

Improving the Biocontrol of Indigenous and Wine Spoilage Microorganisms

The shift in consumer preference to more natural products has had ramifications in the usage of chemical additives by the wine industry. While healthy grapes, good cellar hygiene, and sound oenological practices are the winemaker's battlements against the uncontrolled proliferation of spoilage microbes, uncontrolled microbial growth before, during, or after wine fermentation can alter the chemical composition of the product, thereby detracting from its sensory properties (du Toit and Pretorius, 2000). To provide an added line of defense, approved chemical preservatives such as sulfur dioxide, dimethyl dicarbonate, benzoic acid, fumaric acid, and sorbic acid are introduced to control the growth of unwanted microbial contaminants (Pretorius, 2000). However, excessive use of these preservatives harms drinking quality and is meeting resistance among consumers, whose preferences have shifted to products less heavily preserved with chemicals, less processed, of higher quality, more natural and healthier. In response to this signal from the marketplace, starter strains of *S. cerevisiae* and *O. oeni* that express antimicrobial enzymes or peptides are being developed. The efforts focus on the antimicrobial metabolites (e.g., sulfite, ethanol, acetic acid, medium chain fatty acids, etc.), enzymes (lysozyme, chitinases, and glucanases), and peptides (zymocins and bacteriocins) (Pretorius, 2003, 2004a).

A wine yeast that produces increased SO_2 concentrations could be useful for suppressing the growth of indigenous microbes and provide a means of reducing SO_2 as an antimicrobial and antioxidation agent in wine. Increased SO_2 production has been achieved by decreasing Met10p (sulfite reductase) and Met2p (serine acetyl transferase) activity or increasing Met14p (adenosine 5'phosphosulfates kinase) activity, a process that has been shown to provide flavor stability to beer (Hansen and Kielland-Brandt, 1996a, b; Donalies and Stahl, 2002).

Expressing effective antimicrobial enzymes and peptides in *S. cerevisiae* and *O. oeni* starter strains could alleviate the problem of expense in the use of purified antimicrobial enzymes and bacteriocins. To achieve this, the lysozyme gene of egg white from hens (*HEL1*), the *Pediococcus acidilactici* pediocin gene (*PED1*) and the *Leuconostoc carnosum* leucocin gene (*LCA1*) have been used to engineer bactericidal yeasts (Schoeman et al., 1999; Ibrahim et al., 2001; Van Reenen et al., 2003). The antifungal yeasts, *CTS1*-encoded chitinase and *EXG1*-encoded exoglucanase, have also been expressed in *S. cerevisiae* (Carstens et al., 2003). The main approach in the assembly of zymocidal strains involves the inclusion of a combination of mycoviral killer toxin (zymocin) determinants of *S. cerevisiae* (e.g., a K_1/K_2 double killer) and zymocin-encoding genes from other yeasts (e.g.,

Hanseniaspora, Kluyveromyces, Pichia, Willopsis, Zygosaccharomyces, etc.) into wine yeasts. Although *S. cerevisiae*'s zymocins can only kill in a narrow range, the zymocins of *Pichia anomala, Willopsis saturnus,* and *Kluyveromyces lactis* have been shown to be active on many species common in grape juice, including spoilage yeasts such as *Dekkera bruxellensis* (Yap et al., 2000). The zymocin zygocin, produced by *Z. bailii,* a yeast tolerant to salt, is also active against filamentous fungi. However, it has been shown that heterologous expression of zygocin in *S. cerevisiae* causes cell death (Weiler et al., 2002).

With the overriding proviso of no unforeseen negative side-effects emerging, the ideal would be to incorporate all these antimicrobial activities into selected strains of wine yeast and malolactic bacteria, thereby counteracting all contaminating spoilage bacteria (*Acetobacter, Gluconobacter, Lactobacillus, Pediococcus,* etc.), yeasts (*Brettanomyces/Dekkera, Pichia, Zygosaccharomyces,* etc.), and molds (*Aspergillus, Botrytis, Penicillium, Trichoderma,* etc.) in winemaking.

Improving Wine Processing

In the winemaking stage that follows alcoholic fermentation, the winemaker seeks to produce a product that has clarity and physiochemical stability by means of fining and clarification. But these processes often involve expensive and laborious practices and generate large volumes of lees that need to be disposed of (Pretorius, 2000). Not only does this mean loss of quantity, but the processes sometimes remove important aroma and flavor compounds, leading to loss of quality as well. To minimize these disadvantages, winemakers are adding an array of commercial enzyme preparations (e.g., protein- and polysaccharide-degrading enzymes) to grape must and wine. But these are relatively expensive, adding to the shelf cost of the end product, and yeast and malolactic bacteria that remove proteins and polysaccharides and cause cells to flocculate offer an alternative strategy.

Heat-induced protein haze in wine is caused by thaumatin-like proteins and chitinases. These are PR grape proteins (Pocock et al., 2000) that slowly denature and aggregate, resulting in light-dispersing particles being formed. Proteases have been unsuccessful in degrading these haze proteins (Pocock et al., 2003) during winemaking but several glycoproteins, including yeast invertase and two mannoproteins from *S. cerevisiae* known as haze protection factors, have been found to visibly reduce haze formation (Moine-Ledoux and Dubourdieu, 2003; Dupin et al., 2000).

The main polysaccharides responsible for turbidity, viscosity, and filter stoppages are pectins, glucans, and, to a lesser extent, hemicellulose (mainly xylans), and originate from the grape itself, the fungi on the grape, and microorganisms present during winemaking. Under winemaking conditions, the endogenous pectinase, glucanase, xylanase, and arabinofuranosidase activities of grapes, wine yeasts, and malolactic bacteria are often neither efficient nor sufficient to prevent the occurrence of polysaccharide hazes and filter stoppages (Pretorius, 1997; Van Rensburg and Pretorius, 2000). A pectinolytic wine yeast was developed by co-expressing the *Erwinia chrysanthemi* pectate lyase gene cassette (*PEL5*) and the *Erwinia carotovora* polygalacturonase gene cassette (*PEH1*) in *S. cerevisiae* (Laing and Pretorius, 1992, 1993a, b) or by overexpressing the *S. cerevisiae* polygalacturonase gene (*PGU1*) (Vilanova et al., 2000). Likewise, glucanolytic wine yeasts were developed by expressing glucanase genes in *S. cerevisiae.* These are: *Butyrivibrio fibrisolvens* endo-β-1,4-glucanase gene (*END1*); the *Bacillus subtilis* endo-β-1,3-1,4-glucanase gene (*BEG1*); the *Ruminococcus flavefaciens* cellodextrinase gene (*CEL1*); the *Phanerochaeta chrysosporium* cellobiohydrolase gene (*CBH1*); and the *Saccharomycopsis fibuligera* cellobiase gene (*BGL1*) (Pérez-González et al., 1996; Peterson et al., 1998; Van Rensburg et al., 1994–1998). Xylanolytic yeasts were developed by expressing the following gene cassettes in *S. cerevisiae:* the endo-β-xylanase genes from *Aspergillus kawachii* (*XYN1*) (Crous et al., 1995); *Aspergillus nidulans* (*xlnA* and *xlnB*) (Pérez-González et al., 1996); *Aspergillus niger* (*XYN4* and *XYN5*) (Luttig et al., 1997), and *Trichoderma reesei* (*XYN2*) (La Grange et al., 1996, 2001); the xylosidase genes from *A. niger* (*xlnB*) and *Bacillus pumilus* xylosidase (*XLO1*) (La Grange et al., 1997); and the *A. niger* α-L-arabinofuranosidase gene (*ABF2*) (Crous et al., 1996). Although the levels of secreted enzymes have yet to reach the concentrations needed to remove polysaccharides effectively, a degree of clarification has already been achieved.

Yeast cell flocculation and flotation are two other targets researchers have nominated to improve wine processing and this is in answer to the needs of winemakers producing bottle-fermented sparkling wine and flor sherry when yeast cells are expected to perform two conflicting roles. A high-suspended yeast count during production of base wine for sparkling wine and sherry ensures rapid fermentation. For sparkling wine, the yeast cells must flocculate and settle efficiently once the secondary fermentation in the bottle is complete (Henschke, 1997) to facilitate development of the typical autolysis-mediated fermentation bouquet and mannoprotein-coated bubbles that characterize this wine and to assist in an unproblematic riddling and disgorging procedure. In contrast, sherry production needs yeast flotation and flor (vellum) formation on the surface of the base wine at the end of fermentation so that the biological aging process can take place and the characteristic nutty bouquet be developed (Alexandre et al., 2000; Reynolds and Fink, 2001). It is known that processing sparkling wine and sherry can be simplified by controlling the onset of yeast flocculation or flor formation at the appropriate time (Dequin, 2001) but genetic, physiological, and biochemical factors underlining the adaptive responses of yeast cells to certain physical and chemical signals that lead to filamentation, agglomeration, flocculation, and flotation are not fully understood (Bony et al., 1998; Ishida-Fujii et al., 1998; Gagiano et al., 2002). However, it has now been demonstrated that the expression of *FLO1*, linked to the late-fermentation *HSP30* promoter, can be induced by a heat-shock treatment, and this confirms that controlled flocculation during fermentation is possible (Verstrepen et al., 2001). It is hoped this will lead to the development of wine yeast strains that flocculate only under the stress conditions (e.g., nutrient depletion and the presence of high levels of alcohol) that usually prevail at the end of fermentation. A similar strategy is being followed by expressing the *FLO11* (also known as *MUC1*) gene under the control of the HSP30 promoter to attain control of flor formation in sherry (Verstrepen et al., 2001).

Improving the Sensory Attributes of Wine

In the final product, the most important results of the winemaker's art are appearance, aroma, and flavor—the organoleptic distinction (Pretorius, 2000) and all stages through fermentation to clarification have an impact. The endless variety of wine's sensory properties is a result of the complex interactions of hundreds of metabolites with the human olfactory senses of taste and smell, which are encoded by one of the largest and apparently rapidly evolving gene families in the human genome (de Barros Lopes et al., 2005). *S. cerevisiae* accounts for the major transformations that take place when grape must is turned into wine because it modifies the color, palate, and flavor complexity of wine by helping to extract compounds from solids present in grape must, modifies molecules derived from the grape, and produces yeast metabolites (Lambrechts and Pretorius, 2000). While it is known that *O. oeni* also contributes to the sensory traits of wine during malolactic fermentation, it appears that the complexity of that transformation might be limited.

Several polysaccharide-degrading enzymes lead to the release of color and flavor compounds trapped in the grape skins as they facilitate wine clarification and increase juice yield (Van Rensburg and Pretorius, 2000). In preliminary fermentation trials, some engineered yeast prototypes displayed sensory differences from wines made with the respective parental strains.

Many compounds that potentially impact on aroma and flavor are found in grapes and wine as glycosidically bound aglycons (monoterpenes, norisoprenoids, benzene derivatives, and aliphates) (Jackson, 2003). Although β-glucosidase activity has been demonstrated in *S. cerevisiae* and *O. oeni*, both are limited in their ability to release the disaccharide glycosides (D'Incecco et al., 2004). Preliminary results, however, indicate that the overexpression of the *S. fibuligera* β-glucosidase genes (*BGL1* and *BGL2*) in *S. cerevisiae* does influence aroma profile.

Yeast can influence the concentration of terpenols by two other means: first, strains containing mutations in ergosterol biosynthesis have been shown to produce geraniol, citronelol, and linalool, and second, yeast can convert one terpenol into another (Chambon et al., 1990). Because terpenols are known to have distinct aromas and sensory thresholds, these biotransformations might have a significant effect on wine sensory properties (King and Dickinson, 2000, 2003).

Another focus of the research to improve flavor is on esters because characteristic fruity aromas are caused mainly by the acetate esters of higher alcohols and the C_4–C_{10} ethyl esters (Verstrepen et al., 2003a–c). The enhancement of specific esters that cause desirable outcomes could result from an improved understanding of the formation and further metabolism of esters [acyl alcoholtransferase (ester-synthesizing enzymes) and esterases (ester catabolizing enzymes)] in *S. cerevisiae* and *O. oeni.* Overexpression of the *S. cerevisiae* ester-synthesizing (the *ATF1*- and *ATF2*-encoded alcohol acetyl transferases and *EHT1*-encoded ethanol hexanoyl transferase) and ester-degrading (the *IAH1*- and *TIP1*-encoded esterases) enzyme activities has been shown to have a marked effect on ester formation and the flavor profile of wine (Fujii et al., 1994; Fukuda et al., 1998; Lilly et al., 2000; Yoshimoto et al., 2002; Verstrepen et al., 2003d; Lilly et al., 2005b; Swiegers and Pretorius, 2005). In a recent study, it was found that the overexpression of *ATF1* and *ATF2* increases the concentrations of ethyl acetate, isoamyl acetate, 2-phenylethyl acetate, and ethyl caproate, while the overexpression of *IAH1* resulted in a significant decrease in the concentrations of ethyl acetate, isoamyl acetate, hexyl acetate, and 2-phenylethyl acetate (Lilly et al., 2005b). This study also showed that the overexpression of *EHT1* increases in the concentrations of ethyl caproate, ethyl caprylate, and ethyl caprate, while the overexpression of *TIP1* does not decrease the concentrations of any of the esters. The estery/synthetic fruit flavor was overpowering in the wines fermented with yeast strains in which *ATF1* was overexpressed, but much more subtle in the strain overexpressing *ATF2* (Lilly et al., 2005b). Also, an intense apple aroma was detected in the wines produced by the yeast in which *EHT1* was overexpressed (Lilly et al., 2005b).

While higher alcohols can also add to wine complexity, they are regarded as having a negative influence on quality at excessive levels. Alcohols, which are important precursors to ester formation, are produced either anabolically or catabolically by the breakdown of branched-chain amino acids (Eden et al., 2001; Yoshimoto et al., 2002; Dickinson et al., 2003). Recently, the effect of increased yeast branched-chain amino acid transaminase activity on the production of higher alcohols on the flavor profiles of wine and distillates was investigated (Lilly et al., 2005b). It was found that the overexpression of *BAT1* increased the concentration of isoamyl alcohol and isoamyl acetate and, to a lesser extent, the concentration of isobutanol and isobutyric acid. The overexpression of the *BAT2* gene resulted in a substantial increase in the level of isobutanol, isobutyric acid, and propionic acid production, while the deletion of this gene led to a decrease in the production of these compounds. Sensory analyses indicated that the wines and distillates produced with the strains in which the *BAT1* and *BAT2* genes were overexpressed had more fruity characteristics (peach and apricot aromas) than the wines produced by the wild-type strains. (Lilly et al., 2005a; Swiegers and Pretorius, 2005).

A further negative effect on sensory quality results from high concentrations of ethanol, which can give wine a perceived "hotness" and suppress overall aroma and flavor. Moreover, strict anti-drink driving laws in many countries have led consumers towards a preference for the levels of lower alcohol content often seen in wine made from very ripe grapes. The flow of sugar has been diverted away from ethanol synthesis and into the glycerol pathway by overexpressing the glycerol-phosphate-dehydrogenase genes, *GPD1* and *GPD2*, in wine strains (Michnick et al., 1997; de Barros Lopes et al., 2000) but this increased acetic acid concentrations to unacceptable levels (Remize et al., 1999, 2000). It appears that acetaldehyde dehydrogenase activity encoded by the *ALD6* gene is the main contributor to the oxidation of acetaldehyde during fermentation. A laboratory strain of *S. cerevisiae* overexpressing *GPD2* and lacking *ALD6* had the desired effect of producing more glycerol and less ethanol, without the increase in acetic acid (Eglinton et al., 2002). A "metabolic snapshot" of the glycerol-overproducing ferments demonstrated how seemingly unrelated biochemical pathways could be modified by the change in a single gene (Raamsdonk et al., 2001).

The expression of the *A. niger* glucose oxidase gene (*GOX1*) into *S. cerevisiae* was a second strategy used to decrease ethanol and resulted in a 2% decrease in ethanol concentration (Malherbe et al., 2003). It is thought that the glucose in grape juice was not available for conversion to ethanol because it was converted to D-glucono-δ-lactone and gluconic acid.

Another target for improvement relates to the release of volatile thiols. Several potent thiol compounds provide the basis for the varietal aromas of Sauvignon blanc and Scheurebe wines (Darriet et al., 1995; Tominaga et al., 1998a, b, 2000), and are present in other cultivars, such as Cabernet sauvignon. The sulfur-containing volatiles are present as a cysteine conjugate in grape juice and are released by *S. cerevisiae* during fermentation. However, their release depends on the strain of yeast, which indicates that the concentrations of the thiols in wine can be regulated by modifying the yeast, either through a traditional breeding approach or through genetic engineering (Dubourdieu et al., 2000; Howell et al., 2004; Swiegers and Pretorius, 2005).

The rotten egg aroma of hydrogen sulfide (H_2S) is obviously highly undesirable in wine and is quickly detected. It has been found that modification of genes needed for sulfur metabolism leads to a significant decrease in hydrogen sulfide production (Tezuka et al., 1992; Omura et al., 1992; Spiropoulos and Bisson, 2000). In strains overexpressing *SSU*, a plasma membrane protein specifically for the efflux of sulfite from yeast (Park and Bakalinsky, 2000), the sulfite is secreted before reduction to sulfide (Donalies and Stahl, 2002). Increases in *SSU1* expression also impart SO_2 resistance to *S. cerevisiae* strains. It has been further shown that many wine yeast strains showing a natural resistance to sulfite have undergone chromosome translocation, positioning the *SSU1* gene below a stronger promoter and thus causing its increased expression (Perez-Ortin et al., 2002b).

Medicinal flavors in wine are another consumer disincentive. Free phenolic acids, such as *p*-coumaric and ferulic acids, can be metabolized into 4-vinyl and 4-ethyl derivates by various microorganisms in wine. These volatile phenols contribute to the aroma of the wine, but, if they exceed certain threshold values, they can impart an unwelcome medicinal flavor. Although *S. cerevisiae* possesses a phenolic acid decarboxylase, *POF1 (PAD1)*, it displays low activity. Expressing the phenolic acid decarboxylase or *p*-coumaric acid decarboxylase from *Bacillus subtilis* and *Lb. plantarum*, respectively, gave an approximate twofold increase in volatile phenol formation in a laboratory strain (Smit et al., 2003). Surprisingly, it was also found that the overexpression of the *padc* gene in wine yeasts resulted in wine with elevated levels of favorable monoterpenes. However, no volatile phenols could be detected in wine made with yeasts in which the *PAD1* gene was disrupted and further work might lead to ways through which the concentration of volatile phenols in wine can be controlled.

A further, and profound, influence on sensory quality is the acidity of grape juice and wine, mainly attributed to tartaric and malic acid. Biodeacidification is usually mediated by *O. oeni*, which decarboxylate malic acid to lactic acid and carbon dioxide during malolactic fermentation. There have been many malolactic fermentation sensory studies and much anecdotal evidence that points to changes in the texture and body of wine after malolactic fermentation, with reports indicating a fuller, richer, longer aftertaste (de Revel et al., 1999; Sauvageot and Vivier, 1997). The chemical changes contributing to these favorable mouth-feel properties are poorly understood.

Stuck or sluggish malolactic fermentation often occurs because of nutrient limitation, low temperature, acidic pH, and high alcohol, and SO_2 levels in wine. Several alternatives have been explored, including the use of malate-degrading yeasts. Malo-ethanolic fermentations conducted by *Schizosaccharomyces pombe* caused malate to be converted effectively to ethanol, but produced the side effect of off-flavors (Gallander, 1977). Unlike *S. pombe*, *S. cerevisiae* lacks an active malate transport system, and the efficiency of its malic enzyme is much lower than the *S. pombe* malic enzyme (Volschenk et al., 1997). To engineer a malo-ethanolic wine yeast, the *mae1* malate permease gene and the *mae2* malic enzyme gene from *S. pombe* were co-expressed in *S. cerevisiae*. Similarly, in an endeavor to engineer a malolactic pathway in *S. cerevisiae*, the malolactic genes (*mle*S) from *Lactococcus lactis* were co-expressed with the *S. pombe mae1* permease gene. Malo-ethanolic wine yeast would be favored for low pH wines from the cooler wine-producing regions, while the malolactic wine yeast would provide the best solution for high pH wines from warmer regions (Volschenk et al., 1997). During vinification trials, it was shown that these malo-ethanolic and malolactic wine yeasts were able to degrade all the malic acid in must within 3 days, and no off-flavor was produced. The "malolactic yeast" is the first wine yeast to be commercialized by a yeast-manufacturing company and it has been trialed in 2002/2003 in Moldavia. To the very best of my knowledge, this represents the first large-scale (20,000 l) winemaking trial with a GM wine yeast.

In contrast, the bio-acidification of warmer-region high pH wines can be facilitated by an *S. cerevisiae* strain in which the lacticodehydrogenase-encoding genes from *Lactobacillus casei*, *Rhizopus oryzae*, and bovine muscle were expressed in yeast (Dequin and Barre, 1994; Porro et al., 1995; Skory, 2003). As much as 20% of the glucose was converted into lactic acid in these strains (Dequin et al., 1999).

The desirable "buttery" or "butterscotch" flavor in wines can be imparted by diacetyl (a diketone, 2,3-butanedione), which is one of the most important flavor compounds associated with malolactic fermentation (Davis et al., 1985; Martineau and Henick-Kling, 1995; Ribereau-Gayon et al., 2000). Diacetyl is principally formed during malolactic fermentation by the bacterial metabolism of citric acid (Ramos et al., 1995; Ramos and Santos, 1996; Bartowsky and Henschke, 2000; Bartowsky et al., 2002). The diacetyl pathway has been amenable to genetic manipulation, leading to overproduction of diacetyl in *Lactococcus lactis* (Gasson et al., 1996). With the inactivation of the lactate dehydrogenase gene (*ldh*) in *L. lactis*, there was an accompanying alteration to the metabolic flux, eliminating lactic acid as a metabolic end product and producing ethanol, formate, and acetoin. Acetoin is a degradation product of diacetyl and is considered flavorless in wine because of its high aroma threshold. Overexpression of α-acetolactate synthase (*ilvBN* genes in *L. lactis*) leads to increased production of acetoin (the reaction is driven towards acetoin and away from diacetyl), whereas inactivation of the *aldB* gene (encoding α-acetolactate decarboxylase) results in increased production of α-acetolactate and diacetyl at the expense of acetoin. This latter scenario is quite desirable in the dairy industry. The analogous genes [*alsS* (α-acetolactate synthase) and *alsD* (α-acetolactate decarboxylase)] have been cloned and sequenced from *O. oeni* providing a means of altering diacetyl concentrations in wine (Garmyn et al., 1996). An alternative strategy would be to inactivate the citrate permease gene (*citP*), thus removing the initial substrate for diacetyl. Such a natural mutant of *O. oeni* has been isolated and is commercially available.

An undesirable bitter flavor in wine is a spoilage problem, primarily in the reds. The fermentation of glycerol by some lactic acid bacteria can lead to formation of acrolein, which, when it reacts with phenolic hydroxyl groups, results in bitterness (de Barros Lopes et al., 2005). The manipulation of the pathway of glycerol catabolism, for example, the glycerol dehydratase, might lead to lower formation of acrolein. The presence of certain peptides is also a potential cause of bitterness in wine and bitterness due to proteolytic action and formation of short peptides has been studied extensively in dairy lactic acid bacteria. There appears to be minimal proteolytic activity associated with *O. oeni,* and the peptide transport system of *O. oeni* is poorly understood (de Barros Lopes et al., 2005). Better understanding of potential bitter wine peptides and their formation could reduce the occurrence of bitterness in some wines.

Improving Wine Wholesomeness

Epidemiological studies have led to the widely held view that wine contains protective compounds (e.g., phenolic compounds, flavonoids, resveratrol, salicylic acid, and ethyl alcohol) which, when consumed in moderate quantities, reduce the likelihood of contracting certain diseases (cardiovascular disease, dietary cancers, ischemic stroke, peripheral vascular disease, diabetes, hypertension, peptic ulcers, kidney stones, and macular degeneration) (Armstrong et al., 2001; Finkel, 2003; Howitz et al., 2003; Yang et al., 2003). However, several unwanted compounds (e.g., ethyl carbamate and bioamines) are also present (Pretorius, 2000), therefore generating improved starter strains of *S. cerevisiae* and *O. oeni* that maximize the presence of "beneficial" constituents and minimize the presence of undesirable compounds will be important to enhance the potential health benefits of wine.

In relation to coronary heart disease, it is believed that some grape phytoalexins, including stilbenes such as resveratrol, not only reduce the risk but also act as antioxidants and antimutagens (Armstrong et al., 2001; Finkel, 2003; Howitz et al., 2003; Yang et al., 2003). The conclusion has been drawn that they display anti-inflammatory and cancer chemopreventive properties, as well as the ability to induce specific enzymes that metabolize carcinogenic substances. Resveratrol is synthesized mainly in the skin cells of grape berries; only traces are found in the flesh. As a result, red wines have a much higher resveratrol concentration than whites due to grapeskin contact at the first phase of fermentation. Fermenting white must with wine yeast expressing the *Candida molisciana* β-glucosidase gene (*bgiN*) resulted in an increase in resveratrol in white wine (Gonzalez-Candelas et al., 2000). The suggested mechanism for this was the release of glucose moieties from the glucoside form of resveratrol. In a further attempt to increase the levels of resveratrol in both red and white wine, the phenylpropanoid pathway in *S. cerevisiae* was modified to produce *p*-coumaroyl-coenzyme A, one of the precursors for resveratrol synthesis (Becker et al., 2003). The other substrate, malonyl-coenzyme A, is already found in yeast and is involved in *de novo* fatty acid biosynthesis. It is hypothesized that the production of *p*-coumaroyl-coenzyme A can be achieved by introducing the phenylalanine ammonia-lyase gene (*PAL*), the cinnamate 4-hydroxylase gene (*D4H*), the coenzyme A ligase gene (*4CL9* and *4CL216*) and the grape stilbene synthase gene (*Vst1*) into *S. cerevisiae*. The introduction of the grape stilbene synthase gene (*Vst1*) might then catalyze the addition of three acetate units from malonyl-coenzyme A to *p*-coumaryl-coenzyme A, resulting in the formation of resveratrol. In a preliminary study, the *4CL9, 4CL216*, and *Vst1* genes were cloned under the control of strong yeast promoters of the alcohol dehydrogenase II (*ADH2*) and enolase (*ENO2*) genes and transformed into *S. cerevisiae*. Initially, no resveratrol could be detected in the yeast expressing the coenzyme A ligase and resveratrol synthase genes. It was only after the samples were treated with β-glucosidase to release the glucose moieties from the so-called piceid molecule (the glucoside form of resveratrol) that detectable levels were seen in mass spectrometric assays (Becker et al., 2003). This has encouraged research to co-express the coenzyme A ligase and resveratrol synthase genes together with a glucose-insensitive β-glucosidase gene such as the *BGL1* gene from *S. fibuligera*. However, more investigation at greater depth is needed before it can be recommended as a way to develop resveratrol-producing wine yeast strains.

A suspected mammalian carcinogen, ethyl carbamate, is formed when ethanol reacts with citrulline, urea, or carbamyl phosphate (Ough et al., 1988; Liu et al., 1994; Mira de Orduna et al., 2000a, b). In *S. cerevisiae*, urea is produced by the *CAR1*-encoded arginase during the breakdown of arginine, one of the main amino acids in grape juice. In this process, arginine is converted to ornithine, ammonia, and carbon dioxide, and urea is formed as an intermediate product. Urea is secreted into the wine by certain yeast strains that, depending on fermentation conditions, might be unable to metabolize the external urea further. Although all *S. cerevisiae* strains secrete urea, the extent to which they re-absorb the urea differs (An and Ough, 1993). In a saké strain of *S. cerevisiae* disrupted for the *CAR1* arginase gene, no urea or ethyl carbamate was produced. Low urea-forming yeast strains are selected and commercial preparations of acidic urease are added with some success to reduce the accumulation of urea in wine. To curb the formation of ethyl carbamate, the successive disruption of the *CAR1* arginase gene in an industrial saké yeast successfully eliminated urea accumulation in rice wine, but growth was also impeded, which limits the commercial use of the strain (Kitamoto et al., 1991).

Another attempt to eliminate urea as the precursor of ethyl carbamate from wine involved assembling a urease gene by fusing the α, β, and γ subunits of the *Lactobacillus fermentum* urease operon with the jack bean (*Canavalia ensiformis*) urease linker sequences inserted between both the α and β, and the β and γ subunits, and linking this construct to the yeast phosphoglycerate kinase gene (*PGK1*) promoter (Visser, 1999). However, the secretion of urease peptides was extremely low and did not result in urea being converted into ammonia and carbon dioxide. The absence of biological activity in the recombinant urease was probably the result of a lack of the essential auxiliary proteins, which are present only in urease-producing species, such as the fission yeast *S. pombe*. Without these proteins, *S. cerevisiae* is unable to assemble the various subunits into an active urease so it seems that, for *S. cerevisiae* to express an active urease, accessory genes of *L. fermentum* will also have to be cloned and expressed in addition to the structural urease genes. The addition of urease preparations is an acceptable interim solution.

The role of malolactic fermentation in the formation of ethyl carbamate remains unclear. Lactic acid bacteria vary in their ability to degrade arginine; experiments conducted in a synthetic and a laboratory vinified wine demonstrated a correlation between arginine degradation, citrulline production, and ethyl carbamate formation during malolactic fermentation conducted by an *O. oeni* and *Lb. buchneri* strain (Liu et al., 1994; Mira de Orduna et al., 2000a).The arginine catabolism (*arc*) gene cluster of *O. oeni* has been cloned and characterized, thus providing a basis for manipulating the *arc* genes and reducing the potential of ethyl carbamate production (Divol et al., 2003).

Biogenic amines (e.g., histamine, phenylethylamine, putrescine and tyramine) are known to have undesirable physiological effects when absorbed at high concentrations (McDaniel, 1986; Klijn et al., 1991; Deng et al., 1994). They can trigger hypotension and migraines, lead to histamine toxicity, and produce carcinogenic nitrosamines (Silla Santos, 1996). Wine-associated lactic acid bacteria, including *O. oeni*, have been shown to decarboxylate amino acids to their corresponding amines (Henick-Kling, 1993; Ward et al., 1995; Lonvaud-Funel, 2001; Guerrini et al., 2002). This decarboxylation reaction is purported to favor growth and survival in acidic media because it can provide energy to the lactic acid bacteria. These bacteria vary in their ability to produce the various amines. The *O. oeni* histidine decarboxylase gene (*hdc*) has been cloned and characterized (Coton et al., 1998). A PCR-based test has also been developed to detect amino acid decarboxylating genes in lactic acid bacteria (Le Jeune et al., 1995). Manipulation of the *hdc* gene in *O. oeni* strains with other desirable characteristics would minimize the potential of bioamine formation in wine.

From the studies discussed, it is clear that there is a vast scope to develop strains of wine yeast and malolactic bacteria that could enhance the health benefits associated with wine, while at the same time reducing possible risks and improving the overall quality of wine.

SETTING NEW GOALS FOR THE ELUCIDATION OF CONSUMERS' GUSTATORY AND OLFACTORY CODES—CHALLENGES, OPPORTUNITIES, AND POTENTIAL BENEFITS

Commercial Opportunities in Deciphering Human Gustatory and Olfactory Codes

The importance of consistently producing wine to definable specifications and styles for targeted markets in this modern and economically competitive society was outlined in the first section of this chapter. Specifically designed grapevines and microbial starter strains as well as process management throughout the production stage are crucial in achieving repeatability and control. But there is more to it than that; it is equally important for researchers to provide the means for the wine industry to capture new opportunities in a changing global marketplace by discovering what makes individuals receptive to different tastes and smells (Gilbert and Firestein, 2002; Rønning, 2004). Put differently, the industry needs to understand the expectations and preferences of individuals and populations for wine flavors—and that is why research should be undertaken into the genomics of wine taste and smell so as to benefit from recent scientific advances in this new field of science. Although some of the diversity between individuals and populations is cultural and learnt, it is most likely that there could well be a genetic aspect to defining likes and dislikes for specific tastes and aromas (Pretorius et al., 2005).

Scientists are working to define fully the genetic differences among individuals' sensory perception, and to do that it is necessary to assign particular tastants (taste compounds) and odorants (odor compounds) to their corresponding receptors (sensory organs). This is a challenge because taste and smell receptors constitute the largest, and possibly most diverse, gene family in the human genome, but it is a challenge that will potentially offer great benefits to the wine industry (Firestein, 2001; Dennis, 2004; Pretorius et al., 2005).

While it will be difficult to establish individual and population preferences, the unraveling of the human genome sequence is beginning to provide a scientific understanding of how we smell and taste. At the same time, advances in knowledge about how flavor perception is influenced by factors such as ethnicity, gender, and age can be used to develop an individual's *flavor print* as a means to measure a person's capacity to detect key flavor compounds in wine (Firestein, 2001; Dennis, 2004). This will provide an entrée toward establishing objective measures for predicting flavor preferences in market segments.

Chemical senses exist in all living organisms. Bacteria modify the direction of flagella rotation to change their swim pattern in response to attractants or repellants. Yeasts respond to several external signals such as nutrients and mating-type pheromones, leading to changes in their intracellular metabolism and cell morphology. Insects are known to commit mass suicide by jumping off leaves in response to chemicals produced by parasites, and the use of dogs for tasks such as drug, land mine, and food stuff detection demonstrates their well-known sense of smell. Human beings, although not as dependent on the chemical senses as other creatures, are still able to detect and discriminate between millions of structurally diverse chemicals (Pretorius et al., 2005). The question is: How many genes are required to sense this chemical universe?

Genomics of Taste

Despite the structural diversity of tastants, the gustatory system senses only five types (modalities): sour, salt, bitter, sweet, and umami (savory, the taste of the amino acids, glutamate, and aspartate, and their salts—e.g., MSG) (Finger et al., 2000; Pretorius et al., 2005). Specialized cells named taste receptor (TR) cells mediate recognition of these five types of tastants. These cells are tightly packed into taste buds, mainly on the surface of the tongue. When tastants interact with the proteins in TR cells, they produce an electrical change in the cells that triggers the release of neurotransmitters (Firestein, 2001; Dennis, 2004). This is followed by the activation of nerve fibers and results in an impulse (message) to the brain.

There are two classes of TRs (Zhao et al., 2003). In one, the chemicals that elicit sour and salty tastes, generally ions (e.g., Na^+ and H^+), act directly through ion channels on the cell membrane of these TRs to bring about neuron activation—the message to the brain (Finger et al., 2000). The bitter, sweet, and umami compounds are structurally more diverse and bind directly to the second type of surface TR cells. The proteins in these TRs belong to a family called the *G-protein coupled receptors* (GPCR) (Zhao et al., 2003). These proteins, which span the cell membrane seven times, change shape when a ligand binds to the external regions of the receptor, transmit a signal to the inside of the cell that results in the opening and closing of ion channels in the taste cell, also leading to neuron activation—the message to the brain (Finger et al., 2000; Firestein, 2001; Dennis, 2004).

Ability to taste bitter chemicals has an especially important role in animal survival. Because many natural poisons are bitter, most animals are averse to such compounds. Perception of bitter tastants is performed by a class of GPCRs collectively known as the T2R family. In the human genome, there are 25 functional genes encoding T2R proteins (Mombaerts, 2001). Despite the similarity of their intracellular regions, these proteins show extensive variation in their tastant-binding regions. This variability corresponds well with an ability to detect the chemically diverse ligands associated with bitter tastes.

Surprisingly, the perception of sweet and umami tastants is performed by only three receptors—T1R1, T1R2 and T1R3 (Zhao et al., 2003). Their distinguishing power is achieved by formation of what are called receptor heterodimers (two different proteins attached together). The T1R3–T1R2 dimers sense sweet tastants and have been shown to be able to interact with various sugars, sweet proteins, and artificial sweeteners such as saccharin, glutamate, and aspartate. The umami compounds activate T1R3–T1R1 dimers (Zhao et al., 2003).

Identification of the receptor genes has shown that a single taste bud on the palate contains 50–100 taste cells and possesses receptors to all five taste types (Pretorius et al., 2005). This agrees with sensory and physiology studies that have ruled out the previous understanding that distinct parts of the tongue reacted to specific tastants. Importantly though, the *T1R1* and *T1R2* genes are not expressed in the same cell, ensuring that sweet and umami tastants activate different taste cells and therefore distinguish themselves as distinct tastes or flavors (Zhao et al., 2003). However, we also know now that many bitter receptors (T2Rs) are co-expressed in the same taste cells, which mean they probably act as generalized detectors of bitter tastes. This explains why the human tongue can sense five diverse taste types but it cannot necessarily distinguish between two tastes in the same group, for example, two unrelated compounds that both taste bitter.

Genomics of Smell

Compared with the limitations of the human tongue, our powers of smell—scientifically called olfaction—are greater. We can detect thousands of different odorants that vary considerably in size and structure and we can distinguish and identify these chemicals (Firestein, 2001; Dennis, 2004). This would lead one to expect the aroma genome to require an additional level of complexity.

Odorants are recognized by olfactory nerve cells high in the nasal cavity. When odorants bind to receptors on the surface of these olfactory cells, they trigger a signal that opens ion channels, which in turn send a signal to the olfactory lobe of the brain (Finger et al., 2000). However, the olfactory cells also link with secondary neurons and these transmit information further into the brain, in fact entering the limbic system of the brain that is associated with emotion, sexual behavior, and memory, which explains why aroma has had such a central position in human culture—and romance—throughout history.

The mapping of complete genome sequences has established that mammals have nearly 1000 odor receptor (OR) genes, constituting 1–3% of the total genome (Pretorius et al., 2005). Some of these olfactory receptors are GPCRs, such as the TRs for sweet, umami, and bitter tastes. They also contain similar structural features (e.g., seven transmembrane regions and an intracellular signaling region) but they are extremely diverse, which is consistent with our ability to recognize a great number of smells (Firestein, 2001; Dennis, 2004).

However, they differ from taste cells in that each OR cell expresses only a single OR, which means each odor sends an individual signal to the brain (Zhao et al., 2003). Additionally, although these individual ORs are restricted in their recognition ability, they can bind to multiple odorants—in other words, they can receive combinations of smells. For example, a specific rat receptor, OR-17, recognizes the strongly related seven- to nine-carbon chain aldehydes, interacting strongly with octanal (eight-carbon chain) and to a lesser extent with heptanal (seven-carbon chain) and nonanal (nine-carbon chain), while numerous other related and unrelated odorants are not recognized by OR-17. OR cells have a further advanced capacity—as well as being able to detect a combination of odors, individual odorants can stimulate several different receptor cells (Firestein, 2001; Zhao et al., 2003; Dennis, 2004). This facilitates potentially infinite permutations—each odorant activating a unique combination of olfactory cells—so the olfactory system can, theoretically at least, enable human beings to discriminate among thousands of chemicals.

To add to this complexity, it is now thought some odorants can also inhibit olfactory receptors, rather than activate them. This means that our nasal cavities, lined with OR-carrying minuscule hair-like extensions (called cilia), are a battleground where odors jockey with each other to stimulate or block olfactory receptors (Firestein, 2001; Dennis, 2004). Further, as more odorants are added to a mixture, some will cancel out others or change how we perceive the mixture's smell. On top of all this, it is thought human beings have a physiological limit of about three in the number of odors we can distinguish at once. Therefore, a combination of odors might smell completely different from any of the individual components, and it is impossible to predict what we are going to smell with various combinations of odors (Firestein, 2001; Dennis, 2004).

Genetic Basis for Individual Variation

While we know that individuals differ in the flavors they prefer and that their inability to detect specific smells and tastes can be hereditary, the unknown is how much of this variation depends on changes in the taste and OR areas of the human genome. It is also unknown whether specific populations have particular modifications in their taste and smell repertoires that, if it is the case, could be a key factor in targeting specific wine styles to particular consumer segments, thus opening new markets for the wine industry.

Although smell and taste are ancient senses and have developed to detect thousands of chemicals, humans are less dependent on these senses than animals. These differences are manifested in the genome. For example, human beings have 25 functional bitter TR receptor genes while a mouse has 33. Human beings actually possess these 33 bitter TR sequences, but 8 of them carry mutations rendering them non-functional "pseudogenes" (Firestein, 2001; Mombaerts, 2001; Dennis, 2004). A similar scenario occurs with OR genes. For example, mice, which rely heavily on smell, have about 1200 OR genes, most of which are in full working order. By contrast, nearly two-thirds of the 1000 or so human OR genes have accumulated mutations that render them useless (Mombaerts, 2001). In other words, there has been a decline in the human chemical receptor genome that reflects our decreased dependence on smell and taste for survival and reproduction.

The next question is whether there is any variation between individuals' TR and OR genes and whether this is associated to perceptions and preferences in taste and smell. One source of variation is found in the number of copies of certain receptor genes; for example, one block of OR genes has been shown to be present in 7–11 copies in the genomes of various individuals (Firestein, 2001; Dennis, 2004). Another source of variation is whether a particular genomic sequence encodes a functional receptor or a mutated pseudogene. Mice that show decreased sensitivity to the taste of cyclohexamide have been shown to have a mutation in the mT2R5 bitter receptor gene, making this receptor non-functional. There are similar findings in human beings in the detection of the bitter compound 6-*n*-propyl-2-thiouracil by the hT2R4 receptor. In smells, four ORs have been shown to work in some people but not in others, but the odorants that bind these receptors have not yet been identified.

It has been found that single amino acid changes in receptor proteins can play an important role in creating diversity in smell and taste. In the rat OR-17 receptor, a single alteration in sequence changes its recognition specificity so that heptanal, in place of octanal, is the preferred ligand. Also, mice that differ in their sensitivity to sweet tastants have alterations in the *T1R3* gene.

What emerges from these studies is that human chemical reception is so diverse that it is possible no two individuals have exactly the same repertoire of abilities to taste and smell (Pretorius et al., 2005). This was highlighted in a recent study of the sequence of 51 OR genes in 189 individuals from several ethnic groups (Menashe et al., 2003). Amazingly, each individual had a distinct olfactory pattern, a level of genetic diversity that has been matched only by the human immune system. In addition, some of the variation appeared to be linked to ethnicity with, for example, African Americans having more functional receptors than non-African Americans (Menashe et al., 2003).

To determine how genetic differences influence perception of tastants and odorants, it is necessary to identify the ligands of odor and TRs and define their molecular specificity. This has been hindered by the difficulty in isolating cells bearing a specific receptor. A successful approach has been to express mammalian ORs in yeast because the yeast mating response pathway shares notable similarities with the taste and smell signaling pathways. When the OR expressed in yeast binds to its ligand, the yeast undergoes mating-related changes that can be easily detected. This could ultimately provide a method to define the entire human taste and smell repertoire, and act as a biosensor for the study of wine attributes.

Knowing all this, it is important to remember that, as well as genetic variations, sensory experiences are also dictated by a kaleidoscope of other interactive factors, including vision, emotion, and memory. As an illustration of this complex interactivity, it was found that the odors of a wine are, for the most part, represented by objects that reflect the color of the wine (Morrot et al., 2001). When a white wine was artificially colored red with an odorless dye, it was olfactorily described as a red wine by a panel of tasters. This simple psychophysical experiment demonstrated that because of the visual information, the tasters discounted the evidence of their noses when they spoke about the wine's smell (Morrot et al., 2001). It is therefore clear that there is still much work to do before the complete gustatory and olfactory codes are cracked, but thanks to the advances now being made, it is at least starting to look like a tractable objective (Pretorius et al., 2005).

CONCLUSION

This chapter has shown how biotechnology can support the international wine industry's needs to produce wines that meet the expectations of present and emerging markets around the world for quality, price, health, and environmental compatibility while retaining and building on the image of wine as a desirable consumer product.

Wine is still perceived as a harmonious blend of nature, art, and science but the potential afforded by 21st century grape and wine biotechnology, while providing the grapegrower and the winemaker with the means to meet the challenges of coming decades, invites tension between tradition and innovation (Pretorius and Høj, 2005). The challenge for the researcher is to realize the potential of technological innovation without stripping the ancient art of grapegrowing and winemaking of its charm, mysticism, and romanticism.

Equally challenging for both researcher and industry is the multitude of complex and interconnected agronomic, business, regulatory and social obstacles blocking commercial availability of transgenic grapes, wine yeast, and malolactic bacterial starter strains. But just as the electric light did not come from the continuous improvement of candles, a market-driven wine industry will, despite the abundance of traditional approaches, continue with technological innovation and progress (Pretorius and Høj, 2005). Grape and wine biotechnologists will contribute enormously to this progression, but the commercial uptake of GM technology, because of the complexity of the issues that dominate the international debate, will not depend solely on science providing industry with the capabilities to move ahead.

In this debate, it is my contention that the positioning of GMOs in general as either good or bad is unproductive. It fails to provide an opportunity for informed discussion about the merits of each individual "technology package" or the fact that safety assessments for all "new" products should be conducted on a case-by-case basis (Pretorius and Høj, 2005).

Such merits often depend on the circumstances. In prosperous societies, GM products will be accepted only when there is an obvious direct or indirect benefit of a particular technology package to consumers. Hopefully, progress will enable the GM debate to swing from one of being pro or con GM technology to one in which the debate is concerned with evaluation of individual "technology packages" (Pretorius and Høj, 2005).

Many of the prototype examples of improved grape varieties, yeasts, and microbial starter strains outlined in this chapter will undoubtedly be able to address quality, health, and environmental concerns of consumers and producers. Therefore, the introduction of GM technology into grape and wine production appears to be inevitable.

However, it is hard to predict how long it will take before consumers accept the huge potential benefits offered by grape and wine biotechnology. Historical precedents show how changes that have beneficial impacts on mankind can take many years to win acceptance. A clear and obvious example is the suggestion by Louis Pasteur more than a century ago that milk should be pasteurized. This was subjected to a long period of public argument and it took a couple of decades before this highly beneficial process was accepted throughout the world (Pretorius and Høj, 2005). This example of resistance to technological innovation underlines the well-known observation once made by Machiavelli: "Nothing is more difficult to undertake, more perilous to conduct or more uncertain in its outcome, than to take the lead in introducing a new order of things. For the innovator has for enemies all those who have done well under the old and lukewarm defenders amongst those who may do well under the new."

In light of the arguments and views expressed in this chapter, it is clear that ignoring the obvious potential benefits of biotechnology will disadvantage the international wine industry's efforts to retain and expand the population segments who choose wine as their preferred alcoholic beverage. Likewise, reluctance by consumers to accept biotechnology will deprive them of a further increase in the quality/price ratio of a product of almost universal appeal.

ACKNOWLEDGMENTS

The author thanks Drs Eveline Bartowsky, Miguel de Barros Lopes, Peter Høj, Florian Lopes, Melané Vivier, Maret du Toit, and Pierre van Rensburg for their valuable thoughts and other contributions to this chapter. Research conducted at The Australian Wine Research Institute is supported by Australia's grapegrowers and winemakers through their investment body the Grape and Wine Research and Development Corporation, with matching funds from the Australian Government.

The information in this chapter has been derived from "Grape and Wine Biotechnology: Challenges, opportunities and potential benefits" in *Australian Journal of Grape and Wine Research* 11(2):83–108.

REFERENCES

Alexandre H, Ansannay-Galeote V, Dequin S, Blondin B. 2001. Global gene expression during short-term ethanol stress in *Saccharomyces cerevisiae*. FEMS Microbiol Lett 498:98–103

Alexandre H, Blanchet S, Charpentier C. 2000. Identification of 49-kDA hydrophobic cell wall mannoprotein present in velum yeast which may be implicated in velum formation. FEMS Microbiol Lett 185:147–50

Allen J, Davey HM, Broadhurst D, Heald JK, Rowland JJ, Oliver SG, Kell DB. 2003. High-throughput classification of yeast mutants for functional genomics using metabolic footprinting. Nat Biotechnol 21:692–6

An D, Ough CS. 1993. Urea Excretion and Uptake by Wine Yeasts as Affected by Various Factors. Am J Enol Vitic 44:35–40

Armstrong GO, Lambrechts MG, Mansvelt EPG, Van Velden DP, Pretorius IS. 2001. Wine and Health. S Afr J Sci 97:279–82

Arntzen CJ, Coghlan A, Johnson B, Peacock J. 2003. GM crops: Science, politics and communication. Nat Genet 4:839–43

Arvanitoyannis S. 2003. Genetically engineered/modified organisms in food. Appl Biotechnol Food Sci Policies 1:3–12

Ashe MP, De Long SK, Sachs AB. 2000. Glucose depletion rapidly inhibits translation initiation in yeast. Mol Biol Cell 11:833–48

Axelsson LT. 1993. Lactic acid bacteria: Classification and physiology. In: Salminen, S and von Wright, A. Lactic Acid Bacteria. New York: Marcel Dekker Inc. pp 1–63

Backhus LE, DeRisi J, Brown PO, Bisson LF. 2001. Functional genomic analysis of a commercial wine strain of *Saccharomyces cerevisiae* under differing nitrogen conditions. FEMS Yeast Res 1:111–25

Barnavon L, Doco T, Terrier N, Ageorges A, Romieu C, Pellerin P. 2001. Involvement of pectin methyl-esterase during the ripening of grape berries: Partial cDNA isolation, transcript expression and changes in the degree of methyl-esterification of cell wall pectins. Phytochemistry 58:693–701

Barre P, Vezinhet F, Dequin S, Blondin B. 1993. Genetic improvements of wine yeasts. In: Fleet, GH. Wine Microbiology and Biotechnology. Reading: Harwood Academic Publishers. pp 265–87

Bartowsky E, Costello P, Henschke P. 2002. Management of malolactic fermentation—wine flavour manipulation. Aust Grapegrower Winemaker 461a:7–12

Bartowsky EJ, Henschke PA. 2000. Management of malolactic fermentation for the 'buttery' diacetyl flavour in wine. Aust Grapegrower Winemaker 438a:58–67

Bauer FF, Pretorius IS. 2000. Yeast stress response and fermentation efficiency: How to survive the making of wine. S Afr J Enol Vitic 21:27–51

Becker J, Armstrong G, van der Merwe M, Lambrechts M, Vivier M, Pretorius I. 2003. Metabolic engineering of *Saccharomyces cerevisiae* for the synthesis of the wine-related antioxidant resveratrol. FEMS Yeast Res 4:79–85

Bezier A, Lambert B, Baillieul F. 2002. Study of defense-related gene expression in grapevine leaves and berries infected with *Botrytis cinerea*. Eur J Plant Pathol 108:111–20

Bisson LF, Waterhouse AL, Ebler SE, Walker MA, Lapsley JT. 2002. The present and future of the international wine industry. Nature 418:696–9

Bony M, Barre P, Blondin B. 1998. Distribution of the flocculation protein, Flop, at the cell surface during yeast growth: The availability of Flop determines the flocculation level. Yeast 14:25–35

Boss PK, Davis C, Robinson SP. 1996a. Analysis of the expression of anthocyanin pathway genes in developing *Vitis vinfera* L. cv Shiraz grape berries and the implications for pathway regulation. Plant Physiol 111:1059–66

Boss PK, Davis C, Robinson SP. 1996b. Expression of anthocyanin biosynthesis pathway genes in red and white grapes. Plant Mol Biol 32:847–50

Boss PK, Thomas MR. 2002. Association of dwarfism and floral induction with a grape 'green revolution' mutation. Nature 416:847–50

Bourdineaud J-P, Nehme B, Tesse S, Lonvaud-Funel A. 2003. The *ftsH* gene of the wine bacterium *Oenococcus oeni* is involved in protection against environmental stress. Appl Environ Microbiol 69:2512–20

Breinig F, Tipper DJ, Schmitt MJ. 2002. Kre1p, the plasma membrane receptor for the yeast k1 viral toxin. Cell 108:395–405

Bro C, Regensberg B, Nielsen J. 2003. Yeast functional genomics and metabolic engineering: Past, present and future. In: de Winde, JH. Topics in Current Genetics. Heidelberg: Springer-Verlag. pp 331–60

Capone D, Sefton MA, Pretorius IS, Høj PB. 2003. Flavour scalping by wine bottle closures—the winemaking continues post vineyard and winery. Aust N Z Wine Ind J 18:16–20

Carstens M, Vivier M, Van Rensburg P, Pretorius IS. 2003. Overexpression, secretion and antifungal activity of the *Saccharomyces cerevisiae* chitinase. Ann Microbiol 50:15–28

Chambon C, Ladeveze V, Oulmouden A, Servouse M, Karst F. 1990. Isolation and properties of yeast mutants affected in farnesyl diphosphate synthetase. Curr Genet 18:41–6

Colova-Tsolova V, Perl A, Krastanova S, Tsvetkov IJ, Atanassov AI. 2001. Genetically engineered grape for disease and stress tolerance. In: Roubelakis-Angelakis, KA. Molecular Biology & Biotechnology of the Grapevine. Kluwer Academic Publishers. pp 411–32

Coton E, Rollan G, Lonvaud-Funel A. 1998. Histidine carboxylase of Leuconostoc oenos 9204: Purification, kinetic properties, cloning and nucleotide sequence of the hdc gene. J Appl Microbiol 84:143–51

Coutos-Thevenot P, Poinssot B, Bonomelli A, Yean H, Breda C, Buffard D, Esnault R, Hain R, Boulay M. 2001. *In vitro* tolerance to *Botrytis cinerea* of grapevine 41B rootstock in transgenic plants expressing the stilbene synthase *Vst 1* gene under the control of a pathogen-inducible PR 10 promoter. J Exp Bot 52:901–10

Crous JM, Pretorius IS, Van Zyl WH. 1995. Cloning and expression of an *Aspergillus kawachii* endo-1,4-β-xylanase gene in *Saccharomyces cerevisiae*. Curr Genet 28:467–473.

Crous JM, Pretorius IS, Van Zyl WH. 1996. Cloning and expression of the α-L-arabinofuranosidase gene (*ABF2*) of *Aspergillus niger* in *Saccharomyces cerevisiae*. Appl Microbiol Biotechnol 46: 256–60

Darriet P, Tominaga T, Lavigne V, Boidron J-N, Dubourdieu D. 1995. Identification of a powerful aromatic component of *Vitis vinifera* L. var. Sauvignon wines: 4-mercapto-4-methylpentan-2-one. Flavour Frag J 10:385–92

Davis C, Boss PK, Robinson SP. 1997. Treatment of grape berries, a nonclimacteric fruit with a synthetic auxin, retards ripening and alters the expression of developmentally regulated genes. Plant Physiol 115:1155–61

Davis C, Robinson E. 2000. Differential screening indicates a dramatic change in mRNA profiles during grape berry ripening. Cloning and characterization of cDNAs encodig putative cell wall and stress response proteins. Plant Physiol 122:803–12

Davis C, Robinson SP. 1996. Sugar accumulation in grape berries. Cloning of two putative vacuolar invertase cDNAs and their expression in grapevine tissue. Plant Physiol 111:275–83

Davis CR, Wibowo D, Eschenbruch R, Lee TH, Fleet GH. 1985. Practical implications of malolactic fermentation: A review. Am J Enol Vitic 36:290–301

Day RE. 1998. Presented at the Tenth Australian Wine Industry Technical Conference, Sydney, Australia

de Barros Lopes MA, Bartowsky EJ, Pretorius IS. 2005. The application of gene technology in the wine industry. In: Hui, HY (ed). Handbook of Food Science, Technology and Engineering, Volume 1. CRC Press, Florida/Marcel Dekker, New York, USA, pp. 40-1–40-21.

de Barros Lopes MA, Eglinton JM, Henschke PA, Høj PB, Pretorius IS. 2003. The connection between yeast and alcohol production in wine: Managing the double edged sword of bottled sunshine. Aust N Z Wine Ind J 18:27–31

de Barros Lopes MA, Rehman A-U, Gockowiak H, Heinrich A, Langridge P, Henschke P. 2000. Fermentation properties of a wine yeast over-expressing the *Saccharomyces cerevisiae* glycerol 3-phosphate dehydrogenase gene (*GPD2*). Aust J Grape Wine Res 6:208–15

Delneri D, Brancia FL, Oliver SG. 2001. Towards a truly integrative biology through the functional genomics of yeast. Curr Opin Biotechnol 12:87–91

de Revel G, Martin N, Pripis-Nicolau L, Lonvaud-Funel A, Bertrand A. 1999. Contribution to the knowledge of malolactic fermentation influence on wine aroma. J Agric Food Chem 47:4003–8

Deng WL, Chang HY, Peng HL. 1994. Acetoin Catabolic System of Klebsiella-Pneumoniae Cg43—Sequence, Expression, and Organization of the Aco Operon. J Bacteriol 176:3527–35

Dennis C. 2004. Neuroscience: The sweet smell of success. Nature 428:362–4

Dequin S. 2001. The potential of genetic engineering for improving brewing, wine-making and baking yeasts. Appl Microbiol Biotechnol 56:577–88

Dequin S, Baptista E, Barre P. 1999. Acidification of grape musts by *Saccharomyces cerevisiae* wine yeast strains genetically engineered to produce lactic acid. Am J Enol Vitic 50:45–50

Dequin S, Barre P. 1994. Mixed lactic acid-alcoholic fermentation by *Sacchromyces cerevisiae* expressing the *Lactobacillus casei* L(+)-LDH. Biotechnology 12:173–7

Dickinson JR, Salgado LE, Hewlins MJ. 2003. The catabolism of amino acids to long chain and complex alcohols in *Saccharomyces cerevisiae*. J Biol Chem 278:8028–34

D'Incecco N, Bartowsky EJ, Kassara S, Lante A, Spetolli P, Henschke PA. 2004. Release of Chardonnay glycosidically bound flavour compounds by *Oenococcus oeni* during malolactic fermentation. Food Microbiol 21:257–65

Divol B, Tonon T, Morichon S, Gindreau E, Lonvaud-Funel A. 2003. Molecular characterization of *Oenococcus oeni* genes encoding proteins involved in arginine transport. J Appl Microbiol 94:738–46

Dollinski K, Balakrishnan R, Christie KR, Costanzo MC, Dwight SS, Engel SR, Fisk DG, Hirschman JE, Hong EL, Issel-Tarver L. 2003. *Saccharomyces* genome database. http://genome-www.stamford.edu/Saccharomyces

Donalies UE, Stahl U. 2002. Increasing sulphite formation in *Saccharomyces cerevisiae* by overexpression of *MET14* and *SSU1*. Yeast 19:475–84

Dubourdieu D, Tominaga T, Masneuf I, Peyrot des Gachons C, Murat ML. 2000. Presented at the ASEV 50th anniversary annual meeting, Seattle, Washington, USA

Dupin IVS, Stockdale VJ, Williams PJ, Jones GP, Markides AJ, Waters EJ. 2000. *Saccharomyces cerevisiae* mannoproteins that protect wine from protein haze: Evaluation of extraction methods and immunolocalization. J Agric Food Chem 48:1086–95

du Toit M, Pretorius IS. 2000. Microbial spoilage and preservation of wine: Using weapons from Nature's own arsenal. S Afr J Enol Vitic 21:74–96

Eden A, van Nedervelde L, Drukker M, Benvenisty N, Debourg A. 2001. Involvement of branched-chain amino acid aminotransferases in the production of fusel alcohols during fermentation in yeast. Appl Microbiol Biotechnol 55:296–300

Eglinton JM, Heinrich AJ, Pollnitz AP, Langridge P, Henschke PA, de Barros Lopes MA. 2002. Decreasing acetic acid accumulation by a glycerol overproducing strain of *Saccharomyces cerevisiae* by deleting the *ALD6* aldehyde dehydrogenase gene. Yeast 19:295–301

Eglinton JM, Henschke PA, Høj PB, Pretorius IS. 2003. Winemaking properties and potential of *Saccharomyces bayanus* wine yeast: Harnessing the untapped potential of yeast biodiversity. Aust N Z Wine Ind J 18:16–19

Erasmus DJ, Van der Merwe GK, van Vuuren HJJ. 2003. Genome-wide expression analyses: Metabolic adaptation of *Saccharomyces cerevisiae* to high sugar stress. FEMS Yeast Res 3:375–99

Ferber D. 1999. GM crops in the cross hairs. Science 286:1662–6

Fichtner L, Frohloff F, Burkner K, Larsen M, Breunig KD, Schaffrath R. 2002. Molecular analysis of *KTI12/TOT4*, a *Saccharomyces cerevisiae* gene required for *Kluyveromyces lactis* zymocin action. Mol Microbiol 43:783–91

Fiehn O. 2002. Metabolomics—the link between genotypes and phenotypes. Plant Mol Biol 48:155–71

Fillion L, Ageorges A, Picaud S, Coutos-Thevenot P, Lemoine R, Romieu C, Delrot S. 1999. Cloning and expression of a hexose transporter gene expressed during the ripening of grape berry. Plant Physiol 120:1083–94

Finger TE, Silver WL, Restrepo D. 2000. The Neurobiology of Taste and Smell. Wiley-Liss

Finkel T. 2003. Ageing: A toast to long life. Nature 425:132–3

Firestein S. 2001. How the olfactory system makes sense of scents. Nature 413:211–8

Fleet GH, Heard G. 1993. Yeasts—growth during fermentation. In: Fleet, GH. Wine Microbiology and Biotechnology. Reading: Harwood Academic Publisher. pp 27–54

Fogel S, Welch JW, Cathala G, Karin M. 1983. Gene amplification in yeast: *CUP1* copy number regulates copper resistance. Curr Genet 7:347–55

Ford CM, Boss PK, Høj PB. 1998. Cloning and characterization of *Vitis vinfera* UDP-glucose: Flavonoid 3-O-glucosyltransferase, a homologue of the enzyme encoded by the maize bronze-1 locus that may primarily serve to glucosylate anthocyanidins *in vivo*. J Biochem 273:9224–33

Fujii T, Nagasawa N, Iwamatsu A, Bogaki T, Tamai Y, Hamachi M. 1994. Molecular cloning, sequence analysis, and expression of the yeast alcohol acetyltransferase gene. Appl Environ Microbiol 60:2786–92

Fukuda K, Yamamoto N, Kiyokawa Y, Yanagiuchi T, Wakai Y, Kitamoto K, Inoue Y, Kimura A. 1998.

Balance of activities of alcohol acetyltransferase and esters in *Saccharomyces cerevisiae* is important for production of soamyl acetate. Appl Environ Microbiol 64:4076–8

Gagiano M, Bauer F, Pretorius IS. 2002. The sensing of nutritional status and the relationship to filamentous growth in *Saccharomyces cerevisiae*. FEMS Yeast Res 2:433–70

Gallander JF. 1977. Deacidification of eastern table wines with *Schizosaccharomyces pombe*. Am J Enol Vitic 28:65–8

Garmyn D, Monnet C, Martineau B, Guzzo J, Cavin J-F, Divies C. 1996. Cloning and sequencing of the gene encoding α-acetolactate decarboxylase from *Leuconostoc oenos*. FEMS Microbiol Lett 145:445–50

Gasch AP, Spellman PT, Kao CM, Carmel-Harel O, Eisen MB, Storz G, Botstein D, Brown PO. 2000. Genomic expression programs in the response of yeast cells to environmental changes. Mol Biol Cell 11:4241–57

Gasson MJ, Benson K, Swindell S, Griffin H. 1996. Metabolic engineering of the *Lactococcus lactis* diacetyl pathway. Lait 76:33–40

Giannakis C, Bucheli CS, Skene KGM, Robinson SP, Scott NS. 1998. Chitinase and β-1,3-glucanase in grapevine leaves. Aust J Grape Wine Res 4:14–22

Gilbert A, Firestein S. 2002. Dollars and scents: Commercial opportunities in olfaction and taste. Nat Neurosci 5:1043–5

Godden PW, Francis IL, Field J, Gishen M, Coulter A, Valente P, Høj PB, Robinson E. 2001. Wine bottle closures: Physical characteristics and effect on composition and sensory properties of a Semillon wine 1.—Performance up to 20 months post-bottling. Aust J Grape Wine Res 7:64–105

Goffeau A, Barrell BG, Bussey H, Davis RW, Dujon B, Feldmann H, Galibert F, Hoheisel JD, Jacq C, Johnston M, Louis EJ, Mewes HW, Murakami Y, Philippsen P, Tettelin H, Oliver SG. 1996. Life with 6000 genes. Science 274:546–67

Gonzalez-Candelas L, Gil J, Lamuela-Raventos R, Ramon D. 2000. The use of transgenic yeasts expressing a gene encoding a glycosyl-hydrolase as a tool to increase resveratrol content in wine. Int J Food Microbiol 59:179–83

Grossmann MK, Pretorius IS. 1999. Verfahren zur Identifizierung von Weinhefen und Verbesserung der Eigenschaften von *Saccharomyces cerevisiae*: Eine Übersicht. Weinwissenschaft 54:61–72

Guerrini S, Mangani S, Granchi L, Vincenzini M. 2002. Biogenic amine production by *Oenococcus oeni*. Curr Microbiol 44:374–8

Guzzo J, Jobin M-P, Delmas F, Divies C. 1997. Study on physiology and molecular response to stress in the malolactic bacterium, *Oenococcus oeni*. Revue des Oenologues et des Techniques Vitivinicoles et Oenologiques 86:26–8

Hansen J, Kielland-Brandt MC. 1996a. Inactivation of *MET2* in brewer's yeast increases the level of sulfite in beer. J Biotechnol 50:75–87

Hansen J, Kielland-Brandt MC. 1996b. Inactivation of *MET10* in brewer's yeast specifically increases SO_2 formation during beer production. Nat Biotechnol 14:1587–91

Hauser N, Fellenberg K, Gil R, Bastuck S, Hoheisel J, Perez-Ortin J. 2001. Whole genome analysis of a wine yeast strain. Comp Funct Genomics 2:69–79

Henderson RCA, Cox BS, Tubb R. 1985. The transformation of brewing yeasts with a plasmid containing the gene for copper resistance. Curr Genet 9:133–8

Henick-Kling T. 1993. Malolactic fermentation. In: Fleet, GH. Wine Microbiology and Biotechnology. Camberwell: Harwood Acedemic. pp 289–326

Henschke PA. 1997. Wine yeast. In: Zimmermann, FK and K-D., E. Yeast Sugar Metabolism. Pennsylvania: Technomic Publishing Co. pp 527–60

Hohmann S. 2002. Osmotic stress signalling and osmoadaptation in yeasts. Microbiol Mol Biol Rev 66:300–72

Høj PB, Hayes PF. 1998. Presented at the Tenth Australian Wine Industry Technical Conference, Sydney, Australia

Høj PB, Pretorius IS. 2004. Growing markets and delivering benefit to wine producers and consumers through research and innovation—a perspective and examples from Australia. Aust N Z Wine Ind J 19:51–7

Høj PB, Pretorius IS, Day RE. 2003. Beyond the idea: The importance of industry/researcher communication and enhanced R&D investment as drivers for a 'can do' culture in pursuit of excellence. Aust N Z Wine Ind J 18:18–23

Howell KS, Swiegers JH, Elsey GM, Siebert TE, Bartowsky EJ, Fleet GH, Pretorius IS, de Barros Lopes MA. 2004. Variation in 4-mercapto-4-methyl-pentan-2-one release by *Saccharomyces cerevisiae* commercial wine strains under different wine fermentation conditions. FEMS Microbiol Lett 240:125–129.

Howitz K, Bitterman K, Cohen H, Lamming D, Lavu S, Wood J, Zipkin R, Chung P, Kisielewski A, Zhang L, Scherer B, Sinclair D. 2003. Small molecule activators of sirtuins extend *Saccharomyces cerevisiae* lifespan. Nature 425:191–6

Huh WK, Flavo JV, Gerke LC, Carroll AS, Howson RW, Weissman JS, O'Shea EK. 2003. Global analysis of protein localization in budding yeast. Nature 425:686–91

Ibrahim HR, Matsuzaki T, Aoki T. 2001. Genetic evidence that antibacterial activity of lysozyme is independent of its catalytic function. FEBS Lett 506:27–32

Icco P, Franks T, Thomas MR. 2001. Genetic transformation of major wine grape cultivars of *Vitis vinifera* L. Transgenic Res 10:105–12

Ishida-Fujii K, Goto S, Sugiyama H, Takagi Y, Saiki T, Takagi M. 1998. Breeding of flocculant industrial alcohol yeast strains by self-cloning of the flocculation gene *FLO1* and repeated-batch fermentation by transformants. J Gen Microbiol 44:347–53

Ivorra C, Perez-Ortin JE, del Olmo M. 1999. An inverse correlation between stress resistance and stuck fermentations in wine yeasts. A molecular study. Biotechnol Bioeng 64:698–708

Jackson RS. 1994. Grapevine species and varieties. In: Wine Science: Principles and Application. London: Academic Press. pp 11–31

Jackson RS. 2003. Modern biotechnology of winemaking. In: Sandler, M and Pinder, R. Wine—A Scientific Exploration. United Kingdom: Taylor and Francis. pp 228–59

Kellis M, Patterson N, Endrizzi M, Birren B, Lander ES. 2003. Sequencing and comparison of yeast species to identify genes and regulatory elements. Nature 423:241–54

Kikkert JR, Thomas MR, Reisch BI. 2001. Grapevine genetic engineering. In: Roubelakis-Angelakis, KA. Molecular Biology & Biotechnology of the Grapevine. Kluwer Academic Publishers, Dordrecht, The Netherlands. pp 393–410

Kim J, Alizadeh P, Harding T, Hefner-Gravink A, Klionsky DJ. 1996. Disruption of the yeast *ATH1* gene confers better survival after dehydration, freezing, and ethanol shock: Potential commercial applications. Appl Environ Microbiol 62:1563–9

King A, Dickinson JR. 2000. Biotransformation of monoterpene alcohols by *Saccharomyces cerevisae*, *Torulsapora delbrueckii* and *Kluyveromyces lactis*. Yeast 16:499–506

King A, Dickinson JR. 2003. Biotransformation of hop aroma terpenoids by ale and lager yeasts. FEMS Yeast Res 3:53–62

Kitamoto K, Oda K, Gomi K, Takahashi K. 1991. Genetic-engineering of a sake yeast producing no urea by successive disruption of arginase gene. Appl Environ Microbiol 57:301–6

Klijn N, Weerkamp AH, Devos WM. 1991. Identification of mesophilic lactic-acid bacteria by using polymerase chain reaction-amplified variable regions of 16s ribosomal-RNA and specific DNA probes. Appl Environ Microbiol 57:3390–3

Kobayashi S, Ishimaru M, Ding CK, Yakushiji H, Goto N. 2001. Comparison of UDP-glucose: Flavonoid 3-O-glucosyltransferase (UFGT) gene sequences between white grapes (*Vitis vinifera*) and their spots with red skin. Plant Sci 160:543–50

Kobayashi S, Ishimaru M, Hiraoka K, Honda C. 2002. Myb-related genes of the Kyoho grape (*Vitis labruscana*) regulate anthocyanin biosynthesis. Planta 215:924–33

Lagorce A, Hauser NC, Labourdette D, Rodriguez C, Martin-Yken H, Arroyo J, Hoheisel JD, Francois J. 2003. Genome-wide analysis of the response to cell wall mutations in the yeast *Saccharomyces cerevisiae*. J Biol Chem 278:20345–57

La Grange DC, Claeyssens IM, Pretorius IS, Van Zyl WH. 2001. Degradation of xylan to D-xylose by recombinant *Saccharomyces cerevisiae* co-expressing the *Aspergillus niger* β-xylosidase (*xlnD*) and the *Trichoderma reesei* xylanase II (*xyn2*) genes. Appl Environ Microbiol 67:5512–9

La Grange DC, Pretorius IS, Van Zyl WH. 1996. Expression of the *Trichoderma reesei* β-xylanase gene (*XYN2*) in *Saccharomyces cerevisiae*. Appl Environ Microbiol 62:1036–44

La Grange DC, Pretorius IS, Van Zyl WH. 1997. Cloning of the *Bacillus pumilus* β-xylosidase gene (*xynB*) and its expression in Saccharomyces cerevisiae. Appl Environ Microbiol 47:262–6

Laing E, Pretorius IS. 1992. Synthesis and secretion of an *Erwinia chrysanthemi* pectate lyase in *Saccharomyces cerevisiae* regulated by different combinations of bacterial and yeast promoter and signal sequences. Gene 121:35–45

Laing E, Pretorius IS. 1993a. Co-expression of an *Erwinia chrysanthemi* pectate lyase encoding gene (*pelE*) and an *Erwinia carotovara* polygalacturonase-encoding gene (*peh1*) in *Saccharomyces cerevisiae*. Appl Environ Microbiol 39:181–8

Laing E, Pretorius IS. 1993b. The primary structure and expression of an *Erwinia carotovora* polygalacturonase-encoding gene (*peh1*) in *Escherichia coli* and yeast. J Appl Bacteriol 75:149–58

Lambrechts MG, Pretorius IS. 2000. Yeast and its importance to wine aroma. S Afr J Enol Vitic 21:97–129

Legrand V, Dalmayrac S, Latche A, Pech JC, Bouzayen M, Fallot J, Torregrosa L, Bouquet A,

Roustan JP. 2003. Constitutive expression of *Vr-ERE* gene in transformed grapevines confers enhanced resistance to eutypine, a toxin from *Eutypa lata*. Plant Sci 164:809–14

Le Jeune C, Lonvaud-Funel A, ten Brink B, Hofstra H, van der Vossen J. 1995. Development of a detection system for histidine decarboxylating lactic acid bacteria based on DNA probes, PCR and activity test. J Appl Bacteriol 78:316–26

Lilly M, Styger G, Bauer FF, Lambrechts MG, Pretorius IS. 2005a. The effect of increased yeast branched-chain amino acid transaminase activity and the production of higher alcohols on the flavor profiles of wine and distillates. FEMS Yeast Res (in press).

Lilly M, Bauer FF, Lambrechts MG, Swiegers JH, Pretorius IS. 2005b. The effect of increased alcohol acetyl transferase and esterase activity on flavor profiles of wine and distillates. Appl Environ Microbiol (in press).

Lilly M, Lambrechts MG, Pretorius IS. 2000. Effect of increased yeast alcohol acetyltransferase activity on flavor profiles of wine and distillates. Appl Environ Microbiol 66:744–53

Liu SQ, Pritchard GG, Hardman MJ, Pilone GJ. 1994. Citrulline Production and Ethyl Carbamate (urethane) Precursor Formation from Arginine Degradation by Wine Lactic-Acid Bacteria *Leuconostoc oenos and Lactobacillus buchneri*. Am J Enol Vitcult 45:235–42

Lonvaud-Funel A. 2001. Biogenic amines in wines: Role of lactic acid bacteria. FEMS Microbiol Lett 199:9–13

Lowes KF, Shearman CA, Payne J, MacKenzie D, Archer DB, Merry RJ, Gasson MJ. 2000. Prevention of yeast spoilage in feed and food by the yeast mycocin HMK. Appl Environ Microbiol 66:1066–76

Luttig M, Pretorius IS, Van Zyl WH. 1997. Cloning of two β-xylanase-encoding genes from *Aspergillus niger* and their expression in *Saccharomyces cerevisiae*. Biotechnol Lett 19:411–5

Luyten K, Riou C, Blondin B. 2002. The hexose transporters of *Saccharomyces cerevisiae* play different roles during enological fermentation. Yeast 19:713–26

Malherbe DF, Du Toit M, Cordero Otero RR, Van Rensburg P, Pretorius IS. 2003. Expression of the *Aspergillus niger* glucose oxidase gene in *Saccharomyces cerevisiae* and its potential applications in wine production. Appl Microbiol Biotechnol 61:502–11

Marks VD, Van der Merwe GK, van Vuuren HJJ. 2003. Transcriptional profiling of wine yeast in fermenting grape juice: Regulatory effect of diammonium phosphate. FEMS Yeast Res 3:269–87

Martineau B, Henick-Kling T. 1995. Performance and diacetyl production of commercial strains of malolactic bacteria in wine. J Appl Bacteriol 78:526–36

Martinelli L, Candioli E, Costa D, Minafra A. 2002. Stable insertion and expression of the movement protein gene of grapevine Virus A (GVA) in grape (*Vitis rupestris* S.). Vitis 41:189–93

Martinez-Rodriguez AJ, Polo MC. 2000. Characterization of the nitrogen compounds released during yeast autolysis in a model wine system. J Agric Food Chem 48:1081–5

Martin O, Brandriss MC, Schneider G, Bakalinsky AT. 2003. Improved anaerobic use of arginine by *Saccharomyces cerevisiae*. Appl Environ Microbiol 69:1623–8

Masneuf I, Hansen J, Groth C, Piskur J, Dubourdieu D. 1998. New hybrids between *Saccharomyces* sensu stricto yeast species found among wine and cider production strains. Appl Environ Microbiol 64:3887–92

McDaniel M. 1986. Trained panel evaluation of pinot noir fermented with different strains of malolactic bacteria. Oregon Wine Advisory Board 6

Menashe I, Man O, Lancet D, Giland Y. 2003. Different noses for different people. Nat Genet 34:143–4

Michnick S, Roustan JL, Remize F, Barre P, Dequin S. 1997. Modulation of glycerol and ethanol yields during alcoholic fermentation in *Saccharomyces cerevisiae* strains overexpressed or disrupted for *GPD1* encoding glycerol 3-phosphate dehydrogenase. Yeast 13:783–93

Mira de Orduna R, Liu S, Patchett M, Pilone G. 2000a. Ethyl carbamate precursor citrulline formation from arginine degradation by malolactic wine lactic acid bacteria. FEMS Microbiol Lett 183:31–5

Mira de Orduna R, Liu S, Patchett M, Pilone G. 2000b. Kinetics of the arginine metabolism of malolactic wine lactic acid bacteria *Lactobacillus buchneri* CUC-3 and *Oenococcus oeni* Lo111. J Appl Microbiol 89:547–52

Moine-Ledoux V, Dubourdieu D. 2003. An invertase fragment responsible for improving the protein stability of dry white wines. J Sci Food Agric 79:537–43

Mombaerts P. 2001. The human repertoire of odorant receptor genes and pseudogenes. Ann Rev Genomics Hum Genet 2:493–510

Morita Y, Nakamori S, Takagi H. 2003. L-proline accumulation and freeze tolerance of *Saccharomyces cerevisiae* are caused by a mutation

in the *PRO1* gene encoding gamma-glutamyl kinase. Appl Environ Microbiol 69:212–9
Morrot G, Brochet F, Dubourdieu D. 2001. The color of odor. In: Brain and Language. Academic Press, New York. pp 1–12
Nunan KJ, Davis C, Robinson SP, Fincher GB. 2001. Expression patterns of cell wall-modifying enzymes during grape berry development. Planta 214:257–64
Oliver SG. 1996. From DNA sequence to biological function. Nature 379:597–600
Omura F, Shibano Y, Fukui N, Nakatani K. 1992. Reduction of hydrogen sulfide production in brewing yeast by constitutive expression of *MET25* gene. J Am Soc Brew Chem 53:58–62
Østergaard S, Olsson L, Nielsen J. 2000. Metabolic engineering of *Saccharomyces cerevisiae*. Microbiol Mol Biol Rev 64:34–50
Ough CS, Crowell EA, Gutlove BR. 1988. Carbamyl compound reactions with ethanol. Am J Enol Vitic 39:239–42
Ozcan S, Dover J, Rosenwald AG, Wolfl S, Johnston M. 1996. Two glucose transporters in *Saccharomyces cerevisiae* are glucose sensors that generate a signal for induction of gene expression. Proc Natl Acad Sci USA 93:12428–32
Page N, Gerard-Vincent M, Menard P, Beaulieu M, Azuma M, Dijkgraaf GJ, Li H, Marcoux J, Nguyen T, Dowse T, Sdicu AM, Bussey H. 2003. A *Saccharomyces cerevisiae* genome-wide mutant screen for altered sensitivity to k1 killer toxin. Genetics 163:875–94
Park H, Bakalinsky AT. 2000. *SSU1* mediates sulphite efflux in *Saccharomyces cerevisiae*. Yeast 16:881–8
Paschalidis KA, Aziz A, Geny L, Primikirios NI, Roubelakis-Angelakis KA. 2001. Polyamines in grapevine. In: Roubelakis-Angelakis, KA. Molecular Biology & Biotechnology of the Grapevine. Kluwer Academic Publishers, Dordrecht, The Netherlands. pp 109–51
Patnaik R, Louie S, Gavrilovic V, Perry K, Stemmer WPC, Ryan CM, del Cardayre SB. 2002. Genome shuffling of *Lactobacillus* for improved acid tolerance. Nat Biotechnol 20:707–12
Perez F, Ramirez M, Regodon JA. 2001. Influence of killer strains of *Saccharomyces cerevisiae* on wine fermentation. Anton Leeuw 79:393–9
Pérez-González JA, De Graaf LH, Visser J, Ramon D. 1996. Molecular cloning and expression in *Saccharomyces cerevisiae* of two *Aspergillus nidulans* xylanase genes. Appl Environ Microbiol 62:2179–82
Perez-Ortin JE, Garcia-Martinez J, Alberola TM. 2002a. DNA chips for yeast biotechnology. The case of wine yeasts. J Biotechnol 98:227–41
Perez-Ortin JE, Querol A, Puig S, Barrio E. 2002b. Molecular characterization of a chromosomal rearrangement involved in the adaptive evolution of yeast strains. Genome Res 12:1533–9
Perez-Torrado R, Gimeno-Alcaniz JV, Matallana E. 2002. Wine yeast strains engineered for glycogen overproduction display enhanced viability under glucose deprivation conditions. Appl Environ Microbiol 68:3339–44
Peterson SH, Van Zyl WH, Pretorius IS. 1998. Development of a polysaccharide-degrading strain of *Saccharomyces cerevisiae*. Biotechnol Tech 12:615–9
Pocock KF, Hayasaka Y, McCarthy MG, Waters EJ. 2000. Thaumatin-like proteins and chitinases, the haze-forming proteins of wine, accumulate during ripening of grape (*Vitis vinifera*) berries and drought stress does not affect the final levels per berry at maturity. J Agric Food Chem 48:1637–43
Pocock KF, Høj PB, Adams K, Kwiatkowski MJ, Waters EJ. 2003. Combined heat and proteolytic enzyme treatment of white wines reduces haze forming protein content without detrimental effect. Aust J Grape Wine R 9:56–63
Poole K. 2002. Enhancing yeast performance under oenological conditions by enabling proline utilisation. Ph.D. Thesis. Adelaide: University of Adelaide
Porro DL, Brambilla L, Ranzi BM, Martegani E, Alberghina L. 1995. Development of metabolically engineered *Saccharomyces cerevisiae* cells for the production of lactic acid. Biotechnol Prog 11:294–8
Pratelli R, Lacombe B, Torregrosa L, Gaymard F, Romieu C, Thibaud JB, Sentenac H. 2002. A grapevine gene encoding a guard cell K(+) channel displays developmental regulation in the grapevine berry. Plant Physiol 128:564–77
Pretorius IS. 1997. Utilization of polysaccharides by *Saccharomyces cerevisiae*. In: Zimmermann, FK and Entian, K-D. Yeast Sugar Metabolism. Technomic Publishing Co., Basel, Switzerland. pp 459–501
Pretorius IS. 1999. Engineering designer genes for wine yeasts. Aust N Z Wine Ind J 14:42–7
Pretorius IS. 2000. Tailoring wine yeast for the new millennium: Novel approaches to the ancient art of winemaking. Yeast 16:675–729
Pretorius IS. 2001. Gene technology in winemaking: New approaches to an ancient art. Agric Conspectus Sci 66:27–47
Pretorius IS. 2003. The genetic analysis and tailoring of wine yeasts. In: de Winde, JH. Topics in Current Genetics. Heidelberg: Springer-Verlag. pp 98–142
Pretorius IS. 2004a. The genetic improvement of wine yeasts. In: Arora, DK, Bridge, PD, and Bhatnagar,

D. Handbook of Fungal Biotechnology, Second Edition. New York: Marcel Dekker Inc. pp 209–32

Pretorius IS. 2004b. Tailoring new grapevine varieties for the wine industry. Acenologia 67:1–5

Pretorius IS, Bartowsky EJ, de Barros Lopes MA, du Toit M, Van Rensburg P, Vivier M. 2005. The tailoring of designer grapevines and microbial starter strains for a market-directed and quality-focussed wine industry. In: Hui, HY (ed). Handbook of Food Science, Technology and Food Engineering, Volume 4. CRC Press, Florida/Marcel Dekker, New York, USA. pp. 174-1–174-24.

Pretorius IS, Bauer FF. 2002. Meeting the consumer challenge through genetically customized wine-yeast strains. Trends Biotechnol 20:426–32

Pretorius IS, de Barros Lopes MA, Francis IL, Swiegers JH, Høj PB. 2004b. The genomics of taste and smell—cracking the codes that determine who likes what on the palate and on the nose. Aust N Z Wine Ind J 19:13–18

Pretorius IS, du Toit M, van Rensburg P. 2003. Designer yeasts for the fermentation industry of the 21st century. Food Technol Biotechnol 41:3–10

Pretorius IS, Høj PB. 2005. Grape and wine biotechnology: challenges, opportunities and potential benefits. Aust J Grape Wine Res 11:83–108.

Pretorius IS, van der Westhuizen TJ. 1991. The impact of yeast genetics and recombinant DNA technology on the wine industry. S Afr J Enol Vitic 12:3–31

Pretorius IS, van der Westhuizen TJ, Augustym OPH. 1999. Yeast bioversity in vineyards and wineries and its importance to the South African wine industry. S Afr J Enol Vitic 20:61–74

Querol A, Ramón D. 1996. The application of molecular techniques in wine microbiology. Trends Food Sci Technol 7:73–8

Raamsdonk LM, Teusink B, Broadhurst D, Zhang N, Hayes A, Walsh MC, Berden JA, Brindle KM, Kell DB, Rowland JJ, Westerhoff HV, van Dam K, Oliver SG. 2001. A functional genomics strategy that uses metabolome data to reveal the phenotype of silent mutations. Nat Biotechnol 19:45–50

Ramos A, Lolkema JS, Konings WN, Santos H. 1995. Enzyme basis for pH regulation of citrate and pyruvate metabolism by *Leuconostoc oenos*. Appl Environ Microbiol 61:1303–10

Ramos A, Santos H. 1996. Citrate and sugar cofermentation in *Leuconostoc oenos*, a ^{13}C nuclear magnetic resonance study. Appl Environ Microbiol 62:2577–85

Reiss M. 2002. Labeling GM foods—the ethical way forward. Nat Biotechnol 20:868

Remize F, Roustan JL, Sablayrolles JM, Barre P, Dequin S. 1999. Glycerol overproduction by engineered *Saccharomyces cerevisiae* wine yeast strains leads to substantial changes in by-product formation and to a stimulation of fermentation rate in stationary phase. Appl Environ Microbiol 65:143–9

Remize F, Sablayrolles JM, Dequin S. 2000. Re-assessment of the influence of yeast strain and environmental factors on glycerol production in wine. J Appl Microbiol 88:371–8

Rep M, Krantz M, Thevelein JM, Hohmann S. 2000. The transcriptional response of *Saccharomyces cerevisiae* to osmotic shock. Hot1p and Msn2p/Msn4p are required for the induction of subsets of high osmolarity glycerol pathway-dependent genes. J Biol Chem 275:8290–300

Reynolds TB, Fink GR. 2001. Baker's yeast, a model for fungal biofilm formation. Science 291:878–81

Ribereau-Gayon J, Dubourdieu D, Doneche B, Lonvaud-Funel A. 2000. Lactic acid bacteria. In: Riberau-Gayon, J, Maujean, A, and Dubourdieu, D. Handbook of Enology: The Chemistry of Wine and Stabilization and Treatments. John Wiley & Sons Ltd., New York, pp 107–28

Robert N, Ferran J, Breda C, Coutos-Thevenot P, Boulay M, Buffard D, Esnault R. 2001. Molecular characterization of the incompatible interaction of *Vitis vinifera* leaves with *Pseudomonas syringae* pv. *pisi*: Expression of genes coding for stilbene synthase and class 10 PR protein. Eur J Plant Pathol 107:249–61

Robert N, Roche K, Lebeau Y, Breda C, Boulay M, Esnault R, Buffard D. 2002. Expression of grapevine chitinase genes in berries and leaves infected by fungal or bacterial pathogens. Plant Sci 162:389–400

Rokas A, Williams BL, King N, Carroll SB. 2003. Genomic-scale approaches to resolving incongruence in molecular phylogenies. Nature 425:798–804

Rønning D. 2004. Taste, smell and sound—Future trademarks. Les Novelles16–21

Salmon JM, Barre P. 1998. Improvement of nitrogen assimilation and fermentation kinetics under enological conditions by depression of alternative nitrogen-assimilatory pathways in an industrial *Saccharomyces cerevisiae* strain. Appl Environ Microbiol 64:3831–7

Salzman RA, Tikhonova I, Bordelon BP, Hasegawa PM, Bressan RA. 1998. Coordinate accumulation of antifungal proteins and hexoses constitutes a developmentally controlled defense response during fruit ripening in grape. Plant Physiol 117:465–72

Sauvageot F, Vivier P. 1997. Effects of malolactic fermentation on sensory properties of four Burgundy wines. Am J Enol Vitic 48:187–92

Schoeman H, Vivier MA, du Toit M, Dicks LMT, Pretorius IS. 1999. The development of bactericidal yeast strains by expressing the *Pediococcus acidilactici* pediocin gene (*pedA*) in *Saccharomyces cerevisiae*. Yeast 15:647–56

Shima J, Hino A, Yamada-Iyo C, Suzuki Y, Nakajima R, Watanabe H, Mori K, Takano H. 1999. Stress tolerance in doughs of *Saccharomyces cerevisiae* trehalase mutants derived from commercial baker's yeast. Appl Environ Microbiol 65:2841–6

Shouten A, Wagemakers L, Stefano FL, van der Kaaij RM, van Kan JAL. 2002. Resveratrol acts as a natural profungicide and induces self-intoxication by a specific laccase. Mol Microbiol 43:883–94

Silla Santos M. 1996. Biogenic amines: Their importance in foods. Int J Food Microbiol 29:213–31

Skory CD. 2003. Lactic acid production by *Saccharomyces cerevisae* expressing a *Rhizopus oryzae* lactate dehydrogenase gene. J Ind Microbiol Biotechnol 30:22–7

Smit A, Cordero Otero RR, Lambrechts MG, Pretorius IS, Van Rensburg P. 2003. Enhancing volatile phenol concentrations in wine by expressing various phenolic acid decarboxylase genes in *Saccahromyces cerevisiae*. J Agric Food Chem 51:4909–15

Snow R. 1983. Genetic improvement of wine yeast. In: Spencer, JFT. Yeast Genetics—Fundamental and Applied Aspects. Heidelberg: Springer-Verlag. pp 439–59

Sparvoli F, Martin C, Scienza A, Gavazzi G, Tonelli C. 1994. Cloning and molecular analysis of structural genes involved in flavonoid and stilbene biosynthesis in grape (*Vitis vinifera* L.). Plant Mol Biol 24:743–55

Spiropoulos A, Bisson LF. 2000. MET17 and hydrogen sulfide formation in Saccharomyces cerevisiae. Appl Environ Microbiol 66:4421–6

Stemmer WPC. 1994. DNA shuffling by random fragmentation and reassembly: *In vitro* recombination for molecular evolution. PNAS 91:10747–51

Swiegers JH, Pretorius IS. 2005. Yeast modulation of wine flavor. Adv Appl Microbiol 57:131–175.

Takagi H, Sakai K, Morida K, Nakamori S. 2000. Proline accumulation by mutation or disruption of the proline oxidase improves resistance to freezing and desiccation stresses in *Saccharomyces cerevisiae*. FEMS Microbiol Lett 184:103–8

Tanghe A, van Dijck P, Dumortier F, Teunissen A, Hohmann S, Thevelein JM. 2002. Aquaporin expression correlates with freeze tolerance in baker's yeast, and overexpression improves freeze tolerance in industrial strains. Appl Environ Microbiol 68:5981–9

Tattersall DB, Pocock KF, Hayasaka Y, Adams K, Van Heeswijck R, Waters EJ, Høj PB. 2001. Pathogenesis related proteins—their accumulation in grapes during berry growth and their involvement in white wine heat instability. Current knowledge and future perspectives in relation to winemaking practices. In: Roubelakis-Angelakis, KA. Molecular Biology & Biotexhnology of the Grapevine. Kluwer Academic Publishers, Dordrecht, The Netherlands. pp 183–201

Tezuka H, Mori T, Okumura Y, Kitabatake K, Tsumura Y. 1992. Cloning of a gene suppressing hydrogen sulfide production by *Saccharomyces cerevisiae* and its expression in a brewing yeast. J Am Soc Brew Chem 50:130–3

Tominaga T, Baltenweck-Guyot R, Peyrot des Gachons C, Dubourdieu D. 2000. Contribution of volatile thiols to the aromas of white wines made from several *Vitis vinifera* grape varieties. Am J Enol Vitcult 51:178–81

Tominaga T, Furrer A, Henry R, Dubourdieu D. 1998a. Identification of new volatile thiols in the aroma of *Vitis vinifera* L. var. Sauvignon blanc wines. Flavour Frag J 13:159–62

Tominaga T, Peyrot des Gachons C, Dubourdieu D. 1998b. A new type of flavor precursors in *Vitis vinifera* L. cv. Sauvignon Blanc: *S*-cysteine conjugates. J Agric Food Chem 46:5215–9

Trabalzini L, Paffetti A, Talamo F, Ferro E, Coratza G, Bovalini L, Lusini P, Martelli P, Santucci A. 2003. Proteomic response to physiological fermentation stresses in a wild-type wine strain of *Saccharomyces cerevisiae*. Biochem J 370:35–46

Tsvetkov IJ, Atanassov AI, Tsolova VM. 2000. Gene transfer for stress resistance in grapes. Acta Horticulturae 528:389–96

Van Heeswijck R, Stines AP, Grubb J, Moller IS, Høj PB. 2001. Molecular biology and biochemistry of proline accumulation in developing grape berries. In: Roubelakis-Angelakis, KA. Molecular Biology & Biotechnology of the Grapevine. Kluwer Academic Publishers, Dordrecht, The Netherlands. pp 87–108

Van Reenen CA, Chikindas ML, Van Zyl WH, Dicks LMT. 2003. Characterization and heterologous expression of a class IIa bacteriocin, plantaricin 423 from *Lactobacillus plantarum* 423, in *Saccharomyces cerevisiae*. Int J Food Microbiol 81:29–40

Van Rensburg P, Pretorius IS. 2000. Enzymes in winemaking: Harnessing natural catalysts for efficient biotransformations—a review. S Afr J Enol Vitic 21:52–73

Van Rensburg P, Van Zyl WH, Pretorius IS. 1994. Expression of the *Butyrivibrio fibrisolvens* endo-β-1,4-glucanase gene together with the *Erwinia* pectate lyase and polygalacturonase gene in *Saccharomyces cerevisiae.* Curr Genet 27:17–22

Van Rensburg P, Van Zyl WH, Pretorius IS. 1995. Expression of the *Ruminococcus flavefaciens* cellodextrinase gene in *Saccharomyces cerevisiae.* Biotechnol Lett 17:481–6

Van Rensburg P, Van Zyl WH, Pretorius IS. 1996. Co-expression of a *Phanerochaete chrysosporium* cellobiohydrolase gene and a *Butyrivibrio fibrisolvens* endo-β-1,4-glucanase gene in *Saccharomyces cerevisae.* Curr Genet 30:246–50

Van Rensburg P, Van Zyl WH, Pretorius IS. 1997. Overexpression of the *Saccharomyces cerevisiae* exo-β-1,3-glucanase gene together with the *Bacillus subtilis* endo-β-1,3-1,4-glucanase gene and the *Butyrivibrio fibrisolvens* endo-β-1,4-glucanase gene in yeast. J Biotechnol 55:43–53

Van Rensburg P, Van Zyl WH, Pretorius IS. 1998. Engineering yeast for efficient cellulose degradation. Yeast 4:87–95

Veen M, Stahl U, Lang C. 2003. Combined overexpression of genes of the ergosterol biosynthetic pathway leads to accumulation of sterols in *Saccharomyces cerevisiae.* FEMS Yeast Res 4

Verstrepen KJ, Derdelinckx G, Delvaux FR, Winderickx J, Thevelein JM, Bauer F, Pretorius IS. 2001. Late fermentation expression of *FLO1* in *Saccharomyces cerevisiae.* J Am Soc Brew Chem 59:69–76

Verstrepen KJ, Derdelinckx G, Dufour J-P, Winderickx J, Pretorius IS, Thevelein JM, Delvaux FR. 2003a. The *Saccharomyces cerevisiae* alcohol acetyl transferase gene *ATF1* is a target of the cAMP/PKA and FGM nutrient signalling pathways. FEMS Yeast Res 69:285–96

Verstrepen KJ, Derdelinckx G, Dufour J-P, Winderickx J, Thevelein JM, Pretorius IS, Delvaux FR. 2003b. Flavour-active esters: Adding fruitiness to beer—A practical review. J Biosci Bioeng 96:110–8

Verstrepen KJ, Moonjai N, Derdelinckx G, Dufour J-P, Winderickx J, Thevelein JM, Pretorius IS, Delvaux FR. 2003c. Genetic regulation of ester synthesis in brewer's yeast: New facts, insights and implications for the brewer. In: Smart, K. Brewing Yeast Performance, Second Edition. Oxford, UK: Blackwell Science. pp 234–48

Verstrepen KJ, Van Laere SDM, Vanderhaegen BMP, Dufour J-P, Pretorius IS, Winderickx J, Thevelein JM, Delvaux FR. 2003d. The expression of the yeast alcohol transferase genes *ATF1*, Lg-*ATF1* and *ATF2* control the formation of a broad range of different volatile esters. Appl Environ Microbiol 69: 5228–37

Vilanova M, Blanco P, Cortes S, Castro M, Villa TG, Sieiro C. 2000. Use of a *PGU1* recombinant *Saccharomyces cerevisiae* strain in oenological fermentations. J Appl Microbiol 89:876–83

Visser JJ. 1999. Cloning and Expression of the Lactobacillus Fermentum Acid Ureases Gene in *Saccharomyces cerevisiae.* Stellenbosch, South Africa: Stellenbosch

Vivier M, Pretorius IS. 2000. Genetic improvement of grapevine: Tailoring grape varieties for the third millennium. S Afr J Enol Vitic 21:5–26

Vivier M, Pretorius IS. 2002. Genetically tailored grapevines for the wine industry. Trends Biotechnol 20:472–8

Volschenk H, Viljoen M, Grobler J, Bauer F, Lonvaud-Funel A, Denayrolles M, Subden RE, van Vuuren HJJ. 1997. Malolactic fermentation in grape musts by a genetically engineered strain of *Saccharomyces cerevisiae.* Am J Enol Vitic 48:193–7

Walker GM. 1998. Yeast Physiology and Biotechnology. New York: John Wiley and Sons Ltd

Ward L, Brown J, Davey G. 1995. Detection of dairy *Leuconostoc* strains using the polymerase chain reaction. Lett Appl Microbiol 20:204–8

Weiler F, Rehfeldt K, Bautz F, Schmitt MJ. 2002. The *Zygosaccharomyces bailii* antifungal virus toxin zygocin: Cloning and expression in a heterologous fungal host. Mol Microbiol 46:1095–105

Wibowo D, Eschenbruch R, Davis CR, Fleet GH, Lee TH. 1985. Occurrence and growth of lactic acid bacteria in wine: Review. Am J Enol Vitic 36:302–13

Winzeler EA, Shoemaker DD, Astromoff A, Liang H, Anderson K, Andre B, Bangham R, Benito R, Boeke JD, Bussey H, Chu AM, Connelly C, Davis K, Dietrich F, Dow SW, El Bakkoury M, Foury F, Friend SH, Gentalen E, Giaever G, Hegemann JH, Jones T, Laub M, Liao H, Liebundguth N, Lockhart DJ, Lucau-Danila A, Lussier M, M'Rabet N, Menard P, Mittmann M, Pai C, Rebischung C, Revuelta JL, Riles L, Roberts CJ, Ross-MacDonald P, Scherens B, Snyder M, Sookhai-Mahadeo S, Storms RK, Veronneau S, Voet M, Volckaert G, Ward TR, Wysocki R, Yen GS, Yu KX, Zimmermann K, Philippsen P, Johnston M, Davis RW. 1999. Functional characterization of the *S.*

cerevisiae genome by gene deletion and parallel analysis. Science 285:901–6

Yang S, Kim J, Oh T, Kim M, Lee S, Woo S, Cho H, Choi Y, Kim Y, Rha S, Chung H, An S. 2003. Genome-scale analysis of resveratrol-induced gene expression profile in human ovarian cancer cells using a cDNA microarray. Int J Oncol 22: 741–50

Yap NA, de Barros Lopes M, Langridge P, Henschke PA. 2000. The incidience of killer activity of non-*Saccharomyces* yeasts towards indigenous yeast species of grape must: Potential application in wine fermentation. J Appl Microbiol 89:381–9

Yoshimoto H, Fukushige T, Yonezawa T, Sone H. 2002. Genetic and physiological analysis of branched-chain alcohols and isoamyl acetate production in *Saccharomyces cerevisiae*. Appl Microbiol Biotechnol 59:501–8

Ze-Ze L, Teneiro R, Brito L, Santos MA, Paveia H. 1998. Physical map of the genome of Oenococcus oeni PSU-1 and localization of genetic markers. Microbiology+ 144:1145–56

Ze-Ze L, Teneiro R, Paveia H. 2000. The *Oenococcus oeni* genome: Physical and genetic map of strain GM and comparison with the genome of a 'divergent' strain, PSU-1. Microbiology+ 146:3195–204

Zhang Y-Z, Perry K, Vinci VA, Powell K, Stemmer WPC, del Cardayre SB. 2002. Genome shuffling leads to rapid phenotypic improvement in bacteria. Nature 415:644–6

Zhang Z, Gernstein M. 2003. Reconstructing genetic networks in yeast. Nat Biotechnol 21:1295–9

Zhao GQ, Zhang Y, Hoon MA, Chandrashekar J, Erlenbach I, Ryba NJP, Zuker CS. 2003. The receptors for mammalian sweet and umami taste. Cell 115:255–66

26
Olive Processing

Beatriz Gandul-Rojas and M. Isabel Mínguez-Mosquera

INTRODUCTION

One day the trees went out to anoint a king for themselves. They said to the olive tree, "Be our king." But the olive tree answered, "Should I give up my oil, by which both gods and men are honoured, to hold sway over the trees?" (Judges, IX, 8–9)

THE OLIVE TREE

The olive, *Olea europaea L.*, is the only species of the botanical family Oleaceae with edible fruit. The tree blooms with small white flowers in May, and fruit set begins almost immediately afterward, in May or June. Ripening is complete toward the end of November or December, depending on the variety.

Its cultivation began more than 6000 years ago (making it one of the oldest cultivated plants) in the Middle East, and spread west on both sides of the Mediterranean Sea. It reached Spanish shores with the Roman Empire, but it was the Arabs who promoted cultivation in Andalucía, making Spain the prime producing country in the world. The olive tree was brought from the Mediterranean basin with the discovery of America in 1492. From Seville, the first olive trees were taken to the Antilles and then to the continent. By 1560, olive groves were productive in Mexico and later in Peru, Chile, Argentina, and in the state of California in the United States. In more recent times, olive growing has continued to expand to places far from its origin to southern Africa, China, Japan, and Australia (Civantos, 1997).

There are some 850 million olive trees in the world that occupy an approximate surface of 8.7 million ha of which 95% are concentrated in the countries of the Mediterranean basin. Spain is the leading exporter of olive products, with some 170 million olive trees in more than 2.2 million ha of groves, of which 60% are located in Andalucía. Other countries with substantial populations of olive trees are Italy (125 million), Greece (120 million), Turkey (83 million), Tunisia (55 million), Portugal (49 million), and Morocco (33 million).

Of the annual world production of olive fruit—approximately 10 million tons—10% is consumed as table olives and the rest is processed for olive oil (International Olive Oil Council, IOOC, 2000).

OLIVE FRUIT

The fruit of the olive tree is a fleshy green drupe, oval in shape. In fact, the botanical term drupe is derived from druppa, the Greek and Latin word for ripe olive. Its growth curve is double sigmoidal, in three stages. In the first stage, the fruit weight increases notably; in the second stage, it remains constant; and lastly, there is a phase of rapid growth mainly due to an increase in cell size. The duration of each stage varies depending on variety (Hermoso et al., 1987; Roca and Mínguez-Mosquera, 2003).

Its component parts are the epicarp or epidermis, the mesocarp or flesh, and the endocarp. The olive mesocarp (which is also referred to as pulp) accounts for 70–90% of the olive weight (wet basis); the endocarp (stone or pit) for the other 10–30%, which includes the seed (1–3% of the whole fruit).

Olives differ from all other drupes in chemical composition with a relatively low concentration of sugars, high oil content, and in a characteristic strong bitter taste. Composition basically depends on factors such as cultivar, latitude, climate, state of ripeness, etc. The main constituents of the pulp are water (60–75%) and lipids (oil) (10–25%) (Table 26.1). The 98–99% of the total oil weight consists of a mixture of triglycerides, free fatty acids, waxes, mono- and diglycerides, sterol esters and terpene alcohols, and phospholipids. The fatty acids mainly form esters with glycerol, producing glycerides (mono-, di-, and triglycerides) and phospholipids. Fatty acids also form esters with other linear alcohols (waxes) or terpene alcohols. The fatty acids which are found in larger proportions (such as glycerides) are oleic (C18:1), linoleic (C18:2), palmitic (C16:0), palmitoleic (C16:1), and stearic (C18:0); oleic acid has the highest concentration (55–83% of the oil weight), followed by palmitic (7.5–20%) and linoleic (3.5–21%). Total saturated triglycerides are absent, and the concentration of saturated fatty acids at position two of the glycerine molecule is low.

The olive fruit and processed products, table olive, and olive oil are an important source of linoleic, an essential fatty acid, and monounsaturated fatty acids, so that these products have a high biological and nutritive value.

Carbohydrates and polyols represent 20% of the fresh flesh. The former ones, especially mono- and oligosaccharides, are of great interest in the processing of table olives since it constitutes the available substrate to be fermented by microorganisms. Free sugars (3–4%) include glucose, fructose, and lower levels of sucrose. Small concentrations of xylose and rhamnose are also present. The polyols identified are mannitol and glycerine (Borbolla y Alcalá et al., 1955; Fernández-Bolaños et al., 1982). The fiber fraction (1–4%) (Guillén et al., 1991, 1992) and the pectic substances (0.3–0.6%) (Mínguez-Mosquera et al., 1976, 2002a) constitute a group qualitatively important because of the influence on texture, one of

Table 26.1. Composition of the Green Olive Fruit

Major Components	(g/100 g)
Moisture	60–75
Lipids	10–25
Reducing sugars, soluble	3–6
Nonreducing sugars	≤ 0.3
Polyols	0.5–1.0
Fiber	1–4
Proteins	1–3
Ash	0.6–1
Organic acid and their salts	0.5–1
Phenolic compounds	1–2
Pectic substances	0.3–0.6
Other compounds	3–7
Minor Components	
Chlorophylls (mg/100 g)[a]	1.8–13.5
Carotenoids (mg/100 g)[a]	0.6–2.4
Provitamin A (β-carotene) (μg/100 g)[a]	120–540
Thiamine (μg/100 g)	43–56
Rivoflavin (μg/100 g)	484–550
Minerals (mg/100 g)	
K	283
Ca	51
P	29
Na	8
Mg	14
S	4
Fe	4
Zn	0.5
Cu	0.4
Mn	0.2

Source: Fernández-Díez et al. (1985), Garrido-Fernández et al. (1997).

[a]Data obtained from Mínguez-Mosquera and Garrido-Fernández (1989), Mínguez-Mosquera and Gallardo-Guerrero (1995), Mínguez-Mosquera et al. (1989, 1990a) and Roca and Mínguez-Mosquera (2001a).

the attributes of organoleptic quality in table olives. Protein content is relatively low, between 1% and 3% and all essential amino acids are present although arginine, alanine, aspartic acid, glutamic acid, and glycine predominate (Manoukas et al., 1973). Ash percentage oscillates from 0.6% to 1% and include potassium, followed by calcium, phosphorus, sodium, magnesium, and sulfur and, to a lesser extent iron, zinc, copper, and manganese (Manoukas et al., 1978; Nosti-Vega et al., 1984). Vitamins such as β-carotene and β-cryptoxanthin (provitamin A), thiamine, riboflavin, α-tocopherol, and ascorbic acid are also present (Nosti-Vega et al., 1984; Hassapidou et al., 1994).

Organic acids (oxalic acid, malic acid, and citric acid) and their salts are present in the juice of fruits in concentrations between 0.5% and 1% and are of great importance for table olives because of their buffering capacity during fermentation and storage (Fernández-Díez and González-Pellissó, 1956; Vlahov, 1976).

Olive fruit contains many other substances of a very different nature and chemical structure such as hydrocarbons, terpenes, sterols, alcohols, chlorophyll and carotenoid pigments, tocopherols, polyphenols, and volatile compounds. They are minority compounds but contain the "fingerprint" of the table olive and olive oil. Their composition (global or individual) allows the characterization of the olive cultivars, identification of geographic origin, determination of the system used in olive oil extraction, knowledge of sensory quality, and authentication as opposed to other edible oils, etc. (Mínguez-Mosquera and Garrido-Fernández, 1987; Lanzón et al., 1989; Aparicio et al., 1988; Aparicio and Alonso, 1994; Gandul-Rojas et al., 2000) These compounds also play an important part in the stability and organoleptic characteristics of taste, aroma, and color.

The phenolic compounds (1–2%) are responsible for the characteristic bitterness of olive fruit (Vázquez-Roncero et al., 1974). The main polyphenol in green olives is the oleuropein, a glucoside ester of elenolic acid with 3, 4-dihydroxyphenyl ethanol (hydroxytyrosol). Its content decreases with fruit maturation (Amiot et al., 1986), whereas demethyloleuropein, the acid derivative of oleuropein, becomes the major phenol in black mature olives (Romero et al., 2002). The majority of these compounds are also involved with color changes in olives. Thus, some phenol components are easily oxidized in an alkaline medium to produce the blackening of the Californian-style table olive. Anthocyanins are responsible for the color change during fruit ripening (Maestro-Durán and Vázquez-Roncero, 1976). At the same time as oil content increases during fruit ripening, the color changes from intense green to yellowish green, and finally becomes blackish purple. The green color of the tissue is caused by chlorophyll and carotenoid pigments and their concentrations fall progressively during ripening (Mínguez-Mosquera and Garrido-Fernández, 1989; Minguez-Mosquera and Gallardo-Guerrero, 1995a; Roca and Mínguez-Mosquera, 2001a), giving way to the synthesis of anthocyanin compounds, which appear first as small reddish spots on the skin (turning color or mottled fruit) that cover more and more of the fruit surface in purple until it becomes black at full ripeness. Finally, anthocyanin synthesis invades the interior of the pulp, pigmenting the whole fruit. The "period of ripeness" is the time elapsed from the appearance of the first reddish spots until the definitive coloration of the skin and flesh.

As well as influencing the organoleptic characteristics, the phenol compounds function as antioxidants (Manna et al., 2002) and contribute to virgin olive oil stability by preventing oxidation (Aparicio et al., 1999; Gutiérrez et al., 2001).

Other minor constituents of special importance in the olive fruit are the chloroplastic pigments, chlorophylls and carotenoids, responsible for the green color of table olives and olive oil. The chlorophyll and carotenoid content is subject to wide variations due to the differences among the cultivars, degree of ripeness of the olive fruit, latitude, and environmental conditions. These differences are mainly quantitative since the qualitative composition of the pigments is basically the same in all the olive cultivars and is not modified with the ripening of the fruit. This pattern of chloroplastic pigments is due to chlorophyll a, which is in highest concentration, followed by chlorophyll b, lutein, the major carotenoid β-carotene, and the minority xanthophylls, violaxanthin, neoxanthin, antheraxanthin, and β-cryptoxanthin. The *Arbequina* and *Blanqueta* cultivars are outstanding because in addition to this basic pattern, they exclusively contain other pigments (Gandul-Rojas et al., 1999b; Roca and Mínguez-Mosquera, 2001a).

In addition to a chromatic function, chlorophyll and carotenoid pigments have nutritional benefit (β-carotene and β-cryptoxanthin are precursors of vitamin A) and a biological function. These compounds have considerable potential in preventing damage to human health from mutations, including cancer and other degenerative disorders. The health benefits are derived from the chemoprotective activity against mutagenic agents. Their liposoluble antioxidant nature, capacity to trap reactive oxygen intermediates, and the high affinity of chlorophylls to form molecular complexes with mutagens having polycyclic aromatic structure are responsible for the antimutagenic function (Bendich and Olson, 1989; Dashwood, 1997; Mínguez-Mosquera et al., 2002b).

The pigment content in green olives can vary between 1.8 and 13.5 (mg/100 g fresh pulp) for chlorophylls and 0.6–2.4 (mg/100 g fresh pulp) for

Table 26.2. Typical Chlorophyll and Carotenoid Pigment Composition in Olive Fruit of *Manzanilla, Hojiblanca,* and *Arbequina* Cultivars at Different Degree of Ripeness

	Manzanilla			Hojiblanca			Arbequina		
	YG	TC	B	YG	TC	B	YG	TC	B
Chlorophylls	7.0	3.0	0.5	13.5	7.5	1.1	1.8	1.2	0.2
Carotenoids	1.2	0.6	0.2	2.4	1.7	0.9	0.6	0.6	0.1
β-carotene	0.1	0.1	<0.1	0.5	0.2	0.1	0.1	0.1	<0.1

Source: Mínguez-Mosquera and Garrido-Fernández (1989) and Roca and Mínguez-Mosquera (2001a).

Notes: Degree of ripeness: YG, yellowish green; TC, turning color; B, black. Data expressed in mg/100 g of fresh pulp.

carotenoids in cultivars with high or low pigmentation (Mínguez-Mosquera and Garrido-Fernández, 1989; Roca and Mínguez-Mosquera, 2001a). It is common for all olive cultivars and in general in fruits denominated noncarotenogenic, whose final color is due to the synthesis of compounds of a different nature, that the contents of each of these pigments decrease gradually with the ripening process, although not at the same rate (Table 26.2). The ratio between chlorophylls and carotenoids in the fruit changes with the ripening process due to the greater disappearance rate of chlorophylls as opposed to carotenoids. In all mature fruits, whose purple or black color is due to anthocyanin compounds, chlorophylls as well as carotenoids are still present, although in much lower quantities than that in green fruit.

Olive Cultivars

The earliest olive growers of each zone selected wild olives (acebuche) from woodlands for their productivity, fruit size, oil content, and environmental adaptation. Vegetative propagation maintained the characteristics of cultivars initially selected constituting the first varieties. The repetition of this procedure (spread of cultivars/hybridization, selection of descendance/cloning) was responsible for a great diversity of autochthonous cultivars, products of chance in all the olive-growing zones throughout the world (Barranco, 1997).

Close to 1500 olive cultivars are catalogued throughout the world, although some are the same cultivars under a different denomination. Cultivated olive trees are classified into three categories according to the use given to their fruits: table olive processing, oil extraction, and both purposes, also called double or dual use cultivars.

According to the data reported by the IOOC (2000), a table reports the use and some of the morphologic, agronomic, and technological characteristics of the main olive cultivars (fruit weight, resistance to biotic factors, fat yield, and facility for mechanical harvesting and pitting) (Table 26.3). Principal diseases of olive trees that affect production are repilo, caused by the mushroom *Spilocaea oleagina*; verticilosis, produced by *Verticillium dahliae*; and tuberculosis caused by the bacteria *Pseudomonas syringae pv. savastoni*.

Of the 262 olive cultivars grown in Spain, 24 are classified as main varieties, occupying a considerable geographic area or dominant in at least one region. Only four varieties make up 60% of the olives grown, and one, the *Picual* variety, with more than 700,000 ha cultivated, produces practically half of all Spanish olive oil. Based on the cultivated surface, *Cornicabra*, *Hojiblanca*, and *Lechín de Sevilla* are also important in Spain.

Frantoio cultivar occupies a cultivated surface of 100,000 ha and can be considered the prototype of the Italian cultivars to produce an olive oil of good quality. *Koroneiki* is the main cultivar for oil production in Greece and occupies 50–60% of its cultivated surface. *Ayvalik* cultivar has a high percentage of oil and is considered the Turkish variety with better perspectives for oil production because of the organoleptic quality. *Chemlali of Sfak* is the variety most important in Tunisia and represents 60% of the cultivated surface. The main Portuguese cultivar is *Galega Vulgar* or *Negrihna*, which represents about 80% of its olive production and used primarily for oil production although it is also appreciated as a Greek-style naturally black table olive. In Morocco, the *Picholine marocaine* cultivar occupies 96% of cultivated surface and is a cultivar for dual use, table olives and oil production.

In general, the main technological requirement for processing olives is a facility for mechanical harvesting. For table olives, size and shape, color, texture (delicate pulp and thin epidermis), ease of pit shedding for mechanical pitting, and the pulp-to-pit ratio

Table 26.3. Use and Characteristics of the Main Olive Cultivars

Cultivars	Use	Fw	Fy	Resistance to Biotic Factors			Facility For	
				R	T	V	Mh	Mp
Spanish								
Aloreña	T	High	Low	Medium	Low	Low	High	Medium
Arbequina	O	Medium	High	Low	Low	Medium	Medium	Low
Blanqueta	O	Medium	Low	High	High	Low	Medium	Low
Changlot Real	O	Low	High	Low	Medium	Low	Medium	Low
Cornicabra	O	Low	High	Medium	Medium	Medium	Medium	Low
Empeltre	O	Low	High	Medium	Medium	High	High	Low
Farga	O	Low	High	Medium	High	Medium	Medium	Low
Gordal Sevillana	T	High	Medium	High	Medium	Low	Low	Medium
Hojiblanca	O & T	High	Medium	Medium	Medium	Medium	Medium	Medium
Lechín de Granada	O	Low	High	Medium	Medium	Low	Medium	Low
Lechín de Sevilla	O	Low	Low	High	Medium	Low	Medium	Low
Manzanilla Cacereña	O & T	High	Medium	Low	Low	Medium	High	High
Manzanilla de Sevilla	T	High	High	Medium	Medium	Medium	Medium	High
Morisca	O & T	High	High	Medium	Medium	Low	Low	Low
Morrut	O	Low	High	Medium	Low	Low	High	Low
Picual	O	Low	High	Medium	High	Medium	High	Low
Picudo	O	High	High	Medium	Medium	Low	Medium	Low
Sevillenca	O	Low	High	Medium	Low	Low	High	Low
Verdial de Badajoz	O	High	High	Low	Low	Low	Medium	Low
Verdial de Huevar	O	High	High	Medium	High	Medium	Medium	Low
Villalonga	O & T	High	High	Medium	Medium	Low	High	Medium
Italian								
Ascolana	T	High	Medium	High	High	Low	Low	High
Coratina	O	High	High	Low	Low	Low	Low	Low
Frantoio	O	Low	Low	Medium	Medium	Low	Low	Low
Moraiolo	O	Low	High	Medium	Medium	Low	Low	Low
Nocellara del belice	T	High	Low	Medium	Medium	Medium	Low	High
Nocellara etnea	O & T	High	Medium	Medium	High	Low	Medium	Low
Sant'Agostino	T	High	Medium	High	Medium	Medium	Low	High
Greek								
Chalkidiki	T	High	Low	Low	Low	Low	Low	High
Kalamon	O & T	High	Low	Low	Low	Low	Low	High
Konservolia	O & T	High	Low	Low	Low	Medium	Low	High
Koroneiki	O	Medium	High	High	Medium	Low	Low	Low
Turkish								
Ayvalik	O & T	High	High	Low	Low	Low	High	Medium
Domat	T	High	Low	Low	Low	Low	Low	Medium
Memecik	O & T	High	High	Low	Low	Low	Low	Medium
Memeli	O & T	High	Low	Low	Low	Low	Low	Medium
Tunisian								
Chemlali	O	Medium	High	Low	Medium	Low	Low	Low
Chétoui	O & T	Low	Low	Medium	Low	Low	Low	High
Meski	T	High	Medium	Medium	Low	Low	Low	High
Portuguese								
Carraquenha	O & T	High	Low	Low	Medium	Low	Medium	Low
Galega vulgar	O & T	Low	Medium	Low	Low	High	Medium	High
Marocaine								
Meslala	T	High	Low	High	Medium	Low	High	High
Picholine marocaine	O & T	Low	Low	Medium	Low	Low	Low	High

Source: International Olive Oil Council (IOOC, 2000).

Note: Fw, Fruit weight; Fy, Fat yield; R, repilo, T, tuberculosis; V, verticilosis, Mh, mechanical harvesting, Mp, mechanical pitting; O, Olive oil; T, Table olive. ■ Height □ Medium □ Low For Fw: Height: >4g, Medium: 2–4 g, Low: <2g; for Fy: Height: >22%, Medium: 18–22%, Low: <18%.

are also very important. Ideally the pulp-to-pit ratio must be around 6.5. Higher values can produce a soft final product; lower ratios may be difficult to handle during pitting and stuffing. In this respect, the *Manzanilla* cultivar is considered one of the best by most table olive processors.

The Spanish varieties that best meet this criterion are *Gordal Sevillana* (almost exclusive to the province of Seville), *Manzanilla de Sevilla* (found in the Bajo Guadalquivir valley and in Badajoz, where it is known as Carrasqueña), and *Hojiblanca* (widespread in the provinces of Cordoba and Malaga). Other, less grown varieties are *Aloreña, Morisca,* and *Villalonga.*

Manzanilla de Sevilla is the cultivar, which is internationally most widespread because of high productivity and fruit quality. It has been exported to many other countries including Portugal, the United States, Israel, Argentina, and Australia.

Ascolana is the best Italian cultivar for green table olives and has been exported to many other countries including the United States, Mexico, Israel, and Argentina. On the international market, *Ascolana* cultivar is often confused with the Spanish cultivar, *Gordal Sevillana,* because of its large size although in other characteristics they are really quite different. *Sant'Agostino* and *Nocellara del belice* cultivars are also appreciated as table olives.

The three main Greek olive cultivars used for table olives are *Konservolia, Kalamon,* and *Chalkidiki.* The *Konservolia* cultivar is by far the most important economically, occupying 50–60% of the cultivated surface intended for the production of table olives. *Kalamon* is used mainly for specialities within the Greek-style naturally black olives in brine and *Chalkidiki* is used mainly for Spanish-style green olives in brine.

The most important Turkish cultivars for table olive production are *Domat* (similar to Spanish cultivar called Aloreña) and *Memeli.* In Tunisia, *Meski* is the cultivar more appreciated as table olive. *Carrasquenha* is the preeminent Portuguese cultivar for Spanish-style green table olives and *Meslala* is a marocaine cultivar with minor importance but appreciated as Spanish-style green table olives.

TABLE OLIVE

INTRODUCTION

Table olives are currently the most important fermented vegetable product in the developed world. The first use of olive fruit as table olives is lost in time; the first recorded reference is that of Colmuela, dating from the first century, in the year 54 (Fernández-Díez et al., 1985). Today, Spain is the leading producer and exporter of table olives, at 25% and 50%, respectively, of the world total. In Spain, research on the table olive was begun in 1947, by Borbolla y Alcalá et al. (1955). The hard work of these scientists achieved the conversion of what had been until then in Spain, and more specifically in Andalucía, an industrialized craft to a technological process subject to physicochemical and microbiological rules.

Table olives are defined by the Unified Qualitative Standard Applying to Table Olives in International Trade as "the sound fruits of specific varieties of the cultivated olive tree (*O. europea sativa* Hoffm. Link) harvested at the proper stage of ripeness and whose quality is such that, when they are suitably processed, produce an edible product and ensure its good preservation as marketable goods. Such processing may include the addition of various products or spices of good table quality" (IOOC, 1980).

The three table olive preparation methods or styles of greatest importance in international trade are as follows:

- Spanish (or Sevillian)-style green olives in brine.
- Californian-style black olives in brine.
- Greek-style naturally black olives in brine.

For each style, the fruits are harvested at different degrees of ripeness, yellowish-green, turning color, and black, respectively. The style indicates where the process was first developed and used. Figure 26.1 shows the general scheme of these processes.

The main purpose of table olive processing is the removal, at least partially, of the natural bitterness of the fruit, to make it acceptable as a food or appetizer. In both the Spanish-style and Californian-style processes, the bitterness is removed by means of an alkaline hydrolysis. In the case of the Greek-style process, fruit is placed directly in brine for fermentation and the removal of the bitterness is slow and only partial.

Lactic fermentation gives Spanish-style green olives unique and valued organoleptic characteristics. For Greek-style naturally black olives, the main organisms responsible for the fermentation process are yeasts (González-Cancho et al., 1975). California-style black olives use turning color fruits that are blackened by chemical oxidation in an alkaline medium and do not necessarily require a fermentation process. California-style black olives may

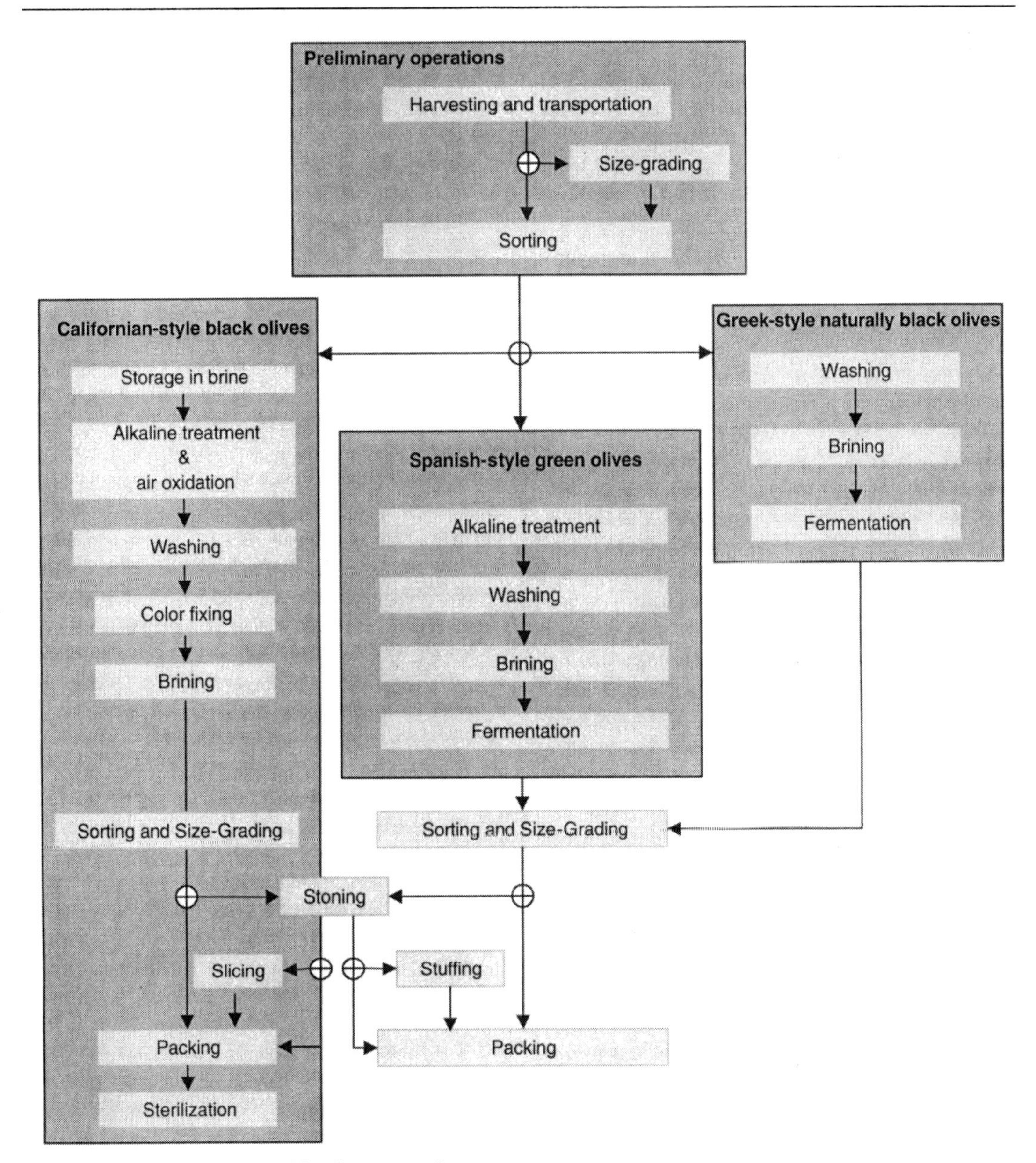

Figure 26.1. Scheme of the table olive processing.

be treated with a solution of sodium hydroxide and oxidized, then washed, placed in brine, packed in cans, and preserved by sterilization. However, some of the crop may be preserved in brine for a variable period prior to the oxidation, in which case a fermentation similar to that of the Greek-style naturally black olives will occur (Fernández-Díez et al., 1985; Garrido-Fernández et al., 1997).

The world production of table olives continues increasing and it has reached a level of 1,748,000 tons in the 2002–2003 season, 30% more than the average of the past four seasons (1998–1999 to 2001–2002), that is, it was ciphered as 1,342,250 tons. The green olives represent more than 45% of the total production, Spain being the main producer. The naturally black olives signify approximately a 35% of the total, Turkey being the greater producer followed by Greece and Italy. The United States maintain the predominance in the production of Californian-style black olives (IOOC, 2003a). Below it is described in greater detail the elaboration process denominated "Spanish-style green olives in brine" since it is considered the most important commercial preparation of the international market.

Green Olive Processing Description

Preliminary Operations

The optimum moment for picking is when the external coloring of the fruit is yellowish green as a result of the exclusive presence of chlorophyll and carotenoid pigments. If the fruits are picked too early, fermentation develops only with difficulty, as the olives are hard and not very palatable; if picked at a later stage of ripeness, when their color begins to become pinkish with the synthesis of anthocyanins, the product is soft and preservation is poor.

Fruits are picked manually to avoid damage, and deposited in specially designed containers permitting good aeration. Normally, the smaller fruits are separated out, as are leaves and twigs. When the fruit arrives at the processing plant, a representative sample is selected to evaluate the percentage of unusable sizes, mean size, size distribution, and percentage of defects.

Alkaline Treatment, Washing, and Brining

Treatment with a dilute solution of sodium hydroxide (lye) is an essential operation in green table olive processing. The main aim is the hydrolysis of the glucoside oleuropein, responsible for the characteristic bitterness, although its action in the fruit components is much more complex, and enables the development of a suitable culture medium for lactic fermentation during brining. The concentration of the alkaline solution is adjusted, so that depending on the ambient temperature, the treatment time is such that penetration of the alkaline solution into the pulp is 2/3 or 3/4 of the distance from skin to pit. This treatment time depends on the characteristics of the cultivar (mainly texture and bitterness), the state of ripeness, and the average fruit size. As a general rule, for the majority of cultivars, this is between 5 and 7 h, except *Gordal* and *Ascolano*, which require 9–10 h with less concentrated lye, because of their texture and bitterness characteristics. A prior classification of the fruit by size is very effective for a more uniform alkaline treatment.

At the end of the alkaline treatment, the lye solution is drained off and the fruit is covered with water in the operation of washing to eliminate the sodium hydroxide solution adhering to the surface and penetrating the pulp. The number and duration of washing varies, although there is normally a first, short washing of 2–3 h, followed by a longer one of 12–15 h. The washing should not be too vigorous, to avoid the excessive loss of hydrosoluble components necessary for the subsequent development of fermentation. Table 26.4 details the characteristics of the alkaline treatment and washing for the main cultivars used in Spain (Rejano, 1997). Because of the scarcity of water and the pollution produced by these wastewaters, the current practice is to perform a single washing of 12–15 h. An excess of sodium salts of organic acids, formed by the reaction of residual sodium hydroxide and acids from the fermentation, is corrected by adding the exact equivalent of a strong acid such as hydrochloric acid (Rejano et al., 1986).

After washing, the fruit is transferred to fermentors and covered with a 10–11°Bé (10%) sodium chloride solution. The concentration of this solution (brine) decreases as the salt penetrates the fruit, with equilibrium established at 5–6% in a few days. If the initial concentration of sodium chloride is excessive, the fruit wrinkles as a result of the higher exterior osmotic pressure, whereas a lower concentration and an equilibrium value of less than 5%, favors the development of certain alterations.

Table 26.4. Characteristics of the Alkaline Treatment and Washing for the Main Cultivars of Table Olives

		Alkaline Treatment		Wash Time (h)	
Cultivar	Temperature (°C)	%NaOH (w/v)	Time (h)	First	Second
Gordal Sevillian	25	2.0	9–11	2–3	12–16
Manzanilla Sevillian	20	2.5	6–7	2–3	12–16
Hojiblanca	15	3.8	6–8	1–5	12–15

Source: Rejano (1997).

Table 26.5. Characteristics of the Fermentation Phases in the Spanish-Style Green Table Olive Processing

			Microorganisms	
Fermentation Phases	Time (days)	pH	Relevant Species	Development
First phase	2–3	10–6	Sporulated Gram-positive bacteria	++
			Gram-negative bacteria	+ + +
			Cocaceae lactic acid bacteria	++
			Lactobacillus	+
Second phase	15–20	6–4.5	Cocaceae lactic acid bacteria: *Pediococcus, Leuconostoc*	+ + +
			Lactobacillus	++
Third phase	30–60	≤4	*Lactobacilus*: *L. plantarum, L. delbrueckii*	+ + +
Conservation	60–120		Consumed fermentable matter	

Source: Fernández-Díez et al. (1985) and Garrido-Fernández et al. (1997).
Note: + + +, abundant; ++, scarce; +, latent.

Fermentation and Preservation. The sugars, vitamins, and amino acids of the fruit pass into the brine by osmosis, and the brine gradually converts into a culture medium suitable for growth of microorganisms. In the first days of brining, due to the residual lye continuing to come out of the pulp, the pH value exceeds 10 units. Through the various stages of fermentation (Table 26.5) (Fernández-Díez et al., 1985), the succession of different species of microorganisms reduce the pH to values of 4 or less, helping the proper long-term preservation of the product. The first phase lasts from brining until the growth of Lactobacilli, which normally occurs after 2–3 days. In this phase, the bacteria detected are sporulated Gram-positive, cocaceae lactic acid, and Gram-negative, mostly belonging to the Enterobacteriaceae family. The second phase begins when the pH of the medium drops to 6.0 and usually lasts 15–20 days. Cocaceae lactic acid bacteria is the characteristic microorganism of this phase, mainly from the genera Pediococcus (homofermentative) and Leuconostoc (heterofermentative). Gram-negative bacteria counts decrease progressively and the total disappearance indicates the end of this phase. The third phase is characterized by an abundant growth of lactobacillus, mainly from the species *Lactobacillus plantarum*, which becomes predominant under normal conditions and is responsible for the typical fermentation of the Spanish-style green table olives. *Lactobacillus delbrueckii* is also identified although its significance is rather limited (González-Cancho, 1963; González-Cancho and Durán-Quintana, 1981).

When the production of acids is stopped by the consumption of fermented matter, preservation begins. At this stage, the salt concentration must be increased to levels of 8.5–9.5% (Rejano, 1997) to prevent the development of a fourth (unwanted) phase of fermentation by bacteria of the genus *Propionibacterium* (González-Cancho et al., 1980). These bacteria consume lactic acid, causing a rise in pH that excludes guaranteeing the proper preservation of the product.

Lastly, the olives are prepared for packing and marketing. First, they are placed on belts where defective fruits are separated manually or automatically using electronic machines. They are then sized-graded by divergent-cable devices. The brines from the different fermentors are mixed, and the acidity and salt concentrations are adjusted to reduce the natural variability. The final olive product, sorted and size graded, is occasionally pitted and stuffed, and packed in small hermetically sealed, tin or glass containers, with a new brine. The packing medium must be kept within the following ranges: pH 3.2–4.1; free acidity 0.4–0.6% as lactic acid, sodium chloride 5–7%, and combined organic acidity or residual lye 0.02–0.07 N.

When the sequence of microorganisms is not correct, the growth of other undesirable ones produces an altered end product, which can be avoided with the proper control of the fermentation process. This involves an initial decrease in pH and maintenance of a suitable temperature (22–25°C) for at least 30 days, using a heat exchanger if necessary. It is also helpful to add a starter culture of lactobacillus (Balatsouras et al., 1983; Ruiz-Barba et al., 1994) or brine from other fermentors in the third phase of fermentation (pH < 4.5 and absence of Gram-negative Bacteria). If necessary, fermentable matter can be added to complete the fermentation process and to achieve a suitable final pH value.

In accordance with the established quality regulations, the table olive must fulfill certain chemical and organoleptic characteristics. The required levels of acidity and salt concentration are adjusted by specific chemical analysis. The type and number of defects are also specified, but there are no objective measures for color, texture, or flavor, and the rules only require that they must be suitable.

Effect of Processing on Components Present in Fresh Fruit

Table 26.6 shows typical values in the final composition of different styles of table olives (Garrido-Fernández et al., 1997). Quantitatively, changes in composition are in the soluble sugars and ash. Soluble sugars disappear and are absent in the final product. During fermentation, sugars pass into the brine and are metabolized by microorganisms. In general, the osmotic exchange between fruit and brine is rapid because of previous lye treatment and washing. The extraction of soluble components is apparently selective with the reducing sugar extraction rate inverse to that of the other organic and inorganic components (Borbolla y Alcalá et al., 1955). Thus, a higher carbohydrate content in brine is obtained with weaker brines, whereas high NaCl brines can provoke surface shriveling and lower sugar diffusion.

The percentage of ash increases as a consequence of alkaline treatment, fermentation, and storage in

Table 26.6. Typical Composition of Table Olives

	Spanish-Style Green Olives[a]	Californian-Style Black Olives[b,c]	Greek-Style Naturally Black Olives[b,d]
Major components (g/100 g)			
Moisture	69[e]	69	60
Lipids	21[e]	21	23
Fiber	1.4[e]	2.4	2.2
Proteins	1.0[e]	1.1	1.2
Ash	4.6[e]	2.0	6.7
Minor components			
Chlorophylls (mg/100 g)	4.2[f]	Nd	Nd
Carotenoids (mg/100 g)	1.0[f]	Nd	Nd
Provitamin A (β-carotene) (μg/100 g)	260[f]	130	Nd
Thiamine (μg/100 g)	0.4[e]	0.2	15
Rivoflavin (μg/100 g)	Nd	96	91
Minerals (mg/100 g)			
K	44[e]	12	29
Ca	54[e]	68	28
P	7[e]	15	3
Na	1435[e]	634	3740
Mg	15[e]	18	8
S	16[e]	8	6
Fe	0.6[e]	7	0.3
Zn	0.3[e]	0.49	0.25
Cu	0.5[e]	0.29	0.06
Mn	0.06[e]	0.14	0.02
Caloric value (cal/100 g)	210[e]	222	250

Note: Nd, not determined.

[a]Typical values for *Manzanilla* cultivar.

[b]Typical values for *Hojiblanca* cultivar.

[c]Data according to Nosti-Vega and Castro-Ramos (1985).

[d]Data according to Nosti-Vega et al. (1984).

[e]Data according to Castro-Ramos et al. (1980).

[f]Data according to Mínguez-Mosquera et al. (1990b).

brine due to the absorption of salts by the flesh. Sodium increases in all types of processing. In Californian-style black olives, there is an increase in iron as a consequence of the use of ferrous salts (gluconate or lactate) in the fixing phase of darkening.

Changes in minority components are very relevant because of the influence on organoleptic characteristics and biological function of the final product. Polyphenols undergo chemical transformations and decrease in concentration in olives during processing. One of the main steps in the processing of Spanish-style green olives is the debittering of fruit under alkaline conditions in which oleuropein is hydrolyzed into hydroxytyrosol (Amiot et al., 1990) and elenolic acid glucoside (Brenes and Castro, 1998). The subsequent lactic acid fermentation does not modify the qualitative phenolic composition (Brenes et al., 1995).

In the Californian-style black olives, during the initial preservation phase, the polyphenols, mainly oleuropein, diffuse from olive flesh into the brine and undergo acid hydrolysis (Brenes et al., 1993). Subsequently, orthodiphenols are oxidized and polymerized during the darkening step (Vincenzo et al., 2001). In Greek-style naturally black olives, the main changes in polyphenols are acid hydrolysis of oleuropein and hydroxytyrosol glucoside and the polymerization of anthocyanin pigments present in the ripe fruit that contributes to color stabilization (Vlahov and Solinas, 1993; Romero et al., 2004).

The optimum final texture is obtained only when the raw material of the correct maturity is used but subsequent treatment during processing can also have an effect. Lye treatment always produces a marked loss of texture, which persists during washing but recovers partially during fermentation due to the effect of salt and the absorption of divalent cations (Mg^{++} or Ca^{++}) from the brine. High sodium hydroxide concentration and temperature can degrade this attribute markedly. An excessive sodium chloride concentration in the initial brine may also cause shriveling that can persist to the final product.

During processing, substantial amounts of cell wall polysaccharides are solubilized. While all noncellulose components appear to be involved in the texture changes, the pectic polysaccharides appear to be most directly affected. Apparently, processing has the greatest impact on neutral sugar-poor less branched polymers. The most important pectic polymer solubilized is uronic acid (Mínguez-Mosquera et al., 1976; Jiménez-Araujo et al., 1994).

Color is one of the organoleptic characteristics most changed during Spanish-style green table olive processing. The chlorophyll and carotenoid pigments responsible for the yellowish-green color of the fresh fruit undergo certain structural transformations as a result of the alkaline treatment as well as the acidity generated in the fermentation medium. The chlorophylls (a and b) initially present in the fresh fruit are totally degraded by different mechanisms. In traditional processing, two mechanisms of chlorophyll degradation coexist (Mínguez-Mosquera et al., 1989). One is enzymatic in origin, produced by the activation of chlorophyllase under alkaline conditions; the other is chemical, arising from the acid pH of the medium. The phytol deesterification reaction catalyzed by chlorophyllase does not affect the properties of the chromophore, altering only chlorophyll solubility. In contrast, the acidity generated during fermentation causes color changes as a consequence of the substitution of Mg^{++} by H^{+} in the chromophore group (porphyrin ring) of chlorophylls and chlorophyllides. This reaction, known as pheophytinization, is responsible for the color change from green to gray brown. Mínguez-Mosquera et al. (1994) established the kinetic parameters governing these reactions, shown in Figure 26.2 in a diagrammatic form.

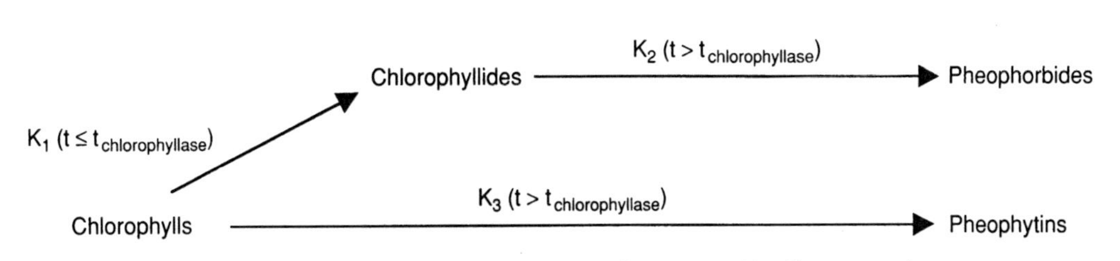

Figure 26.2. Kinetic scheme for the chlorophyll degradation during green table olive processing.

Table 26.7. Composition Changes During Table Olive Processing

	Changes in Composition		
Operation	Spanish-Style Green Olives	Californian-Style Black Olives	Greek-Style Naturally Black Olives
Fermentation in brine		Optional	Fundamental phase
		Slow loss of sugars, tannins and oleuropein; formation of organic acids and aromatic compounds.	
Alkaline treatment (debittering process)	Hydrolysis of oleuropein; formation of organic acids; and parcial desesterification and oxidation of chlorophylls.		
Washing	Loss of sugars, organic acids, and phenols; formation of organic acids.		
Oxidation by air		Polyimerization of phenolic compounds.	
Brining	Diffusion of fermentable Substrates.		
Lactic acid fermentation	Formation of lactic acid and other organic acid; pheophytinization of chlorophyll pigments and isomerization of epoxidized xanthophylls.		
Conservation and storage	Formation of propionic acid and other volatile compounds from lactic acid.	None, under normal conditions.	

Source: Garrido-Fernández et al. (1997).
Note: ▒ Operation not included.

While the enzyme chlorophyllase is active, a part of the chlorophylls are transformed into chlorophyllides; the rest of the chlorophylls remain available for the pheophytinization reaction. This takes place in parallel with the formation of pheophorbides.

The carotenoid pigment fraction is affected only in those components with a molecular structure sensitive to acid medium (epoxidated xanthophylls), which undergoes isomerization, with a decrease in the intensity of yellow coloring (Mínguez-Mosquera and Gandul-Rojas, 1994).

All transformations of chlorophyll and carotenoid pigments are desirable, as they are precisely those that give the highly regarded yellow-golden color of the Spanish-style green table olive. The innovations introduced in the traditional system of processing (elimination of the short washing, addition of acids, reuse of brines, etc.) have changed the previously described mechanism of chlorophyll degradation, with the detection of oxidative reactions that affect the chlorophyll isocyclic ring and production of allomerized chlorophylls (Mínguez-Mosquera and Gallardo-Guerrero, 1995b).

Table 26.7 summarizes the most relevant compositional changes produced during the processing of table olives (Garrido-Fernández et al., 1997).

Alterations During Green Table Olive Processing

Fruits may undergo different kinds of alterations, which may not be a risk to health but depreciate the quality of the final product due to deleterious changes in the olive organoleptic characteristics. Some examples of these alterations, most of them due to the presence of undesirable microorganisms, are given below (Fernández-Díez et al., 1985; Garrido-Fernández et al., 1997; Rejano, 1997).

Fish-Eye and Gas-Pocket

This is a characteristic alteration in the first phase of fermentation and is due to the action of Gram-negative bacteria, producers of carbon dioxide and hydrogen. The accumulation of these gases causes cavities inside the pulp and fissures on the surface, or even blisters under the skin. These changes can be prevented by adjusting the pH at the beginning of fermentation, using a flow of carbon dioxide or acidification of the brine with acetic acid.

Butyric Alteration

This is most often produced during the first two phases of fermentation and is due to the growth of different species of the genus *Clostridium*. The butyric acid produced by these species during fermentative metabolism alters the aroma and flavor of the product. It can be prevented by making sure that the sodium chloride concentration does not fall below 5% and by following good practices of industrial hygiene.

"Zapatería"

This is due to the growth of bacteria of the genera *Clostridium* and *Propionibacterium* during the preservation phase. The volatile compounds from their metabolism (Montaño et al., 1992) give the olive a very unpleasant smell and flavor. The alteration is prevented by raising the salt concentration to inhibit the growth of the responsible microorganisms and by maintaining the pH value below 4.2.

Softening

This appears above all during the preservation phase and is caused by an excessive growth of microorganisms with pectinolytic enzymatic activity (bacilli, yeasts, and molds). It is prevented by maintaining a good anaerobic seal.

Sediment and Gas

These result from the growth of bacteria or yeasts in the packed product under unstable conditions. Using a well-fermented product and adjusting the pH inside the pack to a value of 3.3 units, or stabilizing the product by heat treatment, such as pasteurization, prevents the alteration.

Green Staining

The *Gordal* Spanish cultivar has long presented the problem of the occasional appearance of green staining on the surface of fruits processed as Spanish-style table olives (Mínguez-Mosquera et al., 1995). The visible manifestation of this alteration is due to the presence of copper–chlorophyll derivatives in different degrees of oxidation in the processed fruit (Gandul-Rojas et al., 1999a; Gallardo-Guerrero et al., 1999). The possibility that the copper involved in this alteration is exogenous in origin has been ruled out (Mínguez-Mosquera et al., 1995). The appearance of copper–chlorophyll complexes in olive fruits of the *Gordal* cultivar during industrial processing implies that there has been a strong oxidative disintegration of the chloroplast in at least some fruits, allowing contact between the chlorophyll pigment and the copper of the fruit.

VIRGIN OLIVE OIL

Introduction

The Spanish word "aceite" comes etymologically from the Arab word "azzait," meaning the juice of the olive. The oil obtained from olive fruit has served during centuries as a food, prime material for lighting, medicinal ointment, and a revitalizing liquid for the human beings.

Today, olive oil is exceptional due to its great demand and high profitability, thanks to the successful campaign, which highlights its therapeutic and nutritive properties. It has an encouraging future and an increasing surface area of cultivation. Estimated world production for the 2002–2003 season by the IOOC was 2,515,000 tons. Mediterranean countries produce 98% of the world's olive oil with Spain (36%), Italy (21%), and Greece (17%) being the main producers (IOOC, 2003b).

Virgin olive oil has exceptional organoleptic characteristics and is practically the only vegetable oil that can be consumed raw, conserving intact its nutritive and functional properties. The deterioration in virgin olive oil is almost exclusively due to the faulty handling of the fruit and poorly controlled production process. Independent of the variety, only fruits which have suffered pests or diseases or have fallen on the ground before harvesting inevitably produce altered oil requiring refining due to its high free acidity and/or its disagreeable organoleptic characteristics. Figure 26.3 shows the general scheme of the

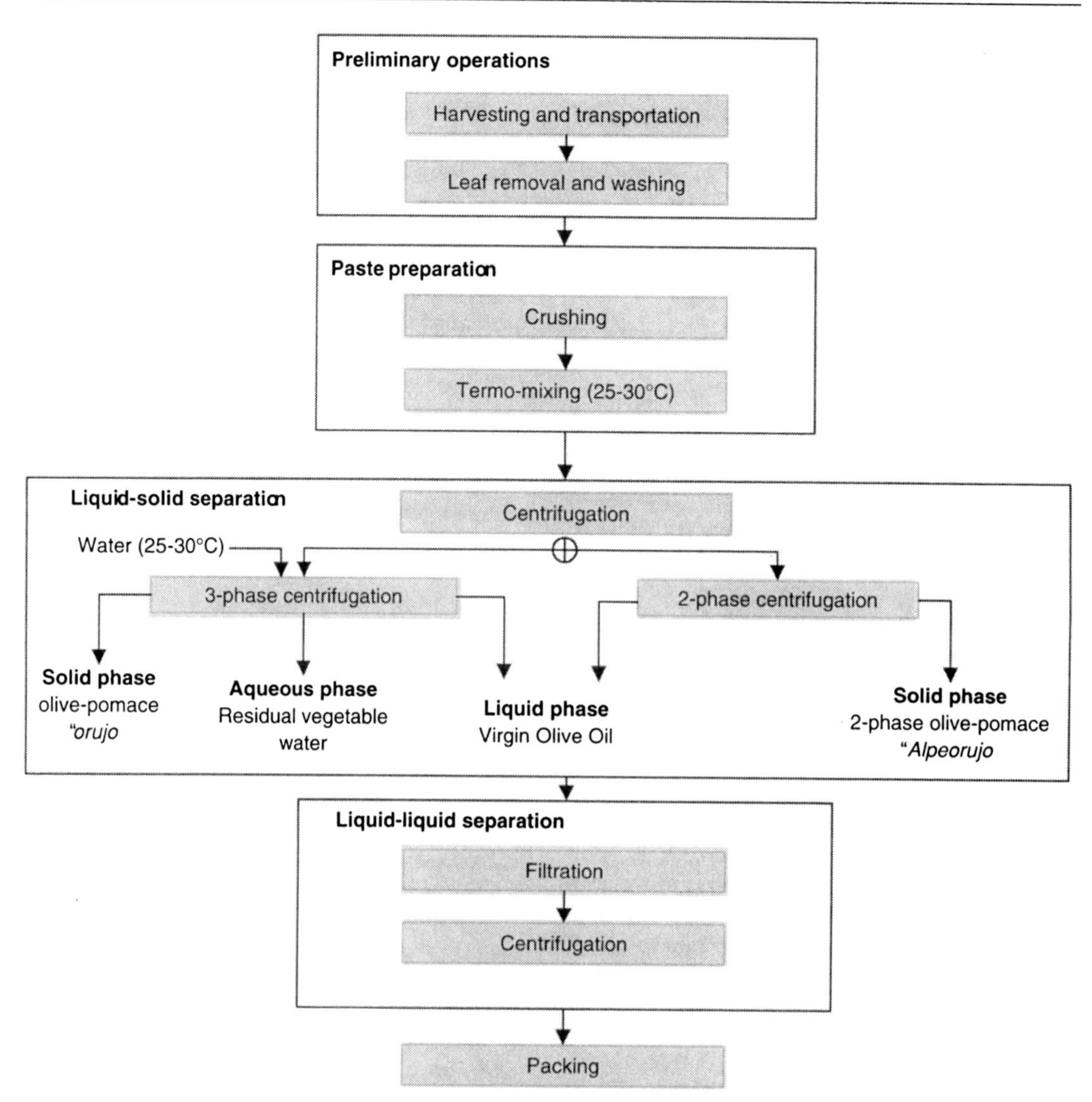

Figure 26.3. Diagram showing olive oil extraction by centrifugation methods.

olive processing to obtain virgin olive oil (Boskou, 1996; Alba, 1997; Giovacchino, 2000).

Processing Description

Preliminary Operations

Harvesting. The time of year in which this operation is carried out, as well as the system used, influences considerably the characteristics of the olive oil obtained. The fruit needs to be harvested at the optimum degree of ripeness taking into account maximum oil content and best characteristics. The factors, which have an impact on the optimum time of harvesting, are (Porras, 1997):

- Resistance to the pull of the stalk of the olive.
- Oil content of the fruit.
- Changes in the oil quality.
- The falling of the fruit.
- Date of the previous harvest.

The resistance to stalk detachment is determined by the force necessary to break the pull of the fruit stalk, which can vary enormously throughout ripening and has a bearing on the performance of the operation, if the harvest is manual. The harvest must coincide with the time when the green fruit disappears from the tree, which is also when the maximum oil content is reached. The organoleptic characteristics of the oil deteriorate when the harvest of the fruit is delayed. If more fruity and aromatic oils are desirable, the harvest can be advanced by a few days. With a substantial percentage of green fruit, a better quality product is obtained, although the production yield of the oil is decreased.

The natural fall of the fruit basically depends on the cultivar, although it can be influenced by climactic conditions and the health of the fruit. The end of the harvest should coincide with the time when the natural fall of the olive starts and should reach a percentage that would have a significant effect on the cost of the harvest.

Lastly, a delay in the time of the harvest also produces certain inhibition in the floral differentiation of the leaf buds that can result in losses in the harvest the following year.

Harvesting that spoils the fruit and produces damage, with bruises, broken branches, or twigs should be avoided. In general, the fruit is brought down manually from the tree, known as handpicking, or mechanically, using a trunk shaker. The fruit is collected in nets of cloth or plastic material that extend below the olive trees, occupying an area larger than the drop zone of the tree. Finally, the contents of these nets are poured into boxes or trailers for transferring to the olive-oil mill. The daily capacity of oil extraction in the mill is normally lower than the rate of receiving the harvested fruit, requiring fruit to be stored in the so-called trojes (or silos) during a variable period of time. During this time, the olives can undergo various changes such as fermentation, enzymatic and microbial lipolysis, autoxidation of fatty acids, etc.; these changes will damage the quality of the oil by increasing acidity, decreasing stability, and degrading organoleptic characteristics. The fundamental objective during this preservation period is to maintain the fruit without changing the oil characteristics and without significantly increasing the cost of the process.

Leaf Removal and Washing. This operation is necessary to prevent contamination from impurities that affect the organoleptic characteristics of the oil, as well to reduce the wear and tear on the mills. The winnowing machine is used as a cleaning system, which separates the impurities of lesser weight than the olive (leaves and twigs) using an adjustable flow of air, and water is used for the wash that eliminates heavier objects such as mud, stones, etc. There are washers that operate by density, adding common salt to the water until the olives start to float, and others that work by trawling, adjusting the circulation rate of the water in such a way that it only carries the fruit.

Paste Preparation

A paste is produced by means of crushing and mixing operations with the objective of separating the olive oil as a continuous oily liquid phase from the rest of the components of the olive, without changing the composition and organoleptic characteristics.

Crushing. This operation breaks down cellular membranes to liberate oil droplets. These droplets group together in variable sized drops that come into contact with the aqueous phase present in the paste, which comes from the vegetation water and the remains of the washing water. This operation is fundamental for the rest of the extraction process and influences the industrial yield and quality of the oil. Currently, the traditional mills with their cone-shaped millstones have been replaced by metallic crushers, generally of the hammer type.

Thermo-Mixing. The olive paste obtained after crushing is mixed slowly with the objective of gathering together the larger sized drops of oil and combining into a continuous oily phase, with the separation from the solid and the aqueous phase (Martínez–Moreno et al., 1957). To improve the yield of this operation, the mixers have a heating system, which helps to decrease the viscosity of the oil and makes outflow easier. The temperature in the thermomixer should not exceed 25–30°C, to prevent loss of aromatic compounds and an increase in the oxidation process.

Liquid–Solid Separation

This stage constitutes the fundamental part of obtaining oil, with the objective of separating liquids contained in the olive paste, that is, the oil and residual water (liquid phase), and the solid mixture (skin, pulp and broken pits), which is known as "olive-pomace." Three separation systems are used in the industry: selective filtration, extraction by pressure, and extraction by centrifugation.

Selective Filtration. After mixing, it is possible to obtain an important part of the freed olive oil contained in the olive paste, by means of cold extraction by selective filtration. In this way, the first oil contained in the fruit is obtained, which has superior characteristics to that obtained later by mechanical extraction. This system is recommended for obtaining high-quality oil but is used little due to high cost (Hermoso et al., 1991).

Extraction by Pressure. The hydraulic press system was used traditionally. It consists, in essence, of placing the paste, once mixed, over some filters made of esparto or synthetic fiber (or a mixture of both), known as pulp mat or olive mat, making up the pressing or load unit. This load, located over a trolley with a central spike, is installed in a container in the press, in such a way that by pressing the load over this container, the filtrate of the liquids contained in the paste is obtained. These are fundamental factors in the execution of the pressing: the paste preparation, its distribution and thickness on the olive mats, the state of the olive mat, specific pressure of the press, and the speed and time of pressing (Martínez-Moreno et al., 1964).

Extraction by Centrifugation. The use of centrifugal force is the modern method to carry out the liquid–solid separation. It is carried out in dynamic phase systems, where the solids are displaced what is long of the draft shaft and continuously removed. The mixed paste is introduced (partially diluted in water) into a horizontal centrifuge called a decanter. This consists, essentially, of a conical-cylindrical revolving rotor and a hollow bore helicoidal scraper, which revolves coaxially in its interior and at a different speed. By the action of the centrifugal force, the solids are pushed to the interior wall of the rotor and are drawn to the furthest point from the axis of spin, that is to say, the space nearest to the wall of the warm gear motor, continuing in this manner until it runs out of liquid. The liquid (oil and aqueous phase) forms concentric rings, which depending on their density move further toward the middle. Two outlets, situated at a different level in the periphery of the decanter, are used as exits for the residual water and oil (Giovacchino and Mascolo, 1988). Therefore, three phases are formed: a solid-olive-pomace or "orujo"—and two liquids–oil and residual vegetable water—and the system is known as three-phase centrifugation.

This system produces a final aqueous phase (residual vegetable water) of approximately 0.7–1.1 l/kg of olives. This effluent contains a high load of contaminants, as assessed by the chemical demand for oxygen and spillage into public waterways negatively affects the flora and fauna (Alba et al., 1995). Government measures compel the olive mills to put a purification system in place. The first system recommended as an emergency measure was the storage of the residual water in pools for natural evaporation. At the same time, exploitation and purification techniques were developed but were not generally accepted by the sector, due to the level of efficiency and installation and operational costs.

Due to this situation, nowadays, the consumption of water tends to be lower to minimize the volume of residual vegetable water produced and thus reducing the environmental problem caused by lack of purification. A variation of the process has been developed called a two-phase centrifugation system, which does not need the fluidity by the addition of extra water. The separation of the olive paste is achieved in only two phases, a solid-olive-pomace and the other liquid–oil. In this system, the vegetable water of the olive is incorporated into the solid phase, which is distinguished from that obtained by the three-phase process for it is a more moist and pliable consistency and is now being called two-phase olive-pomace or "alpeorujo" (Hermoso et al., 1995). The final effluent in this production system is reduced to 0.25 l/kg of olives.

Liquid–Liquid Separation

The oil obtained in the previous stage contains some residual water and a few solids, in the same way that the residual water contains some oil and a large percentage of solids. It is necessary, therefore, to purify the oil (by freeing it from solids and the residual water) as well as the residual water to obtain the little oil that it contains. Before separating the liquids, the solids are eliminated, also known as "fines," by using vibratory sieves or filters, to separate the highest possible quantity of "fines." Finally, the oil separation from the aqueous water is carried out by natural decantation, centrifugation, or a combination of both.

Decantation. The difference in density between the oil (0.915–0.918) and the residual water (1.015–1.086) enables separation by natural decanting into vats or settling pits, which are linked together. It is the oldest method and even today it is used in some oil mills. However, it requires a large space and a long time, during which, the contact of the oil with the aqueous phase can cause fermentation and changes in the oil quality.

Centrifugation. The use of a vertical centrifuge enables continuous and rapid liquid–liquid separation. By the addition of a specific quantity of water, the separation of the oil from the rest of the residual water is achieved and in the same way, but independently, the recovery of the oil fraction from the aqueous phase. The factors to be considered in this operation are the homogeneity and volume of the liquid to be centrifuged, temperature, volume of added water, and time taken between unloading. The oil that comes out of the centrifuge passes into small decanters for degassing and continues to the recipients, where its quality classification is carried out.

Storage

The storage period is limited to a season or part of the following one. Longer periods are only anticipated in industrial mills, which include bottling and marketing operations, as well as the extraction process. During storage, every possible precaution should be taken to prevent three causes of oil spoilage: contact with unsuitable materials, prolonged contact with aqueous impurities, and oxidation. It is advisable to use vats of a capacity of approximately 50 tons, for easier classification and marketing. Vats must be constructed with impermeable and inert materials, so oil will not penetrate or react with the surface since any absorbed oil, which cannot be removed by cleaning, will be altered and will compromise on the further use of the vat. The storage cellar must have minimum luminosity with walls and floors that are easy to clean and are maintained at a temperature between 15°C and 18°C, without abrupt changes in temperature.

Nowadays, the vats which best accomplish these conditions are called "trujales" or traditional underground vats which, thanks to an appropriate cover of vitrified refractory tiles, ensure the preservation of the oils. If surface vats are used, they must be covered and protected from atmospheric agents and variations in temperature. The most ideal material is stainless steel although other materials can be used, provided an inert inner covering is used. Vats must have a conical bottom or a flat incline and a purge tap to facilitate sediment removal.

Composition of Olive Oil

The composition of olive oil basically depends on the factors that influence the composition of the fruit, such as latitude, climate, state of ripeness, etc., although oils are also influenced by the variables associated with processing and storage conditions.

The components in olive oil, as in all vegetable oils, can be subdivided into saponifiable (e.g., triglycerides, free fatty acids, waxes, mono- and diglycerides, sterol esters and terpene alcohols, and phospholipids) and nonsaponifiable (e.g., hydrocarbons, tocopherols, sterols, terpenes, alcohols, pigments, polyphenols, and volatile compounds) fractions. The nonsaponifiable constituents are minority compounds (0.5–1.5% of the total oil weight) but contain the "fingerprint" of the oil; and also play an important part in the stability and organoleptic characteristics of taste, aroma and color (Boskou, 1996; Harwood and Aparicio, 2000).

Saponifiable Fraction

Most of the fatty acids of olive oil are present as triglycerides. The proportion of free fatty acids is small and is directly related to the level of triglyceride hydrolysis, normally associated with the deterioration process.The major olive oil triglycerides are OOO (43.5%), POO (18.4%), OOL (6.8%), POL (5.9%), and SOO (5.1%) (O, oleic; P, palmitic; L, linoleic, and S, stearic). The waxes are esters of fatty acids and primary fatty alcohols. The principal waxes found in virgin olive oil are esters C36 and C38, whereas in olive-pomace oil and in refined olive oil the majority of esters are longer (C40, C42, C44, and C46). The quality regulations of the Economic European Community (EEC) establish a maximum content of waxes of 250 mg/kg for virgin olive oil (nonlampante) and 350 mg/kg for refined oil.

Nonsaponifiable Fraction

Hydrocarbons. Squalene ($C_{30}H_{50}$) is the major hydrocarbon (30–60%) of olive oil and its concentration (2500–9250 mg/kg) is much higher than that in the majority of other vegetable oils (16–370 mg/kg)(Gutfinger and Letan, 1974).

Tocopherols. The tocopherol content can vary between 5 and 300 mg/kg, although in good quality oils it is usually above 100 mg/kg (Maestro-Durán

Table 26.8. Quality Parameters Established by the ECC for Olive Oil and Olive-Pomace Oil

Categories	Acidity (%)	PI[a](mEq O_2/kg)	K_{232}	K_{270}	K_{270}[b]	ΔK	Organoleptic Evaluation[c] Md	Mf
Extra virgin olive oil	≤1.0	≤20	≤2.50	≤0.20	≤0.10	≤0.01	0	>0
Virgin olive oil (fine)	≤2.0	≤20	≤2.60	≤0.25	≤0.10	≤0.01	≤2.5	>0
Ordinary virgin olive oil	≤3.3	≤20	≤2.60	≤0.25	≤0.10	≤0.01	≤6.0[d]	≥0
Lampante virgin olive oil	>3.3	>20	≤3.70	>0.25	≤0.11	–	>6.0	
Refined olive oil	≤0.5	≤5	≤3.40	≤1.20	–	≤0.16		
Olive oil	≤1.5	≤15	≤3.30	≤1.00	–	≤0.13		
Refined olive-pomace oil	≤0.5	≤5	≤5.50	≤2.50	–	≤0.25		
Olive-pomace oil	≤1.5	≤15	≤5.30	≤2.00	–	≤0.20		

Source: Regulation number 2568/91 of the European Communities (EC, 1991)
Note: Commercial categories.
[a]PI, Peroxide index.
[b]Values of K_{270} after alumina treatment.
[c]Median values of defects (Md) and fruity aroma (Mf).
[d]If median of fruity aroma is 0, Md must be ≤2.5.

and Borja-Padilla, 1990; Gutiérrez et al., 1999; Psomiadou et al., 2000); α-tocopherol (vitamin E) represents 90–95% of the total tocopherols, β- and γ-tocopherols are less than 10%,and δ-tocopherol is found in very low amounts. Consumption of 25 g/day of virgin olive oil meets 50% of the recommended daily requirement of vitamin E. Because tocopherols function as antioxidants, they make an important contribution to the stability of virgin olive oil, although are very unstable compounds to thermal treatment (Baldioli et al., 1996; Aparicio et al., 1999). Also, refined olive oils have less protection against oxidation due to the disappearance of practically all the tocopherols during the deodorizing process in the refining.

Aliphatic Alcohols. The most representative of olive oil is linear alcohols with even-numbered carbon atoms: hexacosanol, octacosanol, and tetracosanol. The profile of these compounds is used to detect adulteration with seed oils, which contain a greater proportion of chains with odd-numbered carbon atoms (Morales and León-Camacho, 2000).

Sterols. The total content of sterols is in the order of 800–2600 mg/kg, β-sitosterol being found in the highest proportion (75–90%), followed by Δ^5-avenasterol and campasterol. Sterols are directly related to olive oil quality and are used to detect adulterations with seed oil as well as to characterize the monovarietal virgin olive oils according to geographical origin (Aparicio and Luna, 2002). Table 26.8 shows the limits established by the IOOC (2001) for each one of the olive sterols.

Olive oil contains triterpene alcohols (1000–3000 mg/kg), generally tetracyclic or pentacyclic in structure, which are markers of certain monovarietal olive oils. Two hydroxytriterpenes, erythrodiol and uvaol, have been identified; the presence of these triterpene alcohols together with the wax content allows the detection of adulteration of olive oil with olive-pomace oil (Morales and León-Camacho, 2000). The Commission of the European Communities (EC, 2002) establishes a relative maximum value of 4.5% for these alcohols. To distinguish lampante olive oil from olive-pomace oil, regulations establish that for a content of wax between 300 and 350 mg/kg, it is considered as olive-pomace oil only if the erythrodiol and uvaol is greater than 3.5% or if the total aliphatic alcohol content is more than 350 mg/kg.

Phenolic Compounds. Although water-soluble compounds, a small percentage is transferred from the fruit to the oil during the extraction process.

These phenolic compounds increase the oxidative stability of the olive oil and improve the flavor considerably (Gutiérrez et al., 2001; Tsimidou, 1998). The phenolic profile can also be used to authenticate virgin olive oils of the *Picual* variety (Brenes et al., 2002). The average content (expressed as caffeic acid) is between 50 and 800 mg/kg. The principal polyphenols of olive oil are tyrosol and hydroxytyrosol, both derived from the hydrolysis of oleuropein. Also found are smaller proportions of caffeic, vanillic, siringic, p-coumaric, o-coumaric, protocatechic, synapic, p-hydroxybenzoic, p-hydroxyphenylacetic, and homovanillic acids.

Volatile Compounds. In this group are found all compounds responsible for the characteristic aroma of virgin olive oil. More than a hundred volatile compounds have been identified such as hydrocarbons, alcohols, aldehydes, ketones, esters, acids, phenolic compounds, terpene compounds, and furan derivatives. But there is still no objective method to measure the aroma, which requires evaluation by a panel of specialist tasters (Panel Test).

In good quality oils, the retained volatile compounds come from the actual fruit and are generated by biogenesis mechanisms. Some compounds are originally found in the intact flesh of the olive, while others, called secondary or technological volatiles, are formed during the breaking down of cellular structure as a result of enzyme reactions that take place in the presence of oxygen. The principal biochemistry pathway is by the enzyme lipoxygenase, precursors being the 18-carbon atom polyunsaturated fatty acids, linoleic and linolenic (Olías et al., 1993; Sánchez and Salas, 2000). When the harvest conditions of fruit and production or storage of the virgin olive oil are defective, a whole series of disagreeable volatile compounds responsible for sensory defects (off-flavors) are generated. Very different mechanisms are implicated in the genesis of these undesirable compounds. The majority are related microorganisms (yeast, bacteria, and molds) during storage of fruit before oil extraction, generating volatile compounds responsible for sensory defects such as fusty, musty, or winey-vinegary. During the storage of unfiltered virgin olive oil, sediments are deposited, which can also produce fermentation that generate volatile metabolites responsible for the sensory defect known as muddy sediment.

There are also other chemical mechanisms that produce undesirable volatiles. The main one is autoxidation of fatty acids. This process is generated spontaneously by the action of atmospheric oxygen, producing a series of chain reactions by a free radical mechanism. The products of this oxidative change are responsible for rancidity, as well as a significant deterioration in nutritional quality.

When the intensity of these sensory defects is high, the oils are classified as "lampante" and according to current regulations (EC, 2001), require further refining and are often rejected by the consumer.

Pigments. Virgin olive oil contains two types of pigments, chlorophylls, responsible for the green color and the yellow-colored carotenoids. Due to their lipid soluble nature, these constituents of the olive are transferred to the oil during the extraction process at a time when they are undergoing certain transformation reactions.

The chlorophyll and carotenoid composition of virgin olive oil is subject to wide variations due to the differences between cultivars and the degree of ripeness of the olive fruit, latitude, environmental conditions, processing techniques, and storage conditions (Gandul-Rojas and Mínguez-Mosquera, 1996; Gandul-Rojas et al., 2000; Morello et al., 2003; Salvador et al., 2003). The average content of chlorophyll pigments can vary between 1 and 40 mg/kg and of carotenoids between 2 and 20 mg/kg (Gandul-Rojas et al., 2000; Psomiadou and Tsimidou, 2001).

These differences are mainly quantitative, since the qualitative composition of pigments is basically the same in all the olive cultivars and is not modified with the advance in ripening of the fruit. The color intensity of the olive oil is directly related to the total content of chlorophylls and carotenoids, whereas the color tone depends on the relative proportion between the two pigment fractions. The relationship maintained between the chlorophyll and the carotenoids in the fruit decreases with the ripening process due to the greater rate of chlorophyll disappearance as opposed to carotenoids. The oils obtained from green and mottled olives at the beginning of the season have a greenish color. As the harvesting season of the fruit progresses and the percentage of black olives increases, the color of the olive oils is less intense due to the decrease in the total pigment content of the fruit. At the same time, the color tone becomes more yellowish as a result of an increase in the relative proportion of carotenoids (Mínguez-Mosquera et al., 1991; Gandul-Rojas et al., 2000).

It has been shown that in general, in the ripeness stage at which the fruit is harvested for olive oil, the chlorophyll-to-carotenoid ratio tends to be more or less constant, around 3, independent of the olive cultivar and different pigment content of the fruit. However, this ratio is lowered to 1 with processing into corresponding oils (Roca and Mínguez-Mosquera, 2001b). This change occurs, despite the lipophilic character of these pigments, during the extraction process where only 20% of the chlorophyll content of the fruit is transferred to the oil, while the carotenoid fraction is 50% (Mínguez-Mosquera et al., 1990a). The balance of pigment substances among fruit, oil, and alperujo shows that a large part of the pigmentation stays behind in the vegetable material, which forms part of the alperujo, estimating a retention of 60% chlorophyll and 25% carotenoids. The rest of the pigments (20–25%) are degraded to colorless products (Gallardo-Guerrero et al., 2002).

The changes that the components undergo during olive crushing and paste mixing trigger the release of cytoplasmic components that come into contact with different enzymes and substrates isolated under normal physiological conditions. The modification of pigments is influenced by the release of acid compounds. For the chlorophyll pigment fraction, the most common reaction is pheophytinization and for the carotenoids, the isomerization of the xanthophylls with epoxide groups, such as violaxanthin, neoxanthin, and antheraxanthin, which are transformed into 5,8 furanoids, luteoxanthin + auroxanthin, and neochrome + mutatoxanthin, respectively. β-carotene, lutein, and β-cryptoxanthin are transferred to the oil without transformation (Mínguez-Mosquera et al., 1990a).

Traces of metabolites of chlorophyll catabolism can also be detected, such as 13-OH-pheophytin and 15^1-OH-lactone pheophytin among the oxidation metabolites and chlorophyllides and pheophorbides as intermediaries in the deesterification pathway, initiated by the enzyme chlorophyllase. In the case of cultivars with high chlorophyllase activity, such as *Arbequina*, a significant increase in deesterified chlorophyll derivatives are found (Gandul-Rojas and Mínguez-Mosquera, 1996; Gandul-Rojas et al., 2000).

The small, but always important differences found in chlorophyll and carotenoid metabolism during the ripening of the fruit, depending on cultivar and influenced by endogenous enzyme activities, are going to be reflected in the pigment composition of the corresponding oils. Consequently, the chlorophyll and carotenoid profile of virgin olive oil is used as a parameter of quality and authenticity for this product, since the presence of chlorophyll and carotenoid pigment different from those just described, or the detection of a high level of pigment transformation is indicative of incorrect or fraudulent practices (Gandul-Rojas et al., 1999c).

Two quality parameters are provided to authenticate oils based on the composition of chlorophylls and carotenoids of monovarietal virgin olive oils of Spanish origin, as well as defining the pigment profile specific for virgin olive oil (Gandul-Rojas et al., 2000). On the one hand, it has been shown that there is a constant ratio, of around 1, between the chlorophyll and carotenoid pigment fractions, independent of the cultivar and degree of ripeness. This ratio is maintained within the narrow limits, that is, between 0.5, when the oils are processed from ripe fruits, and 1.4, when oils are obtained from unripe fruits. On the other hand, the carotenoid fraction has a constant relationship between lutein, which is the principal component of this fraction, and the rest of the minority carotenoids. In all virgin olive oils, the minority carotenoid–lutein ratio is around 0.5 and, exceptionally, greater than 1 in the specific case of oils produced from the *Arbequina* cultivar in which this parameter, as well as authenticating the "virgin olive oil" category, allow the differentiation of this variety. In the same way as occurs with the chlorophyll-to-carotenoid ratio, in oils obtained from very ripe fruits the relationship between minor carotenoids and lutein decreases, falling as low as 0.2.

On the other hand, the percentage of lutein, the percentage of violaxanthin, and the total pigment content allow distinguishing between monovarietal virgin olive oils of Spanish origin. For this reason, they have been proposed as suitable qualifying variables in establishing a prediction model, which could indicate the varietal origin of the Spanish virgin olive oil (Gandul-Rojas et al., 2000).

These quality parameters based on relationships between pigments remain stable after 12 months of oil storage at 15°C in the dark, conditions generally used in the industry to store oils that are not going to be marketed immediately after production (Roca et al., 2003). However, it has been shown that during this period, the chlorophyll molecule undergoes specific changes, which implicate alterations in the pigment profile, as compared to recently extracted virgin olive oil. From the quantitative point of view no pigments are lost, although there is an increase in the reactions catalyzed by acids, which started during the extraction of the oil: the pheophytinization reaction of chlorophyll and the isomerization of the 5, 6-epoxide groups of the minority xanthophylls. The color change of the oil during storage is mainly due to the transformation of the chlorophyll (green) into pheophytins (gray brown). Possibly, to a greater or lesser extent the acidic substances that could be transferred from the fruit to the oil will regulate the rate of these reactions.

Likewise, there is a slight increase in the hydroxylation of the C-13^2 of pheophytin a, as well as a small amount of pyropheophytin a, a pigment not present in recently extracted oil. It has been shown that in the previously mentioned storage conditions, the maximum amount of pyropheophytin found in oils is around 3% of the total chlorophyll compounds, and that the ratio of pheophytin a to pyropheophytin a is always more than 20. In this way, these small structural changes of pigment, which are not inherent to the extraction process of the virgin olive oil and therefore not desirable, are indicators that the oil has been stored. Inadequate storage conditions, with high temperatures and/or light, cause marked increases in the content of these compounds (Gallardo-Guerrero et al., 2004). During physical refining of oils through deodorization with nitrogen, the most widespread reaction is pyropheophytinization (Gandul-Rojas et al., 1999a). On the other hand, during classical chemical refining processes to which poor-quality olive oils are subjected, pigments are mostly eliminated by adsorption onto decolorizing earth and/or deodorizing treatment at high temperatures (Usuki et al., 1984).

Quality of Olive Oil

Different definitions of quality exist depending on the use of the olive oil from the established regulations for quality, to the nutritional and therapeutic quality, commercial, culinary, sensory, etc.

The regulated quality is the easiest to define as it is clearly defined by EC Regulation 2568/91 and its later modifications (EC, 1991). Currently two rules exist, which regulate the denominations of olive oils and olive-pomace oil (EC, 2001; IOOC, 2001). In both regulations, there are parameters, known as quality parameters, to evaluate the quality of the oils and these are free acidity, peroxide index, k_{232}, k_{270}, Δk, and organoleptic characteristics.

The degree of free acidity measures the percentage of free fatty acids, expressed as oleic acid. It is an index of the alterations experienced by the fruit and the fermentations that could have taken place during production and storage.

The peroxide index evaluates the state of primary oxidation of oil. The most important causes of a high value of this index are the history of the fruit (i.e., collection from soil, injuries by frosts, etc.) and exposure to factors, which cause oxidation (high temperatures, aeration, and the presence of trace metals) during production and storage.

The k_{232} is a spectrophotometric measurement of the conjugated dienes that is used along with the peroxide index to evaluate the primary oxidation of the oil. This parameter is included only in the EC regulation.

The k_{270} is used to evaluate, spectrophotometrically, the conjugated trienes and to indicate the state of secondary oxidation of the oil. The resulting products of this oxidation (aldehydes and ketones) also absorb at wavelengths of 262, 268, and 274 nm. The Δk coefficient takes into account these absorbencies and is defined as

$$\Delta k = k_{268} - [(k_{262} + k_{274})/2]$$

Organoleptic characteristics of virgin olive oils are determined by a group of trained tasters, according to the rules of sensory analysis (IOOC, 2001). When no defects exist in the flavor (visual together with olfactory-taste-tactile), the tasters evaluate the presence and intensity of positive attributes. The sensory attribute "fruity," responsible for a mixture of olfactory–taste sensations attributed to fresh and healthy fruit at an optimal degree of ripeness is the one which plays an important part in the sensory evaluation since the absence of this attribute excludes the oil from the virgin grade. However, increased intensity of fruity is often accompanied by a perception of high intensities of bitterness and spiciness which makes the oil unacceptable for direct consumption and requires mixing with other virgin oils of less intense flavor. On the new sensory evaluation form, proposed in 1996 by IOOC and adopted by the EC in May 2002, the most frequent negative attributes are scored ("fusty, musty, muddy sediment, and winey-vinegary"). Also included in the "Others" section is all other well-defined negative attributes such as "earthy, esparto, pungent, cucumber, grubby, etc." Only fruity, bitter, and pungent are considered positive attributes. Depending on the median value of the defect perceived with greater intensity and the median value of the fruity attribute, the virgin olive oil is classified into distinct categories.

Color is another organoleptic quality characteristic fundamental to virgin olive oil as it is one of the attributes that most affects the consumer at the time of purchase. The first evaluations of color were based on the visual comparison of oil with the standard solutions of bromothymol blue (BTB), using an adaptation of methods developed for seed oils (Gutierrez et al., 1986; AOCS, 1977, 1988). Photometric evaluations at two wavelengths have been proposed for olive oil (Papaseit et al., 1986) and numerical correlations between the chromatic coordinates and the chlorophyll and carotenoid content (Mínguez-Mosquera et al., 1991), as well as with the BTB indices (Moyano et al., 1999). Despite the importance of this quality attribute, none of the current regulations include the organoleptic quality, color.

According to the International Olive Oil Council, olive oil is "the oil obtained solely from the fruit of the olive tree (*O. europaea sativa* Hoffm. Link) to the exclusion of oils obtained using solvents or by re-esterification processes and of any mixture with oils from other sources." This oil can be produced commercially under different denominations, virgin olive oil being "the oil obtained from the fruit of the olive tree solely by mechanical or other physical means under conditions particularly thermal that do not lead to alteration in the oil and which has not undergone any treatment other than washing, decantation, centrifugation, and filtration."

All the oils obtained in an olive oil mill are considered virgin olive oils and, depending on their quality parameters, are classified into different categories in accordance with the EC and IOOC regulations. In Figure 26.4, the aforementioned classification is

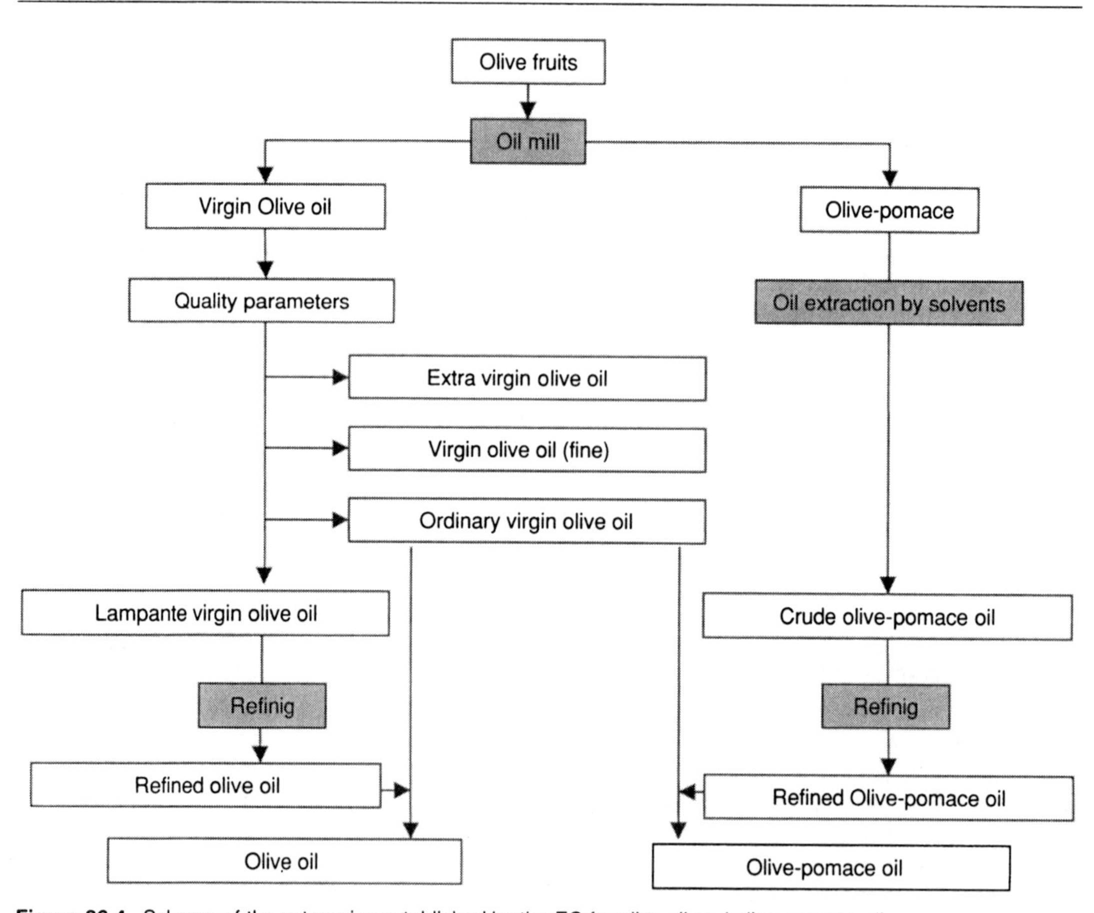

Figure 26.4. Scheme of the categories established by the EC for olive oil and olive-pomace oil.

Table 26.9. Sterol Composition Established by the IOOC for Olive Oil and Olive-Pomace Oil

Sterol	Percentage of Total Sterol Content
β-sitosterol[a]	≥93
Cholesterol	≤0.5
Brasicasterol	≤0.1[b]
Campesterol	≤4.0
Stigmasterol	<Campesterol
Δ^7-Stigmasterol	≤0.5

Source: International Olive Oil Council: Trade standard applying to olive oil and olive-pomace oil. COI/T.15 No2/Rev 10 (IOOC, 2001).

[a]β-sitosterol + Δ^5-avenasterol + $\Delta^{5,23}$stigmastadienol + clerosterol + sitostanol + $\Delta^{5,24}$stigmastadienol.

[b]<0.2 for olive-pomace oil.

shown, and in Table 26.9, the limits established by the EEC for each one of the quality parameters according to the category of the oil is recorded.

REFERENCES

Alba J. 1997. Elaboración de aceite de oliva virgen in El cultivo del Olivo edited by D. Barranco, R. Fernández-Escolar, L. Rallo. Madrid: Junta de Andalucía and Ediciones Mundi-Prensa, pp 509–537.

Alba J., Hidalgo F., Martínez F., Ruiz M.A., Borja R. 1995. Evaluación medioambiental de los sistemas de elaboración de aceite de oliva en Andalucía. Mercacei February–March:20–22.

Amiot M.J., Fleuriet A., Macheix J.J. 1986. Importance and evolution of phenolic compounds in olive during growth and maturation. J. Agric. Food Chem. 34:823–826.

Amiot M.J., Tacchini M., Fleuriet A., Macheix J.J. 1990. The technological debittering process of olives: Characterization of fruits before and during alkaline treatment. Sci. Aliment. 10:619–631.

AOCS. 1977 Official and tentative methods edited by AOCS Technical Committee, Vol. 27. Champaign: Method Cc, pp 13c–50.

AOCS. 1988 Official and tentative methods edited by AOCS Technical Committee, Vol 62. Champaign: Method Cc pp 13d–55.

Aparicio R., Albi T., Cert A., Lanzón A. 1988. SEXIA expert system: Canonical equations to characterize Spanish olive oils by varieties. Grasas y Aceites 39:219–228.

Aparicio R., Alonso V. 1994. Characterization of virgin olive oils by SEXIA expert system. Prog. Lipid Res. 33:29–38.

Aparicio R., Luna G. 2002. Characterization of monovarietal virgin olive Oils. Eur. J. Lipid Sci. Technol. 104:614–627.

Aparicio R., Roda L., Albi M.A., Gutierrez F. 1999. Effect of various compounds on virgin olive oil stability measured by Rancimat. J. Agric. Food Chem. 47:4150–4155.

Balatsouras G., Tsibri A., Dalles T., Doutsias G. 1983. Effects of fermentation and its control on the sensory characteristic of Conservolea cultivar green olives. Appl. Environ. Microbiol. 46:68–74.

Baldioli M., Servili M., Perretti G., Montedoro G.F. 1996. Antioxidant activity of tocopherols and phenolic compounds of virgin olive oil. J. Am. Oil Chem. Soc. 73:1589–1593.

Barranco D. 1997. Variedades y patrones in El cultivo del Olivo edited by D. Barranco, R. Fernández-Escolar, L. Rallo. Madrid: Junta de Andalucía and Ediciones Mundi-Prensa, pp 59–60.

Bendich A., Olson J.A. 1989. Biological actions of carotenoids. FASEB J. 3:1927–1932.

Borbolla y Alcalá J.M.R. de la, Fernández-Díez M.J., González-Pellissó F. 1955. Cambios en la composición de la aceituna durante su desarrollo I. Grasas y Aceites 6:5–22.

Boskou D. 1996. Olive Oil. Chemistry and Technology. Champaign, IL: AOCS.

Brenes M., Castro A. de. 1998. Transformation of oleuropein and its hydrolysis products during Spanish-style green olive processing. J. Sci. Food Agric. 77:353–358.

Brenes M., Garcia A., Rios J.J., Garcia P., Garrido A. 2002. Use of 1-acetoxypinoresinol to authenticate picual olive oils. Int. J. Food Sci. Technol. 37:615–625.

Brenes M., García P., Durán M.C., Garrido A. 1993. Concentration of phenolic compounds change in storage brines of ripe olives. J. Food Sci. 58:347–350.

Brenes M., Rejano L., García P., Sánchez A.H., Garrido A. 1995. Biochemical changes in phenolic compounds during Spanish-style green olive processing. J. Agric. Food Chem. 43:2702–2706.

Castro-Ramos R., Nosti-Vega M., Vázquez-Ladrón R. 1980. Estudio comparativo del valor nutritivo de algunas variedades españolas de la aceituna de mesa. Efecto de la variedad del fruto, del tipo de elaboración y del envasado. Alimentaria XVII, 115:21–24.

Civantos L. 1997. La olivicultura en el mundo y en España in El cultivo del Olivo edited by D. Barranco,

R. Fernández-Escolar, L. Rallo. Madrid: Junta de Andalucía and Ediciones Mundi-Prensa, pp 15–32.

Dashwood R.H. 1997. Chlorophylls as anticarcinogens. Int. J. Oncol. 10:721–727.

EC. 1991. Official Journal of the Commission of the European Communities. Regulation no 2568/91, L248, 11 July, 1991.

EC. 2001. Official Journal of the Commission of the European Communities. Regulation no 1513/2001, L201, 26 July, 2001.

EC. 2002. Official Journal of the Commission of the European Communities. Regulation no 796/2002, L128, 6 May, 2002.

Fernández-Bolaños J., Fernández-Díez M.J., Rivas-Moreno M., Gil-Serrano A. 1982. Azúcares y polioles en aceitunas verdes. II. Identificación y determinación cuantitativa por cromatografía sobre papel. Grasas y Aceites 33:208–211.

Fernández-Díez M.J., Castro-Ramos R., Garrido-Fernández A., González-Cancho F., González-Pellissó F., Nosti-Vega M., Heredia-Moreno A., Mínguez-Mosquera M.I., Rejano-Navarro L., Durán-Quintana M.C., Sánchez-Roldán F., García-García P., Castro Gómez-Millán A. de. 1985 Biotecnología de la aceituna de mesa. Madrid-Sevilla: Consejo Superior de Investigaciones Científica-Instituto de la Grasa.

Fernández-Díez M.J., González-Pellissó F. 1956. Cambios en la composición de la aceituna durante su desarrollo II. Acidez y pH del jugo. Determinación de ácidos oxálico, cítrico y málico. Grasas y Aceites 7:185–189.

Gallardo-Guerrero L., Roca M., Gandul-Rojas B., Mínguez-Mosquera M.I. 2004. Storage influence on the initial content and class of pigments of virgin olive oil. In Proceeding of 3rd International Congress on Pigments in Food, more than colors... edited by L. Dufossé. Quimper: Francia, pp 85–87.

Gallardo-Guerrero L., Roca M., Mínguez-Mosquera M.I. 2002 Distribution of chlorophylls and carotenoids in ripening olives and between oil and alperujo when processed using a two-phase extraction system. J. Am. Oil Chem. Soc. 79:105–109.

Gallardo-Guerrero M.L., Gandul-Rojas B., Mínguez-Mosquera M.I. 1999. Chlorophyll pigment composition in table olives (cv. Gordal) with green staining alteration. J. Food Prot. 62:1167–1171.

Gandul-Rojas B., Gallardo-Guerrero M.L., Mínguez-Mosquera M.I. 1999a. Identification of oxidized chlorophylls and metallochlorophyllic complexes of copper in table olives (cv. Gordal) with green staining alteration. J. Food Prot. 62:1172–1177.

Gandul-Rojas B., Mínguez-Mosquera M.I. 1996. Chlorophyll and carotenoid composition in virgin olive Oils from various Spanish olive varieties. J. Sci. Food Agric. 72:31–39.

Gandul-Rojas B., Roca M., Mínguez-Mosquera M.I. 1999b. Chlorophyll and carotenoid patterns in olive fruits, Olea europaea Cv. Arbequina. J. Agric. Food Chem. 47:2207–2212.

Gandul-Rojas B., Roca M., Mínguez-Mosquera M.I. 1999c. Chlorophyll and carotenoid pattern in virgin olive oil. Adulteration control. In Proceeding of 1st International Congress on Pigments in Food Technology edited by M.I. Mínguez-Mosquera, M. Jarén-Galán, D. Hornero-Méndez. Sevilla, Spain, pp 381–386.

Gandul-Rojas B., Roca M., Minguez-Mosquera, M.I. 2000. Use of chlorophyll and carotenoid pigment composition to determine authenticity of virgin olive oil. J. Am. Oil Chem. Soc. 77:853–858.

Garrido-Fernández A., Fernández-Díez M.J., Adams M.R. 1997. Table Olives. Production and Processing. London: Chapman & Hall.

Giovacchino L. di. 2000. Technological aspect in handbook of olive oil: Analysis and properties edited by J. Harwood, R. Aparicio. Gaithersburg, Maryland: Aspen Publication, pp 17–60.

Giovacchino L. di., Mascolo A. 1988. Incidenza delle tecniche operative nell'estrazione dell'olio d'oliva con il sistema continuo. Nota II. Riv. Ital. Sost. Gras. 65:283–289.

González-Cancho F. 1963 Microorganismos que se desarrollan en el aderezo de aceitunas verdes estilo español. Microbiol. Esp. 16:221–230.

González-Cancho F., Durán-Quintana M.C. 1981. Bacterias cocaceas del ácido láctico en el aderezo de aceitunas verdes. Grasas y Aceites 32:373–379.

González-Cancho F., Nosti-Vega, M., Durán-Quintana M.C., Garrido-Fernández A. 1975. El proceso de fermentación en las aceitunas negras maduras en salmuera. Grasas y Aceites 26:297–309.

González-Cancho F., Rejano L., Borbolla y Alcala J.M.R. de la. 1980. La formación de ácido propiónico durante la conservación de aceitunas verdes de mesa III. Microorganismos responsables. Grasas y Aceites 31:245–249.

Guillén R., Heredia-Moreno A., Felizón B., Jiménez A., Fernández-Bolaños J. 1991. Preparación y caracterización de las fracciones de fibra en aceitunas (variedad Hojiblanca). Grasas y Aceites 42:334–338.

Guillén R. Heredia-Moreno A., Felizón B., Jiménez A., Montaño A., Fernández-Bolaños J. 1992. Fibre fractions carbohydrates in Olea europaea (Gordal and Manzanilla cultivars). Food Chem. 44:173–178.

Gutfinger J., Letan A. 1974. Studies of unsaponifiables in several vegetables Oils. Lipids 9:658–663.
Gutiérrez F., Arnaud T., Garrido A. 2001. Contribution of polyphenols to the oxidative stability of virgin olive oil. J. Sci. Food Agric. 81:1463–1470.
Gutiérrez F., Jiménez B., Ruiz A., Albi M.A. 1999 Effect of olive ripeness on the oxidative stability of virgin olive oil extracted from the varieties picual and hojiblanca and on the different components involved. J. Agric. Food Chem. 47:121–127.
Gutierrez-G., Quijano R., Gutierrez F. 1986 Rapid method to define and classify the color of virgin olive oil. Grasas y Aceites 37:282–283.
Harwood J., Aparicio A. 2000. Handbook of olive oil: Analysis and properties. Gaithersburg, Maryland: Aspen Publication.
Hassapidou M.N., Balatsouras G.D., Manoukas A.G. 1994. Effect of processing upon the tocopherol and tocotrienol composition of table olives. Food Chem. 50:111–114.
Hermoso M., González J., Uceda M., García-Ortiz A., Morales J., Frías L., Fernández A. 1995. Elaboración de aceite de oliva de calidad II. Obtención por el sistema de dos fases. Sevilla, Spain: Junta de Andalucía.
Hermoso M., Uceda M., Frias L., Beltrán G. 1987, Maduración in El cultivo del Olivo edited by D. Barranco, R. Fernández-Escolar, L. Rallo. Madrid: Junta de Andalucía and Ediciones Mundi-Prensa, pp 139–153.
Hermoso M., Uceda M., García-Ortíz A., Morales J., Frías L., Fernández A. 1991. Elaboración de aceite de oliva de calidad. Sevilla, Spain: Junta de Andalucía.
IOOC. 1980. Unified Qualitative Standard Applying to Table Olives in International Trade. Madrid: International Olive Oil Council.
IOOC. 2000. Catálogo mundial de variedades de Olivo. Madrid: International Olive Oil Council.
IOOC. 2001. Trade standard applying to olive oil and olive-pomace oil. COI/T.15 No2/Rev 10. Madrid, Spain: International Olive Oil Council.
IOOC. 2003a. International table olive market: Position and trend. Olivae 99:45–48.
IOOC 2003b. International olive oil market: Position and trend. Olivae 99:42–45.
Jiménez-Araujo A., Labavitch J.M., Heredia-Moreno A. 1994. Changes in the cell wall of olive fruit during processing. J. Agric. Food Chem. 42:1194–1199.
Lanzón A., Cert A., Albi T. 1989. Detección de la presencia de aceite de oliva refinado en el aceite de oliva virgen. Grasas y Aceites 40:385–388.
Maestro-Durán R., Borja-Padilla R. 1990. La calidad del aceite de oliva en relación con la composición y maduración de la aceituna. Grasas y Aceites 41:171–178.
Maestro-Durán R., Vázquez-Roncero A. 1976. Colorantes antociánicos de las aceitunas manzanillas maduras. Grasas y Aceites 27:237–243.
Manna C., D'Angelo S., Migliardi V., Loffredi E., Mazzoni O., Morrica P., Galletti P. Zappia V. 2002. Protective effect of the phenolic fraction from virgin olive oils against oxidative stress in human cells. J. Agric. Food Chem. 50:6521–6526.
Manoukas A.G., Grimanis A., Mazomenos B. 1978. Inorganic nutrients in natural and artificial food of Dacus oleae larvas. (Diptera: Tephritidae). Ann. Zool. Ecol. Anim. 10:123–128.
Manoukas A.G., Mazomenos B., Patrinou M.A. 1973. Amino acid composition of three cultivars of olive fruit. J. Agric. Food Chem. 21:215–217.
Martínez-Moreno J.M., Gómez-Herrera C., Janer del Valle C. 1957. Estudios físico-químicos sobre las pastas de aceitunas molidas. Las gotas de aceite. Grasas y Aceites 8:112–120.
Martínez-Moreno J.M., Gómez-Herrera C., Janer del Valle C., Pereda J. 1964. Estudios físico-químicos sobre las pastas de aceitunas molidas XXI. La prensada como proceso de filtración. Grasas y Aceites 15:299–307.
Mínguez-Mosquera M.I., Castillo-Gómez J., Fernández-Díez M.J. 1976. Variaciones en la composición péctica durante la elaboración, fermentación, y conservación de productos aderezados (pimiento y aceituna). Grasas y Aceites 27:27–32.
Minguez-Mosquera M.I., Gallardo-Guerrero L. 1995a. Disappearance of chlorophylls and carotenoids during the ripening of the olive. J. Sci. Food Agric. 69:1–6.
Mínguez-Mosquera M.I., Gallardo-Guerrero L., Roca M. 2002a. Pectinesterase and polygalacturonase in changes of pectic matter in olives (cv. Hojiblanca) intended for milling. J. Am. Oil. Chem. Soc. 79(1):93–99.
Mínguez-Mosquera M.I., Gallardo-Guerrero M.L. 1995b. Anomalous transformation of chloroplastic pigments in Gordal variety olives during processing for table olives. J. Food Prot. 58:1241–1248.
Mínguez-Mosquera M.I., Gallardo-Guerrero M.L., Hornero-Méndez D., Garrido-Fernández J. 1995. Involvement of copper and zinc ions in "green staining" of table olives of the variety Gordal. J. Food Prot. 58:567–569.
Mínguez-Mosquera M.I., Gandul-Rojas B. 1994. Mechanism and kinetics of the degradation of

carotenoids during the processing of green table olives. J. Agric Food Chem. 42:1551–1554.

Mínguez-Mosquera M.I., Gandul-Rojas B., Gallardo-Guerrero L., Jarén-Galán M. 2002b. Chlorophylls in methods for analysis of functional foods and nutraceuticals edited by W.J. Hurst. Boca Raton, FL: CRC Press LLC, pp 159–218.

Mínguez-Mosquera, M.I., Gandul-Rojas B. Garrido-Fernández J., Gallardo-Guerrero L. 1990a. Pigment present in virgin olive oil. J. Am. Oil Chem. Soc. 67:192–196.

Mínguez-Mosquera M.I., Gandul-Rojas B., Mínguez-Mosquera J. 1994. Mechanism and kinetics of the degradation of chlorophylls during the processing of green table olives. J. Agric Food Chem. 42:1089–1095.

Mínguez-Mosquera M.I., Garrido-Fernández J. 1987. Diferenciación de las variedades de olivo Hojiblanca y Manzanilla según su contenido pigmentario. Grasas y Aceites 38:4–8.

Mínguez-Mosquera M.I., Garrido-Fernández J. 1989. Chlorophyll and carotenoid presence in olive fruit (Olea europaea). J. Agric. Food Chem. 37:1–7.

Mínguez-Mosquera M.I., Garrido-Fernández J., Gandul-Rojas B. 1989. Pigment changes in olives during fermentation and brine storage. J. Agric. Food Chem. 37:8–11.

Mínguez-Mosquera M.I., Garrido-Fernández J., Gandul-Rojas B. 1990b. Quantification of pigments in fermented manzanilla and Hojiblanca olives. J. Agric. Food Chem. 38:1662–1666.

Mínguez-Mosquera M.I., Rejano-Navarro L., Gandul-Rojas B., Sánchez-Gómez A.H., Garrido-Fernández J. 1991. Color-pigment correlation in virgin olive oil. J. Am. Oil Chem. Soc. 68:332–336.

Montaño A., Castro A. de, Rejano L., Sánchez A.H. 1992. Analysis of zapatera olives by gas and high-performance liquid chromatography. J. Chromatogr. 594:259–267.

Morales M.T., León-Camacho M. 2000. Gas and liquid chromatography: Methodology applied to olive oil in handbook of olive oil: Analysis and properties edited by J. Harwood, R. Aparicio. Gaithersburg, Maryland: Aspen Publication, pp 159–207.

Morello J.R. Motilva M.J., Ramo T. Romero M.P. 2003. Effect of freeze injuries in olive fruit on virgin olive oil composition. Food Chem. 81:547–553.

Moyano M.J., Melgosa M., Alba J., Hita E., Heredia F.J. 1999 Reliability of the bromothymol blue method for color in virgin olive oils. J. Am. Oil Chem. Soc. 76:687–692.

Nosti-Vega M., Castro-Ramos R. de. 1985. Composición y valor nutritivo de algunas variedades españolas de aceitunas de mesa. VII. Aceitunas negras oxidadas. Grasas y Aceites 36:203–206.

Nosti-Vega M., Castro-Ramos R. de, Vázquez-Ladrón R. 1984. Composición y valor nutritivo de algunas variedades españolas de aceitunas de mesa. VI. Cambios debidos a los procesos de elaboración. Grasas y Aceites 35:11–14.

Olías J.M., Pérez A.G., Rios J.J., Sanz L.C. 1993. Aroma of virgin olive oil: Biogénesis of the "green" odor notes. J. Agric. Food Chem. 41:2368–2373.

Papaseit T.J. 1986. The color of extra virgin olive oil. A characteristic of quality. Grasas y Aceites 37:204–206.

Porras A. 1997. Recolección in El cultivo del Olivo edited by D. Barranco, R. Fernández-Escolar, L. Rallo. Madrid: Junta de Andalucía and Ediciones Mundi-Prensa, pp 337–363.

Psomiadou E., Tsimidou M. 2001. Pigments in Greek virgin olive oils: Occurrence and levels. J. Sci. Food Agric. 81:640–647.

Psomiadou E., Tsimidou M., Boskou D. 2000. Alpha-tocopherol content of Greek virgin olive oils. J. Agric. Food Chem. 48:1770–1775.

Rejano L. 1997. El aderezo de las aceitunas in El cultivo del Olivo edited by D. Barranco, R. Fernández-Escolar, L. Rallo. Madrid: Junta de Andalucía and Ediciones Mundi-Prensa, pp 565–586.

Rejano L., Castro A. de, González-Cancho F., Durán M.C., Sánchez A., Montaño A., García P., Sánchez F., Garrido A. 1986. Repercusión de diversas formas de tratamiento con ácido clorhídrico en la elaboración de aceitunas verdes estilo sevillano. Grasas y Aceites 37:19–24.

Roca M., Gandul-Rojas B., Gallardo-Guerrero L., Mínguez-Mosquera M.I. 2003. Pigment parameters determining Spanish virgin olive oil authenticity: Stability during storage. J. Am. Oil. Chem. Soc. 80:1237–1240.

Roca M., Mínguez-Mosquera M.I. 2001a. Changes in chloroplast pigments of olive varieties during fruit ripening. J. Agric. Food Chem. 49:832–839.

Roca M., Mínguez-Mosquera M.I. 2001b. Change in the natural ratio between chlorophylls and carotenoids in olive fruit during processing for virgin olive oil. J. Am. Oil Chem. Soc. 78:133–138.

Roca M., Mínguez-Mosquera M.I. 2003. Involvement of chlorophyllase in chlorophyll metabolism in olive varieties with high and low chlorophyll content. Physiol. Plant 117:459–466.

Romero C., Brenes M., García P., Garrido A. 2002. Hydroxytyrosol 4-beta-D-glucoside, and important phenolic compound in olive fruits and derived products. J. Agric. Food Chem. 50:3835–3839.

Romero C., Brenes M., Garcia P., Garcia A., Garrido A. 2004. Polyphenol Changes during fermentation of naturally black olives. J. Agric. Food Chem. 52:1973–1979.

Ruiz-Barba J.L., Cathcarth D.P., Warner P.J., Jiménez-Díaz R. 1994. Use of lactobacillus plantarum LPCOIO, a bacteriocin producer, as starter culture for green Spanish-style olives. Appl. Environ. Microbiol. 60:2059–2064.

Salvador M.D., Aranda F., Gomez-Alonso S., Fregapane G. 2003. Influence of extraction system, production year and area on Cornicabra virgin olive oil: A study of five crop seasons. Food Chem. 80:359–366.

Sánchez J., Salas J.J. 2000. Biogénesis of olive oil aroma in handbook of olive oil: Analysis and properties edited by J. Harwood, R. Aparicio. Gaithersburg, Maryland: Aspen Publication, pp 79–99.

Tsimidou M. 1998 Polyphenols and quality of virgin olive oil in retrospect. Ital. J. Food Sci. 10:99–116.

Usuki R., Suzuki T., Endo Y., Kaneda T. 1984. Residual amounts of chlorophylls and pheophytins in refined edible oils. J. Am. Oil. Chem. Soc. 61:785–788.

Vázquez-Roncero A., Graciani-Constante E., Maestro-Durán R. 1974. Componentes fenólicos de la aceituna I. Polifenoles de la pulpa. Grasas y Aceites 25:269–279.

Vincenzo M., Campestre C., Lanza B. 2001. Phenolic compounds change during Californian-style ripe olive processing. Food Chem. 74:55–60.

Vlahov G. 1976. Gli acidi organici delle olive: II rapportto malico/cítrico quelle indice de maturazione. Ann. Ist. Sper. Elaiot. II:93–112.

Vlahov G., Solinas M. 1993. Anthocyanins polymerization in black table olives. Agric. Med. 123:7–11.

27
Peach and Nectarine

Muhammad Siddiq

INTRODUCTION

Peaches (*Prunus persica*), like apricots, belong to the genus *Prunus* of Rosaceae (rose) family having decorative pink blossoms and a juicy, sweet drupe fruit. They are categorized as "stone fruit," their seed being enclosed in a hard, stone-like endocarp. Peach originated in China, later introduced into Persia, and spread by the Romans throughout Europe. Early Spaniards brought several varieties of peaches to North America. Commercially grown peaches are generally distinguished as clingstone (pit adheres to flesh) or freestone (pit relatively free of the flesh); the famous Elberta peaches belong to the latter type. Nectarine is a smooth-skinned fuzz-less peach; the lack of fuzz is due to a single gene (Anon, 2004a; Rieger, 2004). Peach and nectarine fruits are relatively large in size, ranging from 2 to 3.5 in. in diameter. Peach fruit is pubescent throughout the growing season, and is usually brushed by machine prior to marketing, to remove most of the pubescence (Magness et al., 1971).

World production of peach and nectarine has been on a steady rise since 1997, as shown in Figure 27.1; between 1997 and 2003, peach production saw an increase of about 30%. China is the leading peach and nectarine producing country with about 37% share of the total world production in 2003 (Table 27.1), followed by the United States and Italy. China also leads in the area harvested, followed by Italy and the United States (Table 27.2). However, on a per hectare yield basis (Table 27.3), France is the leader with 18.6 metric tons and the United States a close second (18.5 metric tons). China ranks ninth on per hectare yield basis at 9.2 metric tons, which is just half of both France and the United States. The U.S. commercial peach production is mostly centered in California and in the Southern Atlantic states (South Carolina, Georgia), followed by Michigan, New Jersey, and Pennsylvania. Elsewhere in the world, peaches are cultivated in Southern Europe, Africa, Japan, and Australia. Purple-leaved and double-flowering forms of peaches are cultivated as ornamentals. In China, where the peach flower is largely used in decoration, it is considered a symbol of longevity.

The per capita consumption of fresh and processed peach products in the United States in 2002 was 9.8 lb (Table 27.4). More than half of this consumption was as fresh; this is likely to increase because of the popularity of fresh-cut fruits in recent years.

PRODUCTION AND POSTHARVEST PHYSIOLOGY

The number of peach cultivars commercially harvested far exceeds apple and pear. One of the main

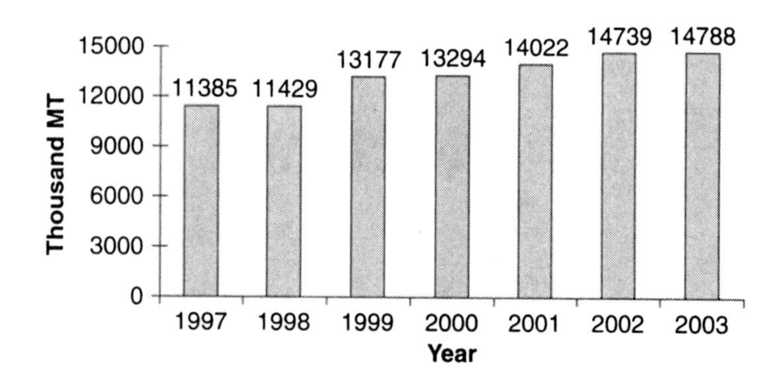

Figure 27.1. World peach and nectarine production (1997–2003). (*Source:* Adapted from FAO (2004), data.)

Table 27.1. Peach and Nectarine: Production in Leading Countries (1997–2003)

	Production (1000 metric tons)						
Countries	1997	1998	1999	2000	2001	2002	2003
China	3,177.1	3,237.2	3,983.4	3,851.9	4,586.2	5,259.8	5,529.4
United States	1,429.0	1,292.3	1,394.2	1,412.4	1,353.2	1,440.2	1,396.7
Italy	1,158.0	1,425.6	1,766.7	1,655.2	1,708.4	1,586.6	1,357.4
Spain	961.9	907.4	982.3	1,129.8	1,082.3	1,247.4	1,284.8
Greece	588.6	527.6	884.2	920.3	927.1	739.6	800.0
Turkey	355.0	410.0	400.0	430.0	460.0	455.0	460.0
Iran	217.8	267.2	317.5	350.0	380.0	385.0	385.0
France	464.0	341.3	477.5	480.7	458.1	464.6	355.8
Chile	242.6	206.3	252.6	260.0	290.0	274.0	275.0
Egypt	377.0	4 29.9	301.2	240.2	247.3	257.0	257.0
Mexico	128.6	116.0	126.1	147.2	175.8	197.9	223.9
Argentina	230.3	261.6	240.0	211.5	259.1	213.3	215.0

Source: FAO (2004).

Table 27.2. Peach and Nectarine: Area Harvested in Leading Countries and World (1997–2003)

	Area Harvested (1000 ha)						
Countries	1997	1998	1999	2000	2001	2002	2003
China	960.4	1,139.9	1,182.5	1,268.5	454.9	549.8	602.7
Italy	95.1	93.8	93.5	93.0	92.8	92.7	97.5
United States	78.4	79.2	78.1	77.2	76.1	76.8	75.5
Spain	70.5	71.2	70.3	72.2	69.1	71.6	70.6
Greece	52.0	52.5	52.5	52.5	52.5	52.5	52.5
Mexico	36.9	39.5	37.3	40.9	39.2	38.6	43.4
Egypt	35.6	34.7	36.1	32.7	33.0	34.0	34.0
Turkey	21.9	23.9	24.1	24.5	25.4	25.4	25.4
Iran	19.5	21.2	24.0	24.5	24.5	25.0	25.0
Argentina	25.3	24.6	23.0	24.0	23.5	23.0	23.0
France	26.7	24.0	23.7	22.8	21.8	20.4	19.1
Chile	17.9	18.2	18.0	17.8	18.1	18.2	18.5
World	1,774.5	1,936.7	1,985.6	2,078.8	1,259.1	1,355.2	1,421.2

Source: FAO (2004).

Table 27.3. Peach and Nectarine: Yield per Hectare, Leading Countries and World Average (1997–2003)

	Yield (metric tons/Ha)						
Countries	1997	1998	1999	2000	2001	2002	2003
France	17.4	14.2	20.1	21.1	21.0	22.7	18.6
United States	18.2	16.3	17.9	18.3	17.8	18.8	18.5
Spain	13.6	12.7	14.0	15.6	15.7	17.4	18.2
Turkey	16.2	17.2	16.6	17.5	18.1	17.9	18.1
Iran	11.2	12.6	13.2	14.3	15.5	15.4	15.4
Greece	11.3	10.0	16.8	17.5	17.7	14.1	15.2
Chile	13.6	11.3	14.0	14.6	16.0	15.1	14.9
Italy	12.2	15.2	18.9	17.8	18.4	17.1	13.9
Argentina	9.1	10.6	10.4	8.8	11.0	9.3	9.3
China	3.3	2.8	3.4	3.0	10.1	9.6	9.2
Egypt	10.6	12.4	8.3	7.3	7.5	7.6	7.6
Mexico	3.5	2.9	3.4	3.6	4.5	5.1	5.2
World (Average)	6.4	5.9	6.6	6.4	11.1	10.9	10.4

Source: FAO (2004).

Table 27.4. Peaches: U.S. Per Capita Consumption (in lb), 1998–2002

			Processed			
Year	Total	Fresh	Total	Canned	Frozen	Dried
1998	8.8	4.8	4.0	3.5	0.5	0.1
1999	9.6	5.4	4.2	3.6	0.6	0.1
2000	9.9	5.4	4.5	3.8	0.6	0.1
2001	9.4	5.2	4.2	3.5	0.6	0.1
2002	9.8	5.3	4.5	3.8	0.6	0.1

Source: Adapted from USDA-ERS (2004).

reasons for this is the ease with which peaches can be bred; in contrast, apple and pear cultivars result mostly from chance selection. Almost all regions of the world, including the United States, have their own breeding programs to produce peach cultivars that adapt to a particular region. Therefore, there is no single cultivar that is dominant worldwide, or in individual peach growing countries (Rieger, 2004).

The peach tree grows better in well-drained, loamy soils (pH 6–7). Peach and nectarine cultivars do not require cross pollination and set satisfactory crops with their own pollen. Peaches have been bred to perform in climates from Canada to the tropics, and have the widest range of chilling sensitivity requirements of any tree crop. Peaches bloom relatively early (e.g., mid-March in Georgia), and are less cold-tolerant than apple or pear. This is the principal reason that they are generally grown in more southern states in the United States than the apples. Frost can be a problem in most peach growing areas of the world as open flowers and young fruits can be seriously damaged or killed by brief exposure to −2°C or below. The U.S. peach production in the southeastern states has suffered a decline in the past several decades due to severe late spring frosts (Rieger, 2004; Logan et al., 2000).

The pruning of peach trees is done extensively to produce quality fruit. Pruning is either done manually or by mechanical hedgers. However, proper manual pruning is widely recommended since it has proven to be more effective (Westwood, 1993). Peaches are heavily thinned for proper size development; this is generally done manually, leaving one fruit per 6 in. of shoot length, which makes it an expensive operation. In some cases, chemical blossom desiccants could be used for flower thinning, but no chemicals are widely used for peach thinning. Miranda-Jimenez and Royo-Diaz (2002) reported that peach tree productivity is improved if trees are pruned early; either in full bloom or soon after the fruit has been set. Chemical thinning could reduce the high cost of manual pruning; however, it distributes the fruit irregularly on the shoots.

Fruit Classification and Maturity

Peaches can be classified according to appearance and sensory characteristics as: round, flat, or beaked; pubescent or smooth-skinned; freestone or clingstone; white-, yellow-, or red-fleshed; sweet, sour, or astringent; and melting-fleshed or non-melting-fleshed (Brovelli et al., 1999). In most cultivars, horticultural maturity for harvesting both peaches and nectarines is determined based on changes in skin ground color from green to yellow. In California, a color chip guide to determine maturity and a two-tier system are used: (1) U.S. Mature, which corresponds to minimum maturity and (2) Well-Mature and/or tree-ripe. Where skin color is masked with a red blush, fruit firmness can be measured to know fruit maturity. Maturity index is the minimum flesh firmness, measured with an 8-mm tip penetrometer, at which stage the fruit can be handled without any damage from bruising (Crisosto and Kader, 2004).

Harvesting and Storage

Harvesting of peaches and nectarines is done manually into bags, baskets, or totes. Picked fruit is then dumped into bins on trailers, transported from orchards to packinghouse, and cooled as soon as possible after harvest. At the packinghouse, the fruit is cleaned and sorted. Attention to details in sorting line efficiency is especially important with peaches, where a range of colors, sizes, and shapes of fruits are encountered. Sizing segregates the fruit by either weight or size. Optimum storage conditions for peaches and nectarines are a temperature of $-1-0$°C and relative humidity of 90–95% with airflow of approximately 50 ft^3/min. Fruit tissue softening is accelerated at elevated temperatures where respiration rate can be as high as 10 times at 20°C than at 0°C. At elevated temperatures, depending on fruit maturity, ethylene production rate increases significantly from <5.0 μl/kg/h at 0°C to as high as 160 μl/kg/h at 20°C (Crisosto and Kader, 2004). Polygalacturonase enzyme is responsible for the softening of fruit tissue as a result of depolymerization of pectic polysaccharide chains during postharvest storage (Tijskens et al., 1998). Togrul and Arslan (2004) reported that the use of carboxymethylcellulose, from sugar beet pulp cellulose as a hydrophilic polymer, in an emulsion coating containing beeswax, triethanolamine, and oleic acid was shown to extend the shelf life of peaches from 12 to 16 days at 25°C and 75% RH.

Chilling Injury

Peaches (most mid- and late-season) and nectarines (some mid- and late-season cultivars) are susceptible to chilling injury during storage. Chilling injury develops faster and more intensely in fruit that is stored at 2.2–7.6°C (peaches) and 2.2–7.8°C (nectarines) than that stored at 0°C or below (Crisosto and Kader, 2004). Brovelli et al. (1998) showed that based on the development of flesh mealiness, the quality of non-melting flesh peaches was not as severely affected as a result of chilling exposure as was the case with melting flesh peaches. This quality of non-melting flesh peaches can be a major advantage for refrigerated storage and long-distance transportation. Modified atmosphere storage (12% CO_2 and 4% O_2) of peaches in unperforated polypropylene bags could be useful as it is associated with lower weight loss, less senescence and chilling injury, reduced incidence of decay, and delayed ripening of the fruit beyond normal shelf-life period (Fernandez-Trujilio et al., 1998).

PROCESSED PRODUCTS

More than 50% of peaches and almost all of nectarines produced are consumed fresh. Canned and frozen peaches are the two major processed peach products. Other products, such as dried peaches, peach jam, jelly, and juice are processed on a much smaller scale (Rieger, 2004). With increasingly new scientific claims of health benefits of phytochemicals, yellow-flesh peaches, which are rich in many such phytochemicals (Gil et al., 2002), have great potential for processing into a variety of new products.

Canned Peaches

About 40% of peaches produced in the United States are processed into canned products, mostly halves and slices. Clingstone peaches are the type most commonly used for commercial canning, as they are exceptional in their ability to retain flavor and consistency. In most cases, peaches are canned within 24 h of delivery to the processing plant, which ensures that the peaches maintain nutritional value and flavor (Anon, 2004b; Rieger, 2004). The optimum maturity of peaches for canning purposes is when the color is orange-yellow and the fruit is still firm. A brief description of processing steps for canning peaches is given below (Lopez, 1981).

1. *Grading*: Since peaches received at the cannery usually include a wide range of sizes, it is necessary to grade fruits using a mechanical grader. Where fruits are to be stored before canning, a pre-grading is done before the fruits have reached canning maturity to minimize bruising.
2. *Halving and Pitting*: Peaches are usually canned as halves or slices. The halving and pitting of peaches is accomplished using automatic machines.
3. *Peeling and Washing*: Clingstone peaches are lye-peeled either by spray or by immersion method. Conditions used for lye-peeling are 5–11% lye solution at a temperature of 102–104°C (215–220°F) for 45–60 s. The treatment time, strength, and temperature of the lye solution depend on the maturity of the fruit. Following the hot lye treatment, fruit pieces are thoroughly sprayed with cold water for complete removal of the peel and the lye residue. Freestone peaches are peeled by (a) steaming, (b) scalding in water, (c) lye-peeling, or (d) a combination of steam and lye.
4. *Quality Grading and Slicing*: An inspection belt is provided for the grading and sorting of poorly sized, off-color, partially peeled, and otherwise imperfect fruit. It is recommended to grade and fill the halves directly from the inspection belt into cans as mechanical grading/filling often results in additional injury to the fruit. Fruits that are not canned as halves go directly from the belt to the slicing machine.
5. *Filling and Syruping*: The peach halves should be filled as rapidly as possible after peeling and grading as extended exposure to the air results in discoloration. Slices should go over an inspection belt for removal of defective pieces. The slices are then discharged to hand-pack fillers. The Standards of Identity for canned peaches specify the cut-out Brix of the syrup (Table 27.6) that correspond to "heavy syrup" or "light syrup," etc. Syruping is accomplished by rotary and straight-line syrupers or by pre-vacuumizing syrupers.
6. *Exhausting and Closing*: The cans are given a steam exhaust for about 6 min at 88–91°C (190–195°F). Cans are closed immediately after exhausting; a gross headspace of 5/16 of an inch is recommended.
7. *Processing and Cooling*: It is recommended to process cans immediately after exhausting and closing of cans so that the heat of the exhaust is not lost before processing begins. Clingstone peaches are processed using continuous reel type cookers at 100°C (212°F) for 20–35 min depending on the diameter of the cans and the type of product—halves or slices. Some processors also use commercial processes of 14–18 min at 116°C (240°F). After processing, the cans are water-cooled immediately to 35–41°C (95–105°F).

Firmness of canned peaches is an important quality attribute. In an attempt to improve the firmness of canned peaches, Wang (1994) treated peach slices by submerging them in pectinase solution containing 100 mg/l $CaCl_2$ under vacuum for 0.5–2 h. This treatment was effective in improving the firmness of peach slices from 7 to 25 J/kg; also, calcium content increased from 280 to 430 mg/kg. Canning is shown to have negative effect on the retention of procyanidins in canned Ross clingstone peaches as well as in the syrup used in the canning (Table 27.5, Hong et al., 2004).

Styles of Canned Clingstone Peaches

The styles of canned clingstone peaches classified as per the U.S. standards are (a) "Halves or Halved" canned peaches—peeled and pitted, cut approximately in half along the suture from stem to apex; (b) "Quarters or Quartered" canned peaches—halved peaches cut into two approximately equal parts; (c) "Slices or Sliced" canned peaches—peeled and pitted peaches cut into sectors smaller than quarters; (d) "Dices or Diced" canned peaches—peeled and pitted peaches cut into approximate cubes; (e) "Whole" canned peaches—peeled, unpitted, whole peaches with or without stems removed; and (f) "Mixed pieces of irregular sizes and shapes" are peeled, pitted, and cut units of canned peaches that are predominantly irregular in size and shape, which do not conform to a single style of halves, quarters, slices, or dices (USDA-AMS, 1985).

Liquid Media and Brix Measurements

Cut-out requirements for liquid media in canned freestone peaches are not incorporated in the grades of the finished product since syrup or any other liquid medium is not a factor of quality for the purpose of these grades. The cut-out Brix measurements for the respective designations are shown in Table 27.6.

Table 27.5. Content of Procyanidin Oligomers in Frozen and Canned Peaches (mg/kg)[a]

Oligomer Fraction	Frozen Peach	Canned Peach			
		0 month	1 month	2 month	3 month
P1	19.59	17.50	17.18	15.77	15.85
P2	39.59	36.50	34.84	29.03	30.55
P3	38.81	34.77	31.33	19.23	19.06
P4	17.81	16.97	10.07	6.26	3.61
P5	12.43	11.85	7.61	2.83	–
P6	10.62	7.87	3.52	–	–
P7	3.94	2.76	–	–	–
P8	1.75	–	–	–	–

Source: Hong et al. (2004).
[a]On wet-weight basis.

Frozen Peaches

About 6–8% of peaches produced are processed as frozen peaches. Processors usually need three different types of peaches: (1) for retail packages of slices, varieties with red color around the pit cavity, such as Rio Oso Gem, are desirable; (2) peaches frozen in bulk for later processing into preserves should have good flavor and should not have red color around the pit cavity for good color/appearance of the jam or marmalade. Fay Elberta peaches picked on the immature side are ideal for preserves; and (3) for the institutional market, mainly pies, a highly flavored, firm-textured variety with resistance to browning is best suited. For freezing peaches, an ideal variety preferably should have the shape, bright color, and firmness of Rio Oso Gem, the flavor of Elberta, and the non-browning characteristic of Sunbeam (Boyle et al., 1977). Fruit preparation steps of washing, peeling, and slicing are the same as discussed under "Canned Peaches."

Air-blast tunnel freezing is the most commonly used system by most processors. In this method, the prepared product ready to be frozen is placed on wire mesh trays and loaded on to racks. The tray racks are moved into the freezing tunnel. Cold air is usually introduced into the tunnel at the opposite end from the one where the product to be frozen enters. The temperature and velocity of the air are of critical importance in the freezing process. The temperature of the air is usually between 0°F and −30°F (−18°C and −34°C, respectively). Air velocity can range from 100 to 3500 ft/min. The length of time a product is subjected to cold air blast in the tunnel depends on the product size (Boyle et al., 1977).

Otero et al. (2000) compared classical methods of freezing and high-pressure-shift freezing (HPSF) with respect to their effect on modification to the microstructure of peach and mango using a histochemical technique. In the HPSF method, samples were cooled under pressure (200 MPa) to −20°C without ice formation, and then pressure was released to

Table 27.6. Cut-Out Brix Levels of Syrups Used in Canned Peaches

Designations	Brix Measurements
"Extra heavy syrup;" or "Extra heavily sweetened fruit juice(s) and water;" or "Extra heavily sweetened fruit juice(s)."	22° or more but less than 35°
"Heavy syrup"; or "Heavily sweetened fruit juice(s) and water"; or "Heavily sweetened fruit juice(s)."	18° or more but less than 22°
"Light syrup;" or "Lightly sweetened fruit juice(s) and water;" or "Lightly sweetened fruit juice(s)."	14° or more but less than 18°
"Slightly sweetened water;" or "Extra light syrup;" or "Slightly sweetened fruit juice(s) and water;" or "Slightly sweetened fruit juice(s)."	10° or more but less than 14°

Source: USDA-AMS (1985).

atmospheric level (0.1 MPa). The super-cooling under high pressure led to uniform and rapid ice nucleation throughout the volume of fruits. The HPSF method prevented quality losses due to freeze-cracking or large ice crystal presence, thus, maintaining the original tissue structure.

Hong et al. (2004) used normal-phase LC-MS to determine the levels and fate of procyanidins in frozen and canned Ross clingstone peaches, as well as in the syrup used in the canning over a 3-month storage period. Retention of these health beneficial compounds was better in frozen peaches as compared to canned peaches (Table 27.5). Storage of canned peaches for 3 months demonstrated a time-related loss in high molecular weight oligomers (P5–P8) and that by 3 months, oligomers larger than tetramers were not observed. After 3 months of post-canning storage, levels of monomers had decreased by 10%, dimers by 16%, trimers by 45%, and tetramers by 80%.

Varietal Types and Styles of Frozen Peaches

Varietal types of peaches that are processed as frozen include: (a) "Yellow freestone"—freestone peaches of the yellow-fleshed varieties, which may have orange or red pigments emanating from the pit cavity, (b) "White freestone"—freestone peaches that are predominately white-fleshed, (c) "Red freestone"—freestone peaches that have substantial red coloring in the flesh, and (d) "Yellow clingstone"—clingstone peaches of the yellow or orange-fleshed varieties. Styles of frozen peaches include: (a) "Halved or halves"—the peaches are cut approximately in half along the suture from stem to apex, (b) "Quartered or quarters"—halved peaches cut into two approximately equal parts, (c) "Sliced or slices"—the peaches are cut into sectors smaller than quarters, (d) "Diced"—the peaches are cut into approximate cube-shaped units, and (e) "Mixed pieces of irregular sizes and shapes"—means peaches cut or broken into pieces of irregular sizes and shapes, and which do not conform to a single style of halves, quarters, or slices (USDA-AMS, 1961).

Dried Peaches

As per the U.S. Standards, dried peaches are the halved and pitted fruit from which greater portion of moisture has been removed. Before packing, the dried fruit is processed to cleanse the fruit and may be sulfured sufficiently to retain a characteristic color. Federal inspection certificates shall indicate the moisture content of the finished product, which shall be not more than 25% by weight. Dried peaches may be processed from freestone or clingstone peach types (USDA-AMS, 1967).

Only 1–2% of peaches produced are processed as dried halves, quarters, or slices. Fruit preparation steps of washing, peeling, and cutting are the same as discussed under "Canned Peaches." To minimize discoloration, cut peaches are dipped in ascorbic acid or other anti-browning solution for 5–10 min (see more detail on anti-browning agents under "Fresh-Cut Peaches and Nectarines"). Prepared fruit is spread in single-layers on trays, usually 3 lb/sq. ft. Peaches are dried to moisture content of about 25%. Generally, forced-draft tunnel dehydrators are used for drying peaches. For drying in a countercurrent tunnel, temperature should not exceed 68°C (155°F). Total drying time (24–30 h) depends on the size of the product and the temperature used. Blanched peaches dry at a faster rate, requiring about 18 h to reach 25% moisture (Brekke and Nury, 1964). In peach-producing developing countries, sun-drying is still the most widely used method due to its low cost; however, quality is not as good as those dried in controlled and sanitary environment.

Hansmann et al. (1998) investigated the drying behavior of clingstone peach halves dehydrated without sulfites, and suggested that enzymatic browning reactions can be controlled during dehydration by selecting dehydration conditions favoring low superficial product temperature and lower water activity (a_w). Also, a peeled fruit dried faster than an unpeeled one. Wang et al. (1996) dried yellow peach fruits cut into halves to a moisture content of 18% with microwaves of 2350 MHz at 0.3–0.45 m/s air velocity and hot air and far ultra-red waves. Microwave dried peaches had better color than those dried by the other two methods.

Many researchers have shown the benefits of osmotic drying before traditional dehydration. Lerici et al. (1988) reported that the osmotically dehydrated products had very good texture and good retention of aroma and color; the a_w was reduced sufficiently to improve shelf life but further processing (e.g., freezing, drying, and pasteurization) was necessary to ensure shelf-stable products. Erba et al. (1994) described "osmodehydro-freezing" as a combined process where osmotic drying is followed by air drying and freezing to prepare reduced-moisture fruit ingredients, free of preservatives, with a natural flavor, color, and texture, and with functional properties suitable for different food applications. Souti et al. (2003) studied suitability of osmotic drying as a method for pre-drying of peaches, and showed that the use of osmotic pre-drying treatment produced dried peaches of better sensory quality than traditionally solar-dried peaches, as judged by a sensory panel.

Peach Jam and Jelly

About 2–3% of peaches are processed into jam, jelly, and juice combined (Rieger, 2004). Jam, jelly, preserves, and marmalades are similar products (all are made from fruit, preserved by sugar and thickened or gelled to some extent). Jam is made from crushed or chopped fruit, holds its shape, but is less firm than jelly. Jelly is a mixture of fruit juice and sugar that is clear and firm enough to hold its shape. Granulated white sugar is most often used to make jelly or jam. Jams and jellies can be made with or without added pectin (Willenberg and Hughes, 2004).

Type of sugar or sweetener and pectin added in jam can affect its textural qualities. Costell et al. (1993) reported that differences in formulations influenced rheological properties of peach jam. Raphaelides et al. (1996) investigated the effects of sugars as present in commercial mixtures on mechanical properties and texture of peach jam. A series of jam samples was prepared using commercial glucose syrups of 38 and 44 dextrose equivalents, isoglucose, maltose syrup, and their mixtures with sucrose. Jam texture was markedly affected by composition of the syrups. Sugar type was important, e.g., monosaccharides and their mixtures with sucrose formed more rigid gels than disaccharides. These researchers concluded that, by carefully selecting blends of glucose syrups with or without sucrose, a range of jam consistencies could be derived. Grigelmo-Miguel and Martin-Belloso (2000) compared the quality of conventional peach jam with those made by total or partial substitution of pectin with added dietary fiber (DF) as a thickener. Peach jam with added DF had similar color but was more viscous than conventional jam. The sensory characteristics of high peach DF jams were similar to conventional jams.

Peach Juice

Peach juice is produced on a much smaller scale as compared to other processed peach products. After cleaning and washing, peaches are heated to about 45°C in large kettles. Stirring with a propeller-type blender and addition of pectinase and hemicullulase enzymes aid in extraction of juice, higher yields, and higher soluble solids contents. Other steps such as filtration, clarification, and pasteurization are similar to those for other juices. Pagan et al. (1997) extracted juice from Caterino variety of peaches by treating pulp with liquefying enzymes at temperatures between 30°C and 70°C. Yield and soluble solids contents of juice were higher with the enzymatic liquefaction method (up to 55°C) than by simple extraction. At temperatures over 55°C, the enzymes used were not effective due possibly to their inactivation. Dogan and Erkmen (2003) investigated the inactivation of *Escherichia coli* by ultra high hydrostatic pressure ranging from 300 MPa to 700 MPa in peach juice and reported that a 12-min treatment at 600 MPa was sufficient to produce a commercially sterile peach juice.

MINIMALLY PROCESSED PRODUCTS

Fresh-Cut Peaches and Nectarines

Fresh-cut produce is one of the fastest growing segments of the food industry in the United States. Retail sales of fresh-cut produce have grown from $5 billion in 1994 to over $10 billion. These figures are projected to reach $15 billion by 2005. Fresh-cut produce is defined as any fresh fruit or vegetable, or any combination thereof that has been physically altered from its original form, but remains in a fresh state (IFPA, 2004). Currently, <1% of peaches and nectarines are processed as fresh-cut; however, both of these fruits have great potential to capture a sizeable share of fresh-cut market owing to the fact that more than half of peaches (Table 27.4) and almost all of nectarines produced are consumed fresh. Fresh-cut produce falls under "Minimal Processing" that has two main purposes: (1) keeping the produce fresh, without losing its nutritional quality and (2) ensuring a product shelf life that is sufficient to make its distribution feasible within the region of its consumption (Laurila and Ahvenainen, 2002).

For processing of fresh-cut peach slices, the optimal ripeness is when flesh firmness reaches 13–27 N (3–6 lb-force). Depending on the cultivar, peach slices retain good eating quality for 2–8 days at 5°C and 90–95% RH. The optimal ripeness for preparing fresh-cut nectarines slices is the partially ripe

(27–49 N or 6–11 lb-force) or ripe (13–27 N or 3–6 lb-force); depending on the cultivar and variety, these slices keep good eating quality at 0°C and 90–95% RH for 2–12 days (Crisosto and Kader, 2004). Gorny et al. (1998) investigated the effects of fruit ripeness and post-cutting storage temperature on the deterioration rate of fresh-cut Flavorcrest peaches and Zee Grand nectarines. They reported that for preparing fresh-cut slices, the optimal ripeness was the ripe stages for peaches and partially ripe or ripe stages for nectarines. While retaining good eating quality, peach and nectarine slices had a shelf life of 6 and 8 days, respectively, at 0°C and 90–95% RH. Gorny et al. (1999) processed slices from 13 cultivars of peaches and 8 cultivars of nectarines using a 2% (w/v) ascorbic acid +1% (w/v) calcium lactate post-cutting dip and reported acceptable sensory color and shelf life of 2–12 days at 0°C for slices from all cultivars, except Cardinal. Controlled atmosphere (CA) storage extended the shelf life by additional 1–2 days for slices from some cultivars.

Some important attributes to be considered for the sensory quality of a fresh-cut product, such as peaches and nectarines, are: (1) color or appearance, (2) texture, (3) flavor, (4) taste, and (5) overall acceptance. However, from consumers' perspective, color or appearance of fresh-cut produce is the single most important factor among all the quality attributes mentioned above (Siddiq et al., 2004). If the color of a fresh-cut product is not acceptable or attractive, the consumer is least likely to purchase it regardless of its excellent texture, flavor, taste, or other quality attributes.

Two enzymes, polyphenol oxidase and peroxidase, have been implicated in color deterioration in cut fruits and vegetables. In addition to visible color changes, these enzymes not only impair the other sensory properties and, hence, the marketability of the product, but also often lower its nutritive value (Vamos-Vigyazo, 1981). Both these enzymes have been widely studied in different fruits and vegetables with special emphasis on their inactivation (Escribano et al., 2002; Nicolas et al., 1994; Siddiq and Cash, 2000; Weemaes et al., 1998). Traditionally, sulfites were used extensively in the food industry to control enzymatic browning. However, since the late 1980s, there has been an effort to avoid the use of sulfiting agents in foods due to safety, regulatory, and labeling issues (Lambrecht, 1995). A number of alternatives to sulfites such as ascorbic acid, citric acid, 4-hexylresorcinol, erythorbic acid, and sodium erythorbate (stereoisomers of ascorbates), benzoic acid, honey, and natural fruit juices (e.g., lemon juice) have been tried with varying success. Chitosan coating (Huaqiang et al., 2004), sodium hexametaphosphate (Pilizota and Sapers, 2004), oxalic acid (Yoruk and Marshall, 2003), and NatureSeal™, a commercially available product containing calcium ascorbate (Arvind et al., 2004), are the other alternatives tried more recently either alone or in conjunction with other inhibitors. In the case of peach and nectarine, post-cutting dips in ascorbate and calcium lactate, or use of modified atmosphere packaging has shown to prolong the shelf life of fresh-cut slices (Gorny et al., 1998).

CHEMICAL COMPOSITION AND NUTRIENT PROFILE

Peaches and nectarines contain significant amounts of some major nutrients as shown in Table 27.7. Peaches and nectarines are good sources of beta-carotene, which is thought to be a powerful anti-aging agent. Palmer-Wright and Kader (1997) studied changes in quality, retinol equivalents, and individual provitamin A carotenoids in fresh-cut Fay Elberta peaches held for 7 days at 5°C in air or CA. They concluded that the limit of shelf life was reached before major losses of carotenoids were observed. Dried peaches, though high in calories, are an excellent source of fiber, most vitamins and minerals. Beta-carotene, along with vitamin C, is important for our immune system as it helps to prevent damage from free radicals. Vitamin C boosts the immune system, promotes healing, and helps prevent cancer, heart disease, and stroke. In addition, peaches are rich in B vitamins, vitamin C, folic acid, calcium, and many other nutrients essential to good health. Peach and its processed products are very low in fat and have no cholesterol.

Peaches and nectarines, especially unpeeled, are a good source of dietary Fibre (DF), which is important to a healthy diet and can help control weight and lower cholesterol levels. Grigelmo-Miguel et al. (1999) investigated insoluble and soluble DF fractions in peach DF concentrates prepared from dried washed peach bagasse or pomace remaining after juice extraction and showed that such concentrates which had low energy value could be an adequate source of DF with an insoluble to soluble DF ratio of 66–34.

Gil et al. (2002) investigated the concentration of total phenolics, total ascorbic acid, beta-carotene, and ascorbic acid equivalent antioxidant capacity

Table 27.7. Composition of Peaches, Their Processed Products and Nectarines (per 100 g Edible Portion)

Nutrient	Unit	Raw Peaches	Canned Peaches[a]	Frozen Peaches[b]	Dried Peaches	Raw Nectarines
Proximate						
Water	g	88.87	84.72	74.73	31.80	87.59
Energy	kcal	39	54	94	239	44
Protein	g	0.91	0.45	0.63	3.61	1.06
Total lipid (fat)	g	0.25	0.03	0.13	0.76	0.32
Fatty acids, total saturated	g	0.019	0	0.014	0.082	0.025
Carbohydrate, by difference	g	9.54	14.55	23.98	61.33	10.55
Fiber, total dietary	g	1.5	1.3	1.8	8.2	1.7
Sugars, total	g	8.39	13.25	22.18	41.74	7.89
Vitamins						
Vitamin A, IU	IU	326	354	284	2163	332
Vitamin C, total ascorbic acid	mg	6.6	2.4	94.2	4.8	5.4
Thiamin	mg	0.024	0.009	0.013	0.002	0.034
Riboflavin	mg	0.031	0.025	0.035	0.212	0.027
Niacin	mg	0.806	0.593	0.653	4.375	1.125
Pantothenic acid	mg	0.153	0.05	0.132	0.564	0.185
Vitamin B-6	mg	0.025	0.019	0.018	0.067	0.25
Folate, total	mcg	4	3	3	0	5
Vitamin E (alpha-tocopherol)	mg	0.73	0.49	0.62	0.19	0.77
Vitamin K (phylloquinone)	mcg	2.6	0	2.2	15.7	2.2
Minerals						
Calcium	mg	6	3	3	28	6
Iron	mg	0.25	0.36	0.37	4.06	0.28
Magnesium	mg	9	5	5	42	9
Phosphorus	mg	20	11	11	119	26
Potassium	mg	190	97	130	996	201
Sodium	mg	0	5	6	7	0
Zinc	mg	0.17	0.09	0.05	0.57	0.17
Copper	mg	0.068	0.052	0.024	0.364	0.086
Manganese	mg	0.061	0.046	0.029	0.305	0.054

Source: USDA (2004).
[a]In light syrup (solids and liquids).
[b]Sweetened.

(AEAC) in a number of cultivars of both yellow-flesh and white-flesh peaches and nectarine; and found a strong correlation (0.93–0.96) between total phenolics and antioxidant activities. Yellow-flesh fruit had significantly higher amounts of total phenolics than white-flesh fruits, and so did peel as compared to flesh regardless of the fruit flesh color (Table 27.8). Asami et al. (2003) reported that lye-peeling of peaches resulted in 21% less loss of total phenolics than manual peeling.

Peach bark has been used as a herbal remedy for a wide variety of ailments. It is said to be "one of the stronger blood moving herbs," and therefore has use in encouraging menstruation in females with delayed menses or congested blood. It also relieves bladder inflammation and urinary tract problems; functions as a mild laxative; has expectorant activity for the lungs, nose, and throat; and relieves chest pain and spasms. The ancient Chinese considered the peach a symbol of long life and immortality (Rieger, 2004).

Table 27.8. Total Phenolics, Total Ascorbic Acid, Beta-Carotene, and Antioxidant Capacity in the Peel and Flesh Tissue of Peaches and Nectarines

	Total Phenolics (mg/kg)		Total Ascorbic Acid (mg/kg)		Beta-carotenes (μg/kg)		AEAC (mg/kg)[a]	
Fruit	Peel	Flesh	Peel	Flesh	Peel	Flesh	Peel	Flesh
Yellow-flesh peaches	485–1,202	172–547	72–181	31–126	2,650–3,350	530–1,680	313–1,107	93–432
White-flesh peaches	670–1,836	228–1,042	112–202	48–65	110–430	40–80	530–1,789	146–1,006
Yellow-flesh nectarine	427–1,403	138–415	78–130	53–61	1,870–3,070	580–1,310	277–981	62–317
White-flesh nectarines	418–2,020	91–901	93–200	42–122	50–570	20–100	230–1,447	46–837

Source: Gil et al. (2002).

[a]Ascorbic Acid Equivalent Antioxidant Capacity.

REFERENCES

Anon. 2004a. Peach. In: The Columbia Electronic Encyclopedia, 6th ed. Columbia University Press, New York (http://www.columbia.edu/cu/cup/).

Anon. 2004b. California Cling Peach Industry. California Cling Peach Board (http://www.calclingpeach.com/html/nav/industry.html).

Arvind AB, Saftner RA, Abbott JA. 2004. Evaluation of wash treatments for survival of foodborne pathogens and maintenance of quality characteristics of fresh-cut apple slices. Food Microbiol 21:319–26.

Asami DK, Hong Y-J, Barrett DM, Mitchell AE. 2003. Processing-induced changes in total phenolics and procyanidins in clingstone peaches. J Sci Food Agric 83:56–63.

Boyle FP, Feinberg B, Ponting JD, Wolford ER. 1977. Freezing Fruits. In: Desrosier ND, Tressler DK, editors. Fundamentals of Food Freezing. Westport: The AVI Publ Co. pp 162–4.

Brekke JE, Nury FS. 1964. Fruits. In: van Arsdel WB, Cople MJ, editors. Food Dehydration: Volume II—Products and Technology. Westport: The AVI Publ Co. p 487.

Brovelli EA, Brecht JK, Sherman WB, Sims CA. 1998. Quality of fresh market melting- and nonmelting-flesh peach genotypes as affected by postharvest chilling. J Food Sci 63:730–3.

Brovelli EA, Brecht JK, Sherman WB, Sims CA, Harrison JM. 1999. Sensory and compositional attributes of melting- and nonmelting-flesh peaches for the fresh market. J Sci Food Agric 79:707–12.

Costell E, Carbonell E, Duran L. 1993. Rheological indices of fruit content in jams: Effect of formulation on flow plasticity of sheared strawberry and peach jams. J Texture Stud 24:375–90.

Crisosto CH, Kader AA. 2004. Apricots, Peach, Nectarine. In: Gross KC, Wang CY, and Saltveit M, editors. The Commercial Storage of Fruits, Vegetables, and Florist and Nursery Stocks—Agriculture Handbook Number 66. Washington, DC: United States Department of Agriculture, Agriculture Research Service.

Dogan C, Erkmen O. 2003. Ultra high hydrostatic pressure inactivation of *Escherichia coli* in milk, and orange and peach juices. Food Sci Technol Int 9:403–07.

Erba ML, Forni E, Colonello A, Giangiacomo R. 1994. Influence of sugar composition and air dehydration levels on the chemical-physical characteristics of osmodehydro-frozen fruit. Food Chem 50:69–73.

Escribano J, Gandý'a-Herrero F, Caballero N, Pedreño MA. 2002. Subcellular localization and isoenzyme pattern of peroxidase and polyphenol oxidase in beet root (*Beta Vulgaris* L.). J Agric Food Chem 50:6123–9.

FAO. 2004. World Primary Crops Data. Food and Agriculture Organization of the United Nations (http://www.fao.org).

Fernandez-Trujilio JP, Martinez JA, Artes F. 1998. Modified atmosphere packaging affects the incidence of cold storage disorders and keeps 'flat' peach quality. Food Res Int 31:571–79.

Gil MI, Tomas-Barberan A, Hess-Pierce B, Kader AA. 2002. Antioxidant capacities, phenolic compounds, carotenoids, and vitamin C contents of nectarine, peach, and plum cultivars from California. J Agric Food Chem 50:4976–82.

Gorny JR, Hess-Pierce B, Kader AA. 1998. Effects of fruit ripeness and storage temperature on the deterioration rate of fresh-cut peach and nectarine slices. HortScience 33:110–3.

Gorny JR, Hess-Pierce B, Kader AA. 1999. Quality changes in fresh-cut peach and nectarine slices as affected by cultivar, storage atmosphere and chemical treatments. J Food Sci 64:429–32.

Grigelmo-Miguel N, Gorinstein S, Martin-Belloso O. 1999. Characterization of peach dietary fiber concentrate as a food ingredient. Food Chem 65:175–81.

Grigelmo-Miguel N, Martin-Belloso O. 2000. The quality of peach jams stabilized with peach dietary fiber. Eur Food Res Technol 211:336–41.

Hansmann CF, Joubert E, Britz TJ. 1998. Dehydration of peaches without sulphur dioxide. Drying Technol 16:101–21.

Hong Y-J, Barrett DM, Mitchell AE. 2004. Liquid chromatography/mass spectrometry investigation of the impact of thermal processing and storage on peach procyanidins. J Agric Food Chem 52:2366–71.

Huaqiang D, Liangying C, Jiahou T, Kunwang Z, Yueming J. 2004. Effects of chitosan coating on quality and shelf life of peeled litchi fruit. J Food Eng 64:355–8.

IFPA. 2004. Fresh-Cut Produce/Fresh-Cut Process. International Fresh-Cut Produce Association, USA (www.fresh-cuts.org).

Lambrecht HS. 1995. Sulfite Substitutes for the Prevention of Enzymatic Browning in Foods. In: Lee CY, Whitaker JR, editors. Enzymatic Browning and Its Prevention. Washington, DC: American Chemical Society. pp 313–23.

Laurila E, Ahvenainen R. 2002. Minimal Processing in Practice. In: Ohlsson T, Bengtsson N, editors. Minimal Processing Technologies in the Food Industry. Cambridge: Woodhead Publishing Ltd. p 223.

Lerici CR, Mastrocola D, Nicoli MC. 1988. Use of direct osmosis as fruit and vegetables dehydration. Acta Aliment Pol 14:35–40.
Logan J, Mueller MA, Searcy MJ. 2000. Microclimates, peach bud phenology, and freeze risks in a topographically diverse orchard. Hort Technol 10:337–40.
Lopez A. 1981. Canning of Fruits—Peaches. In: A Complete Course in Canning, Book II: Processing Procedures for Canned Food Products, 11th ed. Baltimore: The Canning Trade. pp 162–9.
Magness JR, Markle GM, Compton CC. 1971. Food and Feed Crops of the United States. New Jersey Agricultural Experiment Station, Bulletin 828.
Miranda-Jimenez C, Royo-Diaz JB. 2002. Fruit distribution and early thinning intensity influence fruit quality and productivity of peach and nectarine trees. J Am Soc Hort Sci 127:892–900.
Nicolas JJ, Richard-Forget FC, Goupy PM, Amiot MJ, Aubert SY. 1994. Enzymatic browning reactions in apple and apple products. Crit Rev Food Sci Nutr 34:109–57.
Otero L, Martino M, Zaritzky N, Solas M, Sanz PD. 2000. Preservation of microstructure in peach and mango during high-pressure-shift freezing. J Food Sci 65:466–70.
Pagan J, Soliva R, Plque MT, Ibarz A. 1997. Extraction of peach juice using enzymic liquefaction. Aliment Equipos Tecnologia 16(8):65–70.
Palmer-Wright K, Kader AA. 1997. Effect of controlled-atmosphere storage on thc quality and carotenoid content of sliced persimmons and peaches. Postharvest Biol Technol 10:89–97.
Pilizota V, Sapers GM. 2004. Novel browning inhibitor formulation for fresh-cut apples. J Food Sci 69:140–3.
Raphaelides SN, Ambatzidou A, Petridis D. 1996. Sugar composition effects on textural parameters of peach jam. J Food Sci 6:942–6.
Rieger M. 2004. Mark's Fruit Crops Homepage, University of Georgia (http://www.uga.edu/fruit).
Siddiq M, Cash JN. 2000. Physico-chemical properties of polyphenol oxidase from d'Anjou and Bartlett pears (*Pyrus communes L.*). J Food Process Preserv 24:353–64.
Siddiq M, Harte JB, Dolan KD. 2004. Value-added and minimal processing of fresh produce for exports markets. Presented at International Workshop on Intensive Farming and Integrated Resource Management: Traditional and Non-Traditional Approaches, April 28–30, University of Arid Agriculture, Rawalpindi, Pakistan.
Souti M, Sahari MA, Emam-Jomeh Z. 2003. Improving the dehydration of dried peach by applying osmotic method. Iran J Agric Sci 34:283–91.
Tijskens LMM, Rodis PS, Hertog MLATM, Kalantzi U, Dijk C. 1998. Kinetics of polygalacturonase activity and firmness of peaches during storage. J Food Eng 35:111–26.
Togrul H, Arslan N. 2004. Extending shelf-life of peach and pear by using CMC from sugar beet pulp cellulose as a hydrophilic polymer in emulsions. Food Hydrocoll 18:215–26.
USDA. 2004. Nutrient Databse (http://www.nal .usda.gov/).
USDA-AMS. 1961. United States Standards for Grades of Frozen Peaches (http://www.ams.usda .gov/standards/fzpeache.pdf).
USDA-AMS. 1967. United States Standards for Grades of Dried Peaches (http://www.ams.usda .gov/standards/dr-peach.pdf).
USDA-AMS. 1985. United States Standards for Grades of Canned Clingstone Peaches. (http://www .ams.usda.gov/standards/cnpeachc.pdf).
USDA-ERS. 2004. US Per Capita Consumption Data. USDA-Economic Research Service (http://www.ers.usda.gov/).
Vamos-Vigyazo L. 1981. Polyphenol oxidase and peroxidase in fruits and vegetables. CRC Crit Rev Food Sci Nutr 15:49–127.
Wang CY. 1994. Increasing the firmness of canned peach slices. Food Sci (China) 7:34–6.
Wang J, Xu F, Jiang SX. 1996. Effects of microwave drying and pretreatment on the quality of processed yellow peaches. Food Sci (China) 17:39–42.
Weemaes CA, Ludikhuyze LR, Van den Broeck I, Hendrickx ME, Tobback PP. 1998. Activity, electrophoretic characteristics and heat inactivation of polyphenoloxidases from apples, avocados, grapes, pears and plums. Leben Wis Technol 31:44–9.
Westwood MN. 1993. Fruit Growth and Thinning. In: Temperate-Zone Pomology: Physiology and Culture. Portland: Timber Press, Inc. pp 254–74.
Willenberg BJ, Hughes KV. 2004. Jam and Jelly Basics: Tempt Your Taste Buds with Natural Sweets. University of Missouri (http://muextension.missouri .edu).
Yoruk R, Marshall MR. 2003. A survey on the potential mode of inhibition for oxalic acid on polyphenol oxidase. J Food Sci 68:2479–85.

28
Pear Drying

Raquel de Pinho Ferreira Guiné

INTRODUCTION

The drying of fruits allows their better preservation by reducing water content thus inhibiting microbial growth and enzymatic modifications. One of the most important advantages of foodstuff drying is the reduction in size and weight, which facilitates transportation and reduces storage space, and more importantly, avoids the need of using expensive cooling systems and preservation. Finally, it increases food diversity, opening alternatives to the consumption of fresh products, and improving quality of life in rural areas.

The dehydration of foods generally involves a series of interdependent unit operations, such as blanching, pasteurization, and preconcentration, all of which contribute to the overall quality of the final product and to improve the efficiency of the process. Ideally, the process of water removal in dehydration of food products should be reversible. This is not so due to the inevitable loss of nutritional and functional attributes, which in turn depends on the type and extension of dehydration and on the sensitivity of the specific food components (Rizvi, 1986).

Solar drying has been used for centuries, but it is restricted to countries with tropical and semitropical climates. Despite being a slow drying process and requiring much handwork, this is the cheapest of drying methods. However, it has some important disadvantages, like the dependence on natural factors and the need of great exposure areas.

The factors that determine the end of the drying process are mainly the high concentration of sugar, the low moisture content, and the energetic optimization of the process. The desirable amount of water present at the end of the drying varies according to the type of fruit (the smaller the fruit, the less shall be its final moisture content). However, the choice of the final moisture content must take into consideration not only the stability of the fruit, but also the final physical and chemical properties that characterize its quality (Guiné and Castro, 2002b).

The fruits that have been traditionally dried are essentially grapes, figs, plums, apricots, and peaches, but some other species have recently gained increased importance, like apples, mango, pawpaw, pineapple, banana, and pears. The production of dried pears has some relevance in countries like Australia, South Africa, Chile, Argentina, and Portugal (Guiné and Castro, 2003).

The knowledge of the phenomena taking place during drying is crucial to optimize this process, rendering it a more profitable and competitive production method, and offering the consumer products of unquestionable quality. For instance, the study of the drying kinetics is of utmost importance in the design and optimization of dryers (Kiranoudis et al., 1997; Saravia and Passamai, 1997). On the other hand, in the case of fruits, and of pears in particular, the sugar concentrations are relatively high (60–75%, dry basis) and greatly increase as the water evaporation proceeds, offering an additional resistance

To the memory of Professor José Almiro Castro.

to moisture transfer from the fruit (Guiné and Castro, 2002b). In addition, success in food drying and storing (e.g., conditioning and packing) strongly depends on the isothermal relationship between the water activity and water content of the food material being processed. Therefore, the knowledge of such relations, often represented in the form of sorption isotherms, both at the drying and at the storage temperatures, is of unquestionable relevance for the food industry (Weisser, 1986).

This chapter is an overview and presents the personal assessments of the author's work. Dried pears are produced in Portugal in the summer by a traditional solar drying process, but recently many efforts have been made to study pear drying, not only limited to this type of drying. As a matter of fact, many aspects related to the drying process have been investigated to develop reliable process models and new drying techniques applied to this traditional product that is much appreciated and has many economic potentialities.

This work was organized in sections, the first one concerning the shrinking behavior of the pears during drying, namely the evolution of size and density with moisture content. Sorption isotherms of pears are also presented and the experimental data were used to test the suitability of different sorption models. Drying kinetics was also investigated for its importance in the design of alternative driers. From the drying curves, it was possible to examine the dependence of moisture diffusivity on pear composition (in particular, water and sugar concentrations). Finally, commercial drying is also mentioned, the main emphasis being on solar dryers.

SHRINKAGE CHARACTERISTICS

Recently, much attention has been paid to the quality of food products, commonly characterized by a significant number of parameters like texture, taste, color, porosity, and other physical properties including density and specific volume, which are strongly influenced by the processing conditions. In particular during drying, the water present evaporates in a rather important extension, and phenomena such as shrinkage significantly affect physical structure and properties of the food, contributing for its final quality standards (Krokida and Maroulis, 1997).

Shrinking of biological products, as foods, is very much dependent on the internal pressure of the water vapor resulting from the evaporation process, affecting the moisture diffusion and water removal rate, as well as the apparent density. Therefore, the knowledge of the shrinkage behavior assumes an important role in understanding and modeling drying processes and in controlling the characteristics of the food (Zogzas et al., 1994b). Various models have been proposed in the literature to describe shrinkage of the foods in drying, and it has been shown that either volumetric shrinkage or dimensional shrinkage has a strong dependency on the moisture content (Khraisheh et al., 1997).

When studying the physical properties of foods, namely of fruits, it is very important to know with some accuracy estimations for geometrical parameters like shape, size, volume, and surface area, as well as density. Methods based on longitudinal and transversal sections of the material have been developed to assess shape and size, and the results are compiled in the form of charts for some food products, namely apples, peaches, and potatoes (Mohsenin, 1986). The experimental measurements are usually obtained by taking pictures of the object together with a millimeter scale. However, other more sophisticated methods based on computer-aided image processing have also been used (Sabliov et al., 2002). In most cases, fruit shape can be approximated to known geometric forms, allowing the calculation of volume and surface area. In addition, it has been possible to quantify other geometrical parameters associated to shape, like sphericity and roundness (Mohsenin, 1986; Ochoa et al., 2002).

In the present study, the shape of the pears was considered as a combination of a semisphere and a cone, as depicted in Figure 28.1, and thus it was possible, assuming constant shape throughout drying, to estimate their volume and surface area as:

Volume $= \frac{1}{2}$Volume of sphere + Volume of cone

$$V = \frac{1}{2}\left(\frac{4}{3}\pi r^3\right) + \frac{1}{3}\pi r^2 h = \frac{\pi}{3} r^2 (2r + h) \quad (28.1)$$

Surfac area $= \frac{1}{2}$Area of sphere + Area of cone − Area of cone base

$$A_s = \frac{1}{2}(4\pi r^2) + \pi r(\sqrt{r^2 + h^2} + r) - \pi r^2$$
$$= \pi r(2r + \sqrt{r^2 + h^2}) \quad (28.2)$$

where r is the radius of the semisphere and h the height of the cone.

The dimensions r and h were evaluated for pears of the variety D. Joaquina during the drying process from direct readings of pear photographs taken over

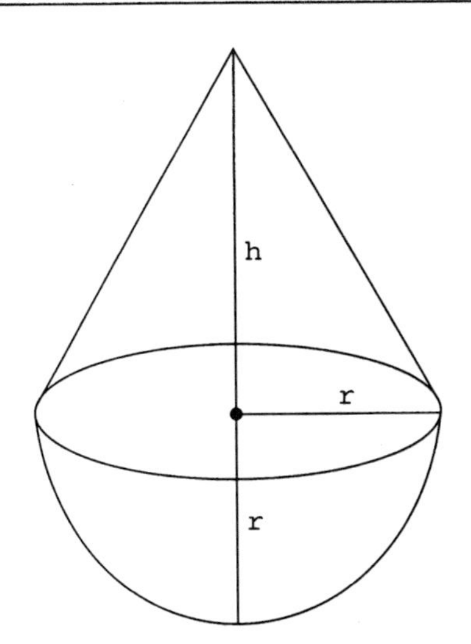

Figure 28.1. Approximation of the pear shape to known geometries.

a millimeter scale paper. Figure 28.2 shows the evolution of the pear dimensions as drying proceeds and consequently the moisture content diminishes. Water content was determined with a Halogen Moisture Analyser (Mettler Toledo HG53).

Both variables (r and h) obey an exponential raise law described by Equations 28.3 and 28.4, where W is the dry basis moisture content (kg of water/kg of dry solids), and r and h are in meters. The quality of the fitting (performed with Sigma Plot 8.0, SPSS, Inc., 2002) is relatively good, as shown by the values of R^2.

$$r = 0.0129 + 0.0268[1 - \exp(-0.1281W)],\quad R^2 = 0.923 \tag{28.3}$$

$$h = 0.0339 + 0.0180[1 - \exp(-0.3598W)],\quad R^2 = 0.951 \tag{28.4}$$

Pear surface area, A_s, was calculated for the assumed geometrical shape [Eq. 28.2] and its evolution throughout drying was also found to follow an exponential law, as a function of dry basis water content (W), expressed by:

$$A_s = 0.0017 + 0.0057[1 - \exp(-0.3081W)],\quad R^2 = 0.987 \tag{28.5}$$

where A_s is expressed in m^2. Figure 28.3 shows the variation of surface area with W.

The volume of the pears was determined not only by using the mean values of the pear dimensions [Eq. 28.1], but also by using liquid displacement (liquid picnometry is currently used for fruits since it is independent of the irregularity of these materials; Mohsenin, 1986). In this work, two different solvents were used (water and toluene) but the results obtained were analogous.

Figure 28.4 shows the plot of the values obtained for pear volume using the two different methodologies against dry basis moisture content, W. The results highlight the good agreement between the two sets of data which were both used to compute the

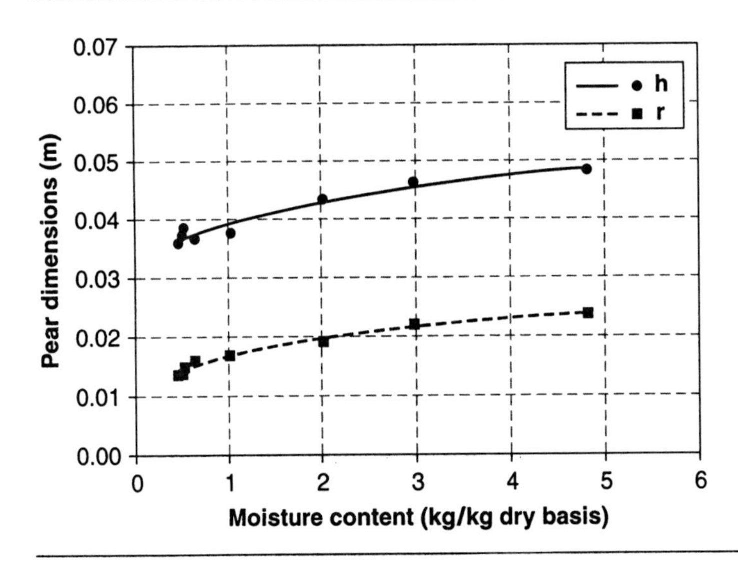

Figure 28.2. Evolution of pear dimensions during drying.

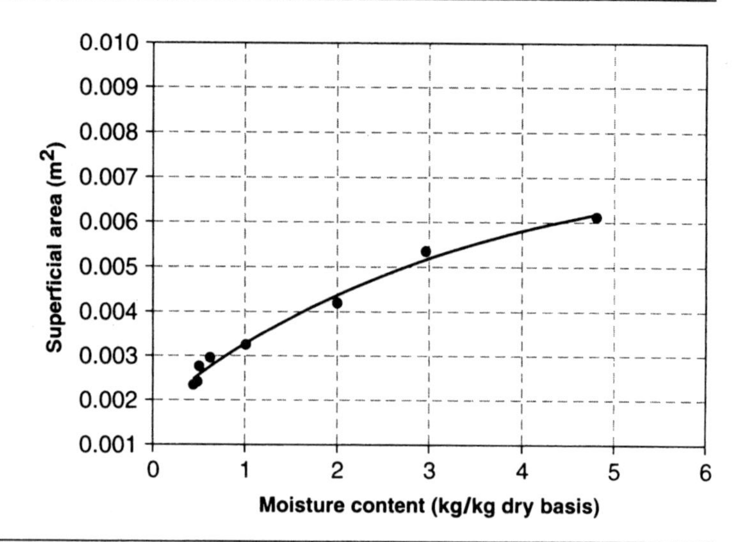

Figure 28.3. Variation of surface area with moisture content.

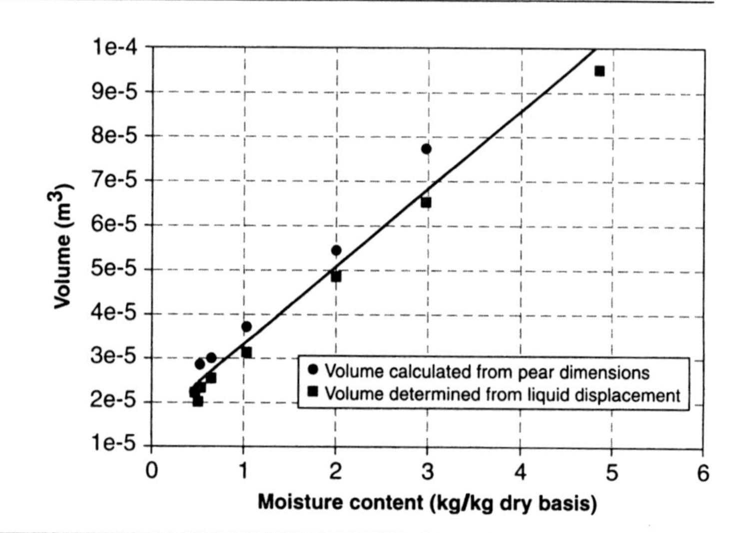

Figure 28.4. Variation of pear volume with moisture content.

following linear relationship between V and W:

$$V = 1.5711\text{e} - 5 + 1.7617\text{e} - 5W,$$
$$R^2 = 0.972 \tag{28.6}$$

where V is the volume (m^3).

In addition, and since both pear volume and weight were known, specific gravity of the pears, d_r, was calculated, being its variation with moisture content presented in Figure 28.5. This variation is an exponential function of the form:

$$d_r = 0.2247 \exp(-2.7168W) + 1.1527 \times \exp(-0.0255W),\ R^2 = 0.898, \tag{28.7}$$

where d_r is dimensionless.

To express the shrinking behavior during drying, a bulk shrinkage coefficient, S_b, has been defined as the ratio of the sample volume at any stage to the initial volume (V/V_0) (Khraisheh et al., 1997). This was determined following two different ways:

a. as the ratio of V/V_0 with both V and V_0 given by Equation 28.6, deriving

$$S_b = \frac{0.89 + W}{0.89 + W_0} \tag{28.8}$$

where W_0 is the initial dry basis water content;

b. by linear regression applied to the experimental values of V/V_0 when plotted against the

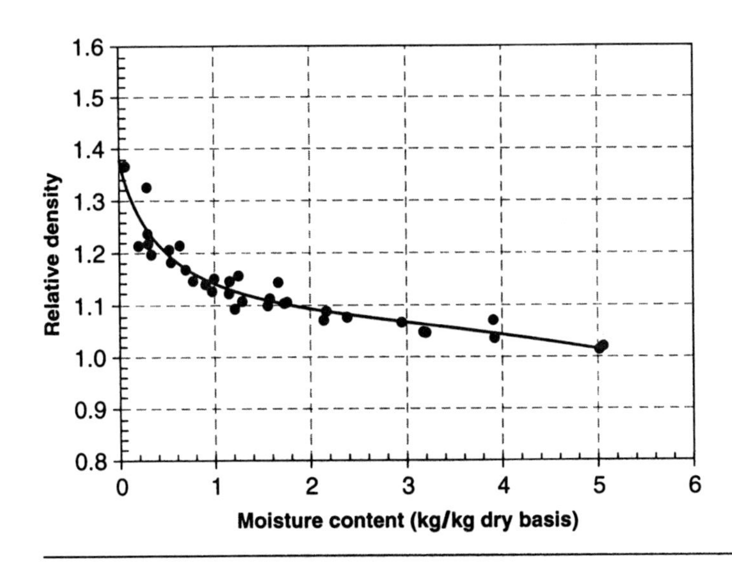

Figure 28.5. Variation of specific gravity with moisture content.

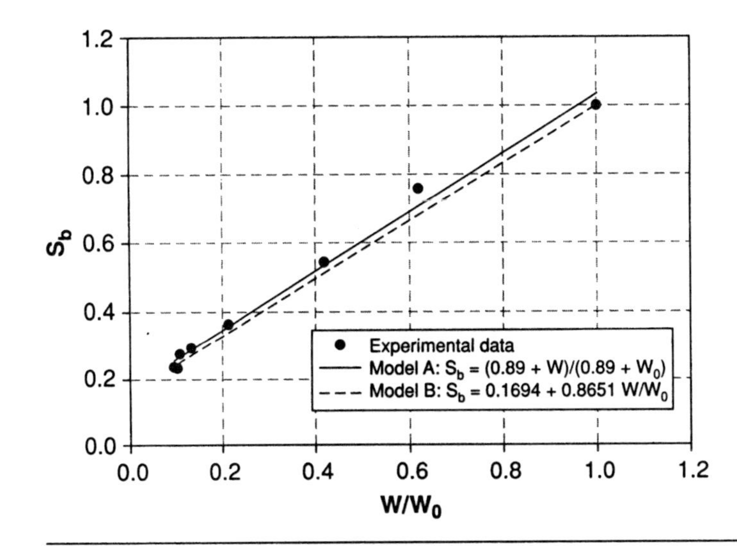

Figure 28.6. Evolution of shrinkage coefficient, S_b, along the drying process.

dimensionless moisture content (W/W_0) (Fig. 28.6), derives

$$S_b = 0.1694 + 0.8651\frac{W}{W_0},\ R^2 = 0.990 \qquad (28.9)$$

Both types of equations have been reported in the literature to compute S_b. Indeed, Equation 28.8 is analogous to the model suggested by Kilpatrick et al. (1955) [$S_b = (0.8 + W)/(0.8 + W_0)$] (Khraisheh et al., 1997), whereas Equation 28.9 is quite similar to the one presented by Raghvan and Silveira (2001) for the microwave drying of strawberries [$S_b = 0.0741 + 0.8121(W/W_0)$].

The good agreement between the experimental points and Equations 28.8 and 28.9, apparent in Figure 28.6, indicates that they are both adequate to correlate the shrinkage coefficient with moisture content.

SORPTION ISOTHERMS

All food materials are characterized by a particular sorption isotherm at a constant temperature. At the end of the drying process, the moisture content of the food becomes stationary, reaching an equilibrium moisture content with the surrounding atmosphere. The moisture sorption isotherms are obtained by plotting the equilibrium moisture content against the corresponding water activity of the food product over a certain range of values, for a constant temperature. Their knowledge is very important for the determination of the optimal storage conditions as well as for the prediction of thermodynamic equilibrium models. Besides, such isotherms also give information on how strongly the water is bound to the material and thus are essential for the prediction of the drying and rehydration rates (Vázquez-Uña et al., 2001; Guiné and Castro, 2002a).

The final water content of the food determines the endpoint of the drying process, which should correspond to the optimum residual moisture content of the final product. In fact, a high water content obtained with a short drying time leads to reduced stability, whereas drying the material to a moisture content below a given optimum value leads to energy wasting with no apparent benefit.

Brunauer et al. (1940) classified adsorption isotherms in terms of the Van der Walls adsorption of gases into five different types (Fig. 28.7). For most foods, sorption isotherms are nonlinear, usually sigmoidal, as type II in Figure 28.7. However, for foods rich in soluble components, such as sugars, the type that best seems to describe the sorption behavior is type III (Rizvi, 1986; Vázquez-Uña et al., 2001).

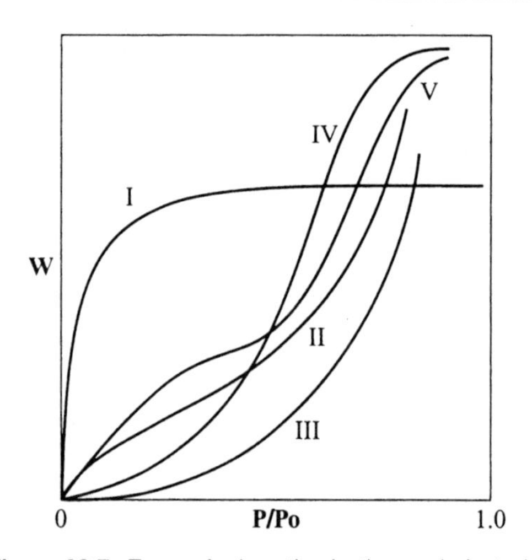

Figure 28.7. Types of adsorption isotherms (adapted from Brunauer et al., 1940).

Many different models can be found in the literature to describe the sorption of water on capillary materials (more than 200; Viswanathan et al., 2003), most of them represented by semiempirical equations with a variable number of parameters. As water is associated with the food matrix in different ways for different activity regions, it is not surprising that one sorption model might not be enough to represent the sorption behavior over the entire water activity range.

Table 28.1 presents eight of the most commonly used sorption isotherm models, where W_e is the equilibrium moisture content (dry basis), W_m the monolayer moisture content (dry basis), a_w the water activity, R the gas constant ($R = 8.31451$ J/mol K), and T the absolute temperature.

Three of these models exhibit an explicit temperature dependency (Chung–Pfost, Halsey, and Henderson), other three include temperature-dependent parameters [Chen, Brunauer–Emmett–Teller (BET), and Guggenheim–Andersen—de Boer (GAB)], and in the remaining (Iglesias–Chirife and Oswin), the temperature dependency has not been considered. Iglesias–Chirife model, corresponding to a type III curve (Fig. 28.7), proved to be quite adequate to describe the sorption behavior of many sugar-rich foods, as most fruits, where the monolayer is completed at a very low moisture content and the dissolution of sugars takes place (Rizvi, 1986).

Generally, higher temperatures result in a reduction of adsorbed molecules and therefore adsorption decreases with increasing temperature, although this dependency is usually small. However, this behavior is not universal as temperature affects many different phenomena and an increase in temperature can in fact increase the rates of adsorption, hydrolysis, and recrystallization processes (Rizvi, 1986). Food rich in soluble solids, such as sugars, seem to exhibit antithetical temperature effects for higher values of a_w due to their increased solubility in water (Rizvi, 1986).

The sorption behavior of pears has been investigated using circular slices of D. Joaquina pears (30 mm diameter and 3 mm thick) dried in a ventilated chamber for constant drying temperatures of 20°C, 25°C, and 30°C (Guiné and Castro, 2002a). The sorptions isotherms were obtained from the measurements of the equilibrium moisture content of the pears (Mettler Toledo, HG53 Halogen

Table 28.1. Most Common Models Found in Literature to Describe Food Sorption Isotherms

Model	Equation	Parameters	References
BET	$\frac{a_w}{W_e(1-a_w)} = \frac{1}{W_m C} + \frac{C-1}{W_m C} a_w$	W_m, $C(T)$	Brunauer et al. (1938)
Chen	$a_w = \exp[-A\exp(-BW_e)]$	$A(T)$, $B(T)$	Chen and Clayton (1971)
Chung–Pfost	$a_w = \exp\left(-\frac{A}{RT}\exp(-CW_e)\right)$	A, C	Chung and Pfost (1967)
GAB	$W_e = \frac{W_m C K a_w}{(1-Ka_w)(1-Ka_w+CKa_w)}$	W_m, $C(T)$, $K(T)$	Guggenheim (1966), Anderson (1946), de Boer (1953)
Halsey	$a_w = \exp\left(-\frac{A}{RT(W_e/W_m)^b}\right)$	W_m, A, b	Halsey (1948)
Henderson	$1 - a_w = \exp(-CTW_e{}^b)$	C, b	Henderson (1952)
Iglesias–Chirife	$\ln[W + (W^2 + W_{0.5})^{1/2}] = Aa_w + B$	A, B	Iglesias and Chirife (1978)
Oswin	$W_e = C\left(\frac{a_w}{1-a_w}\right)^b$	C, b	Oswin (1946)

BET, Brunauer–Emmett–Teller; GAB, Guggenheim–Andersen–de Boer.

Moisture Analyser) and the corresponding water activity (Rotronic Hygrometer). Figures 28.8 and 28.9 confront these data with the models of Table 28.1 for 30°C, as an example. From Figure 28.8, it is clear that the similarity between the BET and GAB models, both quite good in predicting the sorption behavior of the pears for this temperature, contrary to the Chen and Chung–Pfost models that although also similar, are apparently not so adequate to the present case. With regard to Figure 28.9, it can be seen that all models approximately follow the experimental points, with the exception of the Iglesias–Chirife model, that seems to be quite inadequate to reproduce the experimental data in this case, contradicting what would be expected according to Rizvi (1986). However, the quality of fitting of a determined sorption model to the experimental data does not necessarily reveal the nature of the sorption process.

The parameters of the most adequate models were estimated from the experimental data (Guiné and Castro, 2002a) using an orthogonal distance regression algorithm (ODR) with the derivatives approximated by central finite differences, where the unknown errors are taken into account in both dependent and independent variables. The software package used to compute the parameters was ODRPACK, developed by the Center for Computing and Applied Mathematics of the National Institute of Standards and Technology, USA (Boggs, 1992). In Table 28.2, the values of the parameters estimated and the corresponding standard deviation are presented for the models that do not have an explicit dependency on temperature (Chen, BET, GAB, Iglesas–Chirife, and Oswin). In these cases, the experimental data in the form of sets (W, a_w) were treated separately for the three different temperatures studied, and the number of observations varied from 40 to 56.

Similar results are summarized in Table 28.3 for the models that have an explicit dependency on temperature (Chung–Pfost, Halsey, and Henderson). In this case, the experimental data in the form of sets (W, a_w, T) were treated all together, and the number of observations was 138.

To evaluate the influence of temperature on the sorption isotherms, they were predicted by the Halsey model for three different temperatures, and the results are presented in Figure 28.10. The temperatures selected were 20°C, 50°C, and 80°C, as with these intervals the visualization of the differences becomes easier. The selection of the model was based on its nature and its performance. In fact, it should be an explicit temperature-dependent model to allow the easy extrapolation for different temperatures than those studied experimentally, and from the three models available in the present study corresponding to that characteristic (Chung–Pfost, Halsey, and Henderson), the Halsey model is apparently the best one to fit the experimental data (Figs 28.8 and 28.9).

From Figure 28.10, it is apparent that, at constant moisture content, an increase in temperature

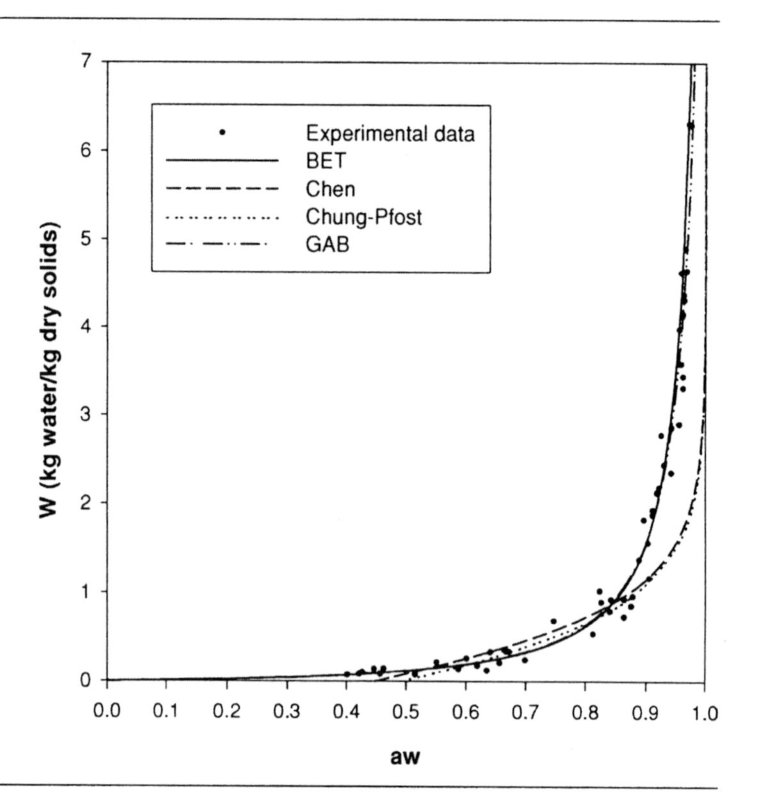

Figure 28.8. Desorption isotherms of pears at 30°C: fitting to BET, Chen, Chung–Pfost, and GAB models.

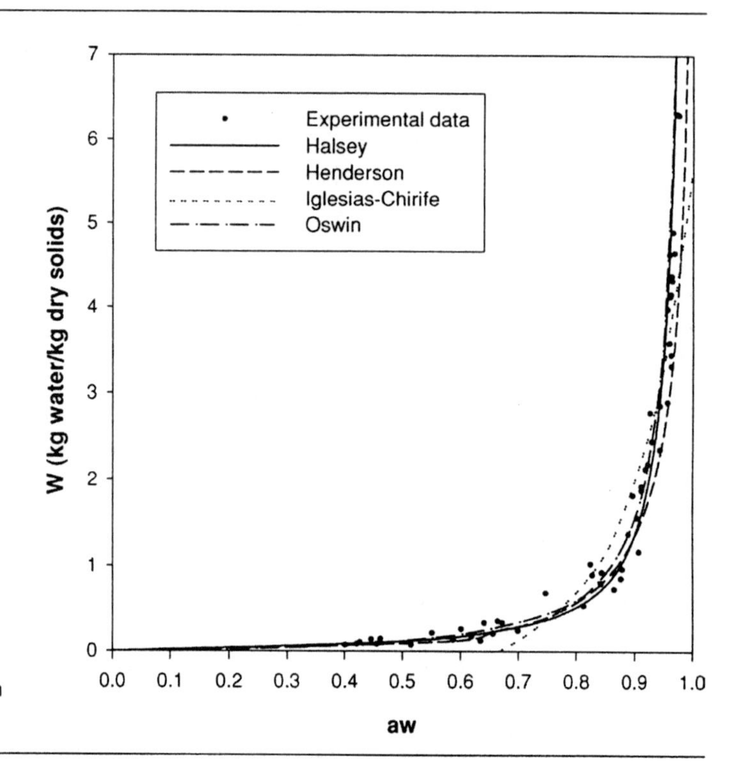

Figure 28.9. Desorption isotherms of pears at 30°C: fitting to Halsey, Henderson, Iglesias–Chirife, and Oswin models.

Table 28.2. Parameter Estimation for the Non-explicit Temperature-Dependent Models

Model	Parameters	Temperature		
		20°C	25°C	30°C
BET	W_m	0.1770 (±0.0182)	0.1819 (±0.0215)	0.1887 (±0.0189)
	C	0.3440 (±0.0757)	0.3434 (±0.0884)	0.4653 (±0.1004)
Chen	A	0.4555 (±0.0251)	0.4433 (±0.0293)	0.8030 (±0.0450)
	B	1.0850 (±0.0931)	1.0942 (±0.1225)	1.7641 (±0.1450)
GAB	W_m	0.1668 (±0.0502)	0.2226 (±0.0994)	0.2372 (±0.0857)
	C	0.3736 (±0.1819)	0.2578 (±0.1596)	0.3379 (±0.1580)
	K	1.0034 (±0.0143)	0.9906 (±0.0192)	0.9880 (±0.0197)
Iglesias–Chirife	A	9.6110 (±0.5629)	9.9273 (±0.6010)	10.2884 (±0.3662)
	B	−7.3664 (±0.4824)	−7.6327 (±0.5082)	−7.8718 (±0.3009)
Oswin	C	0.0960 (±0.0067)	0.0995 (±0.0080)	0.1228 (±0.0079)
	b	1.1941 (±0.0490)	1.1912 (±0.0579)	1.1441 (±0.0468)

BET, Brunauer–Emmett–Teller; GAB, Guggenheim–Andersen–de Boer.

enhances water activity, in agreement to the findings of other authors for most solid fruits (Kiranoudis et al., 1997; Johnson and Brennan, 2000).

DRYING KINETICS

Dehydration is the removal of food moisture which results in an unsaturated gas phase by evaporation due to heat penetration. Thus, this process involves simultaneously heat, mass, and momentum transfer, in a rather complex way, difficult to predict with accuracy. Moreover, in foods such as fruits, with a capillary-porous structure, the food matrix has interstitial spaces, capillaries, and cavities filled with gas, and many different mechanisms of mass transfer have to be considered which, in turn, may act in different combinations. These mechanisms include liquid diffusion due to concentration gradients, liquid transport due to capillary forces, vapor diffusion due to partial vapor pressure gradients, liquid or vapor transport due to the difference in total pressure caused by external pressure and temperature, evaporation and condensation effects, surface diffusion, and liquid transport due to gravity (Rizvi, 1986; Mulet et al., 1989).

Table 28.3. Parameter Estimation for the Temperature-Dependent Models

Model	Parameters	Estimation
Chung–Pfost	A	1717.61 (±81.02)
	C	1.7633 (±0.1117)
Halsey	W_m	0.1281 (±0.0522)
	A	1638.17 (±51.25)
	b	0.7679 (±0.2193)
Henderson	C	0.0067 (±0.0001)
	b	0.4158 (±0.0129)

When developing process models, of utmost importance is the knowledge of the drying kinetics, which accounts for the mechanisms of moisture removal and for the influence of certain variables during the process (Kiranoudis et al., 1997). In air-drying processes, two drying stages are usually present: an initial constant rate period corresponding to pure water evaporation and a falling rate period where the moisture transfer is essentially limited by internal resistances.

Figures 28.11 and 28.12 illustrate the different ways of representing the drying behavior of foods and show respectively, a moisture content versus time plot (commonly referred to as batch drying curve) and the plots of the drying rates versus time and moisture content (i.e., last known as the Krisher rate–moisture curve).

However, it is a common procedure to interconvert between the different types of drying curves, as usually the data recorded from experiments (moisture vs time) are not in the most adequate form for mathematical fitting. In fact, the Krischer rate–moisture curve, which is obtained from the batch drying curve and its differentiation, is actually the most suitable form of treating the drying data in terms of the characteristic curve scaling method (Kemp et al., 2001).

In Figures 28.11 and 28.12, some important points, A to E, are marked in the curves, illustrating the transition between the different stages of the drying process. The induction stage, that goes from A to B, corresponds to the initial unsteady-state heating period, and in most cases, it represents an insignificant

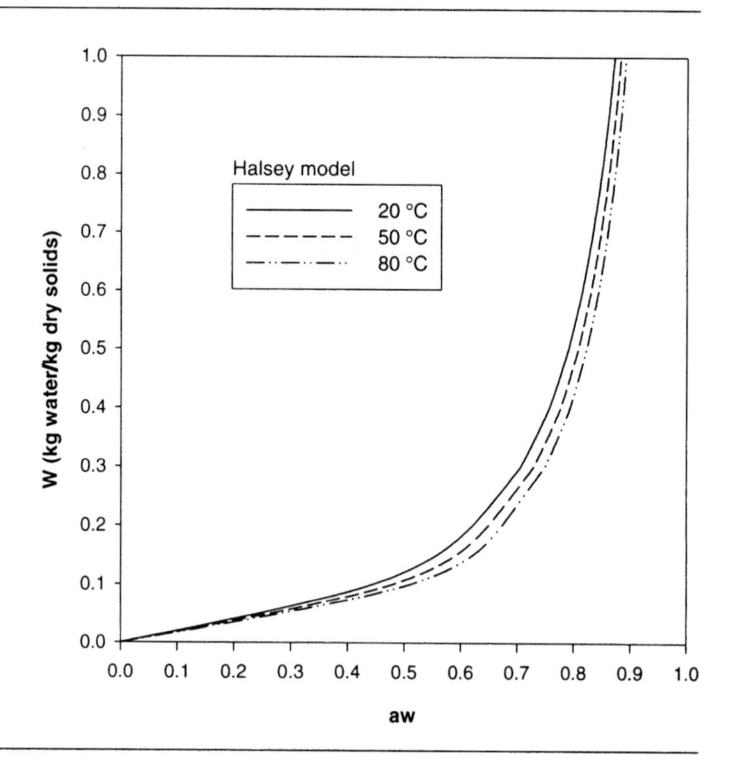

Figure 28.10. Influence of temperature on desorption isotherms of pears using Halsey model.

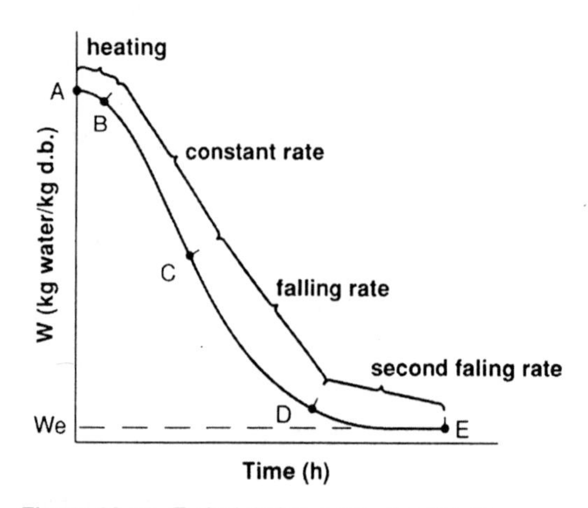

Figure 28.11. Typical drying curve, showing the variation of moisture content with drying time (adapted from Rizvi, 1986).

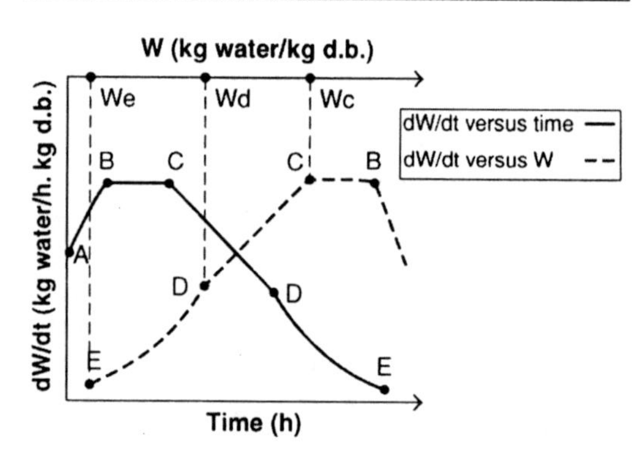

Figure 28.12. Typical drying rate curves, showing the rates of moisture removal as a function of time and moisture content (adapted from Rizvi, 1986).

portion of the total drying cycle. The unhindered drying or constant rate period is represented by *BC*, and during this stage, the drying surface is saturated with water as the rate of migration of water from the inside to the surface equals the rate of evaporation at the surface. The factors that influence the drying rate during this period are the surface area, the temperature and moisture gradients between the surface and the surrounding air, and the heat and mass transfer coefficients. In food systems where moisture transfer is controlled by capillary and gravity forces, the constant rate period is significant and its extension is determined by the food structure and by the mechanisms that govern the internal liquid movement. As drying proceeds, a critical moisture content is reached, W_c, corresponding to point C, where the rate of migration of water from the interior to the surface is reduced to a degree that is no longer sufficient to saturate the entire surface, and this begins to dry. The critical moisture content usually increases with drying rate and thickness of the food. After point C, the surface temperature starts to increase and the hindered drying or falling rate period starts. This generally involves two different stages: the first falling rate period, CD, and the second falling rate period, DE. In the former, the rate of moisture migration to the surface is lower than the rate of evaporation, originating a continuously drier surface until, finally, all the evaporation from the interior of the food is completed (point D). In the second falling rate period, the path for heat and mass transfer becomes longer and more tortuous as the moisture content diminishes. The limiting moisture content, known as the equilibrium moisture content, W_e, is reached when the vapor pressure of the food equals the partial vapor pressure of the drying air, thus ending the drying process (Rizvi, 1986; Brennan et al., 1990).

In typical industrial applications, the kinetic models developed are mostly empirical rather than being mechanistic due to the complexity of the phenomena involved. However, some empirical kinetic models include parameters of phenomenological nature related to the most appropriate driving force, which can lead to rather good prediction of drying kinetics, although they do not have a physical meaning (Kiranoudis et al., 1997).

Drying rate curves were experimentally determined for peeled pears in convective dryers at 30°C, 40°C, and 50°C (Guiné and Castro, 2002b). Every 24 h two samples from each dryer were collected and their moisture contents (Mettler Toledo, HG53 Moisture Analyser) were measured. The corresponding water activity was also analyzed (Rotronic Hygrometer). All experiments were carried out until the moisture content of the pears reached about 20%, since this corresponds to the optimum value for this kind of dried fruit, considering its physical properties and chemical composition (texture, consistency, elasticity, taste, and preservation capacity) (Guiné and Castro, 2002b). From the batch drying curves (dry basis water content vs drying time) obtained with two different fitting methodologies (Figs 28.13 and 28.14), it is possible to verify that they follow the general pattern of a drying curve and that, for higher temperatures, the changes in the water content are sharper, because the evaporation process is highly accelerated.

The data were primarily fitted to cubic polynomials, as suggested by Kemp et al. (2001)

$$W = y_0 + at + bt^2 + ct^3, \qquad (28.10)$$

where W is the dry basis moisture content and t the drying time in hours. The fitting was done with the software Sigma Plot, version 8.0 (SPSS, Inc.).

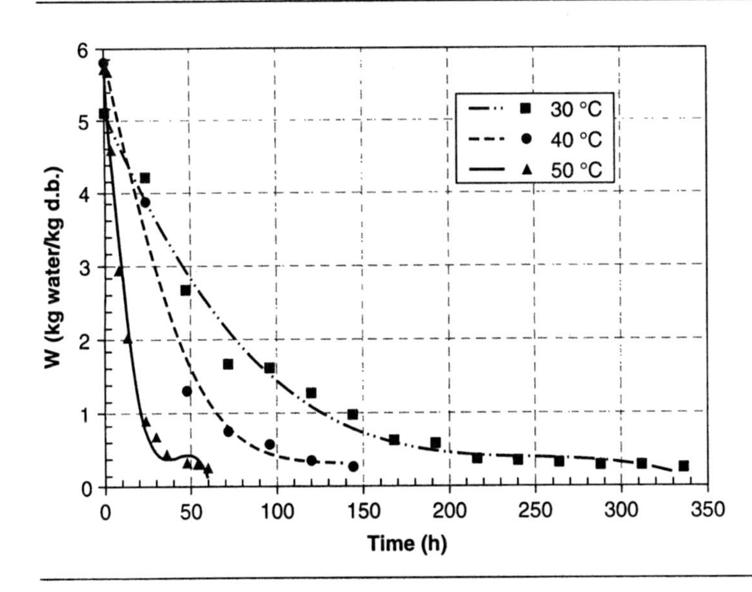

Figure 28.13. Drying curves of pears at 30°C, 40°C, and 50°C: experimental data and predicted by cubic fit.

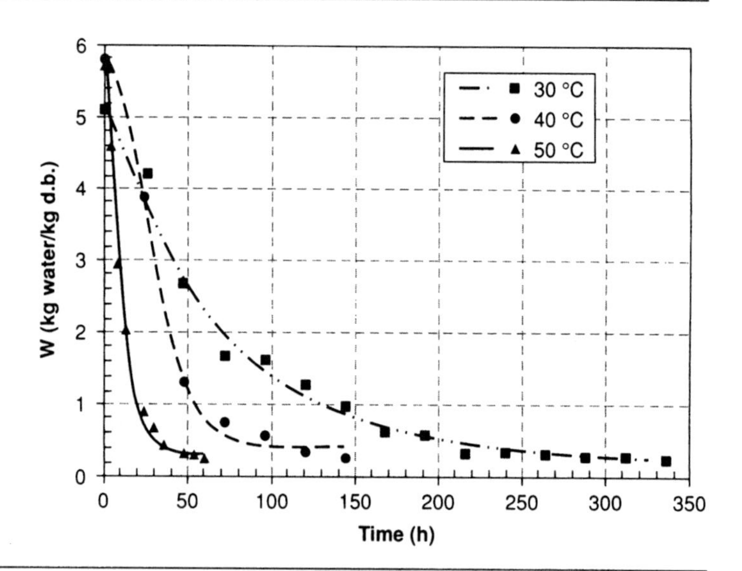

Figure 28.14. Drying curves of pears at 30°C, 40°C, and 50°C: experimental data and predicted by sigmoidal fit.

In addition, a sigmoidal function of the type

$$W = y_0 + \frac{a}{1 + \exp\left[(x_0 - t)/b\right]} \qquad (28.11)$$

was fitted to the experimental data with better results, as can be confirmed from the comparison of Figures 28.13 and 28.14 (where it is evident that the cubic fitting introduces some degree of instability to the final stage of the drying curve) and from the parameter estimation and the corresponding correlation coefficients presented in Table 28.4.

Table 28.4. Results of the Fitting to the Batch Drying Curves

	Temperature		
Parameters	30°C	40°C	50°C
Cubic fit			
y_0	5.1131	5.9685	5.9353
a	−0.0559	−0.1306	−0.4069
b	0.0002	0.0010	0.0098
c	0.0000	0.0000	−0.0001
R	0.9925	0.9912	0.9946
Sigmoidal fit			
y_0	0.2390	0.4199	0.3254
x_0	−87.9603	27.6879	0.7858
a	25.2585	5.9720	10.7263
b	−62.6247	−12.3753	−7.2140
R	0.9941	0.9985	0.9961

The variation of the moisture ratio with the drying time is presented in Figure 28.15 for the three temperatures studied, from which it is clear that the effect of increasing temperature on accelerating the drying process.

A first order kinetics of the form (Togrul and Pehlivan, 2003),

$$\text{MR} = y_0 + a\exp(-bt), \qquad (28.12)$$

where t is expressed in hours, was fitted to the data, using once again Sigma Plot. The results are summarized in Table 28.5, the high values of the correlation coefficient denoting the goodness of the fit.

MOISTURE DIFFUSIVITY

Moisture diffusivity is an unquestionably relevant transport property necessary for correct modeling and understanding of food drying processes. However, its determination is quite complex, not only because of experimental difficulties, but also due to mathematical problems in the solution of the diffusion equation.

As mentioned before, the drying of pears is critically dependent on the temperature as well as on the water and sugar concentrations inside the fruit. In fact, as the drying proceeds, some migration of the sugar occurs along with the water that flows from the inside to the surface of the pears. This combined with the fruit shrinking originates an increase in sugar concentration close to the surface, offering an extra

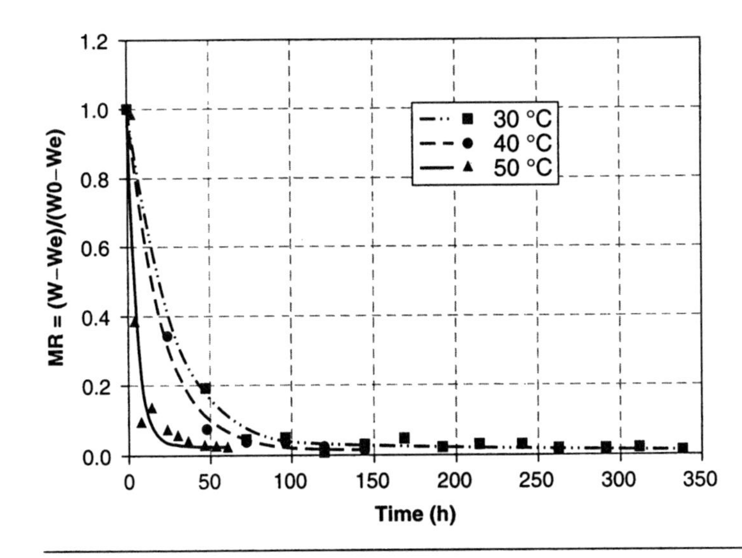

Figure 28.15. Moisture ratio curves for 30°C, 40°C, and 50°C with exponential fit.

resistance to the water diffusion (Guiné and Castro, 2002b).

Despite the complexity of the moisture transfer process and the significant change of the physical characteristics of the fruits along the drying process, it has been reported by many researchers that Fick's law seems to be reasonably good for predicting the moisture distribution inside many food materials during drying (Zogzas et al., 1994a; Bonazzi et al., 1997). For the transient diffusion, assuming an uniform initial distribution of the moisture content as well as a uniform concentration at the surface for $t > 0$, the solution of Fick's law can be approximated by (Konishi et al., 2001)

$$\frac{W - W_e}{W_0 - W_e} = \left(\frac{8}{\pi^2}\right)^3 \exp\left[-D_e t\left(\frac{\pi^2}{4}\right) \times \left(\frac{1}{L_x^2} + \frac{1}{L_y^2} + \frac{1}{L_z^2}\right)\right], \quad (28.13)$$

Table 28.5. Results of the Fitting to the Moisture Ratio Curves

Parameters	Temperature		
	30°C	40°C	50°C
y_0	0.0177	0.0127	0.0287
a	0.9834	0.9901	1.0724
b	0.0383	0.0484	0.2111
R	0.9985	0.9989	0.9616

where D_e is the effective diffusivity, W the dry basis moisture content at time t, W_e the equilibrium moisture content, W_0 the initial moisture content, and L_x, L_y, and L_z represent the half size of the sample in the three spatial dimensions. The evolution of the pear dimensions along drying was studied by measuring the values of the pear radius and height as the moisture content diminished, and the values for L_x, L_y, and L_z were thus calculated from the following equations:

$$L_x = L_y = 0.0129 + 0.0268[1 - \exp(-0.1281W)], \quad R^2 = 0.923 \quad (28.14)$$

$$L_z = 0.0168 + 0.0099[1 - \exp(-0.3285W)], \quad R^2 = 0.945 \quad (28.15)$$

that were obtained by adjusting the experimental data with Sigma Plot 8.0 (SPSS, Inc.), and where L_x and L_y are both equal to the pear radius and L_z is half of the total height, expressed in meters [$L_z = (h + r)/2$ (Fig. 28.1), and the fit of Eq. 28.15 was applied to the experimental values of $(h + r)/2$].

According to Equation 28.13, the values of D_e can be obtained from the slope of a semilog plot of the moisture ratio [$(W - W_e)/(W_0 - W_e)$] versus time, knowing the evolution of pear dimensions.

The dependence of the effective diffusivity on temperature and moisture content is usually represented by a modified Arrhenius relationship of the

Table 28.6. Parameter Estimation for the Diffusivity Models

Parameters	Estimation	
	Equation (28.16)	Equation (28.17)
D_0	3.0450×10^{-3}	4.6246×10^{-3}
B	7.7634×10^{-1}	2.7352×10^{-1}
C	–	8.0543×10^{-1}
E	4.4401×10^{4}	4.5124×10^{4}
Statistic Information		
Number of observations	23	63
$\sum \delta^2(W,T)/\sum \delta^2(W,S,T)$	1.8926×10^{-35}	2.5847×10^{-37}
$\sum \varepsilon^2(D_e)$	1.2785×10^{-17}	1.1249×10^{-17}
Sum of square errors	1.2785×10^{-17}	1.7472×10^{-17}
Residual standard deviation	7.9954×10^{-11}	6.0504×10^{-11}

form (Zogzas and Maroulis, 1996; Ramallo et al., 2001):

$$D_e = D_0(1 + BW)\left[\exp\left(-\frac{E}{RT}\right)\right] \qquad (28.16)$$

where D_0, B, and E are parameters to estimate, T is absolute temperature, W the dry basis moisture content, and R the gas constant ($R = 8.31451$ J/mol K). In the above equation, E represents the activation energy for moisture diffusion (J/mol).

The experimental data obtained for the pears, namely W and a_w and the evolution of the pear dimensions [Eqs 28.14 and 28.15] for the range of the tested moisture contents (between 0.20 and 5.80, dry basis) and temperatures (between 30°C and 50°C), were used to estimate the different values of diffusivity. Equation 28.16 was adjusted to the data in the form of triplets (D_e, W, T), and an orthogonal regression algorithm was applied (explicit ODR with the derivatives approximated by central finite differences), being the results of the estimation listed in Table 28.6. The values of diffusivity obtained with this model vary from 7.86×10^{-11} m^2/s ($W = 0.20$, $T = 30$°C) to 1.11×10^{-9} m^2/s ($W = 5.80$, $T = 50$°C), being the activation energy, E, equal to 44.401 kJ/mol. The latter is within the range of values reported for other food products (15–95 kJ/mol) (Rizvi, 1986; Konishi et al., 2001; Ramallo et al., 2001).

In addition, and since drying of pears is strongly affected by sugar concentration, it was thought more realistic to modify Equation 28.16 to introduce a component that accounts for the dependency of the water diffusivity on sugar concentration, S (kg of sugar/kg dry solids), as

$$D_e = D_0\frac{(1 + BW)}{(1 + CS)}\left[\exp\left(-\frac{E}{RT}\right)\right]. \qquad (28.17)$$

The determination of the parameters of Equation 28.17 was based on the experimental data obtained for the model of Equation 28.16 and also on the corresponding values of the dry basis sugar concentration. The ranges of temperature and moisture concentration used were the same as in the previous case, and the values of the sugar concentration varied from 0.17 to 4.08 (kg/kg dry solids). The same ODR algorithm was used and the data, in the form of sets (D_e, W, S, T), were adjusted to the model, being the estimated parameters also presented in Table 28.6. With this model, the values of diffusivity vary from 1.91×10^{-11} m^2/s ($W = 0.20$, $S = 4.08$, and $T = 30$°C) to 5.34×10^{-10} m^2/s ($W = 5.80$, $S = 0.17$, and $T = 50$°C), and the activation energy is 45.124 kJ/mol. These values are lower than those found by Park et al. (2002) for the osmotic dehydration of pear d'Anjou in the ranges 40–60°C and 40–70°Brix, since the diffusion in osmotic conditions is quite increased in comparison with that in convective drying.

Figures 28.16–28.18 show the variation of diffusivity with the concentrations of water and sugar, as predicted by Equation 28.19, for the temperatures 30°C, 40°C, and 50°C, from which it is apparent that at constant temperature the diffusivity is enhanced as the water concentration increases and the sugar concentration decreases. The diffusivity decreases as drying proceeds, that is, as water contents decreases and sugar concentration increases. This was expected since the presence of dry solids and sugars offers

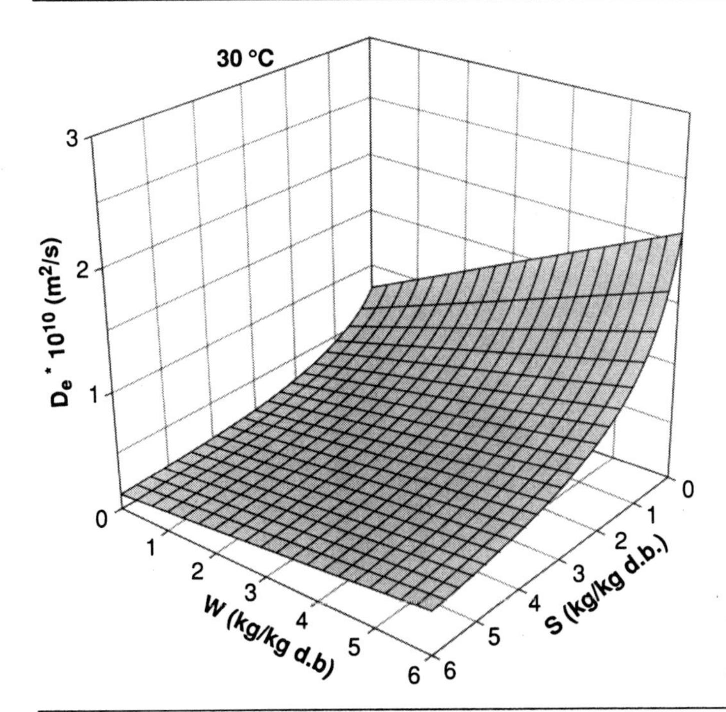

Figure 28.16. Variation of diffusivity with moisture and sugar concentrations at 30°C.

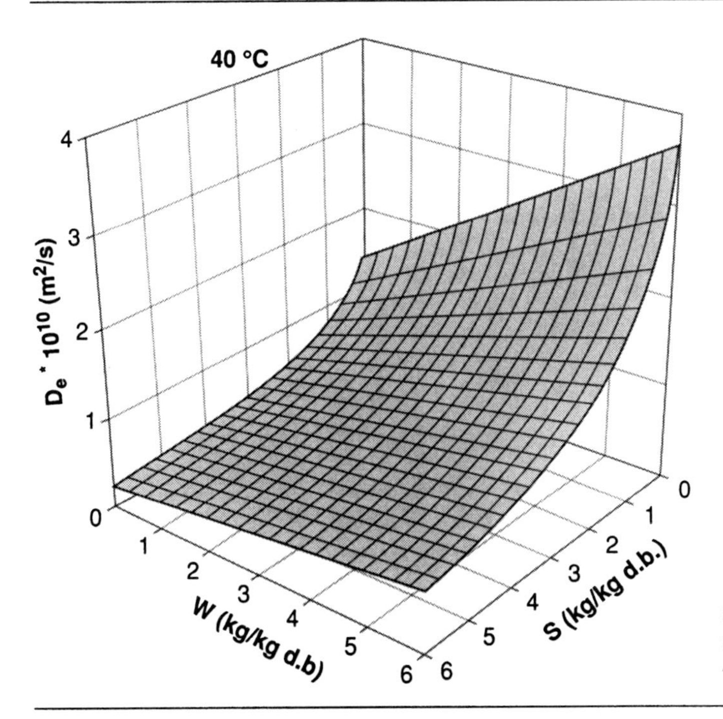

Figure 28.17. Variation of diffusivity with moisture and sugar concentrations at 40°C.

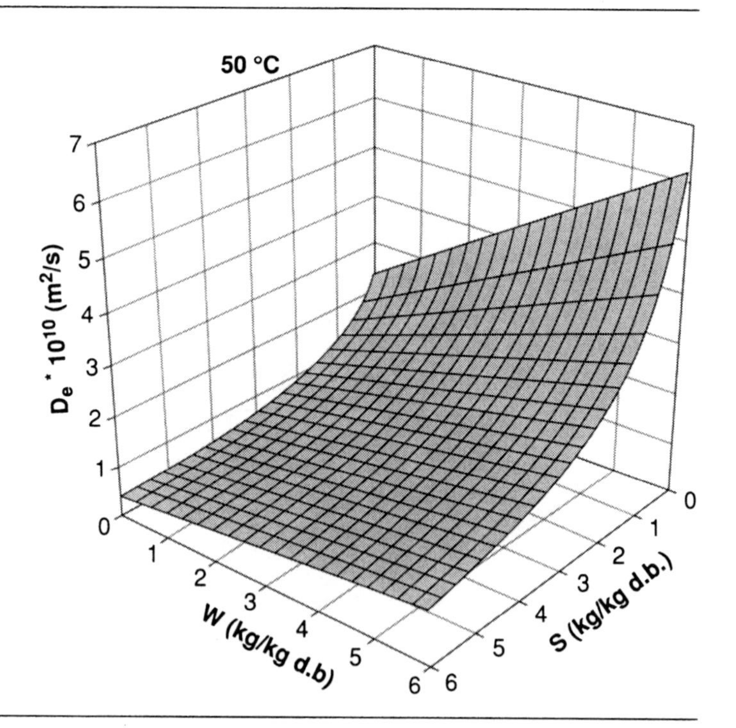

Figure 28.18. Variation of diffusivity with moisture and sugar concentrations at 50°C.

additional resistance to moisture transfer. With regard to the effect of the drying temperature, Figures 28.17 and 28.18 show that although the trends are similar for the different temperatures tested, the influence of this variable is more pronounced for higher temperatures. To better illustrate this effect, both the lowest values of diffusivity (observed for the minimum values of W and maximum values of S) and the highest ones (corresponding to maximum W and minimum S) are presented in Figure 28.19 for the three

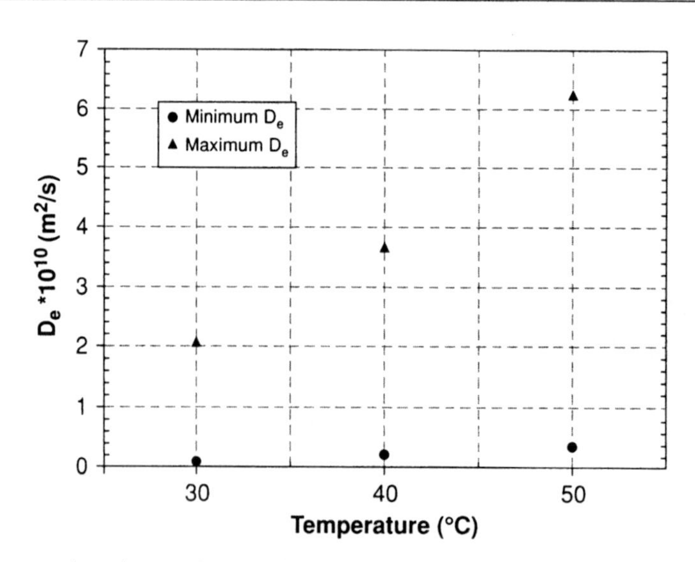

Figure 28.19. Minimum and maximum values obtained for diffusivity at 30°C, 40°C, and 50°C.

Table 28.7. Nutritional Composition of Dried Pears

Property	Variety	
	S. Bartolomeu[a]	D. Joaquina[b]
Water content (g/100 g)	35.2	23.9
Fat (g/100 g)	0.2	0.1
Protein (g/100 g)	1.2	3.7
Sugar (g/100 g)	49.8	62.5
Fiber (g/100 g)	5.0	7.8
Ash (g/100 g)	1.4	1.4
Acidity (cm^3/100 g)	14.2	8.3

[a] Ferreira et al. (1997).
[b] Guiné (2003).

temperatures considered, and it is clearly shown the larger increments of diffusivity as temperature increases.

COMMERCIAL DRYING

Sun-dried pears have been produced in Portugal from the variety S. Bartolomeu, but some other varieties are presently trying to gain importance. The traditional solar drying process involves five different phases, namely a peeling operation, a first drying stage, a barreling phase, a pressing operation, and a second drying stage. In the peeling operation, the skin of the fruit is mechanically removed and then the pears are left at direct sun exposure for 5–6 days. At the end of this first drying, the pears are removed from the sun at the hottest hour of the day and muffled in barrels that are left in the shadow for about 2 days, as this procedure is supposed to increase their elasticity. The pears are then pressed to change their shape from round to flattened, and finally they are subjected to a second sun-drying period of 2–3 days (Ferreira et al., 1997; Guiné and Castro, 2003).

The best pears for drying are sweet small pears, of approximately 4 cm in diameter and 4.5 cm in height, and their state of ripeness is decisive for the quality of the final product. The quality of the final product is usually evaluated in terms of shape, size, and color of the fruits, as well as nutritional composition. Table 28.7 lists typical compositions of pears of both S. Bartolomeu and D. Joaquina varieties. Although both are rich in sugar and fiber (with fiber contents comparable to cereals) and low in protein and fat, D. Joaquina pears are sweeter and less acidic than those from the variety S. Bartolomeu.

In traditional drying, the oxidation of the fruit is favored as to obtain a fruit of reddish color, very much appreciated for its organoleptic characteristics. Alternatively, industrial drying methods making use of different and more efficient driers, either electric or solar, can be applied, but in which the oxidation might not be so intense thus originating fruits less appreciated.

Direct sun drying can be carried out in solar drying stoves, in which the convective airflow is generated by ventilators. In this kind of stoves, the fruits are protected from insects that could damage or even contaminate them, thus compromising the final quality of the product.

The stove (3 m long, 2 m wide, and 2 m high at the center) is a structure of aluminum and glass and is equipped with trays made of nylon nets, as these are less damaging to the pear surface. The tray dimensions are 3 m × 0.6 m and these are located on both sides of the stove, and can be placed in up to three levels, allowing a total drying capacity of 250 kg of fresh pears. Air convection is regulated by a ventilator which provides airflow rates from 1300 to 9000 m^3 of air per hour.

The direct solar dryer is built in wood and plastic and has two stoves, one at the front with a capacity of 60 kg of fresh fruit and the other behind with a capacity of 2000 kg of fresh fruit.

Solar driers in which the product is dried in a chamber away from direct sun exposure are also available. Depending on the air circulation these are available in two versions: natural convection (passive) and forced convection (active). The passive dryer is constituted by a collector, with a layer of dark basaltic stones at the base to absorb the solar energy and accumulate heat, a chamber where the product is dried and a chimney to increase natural convection. The collector, with an area of about 4 m^2, allows an increase in air temperature of approximately 10°C. The airflow is 130 m^3/h and the area available in the chamber is 2.2 m^2, allowing the drying of 12–15 kg of fresh fruit (Leitão, 2003).

The active solar drier, although similar to that of passive solar drier, has a collector equipped with a ventilator, at the entrance, to increase convection. Furthermore, under the drying chamber, there is a furnace that allows wood or crop waste burning, enabling this drier to be used without sun, either when it rains or at night. This particular drier has a collector surface of 5.0 m^2 and increases the air temperature by up to 18°C. The airflow is 570 m^3/h, for a drying surface of 4.3 m^2 corresponding to a capacity of 22–25 kg of fresh fruit. In both driers, passive and active, the product in the drying chambers is protected with nets to avoid insects (Leitão, 2003).

In another model of indirect solar drier, the drying chambers, placed inside a rural construction, are built of wood and possess eight layers of trays each optimizing the capacity of the drier. The air circulating inside the chambers is heated with the solar energy that is collected by the panel installed outside the building. During the night and cloudy days, the air is heated by two electric resistances with 1000 W each. Air convection is regulated by a fan which provides airflow rates from 1000 to 2500 m^3/h.

ACKNOWLEDGMENTS

The information in this chapter has been derived from "Drying Pears," *Food Drying Manual*. © 2003 Raquel Guiné. Used with permission.

REFERENCES

Anderson, R.B. 1946. Modifications of the BET Equation. J. Am. Chem. Soc. 68:686–691.

de Boer, J.H. 1953. The Dynamical Character of Adsorption. Oxford: Clarendon Press.

Boggs, P.T. 1992. User's Reference Guide for ODRPACK Version 2.10. Software for Weighted Orthogonal Distance Regression. Gaithersburg, MD: National Institute of Standards and Technology.

Bonazzi, C.; Ripoche, A.; Michon, C. 1997. Moisture Diffusion in Gelatin Slabs by Modelling Drying Kinetics. Drying Technol. 15(6–8):2045–2059.

Brennan, G.J.; Butters, R.J.; Cowell, D.N.; Liley, V.E.A. 1990. Las Operaciones de la Ingenieria de los Alimentos, 3rd Ed. Spain: Libreia Editorial.

Brunauer, S.; Deming, L.S.; Deming, W.E.; Teller, E. 1940. On a Theory of the Van der Walls Adsorption of Gases. Am. Chem. Soc. J. 62:1723–1732.

Brunauer, S.; Emmett, P.H.; Teller, E. 1938. Adsorption of Gases in Multimolecular Layers. J. Am. Chem. Soc. 60:309–319.

Chen, C.S.; Clayton, J.T. 1971. The Effect of Temperature on Sorption Isotherms of Biological Materials. Trans. ASAE 14:927–929.

Chung, D.S.; Pfost, H.B. 1967. Adsorption and Desorption of Water Vapour by Cereal Grains and their Products. Part I. Heat and Free Energy Changes of Adsorption and Desorption. Trans. ASAE 10:549–551.

Ferreira, D.; Costa, C.A.; Correia, P.; Guiné, R. 1997. Caracterização da Pêra Passa de Viseu. Terra Fértil 3:75–79.

Guggenheim, E.A. 1966. Applications of Satistical Mechanics. Oxford: Clarendon Press.

Guiné, R.P.F. 2003. Avaliação das Propriedades Químicas e Físicas de Pêras Secadas por Diferentes Processos. Actas do 6° Encontro de Química de Alimentos, Lisbon, pp. 195–198.

Guiné, R.P.F.; Castro, J.A.A.M. 2002a. Experimental Determination and Computer Fitting of Desorption Isotherms of D. Joaquina Pears. Food Bioproducts Process.: Trans. IchemE, Part C 80(C3):149–154.

Guiné, R.P.F.; Castro, J.A.A.M. 2002b. Pear Drying Process Analysis: Drying Rates and Evolution of Water and Sugar Concentrations in Space and Time. Drying Technol. 20(7):1515–1526.

Guiné, R.P.F.; Castro, J.A.A.M. 2003. Analysis of Moisture Content and Density of Pears During Drying. Drying Technol. 21(3):581–591.

Halsey, G. 1948. Physical Adsorption on Non-Uniform Surfaces. J. Chem. Phys. 16:931–937.

Henderson, S.M. 1952. A Basic Concept of Equilibrium Moisture. Agric. Eng. 33:29–32.

Iglesias, H.A.; Chirife, J. 1978. An Empirical Equation for Fitting Water Sorption Isotherms of Fruits and Related Products. J. Can. Inst. Food Sci. Technol. 11:12–15.

Johnson, P.N.T.; Brennan, J.G. 2000. Moisture Sorption Isotherm Characteristics of Plantain. J. Food Eng. 44:79–84.

Kemp, I.C.; Fyhr, B.C.; Laurent, S.; Roques, M.A.; Groenwold, C.E.; Tsotsas, E.; Sereno, A.A.; Bonazzi, C.B.; Bimbenet, J.-J.; Kind, M. 2001. Methods for Processing Experimental Drying Kinetics Data. Drying Technol. 19(1): 15–34.

Khraisheh, M.A.M.; Cooper, T.J.R.; Magee; T.R.A. 1997. Shrinkage Characteristics of Potatoes Dehydrated Under Combined Microwave and Convective Air Conditions. Drying Technol. 15(3/4):1003–1022.

Kilpatrick, P.W.; Van Lowe, E.; Arsdel, W.B. 1955. Tunnel Dehydrators For Fruit and Vegetables. Adv. Food Res. 6:313–372.

Kiranoudis, C.T.; Tsami, E.; Maroulis, Z.B.; Marinos-Kouris, D. 1997. Drying Kinetics of Some Fruits. Drying Technol. 15(5):1399–1418.

Konishi, Y.; Horiuchi, J.-I.; Kobayashi, M. 2001. Dynamic Evaluation of The Dehydration Response Curves of food Characterized by a Poultice-Up Process Using a Fish-Paste Sausage. II. A New Tank Model for Computer Simulation. Drying Technol. 19(7):1271–1285.

Krokida, M.K.; Maroulis, Z.B. 1997. Effect of Drying Method on Shrinkage and Porosity. Drying Technol. 15(10):2441–2458.

Leitão, A. 2003. A Utilização da Energia Solar na Conservação de Produtos Horto-Frutícolas. O Segredo da Terra 5:12a–12d.

Mohsenin, N.N. 1986. Physical Properties of Plant and Animal Materials, 2nd Ed. New York: Gordon and Breach Science Publishers.

Mulet, A.; Berna, A.; Rosselló, C. 1989. Drying of Carrots. I. Drying Models. Drying Technol. 7(3):537–557.

Ochoa, M.R.; Kesseler, A.G.; Pirone, B.N.; Márquez, C.A.; Michelis, A. 2002. Volume and Area Shrinkage of Whole Sour Cherry Fruits (*Prunus cerasus*) During Dehydration. Drying Technol. 20(1):147–156.

Oswin, C.R. 1946. The Kinetics of Package Life. III. The Isotherm. J. Chem. Ind. 65:419–423.

Park, K.J.; Bin, A.; Brod, F.P.R.; Park, T.H.K.B. 2002. Osmotic Dehydration Kinetics of Pear D'Anjou (*Pyrus communis* L.). J. Food Eng. 52:293–298.

Raghvan, G.S.V.; Silveira, A.M. 2001. Shrinkage Characteristics of Strawberries Osmotically Dehydrated in Combination with Microwave Drying. Drying Technol. 19(2):405–414.

Ramallo, L.A.; Pokolenko, J.J.; Balmaceda, G.Z.; Schmalko, M.F. 2001. Moisture Diffusivity, Shrinkage, and Apparent Density Variation During Drying of Leaves at High Temperatures. Int. J. Food Prop. 4(1):163–170.

Rizvi, S.S.H. 1986. Thermodynamic Properties of Foods in Dehydration. In: M.A. Rao and S.S.H. Rizvi. Engineering Properties of Foods. New York: Marcel Dekker, pp. 133–214.

Sabliov, C.M.; Boldor, D.; Keener, K.M.; Farkas, B.E. 2002. Image Processing Method To Determine Surface Area and Volume of Axi-Symmetric Agricultural Products. Int. J. Food Prop. 5(3):641–653.

Saravia, L.; Passamai, V. 1997. Relationship Between a Solar Drying Model of Red Pepper and The Kinetics of Pure Water Evaporation (I). Drying Technol. 15(5):1419–1432.

Togrul, I.T.; Pehlivan, D. 2003. Modelling of Drying Kinetics of Single Apricot. J. Food Eng. 58:23–32.

Vázquez-Uña, G.; Chenlo-Romero, F.; Moreire-Martínez, R. 2001. Adsorption and Desorption Isotherms of Lupin (*Lupius albus* L.) at Several Temperatures. CHEMPOR 2001: 8th International Conference of Chemical Engineering, Universidade de Aveiro, Portugal, pp. 1071–1076.

Viswanathan, R.; Jayas, D.S.; Hulasare, R.B. 2003. Sorption Isotherms of Tomato Slices and Onion Shreds. Biosyst. Eng. 86(4):465–472.

Weisser, H. 1986. Influence of Temperature on Sorption Isotherms. In: M.L. Maguer and P. Jelen. Food Engineering and Process Applications: Transport Phenomena, vol. 1. Alberta: Elsevier Applied Science Publishers, pp. 189–200.

Zogzas, N.P.; Maroulis, Z.B. 1996. Effective Moisture Diffusivity Estimation from Drying Data. A Comparison Between Various Methods of Analysis. Drying Technol. 14(7/8):1543–1573.

Zogzas, N.P.; Maroulis, Z.B.; Marinos-Kouris, D. 1994a. Moisture Diffisivity Methods of Experimental Determination. A Review. Drying Technol. 12(3):483–515.

Zogzas, N.P.; Maroulis, Z.B.; Marinos-Kouris, D. 1994b. Densities, Shrinkage and Porosity of some Vegetables During Air Drying. Drying Technol. 12(7):1653–1666.

29
Plums and Prunes

Muhammad Siddiq

INTRODUCTION

Plum is a common name for the tree of many species belonging to the genus *Prunus* of Rosaceae (rose) family and for its drupaceous (fleshy) fruit. Plums are generally cultivated in the temperate zones with numerous varieties and hybrids that are suitable for many soils and regions. Of more than 100 species of plums, 30 are native to North America. It has been cultivated since prehistoric times, longer perhaps than any other fruit except the apple (Anon, 2004a). World production of plums from 1997 to 2003 is shown in Figure 29.1. China is the leading plum-producing country with approximately 42% share of the total world production in 2003 (Table 29.1), followed by Romania and the United States. China also leads in area under plums cultivation accounting for over 57% of the world share (Table 29.2). However, on yield per hectare basis, Chile is the leading country with 17.5 metric tons plum per hectare, which is over four times the world average; in comparison, China's yield per hectare is only 2.9 metric tons (Table 29.3).

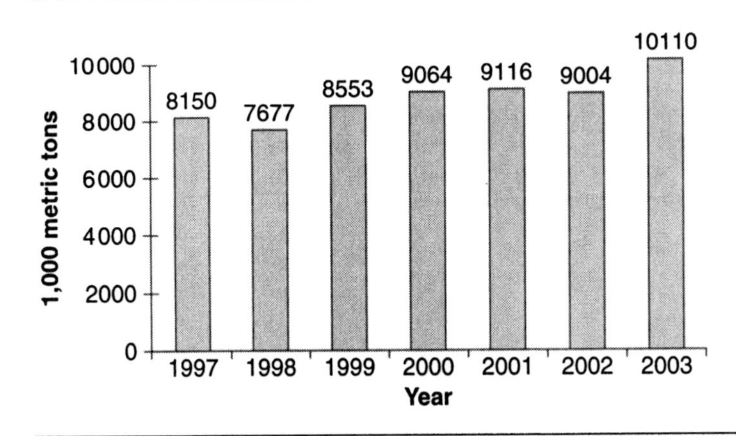

Figure 29.1. World plum production (1997–2003). *Source:* Adapted from FAO (2004) world data.

Table 29.1. Plum Production in Leading Countries (1997–2003)

Countries	Production (1000 metric tons)						
	1997	1998	1999	2000	2001	2002	2003
China	2,956.2	3,161.5	3,919.9	3,942.0	4,060.9	4,397.4	4,234.4
Romania	491.6	404.4	361.3	549.6	557.2	220.6	909.6
United States	840.0	507.0	666.9	819.0	590.8	667.6	725.3
Serbia and Montenegro	471.0	481.0	382.0	362.0	338.0	205.4	577.4
Germany	294.8	338.7	388.1	440.8	388.0	424.5	478.7
France	195.7	205.7	185.0	203.6	271.6	253.2	246.7
Chile	158.8	139.8	186.7	172.0	210.5	230.0	240.0
Turkey	200.0	200.0	195.0	195.0	200.0	200.0	205.0
Spain	158.5	146.5	160.3	168.0	149.7	186.5	196.4
Russian Federation	170.0	105.0	115.0	135.0	125.0	170.0	156.0
Iran	110.3	118.3	133.3	142.6	143.1	145.0	147.0
Italy	114.4	148.8	189.3	179.8	177.4	167.5	127.1

Source: FAO (2004).

Table 29.2. Plums: Area Harvested in Leading Countries and World (1997–2003)

Countries	Area Harvested (1000 ha)						
	1997	1998	1999	2000	2001	2002	2003
China	930.3	1,105.1	1,154.6	1,304.5	1,354.2	1,364.1	1,454.1
Serbia and Montenegro	125.0	125.0	125.0	125.0	125.0	125.0	340.0
Romania	98.6	99.2	95.0	95.7	96.0	87.8	94.5
Germany	57.7	65.0	61.0	61.0	61.0	61.0	68.0
Russian Federation	50.0	50.0	52.0	52.0	53.0	55.0	58.0
United States	52.1	52.4	51.5	51.8	51.5	46.2	45.4
Spain	20.0	20.1	20.0	18.4	17.1	20.4	21.3
France	21.1	20.7	20.3	19.3	19.7	20.0	20.0
Turkey	18.4	18.4	18.4	18.4	18.3	18.3	18.4
Iran	12.5	12.7	13.5	14.9	14.5	14.5	14.5
Italy	12.3	12.2	12.3	12.3	12.2	12.6	14.2
Chile	12.4	13.0	13.1	13.1	13.0	13.0	13.7
World (total)	1,814.6	1,965.8	2,002.9	2,148.0	2,198.7	2,200.2	2,536.4

Source: FAO (2004).

Table 29.3. Plums: Yield per Hectare in Leading Countries and World (1997–2003)

Countries	Yield (metric tons/ha)						
	1997	1998	1999	2000	2001	2002	2003
Chile	12.8	10.7	14.3	13.2	16.2	17.7	17.5
United States	16.1	9.7	13.0	15.8	11.5	14.4	16.0
France	9.3	9.9	9.1	10.6	13.8	12.7	12.3
Turkey	10.9	10.9	10.6	10.6	10.9	10.9	11.2
Iran	8.8	9.3	9.9	9.6	9.9	10.0	10.1
Romania	5.0	4.1	3.8	5.7	5.8	2.5	9.6
Spain	7.9	7.3	8.0	9.1	8.8	9.1	9.2
Italy	9.3	12.2	15.3	14.6	14.6	13.3	8.9
Germany	5.1	5.2	6.4	7.2	6.4	7.0	7.0
China	3.2	2.9	3.4	3.0	3.0	3.2	2.9
Russian Federation	3.4	2.1	2.2	2.6	2.4	3.1	2.7
Serbia and Montenegro	3.8	3.8	3.1	2.9	2.7	1.6	1.7
World (average)	4.5	3.9	4.3	4.2	4.1	4.1	4.0

Source: FAO (2004).

Plums can be divided into three different groups, with the first two being the principal species of commercial plums (Anon, 2004b):

1. European (*Prunus domestica*): Familiar varieties of the European type are Stanley, Reine Claude (Green Gage) and the French and German prune (Fellenburg) types. The European-type plums, purple to black in color, are best for eating fresh and for canning.
2. Japanese (*Prunus salicina*): Examples of Japanese plums are Methley, Shiro, Ozark Premier, Burbank and Elephant Heart. Color of fruit is yellow to crimson.
3. Damson (*Prunus insititia*): Plums in this group produce very tart fruit, which is used chiefly for cooking and preserving. Examples of Damson-type plums are Shropshire and French Damson.

Most of the cultivated plums in the United States are derived from European and Japanese varieties, e.g., *P. salicina*, introduced from Japan in 1870. A wild type red plum (*P. americana*) is found along streams and in thickets from New York to the Rocky Mountains. Its small, sweet fruit has a purple bloom. This plum was utilized by Native Americans, who ate it as raw, cooked, or in dried form; when dried, it was a staple article of diet (Anon, 2004a).

Botanically, all prunes are plums, but not all plums are prunes (Somogyi, 1996); however, in North America, prunes refer to a variety that can be and is normally dried without removing pits. Plums include any variety primarily used for fresh consumption as well as for canning, freezing, and in jams and jellies. Somogyi (1996) indicated that most plum varieties would ferment if dried as a whole fruit (with pits). However, if they are dried after the removal of pit, the product is called "dried plum" and not "prune." California accounts for over 99% of the prunes grown in the United States (see Box 29.1).

Box 29.1. U.S. Prune Facts

- Over 99% of the prunes in the United States are grown in California
- California grows almost 70% of the total world production of dried prunes
- Approximately 1200 growers farm 86,000 bearing and 15,000 non-bearing acres of prunes in California
- In 2000, the crop totaled 214,802 dried tons valued at approximately $166 million
- Approximately 1% of the crop is sold fresh, primarily to the Asian market
- The total cash costs for producing an acre of dried prunes varies from $2740 in the Sacramento Valley to $2840 in the San Joaquin Valley

Source: Anon (2002).

In recent years, per capita consumption of plums and plum products has been on the rise in the United States (Table 29.4). Additionally, use of prunes and prune products as food ingredients has been on a steady increase in the recent years, most likely due to (i) improved production and processing technologies and (ii) discoveries of health benefits of phytochemicals in fruits (like antioxidant benefits of phenolic compounds), which are present in significant quantities in plums and prunes. According to Somogyi (1996), marketers believe that the term "dried plum" has a more positive image to the consumers than "prunes," and that, based on the opinion of the Dried Fruit Association, manufacturers can use the term "dried plums" in labeling prune ingredients.

Table 29.4. Plums: Per Capita Consumption (in Pounds), 1998–2002

			Processed				
Year	Total	Fresh	Total	Canned	Juice	Frozen	Dried
1998	3.1	1.2	1.9	0.1	0.4	0.006	1.4
1999	2.7	1.3	1.4	0.1	0.3	0.004	1.0
2000	2.8	1.2	1.6	0.1	0.3	0.006	1.2
2001	3.0	1.3	1.7	0.1	0.4	0.006	1.2
2002	3.3	1.3	2.0	0.1	0.3	0.003	1.6

Source: Adapted from USDA-ERS (2001).

PRODUCTION AND POSTHARVEST PHYSIOLOGY

Trees of some plum cultivars are capable of self-pollination even if grown as single isolated plant(s); other plum varieties require cross-pollination for fruit set and development. Some popular plum varieties are AU Amber, AU Homeside, AU Producer, AU Roadside, AU Rubrum, Black Ruby, Byron Gold, Crimson, Frontier, Methley, Morris, Ozark Premier, Robusto, Ruby Sweet, Segundo, and Wade. Plum trees usually begin to bear fruits 3–5 years from planting and have a useful life of 15–20 years. Depending on the type of cultivars, average yields can range from 3 to 5 bushels per tree. However, other factors such as fertilizer use, irrigation, pruning and training, and plant protection practices play a big role in yield per tree (DeJong et al., 1992; Keulemans, 1990; Vitanova, 1990). Agricultural practices followed also influence the final quality of fruits, e.g., application of calcium sprays after bloom is reported to improve fruit quality (Wojcik, 2001). Southwick et al. (2000) reported the use of gibberellin at preharvest stage to delay maturity and improve fruit firmness. Common pests and diseases of plum trees include plum curculio, European red mite, brown rot, leaf spot, and black knot (Anon, 2004b).

Visual sign of maturity and harvest date of plums can be estimated by skin color changes, which are cultivar-specific. A color chip guide is used to determine harvest maturity. In California, a two-tier maturity system is currently used to determine maturity: (i) US Mature (minimum maturity) and (ii) California Well Mature. Measurement of fruit firmness is recommended for those plum cultivars where skin ground color is masked by full red or dark color development before maturation. Flesh firmness, measured with a penetrometer, can be used to determine a maximum maturity index, which is the stage at which fruit can be harvested without suffering bruising damage during postharvest handling. For objective measurements of maturity, soluble solids contents and sugar-to-acid ratio provide more reliable markers (Crisosto and Kader, 2000). Slaughter et al. (2003) reported a non-destructive optical method that can be employed successfully using near-infrared spectroscopy to determine total and soluble solids contents in fresh prune. Another non-destructive method, which uses ultrasound technology, was also found useful in evaluating plum quality attributes (Mizrach, 2004).

Plums and fresh prunes are generally machine-picked; however, in some areas, they are hand-picked into bags, and then dumped in bins on trailers that move between tree rows. Each method of harvest has some advantages as well as some disadvantages over the other. At the packinghouse, after washing, sorting is done to eliminate fruit with visual defects and sometimes to divert fruit of high surface color to a high quality pack (sizing segregates fruit by either weight or dimension). Ideal storage conditions for plums are maintaining a temperature of 0°C and 90–95% relative humidity (RH); under such conditions, storage life is from 2 to 4 weeks (Crisosto and Kader, 2000). Martinez-Romero et al. (2003) reported that forced-air cooling after harvesting and before transportation to packinghouse, handling in packinghouse, and during storage helped in maintaining fruit quality and prolong shelf life. Their research showed that forced-air cooling led to a reduction in respiration rate of mechanically damaged plums.

During postharvest storage of fruits, ethylene triggers many, though not all, aspects of fruit ripening. Abdi et al. (1998) reported that within a fruit species, such as plum, high-ethylene producers would soften and ripen at a faster rate than low-ethylene producer cultivars. The ethylene action inhibitor 1-methylcyclopropene has been shown to delay ripening and improves postharvest quality of climacteric fruits (Abdi et al., 1998; Dong et al., 2002). Perez-Vicente et al. (2002) investigated the role of exogenously applied putrescine, a polyamine, during postharvest storage of mechanically damaged plums; exogenous putrescine inhibited and delayed ethylene and CO_2 production rates.

Chilling injury and internal browning are the two physiological disorders reported in plums. Most plum and fresh prune cultivars exhibit flesh translucency associated with flesh browning as chilling injury, or "gel breakdown"—a term used in South Africa for chilling injury (Crisosto et al., 1999; Taylor et al., 1995). Internal browning is a physiological disorder that originates before the harvest and is associated with high temperatures during fruit maturation and delayed harvest. Polyphenol oxidase can initiate flesh discoloration in injured or bruised fruits (Siddiq et al., 1993, 1996). This enzyme not only affects sensory color, but can also be detrimental to nutritional quality.

PRUNE AND PLUM PROCESSED PRODUCTS

Plums have a great potential as a fresh market and/or processing crop, which can be harvested between cherries and apples in many areas, especially in the Unites States. About one-half of the plums are consumed fresh while the rest are processed. The major processed plum products are dried prunes, prune juice, and whole-canned plums. Other processed forms, such as paste, sauce, and plum juice have not been developed and marketed on a scale similar to these products from other fruits such as apples, cherries, citrus, pears, apricots, etc. (Chang et al., 1994; Espie, 1992). However, with new scientific claims of many health benefits of the phytochemicals (plums being rich in many), plums have a great potential for a variety of new processed products than currently available in the market.

Dried Prunes

Dried plums or prunes represent about one-half of the plums and prune products consumed on a per capita basis in the United States (Table 29.4). A flowchart diagram of different processing operations involved in the production of prunes is shown in Figure 29.2. Fruit is harvested at full maturity that allows soluble solids contents to reach at least 22% (22°Brix); other maturity indices include fruit firmness, flesh, and skin color. Modern dehydrators have replaced the old methods of drying prunes in the sun in the United States (Somogyi, 1996). However, in many of the plum-producing developing countries, sun drying still continues to be one of the cheapest sources of drying plums.

Fruits are dried to about 18% moisture, which has sufficiently low-water activity to avoid problems of microbial spoilage allowing long-term storage (Newman et al., 1996). Generally, forced-draft tunnel dehydrators are used for drying plums, with a total drying process time of 24–36 h, depending on the size and soluble solids contents of the prunes. The operating temperature for the tunnel is 62.8–73.9°C (145–165°F)—dry bulb temperature—with wet bulb −9.4°C(15°F) lower than the dry bulb at the cool end. Yield of dried prunes is about 33%. For the prunes that are marketed as dried prunes with pits, dried fruit from above step are rehydrated to

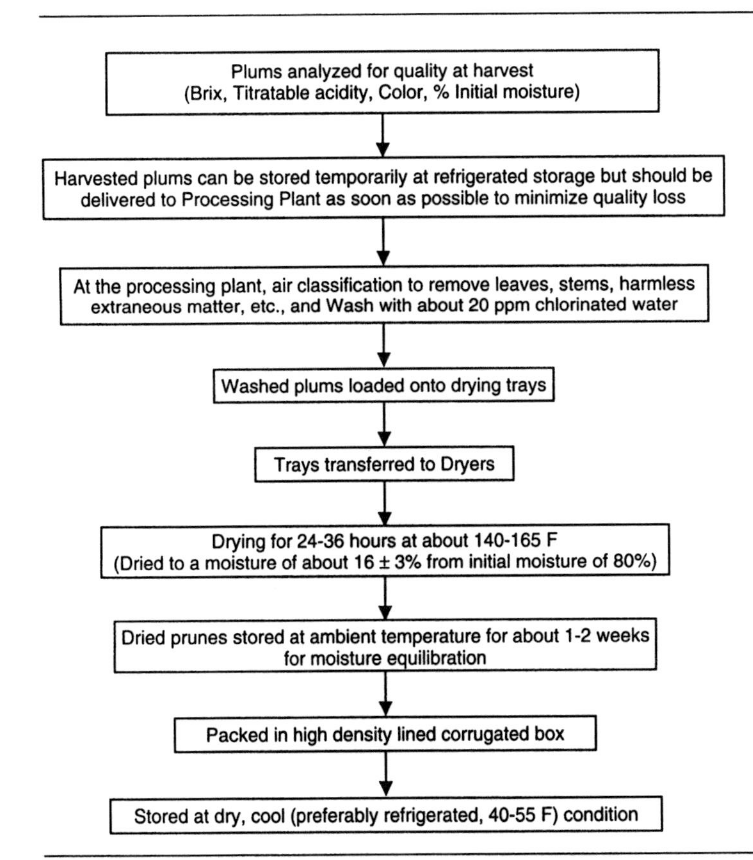

Figure 29.2. Flowchart for drying fresh prune plums into prunes. *Source*: Adapted from Somogyi (1996).

24–30% moisture, sterilized, inspected, and bulk or retail packaged for food service or consumer use, respectively. Potassium sorbate is the most commonly used preservative for prunes when moisture contents of the finished prunes are higher than 25%. Some products made from dried prunes, on a smaller scale, are prune juice, juice concentrate, whole-pitted prunes, canned prunes (pitted), and various forms of dry and low-moisture products (diced prunes, prune bits, prune paste, low-moisture prune granules, low-moisture prune powder, prune fiber, prune fillings, and toppings).

Most of the recent research related to prune processing has been focused on ways to improve efficiency of the drying methods with due consideration given to the retention of best quality characteristics (both chemical and nutritional) in the finished product. Doymaz (2004) studied the effect of dipping treatment on air-drying of plums and observed that dipping of plums in 5% potassium carbonate and 2% ethyl oleate for 1 min was effective in removing the natural wax coating and hence reduced the drying time by about 30% as compared to untreated samples. Cinquanta et al. (2002) reported that physical (abrasion) and chemical (alkaline ethyl oleate dip) pretreatments before drying significantly reduced losses of total phenols. Gabas et al. (2002) studied the rheological properties of prune as a function of drying conditions. Their data showed that prunes exhibited a more pronounced elasticity at low-moisture content and drying temperature. Higher moisture contents and temperature resulted in more viscous and less-rigid prunes. Sabarez et al. (2000) used solid-phase micro-extraction in conjunction with GC-MS to monitor changes in some major volatile flavor compounds under simulated commercial drying conditions (80°C air temperature, 35% RH, 5 m/s air velocity) for d'Agen plums. They observed that aroma profile was significantly modified during drying and substantial loss of the original volatile flavors. Piga et al. (2003) reported that drying had detrimental effect on anthocyanins and ascorbic acid and showed that drying at 85°C doubled the antioxidant activity in plum cultivar they studied.

Sanders (1993) noted a variety of functional attributes of dried plums (Table 29.5) that include natural preservative and a fat substitute, which could be of special benefits owing to more emphasis on low-fat foods and products processed without added preservatives by the consumers.

Dried plum products are suitable for use in many bakery products as summarized in Table 29.6, which in addition to added nutritional benefits can also improve desirable sensory attributes.

Prune Juice

Prune juice is essentially a water extract of dried prunes with a Brix of 18.5°. It is a brownish-to-reddish brown liquid having taste and flavor of prunes. The process involves direct heating of dried prunes in appropriate volumes of water (could be four to five times weight of the fruit) to extract fruit solids without burning and affecting flavor and color. Traditionally, the process can take several hours (1-h boiling and 10-h simmering) under atmospheric cooking. A short duration (10–15 min) pressure-cooking may speed up juice extraction. After extraction, other unit operations of filtration (to remove pits and undissolved solids), pasteurization (88°C for 1 min), and packaging can be similar to processing of other juices. The juice thus made can also be concentrated in a vacuum evaporator. For canning and bottling, the product should be hot filled (88°C) and sealed before processing in boiling water for 20–35 min depending on the can size. As per FDA regulations, prune may contain citrus (lime and lemon juices) and enriched with vitamin C. The federal regulations require that the following label of declaration should appear on the container below the words Prune Juice: "A water extract of dried prunes." Luh (1980) reported that prune juice differed from

Table 29.5. Functional Attributes of Dried Plums

Functional Attribute	Candidates for Reduction/Replacement
Humectancy	Mono- and di-glycerides
Natural color	Caramel color, molasses
Flavor enhancement	Salt, artificial flavors
Natural sweetness	Refined sugars
Natural preservative	Calcium propionate
Fat replacement	Emulsifiers, modified starches, fats

Source: American Society of Bakery Engineers (Sanders, 1993).

Table 29.6. Dried Plum-Based Bakery Ingredients

Product Form	Suggested Uses
Diced dried plums and extruded bits	Breads, muffins, cookies, cakes, fillings
Dried plum paste	Fillings, breads, bagels, cookies
Dried plum juice concentrate	Breads, pastries, cakes, muffins, cookies, fillings
Dried plum powder/granules/flakes	Dry mixes, bagels, reduced-fat/fat-free mixes
Dried plum puree	Reduced fat/fat-free bakery cakes, cookies, muffins, fillings

Source: American Society of Bakery Engineers (Sanders, 1993).

other juices, which are squeezed from fresh fruits, due to this very fact.

PRUNE JUICE CONCENTRATE

Prune juice, processed as above, can be concentrated to 60–72°Brix depending on the intended end use. Concentrate with lower soluble solids is frozen and used for reconstitution into single strength juice. The high Brix (65% soluble solids and above) is shelf-stable/selfpreserving without freezing or added preservatives; the latter quality is of great benefit for shipping long distances, like export market.

For better process efficiency, prune juice before concentration is depectinized by the addition of commercial pectic enzymes (Somogyi, 1996). Juice is concentrated in a high-vacuum evaporator; many types of such evaporators are available commercially. Process is carried at temperature of 48°C or lower. Water in the juice is evaporated, and the juice sugars and other solids are concentrated to the desired level.

Prune juice concentrate offers many benefits and applications in different food systems (Anon, 2000):

1. Extends the shelf life of bread products, serves as an anti-staling agent
2. Sweetens and colors natural baked goods
3. A natural substitute for preservatives
4. A good sugar substitute and natural color/flavor enhancer
5. Can be used as filling for hard candies and chocolate
6. Maintains moisture in chewy cakes and cookies
7. Binding agent in cereal bars

CANNED PRUNES

The dried prunes make an excellent product when canned. Processing consists of washing dried prunes, blanching for 4 min in hot water (to start softening of prunes), filling in cans, and adding syrup. Cans are exhausted for 12–15 min at 76.7°C (170°F) before sealing. Thermal processing is carried out at 100°C (212°F) for 12 min. Canned prunes can be more prone to form hydrogen springer than other canned fruits, which can be minimized with high vacuum as a result of extended exhaustion before can closure (Lopez, 1981).

Canned prunes, which are moist, have no added preservatives, and a long shelf life, can be used for ready-to-eat snacking. Studies on canned prunes in syrup showed that use of syrup previously employed in the osmotic dehydration process did not impair quality of the canned product (Silveira et al., 1984). Bolin et al. (1971) reported that when apple juice was used as stewing medium for canned prunes, a desirable flavor was produced. Flavor did not improve by the addition of citric acid and ascorbic acid; both had an adverse effect instead.

PLUM JUICE

Most of the research on plum juice production was undertaken in the 1970s and 1980s. Plum juice has not been processed to a scale similar to prune juice, probably due to its high acidity. Siddiq et al. (1994) developed a method using pectinase enzymes to press juice from Stanley plums that increased juice yields significantly. Chantanawarangoon et al. (2004) evaluated antioxidant capacity, total phenolics, and total anthocaynins in juices prepared from 10 plum cultivars and suggested that plum juices with high antioxidant capacity, total phenolics, and total anthocyanins could be new healthy beverages for consumers and potential sources of nutraceutical supplements as well.

PLUM PASTE

Plum paste is another product with limited commercial production; however, with recent research on health benefits of polyphenols and dietary fiber, it offers a great potential for increasing production and uses in many food formulations.

Wang et al. (1995) developed a procedure to produce pastes from Stanley plums and studied the effects of processing conditions on chemical, physical, and sensory properties of the pastes. Stanley plums were processed by heat concentration into two pastes of 25°Brix and 30°Brix. Soluble solids of these two pastes were increased to 40°Brix and 45°Brix, respectively, by addition of sugar. Heat concentration resulted in a significant decrease in titratable acidity, total anthocyanins, and total pectin. Pastes showed pseudoplastic behavior within the shear rate range of 20–100 rpm. Sugar addition had a darkening effect on color, but no noticeable effect on rheological properties of the pastes. Sensory evaluation indicated that preference could be adequately predicted by flavor and color under suitable °Brix/acid ratio. Raina et al. (1999) also reported that concentration of plum paste had significant effects on the rheological properties and acidity of the resultant paste. It was not feasible to concentrate the paste beyond 35°Brix. Sweetened paste had higher acceptability scores than unsweetened paste.

Jam and Jelly

Plum jam and jelly has not gained enough consumer acceptances to result in commercial production. However, there is potential for the production of jams and jellies in combination with other fruits, which are lower in phenolic compounds and antioxidant activity. Kim and Padilla-Zakour (2004) investigated the changes in total phenolics, antioxidant capacity, and anthocyanins from fresh fruits (cherries, plums, and raspberry) to processed shelf-stable jams, with the objective of finding whether high sugar and acid levels may provide protection to phenolic compounds. Their results showed that jam processing had no consistent effect on phenolics, although antioxidant capacities generally decreased. Nonetheless, plums can be used as source of phenolics in jams prepared from other fruits that may be low in these phytochemicals.

Fresh-Cut Plums

"Fresh-cut produce" is defined as any fresh fruit or vegetable or any combination thereof that has been physically altered from its original form, but remains in a fresh state. Regardless of commodity, it has been washed, trimmed, peeled, and cut into 100% usable product that is subsequently bagged/packaged to offer consumers with high nutrition, convenience, and value while still maintaining freshness (IFPA, 2004). In recent years, a market for fresh prunes has developed; however, less than 1% of prunes produced in California are marketed through this channel due to volatile market conditions (Anon, 2002). Fresh-cut plum can keep good quality for 2–5 days, depending on the cultivar and ripeness stage (firmness) when stored at 0°C in packages that minimize loss of water (Crisosto and Kader, 2000). Cisneros-Zevallos and Heredia (2004) studied the role of ethylene and methyl jasmonate on the changes in health-promoting antioxidant compounds in different fresh-cut produce that included plums. They found that these two plant hormones combined with wounding (cutting) could be used to enhance the health-promoting antioxidant content of fresh-cut produce.

CHEMICAL COMPOSITION, NUTRIENT PROFILE, AND DIETARY BENEFITS

Plum and prune products, in addition to being low in fat content like other fruits, contain significant amounts of some major nutrients as summarized in Table 29.7. The values shown here are for the fruit grown and processed in the United States; therefore, some differences can be anticipated in composition of plum and prune products in other parts of the world owing to different climatic and soil conditions, agricultural practices, postharvest handling and processing techniques, etc. Additionally, varietal differences can also contribute to variations in the composition of raw and finished products.

Fresh plums are an excellent healthy food due to their low fat and good source of dietary fiber. Prunes are moist and convenient snack, and an easy way to get more natural fiber into the diet. Health experts who have been advising people of all ages to consume more dietary fiber, based on the research findings, suggest that fiber may prevent cancer, diabetes, heart disease, and obesity. Prune fiber mostly consists of soluble fraction (about 80%), mainly pectin, hemicellulose, cellulose, and some lignin. Prunes could be classified as a unique health food that is not only high in dietary fiber, but also exhibits the highest antioxidant activity (as summarized in Table 29.8). Ki and Jong (2000) reported that drying was the mildest processing method for maintaining original levels of dietary fiber in vegetable and fruit products as demonstrated by 7.35% and 3.45% increase in total dietary fiber in prunes and raisins, respectively. Somogyi (1996) gave examples of variations in fiber contents

Table 29.7. Composition of Plums, Prunes, and Their Processed Products (per 100 g Edible Portion)

Nutrient	Plums	Canned[a]	Dried/Prunes	Prune Juice[b]
Proximate				
Water (g)	87.23	83	30.92	81.24
Energy (kcal)	46	63	240	71
Protein (g)	0.7	0.37	2.18	0.61
Total lipid (fat) (g)	0.28	0.1	0.38	0.03
Fatty acids, total saturated (g)	0.017	0.008	0.088	0.003
Carbohydrate, by difference (g)	11.42	16.28	63.88	17.45
Fiber, total dietary (g)	1.4	0.9	7.1	1
Sugars, total (g)	9.92	15.35	38.13	16.45
Vitamins				
Vitamin A (IU)	345	231	781	3
Vitamin C, total ascorbic acid (mg)	9.5	0.4	0.6	4.1
Thiamin (mg)	0.028	0.016	0.051	0.016
Riboflavin (mg)	0.026	0.039	0.186	0.07
Niacin (mg)	0.417	0.297	1.882	0.785
Pantothenic acid (mg)	0.135	0.072	0.422	0.107
Vitamin B-6 (mg)	0.029	0.027	0.205	0.218
Folate, total (μg)	5	3	4	0
Vitamin E (alpha-tocopherol) (mg)	0.26	0.18	0.43	0.12
Vitamin K (phylloquinone) (μg)	6.4	4.3	59.5	3.4
Minerals				
Calcium (mg)	6	9	43	12
Iron (mg)	0.17	0.86	0.93	1.18
Magnesium (mg)	7	5	41	14
Phosphorus (mg)	16	13	69	25
Potassium (mg)	157	93	732	276
Sodium (mg)	0	20	2	4
Zinc (mg)	0.1	0.08	0.44	0.21
Copper (mg)	0.057	0.038	0.281	0.068
Manganese (mg)	0.052	0.032	0.299	0.151

Source: USDA Nutrient Database (USDA, 2004).
[a]In light syrup (solids and liquids).
[b]Canned.

(from 2.04 to 16.1 g/100 g prunes with 23.3–32.4% moisture) reported by various sources and indicated that these variations could not be attributed solely to the natural variations in the composition of fruits. He concluded that the lack of standardized tests, especially for the extraction of soluble fibers, might be responsible for such discrepancies.

Plums are reported to be high in natural phenolic phytochemicals, such as flavonoids and phenolic acids, which function as effective natural antioxidants in human diet and have been reported to reduce the risk of cancer and other chronic diseases (Ames et al., 1995; Block et al., 1992; Machlin, 1995). Cultivar differences are known to affect the phenolic content, as summarized in Table 29.8 for some varieties of plums grown in Michigan (Siddiq et al., 1994). According to Shahidi and Naczk (1995), the presence of phenolic compounds in foods has an important effect on the oxidative stability and microbial safety of these products. In addition, many phenolics in foods possess important biological activity related to their inhibitory effects on mutagenesis and carcinogesis. Therefore, in recent years, a rapid progress has been made on different aspects of polyphenols in food. Kim et al. (2003) demonstrated that antioxidant capacity of plums, expressed as vitamin C equivalent antioxidant capacity, was substantially higher (up to three times) when compared with

Table 29.8. Total Phenolics and Chlorogenic Acid Contents in Different Plum Cultivars

Cultivars	Total Phenolics[a] (μg/g)	Chlorogenic Acid (μg/g)
Beauty	922	103
Pipestone	736	77
La crescent	590	88
Abundance	437	38
Pobeda	353	63
Au Roadside	339	35
Underwood	300	43
Wade	299	33
Shiro	295	46
Stanley	282	75

Source: Siddiq et al. (1994).
[a]As chlorogenic acid.

apples. In another study, Wang et al. (1996) reported 4.4-times higher total antioxidant capacities in plums than apples, the latter is one of the most commonly consumed fruits in our diet.

Prunes are shown to have the highest antioxidant capacity, expressed as oxygen radical absorbance capacity (Table 29.9) that measures the ability to subdue oxygen-free radicals of a food by comparing its absorption of peroxyl or hydroxyl radicals to that of Trolox, a water-soluble vitamin E analog (Keeton et al., 2002). Plums are shown to have a very good free-radical scavenging activity against O_2-derived free radicals, such as hydroxyl and peroxyl radicals (Murcia et al., 2001).

Table 29.9. ORAC Values of Fruits with Antioxidant Potential

Fruits	ORAC Value[a]/100 g
Dried plums	5770
Raisins	2830
Blueberries	2400
Blackberries	2036
Strawberries	1540
Raspberries	1220
Plums	949
Oranges	750
Red grapes	739
Cherries	670
Kiwi fruit	602
Grapefruit, pink	483

Source: Keeton et al. (2002).
[a]ORAC (oxygen radical absorbance capacity).

Plum and prune products, though not a very good source of vitamins based on U.S. RDA requirements, do contain a fair amounts of different vitamins. Dismore et al. (2003) investigated the presence of vitamin K in the U.S. diet containing nuts and fruits and found that with the exception of some berries, green fruits, and prunes, most nuts and fruits are not good sources of this vitamin.

REFERENCES

Abdi N, McGlasson WB, Holfor P, Williams M, Mizrahi Y. 1998. Responses of climacteric and suppressed-climacteric plums to treatment with propylene and 1-methylcyclopropene. Postharvest Biol Technol 14:29–39.

Ames BN, Gold LS, Willett WC. 1995. The causes and prevention of cancer. Proc Natl Acad Sci USA 92:5258–65.

Anon. 2000. Prune Juice Concentrate: A Versatile Ingredient—A Wide Range of Functionality with all the Health Benefits of Dried Plums. Stapleton-Spence Packing Company, San Jose, CA (http://www.stapleton-spence.com).

Anon. 2002. A Pest Management Strategic Plan for California Prune Production. A Report of Center for Integrated Pest Management, North Carolina State University, Raleigh, NC (http://cipm.ncsu.edu/pmsp/pdf/caprune.pdf).

Anon. 2004a. Plum. In: The Columbia Electronic Encyclopedia, 6th Ed. Columbia University Press, New York (http://www.columbia.edu/cu/cup/).

Anon. 2004b. Fact Sheet: Plum Culture. Horticulture Program, University of Rhode Island (http://www.uri.edu/ce/factsheets/sheets/plums).

Block G, Patterson B, Subar A. 1992. Fruit, vegetables, and cancer prevention: A review of the epidemiological evidence. Nutr Cancer 18:1–29.

Bolin HR, Stafford AE, Guadagni DG. 1971. Innovations with canned stewed dried prunes. Canner/Packer 140(7):18.

Chang T-S, Siddiq M, Sinha NK, Cash JN. 1994. Plum juice quality affected by enzyme treatment and fining. J Food Sci 59:1065–69.

Chantanawarangoon S, Kim D-O, Padilla-Zakour OI. 2004. Antioxidant capacity and polyphenolic compounds of plum juices. Presented at the annual meeting of the Institute of Food Technologists, Las Vegas, NV (July 12–17).

Cinquanta L, di Matteo M, Esti M. 2002. Physical pretreatment of plums (*Prunus domestica*). Part II. Effect on the quality characteristics of different prune cultivars. Food Chem 79:233–38.

Cisneros-Zevallos L, Heredia JB. 2004. Antioxidant Capacity of Fresh-Cut Produce May Increase After Applying Ethylene and Methyl Jasmonate. Presented at the annual meeting of the Institute of Food Technologists, Las Vegas, NV (July 12–17).

Crisosto CH, Kader AA. 2000. Plum and Fresh Prune Postharvest Quality Maintenance Guideline. Extension Bulletin. Kearney Agricultural Center, University of California, Davis, CA (http://www.uckac.edu/postharv).

Crisosto CH, Mitchell FG, Ju Z. 1999. Susceptibility to chilling injury of peach, nectarine, and plum cultivars grown in California. HortScience 34:1116–18.

DeJong TM, Day KR, Doyl JF. 1992. Evaluation of training/pruning systems for peach, plum and nectarine trees in California. Acta Hort 322:99–105.

Dismore ML, Haytowitz DB, Gebhardt SE, Peterson JW, Booth SL. 2003. Vitamin K content of nuts and fruits in the US diet. J Am Diet Assoc 103:1650–52.

Dong L, Lurie S, Zhou HW. 2002. Effect of 1-methylcyclopropene on ripening of 'Canino' apricots and 'Royal Zee' plums. Postharvest Biol Technol 24:135–45.

Doymaz I. 2004. Effect of dipping treatment on air-drying of plums. J Food Eng 64:465–70.

Espie M. 1992. Michigan Agricultural Statistics. Michigan Agricultural Statistics Service, Lansing, MI, pp. 29–30.

FAO. 2004. World Primary Crop Data. Food and Agriculture Organization of the United Nations (http://www.Fao.org/).

Gabas AL, Menegalli FC, Ferrari F, Telis-Romero J. 2002. Influence of drying conditions on the rheological properties of prunes. Drying Technol 20:1485–502.

IFPA 2004. Fresh-Cut Produce/Fresh-Cut Process. International Fresh-Cut Produce Association, USA (http://www.fresh-cuts.org).

Keeton JT, Rhee KS, Hafley BS, Nunez MT, Boleman RM, Movileanu I. 2002. Evaluation of Plum/Prune Ingredients as a Component of Meat Products. Part II: Evaluation of Ham and Roast Beef Products Containing Fresh Plum Juice Concentrate, Dried Plum Juice Concentrate, or Spray Dried Plum Powder. Final Report to the California Prune Board, USA, pp. 1–23.

Keulemans J. 1990. Cropping behavior, flowerbud formation, pollination and fruit set of different plum cultivars in Belgium. Acta Hort 283:117–29.

Ki HS, Jong BE. 2000. Total dietary fiber contents in processed fruit, vegetable, and cereal products. Food Sci Biotechnol 9:1–4.

Kim D-O, Jeong SW, Lee CY. 2003. Antioxidant capacity of phenolic phytochemical from various cultivars of plums. Food Chem 81:321–26.

Kim D-O, Padilla-Zakour OI. 2004. Jam Processing Effects on Phenolics and Antioxidant Capacity in Anthocyanins-Rich Fruits: Cherries, Plums and Raspberry. Presented at the annual meeting of the Institute of Food Technologists, Las Vegas, NV (July 12–17).

Lopez A. 1981. Canning of fruits—dried prunes. In: Lopez A, editor. A Complete Course in Canning, Book II: Processing Procedures for Canned Food Products, 11th Ed. The Canning Trade, Baltimore, MD. 178 p.

Luh BS. 1980. Nectars, pulpy juices, and fruit juice blends. In: Woodroof JG, Joslyn MA, editors. Fruit and Vegetable Juice Process Technology. AVI Publishing Co, Westport, CT, pp. 471–76.

Machlin LJ. 1995. Critical assessment of the epidemiological data concerning the impact of antioxidant nutrients on cancer and cardiovascular disease. Crit Rev Food Sci Nutr 35:41–50.

Martinez-Romero D, Castillo S, Valero D. 2003. Forced-air cooling applied before fruit handling to prevent mechanical damage of plums (*Prunus salicina* Lindl.). Postharvest Biol Technol 28:135–42.

Mizrach A. 2004. Assessing plum fruit quality attributes with an ultrasound method. Food Res Int 37:627–31.

Murcia MA, Jimenez AM, Martinez-Tome M. 2001. Evaluation of the antioxidant properties of Mediterranean and tropical fruits compared with common food additives. J Food Prot 64:2037–46.

Newman GM, Price WE, Woolf LA. 1996. Factors influencing the drying of prunes. I. Effects of temperature upon the kinetics of moisture loss during drying. Food Chem 57:241–44.

Perez-Vicente A, Martinez-Romero D, Carbonell A, Serrano M, Riquelme F, Guillen F, Valero D. 2002. Role of polyamines in extending shelf life and the reduction of mechanical damage during plum (*Prunus salicina* Lindl.) storage. Postharvest Biol Technol 25:25–32.

Piga A, Caro A, Corda G. 2003. From plums to prunes: Influence of drying parameters on polyphenols and antioxidant activity. J Agric Food Chem 51:3675–81.

Raina CS, Bawa AS, Ahmed J. 1999. Rheological study and sensory characteristics of plum paste. Indian Food Packer 53(2):15–19.

Sabarez HT, Price WE, Korth J. 2000. Volatile changes during dehydration of d'Agen prunes. J Agric Food Chem 48:1838–42.

Sanders SW. 1993. Dried Plums: A Multi-functional Bakery Ingredient. Bulletin 228. American Society of Bakery Engineers, American Society of Baking, Sonoma, California, pp. 1–23.

Shahidi F, Naczk M. 1995. Food Phenolics—Sources, Chemistry, Effects and Applications. Technomic Publishing Co., Lancaster, PA, pp. 1–5.

Siddiq M, Arnold JF, Sinha NK, Cash JN. 1994. Effect of polyphenol oxidase and its inhibitors on anthocyanin changes in plum juice. J Food Process Preserv 18:75–84.

Siddiq M, Sinha NK, Cash JN. 1993. Characterization of polyphenol oxidase from Stanley plums. J Food Sci 57:1177–79.

Siddiq M, Sinha NK, Cash JN, Hanum T. 1996. Partial purification of polyphenol oxidase from plums (*Prunus domestica* L. cv. Stanley). J Food Biochem 20:111–23.

Silveira ETF, Travaglini DA, Aguirre JM, Mori EEM, Figueiredo IB. 1984. Drying of Carmesim plums. III. Effect of osmotic dehydration of processing and organoleptic properties of the resulting prunes. Boletim Inst de Tecnol de Aliment (Brazil) 21:257–67.

Slaughter DC, Thompson JF, Tan ES. 2003. Nondestructive determination of total and soluble solids contents in fresh prune using near infrared spectroscopy. Postharvest Biol Technol 28:437–44.

Somogyi LP. 1996. Prunes and plums. In: Somogyi LP, Barrett DM, Hui YH, editors. Processing Fruits: Science and Technology, Vol. 2. Technomic Publishing Co., Lancaster, PA, pp. 95–116.

Southwick SM, Moran RE, Yeager JT, Glozer K. 2000. Use of gibberellin to delay maturity and improve fruit quality of 'French' prune. J Hort Sci Biotech 75:591–97.

Taylor MA, Rabe E, Jacobs G, Dodd MC. 1995. Effect of harvest maturity on pectic substances, internal conductivity, soluble solids and gel breakdown in cold stored 'Songold' plums. Postharvest Biol Technol 5:285–94.

USDA. 2004. Nutrient Database (http://www.nal.usda.gov/).

USDA-ERS. 2004. US per Capita Consumption Data. USDA-Economic Research Service (http://www.ers.usda.gov/).

Vitanova IM. 1990. Determination of needs for fertilizers of plum trees. Acta Hort 274:501–8.

Wang H, Cow G, Prior RL. 1996. Total antioxidant capacity of fruits. J Agric Food Chem 44:701–5.

Wang W-M, Siddiq M, Sinha NK, Cash JN. 1995. Effect of processing conditions on the physicochemical and sensory characteristics of Stanley plum paste. J Food Process Preserv 19:65–81.

Wojcik P. 2001. "Dabrowicka Prune" fruit quality as influenced by calcium spraying. J Plant Nutr 24:1229–41.

30

Processing of Red Pepper Fruits (*Capsicum annuum* L.) for Production of Paprika and Paprika Oleoresin

Antonio Pérez-Gálvez, Manuel Jarén-Galán, and M. Isabel Mínguez-Mosquera

INTRODUCTION

CHARACTERISTICS OF RED PEPPER FRUITS AND DESCRIPTION OF VARIETIES

Humans used spices for culinary and medical purposes for centuries, improving the taste and flavor of meals, even adding to them a point of fantasy and mystery, as spices were imported from exotic and strange lands. Spices like cinnamon, vanilla, and ginger were in great demand but so hard to track down and transport since market prices were very high in this lucrative trade. In this sense, pepper fruits from *Capsicum* genus were not a conventional spice. They were introduced in spice trade after the discovery of the New World, although it is thought that peppers made their first appearance around 7000 BC. After its discovery, pepper was no longer a luxury spice that only the rich could afford. This new spice, introduced in Europe, spread rapidly across the world, and was instantaneously incorporated into the cuisines. It is commercially grown in the United States, Brazil, India, Taiwan, South Africa, Zimbabwe, and throughout Europe.

Pepper plant (*Capsicum* genus) belongs to the *Solanaceae*, a large tropical family that includes potato, tomato, tobacco, and petunia. The *Capsicum* genus comprises 22 wild species and 5 domesticated species *Capsicum annuum*, *C. baccatum*, *C. chinense*, *C. frutescens*, and *C. pubescens*. Although several hundred pepper pod types and cultivars are grown worldwide, the most economically important species in the world are *C. frutescens* and *C. annuum*, plants that are used in the manufacture of selected commercial products known for their pungency and color, respectively. *Capsicum frutescens* L. is a short-lived perennial with woody stems that reaches a height of 2 m, has glabrous or pubescent leaves, has two or more greenish white flowers per node, and has extremely pungent fruit. This plant is cultivated in the tropics and in warmer regions of the United States and is used in tabasco sauce. *Capsicum annuum* L. is an herbaceous annual plant that reaches a height of 1 m and has glabrous or pubescent lanceolate leaves, white flowers, and fruits that vary in length, color, and pungency depending on the cultivar. *Capsicum annuum* fruits are spread in every part of the globe. The nonpungent forms are used for direct consumption, fresh or dried, whole or ground. They are also widely used as coloring agents in any of their processed forms, paprika and paprika oleoresin.

The fruits are berries, with different shapes (rounded or elongated) and size (from few g to 250 g) depending on the variety. Seeds are straw colored, and they are placed on the placenta. Generally, shape and size are used to distinguish the varieties. The *C. annuum* sort comprises seven varieties. In all of them, the color is the most important quality of the fruits as they are used for coloring foodstuffs. During ripening of fruits, their color changes from green, due to chlorophylls, to orange and red because of the presence of carotenoids, pigments responsible for fruit color. The carotenoid profile of the ripened fruit includes seven major carotenoid pigments, some of them are only biosynthesized in this fruit (capsanthin and capsorubin) and others with provitamin A activity (β-carotene and β-cryptoxanthin). This profile is kept in all varieties although they present different carotenoid concentrations that directly affect the economical evaluation of fruits and their processed products.

Different plant-plot genealogical selection programs related to productivity and processing have been developed to obtain new cultivars with improved properties, which are related to productivity and processing. To pepper fruits producers, productivity means getting a high fruitful ability (kilograms of fruit per square meter) with a high coloring capacity (American Spice Trade Association [ASTA] degrees, carotenoid concentration). From the processing point of view, the required plant properties include resistance to rot, to avoid losses during the dehydration process, grouped ripeness, and a plant morphology that allows mechanical harvest. With respect to cultivars employed for paprika and paprika oleoresin production, as mentioned before, only the nonpungent varieties are used and they can be subdivided into bell and pimiento types. Bell-type peppers are large fruits with a blocky shape and a thick wall, and they are commercially distributed for direct consumption or cooked. Pimiento types are also large fruits with an elongated heart shape and thick wall, and they are used for processing. Each producer country has developed its own cultivars, which have adapted to weather, land conditions, type of dehydration process to be applied, and, of course, consumer preferences. The available pepper cultivars may be found in different National Plant Variety offices. Some useful links are: International Plant Research http://www.plant.wageningen-ur.nl/; Institut National de la Recherche Agronomique, http://www.inra.fr/; Ministerio de Agricultura Pesca y Alimentación, www.mapya.es/;and the American Seed Trade Association, www.amseed.com.

Harvesting

Conventional pepper cultivars are not suitable for efficient mechanical harvest because of several pepper fruiting characteristics, such as nonuniform ripening, node attachment (pedicel scar) area, fruit orientation (pendant vs erect fruit orientation), branching habit, angle of plant stem laterals, plant height, and uprooting force. Additionally, harvesting must pick up fruits without uprooting the plant; the fruit should be free of trash and without damage. To reach all these objectives, with actual pepper plant features, means that peppers are ordinarily harvested by hand. However, mechanical harvesting is considered essential to expand the production of paprika-type peppers and to reduce harvest labor costs that could account for 35–50% of production costs (Palau and Torregrosa, 1997). Various processors and manufacturing companies have attempted harvest mechanization since the late 60s, without any success. Interest in mechanical harvest renewed in the 90s when the improvement of pepper harvestability through genetic selection was tightly coupled with harvester engineering development (Marshall, 1995). Currently, around 29 harvesting principles are applied to select pepper cultivars in main producer countries with improved harvesting features like Jaranda, Jariza, Negral, and Datler.

Description of Components of Red Pepper Fruits

Up to 40 different compounds of individual phenolics from pericarp and seeds have been observed by chromatographic methods, and the identification and structural elucidation is now under development. Flavonoid content of peppers may reach high levels depending on genotype and maturity, as well as on the growing and processing conditions. Qualitatively, quercetin and luteolin are the main flavonoids present in fresh pepper fruits (Howard et al., 2000). Quantitatively, low (0.1–39.9 mg/kg) or moderate (40–100 mg/kg) flavonoid levels have been described in several red pepper varieties. Assays applied to determine the total soluble phenolics, as the Folin–Ciocalteu assay, provide some data, and concentration of those compounds ranges from 2500 to 5000 mg/kg. The active principle that causes the heat in chili peppers is an alkaloid, which is generically called capsaicin. The word capsaicin actually describes a complex of related components named capsaicinoids by Japanese chemists S. Kosuge and Y. Inagaki in 1964. Capsaicinoids are produced by the glands at the junction of the placenta and the pod wall. The major capsaicinoids and their percentages are capsaicin (69%), dihydrocapsaicin (22%), and three minor-related components: nordihydrocapsaicin (7%), homocapsaicin (1%), and homodihydrocapsaicin (1%).

As in most plants, the major fatty acids accumulated in the pepper are palmitic, oleic, linoleic, and linolenic acid (Pérez-Gálvez et al., 1999b). However, pericarp and seed show different quantitative composition. The pericarp contains the unsaturated acids 18:2 and 18:3 as the major ones and in almost equal amounts, both representing ca. 60% of the total fatty acid content. The saturated-to-unsaturated fatty acids ratio is very similar to that found in the seeds of certain legumes. This fatty acid profile probably constitutes the appropriate membrane lipids for the biosynthesis of carotenoids, particularly when the enzymes involved in the synthesis of carotenoids is highly linked to membranes. As previously mentioned, only saturated fatty acids, oleic and linoleic, take part in the esterification process of xanthophylls, so that it is noteworthy that linolenic is not involved in such process. In the seeds the major acid is 18:2, which accounts for 80% of the total fatty acid content. Taking into account that 30–40% of the weight of paprika are seeds and that the oil content of seeds is ca. 20%, the final fatty acid pattern in paprika resembles that of the seeds, that is, mainly linoleic acid (ca. 70%) and palmitic acid (ca. 12%).

The most striking compounds of red pepper fruits are carotenoid pigments. This fruit contains a wide range of carotenoids, some of them biosynthesized only in the *Capsicum* genus, either free or esterified with fatty acids. This makes the carotenoid profile of red peppers unique. Carotenoids are isoprenoid compounds with a 40-carbon skeleton as basic structure from which the rest of the derivatives originate. The polyene chain in the central part of the molecule is the most characteristic feature of this group of pigments with a delocalized system of double and single bonds that give its physical and chemical properties. In the case of capsicum fruits, this chain is modified by cyclization at both ends giving β-type rings. These carotenoids are named carotenes, while xanthophylls include oxygen functions as hydroxyl, epoxide or keto groups. β-Carotene is the basic carotene from which rest of the xanthophylls are derived. The rest of the pigments are xanthophylls with different oxygenation levels. These include β-cryptoxanthin, zeaxanthin, cucurbitaxanthin A, and violaxanthin, all of which provide the yellow hue, capsanthin and capsorubin. These two keto carotenoids occur only in red peppers and contribute to red color (Levy et al., 1995). Figure 30.1 depicts the structure of the main carotenoids found in red fruits.

This distinction was once used

- to quantify and classify the ratio between those chromatic fractions (Mínguez-Mosquera et al., 1984) and
- as a quality parameter of both processing and storage

This will be further discussed later.

As previously mentioned, xanthophylls are esterified with fatty acids, a process that takes place during the ripening of pepper. Other structural changes that occur include the disappearance of chloroplasts and formation of chromoplasts (Camara and Monéger, 1978). It seems that the metabolic system of the fruit tries to make these pigments more lipid soluble and/or to protect them from enzymatic and nonenzymatic oxidative reactions. Xanthophylls of the yellow fraction are mainly esterified with unsaturated fatty acids like oleic and linoleic, while saturated fatty acids participate in esterification of the red xanthophylls (Biacs et al., 1989). Besides the coloring capacity, some of the carotenoids present in this profile show provitamin A activity, β-carotene and its isomeric forms, and β-cryptoxanthin. This nutritive quality is additionally complemented by antioxidant activities that all carotenoids may perform, once they are ingested and incorporated in human organism. Other antioxidant vitamins described in red peppers are vitamins C and E. Fresh peppers are a good source of vitamin C. They contribute high levels to the Recommended Dietary Allowance (RDA) values for vitamin C. Contents of 70–200 mg/100 g of vitamin C in fresh fruit have been detected that can contribute 124–340% of the RDA of this hydrophilic vitamin (Howard et al., 2000). With respect to the vitamin E content, some reports indicate contents of 0.5 mg/g of dry fruit (Daood et al., 1996).

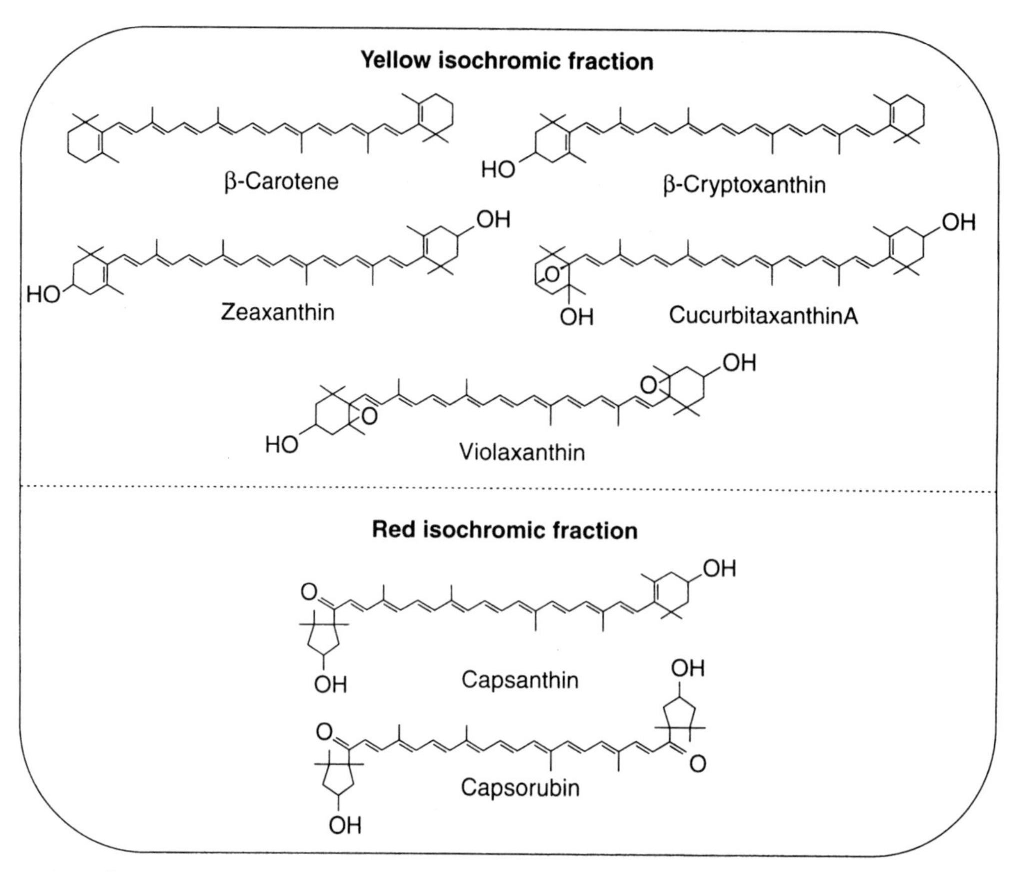

Figure 30.1. Structures of carotenoids described in red pepper fruits (*Capsicum annuum* L.).

INDUSTRIAL PROCESSING OF RED PEPPER, FRESH FRUIT, PAPRIKA, AND PAPRIKA OLEORESIN

FRESH FRUIT

Red pepper fruits are directly consumed, as are most vegetables, by adding to sliced fresh fruits in salads, by cooking, or in the preparation of condiment oils. For direct consumption, the fruits go through the same processing stages as vegetables. Thus, sweet peppers are rinsed and disinfected, generally by dipping the fruits for 3–5 min in hot water (50°C) with the appropriate equipment. After drying in ambient air, the result is a ready-to-package material free of contaminants, dirt, and fungi (Fallik et al., 1999). Fresh fruits are minimally processed for tinning. In this case, they are peeled (by roasting or by using chemicals), washed to remove skin remains, and seeds and stalks are discarded. Finally, the fruits are canned in brine. Nonfood applications consist of medicinal preparations such as a carminative, digestive irritant, stimulant, and tonic. The plants have also been used as folk remedies for dropsy, colic, diarrhea, asthma, arthritis, muscle cramps, and toothache (Bosland, 1994).

PAPRIKA

Processing pepper fruits for obtaining paprika is based on two requirements: dehydration and milling. Dehydration is necessary to extend the shelf life of fruits from days to months and to make them available until the next harvest season. This seasonal fruit has one harvest per year and water content in the fresh fruit is 70–85%, which can result in spoilage, if stored fresh.

Size reduction is necessary to homogenize the color and to facilitate handling and transport. Various cultivars in the batches for processing present different color intensity. Dry fruits are difficult to transport. Size reduction is done by milling or grinding the dry fruits.

Therefore, the production of paprika is based on the above two basic operations. But 60–70% of milled product is used to produce oleoresins (Govindarajan, 1986). Oleoresins are highly concentrated carotenoid oils used by the food industry mainly as a food coloring agent, apart from other applications. This oil is in great demand because of its popularity of use as a natural color in food processing.

Processing Description (Dehydration and Milling)

Preprocessing techniques of raw material before dehydration are not generally used, but recent studies show that some pretreatments of fresh fruits would be advisable to reduce the temperature of processing and to increase rate of mass transfer (Lazarides et al., 1999). Some of them are simple techniques like washing, cutting, and steam blanching. A combination of cutting and steam blanching (for 1 min) increases the drying rate and reduces the processing time and level of thermal stress (Doymaz and Pala, 2002). More sophisticated preprocessing techniques are now under consideration such as osmotic dehydration, which has been successfully applied to pepper fruits. In this case, the fresh material is placed in a dehydration solution consisting of sucrose and sodium chloride. High-intensity electric field pulses that disrupt the cell material and produce pore formation accompany the osmotic process. In this procedure, the temperature applied in hot air dryers could be reduced from 60–80°C to 35°C, without increasing the residence time considerably (Ade-Omowaye et al., 2002).

Dehydration of pepper fruits can be carried out by applying traditional techniques or industrial processes. Traditionally, two conventional dehydration techniques have been used, via direct exposure of fruits to the sunlight in open air or in drying chambers where the heat source is the burning of oak logs. The first form of traditional drying is employed in Spain and Turkey, but now it is not in use because it is both slow and costly. Furthermore, the quality of dry peppers is affected by oxidative degradation promoted by sunlight (Mínguez-Mosquera et al., 1996). Consequently, industrial processes have replaced that dehydration technique. The traditional system applied in La Vera region (Cáceres) is still in use. The drying chambers consist of two floors of cottages where the oak logs are placed on the first floor while the fruits are piled on the second one, the surface of which is a wooden lattice that allows hot air and smoke to circulate from the heat source through the mass of fruits. Fresh peppers are dehydrated in 7–10 days and become dry husks impregnated with a smoke aroma given off by oak logs (Mínguez-Mosquera et al., 1996; Pérez-Gálvez et al., 2001). The flavor features of this dry product (that will be kept in the paprika) have made it to be very highly prized by both the consumer and the food producers who use this smoked paprika as coloring agent and to contribute particular organoleptic characteristics to the end product.

Paprika producers prefer industrial dryers because the dehydration process applied is more uniform and rapid than any of the traditional ways. Industrial techniques considerably reduce the residence time of fruits from days employed in traditional procedures to 4–6 h of processing to complete dehydration of fruits. Additionally, traditional processes enables the fruits to reach a moisture level of about 7–9%, while industrial dryers may reduce moisture of the fruits to as little as 3–4%. Although a batch process may be used with equipment like cabinet dryers, a continuous process is frequently utilized (Chung et al., 1992; Levy et al., 1995). In this case, tunnel dryers are conventionally employed. Fresh fruits are deposited on trays, which are stacked on trucks programmed to move intermittently through an insulated tunnel. The drying element temperature is 60–80°C hot air, which circulates through the tunnel under countercurrent (McGaw, 2001). It is noteworthy that hot air dryers are nowadays complemented with the use of renewable energy like solar irradiation, designs that at present are implemented in some drying equipment (Condori et al., 2001). Another possibility is the use of vacuum dryers that would completely eliminate the thermal stress. Thermal application can be reduced or replaced with a decrease in external pressure (Zhang et al., 2003). Frequently used equipment is a vacuum roller dryer that consists of one or two rollers installed in a vacuum housing. The resulting vapor precipitates in a condenser located between the vacuum chamber and the pump. A screw conveyor removes the product.

Milling is used for the size reduction of solid dry material, thus improving the product's edible quality and suitability for further processing. During this procedure, dry fruits are milled with a certain percentage of fruit seeds. The purpose of this procedure is to homogenize the production and to improve the dark red hue of the dried pericarp, by transforming it to a bright red, an effect from the seed oil. A percentage of about 30–35% fruit seeds is commonly added to dry fruits in this processing step. The seeds included at milling are inert material with regard to their color contribution and act as a dilution factor for the total carotenoid content (Pérez-Gálvez et al., 1999a). However, the addition of seeds has further implications in stabilizing the carotenoid content during storage, a fact to be discussed later.

Common types of mills used are hammer and ball mills. Hammer mills consist of a horizontal or vertical cylindrical chamber lined with a steel breaker plate and contains a high-speed rotor fitted with hammers along its length. The material is broken apart by impact forces as the hammers drive it against the breaker plate. Ball mills consist of a slowly rotating, horizontal steel cylinder, half-filled with steel balls (2.5–15 cm in diameter). The final particle size depends on the speed of rotation and on the size of the balls.

Influence of Processing and Storage in Components Present in Fresh Fruit

Processing. Fruits and vegetables processing must preserve the nutritional and organoleptic qualities of these food items to convince consumers that, although processed, they remain as good as fresh. Vitamins are the most significant nutritional components of vegetables and, therefore, they should be minimally affected by processing conditions. In the case of red peppers, vitamin C and provitamin A carotenoids are the significant valuable components.

Fresh or minimally processed (for direct consumption) red pepper fruits may lose part of their vitamin C content during rinsing and peeling steps due to mechanical injuries, loss of water, and temperature (Lee and Kader, 2000). Processing conditions applied during the dehydration step produce important losses of vitamin C, which are closely joined with the loss of water content. Industrial dehydration using hot air can reduce vitamin C content (ca. 90%). Using inert gas during dehydration may also reduce vitamin C losses (Armes et al., 1999). The loss could be similar in the traditional processing that was described previously. A model traditional dehydration process has shown that vitamin C is very sensitive to the drying process, with a decrease of about 76% after 24 h of drying and remaining at trace levels during the rest of the process (Pérez-Gálvez et al., 2004).

Although vitamin C retention is often used as a marker of processing quality in several fruits and vegetables (Tijskens et al., 1979), red peppers may be considered as an exception. Dehydration conditions must preserve the main quality of pepper fruits, color, and consequently the carotenoid composition responsible for it. Carotenoid pigments are compounds sensitive to light and temperature, the factors that promote oxidation reactions that easily decrease the initial carotenoid amount of fresh fruits. Fruit moisture evolution will set up processing conditions, temperature, and drying time. These variables determine the extent of changes in the fruit carotenoid composition responsible for the final product color (Mínguez-Mosquera et al., 2000).

Additionally, it has been reported that stability of carotenoid concentration in paprika during storage is dependent on the drying conditions. The carotenoid oxidation rate increases as the drying temperatures increase (Levy et al., 1995; Pérez-Gálvez et al., 2001). Therefore, all dehydration systems should have a compromise solution between the drying temperature and the residence time (variables that are indirectly related). Industrial and traditional drying processes fulfill that solution in a different manner, by lowering residence time or temperature, respectively, which will have different consequences in the color quality of dry products obtained.

For economical and processing reasons, industrial processes reach the objective of reducing the residence time of pepper fruits in the dryer by increasing the processing temperature. However, such conditions often provide a blackish red color to dry fruits decreasing their commercial value. This visual drawback is due to the formation of nonenzymatic browning compounds as well as a decrease in the carotenoid concentration in dry fruits with respect to the initial content (Lee and Kim, 1989). Thus, dehydration of Bola and Agridulce varieties in industrial drying tunnels (residence time, 8 h; temperature, 60°C) produce a decrease of ca. 16% on initial carotenoid concentration in both varieties (Mínguez-Mosquera et al., 1994). Finally, it must be borne in mind that when fruits are subjected to a thermal stress, this will reduce the shelf life of paprika.

Mild temperature conditions are applied in the traditional process carried out in the La Vera region. A mean temperature of ca. 40°C is commonly reached during drying which obviously increases the residence time (7–10 days), but carotenoid content is maintained during processing, that is, the color quality remains unaltered. The advantage of this process

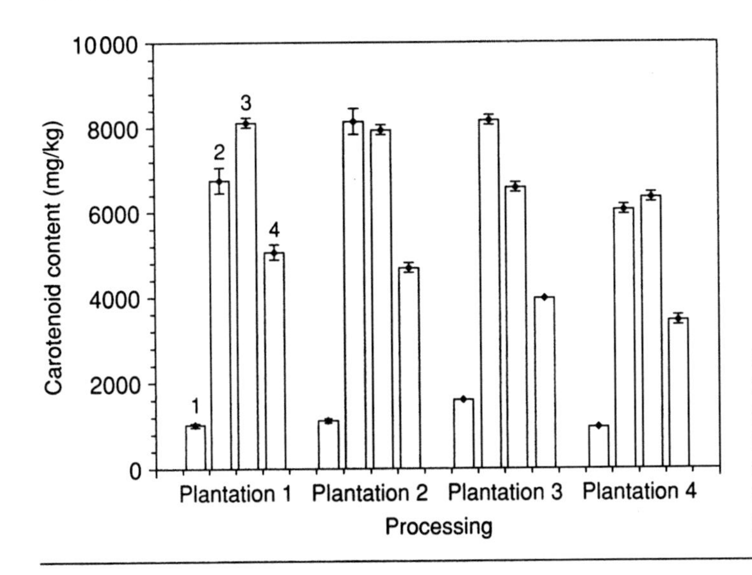

Figure 30.2. Carotenoid content of the red pepper (*Capsicum annuum* L. cv. Jaranda) at each processing step for paprika production: fresh fruit (1 expressed as wet base; 2, expressed as dry base), dehydrated fruit (3), and paprika (4).

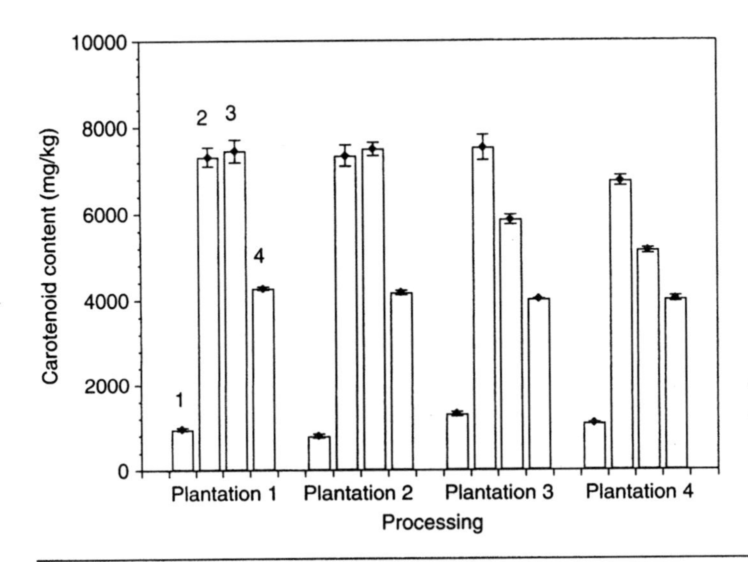

Figure 30.3. Carotenoid content of the red pepper (*Capsicum annuum* L. cv. Jariza) at each processing step for paprika production: fresh fruit (1 expressed as wet base; 2, expressed as dry base), dehydrated fruit (3), and paprika (4).

is emphasized if one considers the fact that overripening of fruits may take place during the dehydration course (Mínguez-Mosquera et al., 1994, 2000; Pérez-Gálvez et al., 2004). Peppers from Agridulce, Bola, Jaranda, and Jariza cultivars subjected to the traditional La Vera dehydration process increase their initial carotenoid content about 15–20%. It should be noted that together with mild conditions, high moisture content of raw material was observed. Water content and gentle temperature allows a continued biosynthesis of carotenoids in the fruits entering the dryer, so metabolic pathways of pigment formation continue to exist. These comments are detailed in Figures 30.2 and 30.3 where the evolution of total carotenoid content during traditional processing of two varieties, Jaranda (Fig. 30.2) and Jariza (Fig. 30.3), commonly grown in La Vera region is shown. The initial carotenoid content (expressed in both wet and dry base) in the experiment rises and falls after dehydration resulting from expression of anabolic or catabolic reactions. Those processes that occur within the appropriate time–temperature regime and ripening stage of the fruits produced a dehydrated product with significant higher carotenoid content in comparison with that of the starting fresh material. On the other hand, processes that yielded a dry product with lower pigment content were not conducted according to fruit conditions (ripening stage, moisture content) together with a more stressed time–temperature regime (Mínguez-Mosquera et al., 2000).

Although the carotenoid profile of pepper fruits is a complex system involving xanthophylls of different structure, it can be easily divided, as stated previously, into two isochromic fractions, red and yellow (R/Y). The ratio between the two fractions (R/Y) should remain invariable if the process does not affect the two fractions, or to the same extent if it affects the initial ratio value. Significant changes in this parameter have been observed during industrial processing of Bola (from 1.88 to 5.36) cultivar (Mínguez-Mosquera and Hornero-Méndez, 1994). In the case of traditional processing of Jaranda and Jariza cultivars, modifications of this parameter are also denoted, as shown in Figures 30.4 and 30.5, respectively. R/Y values resulting at the end of the dehydration step were not homogeneous as they depend on the fraction which is overaccumulated and on the balance between anabolic and catabolic processes that coexist during this stage.

During the milling step, fruits are subjected to heating due to friction and bruising from the mill hammers or balls. In a temperature-sensitive product like pepper fruits, any step increasing the temperature should be minimized. Although operating time of milling takes no more than 3 h, carotenoid content may be affected (Mínguez-Mosquera et al., 2000). In that case, carotenoid content of paprika was lower than that in the previous processing step (dry fruit) even when taking into account the content dilution due to seed addition. Losses during the milling step are approximately 10–15% of carotenoid concentration in dry fruits (see Figs. 30.2 and 30.3). In general, the red-to-yellow ratio in the paprika is higher than that of the fresh fruit. This implies the occurrence of degradative processes affecting especially the yellow fraction. As in the preceding step, the control parameter is temperature that the product reaches during milling.

Storage. Stability of carotenoid content during storage will be a result of previous processing steps. Extent of changes in carotenoid content, especially after thermal dehydration stage, may be an indicative point to predict the stability of paprika at the store. However, an overdrying or excessive milling during paprika production promotes oxidation reactions that not only affect carotenoid concentration but also the rest of the fruit components, especially fatty acids that simply undergo oxidation in processing conditions. These reactions that seriously affect carotenoid profile may be initiated during processing, although their consequences would not be noticeable immediately after processing but during storage (Malchev et al., 1982). Additionally, fatty acid oxidation would give the paprika an undesirable flavor that would make

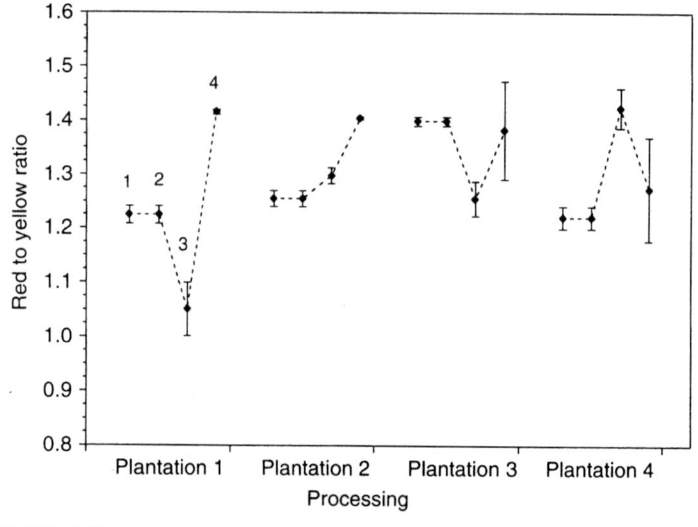

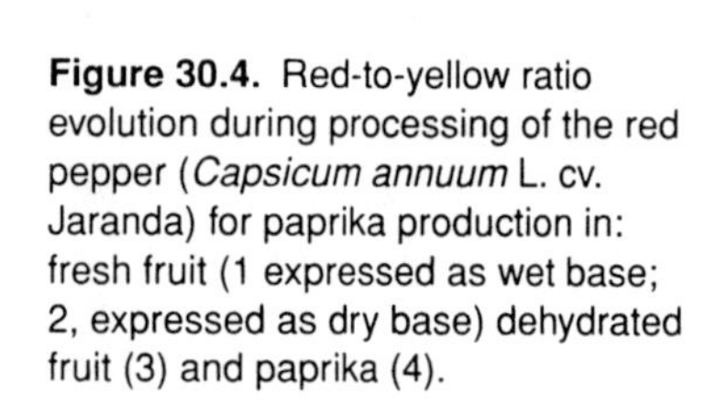
Figure 30.4. Red-to-yellow ratio evolution during processing of the red pepper (*Capsicum annuum* L. cv. Jaranda) for paprika production in: fresh fruit (1 expressed as wet base; 2, expressed as dry base) dehydrated fruit (3) and paprika (4).

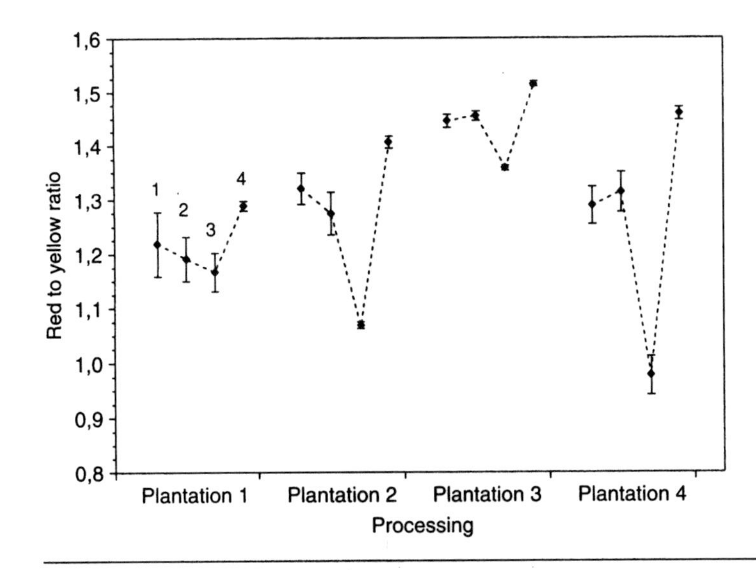

Figure 30.5. Red-to-yellow ratio evolution during processing of the red pepper (*Capsicum annuum* L. cv. Jariza) for paprika production in: fresh fruit (1 expressed as wet base; 2, expressed as dry base) dehydrated fruit (3) and paprika (4).

the product inappropriate for trade and consumption (Biacs et al., 1992). Minimizing these oxidative processes is even more important when considering that during milling the oil content of dry fruits is increased by addition of seeds.

Studies are being conducted on the processes of fatty acid oxidations to prevent negative reactions from affecting the carotenoid content and overall quality of paprika, e.g., the conditions of storage and the effects of the amount of seeds added during milling. It is generally stated that presence of air and light must be avoided during paprika storage. Contact and exposure of the product to air and light could be avoided by vacuum packaging often in red plastic bags. With respect to the seed content, it was initially proposed that addition of seeds would increase paprika stability during storage due to natural antioxidants, mainly tocopherols, which are present in the lipid profile of pepper seeds (Okos et al., 1990). A comparative study of paprika stability with different seed percentage (0%, 20%, 40%, and 60%) showed that high addition of pepper seeds during milling might increase color stability with the drawback of high color dilution, while paprika without added seeds showed the lower carotenoid stability (Pérez-Gálvez et al., 1999a). Similar steadiness was reached when 20% and 40% of seeds were added to dry fruits.

The physical explanation for carotenoid protection by seeds is that solid particles are covered by oil with a thin film, filling the hollow spaces and greatly decreasing the surface exposed to environmental conditions. Paprika processed with high seed content presents a greater oil film protecting the particles that may explain the retardation of pigment degradation.

Chemically, tocopherols of natural occurrence can serve as protective or antioxidative agents. Further, the change in fatty acid composition of dry fruits from the addition of seeds is due to the pericarp of pepper fruits. It has been shown that the pericarp of pepper fruits presents a heterogeneous fatty acid profile with palmitic, linoleic, and linolenic acid as major components, while linoleic acid is the main fatty acid in pepper seed oil. This fatty acid distribution in seed is very similar to that in paprika (Pérez-Gálvez et al., 1999b). Therefore, dry milled fruits without added seeds present a fatty acid mixture, which is very susceptible to oxidation. Adding seeds means not only dilution of color but also dilution of components more prone to oxidation. A good balance between both dilution effects should be ideal for increasing the oxidative stability.

Description of Final Product, Commercial and Industrial Uses, Quality Assurance Methods

This two-step processing of red pepper produces paprika, a fine, brilliant, highly colored powder that may be used directly for consumption in culinary products and commercially distributed in 25–100 g packages. The inclusion of paprika occurs in thousands of recipes. It may be used to improve the presentation or to impregnate the meal with a characteristic flavor.

The food industry makes use of paprika for coloring and flavoring purposes in garnishes, pickles, meats, barbecue sauces, ketchup, cheese, snack food, dips, salads, and sausages. In some cases, the dry fruits are crushed and directly employed for pizza, pasta production, and in preparation of certain flavored oils.

In a product economically evaluated by its coloring capacity, measurement of this organoleptic quality must be the reference and standard procedure for ranking different qualities of paprika. Several standard methods have been developed to achieve the goal of obtaining rapid quantifiable data via easy measurement techniques. A few standard methods for obtaining a "color value" are commonly applied. One of them, "standard," determines the absorbance value of a 1/10,000 solution of a known weight of the oleoresin in acetone at 462 nm, multiplying the result by 66,000 and divided by the sample weight to obtain a standard color value (Guenter, 1948). Another frequently used method is that of the ASTA that consists of a color extraction of a weighed paprika sample with 100 ml acetone during 24 h. An aliquot of that solution (10 ml) is diluted again to 100 ml, and a portion of the diluted solution is used for the spectrophotometric measurement at 460 nm. ASTA units are calculated by multiplying the absorbance by 164 and divided by sample weight (ASTA, 1986).

However, these spectrophotometric methods give only a composite picture, with low information on each carotenoid present in the profile. Specific values of red to yellow pigment fraction and provitamin A content are not available. Recently, a new spectrophotometric method has been developed, which overcome these drawbacks. It is based on measurement at two characteristic wavelengths (472 and 508 nm) for simultaneous quantification of the red and yellow fractions with low error in the determination and lower time analysis in comparison with that of chromatographic methods (Hornero-Méndez and Mínguez-Mosquera, 2001). A global approach to quality should also consider other harmful components that may contribute to degradation of carotenoid fraction, such as peroxide value (PV). Application of spectrophotometric methods using the UV–visible range is currently the most widely used to determine PV in food lipids. However, in food products containing carotenoids, direct application of standard methods is not suitable as carotenoids interfere in the measurement and generate error. This fact has been considered and solved in a recently developed method (Hornero-Méndez et al., 2001). Thus, control of valuable components (carotenoids) and main oxidative components (peroxides) should give better information about sample conditions and how the processing was done.

Paprika Oleoresin

The processing of pepper fruits for obtaining its oleoresin includes the previous dehydration and milling steps to acquire the paprika powder, and extraction of its valuable lipid-soluble components, the carotenoid content, that it is subsequently separated and recovered from the extraction solvent. This procedure is not very selective because it extracts carotenoid and other components of the lipid, which all end up in the final product. All these lipophilic solvents are suitable for this step including both supercritical and organic solvents. Therefore, the extraction methods and equipment are different depending on selection of supercritical gases or liquid solvents.

Organic Solvent Extraction

Organic solvents are classically applied for extraction of oil from oilseeds and those fruits with precious lipophilic components, as in the case of the color content from pepper fruits. Choice of solvent is determined by the efficiency and extraction yield, the recovery of the solvent from the bulky oil, and the ease of removal of all traces of solvent from the product. Additionally, the food laws of the product-consuming countries will restrict the use of a particular solvent. Hence, acetone, dichloromethane, ethyl acetic, and methanol, solvents allowed for extraction purposes with similar good extraction properties, have drawbacks such as low recovery and a source of impurities in the final product. Therefore, hexane, methylene chloride, and ethyl acetate are commonly employed for extraction purposes. In general, they give a good yield with high recovery of color, together with an economic recovery, producing low traces of residues.

Processing Description. The simplest extraction method is the countercurrent extraction either in batch or in continuous process. Batch process uses cylindrical dished-end column percolators with common powder-to-solvent ratio values of 0.6 kg/l. The extraction limit is reached with 5 h of processing. Cofield (1951) describes a continuous countercurrent extractor where the amount of paprika powder-to-solvent volume ratio is 4 kg/l and 3–5 h of processing. The flow of solid material and liquid solvent is made in different forms, such as perforated trays connected to an unbroken loop, screw conveyors, or to an unbroken perforated belt.

The remaining color in the solid material at different times will determine the operating time. Once the extraction limit is reached, defined by economic factors and extraction techniques, the solvent is removed by concentration of the extract in vertical tube evaporators under vacuum. This reduces the temperature of the process, operating in the range of 50–80°C, and recovers the evaporated solvent through condensers to economize the solvent used. The concentration stage leaves about 1–4% of the solvent in the extract; hence, the product must be treated again to remove solvent residues from it to achieve statutory requirements regarding maximum limits of residual solvent in the final oleoresin. This step is done in two different ways depending on the solvent used in the extraction procedure. Thus, the extract is sprayed with steam at a pressure of 160 bars for 15–30 min, if methylene chloride was used, or by indirect heating with water at 100°C, if hexane was the extracting solvent.

Control of the process is set by two factors—extraction limit of coloring content and postprocessing of the extract to reach food law standards with respect to solvent rests. Determining the remaining color in the solid residue at different times controls the first issue. In continuous process, 76% of color is extracted in about 12 min, so raw material is not subjected to heating for an excessive time while a high color extraction has been achieved. It must be remembered that carotenoid content is sensitive to heat, so prolonged processing at high temperatures would decrease both color concentration and stability of the final product. Of course, the remaining 24% of the color content may be recovered but almost 5 h of operating time are required, thus increasing the amount of solvent and thermal stress of raw material. A compromise solution between extraction limit and processing time must be reached. These criteria, scheduled in the extraction step, would affect the final treatment process of extract to remove solvent and solvent residues because legislative standards must be satisfied in any case. Although low-pressure techniques are applied, heating is used to reinforce and improve the final treatment, so again the extract is subjected to thermal stress. Therefore, the selection of solvent and operating conditions at the extraction stage is very important to improve the quality and stability indexes of final product. However, legal restriction limits the solvent used and maximum residue levels in foodstuffs. European Union directives for extraction solvent use and residue levels are 88/344/CEE and 97/60/CE (Council Directive 88/344/EEC, 1988; Directive 97/60/EC, 1997). Maximum residue levels are 50 mg/kg for previously mentioned solvents, except for methylene chloride (10 mg/kg).

Supercritical Fluid Extraction

The increasing restrictions in the use of organic solvents and the fall in the levels of solvent residues in final products are serious limitations for organic solvent extraction procedures. Additionally, lack of selectivity of this process represents a serious disadvantage as markets demand extracts of pure functional components, so producers have to use costly processing steps to remove nonfunctional compounds like fats, waxes, and gums. Both trends may be satisfied with the use of supercritical fluids as extraction solvents. Extraction with supercritical fluids of natural products from plant material is the more extended application of the technique that has been used for tocopherol extraction from wheat germ, caffeine from tea or coffee, and essential oils from a wide range of spices like basil, oregano, turmeric, and paprika (Sugiyama et al., 1985; Jarén-Galán et al., 1999; Marr and Gamse, 2000; Uquiche et al., 2004). A supercritical fluid is a gas above its critical pressure and critical temperature, that is, highly compressed and for most of the gases at near ambient temperatures. Working with supercritical fluids as an extraction solvent has several advantages. The solvency power is retained under those conditions with the particularly interesting property that small changes in operating conditions, temperature or pressure, produce large changes in extraction capacity and selectivity of extracted components. Therefore, this procedure would include at the same time both an extraction step and a fractionation one. Moreover, temperature conditions are reduced to 40°C, gentler than in organic solvent extraction procedures so thermal stress is considerably reduced. Finally, operating steps for removing solvent from extracted material are practically eliminated because decompressing the extract will remove any trace of extraction fluid, which can be recovered and recycled.

Processing Description. Carbon dioxide under supercritical conditions is the gas most applied for extraction purposes because it is inert, nontoxic, highly available, and inexpensive. As discussed above, changing operating conditions generate variations in the kind and amount of extracted compounds. As described by Govindarajan, paprika extraction at 120 bars and 40°C would produce a highly pungent and aromatic oleoresin, while a second extraction at 320 bars and 40°C would produce a highly colored oleoresin free of capsaicinoids and aroma compounds (Govindarajan, 1986). Jarén-Galán et al. reproduced this two-step procedure by increasing the operation pressure, with sweet paprika as raw material (Jarén-Galán et al., 1999). In this case, a β-carotene rich oleoresin was first obtained, whereas a purified carotenoid extract resulted from the second step.

Both examples illustrate the possibilities of selective extraction offered by this technique for the production of purified compounds extract (pure carotenoid oleoresins) and fragmented extracts (β-carotene rich oils, yellow or red oleoresins). As a result, producers can

- diversify their product range,
- satisfy new demands for application in functional foods, and
- supply needs for nonfood industries such as cosmetic and pharmacological items.

However, extraction with supercritical fluids requires more research to determine those operating conditions giving the desired product with higher yields than those currently obtained. Moreover, supercritical extraction equipment is prone to operational problems, requiring higher investments (Marr and Gamse, 2000).

Influence of Processing in Components Present in Fresh Fruit

Processing. Oxidative processes of carotenoids present in paprika oleoresins follow a heterogeneous degradation resulting from different structural features and fatty acid composition of both molecular and substrate environments, reactions that take place in a bulk oily phase. As mentioned before for paprika, the degradation pattern responses to a higher oxidative stability of red pigments attributed to the functional groups taking part in their structure and the saturated molecular environment covering the carotenoid (Pérez-Gálvez et al., 2000; Pérez-Gálvez and Mínguez-Mosquera, 2002). However, processing conditions for obtaining paprika oleoresins are more forceful, with higher temperature regimes to reach an effective extraction and subsequently to remove any trace of solvent residues to accomplish legislative criteria. Thus, processing is a continuous heating line, where the extract is subjected to thermal stress. Those temperature conditions have a direct influence in the degradative pattern described for paprika. Hypothetically, differences in oxidative stability of red and yellow fraction might enlarge, favoring degradation of yellow pigments, but this is not the case. It has been shown that increasing the temperature of processing means a selective reactivity of red carotenoids. Therefore, oxidative stability difference decreases with increasing temperature reaching a zero value and afterward starts to increase in favor of yellow pigments, thereby changing the degradation pattern (Jarén-Galán et al., 1999).

Storage. Stability of Paprika Oleoresins. Main problems during storage of paprika oleoresins can arise from flocculation of gums and waxes that cause cloudiness and changes in density and refractive index of the end product. Additionally, if this sludge is not removed it could interfere in the clarity and solubility of the oleoresin in the food applications. Therefore, physical or chemical refining is often applied to avoid such problems during long storage terms. With respect to the stability of the carotenoid fraction, it will depend on conditions reached during processing, especially during the removing solvent step. Considering that quality criteria is linked to carotenoid content and prediction of its deterioration during storage is important to keep formulation and thus uniformity of production. The industrial sector implied needed a prediction tool that allows it to calculate carotenoid content after a storage period, that is, the content of color in the sample. A thermodynamic study allowed to achieve such a tool (Jarén-Galán et al., 1999).

Description of Final Product, Commercial and Industrial Uses, Quality Assurance Methods

The paprika oleoresin is a red or deep red oily liquid, which depending on features of raw material and dehydration technique applied, may or may not show pungency and smoked flavor. Oleoresins for coloring applications are manufactured from sweet pepper fruits without traditional dehydration techniques, to avoid the presence of smoked aroma. Paprika oleoresins have important applications in the food industry, as trends in the use of natural colorants are more widespread to substitute those of synthetic origin.

Moreover, the oily substrate, solving the carotenoid fraction, is an ideal vehicle to incorporate this product in multiple food formulations. Thus, some applications include processed meats, pasta and pizza, dairy products, soups sauces, and snacks to supply or improve the color of these foodstuffs.

The quality parameter employed is the coloring capacity of the extracted oleoresin measured by same methods applied to paprika evaluation, standard and ASTA. Additionally, another measurement is known as tint determination. This criterion of quality is based on the ratio of absorbance at two wavelengths, and it is an attempt to obtain the ratio between red and yellow carotenoid pigments in a sample (Akira-Mohri et al., 1993).

All the above-mentioned methods meet technical and commercial reasons for oleoresin color evaluation, permitting speed and quantification. However, they showed the same lack, as did paprika color quality control methods. Coloring capacity is evaluated as a composite picture, so valuable information about the nutritional value and the extent of degradation on each compound of the carotenoid profile is missed. As in the case of paprika, an individual carotenoid composition would give information about the influence of processing on the coloring capacity as well as the valuable characteristics on nutritional aspects of the product. However, all producers could not apply quantification of individual carotenoids because the methodology (high pressure liquid chromatography) requires costly equipment and more time for measurement. For that reason, a spectrophotometric method has been developed to determine the concentration of red and yellow fractions together with total carotenoid content in paprika oleoresins, enabling a commercial evaluation of the oleoresin (Mínguez-Mosquera and Pérez-Gálvez, 1998). The measurements could be complemented with information obtained by the method described for paprika (Hornero-Méndez and Mínguez-Mosquera, 2001) and determination of PV (Hornero-Méndez et al., 2001).

REFERENCES

Ade-Omowaye BIO, Rastogi NK, Angersbach A, Knorr D. Osmotic dehydration of bell peppers: Influence of high intensity electric field pulses and elevated temperature treatment. J Food Eng. 2002 54 35–43.

Akira-Mohri I, Kastsumi-Morikawa M, Toshihiko-Matsaya I, Setsuo-Sato M. Natural red coloring matter and its processing method. US Patent 1993 P number 5264212.

Armes MN, Wolf W, Tevini D, Jung G. Studies on inert gas processing of vegetables. J Food Eng. 1999 40 199–205.

ASTA. Official analytical methods of the American spice trade association, 2nd ed. 1986.

Biacs PA, Czinkotai B, Hoschke Á. Factors affecting stability of colored substances in paprika powders. J Agric Food Chem. 1992 40 363–367.

Biacs PA, Daood HG, Pavisa A, Hajda F. Studies on the carotenoid pigments of paprika (*Capsicum annuum* L. var Sz-20). J Agric Food Chem. 1989 37 350–353.

Bosland PW. Chiles: History, cultivation, and uses. In Spices, herbs edible fungi. Ed. Charalambous G. pp 347–366. Elsevier Science BV, London, 1994.

Camara B, Monéger R. Free and esterified carotenoids in green and red fruits of *Capsicum annuum*. Phytochemistry. 1978 17 91–93.

Chung SK, Shin JC, Choi JU. Effect of blanching on dry rates and colour of hot red pepper. Korean Soc Food Nutr. 1992 21 64–69.

Cofield EP. Solvent extraction of oilseed. Chem Eng. 1951 58 127–140.

Condori M, Echazu R, Saravia L. Solar drying of sweet pepper and garlic using the tunnel greenhouse drier. Renewable Energy. 2001 22 447–460.

Council Directive 88/344/EEC. On the approximation of the laws of the Member States on extraction solvents used in the production of foodstuffs and food ingredients. Official Journal L. 1988 157 28.

Daood HG, Vinkler M, Markus F, Hebshi EA, Biacs PA. Antioxidant vitamin content of spice red pepper (paprika) as affected by technological and varietal factors. Food Chem. 1996 55 365–372.

Directive 97/60/EC. Amending for the third time Directive 88/344/EEC on the approximation of the laws of the Member States on extraction solvents used in the production of foodstuffs and food ingredients. Official Journal L. 1997 331 7.

Doymaz I, Pala M. Hot-air drying characteristics of red pepper. J Food Eng. 2002 55 331–335.

Fallik E, Grinberg S, Alkalai S, Yekutieli O, Wiseblum A, Regev R, Beres H, Bar-Lev E. A unique rapid hot water treatment to improve storage quality of sweet pepper. Postharvest Biol Technol. 1999 15 25–32.

Govindarajan VS. Capsicum—Production, technology, chemistry, and quality- part II. Processed products, standards, world production and trade. Crit Rev Food Sci Nutr. 1986 23 207–288.

Guenter E. In The essential oils, Vol I. Van Nostrand, New York, 1948.

Hornero-Méndez D, Mínguez-Mosquera MI. Rapid spectrophotometric determination of red and yellow isochromic carotenoid fractions in paprika and red pepper oleoresins. J Agric Food Chem. 2001 49 3584–3588.

Hornero-Méndez D, Pérez-Gálvez A, Mínguez-Mosquera MI. A rapid spectrophotometric method for the determination of peroxide value in food lipids with high carotenoid content. J Am Oil Chem Soc. 2001 78 1151–1155.

Howard LR, Talcott ST, Brenes CH, Villalon B. Changes in phytochemical and antioxidant activity of selected pepper cultivars (Capsicum species) as influenced by maturity. J Agric Food Chem. 2000 48 1713–1720.

Jarén-Galán M, Nienaber U, Schwartz SJ. Paprika (*Capsicum annuum*) oleoresin extraction with supercritical carbon dioxide. J Agric Food Chem. 1999 47 3558–3564.

Jarén-Galán M, Pérez-Gálvez A, Mínguez-Mosquera MI. Prediction of decoloration in paprika oleoresins. Application to studies of stability in thermodynamically compensated systems. J Agric Food Chem. 1999 47 945–951.

Lazarides HN, Fito P, Chiralt A, Gekas V, Lenart A. Advances in osmotic dehydration. In Processing foods: Quality optimisation and process assessment. Eds. Oliveira ARF and Oliverira JC. CRC Press, London, 1999.

Lee DS, Kim HK. Carotenoid destruction and non-enzymatic browning during red pepper drying as functions of average moisture content and temperature. Korean J Food Sci Technol. 1989 21 425–429.

Lee SK, Kader AA. Preharvest and postharvest factors influencing vitamin C content of horticultural crops. Postharvest Biol Technol. 2000 20 207–220.

Levy A, Harel S, Palevitch D, Akiri B, Menagem E, Kanner J. Carotenoid pigments and β-carotene in paprika fruits (Capsicum Spp.) with different genotypes. J Agric Food Chem. 1995 43 362–366.

Malchev E, Chenov NS, Ioncheva I, Tanchev SS, Kalpakchieva KK. Quantitative changes in carotenoid during the storage of dried red pepper powder. Nahrung. 1982 26 415–418.

Marr R, Gamse T. Use of supercritical fluids for different processes including new developments-a review. Chem Eng Proc. 2000 39 19–28.

Marshall DE. Mechanical pepper harvesting Acta Hort (ISHS). 1995 412 285–292.

McGaw DR, Commissiong E, Holder R, Seepaul N, Maxwell A. Drying of hot peppers Proceedings 1st Nordic Drying Conference—NDC'01. Eds. Alves-Filho O, Eikevik TM and Strømmen I. Trondheim, Norway, June 27–29, 2001.

Mínguez-Mosquera MI, Garrido-Fernández J, Pereda-Marín J. Pimiento pimentonero (*Capsicum annuum*). Relación entre los pigmentos carotenoides rojos y amarillos. Grasas y Aceites. 1984 35 4–10.

Mínguez-Mosquera MI, Hornero-Méndez D. Comparative study of the effect of paprika processing on the carotenoids in peppers (*Capsicum annuum*) of the Bola and Agridulce varieties. J Agric Food Chem. 1994 42 1555–1560.

Mínguez-Mosquera MI, Jarén-Galán M, Garrido-Fernández J. Competition between the processes of biosynthesis and degradation of carotenoids during the drying of peppers. J Agric Food Chem. 1994 42 645–648.

Mínguez-Mosquera MI, Jarén-Galán M, Garrido-Fernández J, Hornero-Méndez D. In Carotenoides en el pimentón. Factores responsables de su degradación. Eds. Mínguez-Mosquera MI, Jarén-Galán M, Garrido-Fernández J and Hornero-Méndez D. CSIC, Madrid, 1996.

Mínguez-Mosquera MI, Pérez-Gálvez A. Color quality in paprika oleoresins. J Agric Food Chem. 1998 46 5124–5127.

Mínguez-Mosquera MI, Pérez-Gálvez A, Garrido-Fernández J. Carotenoid content of the varieties Jaranda and Jariza (*Capsicum annuum* L.) and response during the industrial slow drying and grinding steps in paprika processing. J Agric Food Chem. 2000 48 2972–2976.

Okos M, Csorba T, Szabad J. The effect of paprika seed on the stability of the red colour of ground paprika. Acta Aliment. 1990 19 79–86.

Palau E, Torregrosa A. Mechanical harvesting of paprika peppers in Spain. J Agric Eng Res. 1997 66 195–201.

Pérez-Gálvez A, Garrido-Fernández J, Lozano-Ruíz M, Montero-de-Espinosa V, Mínguez-Mosquera MI. Influencia del secado lento a baja temperatura en el contenido carotenoide de dos variedades de pimiento (*Capsicum annuum* L.). Balance biosintético y/o degradativo en función de las condiciones de procesado. Grasas y Aceites. 2001 52 311–316.

Pérez-Gálvez A, Garrido-Fernández J, Mínguez-Mosquera MI. Participation of pepper seed in the stability of paprika carotenoids. J Am Oil Chem Soc. 1999a 76 1449–1454.

Pérez-Gálvez A, Garrido-Fernández J, Mínguez-Mosquera MI. Effect of high-oleic sunflower seed on the carotenoid stability of ground pepper. J Am Oil Chem Soc. 2000 77 79–83.

Pérez-Gálvez A, Garrido-Fernández J, Mínguez-Mosquera MI, Lozano-Ruíz M, Montero-de-Espinosa V. Fatty acid composition of two new pepper varieties (*Capsicum annuum*, L.) Jaranda and Jariza. Effect of drying process and nutritional aspects. J Am Oil Chem Soc. 1999b 76 205–208.

Pérez-Gálvez A, Hornero-Méndez D, Mínguez-Mosquera MI. Changes in the carotenoid metabolism of Capsicum fruits during application of modelized slow drying process for paprika production. J Agric Food Chem. 2004 52 518–522.

Pérez-Gálvez A, Mínguez-Mosquera MI. Degradation of non-esterified and esterified xanthophylls by free radicals. Biochim Biophys Acta. 2002 1569 31–34.

Sugiyama K, Saito M, Hondo T, Senda M. New double-stage separation analysis method: Directly coupled laboratory-scale supercritical fluid extraction–supercritical fluid chromatography, monitored with a multiwavelength ultraviolet detector. J Chromatogr. 1985 332 107–116.

Tijskens LMM, Koek PC, van der Meer MA, Schijvens EPH, De Witte Y. Quality changes in frozen Brussels sprouts during storage. II. Objective quality parameters: Texture, colour, ascorbic acid content and microbiological growth. J Food Technol. 1979 14 301–313.

Uquiche E, Del Valle JM, Ortiz J. Supercritical carbon dioxide extraction of red pepper (*Capsicum annuum* L.) oleoresin. J Food Eng. 2004 65 55–66.

Zhang M, Li C, Ding X, Cao CW. Optimization for preservation of selenium in sweet pepper under low-vacuum dehydration. Drying Technol. 2003 21 569–579.

31
Strawberries and Raspberries

Nirmal Sinha

Section 1: Strawberries

INTRODUCTION

Strawberry is a member of Rosaceae (Rose) family and *Fragaria* (F) genus. The wild European strawberry is mainly from *Fragaria vesca* L., the cultivated varieties are hybrids from *F. chilosensis* and *F. virginiana*. The strawberry, *F. amanassa*, is cultivated worldwide (Menager et al., 2004) but grows well in a cool, moist climate. Fruits grow on stems in groups of three and are hand harvested at full ripeness. Strawberries are not true berries in the "botanical sense" but an aggregate fruit. Roots produce new plants that bear fruits. The flesh of strawberry is an enlarged receptacle and unlike other berries, seeds (achenes) are attached to the skin surface.

STRAWBERRY VARIETIES

The three major classes of strawberries (Anon, 2004a, b, c; Hanson and Hancock, 2000) are (a) June-bearer or short-day, (b) Ever-bearer, and (c) Day-neutral. The differences are in the day length conditions that stimulate flower bud formation. June-bearers initiate flower buds in the fall when days are relatively short, and bear the following spring. Ever-bearers initiate flowers and fruits during the long days of summer. Day-neutrals can initiate flower buds during any day length. Fruit characteristics such as color, flavor, firmness, size, shape, yield, storage, transport, processing, and end use quality are considered while selecting a particular variety. Some of the key characteristics of selected varieties of strawberries are shown below.

(1) Annapolis: Early to midseason variety, large berries, light-red appearance, soft texture, mild flavor, suitable for freezing, and fresh consumption.
(2) Earliglow: Early to midseason variety, excellent flavor and color, fruit size tends to decrease as the season progresses, suitable for retailing and freezing.
(3) Hood: Developed from ORUS 2315 & Puget Beauty, early to midseason, medium to large fruit, round, conic, glossy, excellent internal and external color, medium firm, high solids and high acidity, low drip loss, suitable for retailing and processing.
(4) Totem: The predominant cultivar in Pacific Northwest, United States, developed from Puget Beauty & Northwest, early to midseason, large fruit, fully red internal and external color, high solids, good flavor, color, and firmness, suitable for retailing and processing.
(5) Jewel: Mid- to late season, developed from *Senga Sengana*, large berry, good flavor, bright and glossy red, good for freezing.
(6) Allstar: Mid- to late season, orange-red color, large conical, medium firm, mild sweet flavor, processing quality fair.
(7) Selva: A California variety, ever-bearer or a Day-neutral cultivar released in 1983, large berry, exceptionally firm but flavor is regarded as fair to poor unless fully ripened, skin is bright red and glossy, dessert and processing quality fair.
(8) Chandler: Ever-bearer, large berry, medium firmness, good dessert and processing quality.
(9) Camarosa: Large conical-shaped fruit, good color, excellent taste and firm texture, good for fresh market, slicing/dicing.
(10) Senga Sengana: Developed in 1954, deep red color, medium size, soft texture, round shape, excellent freeze–thaw stability. This variety is grown in Poland where the majority of strawberries are processed and sold as frozen.

PRODUCTION AND CONSUMPTION

Food and Agriculture Organization (FAO) lists 71 countries where strawberries were grown in 2003 (FAO, 2004). Poland had the highest acreage (about 44,000 ha) under strawberry production, followed by the United States (20,000 ha). The total world production of strawberries in 2003 was approximately 3.2 million metric tons, of which the United States produced about 1.0 million metric ton. Spain was the second leading producer. During 1999–2003, the world production of strawberries expanded only slightly (less than 2%); however, in the United States, production increased by about 14%. Outside the United States, Mexico, Turkey, and Russia have also shown good growth in strawberry production (Table 31.1). In Poland, the production has declined perhaps because of lower yield compared to other major producers. However, Poland is still the leading frozen strawberry exporter, followed by Mexico, China, the United States, Spain, and Germany (FAS-USDA, 2004).

Strawberry is a popular dessert fruit (extensively used in ice cream and other desserts). It is also used in

Table 31.1. Strawberry Production in Leading Countries and World Aggregate

	Production (Mt)				
Country	1999	2000	2001	2002	2003
USA	831,250	862,828	749,520	893,670	944,740
Spain	377,527	343,105	326,000	328,700	262,500
Japan	203,100	205,300	208,600	210,500	208,000
Italy	185,852	195,661	184,314	150,890	154,826
South Korea	152,481	180,501	202,966	209,938	209,938
Poland	178,211	171,314	241,118	154,830	131,332
Russia	115,000	128,000	125,000	130,000	145,000
Turkey	129,000	130,000	117,000	120,000	145,000
Mexico	137,736	141,130	130,688	142,245	150,261
Germany	109,194	104,276	110,130	110,000	95,278
World	3,159,062	3,274,341	3,178,752	3,248,840	3,198,689

Source: FAO (2004).

fruit fillings, jellies and jams, energy bars, breakfast cereals, etc. In the United States, it is the leading berry with per capita consumption of about 3.0 kg. About 75% of strawberries produced in the United States are utilized for fresh market, whereas the remaining 25% are frozen or processed into other products (USDA, 2003).

It is important to note that simply washing with water and freezing fruits do not ensure food safety. In the case of strawberries, hepatitis A (a viral liver disease) outbreak occurred in some parts of the United States in 1997, from consumption of sliced, frozen strawberries. However, washing strawberries with chlorinated water has been reported to significantly cut levels of bacteria, hepatitis A virus, and other viruses that indicate possible contamination by animal or human wastes (Williamson, 1998).

PHYSICOCHEMICAL AND NUTRITIONAL QUALITIES

SUGARS, ORGANIC ACIDS, AND FLAVORS

Flavor of strawberry is related to a balance between sugars and organic acids that are naturally present. Macias-Rodriguez et al. (2002) reported that the main soluble carbohydrates in strawberries are glucose, fructose, and sucrose, followed by myoinositol. However, among individual sugars, sucrose content decreases with storage time (Perez and Sanz, 2001; Castro et al., 2002). Among organic acids, citric acid predominates, although malic acid and ascorbic acid are also present. According to Perez and Sanz (2001), malic acid content decreases sharply during strawberry ripening. Table 31.2 gives quality-related data on Camarosa and Selva strawberries (Castro et al., 2002). The sucrose content of Camarosa strawberries is almost twice that of Selva. However, Selva strawberry has twice as much reducing sugars, glucose, and fructose than the Camarosa strawberry. In both varieties, the predominant sugar is glucose. In terms of acidity, Camarosa has more citric and malic acids than Selva. The ascorbic acid content was similar in both varieties. Sugars contribute to flavor and color development during fruit ripening and are the major nonvolatile flavor components in strawberries (Bood and Zabetakis, 2002). In a study of 24 strawberry cultivars (Azodanlou et al., 2003), correlation between overall acceptance and sensory attributes of aroma, sweetness, and juiciness were shown to be 0.94, 0.87, and 0.49, respectively. Although aroma showed highest correlation ($r = 0.94$) with overall acceptance, sweetness ($r = 0.87$) measured as Brix was also a good indicator of quality.

The fruity, green grass, and other flavor notes of strawberries emanate from esters, alcohols, and carbonyl compounds, which are biosynthesized from amino acid metabolism. Important flavor compounds in strawberries are methyl butanoate, ethyl butanoate, methyl hexanoate, cis-3-hexenyl acetate, and linalool. Menager et al. (2004) showed that in an immature strawberry fruit (cv. Cigaline), C_6 compounds, in particular (E)-hexen-2-al, was the main component; furanones and esters were not detected

Table 31.2. Characteristics of Selected Strawberries (without stem) 48 h after Harvest and Freezer Storage (−18°C) for 6 months

Characteristics	Camarosa		Selva	
	Harvest	Freezer Storage	Harvest	Freezer Storage
Total solids (g/100 g)	9.12 ± 0.20	6.68 ± 0.05	12.69 ± 0.09	7.4 ± 0.59
Sucrose (mg/g)	9.09 ± 0.00	0.00 ± 0.00	5.83 ± 0.20	2.78 ± 0.02
Fructose (mg/g)	30.73 ± 0.03	16.75 ± 0.97	55.12 ± 4.25	32.83 ± 0.36
Glucose (mg/g)	35.37 ± 0.5	21.95 ± 0.67	67.58 ± 1.11	33.78 ± 0.36
Citric acid (mg/g)	7.65 ± 0.25	9.26 ± 0.85	5.19 ± 0.37	6.15 ± 0.14
Malic acid (mg/g)	1.17 ± 0.07	1.18 ± 0.02	ND	4.92 ± 0.45
Ascorbic acid (mg/g)	0.36 ± 0.01	0.06 ± 0.01	0.38 ± 0.05	0.17 ± 0.00
Total phenolics (mg/g)	6.35 ± 0.29	8.32 ± 0.05	9.83 ± 0.18	7.74 ± 0.10
Anthocyanins (mg/100 g)	48.19 ± 1.4	61.36 ± 1.66	29.98 ± 0.98	28.99 ± 0.21
Ash (g/100 g)	0.48 ± 0.01	0.52 ± 0.00	0.33 ± 0.01	0.34 ± 0.02
pH	3.66	3.50	3.73	3.52

ND = Not detected.
Source: Castro et al. (2002).

Table 31.3. Flavor Volatiles in Strawberries (Cv. Cigaline)

Compound	Concentration (μg/kg) At Physiological Maturity (42 days After Anthesis) Mean	Range
Furanones		
Mesifurane	1917	1462–2435
Furaneol	6217	5707–6841
Esters		
Methyl butanoate	755	625–887
Methyl 2-methylbutanoate	209	54–321
Butyl acetate	61	41–90
Isoamyl acetate	56	21–87
Methyl hexanoate	105	23–145
Hexyl acetate	48	41–57
(E)-hex-2-enyl acetate	210	96–321
Terpenes		
(E)-furan linalool oxide	30	15–47
(Z)-furan linalool oxide	37	24–52
C_6 *Compounds*		
Hexanal	43	19–71
(E)-hex-2-enal	370	299–462
Hexanol	211	39–410
Lactones		
γ-decalactone	2887	2540–3279
Carbonyls		
Pentan-2-one	704	558–812
Acids		
2-methylpropanoic acid	2130	1698–2562
Butanoic acid	4103	3905–4560
2-methyl butanoic acid	9810	8750–10,450
Hexanoic acid	12,744	11,240–14,414
Alcohols		
Benzyl alcohol	40	19–61

Source: Menager et al. (2004)

until the fruit was about half red in color. As ripening progressed, C_6 compounds decreased but furanones, acids, lactones, and esters increased. At full maturity, the concentration levels of furaneol and mesifurane were more than that of the other flavor compounds (Table 31.3). Organic cultivation practices had no effect on strawberry volatilities (Hakala et al., 2002).

Color, Phenolic Compounds, and Antioxidant Capacity

The bright red color of fresh strawberries is due to anthocyanins, pelargonidin-3-glucoside, pelargonidin 3-rutinoside, and cyaniding-3-glucoside (Garzon and Wrolstad, 2002; Wang et al., 2002; Kosar et al., 2004). Strawberry varieties differ in their anthocyanin content. For example, the total anthocyanin content of Camarosa strawberries was about 60% more than that of the Selva varieties at the harvest. Freezing strawberries caused little loss in color, in fact the anthocyanin content of Camarosa strawberries increased by 27% after 6 months of frozen storage (Table 31.2). However, maintaining the natural color is a challenge in processed strawberries. About 20% of pelargonidin-3-glucoside was reported to be lost at refrigerated storage in 9 days (Zabetakis et al., 2000). Rwabahizi and Wrolstad (1988) reported lower anthocyanin concentration in strawberry juice clarified by ultrafiltration (possibly due to the removal of high molecular weight constituents) than those processed by conventional filtration. Similarly, strawberry concentrates made from juices clarified by conventional filtration were perceived better in appearance than those clarified by ultrafiltration. As would be expected, there were higher anthocyanin losses (from thermal processing) in strawberry concentrates than in juices (Garzon and Wrolstad, 2002).

Cano et al. (1997) investigated high-pressure treatment (50–400 Mpa) combined with heat (20–60°C) to inactivate color and flavor affecting enzymes polyphenol oxidase (PPO) and peroxidase (POD) in strawberry puree. The high-pressure treatments caused 60% and 25% loss of PPO and POD activities, respectively. However, Rwabahizi and Wrolstad (1988) indicated that browning during concentration and storage of strawberry juice was nonenzymatic. The pH optima of strawberry PPO were reported to be 5.5 with catechol and 4.5 with 4-methylcatechol (Wesche-ebeling and Montgomery, 1990). Since the pH of strawberry is generally below 4.5 (Table 31.2), PPO may not have a significant role in browning of strawberry products. Addition of ascorbic acid was shown to have a negative effect on the color of strawberry syrup (Skrede et al., 1992). It is believed that hydrogen peroxide produced as a result of ascorbic acid degradation affects anthocyanins.

Table 31.4 gives data on phenolic compounds (benzoates, ρ-coumaric acid, ellagic acid, anthocyanins, flavonoids, and myricetin) found in strawberries (Kosar et al., 2004). The concentration of anthocyanins, pelargonidin-3-glucoside was about 17 times higher than cyanidin-3-glucoside, and ellagic acid concentration decreased as the strawberries ripened. In a study on the effects of environmental factors and cultural practices on quality, it was shown that strawberries grown on a hill plasticulture had higher levels of soluble solids, total sugar, ascorbic acid, citric acid, flavonoids, and antioxidant capacities than those grown on a flat matted-row. The concentration of ellagic acid glucoside (believed to be better absorbed than ellagic acid) was slightly higher as well (Wang et al., 2002).

Wang and Lin (2000) reported antioxidant capacity, anthocyanins, and total phenolics of strawberries at different stages of fruit maturity, and in strawberry juices. Table 31.5 shows a typical data on these for Allstar strawberry variety. At full ripeness, antioxidant activity was the highest (12.0 μmole TE/g on wet basis). Wu et al. (2004) analyzed both lipids and water-soluble antioxidants and found more water soluble (35.41 μmole TE/g, as is basis; % moisture = 91.1) than lipid soluble (0.36 μmole TE/g) antioxidant activity in market samples of strawberries. The difference between maximum and minimum values of total antioxidant capacity was 12.51 μmole TE/g, which is close to 12.0 μmole TE/g reported by Wang and Lin (2000).

Nutritional Quality

Besides, phenolic constituents and antioxidant properties, which have created interests in plant products, strawberries are a good source of potassium and vitamin C. However, in fresh strawberries, 26–50% ascorbic acid is lost when cut surfaces are exposed to air for 5 min. Even freezing strawberries did not stop loss of ascorbic acid (Table 31.2). A study of two important strawberry cultivars, Camarosa and Selva, showed that the latter variety not only had higher resistance to thawing, but also exhibited higher contents of ascorbic acid, protein, and total phenolics (Castro et al., 2002).

Table 31.6 shows nutritional values per 100 g in strawberries and its products. Fresh and frozen strawberries and strawberry juices are low calorie products. Even the sugar-infused and dried strawberries, which are used as snacks and as ingredients in various foods, on an ounce (28 g) serving size basis, contribute less than 100 calories. These products are also a good source of dietary fiber.

Table 31.4. Phenolic Compounds in Strawberries (mg/100 g Frozen Fruit)

Compounds	Camarosa	Chandler
ρ-OH-benzoic acid	0.15 ± 0.02	0.26 ± 0.02
ρ-coumaric acid	2.07 ± 0.04	1.38 ± 0.02
Ellagic acid	0.36 ± 0.02	0.42 ± 0.01
Cyanidin-3-glucoside	0.72 ± 0.01	0.95 ± 0.01
Pelargonidin-3-glucoside	11.72 ± 0.12	16.24 ± 0.08
Myricetin	0.69 ± 0.01	0.36 ± 0.03

Source: Kosar et al. (2004)

Table 31.5. Oxygen Radical Absorbance Capacity (ORAC), Anthocyanins, and Total Phenolics in Strawberry (Cv. Allstar) at Different Maturity and in Strawberry Juice

Strawberry	ORAC[a] (μmole of TE/g)	Anthocyanins[b] (mg/100 g)	Total Phenolics[c] (mg/100 g)
50% Red berry			
Wet basis	9.7 ± 0.2	16.2 ± 2.1	91.0 ± 1.9
Dry basis	81.8 ± 1.8	143.4 ± 18.6	916.0 ± 11.5
80% Red berry			
Wet basis	10.4 ± 0.3	23.6 ± 2.3	94.0 ± 0.8
Dry basis	95.4 ± 2.7	216.5 ± 21.1	971.0 ± 6.9
Full Red berry			
Wet basis	12.0 ± 0.5	38.9 ± 1.1	96.0 ± 0.9
Dry basis	118.8 ± 4.9	385.1 ± 10.9	946.0 ± 8.9
Strawberry juice			
Wet basis	12.2 ± 0.3	23.3 ± 1.3	95.0 ± 1.1
Dry basis	120.8 ± 2.9	230.7 ± 12.8	943.0 ± 10.4

Source: Wang and Lin (2000)

[a]μmoles of Trolox equivalent.

[b]As milligrams of pelargonidin-3-glucoside.

[c]As gallic acid equivalent.

Table 31.6. Nutritional Values of Strawberries

Nutrients/100 g	Fresh Strawberry[a]	Frozen Strawberry[a]	Strawberry Juice[a]	Infused Dried Strawberry[b]	Dehydrated Strawberry[c]
Calories (Kcal)	32.0	35.0	30.0	325.0	345.0
Calories from fat (Kcal)	2.7	1.0	3.60	9.36	38.34
Total fat (g)	0.30	0.11	0.40	1.04	4.26
Saturated fat (g)	0.015	0.01	0.02	0.10	NA
Polyunsaturated fat (g)	0.155	0.05	0.19	0.50	NA
Monounsaturated fat (g)	0.043	0.01	0.05	0.10	NA
Trans fat (g)	0.0	0.0	0.0	<0.10	0.0
Cholesterol (mg)	0.0	0.0	0.0	<0.10	0.0
Sodium (mg)	1.0	2.0	1.0	25.0	12.0
Potassium (mg)	153.0	148.0	166.0	382.0	1909.0
Total carbohydrate (g)	7.68	9.13	7.00	82.20	80.70
Total fiber (g)	2.0	2.10	0.10	10.20	6.10
Soluble fiber (g)	0.80	0.65	0.03	3.80	NA
Insoluble fiber (g)	1.20	1.45	0.07	6.40	NA
Sugars (g)	4.66	6.96	6.90	70.30	73.90
Protein (g)	0.67	0.43	0.60	3.16	7.02
Calcium (mg)	16.0	16.0	14.00	160.0	161.0
Vitamin C (mg)	58.8	41.20	28.40	95.0	652.10
Vitamin A (IU)	12.0	45.0	20.0	41.0	311.00
Water (g)	90.95	90.0	91.60	12.0	3.0

NA = Not available.

[a]USDA.

[b]Graceland Fruit Inc., Frankfort, Michigan, U.S.

[c]Esha Nutritional database, Salem, Oregon, U.S.

STRAWBERRY PRODUCTS

Developments in storage, transportation, and processing have made possible the year-round availability of strawberries and strawberry products. As has been indicated before, about 25% of strawberries produced are frozen or processed into other products. There are several processed product options including juice, jelly, jam, fruit fillings, variegates, various dried strawberries, etc. Increasingly, consumers are looking for fresh fruit-like qualities (color, flavor, and texture) and nutritional values in processed products. Beginning with steps involved for frozen strawberries, selected processing methods and products are discussed here.

FROZEN STRAWBERRIES

Strawberries are generally frozen either block or individually quick frozen (IQF) after the removal of stems and cap at the point of production. The preparatory steps after harvesting consist of precooling (about 0–2°C) the harvested fruit to maintain color, texture, and flavor and to remove field heat; air classifying to remove leaves, field debris, etc.; removing berry caps, leaves, etc.; quick rinsing or washing, preferably with about 20 ppm chlorinated water; inspecting and grading for size and defects; quick freezing the fruit individually (at about −40°C) in a blast air freeze tunnel; and packaging and storing under frozen temperatures. Quick freezing helps in minimizing large ice crystal formation, which are believed to cause drip losses on thawing. The IQF strawberries are free flowing, and hold their color and shape better. Thus, they are preferred as raw materials for manufacturing value-added products like freeze dried or infused dried strawberries.

FROZEN SUGAR PACK STRAWBERRIES

In this case, sugar is added to the strawberries after removing stems and caps, and the product is stored frozen at below −18°C. For example, 4 + 1 (80% fruit + 20% sugar) and 7 + 3 (70% fruit + 30% sugar) pack strawberries in a 30 lb pail will contain 24 lb strawberries and 6 lb sugar, and 21 lb strawberries and 9 lb sugar, respectively. These products are used in many applications. However, before using in dairy products such as yogurt and ice cream, they should be pasteurized.

STRAWBERRY PUREE

Purees can utilize fruits, which are not sold as fresh or frozen. They generally form raw material for making fruit fillings, variegates, juices, jams, etc. For making strawberry puree, it is not critical to remove berry cap. The preparatory steps of precooling, washing, and grading are essentially the same as described for frozen strawberries. Subsequently, the berries are cut/chopped and passed through different dimension sieves, depending on whether the product contains seeds or not. For example, to make strawberry puree with and without seeds, the diameter of sieve opening would be, 1.33 ± 0.19 mm and 0.76 ± 0.076 mm, respectively. Then the puree is pasteurized by heating at 88°C for about 2 min and cooled to about 15°C. Following processing, the product is quality analyzed, filled in containers, and stored frozen. The single-pack strawberry puree will have almost the same Brix as that of the starting raw fruit. However, concentrated strawberry purees of about 28 Brix are also available. The puree may be treated with enzymes and filtered before concentration to provide better quality puree for use in jams, juices, etc. Sweeteners such as sucrose can also be added to the puree to adjust Brix as per the end use.

STRAWBERRY JUICE AND CONCENTRATE

Garzon and Wrolstad (2002) described a process, which is similar to commercial production, to manufacture strawberry juice and concentrate. The processing steps are as below (it may be noted that a Bucher press or a centrifuge can also be used for separation of strawberry juice):

(1) Thaw IQF or block frozen strawberries overnight at room temperature.
(2) Crush with a hammer mill (Model D Commuting Machine; W.J. Fitzpatrick Co., Chicago, IL, USA) equipped with a circular pore mesh, 1.27 cm in diameter, at a speed of 182 rpm.
(3) Depectinize in a steam-jacketed kettle at 50°C for 2 h by adding Pectinase, Rapidase® Super BE @ 3 ml/kg (Gist-Brocades Laboratories, Charlotte, NC, USA). Conduct Alcohol test for pectin (1:1 juice–isopropanol + 1% HCl; incubate further until the test for pectin is negative).
(4) Press, after adding 1% rice hull as press aid, at 300 kPa for 30 min using a Willmes press bag, Type 60 (Moffet Co., San Jose, CA, USA).

(5) Filter at 27.6 kPa with 2% diatomaceous earth as filtering aid and using a multipad filtration unit (Strassburger KG, Westhofen, Worms, Germany); the unit equipped with a pad filter SWK supra 2600 (Scott Laboratories, Inc., San Rafael, CA, USA).
(6) Pasteurize at 88°C for 1 min using a tubular heater with screws, Wingear type (Model 200WU; Winsmith Co., Springville, NY, USA).
(7) Concentrate to 65 Brix using Centritherm evaporator, type CT-1B (Alfa Laval, Lund, Sweden). Two passes are needed to achieve the final concentration, and the juice can reach a temperature as high as 80°C.

Strawberry Jelly, Preserve, and Jam

Among various fruit jelly and jams, strawberry jelly (45 parts by weight of fruit juice ingredients to 55 parts of sweeteners, see Code of Federal Regulations 2003: CFR 21, 150.140) and preserves and jams (47 parts by weight of fruit ingredient to each 55 parts by weight of sweeteners, see CFR 21, 150.160) are on the top of the list because of their unique color, flavor, and taste. The soluble solid contents of finished jelly and jam are not less than 65%.

Stabilized Frozen Strawberries

Processed frozen products such as stabilized frozen (strawberry fruit pieces are combined with a syrup matrix containing sweeteners, pectin, starch, and carrageenan or other gums, and heat processed) or infused frozen strawberries for use in ice cream, sorbets, yogurts, and bakery products are made by infusing and pasteurizing whole or sliced strawberries (Sinha, 1998) in sugar syrup or other types of sweeteners. These products typically are about 35 Brix, so that they do not become hard at freezer temperatures. The products are pasteurized and can be added directly to the formulations.

Infused Dried Strawberries

As the name implies, infused dried strawberries are produced by infusing strawberries to a range of sweetness level, prior to drying. These products have found better acceptance in the ingredient market than the traditionally dried strawberries. It has been reported that osmotic treatment of strawberries at atmospheric pressure had a positive effect on flavor (Escriche et al., 2000).

Freeze Dried Strawberries

Freeze dried strawberries have found great success as fruit ingredients in ready-to-eat cereals. Freeze drying retains the typical bright red color of strawberries better than other drying methods. Besides, the freeze dried strawberries have crisp texture and their low bulk density (approximately about 0.1 g/cc) is closer to ready-to-eat cereals. Both whole and sliced strawberries can be processed as freeze dried. However, this product requires special laminated packaging to maintain the crisp texture.

REFERENCES

Anon. 2004a. Strawberry Fact Sheet. California Strawberry Commission (http://www.calstrawberry.com).

Anon. 2004b. Strawberry Varieties. Oregon Strawberry Commission (http://www.oregon-strawberries.org).

Anon. 2004c. Strawberry Variety Information. Shasta Nursery, Inc. (http://www.rootstock.com/variety.html).

Azodanlou R, Darbellay C, Luisier J, Villettaz J, Amado R. 2003. Quality assessment of strawberries (Fragaria Species). J Agric Food Chem 51:715–21.

Bood KC, Zabetakis I. 2002. The biosynthesis of strawberry flavor (II): Biosynthetic and molecular biology studies. J Food Sci 67:2–8.

Cano MP, Hernandez A, De Ancos B. 1997. High pressure and temperature effects on enzyme inactivation in strawberry and orange products. J Food Sci 62:85–8.

Castro I, Goncalves O, Teixeira, JA, Vicente, AA. 2002. Comparative study of Selva and Camarosa strawberries for the commercial market. J Food Sci 67:2132–7.

Code of Federal Regulations (CFR). 2003. CFR 21. (http://www.cfsan.fda.gov).

Escriche I, Chiralt A, Moreno J, Serra JA. 2000. Influence of blanching-osmotic dehydration treatments on volatile fractions of strawberries. J Food Sci 65:1107–11.

FAO. 2004. FAO Crop Database. Food and Agriculture Organization (http://www.faostat.fao.org).

FAS-USDA. 2004. Foreign Agriculture Service, United State Department of Agriculture (http://www.fas.usda.gov).

Garzon GA, Wrolstad RE. 2002. Comparison of the stability of pelargonidin-based anthocyanins in strawberry juice and concentrate. J Food Sci 67:1288–99.

Hakala MA, Lapvetelainen AT, Kallio HP. 2002. Volatile compounds of selected strawberry varieties

analyzed by purge-and-trap headspace GC-MS. J Agric Food Chem 50:1133–42.
Hanson E, Hancock J. 2000. Strawberry varieties for Michigan. Michigan State University Extension (http://www.msue.msu.edu/vanburen/e-839.htm).
Kosar M, Kafkas E, Paydas S, Baser K. 2004. Phenolic composition of strawberry genotypes at different maturation stages. J Agric Food Chem 52:1586–9.
Macias-Rodriguez L, Quero E, Lopez MG. 2002. Carbohydrate differences in strawberry crowns and fruit (Fragaria x ananassa) during plant development. J Agric Food Chem 50:3317–21.
Menager I, Jost M, Aubert C. 2004. Changes in physicochemical characteristics and volatile constituents of strawberry (Cv. Cigaline) during maturation. J Agric Food Chem 52:1248–54.
Perez AG, Sanz, C. 2001. Effect of high-oxygen and high-carbon-dioxide atmospheres on strawberry flavor and other quality traits. J Agric Food Chem 49:2370–5.
Rwabahizi S, Wrolstad RE. 1988. Effects of mold contamination and ultrafiltration on the color stability of strawberry juice and concentrate. J Food Sci 53:857–61, 872.
Sinha NK. 1998. Infused-dried and processed frozen fruits as food ingredients. Cereal Foods World 43:699–701.
Skrede G, Wrolstad RE, Lea P, Enersen G. 1992. Color stability of strawberry and blackcurrant syrup. J Food Sci 57:172–7.
USDA. 2003. Fruit and Tree Nuts Outlook. United States Department of Agriculture, Economic Research Service (http://www.ers.usda.gov/publication/FTS/yearbook03/fts2003.pdf)
Wang SY, Lin HS. 2000. Antioxidant activity in fruits and leaves of blackberry, raspberry, and strawberry varies with cultivar and development stage. J Agric Food Chem 48:140–6.
Wang SY, Zheng W, Galletta GJ. 2002. Cultural system affects fruit quality and antioxidant capacity in strawberries. J Agric Food Chem 50:6534–42.
Wesche-ebeling P, Montgomery MW, 1990. Strawberry polyphenoloxidase: Extraction and partial characterization. J Food Sci 55:1320–24, 1351.
Williamson D. 1998. Chlorinated Water Cuts Strawberry Contamination. http://www.clo2.com/reading/waternews/strawb.html
Wu X, Beecher GR, Holden JM, Haytowitz DB, Gebhardt SE, Prior RL. 2004. Lipophilic and hydrophilic antioxidant capacities of common foods in the United States. J Agric Food Chem 52: 4026–37.
Zabetakis I, Leclerc D, Kajda P. 2000. The effect of high hydrostatic pressure on the strawberry anthocyanins. J Agric Food Chem 48:2749–54.

Section 2: Raspberries

INTRODUCTION

Raspberries (also termed "Brambles") belong to the genus *Rubus* and family Rosaceae (rose). Cultivated raspberries have been derived from the wild red raspberries (*Rubus ideaus*) and black raspberries (*Rubus occidentalis*). These soft and delicate fruits with small seeds and hollow core are made of aggregates of hairy drupelets, which adhere to one another. The origin of red raspberry dates back to 4th century A.D. at Mt. Ida, in the Caucasus Mountains of Asia Minor. The British are credited with popularizing and improving red raspberry cultivation throughout the Middle Ages. The black raspberry is indigenous to North America (Anon, 2004a).

Raspberry plants are a biennial, summer or autumn bearing, and grow on leafy canes (thus called caneberries) in temperate regions of the world. The root system of this plant is perennial and capable of living for several years where there is a good drainage system. All cultivars, especially new plants, are susceptible to root rots. Red and black raspberries dominate commercial production; however, purple and yellow raspberries are also grown for the fresh market.

PRODUCTION AND CONSUMPTION

In recent years, studies linking antioxidative properties having beneficial effect against degenerative diseases have fueled demand for raspberries and other fruits. Raspberry production in the top 10 leading countries of the world is given in Table 31.7. During the last 5 years, the world production has increased by about 9%. Russia leads in raspberry production. In Serbia, where fruits are grown in valleys 400–800 m above sea level, raspberry production has increased by almost 60% during this period. The production has also doubled in Spain. In the United States, raspberry production is concentrated in the states of Oregon and Washington. In Washington State, the leading red raspberry variety is "Meeker," which is a late season, summer fruiting raspberry. In Oregon, "Willamette" an early fruiting, medium-small fruit, typically sold for processing is popular. It is also one of the few cultivars resistant to raspberry bushy dwarf virus infection that causes lower yield and crumbly fruits.

More than 80% of raspberries produced in Oregon and Washington are red raspberries and the remaining black raspberries, Oregon being a leader in

Table 31.7. Raspberries Production in Leading Countries and World Aggregate

Country	Production (Mt)				
	1999	2000	2001	2002	2003
1. Russian Federation	100,000	102,000	90,000	100,000	108,000
2. Serbia and Montenegro	60,000	56,059	77,781	94,366	94,400
3. USA	49,351	51,256	54,885	52,889	53,000
4. Poland	43,195	39,727	44,818	45,026	45,000
5. Germany	35,500	33,700	29,200	29,000	29,000
6. Canada	15,650	16,247	11,658	14,291	13,900
7. Hungry	22,277	19,804	13,306	10,000	10,000
8. France	7020	8743	8549	7999	8000
9. Romania	4287	2390	3990	4000	4500
10. Spain	2000	2500	3200	4500	4500
11. World	381,572	378,787	381,787	407,155	415,836

Source: FAO (2004).

Table 31.8. Red Raspberry and Black Raspberry Production in Oregon and Washington States of United States

	1998	1999	2000	2001	2002
			(1000 lb)		
Red raspberry					
Oregon					
Production	14,200	13,650	14,500	15,900	10,300
Utilization					
Fresh	800	700	1300	1300	1,700
Processed	13,400	12,950	13,200	14,600	8,600
Washington					
Production	60,300	69,350	71,250	75,050	74,100
Utilization					
Fresh	2300	4000	4000	3550	5200
Processed	58,000	65,350	67,250	71,500	68,900
Black raspberry					
Oregon					
Production	2600	2900	3800	3810	2920
Utilization					
Fresh	20	10	30	10	20
Processed	2580	2890	3800	3800	2900

Source: USDA (2003).

production of the latter. In the United States, about 85% of the raspberries produced are processed (Table 31.8). Generally, machine-harvested fruit is used for processing, and handpicked fruits are sold for fresh consumption or to premium quality market. Machine harvesting has become critical to enhance production; however, it requires coordination and synergies of efforts among fruit growers, plant breeders, and other researchers to develop new varieties and use of acceptable insect control practices suitable for machine harvesting of raspberries. Often, the machine-harvested fruits are used for fruit juice and puree/pulp market. There is a growing demand for superior quality whole IQF and frozen fruits in terms of color, shape, size, and freedom from diseases and insects for use as dessert fruit and other applications.

PHYSICOCHEMICAL AND NUTRITIONAL QUALITIES

Raspberries are slightly tart but juicy fruits. Bushway et al. (1992) studied physical, chemical, and sensory characteristics of five red raspberry cultivars (Table 31.9). They indicated that titratable acidity, sucrose, and total sugar could serve as a predictor of flavor in frozen raspberries. The "Newburg" cultivar with highest concentration of sucrose and total sugars was most preferred by panelists, and the traditional red color in frozen raspberry cultivars was liked more than the deeper purple color of cultivar "Boyne."

Typical physicochemical properties of selected raspberry cultivars grown are given in Table 31.10 (Ancos et al., 2000a, b). Cultivars "Autumn Bliss" and "Heritage" are early season (harvested in May), "Zeva" and "Rubi," late season (harvested in autumn, October) berries. Generally, the late season cultivars have higher °Brix, anthocyanins, total phenolics, and ellagic acid content. The main color pigments found in the four raspberry cultivars were cyaniding-based anthocyanins (sophoroside, glucoside, glucorutinoside, and rutinoside) and pelargonidin derivatives (sophroside and glucoside). Late season cultivars showed greater anthocyanin content than early season fruits.

COLOR PIGMENTS OF RASPBERRIES

Boyles and Wrolstad (1993) indicated average Brix of red raspberry cultivars "Willamette" and "Meeker" to be approximately 10.0; their average anthocyanin contents (mg/l cyanidin-3-glucoside) were 620 and 320, respectively. The high pigment concentration of "Willamette" is a desirable characteristic, which increased with ripeness. Approximately, 90–97% of anthocyanins were composed of cyanidin and 3–10% pelargonidin. Cyanidin-3-sophoroside (cyd-3-sop) was the pigment with highest concentration in red raspberry varieties "Willamette" and "Meeker." However, Polish cultivars "Veten" and "Norna" showed much lower concentration of cyd-3-sop.

Processing raspberries can change the quantitative distribution of pigments through partial hydrolysis of glycosidic substituents and/or anthocyanin polymerization. Low total anthocyanin and elevated levels of cyanidin-3-glucoside indicate degradation due to processing and storage. Black raspberries can be distinguished from red raspberries by xylose containing pigments, cyanidin-3-sambubioside, and cyanidin-3-xylosylrutinoside (Torre and Barritt, 1977). Recent studies (Wada and Ou, 2002) report 0.65 mg/g and 5.89 mg/g anthocyanins in red and black raspberries, respectively. These authors indicated that red raspberries anthocyanins were composed of cyanidin 3,5-diglucoside (89.25%) and cyanidin 3-glucoside (10.75%); black raspberries were made of cyanidin3-(6′-p-coumaryl) sambubioside (22%) and cyanidin 3-(6′-p-coumaryl) glucoside (77.0%).

FLAVOR OF RASPBERRIES

Volatile components contributing to the fresh raspberry aroma are α-pinene, citral, β-pinene, phellanderrene, linalool, α-ionone, carryophyllene, and β-ionene. Freezing raspberries for 12 months had

Table 31.9. Sugar and Color Profiles of Selected Red Raspberry Cultivars from Maine in United States

	Cultivars				
Characteristics	Boyne	Festival	Latham	Newburg	Taylor
% Total sugar	6.14	4.84	6.10	7.68	6.30
% Fructose	2.28	2.09	3.71	3.01	2.67
% Glucose	1.93	1.88	2.39	2.43	2.28
% Sucrose	1.93	0.87	0.00	2.24	1.35
% Soluble solids	9.6	10.0	10.6	12.3	11.1
% Acidity (as citric)	1.46	1.85	1.43	1.81	2.09
pH	3.04	3.02	3.05	3.13	2.98
Hunter color					
L	18.25	17.58	21.17	21.27	20.59
a/b	17.97	22.93	22.40	22.46	22.53

Source: Bushway et al. (1992).

Table 31.10. Physicochemical Characteristics of Selected Raspberry Cultivars

Characteristics	Cultivars			
	Autumn Bliss	Heritage	Zeva	Rubi
Brix	9.26	9.50	10.54	10.00
Acidity (% citric)	1.67	1.76	1.75	2.32
pH	3.65	3.87	2.88	2.65
Total solids (%)	15.23	14.69	16.33	17.98
Moisture (%)	84.77	85.31	83.67	82.02
Total anthocyanin (mg/100 g)	31.13	37.04	116.27	96.08
Total phenolics (mg gallic acid/100 g)	121.4	113.7	177.6	155.6
Ellagic acid (mg/100 g)	20.8	21.7	24.4	23.4
Vitamin C (mg/100 g)	30.2	22.0	29.6	31.0
Color				
L	25.89	25.80	18.29	21.26
a	35.03	34.98	33.03	35.10
b	19.05	18.34	17.78	18.63
11. *Flavor volatiles (%)*				
Caryophyllene	37.7	15.0	25.9	20.8
α-ionone	33.5	32.8	43.1	23.9
β-ionone	20.7	19.3	13.3	17.3
α-pinene	2.9	13.2	2.9	17.0
Citral	0.4	0.7	0.7	1.6
β-pinene	0.5	1.2	0.5	1.7
phellandrene	2.4	11.4	12.9	10.4
Linalool	1.8	6.3	0.4	7.1

Source: Ancos et al. (2000a, b).

little affect on flavor volatiles; however, color pigment cyanidin 3-glucoside decreased by about 26.0% in late season raspberry cultivar "Zeva" (Ancos et al., 2000b).

Phenolic Components and Antioxidant Capacity

Liu et al. (2002) reported that the color of raspberry juice correlated with the total phenolic, flavonoid, and anthocyanin content of raspberry. The "Heritage" variety contained the highest total phenolic content (512.7 ± 4.7 mg/100 g), followed by "Kiwigold" (451.1 ± 4.5 mg/100 g), "Goldie" (427.5 ± 7.5 mg/100 g), and "Anne" (359.2 ± 3.4 mg/100 g). Similarly, the "Heritage" had the highest total flavonoids (103 ± 2.0 mg/100 g), followed by "Kiwigold" (87.3 ± 1.8 mg/100 g), "Goldie" (84.2 ± 1.8 mg/100 g), and "Anne" (63.5 ± 0.7 mg/100 g). Raspberry extracts equivalent to 50 mg of "Goldie," "Heritage," and "Kiwigold" fruits inhibited the proliferation of HepG2 human liver cancer cells by 89.4% ± 0.1%, 88% ± 0.2%, and 87.6% ± 1.0%, respectively. Variety "Anne" had the lowest antiproliferation activity of the varieties measured (70.3% ± 1.2%). The antioxidant activity of these raspberry varieties showed significant positive correlation ($P < 0.05$) with the total phenolics and flavonoids found in raspberry. However, there was little significant correlation ($P > 0.05$) between antiproliferative activity and the total phenolics/flavonoids, suggesting that other phytochemicals (such as anthocyanins and ellagic acid) may have a role in the antiproliferative activity of raspberries.

Wang and Lin (2000) measured oxygen radical absorbance capacity (ORAC) of juices from black raspberry and red raspberry at different maturities (Table 31.11). As expected, juice made from ripe fruits had higher antioxidant activity than that from green or pink berries; and black raspberry juice had higher antioxidant activity than red raspberry juice. Moyer et al. (2002) reported very high anthocyanin content of 627, 607, and 464 mg/100 g and ORAC values of 104.6, 146.0, 110.3 μmol TE/g in black raspberries cultivars "Munger," "Jewel," and "Earlysweet," respectively.

Table 31.11. Average Antioxidant Activity (ORAC), Anthocyanin Content, and Total Phenolic Content in Raspberry Juice of Different Maturity Raspberries (on a Fresh Weight Basis)

Raspberry	ORAC (μmol of TE/g)	Anthocyanin (mg/100 g)	Total Phenolics (mg/100 g)
Red raspberry			
Green	16.5 ± 0.8	1.0 ± 0.2	181 ± 5.0
Pink	10.9 ± 0.6	7.2 ± 1.2	99 ± 1.5
Ripe	18.2 ± 0.8	68.0 ± 3.0	234 ± 5.1
Black Raspberry			
Green	33.7 ± 4.0	1.7 ± 0.6	338 ± 7.1
Pink	16.1 ± 0.6	22.8 ± 1.4	190 ± 3.5
Ripe	28.2 ± 1.4	197.2 ± 8.5	267 ± 4.3

Source: Wang and Lin (2000).

Ellagic Acid

Phenolic compound ellagic acid, a dimeric derivative of gallic acid, is suggested as an anticarcinogenic/antimutagenic compound. It is present in plants in the form of hydrolyzable tannins called ellagitannins. Ellagitannins are esters of glucose with hexahydroxydiphenic acid and when hydrolyzed ellagic acid and dilactone of hexahydroxydiphenic acid are produced. Initial studies of ellagic acid in rodents at Ohio State University have shown significant prevention and reduction of certain cancers (Funt et al., 2004). When added to cultured cancer cells *in vitro*, ellagic acid is shown to stop cell division and cancer cells eventually die by apoptosis, while sparing normal cells (Anon, 2004b). However, currently, the role of ellagic acid on human cancer patients is inconclusive. A study at Ohio State University (Anon, 2004c) indicated "Heritage" red raspberry with highest amount of ellagic acid in the pulp among several raspberry cultivars. Raspberry seeds generally had higher ellagic acid content than the pulp (Table 31.12).

Wada and Ou (2002) showed total ellagic acid ranging from 47 mg/g in red raspberries to 90 mg/g in black raspberries grown near Salem, Oregon. They indicated that free ellagic acid level was 40–50% of the total ellagic present. Rommel and Wrolstad (1993) reported an average concentration of ellagic acid and its derivatives in experimental and commercial raspberry juice samples as 30 ppm (0.003%) and 52 ppm (0.052%), respectively. Raspberry juice produced by diffusion extraction (where the berries were exposed to high temperature [63°C] for several hours, thus releasing ellagic acid from the cell walls) contained about twice as much ellagic acid as juice made by high-speed centrifugation. The ellagic acid derivatives

Table 31.12. Ellagic Acid Content of Raspberries (μg/g Dry Weight)

	1997		1998	
	Pulp	Seed	Pulp	Seed
Red cultivars				
Caroline	36.0	173.4	52.5	799.2
Autumn Bliss	22.3	98.9	42.0	263.6
Heritage	40.5	105.8	39.2	467.2
Ruby	39.7	176.4	10.0	85.6
Yellow cultivar				
Anne	11.1	177.7	7.8	60.5
Black cultivar				
Jewel	17.0	240.0		

Source: Anon (2004c).

(4-arabinosylellagic acid, 4-acetylxylosylellagic acid, and 4-acetylarabinosylellagic acid) with the exception of ellagic acid itself remained quite stable with processing and during 6 months of raspberry jam storage. The initial free ellagic acid content of 10 mg/kg increased twofold with processing into jam, and it continued increasing up to 35 mg/kg after 1 month of storage. Thereafter, a slight decrease was observed until 6 months of storage. The increase in ellagic acid was possibly due to the release of ellagic acid from ellagitannins with heat treatment (Zafrilla et al., 2001).

Nutritional Quality

Besides having potential antioxidant and anticancer containing components, raspberries and its products are good sources of dietary fiber and potassium. Typical nutritional profile of raspberries and its products according to National Educational and Labeling Act (NELA) is given in Table 31.13.

RASPBERRY PRODUCTS

Availability of raspberries in IQF, bulk frozen, puree, jam, jelly, juice, concentrated syrups, and dried form enable consumers to enjoy this fruit year round. The postharvest handling, storage, and processing of this fruit generally follow procedures similar to strawberries. Frozen seedless raspberry puree forms a base for many products including stabilized processed products for use in ice cream and sorbet. The raspberry puree, with or without seeds, can be made using sieves of about 1.5–3.2 mm, and 1.1 mm, diameter openings, respectively. Pasteurization and other steps are similar to strawberry puree processing. Single strength seedless raspberry puree is about 10 Brix. However, concentrated raspberry puree of about 28 Brix is also commercially produced for various applications.

Raspberry juice is blended in many beverages where it lends its characteristic flavor, color, and a balance of sweetness and tartness. Essentially similar steps (Boyles and Wrolstad, 1993), as in the case

Table 31.13. Nutritional Values of Raspberries

Nutrients/100 g	Red Raspberry	Black Raspberry	Fresh Raspberry Juice	Infused Dried Red Raspberries[a]	Dehydrated Raspberries
Calories (Kcal)	49.00	73.00	30.0	309	354
Calories from fat (Kcal)	5.00	13.00	0.0	14.0	35.73
Total fat (g)	0.55	1.42	0.00	1.52	3.97
Saturated fat (g)	0.02	NA	0.00	0.2	NA
Polyunsaturated fat (g)	0.31	NA	0.00	0.4	NA
Monounsaturated fat (g)	0.02	NA	0.00	1.0	NA
Trans fat (g)	0.0	0.00	0.00	<0.10	0.00
Cholesterol (mg)	0.0	<0.10	0.00	<0.10	0.00
Sodium (mg)	0.0	0.00	3.00	17.0	0.00
Potassium (mg)	152.00	199.20	153.00	162.0	1097
Total carbohydrate (g)	11.60	15.67	7.14	84.0	83.50
Total fiber (g)	6.80	NA	0.00	16.2	21.70
Soluble fiber (g)	1.22	NA	0.00	2.1	NA
Insoluble fiber (g)	5.58	NA	0.00	14.1	NA
Sugars (g)	4.80	NA	7.14	62.0	61.80
Protein (g)	0.91	1.49	0.31	4.0	6.57
Calcium (mg)	22.0	29.85	18.00	72.0	159.0
Iron (mg)	0.57	0.90	2.60	1.82	4.12
Vitamin C (mg)	25.00	17.91	25.00	98.0	180.50
Vitamin A (IU)	130.0	NA	66.70	76.0	939.0
Water (g)	86.60	81.02	89.40	8.3	3.00

NA: Not available.

Source: Esha Nutritional database, Salem, Oregon, U.S.

[a]Graceland Fruit Inc., Frankfort, Michigan, U.S.

of strawberry juice, are followed to make raspberry juice and concentrates. Raspberry juice concentrates of about 65 Brix, with an acidity of as high as 12%, are commercially available.

Raspberry drying process requires proper handling and use of smaller size IQF fruits because the large fruits have a tendency to breakdown during the process. The drying techniques are essentially similar to that of strawberries and other fruits presented elsewhere in this book.

REFERENCES

Ancos B, Gonzalez EM, Cano MP. 2000a. Ellagic acid, Vitamin C, and total phenolic contents and radical scavenging capacity affected by freezing and frozen storage in raspberry fruit. J Agric Food Chem 48:4565–70.

Ancos B, Ibanez E, Reglero G, Cano MP. 2000b. Frozen storage effects on anthocyanins and volatile compounds of raspberry fruit. J Agric Food Chem 48:873–9.

Anon. 2004a. Blackberries and Raspberries—Rubus spp (http://www.uga.edu/fruit/rubus.htm).

Anon. 2004b. Ellagic Research (http://www.ellagic-research.org/clinical_studies.htm).

Anon. 2004c. Evaluation of Ellagic acid Content of Ohio Berries—Final Report (http://www.ag.ohio-state.edu/~sfgnet/eacid_final.html).

Boyles MJ, Wrolstad RE. 1993. Anthocyanin composition of red raspberry juice: Influences of cultivar, processing, and environmental factors. J Food Sci 58:1135–41.

Bushway AA, Bushway RH, True RH, Work TM, Bergeron D, Handley DT, Perkins LB. 1992. Comparison of the physical, chemical and sensory characteristics of five raspberry cultivars evaluated fresh and frozen. Fruit Variety J 46:229–34.

FAO. 2004. FAO Crop Database. Food and Agriculture Organization (http://www.faostat.fao.org).

Funt RC, Bash WD, Schwartz SJ, Stoner GD. 2004. Research and Education on Phytochemicals and Neutraceutical Foods. Ohio State University (http://www.ag.ohio-state.edu/~sfgnet/ellagic.htm).

Liu M, Li XQ, Webber C, Lee CY, Brown J, Liu RH. 2002. Antioxidant and antiproliferative activities of raspberries. J Agric Food Chem 50:2926–30.

Moyer RA, Hummer KE, Finn CE, Frei B, Wrolstad RE. 2002. Anthocyanins, phenolics, and antioxidant capacity in diverse small fruits: Vaccinium, rubus, and ribes. J Agric Food Chem 50:519–25.

Rommel A, Wrolstad RE. 1993. Ellagic acid content of red raspberry juice as influenced by cultivar, processing, and environmental factors. J Agric Food Chem 41:1951–60.

Torre LC, Barritt BH. 1977. Quantitative evaluation of rubus fruit anthocyanin pigments. J Food Sci 42:488–90.

USDA. 2003. Fruit and Tree Nuts Outlook. United States Department of Agriculture, Economic Research Service (http://www.ers.usda.gov/publication/FTS/yearbook 03/fts2003.pdf).

Wada L, Ou B. 2002. Antioxidant activity and phenolic content of oregon caneberries. J Agric Food Chem 50:3495–500.

Wang SY, Lin H-S. 2000. Antioxidant activity in fruits and leaves of blackberry, raspberry and strawberry varies with cultivar and developmental stage. J Agric Food Chem 48:140–46.

Zafrilla P, Ferreres F, Tomas-Barberan FA. 2001. Effect of processing and storage on the antioxidant ellagic acid derivatives and flavonoids of red raspberry (Rubus idaeus) jams. J Agric Food Chem 49: 3651–55.

32
Tropical Fruits: Guava, Lychee, and Papaya

Jiwan S. Sidhu

INTRODUCTION

Regular and habitual consumption of fruits and vegetables has been strongly associated with reduced risk of many types of cancers such as cancers of the mouth and pharynx, lung, esophagus, stomach, and colon (Slattery et al., 2004). A moderately strong association has also been suggested for the cancers of the breast, pancreas, and bladder (Steinmatz and Potter, 1991). A number of observational studies have also suggested the role of certain dietary nutrients, such as potassium, folic acid, and antioxidants, being abundant in fruits and vegetables to lower the incidence and mortality from cardiovascular diseases (Ness and Powles, 1997; Tribble, 1999; Djousse et al., 2004; Cesari et al., 2004). Increased consumption of fruits and vegetables has also been shown to provide protection against various age-related diseases such as cataract and macular degeneration (Ames et al., 1993).

Exactly what nutrients are responsible for this protection is not known, but certain antioxidant vitamins including vitamins C and E, or β-carotenes are often assumed to be responsible for this benefit (Garcia-Alonso et al., 2004; Einbond et al., 2004; Tylavasky et al., 2004). In addition to these antioxidant vitamins, the role of other phytochemicals present in fruits and vegetables may be equally important (Vinson et al., 2001; Hollman, 2001). As the free oxygen radicals may be involved in several of these pathological conditions, the antioxidant vitamins as well as various phytochemicals present in fruits and vegetables may provide the needed protection against these age-related diseases (Knekt et al., 2002; Hertog et al., 1992; Prior and Cao, 2000). Although our nutrition research knowledge regarding the exact requirements of various antioxidant vitamins and phytochemicals for obtaining maximum health benefits is still rudimentary, it has attracted the attention of agricultural scientists to develop plant genotypes with improved levels of these nutrients in fruits and vegetables. Considering the importance of fruits and vegetables in our diet, three important tropical fruits, namely guava, lychee, and papaya, will be discussed in this chapter.

Section 1: Guava

INTRODUCTION

The common guava (*Psidium guajava* L.) is the most important member of the Myrtaceae family. The genus *Psidium* includes five species, namely *Psidium guianense*, *P.cattleianum*, *P. chinensis*, *P. friedrichsthalianum*, and *P.guajava*. Most of the cultivated guava belongs to the guajava species. It is reported to have originated in Central America but now has thoroughly naturalized throughout the tropics and subtropics (Sidgley and Gardner, 1989). By virtue of its commercial and nutritional values, guava is considered a common man's fruit and can be rightly termed as the "apple of the tropics."

Guava is of commercial importance in about 58 countries, but the leading countries are India, Brazil, Egypt, South Africa, Colombia, USA, Puerto Rico, Jamaica, Taiwan, Sudan, Kenya, Israel, Philippines, Pakistan, Malaysia, Australia, and West Indies (Eipeson and Bhowmik, 1992). Exact figures about guava fruit production are difficult to find, as these are not easily available. Guava fruit production from some of the leading producers is shown in Table 32.1. Among the tropical fruits, guava is becoming popular in the United States and consequently, a large quantity of guava products are imported (Table 32.2).

Table 32.1. Leading Guava Producing Countries of the World

Country	Production, Metric Tons
India	165,000
Mexico	127,000
Pakistan	105,000
Thailand	100,000
Indonesia	56,000
Malaysia	30,000
Colombia	29,000
Egypt	28,000
Brazil	27,000

Source: Anon, 2003.

The guava can be a shrub or tree, commonly multitrunked with wide-spreading branches that reach heights of up to 10 m. Plants are propagated mostly from seeds, but for uniform quality and production, guava should be propagated through root-cutting, grafting, budding, or layering. Seeds normally take 2–3 weeks to germinate under favorable conditions. When seedlings are 5–7 cm tall, these are transferred into another nursery giving larger spacing and allowed to grow for about a year before finally being transplanted into the orchard. Spacing between plants is about 6–8 m depending on the cultivar. Guava trees start bearing fruits from the fourth year onward. Interestingly, the guava tree tends to flower and ripen fruits indiscriminately throughout the year. Guava fruit is a berry, and the fruit consists of a fleshy pericarp and seed cavity.

The fruit generally takes about 17–20 weeks from fruit set to reach full growth, but the fruit is harvested about 2–3 weeks before attaining full maturity, as it continues to undergo changes associated with ripening (Paull and Goo, 1983). The guava tree can easily be grown with less irrigation water in a wide range of soil types; the well-drained light soils are, however, the most suited for guava cultivation. Guava being a hardy plant, requires less irrigation water and is not affected by extremes of hot or cold temperatures, but cannot tolerate frost. The optimum temperature

Table 32.2. The U.S. Guava Imports During 2003

	Guava		
Country	Paste and Puree	Prepared or Preserved	Jams
Brazil	1612	714	340
Colombia	325	274	0
Costa Rica	14	112	271
Dominican Republic	781	1165	33
Ecuador	335	632	0
Fiji	5	0	0
France	26	4	0
India	586	153	0
Malaysia	0	534	0
Mexico	348	761	3
Netherlands	21	0	0
Philippines	25	0	0
South Africa	33	343	0
Thailand	0	121	2
Others	9	10	7

Source: Anon, 2003.

for the growth of guava tree ranges from 23°C to 28°C. The guava fruit has higher nutritional value as a source of vitamin C even in the form of many processed products and has, thus, become an important fruit crop in the domestic economies as well as in the international trade of many tropical countries.

CULTIVARS

The guava tree being hardy, offers a lot of scope for the selection of improved cultivars, and these have appeared in many tropical and subtropical countries. Based on the shape of guava fruit, the varieties are classified in two broad categories, the pyriferum (the pear-shaped guava) and the pomiferum (the round-shaped guava), and these are often called pear guava and apple guava, respectively. The fruits from wild trees range in size from 3 to 8 cm in diameter, but from cultivated trees the fruits attain a size up to 13 cm in diameter with weight around 700 g. Guava fruit has many small hard seeds numbering from 153 to 664 per fruit located in the central core of flesh (Palaniswamy and Shanmugavelu, 1974). Seedless varieties have been developed but the fruit shape and quality is inferior.

Numerous stone cells (scleroids) are present in the fleshy part of the fruit. These stone cells impart a gritty texture to the flesh as well as to the processed juice. The flavor of guava can be described as sweet, musky, strong, and highly aromatic (Wilson, 1980). Based on the color of flesh of ripe fruit, the guava fruit is also classified as pink-fleshed sour type or white-fleshed sweet type. A number of cultivars in both types are being grown commercially for fresh fruit as well as for processing purposes. Pink cultivars such as Beaumont, and Ka Hua Kula in Hawaii, United States; Malberbe and Saxon in South Africa are popularly being cultivated. Among the white cultivars, Paipa in Taiwan; Elisabeth in West Indies; Safeda (most popular), Chittidar, Karela, Lucknow, Sind, Dholka, Nasik, and Habshi, are preferably grown for processing. Some of the other cultivars recommended for cultivation in Florida, United States, are Miami Red, Miami White, Supreme, Red Indian, and Ruby (Malo and Campbell, 1968).

PHYSIOLOGY AND RIPENING

Guava fruit usually takes about 110–150 days from the onset of flowering to reach maturity. More time for maturity is required during rainy season than in winter season. Guava is a climacteric fruit, and exhibits a typical increase in respiration and ethylene production during the ripening period (Brown and Wills, 1983). The peak for ethylene production occurs at the half-ripe stage (usually the fourth day of harvest) and coincides with the peak for rise in respiration rates (Broughton and Leong, 1979). Depending on the cultivar, guava fruits exhibit a change in skin color from green to yellow during ripening. The color becomes yellow due to the loss of chlorophyll. However, some cultivars stay green on maturation. On the other hand, flesh of the mature fruit changes color from white to creamy white, yellowish pink, deep pink, or salmon-red depending on the presence of carotenoids, lycopene, and β-carotene (Wilson, 1980; Wilberg and Rodriguez-Amaya, 1995a, b).

A number of physicochemical changes take place during the development of guava fruit. The fruit weight and volume increases moderately up to the first 50 days after flowering, but rapidly up to 100 days, and very slowly later on until maturity (Yusof and Mohamed, 1987). Sugar content of guava increases during ripening. Among the sugars, fructose increases rapidly but the glucose builds up slowly. The pectin content of guava increases during fruit development but declines in overripe fruits. The differences in the rate of softening between cultivars correlate well with the extent of loss of total pectin content (Chin et al., 1994). Besides pectin, hemicellulose and cellulose are also modified during the fruit ripening process, with a general decline in the level of alkali-soluble hemicelluloses. The activities of the softening enzymes polygalacturonase (PG), pectinesterase (PE), β-galactosidase, and cellulase increase with ripening (El-Buluk et al., 1995).

Guava being a climacteric fruit is highly perishable. The fruit should be harvested only when the green color on the skin starts to fade, as this indicates the onset of ripening. If guava is harvested too green, it will not develop good flavor during postharvest ripening. Fully yellow guavas ripened on the tree have the best flavor, but these are often damaged by insects and birds. Moreover, the fully ripened fruits are too soft and thus difficult to handle and transport to the market. For most cultivars, the fruits are ready for harvest after 120–150 days of flowering and should be harvested only when they turn yellowish green or may be blushed with pink. Guava fruits are usually harvested manually, and the fruits have postharvest life of not more than 10 days (Brown and Wills, 1983).

CHEMICAL COMPOSITION

Guava fruit consists of peel (about 20%), fleshy portion (50%), and seed core (30%). Guava fruit has a low caloric value (275 kJ/100 g) and protein content (1%). The fruit contains 74–83% moisture, 13–26% dry matter (DM), 0.8–1.5% fat, and 0.5–1.0% ash (Mukherjee and Datta, 1967). The DM is made of mainly structural and nonstructural carbohydrates. During the ripening process, the total soluble solids (TSS) and the total soluble sugar contents increase from 10.5% to 12.75%, and from 4.81% to 7.32%, respectively. Ascorbic acid content increases from 118.53 to 199.26 mg/100 g, while the acidity decreases from 0.72% to 0.55%, during this period (Agrawal et al., 2002). Mainly the citric, malic, glycolic, tartaric, and lactic acids contribute toward the acidity of guava fruit (Chang et al., 1971). Guava cultivars are reported to differ in their final sugar contents; fructose varied from 5.6% to 7.7%, glucose from 1.9% to 18.1%, and sucrose from 6.2% to 7.8% (El-Buluk et al., 1996).

The composition of guava varies depending on the cultivar, stage of maturity, and season. Guava is a good source of many important minerals such as phosphorus (23–37 mg/100 g), calcium (14–30 mg/100 g), and iron (0.6–1.4 mg/100 g). It is also a good source of many vitamins like ascorbic acid, niacin, pantothenic acid, thiamine, riboflavin, and vitamin A (Paull and Goo, 1983). White-flesh guava is reported to be a better source of vitamin C (142.6 mg/100 g) than the pink-flesh guava (72.2 mg/100 g) and is also rich in other antioxidants such as phenolics and β-carotenes (Luximon-Ramma et al., 2003). Phenolic compounds decrease with firmness, more rapidly initially in white-fleshed than in pink-fleshed guava, but their content was consistently higher than the latter (Bashir and Abu-Goukh, 2003).

Ascorbic acid content in guava shows a tremendous variation depending on the cultivar, season, and color of flesh. Sachan and Ram (1970) reported 37–1000 mg of vitamin C per 100 g in guava fruits. Pink-flesh guava contained more ascorbic acid than the white-fleshed ones (Kumar and Hoda, 1974). The winter season crop (November–December) contained more vitamin C (325 mg/100 g) than the rainy season crop (July–August) (140 mg/100 g) (Sachan et al., 1969). Vitamin C reaches a maximum level in green but fully ripe fruit and then decreases as the fruit ripens (Agnihotri et al., 1962). Within each guava fruit, the vitamin C content is the highest in the skin but decreases toward the central core (Webber, 1944). Total phenolics, flavonoid, proanthocyanidin, carotenoid, vitamin C contents and antioxidant activities of guava, lychee, and papaya are presented in Table 32.3 (Luximon-Ramma et al., 2003; Setiawan et al., 2001). In another study, the total carotenes and β–carotene contents of ripe papaya were found to be 2464 and 868 μg per 100 g, respectively (Chandrasekhar and Kowsalya, 2002).

Table 32.3. Total Phenolics, Flavonoid, Proanthocyanidin, Carotenoid, Vitamin C Contents, and Antioxidant Activities of Guava, Lychee, and Papaya

Chemical Constituent	Guava, White Flesh (*Psidium guajava*)	Lychee (*Litchi chinensis*)	Papaya (*Carica papaya*)
Total phenolics[a,b]	2473 ± 45	288 ± 17	576 ± 41
Flavonoids[a,c]	209 ± 10	94 ± 6	376 ± 15
Proanthocyanidins[a,d]	263 ± 31	100 ± 9	208 ± 21
Vitamin C[a,e]	1426 ± 26	138 ± 15	929 ± 19
Cryptoxanthin[f,g]	66	–	180
Lycopene[f,g]	1150	–	5750
β-carotene [f,g]	984	–	440
TEAC[h]	17 ± 2	5 ± 1	10 ± 2
FRAP[i]	14 ± 1	3 ± 1	2 ± 0

[a]Data from Luximon-Ramma et al. (2003), mean ± SE with $n = 3$.
[b]μg gallic acid per g fresh weight.
[c]μg quercetin per g fresh weight.
[d]μg cyaniding chloride per g fresh weight.
[e]μg ascorbic acid per g fresh weight.
[f]Data from Setiawan et al. (2001), average of three samples.
[g]μg per 100 g wet weight edible portion.
[h]μmol Trolox per g fresh weight.
[i]μmol Fe(II) per g fresh weight.

Guava fruit is one of the richest sources of pectin ranging from 0.5% to 1.8% and is affected by various factors such as variety, stage of maturity, and crop season. The quality of pectin is judged by its ability to form a gel and is measured in terms of jelly units. Winter season guava fruits are known to contain higher amounts of pectin with more jelly units than the rainy season crop (Dhingra et al., 1983). Half-ripe guava fruits yield pectin having higher jelly units than the unripe ones. On hydrolysis, guava pectin yields 72% D-galacturonic acid, 12% D-galactose, and 4% L-arabinose (Chang et al., 1971).

Guava fruit is also a good source of antioxidant carotenoids (cryptoxanthin, lycopene, and β-carotene) and it contains about 140 μg retinol equivalents/100 g of provitamin A (Setiawan et al., 2001; Goodwin and Goad, 1970). Raw ripe fresh guava is also an excellent source of dietary fiber (12.72 g/100 g) among the commonly consumed foods (Li et al., 2002). Guava fruit is not only rich in dietary fiber but also a good source of natural antioxidant compounds such as polyphenols. The presence of polyphenols gives an astringent taste to fruits. As the fruit matures, the polyphenols decrease considerably, so does the astringency. Pulp and peel fractions are known to contain high contents of dietary fiber (48.55–49.42%, dry basis) and extractable polyphenols (2.62–7.79%, dry basis), thus guava is a good source of both dietary fiber and antioxidant compounds (Jimenez-Escrig et al., 2001). Gorinstein et al. (1999) have compared the total polyphenols and dietary fiber in tropical fruits and persimmon. Guava had a total polyphenol content of 495 mg/100 g fresh fruit, with a high content of gallic acid (374.3 mg/100 g fresh fruit). The guava had a total dietary fiber and soluble dietary fiber contents of 5.6 and 2.70 g/100 g of fresh fruit, respectively. They suggested that in addition to lychee and ripe mango, guava could also be a suitable fruit for the dietary prevention of cardiovascular disease.

The flavor of guava is determined by the type and quantities of sugars, acids, phenolics, volatile, and aroma active compounds present in it. Jordan et al. (2003) have characterized the aromatic profile of fresh guava fruit puree and have identified about 48 components with the predominance of terpenic hydrocarbons and 3-hydroxy-2-butanone. However, using gas chromatograph and mass spectometer (GC–MS), Pino et al. (2002) have characterized about 173 components in aroma concentrate of Costa Rican guava in which (E)-β-caryophyllene, α-terpineol, α-pinene, α-selinene, β-selinene, δ-cadinene, 4,11-selinadiene, and α-copaene were the major constituents. Amounts of aliphatic esters and terpenic compounds were mainly responsible for the unique flavor of this fruit.

POSTHARVEST HANDLING AND STORAGE

Guava fruit has a short shelf life, which varies with the cultivar and the rate of ripening of fruits. It is, therefore, important to find ways of extending the postharvest shelf life of guava fruit to improve their marketability as fresh fruits, as well as to regulate supply to the processing plants. Mature guavas do not keep well and are transported rapidly to the fruit processing factories. Guava fruits are usually shipped in small boxes rather than in bigger crates, as these fruits are easily crushed or bruised and the damaged fruit deteriorates fast. Fully mature fruits can be refrigerated during shipment. Guava, like most other tropical fruits, is highly chill sensitive. In general, a temperature of 8–10°C is considered to be the critical limit for chilling injury for most of the cultivars. Varietal differences in shelf life of guava fruit during cold storage are reported. Allahabad Safeda cultivar stayed acceptable for 6 days compared with 9 days for Chittidar and Sardar cultivars when stored at 18°C ± 2°C and 80–85% relative humidity (Singh et al., 1990).

Storage of fruits under modified atmosphere (MA), as in polybags, or packaging (MAP) or coating in polymeric films prolongs the shelf life (Kader et al., 1989). Guava fruits packed in 300-gauge polyethylene bags can be stored at room temperature for about 10 days (Khedkar et al., 1982). The MA treatments lead to build-up of CO_2 and a depletion of O_2 within the internal atmosphere of the package. In most cases, respiration and ethylene production by the fruit is reduced, and ripening is delayed leading to markedly extended shelf life. Another advantage of MAP is that it reduces the incidence of diseases. In case of guava, ascorbic acid decreases during ripening but MAP minimizes this loss (Mohamed et al., 1994). Cellulose- or carnauba-based emulsions delay color development and suppress the increase in the level of TSS during modified atmosphere coating of mature green guava fruits (McGuire and Hallman, 1995). Coating of guava fruits with 2% or 4% hydroxypropyl cellulose is more effective than the 5% carnauba formulation in retarding softening, but both are effective in preventing water loss.

Other treatments like vacuum infiltration of MAP fruits in 10% $CaCl_2$ at room temperature retards loss of firmness and suppresses the increase in soluble pectin and titratable acidity (TA) but has no effect on the incidence of disease as compared to the control fruits (Lazan and Ali, 1997). A few other postharvest treatments of guava fruit that have been investigated include dipping in sugar syrup (Mudahar and Bhatia, 1983), in metabisulfite solution (Ahlawat et al., 1980), in 1% calcium nitrate (Singh et al., 1981), or in 1000 μl/l cycocel (Tandon et al., 1984). Dipping guava fruits in a solution containing 1000 or 1500 ppm each of 2,4,5-T and maleic hydrazide is reported to extend their shelf life up to 12 days (Garg and Ram, 1974). A minimum loss in weight and spoilage was observed in guava fruits when dipped in 50- or 500-ppm indolebutyric acid solution and stored at room temperature (Garg et al., 1976). Improper storage of guava results in the development of fruit diseases and disorders. The nature and extent of losses due to these diseases depend on the location and storage conditions of these fruits, and this discussion is beyond the scope of this chapter.

PROCESSED PRODUCTS

Guava fruit is among the most important raw materials for commercial processing plants in the tropical and subtropical countries. This fruit is an exceptional source of vitamin C as well as a fairly good source of vitamin A, certain B-complex vitamins, calcium, and phosphorus. Because of its unique and strong flavor, guava fruit lends itself to the production of a number of processed products such as nectar, clarified juice, concentrates, canned, dehydrated powder, jam, jelly, guava cheese, and blends with other juices.

PUREE

Slightly overripe guava fruits having fully yellow skin with brown spots and soft flesh having intense distinct musky odor are the preferred raw material for puree. Guava fruits are inspected, seriously spoiled ones are discarded, and the underripened are kept aside for further ripening. Those fruits that have passed the inspection are washed thoroughly and finally inspected once again before maceration. The washed guavas are processed in a paddle pulper fitted with 0.008 to 0.11 cm screens. This pulp is then passed through a finisher fitted with 0.05 cm screens to remove stone cells (Luh, 1971). It is deaerated and pasteurized for 60 s at 90°C. It can be canned in enameled cans, or preserved by freezing to −29°C and storing at −18°C, or aseptically packaged, or dehydrated. Deaerated aseptically packaged guava pulp retains ascorbic acid much better during storage up to 6 months (Chan and Cavaletto, 1986). Foaming agents such as egg albumen, glycerol monostearate, guar gum, and carboxymethyl cellulose are dissolved in a small amount of water and blended with guava puree and foamed in a mixer using a wire whip. The foam is extruded on to trays and dried in a vacuum shelf drier. The dried product is scraped off the trays and packaged in airtight containers. Guava puree is also a basic starting material for many products such as nectar, juice, cheese, jam, and jelly.

JUICE

Guava juice can be produced either from fresh fruits or from puree. For juice extraction, fully ripe fruits are cut into small pieces followed by the addition of 0.2 g citric acid and 250 ml water/kg. The mix is cooked while stirring constantly, strained through a muslin cloth, and the juice is collected. Juice yield can be increased to more than 80% by using 700 ppm of Pectinex (Ultra SP-L Registered, Novozymes, USA) and pectic enzymes (Chopda and Barrett, 2001). Use of polygalactuonase enzyme results in a small decrease in vitamin C but an increase in acidity, reducing sugar, TSS contents, and a significant increase in volatile compounds was observed in the clarified guava juice (Pong et al., 1996).

The chemical composition of guava juice is more or less similar to that of guava puree. Guava juice has 11% TSS (Brix), 6.64% total sugars, 6.02% reducing sugars, 0.76% acidity (as citric acid), 3.85 pH, 243 mg/100 g ascorbic acid, and 1.02% pectin (Sandhu and Bhatia, 1985). Guava juice can be further processed and utilized for producing concentrates, beverages, jelly, powder, and other products. An acceptable quality whey beverage with 0.5% acidity, 20% TSS using 25% guava fruit juice has been prepared (Gagrani et al., 1987). The shelf life of guava juice has been studied by several researchers. Single-strength guava juice retained higher amounts of ascorbic acid (35%) than did guava juice with 25% added sugar at the end of 270 days of storage (Shaw et al., 1975). Guava juice stored with added sodium citrate retained good color and flavor. Bottled frozen guava juice and acerola juice blends when stored for 8 months retained 90% of the original ascorbic acid (Fitting and Miller, 1959). However, Orr and Miller (1954) have reported that unfrozen bottled guava juice retained 70% of the original ascorbic acid during 11 months of storage compared to 80–85% retention of ascorbic acid in frozen bottled guava juice after 12 months of storage. Guava juice has been shown to blend very well with other fruit juices such as Kinnow mandarin, pear, grape, mango, and pineapple for the preparation of good quality multifruit beverages with 15% juice content to suit the tastes of different consumers (Sandhu and Sidhu, 1992). Ready-to-serve (RTS) beverages made from guava and papaya pulp blend (70:30) have high vitamin C, carotenes, and also have better sensory characteristics due to better consistency and flavor (Tiwari, 2000).

CONCENTRATE

To facilitate overseas shipment and for long-term storage, it is advantageous to produce concentrate from guava puree or clarified guava juice. A falling film evaporator, a rising film evaporator, and a centrifugal evaporator are the equipment required for this purpose. Before concentration, guava puree is treated with pectic enzymes to reduce its viscosity. Depectinized puree can then be easily concentrated to 34 Brix, as it remains flowable in an evaporator (Brekke and Myers, 1978). However, the clarified juice can be concentrated to even 66 Brix (Muralikrishna et al., 1968). The loss of flavor during evaporation is compensated by cutting back the product to a lower concentration, by diluting it with single-strength fresh pasteurized puree or clarified juice. As water is lost during the concentration process, TSS, acidity, sugars, pectin, and ascorbic acid are increased in the finished product but its color changes to brown due to the browning reaction (Sandhu and Bhatia, 1985). Guava juice concentrate is suitable for drying it to guava juice powder as well as for the preparation of RTS beverages. Guava concentrates should be packaged in low oxygen permeable containers to preserve good flavor and color during storage.

NECTAR

Guava nectar, cloudy or clarified, is a very popular beverage that can be prepared from puree, clarified juice, or concentrate with sugar syrup, citric acid and other flavoring additives. Cloudy guava nectar is more popular in most of the countries than the clarified nectar. Guava nectar is prepared using 15% pulp, 14% soluble solids, and 0.25% acidity (Kalra and Tandon, 1984). It can be fortified with vitamin C (100 mg/100 g) and packaged in glass bottles or tetrapacks. Guava nectar can also be diluted (four times its volume with water) to prepare good quality bottled RTS beverage (Jain and Borkar, 1966), or this beverage can also be prepared from fresh or preserved pulp using 1 kg of sugar, 6 l of water, and 20 g of citric acid for every kilogram of pulp.

To improve the consistency of cloudy guava nectar, use of piston-type homogenizer or an ultrasonic homogenizer is essential, but the homogenized nectar should be vacuum deaerated as soon as possible to retain the ascorbic acid and good flavor. Then the nectar is either pasteurized or commercially sterilized for better shelf life. For pasteurization, guava nectar is heated in a plate heat exchanger to 93°C, held for 36 s, cooled down to room temperature, and filled in tetrapacks. For commercial sterilization, guava nectar is heated up to 93°C, held for 45 s, and then filled at a temperature higher than 86°C into enameled tin cans. These cans are sealed, inverted, held for 3 min, cooled to 40°C by spraying water, and then air-cooled to ambient temperature. Under common ambient storage conditions, canned guava nectar has a shelf life of 6 months. During ambient storage, white-fleshed cloudy guava nectar deteriorates in quality due to nonenzymatic browning reactions through the involvement of ascorbic acid and tannins (Chen et al., 1994). However, lowering the pH or the addition of L-cysteine to the nectar formulation effectively reduces the rate of browning (Chen and Wu, 1991).

Canned Guava

Except for guava juice, canned guava is a very popular product in many countries, including India, Pakistan, and Indonesia. The general canning procedure for guava shells is described by Lal et al. (1986). Fully ripe guava fruits are either peeled with a knife or lye-peeled (by dipping for 15 s in boiling solution of 2.5% sodium hydroxide). After rinsing with water, it is dipped in 0.5% solution of citric acid to neutralize the remaining alkali. The fruit is cut into halves or quarters and the seed core is removed to obtain the shells. The shells are firmed by dipping in 2% calcium chloride solution for 1 h. After rinsing, the shells are canned in sugar syrup of 40 Brix containing 0.25% citric acid. The cans are exhausted at 82–100°C for 7–10 min to reach a temperature of 74°C in the center of cans. The cans are sealed, sterilized in boiling water for 20–25 min, and then cooled to room temperature. The suitability of guava cultivars for canning varies; however, Allahabad seedless cultivar from India has been reported to be more suitable (Siddappa, 1982).

Dehydrated Guava Products

Guava slices or chunks can be dehydrated by air-drying, osmotic dehydration, or osmovac dehydration. In the first procedure, guava slices are blanched in boiling water for 4 min, sulfuring is done for 20 min, and then air-dried at 71°C until a final moisture of 6–7% is achieved. This usually takes about 15 h (Campbell and Campbell, 1983). To prevent browning of slices during drying process, there are a number of treatments such as chemical blanching with 0.1% potassium metabisufite (KMS) + 2% $CaCl_2$ at 100°C for 3 min, or sulfiting with 1% KMS for 5 min, or sulfuring with 2 g sulfur/kg of fruit slices for 4 h. Khurdiya and Roy (1974) dried guava slices in a cabinet drier at 60°C ± 5°C for about 18 h to achieve a final moisture content of 3% in the finished product.

Glazed guava slices are prepared by osmotic dehydration technique. Guava slices are heated in an equal amount of 70 Brix sugar syrup containing 0.1% KMS at 90°C for 3 min. After cooling to room temperature, this mixture is allowed to stay overnight. Then these slices are drained, and spread in glycerine-coated trays for drying at 80°C for 1 h. The air temperature is then lowered to 65–70°C for the next 7–8 h. Kannan and Susheela (2002) have recently investigated the storage behavior after osmotic dehydration of guava. Acidity increased and pH decreased in all their samples during storage. Total sugar content decreased slightly, but the ascorbic acid content was reduced significantly by 6 months of storage. As far as appearance, color, texture, and flavor are concerned, Luchnow-49 cultivar was rated the best in sensory quality. Mehta and Tomar (1980a, b) standardized a procedure for the dehydration of guava slices in 70 Brix syrup containing 1000 ppm of sulfur dioxide. The dehydrated guava slices were of very good quality but only about 6% of the original ascorbic acid was retained in the finished product.

Guava fruit powder is prepared by grinding the dehydrated slices. After blanching, the shells are dried at 54°C for about 10–12 h and packed in moisture-impermeable containers. During storage of guava powder for 6 months, a significant decrease in ascorbic acid has been observed with a slight increase in moisture content (CFTRI, 1990). The rate constant of ascorbic acid degradation has been reported to increase with the increasing temperature and water activity (Uddin et al., 2002). Guava powder can be used in the preparation of guava juice, RTS beverage, or milk shake.

Guava Cheese

Guava cheese (guava fruit leather) is prepared from firm ripe fruits. The fruits are washed, sliced, and cooked in equal amounts of water to soften the fruit. The pulp is screened to remove the skins and the seeds. For every kilogram of pulp, 1.25–1.50 kg sugar, 2.2–3.3 g citric acid, and 56 g butter are added. This mixture is cooked to a thick paste. To improve the appearance of the final product, small amounts of permitted color and common salt are added. The hot cheese is spread on a greasy tray and is allowed to set. After cooling, it is cut into small pieces and wrapped in moisture-proof paper (Lal et al., 1986). Guava cheese prepared from pulp can be stored at 4°C for about 4 months without the loss of sensory quality (Singh et al., 1983). Depending on the temperature, pectin and ascorbic contents in guava cheese decrease during storage to a level of 1.24–1.55% and 14.6–41.5 mg/100 g, respectively. This decrease is much less during storage at low temperatures than at room temperature.

Guava cultivars differ in their suitability for cheese preparation. Banarsi Surkha guava cultivar from India was found to produce better quality cheese than the Allahabad Safeda cultivar (Sandhu et al., 2001). They prepared cheese from guava pulp and sugar mixture by drying it in a cabinet drier at 50°C ± 5°C for 4 h to moisture content of about 29.3%. The product that was wrapped in butter paper and packed in polyethylene bags stayed acceptable for 3 months at room temperature. Similarly, Vijayanand et al. (2000) found that guava cheese bars packaged in polyester-polyethylene laminate or pearlized biaxially oriented polypropylene retained their sensory and textural properties for about 3 months of storage at room temperature.

Jelly

For the preparation of jelly, slightly under ripe guava fruits are used. The fruits are crushed, juice is extracted, and allowed to settle overnight. Clear juice is decanted and boiled with sugar. The ratio of sugar to guava juice depends on its pectin content. Usually 0.75 kg sugar/kg of pectin-rich juice and 0.5 kg sugar/kg of low-pectin juice are used. The boiling is continued until the temperature reaches 105°C or it forms a sheet when a small portion is cooled off in a spoon. The hot jelly is filled into glass jars and sealed (Lal et al., 1986). The use of whey in jelly preparation has also been suggested by Joshi et al. (1985). Guava jelly with attractive color, pleasant taste, and aroma has been prepared by blending it with either of the three red grape cultivars, which alone did not produce a good jelly (Aggarwal et al., 1997).

BY-PRODUCTS FROM GUAVA PROCESSING

The use of waste produced from the processing of guava can also be a source of many natural additives and functional food ingredients (Schieber et al., 2001). Extraction of lycopene is one such possibility (Bortlik et al., 2001). Guava is one such fruit whose every part can be utilized for human consumption, either as fresh or in the processed form. Guava is one of the best natural sources of food-grade pectin, which finds uses in various food product formulations as a thickening and gelling agent (Pruthi et al., 1960). Pectin can be obtained from guava fruits by boiling with water and then precipitating with ethanol. Use of sodium hexametaphosphate (0.25–0.75%) or a 1:1 mixture of ammonium oxalate and oxalic acid (0.25–0.75%) increases the yield of high jelly grade pectin (Dhingra and Gupta, 1984). Guava seeds, which are usually discarded after processing, contain 5–13% oil that is rich in essential fatty acids. Guava seed oil is very rich in oleic (54%) and linoleic (29%) acid, making it suitable for use in salad dressing. Prasad and Azeemoddin (1994) have solvent extracted seed oil from processing plant wastes and reported an oil content of 16% in guava seeds. However, linoleic acid was the major fatty acid in guava seed oil (76.4%). Guava seed oil can be easily refined and bleached to produce a light colored bland oil of edible quality. Preparation of protein isolate having good nutritional quality from guava seeds is another possibility. Abd-El-Aal (1992) has obtained guava seed flour and isolate, which were low in sulfur-containing amino acids and lysine but contained other essential amino acids. Due to the susceptibility of mature ripe fruits, a high proportion of damaged (unsaleable) guava fruits are generated during transportation to processing plants, which can be fermented using *Saccharomyces cerevisiae* for ethanol production (Srivastava et al., 1997). A yield of 5.8% (w/v) ethanol has been obtained from such waste products of guava processing plants.

FUTURE RESEARCH NEEDS

Because of its flavor and nutritional quality, guava has the potential to become a commercially important tropical fruit crop not only for fresh consumption but also for processing into value-added products. Unfortunately, despite this, the potential of guava as a commercially important tropical fruit has not been exploited as has been done with mango, papaya, and starfruit. Some of the major constraints that have limited this potential are related to pests and diseases, unavailability of good quality cultivars, and higher susceptibility of guava fruit to damage during transportation and storage. Good management and cultural practices need to be developed to check pests and diseases at the preharvest stage. Most of the guava cultivars are wild and low in quality, and guava trees cannot be grafted as easily as other fruit trees. This problem can be tackled by making use of the recent advances in tissue culture techniques as well as recombinant DNA technology. With the current postharvest technology, the shelf life of guava fruit can only be extended up to 3 weeks. This is not enough if this fruit has to be exported to distant markets. So the existing technology needs to be upgraded to improve the shelf life of fresh guava fruit.

Section 2: Lychee

INTRODUCTION

Lychee, also called Litchi or litchee (*Litchi chinensis* Sonn.), is a subtropical evergreen tree belongs to the family Sapindaceae or soapberry. Lychee is believed to have originated in the Kwantung Province of China where it has been grown for the past 40 centuries (Tao, 1955). Later, it spread to many tropical and subtropical countries. China, Taiwan, Vietnam, Thailand, India, Pakistan, Indonesia, Madagascar, South Africa, and Australia are the major lychee producing countries of the world (Menzel et al., 1988). According to the FAO estimates (FAO, 2001), the top five countries for producing lychee are China (1.2 million metric tons), India (455,000 metric tons), Thailand (85,000 metric tons), Vietnam (50,000 metric tons), and Bangladesh (12,755 metric tons).

The fruit is highly prized in its fresh as well as processed forms. Lychee trees reach a height of 9–12 m and the fruit is borne in bunches. When the fruit is fully ripened, its pericarp is thin, hard, and somewhat warty in texture. Depending on the cultivar, the lychee fruit pericarp may be pale green, bright red, or rose colored. Lychee fruit is round to oval in shape, measures about 2.5–4 cm, and has a large glossy brown seed. Lychee fruit normally contains one seed, but in some cultivars a high proportion of seeds may be abortive. These abortive seeds are small and shriveled. The fruits with such seeds are preferred as they yield higher proportion of flesh and often fetch a higher price (Menzel and Simpson, 1993). Lychee is a nonclimacteric fruit, which is harvested during the summer months (May–July in the Northern Hemisphere, November–February in the Southern Hemisphere). The pulp is white to cream colored, very succulent and has aromatic, sweet, and acidic taste.

Lychee grows well in a variety of soil conditions but a rich, loamy soil or sandy loam is preferred. The lime content of soil is important, as the soil with about 30% lime content is best suited for lychee cultivation. Acidic soil pH favors the growth of mycorhizal fungi on roots, which greatly influences the fruit quality. An ideal climate for lychee cultivation should be free from frost during winter season and hot, dry winds during summer months. The young plants are initially protected from frost during winter months. The dry heat during ripening leads to cracking of fruits and is prevented by frequent irrigation during dry season (Maiti, 1985). Although budding and grafting on seedling rootstock is also practiced, air layering is the most common method of lychee propagation. Plants produced from seed generally do not bear fruits until eighth or ninth year or sometimes not at all, whereas the trees propagated through air layering bear fruits after fourth year, but the latter do not develop a good root system. During air layering, the use of a rooting hormone (200 ppm of α–naphthalene acetic acid) gives higher success in root formation (Singh, 1951).

CULTIVARS

There are many lychee cultivars grown in the world. A cultivar may set fruit with different characteristics in different growing areas and may cause confusion on its identity. Menzel and Simpson (1991) have described the important cultivars being grown in the major lychee producing countries of the world. The major cultivars being grown in China, Australia, Taiwan, and other South-eastern countries have the same original Chinese names, such as, “Haak Yip,” “No Mai Chee,” Kwai Mi,” and “ Wai Chee” (Anon, 1961). In Hawaii, a few other cultivars such as Hak Ip, No Mai T'sz, Brewster, Pat Po Hung, and Groff are also recommended for planting because of their quality and higher yield (Hamilton and Yee, 1970). Brewster (Chen purple) and Mauritius cultivars are also grown in Florida. Among the 12 lychee cultivars grown in India, Dehra Dun, Early Large Red, Kalkattia, and Rose-scented are the most popular ones. Mauritius is the only cultivar commercially grown in South Africa (Knight, 1980). Among the various cultivars grown in Pakistan, Dehra Dun is considered the best (Ahmed, 1961).

PHYSIOLOGY, RIPENING, AND FRUIT CRACKING

The terminal panicles of new shoots in lychee bear two types of flowers, staminate and hermaphrodite. Staminate flowers open before the hermaphrodites. Fruit set is quite low in lychee, as only a small percentage of flowers set fruit. The growth of lychee seeds in fruit occurs initially at a higher rate, followed by membranous mesocarp and aril, which grow fast toward the later stages. Lychee fruits start ripening when the atmospheric temperature is high. Respiration rate of lychee fruits decreases progressively during fruit development, with immature fruits (20 days after anthesis [DAA]) having a rate eight to ten times higher than that of the mature fruits (Akamine and Goo, 1973). Lychee produces low levels of ethylene, most of it comes from the pericarp. Use of low temperatures (5°C) significantly reduces the respiration and ethylene production by lychee fruits, but they increase rapidly above that at harvest when the fruits are transferred to 25°C (Jiang-Ping et al., 1986).

Mature lychee fruits are characterized by a uniform red pericarp color. Four anthocyanin pigments have been found to be associated with the development of red pigments of ripened lychee fruits. Anthocyanin, the main red pigment is synthesized in the pericarp at around 60–80 DAA, when the chlorophyll is simultaneously degraded (Maiti, 1985; Underhill and Critchley, 1992). Total anthocyanin content declines from 1.77 to 0.73 mg/kg fresh weight and individual anthocyanins also decrease during storage period. Decline in anthocyanins is accompanied by an increase in browning. Polymeric pigments gradually increase from 20.9% to 53%. Chlorophyll is still present in the mature lychee fruits, though only in small amounts (Lee and Wicker, 1990). Cyanidin-3-rutinoside is the major anthocyanin found in the skin of lychee fruits, although cyanidin-3-glucoside and malvidin-3-acetylglucoside are also identified. Polymerized anthocyanin pigment is also present and contributes to the brownish-red color of lychee fruit skin (Lee and Wicker, 1991; Rivera et al., 1999).

As the lychee is a nonclimacteric fruit, it does not ripen once harvested, so lychee fruits are allowed to ripen on the tree itself. Since lychee fruit maturity does not change after harvest, the fruit must be harvested at optimal visual appearance and eating quality. While the pericarp color is the most commonly used harvesting index, the relationship between the pericarp color and lychee fruit maturity varies with the cultivar, region of cultivation, and other agricultural practices (Underhill and Wong, 1990). Batten (1989) has suggested a number of maturity criteria for lychee fruits. Lychee fruits are usually analyzed for TSS and TA but TSS (Brix) is not a suitable maturity indicator, whereas TA and the TSS/TA ratio are both good predictors of taste. Dry heat during fruit ripening, especially during windy conditions, is disastrous for a lychee crop. The fruits have been shown to crack prematurely under such desiccating environmental conditions (Maiti, 1985; Underhill and Simons, 1993). Under such conditions of low humidity, frequent irrigations are necessary to maintain the desired humidity conditions during fruit ripening periods.

CHEMICAL COMPOSITION

Although considerable variation is found in the chemical composition of lychee fruits, the most data for moisture ranges from 77% to 83% (Mathew and Pushpa, 1964; Wenkam and Miller, 1965). Lychee is not considered a significant source of proteins, and most analyses show only 0.8–0.9% protein. Similarly, the fat content of lychee fruit is also negligible, less than 1% (Wenkam and Miller, 1965). Sugar content varies considerably among lychee cultivars. The highest sugar content of 20.6% in Kwai Mi is reported by Miller et al. (1957), whereas most Indian cultivars have a sugar content between 10% and 13%, except that of 18% soluble solids in Calcutta Late and Seedless No.1 (Chadha and Rajpoot, 1969). Higgins (1917) observed a total sugar content of 15.3%, consisting of mainly reducing sugars (81.7%) and remaining sucrose (18.3%). The total sugar content of six Indian varieties is reported to vary from 55.92% to 61.37% of the fruit on dry basis, and reducing sugar content of 41.52–43.45% (Mathew and Pushpa, 1964). Thus, more than 70% of the sugars were present as reducing sugars in lychee fruits. However, Chan et al. (1975) have reported a higher proportion of sucrose, roughly 51% of the total sugars. This discrepancy in their findings was attributed to the presence of invertase enzyme in lychee fruit, which if not inactivated prior to analysis, proceeds to catalyze the inversion of sucrose. This invertase enzyme has been characterized by Chan et al. (1975) and found to have an optimum pH of 2.6 considerably lower than other invertases and also lower than the normal pH (4.6) of lychee fruit. The optimum temperature for this enzyme was 55°C.

As abundant varieties of lychee exist, a great variation in acidity might be expected. Fruit acidity in

12 Indian varieties of lychee ranged from 0.20% to 0.64% (Mathew and Pushpa, 1964). Acidity values as high as 1.1% were observed in an unidentified South African variety of lychee (Marloth, 1934). Acidity is known to decrease as the fruit ripens. Following harvest, acidity again decreases during storage (Joubert, 1970). The Brix/acid ratio increases during ripening and storage, and may reach as high as 80:1 before fruit rotting sets in. The predominant nonvolatile acid is malic acid (80% of the total acids) and citric, succinic, levulinic, phosphonic, glutaric, malonic, and lactic acid are the remaining nonvolatile acids in lychee fruit (Chan and Kwok, 1974).

Though considerable variations are reported, lychee fruit is a very good source of ascorbic acid. Wenkam and Miller (1965) found 40.2 mg/100 g of ascorbic acid in Kwai Mi and 80.8 mg/100 g in Brewster. Similarly, 44 mg/100 g in Calcutta Late (Chadha and Rajpoot, 1969) and 90 mg/100 g in an unspecified variety (Thompson, 1955) have been reported in the literature. The ascorbic acid content decreases as the temperature and storage time are increased (Thompson, 1955). Lychee's chief nutritional asset is its high ascorbic acid content, as it is not a significant source of thiamin, riboflavin, calcium, phosphorus, or iron. It is totally lacking in provitamin A. However, variety Kwai Mi was found to be a good source of niacin (Wenkam and Miller, 1965). A small amount of pectin (0.424%) has been reported in lychee fruit.

The volatile profile of lychee fruit has been studied by Johnston et al. (1980). Of the 42 compounds identified by them, β-phenethyl alcohol, its derivatives, and terpenoids comprise the major and characteristic portions of these volatiles. In a more recent study, Chyau et al. (2003) have identified 25 free and glycosidically bound aroma compounds from the lychee juice using GC and GC–MS. Free aroma compounds (FAC) were rich in acetoin, 3-methyl-2-buten-1-ol, and geraniol (874, 445, and 454 mg/kg pulp, respectively); glycosidically bound aroma compounds (GBAC) were rich in geraniol, geranial, neral, and 2-phenylethanol (1162, 125, 60.4, and 54.2 mg/kg pulp, respectively). Total levels of volatile compounds in FAC and GBAC were 2907 and 1576 mg/kg pulp, respectively; and 25 compounds identified in both the fractions were 1 ester, 14 alcohols, 2 aldehydes, 4 acids, 2 ketones, and 2 terpenes. While evaluating aroma, FAC had a fresh, fruity, lychee-like aroma, whereas GBAC was odorless until enzymic hydrolysis. The combination of FAC and hydrolyzed GBAC fractions gave a strong fruity, lychee-like aroma, suggesting that controlled application of glycosidase during lychee processing could enhance the characteristic flavor of lychee juice and other juice-based products.

POSTHARVEST HANDLING AND STORAGE

Maintenance of market quality of fresh lychee is a serious problem when the fruit must be transported great distances. Desiccation accompanied by loss of red shell color and the development of pericarp browning can occur quickly. This browning renders the fruit unsaleable, and is thus of commercial importance in lychee trade. A number of enzymatic and color changes occur during the postharvest storage and transportation of lychee fruit. The decrease in peroxidase (POD) and increase in polyphenyloxidase activities coincided with the onset of discoloration of fruits (Huang et al., 1990). Postharvest physiological characteristics of lychee fruit are described in Table 32.4. Two principal strategies can be employed: reducing the water loss by various methods and the prevention of browning by chemical or physical treatments. If left unpackaged at room temperature, the shelf life of lychee fruit is hardly 72 h or less (Macfie, 1954).

REFRIGERATED STORAGE

Lychee fruit that is quickly cooled to 3°C and then held at low temperature (5°C) tend to be less susceptible to moisture loss and decay. It takes about 2 days for the packaged fruit to reach 5°C in the refrigerated storage (Bagshaw et al., 1994). Cooling methods and shipping containers used affect the lychee fruit quality. Lychee stored without the panicle had higher pulp quality in terms of total TA, pH, and TSS content compared to those stored with the panicles attached (Pornchaloempong et al., 1997). Storage of lychee at 4°C or 10°C increased ethylene production by as much as 8.6 times compared with the control samples stored at 25°C. The green fruit was most responsive to chilling in terms of ethylene production (Chan et al., 1998). Forced-air cooling requires a high-capacity cold room and takes about 12 h. To avoid fruit desiccation, the cold room must have a relative humidity of 95%. Precooling of lychee fruit is most effective, if done before packing (Watkins, 1990). In comparison, hydrocooling is a faster method than the forced-air cooling. It avoids the problem of fruit desiccation and is a relatively less expensive technique. Commercial

Table 32.4. Postharvest Physiological Characteristics of Lychee Fruit

Characteristic	Activity/Concentration	Reference Source
Chlorophyll, mature green fruit		
Chlorophyll a	80 μg/100 mg	Jaiswal et al. (1987)
Chlorophyll b	110 μg/100 mg	Jaiswal et al. (1987)
Total chlorophyll	190 μg/100 mg	Jaiswal et al. (1987)
Chlorophyll, mature red fruit		
Chlorophyll a	25 μg/100 mg	Jaiswal et al. (1987)
Chlorophyll b	14 μg/100 mg	Jaiswal et al. (1987)
Total chlorophyll	40 μg/100 mg	Jaiswal et al. (1987)
Anthocyanins (ACY)		
Cyanidin-3-glucoside	>10% of total ACY content	Prasad and Jha (1978)
Cyanidin-3-galactoside	>10% of total ACY content	Lee and Wicker (1991)
Cyanidin-3-rutinoside	67% of total ACY content	Lee and Wicker (1991)
Pelargonidin-3-glucoside	>10% of total ACY content	Prasad and Jha (1978)
Pelargonidin-3,5-diglucoside	>10% of total ACY content	Prasad and Jha (1978)
Malvidin-3-acetylglucoside	15% of total ACY content	Lee and Wicker (1991)
Sugars		
TSS	13–20 Brix	Nagar (1994)
Fructose	1.6–3.1 g/100 g fruit weight	Chan et al. (1975)
D-Glucose	5.0 g/100 g fruit weight	Chan et al. (1975)
Sucrose	8.5 g/100 g fruit weight	Chan et al. (1975)
Enzyme activity		
Polyphenol oxidase[a]	0.01 ΔOD_{410nm} 0.3 ΔOD_{410nm}	Zauberman et al. (1991)
	0.5 μmol O_2/min/mg protein	Underhill and Critchley (1995)
Peroxidase [a]	0.3 ΔOD_{410nm}/min/g fruit wt	Zauberman et al. (1991)
	0.01 ΔOD_{418nm}/min/mg protein	Underhill and Critchley (1995)
Cellulase	0.18–0.25 mg Glu/h/g fruit wt.	Nagar (1994)
Pectinmethylesterase	1.5–2.0 μequi/min/g fruit weight[b]	Nagar (1994)
	1.0–1.4 μequi/min/g fruit weight[c]	Nagar (1994)
Ethylene production	1–5 μl/kg/h at 25°C	Underhill and Critchley (1993)
Respiration (CO_2)	20 μl/kg/h at 25°C	Nagar (1994)
Ascorbic acid	40–50 mg/100 g aril	Nagar (1994)

Note: All units are as cited by authors. Data relate to activity or concentration prior to storage.
[a]Pericarp PPO and POD activity.
[b]PME activity in the aril.
[c]PME activity in the pericarp.

hydrocooling is being progressively adopted in Australia and Thailand for extending the shelf life of fresh lychee fruits (Bagshaw et al., 1994).

One of the simplest and most commonly used techniques of controlling pericarp browning is to reduce water loss. Traditionally, lychee fruit was stored in woven bags, clay jars, and bamboo baskets. Large bamboo and reed baskets lined with newspapers are still in use in parts of India and Southeast Asia (Singh, 1957). Packing of lychee fruit in plastic containers and over wrapping with a semipermeable membrane reduces fruit desiccation with minimum condensation. This practice when used in combination with refrigerated storage is one of the most effective nonchemical means of controlling pericarp browning (Wara-Aswapati et al., 1990; Wong et al., 1991). Lychee fruits coated with low-pH cellulose formulations designed to lower surface pH to 4.0 prolonged the red color of the fruits throughout the cold and ambient temperature storage (McGuire and Baldwin, 1996).

While refrigerated storage is a simple and highly effective method of increasing the shelf life of fresh lychee fruit, controlling pericarp browning using optimum storage temperatures have not been fully worked out. Lychee is usually stored at 5°C (Jacobi et al., 1993), but there are several reports of fruit being stored at a temperature as low as 0°C for 3 weeks

(Sandhu and Randhawa, 1992). During the refrigerated storage of lychee fruits, usually a relative humidity of 95% is maintained.

CHEMICAL TREATMENTS

A number of chemical treatments have been reported in the literature to increase cell wall strength or to delay fruit senescence, but with little obvious success in retarding the rate of pericarp browning. Patra and Sadhu (1992) investigated the use of calcium nitrate (up to 5%) as a postharvest application to reduce the rate of whole fruit weight loss and to extend the shelf life. In another study, Roychoudhury et al. (1992) reported similar fruit loss during storage using 1% calcium nitrate without any effect on the fruit quality. Both these researchers made no specific reference to pericarp browning in their findings. Recently, a use of 1% calcium nitrate for 5 min was reported to have no effect on the rate of pericarp browning (Duvenhage et al., 1995). Similarly, a number of wax emulsions for coating lychee fruits have not met with any significant success in reducing either the rate of desiccation or the discoloration of pericarp (Bhullar et al., 1983). Underhill and Simons (1993) observed the development of microcracking shortly after harvest. Similar cracking was also noted in wax-coated fruits after 24 h, which led to enhanced desiccation. This may explain the ineffectiveness of current commercial wax coatings to reduce water loss and thereby to inhibit pericarp browning.

Sulfur dioxide is used as an alternative treatment to control physiological browning in fruits but the method of application is critical to treatment success. Burning sulfur powder is the most commonly used method but regulating the dosage is a problem. Fumigation using gaseous sulfur dioxide tends to be more accurate (Duvenhage et al., 1995). The main problem with it is that sulfur dioxide rapidly bleaches the pericarp surface due to the formation of a colorless anthocyanin–SO_3H complex (Zauberman et al., 1991). To overcome this problem, they suggested the immersing of fruit in 1N HCl for 2 min, which leads to complete color recovery within 24–48 h. The SO_2/low-acidity treatment not only leads to a permanent red color of the pericarp but is also very effective in controlling postharvest fungal diseases. In those countries where lychee industry is export oriented, significant advances in the commercialization of SO_2-based technologies have been made. It is, however, unrealistic to consider the use of sulfur dioxide as a long-term commercial solution because of the safety concerns of such chemicals.

To replace the chemical treatments such as SO_2 with more sustainable techniques has increased interest in physical methods. Immersion of lychee fruit in hot water (98°C) for 30 s followed by a low-pH treatment is reported to improve the pericarp color retention during ambient storage (Kaiser, 1995). He suggested that higher temperature inactivated the polyphenoloxidase (PPO), while low pH had an effect on nondegraded anthocyanins. To avoid the need for sulfur dioxide fumigation, the lychee fruits can be sprayed with hot water while undergoing mechanical brushing in a revolving drum, and then dipped in a 4% food grade hydrochloric acid +0.2% prochloraz fungicide solution (Lichter et al., 2000). The quality of lychee, particularly the pericarp color, can be preserved by treating fresh lychee with cold water, then hot water followed by hydrochloric acid, prior to drying the liquid off the surface of the fruit (Moran, 2000). Jiang (2000) has suggested that the anthocyanin-PPO-phenol reactions are involved in lychee pericarp browning. Ascorbic acid content of the pericarp decreased significantly with increasing peel-browning index. As long as ascorbic acid was present, the anthocyanins remained unaltered, but the degradation of anthocyanins started as soon as all the ascorbic acid was consumed; simultaneously, the oxidation products of phenolic extract started to appear.

PROCESSING OF LYCHEE

Although lychee is most desirable as a fresh fruit, but during the peak season it can be processed into canned fruit, juice, and dehydrated products. Dried lychees are known as "Lychee nuts." During drying, the pulp shrivels around the seed and is very pleasant in flavor and develops raisin-like texture. One of the best methods to retain the fresh flavor of lychee is freezing. The fruit can be frozen without peeling or the fruit may be peeled and seeded or left unseeded and frozen in syrup. Hand peeling is commonly used, but Chan and Cavaletto (1973) have successfully employed a combination of a hot-lye dip and mechanical peeling for lychee fruits. Lychee can be fermented for making Chinese medicine or used for making lychee wine, pickles, and preserve in China (Ong and Acree, 1999; Karuwanna et al., 1994.). The lychee fruit can be preserved by canning in syrup. Jelly can also be prepared from lychee fruit (Kuhn, 1962), or it can be used in the preparation of sherbet and ice cream (Shaw et al., 1955). Lychee fruit can also be preserved as a highly flavored squash during the peak season (Sethi, 1985; Jain et al., 1988).

Canning of Lychee

Canning of lychee fruits is another important preservation method for preparing value-added products. The general canning process for lychee fruits involves washing, peeling, pitting, and filling into enameled cans. About 30 Brix syrup is added as a packing medium to improve the flavor, and to lower the pH to 3.8, 0.1–0.2% citric acid is added. The filled cans are exhausted to an internal temperature of 80°C, vacuum sealed, processed in boiling water for 12 min, followed by rapid cooling to room temperature. Alternately, a high-speed spin cooker at 90.6°C for 2–3 min can be used for obtaining a better quality product (Luh et al., 1986). Pink discoloration could be reduced and better texture maintained in canning, if sugar content in syrup (containing 0.2% citric acid) was similar to that of the lychee fruit flesh (Wu and Chen, 1999). Using syrup containing sugar mixtures in the same ratio as found in lychee flesh, canned lychee with higher quality flesh could be produced than by using sucrose alone.

Pink discoloration in canned lychee is a serious problem and has been studied in several laboratories. The sterilization temperature and pH have strong effects on this pink color development (Wu and Fang, 1993). Pink discoloration in canned lychee is due to the hydrolysis of condensed tannins to catechin and leucoanthocyanin, which further degrades to anthocyanin. The cultivar and maturity stage are also reported to influence the discoloration (Hwang and Cheng, 1986). Flavanone-3-hydroxylase and dihydroquercetin-4-reductase from lychee flesh play a key role in the biosynthesis of leucoanthocyanin (Wu, 1992). According to Wu, the flavonones in mature lychee are converted to eriodictyol-containing compounds and then hydrolyzed by flavonone-3-hydroxilase to dihydroquercetin-containing compounds. These compounds are further reduced to leucocyanidin-containing compounds by dihydroquercetin-4-reductase and during heat processing, these are finally converted into cyanidin-containing colored compounds. However, during the storage of canned lychee, leucocyanindin-containing compounds may also develop through some other pathways, in addition to the above (Hwang and Cheng, 1986). To prevent pink discoloration in canned lychee, addition of 0.1–0.15% of citric acid to the packing medium (30 Brix syrup) and restricting the processing time in boiling water to less than 10 min have also been suggested (Chakravorty et al., 1974). Shortening the lapse of time between peeling and heating and immersing the lychee flesh in sodium bisulfite solution prior to heat processing are also known to reduce the extent of pink discoloration (Hwang and Cheng, 1986).

Lychee Juice

Juice from Lychee fruits is used for preparing juice blends or diluted into juice drinks, which are popular among consumers in Taiwan, China, Japan, South Africa, and in many Southeast Asian countries. The fruits for juice extraction are delivered to the factory shortly after harvesting, without much postharvest treatment. Peeling is a necessary step prior to juice extraction to avoid a bitter taste coming from the peels (Redlinghuys and Torline, 1980). As hand peeling is labor intensive and as it delays production, lowers juice quality, and leads to microbial spoilage, now mechanical peeling has become a favorite. Pitting of peeled fruit is not necessary. The peeled fruits are directly fed to a two-stage, paddle-type pulper finisher for juice extraction. By adjusting the clearance between the paddles and screen, the amount of broken seeds is reduced to less than 2%, and these seed fragments can be removed from the juice by the finishing screen. As the acidity is low in lychee fruits, pH of extracted juice is adjusted to 4.0 by adding citric acid prior to pasteurization at 95°C for 30 s. The pasteurized single-strength juice can either be hot-packed in 20-l tin cans or be frozen in 20-l plastic drums. Lychee juice may also be vacuum concentrated to double strength, shipped overseas as frozen concentrate at temperatures lower than −18°C. Taiwan, South Africa, and China are the major producers of lychee juice in the world.

FUTURE RESEARCH NEEDS

Fresh lychee fruit is most valued for its excellent flavor and textural qualities, but the maintenance of market quality of fresh lychees is particularly a serious problem for the fruit handlers when the fruit has to be shipped great distances. Desiccation of fruits with its accompanying loss of red color of fruit skin and the development of pericarp browning can occur very quickly. The development of browning renders the fruit unsaleable and results in commercial loss to the fruit handler. The chemical changes that lead to browning of pericarp need to be further investigated for improving the shelf life of fresh fruit. Pink discoloration in canned lychee is another problem that needs to be tackled to enhance the scope of producing value-added products from lychee. The lychee trees are environmentally exacting, and reach fruiting stage very slowly from seed, and most seedlings do not produce fruits of commercially acceptable quality. Therefore, the development of good quality cultivars and establishment of an industry have been slow to develop. Nevertheless, concerted research efforts with this crop have to be continued, to evolve good cultivars for the successful growth of the lychee processing industry.

Table 32.5. Major Papaya Fruit Producing Countries of the World

Country	Production, 000 Tons
World	5444
Brazil	1450
Nigeria	748
India	644
Mexico	613
Indonesia	470
China	152
Thailand	119

Source: FAO, 2001.

Section 3: Papaya

INTRODUCTION

Papaya (*Carica papaya* L.) is native to Central America but is now distributed throughout the tropical areas of the world. The fruit is mostly consumed fresh, but the immature is cooked or used in preserves, sauces, and pies. A number of acceptable products are being prepared by drying, canning, pickling, and preserving. From the raw fruit latex, papain enzyme is produced, which finds extensive uses in the food and pharmaceutical industry. Latex production is labor intensive, and the papain manufacturing industry is mainly confined to those areas where cheap labor is available. Apart from being an excellent source of ascorbic acid, the fruit is also a good source of provitamin A, some B complex vitamins, and many phytochemicals having antioxidant properties (Murcia et al., 2001; Leong and Shui, 2002). Pureed papaya is a good source of β–carotene and iron for the lactating women (Ncube et al., 2001). Brazil, Nigeria, India, Mexico, Indonesia, China, and Thailand are the leading producers of papaya (Table 32.5). However, USA, Taiwan, Puerto Rico, Peru, Bangladesh, and Australia are also producing sizeable quantities of this fruit (FAO, 2001).

CULTIVARS

Papaya is a rapid-growing, hollow-stemmed and short-lived perennial tree, belonging to the family Caricaceae, which is usually propagated from seeds. Because of open pollination, papaya is a notoriously difficult crop to maintain as a pure or true cultivar. This family includes 4 genera and about 20 species of *Carica* native to tropical and subtropical areas of the world. Papaya attains a height of about 10 m under favorable growing conditions. The plant is dioecious, with either male or female flowers, though trees with hermaphrodite flowers also occur (Samson, 1986). The fruit is a large, fleshy, hollow berry with small numerous seeds and weighs around 0.5–2.0 kg. Now, some cultivars have been developed with seedless fruits. Apart from Waimanalo, Solo is another important commercial variety of papaya, which produces hermaphrodite and female plants. The fruits from this variety have excellent quality for fresh consumption as well as for processing. Hortus Gold of South Africa, Improved Peterson of Australia, Betty, Solo Blue Stem, Red Rock, Cariflora of Florida (Conover et al., 1980), Semank of Indonesia, Sunny Bank, Guinea Gold, Hybrid-5 of Queensland, are other important cultivars. Washington, Honey Dew, Coorg Honey Dew, Pusa Delicious, Pusa Majesty, Pusa Giant, Pusa Nanha, Pusa Dwarf, CO1, CO2, CO4, CO5, and CO6 are grown extensively in India. Pusa Giant is well suited for canning (Muthukrishnan and Irulappan, 1985). Panama Red and Solo No.1 are commercially grown in Taiwan. Solo 62/3, Sunrise, and Solo cultivars are popular in Trinidad. About 10–15% of male trees are planted for pollination of female trees in a dioecious planting.

PHYSIOLOGY AND RIPENING

Papaya trees are fast growing and prolific fruit bearers. The first fruit is ready in 10–14 months from the time the plants are transplanted into the orchard (Sommer, 1985). Most cultivars in India take about 135–155 days from pollination to fruit maturity (Selvaraj et al., 1982a, b), and 168–182 days in Hawaii. However, in a warm, hot climate, "Washington" papaya takes 145–150 days to reach the skin-color-turning stage (Ghanta et al., 1994). The fruit weight (Selvaraj et al., 1982a, b; Ghanta et al., 1994) as well as fruit length (Ong, 1983) shows a typical double sigmoid type of growth curve. Increased fruit numbers could be attributed to improved fruit set and retention induced by the application of boron, and to increased production of indole acetic acid (IAA) induced by zinc application. IAA is known to affect flower production and fruit set in papaya (Kavitha et al., 2000a). The application of these two minerals is also known to affect the biochemical and quality characters of this papaya cultivar (Kavitha et al., 2000b). Treatment with zinc and boron produced higher levels of TSS, total sugars, reducing sugars, and nonreducing sugars (approx. 12.9%, 6.6%, 5.6%, and 1%, respectively). The TA and ascorbic acid in these treated fruits averaged approximately 0.29% and 47.1 mg/100 g. Similarly, uptake of calcium by "Sunset" papaya (*Carica papaya* L.) fruit plays an important role in its ripening process. Mesocarp Ca concentration of about 130 μg/g of fruit weight was associated with slower fruit softening rate than in fruit with a lower Ca concentration (Yunxia et al., 1995).

Yield and quality of papaya fruits are greatly influenced by NPK fertilizer application. Use of N at 200 g, P at 50 g, and K at 100 g per tree was found to be the most effective dose for increasing fruit yield and quality (ascorbic acid, TSS, and sugar contents) of mature papaya fruits (Lavania and Jain, 1995). Application of micronutrients during fruit growth and ripening are known to influence the fruit quality (Chattopadhyay and Gogoi, 1992). Micronutrients (B, ZN, Cu, Fe, and Mn) affected TSS, maximum levels (11.2%) were found in fruits treated with 40 ppm of boron. Treatment with 40 ppm of boron also increased total sugars (7.69% versus 6.6%) and ascorbic acid (65.63 versus 60.84 mg/100 g pulp). Treatment with B, Cu, and Zn (all 40 ppm) reduced TA. Carotene content was found to be higher in treated fruits (2.07–2.33 versus 2.01 mg/100 g). They recommended a combined foliar application of these micronutrients (40 ppm of each) to obtain good quality papaya fruits.

Various hydrolases play an important role in the modification of cell walls and softening of tropical fruits. Lazan and Ali (1993) have reviewed the biochemistry of softening process (depolymerization of pectin and hemicellulose), activity of cell wall hydrolases, role of PG, and β–galactosidase in mango and papaya softening, and the isolation of the β–galactosidase gene from mango and papaya. Mesocarp softening during papaya ripening was impaired by heating at 42°C for 30 min followed by 49°C for 70 min, with the areas of the flesh failing to soften (Paull and Chen, 1990). Disruption of the softening process varied with the stage of maturity and harvest date. The respiratory climacteric and ethylene production were higher and occurred 2 days earlier in the injured fruit than in the noninjured fruit that was exposed to 49°C for only 30 min. Skin degreening and internal carotenoid synthesis were unaffected by the heat treatments. Exposure of ripening fruits to either 42°C for 4 h or 38–42°C for 1 h followed by 3 h at 22°C resulted in the thermotolerance to exposure to the otherwise injurious heat treatment of 49°C for 70 min. Although several physical methods such as reflectance measurement, delayed light emission intensity, and body transmission spectroscopy have been tried to measure fruit maturity, but subjective evaluation based on skin color change is usually used to judge the maturity (Calegario et al., 1997). Once the major accumulation of TSS and DM has occurred after 120 DAA, and papaya fruit does not accumulate starch, commercial fruit quality is ensured if these fruits are harvested at 145 DAA, when the first yellow coloration appears in the peel and the seeds turn black. Softness to touch is another indicator used as a ripening index. Among the physicochemical determinants, pH and TSS (Brix) are very good indicators of ripening of the papaya fruit (Camara et al., 1993). Change of latex color from white to watery, is another index of maturity of papaya fruit (Akamine and Goo, 1971).

The chemical composition of papaya fruit with respect to sugars, organic acids, amino acids, vitamins, and minerals change during ripening. At harvest, water content varied from 87% to 97%, carbohydrates from 2% to 12%. The DM, which was 7% at 15 days after pollination, increased to 13% at harvest. Alcohol-insoluble solids, starch and several minerals decreased, whereas total sugars increased during this time. The total and nonvolatile acidity decreased to a minimum at the fully ripe stage of harvest (Selvaraj et al., 1982a). The organoleptic qualities, volatile profiles, and lipid content of papaya have been shown to be highly dependent on the degree of fruit maturity (Blakesley et al., 1979). Soluble sugars accumulate mainly when papaya fruits are still attached to the plant. Sucrose synthesis still occurs even after harvest, as sucrose-phosphate synthase activity is highly correlated to sucrose content. Sugar content and sweet sensory perception are dissociated, while pulp softening has a strong correlation with sweetness, probably due to the easier release of cellular contents in fully ripe tissues (Gomez et al., 2002).

Ethylene Production

Papaya, being a climacteric fruit, shows a typical respiratory and ethylene production pattern during ripening. Respiration rises to a maximum at the onset of ripening (the climacteric peak) but subsequently declines slowly. Respiratory climacteric is just preceded with a similar pattern of increased ethylene production (Paull and Chen, 1983). The climacteric peaks in four papaya cultivars (Coorg Honey Dew, Pink Flesh Sweet, Sunrise, and Washington) have been observed between 120 and 150 DAA, when the fruit skin color started to turn yellow (Selvaraj et al., 1982b). An increasing trend of mitochondrial protein and RNA content until harvest maturity are associated with an increased synthesis of enzymes responsible for catalyzing the ripening process (Pal and Selvaraj, 1987). Ethylene-forming enzyme (EFE) activity has been observed to be at the maximum in the exocarp of three-fourth ripe papaya fruits, but the EFEs in the mesocarp and endocarp were more heat sensitive than the EFE in the exocarp (Chan, 1991). Slicing and deseeding increases respiration, ethylene production, skin degreening, and flesh softening (Paull and Chen, 1997). Fruit with 60–80% skin yellowing had higher initial ethylene production,and respiration than that of the other ripening stages. Ethylene production and respiration of halved and deseeded fruits declined rapidly within 1 day during storage at 4°C. Fruits with 55–80% skin yellowing and less than 50 N flesh firmness, had more than 50% edible flesh and easily removable seeds. Such fruits were suitable for minimal processing when combined with low storage temperature of 4°C.

CHEMICAL COMPOSITION

Papaya is a common dessert, used in fruit salads and is enjoyed throughout the tropical countries. The raw fruit is also used as a vegetable for cooking. Papaya is a wholesome fruit, rich in sugars, and vitamins C, A, B_1, and B_2. Papaya is second only to mango as a source of β-carotene, a precursor of vitamin A. The physicochemical quality of papaya fruits (such as, mean fruit weight, pulp yield, pulp–peel ratio, TSS as Brix, vitamin C, and total carotenoids) is influenced by various agronomic practices, planting time of the year being an important parameter (Singh and Singh, 1998). September planting produced heavier mean fruit weight (2.30 kg), maximum TSS (11.2 Brix), vitamin C (74.55 mg/100 g) and total carotenoids (1152.50 mg/100 g), higher pulp–peel ratio, than that of the fruits harvested from other months of planting. A typical composition of ripe papaya fruit is presented in Table 32.6.

Table 32.6. Nutritional Value of Papaya (Per 100 g of Edible Portion of Raw Fruit)

Constituent	Content
Water, g	88.7
Food energy, kJ	165
Protein, g	0.6
Fat, g	0.1
Total carbohydrates, g	10
Fiber, g	0.9
Ash, g	0.6
Calcium, mg	20
Phosphorus, mg	16
Iron, mg	0.3
Sodium, mg	3
Potassium, mg	234
Vitamin A, IU	1750
Thiamine, mg	0.04
Riboflavin, mg	0.4
Niacin, mg	0.3
Ascorbic acid, mg	56

Source: USDA, 1968.

Sugars

Among the carbohydrates, sugars are the major constituents of papaya fruit but amounts vary considerably depending on the cultivar and agronomic conditions. Indian cultivars have higher sugar content (10–10.2% TSS) than the papaya cultivars being grown in the United States (5.65–7.1%) (Pal and Subramanyam, 1980; Madhav Rao, 1974). Presence of invertase enzyme in papaya resulted in the discrepancies in nonreducing sugar contents reported by different researchers (Chan and Kwok, 1975). They inactivated this enzyme with microwave heating before sugar extraction and reported papaya fruit to have 48.3% sucrose, 29.8% glucose, and 21.9% fructose of the total sugar content. Frederich and Nichols (1975) have reported that the papaya contained 30 calories, 10 g carbohydrates, 1.1 g fat, 0.6 g protein, and 0.9 g crude fiber per 100 g of fruit flesh. Slightly different values reported by Munsell (1950) for 100 g of papaya fruit flesh are 88.6–89.3% moisture and 0.6–0.7 g crude fiber. In addition to invertase, papaya also contains other enzymes such as papain, esterase, PG, myrosinase, and acid phosphatase, which play important roles in the quality and stability of processed products made from papaya (Jagtiani et al., 1988).

Acids and Volatiles

The pH of papaya pulp ranges between 5.5 and 5.9 and it is low in acidity. The TA, as citric acid, is reported to be 0.099% (Jagtiani et al., 1988). Citric and malic acids are the major acids with smaller quantities of ascorbic acid and α–ketoglutaric acid (Chan et al., 1971). Apart from contributing flavor to the fruit, these acids may be used as a substrate in respiration when sugars have been exhausted. Among the 106 volatile compounds in papaya, linalool is the major constituent having odor characteristic to that of fresh papaya. Linalool was found to be formed by the enzymic activity during cell disruption. The compounds that are significant to papaya flavor include a major volatile component, linalool; several esters, because of their fruity flavors; lactones (γ-hexalactone, γ- and δ-octalactones); and β-ionone, because of their flavor threshold. Another major constituent, benzyl isothiocyanate, has a pungent off-flavor. Butyric, hexanoic, octanoic acids, and their respective methyl esters are the other components responsible for off-flavor in papaya puree (Flath and Forrey, 1977). Among the 18 additional compounds reported by Macleod and Pieris (1983), methyl butanoate was found to be responsible for the sweet odor of some papaya fruits. Volatile aroma compounds emanating from fresh-cut papaya cv. Solo over a 3-day storage period at 20–22°C were found to be linalool (sweet + flowery) and benzaldehyde (almond), with smaller quantities of cis- and translinalool oxides, cyclohexane, hexanoic acid, and benzenemethanol (Mohammed et al., 2001). After this storage period, the relative amounts of these compounds changed by nearly 50%, with benzyl acetate being the dominant aroma volatile component.

Vitamins and Minerals

Ascorbic acid present in fruits is more stable in the acidic medium of natural fruit juice than in vegetables. Furthermore, fruits are always eaten as fresh, thereby avoiding the destruction caused by cooking required for vegetables. During the development of papaya, ascorbic acid increases gradually, reaching the maximum value of 55 mg/100 g at fully ripe stage of maturity (de Arriola et al., 1975). The change in outer color of the skin of fruit is an indicator of ripeness, and this change is considered mainly due to an increase in the carotene content and a decrease in chlorophyll. The total carotenoids content increases many folds from the mature green stage to nearly 4 mg/100 g at the fully ripe stage of maturity (Selvaraj et al., 1982b; Birth et al., 1984). Carotenoids contents differ between the yellow- and red-fleshed papaya fruits (Cynthia et al., 2000). The red-fleshed papaya has 63.5% of the total carotenoids as lycopene, which is absent in yellow-fleshed fruit (Yamamoto, 1964). Munsell (1950) has evaluated two samples of Guatemalan papaya to contain 0.025 and 0.030 mg of thiamin, 0.029 and 0.038 mg of riboflavin, and 0.238 and 0.399 mg of niacin per 100 g of fruit, but Asenjo et al. (1950) found niacin to range between 0.17 and 0.64 mg with an average of 0.320 mg per 100 g of papaya fruit.

Papaya fruits are also good sources of many minerals (K, Mg, and B) in human diet (Hardisson et al., 2001). The most abundant mineral is potassium, which is found combined with various organic acids. Awada and Suehisa (1973) reported the following minerals (in %) in papaya flesh from Hawaii: N, 0.12; P, 0.01; K, 0.21; Ca, 0.03; and Mg, 0.02. Munsell (1950) analyzed two samples of Guatemalan papaya (per 100 g flesh): N, 0.11 and 0.097 g; P, 1.8 and 15.5 mg; Ca, 18.3 and 17.5 mg; Fe, 0.25 and 0.42 mg; and ash, 0.46 and 0.57 g.

Pectin

Reduced firmness or softening of the fruit observed during the ripening process is due to the hydrolytic change of protopectin to pectin. The enzymatic demethylation and depolymerization of protopectin lead to the formation of low molecular weight compounds with less methoxyl groups, which are insufficient to maintain the firmness of fruit. In these textural changes, PG and PME enzymes play an important role (Kertesz, 1951). Loss of firmness is not uniform in papaya fruit, as sometimes the fruit becomes soft before the complete development of TSS (Pelag, 1974). During the ripening process, water-soluble pectin content increases, reaching a maximum 2 days after the fruit starts to ripen. The increase in water-soluble pectin corresponds well with the decrease in protopectin of papaya fruit during ripening (de Arriola et al., 1975). β-galactosidase-I is undetectable in immature fruits but appears to specifically accumulate during ripening (Ali et al., 1998). β-galactosidase-II is present in developing fruits, but its levels seem to decrease during ripening.

β-galactosidase-I seems to be an important softening enzyme, its activity increases four- to eightfold during early ripening stages. β-galactosidase-II may also contribute significantly to the softening of papaya fruit because of its ability to catalyze increased solubility and depolymerization of pectin as well as the alkali-soluble hemicellulose fraction of the cell wall. The degree of methyl esterification of pectin molecules has been reported to be inversely related to the firmness values for green (95.42 N), intermediate (50.70 N), and ripe (9.61 N) papaya fruits (Manrique and Lajolo, 2002). Hemicellulose modification and pectin hydrolysis are involved in the softening of papaya fruit, the latter apparently being more important during the late phase of fruit softening (Paull et al., 1999). The variety of fruit, the growing conditions and the state of development at the time of harvest influence the chemical composition of pectin (Lassoudiere, 1969a, b).

Pigments

During ripening, the color of papaya flesh turns yellow or reddish. The carotenoids content (as β-carotene) showed a five- to tenfold increase in yellow-fleshed cultivars as the green ripe fruit matured to full ripe stage (Selvaraj et al., 1982a, b). However, in red-fleshed cultivars the change of color was due to the marked increase in lycopene content (Bramley, 2000; Sesso et al., 2004). The major difference between yellow- and red-fleshed cultivar is the total absence of lycopene in the yellow-fleshed papaya. Various carotenoids such as β-carotene, lycopene, β-cryptoxanthin, and β-zeacarotene are present in varying amounts in different cultivars (Chan, 1983; Irwig et al., 2002; Cano et al., 1996; Wilberg and Rodriguez-Amaya, 1995a, b; Bhaskarachary et al., 1995; Sugiura et al., 2002). Carotenoids, which are relatively heat stable, showed higher retention than anthocyanins after blanching and drying in fruits (Sian and Soleha, 1991). Levels of carotenoids and anthocyanins decrease progressively in pineapple and papaya as blanching temperature and time increased. Pretreatment with sodium metabisulfite prevented the oxidation of carotenoids but caused bleaching of anthocyanins. Orthophosphoric acid also changed the color intensity of anthocyanins but showed no effect on carotenoids. Carotenoids are more protected in a system in which a higher moisture level is maintained by glycerol and sugar. Anthocyanins, however, are stable only within a certain range of moisture contents.

POSTHARVEST HANDLING AND STORAGE

The increased export of papaya fruit from the growers to the consuming countries has made the storage properties of fresh fruit more important. Papaya fruit is highly perishable and is susceptible to fungal attack during storage and transportation. Under cold storage, apart from the commonly encountered chilling injury (Wills, 1990), papaya fruit also suffers from inability to ripen properly, lack of flesh color development, persistence of green color in skin, loss of firmness, and increased susceptibility of fruit to fungal attack. A range of storage temperatures has been suggested for different cultivars of papaya harvested at the color break stage of maturity. Papaya fruit stored at 30°C has a maximum shelf life of only 7 days (Maharaj, 1988). Papaya at 10–15°C can be stored for 16 days (Aziz-Abou et al., 1975). de Arriola et al. (1980) have recommended 12°C as an optimum temperature for 2 weeks of shelf life. Storage at a temperature lower than 10°C causes chilling injury to papaya fruit, though the tolerance to temperatures below 10°C varies with the maturity of the fruit and the duration of exposure (Ali et al., 1993). Papaya fruits are shown to ripen satisfactorily at temperatures between 20°C and 25°C but storage above 32.2°C leads to delayed coloring and ripening, rubbery pulp texture, latex oozing, and surface bronzing (An and Paull, 1990). The deterioration of cut fruits is not due to injury caused by cold storage but due to the activities of several membrane and cell wall hydrolases, ethylene biosynthetic enzymes, and cell wall polyuronide degradation during low-temperature storage of papaya (Karakurt and Huber, 2003). Paull et al. (1997) have reviewed the available information on the storage and handling of papaya (*C. papaya* L.), with special reference to their effects on fruit quality and losses during marketing covering aspects such as papaya postharvest losses; current handling practices; factors affecting fruit quality (maturity, abrasion/impact, storage temperature, ethylene treatment, ripening conditions, heat treatment, diseases, and nonpathological disorders); market preferences, fruit quality parameters, and observations in the market chain (Fig. 32.1).

Low-oxygen (1–1.5%) conditions during low-temperature storage benefit papaya fruits only slightly. The best atmosphere for maintaining consumer acceptability and market quality of papaya fruit during storage for 14 and 21 days has been suggested to be 1% O_2 and 5% CO_2 (Hatton and

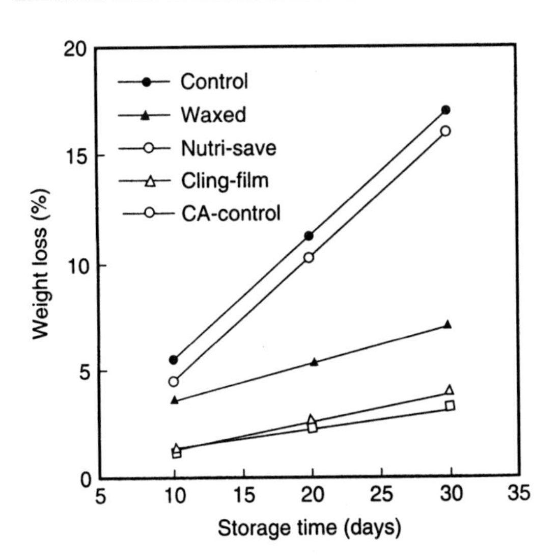

Figure 32.1. Weight loss of papaya during storage at 16°C under modified and controlled atmospheric conditions. (*Source:* Maharaj, 1988.)

Reeder, 1968). Storage of papaya fruit with 1.5–2.0% O_2 and 5% CO_2 at 16°C showed the lowest weight loss, negligible changes in firmness, and a longer time of ripening with slower color development (Maharaj, 1988). He found cling film wrapping of papaya fruits to be very effective in reducing weight loss during storage. Fruit waxing reduced weight loss only by 14–40% compared with that of 90% for plastic cling film. If CO_2 level in controlled atmosphere exceeds 7%, sometimes off-flavors are developed in waxed and wrapped papaya fruits at the fully ripe stage of maturity (Paull and Chen, 1989). Waxing serves two purposes: it reduces the weight loss and improves the fruit appearance to the consumers. Fruits coated with Fresh Mark wax 51V (Fresh Mark Chemical Co., Florida, USA) were successfully stored for 29 days before the onset of fungal infection (Maharaj and Sankat, 1990). Seal packaging is reported to modify both internal O_2 and external (in-package) atmospheres. A significant reduction in internal O_2 and a concomitant decrease in internal ethylene concentration are instrumental in delaying the ripening of sealed papaya fruits (Lazan et al., 1990). Use of methyl jasmonate vapor enhanced postharvest quality of papaya by reducing the loss of firmness, fungal decay and development of chilling injury, and increased retention of organic acids. Modified atmospheric packaging inhibited yellowing, together with the loss of water and firmness (Gonzalez-Aguilar et al., 2003).

Use of chemicals and ionizing irradiations to retard the ripening of fruits are other methods of preservation. Papaya fruits treated with gibberellic acid, vitamin K, silver nitrate, and cobalt chloride showed extended shelf life without any adverse effect on eating quality (Mehta et al., 1986). They attributed this improved shelf life to decreased rate of respiration due to lowered succinate and malate dehydrogenase activity in the TCA cycle. Use of a combination of hot water treatment with irradiation (75 krad) extended the shelf life of papaya fruits by an additional 8 days over the untreated controls, when stored at 20°C (Brodrick et al., 1976). Gamma irradiations alone at 100–150 krad extended the shelf life of "Solo" papaya by an additional 3 days over the controls (Chye et al., 1980). Irradiated fruits exhibited reduced fungal infection, possibly due to the production of chlorogenic acid in the skin of fruit through the increased activity of phenylalanine ammonia lyase enzyme. Although a lower dosage of gamma irradiation extended the shelf life, it decreased the ascorbic acid contents in papaya fruits. On the other hand, higher irradiation dosages caused an excessive softening of fruit and increase in POD activity (Zhao et al., 1996; Paull, 1996). The firmness of irradiated fruits was retained at least for 2 days more than the control, and the irradiated fruits also had a slower rate of firming (D'Innocenzo and Lajolo, 2001). Incidence and severity of peel scald was increased by irradiation, regardless of storage and ripening regime (Miller and McDonald, 1999). However, degree of severity was dependent on fruit maturity at the time of irradiation. Irradiation at quarter yellow stage of maturity causes the most serious incidence and severity of scald. Mature green fruit ripened at 25°C without storage had the lowest incidence of hard areas in the fruit pulp ("lumpy" fruit). Quarter yellow fruits generally were only second to the irradiated mature green fruits stored at 10°C in incidence of lumpiness.

TSS (Brix), cellulase activity, and ethylene production were not affected by irradiation treatment of papaya fruits. The activities of PG, PME, β-galactosidase, cellulase, and 1-aminocyclopropane-1-carboxylate oxidase correlated to changes in firmness. Evidently, irradiation had no direct effect on firmness but acted by altering the ripening-induced synthesis of cell wall enzymes, mainly the PME. POD, PPO, and catalase enzymes are important during ripening as well as during frozen storage of papaya pulp. The POD reactivation in frozen papaya tissues could be important during processing and could lead to undesirable changes in quality, especially in development of off-flavors (Cano et al., 1995). The use of gibberellic acid (150 ppm) was found to be very effective at reducing physiological loss in weight, TSS, and total sugar contents and in maintaining fruit firmness during storage of papaya fruits for 12 days at 30°C ± 2°C and 60–70% relative humidity. Further ripening parameters such as color and total carotenoid content were delayed, thus increasing the shelf life by an additional 4 days over that of control papayas (Ramakrishna et al., 2002).

PROCESSING OF PAPAYA

Papaya trees grow fast, start bearing fruit in less than a year and are prolific bearers of fruits. All these qualities make papaya an important fruit crop for the processing industry. Besides consumption as a fresh fruit, papaya has many applications as a food material. As a consequence, a number of processed food products, such as puree, jam, jelly, candied fruit, juice, mixed beverages, baby foods, nectar, canned slices/chunks, concentrate, powder, dried slices/chunks, have been prepared from papaya on a commercial scale. Some of these products will be discussed in more detail below.

PAPAYA PUREE

Papaya puree is the major semiprocessed product that finds use in juices, nectars, fruit cocktails, jams, jellies, and fruit leather. The initial step for the manufacture of puree is the removal of skins and seeds. A lye-peeling technique involves the use of hot caustic soda solution of 10–20% concentration, followed by water washing (Cancel et al., 1970). Machines have also been developed for the removal of skins and seeds (Brekke et al., 1973; Chan, 1977). Earlier, the processing of papaya into puree was difficult mainly due to product gelation and off-flavor development. The development of undesirable odors due to the presence of butyric, hexanoic, and octanoic acids and their methyl esters was observed in puree prepared by commercial methods. In an improved method for processing puree, acidification and heat inactivation of enzymes prevented the development of these unpleasant odors and flavors (Chan et al., 1973). A number of treatments have been developed to prevent gelation in puree. Gelation develops due to PE activity, which can be prevented by heat treatment of pulp. Increasing the TSS of puree to 26 Brix by adding sucrose (Chang et al., 1965) or by lowering the pH of puree to 3.4 inhibits gelation (Brekke et al., 1973). To inactivate the enzymes and to stabilize the puree against deterioration in quality during frozen storage, the acidified puree is heated at 96°C for 2 min and cooled quickly to 30°C, packed and stored at −23°C or below. Before freezing for storage, the puree is passed through a 0.5-mm screen to remove fruit fibers. Brekke et al. (1972) have developed a process to produce high-quality puree free from off-flavor and gelation problems. Microwave treatment of papaya puree produced a small change in qualitative and quantitative composition of carotenoid pigments, without significant alterations to the original color of the fruit puree (de Ancos et al., 1999). Deoxygenation by glucose oxidase mixed with catalase was found to retain the color, flavor, and ascorbic acid content of aseptically packaged papaya puree after a storage period of 9 months at ambient temperature (Chan and Ramanajaneya, 1992).

Papaya puree can be sold as such or it can be utilized as a raw material for other processed products. When it is sold as puree, it is usually canned or frozen. Puree prepared from varieties grown in Hawaii was reported to contain 11.5–13.5% TSS, 50–90 mg/100 g ascorbic acid, 3.5–3.9 mg/100 g carotenoids, and 84–88% moisture content (Brekke et al., 1973). The degradation of carotenoid pigments and visual color varies linearly with the temperature of processing and is therefore suggested to be used for on-line quality control of papaya puree (Ahmed et al., 2002).

CANNED PAPAYA PRODUCTS

Canned papaya chunks or slices are some of the popular ingredients employed for the preparation of fruit salads. Although fully ripe, soft papaya fruits are ideal for fresh consumption, but they are not suitable for canning purposes. For canning, only the green mature or semiripe papaya fruits are used. Lynch et al. (1959) have described a canning procedure for papaya slices or chunks. Papaya fruits are washed, peeled, and deseeded manually. The peeled fruits are diced in 2-cm cubes. About 300 g of cubes are filled into No.2 cans (307 × 409). Hot 40 Brix syrup containing 0.75% citric acid is poured over these cubes, leaving 0.8-mm headspace. The filled cans are exhausted in steam or hot water at 71°C, and sealed and processed in boiling water for 10 min. For ensuring the inactivation of PE enzyme, Nath and Ranganna (1981) have recommended thermal processing of

3-cm papaya cubes in $2^{1}/_{2}$ size cans (hot filled with syrup at pH 3.8) at 100°C for 16.2 min so as to achieve a F-value of 1.33 or D-value of 2.5 during the canning process. For the establishment of required thermal processing time for canned papaya puree, the use of destruction studies with *Clostridium pasteurianum* conducted at the same temperature used for the inactivation of resistant portion of PE enzyme has been recommended (Dos-Amagalhaes et al., 1996).

Candied Papaya

Fully mature but unripe fruit is hand peeled, deseeded, and cut into 0.5–1.0 cm cubes. These cubes are soaked in 4% brine solution for 2 weeks, after which they are taken out and leached in running tap water to remove the salty taste. The fruit is boiled in 25 Brix syrup for a few minutes and then allowed to stand overnight. The sugar content of syrup is increased by 10% by boiling the mixture of fruit/syrup and allowed to stand overnight. This process is repeated for 5 days until the syrup concentration reaches between 70 and 75 Brix after standing. Other fruit essences such as orange, pineapple, raspberry, and strawberry are added to the syrup and allowed to stand overnight. At this stage a translucent candied papaya is obtained, which can be rolled in powder sugar to prevent stickiness. The candied papaya finds extensive use in baked products, ice cream, and confectionery formulations. The papaya fruit can also be cut into 7.5-cm cubes and pricked with a fork. These pieces are soaked in 1.5% lime solution for 3–4 h and then washed in fresh water. The remaining process of boiling in syrup is the same as explained above. The finished product pieces are crisp, juicy, and almost transparent. This product is packed in sterilized wide-mouthed containers for long-term storage (Kumar, 1952).

Jam and Jelly

Fully mature and ripe but firm fruits are used for the preparation of jam and jelly. The fruits are peeled and seeds and inner white rind are removed. The fruit is cut into thin slices and cooked with little water to soften. The cooked slices are mashed, mixed with equal weight of sugar, and the mixture is cooked. Citric acid at the rate of 5 g/kg of pulp is added to improve the sugar–acid ratio and it also helps in the production of inverted sugars, which prevent sugar crystallization in jam during storage. The cooking of fruit pulp/sugar mixture is continued until it attains a thick consistency, which usually corresponds to a TTS of 65–68 Brix. The jam is filled hot into clean, dry, and sterilized glass jars, sealed airtight and cooled, labeled, and stored. For the preparation of jelly, a clarified fruit extract from papaya pulp is mixed with the sugar, and the mixture is cooked to obtain satisfactory sheeting test or to a temperature of 106.5°C. The product is filled hot in dry, sterilized glass jars, sealed airtight and cooled (Lal and Das, 1956). Recently, reduced calorie tropical mixed fruit jams using low methoxyl pectin, acesulfame-K, and sorbitol have also been made from pineapple, papaya, and carambola mixtures (Abdullah and Cheng, 2001). The most acceptable single fruit was pineapple, followed by papaya, but the papaya had the best color among all these jams.

Juice, Concentrate, and Nectar

Juice and nectar are prepared from papaya puree, either alone or in combination with other fruit juices of different flavors to formulate exotic beverages. A number of formulations have been suggested by various researchers (Benk, 1978; Brekke et al., 1973; Rodriguez and de George, 1972). Nectar is a RTS beverage, which is prepared from thin pulp with sugar and citric acid. The final product has TSS of 15–20 Brix and a mild acidic taste. The pulp is mixed with sugar (1.5–2.0 kg/kg pulp) and citric acid (12.5–17.5 g/kg pulp), color and flavor if required. The nectar is filtered and heated to 85–88°C, filled in plain or lacquered cans and cooled in running water to about 38°C. The cans are allowed to dry under ambient conditions to prevent rusting. Storage time and conditions affect the chemical composition and flavor of nectar, juices, and beverages. The type of container and diffused sunlight do not have significant effect on the darkening reactions (Payumo et al., 1968). However, the storage temperature is an important factor in determining the shelf life and quality of these products. After 50 weeks of storage at 38°C, 100 ppm of Fe and 400 ppm of Sn were found in the enamel-lined canned juices. The metal migration can be stopped at storage temperatures of 13–24°C (Brekke et al., 1976). Ascorbic acid decreases the color darkening of juices stored either at room or at refrigeration temperatures. Readings for color density were three times lower in samples that were refrigerated and/or treated with ascorbic acid than those of untreated controls (Payumo, 1968).

Squash formulations based on mango–papaya juice blends are prepared using sugar, citric acid, and water. Among the mango–papaya blended squash formulations, the one containing 75 parts of mango and 25 parts of papaya was found to be the most acceptable in terms of color, appearance, flavor, and taste attributes even after 90 days of storage at room temperature (Saravana-Kumar and Manimegalai, 2001). Pretreatments given to papaya such as blanching was found to be the most effective in affecting the shelf life quality of papaya squash. Sulfur fumes and citric acid influenced the taste and color of the final product (Sheeja and Prema, 1995). RTS beverage from the guava-papaya blends can be prepared using 15% pulp, TSS of 14 Brix, and 0.3% acidity (as citric acid) and processing at 90°C for 20 min. After a storage period of 6 months at room temperature, the RTS beverage from pure guava had the highest vitamin C content (28.1 mg/100 g), whereas carotene content was highest (441.6 μg/100 g) in pure papaya beverage (Tiwari, 2000). Sensory quality score of guava–papaya (70:30) blend was the highest due to better consistency and flavor, and it also had fair amounts of vitamin C (24.7 mg/100 g) and carotene (303.7 μg/100 g). For the preparation of papaya concentrate, puree is treated with pectolytic enzyme (0.05–0.02%) for about 1–2 h at a temperature of 50–60°C to reduce its viscosity (Sreenath and Santhanam, 1992). The use of hydrolytic enzymes permits preparation of papaya juice free from suspended matter, with a low viscosity and with a flavor similar to that of the fresh fruit (Hermosilla et al., 1991). The depectinized puree is concentrated in a vacuum concentrator up to threefold, packaged, and stored at frozen temperatures. About 5.5% and 14.3% loss of ascorbic acid during pulping and concentrating steps was observed, respectively. About 10–15% losses in carotenoids and 40% in ascorbic acid contents during concentration operations were observed (Chan et al., 1975; Janser, 1997).

Blending of fruit juices could be an economic requisite and blending also helps to improve the appearance, nutrition, and flavor of finished products (Kalra et al., 1991). About 25–33% of papaya pulp (being richer in ascorbic acid than mango) could be incorporated in mango without affecting the quality, nutritive value, and acceptability of the blended beverage. Papaya, being cheaper than mango, could be blended with Dashehari, Chausa, or other varieties of mango in the preparation of economically viable blended beverages. Nectar prepared from papaya alone has an unpleasant aftertaste, whereas adding mango enhances nectar acceptability significantly (Mostafa et al., 1997). A blend of 15% papaya + 15% mango pulp was rated as excellent and was characterized by higher acceptability. In another study, Imungi and Choge (1996) prepared nectar formulations by blending mango and passion fruit purees with papaya puree and pear juice. Based on the cost of ingredients and sensory scores of nectars, blends of passion fruit + papaya (10:90), mango + papaya (10:90), passion fruit + pear (50:50), mango + pear (50:50), and pear + papaya (10:90) were considered as the most acceptable in order of preference.

Dried Papaya Products

A number of low-moisture products such as fruit leather, powder, toffees, chunks, rolls, and slices have been prepared from papaya puree, which finds their places in food commodity market. Siddapa and Lal (1964) have patented a process for drying mixtures of papaya juices, previously concentrated, with sugar and other additives. A procedure for the dehydration of ripe papaya slices after steeping in 70 Brix syrup containing 1000 ppm of SO_2 was standardized to give the best quality product (Mehta and Tomar, 1980a, b). Slices from peeled and deseeded papaya fruit can be dried at 65.5°C to a moisture content of 8–10%. The dried slices are then ground to pass through 20-mesh screen. This product retained red color and much of the papaya flavor, though 5% of the ascorbic acid was lost, which can be minimized by using a vacuum drier. The pulp can also be dried after adding 5–7.5% sugar, 0.5% citric acid, and 0.3% potassium metabisulfite. This mixture is spread on greased trays in 1 cm thickness layer and dried in cabinet drier at 55–60°C. The dried product develops a leathery consistency, is rolled and cut into desirable sizes. This fruit leather has a shelf life of about 8 months when stored at 24–30°C (Ponting et al., 1966). Papaya toffee is a product similar to fruit leather and can be produced from puree. The pulp is first concentrated in a steam-jacketed kettle to a third of its original volume. Then glucose, skim milk powder, margarine, essence, and color are added. The mixture is cooked to reach a final TSS of 82 Brix. Cooked mass is spread on previously greased hard trays to a thickness of 0.33–0.50 cm. After cooling for 2 h, the sheet is cut into toffees and dried at 50–55°C to a final moisture content of 5–6%. The toffees are wrapped and packed in airtight jars or tins (Chan and Caveletto, 1968). Papaya fruit bars when stored at room temperature for 9 months retained 54%, 46%, and 43% of total carotenes, β-carotene, and vitamin C, respectively, and were judged to have superior texture and aroma with fewer physicochemical changes (Aruna et al., 1999). For cheese product containing fruit blends, optimal ratio of papaya puree to pineapple puree was 2:1 with 2% pectin and processed to 77–80 Brix (Barbaste and Badrie, 2000). Sensory analysis indicated a significant preference for the blended fruit cheese. Shelf life of these products at 4–5°C was around 8 weeks.

Most of the dried products prepared from papaya fruit suffer from undesirable darkening effects. To overcome these defects, less severe treatments have been tried. Freeze-drying is one such method that produced good results to reach a moisture content of as low as 3% in the finished papaya powder. Although ascorbic acid does not significantly affect differences during freezing, 15–20% reduction is observed after freeze-drying. Storage of freeze-dried powder in glass jars did not show significant adverse effects on the quality or composition of finished products after 3 months (Salazar, 1968). Carotenoids were found to be most stable in freeze-dried powder at a_w of 0.33 (6–7% moisture content) and were recommended for the storage of freeze-dried papaya (Arya et al., 1983).

A combination of osmotic dehydration and freezing has been investigated for the preservation of papaya slices (Moyano et al., 2002). When fast freezing with liquid nitrogen (−63°C for 10 min) is used then the osmotic drying conditions should be 65 Brix at 20°C for 60 min, and this combination gives a highly acceptable finished product. The ability to predict moisture and sugar contents accurately is useful for producing good quality papaya products by osmotic dehydration. Two models have been developed by Mendoza and Schmalko (2002) to predict the contents of moisture and sugar during osmotic drying of papaya slices. The osmotic (60 Brix, 60°C)—air-drying (60°C) method has been shown to save 8 h in drying papaya cubes from 6.58 to 0.24 kg water/kg DM as compared to the sample air-dried using the air-drying (60°C) method (Kaleemullah et al., 2002). The removal of moisture from papaya cubes increased with the increase of syrup temperature.

Minimally Processed Products

Minimal processing is based on a combination of mild heat treatment (blanching), a_w reduction, pH reduction, and addition of potassium sorbate and sodium metabisulfite. This process is also known as the hurdle technology and has been used for the preservation of fruit slices. Papaya chunks treated with increasing levels of preservatives up to 680 ppm of metabisulfite and 826 ppm of sodium benzoate exhibited good storage stability up to 90 days at 2°C and ambient temperature (Vijayanand et al., 2001). O'Connor-Shaw et al. (1994) studied the shelf life of minimally processed (peeled, deseeded, and diced) honeydew melon, kiwifruit, papaya, pineapple, and cantaloupe when stored at 4°C. Both the length of shelf life and type of spoilage were related to fruit species. Minimally processed fruits had longer shelf life at 4°C than at temperatures recommended for the whole fruit, if these were greater than 4°C. Spoilage of kiwifruit, papaya, and pineapple pieces stored at 4°C was found to be not due to microbial growth. Lopez-Malo et al. (1994) have produced shelf-stable high-moisture minimally processed papaya slices. The moisture and soluble solids contents, pH, and a_w remained almost constant in treated papaya slices during storage. These slices also remained microbiologically stable, due to the sucrose hydrolysis, which reduced a_w in the product. Ascorbic acid decreased during processing and storage at 5°C and 25°C. The total potassium metabisulfite and potassium sorbate concentrations decreased to 62% and 66% and 40% and 60% of their initial concentrations. No differences were observed in sensory properties of samples stored at 5°C or 25°C. Minimally processed papaya slices had good acceptability even after 5 months of storage at 25°C. The use of vacuum osmotic dehydration (VOD) techniques for the production of high-moisture minimally processed papaya has also been reported (Tapia et al., 1999). It was possible to obtain minimally processed papaya (a_w 0.98, pH 3.5) by applying vacuum osmotic drying for just 10 min when sucrose syrup contained 7.5% citric acid, or by applying pulsed VOD treatment (vacuum pulses for less than 15 min followed by osmotic drying for less than 45 min) when the citric acid concentration in sugar syrup was 2.5% or 5%.

BY-PRODUCTS FROM PAPAYA

Papain

Papain is the major by-product from dried latex derived from papaya fruit, which contains a protein-hydrolyzing enzyme. This enzyme has a number of specific technological applications such as in food, meat tenderization, beverages, and animal feeds; pharmaceutical industry; textile industry and detergents; paper and adhesives; medical applications; sewage and effluent treatment; and research and analytical chemistry (Flynn, 1975; Sanchez-Brambila et al., 2002; Kaul et al., 2002). Papain is a hydrolytic enzyme, which digests proteins into amino acids. Papain in solution is easily oxidized by exposure to air, high temperature (above 70°C), or sunlight. Contact with metals such as iron, copper, zinc, and many others also inhibit this enzyme. For improving the stability of papain enzyme, a number of chemicals such as ascorbic acid, sodium ascorbate, erythorbic acid, sodium erythorbate, sodium metabisulfite, 4-hexylresorcinol, TBHQ, rutin, α-tocopherol, trehalose, and sucrose have been tried (Epsin and Islam, 1998). The highest percentage of enzyme activity was retained at 55°C for all chemicals except sucrose and trehalose, which gave their best performance at 40°C. Among the food applications, the use of papain in chill haze removal during beer clarification as well as in the tenderization of meat has shown a steady increase over the past years. There is also a belief in some countries of Asia that eating papaya by pregnant ladies results in abortion (Adebowale et al., 2002). Based on rat feeding studies, they suggested that normal consumption of ripe papaya during pregnancy may not pose any significant risk but unripe or semiripe papaya may be unsafe in pregnancy, as the high concentration of latex produces marked uterine contractions.

Production of papain latex is an economically important alternate product of papaya cultivation. Sun drying of latex is a cause of concern especially in the heavy rainfall areas. Nowadays, vacuum shelf drying is more popular as it yields higher quality product. Papaya cultivars differ in papain yield, Red Panama (Lassoudiere, 1969a, b; Foyet, 1972), CO6 (Balmohan et al., 1992), CO2 (Wagh et al., 1992), and the line CP1512, CP1513, CP4 (Auxcilia and Sathiamoorthy, 1995), and CP5911 (Kanan and Muthuswamy, 1992) are shown to be high yielding. Other factors affecting the papain yield from papaya plants are fruit shape (Lassoudiere, 1969a, b), stage of maturity (Singh and Tripathi, 1957; Bhalekar et al., 1992), season of tapping (Reddy and Kohli, 1992), tapping time of the day (Lassoudiere, 1969a, b; Foyet, 1972), pattern of tapping (Madrigal et al., 1980), and frequency of tapping (Bhutani et al., 1963). Four repeated applications of the plant hormone, Ethephon in coconut oil (at 37 mM), have shown to increase papain yield (Shanmugavelu et al., 1976).

Muthukrishnan and Irulappan (1985) have described the procedure for the collection of latex as well as crude papain manufacturing process using simple equipment. Fruits grown for papain production are thinned on the plant so that each fruit hangs separately for easy collection of latex. Using plastic or stainless steel knives, fruits are lanced and the latex is collected in glass or porcelain containers. Four to five longitudinal cuts made during the morning hours gave the highest yield of latex. Over a period of 2 weeks, this process is repeated three or four times on the untapped portions of the fruits in 3–4 days intervals. The latex, which hardens within 15 min, is then precipitated with alcohol, washed with acetone, and dried in a vacuum drier for obtaining good quality crude papain. The addition of potassium metabisulfite (0.05%) in small quantities to the liquid latex before drying acts as a preservative and improves the quality of crude papain powder. The dried papain is powdered and sieved through a 10-mesh sieve and stored in airtight containers. The yield of crude papain powder obtained from raw green papaya is reported to be usually around 0.025% (Nanjundaswamy and Mahadeviah, 1993) (Fig. 32.2).

Pectin

To make papaya cultivation and papain industry viable, the profitable use of scarred fruit is essential. The quality of such papaya fruits does not appear to be affected adversely but only the fruit appearance seems less attractive to the consumers. The green fruits, whether scarred or not, are rich source of pectin (10% pectin on dry basis), which can be extracted for use in food industry (Das et al., 1954; Varinesingh and Mohammed-Maraj, 1989). Peel is shown to be higher in pectin content than the papaya pulp, and pectin content increases at a higher rate with fruit maturity up to a stage (Paul et al., 1998). The integrated processing of papaya fruits for the production of papain and pectin has been found to be economical (Nanjundaswamy and Mahadeviah, 1993). This process gives a papain yield of 0.25% and a pectin (jelly grade 200) yield of 1% on fresh fruit basis. The variety of the fruit, the growing conditions, and the stage of maturity of fruit are all known to influence the chemical composition of pectin (Lassoudiere, 1969a, b).

FUTURE RESEARCH NEEDS

Papaya fruit is of considerable economic importance, as it enjoys domestic markets in many tropical countries and export markets in the temperate countries. Limited shelf life of papaya fruit under ambient tropical conditions, susceptibility to mechanical

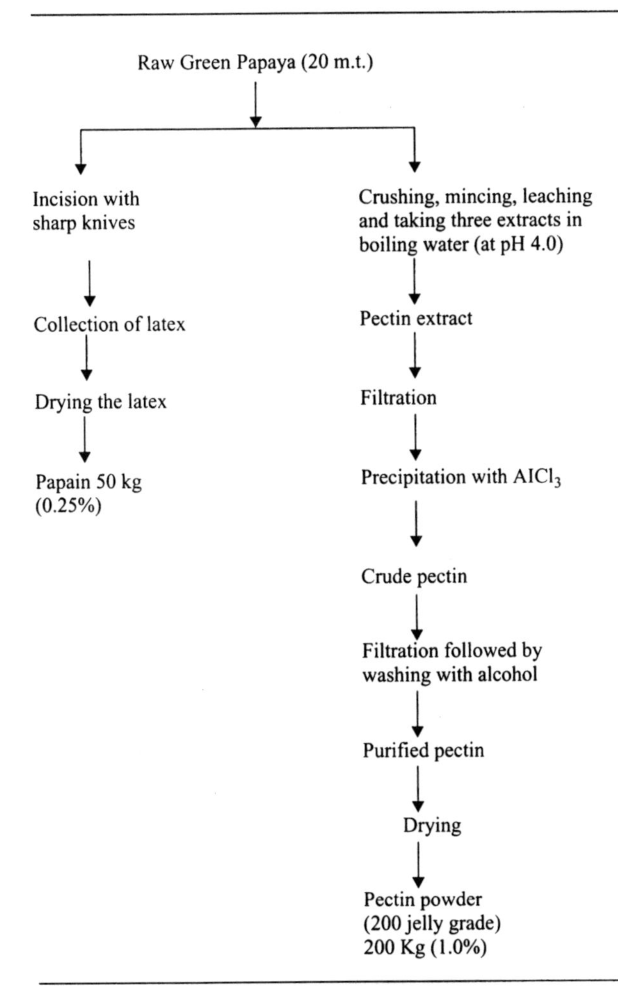

Figure 32.2. Flow sheet of integrated process for the production of papain and pectin. (*Source:* Nanjundaswamy and Mahadeviah, 1993.)

injury during handling and transportation, pest attack, and fungal diseases are some of the constraints that need to be overcome to prevent considerable fruit wastage. Recommended refrigerated storage temperature varies with the variety that need to be optimized to avoid chilling injury and to maintain desirable flavor of papaya fruit. Limited research has so far been reported on the use of modified and controlled atmospheres for papaya fruit storage. Improved methods of postharvest handling, storage, and packaging are necessary to alleviate the chilling injury in papaya fruits.

More research is necessary in tissue culture propagation of papaya rather than depending on the production of plants from seeds. This newer technology raises the prospects for rapid dissemination of high-yielding, disease-resistant cultivars, developed for specific conditions or localities to encourage the production of excellent quality fruits for fresh consumption as well as for processing industry.

REFERENCES

Abd-El-Aal, M.H. 1992. Production of guava seed protein isolates: yield, composition and protein quality. Nahrung 36(1):50–55.

Abdullah, A. and T.C. Cheng. 2001. Optimization of reduced calorie tropical mixed fruit jam. Food Quality and Preference 12(1):63–68.

Adebowale, A., A.P. Ganesan and R.N.V. Prasad. 2002. Papaya (*Carica papaya*) consumption is unsafe in pregnancy: fact or fable? Scientific evaluation of a common belief in some parts of Asia using a rat model. British Journal of Nutrition 88(2):199–203.

Aggarwal, P., G.S. Padda and J.S. Sidhu. 1997. Standardization of jelly preparation from guava:

grape blends. Journal of Food Science and Technology 34(4):335–336.

Agnihotri, B.N., K.L. Kapoor and R. Goel. 1962. Ascorbic acid content of guava fruits during growth and maturity. Science Culture 28:435–437.

Agrawal, R., P. Parihar, B.L. Mandhyan and D.K. Jain. 2002. Physico-chemical changes during ripening of guava fruit (*Psidium guajava* L.). Journal of Food Science and Technology 39(1):94–95.

Ahlawat, V.P., R. Yamdagni and P.C. Jindal. 1980. Studies on the effect of postharvest treatments on storage behavior of guava (*Psidium guajava* L.) cv. Sardar (L49). Haryana Agricultural University Journal of Research 10:242–247.

Ahmed, S. 1961. The Litchi. Agriculture Pakistan 12:769–775.

Ahmed, J., U.S. Shivhare and K.S. Sandhu. 2002. Thermal degradation kinetics of carotenoids and visual color of papaya puree. Journal of Food Science 67(7):2692–2695.

Akamine, E.K. and T. Goo. 1971. Relationship between surface color development and total soluble solids in papaya. HortScience 14:138–139.

Akamine, E.K. and T. Goo. 1973. Respiration and ethylene production during ontogeny of fruit. Journal American Society for Horticultural Science 98:381–383.

Ali, Z., Y.N. Shu, R. Othman, Y.G. Lee and H. Lazan. 1998. Isolation, characterization and significance of papaya β–galactosidases to cell wall modification and fruit softening during ripening. Physiologia Plantarum 104(1):105–115.

Ali, Z.M., H. Lazan, H.S. Ishak and M.K. Selamat. 1993. The biochemical basis of accelerated softening in papaya following storage at low temperature. Acta Horticulturae 343:230–232.

Ames, B.M., M.K. Shigena and T.M. Hagen. 1993. Oxidants, antioxidants and the degenerative diseases of aging. Proceedings of the National Academy of Sciences 90:7915–7922.

An, J. and R.E. Paull. 1990. Storage temperature and ethylene influence on ripening of papaya fruit. Journal of the American Society of Horticulture Science 115:949–953.

Anon. 1961. "Litchi." In: The Queensland Agricultural and Pastoral Handbook, edited by E.T. Hockings, 2nd edition, Vol. II. Brisbane, Australia: S.G. Reid Govt Printer.

Aruna, K., V. Vimala, K. Dhanalakshmi and V. Reddy. 1999. Physico-chemical changes during storage of papaya fruit (*Carica papaya* L.) bar (Thandra). Journal of Food Science and Technology 36(5): 428–433.

Arya, S.S., V. Netesan and P.K. Vijayraghavan. 1983. Stability of carotenoids in freeze dried papaya (*Carica papaya* L.). Journal of Food Technology 18:177–180.

Asenjo, C.F., D.B. Segundo, A.I. Muniz and A.M. Canals. 1950. Niacin content of tropical foods. Food Research 15:465–470.

Auxcilia, J. and S. Sathiamoorthy. 1995. Screening dioecious papaya for latex and proteolytic enzyme activity. South Indian Horticulture 43(1/2):1–4.

Awada, M. and R. Suehisa. 1973. Nutrient removal by papaya fruit. Horticulture Science 5:182–184.

Aziz-Abou, A.B., S.M. El-Nabawy and H.A. Zaki. 1975. Effect of different temperatures on the storage of papaya fruits and respirational activity during storage. Scientia Horticulturae 3:173–177.

Bagshaw, J., S. Underhill and J. Dahler. 1994. Lychee hydrocooling. Queensland Fruit and Vegetable News 16 June, pp. 12–13.

Balmohan, T.N., R. Sankarnarayanan, S. Sathiamoorthy, M. Kulasekaran and R. Armagam. 1992. Evaluation of papaya varieties for papain and fruit yield. National Seminar on Production and Utilization of Papaya. Tamil Nadu Agricultural University, Coimbatore, India, pp. 14 (Abstract).

Barbaste, A. and N. Badrie. 2000. Development of processing technology and quality evaluation of papaya (*Carica papaya*) cheese on storage. Journal of Food Science and Technology 37(3):261–264.

Bashir, H.A. and A.B.A. Abu-Goukh. 2003. Compositional changes during guava fruit ripening. Food Chemistry 80(4):557–563.

Batten, D.J. 1989. Maturity criteria for litchis (lychees). Food Quality and Preference 1(4/5):149–155.

Benk, E. 1978. Exotic fruits as raw material for soft drinks. Naturbrunnen 20:238–240. (German).

Bhalekar, M.N., A.N. Wagh, S.P. Patil and P.N. Kale. 1992. Studies on the effect of age of fruit on yield and quality of crude papain. National Seminar on Production and Utilization of Papaya. Tamil Nadu Agricultural University, Coimbatore, India, pp. 71 (Abstract).

Bhaskarachary, K., D.S. Shankar Rao, Y.G. Deosthale and V. Reddy. 1995. Carotene content of some common and less familiar foods of plant origin. Food Chemistry 54(2):189–193.

Bhullar, J.S., B.S. Dhillon and J.S. Randhawa. 1983. Extending the post harvest life of litchi cultivar 'Seedless Late'. Journal of Research Punjab Agricultural University 20:67–70.

Bhutani, R., J.V. Chankar and P.I.K. Manon. 1963. Papaya. Industrial Monograph, Central Food Technological Research Institute, Mysore, India, pp. 16.

Birth, G.S., G.G. Dull, J.B. Magee, H.T. Chan and C.G. Cavaletto. 1984. An optical method for estimating papaya maturity. Journal of American Society of Horticulture Science 109:62–66.

Blakesley, C.N., J.G. Loots, L.M. Duplessis and G. deBruyn. 1979. Gamma irradiation of subtropical fruits. 2. Volatile components, lipids and amino acids of mango, papaya and strawberry pulp. Journal of Agricultural and Food Chemistry 27(1):42–46.

Bortlik, K., L. Mortezavi and F. Saucy. 2001. A process for extraction of lycopene. European patent application.

Bramley, P.M. 2000. Is lycopene beneficial to human health? Phytochemistry 54(3):233–236.

Brekke, J.E., C.G. Caveletto, T.O.M. Nakayama and R.H. Suehisa. 1976. Effect of storage temperature and container lining on some quality attributes of papaya nectar. Journal of Agricultural and Food Chemistry 24:341–343.

Brekke, J.E., H.T. Chan and C.G. Cavaletto. 1972. Papaya puree: A tropical flavor ingredient. Food Product Development 6(6):36–37.

Brekke, J.E., H.T. Chan and C.G. Cavaletto. 1973. Papaya puree and nectar. Hawaii Agricultural Experiment Station Research Bulletin 170.

Brekke, J.E. and A.L. Myers. 1978. Viscometric behavior of guava puree and concentrates. Journal of Food Science 43:272–273.

Brodrick, H.T., A.C. Thomas, F. Visser and M. Beyers. 1976. Studies on the use of gamma irradiation and hot water treatments for shelf life extension of papayas. Plant Disease Reporter 60:749–754.

Broughton, W.J. and S.F. Leong. 1979. Maturation of Malaysian fruits. III. Storage conditions and ripening of guava (*Psidium guajava* L. var. GU3 and GU4). Mardi Research Bulletin 7:12–26.

Brown, B.I. and R.B.H. Wills. 1983. Post harvest changes in guava fruits of different maturity. Scientia Horticulturae 19:237–243.

Calegario, F.F., R. Puschmann, F.L. Finger and A.F.S. Costa. 1997. Relationship between peel color and fruit quality of papaya (*Carica papaya* L.) harvested at different maturity stages. Proceedings of the Florida State Horticulture Society 110:228–231.

Camara, M.M., C. Diez and M.E. Torija. 1993. Changes during ripening of papaya fruit in different storage systems. Food Chemistry 46(1):81–84.

Campbell, B.A. and C.W. Campbell. 1983. Preservation of tropical fruits by drying. Proceedings of the Florida State Horticulture Society 96:229–234.

Cancel, L.E., I. Hernandez and E. Rodriquez Rosa. 1970. Lye peeling of green papayas. Journal Agricultural University P R 54:19–27.

Cano, M.P., B. de Ancos and G. Lobo. 1995. Peroxidase and polyphenol oxidase activities in papaya during postharvest ripening and after freezing/thawing. Journal Food Science 60(4):815–820.

Cano, M.P., B. de Ancos, M.G. Lobo and M. Monreal. 1996. Carotenoid pigments and color of hermaphrodite and female papaya (*Carica papaya* L.) cv. Sunrise during postharvest ripening. Journal of the Science of Food and Agriculture 71(3):351–358.

Cesari, M., M. Pahor, B. Bartali, A. Cherubini, B.W.J.H. Penninx, G.R. Williams, H. Atkinson, A. Martin, J.M. Guralnik and L. Ferrucci. 2004. Antioxidants and physical performance in elderly persons: The Invecchiare in Chianti (InCHIANTI) study. American Journal of Clinical Nutrition 79(2):289–294.

CFTRI. 1990. Guava in India. Central Food Technological Research Institute, Mysore, India.

Chadha, K.L. and M.S. Rajpoot. 1969. Studies on floral biology, fruit set and its retention and quality of some litchi varieties. Indian Journal of Horticulture 26(3/4):124–129.

Chakravorty, S., R. Rodriguez, S.R. Sampathu and N.K. Saha. 1974. Prevention of pink discoloration in canned litchi (*Litchi chinensis* Sonn.). Journal of Food Science and Technology 11:266–268.

Chan Jr, H.T. 1977. Papaya seed removal. U.S. Patent 4002774.Chan, H.T. 1983. "Papaya." In: Handbook of Tropical Foods, edited by H.T. Chan. New York: Marcel and Dekker, pp. 469–488.

Chan Jr, H.T. 1991. Ripeness and tissue depth effects on heat inactivation of papaya ethylene-forming enzyme. Journal of Food Science 56:996–998.

Chan Jr, H.T. and C.G. Caveletto. 1968. Dehydration and storage stability of papaya leather. Journal of Food Science 43:1723–1725.

Chan Jr, H.T. and C.G. Cavaletto. 1973. Lye peeling of lychee. Hawaii Agricultural Experiment Station Research Report, pp. 215–220.

Chan Jr, H.T. and C.G. Cavaletto. 1986. Effects of deaeration and storage temperature on quality of aseptically packaged guava puree. Journal of Food Science 51:165–167.

Chan Jr, H.T., T.S.K. Chang, A.E. Stafford and J.E. Brekke. 1971. Non volatile acids in papaya. Journal of Agricultural and Food Chemistry 19:263–265.

Chan Jr, H.T., R.A. Flath, R.R. Forrey, C.G. Cavaletto, T.O.M. Nakayama and J.E. Brekke. 1973. Development of off-odors and off-flavors in papaya puree. Journal of Agricultural and Food Chemistry 21(4):566–570.

Chan Jr, H.T., M.T.H. Kuo, C.G. Caveletto, T.O.M. Nakayama and J.E. Brekke. 1975. Papaya puree and concentrate: Changes in ascorbic acid, carotenoids, and sensory quality during processing. Journal of Food Science 40:701–703.

Chan Jr, H.T. and S.C.M. Kwok. 1974. Nonvolatile acids in lychee. Journal of Food Science 39:792–793.

Chan Jr, H.T. and S.C.M. Kwok. 1975. Importance of enzyme inactivation prior to extraction of sugar from papaya. Journal of Food Science 40:770–771.

Chan Jr, H.T., S.C.M. Kwok and C.W.Q. Lee. 1975. Sugar composition and invertase activity in lychee. Journal of Food Science 40:772–774.

Chan Jr, H.T. and K.H. Ramanajaneya. 1992. Enzymatic deoxygenation of aseptically packaged papaya puree during storage. ASEAN Food Journal 7(1):47–50.

Chan, Y.K., Y.H. Yang and N. Li. 1998. Low-temperature storage elicits ethylene production in nonclimacteric lychee (*Litchi chinensis* Sonn.) fruit. HortScience 33(7):1228–1230.

Chandrasekhar, U. and S. Kowalya. 2002. Provitamin A content of selected South Indian foods by high performance liquid chromatography. Journal of Food Science and Technology 39(2):183–187.

Chang, H.T., J.E. Brekke and T. Chang. 1971. Non-volatile organic acids in guava. Journal of Food Science 36:237–239.

Chang, L.W.S., L.L. Morita and N.Y. Yamamoto. 1965. Papaya pectin esterase inhibition by sucrose. Journal of Food Science 30:218–222.

Chattopadhyay, P.K. and S.K. Gogoi. 1992. Influence of micronutrients on fruit quality of papaya (*Carica papaya* L.). Environmental Ecology 10(3):739–741.

Chen, M.L., C.Y. Lee and J.S. Wu. 1994. An evaluation of possible mechanisms for nonenzymatic browning in guava nectar during storage. Food Science Taiwan 21:293–303.

Chen, J.D. and J.S. Wu. 1991. Effect of salt concentration, pH and temperature on activity of pectinesterase from Pear cultivar guava fruit. Food Science Taiwan 18:85–91.

Chin, L.H., Z.M. Ali and H. Lazan. 1994. Comparative softening of guava fruits. Solubilization and depolymerization of cell wall carbohydrates during ripening. Proceedings of the Malaysian Biochemical Society Conference 19:147–150.

Chopda, C.A. and D.M. Barrett. 2001. Optimization of guava juice and powder production. Journal of Food Processing and Preservation 25(6):411–430.

Chyau, C.C., P.T. Ko, C.H. Chang and J.L. Mau. 2003. Free and glycosidically bound aroma compounds in lychee (*Litchi chinensis* Sonn.). Food Chemistry 80(3):387–392.

Chye, S.T., S. Ishak and P. Nitisewojo. 1980. "Effect of gamma irradiation and hot water treatment on papaya." In: Proceedings of the 2nd Symposium of the Federation of Asian and Oceanian Biochemists on Food and Nutritional Biochemistry, edited by H.T. Khor, K.K. Ong and K.C. Oo. Kuala Lumpur, Malaysia: The Malaysian Biochemical Society, pp. 148–156.

Conover, R.A., R.E. Litz and S.E. Malo. 1980. Cariflora-A papaya ring spot virus tolerant papaya for South Florida and the Caribbean. HortScience 21(4):1072–1074.

Cynthia, B., N. Kumar and K. Soorianathasundaram. 2000. Genetic variability and association of economic characters in red fleshed dioecious lines of papaya (*Carica papaya* L.). South Indian Horticulture 48(1–6):11–17.

Das, D.P., G.S. Siddapa and G. Lal. 1954. Effect of extraction of papain on the pectin content of raw papaya. CFTRI Mysore Research Bulletin 3:300–305.

de Ancos, B., M. Pilar-Cano, A. Hernandez and M. Monreal. 1999. Effects of microwave heating on pigment composition and color of fruit purees. Journal of the Science of Food and Agriculture 79(5):663–670.

de Arriola, M.C., J.F. de Calzada, J.F. Menchu, C. Rolz and R. Garcia. 1980. "Papaya." In: Tropical and Subtropical Fruits, edited by S. Nagy and P.E. Shaw. Westport, CT: AVI Publishing Co, pp. 316–340.

de Arriola, M.C., J.F. Menchu and C. Rolz. 1975. Some physical and chemical changes in papaya during its storage. Proceedings of the Tropical Regional American Society for Horticulture Science 19:97–99.

Dhingra, M.K. and O.P. Gupta. 1984. Evaluation of chemicals for pectin extraction from guava (*Psidium guajava* L.) fruits. Journal of Food Science and Technology 21:173–175.

Dhingra, M.K., O.P. Gupta and B.S. Chundawant. 1983. Studies on pectin yield and quality of some guava cultivars in relation to cropping season and fruit maturity. Journal of Food Science and Technology 20:10–14.

D'Innocenzo, M. and F.M. Lajolo. 2001. Effect of gamma irradiation on softening changes and enzyme activities during ripening of papaya fruit. Journal of Food Biochemistry 25(5):425–438.

Djousse, L., D.K. Arnett, H. Coon, M.A. Province, L.L. Moore and R.C. Ellison. 2004. Fruit and vegetable consumption and LDL cholesterol: the National Heart, Lung, and Blood Institute Family Heart Study. American Journal of Clinical Nutrition 79(2):213–217.

Dos-Amagalhaes, M.M., R.M. Tosello and P.R. de Massaguer. 1996. Thermal inactivation of pectinesterase in papaya pulp (pH 3.8). Journal of Food Process Engineering 19(3):353–361.

Duvenhage, J.A., M.M. Moster and J.J. Marais. 1995. Post harvest sulphuring and low pH treatment for retention of red skin color of litchi fruit. South African Litchi Growers Association Yearbook 7:44–46.

Einbond, L.S., K.A. Reynertson, X.D. Luo, M.J. Basile and E.J. Kennelly. 2004. Anthocyanin antioxidants from edible fruits. Food Chemistry 84(1):23–28.

Eipeson, W.E. and S.R. Bhowmik. 1992. Indian fruit and vegetable processing industry-Potential and challenges. Indian Food Packer 46(5):7–12.

El-Buluk, R.E., E.E. Babiker and A.H. Al-Tinay. 1995. Biochemical and physical changes in fruits of four guava cultivars during growth and development. Food chemistry 54:279–282.

El-Buluk, R.E., E.E. Babiker and A.H. Al-Tinay. 1996. Changes in sugar, ash and minerals in four guava cultivars during ripening. Plants Foods for Human Nutrition 49(2):147–154.

Epsin, N. and M.N. Islam. 1998. Stabilization of papain from papaya peels. Food Science and Technology International 4(3):179–187.

FAO. 2001. FAO Food Production Yearbook, Rome, Italy, Vol. 55, pp. 186–187.

Fitting, K.O. and C.D. Miller. 1959. The stability of ascorbic acid in frozen and bottled acerola juice alone and combined with other juices. University of Hawaii Agricultural Experiment Station Technical Bulletin, pp. 443–447.

Flath, R.A. and R.R. Forrey. 1977. Volatile components of papaya (*Carica papaya L.* Solo variety). Journal of Agricultural and Food Chemistry 25:103–109.

Flynn, G. 1975. The Market Potential for Papain, Tropical Products Institute, G-99, London.

Foyet, M. 1972. L'extraction de la papaine. Fruits 27:303–306.

Frederich, C.C. and C.H. Nichols. 1975. Food Values of Portions Commonly Used, 12th Edition. Philadelphia, J.B. Lippincott.

Gagrani, R.L., S.D. Rathi and U.M. Ingle. 1987. Preparation of fruit flavored beverage from whey. Journal of Food Science and Technology 24:93–95.

Garcia-Alonso, M., S. de Pascual-Teresa, C. Santos-Buelga and J.C. Rivas-Gonzalo. 2004. Evaluation of the antioxidant properties of fruits. Food Chemistry 84(1):13–18.

Garg, R.C. and H.B. Ram. 1974. A note on the effect of postharvest treatment with growth regulators on ripening and storage behavior of guava (*Psidium guajava* L.). Progressive Horticulture 6:53–54.

Garg, R.C., H.B. Ram and S.C. Singh. 1976. A note on the effect of postharvest treatment with indolebutyric acid and wax emulsion on the storage behavior of guava (*Psidium guajava* L.). Progressive Horticulture 8:85–86.

Ghanta, P.K., R.S. Dhua and S.K. Mitra. 1994. Studies on fruit growth and development of papaya cv. Washington. Indian Journal of Horticulture 51:246–250.

Gomez, M., F. Lajolo and B. Cordenunsi. 2002. Evolution of soluble sugars during ripening of papaya fruit and its relation to sweet taste. Journal of Food Science 67(1):442–447.

Gonzalez-Aguilar, G.A., J.G. Buta and C.Y. Wang. 2003. Methyl jasmonate and modified atmospheric packaging (MAP) reduce decay and maintain postharvest quality of papaya 'Sunrise'. Postharvest Biology and Technology 28(3):361–370.

Goodwin, T.W. and L.J. Goad. 1970. "Carotenoids and triterpenoids." In: The Biochemistry of Fruits and Their Products, edited by A.C. Hulme, vol. I. New York: Academic Press, pp. 305–368.

Gorinstein, S., M. Zemser, R. Haruenkit, R. Chuthakorn, F. Grauer, O. Martin-Belloso and S. Trakhtenberg. 1999. Comparative content of total polyphenols and dietary fiber in tropical fruits and persimmon. Journal of Nutritional Biochemistry 10(6):367–371.

Hamilton, R.A. and W. Yee. 1970. Lychee cultivars in Hawaii. Proceedings of the Florida State Horticulture Society 83:322–325.

Hardisson, A., C. Rubio, A. Baez, M.M. Martin and R. Alvarez. 2001. Mineral composition of the papaya (*Carica papaya* variety sunrise) from Tenerife island. European Food Research and Technology 212(2):175–181.

Hatton Jr, T.T. and W.F. Reeder. 1968. Controlled atmosphere storage of papayas. Proceedings of the Tropical Regional American Society of Horticulture Science 13:251–256.

Hermosilla, M.J., W.M. Perez, C.J. Moreno and Y. Looze. 1991. Enzymic hydrolysis of papaya pulp. Alimentos 16(4):5–8.

Hertog, M.G.L., P.C.H. Hollman and M.B. Katan. 1992. Content of potentially anticarcinogenic flavonoids of 28 vegetables and 9 fruits commonly consumed in the Netherlands. Journal of Agriculture and Food Chemistry 40:2379–2383.

Higgins, J.E. 1917. The Litchi in Hawaii. Hawaii Agricutural Experiment Station Bulletin 44:1–36.

Hollman, P.C.H. 2001. Evidence for health benefits of plant phenols: Local or systemic effects? Journal of the Science of Food and Agriculture 81:842–852.

Huang, S., H. Hart, H. Lee and L. Wicker. 1990. Enzymatic and color changes during post harvest storage of lychee fruit. Journal of Food Science 55(6):1762–1763.

Hwang, L.S. and Y.C. Cheng. 1986. "Pink discoloration in canned lychees." In: Role of Chemistry in the Quality of Processed Food, edited by O.R. Fennema, W.H. Chang and C.Y. Lii. Westport, CT: Food and Nutrition Press, Inc., pp. 219–228.

Imungi, J.K. and R.C. Choge. 1996. Some physicochemical characteristics of four Kenyan tropical fruits and acceptability of blends of their beverage nectars. Ecology of Foods and Nutrition 35(4):285–293.

Irwig, M.S., A. El-Sohemy, A. Baylin, N. Rifai and H. Campos. 2002. Frequent intake of tropical fruits that are rich in β-cryptoxanthin is associated with higher plasma β-cryptoxanthin concentrations in Costa Rican adolescents. Journal of Nutrition 132(10):3161–3167.

Jacobi, K.K., L.S. Wong and J.E. Giles. 1993. Lychee (*Litchi chinensis* Sonn.) fruit quality following vapor heat treatment and cool storage. Post harvest Biology and Technology 3(2):111–119.

Jagtiani, D.J., H.T. Chan Jr and W.S. Sakai. 1988. "Papaya." In: Tropical Fruit Processing. San Diego: Academic Press, pp. 105–143.

Jain, N.L. and D.H. Borkar. 1966. Preparing beverage from guava. Indian Horticulture 11:5–6.

Jain, S.P., V.K. Tripathi, H.B. Ram and S. Singh. 1988. Varietal suitability of litchi for squash making. Indian Food Packer 42:29–33.

Jaiswal, B.P., N.L. Sah and U.S. Prasad. 1987. Regulation of color break during *Litchi chinensis* Sonn. Ripening. Indian Journal of Experimental Biology 25:66–72.

Janser, E. 1997. Enzyme applications fro tropical fruits and citrus. Fruit Processing 7(10):388–393.

Jiang, Y. 2000. Role of anthocyanins, polyphenol oxidase and phenols in lychee pericarp browning. Journal of the Science of Food and Agriculture 80(3):305–310.

Jiang-Ping, J., S. Mei-xia and L. Pei-man. 1986. The production and physiological effect of ethylene during ontogeny and after harvest of litchi fruit. Acta Phytophysiologica Sinica 72:95–103.

Jimenez-Escrig, A., M. Rincon, R. Pulido and F. Saura-Calixto. 2001. Guava fruit (*Psidium guajava* L.) as a new source of antioxidant dietary fiber. Journal of Agriculture and Food Chemistry 49(11):5489–5493.

Johnston, J.C., R.C. Welch and G.L.K. Hunter. 1980. Volatile constituents of litchi (*Litchi chinensis* Sonn.). J Agric Food Chem 28:859–861.

Jordan, M.J., C.A. Margaria, P.E. Shaw and K.L. Goodner. 2003. Volatile components and aroma active compounds in aqueous essence and pink flesh guava fruit puree (*Psidium guajava* L.) using GC-MS and multidimensional GC/GC-O. Journal of the Agriculture and Food Chemistry 51(5):1421–1426.

Joshi, P.S., P.S. Waghmare, P.N. Zanjad and D.M. Khedkar. 1985. Utilization of curd and whey for preparation of fruit jelly. Indian Food Packer 39:38–40.

Joubert, A.J. 1970. The Litchi. Citrus Subtropical Fruit Research Institute (S. Africa) Bulletin, pp. 389–397.

Kader, A., D. Zagory and E.L. Kerbel. 1989. Modified atmosphere packaging of fruits and vegetables. Critical Reviews in Food Science and Nutrition 28:1–15.

Kaiser, C. 1995. Litchi (*Litchi chinensis* Sonn.) pericarp color retention—alternative to sulphur. South African Litchi Growers Association Yearbook 7:47–52.

Kaleemullah, S., R. Kailappan and N. Varadharaju. 2002. Studies on osmotic-air drying characteristics of papaya cubes. Journal of Food Science and Technology 39(1):82–84.

Kalra, S.K. and D.K. Tandon. 1984. Guava nectars from sulphited pulp and their blends with mango nectar. Indian Food Packer 38:74–76.

Kalra, S.K., D.K. Tandon and B.P. Singh. 1991. Evaluation of mango-papaya blended beverage. Indian Food Packer 45(1):33–36.

Kanan, M. and S. Muthuswamy. 1992. Association of certain biochemical constituents of green papaya fruit with papain production in eight genotypes. National Seminar on Production and Utilization of Papaya. Tamil Nadu Agricultural University, Coimbatore, India, pp. 69 (Abstract).

Kannan, S. and T.A. Susheela. 2002. Effect of osmotic dehydration of guava. South Indian Horticulture 50(1–3):195–199.

Karakurt, Y. and D.J. Huber. 2003. Activities of several membrane and cell-wall hydrolases, ethylene biosynthetic enzymes, and cell wall polyuronide degradation during low-temperature storage of intact and fresh-cut papaya (*Carica papaya* L.) fruit. Post harvest Biology and Technology 28(2):219–229.

Karuwanna, P., M. Boonyaratanakornkit, V. Surojchanamathakul and N. Sarikaputi. 1994. Table wine and fortified wine production from Hong Huay and Thai cultivars of lychee. Food 24(4):264–271.

Kaul, P., H.A. Sathish and V. Prakash. 2002. Effect of metal ions on structure and activity of papain from *Carica papaya*. Nahrung 46(1):2–6.

Kavitha, M., N. Kumar and P. Jeyakumar. 2000a. Role of zinc and boron on fruit yield and associated characters in papaya cv. CO5. South Indian Horticulture 48(1–6):6–10.

Kavitha, M., N. Kumar and P. Jeyakumar. 2000b. Effect of zinc and boron on biochemical and quality characters of papaya cv. CO5. South Indian Horticulture 48(1–6):1–5.

Kertesz, Z.I. 1951. The Pectic Substances. Interscience, New York.

Khedkar, D.M., K.W. Ansarwadkar, R.S. Dabhade and A.L. Ballal. 1982. Extension of shelf life of guava variety L-49. Indian Food Packer 36:49–52.

Khurdiya, D.S. and S.K. Roy. 1974. Studies on guava powder by cabinet drying. Indian Food Packer 28:5–8.

Knekt, P., J. Kumpulainen, R. Jarvinen, H. Rissanen, M. Heliovaara, A. Reunanen, T. Hakulinen, and A. Aromaa. 2002. Flavonoid intake and risk of chronic diseases. American Journal of Clinical Nutrition 76:560–568.

Knight Jr, R. 1980. "Origin and world importance of tropical and subtropical fruit crops." In: Tropical and Subtropical Fruits, edited by S. Nagy and P.E. Shaw. Westport, CT: AVI Publishing Co, pp. 1–120.

Kuhn, G.D. 1962. Some factors influencing the properties of lychee jelly. Proceedings of the Florida State Horticultural Society 75:408–410.

Kumar, V. 1952. Studies in *Carica papaya* L.: Economics of papaya cultivation and fruit utilization with special reference to the production of papain. Indian Journal of Horticulture 9(3):36–39.

Kumar, R. and N. Hoda. 1974. Fixation of maturity standards for guava (*Psidium guajava* L.). Indian Journal of Horticulture 31:140–142.

Lal, G. and D.P. Das. 1956. Studies on jelly making from papaya fruit. Indian Journal of Horticulture 13(1):3–6.

Lal, G., G.S. Siddappa and G.L. Tandon. 1986. Preservation of Fruits and Vegetables. Indian Council of Agricultural research, New Delhi.

Lassoudiere, A. 1969a. Papain—Production, properties and utilization. Fruits 24:503–512.

Lassoudiere, A. 1969b. The papaya crop, packaging for shipment, changes in products for export. Fruits 24:491–502.

Lavania, M.L. and S.K. Jain. 1995. Studies on the effects of different doses of N, P and K on yield and quality of papaya (*Carica papaya* L.). Haryana Journal of Horticulture Science 24(2):79–84.

Lazan, H. and Z.M. Ali. 1993. Cell wall hydrolases and their potential in the manipulation of ripening of tropical fruits. ASEAN Food Journal 8(2):47–53.

Lazan, H. and Z.M. Ali. 1997. "Guava." In: Tropical and Subtropical Fruits, edited by P.E. Shaw, H.T. Chan Jr, S. Nagy and W.F. Wardowski, Vol. III. Gainesville, Florida: Florida Science Publication, pp. 20–41.

Lazan, H., Z.M. Ali and W.C. Sim. 1990. Retardation of ripening and development of water loss stress in papaya fruit seal-packaged with polyethylene film. Acta Horticulturae 269:345–358.

Lee, H.S. and L. Wicker. 1991. Quantitative changes in anthocyanin pigments of lychee fruit during refrigerated storage. Food Chemistry 40(3): 263–270.

Lee, H.S. and L. Wicker. 1990. Anthocyanin pigments in the skin of lychee fruit. Journal of Food Science 56(2):466–468, 483.

Leong, L.P. and G. Shui. 2002. An investigation of antioxidant capacity of fruits in Singapore markets. Food Chemistry 76(1):69–75.

Li, B.W., K.W. Andrews and P.R. Pearsson. 2002. Individual sugars, soluble, and insoluble dietary fiber contents of 70 high consumption foods. Journal of Food Composition and Analysis 15(6):715–723.

Lichter, A., O. Dvir, I. Rot, M. Akerman, R. Regev, A. Wiesblum, E. Fallik, G. Zauberman and Y. Fuchs. 2000. Hot water brushing: An alternative method to SO2 fumigation for color retention of litchi fruits. Post harvest Biology and Technology 18(3): 235–244.

Lopez-Malo, A., E. Palou, J. Welti, P. Corte and A. Argaiz. 1994. Shelf-stable high moisture papaya minimally processed by combined methods. Food Research International 27(6):545–553.

Luh, B.S. 1971. "Tropical fruit beverages." In: Fruit and Vegetable Juice Processing, edited by D.K. Tressler and M.A. Joselyn. Westport, CT: AVI Publishing Co, pp. 341–372.

Luh, B.S., C.E. Kean and J.G. Woodroof. 1986. "Canning of fruit." In: Commercial Fruit Processing, edited by J.G. Woodroof and B.S. Luh, 2nd Edition. Westport, CT: AVI Publishing Co, pp. 286–345.

Luximon-Ramma, A., T. Bahorun and A. Crozier. 2003. Antioxidants actions and phenolic and vitamin C contents of common Mauritian exotic fruits. Journal of the Science of Food and Agriculture (5):496–502.

Lynch, L.J., A.T. Chang, J.C.N. Lum, G.D. Sherman and P.E. Seale. 1959. Hawaii Food Processors Handbook. Hawaii Agricultural Experimental Station, University of Hawaii Circ.

Macfie Jr, G.B. 1954. Packaging and storage of lychee fruits, preliminary experiments. Proceedings of the Florida Lychee Growers Association 1:44–48.

Macleod, A.J. and N.M. Pieris. 1983. Volatile components of papaya (*Carica papaya* L.) with particular reference to glucosinolate products. Journal of Agricultural and Food Chemistry 31:1005–1008.

Madhav Rao, V.N. 1974. Papaya. Farm Information Unit Bull 9, Ministry of Agriculture, New Delhi, India.

Madrigal, L., A. Ortiz, R.D. Cook and R. Fernandez. 1980. The dependence of crude papain on different collection (tapping) procedures for papain latex. Journal of the Science of Food and Agriculture 31:279–283.

Maharaj, R. 1988. The handling and storage of papayas (*Carica papaya* L.) under controlled conditions. M.Sc. thesis, Department of Chemical Engineering, University of West Indies, St. Augustine, Trinidad.

Maharaj, R. and C.K. Sankat. 1990. Storability of papaya under refrigerated and controlled atmosphere. Acta Horticulturae 269:375–386.

Maiti, S.C. 1985. "Litchi." In: Fruits of India: Tropical and Subtropical, edited by T.K. Bose. Calcutta, India: Naya Prokash Publishers, pp. 388–399.

Malo, S.E. and C.W. Campbell. 1968. The Guava. Fruit Crops Fact Sheet No. 4, Florida Agriculture Extension Service, Gainesville, USA.

Mendoza, R. and M.E. Schmalko. 2002. Diffusion coefficients of water and sucrose in osmotic dehydration of papaya. International Journal of Food Properties 5(3):537–546.

Manrique, G.D. and F.M. Lajolo. 2002. FT-IR spectroscopy as a tool for measuring degree of methyl esterification in pectins isolated from ripening papaya fruit. Post harvest Biology and Technology 25(1):99–107.

Marloth, R.H. 1934. The litchi. Farming South Africa 9:423–425, 438.

Mathew, A.B. and M.C. Pushpa. 1964. Organic acids and carbohydrates of litchi. Journal of Food Science and Technology 1:71–72.

McGuire, R.G. and E.A. Baldwin. 1996. Lychee color can be better maintained in storage through application of low-pH cellulose coatings. Proceedings of the Florida State Horticultural Society 109:272–275.

McGuire, R.G. and G.J. Hallman. 1995. Coating guavas with cellulose- or carnauba-based emulsions interferes with post harvest ripening. HortScience 30:294–295.

Mehta, P.M., S. Shiva Raj and P.S. Raju. 1986. Influence of fruit ripening retardants on succinate and malate dehydrogenases in papaya fruit with emphasis on preservation. Indian Journal of Horticulture 43:169–173.

Mehta, G.L. and M.C. Tomar. 1980a. Studies on dehydration of tropical fruits in Uttar Pradesh. III. Papaya (*Carica papaya* L.). Indian Food Packer 34(4):12–16.

Mehta, G.L. and M.C. Tomar. 1980b. Studies on the dehydration of tropical fruits in Uttar Pradesh. II. Guava (*Psidium guajava* L.). Indian Food Packer 34(4):8–11.

Menzel, C.M. and D.R. Simpson. 1991. A description of lychee cultivars. Fruit Varieties Journal 45(1):45–56.

Menzel, C.M. and D.R. Simpson. 1993. "Fruits of tropical climate-Fruits of Sapindaceae." In: Encyclopedia of Food Science, Food Technology and Nutrition. London, edited by R. Macrae, R.K. Robinson and M.J. Saddler. New York: Academic Press, pp. 2108–2112.

Menzel, C.M., B.J. Watson and D.R. Simpson. 1988. The lychee in Australia. Queensland Agriculture Journal 1/2:19–27.

Miller, C.D., K. Benzore and M. Bartow. 1957. Lychee. Fruits of Hawaii, University of Hawaii Press, Honolulu.

Miller, W.R. and R.E. McDonald. 1999. Irradiation, stage of maturity at harvest, and storage temperature during ripening affect papaya fruit quality. HortScience 34(6):1112–1115.

Mohamed, S., K.M.M. Kyi and S. Yusof. 1994. Effects of various surface treatments on the storage life of guava (*Psidium guajava* L.) at 10 degree. Journal of the Science of Food and Agriculture 66:9–11.

Mohammed, M., Y. Wang and S.J. Kays. 2001. Changes in volatile chemistry of fresh-cut papaya (*Carica papaya* L.) during storage. Tropical Agriculture 78(4):268–271.

Moran, I. 2000. Process for preserving product quality of lychee. US Patent No. 6093433.

Mostafa, G.A., E.A. Abd-El Hady and A. Askar. 1997. Preparation of papaya and mango nectar blends. Fruit Processing 7(5):180–185.

Moyano, P.C., R.E. Vega, A. Bunger, J. Garreton and F.A. Osorio. 2002. Effect of combined processes of osmotic dehydration and freezing on papaya preservation. Food Science and Technology International 8(5):295–301.

Mudahar, G.S. and B.S. Bhatia. 1983. Steeping preservation of fruits. Journal of Food Science and Technology 20:77–79.

Mukherjee, S.K. and M.N. Datta. 1967. Physico-chemical changes in Indian guava during fruit development. Current Science 36:674–676.

Munsell, H.E. 1950. Composition of food plants of Central America. VIII. Guatemala. Food Research 15:439–453.

Muralikrishna, M., A.M. Najundaswamy and G.S. Siddappa. 1968. Physico-chemical changes during the concentration of guava juices. Indian Food Packer 22(6):5–7

Murcia, M.A., A.M. Jimenez and M. Martinez-Tome. 2001. Evaluation of the antioxidant properties of Mediterranean tropical fruits compared with common food additives. Journal of Food Protection 64(12):2037–2046.

Muthukrishnan, C.R. and I. Irulappan. 1985. "Papaya." In: Fruits of India- Tropical and Subtropical. Calcutta, India: Naya Prokash, pp. 320–340.

Nagar, P.K. 1994. Physiological and biochemical studies during fruit ripening in litchi (*Litchi chinensis* Sonn.). Postharvest Biology and Technology 4(3):225–234.

Nanjundaswamy, A.M. and M. Mahadeviah. 1993. "Fruit processing." In: Advances in Horticulture, Fruit Crops, edited by K.L. Chadha and O.P. Pareek, Vol. IV. New Delhi, India: Malhotra Publishers, pp. 1865–1927.

Nath, N. and S. Ranganna. 1981. Determination of a thermal process schedule for acidified papaya. Journal of Food Science 46:201–206, 211.

Ncube, T.N., T. Greiner, L.C. Malaba and M. Gebre-Medhin. 2001. Supplementing lactating women with pureed papaya and grated carrots improved vitamin A status in a placebo-controlled trial. Journal of Nutrition 131(5):1497–1502.

Ness, A.R. and J.W. Powles. 1997. Fruits and vegetables, and cardiovascular disease: A review. International Journal of Epidemiology 26:1–13.

O'Connor-Shaw, R.E., R. Roberts, A.L. Ford and S.M. Nottingham. 1994. Shelf life of minimally processed honeydew, kiwifruit, papaya, pineapple and cantaloupe. Journal of Food Science 59(6): 1201–1206, 1215.

Ong, H.T. 1983. Abortion during the floral fruit development in *Carica papaya* in Serdeng, Malaysia. Pertanika 6:105–107.

Ong, P.K.C. and T.E. Acree. 1999. Similarities in the aroma chemistry of Gewuerztraminer variety wines and lychee (*Litchi chinensis* Sonn.) fruit. Journal of Agricultural and Food Chemistry 47(2):665–670.

Orr, K.J. and C.D. Miller. 1954. The Loss of Vitamin C in Frozen Guava Puree and Juice. Hawaii Agricultural Experiment Station Progress Notes, pp. 98–99.

Pal, D.K. and Y. Selvaraj. 1987. Biochemistry of papaya (*Carica papaya* L.) fruit ripening: Changes in RNA, DNA, protein and enzymes of mitochondrial, carbohydrates, respiratory and phosphate metabolism. Journal of Horticulture Science 62:117–124.

Pal, D.K. and M.D. Subramanyam. 1980. Studies on the physico-chemical composition of fruits of twelve papaya varieties. Journal of Food Science and Technology 17(6):254–256.

Palaniswamy, K.P. and K.G. Shanmugavelu. 1974. Physicochemical characters of some guava varieties. South Indian Horticulture 22:8–11.

Patra, D.K. and M.K. Sadhu. 1992. Influence of calcium treatment on shelf life and quality of lychee fruits. South Indian Horticulture 40:252–256.

Paul, P.P., K. Majumder and B.C. Majumdar. 1998. Pectin content in peel and pulp of *Carica papaya* L fruits. Science Culture 64(5/6):127–128.

Paull, R.E. 1996. Ripening behavior of papaya (*Carica papaya* L.) exposed to gamma-irradiation. Post harvest Biology Technology 7(4):359–370.

Paull, R.E. and N.J. Chen. 1983. Post harvest variation in cell wall-degrading enzymes of papaya (*Carica papaya* L.) during ripening. Plant Physiology 72:382–385.

Paull, R.E. and N.J. Chen. 1989. Waxing and plastic wraps influence water loss from papaya fruit during storage and ripening. Journal of the American Society of Horticulture Science 114:937–942.

Paull, R.E. and W. Chen. 1997. Minimal processing of papaya (*Carica papaya* L.) and the physiology of halved fruit. Post Harvest Biology and Technology 12(1):93–99.

Paull, R.E. and N.J. Chen. 1990. Heat shock response in field-grown, ripening papaya fruit. Journal of the American Society of Horticulture Science 115(4): 623–631.

Paull, R.E. and T. Goo. 1983. Relationship of guava (*Psidium guajava* L.) fruit detachment force to the stage of fruit development and chemical composition. HortScience 18:65–67.

Paull, R.E., K. Gross and Q. Yunxia. 1999. Changes in papaya cell walls during fruit ripening. Post Harvest Biology and Technology 16(1):79–89.

Paull, R.E., W. Nishijima, M. Reyes and C. Cavaletto. 1997. Post harvest handling and losses during marketing of papaya (*Carica papaya* L.). Post Harvest Biology and Technology 11(3):165–179.

Payumo, E.M., L.M. Pilac and P.L. Maniguis. 1968. A study of color changes in stored papaya nectar. Philippines Journal of Science 97:127–138.

Pelag, M. 1974. Determination of fresh papaya texture by penetration test. Journal of Food Science 39:701–703.

Pino, J.A., R. Marbot and C. Vazquez. 2002. Characterization of volatiles in Costa Rican guava (*Psidium friedrichsthalianum* (Berg) Niedenzu) fruit. Journal of the Agriculture and Food Chemistry 50(21):6023–6026.

Pong, C.C., Y.H. Lee and C.M. Wu. 1996. Effects of pectinase treatment on guava juice quality and volatile constituents. Food Science Taiwan 23(1): 77–87.

Ponting, J.D., G.G. Watters, R.R. Forrey, R. Jackson and W.L. Stanley. 1966. Osmotic dehydration of fruits. Food Technology 20(10):125–129.

Pornchaloempong, P., S.A. Sargent and C.L. Moretti. 1997. Cooling method and shipping container affect lychee fruit quality. Proceedings of the Florida State Horticultural Society 110:197–200.

Prasad, N.B.L. and G. Azeemoddin. 1994. Characteristics and composition of guava (*Psidium guajava* L.) seed and oil. Journal of the American Oil Chemists Society 71(4):457–458.

Prasad, U.S. and O.P. Jha. 1978. Changes in pigmentation patterns during litchi ripening: flavonoid production. The Plant Biochemical Journal 5: 44–49.

Prior, R.L. and G. Cao. 2000. Antioxidant phytochemicals in fruits and vegetables: diet and health implications. HortScience 35(4):588–592.

Pruthi, J.S., K.K. Mukherji and G. Lal. 1960. A study of factors affecting the recovery and quality of pectin from guava. Indian Food Packer 14:7–10.

Ramakrishna, M., K. Haribabu and K. Purushotham. 2002. Effect of post harvest application of growth regulators on storage behavior of papaya (*Carica papaya* L.) cv. 'CO2'. Journal of Food Science and Technology 39(6):657–659.

Reddy, Y.T.N. and R. Kohli. 1992. Effect of season on papain yield. National Seminar on Production and Utilization of Papaya. Tamil Nadu Agricultural University, Coimbatore, India, pp. 72 (Abstract).

Redlinghuys, H.J.P. and P.A. Torline. 1980. The preparation of litchi juice. South African Food Review 7(1):118–119.

Rivera, L.J., F.C. Ordorica and E.P. Wesche. 1999. Changes in anthocyanin concentration in lychee (*Litchi chinensis* Sonn.) pericarp during maturation. Food Chemistry 65(2):195–200.

Rodriguez, A.J. and L.M.I. de George. 1972. Evaluation of papaya nectar prepared from unpeeled papaya puree. Journal of Agricultural University P R 56:79–80.

Roychoudhury, R., J. Kabir, S.K.D. Ray and R.S. Dhua. 1992. Effect of calcium on fruit quality of litchi. Indian Journal of Horticulture 49:27–30.

Sachan, B.P., D. Pandey and G. Shankar. 1969. Influence of weather on chemical composition of guava fruits (*Psidium guajava* L.) var. Allahabad Safeda. Punjab Horticulture Journal 9:119–122.

Sachan, B.P. and K. Ram. 1970. Ascorbic acid content of different varieties of guava (*Psidium guajava* L.) in Allahabad region. Indian Food Packer 24:6–10.

Salazar, L.A. 1968. Lyophilization of tropical fruits. Chemical Engineering Thesis, University of San Carlos, Guatemala (in Spanish).

Samson, J.A. 1986. Tropical Fruits, 2nd edition. New York, Longman Publishers, pp. 256–269.

Sanchez-Brambila, G.Y., B.G. Lyon, Y.W. Huang, J.R. Franco-Santiago, C.E. Lyon and K.W. Gates. 2002. Sensory and texture quality of canned whelk (*Astraea undosa*) subjected to tenderizing treatments. Journal of Food Science 67(4): 1559–1563.

Sandhu, K.S. and B.S. Bhatia. 1985. Physico-chemical changes during preparation of fruit juice concentrate. Journal of Food Science and Technology 22:202–205.

Sandhu, S.S. and J.S. Randhawa. 1992. Effect of post harvest application of methyl-2-benzimidazole carbamate and in pack fumigant on the cold storage life of litchi cultivars. Acta Horticulturae 269:185–189.

Sandhu, K.S. and J.S. Sidhu. 1992. Studies on the development of multi fruit ready-to-serve beverages. Journal of Plant Science Research 8(1/4):87–88.

Sandhu, K.S., M. Singh and P. Ahluwalia. 2001. Studies on the processing of guava into pulp and guava leather. Journal of Food Science and Technology 38(6):622–624.

Saravana-Kumar, R. and G. Manimegalai. 2001. Formulations of mango-papaya blended squash. South Indian Horticulture 47(1–6):164–165.

Schieber, A., F.C. Stintzing and R. Carle. 2001. By-products of plant food processing as a source of functional compounds-recent developments. Trends in Food Science and Technology 12(11):401–413.

Selvaraj, Y., D.K. Pal, M.D. Subramanyam and C.P.A. Iyer. 1982a. Changes in the chemical composition of four cultivars of papaya (*Carica papaya* L.) during growth and development. Journal of Horticultural Sciences 57:135–143.

Selvaraj, Y., D.K. Pal, M.D. Subramanyam and C.P.A. Iyer. 1982b. Fruit set and the development pattern of fruits of five papaya varieties. Indian Journal of Horticulture 39:50–56.

Sesso, H.D., J.E. Buring, E.P. Norkus and J.M. Gaziano. 2004. Plasma lycopene, other carotenoids, and retinol and the risk of cardiovascular disease in women. American Journal of Clinical Nutrition 79(1):47–53.

Sethi, V. 1985. A simple and low cost preservation of litchi juice. Indian Food Packer 39(4):42–44.

Setiawan, B., A. Sulaeman, D.W. Giraud and J.A. Driskell. 2001. Carotenoid content of selected Indonesian fruits. Journal of Food Composition and Analysis 14(2):169–176.

Shanmugavelu, K.G., R. Chittiraichelum and V.N. Madhav Rao. 1976. Effect of ethephon on latex stimulation in papaya. Journal of Horticulture Science 51:425–427.

Shaw, T.N., S. Sakata, F.P. Boyle and G.D. Sherman. 1955. Hawaii tropical fruit flavors in ice creams, sherbets and ices. Hawaii Agricultural Experimental Station Circular No. 49.

Shaw, W.H., N.A. Sufi and S.I. Zafar. 1975. Studies on the storage stability of guava fruit juice. Pakistan Journal of Scientific and Industrial Research 18:179–183.

Sheeja, N. and L. Prema. 1995. Impact of pre-treatments on the shelf life quality of papaya squash. South Indian Horticulture 43(1/2):49–51.

Sian, N.K. and I. Soleha. 1991. Carotenoid and anthocyanin contents of papaya and pineapple: Influence of blanching and predrying treatments. Food Chemistry 39(2):175–185.

Siddappa, G.S. 1982. Status of fruit and vegetable preservation industry in India and future prospects. Indian Food Industry 1:73–78.

Siddapa, G.S. and G. Lal. 1964 (May 21). Improvement in or relating to the manufacture of fruit juice products. Indian Patent No. 49590.

Sidgley, M. and J.A. Gardner. 1989. International survey of underexploited tropical and subtropical perennials. Acta Horticulturae 250:2–6.

Singh, L.B. 1951. Air layering of litchi without soil or water. Current Science 20:102–104.

Singh, R. 1957. Improvement of packaging and storage of litchi at room temperature. Indian Journal of Horticulture 14:205–212.

Singh, B.P., S.K. Kalra and D.K. Tandon. 1990. Behavior of guava cultivars during ripening and storage. Haryana Journal of Horticulture Science 19:1–5.

Singh, B.P., H.K. Singh and K.S. Chauhan. 1981. Effect of post harvest calcium treatments on the storage life of guava fruits. Indian Journal of Agricultural Sciences 51:44–47.

Singh, R., A.C. Kapoor and O.P. Gupta. 1983. Effect of cultivars, seasons and storage on the nutritive value and keeping quality of guava cheese. Indian Food Packer 37:71–75.

Singh, G. and A.K. Singh. 1998. Physicochemical quality of papaya fruits (*Carica papaya* L.) as influenced by different planting times. Indian Food Packer 52(3):28–32.

Singh, L.B. and R.D. Tripathi. 1957. Studies on the preparation of papain. Indian Journal of Horticulture 14(2):77–80.

Slattery, M.L., K.P. Curtin, S.L. Edwards and D.M. Schaffer. 2004. Plant foods, fiber, and rectal cancer. American Journal of Clinical Nutrition 79(2): 274–281.

Sommer, N.F. 1985. "Post harvest handling systems: Tropical fruits." In: Post Harvest Technology of Horticultural Crops, edited by A.A. Kader, Publ 3311. Davis, CA: Cooperative Extension Service, University of California, pp. 162–169.

Sreenath, H.K. and K. Santhanam. 1992. Comparison of cellulolytic and pectinolytic treatment of various fruit pulps. Chemie Mikrobiologie Technologie der Lebensmittel 14(1/2):46–50.

Srivastava, S., D.R. Modi and S.K. Garg. 1997. Production of ethanol from guava pulp by yeast strains. Bioresource Technology 60(3):263–265.

Steinmatz, K.A. and J.D. Potter. 1991. Vegetables, fruits and cancer. I. Epidemiology. Cancer Causes and Control 2:325–357.

Sugiura, M., M. Kato, H. Matsumoto, A. Nagao and M. Yano. 2002. Serum concentration of β-cryptoxanthin in Japan reflects the frequency of Satsuma mandarin (*Citrus unshiu* Marc.) consumption. Journal of Health Science 48(4): 350–353.

Tandon, D.K., P.G. Adsule and S.K. Kalra. 1984. Effect of certain post harvest treatments on the shelf life of guava fruits. Indian Journal of Horticulture 41:88–92.

Tao, R. 1955. The superior lychee. Proceedings of the Florida Lychee Growers Association 2:73–74.

Tapia, M.S., A. Lopez-Malo, R. Consuegra, P. Corte and J. Welti-Chanes. 1999. Minimally processed papaya by vacuum osmotic dehydration (VOD) techniques. Food Science and Technology International 5(1):41–49.

Thompson, B.D. 1955. A progress report on handling and storage of fresh lychee. Proceedings of the Florida Lychee Growers Association 2:27–28.

Tiwari, R.B. 2000. Studies on blending of guava and papaya pulp for RTS beverage. Indian Food Packer 54(2):68–72.

Tribble, D.L. 1999. Antioxidants consumption and risk of coronary heart disease: Emphasis on vitamin C, vitamin E and beta-carotene: A statement for healthcare professionals from the American Heart Association. Circulation 99:591–595.

Tylavasky, F.A., K. Holliday, R. Danish, C. Womack, J. Norwood and L. Carbone. 2004. Fruits and vegetables intake are an independent predictor of bone size in early pubertal children. American Journal of Clinical Nutrition 79(2):311–317.

Uddin, M.S., M.N.A. Hawlader, D. Luo and A.S. Majumdar. 2002. Degradation of ascorbic acid in dried guava during storage. Journal of Food Engineering 51(1):21–26.

Underhill, S.J.R. and C. Critchley. 1992. The physiology and anatomy of lychee

(*Litchi chinensis* Sonn.) pericarp during fruit development. Journal of Horticultural Science 67(4):437–444.

Underhill, S.J.R. and C. Critchley. 1993. Physiology, biochemical and anatomical changes in lychee pericarp during storage. Journal of Horticultural Science 68:327–335.

Underhill, S.J.R. and D.H. Simons. 1993. The lychee (*Litchi chinensis* Sonn.) pericarp desiccation and the importance of post harvest micro-cracking. Scientia Horticulturae 54(4):287–294.

Underhill, S.J.R. and L.S. Wong. 1990. A maturity standard for lychee (*Litchi chinensis* Sonn.) Acta Horticulturae 16:245–251.

Underhill, S.J.R. and C. Critchley. 1995. Cellular localization of polyphenols oxidase and peroxidase activity in *Litchi chinensis* pericarp. Australian Journal of Experimental Agriculture 34:115–112.

USDA. 1968. Composition of Foods, Raw, Processed and Prepared. Agriculture Handbook 8, Watt and Merrill, USDA.

Varinesingh, P. and R. Mohammed-Maraj. 1989. Solar drying characteristics of papaya (*Carica papaya* L.) latex. Journal of the Science of Food and Agriculture 46:175–179.

Vijayanand, P., K.K.S. Nair and P. Narsimham. 2001. Preservation of pineapple, mango and papaya chunks by hurdle technology. Journal of Food Science and Technology 38(1):26–31.

Vijayanand, P., A.R. Yadav, N. Balasubramanyam and P. Narasimham. 2000. Storage stability of guava fruit bar prepared using a new process. Lebensmittel-Wissenschaft und Technologie 33(2): 132–137.

Vinson, J.A., X. Su, L. Zubik and P. Bose. 2001. Phenol antioxidant quantity and quality in foods: Fruits. Journal of Agriculture and Food Chemistry 49:5315–5321.

Wagh, A.N., S.P. Patil, M.N. Bhalekar, K.N. Wavhal and P.N. Kale. 1992. Evaluation of papaya varieties for yield and quality of crude papain. Maharashtra Journal of Horticulture 6(1):7–10.

Wara-Aswapati, O., J. Sornsrivichai, J. Uthaibutra and C. Oogaki. 1990. Effect of seal packaging by different plastic films on storage life and quality of litchi (*Litchi chinensis* Sonn.) fruits stored at three different temperatures. Japanese Journal of Tropical Agriculture 34:68–77.

Watkins, J.B. 1990. Forced-air cooling. Queensland Department of Primary Industries Information Series, Q188027, Brisbane.

Webber, H.J. 1944. The vitamin C content of guavas. Proceedings. American Society for Horticultural Science 45:87–90.

Wenkam, N.S. and C.D. Miller. 1965. Composition of Hawaii Fruits. Hawaii Agriculture Experiment Station Bulletin, pp. 135–136.

Wilberg, V.C. and D.B. Rodriguez-Amaya. 1995a. HPLC quantitation of major carotenoids of fresh and processed guava, mango and papaya. Lebensmittel-Wissenschaft und Technologie 28(5):474–480.

Wilberg, V.C. and D.B. Rodriguez-Amaya. 1995b. HPLC quantitation of major carotenoids of fresh and processed guava, mango and papaya. Lebensmittel Wissenschaft und Technologie 28(5):474–480.

Wills, R.B.H. 1990. Post harvest technology of banana and papaya in ASEAN. ASEAN Food Journal 5:47–50.

Wilson, W.C. 1980. "Guava." In: Tropical and Subtropical Fruits, edited by S. Nagy and P.E. Shaw. Westport, Connecticut, AVI Publishing Co, pp. 279–299.

Wong, L.S., K.K. Jacobi and J.E. Giles. 1991. The influence of hot benomyl on the appearance of stored lychee (*Litchi chinensis* Sonn.). Scientia Horticulturae 46(3/4):245–251.

Wu, M.C. 1992. Studies on Pink Discoloration of Lychee Flesh in Processing. Ph.D. Dissertation, National Taiwan University, Taiwan, ROC (Chinese).

Wu, M.C. and C.S. Chen. 1999. A research note: effect of sugar types and citric acid content on the quality of canned lychee. Journal of Food Quality 22(4): 461–469.

Wu, M.C. and T.T. Fang. 1993. Prevention of pink discoloration in canned lychee fruit (*Litchi chinensis* Sonn.) Journal of the Chinese Agricultural Chemistry Society 31(5):667–672.

Yamamoto, H.Y. 1964. Comparison of the carotenoids in yellow- and red-fleshed *Carica papaya.* Nature 201:1049–1050.

Yunxia, Q., M.S. Nishina and R.E. Paull. 1995. Papaya fruit growth, calcium uptake, and fruit ripening. Journal of the American Society of Horticulture Science 120(2):246–253.

Yusof, S. and S. Mohamed. 1987. Physico-chemical changes in guava (*Psidium guajava* L.) during development and maturation. Journal of the Science of Food and Agriculture 38:31–35.

Zauberman, G., R. Ronen, M. Akerman, A. Weksler, I. Rot and Y. Fuchs. 1991. Post harvest retention of the red color of litchi fruit pericarp. Scientia Horticulturae 47:89–97.

Zhao, M., J. Moy and R.E. Paull. 1996. Effect of gamma-irradiation on ripening papaya pectin. Post Harvest Biology and Technology 8(3): 209–222.

33
Banana, Mango, and Passion Fruit

Lillian G. Occeña-Po

INTRODUCTION

The dietary message "Diets rich in fruits and vegetables may reduce the risk of some types of cancer and other chronic diseases" endorsed by the Food and Drug Administration (FDA) for food manufacturing use has promoted increased consumption of fruits and fruit-based products, including tropical fruits. Emphasis on value-addition and consumer-driven market has also created interest in the use of tropical fruit products. Further, globalization coupled with the growing demand for tropical fruits has increased the export of these products to the West. This chapter focuses on the major commercial processing of banana, mango, and passion fruit.

BANANA

PRODUCTION, CONSUMPTION, AND STORAGE

In terms of world production, the banana (genus *Musa*) is one of the top three tropical fruits, along with citrus and pineapple. The major banana producers in 2003 were led by India (Fig. 33.1). However, a significant quantity of banana is grown in Africa, Central America, and the Caribbean. A distinction is made between the "dessert" or sweet bananas (*M. acuminata*), which are ripened and best eaten as such, and the "cooking" or starchy bananas and plantains (*M. balbisiana*), which are cooked or processed. Table 33.1 summarizes commercial banana cultivars. In the West, unblemished banana is preferred, almost required. The U.S. consumers prefer a yellow color with green tips, hence the Dwarf Cavendish (*M. cavendeishii*) is the main variety found in the United States. However, this variety is not as sweet and flavorful as the bananas available in the tropics. Blemishes are tolerable to some extent in tropical countries, where degree of ripeness is the main consideration for fresh consumption.

The per capita consumption of bananas in the United States is about 26.7 lbs on a fresh weight basis, down 4.5% from 2000 (USDA, 2003). However, banana still remains the number one fresh fruit consumed by the Americans. Bananas and plantains serve as a staple food in some countries of Africa, Asia, Central and South America. In East Africa, 20 million people are believed to subsist on banana as the principal source of dietary carbohydrate (Karugaba and Kimaru, 1999).

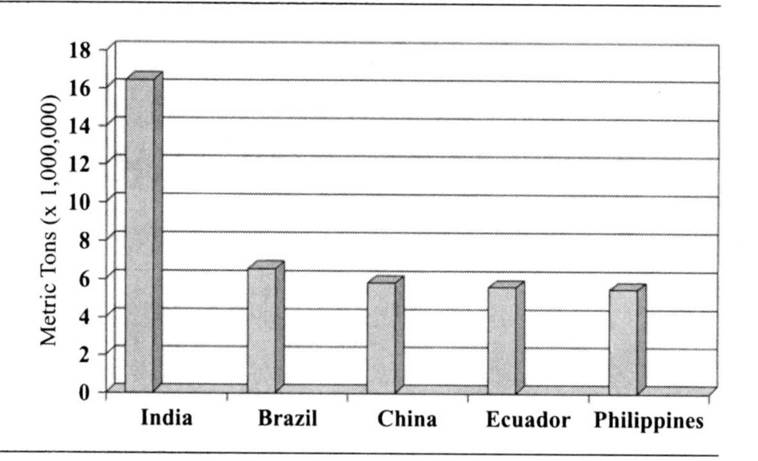

Figure 33.1. Top banana producers (FAO, 2003).

Table 33.1. Commercial Banana Cultivars

Cultivar	Other Terms	Countries	Description
Dwarf Cavendish		Canary Islands, Brazil, Israel, Somalia, South Africa, Guinea, Cameroon, Martinique	Most widespread cultivar; fruit projects from the bunch, so wrapped or cut and boxed for export
Giant Cavendish	Williams Hybrid	Australia, Martinique	Larger fruits than Dwarf
Pisang Masak Hijau (PMH)	Lacatan, Bungulan	Jamaica, Philippines	Distinct from the Philippine Lacatan
Gros Michel	Pisang Ambon, Bluefields	Cameroon, Burma, Thailand, Malaysia, Indonesia, Sri Lanka, Hawaii	Long established in Asia; widely distributed in the Caribbean
Robusta	Valery	Jamaica, Dominica, St. Lucia, St. Vincent, Grenada, Samoa, Fiji	Replacing PMH in Jamaica and Gros Michel in Central America

Source: Knight (1980).

Bananas are cut from the plants when green and are transported by ship to markets. During shipment, they may undergo some ripening. The U.S. Patent 3,950,559 (Kapoor and Turner, 1976) described a method to delay ripening of harvested bananas using gibberellins (A_4 and A_7). This postharvest treatment can be mixed with a fungicide such as benomyl.

The climacteric respiration of banana (60–250 mg/h CO_2/kg fruit), and ripening due to endogenous ethylene, cause a rapid conversion of starch to sugars. Reduced temperatures and modified atmospheres (1–10% O_2, 5–10% CO_2, or combinations of low O_2 and high CO_2) delay the onset of ripening. The optimum storage temperature for banana is about 13°C (Taylor, 2001). Nakasone and Paull (1998) described the treatment with about 100 ppm ethylene for 24 h for ripening to a targeted color. The peel color index of banana, which ranges from 1 to 7 (1 = green; 2 = with trace of yellow; 3 = slightly more green than yellow; 4 = more yellow than green; 5 = green tips at the end; 6 = completely yellow; 7 = yellow peel with brown spots of sugar), is taken into consideration to estimate the degree of ripening. Another measure of ripening is Brix. The ripened banana is about 23°Brix (Sole, 1996).

Processed Banana Products

The year-round supply of bananas makes preservation less economical, and the low acidity (pH 5.1) complicates processing (canning), thus only a very minor proportion of the total world banana production is processed (Nakasone and Paull, 1998). Bananas are extremely susceptible to enzymatic browning by the enzyme polyphenol oxidase, which reacts with various diphenolic compounds in the presence of air. Discoloration also results from the reaction between metal ions (tin and zinc) and vascular bundles loosely attached to the banana skin ("peel rag"). Bananas containing higher starch (green or unripe stage, "cooking" bananas and plantains) are preferred for processing than the ripe ("dessert") bananas.

Banana Chips

Banana chips can be prepared by immersing unripe banana slices in a solution of browning inhibitor (SO_2, citric acid, or aluminum chloride), or alternatively blanched in hot water or steam before frying (Woodroof, 1975). Ammawath et al. (2001) reported the effects of variety and stage of ripeness on sensory characteristics of banana chips. Peeled bananas were sliced (2 mm) and deep-fried in refined palm olein oil. Higher oil absorption and crisp chips were obtained from green bananas (Pisang Abu) with higher carbohydrate content and firmness. They also had lower moisture and water activity (a_w), better aroma, color, texture, flavor, and overall acceptability compared to Pisang Nangka with traces of yellow.

In a storage study (27°C; 2–8 weeks), significantly lower color scores for LDPE-packaged banana chips were observed, along with loss of crispness. Laminated aluminum foil chips possessed significantly higher rancid odor scores, but demonstrated the lowest scores for breaking force, a_w, moisture content, and levels of thiobarbituric acid-reactive substances (Ammawath et al., 2002).

Banana chip processing in the Philippines is by osmotic drying of cooking bananas. Transverse banana slices are soaked in sugar solution, deep fried in coconut oil, and air-dried at room temperature (about 27°C), or cabinet dried (60°C) to allow for moisture equilibration. Finished products are exported in bulk and are either repacked or incorporated in fruit trail mixes or ready-to-eat breakfast cereals. The U.S. Patent 3,510,314 (Lima and Lima, 1970) described a method to make banana chips from unripe bananas by thinly cross-slicing (1/64″–1/32″ thickness) and frying in an edible vegetable oil at 190.6°C (375°F) for about 1 min. Ripe banana flavor was also added to the fried chips.

Mui et al. (2002) reported that banana chips subjected to 90% air-drying followed by 10% vacuum-microwave drying had higher levels of volatile compounds and a crisper texture. Waliszewski et al. (2000) reported that polyphenol oxidase decreased during osmotic dehydration of banana chips and completely inactivated after 60–90 min at 70°C. In an earlier study, Waliszewski et al. (1999) reported that the use of proteolytic enzymes did not help control browning during osmotic dehydration of banana chips.

Banana Flakes and Powder

Ripe fruits are utilized for producing banana flakes and powder, while green and starchy bananas are dehydrated into flour. Banana powder or flour is utilized in infant products, beverages, in baking, and confectionery applications. The varieties Gros Michel, Cavendish, Lady Finger, Nendran, and Plantain are recommended for drying and dehydration (Rodriguez et al., 1975).

Banana flakes (protein 19.7%; fat 7.65%; carbohydrate 60.9%; starch 11.7%; dietary fiber 3.3%; ash content 1.3%; chemical score 80 with lysine as limiting amino acid) were prepared by mixing bananas with full-fat soya flour (60:40 banana:soy, dry basis), antioxidant solution (0.25% ascorbic acid; 0.25% citric acid), sucrose (5%), and sodium chloride (0.1%). The mixture was homogenized, drum dried, and stored at 4°C (Ruales et al., 1990). Drum drying of banana powder is preferred because of larger entrainment and wall losses associated with spray drying (Rodriguez et al., 1975).

Muyonga et al. (2001) reported that predehydration steaming of bananas increased water uptake, density and solubility of banana flour, and resulted in pastes of low bulk density, desirable for weaning and supplementary foods. Chemical composition of banana flour prepared by freeze drying of banana pulp was reported by Mota et al. (2000) for different varieties, such as starch 61–76.5%, amylose 19–23%, protein 2.5–3.3%, moisture 4–6%, lipids 0.3–0.8%, ash 2.6–3.5%, and total fiber 6–15.5%. Gelatinization temperature ranged from 68°C to 76°C depending on the variety, but all increased in viscosity during cooling.

Banana Puree

Banana puree is either canned or frozen and utilized in the preparation of baby foods, beverages, jams, sauces, ice cream, and bakery products. Fresh banana puree, prepared without freezing or heat sterilization, was described by Woodroof (1975) and Lopez (1987). Ripe bananas are peeled, immersed in sodium bisulfite solution (1.5%; 3 min), drained, milled (1/8″ screen), pureed, then pumped through a plate heat exchanger (93.3°C; 1 min), and cooled (37.7°C; 1 min). A finisher (0.033″ screen) removes seeds and any fibrous material. To extend shelf-life, citric acid (to adjust pH to 4.1) and potassium sorbate (200 ppm) were added.

When canning banana puree, the puree is filled into 30-lb cans with plastic film bag liners, sealed, and stored (2–4°C). A process to do away with the acidification step involved extruding ripened, chopped bananas, heating to 121°C by injecting steam, cooling to 2–3°C, filling into fiber containers, and blast freezing (−20°C) (Johnson and Harter, 1981).

In banana puree blanched for 7 min and processed through high hydrostatic pressure (689 MPa; 10 min), the residual PPO activity was <5% (Palou et al., 1999). Both microwave (3 min) and water bath (100°C; 8 min) blanching of banana completely retarded enzymatic browning (Premakumar and Khurdiya, 2002).

Banana ketchup made utilizing banana puree is a popular substitute for tomato ketchup in the Philippines. It is probably better marketed as a brown banana sauce, instead of simulating the red color.

Banana Juice and Essence

Sole (1993) developed a process for banana juice and banana essence. The juice is prepared by mechanical peeling of whole, ripe bananas. The pulp and parenchyma are homogenized and deaerated at a vacuum-holding station (5″ Hg). The mixture is subjected to pectinase digestion (CLAREX L, Miles Lab., Inc.; 48.9°C; 2 h), and the digested puree passed through a screw press to release banana juice. Subsequently, the juice is filtered, pasteurized, cooled, and stored frozen. A banana essence product is also recovered in this operation.

Fermented Banana Products

Anaerobic fermentation of green or cooking banana puree with *Saccharomyces cerevisiae* produces banana wine. Overripe dessert bananas can also be utilized for wine making, but sugar has to be added to increase soluble solids. Beer is brewed from bananas and plantains in Uganda and Tanzania. Juice, squeezed from ripened beer bananas (Bluggoe or Kivuvu), is fermented with sorghum and distilled to produce "waragi" (Nakasone and Paull, 1998; Karugaba and Kimaru, 1999). Banana vinegar is prepared by aerobic fermentation of ripe bananas using *Acetobacter acetii.*

Other Banana Products

Intermediate moisture bananas possess an astringent flavor attributed to tannins located in the latex cells of the fruit, which may be ruptured during the dehydration process (Shewfelt, 1986).

For frozen, green banana slices, the stage of maturity is the key in achieving a firm texture. Canned banana slices are packed in heavy, acidified syrup, but can also be incorporated in canned tropical fruit cocktails/salads, and in other baking and dairy applications.

MANGO

Production and Consumption

The mango (*Mangifera indica* L.) is regarded in tropical countries as the "king of fruits." The world production of mangoes has been led by India for the past 10 years. With the exception of Mexico, Asian countries dominate world production (Figs 33.2 and 33.3). However, more than half of the U.S. imports come from Mexico. Average annual per capita consumption of fresh mangoes in the United States has increased to almost 2 lbs (Fig. 33.4). Generally, varieties with fiber-less flesh, golden yellow color, and pleasant mango flavor are preferred. Table 33.2 features selected mango cultivars.

The mango, a fleshy drupe, is a climacteric fruit where ripeness coincides with the respiratory peak.

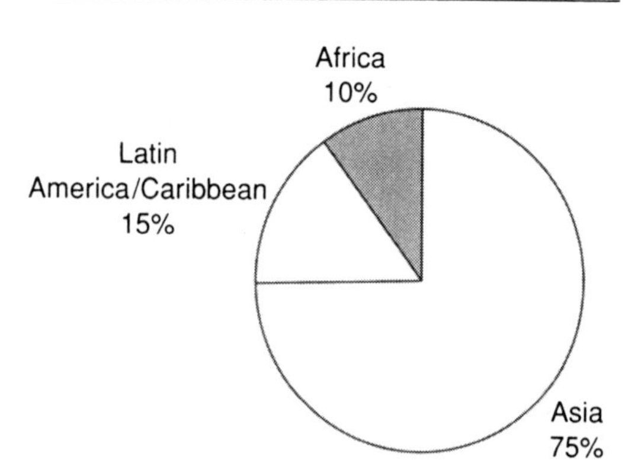

Figure 33.2. World production of mangoes (FAO, 2003).

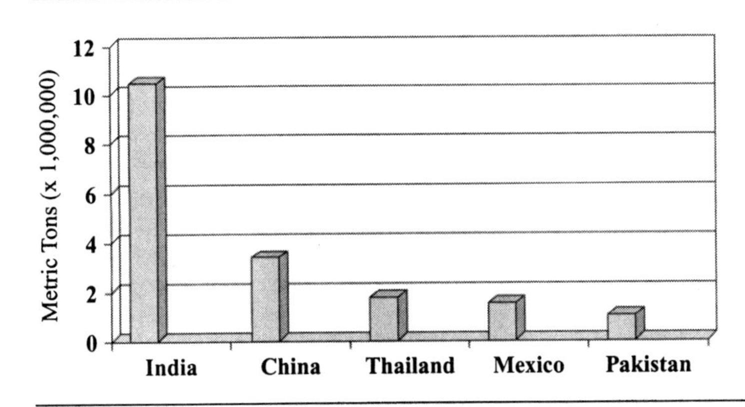

Figure 33.3. Mango producers in the world (FAO, 2003).

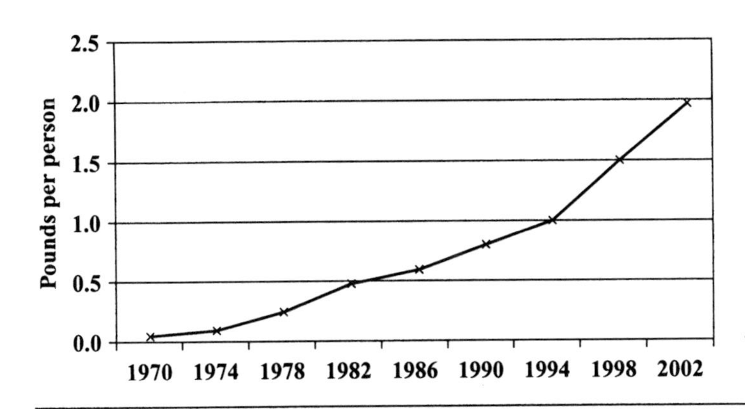

Figure 33.4. Mango consumption in the United States (USDA Fruit and Nuts Outlook, 2003).

Table 33.2. Selected Mango Cultivars

Variety	Country	Characteristics
Alphonso	India	Ovate to oblique; yellow, thin skin; firm to soft flesh; low in fiber; characteristic pleasant aroma
Carabao	Philippines	Long, slender; greenish to bright yellow; lemon yellow, tender flesh; sweet, mild aroma
Kent	Mexico	Oval with round base; greenish-yellow with red or crimson base; thick skin; medium firm texture; little fiber; rich flavor
Kensington	Australia	Round ovate; yellow with orange red; thick skin; soft juicy flesh; moderate to little fiber; characteristic flavor
Kyo Savoy	Thailand	Oblong; greenish-yellow; thin skin; medium-firm flesh; soft texture; no fiber; mild flavor
Neelum	India, China	Oval with flattened or slightly rounded base; bright yellow with no blush and numerous small, white dots; thick, tender, and easily separating skin; soft, melting, and juicy flesh; no fiber; deep yellow, mild, and sweet; delightfully pleasant aroma of good to excellent quality

Source: Knight (1997).

The eating quality is closely associated with the onset of climacteric peak, when fruits change in color, texture, and aroma, and starch is converted into sugars (Hulme, 1971; Beattie and Wade, 2001). Tree-ripened mangoes are reported to be inferior in keeping quality to those which are allowed to ripen after harvest. Mango is hand harvested to minimize quality loss and transported in cushioned bamboo or wooden crates. Although the stage of maturity during harvest is critical, there is no objective method of assessing it. Ripening rate changes with storage temperature, and the quality of mangoes is affected by temperature and relative humidity (RH) during ripening.

Processed Mango Products

Fully ripened mangoes with well-developed flavor, color, and texture are ideal for processing. Figure 33.5 shows various mango products.

Canned Mango Products

Mango Puree. Mango puree is considered the primary mango product in terms of scale of production. It serves as the starting material for mango beverages, jams, and dehydrated products. Either whole or peeled mangoes are processed into puree. Cultivars which differ in their sensory attributes may be blended to compensate for each other. Mangoes are washed, peeled, sliced, and passed through a pulper with a 30-mesh sieve. Seeds, residual peels, and much of the fiber can be separated by centrifugation or pressing through a series of screens. The use of a finishing screen (pore diameter <0.5 mm) further removes fibers (Wu et al., 1993). The pulp is either sweetened (42–43°Brix) or acidified with citric acid (pH 4.0 or lower) (Avena and Luh, 1983). Ascorbic acid can be added for better retention of color, flavor, and carotene. The puree is preheated to 85°C, filled into cans, and processed at 100°C for 45 min (Narain et al., 1998). The physicochemical quality of mango puree (Table 33.3) varies with cultivar, cultural and climatic conditions, ripeness at harvest, postharvest storage, treatment of the fruit, and processing methods (Wu et al., 1993).

Mango Juice. Mango juice is prepared by mixing a measured amount of mango puree with an equal amount of water. The total soluble solids (TSS)

Figure 33.5. Processed mango products sold in the United States.

Table 33.3. Proximate and Physicochemical Composition of Mango and Mango Puree

	Value Ranges (% Fresh Weight)	
Composition	Mango Fruit	Mango Puree
Moisture	72.1–85.5	67–86
Crude protein	0.30–5.42	0.3–1.5
Crude fiber	0.30–2.38	0.05–0.6
Ash	0.29–1.13	0.13–0.61
Total soluble solids	17–24	0.28–0.64
Total sugars	10.5–18.5	12.28–17.54
pH	4.0–5.6	11.27–15.38
TA (as citric)	0.327	0.1–1.1

Source: Hulme (1971), Caygill et al. (1976), Wu et al. (1993), and Narain et al. (1998).

and acidity of the mixture are adjusted as per specifications, and range from 12°Brix to 15°Brix and 0.4–0.5% citric acid, respectively. Mango juice is preheated to 85°C in a heat exchanger, filled into 401 × 411 cans, sealed, and processed at 100°C for 20 min. The cans are cooled, packed, and stored. Mango juice is commonly consumed as a single strength juice, as juice blends, or incorporated in fresh fruit smoothies/shakes.

Cloud stabilization and viscosity of juice can be controlled by the use of pectinolytic enzymes. Enzyme from *Aspergillus oryzae* (10% v/v, 30 ± 2°C for 18 h) was utilized to facilitate mango (Himsagar) juice extraction, decrease viscosity, and increase free flow (Ghosh and Gangopadhyay, 2002). Studies on enzymatic liquefaction of mango pulp with commercial enzymes pectinex and celluclast (0.14 g/l; 30 min; 27–30°C) yielded the highest juice yield from Raspuri variety. Treatment with combined enzymes enhanced low-viscosity juice yield (60–88%) by 8–10% over pectinex alone (Sreenath et al., 1995). Microfiltration after partial enzymatic liquefaction of tropical fruit juices, including mango, pineapple, and passion fruit, resulted in a retentate very similar to the initial juice (Vaillant et al., 2001).

Mango Nectar. The difference between mango juice and nectar is the lower fruit content in the latter. Mango nectars can be made either from fresh mango fruit directly or from canned, aseptically packed, or frozen puree. Mango puree (20–30%) is mixed with other ingredients (water, sugar, citric acid, and carboxymethylcellulose, a stabilizer), filtered through a finishing screen, and processed using any of the following methods: (i) no. 2 vacuum-sealed cans heated in a "spin cooker" (100°C; 3 min; 125 rpm); (ii) flash pasteurized in a plate heat exchanger (80°C), filled into cans, sealed, and processed at 80°C (10 min); or (iii) pasteurized in a plate heat exchanger (95°C; 1 min) and aseptically packed in plastic-lined cartons (Wu et al., 1993). The processing time and temperature are influenced by the viscosity of the nectar, can size, and filling temperature. Mango nectar formulations vary from 20% to 33% pulp content; commercially packed mango nectars contain <25% pulp (Narain et al., 1998).

Mango Squash. Mango nectar and mango squash are similar in composition except for the presence of a preservative (either 350 ppm SO_2 or 0.1% sodium benzoate) in mango squash. A mango squash formulation consists of syrup (68°Brix) blended with mango puree (46 kg), citric acid (2 kg), and potassium metabisulfite (180 g) (Narain et al., 1998). The mixture is packed in sterilized containers. Mango squash is diluted with water (3:1) when served.

Mango Slices and/or Scoops. Canned mango scoop is a Philippine product similar to canned peaches. Carabao variety mangoes are sliced in halves, flesh is scooped out and packed into cans, covered with hot syrup containing citric acid, sealed, and processed.

Mango slices are placed in cans (A2½), packed with syrup (30–40°Brix) containing citric acid (0.20–0.25%), sealed and processed at 100°C for 20–30 min (or 15 psi for 8–10 min; no. 1 cans are processed for 30–32 min at 105°C) (Narain et al., 1998).

Aseptic-Packed Mango Beverages. The tetrapack system is the preferred packaging for mango beverages (Fig. 33.5) because it affords convenience and shelf stability. Aseptic processing involves filling the commercially sterile product into a sterile container and hermetically sealing in a sterile environment. An aluminum foil seal is automatically placed across the top of the filling spot to provide a hermetic and tamper resistant seal. From the filler, the bags are turned over so that the hot product flushes whatever is displaced, and kill any bacteria that may have been in the spout area. The bags are cooled to around 40°C (Shukor et al., 1998). Aseptic bulk packing for the export of mango pulp uses laminated bags of poly/aluminum foil/poly/poly liner and poly/metallized polyester/poly/polyliners for 20 kg bags in boxes and 200 kg bags in drums, respectively (Nanjundaswamy, 1997).

Mango Juice Concentrates. Mango concentrates are utilized in the preparation of juice and energy drinks. Fresh mango puree is pasteurized in a plate heat exchanger (75°C; 1 min), cooled quickly, and centrifuged (5000 rpm). A decanter separates the serum and the pulp portions. The serum is either evaporated or freeze concentrated (45°Brix) prior to mixing with the pulp portion (Wu et al., 1993).

Frozen Mango Products

Mango Puree. Mangoes pass through a cutting mill and screened paddle pulper (0.033″) to strain coarse fibers and residues. The puree is blanched in a heat exchanger (90–93°C; 1 min), rapidly cooled (32–38°C), filled into polyethylene-lined tins (14 kg), and frozen (−23.5°C) (Narain et al., 1998).

For sweetened puree, mango flesh were dipped into cold sucrose syrup (20%) containing ascorbic acid (0.5%), drained and passed through a finisher (0.84 mm screen). The puree was held in a heat exchanger (90.6–93.3°C; 2 min), rapidly cooled (32.3–37.8°C), adjusted to 42.5°Brix, canned, sealed, frozen in a blast freezer, and stored (−17.8°C) (Wu et al., 1993).

Mango Slices. Mango slices are placed in a container, covered with syrup (25–30°Brix), sealed, blast frozen, and stored (≤ −18°C). Calcium pretreatment (2%) prior to freezing firms up the slices (Narain et al., 1998).

Dried Mango Products

Dehydrated Mango Slices. Osmotic dehydration involves dipping mango slices in sugar syrup (70°Brix), draining, dipping in sulfite solution followed by a second draining, then drying in an air convection oven. Vacuum driers can also be used (2–4 mm Hg, 60–66°C) to achieve a final moisture of 2% (Narain et al., 1998). In the Philippines, osmotic-dried mango slices are exported. Cabinet dehydration is preferred, although sun drying is still common in rural areas.

Mango Leather. Fruit leathers ("fruit rolls") in the United States is popular with kids (Sloan, 2004). In India, mango leather ("ampapar") production involves manual extraction of mango puree and straining through bamboo sieves to remove the fiber. The puree is spread on date palm leaf mats and sun dried for about 40 days. Upon drying, another layer is spread over the first, and the process is repeated until a desired thickness of 3 cm is obtained. Slabs of leather are sliced and packed in cellophane, polyethylene, or waxed paper (Narain et al., 1998).

Industrial preparation involves passing the puree through a mechanical finisher with a stainless steel sieve (30–50 mesh). TSS is adjusted to 25–35°Brix. Potassium metabisulfite may be added. The puree is spread evenly on glycine-coated metal trays (≤1 cm per layer) and dried in a forced-air drier (50–80°C; 2–20 h per layer). Mango leather has 15–20% moisture (Woodroof, 1975; Wu et al., 1993).

Mango Powder. Mango powder is utilized either as a beverage base or as a flavoring agent. It is prepared by drying mango puree (≤3% moisture) using any of the following methods: spray drying, freeze drying, foam-mat drying, puff drying, vacuum drying, or drum drying.

Spray-Dried Mango Powder. Mango puree, liquid glucose (5%), and tricalcium phosphate (0.5% anticaking agent) were mixed, heated (30 min; 50°C), homogenized (2800 rpm), pumped to the atomizer (2800 rpm), and spray dried (165°C inlet temperature). The mango powder is cooled, then stored (4°C; 15 min) before transferring to a dehumidified packing room (Wu et al., 1993). Spray-dried mango powder has a nice color but lacks flavor. Hence, sugar, milk solids, and glyceryl monostearate and alginate are added to the mango pulp (Nanjundaswamy, 1997).

Freeze-Dried Mango Powder. Mango puree is poured into freezing trays (12 mm thick) and frozen in a blast freezer (−20°C). The frozen puree is cut into 25 × 25 × 12 mm cubes, further chilled

(−30°C), then held in a freeze-drying chamber (10–11 h; ≤60°C). The dried puree (≤2.4% moisture) is packed in cans flushed with nitrogen (Wu et al., 1993). Freeze-dried mango pulp with added sugar increased shelf-life. There was little loss of flavor, but the cost of production was prohibitive (Nanjundaswamy, 1997).

Mango Flakes. The production of mango drum-dried flakes involves blending mango pulp with small amounts of wheat flour and sugar at certain pH. The homogenized mixture is dried using an atmospheric double drum dryer (50–60 psi). Flakes can be milled and screened into mango powder (Wu et al., 1993). Mango cereal flakes were developed at the Center for Food Technology Research Institute in Mysore, India (Nanjundaswamy, 1997).

Other Mango Products—Jams, Marmalades, Chutneys, and Pickles

Savory jams and marmalades, mixed into dressings, vinaigrettes, marinades, glazes, and cream sauces, are becoming the next generation of "toast toppers" for those avoiding sweets (Sloan, 2004). Mango jams are prepared by combining mango pulp with sugar and citric acid, and heating until viscous.

Immature, green mangoes are made into chutneys, pickles, and refreshing mango beverages. Chutneys are gaining popularity as a dip, meat or fish topping, accompaniment to cheese, and as a sweet-sour flavor enhancer (Sloan, 2004).

PASSION FRUIT

Originally native to Brazil, the leading exporter, passion fruit (*Passiflora edulis*) is widely distributed throughout the tropics. The two major varieties which are processed commercially are the purple passion or granadilla (*P. edulis* Sims) and the yellow passion (*P. edulis* Sims f. *flavicarpa* Deg). Upon ripening, passion fruit falls from the vine and is hand-harvested from the ground. If not processed immediately, they could be stored for 4–5 weeks at 6.5°C (85–90% RH) (Jagtiani et al., 1988; Chan, 1993). Figure 33.6 shows purple passion fruit and juice products. Purple passion fruit is sweeter due to its higher sugar content than the larger and more prolific yellow variety (Chan, 1980; Rutledge, 2001).

The mechanized processing of passion fruit starts with washing. Strong sprays of water are directed on a belt conveyor. Sorting removes rotten and unfit fruits. The rinds are sliced open by serrated, rotating circular knives. The sliced fruit is fed into a continuous basket centrifuge. The juice, pulp, and seeds are forced by centrifugal action through holes in the walls, and the rinds slide up the sloping walls out into a waste conveyor. A screened pulper separates the seeds from the pulp and juice, which is passed through a screened finisher to further remove remaining seed and fiber particles (Woodroof, 1975; Jagtiani et al., 1988; Chan, 1980, 1993).

Another extraction of pulp involves dropping the fruit between two rotating converging cones, where it is caught in the nip and bursts as the cones rotate toward the bottom of the machine. The clearance is reduced to the thickness of the skin of the fruit. The skins are carried through the cones but the pulp drops into a finisher that removes pieces of the skin and seeds (Jagtiani et al., 1988; Rutledge, 2001).

Passion Fruit Juice

Passion fruit is utilized in juice making due to its unique flavor and aroma. However, the aroma components and flavor are extremely heat sensitive, so pasteurization leads to some loss or change of the flavor. Another concern in processing is the high starch content of the juice, which causes the accumulation of gelatinous deposits on the heating surfaces of tubular and plate heat exchangers, affecting efficiency. It also results in localized scorching and deterioration in flavor (Hulme, 1971; Chan, 1980; Jagtiani et al., 1988). The high starch content also results to a thick juice viscosity that could be desirable in some products, but is removed by a decanting centrifugal separation for free-flowing concentrates. The delicate flavor permits its use as pure juice; its strong acidic flavor calls for dilution (1:6) when consumed as nectar. Edwin et al. (2003) investigated methods to increase the pH of the juice from 2.9 to 4.0. Precipitation with $Ca(OH)_2$ and electrodialysis with bipolar membrane were preferred processes based on sensory evaluation.

Passion fruit juice is a refreshing component when blended with banana, mango, papaya, guava, orange, apple, and carrot, and brings out the flavor of these juices. Fresh juice and concentrates can be mixed with tropical fruit cocktail, frozen and heat-processed punches, alcoholic beverages, and serve as a flavoring to cakes, sauces, salads, pies, sherbets, jellies, and jams.

Passion fruit products from different countries include: (i) carbonated drinks and ice cream in Kenya, (ii) Swiss passion fruit-based soft drink called

Figure 33.6. Passion fruit and juice products.

"Passaia" marketed in Western Europe, (iii) blending with milk and alginate in South Africa, and (iv) pulp added to yogurt in Australia. Passion fruit juice boiled down to a syrup is used in making sauce, gelatin desserts, candy, ice cream, sherbet, cake icing/filling, meringue or chiffon pie, and cold fruit soup.

The juice yield of yellow passion fruit averages from 30% to 33%; purple varieties have an average yield of 45–50%. The use of pectinolytic enzymes increased yields by 35% (Jagtiani et al., 1988; Chan, 1993). Physicochemical properties of enzyme-treated juice were similar to untreated juice, except for the increased consistency from the higher content of pectin released during treatment. Alpha amylase has been used to reduce the viscosity of the juice brought about by starch gelatinization. After primary juice extraction, some processors employ an enzymatic process to obtain "secondary" juice from the double juice sacs surrounding each seed.

Heat-Processed Juice

Pasteurization causes the loss of 35% of sensitive volatile flavor components (Nakasone and Paull, 1998). To reduce the effect of heat, juices can be diluted or blended with other fruit juices. An alternative processing utilizes a spin cooker for canned juice where can rotation transfers heat rapidly. Pasteurization to 88°C center temperature is achieved in about $1\frac{3}{4}$ min, and cans are rapidly cooled (Chan, 1980).

Juice processing in Sabah, East Malaysia, involves pasteurization at 62°C (for 20 s). A separator removes starch, and a stripping evaporator removes aroma at 1500 ppm prior to concentration.

Frozen Juice

The extracted passion fruit juice is pasteurized (85°C; 1 min), quick frozen as puree, and kept frozen for processing into frozen sherbet and nectars. The juice can also be frozen directly in 30-lb containers using air-blast freezers (−18°C) for incorporation into juice blends (Jagtiani et al., 1988).

Juice Concentrate

Most volatile flavor components are lost into the distillate stream during concentration. An alternative is to remove volatiles prior to concentration using a spinning cone column consisting of a series of cones (Rutledge, 2001).

Passion fruit juice concentrate was obtained by partial evaporation of the water at a short residence time in the evaporator, to prevent thermal damage of heat-sensitive volatile flavors. The concentrate retained the intensity of the aroma and flavor of the original juice, because aroma evaporated during the concentration process was recovered and incorporated into the concentrate. It had a viscous consistency and characteristic color, and can be stored (−18°C) without deterioration for a year (Rutledge, 2001).

PHYSICOCHEMICAL AND NUTRITIONAL QUALITY OF TROPICAL FRUITS

Tables 33.3–33.6 summarize the nutritional and chemical composition of banana, mango, passion fruit, and processed products.

Sugars and Organic Acids

Mango is one of the few fruits where sucrose concentration exceeds total reducing sugars (Table 33.7). The development of the sweetness in ripe bananas results from the conversion of starch (20–23%) in green bananas to sugars (20%), predominantly sucrose. The conversion of starch to sugar in plantains and cooking bananas is less complete than in sweet varieties. The sweeter taste of purple passion fruit is due to its higher sugar content and sugar/acid ratio. In yellow passion fruit, citric acid predominates, followed by malic acid; purple varieties have lower citric acid, followed by lactic acid (Chan, 1980; Jagtiani et al., 1988). In banana, malic acid predominates, followed by citric and oxalic acids (Shewfelt, 1986).

Phenolic Contents

The astringent flavor of mangoes is imparted by tannins. Smaller fruits were observed to have slightly higher tannin content. Gallic acid, mangiferin, and ellagic acid were reported in ripe mango. Phenolic

Table 33.4. Nutritional Values of Banana (*Musa paradisiaca*) Fruit and Banana Powder

Nutrients/100 g	Banana Fruit	Banana Powder
Calories (Kcal)	89.0	346
Total fat (g)	0.33	1.81
Saturated fat (g)	0.112	0.698
Polyunsaturated fat (g)	0.073	0.337
Monounsaturated fat (g)	0.032	0.153
Cholesterol (mg)	0.0	0.0
Sodium (mg)	1.0	3.0
Potassium (mg)	358.0	1491
Total carbohydrate (g)	22.84	88.28
Total fiber (g)	2.6	9.9
Total sugar (g)	12.23	47.30
Protein (g)	1.09	3.89
Calcium (mg)	5.0	22.0
Iron (mg)	0.26	1.15
Vitamin C (mg)	8.7	7.0
Vitamin A (IU)	64	248

Source: http://www.nal.usda.gov/fnic/foodcomp/cgi-bin/list_nut_edit.pl.

Table 33.5. Nutritional Values of Mango Fruit and Infused-Dried Mango

Nutrients/100 g	Mango Fruit[a]	Infused-Dried Mango[b]
Calories (Kcal)	65.0	331
Total fat (g)	0.27	0.78
Saturated fat (g)	0.066	0.2
Polyunsaturated fat (g)	0.051	0.2
Monounsaturated fat (g)	0.101	0.4
Cholesterol (mg)	0.0	<0.1
Sodium (mg)	2.0	24.0
Potassium (mg)	156.0	182.0
Total carbohydrate (g)	17.0	83.1
Total fiber (g)	1.8	6.1
Total sugar (g)	14.80	75.6
Protein (g)	0.51	0.67
Calcium (mg)	10.0	28.0
Iron (mg)	0.13	0.83
Vitamin C (mg)	27.7	262.0
Vitamin A (IU)	765.0	1554

[a]http://www.nal.usda.gov/fnic/foodcomp/cgi-bin/list_nut_edit.pl.

[b]Data of commercial products; infused with sugar prior to drying, contains added ascorbic acid and high oleic sunflower oil. Courtesy Graceland Fruit Inc., Frankfort, MI, USA.

Table 33.6. Nutritional Values of Passion Fruit and Juices

Nutrients/100 g	Purple Passion Fruit	Purple Variety Juice	Yellow Variety Juice
Calories (Kcal)	97	51	60
Total fat (g)	0.70	0.05	0.18
Saturated fat (g)	0.059	0.004	0.015
Polyunsaturated fat (g)	0.411	0.029	0.029
Monounsaturated fat (g)	0.086	0.006	0.006
Cholesterol (mg)	0.0	0.0	0.0
Sodium (mg)	28	6.0	6.0
Potassium (mg)	348	278	278
Total carbohydrate (g)	23.38	13.60	14.45
Total fiber (g)	10.4	0.2	0.2
Total sugar (g)	11.20	13.40	
Protein (g)	2.20	0.39	0.67
Calcium (mg)	12.0	4.0	4.0
Iron (mg)	1.60	0.24	0.36
Vitamin C (mg)	30.0	29.8	18.2
Vitamin A (IU)	1272	717	2410

Source: http://www.nal.usda.gov/fnic/foodcomp.

Table 33.7. Starch and Sugar Composition of Tropical Fruits

Fruit	Starch (Ripe Fruit) (% Fresh Weight)	Total Sugars (% Fresh Weight)	Reducing Sugars (% Fresh Weight)	Sucrose (% Fresh Weight)
Mango	0.0	14.8	3.50	7.36
Passion fruit		11.2	4.60	3.20
Purple	0.74	17.3		
Yellow	0.06	15.0		
Banana	5.38	20.0	9.84	2.39
Plantain	5–10	17.0		

Sources: Hulme (1971), Chan (1980), Forsyth (1980), Nagy et al. (1993), and http://www.nal.usda.gov/fnic/foodcomp.

Table 33.8. Phenolics and Antioxidant Capacity of Banana and Mango Fruits

Component	Banana	Mango
Moisture (%)	73.5	81.7
Lipophilic—oxygen radical absorbance capacity fluorescent probe (ORACFL) micromole of TE/g	0.66 + 0.14	0.14
Hydrophilic—ORACFL micromole of TE/g	8.13 + 1.02	9.88
Total antioxidant capacity (TAC) micromole of TE/g	8.79	10.02
Total phenolics mg of GAE/g	2.31 + 0.60	2.66
Serving size (g)	118 (one fruit)	165 g (one cup slices)
TAC/serving (micromole of TE)	1037	1653

Source: Wu et al. (2004).

compounds ranged from 31 to 75 mg/100 g flesh in ripe fruits, depending on mango variety and size (Hulme, 1971; Narain et al., 1998). Table 33.8 summarizes the phenolics and antioxidant capacity of banana and mango.

Vitamins and Minerals

Mangoes are rich in provitamin A carotenoids (>50% is β-carotene) and vitamin C. Vitamin B complex is relatively low, except for niacin (Wu et al., 1993). Banana is also considered a source of vitamins A, C, and B, but best known as an excellent source of potassium, containing twice the concentration in other ripe fruits. When dehydrated, the potassium content of banana powder becomes more concentrated. The plantain is relatively rich in ascorbic acid and higher in carotene content than bananas. Vitamin C is also the major vitamin in passion fruits (Tables 33.4–33.6).

Pigments

The major pigments in passion fruit are carotenoids, with β- and γ-carotene, and phytofluene found in purple varieties. Purple varieties have trace quantities of flavones but no anthocyanins (Chan, 1980; Jagtiani et al., 1988; Nagy et al., 1993).

Flavor Components

Monoterpene hydrocarbons constitute the major volatiles in mango. Although no single compound represents a typical mango flavor, 3-carene gives mango flavor and aroma; α-copaene, a sesquiterpene hydrocarbon, also has mango-like aroma. The following volatiles were significant in canned Alphonso mango puree: *cis*- and *trans*-ocimene; butyrolactone and γ-octalactone; furfural, 2-acetyl furan, and 2,5-dimethyl-4-methoxy-2H-furan-3-one. The lactones may contribute to the coconut- or peach-like notes of mangoes, and furfural and 5-methyl furfural contributes the caramel-like note. Methoxy furanone has a berry aroma (Hunter et al., 1974).

Banana fruit has a pleasant flavor. Jordan et al. (2001) identified a total of 43 and 26 volatile flavor compounds in banana essence and fresh banana fruit paste, respectively. These volatile compounds included alcohols, esters, aldehydes, and ketones. In fresh banana paste, aldehydes, E-2-hexenal (floral, herbal; 32.2 ppm), and hexanal (herbal, grassy, green; 21.47 ppm) were found in greatest concentration. In banana essence, E-2-hexenal was not detected; the concentration of hexanal was 31.19 ppm. The concentration of flavor compound isoamyl acetate responsible for overripe sweet banana flavor was 229.73, and 4.85 ppm in banana essence and fresh banana paste, respectively. Liu and Yang (2002) studied changes in banana flavor during ripening at 20°C, 25°C, and 30°C, and found isoamyl alcohol as the most abundant compound at all storage temperatures. Ethanol developed most rapidly at 30°C.

In passion fruits, ethyl butyrate and ethyl hexanoate were associated with the floral mixed-fruit character of the juice. Volatile constituents isolated from yellow varieties were in reversed order of abundance in purple varieties, and further accounts for the subtle flavor differences between the two (Hiu and Scheuer 1961; Chan, 1980). Other flavor compounds reported are 6-(But-2-enylidene)-1,5,5-trimethylcyclohexene, (Z)-Hex-3-enylbutanoate, hexyl butanoate, ethyl(Z)-oct-4-enoate, β-ionone, Edulan I, and (Z)-octa-4,7-dienoate (Chan, 1993).

OPPORTUNITIES FOR TROPICAL FRUITS

TRENDS

Consumer demand for nutritious products, as well as the emphasis on health and wellness, drives food development trends. There is a shift from vitamin fortification to emphasizing fruits/vegetables. Functional beverages and functional juice-based beverages are projected to reach $11.8 billion and $2.17 billion by 2005. The meal-replacement/weight-control beverage segments, and a new generation of 100% organic juice beverages (Sloan, 2003), provide venues for tropical fruits. Fruit smoothies, "bubble teas," probiotics, energy drinks, fortified beverages, and juice drinks have created opportunities for tropical fruit flavors, including Fuze's Mojo Mango; Peach–Mango and Mango–Passion fruit; Oadala's Mango Tango; Sobe's Lizzard Lightning Orange–Mango flavor.

Emerging ethnic influence on American cuisine has product developers exploring concepts, ingredients, and preparation techniques of Asian, Mexican/Tex-Mex, Italian, Middle Eastern, Caribbean, Cajun, Spanish, and Latin cultures. Exotic flavors from freeze-dried fruits and fruit powders deliver more distinctive flavor profiles to sauces and salsas. Ethnic fruit-based appetizers joining the mainstream are Mango Jicama Slaw and Arriba's California Style Berry Mango Salsa (Sloan, 2004). The field of Culinology also creates a venue for tropical fruit products in gourmet TV dinners.

EMERGING TECHNOLOGIES, PRODUCTS, AND PACKAGING

Nonthermal processing technologies are ideal for heat-sensitive tropical fruit juice flavors. High-pressure processing has potential for passion fruit juice. FDA has approved ultraviolet irradiation of juice and juice products. There is a growing market for minimally processed fruits (lightly processed or "fresh cut"), which refers to trimming, peeling, sectioning, slicing, and coring of fruits prior to marketing. They are chilled to optimum temperature levels, until purchased by the consumer.

Mango Treattarome 9830 is a new addition to Treatt's tropical flavor ingredients including passion fruit. It is a water-soluble distillate from tree-ripened mango, containing high levels of fruity esters, lactones, spicy terpenes, and sugary/caramel-like furanones. An intense mango character predominates at 0.1%; fresh, spicy, tropical notes are imparted at $\leq$0.05%. It enhances existing flavors or provides the main body in fruit juice blends, clear beverages, ices, and ice creams.

Infused-dried mango without use of sulfites has been commercialized (Table 33.5). The process involves: (i) removal of the fruit's moisture by immersing in infusion solutions of higher sugar solids (SS) content and (ii) drying to a desired moisture content. The skin serves as a semipermeable membrane allowing the exchange of SS between fruit and the infusion solution (Sinha, 1998). Infused-dried fruits contain about 12% moisture (a_w = 0.55), are shelf stable and retain the qualities of fresh counterparts without using sulfites. They are incorporated in trail mixes, bakery items, cereals, and breakfast/energy bars.

Tetra Pak's new Terra Recart presents possibilities for tropical fruit-based salsas, sauces, purees, whole soft tropical fruits, and ready-to-eat meals/sauces traditionally packed in cans, glass jars, and standup food pouches. Made from plastics, foil, and a paper component, Terra Recart withstands the rigors of retort processing (130°C), 100% humidity for >2 h, and is shelf stable for up to 24 months.

REFERENCES

Ammawath W, Che-Man YB, Yusof S, Rahman RA. 2001. Effects of variety and stage of fruit ripeness on the physicochemical and sensory characteristics of deep-fat fried banana chips. J. Sci. Food Agric. 81(12):1166–1171.

Ammawath W, Che-Man YB, Yusof S, Rahman RA. 2002. Effects of type of packaging material on physicochemical and sensory characteristics of deep-fat fried banana chips. J. Sci. Food Agric. 82(14):1621–1627.

Avena RJ, Luh BS. 1983. Sweetened mango purees preserved by canning and freezing. J. Food Sci. 48:406–410.

Beattie B, Wade N. 2001. Storage, ripening, and handling of fruit. In: David A, Ashurst PR, editors. Fruit processing: nutrition, products, and quality management. Gaithersburg, MD: Aspen Publishers, Inc., pp. 53–81.

Caygill JC, Cooke RD, Moore DJ, Read SJ, Passam HC. 1976. The mango (*Mangifera indica* L.) harvesting and subsequent handling and processing: an annotated bibliography. London: Tropical Products Institute.

Chan HT Jr. 1980. Passion fruit. In: Shaw PE, Chan HT Jr., Nagy S, editors. Tropical and subtropical fruits: composition, nutritive values, properties and uses. Westport, CT: AVI Publishing Co., 568 p.

Chan HT. 1993. Passion fruit, papaya, and guava juices. In: Nagy S, Chen CS, Shaw PE, editors. Fruit juice processing technology. Auburndale, FL: Agscience, Inc, pp. 334–371.

Che-Man YB, Ammawath W, Rahman RA, Yusof S. 2003. Quality characteristics of refined, bleached and deodorized palm olein and banana chips after deep-fat frying. J. Sci. Food Agric. 83(5):395–401.

Edwin VR, Ruales J, Dornier M, Sandeaux J, Persin F, Pourcelly G, Vaillant F, Reynes M. 2003. Comparison between different ion exchange resins for the deacidification of passion fruit juice. J. Food Eng. 57(2):199–207.

Forsyth WGC. 1980. Banana and plantain. In: Nagy S, Shaw PE, editors. Tropical and subtropical fruits. Westport, CT: AVI Publishing, Inc., p 258–272.

Ghosh U, Gangopadhyay H. 2002. Studies on the extraction of mango juice (Himsagar variety) using enzymes from *Aspergillus oryzae*. Indian J. Chem. Tech. 9(2):130–133.

Hiu DN, Scheuer PJ. 1961. The volatile constituents of passion fruit juice. J. Food Sci. 26:557–563.

Hulme AC. 1971. The mango. In: Hulme AC, editor. The biochemistry of fruits and their products. New York, New York: Academic Press, Inc. Vol. 2. pp. 238–252.

Hunter GLK, Bucek WA, Radford T. 1974. Volatile components of canned *Alphonso* mango. J. Food Sci. 39:900–903.

Jagtiani J, Chan HT Jr., Sakai WS. 1988. Tropical fruit processing. San Diego, CA: Academic Press, Inc., 184 pp.

Johnson WP, Harter EH. 1981. U.S. Patent 4,273,792. Banana processing.

Jordan JM, Tandon K, Shaw PE, Goodner KL. 2001. Aromatic profile of aqueous banana essence and banana fruit by gas chromatography-mass spectrometry (GC-MS) and gas chromatography-olfactometry (Gc-O). J. Agric. Food. Chem. 49:4813–4817.

Kapoor JK, Turner JN. 1976. U.S. Patent 3,950,559. Method for delaying ripening of harvested bananas.

Karugaba A, Kimaru G.1999. Banana production in Uganda: an essential food and cash crop. Nairobi: Regional Land Management Unit, RELMA, 72 p.

Knight RJ Jr. 1980. Origin and world importance of tropical and subtropical fruit crops. In: Nagy S, Shaw PE, editors. Tropical and subtropical fruits. Westport, CT: AVI Publishing, Inc., pp. 1–106.

Knight RJ Jr. 1997. Important mango cultivars and their descriptors. In: Litz RE, editor. The mango, botany, production and uses. New York: CAB International, pp. 545–564.

Lima RF, Lima JM. 1970. U.S. Patent 3,510,314. Method of preparing banana chip product.

Liu T-T, Yang T-S. 2002. Optimization of solid-phase microextraction analysis for studying change of headspace flavor compounds of banana during ripening. J. Agric. Food Chem. 50:653–657.

Lopez A. 1987. A complete course in canning. 12th ed. Baltimore, Maryland: The Canning Trade, Inc., 454 p.

Mota RV, Lajota FM, Ciacco C, Cordenunsi BR. 2000. Composition and functional properties of banana flour from different varieties. Starch 52(2/3):63–68.

Mui WWY, Durance TD, Scaman CH. 2002. Flavor and texture of banana chips dried by combinations of hot air, vacuum, and microwave processing. J. Agric. Food Chem. 50(7):1883–1889.

Muyonga JH, Ramteke RS, Eipeson WE. 2001. Predehydration steaming changes physicochemical properties of unripe banana flour. J. Food Process. Preserv. 25(1):35–47.

Nagy S, Chen CS, Shaw PE. 1993. Fruit juice processing technology. Auburndale, FL.: Agscience, 655 p.

Nakasone HY, Paull RE. 1998. Tropical fruits. New York: CAB International, 400 p.

Nanjundaswamy AM. 1997. Processing. In: Litz RE, editor. The mango, botany, production and uses. New York: CAB International, pp. 509–539.

Narain N, Bora PS, Narain R, Shaw PE. 1998. Mango. In: Shaw PE, Chan HT Jr., Nagy S, editors. Tropical and subtropical fruits. Auburndale, FL: Agscience, pp. 1–63.

Palou E, Lopez-Malo A, Barbosa-Canovas GV, Welti-Chanes J, Swanson BG. 1999. Polyphenoloxidase activity and color of blanched and high hydrostatic pressure treated banana puree. J. Food Sci. 64(1):42–45.

Premakumar K, Khurdiya DS. 2002. Effect of microwave blanching on the nutritional qualities of banana puree. J. Food Sci. 39(3):258–260.

Rodriguez R, Raina BL, Pantastico ERB, Bhatti MB. 1975. Quality of raw materials for processing. In: Pantastico ERB, editor. Postharvest physiology, handling and utilization of tropical and subtropical fruits and vegetables. Westport, CT: AVI Publishing Co., Inc., p. 484.

Ruales J, Polit P, Nair BM. 1990. Evaluation of the nutritional quality of flakes made of banana pulp and full-fat soya flour. Food Chem. 36(1):32–43.

Rutledge P. 2001. Production of nonfermented fruit products. In: Arthey D, Ashurst PR, editors. Fruit processing, nutrition, products and quality management. 2nd ed. Gaithersburg, MD: Aspen Publishers, Inc., pp. 96–98.

Shewfelt RL. 1986. Flavor and color of processed fruits. In: Woodroof JG, Luh BS, editors. Commercial fruit processing. 2nd ed. Westport, CT: AVI Publishing Co., Inc., pp. 481–502.

Shukor ARA, Faridah AA, Abdullah H, Chan YK. 1998. Pineapple. In: Shaw PE, Chan HT Jr., Nagy S, editors. Tropical and subtropical fruits. Auburndale, FL: Agscience, Inc., pp. 137–181.

Sinha NK. 1998. Infused-dried and processed frozen fruits as food ingredients. Cereals Foods World 43(9):699–702.

Sloan EA. 2003. Top 10 trends to watch and work on: 2003. Food Technol. 57(4):30–50.

Sloan EA. 2004. Gourmet and specialty food trends. Food Technol. 58(7):27–38.

Sole P. 1993. U.S. Patent RE 34,237. Banana processing.

Sole P. 1996. Bananas (processed). In: Somogyi LP, Barrett DM, Hui YH, editors. Processing fruits: science & technology, Vol. 2. Lancaster, PA: Technomic Pub Co. Inc., p. 367.

Sreenath HK, Krishna KR, Santhanam K. 1995. Enzymatic liquefaction of some varieties of mango pulp. Food Sci. Tech. 28(2):196–200.

Taylor RB. 2001. Introduction to fruit processing. In: Arthey D, Ashurst PR, editors. Fruit processing: nutrition, products, and quality management. Gaithersburg, MD: Aspen Publishers, Inc., pp. 1–17.

USDA. 2003. Fruit and tree nuts outlook – FTS-304 (May 2003). Washington, DC.

Vaillant F, Millan A, Dornier M, Decloux M, Reynes M. 2001. Strategy for economical optimization of the clarification of pulpy fruit juices using cross-flow micro-filtration. J. Food Eng. 48:83 (http://www.elsevier.com/locate/jfoodeng).

Waliszewski KN, Corzo C, Pardio VT, Garcia MA. 1999. Effect of proteolytic enzymes on color changes in banana chips during osmotic dehydration. Drying Technol. 17(4/5):947–954.

Waliszewski KN, Garcia RH, Ramirez M, Garcia MA. 2000. Polyphenol oxidase activity in banana chips during osmotic dehydration. Drying Technol. 18(6):1327–1337.

Woodroof JG. 1975. Other products and processes. In: Woodroof JG, Luh BS, editors. Commercial fruit processing. 2nd ed. Westport, CT: AVI Publishing Co., Inc., pp. 425–478.

Wu JS, Chen H, Fang T. 1993. Mango juice. In: Nagy S, Chen CS, Shaw PE, editors. Fruit juice processing technology. Auburndale, FL: Agscience, Inc., pp. 621–648.

Wu X, Beecher GR, Holden JM, Haytowitz DB, Gebhardt SE, Prior RL. 2004. Lipophilic and hydrophilic antioxidant capacities of common foods in the United States. J. Agric. Food Chem. 52:4026–4037.

WEBSITES

www.nal.usda.gov/fnic/foodcomp
http://apps.fao.org/default.jsp

34
Nutritional and Medicinal Uses of Prickly Pear Cladodes and Fruits: Processing Technology Experiences and Constraints

M. Hamdi

INTRODUCTION

Cactus plants, though native to North America, over the centuries have spread throughout the world: North Africa, Europe, Mediterranean countries, South America, the Middle East, and other countries (Hare and Griffiths, 1907). Even early navigators (1600–1800) used the cladodes of *Opuntia* to treat prophylaxy and scorbut diseases (Diguet, 1928). *Opuntia* is the largest group of cacti and includes several edible kinds which are commonly known as Bunny Ears, Cholla, Prickly Pear, Barbary Fig, Tuna, and Indian Fig.

The Cactaceae family belongs to the order Centrospermae, a group of families whose color is attributed to the presence of betalain pigments (Merin et al., 1987). *Opuntia*, which is one of about 87 genera in the Cactaceae family, includes several species (Yasseen et al., 1996). The genus *Opuntia* includes a number of species that produce vigorous plants in arid and semiarid climates, protect against soil erosion, and stop desert advancement.

Opuntia is a succulent plant with tissues specialized for water storage. Its water-use efficiency makes the *Opuntia* more efficient in converting water into dry matter than most grasses, and some species produce 26 t/ha. year (Yasseen et al., 1996). Retamal et al. (1987a) have estimated the biomass produced by prickly pear in cladodes and in fruit under arid, semiarid, and irrigated conditions of land use (Table 34.1). Indeed, certain highly productive crops of *Opuntia ficus-indica* grown under optimal conditions can have annual aboveground dry matter productivities exceeding 30 tons/ha. year (Garcia de Cortázar and Nobel, 1991). The main producing countries are Mexico and Italy; and some areas of intensive cactus pear plantations have expanded remarkably in the

Table 34.1. Theoretical Biomass Production Under Different Conditions of Land Use (Retamal et al., 1987a)

Production	Regions		
	Arid	Semiarid	Irrigated
Plant material			
Size (m × m)	4 × 4	2 × 2	2 × 1
Plants/ha	625	2500	5000
Biomass (tons/ha)			
Cladodes	2.63	10.5	21
Fruits	0.79	3.15	6.3

past few years, especially in Sicily (Inglese et al., 2002).

Cacti, particularly *O. ficus-indica,* produces the spiny, usually edible fruits and edible pads called nopalitos. *Opuntia* have been used for almost 500 years as a fruit crop, a defensive hedge, a support for cochineal production of dye (carminic acid), and more recently, as a fodder crop and as a standing buffer feed for drought periods; they can also play a key role in erosion control and land rehabilitation, particularly in arid and semiarid zones, and as a shelter, refuge, and feed resource for wildlife (Le Houerou, 2002). The expansion of the prickly pear plantation in arid and semiarid areas could be of interest for stimulating bioindustries in developing countries.

In terms of the future of biotechnology, countries with available raw materials are more highly favored to build strong biotechnology industries (Goma, 1985). *Opuntia* fruits and cladodes are potential sources of foods and functional components such as dietary fiber, natural colorants, and antioxidants. As there have been few industrial considerations of the prickly pear cladodes and fruits, this chapter reviews their promising applications in food and medicine and the technology development possibilities.

COMPOSITION OF PRICKLY PEAR PRODUCTS

The composition of the prickly pear fruits and caldodes depends especially on the species, the horticultural practices, and the level of ripening. Cactus stems, as other vegetables, have low proteins and fats but the crude fiber is higher than that in most other vegetables (Inglese et al., 2002).

The highly gastronomically appreciated prickly pear is a multiseeded berry, with thick, violet pericarp with a number of clefts of small prickles enclosing a red violet or orange yellow, sweet, lucsius pulp, intermixed with a number of small, black, shiny seeds. The pear-shaped fruit weighs approximately 30–70 g, and the overall composition of the pears fruit is 48% peel, 45% strained pulp, and 7% seeds (Merin et al., 1987). Twenty-two *Opuntia* clones were selected for increased cold hardiness, fruit yield, and fruit quality, i.e., pH, sugar content, and seed content. Chilean varieties were most promising for high sugar content and low seed weight per fruit. Mexican varieties with high yields did not contain high sugar (Parish and Felker, 1997).

The pH of the pulp is slightly acidic, e.g., 5.75 with low acidity (0.18% expressed as citric acid). Total solids (14.2 Brix) are comparable to the common fruit pulps, such as apricots, apples, cherries, and plums, and are greater than those of strawberries, raspberries, and peaches (Sawaya et al., 1983). Proximate analysis of the pulp (Table 34.2) showed that the percentages of protein, crude fat, and fiber

Table 34.2. Characteristics of Prickly Pear Pulp (Sawaya et al., 1983)

pH	5.75
Moisture	85.6%
Acidity (expressed as citric acid)	0.18%
Total solids	14.5%
Total sugars	12.8%
Crude proteins (N × 6.25)	0.21%
Pectin	0.19%
Ca	276 mg/l
Mg	277 mg/l
Na	8 mg/l
K	1,610 mg/l
P	154 mg/l
Fe	1.5 mg/l

were relatively low. Volatile compounds of prickly pear juice are so complex that the distillates contained at least 50 elements including alcohols, aldehydes, lactones, hydrocarbons, and esters (Di Cesare and Nani, 1992). The content of Vitamin C in the pulp leaves represent about half the concentration of Vitamin C of some of the Vitamin-C-rich fruits such as oranges and lemons, and are one and half times higher than that of cherries (Sawaya et al., 1983). At harvest, the concentration of reducing sugars is 90% of that of fully ripe fruits and should not be less than 13% by fresh weight. Sugar analysis revealed glucose and fructose in the ratio of 60:40 at a 12.8% content in the pulp on fresh weight basis. The Argentine varieties had the greatest fruit pulp firmness (about 2 kg) and sugar contents (13.4–15.2%) but had a lower percentage pulp (40–47%) than that of the North American materials (Felker et al., 2005). Minerals (Ca, Mg, K, Na, P, and Fe) were comparable to those of other fruit pulp such as apricots, pineapples, and strawberries. The pulp was rich in potassium, fair in calcium, magnesium, and phosphorus, and low in sodium and iron (Sawaya et al., 1983; Inglese et al., 2002; Jaramillo et al., 2003). Some beneficial effects of *O. ficus-indica* fruits can be attributed to their biochemical activities associated with some enzymes as an acid β-fructofuranosidase (Ouelhazi et al., 1991), peroxidase (Padiglia et al., 1995), and pectinesterase (Contreras-Esquivel et al., 1999).

Fresh cladodes have high moisture content (92%), due to their crassulacean acid metabolism (Brulfert et al., 1984) with stem tissues containing numerous mucilaginous cells that store large volumes of water (Amin et al., 1970; Merin et al., 1987). Fresh cladodes comprise 5.1% carbohydrates (with lipids 0.2%), minerals (1.9%), and protein (0.8%). *Opuntia vulgaris* is a source of Nopal gum (Bonnassieux, 1988). Phytochemical investigation of *Opuntia* reveals 17 amino acids, 8 of which are essential. Nopal is full of vitamins and minerals, such as the B vitamins, calcium, magnesium, iron, and fiber (Jaramillo et al., 2003).

Particular considerations were made for lipids, polymers, and flavonoids contained in the *Opuntia* products, as reported by some published works.

Lipids

The chemical investigation of cactus pear oil promotes the industrial utilization of the fruit as a raw material of oils and functional foods (Ramadan and Mörsel, 2003a). The seed oil has been proposed for human and/or animal consumption (Sawaya and Khan, 1982). Seeds and pulp of *O. ficus-indica* were compared in terms of fatty acids, lipid classes, sterols, fat-soluble vitamins, and β-carotene.

Total lipids (TL) in lyophilized seeds and pulp were 98.8 (dry weight) and 8.70 g/kg, respectively. High amounts of neutral lipids were found (87.0% of TL) in seed oil, whereas glycolipids and phospholipids occurred at high levels in pulp oil (52.9% of TL). In both oils, linoleic acid was the dominating fatty acid, followed by palmitic and oleic acids, respectively. Trienes, γ- and α-linolenic acids were estimated in higher amounts in pulp oil, whereas α-linolenic acid was only detected at low levels in seed oil. Neutral lipids were characterized by higher unsaturation ratios, while saturates were in higher levels in polar lipids. The sterol marker, β-sitosterol, accounted for 72% and 49% of the total sterol content in seed and pulp oils, respectively. Vitamin E level was higher in the pulp oil than in the seed oil, whereas γ-tocopherol was the predominant component in seed oil and δ-tocopherol was the main constituent in pulp oil. β-Carotene was also higher in pulp oil than in seed oil (Ramadan and Mörsel, 2003a). Over 50% of fruit (peel and seeds) by-products used as animal feed contain oil. Oil from seeds was 13.6% by weight. This oil has a high degree of unsaturation, 82% (linoleic acid 73.4%, palmitic 12%, oleic 8.8%, and stearic acid 5.8%). Compositions and concentrations of fatty acids, lipid classes, sterols, fat-soluble vitamins, and β-carotene were determined in extracted lipids from prickly pear peel. TL recovered were found to be 36.8 g/kg (on dry weight basis). Recovered lipids were characterized by a high percentage of unsaponifiables (12.8% TL) and found to be a rich source of vitamin E and sterols. The level of neutral lipids was the highest, followed by glycolipids and phospholipids. Among the TL, linoleic acid was the dominating fatty acid, while oleic and palmitic acids were estimated to be in relatively equal amounts. (Ramadan and Mörsel, 2003b).

Polymers

Large quantities of complex carbohydrates, especially mucilage (Inglese et al., 2002; Saenz et al., 2004) and others as cellulose (Malainine et al., 2003), pectin (Forni et al., 1994; Contreras-Esquivel et al., 1999; Majdoub et al., 2001), xylan (Habibi et al., 2002), and arabinogalactan (Habibi et al., 2004a) were extracted from *Opuntia* products and investigated.

Opuntia genus is widely known for its mucilage production, and the mucilage composition and content vary with the time of year and species (Inglese et al., 2002). The mucilage extracted from the cladodes (modified stems) of *O. ficus-indica* contains residues of D-galactose, D-xylose, L-arabinose, L-rhamnose, and D-galacturonic acid (McGarvie and Parolis, 1979; Saenz et al., 2004).

The prickly pear pectin yield was 0.12% on fresh weight. This pectin had a galacturonic acid content of 64%, a low degree of methoxylation (10%), a high acetyl content (10%), and neutral sugar content (51% galacturonic). It might be related to similar polygalacturonides present in the mucilages of other Cactaceae (Forni et al., 1994). Pectic polysaccharides were solubilized from *O. ficus-indica* fruit skin by sequential extraction with water at 60°C and EDTA solution at 60°C prior to removal of the mucilage with water at room temperature. The sugar analysis supported by ^{1}H and ^{13}C NMR spectroscopy showed that the water-soluble fraction and the EDTA-soluble fraction consisted of a disaccharide repeating unit $\rightarrow$ 2)- α-L-Rha*p*-(1 $\rightarrow$ 4)-. α-D-Gal*p*A-(1 $\rightarrow$ backbone, with side chains attached to O-4 of the rhamnosyl residues (Habibi et al., 2004b). Pectinesterase activity was detected in the extracts from prickly pear peels (Contreras-Esquivel et al., 1999). The water-soluble fraction of peeled prickly pear nopals called native sample contain mainly two components: pectic polysaccharide with high average molar mass (M_w of 13×10^6 g/mol) and protein with low molecular weight (Majdoub et al., 2001).

The cold-water extract from the skin of *O. ficus-indica* fruits contains a polysaccharide, which can be classified under the type I of the arabinogalactan family (Habibi et al., 2004a).

Xylans were isolated from the pericarp of prickly pear seeds of *O. ficus-indica* by alkaline extraction, fractionated by precipitation, and purified. Sugar analysis and NMR spectroscopy showed that, on an average, the water-soluble xylans have one nonreducing terminal residue of 4-*O*-methyl-D-glucuronic acid for every 11–14 xylose units, whereas in the water-insoluble xylans, the xylose units varied from 18 to 65 residues for one nonreducing terminal residue of 4-*O*-methyl-D-glucuronic acid (Habibi et al., 2002).

Flavonoid and Carotenoid Contents

The antioxidant capacity of cactus fruits may be attributed to their flavonoid, ascorbic acid, and carotenoid contents, and the cactus fruits are a rich source of natural antioxidants for foods (Kuti, 2004). The prickly pear juice contains betanine, isobetanine, and betalainic glucoside, which are responsible for the redbeet color of the fruit (Forni et al., 1992), and the structure details of indicaxanthin, the yellow betaxanthin (Piatelli et al., 1964; Forni et al., 1992). The stability of the red color of betacyanine was determined in fruit juice at temperatures up to 90°C. Degradation occurred with rates being slower at higher pigment concentration. The presence of oxygen has only a marginal effect on the thermostability of the dye, and thus autoxidation was not a major chemical mechanism responsible for decoloration (Merin et al., 1987). The qualitative and quantitative betalain pigment contents of the two cultivars of prickly pear fruits grown in southeastern Spain were evaluated. Two main pigments were obtained, which were identified as indicaxanthin (484 nm) and betanin (535 nm). Both the pigments showed a yield of around 20–30 mg/100 g of fresh pulp. When the influence of temperature (25–90°C) on betacyanin pigment stability was investigated, the results revealed a substantial degree of thermodegradation at temperatures higher than 70°C (Fernández-López and Almela, 2001).

The carotenoids cryptoxanthin, carotene, and lutein were identified in the cladodes, the latter having the highest concentration. Mucilage presentin the stems decreased the extractability of the carotenoids. Thermal treatmentsincreased the extractability of these pigments, and the antioxidant activity was related to the carotenoids concentration. Total phenolic content decreased after the thermal treatments; however, this result had little effect on the antioxidant activity (Jaramillo et al., 2003).

USES OF PRICKLY PEAR CLADODES IN ANIMAL FOOD

Under domesticated cultivation, the prickly pear yields pads used as a vegetable for animal consumption. *Opuntia ficus-indica* could safely substitute grass hay to a level of 60%, and it has a substantial contribution in satisfying the water requirement of sheep (Tegegne, 2002a). The Food and Agriculture Organization of the United Nations calculated the biological value of prickly pear cactus protein to be 72.6, relative to egg protein (Teles et al., 1984).

Experiments carried out by the Tunisian government and the United Nations (1962) showed that cactus forage was 2–3 times cheaper than conventional forage. The prickly pear cactus yields 0.075

Fodder Units (FU)/kg of fresh cladodes with 10% dry matter (Theriez, 1966). Compared with ruminant nutrient requirements, *O. ficus-indica* was moderate in CP, Ca, Mg, and P, and low in Na and K contents. It was highly digestible (Tegegne, 2002b). *Opuntia ficus-indica* f. *inermis* pads have been used as alternative energy supplements for growing Barbarine lambs, which were given straw-based diets (Ben Salem et al., 2004). A daily ration of 6 kg cladodes and 1.5 kg straw corresponds to 1.2 FU feed for an ewe weighing 45 kg and producing 2 l of milk/day (Arces, 1941). The addition of straw (0.5 kg) to the cladodes (5 kg) feed prevents diarrhea. In an ensilage of 84% minced cladodes plus 16% gives a product lucerne, straw hay gives a product of high quality, which maintains a good level of milk production by cows in South Africa (Le Houerou, 1965).

BENEFITS TO HEALTH AND POTENTIAL USES OF *OPUNTIA* PRODUCTS IN MEDICINE

The chemical composition of *Opuntia* fruits and cladodes, which depend on the age, define their uses in medicine and nutrition. The cladodes of the plant are a good source of fiber, an important element for the human diet, and are of considerable potential for medical use (Saenz, 2000). *Opuntia ficus-indica* cladodes are used in traditional medicine of many countries for their cicatrisant activity. The major components of cladodes are carbohydrate-containing polymers, which consist of a mixture of mucilage and pectin. The treatment with *Opuntia ficus-indica* cladodes provokes an increase in the number of secretory cells. Probably, the gastric fibroblasts are involved in the antiulcer activity (Galati et al., 2002a). In Sicily folk medicine, *O. ficus-indica* cladodes are used for the treatment of gastric ulcer, and probably, the mucilage of *O. ficus-indica* is involved (Galati et al., 2002a). The wide range of buffering activity of the edible young cladodes could partially explain the therapeutic effect attributed to nopalitos in gastrointestinal disorders (Corrales-García et al., 2003). The increase of the diuresis is more marked with the prickly pear cladode, and it is particularly significant during the chronic treatment (Galati et al., 2002b).

The regular ingestion of *O. robusta* enables to significantly reduce *in vivo* oxidation injury in a group of patients suffering from familial hypercholesterolemia (Budinsky et al., 2001).

The young, rapidly growing flattened stems or cladodes of the prickly pear cactus (*Opuntia* spp.) are commonly consumed in Mexico as a vegetable and are shown to reduce blood glucose levels (Guevara et al., 2001). The hypoglycemic activity of a purified extract from prickly pear cactus (*O. fuliginosa* Griffiths) was evaluated on streptozotocin-induced diabetic rats. Blood glucose and glycated hemoglobin levels were reduced to normal values by a combined treatment of insulin and *Opuntia* extract (Trejo-González et al., 1996). The symptoms of the alcohol hangover are largely due to the activation of inflammation. An extract of the *O. ficus-indica* plant has a moderate effect on reducing the hangover symptoms, apparently by inhibiting the production of inflammatory mediators (Wiese et al., 2004).

The fresh fruits have long been utilized as a source of carbohydrates and vitamins in human nutrition (Duisberg and Hay, 1971). Because of the high content of insoluble fiber (cellulose and lignin), many people use *Opuntia* products to aid digestion and monitor regularity (Jaramillo et al., 2003). The flavonoids quercetin, (+)-dihydroquercetin, and quercetin 3-methyl ether are the active antioxidant principles in the fruits and stems of *O. ficus-indica* var. *saboten* exhibiting neuroprotective actions against the oxidative injuries induced in cortical cell cultures. Furthermore, quercetin 3-methyl ether appears to be the most potent neuroprotectant of the three flavonoids isolated from this plant (Dok-Go et al., 2003). The consumption of cactus pear fruit positively affects the body's redox balance, decreases oxidative damage to lipids, and improves antioxidant status in healthy humans (Tesoriere et al., 2004). Moreover, the fruit infusion shows also antiuric effect (Galati et al., 2002b).

PRICKLY PEAR FRUITS AND CLADODES PROCESSING

The use of prickly pear cladodes and fruits as raw materials in bioindustries can be developed especially when their annual production becomes abundant and the harvest costs are reduced (Hamdi, 1997). Moreover, the *Opuntia* products processing is limited by the cactus plantation area and the technological aspects, especially the fast postharvest deterioration of cladodes and fruits (Fig. 34.1).

Postharvest Changes and Shelf Life of Prickly Pear Fruits and Cladodes

Postharvest deterioration of fruits, cladodes, and nopalitos tender young cladodes result from

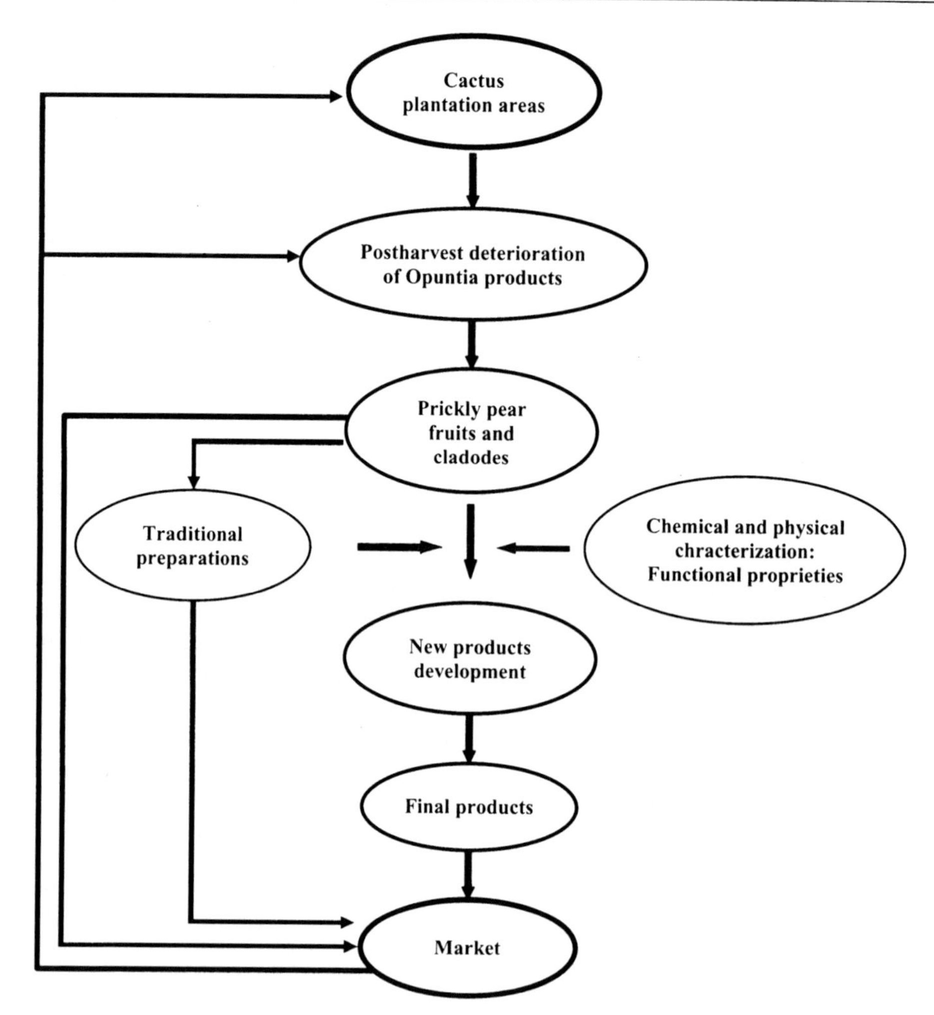

Figure 34.1. General approaches for promising the development of products from *Opuntia ficus-indica* harvests.

physiological disturbance inducing surface softening, causing microbial infections and reactions.

Shelf Life of Prickly Pear Fruits

Fruits should be harvested early in the morning, when their internal temperature is not higher than 25°C, and when the glochids are still wet and adhering to the pell. Postharvest changes of internal quality characteristics, such as pH, titrable acidity, soluble solids, acetaldehyde concentration, and ethanol concentration in the flesh are low, whereas ascorbic acid concentration may decrease, depending on storage conditions (Inglese et al., 2002). Summer ripening cactus pear fruits were more susceptible to chilling injury and to weight loss but less sensitive to decay than autumn fruit (Schirra et al., 1999).

Keeping the fruits in ventilated cold rooms at 6–8°C and 90–95% relative humidity will prevent chilling injury and reduce respiration (Inglese et al., 2002). Wrapping with polyolefinic film retains fruit freshness and greatly reduces fruit weight loss during 4 weeks of storage at 9°C and subsequent marketing (Piga et al., 1997).

Physicochemical and sensorial parameters did not show significant changes in 4°C-stored fruits of cactus pear fruits until day 8, while those kept at 15°C experienced a dramatic increase in acidity, ethanol accumulation, strong off-flavor development, and loss of overall fruit freshness and firmness. Microbiological growth rate was much higher at 15°C than at 4°C. Visible manifestation of molds was detected on the 4°C-stored fruits on day 11, while in the 15°C-stored fruits, fungi colonization had already started on the day 4 (Piga et al., 2000). Yet, in a study of ripeness and storage conditions of prickly pear fruits (var. Rossa and Gialla) the prickly pear fruit could be preserved for 30 days at 5°C. Storage at 5°C under 2% carbon dioxide and 2% oxygen increased the preservation up to 45 days (Testoni and Eccher, 1990).

Opuntia ficus-indica fruits were manually peeled, then placed in plastic boxes sealed with a film with high permeability to gases, and kept at 4°C for 9 days. After 3, 6, and 9 days, chemical, physical, microbiological and sensorial parameters, total phenols, vitamin C, and antioxidant capacity were determined. In-package gas concentrations were measured almost daily. Vitamin C and antioxidant capacity remained unchanged, while polyphenols decreased after 6 days in storage. Of the chemical parameters, only pH and acidity changed significantly, without however, adversely affecting the sensorial properties. Microbiological growth was limited and fungal colonies were never visually detected (Piga et al., 2003). Minimally processed cactus pear fruit had longer shelf life at 4°C than at temperatures recommended for whole fruit when these were greater than 4°C. The packaging of processed cactus pear fruit in modified atmospheres during storage resulted in a homogeneous bacterial population, compared to that isolated from fruit stored in air, and favored the growth of *Leuconostoc mesenteroides* (Corbo et al., 2004).

Shelf Life of Prickly Pear Cladodes

Cladodes for nopalitos are harvested by hand, gripping the bottom of pad, and twisting at more than 90° until it snaps off the mother plant. Storing at low temperatures extends the shelf life of nopalitos and maintains their vitamin contents especially under modified atmospheres, which implies low oxygen availability for oxidation and browning (Inglese et al., 2002). The chilling injury causes damage of nopalitos when conditions favor cactus physiological disorders and microbial growth. Indeed, as mentioned by Guevara et al. (2001), the nopalitos are highly perishable with a storage life of 1 day at room temperature and 6 days when packaged in polyethylene bags and stored at 5°C.

Cladodes of *Nopalea cochenillifera* were found to be suitable for harvest at the end of the rapid growth phase, at which time they were 11–13 cm long, weighed about 40 g, and had a chemical composition similar to that of certain *Opuntia* species used as a vegetable. With respect to both acidity and weight loss, the most favorable treatment for long-term storage is wrapping the cladodes in PVC film and storing them at either 12°C or 20°C (Nerd et al., 1997). Modified atmosphere packaging (MAP) of nopalitos at 5°C, where O_2 concentration was decreased by up to 8.6 kPa and CO_2 concentration was increased by up to 6.9 kPa, for up to 30 days, increased significantly the storage life and decreased losses in texture, weight, chlorophyll content, crude fiber content, and color. MAP also decreased chlorophyllase activity and caused the least increase in yeast and molds, and total aerobic mesophiles (AeM) counts, but slightly increased the total anaerobic mesophiles (AnM) counts (Guevara et al., 2001).

The effects of passive and semiactive MAP were tested on the postharvest life and quality of flattened stems or cladodes of the prickly pear cactus stored at 5°C. Passive MAP and semiactive MAP with 20 kPa CO_2 significantly decreased the losses in the quality deterioration and also decreased the microbial counts (total AeM, mold, and yeasts) but slightly increased the total AnM counts. The microorganisms identified were *Pseudomonas*, *Leuconostoc*, *Micrococcus*, *Bacillus*, *Ruminicoccus*, *Absidia*, *Cladosporium*, *Penicillium*, and *Pichia*. Therefore, fresh prickly pear cactus stems can be stored for up to 32 days in MAP with $\leq$20 kPa CO_2, without significant losses in quality nor any significant increase in microbial counts (Guevara et al., 2003).

Foods and Functional Foods From *Opuntia* Products

Prickly pear is a classical food in Mexico of nomadic tribes of the desert, and the civilized centers of the pre-Columbia Anauac. The Papago and Pima Indians in the southwestern United States are also known to have used prickly pear pads and fruits as food. Nopales could also be added to a tossed salad or stirred into any casserole. Panamint Indians dry the gray-green pads, buds, flowers, and fruits in the sun and may store them until needed for cooking by boiling (Teles et al., 1984). The preparation of juices, marmalades, gels, liquid sweeteners, dehydrated foods, and other products from *Opuntia* products are reported (Saenz, 2000; Saenz et al., 2002a).

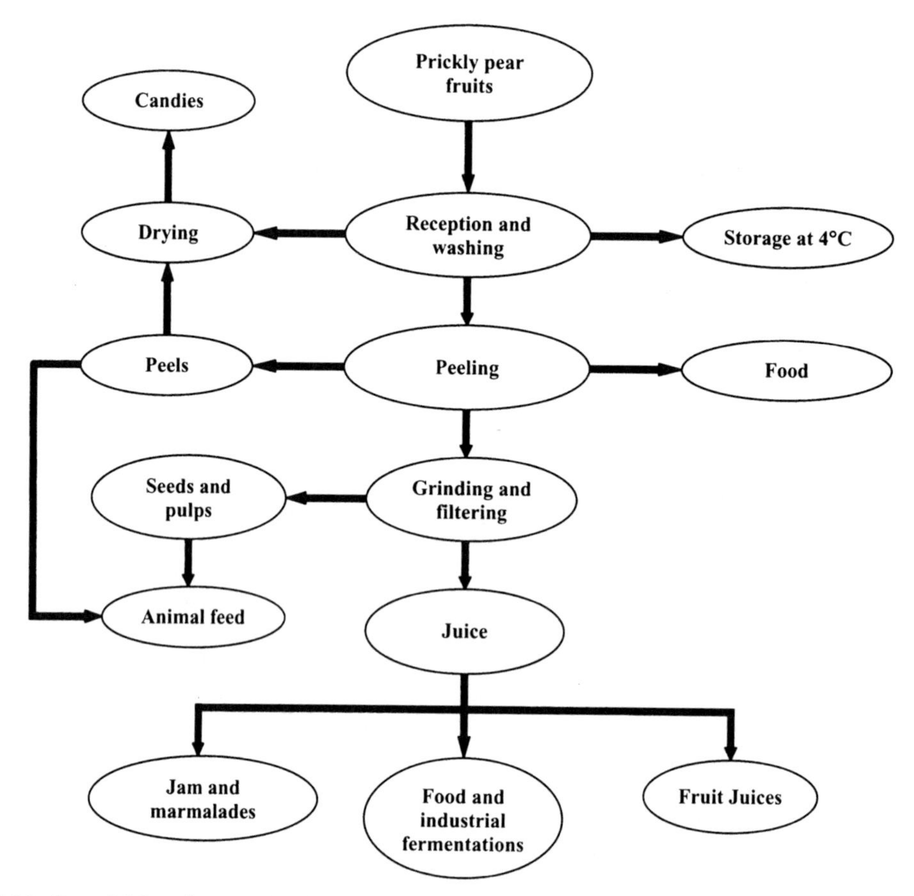

Figure 34.2. Potential flow diagrams for *Opuntia ficus-indica* fruits processing.

Fruit Products

The fruit is an excellent dietetic food because it aids in digestion and can cure diarrhea and constipation (Paris and Moyse, 1940). However, when used alone, it becomes a constipating food (Gobert, 1940). Figure 34.2 summarizes the traditional preparations and potential industrial products from the prickly pear fruits.

Traditionally, the prickly pear fruits are cut manually and dried under the sun. Dried slices of prickly pear fruit can be used as human food (Ewaidah and Hassan, 1992). Food additives did not affect their organoleptic quality and allow direct preservation of fruits. The dried product (candy) of high quality from the peel of cactus pear fruit was tested for yeast and bacterial activity and storage life. It had a good flavor, texture, and appearance, with wide appeal, and stored satisfactorily for up to 5 months (Mnkeni and Brutsch, 2002). The prickly pear fruits are sufficiently dried in a convective solar oven between 32°C and 36°C of ambient air temperature, 50–60°C of drying air temperature, 23–34% of relative humidity, 0.0277–0.0833 m^3/s of drying air flow rate, and 200–950 W/m^2 of solar radiation (Lahsasni et al., 2004).

Traditionally, the prickly pear juice has been used to prepare such commercial foods like Tuna honey, Melcocha pastry, and Tuna Queso cake (Bonnassieux, 1988). The economics of jam manufacture can be enhanced, especially if efficient mechanical devices to replace handpeeling of the fruits are used (Sawaya et al., 1983). The low pectin content suggests that additional pectin should be added to ensure gelation of the jam. Citric acid and a combination of citric and tartaric acids (1:1) were preferred over several other natural acids used as acidifying agents. The sensory evaluation results showed

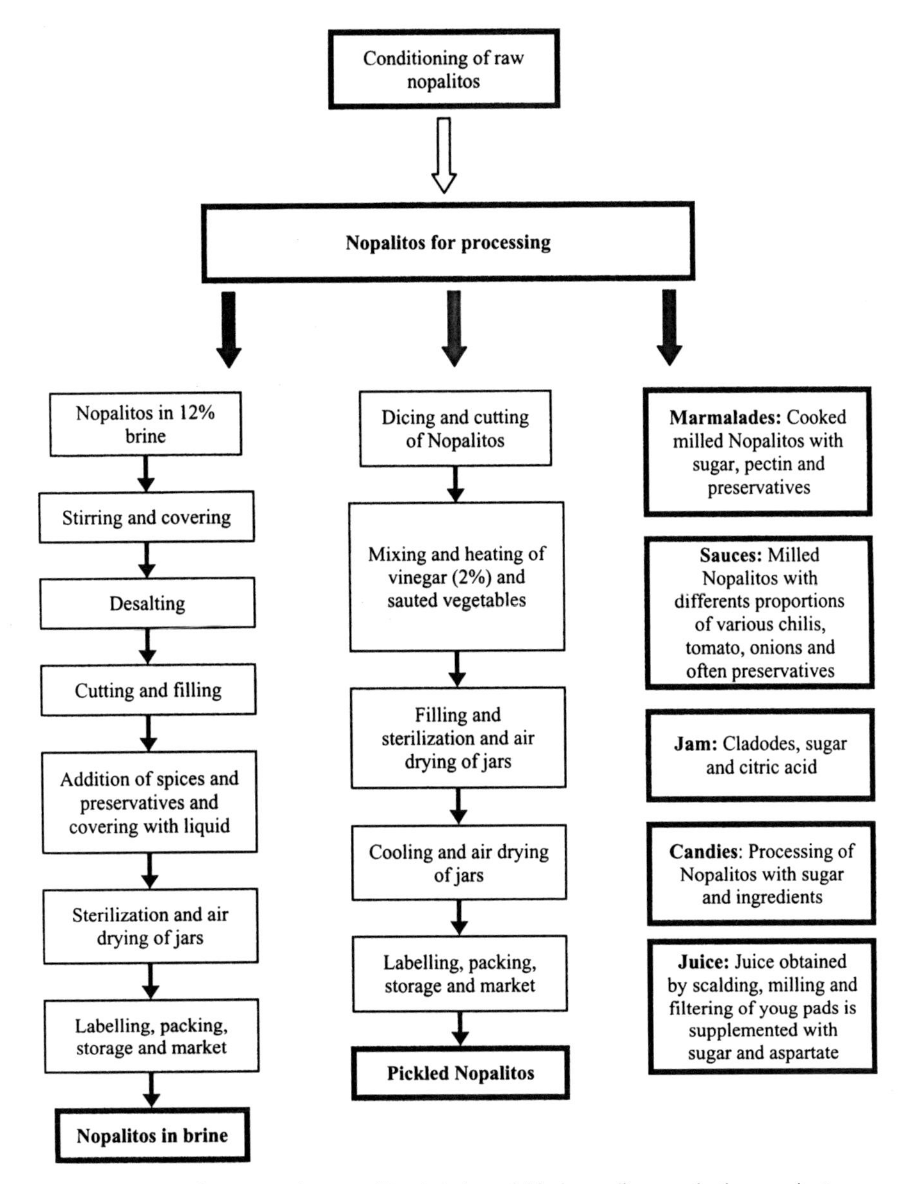

Figure 34.3. Flow diagrams for processing nopalitos in brine, pickled nopalitos, and other products prepared with mille nopalitos (adapted from Saenz et al., 2002a).

that the jam was acceptable with or without any flavor added. Blanching resulted in no significant difference to the sensory quality of the jam (Sawaya et al., 1983). The sensory evaluation of the color acceptability of concentrated juice of prickly pears (*Opuntia ficus indica*) indicated that reconstituted juices were more acceptable at shorter storage times than the respective concentrated juices (Saenz et al., 1993).

Nopalitos Products

The fresh product is marketed directly after harvesting and cleaning. The cleaned pads are cut into small pieces and preserved by various methods (Fig. 34.3). The minimally processed nopalitos washed with cold, clean water to eliminate adhering pectines and mucilages can be used for various products, e.g., nopalitos in salt brine with 2% NaCl, pickled nopalitos preserved in vinegar, and marmalades prepared using milled nopalitos with various concentrations of sugar, pectin, and preservatives (Saenz et al., 2002a).

Mucilage as Functional Foods

The functional components from cactus pear open new possibilities for adding value to a crop of the arid and semiarid regions of the world. Mucilage, a complex carbohydrate with a great capacity to absorb water, should be considered a potential source of gelatinous colloid, dietary fiber, and thickening agents (Amin et al., 1970; Saenz, 2002; Saenz et al., 2004). Fractionation studies indicate that the mucilage of *O. ficus-indica* is essentially homogeneous (McGarvie and Parolis, 1979). The rheological properties of aqueous solutions of the mucilage isolated from *O. ficus-indica* have been examined. Steady-shear viscosities in a range of shear rate from 1 to 300 s^{-1} were observed as a function of mucilage concentration, temperature, pH, and ionic strength. Mucilage aqueous solutions showed non-Newtonian shear-thinning behavior with high elastic properties, similar to the high elastic synthetic polymers like polyisobutylene (Medina-Torres et al., 2000).

The mechanical properties of gels formed by either mixtures of mucilage gum obtained from *O. ficus-indica* and K-carrageenan or i-carrageenan have been examined using dynamic shear and uniaxial compression measurements. A total polymer concentration of 2% (w/w) was used, the proportion of mucilage gum varying from 0% to 80% (w/w), and KCl or $CaCl_2$ in the range from 12 to 60 mM. For the mixed gels of i-carrageenan and mucilage gum, no enhancement of the mechanical properties with respect to the pure i-carrageenan was observed. For the K-carrageenan/mucilage gum system, at an 80:20 ratio and 12 mM KCl, higher gel rigidity than the pure K-carrageenan gels was observed, and failure stress and strain increases as mucilage gum concentration increases. The main feature of the mucilage is to enhance elasticity on the final gels (Medina-Torres et al., 2003). The use of prickly pear cactus mucilage (*Opuntia ficus indica*) as an edible coating to extend the shelf life of strawberries leads to increased shelf life (Del-Valle et al., 2005). The addition of cactus pear flour in oatmeal cookies considerably increases the dietary fiber content in this type of product and could comprise a part of the daily intake requirements of this nutrient (Saenz et al., 2002b).

FERMENTATION PROCESSING OF *OPUNTIA* PRODUCTS

Prickly pear juice, which contains high concentrations of glucose and fructose, offers a potential substrate for fermentation processes.

Production of Single-Cell Protein

The protein production with *Candida utilis*, using prickly pear juice in batch and in continuous culture was studied (Paredes-Lopez et al., 1976). In batch culture, the maximum specific growth (μ_m) rate and the substrate yield coefficient (Ys) varied according to sugar concentration. At 1% sugar, μ_m and Ys were 0.47/h and 42.6%, respectively, the best yields occuring in a chemostat at pH 3.5–4.5 at 30°C. A beneficial effect on Ys as observed when the dilution rate (D) was increased. At a D of 0.55/h, the productivity was 2.38 g/l.h. The maintenance coefficient was 0.09 g of sugar/g of biomass. Increases of D produced higher protein content of the biomass.

Since the prickly pear juice contains fermentable sugars (glucose and fructose), it could find use as a raw material for baker's yeast production. The mineral and vitamin contents of prickly pear juice also favor good growth of yeasts.

Fermentation Substrate for Wine and Alcohol Production

The prickly pear juice is used in Mexico to produce the traditional fermented drinks Tequila, Colonche, and Pulque (Bonnassieux, 1988). The transformation of fruit juice into wine is essentially a microbial process. A pure yeast culture, inoculated in a pasteurized fruit must digest the available sugars into ethanol and carbon dioxide. Fermentation has a relatively low capital and operating costs, and it is a simple technology. The effect of fermentation on foods produces few deleterious changes to the nutritional and sensory properties. The aroma of fermented foods is due to the large number of volatile chemical components like amines, fatty acids, aldehydes, esters, and ketones, and products from their interactions of these compounds during fermentation and maturation.

The sugar and energy conversions by *Saccharomyces cerevisiae* into ethanol were 99% and 32%, respectively (Retamal et al., 1987b). The final ethanol concentration obtained on the twice diluted prickly pear juice was 5.45% v/v. These experiments were used to determine the optimal area of prickly pear crop necessary to supply an industrial fermentation plant processing 50,000 l/day of ethanol for 5 months of the year. In Spain, and with production of 1500 l of ethanol/ha, the theoretical radius of the cultivated land would be approximately 4 km.

The ethanol production by *S. cerevisiae* using prickly pear cladodes as a substrate (Retamal et al., 1987b) was useful when enzymatic hydrolysis using cellulase followed by acid hydrolysis of the substrate considerably improved the release of sucrose. The highest ethanol concentration obtained by direct fermentation was 0.86% (v/v). This is not economical because final ethanol concentration did not reach 7–10% (v/v).

Prickly Pear Juice Substrate for Microbial Metabolites Production

The use of agricultural by-products as substrates for the production of red pigments by *Monascus* is interesting and economically feasible. The juice from the prickly pear fruit can be used as a medium for red pigment production by *Monascus purpureus*. The red-orange pigment present in the crude prickly pear juice was negligible in comparison to red pigments produced by *M. purpureus*. Optimization of a growth medium based on the prickly pear juice showed that the pigment formation occurred when juice was diluted fourfold, and the addition of monosodium glutamate (MSG) was necessary to obtain high yields. The higher yields of 32.8 U red pigment were obtained with initial pH of 5.2 and 10 g/l of MSG. The influence of gaseous environments by varying the pressure of oxygen on *M. purpureus* growth on diluted prickly pear juice, sugar consumption, ethanol formation, and then on red pigment production from ethanol has been investigated (Hamdi et al., 1996).

Unusual lipids such as d-linolenic acid or other polyunsaturated fatty acids undoubtedly can be produced economically by various microorganisms with yields of oil production of 40–65% of the biomass by certain oleaginous yeasts: *Lipomyces*, *Rhodotorula*, and *Candida* (Ratledge, 1982). An unsaturated fatty acid auxotroph derived from the oleaginous yeast *Cryptococcus curvatus,* and named UfaM3, was defective in the conversion of stearic to oleic acid. It was cultivated on diluted (25%) prickly pear juice in batch culture. The carbon-to-nitrogen ratio (C:N ratio) of the juice (44% glucose and 56% fructose content) was about 50 g/g. Differential utilization of glucose and fructose was observed. The efficiency of substrate conversion was 0.50 g/g for biomass and 0.21 g/g for lipids. The quality of lipids produced by UfaM3 approached cocoa butter composition. The extracted lipids were mainly oleic (C 18:1) and palmitic (C16:0) acids (Hassan et al., 1995).

The growth of *Kluyveromyces marxianus* on five agroindustrial residues proved the feasibility of using cassava bagasse and giant palm bran (*O. ficus-indica*) as substrates to produce fruity aroma compounds by the yeast culture (Medeiros et al., 2000).

FUTURE PROSPECTS

The prickly pear cladodes and fruits are a potential food reserve for humans and animals, especially in arid and semiarid regions of the world. However, some constraints limiting the expansion of the plantation of cactus pear are low productivity and quality, the poor value of crops, and the modest consumption. Much research still needs to be done to promote the commercial potentials for foods, cosmetics, and medicines.

REFERENCES

Amin E.S., Awad O.M. and El-Sayed M.M. 1970. The mucilage of *Opuntia ficus-indica* mill. Carbohydrate Research 15, 159–161.

Arces J. 1941. Un aliment du bétail: le cactus. Doc. Rens. Agric. Bull. No 53. G.G. XII. Alger. 4 p.

Ben Salem H., A. Nefzaoui A. and Ben Salem L. 2004. Spineless cactus (*Opuntia ficus indica* f. *inermis*) and oldman saltbush (*Atriplex nummularia* L.) as alternative supplements for growing Barbarine lambs given straw-based diets. Small Ruminant Research 51, 65–73.

Bonnassieux M.P. 1988. Les fruits commestibles du monde (Editor C. Doremus). Bordas, Paris.

Brulfert J., Guerrier D. and Queiroz O. 1984. Rôle de la photopériode dans l'adaptation à la sécheresse: cas d'une plante à métabolisme crassulacèen, l'*Opuntia ficus-indica* Mill. Bulletin de la Societe Botany Francaise 131, 677–682.

Budinsky A., Wolfram R., Oguogho A., Efthimiou Y., Stamatopoulos Y. and Sinzinger H. 2001. Regular ingestion of Opuntia robusta lowers oxidation injury. Prostaglandins, Leukotrienes and Essential Fatty Acids 65, 1, 45–50.

Contreras-Esquivel J.C., Correa-Robles C., Aguilar C.N., Rodríguez J., Romero J. and Hours R.A. 1999a. Pectinesterase extraction from Mexican lime (*Citrus aurantifolia* Swingle) and prickly pear (*Opuntia ficus indica* L.) peels. Food Chemistry 65, 2, 153–156.

Corbo M.R., Altieri C., D'Amato D., Campaniello D., Del Nobile M.A. and Sinigaglia M. 2004. Effect of temperature on shelf life and microbial population of lightly processed cactus pear fruit. Postharvest Biology and Technology 31, 1, 93–104.

Corrales-García J., Peña-Valdivia C.B., Razo-Martínez Y. and Sánchez-Hernández M. 2003. Acidity changes and pH-buffering capacity of nopalitos (*Opuntia* spp.). Postharvest Biology and Technology 32, 2, 169–174.

Del-Valle V., Hernández-Muñoz P., Guarda A. and Galotto M.J. 2005. Development of a cactus-mucilage edible coating (*Opuntia ficus indica*) and its application to extend strawberry (*Fragaria ananassa*) shelf-life. Food Chemistry 91(4):751–756.

Di Cesare, L.F. and Nani, R. 1992. Analysis of volatile constituents of prickly pear juice (Opuntia ficus indica var. Fructa sanguineo). Fruit Process 2:6–8.

Diguet L. 1928. Les Cactées utiles du Mexique, 8th ed. Paris, 552 p.

Dok-Go H., Heun Lee K., Kim H.J., Lee E.H., Lee J., Song Y.S., Lee Y.H., Jin C., Lee Y.S. and Cho J. 2003. Neuroprotective effects of antioxidative flavonoids, quercetin, (+)-dihydroquercetin and quercetin 3-methyl ether, isolated from *Opuntia ficus-indica* var. *saboten.* Brain Research 965, 1–2, 130–136.

Duisberg P.C. and Hay J.L. 1971. Economic botany of arid regions. In food, fiber and their lands. The University of Arizona Press, Tucson, AZ, 252 p.

Ewaidah E.H. and Hassan P.H. 1992. Prickly pear sheets: A new fruit product. International Journal of Food Science Technology 27, 353–358.

Felker P., del S. Rodriguez C., Casoliba R.M., Filippini R., Medina D. and Zapata R. 2005. Comparison of *Opuntia ficus indica* varieties of Mexican and Argentine origin for fruit yield and quality in Argentina. Journal of Arid Environments 60, 3, 405–422.

Fernández-López J.A. and Almela L. 2001. Application of high-performance liquid chromatography to the characterization of the betalain pigments in prickly pear fruits. Journal of Chromatography A 913, 1–2, 415–420.

Forni E., Penci M. and Polesello A. 1994. A preliminary characterization of some pectins from quince fruit (*Cydonia oblonga* Mill.) and prickly pear (*Opuntia ficus indica*) peel. Carbohydrate Polymers 23, 4, 231–234.

Forni E., Pollesello A., Montefiorni D. and Mastrelli A. 1992. High performance liquid chromatographic analysis of the pigments of blood-red prickly pear. Journal of Chromatography 593, 177–183.

Galati E.M., Pergolizzi S., Miceli N., Monforte M.T. and Tripodo M.M. 2002a. Study on the increment of the production of gastric mucus in rats treated with *Opuntia ficus indica* (L.) Mill. Cladodes. Journal of Ethnopharmacology 83, 3, 229–233.

Galati E.M., Tripodo M.M., Trovato A., Miceli N. and Monforte M.T. 2002b. Biological effect of *Opuntia ficus indica* (L.) Mill. (Cactaceae) waste matter. Journal of Ethnopharmacology 79, 1, 17–21.

Garcia de Cortázar V. and Nobel P.S. 1991. Prediction and measurement of high annual productivity for *Opuntia ficus-indica.* Agricultural and Forest Meteorology 56, 3–4, 261–272.

Gobert E.G. 1940. Usages et rites alimentaires des tunisiens. Archives de l'Institut Pasteur de Tunis 29, 475—485.

Goma G. 1985. Substrates for bioconversion. In Proceeding of the Seventh International Biotechnology Symposium: Biotechnology and Bioprocess Engineering. New Delhi, India, February 19–25. 1984. Tarun K. Ghose (Ed). IIT Delhi.

Guevara J.C., Yahia E.M. and Brito de la Fuente E. 2001. Modified atmosphere packaging of prickly pear cactus stems (*Opuntia* spp.). Lebensmittel-Wissenschaft und-Technologie 34, 7, 445–451.

Guevara J.C., Yahia E.M., Brito de la Fuente E. and Biserka S.P. 2003. Effects of elevated concentrations of CO_2 in modified atmosphere packaging on the quality of prickly pear cactus stems (*Opuntia* spp.). Postharvest Biology and Technology 29, 2, 167–176.

Habibi Y., Mahrouz M., Marais M.F. and Vignon M.R. 2004a. An arabinogalactan from the skin of *Opuntia ficus-indica* prickly pear fruits. Carbohydrate Research 339, 6, 1201–1205.

Habibi Y., Heyraud A., Mahrouz M. and Vignon M.R. 2004b. Structural features of pectic polysaccharides from the skin of *Opuntia ficus-indica* prickly pear fruits. Carbohydrate Research 339, 6, 1119–1127.

Habibi Y., Mahrouz M., Marais M.F. and Vignon M.R. 2002. Isolation and structure of D-xylans from pericarp seeds of *Opuntia ficus-indica* prickly pear fruits. Carbohydrate Research 337, 17, 1593–1598.

Hamdi M. 1997. Prickly pear fruit and cladodes a potential raw material for bioindustries: A review. Bioprocess Engineering 17/6, 387–391.

Hamdi M., Blanc P.J., and Goma G. Effect of aeration conditions on the production of red pigment by *Monascus purpureus* growth on prickly pear juice. Process Biochem. 36 (1996) 534–547.

Hare R.F. and Griffiths G. 1907. New Mexico agricultural experiment station. Bulletin 1964, 88–95.

Hassan M., Blanc P.J., Pareilleux A. and Goma G. 1995. Production of cocoa butter equivalents from prickly pear juice fermentation by an unsaturated fatty acid auxotroph of *Cryptococcus curvatus* grown in batch culture. Process Biochemistry 30, 7, 629–634.

Inglese P., Basile F. and Schirra M. 2002. Cactus pear fruit production. In Cacti Biology and Uses (Editor P.S. Nobel) University of California Press, Ltd., 280 p.

Jaramillo-Flores M.E., González-Cruz L., Cornejo-Mazón M., Dorantes-Alvarez L., Gutiérrez-López G.F. and Hernández-Sánchez H. 2003. Effect of thermal treatment on the antioxidant activity and content of carotenoids and phenolic compounds of cactus pear cladodes (Opuntia ficus-indica). Food Science and Technology International 9, 4, 271–278.

Kuti J.O.2004. Antioxidant compounds from four *Opuntia* cactus pear fruit varieties. Food Chemistry 85, 4, 527–533.

Lahsasni S., Kouhila M., Mahrouz M. and Jaouhari J.T. 2004. Drying kinetics of prickly pear fruit (*Opuntia ficus indica*). Journal of Food Engineering 61, 2, 173–179.

Le Houerou H.N. 1965. Les cultures fourragères en Tunisie. Doc. Tech. No 13. INRAT. Tunisie, 81 p.

Le Houerou H.N. 2002. Cacti (Opuntia spp) as a fodder crop for marginal lands in the mediterranean basin. ISHS Acta horticultura 581: In Proceeding of the IV International Congress on Cactus Pear and Cochineal (Editors A. Nefzaoui, P. Inglese) Hammamet, 22–28 October 2002, Tunisia.

Majdoub H., Roudesli S., Picton L., Le Cerf D., Muller G. and Grisel M. 2001. Prickly pear nopals pectin from *Opuntia ficus-indica* physico-chemical study in dilute and semi-dilute solutions. Carbohydrate Polymers 46, 1, 69–79.

Malainine M.E., Dufresne A., Dupeyre D., Mahrouz M., Vuong R. and Vignon M.R. 2003. Structure and morphology of cladodes and spines of *Opuntia ficus-indica*. Cellulose extraction and characterisation. Carbohydrate Polymers 51, 1, 77–83.

McGarvie D. and Parolis H. 1979. The mucilage of Opuntia ficus-indica. Carbohydrates Research 69, 1, 171–179.

Medeiros A.B.P., Pandey A.R., Freitas J.S., Christen P. and Soccol C.R. 2000. Optimization of the production of aroma compounds by *Kluyveromyces marxianus* in solid-state fermentation using factorial design and response surface methodology. Biochemical Engineering Journal 6, 1, 33–39.

Medina-Torres L., Brito-De La Fuente E., Torrestiana-Sanchez B. and Alonso S. 2003. Mechanical properties of gels formed by mixtures of mucilage gum (*Opuntia ficus indica*) and carrageenans. Carbohydrate Polymers 52, 2, 143–150.

Medina-Torres L., Brito-De La Fuente E., Torrestiana-Sanchez B. and Katthain R. 2000. Rheological properties of the mucilage gum (*Opuntia ficus indica*). Food Hydrocolloids 14, 5, 417–424.

Merin U., Gagel S., Popel G., Bernstein S. and Rosenthal I. 1987. Thermal degradation kinetics of prickly pear fruit red pigments. Journal of Food Science 52, 485–486.

Mnkeni A.P. and Brutsch M.O. 2002. Production of an edible dried product (candy) of high quality from the peel of cactus pear fruit in the eastean cape province of South Africa using a simple portable solar drier. ISHS Acta horticultura 581: In Proceeding of the IV International Congress on Cactus Pear and Cochineal (Editors A. Nefzaoui, P. Inglese) Hammamet, 22–28 October 2002, Tunisia.

Nerd A., Dumoutier M. and Mizrahi Y. 1997. Properties and postharvest behavior of the vegetable cactus *Nopalea cochenillifera*. Postharvest Biology and Technology 10, 2, 135–143.

Ouelhazi N.K., Ghrir R., Di K.H. and Lederer F. 1991. Invertase from *Opuntia ficus-indica* fruits. Phytochemistry 31, 1, 59–61.

Padiglia A., Cruciani A., Pazzaglia G., Medda R. and Floris G. 1995. Purification and characterization of *Opuntia* peroxidase. Phytochemistry 38, 2, 295–297.

Paredes-Lopez O., Camargo-Rubio E. and Ornelas-Vale A. 1976. Influence of specific growth rate on biomass yield, productivity, and composition of *Candida utilis* in batch and continous culture. Applied Environmental Microbiology 31, 487–491.

Parish J. and Felker P. 1997. Fruit quality and production of cactus pear (*Opuntia* spp.) fruit clones selected for increased frost hardiness. Journal of Arid Environments 37, 123–143.

Paris R. and Moyse H. 1940. Matière médicale. Tome 3 (Editors Masson & Cie) 509 p.

Piatelli M., Minale L. and Prota G. 1964. Isolation, structure and absolute configuration of indicaxanthin. Tetrahedron 20, 2325–2328.

Piga A., D'Aquino S., Agabbio M., Emonti G. and Farris G.A. 2000. Influence of storage temperature on shelf-life of minimally processed cactus pear fruits. Lebensmittel-Wissenschaft und-Technologie 33, 1, 15–20.

Piga A., Del Caro A., Pinna I. and Agabbio M. 2003. Changes in ascorbic acid, polyphenol content and antioxidant activity in minimally processed cactus pear fruits. Lebensmittel-Wissenschaft und-Technologie 36, 2, 257–262.

Piga A., D'hallewin G., D'Aquino S., Agabbio M., Emonti G. and Farris G.A. 1997. Influence of film

wrapping and UV irradiation àn cactus pear quality after storage. Packaging Technology and Science 10, 59–68.
Ramadan M.F. and Mörsel J.T. 2003a. Oil cactus pear (*Opuntia ficus-indica* L.). Food chemistry 82, 3, 339–345.
Ramadan M.F. and Mörsel J.T. 2003b. Recovered lipids from prickly pear [*Opuntia ficus-indica* (L.) Mill] peel: A good source of polyunsaturated fatty acids, natural antioxidant vitamins and sterols. Food Chemistry 83, 3, 447–456.
Ratledge C. 1982. Micrbiol oil and fats: An assessment of their commercial potential. Progress in Industrial Microbiology 16, 119–206.
Retamal N., Duran J.M. and Fernandez J. 1987a. Seasonal variations of chemical composition in prickly pear *Opuntia ficus-indica* L. Miller. Journal of the Science of Food and Agriculture 38, 303–311.
Retamal N., Duran J.M. and Fernandez J. 1987b. Ethanol production by fermentation of cladodes and fruits of prickly pear Cactus *Opuntia ficus-indica* L. Miller. Journal of the Science of Food and Agriculture 38, 303–311.
Saenz C. 2000. Processing technologies: An alternative for cactus pear (*Opuntia* spp.) fruits and cladodes. Journal of Arid Environments 46, 3, 209–225.
Sáenz C. Cactus pear fruits and cladodes: A source of functional components for foods. ISHS Acta horticultura 581: In Proceeding of the IV International Congress on Cactus Pear and Cochineal (Editors A. Nefzaoui, P. Inglese) Hammamet, 22–28 October 2002, Tunisia.
Saenz C., Corrales-Garcia J. and Aquino-Peres G. 2002a. Nopalitos, mucilage, fiber and cocheneal. In Cacti biology and uses (Editor P.S. Nobel) University of California Press, Ltd., 2002, 280 p.
Saenz C., Estevez A.M., Fontanot M. and Pak N. 2002b. Oatmeal cookies enriched with cactus pear flour as dietary fiber source: physical and chemical characteristics. ISHS Acta horticultura 581: In Proceeding of the IV International Congress on Cactus Pear and Cochineal (Editors A. Nefzaoui, P. Inglese) Hammamet, 22–28 October 2002, Tunisia.
Sáenz C., Sepúlveda E., Araya E. and Calvo C. 1993. Colour changes in concentrated juices of prickly pear (*Opuntia ficus indica*) during storage at different temperatures. Lebensmittel-Wissenschaft und-Technologie 26, 5, 417–421.
Sáenz C., Sepúlveda E. and Matsuhiro B. 2004. Opuntia spp mucilage's: A functional component with industrial perspectives. Journal of Arid Environments 57, 3, 275–290.
Sawaya W.N. and Khan P. 1982. Chemical characterization of pricly pear seed oil, *Opuntia ficus-indica*. Journal of Food Science 47, 2060–2061.
Sawaya W.N., Khatchadourian H.A., Safi W.M. and Al-Muhammad H.M. 1983. Chemical characterization of prickly pear pulp, *Opuntia ficus-indica*, and the manufacturing of prickly pear jam. Journal of Food Technology 18, 183–193.
Schirra M., Inglese P. and La Mantia T. 1999. Quality of cactus pear [*Opuntia ficus-indica* (L.) Mill.] fruit in relation to ripening time, $CaCl_2$ pre-harvest sprays and storage conditions. Scientia Horticulturae 81, 11, 425–436.
Tegegne F. 2002a. In vivo assessment of the nutritive value of cactus pear as a ruminant feed. ISHS Acta horticultura 581: In Proceeding of the IV International Congress on Cactus Pear and Cochineal (Editors A. Nefzaoui, P. Inglese) Hammamet, 22–28 October 2002, Tunisia.
Tegegne F. 2002b. Fooder potential of Opuntia ficus indica. ISHS Acta horticultura 581: In Proceeding of the IV International Congress on Cactus Pear and Cochineal (Editors A. Nefzaoui, P. Inglese) Hammamet, 22–28 October 2002, Tunisia.
Teles F.F.F., Stull J.W., Brown W.H. and Whiting F.M. 1984. Amino and organic acids of the prickly pear cactus *Opuntia ficus-indica* L. Journal of the Science of Food and Agriculture 35, 421–425.
Tesoriere L., Butera D., Pintaudi A.M., Allegra M. and Livrea M.A. 2004. Supplementation with cactus pear (*Opuntia ficus-indica*) fruit decreases oxidative stress in healthy humans: A comparative study with vitamin C[1,2,3]. American Journal of Clinical Nutrition 80, 2, 391–395.
Testoni A. and Eccher Z.P. 1990. Conservazione del fico d'Indica in atmosfera normale e controllata. Annali Instituto Valorrzzazione Tecnologia Prodotti Agricoli XXI, Milan, pp. 131–137.
Theriez M. 1966. Recherches sur la digestibilité de *Opuntia ficus-indica* en Tunisie. INRAT. Tunis (umpublished).
Trejo-González A., Gabriel-Ortiz G., Puebla-Pérez A.M., Huízar-Contreras M.D., Munguía-Mazariegos M.R., Mejía-Arreguín S. and Calva E. 1996. A purified extract from prickly pear cactus (*Opuntia fuliginosa*) controls experimentally induced diabetes in rats. Journal of Ethnopharmacology 55, 1, 27–33.
Wiese J., McPherson S., Odden M.C. and Shlipak M.G. 2004. Effect of *Opuntia ficus indica* on Symptoms of the Alcohol Hangover. Archives of Internal Medicine 164, 1334–1340.
Yasseen M.Y., Barringer S.A. and Walter E.S. 1996. A note on the uses of *Opuntia*spp. in Central/North America. Journal of Arid Environments 32, 3, 347–353.

35
Speciality Fruits Unique to Hungary

Mónika Stéger-Máté

INTRODUCTION

Hungary is located in Central Europe, at the western-central half of the Carpathian Basin; its territory is 90,036 km^2. It is located midway between the equator and the North Pole and the northern latitudes of 45.7° and 48.5°. The area is dominated by flatlands with elevations reaching 200–400 m but only 4% above 400 m. Hungary is ideal for fruit production because of its location in the northern temperate zone and within the influence of the ocean and the Mediterranean Sea (Czellár and Somorjai, 1996). The continental climate of Hungary with mean annual temperatures of 9–10°C and precipitations of 500–800 mm enable production of all fruit varieties native to the temperate zone (Kállay, 1999). The distribution of the total fruit production area (approximately 90,000 ha) is as follows: 48% pomaceous, 40% stone fruits, 7% berries, and 5% shelled fruits (KSH, 2002).

POME FRUITS

Hungary provides an excellent environment for the production of pomes. The most significant pomes are apples and pears. Quince and naseberry can also be found; however, these are mainly produced in household plots.

APPLE (*MALUS DOMESTICA* BORKH.)

Production

Apples are produced and processed in the highest quantity of pome fruits. Annually 58–60 million tons are produced worldwide, with Europe producing 28–30%. The largest apple producing countries of Europe are Turkey, Italy, France, Germany, and Poland. Apples are also the predominant fruit of Hungary.

Total production was more than 500,000 tons in 2001 and 2002 (Table 35.1); meanwhile, producing areas were between 40,000 and 45,000 ha (KSH, 2002).

Varieties

Hungarian apple production is in transition because of the change in varieties that are grown. Before the 1990s, *Jonathan* (50–80%), *Golden Delicious,* and *Starking* were the most common varieties. At present, varieties that are hardy and easy to store and process are being selected for production. The

Table 35.1. Fruit Production Trends (1000 tons)

	World			Europe			Hungary		
	2000	2001	2002	2000	2001	2002	2000	2001	2002
Apple	59,963	60,965	57,095	15,679	17,984	15,820	711	564	510
Pear	16,627	17,044	17,115	3641	3691	3626	40	27	27
Plum	8223	8760	9315	2408	2659	2661	75	85	75
Sour cherry	884	996	883	566	665	651	38	46	47
Cherry	1769	1839	1787	884	926	921	16	16	15
Peach	13,457	13,470	13,815	4674	4548	4640	70	50	29
Apricot	2742	2741	2708	811	855	811	20	20	12
Raspberry	355	364	414	298	306	342	7	7	10
Strawberry	3110	3175	3238	1425	1339	1276	6	8	9
Currant	613	680	653	609	675	648	9	11	12

Source: Horticultural Bulletin (2002).

importance of *Jonathan* has decreased, while *Idared, Jonagold, Starking,* and *Golden* varieties have become popular.

Composition

Apples constitute 90% water and contain small amounts of energy, proteins, and carbohydrates. Apples are rich in certain vitamins (B_1, B_2, B_6, and folic acid), minerals (magnesium, iron, and potassium), organic acids, and dietary fiber (Tables 35.2 and 35.3). Sugar content and organic acids may vary according to apple varieties. Apples contain twice as much fructose (8.2 g/100 g) as glucose (Barta, 2000) and are high in fiber (1.3%). In addition, the Hungarian climate produces an apple with a unique flavor that is desirable.

Processing

One of the most significant advantages of apples is that there are many different ways in which apples can be used, processed, and sold.

Processing and utilization opportunities are as follows:

- Fresh apple: immediate use, storage
- Semiprocessed products: aseptic pulp, concentrate (70–72 Brix), semiconcentrate (40–45 Brix), pulp concentrate, opalescent concentrate
- Juices: clarified filtered apple juice, apple juice with fiber, opalescent apple juice, nectar, syrup, infant drinks
- Products preserved by heat: canned apples, apple pudding, jams, jellies, marinated apples, apple strudel, apple sauce, and infant foods

Table 35.2. Vitamin Compositions of Fruit Species (In 100 g Fruit)

	Carotene (mg)	Tocopherol (E) (mg)	Thiamin (B_1) (μg)	Riboflavin (B_2) (μg)	Pyridoxine (B_6) (mg)	Folacin (μg)	Ascorbic Acid (mg)
Apple	0.05	0.6	50	50	0.07	6.0	5
Pear	0.03	0.4	30	30	0.01		5
Plum	0.2	0.8	50	20	0.04	1.9	6
Sour cherry	0.3		50	20	0.05		10
Cherry	0.08	0.3	50	20	0.02	5.3	8
Peach	0.4	0.6	20	20	0.07	2.5	7
Apricot	1.8	0.5	20	30	0.06	3.0	10
Raspberry	0.08	1.4	20	30	0.05		30
Blackberry	0.3		40	40	0.05		20
Strawberry		1.2	30	70	0.06		40
Red currant	0.04	0.2	40	30	0.02		30
Black currant	0.1	1.0	60	10	0.02		160
Grape	0.3		50	50	0.07	5.2	5

Source: Bíró and Lindner (1999).

Table 35.3. Mineral Compositions in Fruit Species (In 100 g Fruit)

	K	Ca	Mg	P	Fe	Cu	Zn
				mg/100 g			
Apple	112	5.5	6	8	0.3	0.028	0.046
Pear	100	15.7	10	20	0.2	0.050	0.073
Plum	240	16.0	16	30	0.2	0.029	0.071
Sour cherry	186	31.3	15	50	0.6	0.057	0.142
Cherry	174	16.3	16	20	0.3	0.040	0.110
Peach	183	5.7	10	12	0.3	0.034	0.090
Apricot	226	13.8	14	20	0.3	0.032	0.163
Raspberry	172	27.3	24	45	0.4	0.095	0.214
Blackberry	160	52	22	35	1.3		
Strawberry	145	28.1	18	35	0.3	0.029	0.063
Red currant	316	56.8	10	35	4.5	0.024	0.033
Black currant	187	39.8	17	35	4.5		
Grape	195	28.2	14	75	0.7	0.042	0.136

Source: Bíró and Lindner (1999).

- Dried products: dehydrated apple (slices, cubed), powder, flakes, instant powder, apple chips, and crystallized apple
- Fermented products: cider, sparkling cider, distilled products, calvados, and vinegar
- Others: pectin, flavoring

Apples are mainly processed as juice or concentrate. Hungary's special apple product is the Szabolcs Almapálinka, which has had protected origin status since 2000, according to the Hungarian Council of Origin Protection. Products with "szabolcsi" denomination must be made of apples produced in Szabolcs-Szatmár-Bereg county, and processed using the traditional method in a small-scale copper distillery. The product is clean, colorless, with a pleasant apple taste and aroma, and 40–50 v/v% alcohol content. Approximately 1 million liters of this distilled product is produced annually in the Szabolcs-Szatmár-Bereg region (Drexler and Páll, 2001).

PEAR (*PYRUS COMMUNIS* L.)

Production

Pears are one of the most valuable fruits, rich in flavors and aroma and very popular in both fresh and processed forms. The world's pear production was about 17 million tons in the past years, 20–22% of which was produced in Europe. The most significant pear producing countries are China, USA, Italy, and Spain. In Hungary, pears are produced on 2000–2500 ha (KSH, 2002); the annual production can be seen in Table 35.1.

Varieties

There are a large number of pears, which are harvested from the middle of June until the end of October. Varieties can be divided into three categories:

- Early summer varieties (harvested in July): small-sized fruit, good for raw fresh consumption, tend to overripen, e.g., *Kornélia, Mirandino Rosso, Napoca.*
- Summer varieties (harvested in August): short storage period (1–2 months), e.g., *Clapp's favourite, Jules Guyot dr, Williams.*
- Autumn varieties (harvested at the beginning or in the middle of September): when harvested at optimal maturity, they can be stored for 4–5 months, e.g., *Conference, Beurré Bosc.*
- Winter varieties (harvested in late September, October): excellent for storage, even for 5–7 months, e.g., *Packham's Triumph, Beurré d'Hardenpont, Olivier de Serres.*

The most widespread varieties in Hungary are *Beurré d'Hardenpont, Bosc Kobak,* and *Williams* (Göndör, 1998).

Composition

Pears are high in minerals and fiber (2.6%), and contain substantial amounts of tocopherols, carotene, and vitamins B_1 and B_2 (Tables 35.2 and 35.3). Fructose is the predominant carbohydrate, with a level three times higher than glucose.

Processing

Only a small part of the annual pear production is sold as fresh fruit (primarily the *Bosc Kobak* variety). The market for processed pears is of far greater importance. Varieties intended for processing should be tasty and rich in aroma, with flesh that does not tend to turn brown, for example, *Williams, Packham's Triumph,* and *Conference* varieties (Göndör, 1997).

Pears can be processed as canned fruit, juice, nectar, puree, jam, dried fruit, or pálinka (an alcoholic drink made by distillation). Because of their high fructose and low glucose content, pears are a suitable raw material for lowered sugar products. Sucrose can be substituted with high fructose content sweeteners, e.g., Jerusalem artichoke concentrate, without changing the carbohydrate content of the fruit (Barta, 1993).

One of Hungary's unique products is the spicy Halasi Kiffer körtepálinka (a spirit) with intense aroma and an alcohol content of at least 40 v/v%. This product is exported to the United States and Italy as well (Edelényi, 2001).

STONE FRUITS

Among stone fruits, plums are produced in the largest volume in Hungary but the production of sour cherries and peaches is substantial as well. However, cherries and apricots are of less importance in Hungary.

Plums (*Prunus domestica* L.)

Production

Plums have an important role in both production and consumption worldwide. Plums are usually second behind apple production in temperate zones. Both fresh and processed plums are popular in Hungary, because of the unique characteristics (sweetness, taste, flavor) of the fruit and the long harvest period (from the middle of July until the end of September). The world's annual plum production exceeds 9 million tons, approximately one third in Europe, primarily in Germany and France.

In Hungary, plums are produced on 6800 ha (KSH, 2002), second to that for apples.

Varieties

Varieties may differ in each country depending on ecological conditions and utilization. The most widespread varieties are—in Hungary as well—the ones that belong to the European plum species (*Prunus domestica* L.). These are elongated, waxy, and dark blue, although there is a great diversity within the species. Fruits may vary in size from small to large, in shape from round to oblong, in skin color from blue to greenish yellow, and in the flesh color from greenish yellow to gold. Greengage varieties are mainly round and are in a variety of colors.

In Mediterranean and subtropical areas, Japanese type plums (*Prunus salicina* L.) are produced. This species originated in China, thus are called Chinese and Chinese–Japanese plums. China produces one fourth of the world's plums. The fruit is usually round and large with a low proportion of stone (2–3%) to flesh.

In Hungary, Cherryplum (*Prunus cerasifera* E.) is primarily used as rootstock and as the pollen donor for Japanese-type plums (Szabó, 1997a).

Besztercei plums variety is still important in Hungarian plum production, with 40% of the total commercial production. The varieties produced in Hungary have expanded in recent years. Plum varieties produced in Hungary are as follows:

- European varieties: *Ageni, Besztercei* (in Hungary, only its clones are allowed for propagation), *Bluefre, Cacanska lepotica, Cacanska rana, Cacanska rodna, Debreceni muskotály, President, Silvia, Stanley*
- Greengage varieties: *Althann ringló, Sermina, Zöld ringló*
- Japanese varieties: *Nagrada, Oilnaja* (Szabó, 1997a)

Composition

Among fruits, plums are highest in carbohydrates, mainly sucrose and glucose, but plums also contain substantial amounts (3 g/100 g) of sorbitol (Bíró and Lindner, 1999) that has a slight laxative effect.

Its vitamin and mineral contents are remarkably balanced. Plums contain high quantities of vitamins E, B_1, and B_6 as compared with other fruits. As far as minerals are concerned potassium (240 mg/100 g), magnesium (16 mg/100 g), and zinc (0.07 mg/100 g) contents are outstanding (Tables 35.2 and 35.3).

Processing

Varieties that are harvested early, European ones (*Cacanska rana* and *lepotica*), greengage varieties, and Japanese types are consumed primarily as fresh fruit. Varieties that ripen later and plums high in dry matter are suitable for processing. *Besztercei* and *Stanley* are excellent raw materials for high-quality canned and frozen plums. Due to the high sugar content, *Besztercei* is good for jam and spirit (pálinka) production. Moreover, its small size makes it a popular raw material for plum dumplings. For dried plum production, *Ageni* and its derivatives are used worldwide. *Stanley, Bluefre,* and *President* varieties can also be processed as delicious dried plum, jam, or canned fruit (Szabó, 1997a). Greengage type plums are consumed in fresh and canned form as well. Besides consumption and exportation of fresh plums, Hungary plays an important role in the processed plum market. Plum products are considered traditional Hungarian goods that are popular all over the world. Canned plum products are important, particularly products made with rum or vinegar, and those stuffed with garlic.

The Szatmár region—located in Northeast Hungary, beside the river Tisza—is known as "plum-country." Tarpa is a significant village in this area, where a famous and traditional Hungarian product—sugar-free plum jam—is produced. The raw material is from a small variety called "*nemtudom szilva*" found only in this region and having a very sweet taste and rich in aroma. This plum jam is made by long, gentle cooking to form a concentrate, while nothing is added, not even sugar.

Another well-known product of this region is Szatmári szilvapálinka (a spirit), which has a protected denomination of origin. It is also made of the "*nemtudom szilva*" variety, with the traditional method of aging in oak barrels for 1–3 years (www.tarpa.de/magyar/szilva.htm).

Distilled plum products, with a protected denomination of origin, are also produced in the Békés region, Békési szilvapálinka.

Sour Cherry (*Prunus cerasus* L.)

Production

The Hungarian climate is excellent for sour cherry production; currently it is our number one unique fruit. It is the only fruit, which is known as a popular fruit of our predecessors, even before the Hungarian conquest (Rapaics, 1940). Hungary is considered a significant sour cherry producing country because of its traditions, varieties, technology, and market opportunities (Nyéki et al., 2003).

The world's sour cherry production is above 900,000 tons per year. More than half are produced in Europe, primarily in Poland and Turkey. Hungary is a significant sour cherry producer, and is number one in breeding improvement (Apostol, 1995).

Varieties

Hungary's own sour cherry selection is the largest in the world, from light-colored to deep claret-colored varieties. Hungarian varieties are produced in the world's leading sour cherry producing countries such as the USA (Michigan), Germany, and Poland. Hungarian varieties are unique and suitable for preservation, because of the high acid–sugar ratio, and are good for raw consumption because of the lack of bitter substances (Apostol, 1998). From the end of the 19th century, *Pándy meggy* and *Cigánymeggy* varieties dominated in sour cherry production. The quality of *Pándy meggy* is considered a standard; thus, it is a unique Hungarian variety (Nyéki et al., 2003). However, due to production and propagation difficulties nowadays only certified clones can be used. *Pándy meggy* was substituted by varieties improved in the past three decades (e.g., *Érdi bőtermő, Érdi jubileum, Maliga emléke*) and by the ones selected in Northeast Hungary (e.g., *Újfehértói fürtös, Kántorjánosi 3, Debreceni bőtermő*) (Nyéki et al., 2003). *Korai pipacsmeggy, Meteor korai,* and *Csengődi* varieties are produced to expand the selection of sour cherries (Harsányi and Mády, 2003).

Composition

The acid–sugar ratio is 1 : 6, providing a refreshing sweet taste. As far as vitamins are concerned, carotene and vitamins B_1 (50 μg/100 g) and B_6 are outstanding. Its minerals are valuable too. The level of calcium (31.3 mg/100 g), magnesium (15 mg/100g), and phosphorus (50 mg/100 g) is remarkable compared with other fruits. In addition, sour cherries contain high amounts of iron (0.6 mg/100 g), copper (0.057 mg/100g), and zinc (0.142 mg/100g) (Tables 35.2 and 35.3).

Processing

Hungarian sour cherry varieties are unique, since they are suitable for processing and fresh consumption as well.

Sour cherries, intended for processing, are mainly used by the canning industry for canned fruit and jam production (about 60%) and about 20% are processed by the refrigeration and the soft drink industry

(Kállay, 1999). *Meteor korai* is the earliest variety and is only used for fresh consumption. Meanwhile, *Érdi nagygyümölcsű* variety is good for juice, syrup, and jam production; moreover, it is an excellent raw material for canned fruit, due to its large size and slightly acidic, staining juice. Deep claret-colored varieties, possessing staining juice (*Cigánymeggy* clones, *Pándy meggy* clones, *Csengődi*) are significant, in terms of juice, concentrate, syrup, and production of food colorants.

Pipacsmeggy varieties, that have white or pale-pink fruit flesh, nonstaining juice, and sourish-sweet taste, are primarily used for confectionary application (for cakes) and are sold in Western Europe, mainly in Germany (Apostol, 2000).

Sour cherries can also be processed as dried fruit, fruit tea, crystallized fruit, spirits, and as a special Hungarian confectionary product called "konyakos-meggy" (sour cherry bonbon with alcohol).

CHERRIES (*PRUNUS AVIUM* L.)

Production, Varieties

In contrast with other countries, in Hungary, cherries and sour cherries are known as distinct species. In Hungary, cherries are produced on approximately 12,000 ha, or one-tenth of the sour cherry producing area (KSH, 2002). For a long time *Germersdorfi óriás* was the main variety in Hungary but now the selection of varieties has widened.

Varieties produced in Hungary are as follows:

- Selected regional varieties: *Pomázi hosszúszárú, Szomolyai fekete, Solymári gömbölyű*
- Selected clones of widespread varieties: *Münchebergi korai, Germersdorfi óriás* (only its clones can be propagated in Hungary), *Hedelfingeni óriás*
- Varieties improved in Hungary: *Margit, Linda, Katalin, Kavics, Alex, Vera*
- Acclimatized varieties: *Bigarreau Burlat, Valerij Cskalov, Vega, Van, Stella* (Apostol and Brózik, 1998)

According to fruit-flesh quality, cherries can be classified as crispy, stiff, or mellow; the latter being well known for its heart shape (G. Tóth, 1997b).

Composition, Processing

Because of its early ripening and delicious taste, it is the most significant summer delicacy. The majority of this fruit is consumed fresh around the world. The amount of magnesium, vitamin B_1, and folic acid is remarkable (Bíró and Lindner, 1999).

The direction of cherry processing primarily depends on fruit color, flesh stiffness, ripening period, and composition. Red and claret-colored early ripening varieties (*Münchenbergi korai, Biggareau Burlat, Valerij Cskalov*) are consumed fresh. Crispy and claret-colored varieties (*Solymári gömbölyű, Vega, Linda, Germersdorfi clones, Van*) are mainly exported or used for canned cherry production. Black varieties with staining juice (e.g., *Szomolyai fekete*) are suitable for juice, concentrate, and food coloring (G. Tóth, 1997b). Dark purple colored, sweet tasting varieties are purchased by the refrigeration industry as well. In addition, cherries are used as a raw material for confectionary products, crystallized fruit, and distilled spirit products (cseresznyepálinka).

PEACHES (*PRUNUS PERSICA* L.)

Production

Peaches are considered the queen of fruits. Production began some 4000 years ago, and only *Prunus persica* varieties have become widespread. The majority of them are produced between the northern latitudes of 30° and 45°; however, peach production is rapidly growing in tropical and subtropical areas (Szabó, 1998). Peaches were acclimatized in Hungary, thus these climatic conditions are not optimal for peach production. However, under the microclimatic circumstances of the Carpathian Basin, peaches are superior in both taste and aroma compared to the larger fruits produced in the Mediterranean countries. Peaches are produced on some 7000 ha in Hungary (KSH, 2002).

Varieties

More than half of the varieties produced in Hungary came from the United States and 40% of these originated in California. Only one tenth of Hungary varieties were derived from Hungarian breeding (Horváth, 2003).

Peach varieties produced in Hungary are as follows:

- yellow fruit-flesh, tomentose varieties: *Dixired, Early Redhaven, Redhaven, Springcrest, Suncrest*
- white fruit-flesh, tomentose varieties: *Champion, Michelini, Nektár H, Springtime*

- nectarines: *Andosa, Fantasia, Flavortop, Independence*
- industrial varieties: *Babygold 5, Babygold 6, Babygold 7, Baladin, Loadel* (Szabó, 1998)

All categories contain white- and yellow-flesh varieties and both freestone and clingstone peaches.

Composition

Peaches are a very delicious and refreshing fruit eaten fresh, processed (canned), and as juice. The energy, carbohydrate, and protein contents are low compared with other stone fruits. Sucrose is the most significant carbohydrate, which is 7–8 times higher than glucose and fructose (Bíró and Lindner, 1999). As far as acids are concerned, malic acid is predominant, but citric acid and succinate acid can also be found in peaches. Industrial varieties are usually less acidic than varieties eaten fresh. Nectarines are higher in acid.

The vitamin and mineral contents of peaches are insignificant, except for high carotene content in yellow-flesh varieties (Tables 35.2 and 35.3).

Processing

Most of the annual peach production is consumed as fresh fruit (50–70%), a smaller amount (10–20%) is processed by the canning industry as jam and canned fruits; however, there is an increasing demand for peach juices and nectars made of peach pulp. Only varieties that do not turn brown or red around the stone are used for canning. Due to the improvements in breeding, industrial varieties are excellent for processing. These varieties do not soften, even when ripe, but remain hard and elastic. Most industrial varieties have yellow flesh and are clingstone, though some are freestone and semiclingstone types.

Peaches are a popular raw material for infant foods and drinks as well as for spirits.

APRICOTS (*PRUNUS ARMENIACA* L)

Production

Apricot varieties belong mainly to *Prunus armeniaca* L. species. In the past years, apricot production has increased worldwide, with Asia taking the lead. Africa, North and Central America are also involved in apricot production. The United States produces 90% of the total produce of North and Central America (Kerek and Nyujtó, 1998).The most ideal conditions are the warm Mediterranean climates in the temperate zone.

In Hungary, apricots are produced on some 5700 ha (KSH, 2002); production yields are shown in Table 35.1.

Varieties

Since apricots cannot adapt easily to different conditions, one or two dominant varieties are grown locally. Selected varieties should guarantee a high yield and be suitable for fresh consumption, as well as for processing.

In Hungarian apricot production, *Magyar kajszi* varieties are predominant, but only the clone *Gönci magyar kajszi* can be propagated. The production of giant apricots is significant, whereas rose apricot production has declined. The most substantial varieties produced in Hungary are *Bergeron, Ceglédi arany, Ceglédi bíbor, Ceglédi Piroska, Ceglédi óriás, Gönci magyar kajszi, Mandulakajszi,* and *Pannonia* (Szabó, 1997b). *Harmat* and *Korai zamatos* varieties have been improved in the past few years. Furthermore, the Canadian frost resistant *Harcot* variety and the late ripening Romanian *Sulmona* and *Comondor* varieties are also promising in Hungary (Szalay, 1998).

Composition

Due to its valuable nutrients, apricot is an important supplement to our diet. The balanced sugar–acid proportion and intense aroma contribute to its popularity. Amounts of carotene (1.8 mg/100 g) are significant and are higher in apricots than that in other fruits grown in Hungary. As far as minerals are concerned, potassium (226 mg/100 g) and zinc (0.16 mg/100 g) are significant. Due to its high fiber content (0.8 g/100 g), apricots have nutritional benefits (Tables 35.2 and 35.3).

Processing

Apricots can be processed in many ways, as canned fruit, jam, pulp juice, nectar, infant drinks, dried fruit, and pálinka.

Early ripening varieties (*Ceglédi Piroska, Ceglédi óriás*) are primarily used for fresh consumption, due to their clingstone characteristic. Varieties, which ripen later, are good for fresh consumption and processing as well (Balla and Koncz, 1996; Fekete et al., 1997). *Gönci magyar kajszi, Pannónia,* and *Ceglédi arany* are excellent for canned fruit and jam production. The fruit of *Mandulakajszi,* that looks like an almond, and the late ripening *Bergeron* variety are very good raw material for processing (Szalay, 1998).

Hungarian apricot producing regions have their own special products that are well-known worldwide. Kecskemét apricot jam, which possesses a unique aroma and taste, is made of *Magyar kajszi*, produced in the Kecskemét region. Distilled spirit products such as Gönci barackpálinka and Kecskeméti barackpálinka are also very popular; the latter has had a protected denomination of origin since 2000; therefore, it is protected in 17 countries, according to the Lisbon Agreement (Gerencsér, 2001).

BERRIES

Among berries in Hungary, raspberries and strawberries are significant, while red and black currants are less important. Gooseberries are mainly grown in household gardens.

Raspberries (*Rubus idaeus* L.), Bramble

Production

Rasberry production is greater than that of other berries. Raspberries can be used in many ways: fresh, frozen, juice, juice concentrate, drinks, and jams. The majority of European varieties and some early American varieties were derived from *Rubus ideaus* L., and new American varieties were derived from *Rubus strigous* (Dénes, 1997a).

Some 80–85% of the world's raspberry production is in Europe (Table 35.1), although Canada, the United States, New Zealand, and Australia produce substantial amounts as well. In recent years, raspberry production has stagnated or decreased in Europe because of labor costs associated with handpicking. However, in Eastern European countries including Hungary and (Vinic, 2002), especially Poland, which has become the main producer, raspberry production is increasing.

Varieties

Most countries that produce substantial amounts of raspberries develop varieties to suit their ecological conditions. In England, mainly the Malling series varieties are produced, but other countries—Switzerland, Poland, the United States, and Australia—have developed varieties suitable for the environment and contributing to a greater selection.

In Hungary, first the *Nagymarosi* variety was predominant; then came the *Malling Promise* and *Malling Exploit* varieties (Dénes, 1997a).

New varieties have been produced that meet special requirements, for example, yellow and white in addition to red varieties are now available. Beside traditional varieties that produce only once a year, there are commercial varieties that produce twice a year (autumn varieties) or continuously. In the improvement of bramble varieties, the main target is to develop and select thornless varieties, like the Scottish *Loch Ness* variety.

The main Hungarian raspberry and bramble varieties are as follows:

- summer raspberry: *Fancsalszki egyszertermő, Fertődi zamatos, Malling Exploit, Malling Promis, Nagymarosi, Willamette, Tulameen*
- autumn raspberry: *Autumn Bliss, Fertődi kétszertermő, Golden Bliss, Zeva Herbsternte*
- bramble: *Dirksen, Thornfree, Loch Ness, Hull*

Composition

Raspberries and brambles are remarkably valuable regarding their composition (Tables 35.2 and 35.3). Both are low in energy and carbohydrate content with substantial amounts of fiber (4–5.6 g/100 g). As far as vitamins are concerned, the tocopherol content of raspberry (1.4 mg/100 g) is outstanding compared to other fruits. They are also rich in minerals. Raspberry contains high amounts of potassium, calcium, and magnesium (24 mg/100 g). Moreover, the level of microelements, particularly zinc (0.214 mg/100 g), is outstanding.

In the case of bramble, calcium (52 mg/100 g), magnesium (22 mg/100g), and iron (1.3 mg/100 g) need to be emphasized. Both fruits contain significant level of flavonoids, primarily anthocyanins that are responsible for the color. Flavonoids have an antioxidant effect and also have an important role in the prevention of several diseases (e.g., heart attack, cancer, thrombosis).

Processing

Raspberries are soft and deteriorate easily, and therefore fresh raspberry consumption is only possible with household production. However, there are new packaging techniques for the storage of fresh raspberries, e.g., modified atmosphere packaging (Jacxens et al., 2001a).

As the fruit of bramble is harvested with the peduncle, it can be transported more easily.

The refrigeration industry purchases substantial amounts of both fruits. The frozen rolling raspberry is an important Hungarian export product. Mainly *Willamette*, *Fertődi zamatos*, and *Autumn Bliss* varieties are used for freezing, since the fruit does not tend to crumble (Dénes, 1997a). Fruits are required to keep their texture and color even after melting. The canning industry prefers dark colored varieties with intense aroma for producing concentrate. Raspberry concentrate is the raw material of nectars, drinks, and syrups; meanwhile, bramble is used for mixed fruit drinks or applied as a natural coloring agent. Both fruits are excellent for jam and jelly production. These can be made of fresh fruit and fruit pulp preserved by aseptic technology. Raspberries and brambles are also raw materials for dairy fruit products (e.g., yoghurt, cottage-cheese cream) and confectionery goods such as cake decorating and jam fillings for baking.

STRAWBERRIES (*FRAGARIA ANANASSA* L.)

Production

Strawberries are one of the most popular fruits. Since the Second World War, the world's strawberry production has been increasing continuously. Almost one third of the world's production is from the United States, but strawberries are also produced in significant amounts in Spain, Italy, Poland, and Japan. Approximately one third of the world's production is in Europe. In the past few years, domestic production was between 11,000 and 14,000 tons annually (Table 35.1).

Varieties

Strawberries are a cosmopolitan species, which is produced almost everywhere except in the tropics. Varieties have been developed for different climatic conditions.

These varieties can be divided into three main groups: traditional, continuously yielding, and day neutral varieties.

German and Dutch varieties are predominant in Hungary. The *Gorella* variety has been planted for a long period, but *Cambridge Rival* and *Korona* are also significant. Nowadays, *Elsanta* is the most widespread variety (Dénes, 1997b). Among Hungarian varieties, *Fertődi 5* and *Kortes* are produced.

Composition

Strawberries are a very popular fruit, because it ripens early, making it one of the first fruits to be harvested after winter. It is mainly consumed in fresh form. Strawberries are high in water and low in energy and carbohydrates contents. Among the vitamins, tocopherols (1.2 mg/100 g), vitamin C (40 mg/100 g), and vitamin B_2 (70 μg/100 g) are outstanding. The mineral content is also substantial, especially potassium, calcium, magnesium, and phosphorus (Tables 35.2 and 35.3).

Processing

Strawberries are primarily consumed fresh, frozen and in several heat-preserved forms. According to recent research, high oxygen packaging can increase the shelf life of fresh strawberries (Jacxsens et al., 2001b). As far as preserved foodstuffs are concerned, frozen strawberry production and export is considered significant. The industry produces preserved aseptic pulp, which can be the raw material of jams, jellies, and dairy products. Strawberry concentrate is used for syrup, juice, nectar, and jelly production.

CURRANTS (*RIBES NIGRUM, RUBRUM* L.)

Production

Currants (red and black) are also considered as important berries. The world's annual currant production is more than 600,000 tons, and 99% of that is produced in Europe. The main currant producing countries are Poland and Germany. Hungarian production is approximately 12,000–14,000 tons (Table 35.1); black and red currant producing areas are greater than 2000 ha (KSH, 2002).

Varieties

Red and white currants produced in Hungary belong to *Ribes petreanum W.* or *Ribes rubrum L.* species; meanwhile, black currants are derived from the *Ribes nigrum L.* species.

Red currants can be produced on larger areas than the black ones. Hungarian large-scale production of red currants began with the importation of the *Jonkheer van Tets* variety. Hungarian varieties such as *Fertődi hosszúfürtű* were not widespread at that time.

Since Hungary is located at the southern border of black currant production, sprouts budding too early in the spring often suffer from frost damage. Therefore, new frost-resistant varieties need to be developed.

The main Hungarian currant varieties are as follows:

- Red currant: *Fertődi hosszúfürtű, Jonkheer van Tets, Rondom,* and *Red Lake*
- White currant: *Blanka*
- Black currant: *Altajszkaja deszartnaja, Fertődi 1, Hidas bőtermő, Silvergieter F. 59, Titania, Triton,* and *Wellington XXX* (Harsányi and Mády, 2003)

Composition

The nutrient value of currants is significant. Both are low in energy and quite acidic, with black currant being one of the most acidic fruits. Their fiber content is outstanding (3.8–4.3 g/100 g) when compared with other fruits.

As far as vitamins are concerned, the level of vitamin C is significant (160–180 mg/100 g), particularly in the case of black varieties (Stéger-Máté et al., 2002). Both fruits possess remarkable amounts of biotin (2.4–4.2 mg/100 g) (Bíró and Lindner, 1999). In addition, black currant varieties contain substantial amounts of tocopherols and vitamin B_1. Their iron content is very high (4.5 mg/100 g), compared with other fruits, as is potassium and calcium (Tables 35.2 and 35.3). Black currants contain high amounts of flavonoids, mainly anthocyanins, which have an antioxidant effect, offering protection against the development of chronic diseases.

Processing

The majority of currants are processed by refrigeration and heat preservation. Black currants, which are acidic and aromatic, are used by the industry. Currants are used for concentrate, the raw material for juices, nectars, and syrups. Aseptic black currant pulp is used for jams, jellies, and dairy products. It is also used for fruit wine making.

Currants are also used as a coloring agent. Coloring effect is due to the presence of anthocyanins, which are extracted from the concentrate or fruit.

Red currants are reprocessed in large quantities both by the canning and the cooling industries. The refrigeration industry usually purchases handpicked fruits and freezes only the berries, without peduncle.

Grape (*Vitis vinifera* L.)

Production

Grape production and wine making originated in Asia (Armenia, Azerbaijan, and Iran). In those regions, small-berry *Vitis sylvestris* varieties were available and were produced 4000–5000 years before Christ. The production over 2000–3000 years led to the improvement of *Vitits vinifera* L., with bigger berries and bunches. In taxonomy, it is an individual species.

Nowadays, grapes are produced on about 8 million ha, with 70% being processed in Europe. Europe also produces the largest volume, about 60%, ahead of America and Asia. However, Italy, France, and the United States are the most significant grape producing countries. Also, Italy, Turkey, and the United States lead in the production of dessert grapes, whereas the United States and Iran lead in raisin production.

In wine production, Europe and some countries like Italy, France, and Spain are outstanding (Kozma, 2000; http://mail.bbkvtar.hu).

In 2002, Hungarian vineyards covered about 83,000 ha, with 501,000 tons of grape production, 23,000 tons of which were used for raw consumption, and the rest for winemaking. That year's wine production was 333 million liters (KSH, 2003).

Varieties

There are two main groups of grape varieties: white and red. In Hungary, the most widespread variety, in terms of high-quality white wine raw materials, is the *Olasz Rizling* (produced on some 10% of the vineyards). It is followed by *Rizlingszilváni* (about 4%), *Chardonnay, Muscat Ottonel, Hárslevelű, Furmint, Rajnai Rizling,* and *Leányka.* Regarding the high-quality red wine raw materials, *Kékfrankos* leads, produced in about 5% of the vineyards, ahead of *Zweigelt, Kékoportó, Merlot, Cabernet Franc, Cabernet Sauvignon,* and *Pinot noir* (www.pannonwine.hu).

Wine quality is primarily determined by the quality of the grape, dependent on technological ripeness. For table wines, the minimum requirement is 15 MF (i.e., 15 g sugar/100 g grape). For high-quality wines, the minimum is 17 MF. In the case of varieties that are rich in aromas and those used for sparkling wine production, the optimal harvest occurs before full ripeness. On the other hand, special high-quality wines are usually made of overripe grapes exceeding 19 MF (www.boraszat.hu).

The famous Tokaj region—possessing 5500 ha of vineyard—provides unique conditions for wine making. The traditional aszu-wine has been produced for more than 350 years. The climatic and soil circumstances of the region enable the production of high-extract white wines that are rich in minerals. Tokaj is a closed or restricted wine district;

therefore, only four wine varieties (*Furmint, Hárslevelű, Sárgamuskotály,* and *Oremus*) are allowed for wine making. To protect the unity of the region and the wine growing traditions of the past thousand years, the UNESCO declared the Tokaj wine district as a part of the World's Heritage.

Composition

The fruit of the grape contains important and valuable nutrients. Its vitamin—B_1, B_2, B_6, and folic acid—contents are remarkable. As far as minerals are concerned, the level of potassium, calcium, magnesium, and microelements is substantial (Tables 35.2 and 35.3). Both the fruit flesh and the skin of red grapes contain high amounts of flavonoids, such as anthocyanides, flavonols, and tannins, responsible for the red color of the berries. Some of these latter compounds have antioxidant effect. Furthermore, the grape seed oil is rich in valuable unsaturated fatty acids (linoleic acid, oleic acid, and palmic acid).

Processing

Only a small proportion of the grape is consumed fresh and the majority is processed. This usually means fermentation for making table wines, high quality white and red wines, vermouths and sparkling wines. If not fermented the grape is utilized as must, condensed must, or juice for beverage production. It is also processed as canned fruit or jelly in small quantities. Raisin and dried grape can be obtained and also available as seedless. More than half of the world's raisin production is concentrated in California, where it is sold not only in dried form but also as pulp and raisin juice concentrate. This concentrate is prepared by the multiple aqueous extraction of the raisin, to be condensed up to 70 Brix. Meanwhile, the pulp, with 100% fruit content, is made by the grinding and mashing of the raisin. Both are used in the baking and confectionery industries for biscuits, ice creams, mueslis, and sauces. Due to its high tartaric acid content, raisin has a strong antimold effect, and its high fiber content prohibits moisture accumulation, improving the shelf life of breads and bakery products (www.rac-intl.org/hungary/).

The by-products of wine making can be used for tartaric acid, grape seed oil, yeast, distilled spirit drink, and vinegar production.

OTHER FRUITS (WILD, SPECIAL)

In Hungary, many traditional wild fruits can be found. These have been gathered for ages and play an important role in folk medicine. Elderberries, rose hips, and dogwood berries can be found everywhere in the country; wild-growing blackthorn, rowanberry, hawthorn species, and blackthorn also grow in Hungary.

The most significant special fruits are elderberries, rose hips, and sea blackthorn; these are produced on plantations as well.

The *Haschberg* elderberry variety is rich in vitamin C (131 g/100 g) and in micro- and macroelements (Szabó et al., 2002).

This variety, produced under Hungarian conditions, is richer in nutrients as compared to varieties grown in other European countries. In addition, elderberries are high in anthocyanin content that is used as a coloring agent. In the case of *Haschberg* variety, produced in Hungary, the anthocyanin content is between 8000 and 12,000 mg/l (Stéger-Máté et al., 2001).

REFERENCES

Anon. 2002. The grape varieties. (in Hungarian) www.pannonwine.hu.

Anon. 2002. The technology of winemaking. (in Hungarian) www.boraszat.hu

Anon. 2002. The winemaking of the world and Hungary. (in Hungarian) http://mail.bbkvtar.hu/bor/bor.html.

Anon. 2003. Raisin production in California. (in Hungarian) www.rac-intl.org/hungary.

Anon. 2003. Tarpa. The plum spirit. (in Hungarian) www.tarpa.de/magyar/szilva.htm.

Apostol, J. 1995. Results of sour cherry and sweet cherry breeding and production in Hungary. *Acta Horticulture*. 40(3):26–29.

Apostol, J. 1998. Sour cherry. (in Hungarian) In: Soltész, M. (ed.) Gyümölcsfajta-ismeret és –használat (Knowledge and Use of Fruit Variety) Mezőgazda Kiadó. Budapest, Hungary. 288–308.

Apostol, J. 2000. Korai pipacsmeggy for confectioners. (in Hungarian) *Kertészet és Szőlészet*. 49(2):4.

Apostol, J. and Brózik, S. 1998. Cherry (in Hungarian) In: Soltész, M. (ed.) Gyümölcsfajta-ismeret és –használat (Knowledge and Use of Fruit Variety) Mezőgazda Kiadó. Budapest, Hungary. 309–329.

Balla, Cs. and Koncz, K. 1996. Observation of the ripening process of apricots reflected as coloring. *Élelmiszerfizikai Közlemények*. IX. 1996/1–2. 23–34.

Barta, J. 1993. Jerusalem artichoke as a multipurpose raw material for food products of high fructose or inulin content. In: Fuch, A. (ed.) Inulin and

Inulin-Containing Crops Studies in Plant Science, Vol. 3. Elsevier Science Publishers, Amsterdam. 323–339.

Barta, J. 2000. The composition, storage and processing of the Jerusalem artichoke tubers. (in Hugarian) In: Angeli, I., Barta, J. and Molnár, L.A. (eds.) gyógyító csicsóka. (The Medicinal Jerusalem Artichoke). Mezőgazda Kiadó. Budapest, Hungary. 77–147.

Bíró, Gy. and Lindner, K. 1999. Nutrition Tables. (in Hungarian) Medicina Könyvkiadó Rt. Budapest, Hungary.

Czellár, K. and Somorjai, F. 1996. Hungary. (in Hungarian) Medicina Könyvkiadó Rt. Budapest, Hungary. 14–17.

Dénes, F. 1997a. Raspberry, blackberry. (in Hugarian) In: G. Tóth, M.(ed.) Gyümölcsészet (Fruit Cultivation) Primom Vállalkozásélénkítő Alapítvány. Nyíregyháza, Hungary. 295–316.

Dénes, F. 1997b. Strawberry. (in Hungarian) In: G. Tóth, M. (ed.) Gyümölcsészet (Fruit Cultivation). Primom Vállalkozásélénkítő Alapítvány. Nyíregyháza, Hungary. 275–294.

Drexler, Gy. and Páll, I. 2001. Drinks. Szabolcsi almapálinka. (in Hungarian) In: Farnadi, É. (ed.) Hagyományok Ízek Régiók I. (Traditions, Tastes, Regions I.) Kesztler Marketing Kft. Budapest, Hungary. 264–266.

Edelényi, M. 2001. Drinks. Halasi körtepálinka. (in Hungarian) In: Farnadi, É. (ed.) Hagyományok Ízek Régiók I. (Traditions, Tastes, Regions I.) Kesztler Marketing Kft. Budapest, Hungary. 110–111.

Fekete, A., Felföldi, J. and Balla, Cs. 1997. Quality control of fresh apricots by measuring physical properties. Gyümölcs-zöldség postharvest és logisztikai konferencia. Budapest, Hungary. 167–180. (1997).

Gerencsér, E. 2001. Drinks. Kecskeméti barackpálinka (in Hungarian) In: Farnadi, É. (ed.) Hagyományok Ízek Régiók I. (Traditions, Tastes, Regions I.) Kesztler Marketing Kft. Budapest, Hungary. 112–113.

Göndör, M. 1997. Pear. (in Hungarian) In: G. Tóth, M. (ed.) Gyümölcsészet (Fruit Science). Primom Vállalkozásélénkítő Alapítvány. Nyíregyháza, Budapest. 111–136.

Göndör, M. 1998. Pear. (in Hungarian) In: Soltész, M. (ed.) Gyümölcsfajta-ismeret és –használat, (Knowledge and Use of Fruit Variety). Mezőgazda Kiadó. Budapest, Hungary. 156–182.

G. Tóth, M. 1997a. Sour cherry. (in Hungarian) In: G. Tóth, M. (ed.) Gyümölcsészet. (Fruit Cultivation) Primom Vállalkozásélénkítő Alapítvány. Nyíregyháza, Hungary. 257–272.

G. Tóth, M. 1997b. Cherry. (in Hungarian) In: G. Tóth, M. (ed.) Gyümölcsészet (Fruit Cultivation). Primom Vállalkozásélénkítő Alapítvány. Nyíregyháza, Hungary. 237–256.

Harsányi, J. and Mády, R. 2003. Grape and fruit varieties. Catalogue. (in Hungarian). Országos Mezőgazdasági Minősítő Intézet. Budapest, Hungary. 13–32.

Horticultural Bulletin. 2002. Annual Report of Hungarian Horticultural Sector. (in Hungarian) Magyar Zöldség-Gyümölcs Szakmaközi Szervezet és Terméktanács. Budapest, Hungary.

Horváth, Cs. 2003. Peach demonstration in Tordas. (in Hungarian) *Kertészet és Szőlészet*. 53 (31):15.

Jacxsens, L., Devlieghere, F., Siró, I. and Debevere, J. 2001. Survival of acid resistance pathogens on packaged strawberry fruit. 21–22. June 2001. Luik, Belgium (Abstract, poster).

Jacxens, L., Devlieghere, F., van der Steen, C., Siró, I. and Debevere, J. 2001a. Application of ethylene adsorbenrs in combination with high oxygen atmospheres for the storage of strawberries and raspberries. Proceedings of 8th International Controlled Atmosphere Research Conference, July 2001. Rotterdam, The Netherlands. ISHS, Leuven, Belgium. 8–13.

Jacxsens, L., Devlieghere, F., Siró, I. and Debevere, J. 2001b. Survival of acid resistance pathogens on packaged strawberry fruit. On Sixth Conference in Food Microbiology 21–22 June 2001. Luik, Belgium (Abstract, poster).

Kállay, E. 1999. Actual situation of cherry production in Hungary and tasks to promote it. (in Hungarian) *Kertgazdaság*. 31(1):77–85.

Kerek, M. and Nyujtó, F. 1998. Apricot. (in Hungarian) In: Soltész, M. (ed.) Gyümölcsfajta-ismeret és –használat (Knowledge and Use of Fruit Variety). Mezőgazda Kiadó. Budapest, Hungary. 234–257.

Kozma, P. 2000. Growing of the grape. (in Hungarian). A szőlő és termesztése I. Akadémiai Kiadó. Budapest, Hungary.

KSH. 2002. Fruit plantations in Hungary, 2001. (in Hungarian). Központi Statisztikai Hivatal (Hungarian Central Statistical Office). Budapest, Hungary.

KSH. 2003. Statistical yearbook of Hungary, 2002. (in Hungarian). Központi Statisztikai Hivatal (Hungarian Central Statistical Office). Budapest, Hungary. ISBN 963 215 554 8.

Nyéki, J., Soltész, M. and Papp, J, et al. 2003. Fruits and fruits-products as "Hungaricums". (in Hungarian) *Kertgazdaság*. 35(2):66–74.

Rapaics, R. 1940. The Hungarian fruit (in Hungarian) Királyi Magyar Természettudományi Társulat. Budapest, Hungary.

Stéger-Máté, M., Horváth, E. and Barta, J. 2001. Evaluation of elder (*Sambucus nigra* L.) varieties and candidates for the canning industry. *International Journal of Horticultural Science.* 7(1):102–107.

Stéger-Máté, M., Horváth, E. and Sipos, B. 2002. The examination of the composition and processing possibilities of the different current varieties I. (in Hungarian) *Ásványvíz, üdítőital, gyümölcslé.* 3(1):3–8.

Szabó, Z. 1997a. Plum. (in Hungarian) In: G. Tóth, M. (ed.) Gyümölcsészet (Fruit Cultivation). Primom Vállalkozásélénkítő Alapítvány. Nyíregyháza, Hungary. 211–236.

Szabó, Z. 1997b. Apricot. (in Hungarian) In: G. Tóth, M. (ed.) Gyümölcsészet (Fruit Cultivation). Primom Vállalkozásélénkítő Alapítvány. Nyíregyháza, Hungary. 195–210.

Szabó, Z. 1998. Peach. (in Hungarian) In: Soltész, M. (ed.) Gyümölcsfajta-ismeret és –használat. (Knowledge and Use of Fruit Variety) Mezőgazda Kiadó. Budapest, Hungary. 200–233.

Szabó, N., Stéger-Máté, M., Stefanovits-Bányai, É. and Sipos, B. 2002. Microelement content of elderberry candidates. 10th International Trace Element Symposium, Proceedings, Budapest, Hungary. 269–280.

Szalay, L. 1998. Our favourite is the apricot. (in Hungarian) *Kertbarát.* 21(7):19.

Vinic. 2002. Strawberry is in full swing. (in Hungarian) *Magyar Mezőgazdaság.* 57(12): 25–27.

Index

B

C

D

G

H

I

J

K

L

N

O

P

Q

R

S

T

U

V

W

X

Y

Z

Printed in the United States
125921LV00001BA/1-42/P

9 780813 819815